IRRIGATION WATER RESOURCES *AND* WATER POWER ENGINEERING

AF662202

By

Dr. P.N. MODI

B.E., M.E., Ph.D.

Former Professor of Civil Engineering

M.R. Engineering College, (Now M.N.I.T), Jaipur

Presently Principal, Kautilya Institute of Technology and Engineering, Jaipur

STANDARD BOOK HOUSE

unit of: **RAJSONS PUBLICATIONS PVT. LTD.**

1705-A, Nai Sarak, PB.No. 1074, Delhi-110006 Ph.: +91-(011)-23265506
Show Room: 4262/3, First Lane, G-Floor, Gali Punjabian, Ansari Road, Darya Ganj,
New Delhi-110002 Ph.: +91-(011) 43751128 Tel Fax : +91-(011)43551185,
Fax: +91-(011)-23250212

E-mail: sbh10@hotmail.com www.standardbookhouse.in

Published by:
RAJINDER KUMAR JAIN
Standard Book House
Unit of: Rajsons Publications Pvt. Ltd.
1705-A, Nai Sarak, Delhi - 110006
Post Box: 1074
Ph.: +91-(011)-23265506 Fax: +91-(011)-23250212

Showroom:
4262/3, First Lane, G-Floor, Gali Punjabian
Ansari Road, Darya Ganj
New Delhi-110002
Ph.: +91-(011)-43751128, +91-(011)-43551185
E-mail: sbh10@ hotmail.com
Web: www.standardbookhouse.in

First Edition	:	1988
Second Edition	:	1990
Third Revised and Enlarged Edition	:	1995
Fourth Revised and Enlarged Edition	:	2000
Fifth Revised and Enlarged Edition	:	2004
Sixth Revised and Enlarged Edition	:	2007
Seventh Revised and Enlarged Edition	:	2009
Eighth Revised and Enlarged Edition	:	2012
Ninth Revised and Enlarged Edition	:	2014
Tenth Revised and Enlarged Edition	:	2017
Eleventh Revised and Enlarged Edition	:	2020

© *Publishers*

All rights are reserved with the Publishers. This book or any part thereof, may not be reproduced, represent, photocopy in any manner without the prior written permission of the Publishers.

***Price*: Rs. 720.00**

ISBN: 978-81-89401-29-0

Typeset by:
N.D. Enterprises, Delhi.

Printed by:
R.K. Print Media Company, New Delhi

Preface to Eleventh Edition

In the Eleventh Edition the book has been thoroughly revised and enlargeld. Additional illustrative examples and problems with answers have been given in the various chapters. A large number of additional Multiple Choice Questions with answers have been given.

The author thank all the learned Professors as well as the students of the various Universities and Engineering Collage for thier appreciation of the book.

The author thank the Publishers Shri Rajindra Kumar Jain and Shri Sandeep Jain for bringinig this edition of the book with nice get up.

2020 **Dr. P.N. MODI**

Preface to Sixth Edition

In the sixth edition, the book has been thoroughly revised and enlarged. In this edition the book has been brought out in A-4 size thereby general get-up of the book has been considerably enhanced. The soil group of India have been discussed in detail in Chapter 3. The method for determining the wave height as per I.S.S. for computing the wave pressure on gravity dams has been indicated in Chapter 9. The revised recommendation for providing the free board for gravity dams as per IS: 6512-1984 has been given in Chapter 9. Similarly the revised recommendation for providing the free board for earth dams as per IS: 10635-1983 has been given in Chapter 11. The criteria for the design of weirs and barrages have been discussed in detail in Chapter 13. At the end of each chapter number of solved illustrative examples as well as number of unsolved problems with their answers have been added. Further a number of additional multiple choice questions have been given in the relevant Appendix. A new Appendix indicating the cause of earthquake and earthquake intensity scales has been added.

The author thank all the learned Professors as well as the students of the various Universities for their appreciation of the book and also for their valuable suggestions which have been incorporated in this edition of the book.

The author thank the publishers Shri Rajinder Kumar and Shri Sandeep Jain for bringing out the book with nice get-up.

2007 **P.N. MODI**

Preface to First Edition

The present book is the outcome of the author's long association of more than two decades with he subject of Irrigation and Water Resources Engineering as a teacher both at Under Graduate as well as Post Graduate Level. In this book an attempt has been made to present the subject matter of Irrigation and Water Resources Engineering in such a manner that it may be easily understood by the students. The basic principles involved in the designs of various irrigation works have been thoroughly explained. The book covers the complete syllabii of this subject for the students studying at first degree course of the various Indian Universities. Some advanced topics included in this book will be useful for the students studying at Post Graduate Level.

In this book a number of design problems have been solved in detail and these designs are based on the latest IS Codes. The treatment of earthquake forces acting on gravity dams has been thoroughly explained. The topics such as planning for water resources development and flood control have been discussed in detail. The subject matter has been supported by very good diagrams and selected multiple choice questions have been given in one of the appendices at the end of the book.

The book will be quite useful for the various competitive examinations such as Engineering Services and ICS Examinations and being self contained it will be equally suitable for the students preparing for AMIE Examinations.

The author is grateful for encouragement and help received by him from his collegues in the Civil Engineering Department. The cooperation of all the family members of the author is bringing out this book is worth appreciation. The excellent diagrams prepared by Shri Chiranji Lal are thankfully acknowledged. The full cooperation and understanding of our publisher Shri Rajinder Kumar Jain is greatly appreciated.

1988 **P.N. MODI**

Contents

CHAPTER

1 Introduction

1.1 DEFINITION

Irrigation may be defined as the process of supplying water to land by artificial means for the purpose of cultivation. Ordinarily water is supplied to land by nature through rain but generally it is not enough for the proper growth of the plants. As such the basic objective of irrigation is to supplement the natural supply of water to land so as to obtain an optimum yield from the crop grown on the land.

In order to achieve this objective of irrigation, an irrigation system is required to be developed, which involves planning, designing, construction, operation and maintenance of various irrigation works viz., a source of water supply, a distribution system for carrying water from the source to the agricultural land and its application on the land, and various other associated works. An irrigation engineer is associated with every phase of an irrigation system right from its inception and hence it is primarily his responsibility to develop an efficient and economic irrigation system to suit the natural conditions and the potential of the region. Moreover, an irrigation engineer must also be well versed with the agricultural aspects to optimise the irrigation system. The technology adopted by an irrigation engineer to develop a total irrigation system is termed as irrigation engineering.

1.2 NECESSITY OF IRRIGATION

The yield from an agricultural land depends upon several factors but the most important factor is that the crops get adequate water at the various stages of the growth of the plants. The natural supply of water to the agricultural land for the production of crop is through rain. Thus, if the normal rainfall at any place is adequate to meet the total water requirements of the crops grown and the time interval of the rainfall is such that water is available whenever the plants need it, then irrigation is not required. Such ideal conditions exist only for some small regions of the world where the normal rainfall is sufficient to fulfil the water requirements of the crops grown. However, for most of the other regions of the world crop production is not possible without the help of irrigation. Thus, the factors which necessitate irrigation are as described below.

1. Inadequate Rainfall. When the rainfall at a place is inadequate to meet the crop requirements, then it would be necessary to use irrigation. In such cases water may be conveyed from the places where it is available in abundance to the places of deficiency. In India, Indira Gandhi Canal is one such example in which water is conveyed from the river Sutlej to the Thar desert in the north-west part of the country where the average annual rainfall at many places is less than 100 mm.

2. Uneven Distribution of Rainfall. The total rainfall in a region may be adequate but it may be unevenly distributed over time as well as place. The entire rainfall for any region may occur only during some period of the year and for the rest of the period there may be no rains. In addition to the uneven distribution of the rainfall in a year, the rainfall may also show considerable variation from year to year. Further in some part of the region there may be excessive rains and the remaining part may have very little or almost no rains. The rainfall in India has such characteristics. Almost the entire rainfall for most part of the country is received from the monsoon during the four months from June to September and the rainfall during the winter is scanty. The only exception to this is the south-eastern part of the country where a considerable rainfall is also during the winter season from October to January. Further the average annual rainfall in the country varies from less than 100 mm in the Thar desert in the north-west to more than 2500 mm in Assam and other states in the east. With such an uneven distribution of rainfall, it would be necessary to use irrigation. The water may be collected during the period of rainfall and it may be utilised for irrigation when there are no rains. Most of the irrigation projects in India are based on this premise.

3. Growing a Number of Crops during a Year. The rainfall in a region may be sufficient to grow only one crop in a year for which no irrigation may be required. However, in the same region if more number of crops are to be grown during the same year it would be possible only if irrigation facilities are available. Moreover, some crops such as sugarcane have a much longer period of maturity for which irrigation is invariably needed.

4. Growing Superior Crops. Certain inferior or low-priced crops require less water and hence for growing such crops only the rain water may be sufficient and no irrigation is therefore needed. However, there are several superior or high-priced crops which need frequent application of large quantity of water and for growing such crops irrigation is necessary.

1.3 TOTAL PLANNING CONCEPT OF IRRIGATION PROJECT—SCOPE OF IRRIGATION ENGINEERING

In order to maximize the benefits from any irrigation project it is essential to adopt the concept of total and comprehensive planning which involves the consideration of all the aspects of the project. Broadly the various aspects of an irrigation project may be divided into the following three groups.

1. Engineering Aspect.
2. Agricultural Aspect.
3. Management Aspect.

1. Engineering Aspect. The engineering aspect of an irrigation project involves the development of a source of water for irrigation and the arrangement for the conveyance of water from the source right upto the agricultural fields.

The development of the source of water constitutes the *first phase* of an irrigation project. The water for irrigation may be derived from the surface water sources such as rivers, lakes, ponds, tanks etc., and it is known as *surface water*. It may also be derived from the ground water storage which is known as *ground water*. For obtaining water for irrigation from rivers two methods are commonly adopted. In one of the methods a storage reservoir is created by constructing a dam across the river and the water stored in the reservoir is used for irrigation through a canal system. In the second method the water is not stored but it is diverted from the river into the canal system by constructing a suitable diversion work such as a weir or a barrage. The first method is adopted when the minimum or the dry weather flow in the river is much less than the demand for water during that period, or when it is a non-perennial river. The second method may however be adopted only in the case of a perennial river having minimum or dry weather flow greater than the demand during that period. When the storage of water is needed the irrigation project becomes quite costly due to the high cost of construction of dam. In such cases the project is usually planned as a multipurpose project involving other uses such as power generation, flood control etc., so as to render it economically feasible.

In places where ample ground water storage exists and the water table is fairly high, wells can be dug, bored or sunk from which water can be lifted and used for irrigation. Often a conjunctive use of surface and ground water is made for irrigation purposes, which in many cases becomes necessary to lower the water table in the region.

The *second phase* of an irrigation project comprises the development of a system for the conveyance of water for irrigation. For the conveyance of water from a storage reservoir or a diversion work to the agricultural fields, a net-work of irrigation canals is provided which comprises main canals, branch canals, major distributaries (or majors) and minor distributaries (or minors). From the canals the water is led to the agricultural fields through water courses or field channels. The water courses take off from the major or minor distributaries through various types of outlets. In order to regulate the flow of water in canals, several irrigation structures such as head regulators, cross regulators, falls etc., are provided. In addition to these structures certain cross drainage works are also required to be provided at places where the canals cross the rivers and natural drains.

In the case of wells a net-work of canals is usually not required because the wells are mostly located close to the agricultural fields and hence the water lifted from the wells is conveyed to the fields through water courses or pipes.

2. Agricultural Aspect. The agricultural aspect of an irrigation project involves the timely and systematic application of irrigation water to the agricultural fields by choosing suitable irrigation methods. In addition to the application of water to the fields the agricultural aspect also involves the following factors:

(i) Proper levelling and shaping of the agricultural fields.

(ii) Soil investigation and classification of the agricultural land.

(iii) Consolidation of the small and scattered holdings of the farmers.

(iv) Provision of field channels.

(v) Choosing the proper cropping pattern to suit the local soil and climatic conditions.

(vi) Introduction of improved hybrid and high-yielding varieties of crops.

(vii) Assessment of the water requirement of the crops and distribution of water according to the needs of the crops grown.

(viii) Conservation of soil against erosion.

(ix) Provision of drainage system to control the water logging of the agricultural land.

(x) Reclamation of saline and alkaline land to make it suitable for cultivation.

(xi) Plan the development of ancillary inputs such as fertilizers.

All the above noted factors of the agricultural aspect of an irrigation project are included in the general development of the command area (or ayacut) of the project. The concept of the command area development is relatively new in India, but it is now generally accepted as a necessary feature in all the major and medium sized irrigation projects. It aims at the overall development of the command areas of the irrigation projects and thus it embodies the concept of total planning. Prior to the introduction of this concept the responsibility of an irrigation engineer was limited to the maintenance and upkeep of the various irrigation structures such as headworks, canals etc., the supply and distribution of water at outlets and the assessment and collection of the water charges from the farmers. However, with the introduction of the command area development (CAD) concept of total planning, the scope of the irrigation engineering has considerably widened. Further an irrigation engineer has also to be now well versed with the agricultural aspect in addition to the other aspects.

3. Management Aspect. The management aspect of an irrigation project deals with the successful implementation and efficient management of both the engineering as well as the agricultural aspects of the project. Some of the points to be considered in this regard are as follows.

(i) It may be emphasized that the cultivation should be carried out in a scientific manner with due control on all the inputs so as to obtain the maximum yield.

(ii) The farmers should be properly trained and educated through courses organised regularly as a part of the management.

(iii) The distribution of water to the farmers should be managed properly by using some kind of rotation system such as *wara bandi*.

(iv) In order to increase the efficiency of irrigation the agricultural land holdings of the small farmers should be consolidated. For this purpose cooperative societies of the farmers may be formed which would solve many operational problems.

(v) Proper arrangement for the measurement of water used by the farmers should be made to facilitate the collection of the water charges from the farmers. However, any other suitable method for the assessment of the water charges to be recovered from the farmers may also be adopted.

(vi) The ill effects of irrigation should be as far as possible eliminated by adopting suitable measures in the management of the project.

1.4 BENEFITS OF IRRIGATION

There are several benefits of irrigation which are narrated below.

1. Increase in Crop Yield. The yield of almost all the crops increases by providing a timely supply of an optimum quantity of water. Such controlled supply of water is possible only through irrigation. Thus, irrigation leads to increased crop yield. This has a special significance for an agricultural country like India for which, with the increase in the crop yield, it would be possible to achieve self sufficiency in the production of food grains.

2. Protection from Famine. The provision of adequate irrigation facilities in any region ensures protection against failure of crops or famine from droughts. However, in the regions where no irrigation facilities exist and the farmers have to depend only on rains for growing crops, the failure of rains may result in famine. In such areas during the period of famine if the construction of irrigation works is undertaken it would provide help in two ways. During the construction of the irrigation works employment is provided to a large number of people who are rendered jobless due to famine. Hence these people get some means of living and consequent relief against famine. With the construction of these works, irrigation facilities would be made available in these areas also, which would provide protection against famine in future.

3. Cultivation of Superior Crops. When regular supply of water is assured for irrigation, the cultivators can grow certain superior or high-priced crops such as wheat, sugarcane, rice, tobacco, cotton, oil seeds etc., in place of inferior or low priced crops such as bajra (spiked millet), ragi etc., which can be matured on rainfall alone. Further the cultivators also have the incentive to improve the method of cultivation and use improved seeds and manure as with an assured water supply they are assured of better crop.

4. Elimination of Mixed Cropping. In the areas where irrigation facilities do not exist, generally mixed cropping is adopted by the farmers. Mixed *cropping* means sowing of two or more crops together in the same field. The reason for adopting the mixed cropping is that if the weather conditions are not favourable to one of the crops, they may be tolerably suitable for the other, and thus the farmer gets at least some yield from the field. Thus, mixed cropping is a sort of insurance for the farmer against weather. But when irrigation facilities are made available, the main advantage of mixed cropping disappears and hence it is either totally eliminated or considerably reduced.

Mixed cropping has certain drawbacks on account of which as far as possible it should not be adopted. Since every crop requires special preparation of the field, manuring, watering etc., if two or more crops are sown together all these conditions cannot be accomplished to suit the special needs of the either crop. Moreover, at the time of harvesting inspite of enough care, the crops would get intermixed, thereby reducing the purity and the value of the product in the market.

5. General Prosperity of the Farmers. In the regions where adequate irrigation facilities are made available, two crops are generally grown and in places even three crops can be grown. Also the farmers can grow superior or high priced crops. Moreover, the value of agricultural land and the other immovable property is increased with the advent of irrigation facilities. Thus, on account of all these factors the farmers become prosperous and their living standard is raised.

6. Increase in the Wealth of the Country. From the farmers who utilise water for irrigation from any of the irrigation projects, a tax is collected by the state in lieu of the benefits derived by them from these projects. Thus, almost all the irrigation projects bring revenue to the state. The revenue so collected by the state may be utilised either for some new irrigation projects or for other development works which in general lead to increase in the wealth of the country. Moreover, the self sufficiency achieved by the country in the production of food grains due to the provision of the irrigation facilities, would save a lot of foreign exchange otherwise required for importing food grains.

7. Generation of Hydro-electric Power. Major river valley projects are usually planned to provide hydroelectric power together with irrigation. However, relatively small quantity of hydro-electric power may also be generated at comparatively small cost on projects, which are primarily planned for irrigation alone. Such generation of hydro-electric power is possible at falls on the irrigation canals. For example, Ganga and Sharda canals in Uttar Pradesh, initially constructed for irrigation are now having a generating capacity of about 80,000 kilo-watts of hydro-electric power.

8. Domestic and Industrial Water Supply. The irrigation canals may also be utilised for domestic and industrial water supply for the nearby areas where no other source of water supply exists. The canals also provide facilities for bathing, cattle watering, swimming, boating and other recreations.

9. Inland Navigation. The network of irrigation canals can also be used for inland navigation. These inland water-ways may constitute the most economical means for transportation of goods as well as human beings. This would supplement the other means of transportation.

10. Additional Facilities of Communication. Almost all the irrigation canals are provided with unsurfaced roads primarily for purposes of inspection and maintenance. These roads though having an earthen surface, provide a good pathway to the villagers for walking, cycling, riding and sometimes even for motoring. In some cases these are the only roads available for the residents of nearby small villages. Further, sometimes even the embankments of the canals may also be used as pathways.

11. Canal Plantations. Different types of trees are usually planted along the canal banks and also in any open land available near the canal. These trees would provide timber and other associated products and also check soil erosion.

12. Increase in Ground Water Storage. In the areas where irrigation facilities are provided, due to constant percolation of a portion of water flowing in the canals and also that is supplied to the field, the ground water storage is increased and consequently the ground water table is raised. Such increase in the ground water storage is however beneficial only so long as the water table remains well below the root zone of the crops normally grown.

13. Overall Development of the Country. The provision of irrigation facilities leads to an overall development of the country. This is of special significance for an agricultural country like India in which majority of people live in villages and has agriculture as their main profession. The supply of irrigation water leads to improved agricultural pattern, increased crop yield and increased employment; and thus brings about significant changes in the social and economic pattern of society. Moreover, with the increase in the agricultural revenue of the state, several industries may be set up and also other development works may be undertaken to provide better amenities for its people.

1.5 ILL-EFFECTS OF IRRIGATION

If the irrigation water is used judiciously with proper scientific consideration then there would be no ill-effects of irrigation. It is only the excess irrigation and unscientific use of irrigation water that may give rise to the following ill-effects.

1. Waterlogging. The abundant supply of irrigation water tempts the cultivators to use more water than the requirement of the crop grown. The excess water supplied to the field would percolate into the soil. Due to constant percolation of large quantity of water, the ground water table would be raised so much that it may completely saturate the root zone of the crops grown. With complete saturation of the soil pores, the normal circulation of air in the root zone of the crops is cut off. The agricultural land rendered in this state is said to be waterlogged and this phenomenon is known as waterlogging. The fertility of a waterlogged soil is lost resulting in a considerably reduced crop yield. Further in course of time such land turns saline or alkaline and ultimately becomes unfit for cultivation.

2. Breeding Places for Mosquitoes. The excess application of irrigation water may also lead to the formation of stagnant pools of water in the pits and depressions existing in the area. These pools of water would be breeding places for mosquitoes, which may result in the outbreak of malaria in the area.

3. Unhealthy Climate. Due to intense irrigation the climate of the area becomes quite cold and damp, which is an unhealthy climate for the human beings. In such a climate the body resistance to disease is decreased. Further such a climate is susceptible to outbreak of various diseases.

4. Pollution of River Flows and Groundwater. About one-half to two-third of the water applied on the irrigated lands is lost to the atmosphere by evapotranspiration and the remaining water appears as irrigation return flow. The return flow will either be surface runoff and shallow horizontal subsurface flow which may join a river, or will move vertically through the soil and may join the groundwater reservoir. The return flow has a higher concentration of salts including nitrates and phosphates because it contains most of the salts left behind by the water returned to atmosphere as vapour through the phenomena of evaporation and transpiration. Moreover several harmful toxic chemicals used as pesticides are contained by dissolution in the return flow. Also as the water moves through the soil, it may pick up additional salts from the soil by dissolution. Due to the addition of such a contaminated irrigation return flow the water in a river or groundwater reservoir may be polluted and rendered less fit for irrigation, drinking purposes and other beneficial uses. However, the return flow will affect the receiving water in proportion to its discharge and the relative quality of the receiving water.

From the study of ill-effects of irrigation it is obvious that these are mainly the side effects of an incomplete planning of an irrigation project. As such all the ill-effects of irrigation may either be altogether eliminated or effectively controlled by exercising economical use of water and also simultaneously developing all the aspects of an irrigation project.

1.6 HISTORY OF IRRIGATION DEVELOPMENT IN INDIA

Irrigation is an age-old art, as old as civilization. In India also the irrigation has been practised since pre-historic times. Historical records bear testimony to the existence of a number of old irrigation works in different parts of the country. The earliest form of irrigation practised in India was flood or inundation irrigation, mostly in the flood plains of rivers like the Indus and the Ganga and their tributaries in North India and the deltaic tracts of various rivers of Central and South India. In the southern peninsula where the country is mostly undulating, large tanks were built from very early times for collection of rain water during the rainy season to be used later on for irrigation purposes during the dry season. Further large number of wells have also been in use for irrigation since ancient times in various parts of the country.

Ancient Hindu rulers as well as the Muslims rulers took keen interest in the construction of irrigation works. Numerous references to wells of different types, canals, tanks and reservoirs are found in old records and also ruins of ancient irrigation works mostly tanks and reservoirs are found scattered all over India. Some of these ancient irrigation works which have been properly maintained are still in use.

With the advent of British rule irrigation attained great importance. The first efforts of irrigation development under the British rule began in the beginning of the nineteenth century (1800–1836), which were directed towards the improvement and utilization of old indigenous works, rather than the construction of new irrigation works. Three important irrigation works—Western Yamuna Canal, Eastern Yamuna Canal and Cauvery Delta System were renovated and opened for use. Encouraged by these three canal systems the British rulers ventured on new major irrigation projects and during the middle of the nineteenth century (1836–1866) four important irrigation works—Upper Ganga Canal, Upper Bari Doab Canal and the Krishna and the Godavari Delta systems—were taken up. The other irrigation works constructed during the 19th century are Agra canal, Betwa canal, Periyar canal system, Khadakwasla Storage Dam, Sone canal system etc.

The beginning of the 20th century was marked by an important event in the history of irrigation in India, namely the appointment of the First Indian Irrigation Commission. Subsequent to the submission of the report of the First Irrigation Commission in the early 20th century some more projects such as Godavari Canal, Sarda Canal, Krishnaraja Sagar, Nizam Sagar and a few others were taken up. These irrigation projects are in addition to the number of smaller irrigation works taken up all over the country to irrigate small areas.

In the early forties just before independence undivided India was a large irrigated country in the world. With partition nearly one third of the irrigated area of the country went to Pakistan. Thus, at the time of Independence the total gross irrigated area that remained in India after partition was 22.6 million hectares. Soon after independence India was facing severe food shortage and hence much attention was paid to irrigation to increase agricultural production in the plans that were formulated. Accordingly a large number of irrigation and multi-purpose projects were taken up during the various Five Year Plans. Some of the major projects taken up since Independence are Bhakra–Nangal, Gandak, Hirakud, Chambal, Nagarjunasagar, Damodar Valley, Koshi, Mahi, Indira Gandhi Nahar, Ramganga etc. With the completion of the various projects the gross irrigated area has increased from 22.6 million hectares at the beginning of 1951 to 63.4 million hectares upto the end of 1983. Several other projects are in progress and with their completion more and more area will be provided with the irrigation facilities.

1.7 TYPES OF IRRIGATION SYSYTEMS

The various irrigation systems may be classified in different ways as indicated below.

1. Classification of Irrigation Systems on the Basis of the Availability of Irrigation Water. On the basis of the availability of the irrigation water, the irrigation systems may be classified in the following two classes.

(i) Flow irrigation system.

(ii) Lift irrigation system.

Flow irrigation system is that system of irrigation in which the irrigation water is available at such a level that it is conveyed to the land to be irrigated by gravity flow. The irrigation practised in this system of irrigation is known *as flow irrigation.* The flow irrigation system can be further classified on the basis of the source from which the irrigation water is obtained as follows.

(a) Direct irrigation system.

(b) Reservoir or Tank or Storage irrigation system.

In *direct irrigation system,* the irrigation water is obtained directly from a river without creating any storage of water. Thus, in this system of irrigation only a weir or a barrage is constructed across a river to raise the level of water in the river and facilitate its diversion to irrigation canals.

In *reservoir or tank or storage irrigation system,* the irrigation water is obtained through canals from a storage reservoir or a tank created by constructing a dam or a bund across a river.

Lift irrigation system is that system of irrigation in which the irrigation water is available at a level lower than that of the land to be irrigated, and hence the water is lifted up by pumps or other mechanical devices for lifting water and then conveyed to the land to be irrigated by gravity flow. The irrigation practised in this system of irrigation is known as *lift irrigation.* Irrigation from wells is an example of lift irrigation system. In some cases the irrigation water may also be required to be lifted up from a canal or any other source of water when the level of water is lower than that of the land to be irrigated.

2. Classification of Irrigation Systems on the Basis of the duration of Irrigation. On the basis of the duration of the irrigation practised, the irrigation systems may be classified in the following two classes.

(i) Inundation irrigation system.

(ii) Perennial irrigation system.

Inundation irrigation system is that system of irrigation in which a large quantity of water flowing in a river during floods is allowed to flood or inundate the land to be cultivated, thereby causing a thorough saturation of the land. The excess water is then drained off and the land is used for cultivation. Generally the moisture stored in the soil by inundation is sufficient to bring the crop to maturity, but in some cases it may be required to be supplemented by occasional minor waterings. The inundation irrigation system is therefore basically a flow irrigation system. This method of irrigation of inundating the land during floods, as practised in an inundation irrigation system, is known as *inundation irrigation* or *flood irrigation.* The inundation irrigation is commonly practised in delta regions where during floods the water may overflow the river banks and flood the adjoining land. However, the inundation irrigation may also be practised by conveying water from a river during floods, through artificially built inundation canals, to the land to be inundated.

Perennial irrigation system is that system of irrigation in which irrigation water is supplied as per the crop requirement at regular intervals, throughout the period from sowing to harvesting of the crop. The irrigation practised in a perennial irrigation system is known as *perennial irrigation.* The water for this system of irrigation may be obtained from rivers or wells or any other perennial sources of water. As such a perennial irrigation system may be a flow irrigation system or a lift irrigation system or a combination of both these systems. The perennial irrigation system is the most commonly adopted irrigation system.

REVIEW QUESTIONS

1.1 Define irrigation.

1.2 Enumerate and explain the factors which necessitate irrigation.

1.3 What do you understand by total planning concept of irrigation project?

1.4 Discuss briefly the benefits as well as the ill effects of irrigation.

1.5 Write a brief note on 'History of development of irrigation in India'.

1.6 What are the different types of irrigation systems ? Discuss each of these systems briefly.

1.7 Explain the terms

(i) Flow irrigation

(ii) Lift irrigation

(iii) Inundation irrigation; and

(iv) Perennial irrigation.

CHAPTER

2 Method of Irrigation

2.1 CLASSIFICATION OF IRRIGATION METHODS

Irrigation methods are commonly designated according to the manner in which water is applied to the land (or field) to be irrigated. The water may be applied to the land to be irrigated either by spreading it on the land surface, or by spraying it over the land surface, or by applying it beneath the land surface. As such the various methods of irrigation may be classified in the following three categories.

(i) Surface irrigation methods.

(ii) Sprinkler irrigation methods.

(iii) Sub-irrigation methods, or sub-surface irrigation methods.

All these methods of irrigation are in general used for the perennial irrigation system. Further each of the methods of application of water to the land to be irrigated represents the type of irrigation practised. The choice of the method of irrigation is affected by several factors as indicated in the next section which is followed by the sections in which the various methods of irrigation are described.

2.2 FACTORS AFFECTING THE CHOICE OF THE METHOD OF IRRIGATION

Certain methods of irrigation are of general applicability and are used in almost all irrigation regions, but there are other methods of limited applicapability which can be used only when special conditions prevail. As such a proper selection of the method of irrigation is very essential and is based on the following factors.

(i) Soil characteristics of the land to be irrigated.

(ii) Topography of the country—slope of land surface, roughness of the surface, etc.

(iii) Size of the stream supplying irrigation water to the land to be irrigated.

(iv) Available water supplies and the rate of advance of irrigating water.

(v) Length of run and time required for wetting the total area of the land to be irrigated.

(vi) The water requirements of the crops grown and the growth habits of the plants.

(vii) Rate of infiltration of the soil.

(viii) Depth of the root zone of the plants.

(ix) Depth of the water table.

(x) Possible erosion hazard.

(xi) Amount of water to be applied during each irrigation.

Further the method of irrigation should be so selected that the following objectives are fulfilled while providing water to the land to be irrigated.

(i) Adequate amount of water is stored in the root zone of the plants.

(ii) Uniform application of water is made possible.

(iii) As far as possible minimum soil erosion takes place.

(iv) There is minimum wastage of water.

(v) Reuse of water is made possible.

(vi) Minimum land is utilized for field channels, borders, etc., so that as far as possible maximum land is made available for cultivation.

(vii) The method properly fits to the boundaries of the land to be irrigated.

2.3 SURFACE IRRIGATION METHODS

In the surface irrigation methods the irrigation water is applied by spreading it in sheet or small streams on the land to be irrigated. The various methods of surface irrigation are classified as explained below.

Surface Irrigation Methods

- Flooding methods
  - Uncontrolled or wild flooding
  - Controlled flooding
    - Flooding from field channels
    - Border strip method
    - Check basin method
    - Ring basin method
- Furrow method
- Contour farming

All the above noted methods of surface irrigation are adopted for the perennial irrigation system. The inundation irrigation system however employs only the uncontrolled or wild flooding method of irrigation. Brief descriptions of the above noted surface irrigation methods are given below.

2.3.1 Uncontrolled or Wild Flooding

In this method of irrigation water is applied by spreading it over the land to be irrigated without any prior preparation of the land and without enforcing any control in the form of levees etc. to guide the flow of water or otherwise restrict the movement on the land. During the season of high stream flow, water is brought from the natural stream or river to the field in permanent supply channels and it is allowed to spread on the land along the natural slope of the land without any control. Alternatively the water from the supply channels may be distributed to field channels built across the field from which water is allowed to spread on the land without any control. This method of irrigation is suitable for smooth and flat land for which no prior preparation of the land would normally be required and also better water application efficiency would be achieved. Moreover, this method of irrigation is well adapted to the lands that have such irregular surfaces that the other flooding methods are impractical. However, this method of irrigation involves a wasteful use of water and hence it is practised only where water is available for irrigation in abundance and it is inexpensive.

2.3.2 Controlled Flooding

In controlled flooding the irrigation water is applied by spreading it over the land to be irrigated with proper control being enforced on the flow of water as well as on the quantity of water applied. Further for all the methods of irrigation by controlled flooding prior preparation of the land is essential. Each of these methods of irrigation by controlled flooding is described below.

2.3.3 Flooding from Field Channels

In this method the land to be irrigated is divided into small strips by a series of field channels (or laterals) which are supplied water from the supply channels. The supply channels are located at the higher edges of the field and are aligned along the general slope of the land. The laterals may be aligned either along the contour lines; or at right angles to the contour lines; or at right angles to the sides of the land as shown in Fig. 2.1. When the laterals are aligned at right angles to the contour lines then the land on both the sides of the channels can be irrigated, but in the other two cases irrigation would be possible only on one side of the channel in the direction of the general slope of the land. Further when the laterals are aligned along the contour lines this method is known as irrigation by the contour laterals. From the laterals, water is applied to the land through the feeding points provided at some interval along their lengths. The excess water from the laterals is carried through waste channels and it may be used to irrigate the lower areas.

This method of irrigation can be used on relatively steep slopes as well as on flat lands. Further no major land preparations are required in this case but careful location and proper spacing of the laterals is essential. The proper arrangement and spacing of the laterals mainly depends on the topography of the land to be irrigated and the type of soil. Lands having steep slopes require closer spacing of laterals than those having flat slopes. Similarly porous soils require closer spacing of laterals than tight soils. Thus, in general the spacing of laterals may vary from 15 to 60 metres. However, in some cases the spacing of laterals may even be less than 15 metres or more than 60 metres. The lands best suited for irrigation by flooding from field channels are those having medium type soil and slopes in the range of 1 in 100 to 1 in 200.

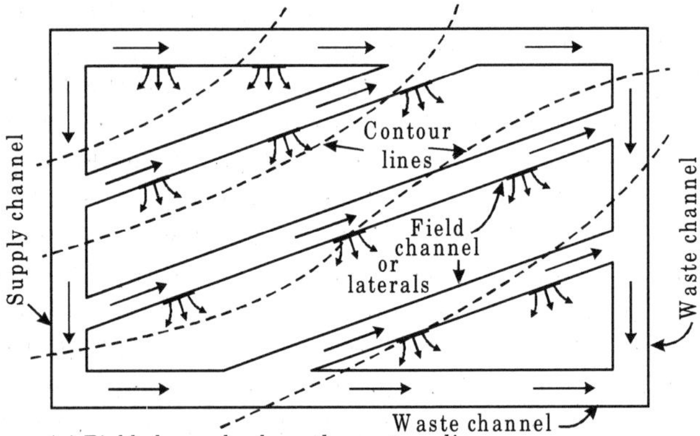

(a) Field channels along the contour lines

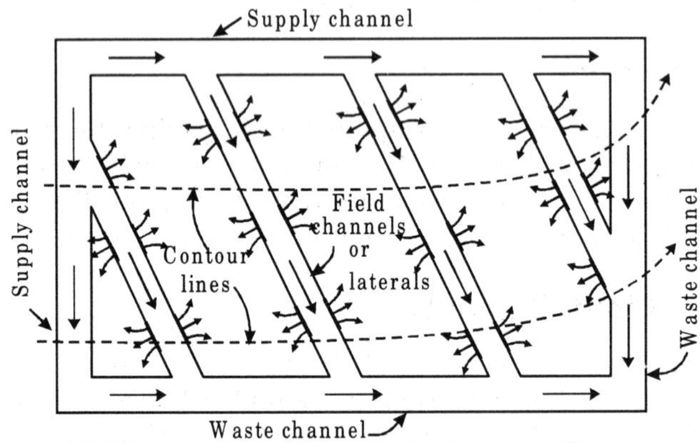

(b) Field channels at right angles to the contour lines

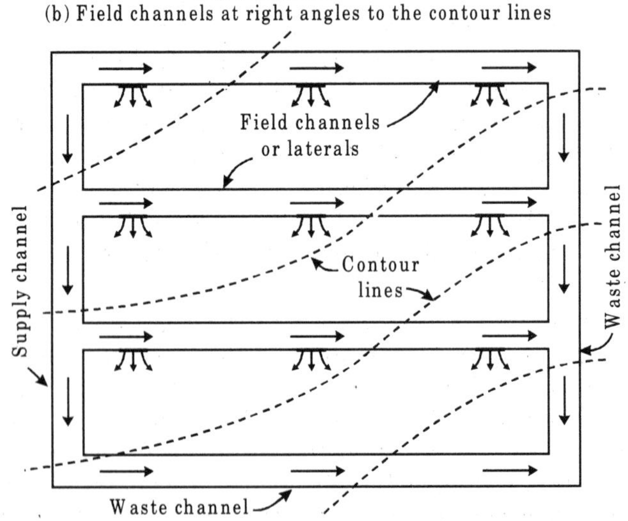

(c) Field channels at right angles to the sides of the field

Fig. 2.1 *Flooding from feld channels*

2.3.4 Border Strip Method

In the border strip method the land to be irrigated is divided into a series of long narrow strips separated from each other by low levees or borders (low flat dikes). Each strip is irrigated independently by supplying water at its upper end from a supply channel or an underground pipe. The strips have a uniform gentle slope in the longitudinal direction but have no cross slope (i.e., slope along the width of the strip), so that the irrigation water applied to each strip is uniformly spread over its entire width without getting accumulated on either side as it flows down the slope. On a relatively flat land the strips are laid along the general slope of the land

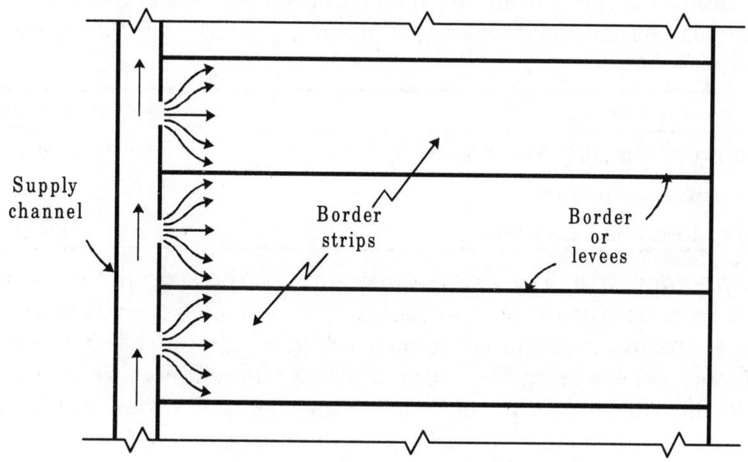

(a) Border strips with supply channel

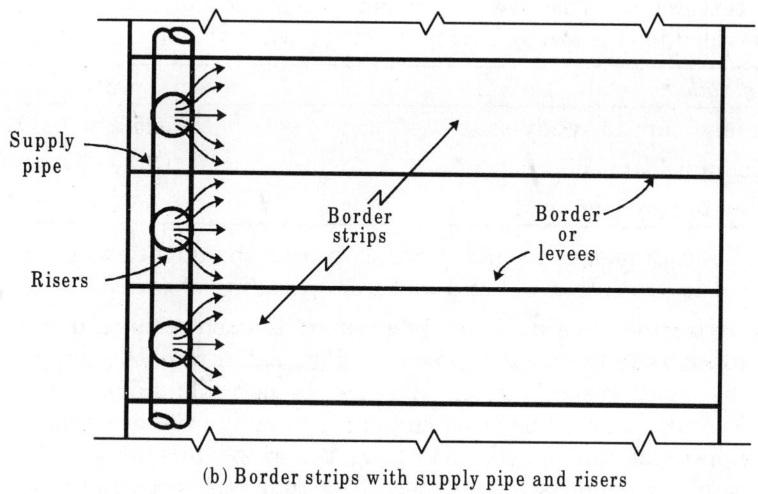

(b) Border strips with supply pipe and risers

Fig. 2.2 *Border strip method*

and hence these are straight and parallel to each other. However, when the land slopes are excessive the strips are laid across the general slope of the land and these are aligned to approximately follow the different contours at a vertical interval of 0.30 to 0.60 metres. Thus, in the later case the strips are curved and are in the form of series of steps formed in successive elevations around the slope. These strips are usually known as contour strips to distinguish them from the straight strips of the former case. The design criteria of the strips is same for both the cases which is as follows.

(i) *Width of border strip.* The width of the border strip depends on the size of irrigation stream and degree of land levelling practicable and it usually varies from 3 to 15 metres. When the size of irrigation stream available is small, the width of the strip is reduced, but it is usually not less than 3 metres because with the reduction in the width of the strip too many levees will have to be formed which will occupy considerable area of the land.

(ii) *Length of border strip.* The length of the strip depends on the infiltration rate of the soil, the slope of the land and the size of the irrigation stream available. For moderate slopes and small to moderate size irrigation streams, the following lengths of the strips may be adopted for different types of soils.

Type of soil		*Length of strip*
(i) Sandy and sandy loam soils	:	60 to 120 metres
(ii) Medium loam soils	:	100 to 120 metres
(iii) Clay loam and clay soils	:	150 to 300 metres

(iii) *Slope of border strip.* The longitudinal slope of the strip mainly depends on the type of soil and it should be neither excessive nor too flat. This is so because excessive slopes will make the water to flow to the lower end of the strip quickly, causing insufficient irrigation at the upper reaches, deep percolation losses at the lower reaches and soil erosion in the strip. On the other hand too flat slopes will result in very slow movement of water in the strip, causing deep percolation losses at the upper reaches and inadequate irrigation at the lower reaches. In general for course textured soils with high infiltration rates relatively steeper slopes may be provided as compared to fine textured soil with low infiltration rates. The safe limits of the longitudinal slopes which may be provided for the strips having different types of soil are as follows:

Type of soil		*Slope*
(i) Sandy loam to sandy soils	:	0.25% to 0.60%
(ii) Medium loam soils	:	0.20% to 0.40%
(iii) Clay to clay loam soil	:	0.05% to 0.20%

For the initial 6 to 12 metres length of the strip from the upper end the slope is not provided but it is made level to promote rapid and uniform spreading of water.

(iv) *Size of irrigation stream.* Water is supplied to each strip at upper end from a supply channel or an underground pipe as shown in Fig. 2.2. From the supply channel water is discharged into the strips through gated openings or portable siphons of suitable material or plastic piping. Generally it would be possible to irrigate only from one side of the supply channel but when the supply channel is aligned along the ridge line or if it is aligned along the predominant slope of the land as in the case of contour strips it can be used for irrigation on both the sides. From the underground pipe water is discharged into the strips through risers provided with light weight portable pipes. The risers are so located that each riser can irrigate two strips or more on either side of the pipe line. The water flows slowly along the strip towards its lower end in the form of a sheet of water confined by levees without overtopping them and it infiltrates into the soil as it advances. When the advancing water front either reaches the lower end of the strip or a few minutes before or after that the supply of water to the strip is turned off.

The size of the irrigation stream needed depends on the infiltration rate of the soil and the width of the border strip. Coarse textured soils with high infiltration rates require large streams to spread water over the entire strip rapidly and avoid excessive losses due to deep percolation at the upper reaches. On the other hand fine textured soils with low infiltration rates require smaller streams to avoid excessive losses due to runoff at the downstream end and deep percolation at the lower reaches. The following Table gives some values of streams sizes for different types of soils and slopes:

S.No.	Soil type	Longitudinal slope (%)	Flow per metre width of strip in litres per secon
1.	Sandy soil	0.20 to 0.40	10 to 15
	(Infiltration rate 25 mm/ hour)	0.40 to 0.65	7 to 10
2.	Loamy sand	0.20 to 0.40	7 to 10
	(Infiltration rate 18 to 25 mm/hour)	0.40 to 0.60	5 to 8
3.	Sandy loam	0.20 to 0.40	5 to 7
	(Infiltration rate 12 to 18 mm/hour)	0.40 to 0.60	4 to 6
4.	Clay loam	0.15 to 0.30	3 to 4
	(Infiltration rate 6 to 8 mm/hour)	0.30 to 0.40	2 to 3
5.	Clay	0.10 to 0.20	2 to 4
	(Infiltration rate 2 to 6 mm/hour)		

By varying the size of irrigation stream it is possible to vary the depth of water required to be applied. Thus, a larger stream is used to apply water at a shallower depth and a smaller stream is used to apply water at a greater depth. This is so because the amount of water

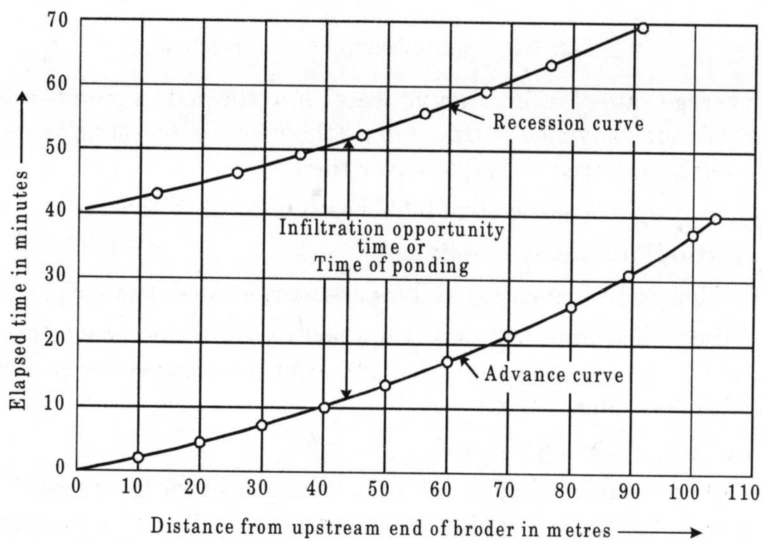

Fig. 2.3 *Advance and recession curves obtained in the field test of a border strip*

entering the soil is related to infiltration opportunity time or time of ponding. The *infiltration opportunity time* is the difference between the time the water front reaches a particular point along the border strip and the time at which the tail water recedes from the same point when the supply of water to the strip is turned off. Thus, infiltration opportunity time at any point along the border strip is the vertical distance (in time scale) between the advance and recession curves as shown in Fig. 2.3. If the advance and recession curves are parallel to each other than uniform distribution of water throughout the border strip is ensured.

2.3.5 Design of Border Strip Irrigation

Consider a border strip as shown in Fig. 2.4 which is irrigated by flooding a thin sheet of water over it. Let q be the quantity of water supplied to the strip in cumec per unit width of the strip; A be the area of the strip in m^2 covered with water in time t seconds after the supply was

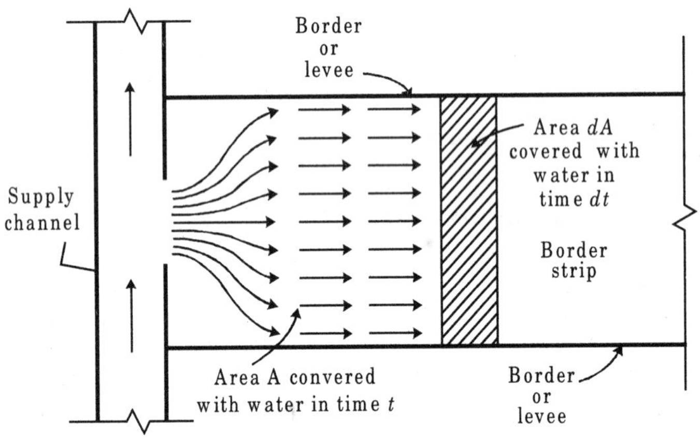

Fig. 2.4 *Design of border strip irrigation*

started; I be the average rate of infiltration of water into the soil in metres per second; x be the distance in metres of water advance in time t; b be the width of the strip in metres and y be the average depth of water in metres as it flows over the strip.

The water that is supplied to the strip is disposed of in the following two ways:

(i) Part of it percolates into the soil; and

(ii) Part of it flows down the strip and covers more area of the strip.

During time t the total volume of water supplied to the strip is (qbt); the volume of water that percolates into the soil over the area A is (IAt); and the volume of water that flows to cover the area A is (AY). It thus follows that

$$qbt = IAt + AY \qquad ...(i)$$

Further as shown in Fig. 2.4 if in time dt the water advances over the area dA, then during time $(t + dt)$ the total volume of water supplied to the strip is $qb\ (t + dt)$; the volume of water that percolates into the soil over the area $(A + dA)$ is $I\ (A + dA)\ (t + dt)$; and the volume of water that flows to cover the area $(A + dA)$ is $(A + dA)y$. It, thus, follows that

$$qb(t + dt) = I(A + dA)(t + dt) + (A + dA)Y$$

which may be approximately expressed as

$$qb(t + dt) = IA(t + dt) + (A + dA)Y \quad ...(ii)$$

By subtracting Eq. (i) from Eq (ii) we obtain

$$qbdt = IAdt + ydA \quad ...(iii)$$

Since $A = (bx)$ and $dA = (bdx)$, by substitution in Eq. (iii) we get

$$qbdt = Ibxdt + ybdx$$

or

$$qdt = Ixdt + ydx$$

or

$$dt = \frac{ydx}{q - Ix} \quad ...(iv)$$

Considering y, q and I as constant and integrating the above Eq. (iv) we get

$$t = -\frac{y}{I}\log_e(q - Ix) + C$$

Since at $t = 0, x = 0$

$$C = \frac{y}{I}\log_e q$$

$\therefore$

$$t = -\frac{y}{I}\left[\log_e(q - Ix) - \log_e q\right]$$

or

$$t = -\frac{y}{I}\log_e\left(\frac{q - Ix}{q}\right)$$

or

$$t = \frac{y}{I}\log_e\left(\frac{q}{q - Ix}\right) \quad ...(2.1)$$

If Q is the discharge in cumec for the entire strip, then since $Q = qb$ and $A = bx$, Eq. 2.1 may be expressed as

$$t = -\frac{y}{I}\log_e\left(\frac{Q}{Q - IA}\right) \quad ...(2.1\text{ a})$$

or

$$t = 2.303\,\frac{y}{I}\log_{10}\left(\frac{Q}{Q - IA}\right) \quad ...(2.1\text{ b})$$

Let

$$\log_{10}\left(\frac{Q}{Q - IA}\right) = \phi = \frac{It}{2.303y}$$

Then

$$A = \frac{\left(10^{\phi} - 1\right)Q}{10^{\phi} I} \quad ...(2.1\text{ c})$$

If it is considered that

$$\frac{10^{\phi}-1}{10^{\phi}} \simeq 1$$

Then $$A = \frac{Q}{I};$$

or $$x = \frac{q}{I} \text{ or } x = \frac{q}{I} \qquad ...(2.1\text{ d})$$

By introducing Eq. 2.1(d) in Eq. 2.1 or 2.1 (a) or 2.1 (b), we get $t = \infty$ (infinity). Further Eqs. 2.1 and 2. 1 (a) or 2.1 (b) cannot be solved for a value of $x > \frac{q}{I}$ or $A > \frac{Q}{I}$, and it is only for a value of $x < \frac{q}{I}$ or $A < \frac{Q}{I}$ that Eqs. (2.1) and (2.1 a) or (2.1 b) give a finite value of t. It may therefore be concluded that Eq. (2.1 d) gives a maximum value of the area or the length of the strip having rate of infiltration I, which can be irrigated by a stream of discharge Q for the entire strip or q per unit width of the strip.

From Eq. 2.1 the following expression for x may be obtained

$$x = \frac{q}{I}\left[1-e^{(D/y)}\right] \qquad ...(2.2)$$

Further if t is the time required for the infiltration of water to the desired depth D of the root zone, then

$$D = It \qquad ...(2.3)$$

By introducing Eq. (2.3) in Eq. (2.2) we get

$$x = \frac{q}{I}\left[1-e^{(D/y)}\right] \qquad ...(2.4)$$

From Eq. (2.4) for the known values of I, q, D and y and length of run of border x can be determined, and for the known values of x, I, D and y the flow rate q can be determined.

Average depth of surface flow y

The average depth of flow y is considered to be equal to a constant times the normal or uniform depth y_0. The value of the constant is found to be 0.665 and hence

$$y = 0.665\, y_0 \qquad ...(2.5)$$

Further from Manning's equation the value of the normal or uniform depth y_0 is given as

$$y_0 = \left(\frac{nq}{\sqrt{S}}\right)^{0.6} \qquad ...(2.6)$$

in which S is the slope of the border strip; and n is roughness coefficient.

The value of n to be used in Eq. (2.6) has been estimated to be given by the following empirical formula.

$$n = 0.55425 - 0.0001386q' \quad ...(2.7)$$

in which

q' = inflow per unit width or border strip in cm³/minute/cm.

Time of application of water

Knowing I and D, the time for which the irrigation is to be continued (or water is to be applied) for each strip is given by Eq. (2.3) as

$$t = \frac{D}{I} \quad ...(2.8)$$

2.3.6 Check Basin Method

It is the most common method of irrigation used in India as well as in many other countries of the world. This method is also known as the method of irrigation by plots. In this method the land to be irrigated is divided into small plots or check basins surrounded by checks or levees (or low dikes or bunds) as shown in Fig. 2.5. Each plot or basin has a nearly level surface. The irrigation water is applied by filling the plots will water upto the desired depth without overtopping the levees and the water is retained there to allow it to infiltrated into the soil. The

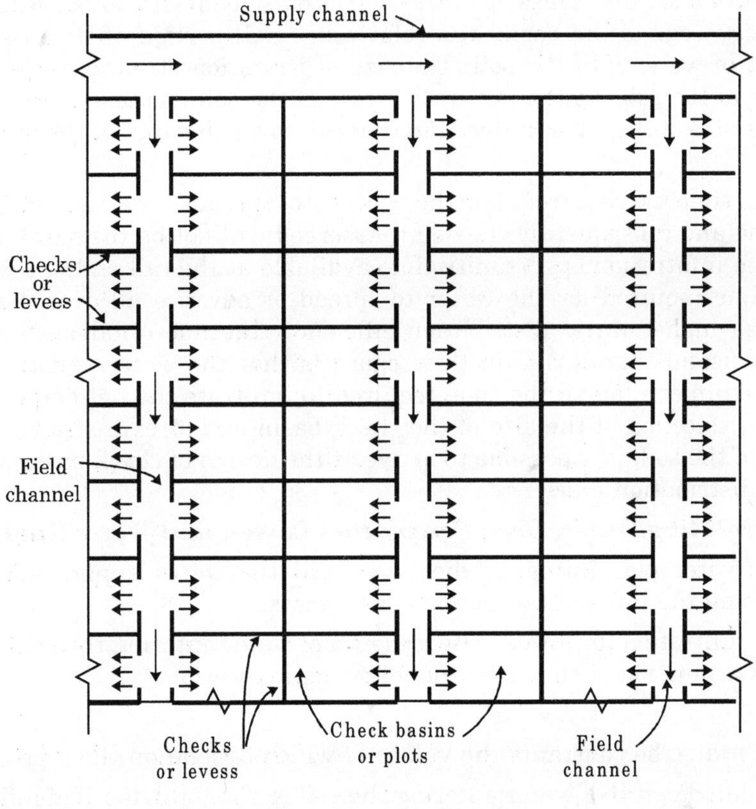

Fig. 2.5 *Check basin method*

levees may be temporary for a single irrigation as in the pre-sowing irrigation of seasonal crops, or may be for a cropping season to permit number of irrigations to be carried out as per the requirements of the crops. The levees may also be constructed semi-permanently for repeated used as in the case of paddy fields. The size of the levees depends on the depth of water to be impounded as well as on the stability of the soil when wet.

Water is conveyed to the land to be irrigated by a system of supply channels and lateral field channels. The supply channel is aligned on the upper side of the field and there is usually one lateral field channel for every two rows of plots as shown in Fig. 2.5. Water from the lateral field channel is supplied to the plots and the supply is cut-off when sufficient quantity of water has been admitted to the plots. Water is retained in the basin until it soaks into the soil.

Size and shape of basins. The size of the check basins depends on the infiltration rate of the soil. In general the size of the basins may vary from one metre square, used for growing vegetables and other intensive cultivation, to as large as one or two hectares or more, used for growing rice under wet land conditions. However, for medium soils plots of area 0.04 to 0.05 hectare may be suitable. Further oblong or rectangular plots are preferable to square plots. As such for land which is level or is very gently sloping, the basins are rectangular in shape. However, for sloping land the basins or plots are prepared by constructing the levees along contours having vertical intervals of 60 to 120 mm and connecting them with cross levees at convenient places. These are called contour levees or checks. In this case the plots or basins have some odd shapes depending on the configuration of the contours of the land.

Size of irrigation stream. The size of the irrigation stream should be such that it is sufficient to provide the coverage of the basin in a relatively small portion of time required to apply the desired amount of water into the soil. The size of irrigation stream, however, depends on the infiltration rate of the soil. As the infiltration rate of the soil increases, the stream size must be increased or the size of the basin reduced in order to cover the area of the entire basin in a short period of time.

Since some time is required for the water to spread over the basin, the infiltration opportunity time and consequently the depth absorbed will not be the same over the entire area of the basin. The infiltration opportunity time available at the tail end of the basin is less to an extent of the time required for the water to spread or advance all over the basin. As such in order to attain a higher water distribution efficiency the time of advance should be as far as possible less. A useful thumb rule in this regard is that the water spread in the entire basin should be covered in one fourth the time required to infiltrate the net depth of irrigation water. Thus, by careful selection of the size of the check basin and stream size to suit the infiltration characteristics of the soil, it is possible to arrive at the design of check basin which will give high value of water distribution efficiency.

Volume of Percolating Water, Deep Percolation Losses, and Water Application Efficiency

The volume of water percolating in the basin and the water application efficiency may be calculated by using the infiltration functions as follows.

The rate of infiltration (or intake rate or depth of water entering the soil per unit time) I can be expressed as a function of time t by the following expression.

$$I = Kt^n \qquad \text{...(i)}$$

in which K and n are the constants the values of which depend on the type of soil.

The accumulated depth of water entering the soil or the total depth of infiltration during the time of ponding t_d is then given by

$$I_c = \int_0^{t_d} Kt^n dt = \frac{Kt_d^{(n+1)}}{(b+1)} \quad \text{...(ii)}$$

If the volume of water absorbed during the time of advance is neglected then the total volume of water absorbed over a length L for a unit width of check basin is given by

$$V_1 = \frac{K}{(n+1)}\int_0^L t_d^{n+1} dL = \frac{KLt_d^{(n+1)}}{(n+1)} \quad \text{...(iii)}$$

Further if t_L is the time of advance or the time to cover the entire length L of the basin then the total time T that elapsed since the irrigation started may be expressed as

$$T = (t_L + t_d)$$

The time of advance t_L may be determined by the following expression in which a and b are the constants determined for a particular stream size and soil characteristics.

$$L = at_L^b \quad \text{...(iv)}$$

The total volume of water V_2 absorbed over a length L for a unit width of check basin during the total time T may be obtained as

$$V_2 = \frac{KLT^{(n+1)}}{(n+1)} \quad \text{...(v)}$$

The difference between V_2 and V_1 may be considered as the loss of water due to deep percolation (or deep percolation loss) which is usually expressed as a fraction of the total volume of water as

$$P = \frac{V_2 - V_1}{V_2}$$

or

$$P = \frac{T^{n+1} - t_d^{n+1}}{T^{n+1}}$$

or

$$P = \left[1 - \left(\frac{t_d}{T}\right)^{n+1}\right] \quad \text{...(2.9)}$$

The water application efficiency η_a may thus be expressed as

$$\eta_a = \frac{V_1}{V_2} \times 100 = \left(\frac{t_d}{T}\right)^{n+1} \times 100 \quad \text{...(2.10)}$$

or

$$\eta_a = \left(\frac{t_d}{t_L + t_d}\right)^{n+1} \times 100 \quad \text{...(2.11)}$$

or

$$\eta_a = (1 - P) \times 100 \quad \text{...(2.12)}$$

The expression derived for percolation losses can be used for the design of the check system. For a given stream size and plot size the percolation losses in percentage may be calculated. If these losses are high either the size of the check basin is to be reduced or the size of the irrigation stream is to be increased. However, smaller size of plots will involve more ridges and channels resulting in wastage of land. On the other hand larger stream size may cause the erosion of soil. Thus, a compromise between the area of land occupied by the ridges and channels and soil erosion has to be made to obtain the most suitable design.

Further knowing the percolation losses (or water application efficiency), size of plot and depth of water to be applied, the total volume of water to be applied can be calculated and knowing the stream size the total time of irrigation can be calculated as expressed below.

Volume of water to be applied

$$= \frac{d \times L \times B}{\eta_a} \qquad ...(2.13)$$

Time of irrigation

$$= \frac{d \times L \times B}{\eta_a \times Q} \qquad ...(2.14)$$

where d = depth of water to be applied

L = length of the basin

B = width of the basin

η_a = water application efficiency; and

Q = flow rate (or size) of irrigation stream.

2.3.7 Ring Basin Method

The ring basin method of irrigation is a special form of check basin method of irrigation, which is used for the irrigation of orchards (enclosures with fruit trees). In this method generally for each tree a separate basin is made which is usually circular in shape and hence it is known as

Fig. 2.6 *Ring basin method*

ring basin (Fig. 2.6). However, in some cases basins of larger size are made to include two to five or more trees in one basin, water is supplied to the basins from a supply channel through small field channels connecting the basins with the supply channel. In most of the cases each basin is provided with a separate field channel so that water is directly supplied to each basin from the supply channel. However, in some cases all the basins are interconnected so that from the supply channel water is conveyed to one of the basins and by flowing from one basin into another the water is supplied to all the basins.

2.3.8 Furrow Method or Furrow Irrigation

In the furrow method of irrigation water is applied to the land to be irrigated by a series of long, narrow field channels called furrows which are dug in the land at regular intervals as shown in Fig. 2.7. The water flowing in the furrows infiltrates into the soil and spreads laterally to

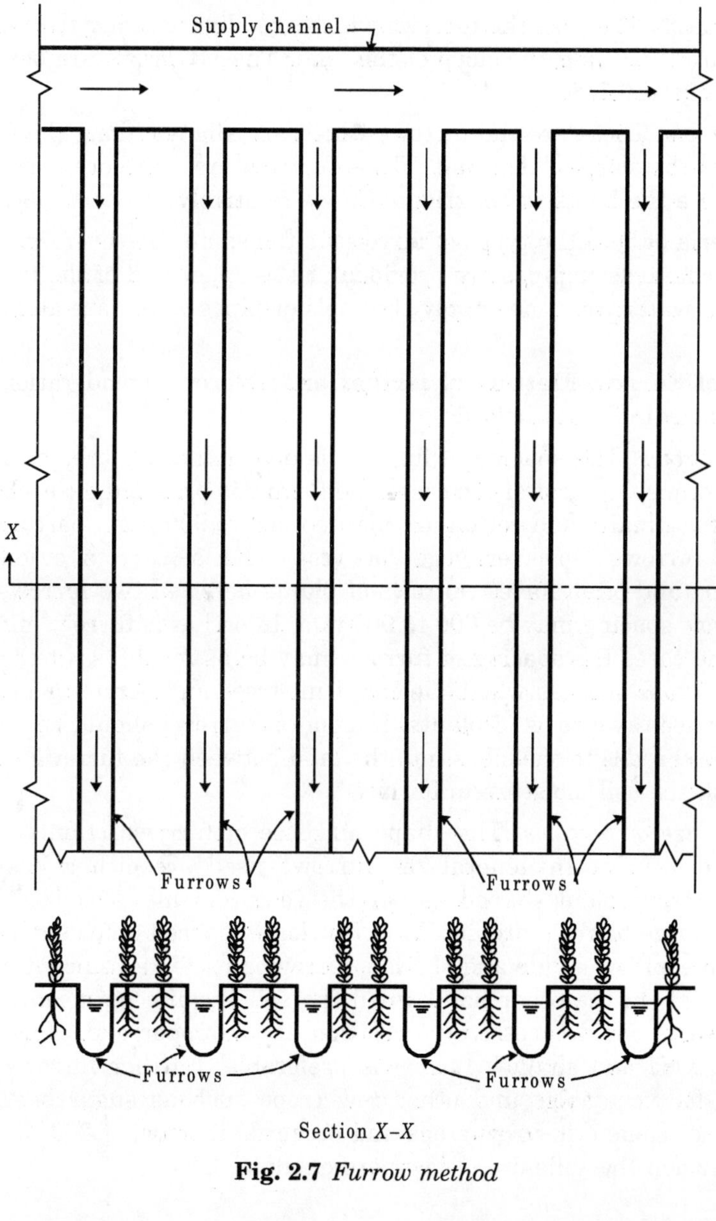

Fig. 2.7 *Furrow method*

irrigate the land between the furrows. Thus, whereas in each of the methods of irrigation by flooding described earlier almost the entire land is wetted, in the furrow method only a part of the land varying from one-half to one-fifth is wetted which results in reducing the evaporation losses. The furrow method of irrigation is commonly used for row crops such as maize, cotton, potatoes, sugar cane, sugar beat, groundnut, tobacco etc.

Types of Furrows. The furrows may be classified on the basis of their alignment into the following two types:

(i) Straight furrows

(ii) Contour furrows

(i) *Straight furrows.* These are the furrows which are aligned along the more or less parallel straight lines and are laid along the slope of the land. These furrows are best suited where the land slopes do not exceed 0.75%.

(ii) *Contour furrows.* These are the furrows which are aligned along the contours and hence these are laid across the slope of the land. These furrows are therefore curved in plan and are usually adopted where the lands to be irrigated have relatively steeper slopes.

The design criteria of both the types of furrows is the same. However, in the case of straight furrows the supply channels or pipes are provided at the upper end of the field, but the contour furrows are supplied water from the supply channels or pipes which are aligned along the slope of the land.

Design Features of Furrow Method of Irrigation. Different considerations for the design of furrow irrigation are as indicated below.

(i) *Spacing of furrows.* The spacing of furrows is determined by the spacing desired for the rows of the plants because in most of the cases one furrow is provided for each row of the plants. The crops like potatoes, maize and cotton are planted 600 to 900 mm apart and have one row of plants between two furrows. However, vegetable crops such as carrots, onions, etc., are spaced 300 to 400 mm apart and often have two rows of plants between two furrows. Thus, in each of these cases the furrow spacing may be 600 to 900 mm. In orchards furrows may be spaced 1 to 2 m apart and in some cases the spacing of furrows may be as much as 3 to 4 metres. Moreover, certain wide spaced crops such as water melons, fruit trees and berries generally require more than one furrow between two rows of plants. In general furrows should be spaced close enough to ensure that water spreads to the sides into the land between the furrows and the root zone of the crop to replenish the soil moisture uniformly

(ii) *Shape and size of furrows.* The shape and size of furrows mainly depend on the soil conditions and type of crop. In general the furrows used are either V shaped, U shaped, parabolic shaped or trapezoidal shaped as per the requirements of soil stability. Further for widely spaced crops large furrows are used and for closely spaced crops small furrows are used. For soils of low permeability a wide and shallow furrow is preferred since it gives more area for the water to infiltrate. On the other hand when the soil is quite permeable, narrow and deep furrows may be used to avoid excessive percolation at the upper end. However, in general in furrow irrigation a wide and shallow furrow is preferable. Further furrows of 75 to 125 mm depth are suitable for vegetables and other row crops such as sugar beet, potatoes, cotton, tobacco, etc., while for some other row crops and orchards furrows of 200 to 300 mm depth are required especially when the soils are of low permeability.

(iii) *Furrow length.* The furrows should neither be too long nor too short. This is so because if a furrow is too long water may infiltrate to a greater depth at the upper end of the furrow by the time it reaches the lower end. This results in excessive deep percolation losses and over irrigation at the upper end of the furrow. On the other hand short furrows require supply channels to be spaced too close with consequent loss of land and increase in labour requirement. The furrow length is considerably affected by the hydraulic conductivity of the soil. Thus, furrows must be shorter for porous sandy soils than for tight clayey soils or heavy soils. The furrow length is also influenced by the slope, rate of advance and depth of application besides the type of soil. The stream size, slope and furrow length should be so adjusted that the deep percolation are a minimum. The following Table may serve as a rough guide for selecting the maximum furrow length.

Furrow slope %	*Furrow length metres*						
	Soil type						
	Clay		Loams		Sands		
	Net depth of water application, mm						
	75	150	50	100	50	75	100
0.05	300	400	120	270	60	90	150
0.10	350	440	180	330	90	120	190
0.20	370	470	220	370	120	190	250
0.30	390	500	280	400	150	225	280
0.50	380	500	280	370	120	190	250
1.00	270	400	250	300	90	150	220

From the Table it may be observed that with flat sloped upto 0.3% the furrow length increases with the increase in the furrow slope but for slopes steeper than 0.3%, furrows of shorter lengths are used. Further the furrow lengths can be increased as the average depth of water to be applied increases.

The time of advance should be such that minimum deep percolation losses are caused at the upper end of the furrow. It is thus suggested that the time taken for the stream to reach the lower end of the furrow should be limited to about one fourth of the time required to infiltrate the required depth of irrigation water.

(iv) *Furrow slope.* The slope of the furrow controls the velocity of flow of water in the furrow. Thus, steeper slopes lead to higher velocities of flow of water in the furrows which may result in more erosion of soil in the furrow. Moreover, as the furrow slope increases, the rate of infiltration slows down and the side spread of water into the land decreases, with the result that wastage of water may occur at the end of the furrow. If the furrow slopes are too less than proper surface drainage may not take place with the result that excessive deep percolation losses may occur. As such proper furrow slope needs to be provided. In general the ranges of slopes as recommended for the border strips in different soils apply to furrows also.

Further a minimum furrow slope of 0.05% is needed to ensure surface drainage. However, in soils having very low infiltration rates, the furrows are usually level without any slope being provided along their lengths. These furrows are known as level furrows in which the water is ponded until it is absorbed by the soil.

(v) *Furrow stream size.* Water is supplied to the furrows from a supply channel or a concrete pipe placed underground at the upper end of the field. From the supply channel water is supplied to the furrows either through small opening made by cutting the bank of the channel, or through small portable siphons. For siphons about 1.2 m long curved pipes of small diameter are used which are made of light weight plastic, aluminium, galvanized iron, or rubber. From the concrete pipe water is supplied to the furrows through vertical riser pipes. Sometimes aluminium pipe provided with gated openings and placed on the ground surface is used for supplying water to the furrows. Small, easily adjustable gates in the pipe facilitate control of quantity of water supplied to the furrows. A tail channel at the end of the furrows collect water for reuse at lower levels.

The stream size of the furrow should be such that it will not cause erosion and most uniform irrigation would be obtained. The size of the furrow stream usually varies from 0.5 to 2.5 litres per second. However, the maximum non-erosive flow rate in furrows is estimated by the following empirical equation

$$q_m = \frac{0.60}{s} \quad ...(2.15)$$

in which

q_m = maximum non-erosive stream size in litres per second; and

s = slope of furrow expressed as percent.

After the water has reached the lower end of the furrow, the stream size is reduced. The reduction in the stream size is known as cut back and the stream of reduced size is known as cut back stream. The cut back stream flows until the required amount of water has been applied. By this arrangement higher water application efficiency may be obtained. The stream size can be conveniently reduced by reducing the number of siphons or by reducing the size of the opening in the gated pipe.

With a stream of known size the average depth of water applied during an irrigation can be calculated from the following expression

$$d = \frac{q \times 3600 \times t}{W \times L} \quad ...(2.16)$$

in which

d = average depth of water applied in mm

q = stream size in litres per second

t = duration of irrigation (elapsed time) in hours

W = furrow spacing in metres; and

L = furrow length in metres.

Determination of Infiltration Rate and Time required to infiltrate the given depth of water. The infiltration rate or intake rate in furrows is determined either by ponding method or by the inflow-outflow method (or volume balance method). Out of the two methods the inflow-outflow method gives a correct assessment of furrow infiltration and the same is described below.

In the inflow-outflow method a 30 m reach is selected and water is allowed to flow in three consecutive furrows. The inflow and outflow rates are measured for this reach of the central furrow at different instants of time. The outer furrows are known as guard furrows as they help in obtaining a representative value of the furrow infiltration in the central furrow. Now if Q_1

and Q_2 are the inflow and outflow rates in litres per minute at any instant of time t in minutes since the flow began then the quantity of water absorbed per unit time is

$$I = (Q_1 - Q_2) \text{ litres/minute} \quad \text{...(i)}$$

per 30 m length of furrow.

Different values of I are obtained for different values of t and the same are plotted on log-log scale, which gives a straight line plot. From this plot the following equation may be obtained for the infiltration rate

$$I = Kt^n \quad \text{...(ii)}$$

in which

I is infiltration rate in litres/minute per 30 m length of furrow

t is time in minutes; and

K and n are the constants.

The accumulated infiltration in time T is given by

$$V = \int_0^T I dt = \int_0^T Kt^n dt = \frac{KT^{(n+1)}}{(n+1)} \quad \text{...(iii)}$$

Further if D mm is the cumulative or total depth of infiltration then

$$V = \frac{D}{100} \times W \times 30 \times 1000 = D \times W \times 30 \quad \text{...(iv)}$$

in which

W furrow spacing in metres.

Thus, equating the two values of V, we get

$$D \times W \times 30 = \frac{KT^{(n+1)}}{(n+1)}$$

or

$$D = \frac{KT^{(n+1)}}{30 \times W \times (n+1)} \quad \text{...(2.17)}$$

From Eq. (2.17) the time T to infiltrate the given depth D is obtained as

$$T = \left[\frac{30 \times D \times W \times (n+1)}{K}\right]^{\left(\frac{1}{n+1}\right)} \quad \text{...(2.18)}$$

Further the instantaneous rate of infiltration i in millimetre per hour may be obtained as follows:

From Eq. (i), we have

$$I = (Q_1 - Q_2) \text{ litres/minute}$$

$$\text{Area covered} = (30 \times W) \text{ m}^2$$

Thus, $$I = \left(\frac{i}{1000} \times 30 \times W\right) \text{ m}^3/\text{ hour}$$

or $$I = \left(\frac{i}{1000} \times 30 \times W\right) \times \frac{1000}{60} \text{ litres/minute}$$

Equating the two values of I, we get

$$\left(\frac{i}{1000} \times 30 \times W\right) \times \frac{1000}{60} = (Q_1 - Q_2)$$

or $$i = \left(\frac{Q_1 - Q_2}{W}\right) \times 2 \text{ mm/hour} \qquad ...(2.19)$$

Advantages of Furrow Method of Irrigation

The furrow method of irrigation has the following advantages:

(i) In this method of irrigation only a part of the land varying from one-half to one-fifth is wetted, which results in reducing evaporation losses, lessening the puddling of heavy soils and making it possible to cultivate the soil soon after irrigation.

(ii) It is suitable for row crops such as maize, cotton, potatoes, sugar cane, sugar beet, groundnut, tobacco etc.

(iii) It is especially suitable for crops such as maize etc., which are subject to injury if allowed to come in contact with water ponded on the land in any of the methods of irrigation by flooding.

(iv) In this method the requirements of labour for land preparation and irrigation are very much reduced as compared to the various methods of irrigation by flooding.

(v) In the furrow method there is no wastage of land in field channels as compared to the checks or levees method of irrigation.

Corrugation Method

Corrugations are in general similar to furrows but the main difference between the two being that whereas the furrows are the channels of relatively larger cross-section, the corrugations are the channels of smaller cross-section. Further as indicated earlier the furrows may be laid along the slope of the land only if the land slopes are relatively flat and for lands with steeper slopes contour furrows are used but the corrugations may be laid along the slope of the land even for lands with steeper slopes. As such corrugations may also be known as small furrows used for irrigating the lands with steeper slopes. Corrugations have either V- or U-shaped cross-section and are made about 60 to 100 mm deep and only slightly wider than their depth. The spacing of the corrugations may be 400 to 900 mm. This method of irrigation is commonly used for close-growing crops such as small grains and for pasture growing on lands having steep slopes.

Contour Farming

Contour farming is practised in hilly regions where the lands to be irrigated have steep slopes. It also controls erosion due to rainfall. In this method the land to be irrigated is divided into a series of strips usually known as terraces or benches which are aligned to approximately follow the different contours at a vertical interval of 300 to 600 mm. Each of these strips forms an independent portion of the land to be irrigated. Thus, in this case as shown in Fig. 2.8, a series of steps is formed in successive elevations around the slope. The strips are level in the direction of the original land slope but are given a gentle slope along their length to ensure efficient irrigation and drainage of excess water at non-erosive velocities. At the outer end of

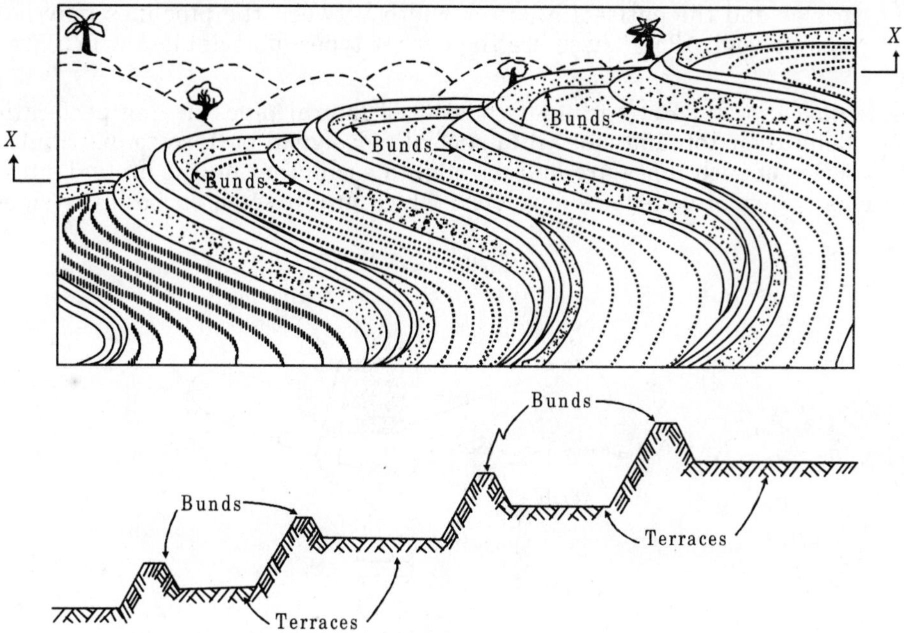

Fig. 2.8 *Contour fanning*

each strip following its boundary configuration a low earthen bund or a dike is provided. The height of the bund should be sufficient to safely contain both irrigation water and the rainfall runoff. For applying the irrigation water to the strips any of the methods such as border strips, furrows and check basins can be used.

2.4 SPRINKLER IRRIGATION METHOD

In the sprinkler irrigation method the irrigation water is applied to the land in the form of spray, somewhat as in ordinary rain. As such the sprinkler irrigation is also some times known as overhead irrigation. The sprinkler irrigation can be used for all the crops except rice and jute (for which standing water is required) and for almost all the soils except very heavy soils with very low infiltration rates.This method of irrigation is best suited for very light soils as deep percolation losses are avoided. Further the method is flexible to suit undulating topography and hence land levelling is not necessary.

The components of a sprinkler irrigation system are main pipelines (or mains), submain pipelines (or sub-mains), lateral pipelines (or laterals), riser pipes (or risers), sprinklers and a pumping unit. The pump lifts water from the source and supplies it through the pipes to the sprinklers which develop the required spray of water. Three types of sprinklers as indicated below are generally used.

(i) Fixed nozzle pipe

(ii) Perforated pipe

(iii) Rotating sprinkler.

The *fixed nozzle pipe* type of sprinkler consists of the lateral pipes having a line of small holes drilled at the top at some regular interval along their lengths and on each of these holes small nozzles are fitted. A series of such pipes are installed parallel to each other at a distance of about 15 metres apart and are supported on rows of posts. The spray of water is developed through the nozzles and the entire 15 metres width between the pipelines may be irrigated by turning the pipes through 135°. These are the earlier types of sprinklers which are not common nowadays.

The *perforated pipe* type of sprinkler consists of lateral pipes having perforations or small holes drilled along their top and sides in a specially designed pattern to distribute water fairly uniformly. A series of such pipes are laid parallel to each other on the ground surface. Through the perforations sprays of water are developed which are directed on both the sides of the pipe

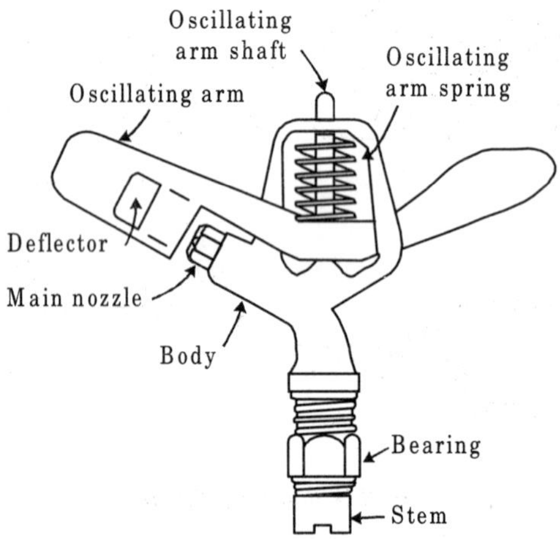

(a) Single-nozzle rotating sprinkler

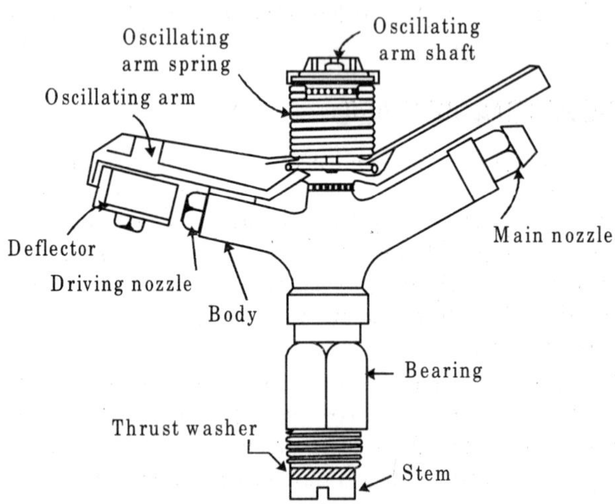

(b) Twin-nozzle rotating sprinkler

Fig. 2.9 *Types of rotating sprinklers*

and can cover a strip of land from 6 to 15 metre wide. In this case the minimum rate of application of water is about 20 mm per hour. Since the water is applied at a relatively higher rate, these sprinklers are suitable for soils having moderately high infiltration rates. The operating pressures for these sprinklers are in the range of 0.049 to 0.245 N/mm^2 [0.5 to 2.5 kg(f)/cm^2]. The pressure in this case being low the water may be supplied from an overhead tank instead of a pump. These sprinklers are used more extensively in orchards and nurseries.

The *rotating sprinkler* consists of one or two nozzles mounted on a body which is rotated slowly about vertical axis by the action of a deflecting vane connected to it (Fig. 2.9). The jet of water issuing from one of the nozzles impinges on the vane and thrusts it aside. The motion of the vane is restrained by light spring which tends to bring it back to original position, but the return of the vane is terminated by a stop on the body which thus rotates through a small angle by the impulse. The vane again intercepts the jet and the cycle repeats. The spray of water is developed through the nozzles and it is sprinkled over a circular area of the land to be irrigated which is covered by each sprinkler as it rotates.

The rotating sprinklers are used very extensively for irrigating almost all the types of crops. The advantage of rotating sprinkler over other types is that even with relatively large nozzle openings provided for a rotating sprinkler the water is applied at a slower rate. Further with large nozzle openings the problem of clogging caused by the water containing silt and debris would be minimized. The water application rates less than 2.5 mm per hour are possible with these sprinklers. The slow rate of water application is desirable on soils having low infiltration rates. Single nozzle sprinklers are used for low water application rates. However, sprinklers most commonly used have two nozzles, one to apply water at a considerable distance from the sprinkler and the other to cover the area near the sprinkler centre (Fig. 2.9). As such although two nozzle sprinklers cannot apply water at as low a rate as single-nozzle sprinklers, but they provide uniform application of water. The pressure for rotating sprinklers normally range from about 0.2 N/mm^2 [2 kg(f)/cm^2] for the smaller sprinklers to more than 0.7 N/mm^2 [7 kg(f)/cm^2] for the larger sprinklers.

The rotating sprinklers are placed on the riser pipes and are located just above the crops to be irrigated. As such the height of the riser pipes depends upon the maximum height of the crop. However, the minimum height of the riser pipe is 0.3 m when the riser pipe is of 25 mm diameter and 0.15 m when it is of 18 to 20 mm diameter. The riser pipes along with the sprinklers are fixed at regular intervals along the length of the lateral pipes and their spacing is so adjusted that the water spread areas of the adjacent sprinklers partially overlap with each other in order to achieve uniform application of water. The overlap required to be provided increases with the increase in the wind velocity as shown in the Table 2.1. The lateral pipes are usually either laid on the ground surface or burried below the plough depth. However, the lateral pipes may also be mounted on posts above the crop height. In the later case the lateral pipes may be rotated through 90° to irrigate a rectangular strip of the land.

The required discharge of each sprinkler is a function of the water application rate, the spacing of the sprinklers along the lateral pipelines and the spacing of the lateral pipelines along the main pipeline. It may be determined by the following formula:

$$q = \frac{S_1 \times S_m \times I}{3600}$$

in which

q = required discharge of each sprinkler in litres/second

S_1 = spacing of sprinklers along the lateral pipelines in metres

S_2 = spacing of laterals along the main pipeline in metres

I = optimum water application rate in mm/hour.

TABLE 2.1 Maximum spacing of sprinklers

S.No.	*Average wind speed*	*Spacing*
1.	No wind	65% of the diameter of the water spread area of the sprinklers
2.	0 to 6.5 km/hr	60% —Do—
3.	6.5 to 13 km/hr	50% —Do—
4.	Above 13 km/hr	30% —Do—

Based on the portability, sprinkler irrigation systems are classified into the following types:

(i) *Portable system.* A portable system has the main pipelines, submain pipelines, the lateral pipelines with sprinklers and the pumping unit portable. It is designed to be moved from one agricultural field to another or to different pump sites in the same field. The portable system may be designed to be moved manually or by mechanical power. The system operated manually is known as *hand move system* in which the main and the lateral pipelines are moved manually. On the other hand the system using mechanical power for moving the lateral pipeline is known as *wheel move system* in which the entire lateral pipeline is mounted on wheels and it is moved as a single unit instead of moving separate pipe lengths by disassembling the lateral pipeline.

(ii) *Semi-portable system.* A semi-portable system is similar to a fully portable system except that the location of the water source and the pumping unit are fixed. As such this system may be used on more than one agricultural field where there is an extended main pipeline or there are additional pumping units.

(iii) *Semi-permanent system.* A semi-permanent system has the portable lateral pipelines but the main and the submain pipelines are permanently laid and are usually burned. Further the source of water as well as the pumping unit are also fixed.

(iv) *Solid-set system.* A solid-set system has enough laterals to eliminate their movement. The laterals are placed in the field at the beginning of the crop season and are kept there for the entire crop season. As such this system is used for those crops which require short and frequent irrigations.

(v) *Permanent system.* A fully permanent system consists of permanently laid main, submain and lateral pipelines and the source of water as well as the pumping unit are also fixed. The main, submain and lateral pipelines are usually burried below the plough depth. The sprinklers are also permanently located on each riser. Sprinkler installations in orchards are usually of the permanent type. This system is however costly.

Advantages of Sprinkler Irrigation

1. Sprinkler irrigation can be efficiently used for a wide range of topographic conditions, soils and crops.

2. With the use of sprinkler irrigation since surface runoff is eliminated, erosion of soil can be controlled.

3. Uniform application of water is possible with sprinkler irrigation.

4. In the sprinkler irrigation method a better control on irrigation can be enforced and also light irrigation is possible which is required for seedlings and plants which are very young.

5. For sprinkler irrigation no land preparations are required thus labour cost is reduced.

6. The sprinkler irrigation does not require borders, field channels, etc. and hence more land is available for cropping.

7. For sprinkler irrigation even small streams of irrigation water can be used efficiently.

8. The time and amount of application of fertilizer can be better controlled in sprinkler irrigation.

9. The protection to the crop against frost can be provided by sprinkler irrigation. The spray of irrigation water makes the ice formed on the plants to melt and thus damage to the crop due to frost is prevented. Crops have been saved in temperatures as low as 9°C by sprinkler irrigation.

10. With sprinkler irrigation in semi arid and humid regions about 80% water application efficiency (defined in Chapter 4) is possible. However, the efficiency decreases by 5% for each 8 km/hr wind velocity above a base velocity of 8 km/hr and decreases further by 5% for each 5% slope of the land above a base of 12% slope of the land.

11. The sprinkler irrigation method is especially adaptable to more humid regions be cause by this method of irrigation the increase in the humidity of the atmosphere is much less as compared to the other methods of irrigation.

12. The pumping unit of the sprinkler irrigation system will act as standby drainage pumping unit.

Limitations of Sprinkler Irrigation

1. Wind may distort the sprinkler pattern, thus resulting in non-uniform application of irrigation water.

2. The first investment involved in the sprinkler irrigation method is high.

3. A constant water supply is needed for the most economical use of the equipment of the sprinkler irrigation system.

4. Water must be clean and free from sand etc., to prevent the clogging of the sprinklers.

5. The power requirement is high since continuous pumping of water is required in this method of irrigation.

6. Heavy soils with poor intake cannot be irrigated efficiently by the sprinkler irrigation method.

2.5 SUB-IRRIGATION METHODS OR SUB-SURFACE IRRIGATION METHODS

In the sub-irrigation methods water is applied below the ground surface so that it is supplied directly to the root zone of the plants. The main advantages of these methods of irrigation are that the evaporation losses are considerably reduced and the hindrance caused to cultivation by the presence of borders, pipes and field channels in the other methods of irrigation is eliminated. The sub-irrigation methods are classified as follows.

(i) Natural sub-irrigation

(ii) Artificial sub-irrigation

(iii) Drip or trickle irrigation.

(i) **Natural sub-irrigation.** In natural sub-irrigation water is supplied to the plants from below the surface by controlling the level of the water table. The method is so called because ordinarily under the favourable geological and topographical conditions the water seeping from earthen canals, drains, rivers, etc., results in raising the water table to such a level that water is made available in the root zone of the plants and hence natural irrigation occurs. However, in order to ensure requisite supply of water to the crops grown by adopting this method of irrigation the water table is required to be maintained at the desired level by artificial means. For this purpose a series of parallel channels with almost vertical sides are excavated in the land to be irrigated. The channels are spaced 15 to 100 metres apart depending on the permeability of the sub-soil and the topography of the land to be irrigated. The depth of the channels varies from 0.3 to 1 metre and their width varies from 0.25 to 0.5 metre. The channels have relatively flat bed-slopes. Water is made to flow through these channels at a slow rate so that these remain filled with water, thus, allowing it to seep into the ground. The flow of water in the channels is so regulated that the seepage raises the water table high enough to produce moist conditions in the top soil by capillary movement and thus adequate supply of moisture is maintained in the root zone. Proper drainage of excess water is however needed to prevent water fogging of the land. Moreover, the water is reduced or discontinued and the top soil is allowed to dry just before harvesting.

However, in this method of irrigation since all water movement in the process of supply to the plant is upwards from the water table, there is also an upward movement of unwanted salts within the soil. As such there is possibility of a build-up of harmful salts close to or on the land surface. Provision is therefore required to be made for periodic leaching of the soil by heavy applications of water to the land surface and for the drainage to remove the salts thus leached.

This method of irrigation is excellent from both water distribution and labour saving points of view, provided it is employed where suitable. The conditions required to allow the use of this method of irrigation are as follows.

1. The soil in the root zone should be highly permeable loam or sandy loam.

2. At a depth of 2 to 3 metre below the surface level there must be an impermeable sub-stratum or free water table.

3. The land to be irrigated must have uniform topographic conditions and it must be flat and near level. However, lands with moderate slopes can also be irrigated by this method.

4. There must be abundant water supply of excellent quality and very low salinity.

Since this method of irrigation requires an unusual combination of natural conditions which are very rarely found, it can be used in only a few areas. Moreover, if this method is used under conditions which are not favourable to it then this may prove to be the worst irrigation method.

(ii) **Artificial Sub-irrigation.** In artificial sub-irrigation water is applied beneath the land surface by a network of burried perforated pipes through which water is passed under pressure to percolate into the soil. This method of irrigation will function effectively only if the soil in the root zone has high horizontal permeability to permit free lateral movement of water and low vertical permeability to prevent deep percolation of water. For uniform distribution of water percolating into the soil, the pipes are required to be very closely spaced at a spacing as low as 0.45 m. Further in order to avoid interference with cultivation the pipes have to be buried not less than about 0.4 m below the ground surface. This method of irrigation is not used widely because of the following shortcomings:

1. This method is highly expensive because of high cost of the pipes as well as tremendous installation cost involved for laying the network at the desired depth.

2. In most of the cases the actual water distribution will be poor because rootlets tend to grow close to the pipes and thus clog part of the pipe perforations.

3. In order to avoid clogging of the small perforations the water must be free from silt, sand and debris. Moreover, the water used must be very low in salinity as this method affords no leaching.

(iii) **Drip or Trickle Irrigation.** Drip or trickle irrigation is one of the latest methods of irrigation which is becoming increasingly popular in areas with water scarcity and salt problems. In this method irrigation is accomplished by using small diameter plastic pipes with drip nozzles commonly called emitters or drippers to deliver water to the land surface near the base of the plants. In this case the water is applied at a very low rate varying from about 2 to 10 litres per hour to keep the soil moisture within the desired range for plant growth.

The main components of a drip irrigation system are a pumping unit, main pipelines, submain pipelines, lateral pipelines and emitters. All the pipes are usually made of black PVC (Poly Vinyle Chloride) tubings. The emitters are also usually made of PVC material. The emitters are so designed that no jet of water is issued by them but they allow water to emit or trickle through them. Further the emitters are provided at regular intervals on the lateral pipelines and their spacings are adjusted to produce either a wetted strip along the row of the crop or a wetted bulb of soil at every plant. There is usually one lateral pipeline for each row of the crop.

The main advantage of drip irrigation over other methods of irrigation is the excellent control of water application which it provides. As such by drip irrigation efficiency as high as 90% or more can be achieved which can hardly be achieved by surface irrigation methods or sprinkler irrigation method. Moreover, in the case of drip irrigation the evaporation losses from the land surface are minimal and the deep percolation of water can be almost entirely avoided. It has also been found that the drip irrigation method is particularly advantageous when the water is saline. This is so because the unwanted salts move to the outer edge of the wetted zone while within the region of the root zone relatively less saline water is made available.

However, the main drawbacks of this method of irrigation are that the cost of drip irrigation equipment is considerably higher than that of sprinkler irrigation and also the problem of blockage of the outlets exists.

A comparison of furrow, sprinkler and drip irrigation methods is given in a tabular form in Table 2.2.

TABLE 2.2 Comparison of furrow, sprinkler and drip irrigation methods

S. No.	*Characteristic*	*Furrow irrigation*	*Sprinkler irrigation*	*Drip irrigation*
1.	Water requirement	Higher water requirement as flow in furrows entail considerable water losses to the extent of 55 per cent, occur	Low water requirement, sixty five per cent of that required in surface irrigation methods. Total loss is estimated to be about 18 per cent	Least water requirement, since water is delivered directly to the plant. Water consumption is about 20–30 per cent of the conventional methods
2.	Evaporation loss	Less water lost by evaporation	Higher evaporation loss	Evaporation loss almost negligible
3.	Suitability for soils	Suitable for all soils except for very light soils having high permeability	Suitable for all types of soils, especially for shallow coarse textured, highly permeable soils requiring light irrigation at frequent intervals	Suitable for all types of soils, especially for coarse textured soils with scarce water resources. Crops can be grown over saline soils also
4.	Soil surface wetting	Wets 20–25 per cent of irrigated soil surface	Entire surface area is wetted	Least wetting of soil surface due to direct water supply to the plants
5.	Surface temperature and its effect on crop period.	Soil temperature is reduced owing to flooding of entire furrows line. This delays the maturity of crop and hence the crop period is extended.	Entire surface temperature is reduced, resulting in comparatively longer crop period due to delayed maturity	Surface temperature is kept high, the moisture availability being restricted to root zone This hastens the ripening or maturity of the produce and hence the crop period is reduced
6.	Rate of growth	Comparatively low rate of vegetative growth	Better and quicker germination of seeds. Plants grow more vigorously. Improved response of crops as compared to surface irrigation	Highest rate of vegetative growth. Much quicker germination of seeds and rate of plant development
7.	Suitability for crops	Suitable for all cultivated crops planted in rows and trunk crops. Not suitable for shallow rooted crops requiring light irrigation. Also not suitable for seed germination	Widespread application. Especially suitable for tea and coffee gardens, young seedlings and plants	Suitable for close growing crops such as vegetables and also for orchards

Contd.

Contd. Table 2.2

S. No.	*Characteristic*	*Furrow irrigation*	*Sprinkler irrigation*	*Drip irrigation*
8.	Adjustment of water application	Regulation of water application is almost absent	Sprinkler pressure controls the amount of water applied and the rate of rotation of the sprinkler	Adjustment of water application very easy by using nozzles of different sizes
9.	Automation of water supply	Automation is absent. Higher labour cost is involved in guiding and distributing water to the furrows	Complete automation is rare	Economical and convenient water control. Automatic valve at head terminates supply after the desired supply is discharged. Water meter is also installed in the head to know the discharge
10.	Uniformity of irrigation	Non-uniform distribution of water along the length of the furrow. The uniformity of irrigation may be as low as 25 per cent. Natural conditions such as texture and structure of the soil are the deciding factors of the uniformity over which the irrigationist has no or little control	The uniformity of irrigation may be about 50 per cent. Intensity of wind is the controlling factor.	The uniformity of irrigation in the drip irrigation is above 90 per cent
11.	Fertiliser application	Localised application of fertiliser is absent	Fertiliser is applied through irrigation water by means of fertiliser applicator. Localised application is absent	Fertiliser application is localised, as it is delivered to the plant root only through irrigation water. Thus, the fertiliser application is best regulated, besides minimising its requirement. However application of insoluble and slightly soluble fertiliser such as superphosphate or sulphate of potash through the flow is not readily possible in sprinkler and drop irrigation systems

Contd.

Contd. Table 2.2

S. No.	*Characteristic*	*Furrow irrigation*	*Sprinkler irrigation*	*Drip irrigation*
12.	System and component parts	Furrows change with the type of crops grown. Furrows made on gradient involve skill and labour. However no piping system is involved	Sprinkling system may be stationary or portable. The system consists of main pipes and laterals besides risers. It has movable components like nozzles, etc. The risers constitute obstruction to cultivation operations. However the spacing of laterals and nozzles is easy to vary	The drip irrigation system is generally stationary. The pipe lines are made from plastic materials and thus cheaper. The layout is free from any movable component. The spacing of laterals and nozzles is easy to vary to suit crop and soil requirement. Risers are absent and as such there is no obstruction to cultivation operations by machinery
13.	Effect of wind velocity	In level furrow method, strong wind (10 to 12 km per hour), blowing in opposite direction interferes with the application of irrigation water	Strong wind interferes with application of water, particularly when it flows in the opposite direction. High wind velocity also causes high evapotranspiration	Not much affected by wind conditions
14.	Soil erosion	Soil erosion near upper end of the field results in furrows of long length	Soils erosion in controlled	No soil erosion or tail water loss
15.	Weed control	Wild weed and pest growth	Weed and pest growth, and diseases are less. Cost of their removal is less	Weed, pest and insect troubles are least since only selected points on the land are wet. Hence the problems of weed control and the preservation of soil nutrients for the crops are easy
16.	Percolation of water and leaching of salts and nutrients	Percolation losses are very high, especially in sandy soils. Thus, the salt nutrients and fertilisers are leached down to greater depths and are not available for plant growth	Percolation of water and leaching down of salts, fertilisers etc. is less. Leaching takes place in basic soils only	Minimum percolation reduces leaching down of fertilisers and nutrients.

Contd.

Contd. Table 2.2

S. No.	*Characteristic*	*Furrow irrigation*	*Sprinkler irrigation*	*Drip irrigation*
17.	Land levelling requirement	Levelling of fields is very essential in straight graded furrows	No land levelling is required	Land levelling is not required
18.	Moisture distribution	Moisture distribution is normal to the furrow shape	Moisture distribution is two dimensional	Moisture distribution is two dimensional for soils of low permeability and for soils with an impervious sub soil layer. The moisture distribution in soils of high permeability (sandy soils) is mainly vertical i.e., unidirectional
19.	Land utilisation	About 7.5 per cent of land is lost in field channels	More land is available for cultivation as water courses are not required	Makes available maximum land for cultivation
20.	Crop yield	The yield is less as compared to other method.	Higher yield, though less than drip irrigation system	Higher yield if crops in this system are possible such as tomato, beet root, lady finger, etc. and for other crops, the yield will beat least as good as in other methods of irrigation
21.	Cost	Lowest initial investment but high operation labour charges	High initial cost and operation charges. Steel or light weight aluminium pipes are costly as they have to withstand high operating pumping pressures	Low initial and operating charges as compared to sprinkler irrigation. However the plastic pipes have low durability and therefore need frequent replacement (4-5 years).

SOLVED ILLUSTRATIVE EXAMPLES

Illustrative Example 2.1 *For a depth of water application of 50 mm and average intake rate of 50 mm per how determine:*

(a) the optimum length of each border strip if the discharge of the source of water is 20 litres per second and width of each border strip is 10 m

(b) the inflow rate if the length of each border strip is 200 m and its width is 10 m;

(c) the width of each border strip if the discharge of the source of water is 20 litres per second and its length is 200 m;

(d) Time of irrigation in each case.

Slope of the border strip is 0.3%

Solution

Given $$D = \frac{50}{1000} = 0.05 \text{ m}$$

$$I = \frac{50}{1000 \times 3600} = 1.389 \times 10^{-5} \text{ m/s}$$

(a) $$q = \frac{20}{10 \times 1000} = 2 \times 10^{-3} \text{ m}^3\text{/s per m width}$$

From Eq. (2.6), we have

$$y_0 = \left(\frac{nq}{\sqrt{s}}\right)^{0.6}$$

From Eq. (2.5), we have

$$y = 0.665\, y_0$$

Also from Eq. (2.7), we have

$$n = 0.55425 - 0.0001386\, q'$$

$$q' = \frac{2 \times 10^{-3} \times 60 \times 10^6}{10^2} = 1.2 \times 10^3 \text{ cm}^3\text{/min/cm}$$

Thus, by substitution, we get

$$n = (0.55425 - 0.0001386 \times 1.2 \times 10^3)$$
$$= 0.388$$

$$y_0 = \left(\frac{0.388 \times 2 \times 10^{-3}}{\sqrt{0.003}}\right)^{0.6} = 0.078 \text{ m}$$

$$y = (0.665 \times 0.078) - 0.052 \text{ m}$$

From Eq. (2.4), we have

$$x = \frac{q}{I}\left[1 - e^{(-D/y)}\right]$$

Thus, by substitution, we get

$$x = \frac{2 \times 10^{-3}}{1.389 \times 10^{-5}}\left[1 - e^{\left(-\frac{0.05}{0.052}\right)}\right]$$

$$= 89.35 \text{ m } 90 \text{ m} \simeq 90 \text{ m}$$

(b) From Eq. (2.4), we have

$$x = \frac{q}{I}\left[1 - e^{(-D/y)}\right]$$

or
$$q = xI\left[\frac{e^{(D/y)}}{e^{(D/y)-1}}\right]$$

Given $x = 200$ m; $I = 1.389 \times 10^{-5}$; and $D = 0.05$ m

Thus, by substitution, we have

$$q = 200 \times 1.389 \times 10^{-5}\left[\frac{e^{(0.05/y)}}{e^{(0.05/y)-1}}\right] \qquad ...(i)$$

This equation can be solved by trial and error as indicated below

First trial

Assume $q = 5$ lps per m width

or $q = 5 \times 10^{-3}$ m³/s per m width

Then
$$q' = \frac{5\times10^{-3}\times60\times10^{6}}{10^{2}}$$
$$= 3 \times 10^{3} \text{ cm}^3\text{/min/cm}$$

From Eq. (2.7), we have

$$n = 0.55425 - 0.0001386\, q'$$
$$n = (0.55425 - 0.0001386 \times 3 \times 10^{3})$$
$$= 0.13845$$

From Eq. (2.6), we have

$$y_0 = \left(\frac{nq}{\sqrt{s}}\right)^{0.6}$$

$$\therefore \quad y_0 = \left(\frac{0.13845\times5\times10^{-3}}{\sqrt{0.003}}\right)^{0.6} = 0.072\text{ m.}$$

From Eq. (2.5), we have

$$y = 0.665\, y_0$$
$$\therefore \quad y = (0.665 \times 0.072) = 0.0479 \text{ m}$$

By substituting the value of y in Eq. (i), we get

$$q = 200 \times 1.389 \times 10^{-5}\left[\frac{e^{(0.05/0.0479)}}{e^{(0.05/0.0479)}-1}\right]$$
$$= 0.0043 \text{ m}^3\text{/s per m width}$$
$$= 4.3 \text{ lps per m width}$$

The assumed value of q is higher than the calculated value.

Second trial

Assume $q = 3.3$ lps per m width

or $q = 3.3 \times 10^{-3}$ m³/s per m width

Then
$$q' = \frac{3.3 \times 10^{-3} \times 60 \times 10^6}{10^2}$$
$$= 1.98 \times 10^3 \text{ cm}^3\text{/min/cm}$$

From Eq. (2.7), we have
$$n = 0.55425 - 0.0001386\, q'$$
$$n = (0.55425 - 0.0001386 \times 1.98 \times 10^3)$$
$$= 0.27982$$

From Eq. (2.6), we have
$$y_0 = \left(\frac{nq}{\sqrt{s}}\right)^{0.6}$$

$$\therefore \quad y_0 = \left(\frac{0.27982 \times 3.3 \times 10^{-3}}{\sqrt{0.003}}\right)^{0.6} = 0.086 \text{ m.}$$

From Eq. (2.5), we have
$$y = 0.665\, y_0$$
$$\therefore \quad y = (0.665 \times 0.086) = 0.0572 \text{ m}$$

By substituting the value of y in Eq. (i), we get
$$q = 200 \times 1.389 \times 10^{-5}\left[\frac{e^{(0.05/0.0572)}}{e^{(0.05/0.0572)} - 1}\right]$$
$$= 0.00477 \text{ m}^3\text{/s per m width}$$
$$= 4.77 \text{ lps per m width}$$

The assumed value of q is less than the calculated value.

Third trial

Assume $q = 4.5$ lps per m width

$q = 4.5 \times 10^{-3}$m³/s per m width

Then
$$q' = \frac{4.5 \times 10^{-3} \times 60 \times 10^6}{10^2}$$
$$= 2.7 \times 10^3 \text{ cm}^3\text{/min/cm}$$

From Eq. (2.7), we have
$$n = 0.55425 - 0.0001386\, q'$$
$$\therefore \quad n = (0.55425 - 0.0001386 \times 2.7 \times 10^3)$$
$$= 0.18003$$

From Eq. (2.6), we have
$$y_0 = \left(\frac{nq}{\sqrt{s}}\right)^{0.6}$$

$$\therefore \qquad y_0 = \left(\frac{0.18003 \times 4.5 \times 10^{-3}}{\sqrt{0.003}}\right)^{0.6} = 0.0798$$

From Eq. (2.5), we have

$$y = 0.665\, y_0$$

$$\therefore \qquad y = (0.665 \times 0.0798) - 0.05307 \text{ m}$$

By substituting the value of y in Eq. (i), we get

$$q = 200 \times 1.389 \times 10^{-5}\left[\frac{e^{(0.05/0.05307)}}{e^{(0.05/0.05307)}-1}\right]$$

$$= 0.00455 \text{ m}^3\text{/s per m width}$$

$$= 4.55 \text{ lps per m width}$$

The assumed value of q is close to the calculated value and hence the inflow rate may be taken as 4.5 lps per m width.

Since the width of each border strip is 10 m, the total inflow rate for each border strip is (4.5 × 10) = 45 lps.

(c) In this case also firstly the value of q is obtained, for which the same procedure of trial and error is adopted as in the previous case. Since in this case the length of each border strip is same equal to 200 m as in the previous case, the value of q will be obtained as 4.5 lps per m width.

The total discharge available from the source is

$$Q = 20 \text{ lps}$$

∴ Width of each border strip

$$= \frac{20}{4.5} = 4.45 \text{ m}$$

(d) Time of irrigation in each case is given by Eq. (2.8) as

$$t = \frac{D}{I} = \frac{50}{50} = 1 \text{ hour}$$

Illustrative Example 2.2 *Find the time required to irrigate a border strip of length 203 m and width 10 m by a stream of discharge 0.043 cumec. Average depth of flow is expected to be 63.5 mm and average infiltration rate for the soil may be taken as 50 mm/hour.*

Also determine the maximum length of the strip which can be irrigated in this case.

Solution

From Eq. (2.2), we have

$$x = \frac{q}{I}\left[1 - e^{(-It/y)}\right]$$

Given $\qquad x = 203$ m

$$q = \frac{0.043}{10}$$

$= 0.0043$ cumec per m width

$= 0.0043 \times 3600$ m^3/hour per m width

$I = 50$ mm/hour

$= 50 \times 10^{-3}$ m/hour

$y = 63.5$ mm

If t is the time in hours, required to irrigate the strip, then by substituting the given values with appropriate units in the above equation, we get

$$203 = \frac{0.0043 \times 3600}{50 \times 10^{-3}}[1 - e^{(-50t/63.5)}]$$

Solving the above equation for t, we get

$t = 1.354$ hours

From Eq. (2.1 d) the maximum length of the strip which can be irrigated is given by

$$x_{max} = \frac{q}{I}$$

Thus, by substitution, we get

$$x_{max} = \frac{0.0043 \times 3600}{50 \times 10^{-3}}$$

$= 309.6$ m

Alternatively from Eq. (2.1 b), we have

$$t = 2.303\ \frac{y}{I}\ \log_{10}\left(\frac{Q}{Q - IA}\right)$$

$y = 63.5$ mm $= 0.0635$ m;

$I = 50$ mm/hour $= 0.05$ m/hour;

$Q = 0.043$ cumec $= 0.043 \times 3600$ m^3/hour

$A = (203 \times 10) = 2030$ m^2

Thus, by substitution, we get

$$t = \frac{2.303 \times 0.0635}{0.05} \times \log_{10}\left(\frac{0.043 \times 3600}{0.043 \times 36000.05 \times 2030}\right)$$

$= 1.354$ hours

From Eq. (2.1 d) the maximum area of the strip which can be irrigated is given by

$$A_{max} = \frac{Q}{I}$$

Thus, by substitution, we get

$$A_{max} = \frac{0.043 \times 3600}{0.05}$$

$$= 3096 \text{ m}^2$$

The width of the strip = 10 m

∴ The maximum length of the strip which can be irrigated is

$$x_{max} = \frac{3906}{10} = 309.6 \text{ m}$$

Illustrative Example 2.3 *The following data pertain to a check basin system :*

(i) size of check basin = 10 m × 40 m;

(ii) cumulative infiltration $I_c = 9.951^{0.43}$, in which I_c is in mm and t is time in minutes;

(iii) advance function

(a) $L = 5.6\ t_i^{0.63}$ for a stream size of 1.7 lps/m

(b) $L = 9.0\ t_L^{0.63}$ for a stream size of 3.4 lps/m

(iv) depth of irrigation = 75 mm

Calculate (A) deep percolation losses; (B) water application efficiency; and (C) time of irrigation.

Solution

(A For stream size of 1.7 lps/m, the time to cover the entire area of the check basin for length 40 m is obtained as

$$t_L = \left(\frac{40}{5.6}\right)^{\frac{1}{0.63}} = 22.66 \text{ minutes.}$$

Time to infiltrate the depth I_c = 75 mm is given as

$$t_d = \left(\frac{75}{9.95}\right)^{\frac{1}{0.43}} = 109.67 \text{ minutes}$$

Thus, total time T is obtained as

$$T = t_L + t_d$$

$$= (22.66 + 109.67) - 132.33 \text{ minutes}$$

Thus, from Eq. (2.9), deep percolation loss is obtained as

$$P = \left[1 - \left(\frac{t_d}{T}\right)^{n+1}\right]$$

$$(n + 1) = 0.43$$

$$\therefore \quad P = \left[1 - \left(\frac{109.67}{132.33}\right)^{0.43}\right] = 0.08 \text{ or } 8\%$$

From Eq. (2.12), water application efficiency is obtained as

$$\eta_a = (1 - P) \times 100$$
$$= (1 - 0.08) \times 100 = 92\% \text{ or } 0.92$$

From Eq. (2.14), time of irrigation is obtained as

$$\text{Time of irrigation} = \frac{d \times L \times B}{\eta_a \times Q}$$

$$Q = (1.7 \times B) = (1.7 \times 10)\text{lps} = (1.7 \times 10) \times 10^{-3} \text{ m}^3/\text{s}$$

$$\therefore \text{ Time of irrigation} = \frac{75 \times 10^{-3} \times 40 \times 10}{0.92 \times (1.7 \times 10) \times 10^{-3}} \times \frac{1}{60}$$

$$= 31.97 \text{ minutes.}$$

(b) For stream size of 3.4 lps/m, the time to cover the entire area of the check basin for length 40 m is obtained as

$$t_L = \left(\frac{40}{90}\right)^{\frac{1}{0.63}} = 10.67 \text{ minutes}$$

Time to infiltrate the depth I_c %= 75 mm is given as

$$t_d = \left(\frac{75}{9.95}\right)^{\frac{1}{0.43}} = 109.67 \text{ minutes}$$

Thus, total time T is obtained as

$$T = t_L + t_d$$
$$= (10.67 + 109.67) - 120.34 \text{ minutes}$$

Thus, from Eq. (2.9), deep percolation loss is obtained as

$$P = \left[1 - \left(\frac{t_d}{T}\right)^{n+1}\right]$$

$$(n + 1) = 0.43$$

$$\therefore \quad P = \left[1 - \left(\frac{109.67}{120.34}\right)^{0.43}\right] = 0.04 \text{ or } 4\%$$

From Eq. (2.12), water application efficiency is obtained as

$$\eta_a = (1 - P) \times 100$$
$$= (1 - 0.04) \times 100 = 96\% \text{ or } 0.96$$

From Eq. (2.14), time of irrigation is obtained as

$$\text{Time of irrigation} = \frac{d \times L \times B}{\eta_a \times Q}$$

$$Q = (3.4 \times B) = (3.4 \times 10) \text{ 1ps} = (3.4 \times 10) \times 10^{-3} \text{ m}^3/\text{s}$$

$$\therefore \text{ Time of irrigation } = \frac{75 \times 10^{-3} \times 40 \times 10}{0.96 \times (3.4 \times 10) \times 10^{-3}} \times \frac{1}{60}$$

$$= 15.32 \text{ minutes.}$$

Illustrative Example 2.4 *In irrigation of furrow 120 m long observations on advance time were taken as shown below:*

Distance (m)	*Time of advance (min.)*
0	*0*
30	*26*
60	*58*
90	*98*
120	*142*

The stream size used was 25 lpm. The infiltration characteristics of the furrow indicated the following infiltration equation

$I = 40t^{-0.5}$ litres per min per 30 m length

For 18 minutes after the water reached the end of the furrow, irrigation was continued with original stream size and then the stream size was cut back. Calculate the cut back stream size.

If the crop needs 40 mm of water calculate the time of irrigation and the average depth of water applied. Assume furrow spacing of 1 m.

Solution

Using the given infiltration equation for different opportunity times the following table may be prepared.

Distance (m)	*Time of advance (min)*	*Opportunity time (min)*	*Furrow infiltration (lpm/30 m)*	*Average infiltration (lpm/30 m)*	*Cumulative infiltration (lpm)*
0	0	160	3.16	—	—
30	26	134	3.46	3.31	3.31
60	58	102	3.96	3.71	7.02
90	98	62	5.08	4.52	11.54
120	142	18	9.43	7.26	18.80

Thus, it is observed that at the end of 160 minutes the infiltration rate for the entire furrow is 18.8 lpm and hence the stream size entering the furrow should also be equal to 18.8 lpm or slightly less than this. Time required to put 40 mm of water into the soil is given by Eq. (2.18), as

$$T = \left[\frac{30 \times D \times W \times (n+1)}{K}\right]^{\left(\frac{1}{n+1}\right)}$$

$$n = -0.5, K = 40 = 40 \text{ mm}, W = 1 \text{ m}$$

Thus, by substitution, we get

$$T = \left[\frac{30 \times 40 \times 1 \times 0.5}{40}\right]^{\left(\frac{1}{0.5}\right)} = 225 \text{ minutes}$$

Average depth of water applied may be calculated by taking the stream size as 25 lpm for 160 minutes and (i) 18.8 lpm, (ii) 17 1pm for the remaining period of (225 – 160) = 65 minutes as indicated below:

(i) When cut back stream size = 18.8 lpm

Average water depth applied in mm

$$= \left(\frac{25 \times 160 + 18.8 \times 65}{1000}\right) \times \frac{1}{120 \times 1} \times 1000$$

$$= 43.52 \text{ mm}$$

(ii) When cut back stream size = 17 1pm

Average water depth applied in mm

$$= \left(\frac{25 \times 160 + 17 \times 65}{1000}\right) \times \frac{1}{120 \times 1} \times 1000$$

$$= 42.54 \text{ mm}$$

Illustrative Example 2.5 *Furrows 120 m long and 0.9 m apart have a slope of 0.35 percent and are irrigated by an initial stream equal to non-erosive stream for 1 hour 30 minutes. The stream size is then reduced to one-third, which is continued for 40 minutes. Determine the average depth of irrigation.*

Solution

From Eq. (2.15), the maximum non-erosive flow rate is given as

$$q_m = \frac{0.60}{S}$$

$$= \frac{0.60}{0.35} = 1.714 \text{ lps}$$

After 1 hour 30 minutes the stream size becomes

$$q = \frac{q_m}{3} = \frac{1.714}{3}$$

$$= 0.571 \text{ lps}$$

The average water depth applied in mm

$$= \left(\frac{1.714 \times 5400 + 0.571 \times 2400}{1000}\right) \times \frac{1}{120 \times 0.9} \times 1000$$

$$= 98.39 \text{ mm}$$

REVIEW QUESTIONS

2.1 Discuss briefly the factors affecting the choice of the method of irrigation.

2.2 Discuss in brief the flooding methods of irrigation.

2.3 Describe in detail the border strip method of irrigation.

2.4 Find the time required to irrigate a border strip of length 150 m and width 10 m by a stream of discharge 0.03 cumec. Average depth of flow is expected to be 65 mm and average infiltration rate for the soil may be taken as 45 mm/hour. Also calculate the maximum length of the strip which can be irrigated in this case. [*Ans.* 1.417 hours; 240 m]

2.5 Derive an expression for the time taken to irrigate a given area in border strip method for a given discharge and a given rate of infiltration of water in the soil.

2.6 Write a note on furrow method of irrigation. Indicate the advantages of this method of irrigation.

2.7 Explain the terms (i) Contour farming; (ii) Wild flooding; (iii) Corrugations.

2.8 Describe in detail sprinkler method of irrigation. Indicate the advantages and limitations of this method.

2.9 Discuss the various sub-surface irrigation methods. Indicate their limitations.

2.10 How length and slope of a furrow are decided?

2.11 What do you understand by a cut back stream in the case of furrow irrigation?

CHAPTER

3 Soil Water and Plant Relationship

3.1 INTRODUCTION

The aim of irrigation practice is to ensure that the plants have an adequate supply of water in their root zone for achieving optimum yield of crops without damaging the quality of the soil. This necessitates a clear understanding of the interaction among the soil, water and plants in the root zone of the plants. The root zone of the plants of the most of the crops extends upto one to two metres below the land surface. As such the soil properties within the top one to two metres of the soil depth are of main interest to the agriculturists. However, the soil layers down may also need to be investigated to determine their infiltration characteristics.

The soil provides a structural base to the plants and allows the root system to get firmly embedded in the soil. Further the soil also provides water as well as the nutrients required by the plants for their growth. As such the soil-water-plant relationships which are of special significance from the point of view of irrigation include the capacities of the soils to retain water, use of this water by the plants and the movement of the excess water through the soils. The water retained by the soil is used by the plants through their roots. Water is required by plants in small quantities for metabolism and transportation of plant nutrients, and in much larger quantities in the physiological process of transpiration which protects the plants against the injurious effects of high temperatures.

The nutrients required by the plants are normally present in the soil in the form of salts. However, if the soil is deficient in any of these nutrients it may be supplemented by the required type of fertilizers. The salts present in the soil are dissolved in the water retained by the soil. The roots of the plants absorb this salt solution by the osmotic action which induces a pressure from a weaker or dilute salt solution towards a stronger or concentrated salt solution. Within the plants the salt solution is relatively concentrated due to the loss of water from their leaves by transpiration. It is therefore essential that sufficient water is retained by the soil to make the salt solution dilute enough to enable the plants to readily use it. Since soil is capable of retaining a limited amount of water and that too goes on depleting due to evaporation and transpiration, it needs to be replenished by irrigation at some interval.

In addition to water, air must also be present in the root zone for the respiration of the microorganisms to occur and to provide a favourable environment for root development and the absorption of nutrients. As such the excess water must be drained out from the soil to enable free circulation of air in the root zone.

For proper growth of the plants requisite quantites of water, air and nutrients must be available in the soil. Therefore while practicing irrigation it must be ensured that these quantites are available in correct proportions.

3.2 COMPOSITION OF SOIL, SOIL TEXTURE AND SOIL STRUCTURE

Soil is a complex mass of mineral and organic particles. The mineral particles of the soil are formed by the weathering and erosion of rocks. They are mainly composed of silica and silicates incorporating other chemicals such as potassium, calcium, phosphorous etc. The organic matter is formed from the residues of plants and animals. The minerals and the organic matter provide the nutrients required for the plants. The organic matter also provides the micro-organisms needed for the plant life.

A soil is usually classified according to the size of its particles. The material having particles of diameter between 20 mm and 2 mm is called *gravel,* while that having particles smaller than 2 mm in diameter is called *fine earth.* The term soil in general refers to fine earth, which has three components viz., sand, silt and clay. According to International Soil Classification the range of diameters of particles of each of these components of the soil are as indicated below.

Type of soil	*Range of particle diameter in millimetre*
Coarse sand	2 to 0.2
Fine sand	0.2 to 0.02
Silt	0.02 to 0.002
Clay	Less than 0.002 m

Soil Texture. Soil texture refers to the composition of the soil and it is reflected by the particle size, shape and gradation. Generally the soils occurring in nature are a combination of and, silt and clay. The relative proportions of sand, silt and clay in a soil mass determines the soil texture. According to textural gradations the soils may be broadly classified as (i) 'open' or 'light' textured soils, (ii) 'medium' textured soils, and (iii) 'tight' or 'heavy' textured soils. The 'open' or 'light' textured soils contain very low content of silt and clay, and hence these soils are coarse or sandy. The 'medium' textured soils contain sand, silt and clay in sizable proportions. In general loam is a soil which has all the three major size fractions in sizable proportions, and hence the loams are medium textured soils. The 'tight' or 'heavy' textured soil contain high content of clay. Thus, clayey soils are tight or heavy textured soils. Figure 3.1 presents the textural classification chart for 12 main textural classes of soils which is given by United States Department of Agriculture (USDA).

Soil Structure. In almost all the soils, groups of individual soil particles can adhere to form aggregates. The arrangement of individual soil particles and the aggregates with respect to each other into a pattern is called soil structure. The studies of soil structure have shown that soil is a porous medium, the interstices of which consist of irregularly shaped pores of various sizes.

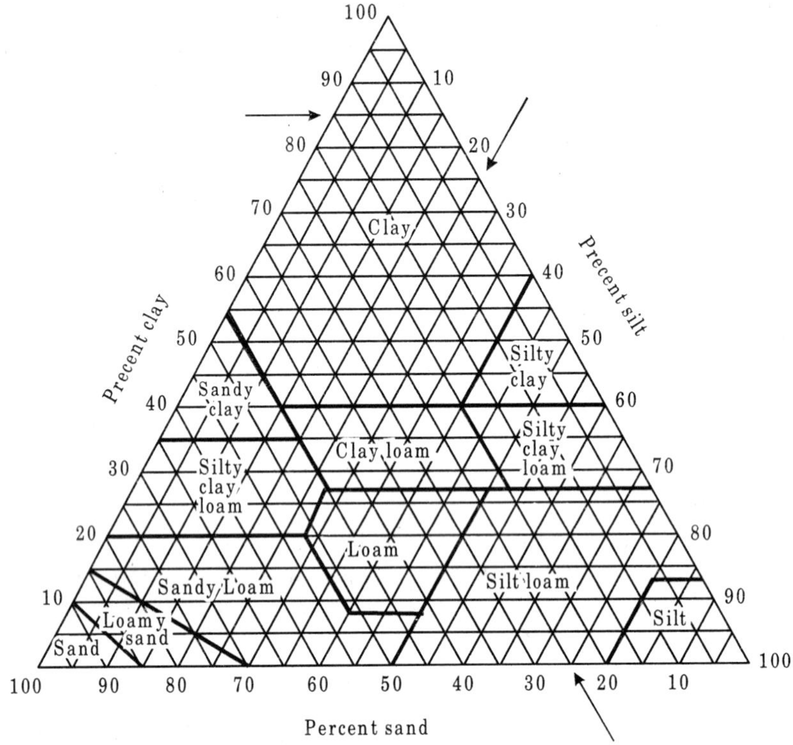

Fig. 3.1 *Soil textural classification chart*

Soil structures may be classified as (i) simple structure, and (ii) compound structure. Simple structure is the one in which natural cleavage planes are absent or indistinct. Simple structure may be a single grained structure or a massive structure. Single grained structure occurs only in sands and silts with low organic matter content. Massive structure is a coherent structure and its example is kaolinite clays. Compound structure is the one in which natural cleavage planes are distinct. Montmorillonite clays are the examples of compound structure.

Both texture and structure of a soil have pronounced effect on the properties of soil influencing irrigation.

3.3 WATER HOLDING CAPACITY OF SOIL

Water holding capacity of the soil is one of the dominant factors influencing irrigation. The water holding capacity of a soil mainly depends on its porosity. The porosity of a soil is defined as the ratio of the volume of pores in the soil mass to its total volume, and it is expressed as percentage. In general there are two types of soil pores viz., (i) capillary or small pores, and (ii) non-capillary or large pores. The capillary pores hold tightly by capillarity a large amount of water held by the soil at saturation and prevent it from being drained off under gravity. On the other hand the non-capillary pores do not hold water tightly and hence a large amount of water held by the soil at saturation is drained off under gravity. Thus, capillary pores induce greater water holding capacity while non-capillary pores induce drainage and aeration. The relative magnitudes of these types of pores in a soil depend on its texture and structure. Thus, a sandy soil has more non-capillary pores which result in better drainage and aeration but low water holding capacity. On the other hand a clayey soil has more capillary pores which result in better water holding capacity but poor drainage and aeration.

The water held by the soil is extracted by the roots of the plants for being used by the plants. In general the extraction of water from the soil by the roots of the plants is resisted by some forces, but the resisting forces are more in clayey soils than in sandy soils. Thus, water cannot be easily extracted by the roots of the plants in clayey soils although large amount of water is held by these soils. On the other hand relatively less amount of water is held by sandy soils, but water can be easily extracted from these soils by the roots of the plants.

Thus, an ideal soil for irrigation is that which has its pore space almost equally divided between capillary and non-capillary pores. Such a soil has enough small pores to provide adequate water holding capacity and also enough large pores to permit adequate drainage and aeration, and easy extraction of water by the roots of the plants. The loams are therefore ideal soils as they possess good water holding capacity, have good drainage and aeration, and allow extraction of water by the roots of the plants without much resistance.

3.4 SOIL GROUPS OF INDIA

The Indian soils may be divided into four major groups viz., (i) alluvial soils, (ii) black soils, (iii) red soils, and (iv) laterite soils. In addition to these four groups there exist another group of soils which includes forest soils, desert soils, and saline and alkali soils.

3.4.1 Alluvial Soils

Alluvial soils are formed by successive deposition of silt transported by rivers during floods, in the flood plains and along the coastal belts. The silt is formed from the weathering of the rocks by river water in the hilly terrain through which it flows. These soils form the largest and the most important group of soil in India. The alluvial soils occur in the Indo-Gangetic plains and Brahamputra plains in north India, and also in the plains of various rivers in other parts of the country. These are in general deep soils, that is having more than 1 metre depth above a hard stratum, but the properties of these soils occurring in different parts of the country vary mainly because the parent material from which they have been derived are different. These soils vary from clayey loam to sandy loam. The values of pH for these soils usually range between 7.0 and 9.0, and hence these soils may be neutral or alkaline in character. The water holding capacity of these soils is fairly good and they give good response to irrigation.

3.4.2 Black Soils

Black soils have evolved from the weathering of rocks such as basalts, traps, granites and gneisses. These soils occur chiefly in the states of Andhra Pradesh, Gujrat, Madhya Pradesh, Tamil Nadu, Maharashtra and Karnataka. The colour of these soils ranges from dark brown to black. Further irrespective of the nature of the parent rock from which black soils have developed, they do not differ much in general physical and chemical properties. These soils are highly argillaceous and very fine-grained. Thus, these are heavy textured soils and their clay content varies from 40 to 60 percent. The values of pH for these soils vary from 8.0 to 9.0 or higher in different states, and hence these soils are alkaline in character. A special feature of the black soils is that they are plastic and sticky when wet, and very hard when dry. These soils possess a high water holding capacity but poor drainage. Black soils are sub-divided as (i) shallow black soils which have a depth of 0.3 metre or less, (ii) medium black soils which are

0.3 metre to 1.0 metre in depth, and (iii) deep soils which are over 1 metre deep. Deep black soils are also referred to as black cotton soils since cotton is the most important crop in these soils.

3.4.3 Red Soils

Red soils are formed by the weathering of igneous and metamorphic rocks comprising gneisses and schists. These soils mostly occur in Tamil Nadu, Karnataka, Maharashtra, Andhra Pradesh, Madhya Pradesh and Orissa. They also occur in Bihar, West Bengal and some parts of Uttar Pradesh. These soils are in general light textured loams, but the properties of these soils vary from place to place. These are in general deep soils with values of pH ranging between 5.0 and 8.0, and hence in most of the cases these soils are acidic in character. The red soils have low water holding capacity. These soils react well to the application of irrigation water and on account of low water holding capacity they are well drained.

3.4.4 Laterite Soils and Forest Soils

Laterite soils are derived from the weathering of the laterite rocks. These soils occur mostly in Karnataka, Kerala, Madhya Pradesh, the Eastern Ghat region, Orissa, Maharashtra, Malabar and in some parts of Assam. These soils are reddish or yellowish-red in colour. The laterite soils have values of pH between 5.0 and 6.0 and hence these soils are acidic in character. These soils have low clay content and hence possess good drainage characteristics. Forest soils are formed by the deposition of organic matter derived from forest growth.

3.4.5 Desert Soils

These soils are found in the arid areas in the north-western region in the states of Rajasthan, Haryana and Punjab and are lying between the Indus river on the west and the range of Aravali Hills on the east. These soils are blown in from the coastal region and Indus valley, and are also derived from disintegration of rocks in the adjacent areas. These are light textured sandy soils of depth extending beyond 0.5 metre and react well to the application of irrigation water. However, these soils have a fairly high values of pH, and some of these soils contain a high percentage of soluble salts.

3.4.6 Saline and Alkali Soils

These are the soils which have appreciable concentration of soluble salts and exchangeable sodium content. These salts usually appear in the form of a white efflorescent crust on the surface of the soil. These soils are formed due to inadequate drainage of the irrigated lands, and hence these are found among the groups of alluvial, black and red soils. These soils are not suitable for cultivation, unless these are reclaimed by adopting suitable methods.

3.5 CLASSIFICATION OF SOIL WATER

The water added to a soil mass during irrigation or otherwise is held in the pores of the soil which is termed as soil water or soil moisture. The soil water may exist in the soil in various forms, on the basis of which it may be classified in the following three categories.

(i) Gravitational water

(ii) Capillary water

(iii) Hygroscopic water.

3.5.1 Gravitational Water

It is that water which is not held by the soil but drains out freely under the influence of gravity. With the addition of water to the soil during irrigation or otherwise, the water content of the soil is raised to a state of saturation corresponding to point A on the water-content (or moisture-content) versus time curve shown in Fig. 3.2. At this point the soil pores are completely filled

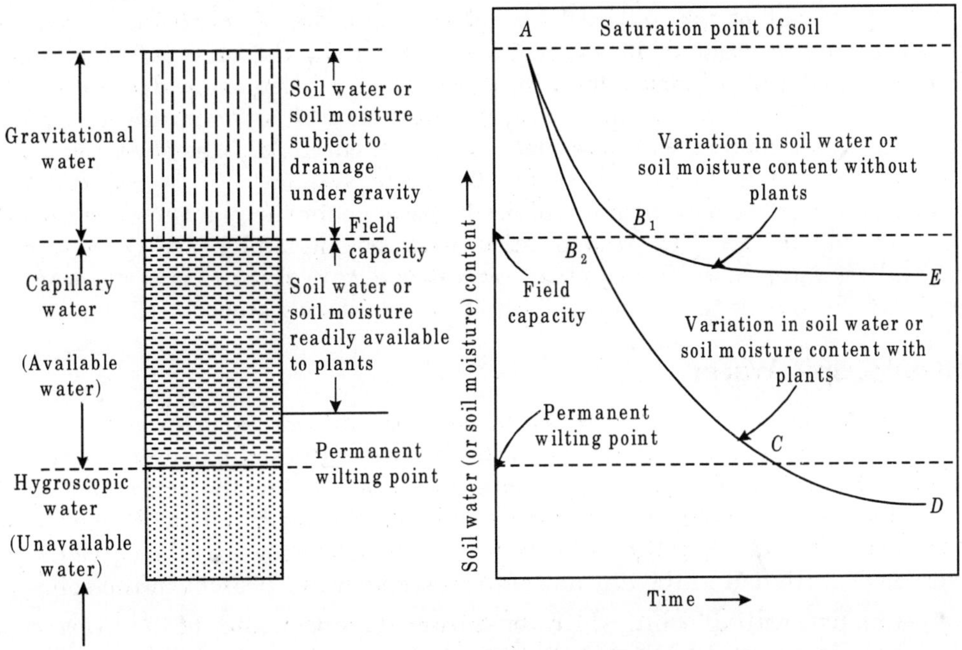

Fig. 3.2 *Classification of soil water*

with water and the soil contains the maximum possible water content, which thus constitutes the upper limit of the gravitational water. Some of this water is held very loosely by the soil and readily moves downward under the pull of gravity, unless it is prevented by an impermeable barrier such as hard pan or a high water table. As the loosely held water drains from the soil the water content of the soil is reduced as shown by the curve AB_1 (or AB_2) in Fig. 3.2, until it reaches a value corresponding to point B_1 (or B_2) beyond which free drainage of water from the soil ceases. The reduction in the water content of the soil will be along the curve AB_1 if there are no plants growing on the soil, and it will be along the curve AB_2 if there are plants growing on the soil which will utilize some water through transpiration. The water content between the points A and B_1 (or B_2) thus represents the gravitational water. Most of the gravitational water is not available to the plants for their use because it drains out rapidly from the root zone. The water content of the soil corresponding to point B_1 (or B_2) is termed as *field capacity* which thus constitutes the lower limit of the gravitational water.

3.5.2 Capillary Water

The water content retained in the soil after the gravitational water has drained off from the soil is known as *capillary water,* which is held in the soil by surface tension as a continuous film around the soil particles and in the capillary pores between the soil particles. The capillary water is thus held in the soil against the force of gravity. The plant roots gradually absorb the capillary water which thus constitutes the principal source of water for plant growth. The capillary water is also therefore designated as available water.

As shown in Fig. 3.2 the field capacity forms the upper limit of the capillary water which is gradually reduced due to evaporation and transpiration. If no plants are growing on the soil then the water content of the soil will be reduced only due to evaporation and hence the variation in the water content of the soil will take place along the curve AB_1E as shown in Fig. 3.2. On the other hand if the plants are growing on the soil then the water content of the soil will be reduced at a faster rate because of evaporation as well as transpiration and hence the variation in the water content will take place along the curve AB_2D as shown in Fig. 3.2. The plant roots will continue to extract water from the soil until it is reduced to a value corresponding to point C below which it will not be possible for the plant roots to extract enough water from the soil to meet the transpiration requirement, and hence the plants will wilt permanently. The water content in the soil corresponding to point C is termed as *permanent wilting point* which constitutes the lower limit of the capillary water.

3.5.3 Hygroscopic Water

The *hygroscopic water* is that water which is absorbed by the particles of dry soil from the atmosphere and is held as a very thin film on the surface of the soil particles due to adhesion or attraction between surface of particles and water molecules. Below the permanent wilting point the soil contains only hygroscopic water. Since hygroscopic water is held with considerable force, it cannot be removed easily from the soil particles. As such as shown by the curve CD in Fig. 3.2 there is very little variation in the amount of hygroscopic water present in the soil.

At the permanent wilting point, which constitutes the upper limit of the hygroscopic water, the plant growth ceases, and after this only small amount of water can be extracted from the soil by the plant roots which is enough only to maintain life of the plants. This continues till the water content of the soil is reduced to such an extent that ultimate wilting of plants occurs, that is the plants are completely wilted and are dead. The water content of the soil corresponding to ultimate wilting is known as *ultimate wilting point.* It is thus observed that only very little amount of hygroscopic water lying between permanent and ultimate wilting points can be extracted just for the survival of the plants. Moreover, below ultimate wilting point the hygroscopic water present in the soil cannot be extracted by the plant roots. As such hygroscopic water is in general not available for plant use.

The soil water may also be classified on the basis of its availability to the plant roots or otherwise as unavailable, available and gravitational water. Unavailable water is held too tightly by adhesive forces and is generally not available to plant roots. As described earlier most of the hygroscopic water cannot be extracted by plant roots and hence it is considered as unavailable water. Gravitational water drains quickly from the root zone under normal drainage conditions. Available water is the difference between gravitational and unavailable water. The capillary water held by the soil can be easily extracted by the plant roots and hence it is considered as available water.

3.6 SOIL MOISTURE TENSION

Soil moisture tension is defined as the force per unit area that must be exerted in order to extract water from the soil. Thus, it is a measure of the tenacity with which water is retained in the soil. Soil moisture tension is usually expressed in terms of atmospheres. Several other terms such as capillary potential, capillary tension, soil pull, and the force of suction are often used with the same meaning as that of soil moisture tension. For a soil of given texture and structure, soil moisture tension is inversely proportional to its moisture content. Thus, as indicated earlier at the permanent wilting point the moisture content of the soil is very low and it is held by the soil with such a tenacity that the corresponding soil moisture tension for almost all the soils is in the range of 7 to 32 atmospheres and hence it cannot be extracted by the roots of the plants. A knowledge of the amount of moisture held by the soil at various tensions is required in order to determine the amount of moisture that would be available to the plants and also the amount of water that must be used for irrigation. Soil moisture tensions for different soil moisture contents may be measured in the laboratory with the help of instruments such as centrifuge, pressure membrane (or pressure plate), suction plate, tensiometer etc. The results so obtained are presented as the plots of soil moisture content expressed as percentage of weight of dry soil versus soil moisture tension.

3.6.1 Soil Moisture Stress

Soil moisture stress is defined as the sum of the soil moisture tension and osmotic pressure of soil solution. In many irrigated soils, the soil solution contains an appreciable amount of salts. Salts in soil water increase the force that must be exerted to extract water and thus affect the amount of water available to plants. The increase in the force (or tension) caused by salts is from osmotic pressure. If two salt solutions differing in concentration are separated by a membrane impermeable to the dissolved substance, such as a cell membrane in a plant root, water moves from the solution of lower concentration to the one of higher concentration. The force with which water moves across such a membrane is called osmotic pressure and is measured in atmospheres. The osmotic pressure developed by the soil solution retards the uptake of water by plants. Thus, plants growth is a function of both soil moisture tension as well as osmotic pressure or in other words it is a function of soil moisture stress. For example, plants growing in a soil in which the soil moisture tension is, say, I atmosphere, apparently can extract enough moisture for good growth. But if the osmotic pressure of the soil solution is, say 10 atmospheres, the total soil moisture stress is 11 atmospheres and the plants cannot extract enough water for good growth. Thus, for good growth of plants in soils having appreciable salts, the osmotic pressure of the soil solution must be maintained as low as possible by controlled leaching so that in the root zone soil moisture stress is maintained in a range that will provide adequate moisture to the plants.

3.7 SOIL MOISTURE CONSTANTS

Certain soil moisture contents are of particular significance in relationship between soil moisture and plants. These soil moisture contents are often called soil moisture constants and the same are described below.

3.7.1 Saturation Capacity

Saturation capacity is defined as the total water content of a soil when all the pores of the soil are filled with water. This is also known as the maximum water holding capacity of the soil. At saturation capacity soil moisture tension is almost equal to zero as it is equal to the surface tension at free water surface.

3.7.2 Field Capacity

Field capacity is defined as the maximum amount of moisture which can be held by a soil against gravity. Thus, immediately after the gravitational water has drained off from a saturated soil, the moisture content held by the soil is the field capacity of the soil. The field capacity is usually expressed as the weight of the maximum amount of moisture held by the soil against gravity per unit weight of the dry soil and is given as a percentage. At field capacity the large or non-capillary pores of the soil are filled with air and the small or capillary pores are filled with water. Further as indicated in Fig. 3.2 the field capacity is the upper limit of the capillary water or the moisture content available to the plant roots.

The soil moisture tension at field capacity varies from soil to soil, but it is normally between 1/10 to 1/3 atmospheres. At field capacity for sandy soils the soil moisture tension tends to be near 1/10 atmospheres while for clayey soils it tends to be near 1/3 atmospheres.

In coarse textured sandy soils the field capacity may usually be achieved in about 1 to 3 days after the soil has been thoroughly wetted by irrigation. In medium textured sandy silt and sandy clay soils about 4 to 8 days may be required for the moisture content to reach field capacity. In fine textured soils containing large proportions of clay, somewhat longer periods are required for the moisture content to reach field capacity. However, a soil will come to field capacity more quickly when plants are growing on the soil than when there are no plants growing on the soil.

The field capacity of a soil may be determined as defined below.

(i) The soil in situ is wetted to saturation capacity.

(ii) It is allowed to drain freely under the action of gravity for two to three days. During this period the soil is covered with polythene sheet or a thick layer of mulch in order to stop evaporation.

(iii) A sample of the drained moist soil is taken and weighed.

(iv) The sample of the soil is then dried in an oven at 105°C. Higher temperatures are avoided as these may cause loss of organic matter of the soil and hence show a greater loss of weight than the actual resulting from the removal of only the moisture from the soil.

(v) The oven dried sample is then weighed.

(vi) The difference in the weight of the moist and the dried sample of the soil is determined which gives the weight of the maximum moisture content held by the soil sample. The field capacity of the soil is obtained by using the following expression.

$$\text{Field capacity} = \frac{\text{Weight of maximum moisture content held in the soil sample}}{\text{Weight of the dried soil sample}} \times 100$$

3.7.3 Moisture Equivalent

Moisture equivalent is defined as the percentage of moisture retained in an initially saturated sample of soil 10 mm thick after being subjected to a centrifugal force of 1000 times gravity for a period of 30 minutes. The moisture equivalent may vary from about 5% for coarse sandy soils to more than 50% for heavy fine textured clay soils. However, for some clean sands, the moisture equivalent may be as low as 2%. The moisture equivalent of a soil can be quickly determined in a laboratory and it is used as an approximate measure of field capacity. For medium textured soils the values of the field capacity and moisture equivalent are nearly equal. For sandy soils the field capacity exceeds the moisture equivalent. For very clayey soils the field capacity is generally lower than the moisture equivalent.

3.7.4 One-third Atmosphere Moisture Point

It is the percentage of moisture retained in a soil sample when placed on a porous plate and subjected to a pressure of one-third atmosphere. It also provides a good estimate of field capacity alike moisture equivalent. Moreover this method is less costly.

3.7.8 Permanent Wilting Point

The permanent wilting point is defined as the amount of moisture content present in a soil when the plants become permanently wilted. A plant is considered to be permanently wilted when it does not regain turgidity even after being placed in a saturated atmosphere, but it may however regain turgidity if water is added to the soil. At the permanent wilting point the films of water around the soil particles are held so tightly that the plant roots cannot extract enough moisture at sufficiently rapid rate to satisfy transpiration requirements thus resulting in the wilting of the plants. As such the permanent wilting point is at the lower end of the available moisture range as shown in Fig. 3.2. Some of the terms such as wilting coefficient and wilting point are generally used with the same meaning as that of permanent wilting point. The permanent wilting point is usually expressed as the weight of the moisture held by the soil per unit weight of the dry soil when the plants are permanently wilted. Moreover, if the permanent wilting point is expressed in percentage then it is usually termed as permanent wilting percentage. Permanent wilting points differ widely for different soils but have approximately the same values for different plants grown on the same soil. The value of permanent wilting percentage may be as low as 2% for light sandy soils and it may be as high as 30% for heavy clay soils.

The soil moisture tension of a soil at the permanent wilting point ranges from 7 to 32 atmospheres, depending on soil texture, the kind and condition of the plants, the amount of soluble salts in the soil solution and to some extent on the climatic environment. When permanent wilting point is reached, there is little change in the moisture content of the soil with rather large changes of soil moisture tension, and hence in reality whether the soil moisture tension is 10 to 20 atmospheres makes very little difference. Therefore when the soil is at permanent wilting point the soil moisture tension is generally considered to be 15 atmospheres.

The most common method of determining the permanent wilting point is to grow indicator plants in containers, usually in small cans, holding about 600 grams of soil. Sunflower plant is commonly used as the indicator plant. The plants are allowed to wilt and are then placed overnight in a chamber having saturated atmosphere to test them for permanent wilting. If permanent wilting has occurred then the plants will not regain turgidity even after being placed in a saturated atmosphere for such a long duration. The moisture content present in the soil is then determined which gives the permanent wilting point. This method for determining the permanent wilting point however involves a longer time. As such for quick results an alternative method may be adopted for the determination of permanent wilting point in which pressure membrane apparatus is used. In this method a soil sample is subjected to a pressure of 15 atmospheres, which is the usually assumed value of soil moisture tension for soil at permanent wilting point as mentioned earlier. The residual moisture content of the soil is then determined which gives the permanent wilting point. Further as an approximation, the permanent wilting percentage can be estimated by dividing the field capacity by a factor varying from 2.0 to 2.4, depending upon the amount of silt in the soil. For soils of high silt content 2.4 should be used.

3.7.9 Temporary Wilting

Temporary wilting of plants may occur on a hot windy day but the plants recover from wilting in the cooler portion of the day or during the night without any addition of water to the soil. Thus, temporary wilting of plants may occur even without much reduction in the soil moisture content.

3.7.10 Ultimate Wilting

When ultimate wilting occurs the plants are completely wilted, that is the plants die away and they cannot recover from wilting even after the addition of water to the soil. As mentioned in section 3.5 the moisture content present in the soil when ultimate wilting of plants occurs is termed as *ultimate wilting point* and the same when expressed as percentage is termed as *hygroscopic coefficient.* The hygroscopic coefficient is about 67% of the permanent wilting percentage. Although the difference in the amount of moisture content of the soil between the permanent wilting point and the ultimate wilting point is small, there is large difference in the values of soil moisture tensions at these points. At ultimate wilting point soil moisture tension may be as high as 60 atmospheres.

3.7.11 Available Moisture

The difference in moisture content of the soil between the field capacity and the permanent wilting point is termed as the available moisture. As described in Article 3.5, this represents the moisture which is stored in the soil in the form of capillary water for being used subsequently by the plants.

3.7.12 Readily Available Moisture

It is that portion of the available moisture which is most easily extracted by plant roots. Only about 75% of the available moisture is usually readily available, because the soil moisture content near the permanent wilting point cannot be easily extracted by the plant roots and hence it is not readily available for the plant use.

The average values of field capacity, permanent wilting point and available moisture (in percent) for the major soil types are as given below.

Percentage of dry weight of soil

Soil	*Field Capacity*	*Permanent wilting point*	*Available moisture*	*Dry density* (kg/m^3)
Sand	5	2	3	1500
Sandy loam	12	5	7	1400
Loam 1	18	10	8	1350
Silt loam	24	15	9	1300
Clay loam	30	19	11	1300
Clay	40	24	16	1200
Peat*	140	75	65	400

* Soil comprising dark fibrous or spongy textured vegetable matter formed from the decomposition of plants.

3.8 DEPTH OF WATER (OR MOISTURE) HELD BY SOIL IN ROOT ZONE AND AVAILABLE TO PLANTS

The water (or moisture) held by soil in root zone may be expressed in terms of depth of water (or moisture) which may be determined as indicated below.

Let d be the depth of root zone, w_s, be the specific weight of soil and w be the specific weight of water.

If unit area of soil is considered then the weight of soil per unit area and depth d is equal to $(w_s \times 1 \times d)$.

Further the weight of water (or moisture) held by the soil per unit area is equal to [$w \times 1 \times$ depth of water (or moisture)].

Field capacity

$$= \frac{\text{Weight of water (or moisture) held by soil per unit area}}{\text{Weight of soil per unit area}}$$

$$= \frac{w \times 1 \times \text{depth of water (or moisture)}}{w_s \times 1 \times d}$$

∴ Depth of water (or moisture) held by soil at field capacity

$$= \frac{w_s}{w} \times d \times \text{ (Field capacity)}$$

$$= S \times dx \text{ (Field capacity)}$$

in which S is the specific gravity of soil.

Similarly the depth of water (or moisture) held by soil at permanent wilting point

$$= \frac{w_s}{w} \times d \times \text{ (Permanent wilting point)}$$

$$= S \times d \times \text{ (Permanent wilting point)}$$

∴ Depth of available water (or moisture)

$$= \frac{w_s}{w} \times d \times \text{ (Field capacity – Permanent wilting point])}$$

$$= S \times d \times \text{ (Field capacity – Permanent wilting point)}$$

Further depth of available water (or moisture) per metre depth of soil

$$= \frac{w_s}{w} \text{ (Field capacity – Permanent wilting point)}$$

$$= S \text{ (Field capacity – Permanent wilting point)}$$

However, as described earlier the plants can easily extract only the readily available moisture. Hence, the readily available moisture

$$= \frac{w_s}{w} \times d \times \left(\text{Field capacity} - \left(\begin{array}{c} \text{Weight of moisutre content} \\ \text{held by soil at lower limit of} \\ \text{readily available moisture per unit} \\ \text{weight of dry soil} \end{array} \right) \right)$$

$$= S \times d \times \left(\text{Field capacity} - \left(\begin{array}{c} \text{Weight of moisutre content} \\ \text{held by soil at lower limit of} \\ \text{readily available moisture per unit} \\ \text{weight of dry soil} \end{array} \right) \right)$$

The readily available moisture per metre depth of soil may also be determined in the same manner as shown above.

3.9 EXTRACTION PATTERN OF SOIL MOISTURE IN ROOT ZONE BY PLANT ROOTS

The moisture extraction pattern shows the relative amounts of moisture extracted by plant roots from different depths in the root zone. Figure 3.3 shows the moisture extraction pattern of most of the crop plants growing in deep uniform soils with adequate supply of soil moisture. It

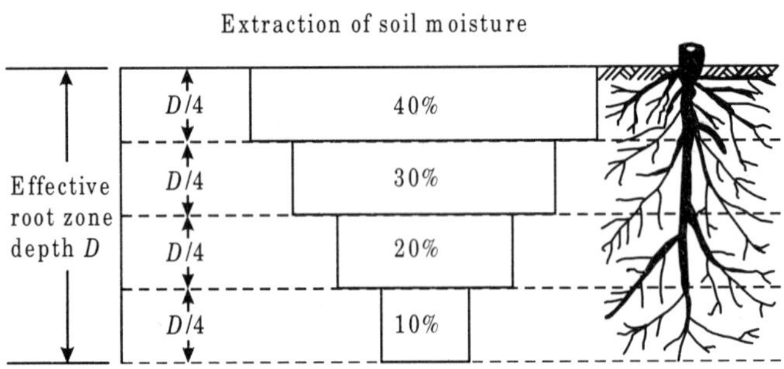

Fig. 3.3. *Extraction pattern of moisture from soil by plant roots*

may be observed that about 40 per cent of the total moisture used is extracted from the first quarter of the root zone, 30 per cent from the second quarter, 20 per cent from the third quarter and only 10 per cent from the last quarter. The larger extraction of moisture from the upper layers is due to the fact that in a uniform soil generally greater root development takes place in the upper layers of the soil than elsewhere. However, in a layered soil, the moisture available per unit depth varies and hence the extraction pattern and the root development vary from the one in a uniform soil.

3.10 ELEMENTS ESSENTIAL FOR PLANT GROWTH

For proper growth of plants various elements are essential. These elements may be classified according to their agricultural supply and/or sources from which these are derived in the following four groups.

Group I. Water and air elements. These elements include oxygen, hydrogen and carbon. The plants obtain these elements principally from water and carbon dioxide. However, certain quantities of these elements may also be taken up with other elements in such forms as bicarbonate ions, sulphate ions, etc.

Group II. Fertilizer elements. These elements include nitrogen, phosphorous, and potassium. These elements are important because they are used in large quantities by plants. The plants derive these elements from the soil. However, the soils are usually deficient in these elements. As such various fertilizers and/or manures are required to be added to the soils to provide these elements to the plants. Often sulphur is also included in this group of elements because it is closely related to the mineral supply.

Group III. Lime elements. These elements include calcium and magnesium. They are so named because they are the important elements in limestone which are acidic in nature. However, in irrigated soils generally there are ample quantities of both these elements.

Group IV. Minor elements. These elements include iron, manganese, boron, zinc, copper and molybdenum. These elements are also very important but they are needed in relatively small quantities by plants. As such these are also termed as trace elements.

3.11 NON-ESSENTIAL ELEMENTS OF INTEREST IN PLANT NUTRITION

There are several elements which have not been established as essential for plant growth but are still of great interest. These elements may be classified in five groups according to their significance in the plant growth or otherwise as defined below.

The first group of elements includes sodium, silicon, cobalt, titanium and nickel. Some of these elements commonly occur in plants and are beneficial in the growth of certain plants. For example sodium has been found to be beneficial in the growth of sugar beets and some other plants. It has also been found to be a substitute in part for potassium in the nutrition of the several plants. Silicon has been reported to be beneficial in strengthening the straw of small grain plants and is of further interest because of its very wide occurrence. Cobalt, nickel and titanium have been reported to be stimulatory to the growth of several plants under certain conditions.

The second group of elements includes chlorine, iodine, cobalt and sodium. These elements are not considered essential for plant growth but they are of interest because they are essential for man and animals, and the health of man and animals is related to the concentration of these elements in plants used for food or feed.

The elements included in the third group are selenium, lead, thallium, arsenic, molybdenum and fluorine. These elements if present in plants are injurious to the animals eating them.

The fourth group of elements includes sodium, chlorine, fluorine, nickel, lead, arsenic, selenium, aluminium, manganese and chromium. These elements if present in soil in sufficient quantities are very toxic to plants.

The fifth group of elements includes lithium, strontium, tin, radium, beryllium, vanadium, barium, mercury, silver and bromine. These elements also occur commonly in plants and some of which are reported to stimulate the growth of plants. However, it is probable that these elements are simply taken from the soil as impurities in the plant food.

3.12 MAINTAINING SOIL FERTILITY

For optimum yield of a crop it is essential that the soil on which it is grown is fertile. A soil is said to be fertile if it has ample storage of the nutrients required by the crop for its growth. However, if the land is used repeatedly for growing crops then since nutrients are consumed the soil becomes deficient in the nutrients and ultimately the soil fertility and hence the crop yield decreases considerably unless some methods are adopted to maintain the soil fertility. The various methods adopted to maintain the fertility of a soil are as follows.

3.12.1 By Keeping the Land Fallow

If the land is kept fallow, i.e., it is left uncultivated for one or more crop seasons then the soil is allowed to recoup in respect of the nutrients which are deficient and thus regain its fertility.

3.12.2 Addition of Manure and Fertilizers

If the soil is deficient in some nutrients then by adding manure as well as fertilizers the deficiency is removed and the fertility of the soil is improved.

3.12.3 Crop Rotation

Crop rotation means changing the crops to be grown every year in the same field. In other words it is a process of growing different crops in rotation in the same field. The soil fertility is maintained by crop rotation as explained below.

If the same crop is grown every year in any field then since the same type of nutrients are consumed the soil becomes deficient in these nutrients. On the other hand if different crops are grown in the same field then since different crops require different nutrients and in different proportions a balanced utilization of the nutrients results and the soil does not become deficient in particular type of nutrients. Moreover, if a soil becomes deficient in some of the nutrients by a certain crop grown in the field then it is allowed to recoup when next time a different crop is grown in the same field.

Different crops are affected by different types of diseases. Similarly different types of pests (destructive insects) thrive on different types of crops. As such if same crop is grown every year in the same field then crop diseases as well as pests multiply at an alarming rate. However, by adopting crop rotation the crop diseases as well as the growth of pests is controlled.

Different crops have different depths of the root zone. Thus, by combining deep rooted crops and shallow rooted crops in the rotation of crops the optimum utilization of the nutrients available in the soil is made.

A leguminous crop such as gram if introduced in rotation will increase nitrogen content of the soil thus increasing its fertility.

In addition to the soil fertility being maintained by crop rotation it is also beneficial to the cultivators because rotation of cash crops, fodder crops and soil renovating crops can be made. A larger number of crop rotations are being practised in our country, out of which the common crop rotations are as listed below.

(i) Wheat—Great Millet—Gram

(ii) Rice—Gram

(iii) Cotton—Wheat—Gram
(iv) Cotton—Wheat—Sugarcane
(v) Cotton—Great Millet—Gram.

3.12.4 Mixed or Inter-cropping

Mixed or inter-cropping is defined as the growing of two or more crops together in the same field during the same crop season. For example, wheat and mustard or gram and barley may be grown simultaneously in the same field. In mixed or inter-cropping the optimum utilization of the nutrients available in the soil is made. However, as indicated in Chapter 1, there are certain drawbacks of mixed cropping.

3.13 QUALITY OF IRRIGATION WATER

Generally irrigation water is not intended to provide any nutrient, trace element or any other element needed for the growth of the plants. However, except rain water which is near pure water, the irrigation water obtained from any other source contains some silt content as well as certain salts dissolved in it. The availability of some of these ingredients in irrigation water would supplement and even replenish the nutrients taken by the plants from the soil. However, some of the ingredients present in irrigation water may be injurious to the plant growth, which would render the water unsuitable for irrigation. As such the quality of irrigation water depends primarily on its silt content and salt constituents. The effect of each of these factors on the quality of irrigation water is discussed below.

3.13.1 Effect of Silt

The effect of silt on the quality of irrigation water depends on the nature of the silt material and the characteristics of the soil receiving the water. Thus, if the silt contains a larger content of plant nutrients and/or it comes from the fertile lands then it is quite beneficial, particularly for virgin soils of the agricultural land which has a low content of plant nutrients and a very low water holding capacity. The gradual accumulation of silt on this type of soil through irrigation water improves the soil characteristics considerably. However, in many cases the silt carried by the irrigation water may be injuries. This is especially so when the silt is not rich in plant nutrients and it is deposited on the surface of rather fertile land. Moreover, in some cases the silt accumulation may reduce the permeability of soils and make irrigation more difficult.

3.13.2 Effect of Salts

Salts of heavy elements such as lead, zinc, selenium, arsenic etc., are injurious to the plant growth even if they are present in low concentrations. However, under normal conditions the solubilities of these salts are quite low and hence the concentrations of these salts are usually less than those which may be harmful for the plants. On the other hand certain salts, primarily the chlorides, sulphates and borates of sodium, potassium, calcium and magnesium, have solubilities significantly larger than the tolerance limit of the plants. As such the quality of irrigation water is determined by the concentrations of these salts in water. The concentration of dissolved salts in water or salinity of water is determined by measuring the electrical conductivity (EC) of water and it is usually expressed in the units millimhos per centimetre or micromhos per centimetre.* Moreover the concentration of dissolved salts in water may also be

* Electrical conductivity is also expressed in the units milli siemen per metre and the following relation exists between the two units of electrical conductivity. 1 milli siemen per m = 10 micromhos per cm.

expressed in terms of parts per million (ppm); or in terms of milliequivalents per litre $\left(\frac{meq}{l}\right)$. The following relationships may be employed for evaluating the salt concentration from one unit to other.

(i) parts per million is numerically equal to milligram per litre $\left(\frac{mg}{l}\right)$, i.e., 1 ppm $= \left(\frac{1mg}{l}\right)$.

(ii) milliequivalent per litre is equal to parts per million divided by equivalent weight, i.e., $\left(\frac{meq}{l}\right) = \left(\frac{\text{ppm}}{\text{Eq.wt.}}\right)$

(iii) parts per million is equal to $\left(\frac{1}{640}\right)$ millimhos per centimetre ; i.e., 1 ppm $= \left(\frac{1}{640}\right)$ millimhos per cm.

Waters have been divided into four classes depending on their soluble salt content or salinity as assessed using the electrical conductivity value as given below.

Classification of water	*Electrical conductivity* $\frac{micromhos}{cm \text{ at } 25° C}$ *(EC)*
C1 Low salinity water	0 < EC < 250
C2 Medium salinity water	250 < EC < 750
C3 High salinity water	750 < EC < 2250
C4 Very high salinity water	2250 < EC < 5000

In general the irrigation water must have low concentration of dissolved salts or low salinity, because a high concentration of dissolved salts may result in dehydration of plants due to osmotic effects. As such water with high concentration of salts or high salinity is usually not suitable for irrigation. However, the suitability of water for irrigation is also influenced by the constituents of the soil which is to be irrigated and the crops to be grown. This is so because water with certain ingredients may be harmful for irrigation on a particular soil but the same water may be tolerable or even useful for irrigation on some other soil. Similarly some crops are salt-sensitive while others are salt-tolerant. Thus, waters of low salinity can be used for irrigating most crops on most soils but as the salinity level increases, the water becomes less suitable for the salt-sensitive crops and for use on heavy soils having low permeability and plenty of soluble salts. High salinity water can be used only for salt-tolerant crops on well-drained permeable soils.

Early studies on the quality of irrigation water were based primarily on the estimates of total quantities of soluble salts. However, this criterion of determining the quality of irrigation water is not reliable, because some salts may be more harmful to the plants than others.

Moreover, a high concentration of a particular salt in irrigation water may ultimately destroy the soil structure due to continuous exchange of ions between the salt contents of irrigation water and soil. For example, if sodium ions predominate in irrigation water, they will tend to replace calcium and magnesium ions from the clay and organic matter in the soil. The presence of calcium and magnesium keeps the soil in coagulated clod form since the silicates and aluminates of calcium and magnesium are insoluble in water and they possess cementing properties, but the exchange of calcium and magnesium ions by sodium produces soluble sodium silicates due to which the cementing property of the soil is lost. The clods then crumble and fine silica grains are released which clog the soil resulting in the reduction of its permeability and destruction of the structure of soil. Since this action depends on the exchange of calcium and magnesium ions by sodium ions, the quality of irrigation water will depend on the amount of exchangeable sodium ions available in water. The sodium hazard of water is thus assessed in terms of exchangeable sodium ions, the assessment of which is made by determining the exchangeable sodium ratio. The *exchangeable sodium ratio* (ESR) is defined as the concentration of exchangeable sodium ions divided by the sum of the concentrations of exchangeable calcium, magnesium and potassium ions, that is,

Exchangeable Sodium Ratio,

$$ESR = \frac{Na^+}{Ca^{++} + Mg^{++} + K^+} \quad ...(3.1)$$

The concentrations of all the elements are expressed in milliequivalents per litre (*meq/l*) which is equal to the concentrations in parts per million (ppm) divided by the equivalent weight of the element.

In most of the waters used for irrigation the concentration of potassium is relatively small and hence often the concentration of potassium is not considered. The expression for ESR then becomes

$$ESR = \frac{Na^+}{Ca^{++} + Mg^{++}} \quad ...(3.2)$$

Another parameter commonly used for the assessment of sodium hazard is *exchangeable sodium percentage* (ESP) which is related to ESR by the expression

$$ESP = 100 \times \frac{ESR}{1 + ESR} \quad ...(3.3)$$

Introducing the values of ESR given by Eqs (3.1) and (3.2), we obtain

$$ESP = 100 \times \frac{Na^+}{Ca^{++} + Mg^{++} + K^+ + Na^+} \quad ...(3.4)$$

or

$$ESP = 100 \times \frac{Na^+}{Ca^{++} + Mg^{++} + Na^+} \quad ...(3.5)$$

The assessment of sodium hazard may also be made on the basis of another more rational parameter termed as *sodium adsorption ratio* (SAR) which is defined as

Sodium Adsorption Ratio

$$SAR = \frac{Na^+}{\left[\frac{1}{2}(Ca^{++} + Mg^{++})\right]^{1/2}} \quad ...(3.6)$$

The concentrations of all the elements in this case are also expressed in milliequivalents per litre. An empirical relationship between SAR and ESP has been developed as follows.

$$ESP = \frac{100(-0.0126 + 0.01475\ SAR)}{1+(-0.0126 + 0.01475\ SAR)} \quad ...(3.7)$$

Equation 3.7 may be used to determine ESP for a known value of SAR.

Waters have also been divided into four classes depending on their sodium hazards as water with low sodium hazard or low-sodium water (S_1), water with medium sodium hazard or medium—sodium water (S_2), water with high sodium hazard or high-sodium water (S_3), and water with very high sodium hazard or very high-sodium water (S_4). Low-sodium water can be used for irrigation on almost all soils, with little danger of the development of a sodium problem. Medium-sodium water is suitable for only coarse-textured (sandy) or organic soils, as this water may present a moderate sodium problem in fine textured (clay) soils unless there is gypsum in the soil. High-sodium water may produce some sodium problems in most soils and will require special management : good drainage, high leaching and addition of organic matter. However, with high-sodium water also there may not be serious sodium problem if there is plenty of gypsum already present in the soil; or if gypsum or some similar material is added to the soil. Very high-sodium water is generally unsatisfactory for irrigation except on low-salinity or medium-salinity levels where the use of gypsum or some similar material makes its use possible.

From the foregoing discussion it is evident that in judging the suitability of irrigation water, both salinity hazard and sodium hazard (or SAR) should be kept in view. The classification of irrigation waters based on salinity and sodium hazards is shown in Fig. 3.4, which also shows how SAR limits change in the quality of irrigation water with changing salinity for given sodium hazards.

The quality of irrigation water is also significantly affected by the *boron* content of water. Some of the crops such as beans are very sensitive to an excess of boron, while others such as sugar beets will tolerate large quantities of boron. However, the presence of boron beyond 2 ppm in irrigation water renders it toxic for most of the crops.

Irrigation waters rich in *bicarbonate* content tend to precipitate calcium and magnesium in the soil as their carbonates which are insoluble under most conditions. The reactions involved can be represented by the following relations.

$$Ca + 2HCO_3 \Leftrightarrow CaCO_3 \downarrow + H_2O + CO_2 \uparrow$$
$$Mg + 2HCO_3 \Leftrightarrow MgCO_3 \downarrow + H_2O + CO_2 \uparrow$$

This reduces soluble calcium and magnesium and thus increases the sodium adsorption ratio or the sodium hazard. In this connection a term *residual sodium carbonate* (RSC) has been introduced which is defined as follows:

$$RSC = (CO^{--}_3 + HCO_3^-) - (Ca^{++} + Mg^{++}) \text{ (in } meq/l\text{)}$$

If RSC > 2.5 the water is not suitable for irrigation, if 1.25 < RSC < 2.5 the water is considered as marginal and if RSC < 1.25 the water is safe for irrigation.

Another parameter used for the assessment of the quality of irrigation water is pH which is defined as the logarithm of the reciprocal of hydrogen ion (H^+ ion) concentration in water. It is a measure of acidity and alkalinity of water. In general water (H_2O) is a combination of positively

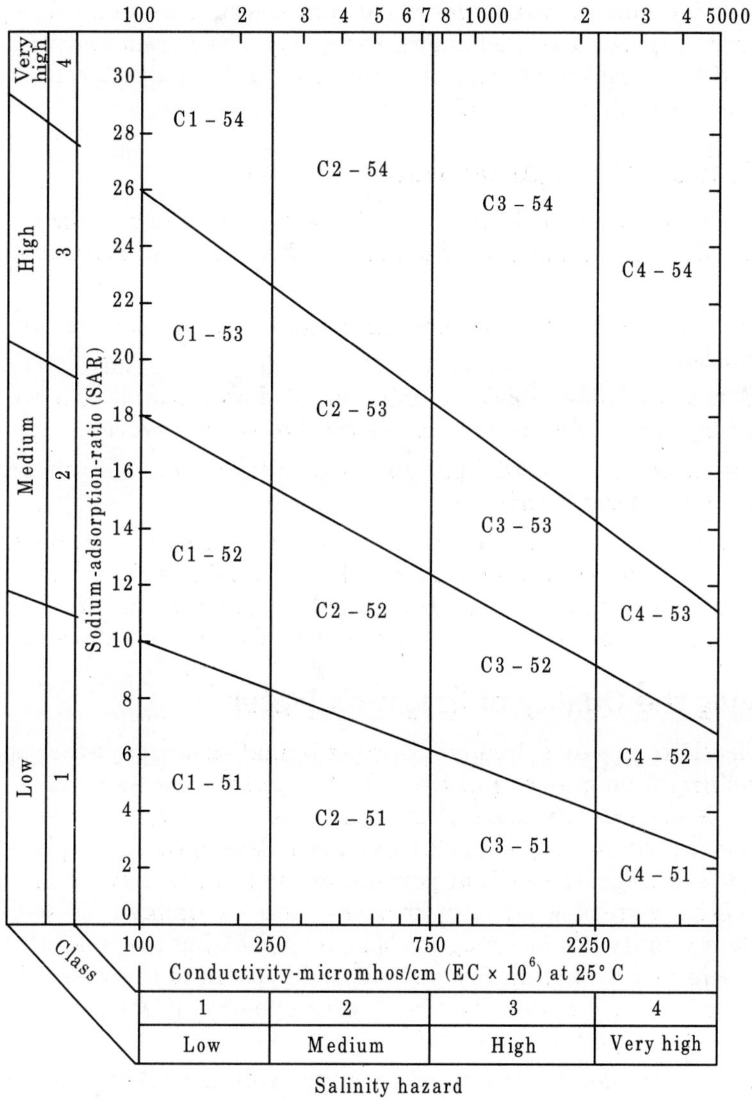

Fig. 3.4 *Diagram for determining the quality rating of irrigation water from its sodium adsorption ratio and electrical conductivity.*

charged hydrogen ions (H^+ ions) and negatively charged hydroxyl ions (OH^- ions). In pure water the concentration of H^+ ions is equal to that of OH^- ions. When some substance is dissolved in pure water the solution formed ionizes (i.e., splits up into H^+ ions and OH^- ions) and the balance between the concentration of H^+ ions and OH^- ions is disturbed. If the concentration of H^- ions is in excess of the concentration of OH^- ions the water solution becomes acidic and if the concentration of OH^- ions is in excess of the concentration of H^+ ions the water solution becomes alkaline. For pure or neutral water (i.e., neither acidic nor alkaline) concentration of H^+ ions is such that it is expressed by a pH value of 7. The value of pH for an acidic water solution will be less than 7 and that for an alkaline water solution will be more than 7. A pH value of 7.5 to 8.0 usually indicates the presence of carbonates of calcium and magnesium, and a pH value of 8.5 or above usually indicates appreciable exchangeable sodium. The parameter pH is also used for the assessment of the characteristics of soils. As mentioned later for different soils the pH of the saturated-soil pastes have different values.

3.13.3 Classification of Irrigation Water

On the basis of suitability of water for irrigation it may be classified in three categories as class I, class II and class III. The standard for each of the three classes of irrigation water are given in Table 3.1.

Class I—waters are considered to be excellent to good and are suitable for most crops under most growing conditions.

Class II— waters are considered good to injurious and are suitable only with permeable soils and moderate leaching. These waters may be harmful to some sensitive crops.

Class III— waters are considered injurious to unsatisfactory and are unsuitable for most crops except a few most tolerant varieties.

Further if a water falls in Class III on any basis, i.e., electrical conductivity, salt content percentage of sodium or boron content, it should be considered as unsuitable under most conditions. However, if the salts present are largely sulphates, the limiting values for total dissolved salts in each class given in Table 3.1 may be raised by 50 per cent.

3.13.4 Improving the Quality of Irrigation Water

Since in most areas there is only a limited choice of irrigation water, either the water available must be used or no irrigation may be practised. If the quality of water is poor, its prolonged use for irrigation may deteriorate the agricultural land. As such it is necessary to improve the quality of irrigation water. A simple device has been developed for improving the quality of waters having high exchangeable sodium percentages but which have relatively low total salt contents. Since these waters gradually increase the exchangeable sodium ions in soil, improvement in their quality can be brought about by increasing the concentration of calcium in the water. This is achieved by adding finely powdered gypsum to the water. Use of such treated water has improved the infiltration rate and permeability of soils previously damaged by irrigation with untreated high sodium water.

Study of water quality along a length of an intensively used stream often shows that poor quality of water is sometimes created by the entrance of a few small streams of high salt content. If these streams can be diverted the quality of water would be considerably improved, though there will be some reduction in the quantity of water available for irrigation.

3.14 SALINE, SALINE-ALKALI AND ALKALI (OR NON-SALINE ALKALI) SOILS

Soils may be classified as saline, saline-alkali and alkali (or non-saline alkali) soils on the basis of their soluble salt concentration and exchangeably sodium content. These soils are formed from the normal (i.e., non-saline) soils through the accumulation of salts from applied irrigation water or by upward moving ground water or by a combination of both these processes. The problem with these soils is that their crop yield is considerably low and gradually these soils become unsuitable for cultivation, unless remedial measures are taken for their reclamation. The formation of these soils may be prevented through proper planning and management of irrigation, but where this problem already exists the soil must be reclaimed before cultivation can be practised satisfactorily. The characteristics of these soils as well as the methods of their reclamation are described below.

3.14.1 Saline Soils

Saline soils are the soils which contain soluble salts in such quantities that they interfere with the growth of most of the plants. The electrical conductivity of the saturation extract* of saline soils is greater than 4 millimhos per centimetre (at 25 °C) and the exchangeable sodium percentage is less than 15. The pH of the saturated-soil paste of a saline soil is usually less than 8.5.

The soluble salts present in the saline soils are mainly chlorides, sulphates and sometimes nitrates of sodium, calcium, magnesium and potassium. Small amounts of bicarbonates of these elements may also occur but the soluble carbonates are usually absent. In addition to the readily soluble salts, saline soils may also contain relatively insoluble salts such as calcium sulphate (gypsum) and calcium and magnesium carbonates. The salts present in saline soils usually appear in the form of a white efflorescent crust on the surface of soil and hence the saline soils are also commonly termed as 'white alkali' soils.

In order to reclaim a saline soil it is necessary to reduce the soluble salt concentration to acceptable limits and this can be done by leaching, provided there is adequate drainage either natural or artificial.

TABLE 3.1 Standards for irrigation waters

Class of Water	*Electrical condutivity in micro-mhos/cm*	*Total dissolved salts (TDS) in ppm*	*Exchangeable Sodium percentage (ESP)*	*Chlorides in ppm*	*Sulphates in ppm*	*Boron in ppm*	*Remarks*
I	0–1000	0–700	0–60	0–142	0–192	0–0.5	Excellent to good for irrigation
II	1000–3000	700–2000	60–75	142–355	192–480	0.5–2.0	Good to injurious suitable only with permeable soils and moderate leaching. Harmful to more sensitive crops
III	Over 3000	Over 2000	Over 75	Over 355	Over 480	Over 2.0	Unfit for irrigation

* The saturation extract is the solution extracted by suction from soil brought to saturation by stirring in distilled water until the soil flows slightly without free water standing on the surface.

3.14.2 Saline-Alkali Soils

These are the soils for which the electrical conductivity of saturation extract is greater than 4 millimhos per centimetre (at 25°C) and the exchangeable sodium percentage is greater than 15. The pH of the saturated-soil paste of saline-alkali soil is generally about 8.5.

These soils can also be reclaimed by leaching. However, before leaching is carried out the exchangeable sodium ions present in the soils must be replaced by calcium or other suitable ions and then the replaced sodium ions and soluble salts are leached out. This is so because if soluble salts are removed first leaving behind the exchangeable sodium ions, the soil structure may deteriorate due to dispersion of soil particles resulting in the reduction of permeability of soil.

3.14.3 Alkali (or Non-saline Alkali) Soils

These are the soils for which the electrical conductivity of saturation extract is less than 4 millimhos per centimetre (at 25°C) and the exchangeable sodium percentage is greater than 15. The pH of the saturated-soil paste of an alkali soil is generally between 8.5 and 10. These soils are also termed as 'black alkali' soils because a black crust forms on the surface of these soils if organic matter is present.

The alkali soils can be reclaimed by reducing the exchangeable sodium percentage and removing the released sodium salts. However, the main difficulty with alkali soils is that they are usually almost impermeable, so that it is very difficult to get water and salts into them, and once in, it is often more difficult to get them out. To replace the sodium, chemicals such as gypsum and sulphur are introduced to increase the concentration of soluble calcium so that it replaces the exchangeable sodium. These chemicals are mixed into the soil mechanically or washed in, if intake rates and permeability can be improved. Soil permeability can be improved temporarily by soil cultivation and more permanently by converting the alkali soil into a saline-alkali soil. This can be done by irrigating with saline water, since the increase in soluble salt improves the soil structure and hence soil permeability. Once the soil becomes permeable the chemicals can be introduced to replace the sodium. Having replaced the sodium, it can be removed by controlled leaching and drainage.

SOLVED ILLUSTRATIVE EXAMPLES

Illustrative Example 3.1 *The analysis of a farm soil sample indicated that it has 30% clay, 30% silt and 40% sand. Corresponding to USDA classification of soils how would you classify this soil?*

Solution From Fig. 3.1, trace the intersection of line in the direction of arrows following the starting point of 40% sand, 30% clay and 30% silt. The intersecting point is in the region of clay loam. Hence the soil is classified as clayey loam.

Illustrative Example 3.2 *The root zone of a certain soil has a field capacity of 30% and permanent wilting percentage is 10%.*

(i) What is the depth of moisture in the root zone at field capacity and permanent wilting point ?

(ii) How much water is available if the root zone depth is 1.2 m. The dry weight of the soil is 13.73 Wm^3 *[1400 kg(f)/m*^{3}*].*

Solution

(i) The depth of moisture in root zone at field capacity per metre depth of soil

$$= \text{Sp. gr. of soil} \times \text{Field capacity}$$

$$= \left(\frac{13.73\times10^3}{9810}\times\frac{30}{100}\right)$$

$$= 0.42 \text{ m/m} = 420 \text{ mm/m}$$

The depth of moisture in root zone at permanent wilting point per metre depth of soil

$$= \text{Sp. gr. of soil} \times \text{Permanent wilting point}$$

$$= \left(\frac{13.73\times10^3}{9810}\times\frac{10}{100}\right)$$

$$= 0.14 \text{ m/m} = 140 \text{ mm/m}$$

(ii) Depth of available water per metre depth of soil

$$= \frac{w_s}{w} \text{ (Field capacity} - \text{Permanent wilting point)}$$

$$= \frac{13.73\times10^3}{9810}\left(\frac{30}{100}-\frac{10}{100}\right)$$

$$= 0.28 \text{ m/m} = 280 \text{ mm/m}$$

Total water available in the root zone = 280 ×1.2 = 336 mm

Illustrative Example 3.3 *Find the field capacity of a soil for the following data:*

(*i*) *Depth of root zone = 2 m*

(*ii*) *Existing water content = 5%*

(*iii*) *Dry density of soil = 1500 kg/m*^{3}

(*iv*) *Water applied to soil = 600 m*^{3}

(*v*) *Water lost due to evaporation and deep percolation = 10%*

(*vi*) *Area of land irrigated = % 900 m*^{2}.

Solution

Total water applied = 600 m^3

Loss of water = 10%

∴ Water retained in the soil

$$= 600\times\frac{90}{100}$$

$$= 540 \text{ m}^3$$

$$= 540 \times 9810 \text{ N}$$

Total dry weight of soil

$$= 900 \times 2 \times 1500 \times 9.81 \text{ N}$$

∴ % of water retained in the soil

$$= \frac{540 \times 9810}{900 \times 2 \times 1500 \times 9.81} = 0.2 \text{ or } 20\%$$

Existing water content = 5%

∴ Field capacity of soil

= 5% + 20%

= 25%

Illustrative Example 3.4 *A loam soil has field capacity 25% and permanent wilting percentage 10%. The dry unit weight of soil is 14.72 kN/m³ [1500 kg (f)/m³]. If the depth of the root zone is ft 75 m, determine the storage capacity of the soil. Irrigation water is applied when moisture content drops to 14%. If water application efficiency is 75%, determine the water depth required to be applied in the field.*

Solution

Maximum storage capacity of soil

= Available moisture

$$= \frac{w_s d}{w} \text{ (Field capacity – Permanent wilting percentage)}$$

$$= \frac{14.72 \times 10^3 \times 0.75}{9810}(0.25 - 0.10)$$

= 0.169 m = 169 mm

Depth of irrigation water

$$= \frac{w_s d}{w} \text{ (Field capacity – Minimum moisture content at which irrigation is required)}$$

$$= \frac{14.72 \times 10^3 \times 0.75}{9810}[0.25 - 0.14]$$

= 0.124 m = 124 mm

∴ Depth of irrigation water to be applied in the field

$$= \frac{124}{0.75} = 165 \text{ mm}$$

Illustrative Example 3.5 *The field capacity of a certain soils is 18.3% and its specific gravity is 1.25. A wet sample of the soil taken before irrigation weighs 153 gm and its weight after drying in the oven is 138 gm. What depth of water must be applied to irrigate the soil to a depth of 1.2 m?*

Solution

The moisture content before irrigation

$$= \frac{153 - 138}{138} \times 100 = 10.9 \text{ \% by weight}$$

∴ Depth of water required to be applied to bring the moisture upto its field capacity

$$= 1.25 \times 1.2 \left[\frac{18.3}{100} - \frac{10.9}{100}\right]$$

$$= 0.111 \text{ m} = 111 \text{ mm}$$

Illustrative Example 3.6 *The field capacity of a certain soil is 12.6% and its specific gravity is 1.42. If the moisture content present in the soil before irrigation is 8.2% how deep the soil profile will be wilted with an application of 50 mm of irrigation water.*

Solution

Depth of irrigation water applied

$$= S \times d \text{ (Field capacity – Moisture content before irrigation)}$$

Thus, by substitution we have

$$\frac{50}{1000} = 1.42 \times d\left[\frac{12.6}{100} - \frac{82}{100}\right]$$

$$\therefore \quad d = \frac{50}{1000} \times \frac{1}{1.42} \times \frac{100}{4.4}$$

$$= 0.80 \text{ m} = 800 \text{ mm}$$

Illustrative Example 3.7 *Determine the net depth of irrigation required to irrigate a field 1000 m long and 10 m wide from a source supplying water at the rate of 30 000 litres per hour in clay loam soil in the moderate climate. The field capacity of the soil is 27%, depth of root zone is 1 m, specific gravity of soil is 1.35 and irrigation is started when 50% of the available moisture has been used. Also determine the time required to irrigate the field.*

Solution

Depth of irrigation water required to be applied

$$d' = S \times d \text{ (Field capacity – Moisture content before irrigation)}$$

or

$$d' = 1.35 \times 1 \times \left[\frac{27}{100} - \frac{0.5 \times 27}{100}\right]$$

$$= 0.182 \text{ m} = 182 \text{ mm}$$

If irrigation water is applied at the rate of q cubic metre per hour for t hours over area of a hectare and d' is the depth of water required to be applied in metre, then we have

$$q \times t = a \times 10^4 \times d'$$

Thus, by substitution we get

$$\frac{30000}{1000} \times t = \frac{1000 \times 10}{10^4} \times 10^4 \times 0.182$$

$$\therefore \quad t = 60.7 \text{ hours}$$

Illustrative Example 3.8 *Determine the area which can be irrigated using surface method of irrigation with a source of warter having a discharge of10000 litres per minute. The available moisture holding capacity of the soil is 180 mm per metre and depth of root zone is 1.1 m.*

Irrigation is done when 40% of available moisture is used. The total losses during irrigation are 35% and peak daily moisture use is 4.5 mm.

Solution

Effective discharge of the source for irrigation

$= (10\ 000 \times 0.65) = 6500$ litres/minute

Depth of water available for the use of plants

$= (180 \times 0.40) = 72$ mm/m

Depth of root zone $= 1.1$ m

$\therefore$ Depth of water available for the entire root zone

$= (72 \times 1.1) = 79.2$ mm

Peak daily moisture use $= 4.5$ mm

$\therefore$ Watering interval $= \dfrac{79.2}{4.5} = 17.6$ days

$\simeq$ 17 days (say)

Thus, if a is the area in hectares which can be irrigated then we have

$$\frac{6500}{1000} \times 17 \times 24 \times 60 = a \times 10^4 \times \frac{79.2}{1000}$$

$\therefore$ $a = 201$ ha.

Illustrative Example 3.9 *The following data pertains to healthy growth of a crop:*

(i) Field capacity of soil = 30%

(ii) Permanent wilting percentage = 11%

(iii) Density of soil = 1300 kg/m^3

(iv) Effective depth of root zone = 700 mm

(v) Daily consumptive use of water for the given crop = 12 mm

For healthy growth moisture content must not fall below 25% of the water holding capacity between the field capacity and the permanent wilting point. Determine the watering interval in days.

Solution Maximum storage capacity of soil

$$= \frac{w_s}{w} d \text{ (Field capacity – Permanent wilting percentage)}$$

$w_s = 1300$ kg/m^3; $w = 1000$ kg/m^3; $d = 700$ mm $= 0.7$ m;

Field capacity $= \dfrac{30}{100} = 0.30$; and

Permanent wilting percentage = 11% = 0.11

Thus, by substitution, we get

Maximum storage capacity of soil

$$= \frac{1300}{1000} \times 0.7(0.30 - 0.11)$$

$= 0.1729$ m $= 172.9$ mm

Since for healthy growth moisture content must not fell below 25% of the maximum water holding capacity of soil.

$$\text{Available moisture} = 172.9 \times \frac{75}{100} = 129.675 \text{ mm}$$

Daily consumptive use of water = 12 mm

$$\therefore \quad \text{Watering interval} = \frac{129.675}{12}$$

$$= 10.81 \text{ days} = 10 \text{ days}$$

REVIEW QUESTIONS

3.1 Explain how soil properties affect the irrigation requirements.

3.2 Describe briefly the various soil groups of India.

3.3 Describe with the help of sketch various forms of soil moisture. Which of these soil moistures is mainly available for utilisation by the plants?

3.4 Explain the terms:

(i) Saturation capacity; (ii) Field capacity;

(iii) Moisture equivalent; (iv) Wilting point;

(v) Available moisture; (vi) Readily available moisture.

3.5 The analysis of a farm soil sample indicated that it has 25% clay, 35% sand and 40% silt. Corresponding to USDA classification of soils how would you classify this soil?

[*Ans.* Loam]

3.6 The root zone of a certain soil has a field capacity of 25% and permanent wilting percentage is 8%.

(i) What is the depth of moisture in the root zone at field capacity and permanent wilting point ?

(ii) How much water is available if the root zone depth is 1.1 m? The dry weight of the soil is 13.75 kN/m^3. [*Ans.* (i) 0.35 m/m, 0.112 m/m; (ii) 0.262 m]

3.7 Explain the terms (i) Soil moisture tension and (ii) Soil moisture stress.

3.8 Find the field capacity of a soil for the following data:

(i) Depth of root zone = 2 m

(ii) Existing water content = 6%

(iii) Dry density of soil = 1400 kg/m^3

(iv) Water applied to soil = 500 m^3

(v) Water lost due to evaporation and deep percolation = 10%

(vi) Area of land irrigated = 1000 m^2 [*Ans.* 22%]

3.9 A loam soil has field capacity 27% and permanent wilting percentage 12%. The dry weight of the soil is 13.73 kN/m^3 (1400 kg(f)/m^3). If the depth of root zone is 1 m, determine the storage capacity of the soil. Irrigation water is applied when moisture content drops to 15%. If water application efficiency is 75%, determine the water depth required to be applied in the field. [*Ans.* 0.210 m; 0.224 m]

CHAPTER

4 Water Requirements of Crops

4.1 INTRODUCTION

Water requirement of a crop is the total quantity of water required by the crop from the time it is sown to the time it is harvested. Different crops require different quantities of water. Since the growing crops use water continuously, it is essential to maintain the quantity of readily available moisture in the soil by irrigation. As such the total quantity of water required by a crop is so distributed that a part of it is applied each time at a more or less fixed interval throughout the period of growth. The quantity of water applied at each irrigation should be such that water sufficient to meet the needs of the crop for a period between two successive irrigations is stored in the soil. Therefore, in addition to the total quantity of water required by a crop, it is also essential to determine the frequency of irrigation as well as the quantity of water required to be applied during each application.

4.2 LIMITING SOIL MOISTURE CONDITIONS

The growth of most crops is retarded by either excessive or deficient amounts of soil moisture content. The excessive moisture content in the soil results in filling the soil pore spaces completely with water, thus driving out air from the soil in the root zone. Since for satisfactory plant growth the presence of air in the root zone is essential, the absence of the same retards the plant growth. On the other hand soils having deficient amounts of moisture hold it so tightly that plants are required to expend extra energy to obtain it. Moreover, if the rate of intake by the plants is not sufficient to

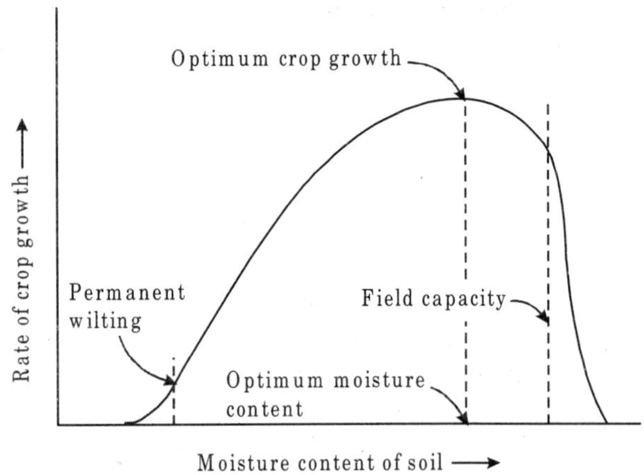

Fig. 4.1 *Limiting soil moisture conditions*

maintain turgidity of the leaves, permanent wilting follows. In between these two extreme soil moisture conditions, there is a moisture content designated as optimum moisture content at which plants grow most rapidly, resulting in the optimum growth of crops. For most crops the optimum moisture content is somewhat less than the field capacity of the soil as shown in Fig. 4.1 which shows the various limiting soil moisture conditions.

4.3 DEPTH OF WATER APPLIED DURING IRRIGATION AND FREQUENCY OF IRRIGATION

During irrigation the amount of water applied should be such that the moisture content of the soil is raised to its field capacity. After the irrigation is discontinued the moisture content of the soil is gradually reduced due to transpiration by the plants growing on the soil and evaporation from the soil surface. The total water (or moisture) lost from the soil due to transpiration and evaporation is termed as evapotranspiration or consumptive use. The moisture content of the soil should however not be allowed to get depleted below the amount of the readily available moisture. As such when the soil moisture content reaches the lower limit of the readily available moisture, it should be replenished by irrigation and again raised to the field capacity of the soil as shown in Fig. 4.2. The minimum depth of water d_w to be applied during irrigation is thus given by

$$d_w = \frac{w_s}{w} \times d \left(\text{Field capacity} - \left\{ \begin{array}{c} \text{Weight of moisture content} \\ \text{held by soil at lower limit of} \\ \text{readily available moisture per} \\ \text{unit weight of dry soil} \end{array} \right\} \right)$$

or $$d_w = \frac{w_s}{w} \times d \text{ (Weight of readily available moisture per unit weight of dry soil)}$$

or d_w = Depth of readily available moisture in the root zone of the soil

in which w_s is the specific weight of soil, w is the specific weight of water and d is the depth of root zone.

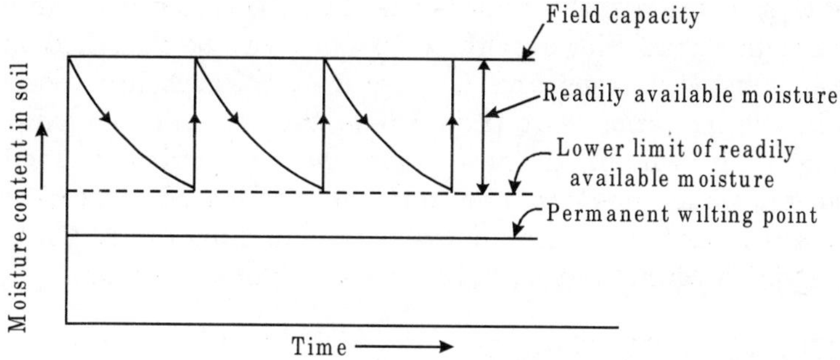

Fig. 4.2 *Frequency of irrigation*

The frequency of irrigation depends on the amount of readily available moisture in the root zone of the soil and the rate of consumptive use. Thus, if C_u is the rate of consumptive use expressed in terms of depth of moisture lost from the soil per day then the frequency of irrigation f_w is given by

$$f_w = \frac{d_w}{C_u} \text{ (days)}$$

4.4 CROP SEASONS AND CROPS OF INDIA

In general there are two main crop seasons of India viz., *Rabi* and *Kharif.* The crops grown during these crop seasons are designated as Rabi crops and Kharif crops. Rabi crops are also known as winter crops and Kharif crops are known as monsoon crops. Normally, Rabi crops are sown in the month of October and are harvested by the end of March, while Kharif crops are sown in the month of April and are harvested by the end of September. However, the sowing and harvesting times of the crops grown during the two crop seasons may vary from place to place in the country. Moreover, there is usually considerable overlapping of the crops grown in the two seasons. Table 4.1 gives some of the important crops grown during these crop seasons and their sowing and harvesting times.

There are certain crops which have a longer period between their sowing and harvesting times which extends from one crop season to the other. For example, sugarcane is sown sometimes during February to March and it is harvested sometimes during November to March next year. Thus, sugarcane takes almost a full year for maturity covering both the crop seasons and hence it is classified as a perennial crop. Similarly cotton also requires relatively longer time for maturity. Cotton is sown in May or early June and harvested in December or January next year. Since cotton requires about eight months for maturity, it is classified as eight month crop.

Sometimes in between the Rabi and Kharif crops intermediate crops are grown, in which case the crops may be classified as *Hot weather crops, Kharif (or monsoon) crops* and *Rabi (or winter) crops.* Hot weather crops are sown in February and harvested in May or early June, Kharif crops are sown in June and harvested in September or early October and Rabi crops are sown in October and harvested in February. Often the Hot weather crops and the Kharif crops are combinedly designated as *Summer crops.*

The above noted classification of the crops is based on the crop season, that is, on the sowing and harvesting times of the crops. However the crops may also be classified on the basis of their irrigation requirements as *Dry crops* and *Wet crops.* Dry crops are those crops which are ordinarily grown without irrigation, but they utilize the moisture stored in the soil during rainfall. The dry crops are grown only in those areas where the irrigation facilities are normally not available. However, sometimes dry crops are also irrigated, especially in the years of deficient rainfall, in which case these are known as irrigated dry crops. On the other hand wet crops are those crops which cannot normally be grown without irrigation.

4.5 CROP PERIOD AND BASE PERIOD

Crop period is defined as the total time that elapses between the sowing of the crop and its harvesting. Thus, crop period represents the total time during which the crop remains in the field. On the other hand *base period* is defined as the total time between the first watering done for the preparation of the land for sowing of a crop and the last watering done before its harvesting. It is thus evident that the crop period is slightly more than the base period for any crop. Both the crop period and the base period are expressed in days. The consideration of base period is essential for determining the total water requirement of a crop.

4.6 DUTY OF WATER AND DELTA

Duty of water is the relation between the area of the land irrigated and the quantity of water required to be supplied for growing a crop. It is usually defined as the area of land in hectares which can be irrigated for growing any crop if one cumec (one cubic metre per second) of water is supplied continuously to the land for the entire base period of the crop. For example, if 1500 hectares of land can be irrigated for growing any crop by one cumec of water supplied continuously for the entire base period of the crop then the duty of water for this crop is 1500 hectares per cumec. Since each crop has different water requirement the duty of water varies with crop. Further duty of water is said to be high or low according to the area of the land irrigated per cumec of water is large or small.

The duty of water expressed in hectares per cumec is convenient in the case of direct flow irrigation from canals, because by knowing the duty of water and the area of the land to be irrigated for growing crop, the required discharge for the canal can be determined. The duty of water expressed in this manner is usually termed as *flow duty of water*. However, the duty of water may also be expressed in the following different ways which are convenient in some cases.

The duty of water may be expressed in terms of the total area of land in hectares which may be irrigated for growing a crop per million cubic metre of water stored in a tank or reservoir. This mode of expressing the duty of water is specially suitable for tank irrigation because in such cases the areas which can be irrigated are dependent on the amount of water stored in the tanks supplying them. The duty of water expressed in this manner is usually termed as *quantity duty of water.*

In the case of a well (or tube well) the area which can be irrigated by the well (or tube well) annually is called the *duty of the well (or tube well).* On account of reduced transmission losses the duty of a well (or tube well) is usually higher than that of a canal irrigation. It has been ascertained that from a well with a lift of 7.5 m and using 'churus' as the lifting device about 1.5 hectares can be irrigated each year.

The duty of water may also be expressed in terms of the depth of water which is termed as delta and is defined below.

Delta is defined as the total depth of water over the irrigated land required by a crop grown on it during the entire base period of the crop. It is denoted by a symbol Δ (Greek 'delta'). The delta for any crop may be determined by dividing the total quantity of water in hectare-metres required by the crop for its growth by the area of the land in hectares over which the crop is growing. Thus, in other words delta may be defined as the total water required by a crop for its growth in hectare-metre per hectare. Since total water required by a crop is to be applied in stages and is distributed over the base period of the crop, the delta for the crop may be conveniently subdivided into the depths of water to be applied during each watering depending upon the crop requirements. The duty of water expressed in terms of delta is reckoned high or low according as the delta is less or more. Table 4.1 gives the average delta for the principal crops of India.

Table 4.1 Data for Principal Crops of India

Crop	*Sowing time*	*Harvesting time*	*Average delta* (Δ) *mm*
(A) Kharif Crops			
Early Rice	June–July	Sept.–Oct.	900
Transplanted Rice	July	November	1250
Maize	June	Sept.–Oct.	450
Bajra (Spiked Millet)	July–Aug.	Oct.–Nov.	300
Jowar (Great Millet)	June–July	Sept.–Oct.	300
Arhar	July–Aug.	March–April	300
Pulses (Moong, Moth, Urd)	July–Aug.	Oct.–Nov.	300
Groundnut	May	Nov.–Dec.	450
Til	July–Aug.	Oct.–Nov.	–
(B) Rabi Crops			
Wheat	Oct.–Nov.	March–April	375
Gram	Sept.–Oct.	March–April	300
Barley	Oct.–Nov.	March–April	450
Peas	Oct.–Nov.	March–April	500
Mustard	October	Feb.–March	450
Potato	October	February	750
Tobacco	Feb.–March	June	600
(C) Eight Month Crops			
Cotton	May–June	Dec.–Jan.	450
(D) Perennial Crops			
Sugarcane	Feb.– March	Dec.–March next year	900

In general the duty of water is expressed to include only the water applied through irrigation. However, if during the base period of a crop there is rainfall then the quantity of water required to be supplied through irrigation is correspondingly reduced. Thus, in this case duty of water may be expressed by taking into account either the total quantity of water, that is, rain plus the water supplied through irrigation, or only the water supplied through irrigation. The duty of water may therefore be expressed in either of the two ways viz., (i) duty of water inclusive of rain, and (ii) duty of water exclusive of rain. When rainfall is included in the duty of water this must be specifically stated, otherwise the duty of water refers only to the water supplied through irrigation. For example, if a crop requires a total depth of 1200 mm of water and during the base period of the crop 350 mm of effective or useful rainfall can be counted then the duty of water for the crop can be expressed either as (i) duty of water inclusive of rain is 1200 mm or (ii) duty of water exclusive of rain is 850 mm. The effective or usefull, rainfall refers to that quantity of rain water which is actually available for the use of crop.

In order to state the duty of water in a precise way it is essential to state (i) the base of duty of water, and (ii) the place of measurement of duty of water.

(i) *Base of duty of water.* The base of duty of water is defined as the period to which the stated duty of water has reference. The total quantity of water required by a crop is applied by irrigation through a number of waterings at certain interval. The quantity of water required to be applied during each watering however varies and it is required to be applied in a certain specified duration. The term duty of water is also frequently used to indicate the quantity of water required to be applied during each watering and it therefore refers to the period in which water is applied which is known as the base of the duty of water. When duty of water is expressed for each watering its base must also be stated. The duty of water expressed for each watering gives information about the actual rate at which water is applied on any day, or series of days. Similar information is however not given by the duty of water expressed for the entire base period of a crop because it only gives an average rate of water application during this period. Moreover, when the duty of water is expressed for the entire base period and if its base is not mentioned it generally means that the duty of water refers to the entire base period.

(ii) *Place of measurement of duty of water.* The duty of water varies with the place of its measurement. This is so because due to losses of water in transmission between the head of a canal system and the field by evaporation and percolation, the quantity of water delivered on the field is different from that enters the canal system at the head, and therefore the duty of water 'on the field' is different from that 'at the head ' of the canal system. For example, if 0.25 cumec of water is lost in transmission between the head of the canal system and the field and if the duty of water on the field is 1500 hectares per cumec, then the duty of water at the head of the canal system will be (1500/1.25) = 1200 hectares per cumec, because for delivering 1 cumec of water at the field 1.25 cumec of water must enter the canal system at the head. It, therefore, indicates that the duty of water increases as one moves from the head of the canal system to the field, and hence the place at which duty of water is measured must be specified. However, if the place at which the duty of water is measured is not mentioned then generally it means that the duty of water refers to that on the field.

Relation between duty of water in hectares per cumec and delta. Let D be the duty of water on the field in hectares per cumec, Δ be the delta or the total depth of water in metres supplied to a crop growing on the field during the entire base period and B be the base period of the crop in days.

For a field of area D hectares corresponding to the depth of water Δ metres the total quantity of water supplied for growing a crop on the field

$$= D \times \Delta \text{ hectare-metre}$$

$$= D \times \Delta \times 10^4 \text{ cubic metre} \quad \ldots\text{(i)}$$

Further for the same field of area D hectares for growing a crop on it if water is supplied at the rate of 1 cumec for the entire base period of B days, then the total quantity of water supplied to the field

$$= 1 \times B \times 24 \times 60 \times 60 \text{ cubic metre}$$

$$= 8.64 \times 10^4 \times B \text{ cubic metre} \quad \ldots\text{(ii)}$$

Equating (i) and (ii), we get

$$D \times \Delta \times 10^4 = 8.64 \times 10^4 \times B$$

or

$$D = \frac{8.64 \times B}{\Delta} \quad \ldots(4.1)$$

Equation (4.1) gives a relation between duty of water, delta and base period of a crop.

4.7 FACTORS AFFECTING DUTY OF WATER

The duty of water mainly depends on the following factors.

(1) Type of crop.

(2) Climatic condition of the area.

(3) System of irrigation.

(4) Method of irrigation.

(5) Quality of irrigation water.

(6) Method of cultivation.

(7) Time of irrigation and frequency of cultivation.

(8) Type of soil and sub-soil of the irrigated field.

(9) Type of soil and sub-soil of the area through which canal passes.

(10) Canal conditions.

(11) Method of assessment of irrigation water rate.

(12) Skill of the cultivator.

(13) Topography of land.

(14) Base period of crop.

1. **Type of Crop.** Different crops require varying quantities of water and therefore duty of water varies from crop to crop. The crops which require large quantity of water have lower duty of water than for the crops which requires less quantity of water.

2. **Climatic Condition of the Area.** The water requirement of a crop varies with the climatic condition of the area and hence it also affects the duty of water. The climatic conditions which affect the duty of water are (i) termperature, (ii) wind velocity, (iii) humidity and (iv) rainfall. If the temperature is high, the loss of water due to evapotranspiration will be more and hence the duty of water will be less. Similarly higher wind velocity will result in greater loss of water due to evapotranspiration and hence lower duty of water. On the other hand higher is the humidity higher will be the duty of water since the loss of water due to evapotranspiration will be less. Further if during the base period of a crop there is rainfall then since less quantity of irrigation water will be required to be supplied the duty of water will increase.

3. **System of Irrigation.** In the perennial irrigation system the soil of the irrigated area remains continuously wet and hence less quantity of water is required for initial saturation of the soil. Further in the areas where perennial irrigation is practised the water table is generally high and hence the loss of water due to deep percolation is less. Also in the perennial system of irrigation a better control on the use of irrigation water may be enforced. On the other hand in the inundation irrigation system there is a wasteful use of water. As such for the perennial irrigation system the duty of water is higher than for the inundation irrigation system. Similarly for the flow irrigation system the duty of water is low due to the transmission losses in the network of the canals. However, for tank irrigation system the duty of water may be high if the land to be irrigated is in the close vicinity of the tank and through proper control efficient use of water is made. On the other hand for the lift irrigation system the duty of water is generally high because the land to be irrigated is mostly near the well and hence the transmission losses are considerably reduced.

4. **Method of Irrigation.** The method of irrigation or the mode of applying water to the fields affects the duty of water because the water application efficiency varies with the method of irrigation. Thus, among the various surface irrigation methods the duty of water is higher for furrow method than for any of the flooding methods. This is so because in furrow method since water is not applied to the entire land surface the water losses are considerably reduced. Further as compared to the surface irrigation methods the duty of water is high for the sprinkler irrigation method as well as for the sub-surface irrigation methods, and generally for the drip irrigation method the duty of water is the highest.

5. **Quality of Irrigation Water.** If the irrigation water contains an appreciable amount of harmful salts and alkalies dissolved in it then it is required to be applied in large quantity so that the salts are leached off. This however results in a lower duty of water due to the wastage of considerable amount of water. On the other hand if the water is of better quality and contains useful salts dissolved in it then the wasteful use of water is avoided and it results in a higher duty of water.

6. **Method of Cultivation.** If the land is properly ploughed upto the required depth and made quite loose before irrigating, the soil will have high water retaining capacity in the root zone of the plants. This will reduce the number of waterings and hence result in a higher duty of water. Further the use of modern methods of cultivation gives higher duty of water than the old conventional methods of cultivation.

7. **Time of Irrigation and Frequency of Cultivation.** In the initial stages the land to be cultivated may not be properly levelled and hence more than the required quantity of water may be applied, which will result in a lower duty of water. The land slope may gradually improve with time then only the required quantity of water will be applied and hence the duty of water will be high. Further gradual rise of water table with time due to continuous irrigation will make water available in the root zone of the plants, thus relatively less quantity of water will be required to be applied which will result in a higher duty of water.

Frequent cultivation of land reduces the loss of moisture through weeds and evaporation from soil and hence results in a higher duty of water. Moreover, frequent cultivation helps to maintain the soil structure in good condition so that the soil is in good tilth having higher water retaining capacity and hence results in a higher duty of water.

8. **Type of Soil and Sub-soil of the Irrigated Field.** If the soil and sub-soil of the field to be irrigated is coarse grained then due to high percolation loss the duty of water will be low. However, the presence of an impervious layer below the root zone of the plants i.e., at a depth of 1 to 2 metres below the surface will reduce percolation of water and hence the duty of water will be high.

9. **Type of Soil and Sub-soil of the Area through which Canal Passes.** If the canal is unlined and it passes through coarse grained soil then since there will be greater percolation loss the duty of water will be low. On the other hand if an unlined canal is passing through fine grained soil then the percolation loss will be less and hence the duty of water will be high.

10. **Canal Conditions.** In an earthen canal, the percolation loss will be high which will result in a low duty of water. However, if the canal is lined then the percolation loss will be less and hence the duty of water will be high. The canals with good maintenance will have a higher duty of water than those having poor maintenance. Further if a canal is so aligned that the land to be irrigated is concentrated along it then since the transmission losses will be less the duty of water will be high. However, the dispersion of the fields to be irrigated with respect to the canal will increase the transmission losses and hence the duty of water will be low.

11. **Method of Assessment of Irrigation Water.** The assessment of irrigation water on volumetric basis prevents wastage of water and over irrigation by the farmers, thus leading to economy of water and a higher duty of water. However, if the assessment is made on flat rate basis or on the basis of area under cultivation then the farmers are tempted to use more water which leads to wastage of water and a lower duty of water.

12. **Skill of Cultivators.** The judicious use of water by the cultivators can save large quantity of water, which can be used to irrigate more area of the land thus resulting in a higher duty of water. However, the economical use of water will mainly depend on the skill of the cultivators and therefore if the cultivators are not skilled they should be trained to use irrigation water with utmost care and to avoid its wastage.

13. **Topography of Land.** If the land to be irrigated is properly levelled then uniform application of water will be possible which will result in economical use of water and hence a higher duty of water. On the other hand if the land is not levelled then the lower portion will receive more water than the higher portion, resulting in a wasteful use of water and hence the duty of water will be low.

14. **Base Period of Crop.** In general when the base period of a crop is long, more water may be required thus resulting in a lower duty of water. However, the requirement of water is not directly proportional to the base period of the crop.

4.8 METHODS OF IMPROVING DUTY OF WATER

The duty of water may be improved by counteracting all the factors which tend to lower the duty of water. Thus, duty of water may be improved by taking the following measures.

(1) The land should be properly ploughed to the required depth and levelled before sowing the crop.

(2) The land should be cultivated frequently because frequent cultivation reduces loss of moisture from the soil especially when the ground water is within capillary reach of the ground surface.

(3) Suitable methods of irrigation should be used.

(4) The canals should be lined. This will reduce the transmission losses.

(5) The canal should be close to the land to be irrigated by it so that the idle length of the canal is minimum and the transmission losses are reduced.

(6) The rotation of crops must be practised.

(7) Volumetric assessment of irrigation water should be made so that efficient and economical use of water is made by the cultivators.

(8) Good quality of water should be used for irrigation. The canal water should not be allowed to get polluted. Canal alignment through the soil which may dissolve harmful salts in the canal water should be avoided.

(9) The alignment of the canal either in sandy soil or in fissured rock should be avoided. However, if it cannot be avoided then the canal should invariably be lined.

(10) The cultivators must be trained to use irrigation water efficiently and economically.

(11) If possible instead of a single big size canal with higher full supply level, parallel canals of smaller size with lower full supply level should be constructed so that the loss of water due to percolation is reduced.

(12) The canals should be properly maintained so that the loss of water is minimised.

(13) The land should be redistributed among the cultivators by grouping together their small holdings so that irrigation water may be used more efficiently.

(14) Measures should be taken to increase water holding capacity of the soil and to conserve the soil moisture.

(15) The canal administrative staff should be well versed with the economical use of irrigation water. Further the operation of the canal system should be such that the cultivators both at the head as well as the tail end of the canal get water in sufficient quantity as and when they need it.

(16) Cultivators must be encouraged to adopt modern methods of cultivation instead of obsolete conventional methods.

4.9 COMMANDED AREAS AND INTENSITY OF IRRIGATION

A commanded area is defined as the area which can be irrigated (or commanded) by a canal system. The commanded areas may be classified as follows.

4.9.1 Gross Commanded Area (G.C.A.)

The gross commanded area is defined as the total area which can be irrigated by a canal system on the presumption that unlimited quantity of water is available. A canal is usually aligned along water-shed in between two drainage valleys, so that water can flow from it on both sides under gravity to the maximum possible area. However, the area to which water can flow from a canal will be restricted by the drainage boundaries. Thus, the gross commanded area is the total area lying between the drainage boundaries which can be irrigated by a canal system. The entire gross commanded area is however not culturable or cultivable because it also includes unculturable or uncultivable areas such as areas of habitation, roads, ponds, hillocks, barren land etc.

4.9.2 Culturable Commanded Area (C.C.A.)

The culturable commanded area is that portion of the gross commanded area which is culturable or cultivable. Thus, culturable commanded area may be obtained by subtracting the unculturable area from the gross commanded area. Thus, C.C.A. = G.C.A.– unculturable area.

The entire culturable commanded area is however never put under cultivation during any crop season due to the following reasons.

(i) The required quantity of water, manure, fertilisers etc. may not be available to cultivate the entire culturable commanded area.

(ii) If the land is kept fallow i.e., without cultivation for one or more crop seasons the fertility of the soil increases.

(iii) A portion of the land is required to be left uncultivated to serve as a pasture for the cattle.

(iv) The area under cultivation at any instant of time needs to be restricted to protect the land from waterlogging especially in areas where water table is high.

Thus, during any crop season only a part of the culturable commanded area is put under cultivation and hence the culturable commanded area may be subdivided into the following two categories.

(i) Culturable cultivated area

(ii) Culturable uncultivated area.

(i) **Culturable Cultivated Area.** It is that portion of the culturable commanded area which is actually cultivated during a crop season.

(ii) **Culturable Uncultivated Area.** It is that portion of tbe culturable commanded area which is not cultivated during a crop season.

4.9.3 Intensity of Irrigation

The intensity of irrigation is defined as the percentage of the culturable commanded area proposed to be irrigated annually. Usually the area irrigated during each crop season is expressed as a percentage of the culturable commanded area which represents the intensity of irrigation for the crop season. By adding the intensities of irrigation for all the crop seasons the yearly intensity of irrigation may be obtained. As such the yearly intensity of irrigation may often exceed 100%. For example, if the intensity of irrigation for Rabi is 50% and that for Kharif is 60% then the yearly intensity of irrigation will be 110%.

By knowing the intensity of irrigation for a crop season the *area to be irrigated* during that crop season can be determined. Further by knowing the areas over which different crops are grown during any crop season as well as the duty of water for each of these crops, the water requirement for the crop season can be determined. The water requirement for each crop season is required to be determined because different crops are grown during each crop season and their water requirements are different. Moreover, the canal for supplying irrigation water is designed to fulfill the water requirement for that crop season which needs maximum water during the crop period.

4.10 OTHER TERMS RELATED TO WATER REQUIREMENTS OF CROPS

There are certain terms related to the water requirements of the crops which are defined below.

4.10.1 Paleo Irrigation (or Paleo)

It is defined as the watering done prior to the sowing of a crop. This is done to prepare the land for sowing and to add sufficient moisture to the soil which would be required for the initial growth of the crop.

4.10.2 Kor Watering, Kor Depth and Kor Period

The total quantity of water required by a crop is applied through a number of waterings at certain interval during the base period of the crop. However, the quantity of water required to be applied during each of these waterings is not same. In general for all the crops during the first watering after the plants have grown a few centimetres high, the quantity of water required is more than that during the subsequent waterings. The first watering after the plants have grown a few centimetres high is known as *kor watering* and the depth of water applied during this watering is known as *kor depth.* The kor watering must be done in a limited period which is known as *kor period.* Since during kor watering certain quantity of water is required to be applied in a relatively short duration, the discharge capacity of the canal supplying irrigation water has to be maximum during this period which is indicated in the following example.

Wheat requires a total water depth of 375 mm and its base period is 182 days. Then using Eq. (4.1) the average duty of water for wheat for the entire base period is obtained as 4193 hectares per cumec. Thus, for irrigating 1000 hectares of land with wheat crop growing on it, the canal should have a discharge capacity of (1000/4193) = 0.238 cumec. Further for wheat the kor depth required is 135 mm and its kor period varies from three to eight weeks. Using Eq. (4.1) for a period of n weeks ($8.64 \times 7n/0.135$) = $448n$ hectares per cumec. Thus, for a kor period of three weeks the duty of water will be (448 × 3) = 1344 hectares per cumec and for irrigating 1000 hectares of the same land the canal should have the discharge capacity of (1000/1344) = 0.744 cumec. Similarly for a kor period of eight weeks the duty of water will be (448 × 8) = 3 584 hectares per cumec and again for irrigating 1000 hectares of the same land the canal should have a discharge capacity of (1000/3584) = 0.279 cumec. It may therefore be noted that for kor watering even for a relatively long kor period the required discharge is more than the average value of the discharge determined for the entire base period. As such the canal supplying irrigation water must be designed to meet the peak demand of kor watering and not merely the average demand. Further as indicated above in the two cases of kor watering with different kor periods although the total quantity of water to be supplied in each case is the same, more discharge will be required when the kor period is of smaller duration.

4.10.3 Outlet Factor

The irrigation water is supplied to the land to be irrigated by field channels (or water courses) which in the case of a flow irrigation system are supplied water from canals through the outlets. The duty of water at the outlet is known as the *outlet factor*. In other words outlet factor is the duty of water at the head of a field channel and as indicated earlier it is always less than the duty of water at the field on account of the transmission losses.

4.10.4 Capacity Factor

A canal is designed for a certain maximum discharge capacity, but it need not carry the same discharge at all the times. As such the ratio of the mean supply discharge of a canal for a certain duration to its maximum discharge capacity is defined as *capacity factor*.

4.10.5 Time Factor

Normally canal supplying irrigation water must run on all the days during each watering period but often on account of some unavoidable circumstances the canal may have to be closed for some days during the watering period. Thus, the ratio of number of days the canal has actually run during a watering period to the total number of days of the watering period is known as *time factor*. For example, if the total number of days of a watering period is 15 but the canal has actually run only for 9 days during this watering period then the time factor is equal to (9/15).

The canal may have to be closed for some days during any watering period due to some emergency arising out of either a breach in the canal bank or other major damage caused to the canal when it is running and which may need immediate repairs. In order to allow for any such eventuality the discharge capacity of the canal will have to be increased so that the required quantity of water may be made available to the cultivators for irrigation even if the canal actually runs for less number of days during any watering period. The increased discharge capacity of the canal is obtained by dividing the required discharge by the time factor.

4.10.6 Cumec Day

The total quantity of water flowing for one day at the rate of 1 cumec is known as **cumec-day.** One cumec-day is equal to 8.64 hectare-metres. Thus, cumec-day is a unit for measurement of quantity of water and it is used when relatively large quantity of water is involved.

4.10.7 Crop Ratio

Crop ratio is defined as the ratio of the areas of the land irrigated (or anticipated to be irrigated) during the two main crop seasons viz., Rabi and Kharif. Thus, it is also sometimes called *Rabi-Kharif ratio.* For example, if the area irrigated under Rabi crops is 500 hectares and that under Kharif crops is 1000 hectares then the crop ratio is 1:2. The crop ratio is selected in such a manner that the discharge requirement of the canal supplying irrigation water for both Rabi and Kharif crops may be nearly the same so that the canal section is put to the most efficient use. The crop ratio usually adopted varies from 1:1 to 1:2 for different states.

4.10.8 Overlap Allowance

It may so happen that the crop of one season may sometimes overlap the next crop season for some period. During such a period of overlapping irrigation water is required to be supplied simultaneously to the crops of both the seasons. Thus, there is extra demand of water during this period and to cater for this the canal discharge will have to be increased by some amount. The extra discharge of the canal required for this purpose is known as *overlap allowance.* The overlap allowance is usually expressed in percentage and it may be of the order of 5 to 10 per cent.

4.11 CONSUMPTIVE USE OF WATER-EVAPOTRANSPIRATION

Consumptive use of water or evapotranspiration is defined as the total quantity of water used by the vegetative growth of a given area in transpiration and building of plant tissue, and that evaporated from the adjacent soil in the area in any specified time. It therefore includes the water removed from the soil by transpiration and evaporation. *Transpiration* is the process in which the water that enters the plant roots and is used in building plant tissue, finally passes into the atmosphere in the vaporous form through the leaves of the plants. *Evaporation* is the process in which water from the adjacent soil passes into the atmosphere in the vaporous form. Moreover, the water deposited on the plant leaves in the form of dew, or a portion of rain water as well as sprinkler irrigation water which is intercepted by the plant leaves and subsequently evaporated without entering the plant system also constitutes a part of the consumptive use of water.

Consumptive use is expressed in hectare-metres per hectare or depth of water in metres for specified periods such as days, months or crop growing season. Consumptive use of water for each crop can be determined which gives the water requirement of the crop. The water requirement of a crop is also expressed in terms of duty of water as defined earlier. However, these days the consumptive use of water is widely used to indicate the total water requirement of a crop instead of duty of water. This is so because the duty of water includes in addition to the consumptive use of water for a crop, the water lost by evaporation and seepage from canals and that lost through deep percolation.

The value of the consumptive use of water varies from crop to crop and also for the same crop it varies with time as well as place. There are several factors as described below which operate individually or combinedly to influence the consumptive use of water.

Factors affecting consumptive use of water :

The various factors affecting the consumptive use of water are as follows.

(1) Evaporation from the soil
(2) Temperature
(3) Wind velocity
(4) Relative humidity of air
(5) Precipitation
(6) Day time hours
(7) Intensity of sunlight
(8) Soil type and topography
(9) Type of crop
(10) Cropping pattern
(11) Length of growing season
(12) Stage of the growth of the plant
(13) Amount of foliage of plants
(14) Nature of leaves of plants
(15) Method of irrigation
(16) Quantity of irrigation water applied
(17) Quantity of readily available soil moisture.

4.11.1 Potential Evapotranspiration (PET) and Actual Evapotranspiration (AET)

For a given set of climatic conditions, evapotranspiration obviously depends on the availability of water. If sufficient moisture is always available to completely meet the needs of the plants the resulting evapotranspiration is called *Potential evapotranspiration* (PET). Potential evapotranspiration no longer critically depends on soil and plant factors but depends essentially on climatic factors. The real evapotranspiration occurring in a specific situation is called *acutal evapotranspiration* (AET). At the field capacity, since the water supply to the plants is adequate AET will be equal to PET, i.e., the ratio (AET/PET) is equal to 1. With the reduction in the available moisture (as the soil dries out) the ratio (AET/PET) would be less than unity. The decrease in the ratio (AET/PET) with available moisture depends on the the type of soil and the rate of drying. In general for clayey soils the ratio (AET/PET) ≃ for nearly 50% drop in the available moisture. It is, however, evident that when the soil moisture reaches the permanent wilting point, the AET reduces to zero.

Methods of Determining the Consumptive use of Water

The various methods adopted for determining the consumptive use of water may be broadly classified under the following two categories.

(a) Direct measurement of consumptive use of water

(b) Use of empirical formulae for determining consumptive use of water.

(a) **Direct measurement of consumptive use of water**

The consumptive use of water can be measured by the following methods.

(i) Soil moisture studies on plots

(ii) Tank or Lysimeter method

(iii) Field experimental plots

(iv) Integration method

(v) Inflow and outflow studies for large areas.

(i) *Soil moisture studies on plots.* This method is usually employed to determine the consumptive use of water in irrigated field plots in which the soil is fairly uniform and the depth of ground water is such that it does not affect the soil moisture fluctuations within the root zone. Soil samples are taken before and after irrigation with some samples between irrigations. Usually a large number of soil samples must be taken at different depths in the roots zone. The moisture content of the soil samples is determined by standard laboratory methods and it is expressed as a percentage of the oven dry weight of the soil. From the moisture percentage thus obtained the quantity of water in hectare-metres per hectare (or metres) removed from the soil evaporation and transpiration is detetmined by using the formula

$$D = \frac{pSd}{100} \qquad \text{...(4.2)}$$

in which p represents the moisture percentage by weight ; S the apparent specific gravity of the soil ; d the depth of soil in metres ; and D the depth of water in metres removed from the soil.

The total quantity of water removed from the soil during each period is computed which gives the value of the evapotranspiration for that period from which evapotranspiration for a period of 30 days is determined. The value of the 30 days period evapotranspiration are plotted against time and a use-of-water curve for the crop season is obtained. The average use of water for each month may be obtained directly from this curve.

(ii) *Tank or Lysimeter method.* Tanks are watertight cylindrical containers which are open at one end and are set into the ground with their rim approximately flush with the surface. The tanks are 1 to 3 metres in diameter and 2 to 3 metres deep. Usually larger tanks are preferred because they permit more nearly natural root development for the plants grown in the tanks. The consumptive use of water is determined by measuring the quantity of water required to be supplied to maintain satisfactory growth of the plants within the tank.

Lysimeters are similar to the tanks with the difference that lysimeters have pervious bottoms. In the case of lysimeters also the plants are grown in the same manner as in the case of tanks but owing to their pervious bottom a part of the water applied is allowed to drain through it and the same is collected in a pan kept below it. The consumptive use of water in this case is therefore given by the difference between the water applied and that draining through the pervious bottom and collected in the pan. The advantage in the case of lysimeter is that gravity flow is permitted and hence conditions more close to the natural ones are developed.

(iii) *Field experimental plots.* In this method irrigation water is applied to selected field plots in such a way that there is neither runoff nor deep percolation but the quantity of water applied is sufficient for satisfactory growth of the crops grown. The consumptive use of water is then given by the quantity of water applied. However if there is some runoff resulting from the application of the irrigation water then the same should be measured and subtracted from the quantity of water applied to obtain the consumptive use of water. Further since deep percolation is not susceptible to precise measurement, it may introduce error in the result and in order to avoid this irrigation water is applied in small quantities.

(iv) *Integration method.* In this method the consumptive use of water is determined by the summation of the products of (a) consumptive use of water for each crop times its area, plus (b) consumptive use of water for natural vegetation times its area, plus (c) evaporation from water surface times water surface area, plus (d) evaporation from bare land times its area. For successful application of this method it is necessary to know the consumptive use of water and the areas of various classes of agriculture crops and natural vegetation, as well as the rate of evaporation from the water surface and bare land and their areas. Thus, in this method the annual consumptive use of water for the entire area is found in hectare-metres.

(v) *Inflow and outflow studies for large areas.* In this method the consumptive use of water is determined by measuring the inflow to and the outflow from a selected area. Thus, if U is the consumptive use of water for a certain area then its value is given by

$$U = (I + P) + (G_s - G_e) - O \qquad \ldots(4.3)$$

in which

I = Total inflow into the area during a 12 months year

P = Yearly precipitation of the area

G_s = Ground water storage of the area at the beginning of the year

G_e = Ground water storage of the area at the end of the year

O = Yearly outflow from the area

All the above noted quantities of water are measured in hectare-metres. Further in this method also annual consumptive use of water for large areas or valleys is found.

(b) **Use of empirical formulae for determining consumptive use of water**

Several reserchers have studied the influence of the various climatological parameters on the evapotranspiration and developed a number of formulae for determining the potential evapotranspiration (PET). Some of these formulae are purely empirical but some are supported by theoretical analysis. The following are some of the methods used commonly for determining evapotranspiration or consumptive use of water using these formulae.

(i) Modified Penman method

(ii) Jensen – Haise method

(iii) Hargreaves method

(iv) Thornthwaite method

(v) Blaney – Criddle method

(vi) Hargreaves class A pan evaporation method.

(i) *Modified Penman method.* Based on the concept that evapotranspiration is intimately connected to incoming solar energy, Penman developed a theoretical formula for potential evapotranspiration which in its modified form is as follows.

$$E_{t_p} = WR_n + (1- W) f(U) (e_s - e_a) \quad ...(4.4)$$

in which

E_{t_p} = potential evaporrtranspiration for reference crop alfalfa in mm/day

W = a weighting or weightage factor which is dimensionless

R_n = net radiation in mm of evaporable water per day

$f(U)$ = a function of wind velocity

e_s = saturation vapour pressure in mbar at the mean daily temperature, and

e_a = actual mean vapour pressure of the air in mbar.

The weighting factor W is given by the following expression

$$W = \frac{\Delta}{\Delta+\gamma} \quad ...(4.5)$$

in which

Δ = slope of saturation vapour pressure *v/s* temperature curve (de_s/dt) in mbar/°C at the mean daily temperature; and

γ = psychrometric constant.

For a known value of mean daily temperature t°C the value of Δ may be obtained from Table 4.2 which also gives the value of saturation vapour pressure e_s. However, for $t \geq 23$°C, the value of Δ in mbar/°C may also be computed by the following expression.

$$\Delta = 33.86\ [0.05904\ (0.00738t + 0.8072)7 - 0.0000342] \quad ...(4.6)$$

Table 4.2 Saturation vapour pressure e_s mbar and Δ mbar/°C

Temp. t°C	0	1	2	3	4	5	6	7	8	9	10	11	12	13	14
e_s mbar	6.1	6.6	7.1	7.6	8.1	8.7	9.3	10.0	10.7	11.5	12.3	13.1	14.0	15.0	16.1
Δ mbar/°C	0.4	0.5	0.5	0.5	0.5	0.6	0.6	0.7	0.7	0.8	0.8	0.8	0.9	1.0	1.1

Temp. t°C	15	16	17	18	19	20	21	22	23	24	25	26	27	28	29
e_s mbar	17.0	18.2	19.4	20.6	22.0	23.4	24.9	26.4	28.1	29.8	31.7	33.6	35.7	37.8	40.1
Δ mbar/°C	1.1	1.2	1.2	1.2	1.4	1.4	1.5	1.5	1.7	1.7	1.9	1.9	2.1	2.1	2.3

Temp. t°C	30	31	32	33	34	35	36	37	38	39	40	45	50	55	60
e_s mbar	42.4	44.9	47.6	50.3	53.2	56.2	59.4	62.8	66.3	69.9	73.8	95.8	123.4	157.5	199.3
Δ mbar/°C	2.5	2.5	2.7	2.7	2.9	3.1	3.2	3.4	3.5	3.6	3.9	4.4	5.5	6.8	8.4

The value of psychrometric constant γ is given by the following expression

$$\gamma = C_p \frac{P}{(0.622\lambda)} \quad ...(4.7)$$

in which

$$C_p = \text{a constant of value } 0.240$$

$$P = (1013 - 0.1055\ EL) \text{ mbar} \qquad \text{...(4.8)}$$

in which

EL = altitude of the place in metres

λ = latent heat of water in cal/gm, which is estimated by

$$\lambda = (595.9 - 0.55t) \qquad \text{...(4.9)}$$

in which t is mean daily temperature in °C

The value of weighting factor W may, however, be obtained directlly from Table 4.3 for a known value of mean daily temperature t°C and the altitude of the place in metres.

The net radiation R_n is given by the followng expression

$$R_n = R_{ns} - R_{nl} \qquad \text{...(4.10)}$$

in which

R_{ns} = the net shortwave solar radiation; and

R_{nl} = the net longwave solar radiation.

The net shortwave solar radiation R_{ns} is given by the following expression

$$R_{ns} = R_a\,(1 - r)\left(a + b\frac{n}{N}\right) \qquad \text{...(4.11)}$$

in which

R_a = extra-terrestrial radiation or incident solar radiation outside the atmosphere on a horizontal surface expressed in mm of evaporable water per day, (it is a function of the latitude of the place and period of the year as indicated in Table 4.4);

r = reflection coefficeient (or albedo), usual ranges of values of r are given below for different surfaces;

a = a constant depending upon the latitude ϕ of the place and is given by $a = (0.29 \cos \phi)$;

b = a constant with its value ranging from 0.50 to 0.54 and average value being taken as 0.52;

n = actual duration of bright sunshine in hours; and

N = maximum possible hours of bright sunshine, (it is a function of latitude of the place and period of the year as indicated in Table 4.5).

Values of albedo *r*

Surface	Range of r values
Close ground crops	0.15 – 0.25
Bare lands	0.05 – 0.45
Water surface	0.05
Snow	0.45 – 0.90

Table 4.3 Values of Weighting factor *W* for different temperatures and altitudes

Temp.°C	2	4	6	8	10	12	14	16	18	20	22	24	26	28	30	32	34	36	38	40
Altitude m	*Values of W*																			
0	0.43	0.46	0.49	0.52	0.55	0.58	0.61	0.64	0.66	0.68	0.71	0.73	0.75	0.77	0.78	0.80	0.82	0.83	0.84	0.85
500	0.44	0.48	0.51	0.54	0.57	0.60	0.62	0.65	0.67	0.70	0.72	0.74	0.76	0.78	0.79	0.81	0.82	0.84	0.85	0.86
1000	0.46	0.49	0.52	0.55	0.58	0.61	0.64	0.66	0.69	0.71	0.73	0.75	0.77	0.79	0.80	0.82	0.83	0.85	0.86	0.87
2000	0.49	0.52	0.55	0.58	0.61	0.64	0.66	0.69	0.71	0.73	0.75	0.77	0.79	0.81	0.82	0.84	0.85	0.86	0.87	0.88
3000	0.52	0.55	0.58	0.61	0.64	0.66	0.69	0.71	0.73	0.75	0.77	0.79	0.81	0.82	0.84	0.85	0.86	0.87	0.88	0.89
4000	0.54	0.58	0.61	0.64	0.66	0.69	0.71	0.73	0.75	0.77	0.79	0.81	0.82	0.84	0.85	0.86	0.87	0.89	0.90	0.90

Table 4.3 (a) Values of weightingfactor (1 – *W*) for different temperature and altitudes

Temp. °C	2	4	6	8	10	12	14	16	18	20	22	24	26	28	30	32	34	36	38	40
Altitude m	*Values of (1 – W)*																			
0	0.57	0.54	0.51	0.48	0.45	0.42	0.39	0.36	0.34	0.32	0.29	0.27	0.25	0.23	0.22	0.20	0.18	0.17	0.16	0.15
500	0.56	0.52	0.49	0.46	0.43	0.40	0.38	0.35	0.33	0.30	0.28	0.26	0.24	0.22	0.21	0.19	0.18	0.16	0.15	0.14
1000	0.54	0.51	0.48	0.45	0.42	0.39	0.36	0.34	0.31	0.29	0.27	0.25	0.23	0.21	0.20	0.18	0.17	0.15	0.14	0.13
2000	0.51	0.48	0.45	0.42	0.39	0.36	0.34	0.31	0.29	0.27	0.25	0.23	0.21	0.19	0.18	0.16	0.15	0.14	0.13	0.12
3000	0.48	0.45	0.42	0.39	0.36	0.34	0.31	0.29	0.27	0.25	0.23	0.21	0.19	0.18	0.16	0.15	0.14	0.13	0.12	0.11
4000	0.46	0.42	0.39	0.36	0.34	0.31	0.29	0.27	0.25	0.23	0.21	0.19	0.18	0.16	0.15	0.14	0.13	0.12	0.11	0.10

Table 4.4 Extra Terrestrial Radiation Ra expressed in equivalent evaporation in mm/day

Latitude (degree)	*Jan.*	*Feb.*	*Mar.*	*Apr.*	*May.*	*Jun.*	*Jul.*	*Aug.*	*Sep.*	*Oct.*	*Nov.*	*Dec.*
					Northern Hemisphere							
50	3.8	6.1	9.4	12.7	15.8	17.1	16.4	14.1	10.9	7.4	4.5	3.2
48	4.3	6.6	9.8	13.0	15.9	17.2	16.5	14.3	11.2	7.8	5.0	3.7
46	4.9	7.1	10.2	13.3	16.0	17.2	16.6	14.5	11.5	8.3	5.5	4.3
44	5.3	7.6	10.6	13.7	16.1	17.2	16.6	14.7	11.9	8.7	6.0	4.7
42	5.9	8.1	11.0	14.0	16.2	17.3	16.7	15.0	12.2	9.1	6.5	5.2
40	6.4	8.6	11.4	14.3	16.4	17.3	16.7	15.2	12.5	9.6	7.0	5.7
38	6.9	9.0	11.8	14.5	16.4	17.2	16.7	15.3	12.8	10.0	7.5	6.1
36	7.4	9.4	12.1	14.7	16.4	17.2	16.7	15.4	13.1	10.6	8.0	6.6
34	7.9	9.8	12.4	14.8	16.5	17.1	16.8	15.5	13.4	10.8	8.5	7.2
32	8.3	10.2	12.8	15.0	16.5	17.0	16.8	15.6	13.6	11.2	9.0	7.8
30	8.8	10.7	13.1	15.2	16.5	17.0	16.8	15.7	13.9	11.6	9.5	8.3
28	9.3	11.1	13.4	15.3	16.5	16.8	16.7	15.7	14.1	12.0	9.9	8.8
26	9.8	11.5	13.7	15.3	16.4	16.7	16.6	15.7	14.3	12.3	10.3	9.3
24	10.2	11.9	13.9	15.4	16.4	16.6	16.5	15.8	14.5	12.6	10.7	9.7
22	10.7	12.3	14.2	15.5	16.3	16.4	16.4	15.8	14.6	13.0	11.1	10.2
20	11.2	12.7	14.4	15.6	16.3	16.4	16.3	15.9	14.8	13.3	11.6	10.7
18	11.6	13.0	14.6	15.6	16.1	16.1	16.1	15.8	14.9	13.6	12.0	11.1
16	12.0	13.3	14.7	15.6	16.0	15.9	15.9	15.7	15.0	13.9	12.4	11.6
14	12.4	13.6	14.9	15.7	15.8	15.7	15.7	15.7	15.1	14.1	12.8	12.0
12	12.8	13.9	15.1	15.7	15.7	15.5	15.5	15.6	15.2	14.4	13.3	12.5
10	13.2	14.2	15.3	15.7	15.5	15.3	15.3	15.5	15.3	14.7	13.6	12.9
8	13.6	14.5	15.3	15.6	15.3	15.0	15.1	15.4	15.3	14.8	13.9	13.3
6	13.9	14.8	15.4	15.4	15.1	14.7	14.9	15.2	15.3	15.0	14.2	13.7
4	14.3	15.0	15.5	15.5	14.9	14.4	14.6	15.1	15.3	15.1	14.5	14.1
2	14.7	15.3	15.6	15.3	14.6	14.2	14.3	14.9	15.3	15.3	14.8	14.4
0	15.0	15.5	15.7	15.3	14.4	13.9	14.1	14.8	15.3	15.4	15.1	14.8
					Southern Hemisphere							
50	17.5	14.7	10.9	7.0	4.2	3.1	3.5	5.5	8.9	12.9	16.5	18.2
48	17.6	14.9	11.2	7.5	4.7	3.5	4.0	6.0	9.3	13.2	16.6	18.2
46	17.7	15.1	11.5	7.9	5.2	4.0	4.4	6.5	9.7	13.4	16.7	18.3
44	17.8	15.3	11.9	8.4	5.7	4.4	4.9	6.9	10.2	13.7	16.7	18.3
42	17.8	15.5	12.2	8.8	6.1	4.9	5.4	7.4	10.6	14.0	16.8	18.3
40	17.9	15.7	12.5	9.2	6.6	5.3	5.9	7.9	11.0	14.2	16.9	18.3
38	17.9	15.8	12.8	9.6	7.1	5.8	6.3	8.3	11.4	14.4	17.0	18.3
36	17.9	16.0	13.2	10.1	7.5	6.3	6.8	8.8	11.7	14.6	17.0	18.2
34	17.8	16.1	13.5	10.5	8.0	6.8	7.2	9.2	12.0	14.9	17.1	18.2

Contd.

Contd.

Latitude (degree)	*Jan.*	*Feb.*	*Mar.*	*Apr.*	*May.*	*Jun.*	*Jul.*	*Aug.*	*Sep.*	*Oct.*	*Nov.*	*Dec.*
32	17.8	16.2	13.8	10.9	8.5	7.3	7.7	9.6	12.4	15.1	17.2	18.1
30	17.8	16.4	14.0	11.3	8.9	7.8	8.1	10.1	12.7	15.3	17.3	18.1
28	17.7	16.4	14.3	11.6	9.3	8.2	8.6	10.4	13.0	15.4	17.2	17.9
26	17.6	16.4	14.4	12.0	9.7	8.7	9.1	10.9	13.2	15.5	17.2	17.8
24	17.5	16.5	14.6	12.3	10.2	9.1	9.5	11.2	13.4	15.6	17.1	17.7
22	17.4	16.5	14.8	12.6	10.6	9.6	10.0	11.6	13.7	15.7	17.0	17.5
20	17.3	16.5	15.0	13.0	11.0	10.0	10.4	12.0	13.9	15.8	17.0	17.4
18	17.1	16.5	15.1	13.2	11.4	10.4	10.8	12.3	14.1	15.8	16.8	17.1
16	16.9	16.4	15.2	13.5	11.7	10.8	11.2	12.6	14.3	15.8	16.7	16.8
14	16.7	16.4	15.3	13.7	12.1	11.2	11.6	12.9	14.5	15.8	16.5	16.6
12	16.6	16.3	15.4	14.0	12.5	11.6	12.0	13.2	14.7	15.8	16.4	16.5
10	16.4	16.3	15.5	14.2	12.8	12.0	12.4	13.5	14.8	15.9	16.2	16.2
8	16.1	16.1	15.5	14.4	13.1	12.4	12.7	13.7	14.9	15.8	16.0	16.0
6	15.8	16.0	15.6	14.7	13.4	12.8	13.1	14.0	15.0	15.7	15.8	15.7
4	15.5	15.8	15.6	14.9	13.8	13.2	13.4	14.3	15.1	15.6	15.5	15.4
2	15.3	15.7	15.7	15.1	14.1	13.5	13.7	14.5	15.2	15.5	15.3	15.1
0	15.0	15.5	15.7	15.3	14.4	13.9	14.1	14.8	15.3	15.4	15.1	14.8

The net longwave solar radiation R_{nl} is given by the following expression

$$R_{nl} = f(t) \cdot f(e_a) \cdot f\left(\frac{n}{N}\right) \qquad ...(4.12)$$

in which

$$f(t) = \sigma\, T_a \qquad ...(4.13)$$

where σ = Stefan – Boltzman constant the value of which is about 2.0×10^{-9} mm/day; and

T_a = mean air temperature in °K, i.e., equal to (273 + t°C)

$$f(e_a) = \left[0.34 - 0.044\sqrt{e_a}\right] \qquad ...(4.14)$$

in which e_a is actual mean vapour pressure of air in mbar as defined earlier

$$f\left(\frac{n}{N}\right) = \left[0.10 + 0.90\frac{n}{N}\right] \qquad ...(4.15)$$

in which n and N are same as defined earlier.

The function of wind velocity $f(U)$ in Eq. (4.4) is given by the following expression

$$f(U) = 0.27\left(1 + \frac{U_2}{100}\right) \qquad ...(4.16)$$

in which

U_2 = mean wind velocity in km/day at a height of 2 m above the ground.

If wind velocity is available at a height other than 2 m, then the following equation may be used to estimate the wind velocity at 2 m height from a wind velocity at height z

$$U_2 = U_z \left(\frac{2}{z}\right)^{0.2} \qquad ...(4.17)$$

in which

U_z = wind velocity at height z; and

z = height above ground in m at which wind velocity is measured.

Further if e_s is known then e_a can be calculated by the following expression

$$e_a = e_s \times \frac{RH_{\text{mean}}}{100} \qquad ...(4.18)$$

in which

RH_{mean} = mean relative humidity in percent.

The Penman Eq. (4.4) gives the value of E_{t_p} for the conditions such that the incident solar radiations R_s is medium to high, maximum relative humidity RH_{max} is medium to high and moderate day time wind velocity U_{day} is about double the night time wind velocity U_{night}. However, these conditions may not always be met. As such a correction to the Penman equation is required and the equation is expressed as follows

$$E_{t_p} = c\,[WR_n + (1 - W)\,f(U)\,(e_s - e_a)] \qquad ...(4.19)$$

in which c is known as adjustment factor. Table 4.6 presents the values c for different conditions of RH_{max}, R_s, U_{day} and ($U_{\text{day}}/U_{\text{night}}$). The incident solar radiation R_s is given by the following expression

$$R_s = R_a\left(a + b\frac{n}{N}\right) \qquad ...(4.20)$$

in which all the terms are same as defined in Eq. (4.11).

(ii) Jensen–Haise method. Jensen and Haise used observations of consumptive use from the western United States and developed the following linear relationship for estimating the potential evapotranspiration

$$E_{t_p} = C_t\,(t - t_x)\,R_s \qquad ...(4.21)$$

in which

E_{t_p} = potential evapotranspiration for reference crop alfalfa in mm/day

C_t = a temperature coefficient

t = mean daily temperature in °C

t_x = the intercept on the temperature axis, and

Table 4.5 Mean daily duration of maximum possible hours *N* of bright sunshine for each month of the year at different lattiudes

Latitude (degree) North South	*Jan. Jul.*	*Feb. Aug.*	*Mar. Sep.*	*Apr. Oct.*	*May. Nov.*	*Jun. Dec.*	*Jul. Jan.*	*Aug. Feb.*	*Sep. Mar.*	*Oct. Apr.*	*Nov. May.*	*Dec. Jun.*
0	12.12	12.12	12.12	12.12	12.12	12.12	12.12	12.12	12.12	12.12	12.12	12.12
5	11.87	11.96	12.08	12.22	12.35	12.41	12.38	12.28	12.16	12.02	11.90	11.83
10	11.61	11.81	12.06	12.35	12.57	12.70	12.64	12.45	12.17	11.91	11.67	11.55
15	11.34	11.66	12.04	12.47	12.82	13.00	12.92	12.62	12.22	11.81	11.44	11.25
20	11.07	11.50	12.01	12.60	13.07	13.32	13.22	12.81	12.26	11.70	11.20	10.94
25	10.78	11.33	11.97	12.74	13.34	13.66	13.53	13.02	12.31	11.58	10.94	10.62
30	10.45	11.14	11.97	12.88	13.65	14.05	13.88	13.23	12.35	11.47	10.67	10.26
35	10.09	10.95	11.95	13.06	13.98	14.47	14.27	13.47	12.42	11.33	10.36	9.86
40	9.68	10.71	11.91	13.25	14.36	14.96	14.71	13.76	12.48	11.18	10.00	9.39
45	9.19	10.45	11.87	13.48	14.82	15.55	15.25	14.09	12.55	11.01	9.60	8.85
50	8.61	10.13	11.84	13.78	15.38	16.29	15.91	14.48	12.66	10.80	9.07	8.17
55	7.83	9.73	11.79	14.10	16.14	17.28	16.78	14.99	12.76	10.55	8.45	7.28
60	6.79	9.21	11.74	14.62	17.10	18.70	18.01	15.67	12.92	10.22	7.60	6.04

Table 4.6 Adjustment factor C in Penman equation

	RH_{max} = 30%				RH_{max} = 60%				RH_{max} = 90%			
R_s* mm/day	3	6	9	12	3	6	9	12	3	6	9	12
u_{day} m/s	u_{day}/u_{night} = 4.0											
0	0.86	0.90	1.00	1.00	0.96	0.98	1.05	1.05	1.02	1.06	1.10	1.10
3	0.79	0.84	0.92	0.97	0.92	1.00	1.11	1.19	0.99	1.10	1.27	1.32
6	0.68	0.77	0.87	0.93	0.85	0.96	1.11	1.19	0.94	1.10	1.26	1.33
9	0.55	0.65	0.78	0.90	0.76	0.88	1.02	1.14	0.88	1.01	1.16	1.27
	u_{day}/u_{night} = 3.0											
0	0.86	0.90	1.00	1.00	0.96	0.98	1.05	1.05	1.02	1.06	1.10	1.10
3	0.76	0.81	0.88	0.94	0.87	0.96	1.06	1.12	0.94	1.04	1.18	1.28
6	0.61	0.68	0.81	0.88	0.77	0.88	1.02	1.10	0.86	1.01	1.15	1.22
9	0.46	0.56	0.72	0.82	0.67	0.79	0.88	1.05	0.78	0.92	1.06	1.18
	u_{day}/u_{night} = 2.0											
0	0.86	0.90	1.00	1.00	0.96	0.98	1.05	1.05	1.02	1.06	1.10	1.10
3	0.69	0.76	0.85	0.92	0.83	0.91	1.99	1.05	0.89	0.89	1.10	1.14
6	0.53	0.61	0.74	0.84	0.70	0.80	1.94	1.02	0.79	0.92	1.05	1.12
9	0.37	0.48	0.65	0.76	0.59	0.70	0.84	0.95	0.71	0.81	0.96	1.06
	u_{day}/u_{night} = 1.0											
0	0.86	0.90	1.00	1.00	0.96	0.98	1.05	1.05	1.02	1.06	1.10	1.10
3	0.64	0.71	0.82	0.89	0.78	0.86	0.94	0.99	0.85	0.92	1.01	1.05
6	0.43	0.53	0.68	0.79	0.62	0.70	0.84	0.93	0.72	0.82	0.95	1.00
9	0.27	0.41	0.59	0.70	0.50	0.60	0.75	0.87	0.62	0.72	0.87	0.96

$*R_s = R_a \left(a + b\frac{n}{N}\right)$

R_s = incident solar radiation in mm of evaporable water per day.

The value of C_t is given by the following expression

$$C_t = \frac{1}{C_1 + C_2 C_H} \qquad \text{...(4.22)}$$

in which

$$C_1 = 38 - \left(2°C \times \frac{EL}{305}\right)$$

$$C_2 = 7.6\ °C$$

$$C_H = \frac{50 \text{ mbar}}{e_2 - e_1}$$

where

EL = altitude of the place in metres; and

e_2 and e_1 = saturation vapour pressures of water in mbar at the mean daily maximum and minimum temperatures respectively, for the warmest month of the year in a given area.

The value of t_x is given by the following expression

$$t_x = -2.5 - 0.14\,(e_2 - e_1)\ °\text{C/m bar} - \frac{EL}{550} \qquad \text{...(4.23)}$$

in which the various terms are same as defined earlier.

The value of R_s is given by Eq. (4.20).

(iii) Hargreaves method. On the basis of data obtained from grass lysimeters, Hargreaves developed the following expression

$$E_{t_g} = 0.0135\,(t + 17.78)\,R_s \qquad \text{...(4.24)}$$

in which

E_{t_g} = reference crop potential evapotranspiration in mm/day, the reference crop being well watered grass;

t = mean daily temperature in °C; and

R_s = incident solar radiation in mm of evaporable water per day.

The value of R_s is given by Eq. (4.20).

Hargreaves indicated that a reference crop alfalfa would have a consumptive use about 20 percent greater than for grass, thus E_{tp} can be obtained from E_{tg} by the following expression

$$E_{tp} = 1.2\,E_{t_g} \qquad \text{...(4.25)}$$

(iv) Thornthwaite method. Thornthwaite presented an empirical formula based on temperature for the estimation of potential evaporation. His formula is as follows

$$E = 16.0\left(\frac{10t}{I}\right)^a \qquad \text{...(4.26)}$$

in which

E = potential evaporation in mm/month;

t = mean monthly temperature in °C;

I = 'heat index' for the 12 months in a year, given by

$I = \Sigma_i = \Sigma(T/5)^{1.514}$

(i) being 'heat index' for each month in the year); and

$a = 6.75 \times 10^{-7} I^3 - 7.71 \times 10^{-5} I^2 + 1.792 \times 10^{-2} I + 0.49239$

Evapotranspiration comprises evaporation from the soil surface and transpiration through the plants via leaves. If the crop completely covers the ground surface, evaporation from the soil surface would be negligible and the evaporation takes place entirely through the plants, and if the roots can absorb water at a sufficiently high rate, the vapour transfer is controlled by the climate alone. This rate of moisture use is referred to as the *Potential evaporation rate* or *Potential evaporation.* Equation (4.26), however, gives the 'unadjusted' potential evaporation. This is converted to the *Thornthwaite potential evaporation* E_{t_0} by correcting it for the duration of sunlight and for this the value of E obtained by Eq. (4.26) is multiplied by the adjustment factor given in Table 4.7 for the month and latitude.

(v) *Blaney – Criddle method.* Blaney and Criddle (1950) developed a simplified formula in which the consumptive use of water is correlated with the temperature and daytime hours. By multiplying the mean monthly temperature t by the mean monthly percentage p of the maximum possible daytime hours of the year, a monthly consumptive use factor f is obtained as $f = (pt/100)$. The value of p depends on the latitude of the place and the period of the year and it may be obtained from Table 4.8. It is assumed that the consumptive use varies directly with this factor when an ample water supply is available. Thus, if u is the monthly consumptive use of water for a crop then

$$u = kf = k\left(\frac{pt}{100}\right) \qquad \text{...(4.27)}$$

in which k is the monthly consumptive use coefficient.

Equation (4.27) is, however, applicable in English system of units which gives the monthly consumptive use of water u in inches and the mean monthly temperature t is in degrees Fahrenheit. Thus, if Eq. (4.27) is expressed in SI/metric system of units then it reduces to

$$u = kf = kp\ (0.46t + 8.13) \qquad \text{...(4.27a)}$$

in which u is monthly consumptive use of water in millimetres; t is mean monthly temperature in degrees celcius or centigrade (°C) and p and k are same as defined earlier. The value of k is same in both the cases. Further the consumptive use of water for the entire crop season (also called seasonal consumptive use of water) or the consumptive use of water for any given period is given by the sum of the monthly consumptive use values. Thus, if U is the consumptive use of water for the entire crop season or for any given period then

$$U = \Sigma kf = \Sigma kp\ (0.46t + 8.13) \qquad \text{...(4.28)}$$

Table 4.7 Adjustment factor for thornthwaite method

Latitude (degree)	Jan.	Feb.	Mar.	App.	May	June	July	Aug.	Ssp.	Oct.	Nov.	Dec.
	Month											
N												
0	1.04	0.94	1.04	1.01	1.04	1.01	1.04	1.04	1.01	1.04	1.01	1.04
5	1.02	0.93	1.03	1.02	1.06	1.03	1.06	1.05	1.01	1.03	0.99	1.02
10	1.00	0.91	1.03	1.03	1.08	1.06	1.08	1.07	1.02	1.02	0.98	0.99
15	0.97	0.91	1.03	1.04	1.11	1.08	1.12	1.08	1.02	1.01	0.95	0.97
20	0.95	0.90	1.03	1.05	1.12	1.11	1.14	1.11	1.02	1.00	0.93	0.94
25	0.93	0.89	1.03	1.06	1.15	1.14	1.17	1.12	1.02	0.99	0.91	0.91
30	0.90	0.87	1.03	1.08	1.18	1.17	1.20	1.14	1.03	0.98	0.89	0.88
35	0.87	0.85	1.03	1.09	1.21	1.21	1.23	1.16	1.03	0.97	0.86	0.85
40	0.84	0.83	1.03	1.11	1.24	1.25	1.27	1.18	1.04	0.96	0.83	0.81
42	0.82	0.83	1.03	1.12	1.26	1.27	1.28	1.19	1.04	0.95	0.82	0.79
44	0.81	0.82	1.02	1.13	1.27	1.29	1.30	1.20	1.04	0.95	0.80	0.76
46	0.79	0.81	1.02	1.13	1.29	1.31	1.32	1.22	1.04	0.94	0.79	0.74
48	0.76	0.80	1.02	1.14	1.31	1.33	1.34	1.23	1.05	0.93	0.77	0.72
50	0.74	0.78	1.02	1.15	1.33	1.36	1.37	1.25	1.06	0.92	0.76	0.70
S												
5	1.06	0.95	1.04	1.00	1.02	0.99	1.02	1.03	1.00	1.05	1.03	1.06
10	1.08	0.97	1.05	0.99	1.01	0.96	1.00	1.01	1.00	1.06	1.05	1.10
15	1.12	0.98	1.05	0.98	0.98	0.94	0.97	1.00	1.00	1.07	1.07	1.12
20	1.14	1.00	1.05	0.97	0.96	0.91	0.95	0.99	1.00	1.08	1.09	1.15
25	1.17	1.01	1.05	0.96	0.94	0.88	0.93	0.98	1.00	1.10	1.11	1.18
30	1.20	1.03	1.06	0.95	0.92	0.85	0.90	0.96	1.00	1.12	1.14	1.21
35	1.23	1.04	1.06	0.94	0.89	0.82	0.87	0.94	1.00	1.13	1.17	1.25
40	1.27	1.06	1.07	0.93	0.86	0.78	0.84	0.92	1.00	1.15	1.20	1.29
42	1.28	1.07	1.07	0.92	0.85	0.76	0.82	0.92	1.00	1.16	1.22	1.31
44	1.30	1.08	1.07	0.92	0.83	0.74	0.81	0.91	0.99	1.17	1.23	1.33
46	1.32	1.10	1.07	0.91	0.82	0.72	0.79	0.90	0.99	1.17	1.25	1.35
48	1.34	1.11	1.08	0.90	0.80	0.70	0.76	0.89	0.99	1.18	1.27	1.37
50	1.37	1.12	1.08	0.89	0.77	0.67	0.74	0.88	0.99	1.19	1.29	1.41

Table 4.8 Mean monthly percentage p of the maximum possible daytime hours of the year

Latitude (degree) North South	*Jan. Jul.*	*Feb. Aug.*	*Mar. Sep.*	*Apr. Oct.*	*May. Nov.*	*Jun. Dec.*	*Jul. Jan.*	*Aug. Feb.*	*Sep. Mar.*	*Oct. Apr.*	*Nov. May.*	*Dec. Jun.*
60	4.67	5.65	8.08	9.65	11.74	12.39	12.31	10.70	8.57	6.98	5.04	4.22
58	5.01	5.81	8.12	9.55	11.46	12.00	11.98	10.55	8.55	7.10	5.31	4.56
56	5.30	5.95	8.15	9.45	11.22	11.67	11.69	10.40	8.53	7.21	5.54	4.89
54	5.55	6.08	8.18	9.36	11.03	11.38	11.43	10.26	8.51	7.30	5.74	5.18
52	5.77	6.19	8.21	9.29	10.85	11.13	11.20	10.12	8.49	7.39	5.93	5.43
50	5.98	6.30	8.24	9.24	10.68	10.91	10.99	10.00	8.46	7.45	6.10	5.65
48	6.17	6.41	8.27	9.18	10.53	10.71	10.80	9.89	8.44	7.51	6.23	5.86
46	6.34	6.50	8.29	9.12	10.39	10.54	10.64	9.79	8.42	7.57	6.36	6.04
44	6.49	6.58	8.30	9.06	10.26	10.38	10.49	9.70	8.41	7.63	6.49	6.21
42	6.63	6.65	8.31	9.00	10.14	10.22	10.35	9.62	8.40	7.69	6.62	6.37
40	6.76	6.72	8.33	8.95	10.02	10.08	10.22	9.54	8.39	7.75	6.72	6.52
38	6.87	6.79	8.34	8.90	9.92	9.95	10.10	9.47	8.38	7.80	6.82	6.66
36	6.99	6.85	8.35	8.85	9.82	9.82	9.99	9.40	8.37	7.85	6.92	6.79
34	7.10	6.91	8.36	8.80	9.72	9.70	9.88	9.33	8.36	7.90	7.02	6.92
32	7.20	6.97	8.37	8.76	9.62	9.59	9.77	9.27	8.34	7.95	7.11	7.05
30	7.30	7.03	8.38	8.72	9.53	9.49	9.67	9.22	8.33	7.99	7.19	7.15
25	7.53	7.14	8.39	8.61	9.33	9.23	9.45	9.09	8.32	8.09	7.40	7.42
20	7.74	7.25	8.41	8.52	9.15	9.00	9.25	8.96	8.30	8.18	7.58	7.66
15	7.94	7.36	8.43	8.44	8.98	8.80	9.05	8.83	8.28	8.26	7.75	7.88
10	8.13	7.47	8.45	8.37	8.81	8.60	8.86	8.71	8.25	8.34	7.91	8.10
5	8.32	7.57	8.47	8.29	8.65	8.41	8.67	8.60	8.23	8.42	8.07	8.30
0	8.50	7.66	8.49	8.21	8.50	8.22	8.50	8.49	8.21	8.50	8.22	8.50

Table 4.8 (a) Mean dsily percentage p of the maximum possible daytime hours of the year

Latitude (degree) North South	*Jan. Jul.*	*Feb. Aug.*	*Mar. Sep.*	*Apr. Oct.*	*May. Nov.*	*Jun. Dec.*	*Jul. Jan.*	*Aug. Feb.*	*Sep. Mar.*	*Oct. Apr.*	*Nov. May.*	*Dec. Jun.*
60	0.15	0.20	0.26	0.32	0.38	0.41	0.40	0.34	0.28	0.22	0.17	0.13
58	0.16	0.21	0.26	0.32	0.37	0.40	0.39	0.34	0.28	0.23	0.18	0.15
56	0.17	0.21	0.26	0.32	0.36	0.39	0.38	0.33	0.28	0.23	0.18	0.16
54	0.18	0.22	0.26	0.31	0.36	0.38	0.37	0.33	0.28	0.23	0.19	0.17
52	0.19	0.22	0.27	0.31	0.35	0.37	0.36	0.33	0.28	0.24	0.20	0.17
50	0.19	0.23	0.27	0.31	0.34	0.36	0.35	0.32	0.28	0.24	0.20	0.18
48	0.20	0.23	0.27	0.31	0.34	0.36	0.35	0.32	0.28	0.24	0.21	0.19
46	0.20	0.23	0.27	0.30	0.34	0.35	0.34	0.32	0.28	0.24	0.21	0.20
44	0.21	0.24	0.27	0.30	0.33	0.35	0.34	0.31	0.28	0.25	0.22	0.20
42	0.21	0.24	0.27	0.30	0.33	0.34	0.33	0.31	0.28	0.25	0.22	0.21
40	0.22	0.24	0.27	0.30	0.32	0.34	0.33	0.31	0.28	0.25	0.22	0.21
35	0.23	0.25	0.27	0.29	0.31	0.32	0.32	0.30	0.28	0.25	0.23	0.22
30	0.24	0.25	0.27	0.29	0.31	0.32	0.31	0.30	0.28	0.26	0.24	0.23
25	0.24	0.26	0.27	0.29	0.30	0.31	0.31	0.29	0.28	0.26	0.25	0.24
20	0.25	0.26	0.27	0.28	0.29	0.30	0.30	0.29	0.28	0.26	0.25	0.25
15	0.26	0.26	0.27	0.28	0.29	0.29	0.29	0.28	0.28	0.27	0.26	0.25
10	0.26	0.27	0.27	0.28	0.28	0.29	0.29	0.28	0.28	0.27	0.26	0.26
5	0.27	0.27	0.27	0.28	0.28	0.28	0.28	0.28	0.28	0.27	0.27	0.27
0	0.27	0.27	0.27	0.27	0.27	0.27	0.27	0.27	0.27	0.27	0.27	0.27

Moreover, the consumptive use of water for the entire crop season or for any given period may also be expressed as

$$U = KF = K\Sigma p\,(0.46t + 8.13) \qquad ...(4.28a)$$

in which K is the seasonal consumptive use coefficient or the consumptive use coefficient for any given period and F is the sum of the monthly consumptive use factors for the entire crop season or for any given period.

Table 4.9 Correction factors for modifies blaney–Criddle formula

Correction factor	*a*	*b*	*a*	*b*	*a*	*b*
Relative humidity (RH)	*Low < 20%*		*Medium 20 – 50%*		*High > 50%*	
Sunshine ratio (n/N) Low (0.3 – 0.6)						
Wind velocity Low (0 – 2 m/s)	1.15	2.00	1.05	2.00	0.80	1.45
Wind velocity Medium (2 – 5 m/s)	1.28	1.80	1.15	1.85	0.88	1.55
Wind velocity High (> 5 m/s)	1.40	1.60	1.25	1.70	0.98	1.65
Sunshine ratio (n.N) Medium (0.6– 0.8)						
Wind velocity Low (0 – 2 m/s)	1.35	2.30	1.20	2.20	0.97	1.80
Wind velocity Medium (2 – 5 m/s)	1.55	2.05	1.38	2.15	1.06	1.75
Wind velocity High (> 5 m/s)	1.73	1.80	1.52	2.10	1.16	1.70
Sunshine ratio (n/N) High (> 0.8)						
Wind velocity Low (0 – 2 m/s)	1.55	2.60	1.37	2.40	1.14	2.15
Wind velocity Medium (2 – 5 m/s)	1.82	2.30	1.61	2.50	1.22	1.95
Wind velocity High (> 5 m/s)	2.06	2.00	1.82	2.55	1.31	1.70

Doorenbos and Pruitt, however, rejected the use of crop coefficient k or K in the Blaney–Criddle formula because depending on local conditions wide varieties of the values of crop coefficients have been developed from which the selection of the correct value of the crop coefficient is rather difficult. Moreover, once potential evapotranspiration E_{t_p} has been determined by any of the standard methods then as indicated later, one set of crop coefficient k_c can be used to determine the consumptive use of water or evapotranspiration for the specific crop. Thus, in 1975 Doorenbos and Pruitt modified the Blaney–Criddle method and proposed the following expression for the potential evapotranspiration E_{t_p}

$$E_{t_p} = f = p\,(0.46t + 8.13) \quad (4.29)$$

in which

E_{t_p} = potential evapotranspiration in mm/day

p = mean daily percentage of the maximum possible day time hours of the year, (its value may be obtained for a given month and latitude from Table 4.8 (a); and

t = mean monthly temperature in °C.

In this case the potential evapotranspiration E_{t_p}, however, does not pertain to any reference crop as in the case of the other methods described earlier.

Further it was indicated by Doorenbos and Pruitt that in this case the three important climatic factors viz., relative humidity, sunshine ratio (or cloudiness) and day time wind speed, which are not included, have a significant influence on the value of potential evapotranspiration E_{t_p}. Thus, Doorenbos and Pruitt proposed correction factors for the value of E_{t_p} given by Eq. (4.29). These correction factors are in the form of the constants of a linear correlation function, and the corrected formula is as given below.

$$E = ap\,(0.46t + 8.13) - b \quad \ldots(4.29a)$$

The values of the correction factors a and b are given in Table 4.9 for three levels of each of the three climatic factors explained earlier.

Crop Coefficient k_c and its Determination

From the potential evapotranspiration E_{t_p} or E_{t_0} determined by any of the five methods described earlier the evapotranspiration or consumptive use of water for the specific crop may be estimated by the use of a crop coefficient as follows.

$$E_{t_c} = k_c\,E_{t_p} \quad \ldots(4.30)$$

in which

E_{t_c} = evapotranspiration or consumptive use of water for the crop; and

k_c = crop coefficient.

Since the evapotranspiration or consumptive use of water for a crop varies with the stage of growth of the crop, the values of k_c for different stages of growth are needed to be determined. The growing season of a crop is usually divided into four stages viz., initial stage, development stage, mid-season stage and late-season stage. The values of k_c for the different stages may be determined as discussed below.

(1) Establish planting or sowing date of the crop from local information or from practices in similar climatic zones.

(2) Determine total growing season of the crop and duration of each of the four stages from local information.

(3) *Initial stage.* Assume irrigation and/or rainfall frequency and for predetermined E_{t_p} or E_{t_0} value obtain k_c as shown in Fig. (4.3). Usually irrigation and/or rainfall frequency of 7 days

is assumed. Plot the value of k_c so obtained as a horizontal straight line for the entire duration of the initial stage.

(4) *Mid-season stage.* For given climatic condition (humidity and wind velocity) select k_c value for the mid-season stage of the crop from Table 4.10 and plot the same as horizontal straight line for the entire duration of the mid-season stage.

(5) *Late-season stage.* For time of full maturity (or harvest within a few days) select k_c value for the late-season stage of the crop from Table 4.10 for given climatic condition (humidity and wind velocity) and plot this value of k_c at the end of the growing season or full maturity. Assume straight line between k_c values at the end of mid-season period and at the end of growing season.

Table 4.10 Values of crop coefficient k_c for some of the common crops

Crop	*Humidity*	*RH_{min} > 70%*		*RH_{min} < 20%*	
	Wind m/s	*0 – 5*	*5 – 8*	*0 – 5*	*5 – 8*
	Crop stage				
Barley	M	1.05	1.1	1.15	1.2
	H	0.25	0.25	0.2	0.2
Carrot	M	1.0	1.05	1.1	1.15
	H	0.7	0.75	0.8	0.85
Maize (grain)	M	1.05	1.1	1.15	1.2
	H	0.55	0.55	0.6	0.6
Cotton	M	1.05	1.15	1.2	1.25
	H	0.65	0.65	0.65	0.7
Cabbage, Cauliflower	M	0.95	1.0	1.05	1.1
	H	0.80	0.85	0.9	0.95
Grain	M	1.05	1.1	1.15	1.2
	H	0.3	0.3	0.25	0.25
Lentil	M	1.05	1.1	1.15	1.2
	H	0.3	0.3	0.25	0.25
Lettuse	M	0.95	0.95	1.0	1.05
	H	0.9	0.9	0.9	1.0
Melons	M	0.95	0.95	1.0	1.05
	H	0.65	0.65	0.75	0.75
Millet	M	1.0	1.05	1.1	1.15
	H	0.3	0.3	0.25	0.25
Onion (dry)	M	0.95	0.95	1.05	1.1
	H	0.75	0.75	0.8	0.85
Onion (green)	M	0.95	0.95	1.0	1.05
	H	0.95	0.95	1.0	1.05
Groundnut	M	0.95	1.0	1.05	1.1
	H	0.55	0.55	0.6	0.6
Peas	M	1.05	1.1	1.15	1.2
	H	0.95	1.0	1.05	1.1
Potato	M	1.05	1.1	1.15	1.2
	H	0.7	0.7	0.75	0.75
Radishes	M	0.8	0.8	0.85	0.9
	H	0.75	0.75	0.8	0.85

Contd.

Contd.

Crop	*Humidity*	*RH_{min} > 70%*		*RH_{min} < 20%*	
	Wind m/s	*0 – 5*	*5 – 8*	*0 – 5*	*5 – 8*
Safflower	M	1.05	1.1	1.15	1.2
	H	0.25	0.25	0.2	0.2
Sorghum	M	1.0	1.05	1.1	1.15
	H	0.5	0.5	0.55	0.55
Soybeans	M	1.0	1.05	1.1	1.15
	H	0.45	0.45	0.45	0.45
Sunflower	M	1.05	1.1	1.15	1.2
	H	0.4	0.4	0.35	0.35
Tomato	M	1.05	1.1	1.2	1.25
	H	0.6	0.6	0.65	0.65
Wheat	M	1.05	1.1	1.15	1.2
	H	0.25	0.25	0.2	0.2

M–mid-season stage (from time of attainment of full ground cover to time of start of maturing).
H–late-season stage (from end of mid-season to full maturity or harvest).

(6) *Development stages.* Assume straight line between *kc* values at the end of the initial stage and the beginning of the mid-season stage.

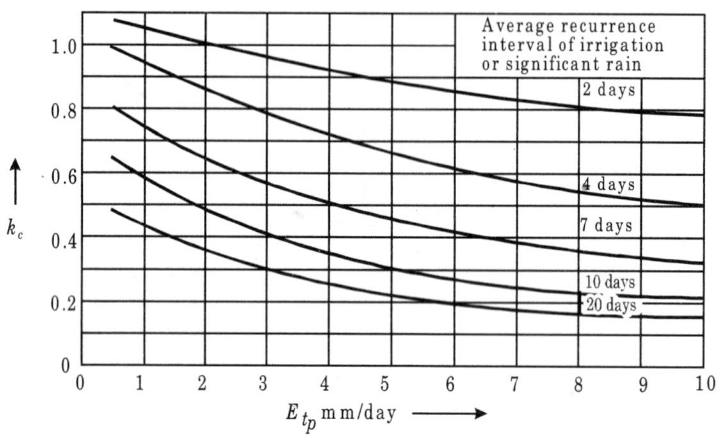

Fig. 4.3 *Average value of k_c for initial stage of crop growth as related to E_{tp} and frequency of irrigation and/or significant rain*

A crop coefficient graph is thus prepared for the entire growing season of the crop, from which the k_c values can be obtained for different periods of the growing season of the crop.

The method described above for determining the k_c values for different periods of the growing season of the crop and the E_{t_c} values for the crop is indicated in the Illustrative Example 4.19.

(*vi*) *Hargreaves class A pan evaporation method.* In this method evapotranspiration or consumptive use of water for a crop is related to Class *A* pan evaporation by the following expression

$$E_{t_c} = K E_p \qquad \text{...(4.31)}$$

in which

E_{t_c} = evapotranspiration or consumptive use of water for the crop

K = crop coefficient, or crop factor or crop consumptive use coefficient, and

E_p = class A pan evaporation

The value of crop coefficient K depends on several factors and it is different for different crops and also it is different for the same crop at different places. Further the value of K also varies with the growth of the crop and it is different at different stages of the growth of the same crop. The values of K obtained by Hargreaves for a range of crops at various stages of growth are given in Table 4.11. These values of the crop coefficient are suggested as representing average conditions and as a cautious initial estimate, these values can be used for any climate. However, on the basis of research the values of K for certain crops at various stages of growth have been determined for several countries. Tables 4.12 and 4.13 give the values of K for certain crops grown in different parts of India and U.S.A. respectively. Further where specific data are not available, certain average values of K have been recommended for different groups of crops as given in Table 4.14. For this the crops have been divided into 8 groups as following.

Table 4.11 Values K given by Hargreaves for different Ccrops grown in U.S.A.

	Percentage of crop growing season										
	0	*10*	*20*	*30*	*40*	*50*	*60*	*70*	*80*	*90*	*100*
Alfalfa	0.55	0.60	0.70	0.80	0.90	0.95	0.95	0.95	0.90	0.80	0.65
Beans	0.20	0.30	0.40	0.65	0.85	0.90	0.90	0.80	0.60	0.35	0.20
Citrus and Avocados	0.50	0.45	0.45	0.45	0.45	0.45	0.50	0.55	0.60	0.55	0.50
Corn	0.20	0.30	0.50	0.65	0.80	0.90	0.90	0.85	0.75	0.60	0.50
Cotton	0.10	0.20	0.40	0.55	0.75	0.90	0.90	0.85	0.75	0.55	0.35
Fruit, deciduous	0.20	0.30	0.50	0.65	0.70	0.75	0.70	0.60	0.50	0.40	0.20
Fruit with cover	Average about 1.00 for periods of rapid growth of cover crop										
Grain Sorghum	0.20	0.35	0.55	0.75	0.85	0.90	0.85	0.70	0.60	0.35	0.15
Grain, Spring	0.15	0.20	0.25	0.30	0.40	0.55	0.75	0.85	0.90	0.90	0.30
Grain, Winter	0.15	0.25	0.35	0.40	0.50	0.60	0.70	0.80	0.90	0.90	0.30
Grapes	0.15	0.15	0.20	0.35	0.45	0.55	0.55	0.45	0.35	0.25	0.20
Ladino Clover	Average about 0.95 for maximum growth										
Nuts, Walnuts	0.30	0.35	0.55	0.70	0.75	0.75	0.75	0.65	0.55	0.30	0.15
Nuts, Pecan	0.35	0.45	0.55	0.75	0.75	0.65	0.50	0.45	0.40	0.35	0.30

Contd.

Contd.

	0	*10*	*20*	*30*	*40*	*50*	*60*	*70*	*80*	*90*	*100*
Peanuts	0.15	0.25	0.35	0.45	0.55	0.60	0.65	0.65	0.60	0.45	0.30
Potatoes	0.20	0.35	0.45	0.65	0.80	0.90	0.95	0.95	0.95	0.90	0.90
Rice	0.80	0.95	1.05	1.15	1.20	1.30	1.30	1.20	1.10	0.90	0.50
Sugar beets	0.25	0.45	0.60	0.70	0.80	0.85	0.90	0.90	0.90	0.90	0.90
Sugar cane	Varies from 0.55 to 1.00 depending upon rate and stage of growth										
Vegetable, deep rooted	0.20	0.20	0.25	0.35	0.50	0.65	0.70	0.60	0.45	0.35	0.20
Vegetable, shallow rooted	0.10	0.20	0.40	0.50	0.60	0.60	0.60	0.55	0.45	0.35	0.30

Table 4.12 Values of K for certain crops grown in different parts of India

% of crop growing season	*Wheat Ludhiana (India)*	*Wheat Poona (India)*	*Cotton Poona (India)*	*Maize Ludhiana (India)*
0	0.14	0.30	0.22	0.40
5	0.17	0.40	0.22	0.42
10	0.23	0.51	0.23	0.47
15	0.33	0.62	0.24	0.54
20	0.45	0.73	0.26	0.63
25	0.60	0.84	0.35	0.75
30	0.72	0.92	0.58	0.85
35	0.81	0.96	0.80	0.96
40	0.88	1.10	0.95	1.04
45	0.90	1.10	1.03	1.07
50	0.91	1.00	1.08	1.09
55	0.90	0.91	1.08	1.10
60	0.89	0.80	1.07	1.11
65	0.86	0.65	1.05	1.10
70	0.83	0.51	1.00	1.07
75	0.80	0.40	0.93	1.04
80	0.76	0.30	0.85	1.00
85	0.71	0.20	0.73	0.97
90	0.65	0.12	0.62	0.89
95	0.58	0.10	0.50	0.81
100	0.51	0.10	0.40	0.70
Seasonal value of K	0.61	0.61	0.68	0.86

Table 4.13 Values of *K* for certain crops grown in different parts of U.S.A

% of crop growing season	*Sugarcane Hawaii (U.S.A.)*	*Rice Los Banos (U.S.A.)*	*Maize Alabama (U.S.A.)*	*Jowar Alabama (U.S.A.)*
0	0.34	1.00	0.40	0.42
5	0.37	1.02	0.40	0.44
10	0.40	1.03	0.43	0.46
15	0.44	1.05	0.46	0.48
20	0.50	1.07	0.52	0.50
25	0.60	1.09	0.59	0.52
30	0.72	1.11	0.67	0.56
35	0.86	1.13	0.76	0.59
40	0.93	1.16	0.85	0.64
45	0.98	1.18	0.93	0.71
50	1.02	1.20	1.02	0.79
55	1.05	1.21	1.09	0.92
60	1.07	1.22	1.14	1.01
65	1.10	1.22	1.19	1.07
70	1.13	1.21	1.21	1.09
75	1.16	1.19	1.23	1.09
80	1.19	1.16	1.22	1.05
85	1.20	1.10	1.16	0.99
90	1.20	1.03	1.06	0.91
95	1.19	0.96	0.95	0.82
100	1.19	0.86	0.75	0.70
Seasonal Value of *K*	0.89	1.10	0.86	0.75

Table 4.14 Average values of *K* for different groups of crops

% of crop growing season	*Group A*	*Group B*	*Group C*	*Group D*	*Group E*	*Group F*	*Group G*	*Rice*
0	0.20	0.15	0.12	0.08	0.90	0.60	0.50	0.80
5	0.20	0.15	0.12	0.08	0.90	0.60	0.55	0.90
10	0.36	0.27	0.22	0.15	0.90	0.60	0.60	0.95
15	0.50	0.38	0.30	0.19	0.90	0.60	0.65	1.00
20	0.64	0.48	0.38	0.27	0.90	0.60	0.70	1.05
25	0.75	0.56	0.45	0.33	0.90	0.60	0.75	1.10

Contd.

Contd.

% of crop growing season	*Group A*	*Group B*	*Group C*	*Group D*	*Group E*	*Group F*	*Group G*	*Rice*
30	0.84	0.63	0.50	0.40	0.90	0.60	0.80	1.14
35	0.92	0.69	0.55	0.46	0.90	0.60	0.85	1.17
40	0.97	0.73	0.58	0.52	0.90	0.60	0.90	1.21
45	0.99	0.74	0.60	0.58	0.90	0.60	0.95	1.25
50	1.00	0.75	0.60	0.65	0.90	0.60	1.00	1.30
55	1.00	0.75	0.60	0.71	0.90	0.60	1.00	1.30
60	0.99	0.74	0.60	0.77	0.90	0.60	1.00	1.30
65	0.96	0.72	0.58	0.82	0.90	0.60	0.95	1.25
70	0.91	0.68	0.55	0.88	0.90	0.60	0.90	1.20
75	0.85	0.64	0.51	0.90	0.90	0.60	0.85	1.15
80	0.75	0.56	0.45	0.90	0.90	0.60	0.80	1.10
85	0.60	0.45	0.36	0.80	0.90	0.60	0.75	1.00
90	0.46	0.35	0.28	0.70	0.90	0.60	0.70	0.90
95	0.28	0.21	0.17	0.60	0.90	0.60	0.55	0.80
100	0.20	0.20	0.17	0.60	0.90	0.60	0.50	0.20

(i) *Group A.* The important crops included in this group are: Sugar beets, Maize, Cotton, Jowar, Beans, Peas, Potatoes, etc. The value of *K* in this case varies from 0.2 to 1.00.

(ii) Group *B*. This group includes deciduous fruits and some field crops. Some of the important crops of this group are: Tomatoes, Hybrid Walnuts, Plumes, Olives and some crops of group A which fail to produce maximum vegetative cover and maximum growth ratio. The value of *K* in this case varies from 0.15 to 0.75.

(iii) *Group C.* It includes crops such as Melons, Onions, Carrots, Hops, Grapes, etc. The value of *K* in this case varies from 0.12 to 0.6.

(iv) *Group D.* The important crops included in this group are: Wheat, Barley, Celery, Flax and other small grains. The value of *K* in this case varies from as low as 0.08 to as high as 0.9. Further the maximum value of *K* usually occurs at about 75 to 80% of the crop growing season, whereas in the case of all other groups the maximum value of *K* occurs at about 50 to 60% of the crop growing season.

(v) *Group E.* The important crops included in this group are: Pastures, Orchard with cover crop, Plantain etc. The value of *K* in this case is more or less constant being 0.9 for the entire growing season. However, in actual practice the value of *K* may vary from 0.7 to 1.1.

(vi) *Group F.* This group includes citrus crops such as Oranges, Grapes, etc. The value of *K* in this case is also constant being 0.6 for the entire growing season.

(vii) *Group G.* The main crops included in this group are: Sugarcane, Alfalfa etc. The value of *K* in this case varies from 0.50 to 1.00.

(viii) *Rice or Paddy.* In this case the value of *K* increases from 0.8 to 1.3 with the crop growth and then again decreases.

The values of K given in Table 4.14 are only the average values. As such care must be taken while using these values of K and if local values of K are available then the same should only be used. Further the factors which increase or decrease the evapotranspiration may be taken into consideration. For example taller and more uneven vegetation tends to result in greater turbulence and more efficient utilisation of radiation in the production of water use. Also dark green vegetation produces higher rate of absorption of solar energy and hence, its evapotranspiration rate will be higher than that for light green vegetation. For this reason, the plant diseases, causing yellowing of the leaves of the plants, greatly reduce evapotranspiration.

The value of pan evaporation E_p may be determined by directly measuring the quantity of water evaporated from the standard Class A Pan. The Class A Pan used in India is as per IS: 5973–1970. It is circular 1225 mm in diameter and 255 mm deep. It is made from 1 mm copper sheet tinned inside and painted white outside or from fibre glass reinforced polyster and at the top covered with a wire mesh. A stilling well provides an undisturbed water surface for measurement. The pan rests on a white painted wooden grill about 100 mm above the ground. The pan is filled within 50 mm of the rim. Evaporation measurements are made by hook gage or refilling to a fixed point.

The value of pan evaporation E_p may also be determined by using the formula given by Christiansen which is as follows:

$$E_p = 0.473\, R.C_t\, .C_w.C_h.C_s.C_e \qquad ...(4.32)$$

where R = Extra-terrestrial radiation in the same units as E_p and its value may be obtained from Table 4.4

C_t = Coefficient for temperature, given by

$$C_t = 0.393 + 0.027\,96\, T_c + 0.000\,118\,9\, T_c^2 \qquad ...(4.32a)$$

(T_c being mean temperature in °C)

C_w = Coefficient for wind velocity, given by

$$C_w = 0.708 + 0.003\,4\, W - 0.000\,0038\, W^2 \qquad ...(4.32b)$$

(W being the mean wind velocity in km/day at 0.5 m above ground surface)

C_h = Coefficient for relative humidity, given by

$$C_h = 1.250 - 0.0087\, H + 0.75 \times 10^4\, H^2 - 0.85 \times 10^{-8}\, H^4 \qquad ...(4.32\text{ c})$$

(H being the mean % relative humidity at noon or average relative humidity for 11 and 18 hours)

C_s = Coefficient for percent of possible sunshine, given by

$$C_s = 0.542 + 0.008\, S - 0.78 \times 10^{-4}\, S^2 + 0.62 \times 10^{-6}\, S^3 \qquad ...(4.32d)$$

(S being mean sunshine percentage) and

C_e = Coefficient for elevation, given by

$$C_e = 0.97 + 0.009\,84\, E \qquad ...(4.32e)$$

(E being the elevation in 100 metres).

4.12 IRRIGATION EFFICIENCIES

It is essential that efficient use of irrigation water is made, but even in the best method of irrigation because of various losses of water the entire quantity of water applied during irrigation is not available for use. The ratio of the water available for use to the water applied is in general defined as irrigation efficiency. However, various types of irrigation efficiencies are defined as explained below which are usually expressed as percentage.

1. **Water Conveyance Efficiency** (η_c). It is defined as the ratio of the quantity of water delivered to the fields or the irrigated land to the quantity of water diverted into the canal system from the river or reservoir. Thus, if W_f is the quantity of water delivered to the field and W_r is the quantity of water diverted into the canal system from the river or reservoir then

$$\eta_c = \frac{W_f}{W_r} \times 100 \quad ...(4.33)$$

The water conveyance efficiency therefore accounts for the water losses which occur in conveyance from the point of diversion into the canal system to the fields.

2. **Water Application Efficiency** (η_a). It is defined as the ratio of the quantity of water stored in the root zone of the plants to the quantity of water delivered to the field. Thus, if W_s is the quantity of water stored in the root zone and W_f is the quantity of water delivered to the field then

$$\eta_a = \frac{W_s}{W_f} \times 100 \quad ...(4.34)$$

The water application efficiency therefore accounts for the water losses which occur during the application of irrigation water to the field. The common sources of loss of water during its application to the field are surface runoff from the field and deep percolation. Thus, if R_f is the quantity of water lost as surface runoff from the field and D_f is the quantity of water lost due to deep percolation to a level far below the root zone, then

$$W_f = W_s + R_f + D_f$$

and

$$\eta_a = \frac{W_f - (R_f + D_f)}{W_f} \times 100 \quad ...(4.35)$$

The water application efficiency varies with the irrigation method. For example in the case of sprinkler irrigation method the water application efficiency may be as high as 80% while in the case of a surface irrigation method it may not exceed 60%.

3. **Water Use Efficiency** (η_u). It is defined as the quantity of water used beneficially including the water required for leadhing to the quantity of water delivered. Thus, if W_u is the quantity of water used beneficially and W_f is the quantity of water delivered to the field then

$$\eta_u = \frac{W_u}{W_f} \times 100 \quad ...(4.36)$$

4. **Water Storage Efficiency** (η_s). It is defined as the ratio of the quantity of water stored in the root zone during irrigation to the quantity of water needed to bring the moisture content of the soil to the field capacity. Thus, if W_s is the quantity of water stored in the root zone during irrigation and W_n is the quantity of water needed to bring the moisture content of the soil to the field capacity (i.e., W_n = Field capacity – Available moisture in the soil prior to irrigation), then

$$\eta_s = \frac{W_s}{W_n} \times 100 \quad ...(4.37)$$

Water storage efficiency becomes significant when insufficient quantity of water is stored in the root zone during irrigation. Further the presence of excess salts in the soil would require that water storage efficiency should be high in order to keep the salts washed out of the soil.

5. **Water Distribution Efficiency** (η_d). It is determined from the following expression

$$\eta_d = \left[1 - \frac{y}{d}\right] \times 100 \qquad \text{...(4.38)}$$

in which y is the average numerical deviation in depth of water stored from the average depth of water d stored in the root zone during irrigation.

Water distribution efficiency evaluates the degree to which water is uniformly distributed throughout the root zone during irrigation and hence it is also known as *uniformity coefficient*. Further higher is the value of η_d more uniformly is the water distributed in the root zone which in turn will result in a better crop response. For example, a water distribution efficiency of 80% means that as compared to average depth of water application 10% is the excess depth of water applied and 10% is the deficient depth of water applied.

The water distribution efficiency provides a measure for comparing different methods of irrigation i.e., sprinkler irrigation method may be compared to surface irrigation method, or one sprinkler irrigation method may be compared to the other sprinkler irrigation method; or one surface irrigation method may be compared to the other surface irrigation method.

6. **Consumptive Use Efficiency** (η_{cu}). It is defined as the ratio of the normal consumptive use of water to the net amount of water depleted from the root zone. Thus, if W_{cu} (or C_u or E_{t_c}) is the normal consumptive use of water or evapotranspiration and W_d is the net amount of water depleted from the root zone then

$$\eta_{cu} = \frac{W_{cu}}{W_d} \times 100 \qquad \text{...(4.39)}$$

The consumptive use efficiency therefore accounts for the loss of water by deep percolation and any excessive evaporation following an irrigation.

4.13 IRRIGATION REQUIREMENTS OF CROPS

The irrigation requirements of the various crops may be expressed in different ways as described below.

1. **Consumptive Irrigation Requirement (CIR).** It is defined as the amount of irrigation water that is required to meet the evapotranspiration needs of a crop during its full growth. However, if during the growth period of a crop rain occurs then since a part of it will be retained by the soil in the root zone and the same will be available to meet a part of the evapotranspiration requirements of the crop, the quantity of irrigation water required to be applied will be correspondingly reduced. This part of the rainfall is known as *effective rainfall* and its value may be determined as indicated later. Thus, if E_{t_c} or C_u is the evapotranspiration or consumptive use of water for a crop and R_e is the effective rainfall during the growth period of the crop then

$$\text{CIR} = E_{t_c} - R_e \text{ (or } C_u - R_e) \qquad ...(4.40)$$

The irrigation water in addition to satisfying the consumptive use or evapotranspiration requirements, has also to satisfy other requirements such as leaching, presowing and nursery requirements (for some crops like tobacco, vegetables, etc.). Thus, net irrigation requirement is defined as follows.

2. **Net Irrigation Requirement (NIR).** It is defined as the amount of irrigation water required to be delivered at the field to meet the evapotranspiration needs of a crop as well as the other needs such as leaching, presowing requirement and nursery water requirement (if any). Thus, we have

$$\text{NIR} = \text{CIR} + \text{LR} + \text{PSR} + \text{NWR} \qquad ...(4.41)$$

where

LR = Leaching Requirement

PSR = Presowing Requirement, and

NWR = Nursery Water Requirement.

3. **Field Irrigation Requirement (FIR).** It is defined as the amount of water required to meet the 'net irrigation requirements' plus the amount of water lost as surface runoff and through deep percolation. As described in the previous section the water application efficiency η_a accounts for the loss of irrigation water by surface runoff and through deep percolation and hence

$$\text{FIR} = \frac{\text{NIR}}{\eta_a} \qquad ...(4.42)$$

4. **Gross Irrigation Requirement (GIR).** It is defined as the amount of water required to meet the field irrigation requirements plus the amount of irrigation water lost in conveyance through the canal system by evaporation and by seepage. Again as described in the previous section the water conveyance efficiency η_c accounts for the conveyance losses and hence

$$\text{GIR} = \frac{\text{FIR}}{\eta_c} \qquad ...(4.43)$$

Determination of Effective Rainfall R_e

Effective rainfall can be determined by the (evapotranspiration/precipitation) ratio method. In this method for the known mean monthly rainfall and average monthly evapotranspiration E_{t_c} of crop the corresponding effective rainfall R_e can be obtained from Table 4.15.

The effective rainfall so obtained from this table is for an effective storage of 75 mm and for other values of the effective storage, this effective rainfall is multiplied by a correction factor called storage factor which is obtained for the known effective storage from Table 4.15 (a)

The consideration of the effective rainfall is based on the fact that frequent light rains intercepted by plant foliage with full ground cover are close to 100% effective. For heavy rains due to runoff, deep percolation and evaporation the effectiveness is reduced and it is close to 60%.

Table 4.15 Values of effective monthly rainfall R_e (mm)

Average monthly E_{tc} (mm) \ Mean monthly rainfall (mm)	12.5	25.0	37.5	50.0	62.5	75.0	87.5	100.0	112.5	125.0	137.5	150.0	162.5	175.0	187.5	200.0
25	8	16	24													
50	8	17	25	32	46											
75	9	18	27	34	41	48	56	62	69							
100	9	19	28	35	43	52	59	66	73	80	87	94	100			
125	10	20	30	37	46	54	62	70	76	85	92	98	107	116	120	
150	10	21	31	39	49	57	66	74	81	89	97	104	112	119	127	133
175	11	23	32	42	52	61	69	78	86	95	103	111	118	126	134	141
200	12	24	33	44	54	64	73	82	91	100	109	117	125	134	142	150
225	12	25	35	47	57	68	78	87	96	106	115	124	132	141	150	159
250	13	25	38	50	61	72	84	92	102	112	121	132	140	150	158	167

Table 4.15 (a) Values of storage factor

Effective storage (mm)	20	25	37.5	50	62.5	75	100	125	150	175	200
Storage factor	0.73	0.77	0.86	0.93	0.97	1.00	1.02	1.04	1.06	1.07	1.08

The method described above for determining the effective rainfall R_e is indicated in the Illustrative Example 4.20.

Presowing Requirement (PSR)

Presowing irrigation is important for field preparation as availability of moisture is essential for good germination of seeds. The amount of water required for presowing can be computed on the basis of the requirement to replenish the moisture to field capacity. For this suitable root zone depth is assumed and knowing the available moisture holding capacity (AMHC) of the soil the presowing requirement may be computed as follows:

$$PSR = (AMHC) \times D_e$$

where

PSR = presowing requirement of water

AMHC = available moisture holding capacity of soil; and

D_e = reasonable depth of root zone.

For example, if root zone depth is assumed as 800 mm and AMHC is 130 mm/m, then

$$PSR = \frac{130}{1000} \times 800 = 104 \text{ mm}$$

Nursery Water Requirement (NWR)

The water requirement for nursery is required to be considered in the case of those crops which are sown on nursery beds and are transplanted within few days after sowing when the plants are a few cm tall. The water required for this purpose may be expressed in m^3 per unit area per day. The total water required may be computed for the area used for this purpose and the total nursery period in days. For example, in the case of Tobacco it is usually sown on nursery beds and transplanted to the fields in about 40 to 60 days after sowing when the plants are about 150 mm tall. The recommended water requirement during this period is 3 to 5 litres per m^2. The total water required for the nursery may then be expressed as depth of water in mm for the entire area of the crop.

Leaching Requirement (LR)

In order to remove or leach out excess salts from the soil it is usually essential that large depths of water be applied to the land and be made to percolate through the soil. This process of removing excess salts from the soils is known as leaching. Leaching can be practised during, before or after the crop season, provided that salt accumulation in the soil does not exceed the crop tolerence level.

Leaching requirement (LR) is the minimum amount of irrigation water supplied that must be drained through the root zone to control soil salinity at the given specific level. For sandy loam to clay loam soils with good drainage and where rainfall is low the leaching requirement can be obtained from the following expression.

For surface irrigation methods including sprinkler

$$LR = \frac{EC_w}{5EC_e - EC_w} \quad ...(4.44)$$

For drip irrigation and high frequency sprinkler irrigation (nearly daily)

$$LR = \frac{EC_w}{2\text{ Max. }EC_e} \quad ...(4.45)$$

in which

LR = leaching requirement expressed as a fraction of total irrigation water required for plant consumption and for leaching

EC_w = electrical conductivity of the irrigation water, m mhos/cm

EC_e = electrical conductivity of the soil saturation extract for a given crop appropriate to the tolerable degree of yield reduction as given in Table 4.16, and

Max. EC_e = maximum tolerable electrical conductivity of the soil saturation extract for a given crop as given in Table 4.16.

The above noted expression for *LR* corresponds to leaching efficiency *Le* = 100%. However, for any other value of *Le* the value or *LR* obtained by this expression is to be divided by *Le*. The leaching efficiency varies with the type of soil and particularly with the internal drainage properties of the soil and field. Thus *Le* may be as low as 30% for cracking and swelling heavy clays and may be almost 100% for sandy soils and it must be measured at a most early date for the area under investigation.

Table 4.16 Crop salt tolerance levels for different crops

Crop	*Values of EC_e in m mhos/cm for Yield Potential*				*Max. EC_e (m mhos/cm)*
	100%	*90%*	*75%*	*50%*	
Field Crops					
Barley	8.0	10.0	13.0	18.0	28
Beans (field)	1.0	1.5	2.3	3.6	7
Maize (corn)	1.7	2.5	3.8	5.9	10
Cotton	7.7	9.6	13.0	17.0	27
Groundnut	3.2	3.5	4.1	4.9	7
Rice (paddy)	3.0	3.8	5.1	7.2	12
Sorghum	4.0	5.1	7.2	11.0	18
Soybean	5.0	5.5	6.2	7.5	10
Sugarbeet	7.0	8.7	11.0	15.0	24
Wheat	6.0	7.4	9.5	13.0	20
Vegetable Crops					
Beans	1.0	1.5	2.3	3.6	7
Cabbage	1.8	2.8	4.4	7.0	12
Carrot	1.0	1.7	2.8	4.6	8

Contd.

Contd.

Crop	*Values of EC_e in m mhos/cm for Yield Potential*				*Max. EC_e (m mhos/cm)*
	100%	*90%*	*75%*	*50%*	
Onion	1.2	1.8	2.8	4.3	8
Potato	1.7	2.5	3.8	5.9	10
Radish	1.2	2.0	3.1	5.0	9
Tomato	2.5	3.5	5.0	7.6	13
Forage Crops					
Alfalfa	2.0	3.4	5.4	8.8	16
Bermuda grass	6.9	8.5	10.8	14.7	23
Fruit Crops					
Almond	1.5	2.0	2.8	4.1	7
Apple, pear	1.7	2.3	3.3	4.8	8
Date, palm	4.0	6.8	10.9	17.9	32
Grape	1.5	2.5	4.1	6.7	12
Lemon	1.7	2.3	3.3	4.8	8
Orange	1.7	2.3	3.2	4.8	8
Straw berry	1.0	1.3	1.8	2.5	4

The total depth of irrigation water needed to satisfy both plant consumptive use E_{t_c} and leaching requirement LR may be determined as follows.

If D_{iw} is total irrigation depth required to satisfy the consumptive use depth D_{CW} and leaching depth D_{lw}, then

$$D_{iw} = D_{cw} + D_{lw}$$

but

$$D_{cw} = (E_{t_c} - R_e)$$

and

$$\text{LR} = \frac{D_{lw}}{D_{iw}}$$

or

$$D_{lw} = (LR) \times D_{iw}$$

Thus, by introducing these expressions, we get

$$D_{iw} = (E_{t_c} - R_e) + (LR) \times D_{iw}$$

$$\therefore \quad D_{iw} = \frac{(E_{t_c} - R_e)}{(1 - LR)} \quad \ldots(4.46)$$

Frequency of Irrigation

Frequency of irrigation may be defined as the interval in days between the two successive waterings. It is generally determined by the allowable soil moisture depletion method, according to which irrigation water is to be applied before the soil moisture has been depleted below the permissible lower limit. The soil moisture may be allowed to be depleted upto the lower limit of the readily available moisture. As mentioned in Chapter 3, readily available moisture is usually taken as about 75% of the available moisture. However, for healthy growth of the plants, irrigation water is usually applied when the soil moisture has been depleted to only about 50% of the available moisture, i.e., well before the lower limit of the readily available moisture is reached. During each watering the effective storage available in the root zone for irrigation water is thus equal to the amount of moisture allowed to be depleted from the soil. The frequency of irrigation is then equal to the water stored in the root zone during each watering divided by the evapotranspiration or consumptive use of water per day.

4.15 ASSESSMENT OF IRRIGATION WATER CHARGES

For supplying irrigation water to the cultivators irrigation projects are undertaken by the Government. The irrigation projects are financed by the Government from its own resources and/or by obtaining loans from the financial institutions. The loans obtained for these projects are required to be repaid along with interest if any. Moreover, certain expenditure has to be incurred on the maintenance and operation of the projects. As such some return on the capital investment on the project should be obtained to facilitate the repayment of the loan as well as to meet the other expenditures. Further a part of the revenue so realized may also be used for undertaking other new projects. The return is, however, obtained by imposing charges for the irrigation water supplied to the cultivators. The imposition of such water charges on the cultivators is justified because they derive considerable benefits by way of (i) increased production with insurance against crop failure, (ii) use of improved seeds, fertilizers, pesticides, etc., with the availability of irrigation facilities, (iii) increased cost of land with the advent of irrigation project. The fixation of the charges for the irrigation water supplied to the cultivators is also termed as assessment of irrigation water charges. The responsibility for the assessment of the water charges is, in most cases, that of the engineer incharge. Therefore a knowledge of the methods of assessment is essential for an irrigation engineer.

Methods of Assessment

There are five methods of assessment of irrigation water charges which are mentioned below.

(i) Assessment on area basis or crop rate basis

(ii) Volumetric assessment

(iii) Assessment on seasonal basis

(iv) Assessment on composite rate basis or consolidated assessment

(v) Permanent assessment or betterment levy

(i) *Assessment on area basis or crop rate basis.* In this method the water charges are fixed per unit area of the land irrigated for each of the different crops grown. The rate of water charges for the different crops is fixed on the basis of the following factors.

(a) Cash value of crop and hence its paying capacity.

(b) Water requirement of the crop.

(c) Time of demand for water in relation to the available supplies in the rivers.

The rates of water charges for a few main crops are given in Table 4.16.

Table 4.16 Rates of water charges for main crops

Crop	*Delta in mm*	*Yield per hectare*	*Rate of water charges per hectare in*	
			Punjab	*U.P.*
Wheat	375	13.54 kN (1380 kg)	₹ 16.00	₹ 15.00
Maize	450	13.54 kN (1380 kg)	₹ 16.00	₹ 11.00
Cotton	450	6.28 kN (640 kg)	₹ 16.75	₹ 15.00
Rice	900	10.30 kN (1050 kg)	₹ 24.00	₹ 39.50
Sugarcane	900	29.43 kN (3000 kg)	₹ 41.00	₹ 74.25

In this method of assessment, irrigated area showing crops is recorded both at the time of sowing and maturity. At the end of the crop period a 'demand statement' is prepared for each cultivator to enable him to deposit the water charges.

This method of assessment has certain drawbacks as noted below.

(a) Since the water charges are not made on the basis of the actual quantity of water used by the cultivators but are made on the basis of the area irrigated, the cultivators make a wasteful use of water. In other words this method lacks incentive to the cultivators for economical use of water.

(b) The cultivators near the head reaches of the canal use more water than their due share, and hence deprive the cultivators near the tail end of the canal of their due share of water which results in mal-distribution of water.

(c) Cultivators are tempted to over irrigate a limited area rather than bringing more area under irrigation by judicious application of water.

However, inspite of the various drawbacks, because of the convenience, almost all the canal irrigated lands in our country are assessed for irrigation water charges by this method. Moreover, the method is not only adopted in flow irrigation system but also adopted in lift irrigation system.

(ii) *Volumetric Assessment*. In this method the water charges are fixed on the basis of the volume of water supplied to the cultivators. For measuring the quantity of water supplied, numerous water meters are required to be installed at all the outlets in the canal system. Alternatively certain modular outlets may be installed to supply the measured quantity of water. Thus, this is an ideal method in which the cultivators have an incentive for economic use of water and the short falls in supplies can be distinctly located. However, there are several short comings of this method as described following.

(a) Heavy initial cost is involved in the installation of the devices for the measurement of water and the same are difficult to maintain.

(b) A whole time staff is required to keep records of the water meters.

(c) The meters may not work on small heads at the outlet.

(d) Silt and debris may obstruct the flow through the metered outlets.

(e) The cultivators may pilfer water by unauthorized means.

(f) The internal distribution of water supplies and apportionment of dues between petty cultivations may be difficult.

On account of these shortcomings this method has found little practical use in our country. Moreover, with the present stage of development and enlightenment of the cultivators in our country this method is unlikely to work. However, this method has found use in tubewell irrigation system where water is sold out on a volumetric basis by litres per rupee or by unit rate of electric power consumed, as a tubewell operator is there to keep the full record. Further a feasible proposal for the use of this method in canal irrigation system is to sell out water on volumetric basis to cooperative societies or panchayats of cultivators who should be made responsible both for internal distribution as well as collection and payment of water charges.

(iii) *Assessment on seasonal basis*. In this method the water charges are fixed for a certain crop season on the basis of the crops grown in that season and the water consumed. The value of the crop is not considered in fixing the water charges. However, the water charges varies from season to season. For example, the charges for the water supplied during summer will be high and for the water supplied during monsoon will be less. This method is however, not in vogue in our country.

(iv) *Assessment on composite rate basis or consolidated assessment*. In addition to the water charges the cultivators are also required to pay a tax on the land owned by them which is known as the land revenue. In this method of assessment both the water charges and the land revenue are combined to obtain a composite rate which is collected from the cultivators. Since in this method the cultivators pay a consolidated levy this method is also known as consolidated assessment. However, this method is also not in common use except in some states where it is preferred to collect both the revenues together.

(v) *Permanent assessment or betterment levy.* The cultivators in certain areas may have their own source for the supply of irrigation water and hence under ordinary conditions they may not use water from the canals provided in these areas. However, during the drought periods these cultivators may use water for irrigation from the canals. As such by virtue of irrigation facilities provided an insurance against the drought exists for the cultivators of these areas. Moreover, with the provision of irrigation facilities the cost of the land in these areas would appreciate considerably. Therefore certain fixed levy is imposed on the cultivators of these areas which they have to pay every year. Such a levy is known as *betterment levy*. In lieu of the betterment levy charged from the cultivators they are authorised to use water from the canals during the drought period without extra payment. Further since the betterment levy is permanently imposed on the cultivators, this method is also known as permanent assessment method.

SOLVED ILLUSTRATIVE EXAMPLES

Illustrative Example 4.1 *Find the delta for a crop if the duty for a base period of 120 days is 1500 hectares/cumec.*

Solution

From Eq. (4.1)

$$\Delta = \frac{8.64B}{D} = \frac{8.64 \times 120}{1500}$$

$$= 0.69 \text{ m} = 690 \text{ mm}$$

Illustrative Example 4.2 *A crop requires 900 mm of water for a base period of 120 days. Find the duty of water.*

Solution

From Eq. (4.1)

$$\Delta = \frac{8.64B}{\Delta}$$

$$= \frac{8.64 \times 120}{0.9}$$

$$= 1152 \text{ hectares/cumec.}$$

Illustrative Example 4.3 *Water is released at the rate of 12 cumec at the head of a canal. If duty at the field is 1250 hectares/cumec and loss of water in transit is 25%, find the area of the land that can be irrigated.*

Solution

Discharge at the canal head

$$= 12 \text{ cumec}$$

Loss of water in transit

$$= 25\%$$

∴ Discharge at the field

$$= 12 \times 0.75$$

$$= 9 \text{ cumec}$$

Duty at the field = 1250 hectares/cumec

∴ Area irrigated = 1250×9

$$= 1.125 \times 10^4 \text{ hectares.}$$

Illustrative Example 4.4 *The gross commanded area for an irrigation canal is 20 000 hectares out of which 75% is culturable commanded area. The intensity of irrigation is 40% for rabi and 10% for rice. If kor period is 4 weeks for rabi and 2.5 weeks for rice, determine the outlet discharge. Outlet factors for rabi and rice may be assumed as 1800 hectares/cumec and 775 hectares/cumec. Also calculate delta for each case.*

Solution

Gross commanded area

$$= 20\,000 \text{ hectares}$$

Culturable commanded area

$$= 20\,000 \times 0.75$$

$$= 15\,000 \text{ hectares}$$

Area to be irrigated in rabi at 40% intensity

$$= 15\,000 \times 0.40$$

$$= 6\,000 \text{ hectares}$$

Outlet discharge $= \dfrac{6\,000}{1\,800}$

$= 3.33$ cumec

Area to be irrigated for rice with 10% intensity

$= 15\,000 \times 0.10$

$= 1\,500$ hectares

Outlet discharge $= \dfrac{1\,500}{775}$

$= 1.94$ cumec

From Eq. (4.1)

$$\Delta = \frac{8.64B}{D}$$

Thus, for rabi $\Delta = \dfrac{8.64 \times (4 \times 7)}{1\,800}$

$= 0.134 \text{ m} = 134 \text{ mm}$

Similarly for rice $\Delta = \dfrac{8.64 \times (2.5 \times 7)}{775}$

$= 0.195 \text{ m} = 195 \text{ mm}.$

Illustrative Example 4.5 *A water course commands an irrigated area 1 000 hectares. The intensity of irrigation of rice in this area is 70%. The transplantation of rice crop takes 15 days and during the transplantation period the total depth of water required by the crop on the field is 500 mm. During the transplantation period, the useful rain falling on the field is 120 mm. Find the duty of irrigation water for the crop on the field during transplantation, at the head of the field and also at the head of the water course assuming losses of water to be 20% in the water course. Also calculate the discharge required in the water course.*

Solution

Depth of water required on the field during transplantation

$= 500$ mm

Useful rainfall during this period

$= 120$ mm

∴ Depth of water required to be supplied by the water course

$= (500 - 120)$

$= 380 \text{ mm} = 0.38 \text{ m}$

From Eq (4.1)

$$\Delta = \frac{8.64B}{\Delta}$$

∴ Duty of water on the field is

$$D = \frac{8.64 \times 15}{0.38}$$

$= 341$ hectares/cumec

Since the losses of water in the water course are 20%, a discharge of 1 cumec at the head of the water course will be reduced to 0.8 cumec at the head of the field, and hence will irrigate

$$341 \times 0.8 = 272.8 \text{ hectares}$$

$\therefore$ Duty of water at the head of the water course

$= 272.8$ hectares/cumec

Total area under rice plantation

$= 1000 \times 0.7 = 700$ hectares

$\therefore$ Discharge at the head of water course

$$= \frac{700}{272.8}$$

$= 2.57$ cumec.

Illustrative Example 4.6 *Following Table gives the necessary data about the crop, duty of water and the area under each crop commanded by a canal taking off from a storage reservoir. Taking a time factor for the canal to be (12/20), calculate the discharge required at the head of the canal. If the capacity factor is 0.8, determine the design discharge.*

Crop	*Base period (days)*	*Area (hectares)*	*Duty of water at the head of the canal (hectares/cumec)*
Sugarcane	*320*	*900*	*580*
Overlap for sugarcane in hot weather	*90*	*150*	*580*
Wheat (Rabi)	*120*	*750*	*1 600*
Bajra (Kharif)	*120*	*600*	*2 000*
Vegetables (Hot weather)	*120*	*320*	*600*

Solution

Discharge required for sugarcane

$$= \frac{900}{580} = 1.552 \text{ cumec}$$

Discharge required for overlapping sugarcane

$$= \frac{150}{580} = 0.259 \text{ cumec}$$

Discharge required for wheat

$$= \frac{750}{1600} = 0.469 \text{ cumec}$$

Discharge required for Bajra

$$= \frac{600}{2\,000} = 0.3 \text{ cumec}$$

Discharge required for vegetables

$$= \frac{320}{600} = 0.533 \text{ cumec}$$

Since sugarcane has a base period of 320 days, it will require water during Rabi, Kharif and hot weather. Thus,

Discharge required in Rabi

$$= 1.552 + 0.469$$
$$= 2.021 \text{ cumec}$$

Discharge required in Kharif

$$= 1.552 + 0.3$$
$$= 1.852 \text{ cumec}$$

Discharge required in hot weather

$$= 1.552 + 0.259 + 0.533$$
$$= 2.344 \text{ cumec}$$

Out of the three demands the maximum demand is 2.344 cumec which is during the hot weather.

The *time-factor* is the ratio of the number of days the canal has actually run to the number of days the canal was supposed to run. In this case time factor is (12/20).

∴ Discharge required at the head of the canal

$$= 2.344 \times 2012$$
$$= 3.907 \text{ cumec}$$

$$\text{Design discharge} = \frac{\text{Mean discharge required}}{\text{Capacity factor}}$$

$$= \frac{3.907}{0.8} = 4.884 \text{ cumec}$$

Illustrative Example 4.7 *The base period, intensity of irrigation and duty of water for various crops under a canal system are given in the follwoing Table. Determine the reservoir capacity if the culturable commanded area is 40000 hectares, canal losses are 20% and reservoir losses are 10%.*

Crop	*Base Period (days)*	*Duty of water at the field (hectares / cumec)*	*Intensity of Irrigation (percentage)*
Wheat	*120*	*1800*	20
Sugarcane	*360*	*1700*	20
Cotton	*180*	*1400*	10
Rice	*120*	*800*	15
Vegetables	*120*	*700*	15

Calculations are done in the following table:

Crop	*Duty of water at the field (hectares/cumec)* D	*Base period (days)* B	*Delta (m)* $\Delta = \frac{8.64B}{D}$	*Area under each crop (hectares)* A	*Volume of water required (hectare-m)* $V = A \times \Delta$
Wheat	*1800*	*120*	*0.576*	*8000 (40 000 × 0.2)*	*4 608*
Sugarcane	*1700*	*360*	*1.830*	*8 000 (40 000 × 0.2)*	*14 640*
Cotton	*1400*	*180*	*1.111*	*4 000 (40 000 × 0.1)*	*4 444*
Rice	*800*	*120*	*1.296*	*6 000 (40 000 × 0.15)*	*7 776*
Vegetables	*700*	*120*	*1.481*	*6 000 (40 000 × 0.15)*	*8 886*
				Total	***40 354***

Solution

Total volume of water required by the crops

$$= 40354 \text{ hectare-metres}$$

∴ Required capacity of the reservoirs

$$= \frac{40359}{0.3 \times 0.9}$$

$$= 5.605 \times 10^4 \text{ hectare-metres.}$$

Illustrative Example 4.8 *After how many days will you order irrigation in order to ensure healthy growth of crops, if*

(*i*) *Field capacity of soil* = *29%*

(*ii*) *Permanent wilting percentage* = *11%*

(*iii*) *Density of soil* = *1 300 kg/m*^{3}

(*iv*) *Effective depth of root zone* = *700 mm*

(*v*) *Daily consumptive use of water for the given crop* = *12 mm*

For healthy growth moisture content must not fall below 25% of the water holding capacity between the field capacity and the permanent wilting point.

Solution

Available moisture

= Field capacity—Permanent wilting percentage

= 29 – 11 = 18%

Readily available moisture is 75% of the available moisture.

∴ Readily available moisture

$$= 18 \times 0.75 = 13.5\%$$

$\therefore$ Minimum moisture content at which irrigation is required

$$= 29 - 13.5 = 15.5\%$$

i.e., when irrigation water is applied, the moisture content is raised from 15.5% to 29%.

Depth of water stored in the root zone during each watering

$$\frac{w_s d}{w} = \text{(Field capacity – Minimum moisture content at which irrigation is required)}$$

$$= \frac{1300 \times 0.7}{1000}(0.29 - 0.155)$$

$$= 0.123 \text{ m} = 123 \text{ mm}$$

Thus, depth of water available for evapotranspiration

$$= 123 \text{ mm}$$

Daily consumptive use of water for the given crop

$$= 12 \text{ mm}$$

$\therefore$ Watering interval

$$= \frac{12.3}{1.2} \text{ 10 days.}$$

Illustrative Example 4.9 *Work out the irrigation schedule based on the soil moisture concept given the following information. Also extract the data on the total depth of irrigation water required and the respective dates of irrigation water supply:*

(a) The crop is grown in an appropriate soil with no restrictive layers within the top 1.5 m depth of soil.

(b) Normal root zone depth of the crop is 1.2 m.

(c) Bulk density of the soil is 1.35.

(d) Field capacity is 18% and permanent wilting point is 7%.

(e) Moisture level in the soil is to be maintained at not less than one-third of available retention. Irrigation will then be done over a duration of 2 days at a uniform rate of supply and at a uniform rate of advance to fully and just compensate for the depletion.

(f) No extra water is ever required for leaching.

(g) Sowing is done on 1 November when the soil moisture is left just at field capacity in the entire root zone.

(h) For the crop, at the location, the average evapotranspiration rates are:

1 November	–	*30 November*	:	*1.1 mm/day*
1 December	–	*31 December*	:	*1.7 mm/day*
1 January	–	*31 January*	:	*2.4 mm/day*
1 February	–	*28 February*	:	*1.5 mm/day*
1 March	–	*25 March*	:	*3.5 mm/day*

(*i*) *Harvesting is done on or after 26th March.*

(*j*) *There is expected an effective rainfall of 24 mm during 4 January to 19 January, both days inclusive, with uniform intensity.*

(*k*) *By the end of the crop growth season, only the minimum water needed be left unused in the root zone.*

Solution

On 1 November moisture in root zone

$$= \left(1.35 \times 1.2 \times \frac{18}{100}\right)$$

$$= 0.2916 \text{ m} = 291.6 \text{ mm}$$

During 30 days of November evapotranspiration

$$= (1.1 \times 30) = 33 \text{ mm}$$

Thus, on 1 December moisture in root zone

$$= (291.6 - 33) = 258.6 \text{ mm}$$

During 31 days of December evapotranspiration

$$= (1.7 \times 31) = 52.7 \text{ mm}$$

Thus, on 1 January moisture in root zone

$$= (258.6 - 52.7) = 205.9 \text{ mm}$$

During 3 days of January evapotranspiration

$$= (3 \times 2.4) = 7.2 \text{ mm}$$

Thus, on 4 January moisture in root zone

$$= (205.9 - 7.2) = 198.7 \text{ mm}$$

During 16 days of January (4 January to 19 January) evapotranspiration

$$= \left(2.4 - \frac{24}{16}\right) \times 16 = 14.4 \text{ mm}$$

Thus, on 20 January moisture in root zone

$$= (198.7 - 14.4) = 184.3 \text{ mm}$$

During 12 days of January (20 January to 31 January) evapotranspiration

$$= (12 \times 2.4) = 28.8 \text{ mm}$$

Thus, on 1 February moisture in root zone

$$= (184.3 - 28.8) = 155.5 \text{ mm}$$

During 28 days of February evapotranspiration

$$= (28 \times 1.5) = 42 \text{ mm}$$

Thus, on 1 March moisture in root zone

$$= (155.5 - 42) = 113.5 \text{ mm}$$

Minimum moisture level to be maintained

$$= \frac{1}{3} \text{ available moisture}$$

$$= \frac{1}{3}\left(1.35 \times 1.2 \times \frac{18-7}{100}\right)$$

$$= 0.0594 \text{ m} = 59.4 \text{ mm}$$

Thus, in March irrigation must be done when the available moisture just falls below 59.4 mm.

∴ Allowable depletion of moisture in March

$$= (113.5 - 59.4) = 54.1 \text{ mm}$$

At an evapotranspiration rate of 3.5 mm/day this depletion would occur in

$$\frac{54.1}{3.5} = 15 \text{ days}$$

Evapotranspiration in 15 days of March

$$= (15 \times 3.5) = 52.5 \text{ mm}$$

Thus, on 16 March moisture in root zone

$$= (113.5 - 52.5) = 61 \text{ mm}$$

On 26 March moisture to be left in root zone should be equal to the minimum moisture content, i.e., 59.4 mm.

∴ Moisture needed on 16 March to take care of the evapotranspiration from 16 March to 25 March and leave a balance of 59.4 mm on 26 March

$$= [59.4 + (3.5 \times 10)] = 94.4 \text{ mm}$$

∴ Irrigation needs on 16 and 17 March (i.e., 2 days)

$$= (94.4 - 61) = 33.4 \text{ mm}$$

Moisture on 16 March

$$= \left(61.0 + \frac{33.4}{2}\right) = 77.7 \text{ mm}$$

Moisture on 17 March

$$= \left(77.7 - 3.5 + \frac{33.4}{2}\right) = 90.9 \text{ mm}$$

On 18 March moisture in root zone

$$= (90.9 - 3.5) = 87.4 \text{ mm}$$

During 8 days of March (18 March to 25 March) evapotranspiration

$$= (3.5 \times 8) = 28 \text{ mm}$$

Thus, on 26 March moisture in root zone

$$= (87.4 - 28) = 59.4 \text{ mm}$$

which is alright.

Hence irrigation required on 16 and 17 March uniformly @ 16.7 mm/day.

Illustrative Example 4.10 *Determine the consumptive use requirement for a certain crop with the climatic and other data given in the following Table. Also calculate the filed irrigation requirement if the water application efficiency is 75%. Use Blaney–Criddle formula.*

Month	*Mean monthly temperature* $t°C$	*Monthly percent of day-time hours of the year* p	*Monthly consumptive use coefficient* k	*Mean monthly effective rainfall in mm* R_e
April	25	8.60	0.60	—
May	27	9.29	0.65	—
June	28	9.18	0.70	52.3
July	29	9.39	0.75	74.6
August	29	9.04	0.75	62.8
September	27	8.31	0.65	31.2
October	24	8.10	0.60	25.3

Solution

Month	$t°C$	p	k	$u = kp\,(0.46\,t + 8.13)$ *(mm)*	R_e *(mm)*
April	25	8.60	0.60	101.29	—
May	27	9.29	0.65	124.09	—
June	28	9.18	0.70	135.01	52.3
July	29	9.39	0.75	151.20	74.6
August	29	9.04	0.75	145.57	62.8
September	27	8.31	0.65	111.00	31.2
October	24	8.10	0.60	93.17	25.3
				$\Sigma u = 861.33$	$\Sigma R_e = 246.2$

Total consumptive use requirement

$$U = \Sigma u = 861.33 \text{ mm}$$

Consumptive irrigation requirement

$$\text{CIR} = U - \Sigma R_e$$

$$= (861.33 - 246.2) = 615.13 \text{ mm}$$

Field irrigation requirement

$$\text{FIR} = \frac{\text{NIR}}{\eta_a} = \frac{\text{CIR}}{\eta_a}$$

$$= \frac{U - \Sigma R_e}{\eta_a}$$

where η_a is the water application efficiency.

$$\therefore \qquad \text{FIR} = \frac{615.13}{0.75} = 820.17 \text{ mm.}$$

Illustrative Example 4.11 *Calculate the potential evapotranspiration for an area in the month of September by Penman's method. The following data are available:*

Latitude ϕ = *6°N, Altitude EL = 1500 m a SL,*

Max. Temperature = 28.93°C, Min. Temperature = 13.44°C,

Mean humidity RH_{mean} = *62.40%*

Actual sunshine hours n = 6.38 hours

Wind speed at 4 m height = 5.5 m/s

Solution

Mean temperature $t = \dfrac{28.93 + 13.44}{2} = 21.19°\text{ C}$

From Table 4.2, for t = 21.19°C

$$e_s = 25.19 \text{ mbar}, \Delta = 1.5 \text{ mbar/°C}$$

From Eq. (4.9),we have

$$\lambda = (595.9 - 0.55T)$$

or $$\lambda = (595.9 - 0.55 \times 21.19) = 584.25 \text{ cal/gm}$$

From Eq. (4.8), we have

$$P = (1013 - 0.1055EL)$$

or $$P = (1013 - 0.1055 \times 1500) = 854.75 \text{ mbar}$$

From Eq. (4.7), we have

$$\gamma = C_p \frac{P}{(0.622)\lambda}; C_p = 0.240$$

$$\therefore \quad \gamma = \frac{0.240 \times 854.75}{(0.622 \times 584.25)} = 0.564$$

From Eq. (4.5), we have,

$$W = \frac{\Delta}{\Delta + \gamma}$$

$$\therefore \quad W = \frac{1.5}{1.5 + 0.564} = 0.73$$

Alternatively the value of W may be obtained directly from Table 4.3.

From Table 4.4 for latitude ϕ = 6°N during the month of September extra terrestrial radiation

$$R_a = 15.3 \text{ mm/day}$$

From Table 4.5 maximum possible hours of sunshine

$$N = 12.162 \text{ hours}$$

From Eq. (4.11), we have,

$$R_{ns} = R_a(1 - r)\left(a + b\frac{n}{N}\right)$$

For close ground crops $r = 0.25$,

$$a = 0.29 \cos \phi$$

or
$$a = 0.29 \cos 6° \approx 0.25$$

$$b = 0.52$$

$$\frac{n}{N} = \frac{6.38}{12.162} = 0.525$$

Thus, by substitution we get,

$$R_{n_s} = 15.3\,(1 - 0.25)\,(0.25 + 0.52 \times 0.525)$$
$$= 6.00$$

From Eq. (4.12) we have,

$$R_{nl} = f(t) \cdot f(e_a) \cdot f\left(\frac{n}{N}\right)$$

From Eq. (4.13) we have,

$$f(t) = \sigma T_a^4,\ T_a = (273 + 21.19) = 294.19\ °K$$

$$\therefore \quad f(t) = 2.0 \times 10^{-9} \times (294.19)^4$$
$$= 14.98$$

From Eq. (4.14) we have,

$$f(e_a) = \left[0.34 - 0.044\sqrt{e_a}\right]$$

From Eq. (4.18) we have

$$e_a = \frac{e_s \times RH_{\text{mean}}}{100}$$

or
$$e_a = \frac{25.19 \times 62.4}{100} = 15.72 \text{ mbar}$$

$$\therefore \quad f(e_a) = \left[0.34 - 0.044\sqrt{15.72}\right]$$
$$= 0.166$$

From Eq. (4.15) we have

$$f\left(\frac{n}{N}\right) = \left(0.10 + 0.90\frac{n}{N}\right)$$

$$\therefore \quad f\left(\frac{n}{N}\right) = (0.10 + 0.90 \times 0.525)$$
$$= 0.573$$

Thus, by substitution we get,

$$R_{nl} = (14.98 \times 0.166 \times 0.573)$$

$$= 1.425$$

Thus, from Eq. (4.10) we have

$$R_n = R_{ns} - R_{nl}$$
$$= (6.00 - 1.425) = 4.575$$

From Eq. (4.16) we have

$$f(U) = 0.27\left(1+\frac{U^2}{100}\right)$$

From Eq. (4.17) we have

$$U_2 = U_z\left(\frac{2}{z}\right)^{0.2}$$

$$U_z = U_4 = \frac{5.5\times3600\times24}{1000} = 475.2 \text{ km/day}$$

$$\therefore \quad U_2 = 475.2\left(\frac{2}{4}\right)^{0.2} = 413.42 \text{ km/day}$$

Thus, by substitution we get,

$$f(U) = 0.27\left(1+\frac{413.42}{100}\right) = 1.39$$

Also $\quad (e_s - e_a) = (25.19 - 15.72) = 9.47$

From Eq. (4.4), we have

$$E_{t_p} = W_{Rn} + (1 - W)\, f(U)\, (e_s - e_a)$$

Thus, by substitution, we get

$$E_{t_p} = 0.73 \times 4.575 + (1 - 0.73) \times 1.39 \times 9.47$$
$$= 6.894 \text{ mm/day}$$

Illustrative Example 4.12 *Solve Illustrative Example 4.11 by Jensen–Haise method with the following additional data.*

Hotest month March during which

Mean Max. Temperature = 33.04°C

Mean Min. Temperature = 14.31°C

Solution

From Table 4.2 corresponding to mean max. temperature equal to 33.04°C and mean min. temperature equal to 14.31°C the values of saturation vapour pressures are

$$e_2 = 50.416 \text{ mbar}$$

and $\quad e_1 = 16.379$ mbar

Mean temperature t = 21.19°C

Altitude EL = 1500 ma SL

From Eq. (4.23), we have

$$t_x = -2.5 - 0.14\,(e_2 - e_1) - \frac{El}{550}$$

or

$$t_x = -2.5 - 0.14\,(50.416 - 16.379) - \frac{1500}{550}$$

$$= -9.992°\text{C}$$

From Eq. (4.22), we have

$$C_t = \frac{1}{C_1 + C_2 C_H}$$

in which

$$C_1 = 38 - \left(2°C \times \frac{EL}{305}\right)$$

$$C_1 = 38 - \left(2 \times \frac{1500}{305}\right) = 28.164$$

$$C_2 = 7.6$$

$$C_H = \frac{50}{(e_2 - e_1)}$$

or

$$C_H = \frac{50}{(50.416 - 16.379)} = 1.469$$

Thus, by substitution, we get

$$C_t = \frac{1}{28.164 + (7.6 \times 1.469)} = 0.0254$$

From Eq. (4.20), we have

$$R_s = R_a\left(a + b\frac{n}{N}\right)$$

$$R_a = 15.3,\ a = 0.25,\ b = 0.52, = 0.525$$

$\therefore$

$$R_s = 15.3\,(0.25 + 0.52 \times 0.525)$$

$$= 8.002$$

From Eq. (4.21), we have

$$E_{t_p} = C_t\,(t - t_x)\,R_s$$

Thus, by substitution, we get

$$E_{t_p} = 0.0254\,(21.19 + 9.992) = 8.002$$

$$= 6.338 \text{ mm/day}$$

Illustrative Example 4.13 *Solve illustrative Example 4.11 by Hargreaves method.*

Solution

From Eq. (4.24), we have

$$E_{t_g} = 0.0135\,(t + 17.78)\,R_s$$

$$t = 21.19°\text{C and } R_s = 8.002$$

Thus, by substitution, we get

$$E_{t_g} = 0.0135\,(21.19 + 17.78) \times 8.002$$

$$= 4.21 \text{ mm/day}$$

Further from Eq. (4.25), we have

$$E_{t_p} = 1.2\,E_{t_g}$$

$$\therefore \quad E_{t_p} = 1.2 \times 4.21$$

$$= 5.052 \text{ mm/day}$$

Illustrative Example 4.14 *Solve Illustrative Example 4.11 by Blaney–Criddle method.*

Solution

From Eq. (4.29), we have

$$E_{t_p} = p\,(0.46t + 8.13)$$

For latitude ϕ = 6°N, during the month of September, the value of p is obtained from Table 4.8 (a) as

$$p = 0.28$$

Also monthly mean temperature t = 21.19°C

Thus, by substitution, we get

$$E_{t_p} = 0.28\,[(0.46 \times 21.19) + 8.13]$$

$$= 5.006 \text{ mm/day}$$

Further for medium values of sunshine ratio, wind velocity and relative humidity, the values of correction factors a and b are obtained from Table 4.9 as

$$a = 1.38 \text{ and } b = 2.15$$

Thus, from Eq. (4.29 a) we obtained the corrected value of potential evapotranspiration as

$$E_{t_p} = ap\,(0.46t + 8.13) - b$$

$$= 1.38 \times 08\,[(0.46 \times 21.19) + 8.13] - 2.15$$

$$= 4.758 \text{ mm/day}$$

Illustrative Example 4.15 *Compute the evapotranspiration by Thornthwaite method for July in a region at 44°N, with the following average monthly temperatures:*

Month	: *J*	*F*	*M*	*A*	*M*	*J*	*J*	*A*	*S*	*O*	*N*	*D*
Temperature – T°C	: *2.2*	*1.0*	*5.2*	*9.5*	*13.3*	*18.0*	*22.5*	*21.6*	*16.3*	*10.0*	*4.3*	*–1.0*

Solution

The heat index i is given by

$$i = (T/5)^{1.514} \quad \text{...(1)}$$

Using Eq. (1) the values of heat index i are computed for each month and the same are as given below:

Month	: J	F	M	A	M	J	J	A	S	O	N	D
$= (T/5)^{1.514}$	: 0	0.09	1.06	2.64	4.40	6.95	9.75	9.16	5.98	2.86	0.80	0

Note: For values of T less than 0, i is taken as zero.

Thus, $\Sigma i = 43.69 = I$

From Eq. (4.19), we have

$$a = 6.75 \times 10^{-7} (43.69)^3 - 7.71 \times 10^{-5} (43.69)^2 + 1.792 \times 10^{-2} (43.69) + 0.492\ 39$$

or $a = 1.184$

From Eq. (4.17), we have

$$E_t = 16.0 \left(\frac{10 \times 22.5}{43.69}\right)^{1.184}$$

or $$E_t = 16.0\ (5.15)^{1.184} = 111.40 \text{ mm/month}$$

This is, however, unadjusted value of potential evapotranspiration.

From Table 4.10 the adjustment factor for July at 44°N is 1.30 and hence Thornthwaite potential evapotranspiration

$$= (1.3 \times 111.4)$$
$$= 144.82 \text{ mm/month} \simeq \text{mm/month}$$

Table : Calculations for determining irrigation requirements for wheat crop

Dates	*No. of days upto midpoint of time interval*	*% of crop growing season =* $Col.(2) \times \frac{100}{135}$	*Coefficient* *K*	*Pan Evaporation* E_p *(mm)*	*Evapo-transpiration* $E_t = KE_p$ *(mm)*	*Effective Rainfall* R_e *(mm)*	*N.I.R.* $= E_t - R_e$ *(mm)*
(1)	*(2)*	*(3)*	*(4)*	*(5)*	*(6)*	*(7)*	*(8)*
Nov 1–30	15	11	0.16	130	20.8	5	15.8
Dec. 1–31	46	34	0.45	110	49.5	16	33.5
Jan. 1–31	77	57	0.73	90	65.7	32	33.7
Feb. 1–28	106	79	0.90	150	135.0	27	108.0
March 1–15	128	95	0.60	100	60.0	Nil	60.0

Net irrigation requirement = Σ N.I.R. = 251.0 mm

$$\text{Field irrigation requirement F.I.R.} = \frac{\text{N.I.R.}}{\eta_a}$$

$$= \frac{251.0}{0.70} = 358.6 \text{ mm}$$

$$\text{Gross irrigation requirement G.I.R} = \left(\frac{\text{F.I.R.}}{\eta_c} + 70\right) \text{ mm}$$

$$= \left(\frac{358.6}{0.75} + 70\right) \text{mm} = 548.1 \text{ mm}$$

Illustrative Example 4.16. *For use in Hargreaves method, in respect of Group D wheat crop, the following K values are suggested:*

% of crop growing season K	*0*	*5*	*10*	*15*	*20*	*25*	*30*	*35*	*40*	*45*	*50*
	0.08	*0.08*	*0.15*	*0.19*	*0.27*	*0.33*	*0.40*	*0.46*	*0.52*	*0.58*	*0.65*
% of crop growing season K	*55*	*60*	*65*	*70*	*75*	*80*	*85*	*90*	*95*	*100*	
	0.71	*0.77*	*0.82*	*0.88*	*0.90*	*0.90*	*0.80*	*0.70*	*0.60*	*0.60*	

Wheat is grown between 1 November and 15 March. The effective rainfall in the individual months is taken as follows: Nov–5 mm; Dec–16 mm; Jan–32 mm; Feb–27 mm; March-nil. Percolation losses are nil throughout. Field Irrigation Efficiency is 70% and Gross Irrigation Efficiency is 75%. The mean Class A pan evaporation through the respective months are: Nov–130 mm; Dec–110 mm; Jan–90 m; Feb–150 mm; March–100 mm. Compute the gross irrigation requirement for the whole crop period including 70 mm of net presowing requirement at the site.

Solution

The computations are shown in the Table given on page 142.

Illustrative Example 4.17. *The discharge available from a tube well is 140 $m^3/hour$. Assuming 10 hours of pumping per day for 350 days in a year, estimate the culturable area that this tube well can command. The intensity of irrigation is 45% and average water depth required for the rabi and kharif crop is 500 mm.*

Solution

Total quantity of water available from the tube well per year

$$= (140 \times 10 \times 350) \text{ m}^3 \quad \ldots\text{(i)}$$

If x is the culturable area in hectares which this tube well can command then the total quantity of water required

$$= \left(x \times 10^4 \times \frac{500}{1000}\right) \text{m}^3 \quad \ldots\text{(ii)}$$

Equating Eq. (i) and (ii), we get

$$\left(x \times 10^4 \times \frac{500}{1000}\right) = (140 \times 10 \times 350)$$

or $$x = \frac{140 \times 10 \times 350 \times 1000}{10^4 \times 500}$$

∴ $x = 98$ hectares

The intensity of irrigation is 45% and hence the area actually cultivated during rabi or kharif season = (98 × 0.45) = 44.1 hectares.

Illustrative Example 4.18. *Following particulars were recorded from an agricultural land growing wheat:*

(i) Field capacity	*: 20%*
(ii) Wilting point	*: 8%*
(iii) Permissible depletion of available moisture content	*: 50%*
(iv) Root zone depth	*: 2 m*
(v) Dry unit weight of soil	*: 1400 kg/m³*
(vi) Effective rainfall	*: 20 mm*
(vii) Daily consumptive use	*: 5.6 mm*

Find the net and field irrigation requirement if loss in field application is 10%. What is the frequency of irrigation?

Solution

Available moisture

= Field capacity – Permanent wilting percentage

= (20 – 8) = 12%

Permissible depletion of available moisture content = 50%

∴ Readily available moisture

= (12 × 0.5) = 6%

∴ Minimum moisture content at which irrigation is required

= (20 – 6) = 14%

i.e., when the moisture content is reduced to 14% below field capacity, irrigation water is required to be applied and the moisture content is raised from 14% to 20%.

The net irrigation required is equal to the depth of water stored in the root zone when the moisture content is raised from 14% to 20% minus the effective rainfall.

Depth of water stored in the root zone when the moisture content is raised from 14% to 20%

$$= \frac{w_s d}{w}[0.20 - 0.14]$$

$$= \frac{1400 \times 2}{1000} \times 0.06$$

$$= 0.168 \text{ m} = 168 \text{ mm}$$

Effective rainfall = 20 mm

∴ Net irrigation requirement (NIR)

= (168 – 20) = 148 mm

Loss in field application = 10%, i.e., water application efficiency = 90% = 0.9

∴ Field irrigation requirement (FIR)

$$= \frac{148}{0.9} = 164.4 \text{ mm}$$

The depth of water stored in the root zone when the moisture content is raised from 14% to 20% is equal to the depth of water available for consumptive use or evapotranspiration.

Thus, depth of water available for consumptive use or evapotranspiration

$$= 168 \text{ mm}$$

Daily consumptive use for wheat

$$= 5.6 \text{ mm}$$

∴ Frequency of irrigation $= \frac{168}{5.6} = 30$ days

Illustrative Example 4.19 *The values of potential evapotranspiration (PET) for reference crop alfalfa at a place having latitude 6°N, altitude (or elevation) 1500 maSL as computed by Modified Penman method for each month of the year are noted below:*

Month :	*J*	*F*	*M*	*A*	*M*	*J*	*J*	*A*	*S*	*O*	*N*	*D*
E_{tp} *(mm/day)* :	*7.18*	*7.40*	*7.81*	*7.15*	*6.65*	*6.54*	*6.52*	*6.92*	*6.91*	*6.61*	*6.76*	*6.60*

Determine the evapotranspiration or consumptive use requirement E_{tc} *for the crop maize (or corn) with the following data.*

Planting date July 6.

Initial stage duration 30 days (in July and Aug.) Development stage duration 50 days (in Aug. and Sept.) Mid-season stage duration 60 days (in Sept., Oct., and Nov.) Late-season stage duration 40 days (in Nov., Dec., and Jan.). Each month may be assumed to be of 30 days.

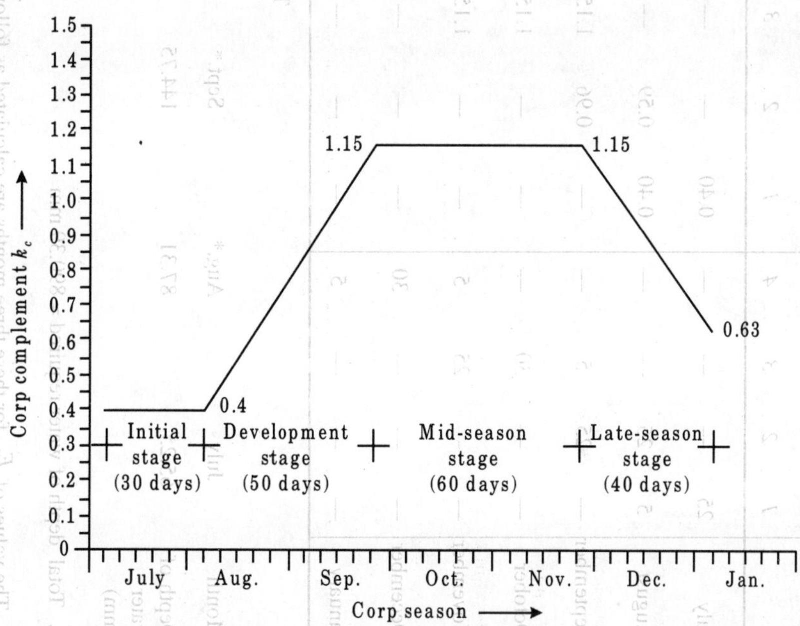

Fig. Ex. 4.19

Table

Month	*Length of each stage in days*				k_c				$E_{t_c} = k_c \times E_{t_p}$ (mm/day)				*Total E_{t_c} (in mm) = (No. of days) × E_{t_c} (in mm/day)*			
	1	*2*	*3*	*4*	*1*	*2*	*3*	*4*	*1*	*2*	*3*	*4*	*1*	*2*	*3*	*4*
July	25	—	—	—	0.40	—	—	—	2.61	—	—	—	65.25	—	—	—
August	5	25	—	—	0.40	0.59	—	—	2.77	4.08	—	—	13.85	102.00	—	—
September	—	25	5	—	—	0.96	1.15	—	—	6.63	7.95	—	—	165.75	39.75	—
October	—	—	30	—	—	—	1.15	—	—	—	7.60	—	—	—	228.00	—
November	—	—	25	5	—	—	1.15	1.12	—	—	7.77	7.57	—	—	194.25	37.85
December	—	—	—	30	—	—	—	0.88	—	—	—	5.81	—	—	—	174.30
January	—	—	—	5	—	—	—	0.63	—	—	—	4.52	—	—	—	22.60

Total Water Required

Month	July	Aug.*	Sept.*	Oct.	Nov.*	Dec.	Jan.
Depth of water (mm)	65.25	87.31	144.75	228.00	168.18	174.30	22.60

∴ Total depth of water required = 890.39 mm

* The values of E_{t_c} for these three months are calculated as follows:

Aug. $\frac{13.85 \times 5 + 102.00 \times 25}{30} = 87.31$ mm;

Sept. $\frac{165.75 \times 25 + 39.75 \times 5}{30} = 144.75$ mm;

Nov. $\frac{194.25 \times 25 + 37.85 \times 5}{30} = 168.18$ mm

Solution

For the initial stage the following values of k_c are obtained from Fig. 4.3 for irrigation frequency of 7 days

Month	E_{tp}	k_c
July	6.52	0.40
August	6.92	0.39

Thus, for the initial stage average value of k_c is obtained as

$$k_c = \frac{25 \times 0.40 + 5 \times 0.39}{30} = 0.398 \approx 0.40$$

As shown Fig. Ex. 4.19 this value of k_c is plotted as horizontal straight line for the entire duration of the initial stage.

For the mid-season stage the value of k_c is obtained as 1.15 from Table 4.10, and this value of k_c is plotted as a horizontal straight line for the entire duration of this stage.

Thus, for development stage the value of k_c increase from 0.40 to 1.15. Again from Table 4.10, at the end of the growing season the value of k_c is obtained as 0.6.

Thus, during the late season stage the value of k_c reduces from 1.15 to 0.6.

The variation of the crop coefficient k_c for the entire growing season of the crop maize (or corn) is as shown in Fig. Ex. 4.19.

From Eq. (4.30), we have

$$E_{t_c} = k_c E_{t_p}$$

The value of k_c for each stage of growth during different months are obtained from the (Fig. Ex. 4.19) and the same are as indicated in the Table Ex. 4.19 which also shows the total depth of water required at different stages of growth.

Illustrative Example 4.20 *The mean monthly rainfall for a place with latitude 6°N and altitude EL = 1500 maSL are as follows:*

Month	*J*	*F*	*M*	*A*	*M*	*J*	*J*	*A*	*S*	*O*	*N*	*D*
Mean monthly rain fall (mm)	*24.28*	*14.35*	*62.55*	*124.57*	*139.90*	*62.00*	*25.41*	*49.48*	*89.20*	*94.60*	*53.80*	*19.40*

The depth of the root zone = 1.2 m; available moisture holding capacity = 160 mm/m. Irrigation is to be provided when the moisture is depleted to 50 % of the available moisture holding capacity.

Determine the effective rainfall for the crop corn for which the E_{t_c} *values are as obtained in the Illustrative Example 4.19. Also calculate the irrigation water required for the same crop.*

Solution

For the known mean monthly rainfall and E_{t_c} the values of effective rainfall are obtained from Table 4.15 and the same are given in the table given on next page. These values of the effective rainfall are to be multiplied by storage factor the value of which may be calculated as follows.

The effective storage in this case

$$= 1.2 \times 160 \times \frac{50}{100} = 96 \text{ mm}$$

Thus, from Table 4.15 (a) the value of storage factor is obtained as

$$= \left[1.00 + \frac{(1.02 - 1.00)(96 - 75)}{100 - 75}\right] = 1.017$$

Month	*Mean monthly rainfall (mm)*	E_{tc} *(mm)*	*Effective rainfall from Table (4.15) (mm)*	*Effective rainfall* R_e *(mm)*	*Irrigation water required (mm)*
July	25.41	65.25	17.89	18.19	47.06
August	49.48	87.31	34.20	34.78	52.47
September	89.20	144.75	66.25	67.38	77.37
October	94.60	228.00	83.76	85.18	142.82
November	53.80	168.18	44.22	44.97	123.21
December	19.40	174.30	17.58	17.88	156.42
January	24.28	22.60	15.54	15.80	6.80
				Total	606.15 mm

Thus, total irrigation water required for this crop = 606.15 mm.

Illustrative Example 4.21. *The total evapotranspiration requirement* E_{tc} *for maize crop is 890.39 mm and effective rainfall* R_e *= 284.18 mm. If the available irrigation water has electrical conductivity* EC_w *= 2.7 m mhos/cm, irrigation is done by surface method and soil is slightly layered, medium textured with measured leaching efficiency* L_e *= 70%, determine the leaching requirement LR for 100%, 90% and 75% yield potential. Also calculate the total depth of irrigation water required to meet evapotranspiration and leaching requirements in each of the three cases.*

Solution

From Table 4.16 for the maize crop the values of EC_e for 100%, 90% and 75% yield potential are obtained as noted below.

Yield Potential	100%	90%	75%
EC_e	1.7	2.5	3.8

From Eq. (4.44) the leaching requirement LR is given as

$$\text{LR} = \frac{EC_w}{5EC_e - EC_w} \times \frac{1}{L_e}$$

Thus, by substitution, we get

For 100% yield potential

$$\text{LR} = \frac{2.7}{(5 \times 1.7) - 2.7} \times \frac{1}{0.7} = 0.67$$

For 90% yield potential

$$\text{LR} = \frac{2.7}{(5\times2.5)-2.7}\times\frac{1}{0.7} = 0.39$$

For 75% yield potential

$$\text{LR} = \frac{2.7}{(5\times3.8)-2.7}\times\frac{1}{0.7} = 0.23$$

From Eq. (4.46), the total depth D_{iw} of irrigation water required to meet evapotranspiration and leaching requirements is given as

$$D_{iw} = \frac{(Et_c - R_e)}{(1-\text{LR})}$$

Thus, by substitution, we get

For 100% yield potential

$$D_{iw} = \frac{(890.39-284.18)}{(1-0.67)} = 1837 \text{ mm}$$

For 90% yield potential

$$D_{iw} = \frac{(890.39-284.18)}{(1-0.39)} = 994 \text{ mm}$$

For 75% yield potential

$$D_{iw} = \frac{(890.39-284.18)}{(1-0.23)} = 787 \text{ mm}$$

Illustrative Example 4.22. *A stream of 150 litres per second was delivered from a canal and 110 litres per second were delivered to the field. An area of 2.2 hectares was irrigated in eight hours. The effective depth of root zone was 1.5 m. The runoff loss in the field was 445 m^3. The depth of water penetration varied linearly from 1.5 m at the head end of the field to 1.1 m at the tail end. Available moisture holding capacity of the soil is 200 mm per metre depth of soil. Determine the water conveyance efficiency, water application efficiency, water storage efficiency, and water distribution efficiency. Irrigation was started at a moisture extraction level of 50%.*

Solution

From Eq. (4.33), water conveyance efficiency

$$\eta_c = \frac{W_f}{W_r} \times 100$$

$$= \frac{110}{150}\times100 = 73.33\%$$

From Eq. (4.34), water application efficiency

$$\eta_a = \frac{W_s}{W_f}\times100$$

Water delivered to the plot

$$= \frac{110 \times 60 \times 60 \times 8}{1000} = 3168 \text{ m}^3$$

$$\text{Runoff loss} = 445 \text{ m}^3$$

∴ Water stored in the root zone

$$W_s = (3168 - 445) = 2723 \text{ m}^3$$

$$\therefore \quad \eta_a = \frac{2723}{3168} \times 100 = 86\ \%$$

From Eq. (4.37) water storage efficiency

$$\eta_s = \frac{W_s}{W_n} \times 100$$

Water holding capacity of the soil

$$= 200 \times 1.5 = 300 \text{ mm}$$

Moisture required in the root zone

$$= \left(300 - \frac{300 - 50}{100}\right) = 150 \text{ mm}$$

$$= \frac{150}{1000} \times 100 = 2.2 \times 10^4 = 3300 \text{ m}^3$$

$$\therefore \quad \eta_s = \frac{2723}{3300} \times 100 = 82.52\ \%$$

From Eq. (4.38), water distribution efficiency

$$\eta_d = \left(1 - \frac{y}{d}\right) \times 100$$

Average depth of water stored in the root zone is

$$d = \frac{1.5 + 1.1}{2} = 1.3 \text{ m}$$

Numerical deviation from depth of penetration

at the head end of the field = (1.5 – 1.3) = 0.2 m

at the tail end of the field = (1.3 – 1.1) = 0.2 m

∴ Average numerical deviation in depth of water stored is

$$y = \frac{0.2 + 0.2}{2} = 0.2 \text{ m}$$

$$\therefore \quad \eta_d = \left(1 - \frac{0.2}{1.3}\right) \times 100 = 84.62\ \%$$

Illustrative Example 4.23 *A field of 4 hectares has an average root zone depth of 1.0 m, a field capacity of 18 % and permanent wilting percentage of 10 % (both by weight). Assume that it is desirable to irrigate when 60 % of available moisture has been extracted. The field is irrigated by a sprinkler system which delivers 300 m³/hour over a period of 12 hours. What is the water application efficiency? Density of soil is 1400 kg/m³.*

Solution

Available moisture

$$= \text{Field capacity} - \text{Permanent wilting percentage}$$

$$= (18 - 10) = 8\ \%$$

Moisture extracted from the soil

$$= (0.6 \times 8\ \%) = 4.8\ \%$$

Minimum moisture content at which irrigation is required

$$= (18.0 - 4.8) = 13.2\ \%$$

∴ Depth of water stored in the root zone during the watering

$$= \frac{1400}{1000} \times 1 \times (18.0 - 13.2) \times \frac{1}{100}$$

$$= 0.0672 \text{ m} = 67.2 \text{ mm}$$

The sprinkler system delivers water at the rate of 300 m³/hour. Thus, in 12 hour period it delivers

$$300 \times 12 = 3600 \text{ m}^3$$

over an area of 4 hectares.

Thus, water is delivered at the rate of (3600/4) = 900 m³/heactare.

Since 1 m³ /hectare = 0.10 mm depth of water, the depth of water applied = (900 × 0.10) = 90 mm

∴ Water application efficiency

$$\eta_a = \frac{67.2}{90} \times 100 = 74.7\ \%$$

Illustrative Example 4.24 *A certain crop is grown on an area of 3000 hectares which is fed by a canal system. The data pertaining to irrigation are as follows:*

Field capacity of soil	*= 26%*
Optimum moisture	*= 12%*
Permanent wilting point	*= 10%*
Effective depth of root zone	*= 0.8 m*
Apparent relative density of soil	*= 1.4*

If the frequency of irrigation is 10 days and the overall irrigation efficiency is 22%, find (i) the daily consumptive use and (ii) the water discharge in m^3/s required in the canal feeding the area.

Solution

Since optimum moisture is 12%, irrigation water is required to be applied when the moisture content is reduced to 12% below field capacity.

Depth of water stored in the root zone when the moisture content is raised from 12% to 26%

$$= \frac{w_s \times d}{w}(0.26 - 0.12)$$

$$= (1.4 \times 0.8 \times 0.14) = 0.1568 \text{ m}$$

∴ Daily consumptive use C_u

$$= \frac{0.1568}{10}$$

$$= 0.015\ 68 \text{ m} = 15.68 \text{ mm}$$

Discharge required in the canal

$$= \frac{0.01568 \times 3000 \times 10^4}{0.22 \times 24 \times 3600}$$

$$= 24.75 \text{ m}^3/\text{s}$$

Illustrative Example 4.25 *A farmer wishes to have his own pump set for the following cropping pattern to be followed in five hectares of his land. Calculate the right size of pump he should have, litres / second.*

Season	*Crop*	*Area to be irrigated , ha*	*Intensity irrigation, cm*	*Rotation period days*
Rabi	*Wheat*	*2.0*	*7.5*	*12*
	Cotton	*0.4*	*7.5*	*20*
	Vegetables	*0.4*	*7.5*	*10*
	Mustard	*2.2*	*5.0*	*40*

For each crop, duration of pumping hours per day is 10.

Solution

From Eq. (4.1) duty is given as

$$D = \frac{8.64B}{\Delta}$$

Thus, for wheat duty

$$D = \frac{8.64 \times 12}{0.075} = 1382.4 \text{ ha/cumec}$$

∴ Discharge required for wheat

$$= \frac{\text{Area under irrigation}}{\text{Duty}}$$

$$= \frac{2}{1382.4}$$

$$= 1.4468 \times 10^{-3} \text{ m}^3/\text{s} = 1.4468 \ l/\text{s}$$

For cotton duty

$$D = \frac{8.64 \times 20}{0.075} = 2304 \text{ ha/cumec}$$

∴ Discharge required for cotton

$$= \frac{0.4}{2304}$$

$$= 0.1736 \times 10^{-3} \text{ m}^3/\text{s} = 0.1736 \ l/\text{s}$$

For vegetables duty

$$D = \frac{8.64 \times 10}{0.075} = 1152 \text{ ha/cumec}$$

∴ Discharge required for vegetables

$$= \frac{0.4}{1152}$$

$$= 0.3472 \times 10^{-3} \text{ m}^3/\text{s} = 0.3472 \ l/\text{s}$$

For mustard duty

$$D = \frac{8.64 \times 40}{0.050} = 6912 \text{ ha/cumec}$$

∴ Discharge required for mustard

$$= \frac{2.2}{6912}$$

$$= 0.3183 \times 10^{-3} \text{ m}^3/\text{s} = 0.3183 \ l/\text{s}$$

Total discharge required

$$= (1.4468 + 0.1736 + 0.3472 + 0.3183) \ l/\text{s}$$

$$= 2.2859 \ l/\text{s}$$

Since pump is operating daily for 10 hours only, the pump should have a capacity

$$= \frac{2.2859 \times 24}{10}$$

$$= 5.486 \ l/\text{s} \simeq 5.5 \ l/\text{s} = 19{,}800 \ l/\text{hour}$$

REVIEW QUESTIONS

4.1 Explain how frequency of irrigation is determined.

4.2 Write a brief note on crop seasons and crops of India.

4.3 Explain the terms 'crop period' and 'base period'.

4.4 Explain the terms 'duty' and 'delta'. Derive a relationship between the two for a given base period.

4.5 What are the factors affecting duty? How can duty be improved?

4.6 What do you understand by gross commanded area (G.C.A) and culturable commanded area (C.C.A)?

4.7 Explain the terms

(i) Intensity of irrigation
(ii) Kor watering
(iii) Kor depth
(iv) Kor period
(v) Outlet factor
(vi) Capacity factor
(vii) Time factor
(viii) Crop ratio
(ix) Overlap allowance.

4.8 What is consumptive use of water? Describe any two methods for determining the consumptive use of water.

4.9 What do you understand by potential evapotranspiration (PET)? How does it differ from actual evapotranspiration (AET)?

4.10 Discuss briefly the factors affecting the consumptive use of water.

4.11 Explain the terms

(i) Net irrigation requirement (NIR);
(ii) Field irrigation requirement (FIR); and
(iii) Gross irrigation requirement (GIR).

4.12 Following data refer to an agricultural land growing rice:

(i) Period of growth 16th July to 31st July.
(ii) USBR Class A type Pan evaporation = 6 mm/day
(iii) Crop coefficient (or Crop index) $K = 1.2$
(iv) Percolation loss = 5 mm/day
(v) Effective rainfall during the period = 60 mm

Calculate values of NIR, FIR and GIR. Assume 20% loss of water in application and 25% in conveyance. [*Ans.* 135.2 mm; 169 mm; 225.33 mm]

4.13 A water course has a culturable commanded area of 1000 hectares. The intensity of irrigation for crop A is 40% and for crop B is 45%, both the crops being rabi crops. Crop A has a kor depth of 150 mm and kor period 3 weeks; and crop B has a kor depth of 100 mm and kor period 2 weeks. Calculate the discharge of the water course.

[*Ans.* 0.703 cumec for the first two weeks of the Kor period of both the crops A and B and 0.331 cumec for the third week of the Kor period of the crop A]

4.14 An area of 20 000 hectares has to be irrigated for rice having field delta as follows:

Month	June	July	August	September	October
Delta (mm)	200	300	400	350	200

Determine the design capacity of a canal for supplying water to the above crop, assuming no loss of water in conveyance. [*Ans.* 29.87 cumec]

4.15 How many hours should a tube well, having a discharge of 550 m^3/hour, run if it has to irrigate 140 hectares of wheat and 80 hectares of sugarcane during a year? Make suitable assumptions regarding delta for each crop. [*Ans.* 2264 hours]

4.16 Explain the following irrigation efficiencies:

(i) Water conveyance efficiency
(ii) Water application efficiency
(iii) Water use efficiency
(iv) Water storage efficiency
(v) Water distribution efficiency
(vi) Consumptive use efficiency.

4.17 Discuss the various methods of assessment of irrigation water charges.

4.18 The discharge at an outlet is 0.2 m^3/s. Average losses from outlet to field are 10% of water flowing through the outlet. If kor period and kor depth for wheat and rice are 3 weeks, 120 mm and 2 weeks, 250 mm, calculate how much area can be irrigated for each crop.

[*Ans.* 272.16 hectares of wheat; 87.09 hectares of rice]

4.19 Calculate the reference crop evapotranspiration from an area near New Delhi in the month of October by Penman method with the following data:

(i) Latitude : 28°4' N
(ii) Elevation : 230 m above MSL
(iii) Mean air temperature : 20°C
(iv) Mean relative humidity : 77 %
(v) Mean observed sunshine hours : 9 hours
(vi) Wind velocity at 2 m height : 87 km/day
(vii) Reflection coefficient *r* : 0.25 [*Ans.* 3.575 mm/day]

CHAPTER

5 Hydrology

5.1 DEFINITION

Hydrology may be defined as the science that deals with the processes governing the depletion and replenishment of the water resources of the land areas of the earth. In its broadest sense, it is the study of water in all its phases and includes hydraulics, the physics and chemistry of water, meteorology and various other allied natural sciences.

5.2 HYDROLOGIC CYCLE

The hydrologic cycle is the descriptive term applied to the general circulation of water from the oceans to the atmosphere, to the ground and back to the oceans again. Figure 5.1 shows the various phases of hydrologic cycle. The cycle may be considered to begin with the water of the oceans. Water from the ocean surface is evaporated into atmosphere. The vapour is condensed by various processes and falls to the earth as precipitation. Some of this precipitation falls directly on the ocean, and some falls on the land surfaces. A portion of that falling on the land is retained temporarily in the soil, in surface depressions and on vegetation and on other objects until it is returned to the atmosphere by evaporation and transpiration. Another portion runs off from the ground surface into the streams and is returned to the ocean. Still another portion percolates into the ground and joins the ground water which also slowly finds its way to the streams as ground water flow. However, some portion of the ground water which percolates the great depths appears after long intervals as springs, artesian wells and geysers. Further the entire quantity of water that reaches the stream does not flow directly to the oceans, because throughout its travel a portion of it returns to the atmosphere by evaporation and transpiration and some portion seeps into the ground. The hydrologic cycle provides the ground work upon which the science of hydrology is based.

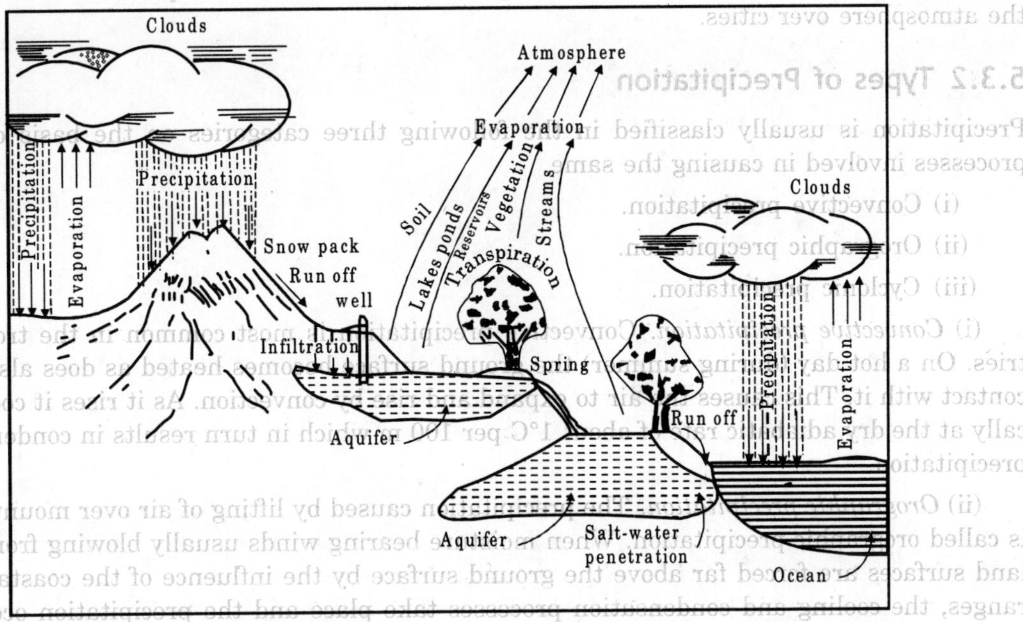

Fig. 5.1 *The hydrologic cycle*

5.3 PRECIPITATION

Precipitation is one of the main phases of the hydrologic cycle. It includes all moisture that reaches the earth's surface in liquid or solid form due to the condensation of the atmospheric vapour. The principal forms of the precipitation are rain, snow, sleet (snow mixed with rain), hail (frozen rain drops), mist, dew and frost (frozen dew). Out of the various forms of precipitation, a major part of precipitation occurs in the form of rain and a minor part occurs in the form of snow. The contribution of the other forms of precipitation is very little and hence it is generally ignored.

5.3.1 Mechanism of Precipitation

Water vapour from which precipitation is derived is present in the atmosphere as a gas in varying amounts at all times. The absolute amount of vapour in the air depends upon the temperature of the air and upon whether it has passed over features of the earth from which moistures may be derived. A general relation between temperature and moisture is that the moisture carrying capacity of the atmosphere doubles with an increase in temperature of about 11°C.

For the precipitation to take place, it is first necessary for a portion of the atmosphere to be cooled until the air is saturated with water vapour. Thereupon condensation occurs resulting in the formation of minute droplets, which are visible collectively in the form of clouds or fog. These droplets will not in themselves form into rain drops and may remain indefinitely as droplets in a state of colloidal suspension. Rain drops or other forms of precipitation result when the minute droplets of moisture collect on particles in the air termed as hygroscopic nuclei. The most common hygroscopic nuclei are salt particles from oceans and products of combustion in the atmosphere over cities.

5.3.2 Types of Precipitation

Precipitation is usually classified in the following three categories on the basis of the basic processes involved in causing the same.

(i) Convective precipitation.

(ii) Orographic precipitation.

(iii) Cyclonic precipitation.

(i) *Convective precipitation.* Convective precipitation is most common in the tropical countries. On a hot day (during summer) the ground surface becomes heated as does also the air in contact with it. This causes the air to expand and rise by convection. As it rises it cools dynamically at the dry adiabatic rate of about 1°C per 100 m which in turn results in condensation and precipitation.

(ii) *Orographic precipitation.* The precipitation caused by lifting of air over mountain barrier is called orographic precipitation. When moisture bearing winds usually blowing from oceans to land surfaces are forced far above the ground surface by the influence of the coastal mountain ranges, the cooling and condensation processes take place and the precipitation occurs on the wind-ward side of the mountains.

(iii) *Cyclonic precipitation.* Cyclonic precipitation results from lifting of air masses converging into a low pressure area or cyclone. The cyclonic precipitation may be classified as (a) frontal precipitation; and (b) non-frontal precipitation.

(a) *Frontal precipitation.* A surface separating the warm air mass and the cold air mass is called a frontal surface or a front which may be subdivided as warm front and cold front. A warm front is the one in which warm air replaces cold air and in a cold front cold air replaces warm air.

Frontal precipitation results from lifting of warm air over cold air and it may be subdivided as warm front precipitation and cold front precipitation. In warm front precipitation the warm air moves upwards over a relatively stationary wedge of cold air. In this case the precipitation is spread over a large area which may extend 300 to 500 kilometres ahead of the front and the precipitation is generally light to moderate and nearly continuous until after the passage of the warm front. On the other hand in cold front precipitation the warm air is forced upwards by an advancing wedge of cold air. In this case, the precipitation occurs on a small area which may extend only 100 to 150 kilometres ahead of the front and the precipitation is relatively more intense.

(b) *Non-frontal precipitation.* Non-frontal precipitation occurs when there is low pressure (or barometric depression) caused in any region. In this case air from an adjacent high pressure area flows into the area of low pressure which causes the lifting of the air of the low pressure area to high altitudes where it cools down and results in condensation and precipitation.

The above noted general causes of precipitation may however combine in various ways and hence the resulting precipitation cannot be identified as being of any one type.

5.4 MEASUREMENT OF PRECIPITATION

For almost all the hydrological designs it is necessary to have the records of precipitation for a long period. Such records may be developed by the measurement of the precipitation occurring at any place from time to time. All forms of precipitation are measured on the basis of the vertical depth of water or water equivalent (in the case of snow) which would accumulate on a level surface if all the precipitation without any loss remained where it fell. The precipitation is therefore expressed in millimetres (or centimetres). Many types of gages have been developed for the measurement of rain and snow which constitute the major part of precipitation. Some of these gages which are commonly used are described below. Since the amount of precipitation varies from place to place, it would be necessary to instal the gages for the measurement of precipitation at various key points in the area. Especially in the drainage basin of a stream or river a network of such gage stations evenly distributed over the entire area of the drainage basin should invariably be provided. In the measurement of precipitation it is, however, assumed that the observations of precipitation made at any gage station is representative precipitation of certain area around the gage station where the measurement is made. The Indian Standard, IS: 4987–1991 gives detailed specifications for establishing a network of gage stations for the measurement of rainfall.

(A) Measurement of Rainfall

For the measurement of rainfall rain gages are used which may be classified as follows:

(i) Non-recording type rain gages.

(ii) Self-recording type or automatic recording type rain gages.

(i) *Non-recording type rain gages*. As the name indicates these rain gages do not record the rainfall directly but only collect the rain water which when measured gives the total amount of rainfall at the rain gage station during the measuring interval. In our country until about 1969, for the measurement of rainfall at the various rain gage stations, the non-recording type rain gage extensively used by the Indian Meteorological Department (IMD) is the Symon's rain gage. However, since 1969 the Indian Meteorological Department (IMD) started using another non-recording type rain gage which has been standardized by the Indian Standard Institution (ISI) in collaboration with the Indian Meteorological Department (IMD). The Indian Standard, IS: 5225–1992 gives details of this newly standardized non-recording type rain gage, known as *Standard non-recording type rain gage*, which is an improvement over the Symon's rain gage. Both the Symon's rain gage and the Standard non-recording type rain gage are described below.

Symon's rain gage. As shown in Fig. 5.2, the Symon's rain gage consists of a cylindrical metal case of internal diameter 127 mm (5 inches) with its base enlarged to 203.2 mm (8 inches) diameter. At the top of the case a funnel is fixed which is provided with a brass rim measuring exactly 127 mm (5 inches) inside diameter. The funnel shank is inserted in a glass bottle placed inside the case. The case of the rain gage is fixed in masonry or concrete foundation block 600 mm × 600 mm × 600 mm which is sunk into the ground such that the funnel rim is exactly 304.8 mm (12 inches) above the ground level. The rain water enters the bottle through the funnel and gets collected in the bottle.

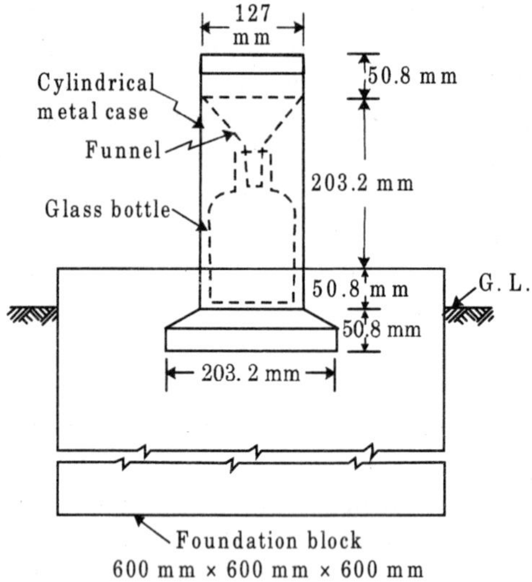

Fig. 5.2 *Symon's rain gage*

Standard non-recording type rain gage. As shown in Fig. 5.3, the Standard non-recording type rain gage consists of a collector with a gun metal or aluminium alloy rim, and a receiver consisting of a base and a bottle. The collector is exposed above ground level while the receiver is fixed partially below ground level. Both the collector and the base are made of fibre glass reinforced polyester. The collector has a deep set funnel and the complete rain gage has a slight taper with the narrower portion at the top. The collector and the base are locked to each other by means of two complementary locking rings, one fixed inside the collector at its lower end and the other fixed at the top end of the base. The collectors having apertures of either 100 cm^2 or 200 cm^2 area at the top are used. The collector intercepts the rainfall and the rain water entering the collector is led through the funnel into the bottle where it is stored. The bottles used in this rain gage are made of polythene and have capacities of 2, 4 and 10 litres. With different combinations of aperture size of the collector and the bottle capacity the rain gage may have measuring capacities of 100, 200, 400 and 1000 mm of rainfall. For example, a rain gage having collector of aperture area 200 cm^2 and bottle of capacity 2 litres will have a measuring capacity of $\left(\frac{2\times10^3}{200}\right)$ = 10 cm or 100 mm of rainfall. A rain gage of measuring capacity of 200 mm or rainfall with collector of aperture area 200 cm^2 and bottle of capacity 4 litres is most widely used. However, if a bottle of 2 litres capacity is used then as shown in Fig. 5.3, an additional cylinder is used to collect the overflow, if any, from the bottle.

The rain water collected in the bottle of a non-recording type rain gage is measured with the help of a standard measuring glass supplied with each rain gage which indicates the millimetres of rain that has fallen at the rain gage station. The Indian Standard, IS: 4849–1992 gives detailed specifications for the measuring glasses to be used with the rain gages having collectors of different aperture areas. At each rain gage station the observations for the rainfall are taken daily at 8.30 a.m. (IST). However, if the rainfall is likely to exceed the capacity of the bottle then a few intermediate observations are also taken. The sum of the observations taken will represent the total rainfall of the past 24 hours of the day on which the observation at 8.30 a.m. is taken. The detailed specifications for the installation of non-recording type rain gage and measurement of rain may be obtained from the Indian Standard, IS: 4986–1990.

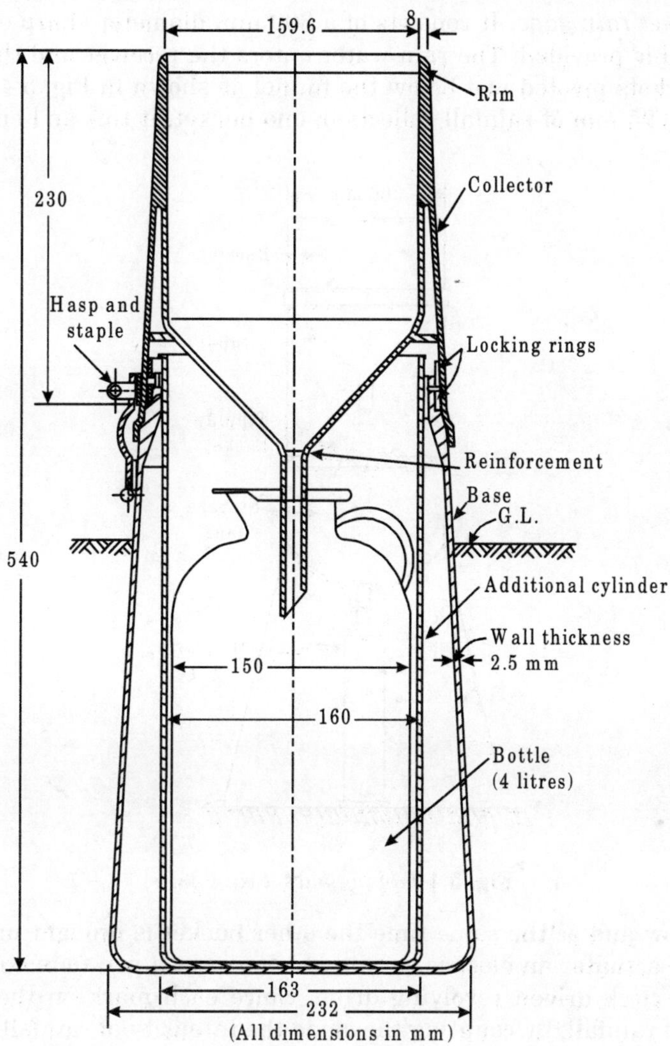

Fig. 5.3 *IMD's Standard non-recording type rain gage*

(ii) *Self-recording type or automatic recording type rain gages*. These rain gages automatically record the intensity of rainfall and the time of its occurrence in the form of a pen trace on a clock driven chart, from which the total amount of rainfall for the desired duration may also be determined. These gages may be operated over extended periods without attention and the recorded observations may be collected from the gages each time after a certain fixed duration. The most widely used self-recording type rain gages are as given below.

(a) Tipping bucket rain gage.

(b) Weighing type rain gage.

(c) Float type rain gage.

Each of these rain gages is described below.

(a) *Tipping bucket rain gage.* It consists of a 300 mm diameter sharp edged receiver at the end of which a funnel is provided. The rain water enters the receiver and the funnel conducts it to a pair of small buckets pivoted just below the funnel as shown in Fig. 5.4. The buckets are so designed that when 0.25 mm of rainfall collects in one bucket, it tips and empties its water into

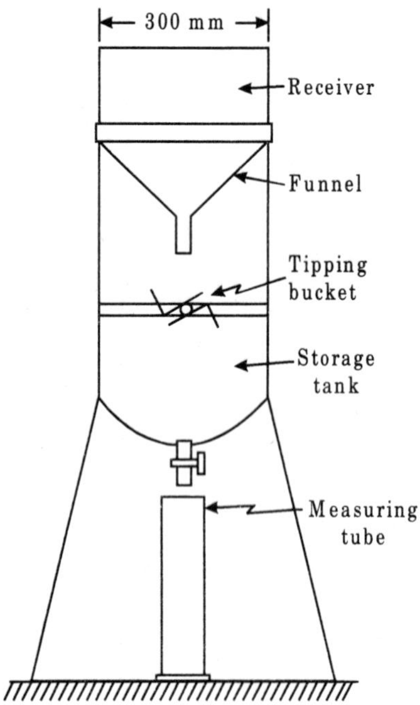

Fig. 5.4 *Tipping bucket rain gage*

the storage tank below and at the same time the other bucket is brought under the funnel. The tipping of the bucket actuates an electric circuit which causes a pen to make a mark on a record sheet mounted on a clock driven revolving drum. Since each mark on the record sheet corresponds to 0.25 mm of rainfall, by counting the same the intensity of rainfall may be determined. The total rainfall as determined from the records at the end of the day may also be checked by measuring the rain water collected in the storage tank in the same manner as in the case of a non-recording type rain gage. The movement of the tipping bucket may also be transmitted electronically to the control room so that the rainfall is directly recorded in the control room without any manual assistance. As such these rain gages are quite suitable for the measurement of rainfall in hilly and other inaccessible areas where manual recording of rainfall may not be possible.

(b) *Weighing type rain gage.* In this rain gage the rain water passes through a funnel into a bucket, which is supported on the weighing platform of a spring or lever balance as shown in

Fig. 5.5. The increase in the weight of the bucket due to the addition of the rain water causes the platform to move. The movement of the platform is transmitted through a system of links and levers to a pen which makes trace of accumulated amounts of rainfall on a suitably graduated chart wrapped round a clock driven revolving drum. The mechanism is arranged to

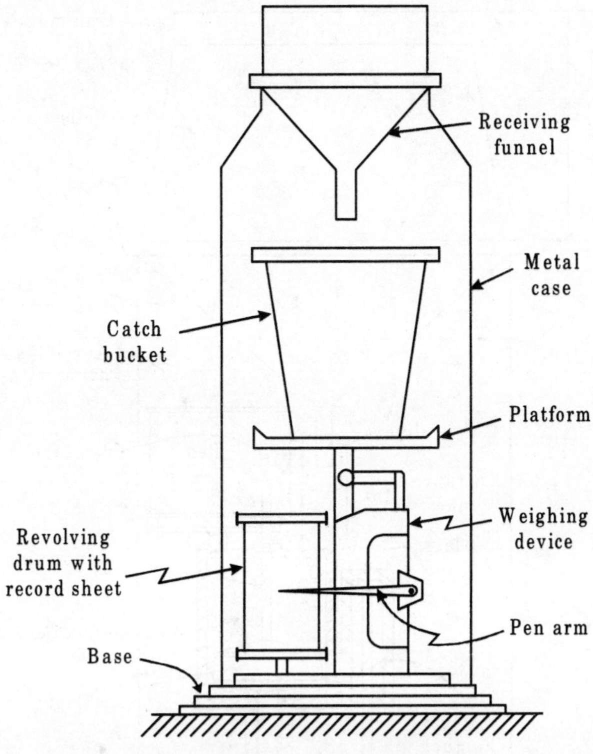

Fig. 5.5 Weighing type rain gage

reverse the travel of the pen after a certain amount of rainfall (say 150 mm) has accumulated and reverse again after another equal amount, so that the gage may operate unattended for a week at a time, except in regions of very intense rainfall, where the total rainfall may exceed the capacity of gage (usually 300 mm). The rainfall record produced by this gage is in the form of a mass curve of rainfall in which as shown in Fig. 5.6. The total rainfall is plotted with respect to time. The slope of the curve gives the intensity of the rainfall.

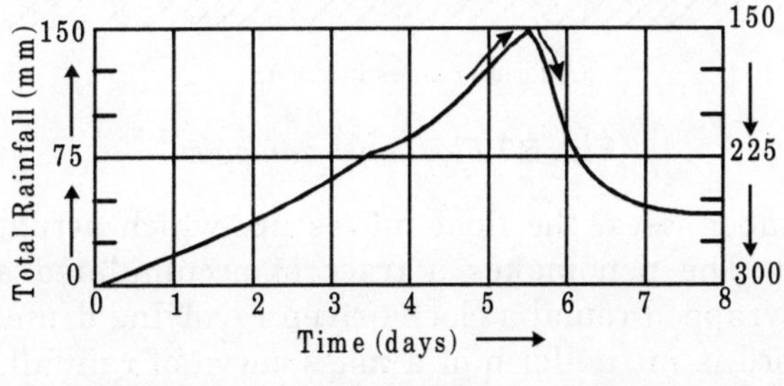

Fig. 5.6 *Rainfall record sheet of a weighing type rain gage*

(c) *Float type rain gage*. In this rain gage the rain water after passing through the funnel enters a chamber which contains a float as shown in Fig. 5.7. As the level of the rain water

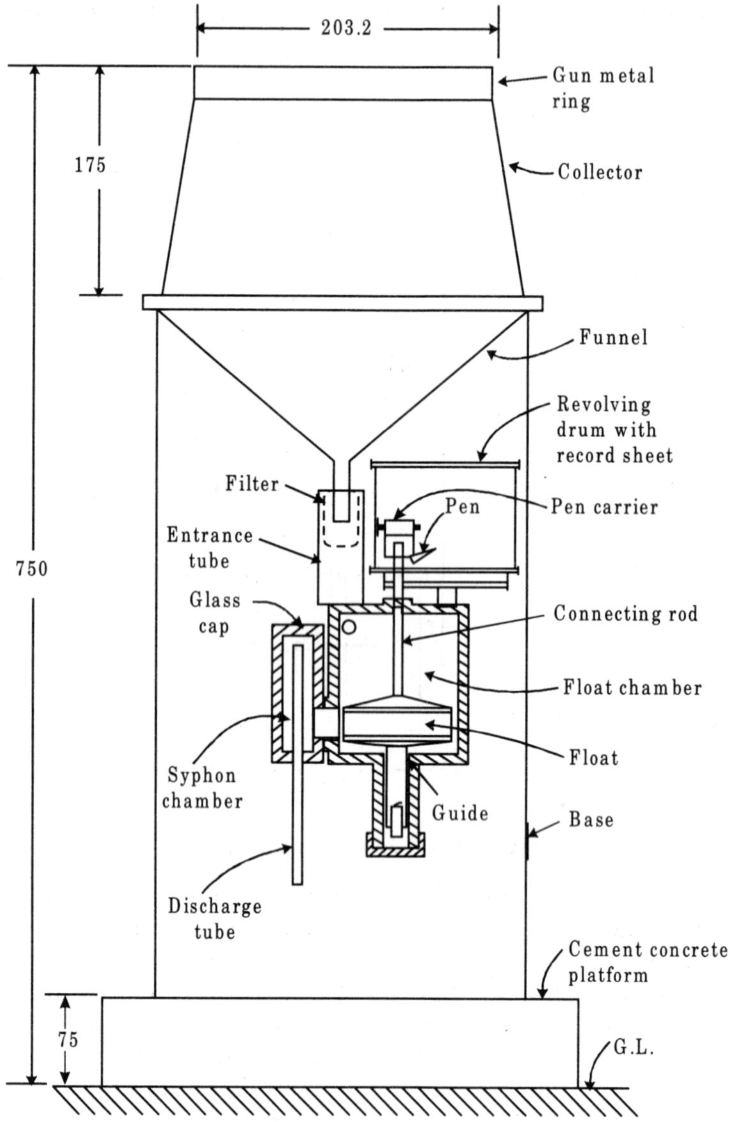

Fig. 5.7 *Float type rain gage*

collected in the float chamber rises, the float moves up, which actuates a pen connected to it through a connecting rod. The pen makes a trace of accumulated amounts of rainfall on a suitably graduated chart wrapped round a clock driven revolving drum. Thus, in the case of this gage also the rainfall record is in the form of a mass curve of rainfall, which is same as in the case of weighing type rain gage. When the float chamber gets completely filled it is automatically emptied by means of a syphon as shown in Fig. 5.7. As such this type of rain gage is also known as natural-syphon type rain gage. In our country this type of rain gage is adopted by the Indian Meteorological Department (IMD) as the *Standard self recording-type rain gage*. The Indian Standard, IS : 5235–1992 gives the details of the Standard self-recording type rain gage.

The self-recording type rain gages are installed on a concrete or masonry platform 450 mm square with the rim of the receiver being kept at a height of exactly 750 mm above the ground level. The self-recording type rain gage is generally used in conjunction with a non-recording type rain gage so that the readings of the self-recording type rain gage can be checked and if necessary adjusted.

In addition to the above noted gages, in the modern times the use of radar as an aid in the measurement of rainfall is being made. However, the main use of radar is in the determination of the areal extent, orientation and movement of rain storms.

(B) Measurement of Snowfall

For the measurement of snowfall, storage gages are used. In principle these gages are same as the non-recording type and the self-recording type rain gages described earlier, but they are designed to hold much greater amounts of precipitation. Self recording type gages designed to operate for 30 to 60 days without servicing and non-recording type gages with sufficient capacity for an entire season are now in use. A non-recording type snowfall measuring gage used by U.S. Weather Bureau is made from 300 mm diameter thin-walled pipe. The gage is made in 1.5 m long sections so that any height in multiples of 1.5 m may be obtained. To permit free fall of snow inside the gage, the upper portion of the gage is in the shape of a frustum of a cone. The equivalent depth of water contained in any snowfall may be found by melting the snow collected in the gage and measuring the quantity of water formed. The snow may be melted by adding an adequate quantity of warm water, accurately measured previously with the measuring glass. The total water content in the gage is then measured using the measuring glass. The measurement thus obtained would give the total of the precipitation and the added warm water. By subtracting from the total the amount of the added warm water, the amount of precipitation is determined. The snow can also be melted in the gage as it gets collected there by a heating system fitted to the gage or by placing in the gage certain chemicals such as calcium chloride, ethylene glycol, etc. This arrangement for the melting of snow would, however, provide gages with capacity for large amounts of snowfall without making the gage excessively large.

For the measurement of snowfall another method is also used which is known as *pole platform method*. The Indian Standard, IS : 4986–1990 gives the details about the pole platform method which is described below.

Pole Platform Method. A concrete platform 2 m × 2 m and 75 mm thick is made at the ground level at a place where there is least drifting due to wind. At the centre of the platform a scale is permanently fixed. The scale consists of a 50 mm square wooden pole projecting 3 m above the level of the platform and 1 m grouted below it. The scale is graduated in millimetres and the graduations are painted on all the four sides of the pole. Another scale in the form of a wooden stick of 25 mm square with similar graduations is provided to the gage station. While taking observation, the observer first reads the scale at the centre for the depth upto which it is submerged in snow. With the help of the other scale the depth of snow at each of the four corners of the platform is measured. The mean of the five readings is taken as the depth of snowfall. The mean depth of snowfall in millimetres divided by 10 approximately gives the water equivalent of snow in millimetres. At the end of each observation, the observer clears the platform of the snow so that the next snowfall measurement is not vitiated by the snow accumulation which has already been measured.

(C) Rain Gage Network

Since the amount of rain fall varies from place to place, it would be necessary to instal the rain gages at various points. Especially in the drainage basin (or catchment area) of a stream a network of rain gage stations evenly distributed over the entire area of the drainage basin should invariably be provided. A network of the rain gages should be so planned that it yields a representative picture of the areal distribution of rainfall, for which it is obvious that the number of rain gages installed should be as large as possible, or the area per gage should be small. However, the cost of installing and maintaining a network of rain gages and accessibility of the gage site to an observer are also important considerations. For general hydrometeorological purposes the following minimum densities of rain gage networks have been recommended by World Meteorological Organization (WMO).

(i) For flat regions of temperate, mediterranean and tropical zones-1 station for 600–900 km^2.

(ii) For mountainous regions of temperate, mediterranean and tropical zones-1 station for 100–250 km^2

(iii) For small mountainous islands with irregular rainfall-1 station for 25 km^2.

(iv) For arid and polar zones-1 station for 1500–10000 km^2.

Further 10 percent of rain gage stations should be equipped with self-recording type rain gages to know the intensities of rainfall.

From practical considerations for Indian conditions, the Indian Standard IS: 4987–1991 recommends the following minimum densities of rain gage networks:

(i) In plains-1 station for 520 km^2.

(ii) In regions of average elevation 1000 m above sea level-1 station for 260–390 km^2.

(iii) In areas predominently hilly and where heavy rainfall is expected-1 station for 130 km^2.

Further in order to know the intensities of rainfall it is desirable to have at least 10 percent of the rain gage stations equipped with self recording type rain gages.

(D) Optimum Rain Gage Network Design

The aim of optimum rain gage network design is to obtain the average value of the rainfall of an area with sufficient accuracy for practical purposes. When the mean areal depth of rainfall is calculated by the simple arithmetic average, the optimum number of rain gage stations to be established in a given basin is given by the following equation

$$N = \left(\frac{C_v}{p}\right)^2 \qquad \ldots(5.1)$$

in which

N = optimum number of rain gage stations to be established in the basin

C_v = coefficient of variation of the rainfall of the existing rain gage stations (say, n) ; and

p = desired degree of percentage error in the estimate of the average depth of rainfall over the basin.

The number of additional rain gage stations ($N - n$) should be distributed in the different zones caused by isohyets (which are the contours of equal rainfall) in proportion to their areas, i.e., depending upon the spatial distribution of the existing rain gage stations and the variability of the rainfall over the basin.

The method described above for the design of optimum rain gage network is indicated in Illustriative Example 5.4.

(E) Estimation of Missing Rainfall Data

A rain gage station may have a short break in its records because of absence of the observer or because of instrumental failure. In order to complete the record of this station it is necessary to estimate the missing rainfall data which may be done by the *normal ratio method* as explained below.

In this method a number of rain gage statins (say m) as close to and as evenly spaced around the rain gage station with the missing rainfall data as possible are selected as index stations. The amount of rainfall at each of the index stations is multiplied (or weighted) by the ratio of average (or normal) annual rainfall of the station with missing rainfall data and the average (or normal) annual rainfall of the index station. The average of these values then gives the value of the missing rainfall.

For example, let rain gage station X became inoperative during a certain period. During this period the total rainfall recorded at the m index stations 1, 2, 3, 4, ..., m were P_1, P_2, P_3, P_4, ..., P_m, respectively. If N_1, N_2, N_3, N_4......, N_m are the average annual rainfalls at the index stations 1, 2, 3, 4, ..., m, respectively and N_x is the average annual rainfall at the station X, then the rainfall P_x at the station X during this period is given by

$$P_x = \frac{1}{m}\left[\frac{N_x}{N_1}\times P_1 + \frac{N_x}{N_2}\times P_2 + \dots\dots + \frac{N_x}{N_m}\times P_m\right] \quad \dots(5.2)$$

However, if the average annual rainfall at the index stations is within about 10% of the average annual rainfall at the station X with missing rainfall data then the rainfall P^x is given by a simple arithmetic average as given below.

$$P_x = \frac{1}{m}\left[P_1 + P_2 + P_3 \dots\dots\dots + P_m\right] \quad \dots(5.3)$$

(F) Double-Mass Curve Technique–Test for Consistency of Rainfall Data

If the conditions affecting the rainfall catch at any rain gage station have undergone significant change during the period of record, the rainfall data of that rain gage station would not be consistent. Some of the common causes which may lead to inconsistency in the rainfall data of any rain gage station are (i) shifting of the rain gage station to a new location; (ii) the neighbourhood of the rain gage station undergoing a marked change due to coming up of a new building, fence, planting of trees or cutting of near by forest; (iii) change in the surroundings due to natural calamities such as forest fires, land slides, etc.; and (iv) change in the observational procedure and/or occurrence of observational error from a certain date. Often such changes are not disclosed in the published records of the rainfall data, and hence it becomes necessary to test the consistency of the rainfall data of a rain gage station. The consistency of the rainfall data of a rain gage station (say X) is tested and the necessary adjustment of the rainfall data made by the use of *double-mass curve technique.* In this method out of the various rain gage stations in the neighbourhood of the rain gage station X at least 10 or more rain gage stations are selected as base stations. The annual (or seasonal) rainfall of the rain gage station

X and the average annual (or seasonal) rainfall of the base stations for the corresponding years covering a large number of consecutive years of record are arranged in the reverse chronological order (i.e., the rainfall of the latest year as the first entry and the rainfall of the oldest year as the last entry in the list). The accumulated rainfall of the rain gage station X (i.e., ΣP_x) and the accumulated values of the average rainfall of the base stations (i.e., ΣP_{av}) are calculated starting from the latest year. Values of ΣP_x are plotted against ΣP_{av} for the consecutive years. From the resulting plot of double-mass curve, the year in which a change in the conditions has occurred at rain gage station X is indicated by a distinct change in the slope of the curve. The rainfall values of the rain gage station X recorded during the years prior to the change may be adjusted (or corrected) to make them comparable with those recorded during the years after the change by the following relation

$$P_{cx} = P_x \times \frac{M_c}{M_a} \qquad \text{...(5.4)}$$

in which

P_{cx} = adjusted or corrected value of rainfall at the rain gage station X

P_x = originally recorded value of rainfall at the rain gage station X at any time prior to the change

M_c = slope of the double-mass curve for the years after the change, and

M_a = slope of the double-mass curve for the year prior to the change.

The method described above for testing the consistency of rainfall data of a rain gage station is described in Illustrative Example 5.5.

5.5 MEAN ANNUAL RAINFALL, MINIMUM AND MAXIMUM ANNUAL RAINFALL

The rainfall recorded on each day at a rain gage station being the amount of rainfall that has accumulated in the rain gage during a period of 24 hours, it is known as *daily rainfall*. By adding the values of the daily rainfall recorded for various periods such as week, month and year, weekly, monthly and annual (or yearly) rainfall may be obtained. From the rainfall data so obtained for a rain gage station the variation of rainfall during the same month in different years as well as the year to year variation of the rainfall may be known. Further from the rain fall data obtained for a number of consecutive years the mean annual rainfall may be determined. The mean annual rainfall is determined by taking simple average of the annual rainfall of several consecutive years recorded at a rain gage station. Since the annual rainfall varies at the same rain gage station, rainfall record for a large number of years is required to obtain a reliable estimate of the mean annual rainfall at any place. According to Alexander Binnie the mean annual rainfall worked out from 35 years of rainfall record is within 1.8% of the true mean value and hence in India a rainfall record of 35 consecutive years is usually considered to determine the mean annual rainfall. However, if the rainfall record is for a period less than 35 years, then the error in the computed mean annual rainfall will be as given below in tabular form.

Period (years)	5	10	15	20	25	30	35
Error in mean annual rainfall (%)	14.9	8.2	4.8	3.3	2.8	2.3	1.8

For catchments (or drainage basins) having different layouts, IS: 4987–1991 has recommended the following number of years of rainfall record to be considered for determining the mean annual rainfall.

Catchment layout	Island	Shore	Plain	Mountains
No. of years of rainfall record	30	40	40	50

On the other hand according to U.S. Weather Bureau rainfall record for 40 years should be considered to give results within 1.5% of the true mean.

From the rainfall data for a number of consecutive years the minimum and the maximum annual rainfall may also be determined. The information about the average annual rainfall and minimum and maximum annual rainfall at a place is very useful for planning water resources projects because it gives an idea of the probable amount of water that may be available at that place.

5.6 AVERAGE DEPTH OF RAINFALL OVER AN AREA

For the analysis of most of the hydrologic problems, it is necessary to determine the average (or mean) depth of rainfall over an area. For a small area the rainfall recorded at a single rain gage station located in that area may be taken as the average depth of rainfall over that area. For large areas there will be a network of rain gage stations as recommended by the Indian Standard IS: 4987–1991. As the rainfall over a large area is not uniform, the average depth of rainfall over the area is determined. The following three methods are commonly used for determining the average depth of rainfall over an area.

(i) Arithmetic mean method

(ii) Thiessen polygon method

(iii) Isohyetal method.

(i) *Arithmetic mean method.* In this method the average depth of rainfall over an area is obtained by dividing the sum of the depths of rainfall recorded at all the rain gage stations located in the area by the number of stations. Thus, if P_1, P_2, P_3,.......P_n are the depths of rainfall recorded at the n rain gage stations distributed over an area then the average depth of rainfall P for the entire area is given as

$$P = \frac{P_1 + P_2 + P_3 + \dots\dots\dots + P_n}{n} = \frac{\sum_{i=1}^{i=n} P_i}{n} \qquad \dots(5.5)$$

The results obtained by this method will be correct only if the rain gage stations are uniformly distributed over the area and the rainfall varies in a regular manner. This is so because in this method every rain gage station has equal weightage regardless of its location.

(ii) *Thiessen polygon method.* In this method the adjacent rain gage stations are joined by straight lines thus dividing the entire area into a series of triangles as shown in Fig. 5.8. On each of these lines perpendicular bisectors are erected, thereby forming a series of polygons, each containing only one rain gage station. The entire area within any polygon is nearer to the rain gage station contained therein than to any other station, and hence it is assumed that the rainfall recorded at that station is the representative rainfall for that area. If P_1, P_2, P_3,...... P_n

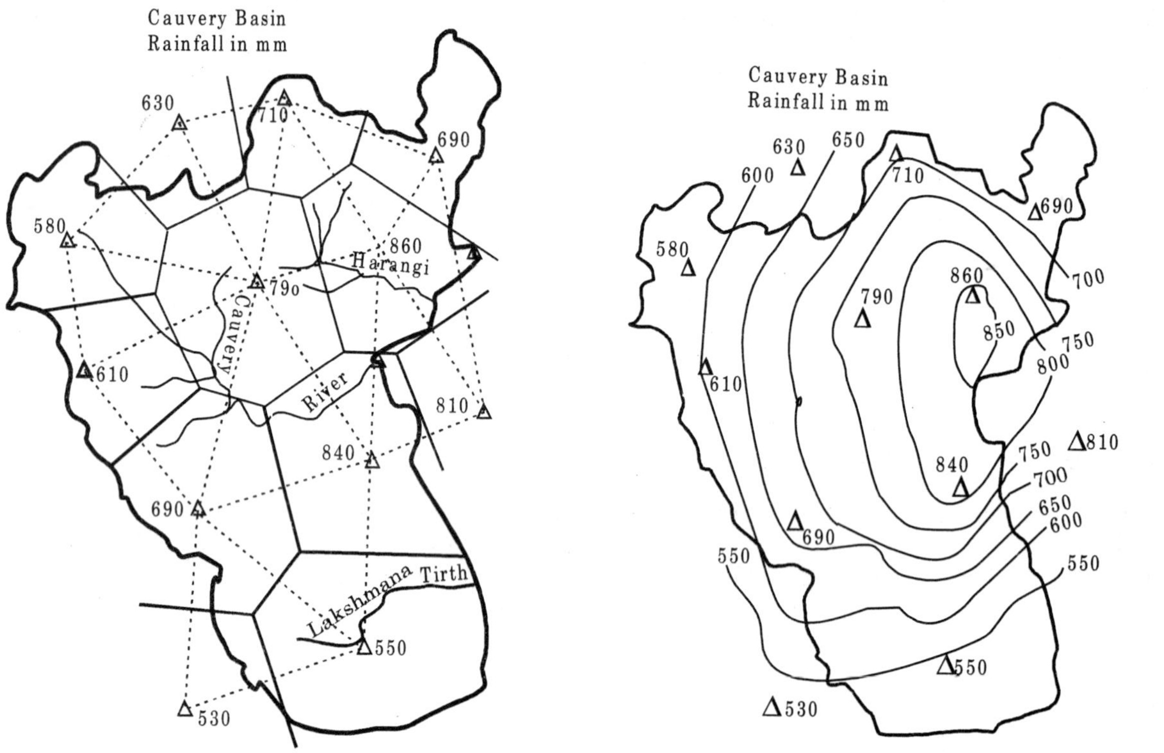

(a) Thiessen Polygon method

(b) Isohyetal method

Fig. 5.8

represent the depths of rainfall recorded at the rain gage stations enclosed by polygons, the areas of which are respectively A_1, A_2, A_3, ... A_n, then the average depth of rainfall P for the entire area A is given as

$$P = \frac{P_1 A_1 + P_2 A_2 + P_3 A_3 + ...P_n A_n A}{A}$$

$$= \frac{\sum_{i=1}^{i=n} P_i A_i}{A}$$

where

$$A_1 + A_2 + A_3 + ... + A_n = A \qquad ...(5.6)$$

In the Theissen polygon method each rain gage station is given weightage according to its position with respect to the boundary of the area under consideration, and hence this method is better than the arithmetic mean method.

(iii) *Isohyetal method.* Isohyets are the contours of equal rainfall. In the isohyetal method isohyets are drawn on the map of the area under consideration. For drawing the isohyets the depth of rainfall recorded at each of the rain gage stations are noted on the map of the area, at the respective stations. Then assuming a linear variation of rainfall between the two stations, the probable positions of rainfall values at some interval are interpolated between the stations. The points with equal values of rainfall are connected by smooth curves which forms the isohyetal pattern for the area as shown in Fig. 5.8 (a). The area between the two adjacent isohyets is measured with the help of planimeter and for this area the rainfall is presumed to be equal to the mean of the two isohyet values. The average depth of rainfall P for the entire area A may be found by the following equation.

$$P = \frac{\Sigma(\text{Area between two adjacent isohytes} \times \text{Mean of the two isohyet values})}{A} \quad ...(5.7)$$

The isohyetal method is the most accurate method for computing the average depth of rainfall. This is so because the isohyetal method takes into account the actual spatial relationship of the rain gage stations.

5.7 EVAPORATION, TRANSPIRATION AND EVAPOTRANSPIRATION

Evaporation is the process by which water is converted into gaseous state and is returned to the atmosphere as vapour. The water is continuously evaporating from the water surfaces of natural and artificial reservoirs, lakes etc., snow and soil surfaces and also from the surfaces of vegetation. As such a part of the precipitation reaching the earth's surfaces is lost through evaporation. Evaporation depends upon several factors such as temperature, wind velocity, humidity, salinity of water and surface area. The rate of evaporation increases with the increase of temperature as well as with the increase of wind velocity, but it decreases with the increase of humidity as well as with the increase of dissolved salts in water. Further the evaporation increases with the increase in the open surface area of water and soil.

In order to determine the quantity of water lost through evaporation from the water surfaces of natural and artificial reservoirs, lakes etc., several methods have been developed, but those commonly adopted are (i) the use of empirical formulae, and (ii) the direct measurement with the help of evaporation pans.

A large number of empirical formulae have been developed for determining the quantity of water lost through evaporation but the one which is widely used is the Lake Hefner formulae which is

$$E = 6.9317\,(e_s - e_2)\,V_4 \quad ...(5.8)$$

and

$$E = 5.5046\,(e_s - e_8)\,V_8 \quad ...(5.8a)$$

where E is evaporation in mm/day ; e_s is vapour pressure at water surface in N/mm^2; e_2 and e_8 are vapour pressures in N/mm^2 at the heights of 2 m and 8 m respectively above the water surface ; and V_4 and V_8 are wind velocities in km/hour at heights of 4 m and 8 m respectively above the water surface.

For the direct measurement of the quantity of water lost through evaporation various types of evaporation pans have been designed, but the most commonly employed pans are (a) USWB Class A Land Pan, (b) Colorado Sunken Pan, (c) Floating Pan. The first two types of evaporation pans are installed in the vicinity of the reservoir or lake for which the loss of water through evaporation is to be determined, while the third type of the evaporation pan is held in floating state on the water surface in the reservoir or lake. The quantity of water lost through evaporation from the pan is measured. Since the evaporation from pan is greater than that from the adjoining reservoir or lake, the observed value of the water lost through evaporation from the pan is multiplied by a predetermined coefficient of the pan to obtain the actual quantity of water lost through evaporation from the adjoining reservoir or lake.

Transpiration is the process by which the moisture absorbed by the roots of the plants and circulated through the plant body, returns back to the atmosphere as water vapour breathed out by the stomata of the leaves of the plants. Since transpiration is essentially the evaporation of water from leaf cells, the rate of transpiration is also influenced by all the factors which affect the evaporation. Thus, high temperature, low humidity and strong wind all tend to increase transpiration. However, in addition to these factors transpiration is also affected by a few other factors such as the available sun-shine, available soil moisture, type and extent of plant growth etc.

It is not possible to measure the quantity of water loss through transpiration from a large area under natural conditions, and hence the measurement of the transpiration losses is restricted to studies of small samples under laboratory conditions. One such practical method for measuring the transpiration loss is with the help of instruments called potometers and phytometers. A *potometer* is a small vessel containing water and sealed so that the only escape of moisture is by transpiration from a small plant with its cut end inserted in the water. A *phytometer* is a large vessel filled with soil in which one or more plants are rooted. The soil surface is sealed to prevent evaporation so that the only escape of moisture is by transpiration. Thus, in each of the above devices the loss in the weight of plant and vessel is a measure of the transpiration loss for the specimen tested.

The land surfaces which are partially covered with vegetation, loose moisture simultaneously through evaporation from the land surface and through transpiration from vegetation. As such the common practice is to estimate the combined loss of moisture from the land surface through evaporation as well as through transpiration. This combined loss of moisture from the land surface is known as evapotranspiration or consumptive use of water. The term *evapotranspiration* is defined as the total quantity of water used by the vegetative growth of a given area in transpiration and building of plant tissue, and that evaporated from the adjacent soil in the area in any specified time. In the above definition the term vegetative growth includes the natural vegetation as well as the crops grown on the area. The various factors affecting evapotranspiration as well as the methods commonly adopted to determine the same are described in Chapter 4.

5.8 INFILTRATION

Infiltration is the process by which water enters the surface strata of the soil and moves downwards to join the ground water. This water first replenishes the soil moisture deficiency and the excess water moves downwards to become a part of the ground water. The rate at which the water actually infiltrates through a soil at a given time is known as *infiltration rate.* Further the maximum rate at which a soil in any given condition is capable of absorbing water is called its *infiltration capacity.* Initially when the soil is dry and the rainfall occurs some rain water is consumed in wetting the soil particles and hence the infiltration rate is high. But as

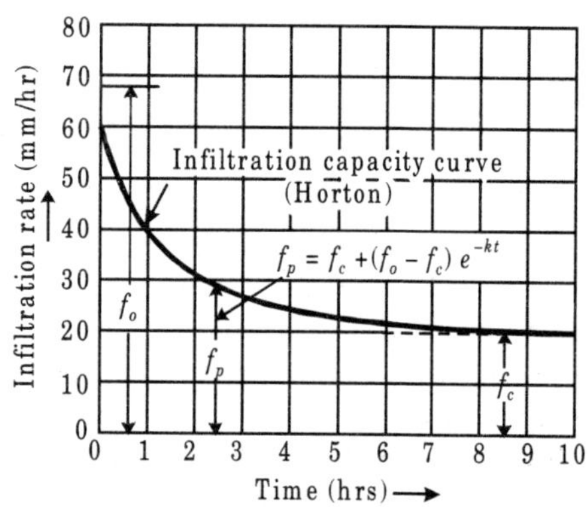

Fig. 5.9 *Infiltration capacity curve (Horton)*

the rain continues, the capacity of the soil to absorb water progressively decreases. Thus, if infiltration rate is plotted with respect to time then as shown in Fig. 5.9 a falling curve is obtained which is known as infiltration capacity curve. Horton found that infiltration capacity curves approximate the form

$$f_p = f_c + (f_o - f_c)\, e^{-kt} \qquad \text{...(5.9)}$$

where f_p is infiltration capacity in mm/hr at any time t ; f_o is initial infiltration capacity in mm/hr ; f_c is final constant infiltration capacity in mm/hr at saturation; t is time in hours from the beginning of rainfall ; and k is an empirical constant.

The infiltration takes place at capacity rates only when the intensity of rainfall equals or exceeds f_p, i.e., $f = f_p$ when $i \geq f_p$; but when $i < f_p$, $f < f_p$ and actual infiltration rates are approximately equal to the rainfall rates.

The infiltration depends upon the intensity and duration of rainfall, climatic condition, soil characteristics, vegetal cover, initial soil moisture content, entrapped air and depth of the ground water table.

For determining infiltration capacity of a soil the following two methods are adopted: (i) by infiltrometer, (ii) by analysis of rainfall and runoff data. The first method is commonly used and the same is briefly discussed below.

In general there are two types of infiltrometers (a) Ring type and tube type infiltrometers, and (b) Rain simulators.

Ring type infiltrometer consists of two shallow concentric rings of sheet metal, usually ranging from 225 mm to 900 mm in diameter. The rings are placed with their lower edges a few centimetres below the ground surface and with the upper portion projecting above the ground surface. Water is then applied in the inner ring as well as in the annular space between the two rings and is always kept at the same level in both. The function of the outer ring is to prevent the water within the inner ring from spreading over a large area after penetrating below the bottom of the ring. The infiltration capacity of the soil is determined directly as the rate at which water must be added to maintain a constant level.

Tube type infiltrometer consists of a single tube about 225 mm in diameter and 450 mm to 600 mm long, which is placed in the ground upto a depth at least equal to that to which the water penetrates during the experiment. Thus, in this case no spreading of water can occur. Again the rate at which water must be added to maintain a constant depth within the tube is then observed which directly gives the infiltration capacity of the soil.

In the case of rain simulators, on a small area of land ranging from 0.1 m^2 to 50 m^2, water is applied by sprinkling at a uniform rate that is in excess of infiltration capcity. The resulting surface runoff is measured and by subtracting it from the rate of application of water, the infiltration capacity of the soil is determined.

Infiltration Indices

As explained in section 5.11 the concept of infiltration can be applied to the rational computation of surface runoff. For this purpose certain infiltration indices have been devised as indicated below.

ϕ-index. ϕ-index may be defined as the average rainfall intensity above which the volume of rainfall equals the volume of runoff (Fig. 5.10). Since the difference between the rainfall and runoff volumes is equal to the sum of the volume of water intercepted, depression storage and that infiltrates into the soil, the ϕ-index represents combined effect of interception, depression

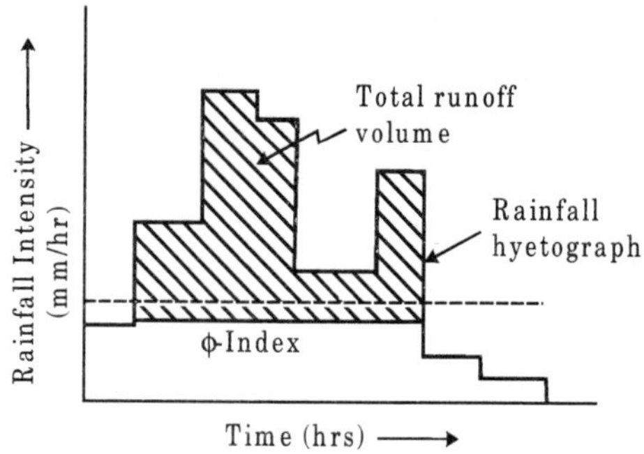

Fig. 5.10 *Diagrammatic representation of φ-index*

storage and infiltration, all of which are combinedly designated as basin recharge. Further the φ-index is based on the assumption that for a specified storm with given initial conditions, the rate of basin recharge remains constant throughout the storm period.

W-index. W-index is defined as the average rate of infiltration during the time rainfall intensity exceeds the infiltration capacity. The value of W-index may thus be obtained from the following equation

$$W\text{-index} = \frac{F_i}{t_f} = \frac{1}{t_f}(P - Q_s - S_e) \qquad \text{...(5.10)}$$

where F_i is the total infiltration, t_f is the time during which rainfall intensity exceeds infiltration capacity (it is mostly same as duration of rainfall), P is the precipitation, Q_s is the surface runoff, and S_e is the effective surface retention by interception and depression storage. Thus W-index is equivalent to f-index minus the retention by interception and depression storage.

W_{min}-index. With very wet conditions, when the infiltration capacity is essentially equal to the final infiltration capacity f_c and the retention rate is at a minimum, the values of W and f indices are almost indentical. Under these conditions, W-index becomes the W_{min}-index by definition.

f_{ave}-index. f_{ave}-index is defined as the average rate of infiltration throughout the storm for the period $i > f$.

5.9 RUNOFF

Runoff is that part of precipitation as well as any other flow contribution, which is transmitted through natural surface channels or streams or rivers. In the general sense runoff includes (i) surface runoff or overland flow received in the stream immediately after a heavy rain; (ii) inter-flow which is a portion of soil moisture that flows laterally through the upper soil layers and joins the stream before joining the ground water; (iii) delayed runoff or ground water flow that enters the stream after passing through deeper portions of the earth ; and (iv) other delayed runoff that has been temporarily detained as snow cover or stored in natural lakes and swamps. Thus, runoff is the total quantity of water received by a stream from its drainage basin or catchment area. The runoff is generally classified as direct runoff and base flow (or base runoff). The direct runoff comprises the overland flow and the inter-flow which are usually grouped together, while the ground water flow that enters a stream is termed as base flow (or base runoff).

The runoff is generally considered in terms of the total flow carried by a stream during a month, season or year and accordingly it is termed as monthly, seasonal or annual runoff. The runoff is expressed either in millimetres or centimetres or metres of water depth on the entire drainage basin, or in hectare-metre or cubic metre per unit area of the drainage basin. Further the terms stream flow, discharge of a stream and rate of runoff are generally used to mean one and the same thing.

5.10 FACTORS AFFECTING RUNOFF

The runoff from the drainage basin or catchment area of a natural stream depends on several factors as noted below.

1. **Characteristics of Precipitation.** The pertinent characteristics of precipitation which may considerably affect the runoff are the type of precipitation, its intensity, extent and duration. The precipitation may occur either in the form of rain or snow. If the precipitation occurs in the form of rain, it will immediately produce the runoff, but if the precipitation occurs in the form of snow, it will produce the runoff much later, only when the snow will melt due to rise in the temperature. The runoff increases with the increase in the intensity of the rainfall. This is so because an intense rainfall in a short duration results in less evaporation and infiltration losses. On the other hand a mild rainfall extending over a longer duration would produce less runoff. Further a greater area extent of the rainfall over the drainage basin and also the rainfall for a longer duration would produce greater runoff.

2. **Characteristics of Drainage Basin.** The runoff is considerably affected by the characteristics of the drainage basin such as size, shape, surface, orientation, altitude, topography and geology of the drainage basin.

In general larger the drainage basin greater will be the runoff, but it may not be always true. This is so because for a larger drainage basin it is very seldom that the precipitation of same intensity occurs over the entire basin. Further a larger drainage basin may have larger storage effect resulting in the reduction of the runoff. On the other hand even for a smaller drainage basin with intense rainfall, uniformly spread over the entire basin, greater runoff rate may be produced.

A drainage basin may be either fan-shaped or fern-shaped. A fan-shaped drainage basin produces a greater runoff rate than a fern-shaped drainage basin. This is so because in a fan-shaped drainage basin all the tributaries of a stream are of comparable length and hence the runoff from each tributary reaches almost simultaneously. However, with a fern-shaped drainage basin reverse is the case.

The surface of the drainage basin also has substantial effect on the runoff. For example, a bare surface gives more runoff as compared with a surface having vegetal covering. Similarly a paved surface also produces a greater runoff.

For the drainage basin located on the windward side of the mountains heavy precipitation occurs and hence greater runoff is produced. On the other hand if the drainage basin is located on the leeward side it receives less precipitation and hence less runoff is produced. Similarly the drainage basins located at higher altitude receive more precipitation and hence yields greater runoff.

The topography of the drainage basin also has significant effect on the runoff. A steeply sloping drainage basin produces greater runoff because a steep slope helps quicker drainage and there being virtually no depression storage, less opportunity for evaporation and infiltration losses is provided. On the contrary a drainage basin with mild slope or with undulating land will yield less runoff.

The runoff is also significantly affected by the geology of the drainage basin. A drainage basin with pervious soil will yield less runoff due to larger infiltration losses, while the one with impervious soil or having a hard stratum will produce large runoff due to less infiltration losses. Further if a drainage basin comprises unfissured rock it will produce greater runoff, but the runoff will be considerably reduced if the rock is fissured.

3. **Meteorological Characteristics.** The runoff is significantly affected by the meteorological characteristics such as temperature, humidity, wind velocity, pressure variation etc. Higher temperature and higher wind velocity cause more losses due to evaporation and transpiration and hence reduce the runoff. On the other hand greater humidity decreases evaporation and transpiration losses, resulting in greater runoff. The variation of the pressure in the atmosphere helps the movement of storms causing the precipitation. If the direction of the storm is same as the direction of flow of water in the drainage basin, then water will remain in the basin for lesser period and hence more runoff will be produced in a short duration as compared to the case when the storm is moving in the opposite direction.

4. **Storage Characteristics.** The storage characteristics of a drainage basin also have significant effect on the runoff. If a drainage basin has a larger number of natural depressions, pools, lakes etc., which will store a part of the precipitation and the same will not appear as runoff, then the runoff at the out-flow point of the basin will be reduced. Similarly the creation of artificial reservoirs or tanks will also increase the storage capacity of a drainage basin, resulting in the reduction of the runoff. However, if from these surface storages of the basin, water is released subsequently, it will appear in the form of delayed runoff. Further for drainage basins having pervious deposits, large ground water may be created, which may also appear in the form of delayed runoff.

5.11 METHODS OF DETERMINATION OF RUNOFF

The runoff from catchments of rivers may be determined either by direct measurement of discharge through rivers or it may be estimated in an indirect manner.

The discharge through a small stream or river may be measured by providing a weir across the river. However, for measuring discharge through a large river, velocity area method is commonly adopted. In this method the cross-section of the river is divided into a number of segments of widths l_1, l_2, l_3, etc. The mean depths of flow d_1, d_2, d_3, etc., of the segments are measured along the centre lines by soundings. The mean velocities of flow V_1, V_2, V_3, etc., through the segments are measured with the help of a current meter by inserting it along the centre line of each of the segments to a depth equal to 0.6 times the mean depth below the free water surface. The total discharge through the river is obtained by adding the discharge through each of the segments. Thus,

$$Q = [l_1 d_1 V_1 + l_2 d_2 V_2 + l_3 d_3 V_3 + ...] = \Sigma\,(ld\,)V$$

A greater accuracy in the computation may be obtained by dividing the cross section of the river into segments of equal widths and then using either Simpson's rule or Weddle's rule.

Further the elevation of the water surface at any section in the river measured above an arbitrary datum is known as stage. Corresponding to the measured discharge of the river the stage is also recorded with the help of a gage provided at the section of the river where the discharge is measured. By plotting the stage against the corresponding discharge a stage discharge curve may be prepared for a river which may be used for determining the discharge corresponding to a known stage.

The runoff may be indirectly estimated by the following methods:

(a) by empirical formulae, curves and tables relating rainfall and runoff

(b) by infiltration method

(c) by rational method

(d) by unit hydrograph.

The first three methods are discussed below and the method of unit hydrograph is discussed in section 5.14.

(a) **Empirical Formulae, Curves and Tables Relating Rainfall and Runoff.** Several empirical formulae, curves and tables relating the rainfall and runoff have been developed which may be used for the estimation of runoff. Some of these are as described below.

(i) *Binnie's percentage.* In India the earliest efforts in estimating runoff from the rainfall were those of Alexander Binnie. He made observations on two rivers in Madhya Pradesh and established certain percentages of runoff from rainfall which are given below :

Annual rainfall (mm)	*Runoff %*
500	15
600	21
700	25
800	29
900	34
1000	38
1100	40

A graphical plot of runoff v/s rainfall has also been given. However, the main drawback in the use of these percentages is that the conditions of catchment and other geological and hydrological factors are not taken into account.

(ii) *Strange's table and curves.* W.L. Strange was the next to evolve some ratios between rainfall and runoff. His results are based on data of Maharashtra and are given in Table 5.1

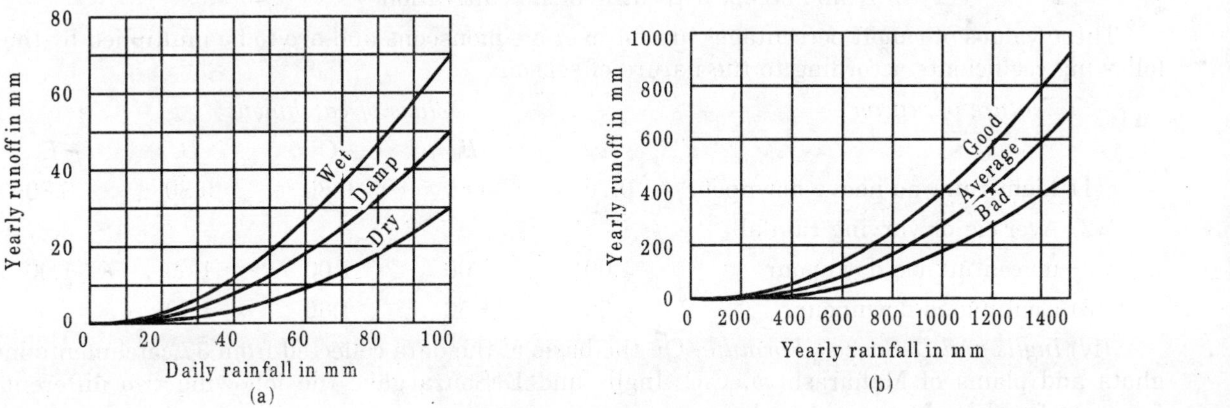

Fig 5.11 *Strange's runoff curves*

along with curves as shown in Fig. 5.11. Strange's tables and curves give runoff for daily rainfall for good, average and bad catchments and three surface conditions viz., dry, damp and wet prior to rain. Since in this case geological conditions of the catchment is taken into account, it is an improvement on Binnie's percentages.

Table 5.1 Daily Runoff percentage and yield according to Strange

Daily rainfall (mm)	*Runoff percentage and yield when the original state of ground was*					
	Dry		*Damp*		*Wet*	
	Percentage (mm)	*Yield (mm)*	*Percentage*	*Yield (mm)*	*Percentage*	*Yield*
6.25	...	...	...	...	8	0.50
12.50	...	...	6	0.75	12	1.50
18.75	...	...	8	1.50	16	3.00
25.00	3	0.75	11	2.75	18	4.50
31.25	5	1.56	14	4.38	22	6.88
37.50	6	2.25	16	6.00	25	9.38
43.75	8	3.50	19	8.31	30	13.10
50.00	10	5.00	22	11.00	34	17.00
62.50	15	9.38	29	18.13	43	26.88
75.00	20	15.00	37	27.75	55	41.25
100.00	30	30.00	50	50.00	70	70.00

Note. For good and bad catchment, add or deduct upto 25% of yield.

(iii) *Barlow's Table.* T.G. Barlow carried out studies of catchments mostly under 130 square kilometres in Uttar Pradesh. He divided catchments into five classes and assigned values of runoff in percentage for each class as given below.

Class	*Description of catchment*	*Runoff percentage*
A	Flat, cultivated and black cotton soils	10
B	Flat, partly cultivated stiff soils	15
C	Average catchment	20
D	Hills and plains with little cultivation	35
E	Very hilly and steep, with little or no cultivation	45

These values of runoff percentages are for average monsoons and are to be multiplied by the following coefficients according to the nature of season.

Nature of season	*Class of catchment*				
	A	B	C	D	E
(1) Light rain, no heavy downpour	0.70	0.80	0.80	0.80	0.80
(2) Average or varying rainfall, no continuous downpour	1.00	1.00	1.00	1.00	1.00
(3) Continuous downpour	1.50	1.50	1.60	1.70	1.80

(iv) *Inglis and DeSouza's Formula.* On the basis of the data collected from 37 catchments in ghats and plains of Maharashtra, C.C. Inglis and DeSouza gave the following two different formulae for ghat areas and for plains.

For ghat areas

$$R = 0.85\,P - 305 \qquad ...(5.11)$$

For plains

$$R = \frac{(P-178)P}{2540} \qquad ...(5.12)$$

where R is average annual runoff in millimetres and P is average annual rainfall in millimetres over the entire drainage basin.

(v) *Lacey's Formula.* G. Lacey gave the following formula

$$R = \frac{P}{1+\dfrac{3048}{P}\left(\dfrac{F}{S}\right)} \qquad ...(5.13)$$

where R is average annual runoff in millimetres; P is average annual rainfall in millimetres over the entire drainage basin ; F is a monsoon duration factor and S is a catchment (or drainage basin) factor.

Lacey gave the following values for the catchment factor S corresponding to Barlow's five classes of catchments.

Class of catchment	*Value of S*
A	0.25
B	0.60
C	1.00
D	1.70
E	3.45

Further Lacey also divided the monsoon into three classes depending upon its duration and gave the following values of (F/S) for the Barlow's five classes of catchments.

Class of monsoon	*Values of (F/S) for the catchments of type*				
	A	*B*	*C*	*D*	*E*
(1) Very short	2.00	0.83	0.50	0.23	0.14
(2) Standard length	4.00	1.67	1.00	0.58	0.28
(3) Very long	6.00	2.50	1.50	0.88	0.48

(vi) *Khosla's Formula.* A.N. Khosla considered mean annual temperature as a measure of the various factors affecting losses by evaporation, transpiration, sunshine and wind velocity and hence gave the following formula

$$R = P - \frac{T}{0.208} \qquad ...(5.14)$$

where R is average annual runoff in millimetres; P is average annual rainfall in millimetres over the entire drainage basin ; and T is mean annual temperature in °C for the entire drainage basin.

(vii) *Formula for some of the drainage basins in India.* For some of the drainage basins in India empirical formulae have been developed which are as follows:

Ganga basin	$R = 4.902\,5\, P^{0.64}$	...(5.15)
Yamuna basin, (Delhi)	$R = 0.111\,2\, P^{1.1}$	...(5.16)
Rihand basin, (U.P.)	$R = P - 1.615\,0\, P^{0.86}$	...(5.17)
Chambal basin, (Rajasthan)	$R = 120\, P - 49\,450$	...(5.18)
Tawa basin, (M.P.)	$R = 90.5\, P - 48\,000$	...(5.19)
Tapti basin, (Gujarat)	$R = 435\, P - 172\,000$	...(5.20)

where R is average annual runoff in millimetres and P is average annual rainfall in millimetres over the entire drainage basin.

(b) Infiltration Method. In this method the runoff may be estimated either by the use of the infiltration capacity curve, or by the use of filtration indices. In the first method as shown in Fig. 5.12 the infiltration capacity curve for a given soil and moisture conditions is subtracted from the curve of rainfall pattern to derive the excess rainfall which represents the runoff. In the second method by subtracting the predetermined values of the infiltration indices from the rainfall intensity the excess rainfall representing the runoff may be obtained.

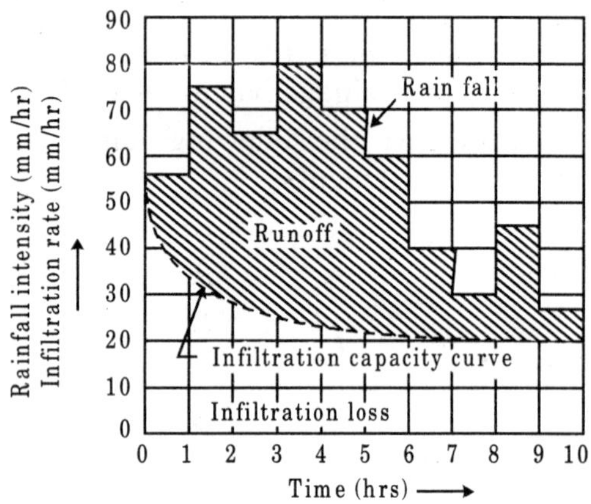

Fig. 5.12 *Determination of runoff by using infiltration capacity curve*

The infiltration capacity method is rational and would give good results if there is high degree of areal uniformity in rainfall pattern and in infiltration capacity, and if reasonably accurate infiltration curves are available. On the other hand the infiltration index methods are largely empirical and the derived values of the indices are applicable only when the rainfall characteristics and initial soil moisture conditions are identical to those for which these are derived. However, the infiltration index methods are best suited for estimating runoff which would be produced by a rainfall occurring on a drainage basin with its soils being saturated either by a preceding rainfall, or shortly after the starting of the rainfall of higher intensity and longer duration, so that the rate of infiltration might be assumed to have reached a final constant rate prior to or shortly after the starting of the rainfall.

(c) Rational Method. In the rational method the basic equation which correlates runoff and rainfall is as follows

$$Q = CIA \qquad ...(5.21)$$

where Q is flood flow in hectare-metre per hour (or cubic metre per hour); I is intensity of rainfall in metres per hour; A is area of the drainage basin in hectares (or square metres); and C is a runoff coefficient.

The value of the runoff coefficient C depends on the characteristics of the drainage basin such as soil type, vegetation, geological features etc. For different types of drainage basins the values of C are as given below.

Types of drainage basin	*Value of C*
Rocky and impermeable	0.8 — 1.0
Slightly permeable, bare	0.6 — 0.8
Cultivated or covered with vegetation	0.4 — 0.6
Cultivated absorbent soil	0.3 — 0.4
Sandy soil	0.2 — 0.3
Heavy forest	0.1 — 0.2

In the rational method the entire area of the drainage basin lying on the upstream of the point where the runoff is estimated, is divided into a number of sub-areas such that the time taken for the flow to reach the reference point (i) from the extreme dividing line of the nearest sub-area is one hour; (ii) from the extreme dividing line of the next sub-area is two hours and so on. Then knowing the intensity and duration of the rainfall and the values of the runoff coefficient for each sub-area at different times, and using the rational formula the runoff contributed by each sub-area at different times can be obtained. By adding the runoff contributed by each of the sub-areas during every hour, the direct runoff obtained from the entire drainage basin during every hour at the reference point is determined.

5.12 ESTIMATION OF MAXIMUM OR PEAK FLOOD DISCHARGE

While estimation of runoff assists in ascertaining the quantity of water available for utilization directly or through storage, the estimation of maximum or peak flood discharge helps in knowing the maximum quantity of water ever to be expected and is useful in designing the structures required for the disposal of the surplus or excess water. The estimation of the maximum or peak flood discharge can be made by the following methods:

(a) by physical indication of past floods
(b) by rating curve
(c) by empirical formulae and envelope curves
(d) by rational method
(e) by flood frequency studies
(f) by unit hydrograph.

The first five methods are discussed below and the method of unit hydrograph is discussed in section 5.14.

(a) Estimation of maximum flood discharge from physical indication of the past floods. The maximum flood discharge may be approximately estimated by enquiring from the residents about the marks that the highest flood in their memory in the past about 35 years may have left on the river banks or bathing ghats or any other ancient monuments situated on the river bank. By noting the high water marks along the banks of the river the cross-sectional area and wetted perimeter of the flow section as well as the water surface slope may be computed and using the Manning's formula, with suitably assumed value of roughness coefficient, the corresponding maximum value of the flood discharge may be determined. This method of determining the discharge through a river is known as slope-area method.

(b) Estimation of maximum flood discharge from rating curve. During high floods it is often impossible to make the discharge measurement, but the stage or the elevation of the water surface corresponding to the flood may be determined by noting the elevation of the high water marks on the banks of the river. Corresponding to this value of stage the maximum or peak value of flood discharge may be determined by extrapolation from the stage-discharge curve or the rating curve already prepared as indicated in section 5.11.

(c) Empirical formulae and envelope curves for estimation of maximum flood discharge. Several empirical formulae have been developed for estimating the maximum or peak value of the flood discharge. In these formulae the maximum flood discharge Q of a river is expressed as a function of the catchment area A. Most of these formulae may be written in a general form as

$$Q = CA^n$$

where C is coefficient and n is index, the values of which depend on the size and location of the catchment. Some of the formulae which are commonly used in our country are as indicated below.

(i) *Dicken's Formula*

$$Q = CA^{3/4} \qquad \text{...(5.22)}$$

where Q is maximum flood discharge in cumec; A is area of catchment in square kilometre ; and C is coefficient the values of which for different regions are as follows.

Region	*Value of C*
Northern India	11.4
Central India	13.9 to 19.5
Western Ghats	22.2 to 25

The maximum value of C is 35.

Dicken's formula is generally applicable for moderate size catchments of rivers of North and Central India.

(ii) *Ryve's Formula*

$$Q = CA^{2/3} \qquad \text{...(5.23)}$$

where Q is maximum flood discharge in cumec; A is area of catchment in square kilometre; and C is coefficient the values of which for different regions are as follows:

Region	*Value of C*
1. Area within 24 km from the coast	6.75
2. Area between 24 km to 161 km from the coast	8.45
3. Limited areas near hills	10.00

Ryve's formula is generally applicable for the catchments of rivers of South India.

(iii) *Inglis Formula*

$$Q = \frac{124A}{\sqrt{A + 10.4}} \simeq 124\,A^{1/2} \qquad \text{...(5.24)}$$

where Q is maximum flood discharge in cumec ; A is area of catchment in square kilometre.

Inglis' formula is derived by using the data of rivets of Maharashtra, where it is commonly used.

(iv) *Ali Nawaz Jung Bahadur's Formula*

$$Q = C\,[0.386\,A]^{[0.925-(1/14)\log 0.386\,A]} \qquad ...(5.25)$$

where Q is maximum flood discharge in cumec; A is area of catchment in square kilometre ; and C is a coefficient the value of which varies from 48 to 60. The maximum value of C is 85.

This formula is derived on the basis of the data of rivers of former Hyderabad State. However, the formula is applicable with lower values of C for the catchments of rivers of South India and with higher values of C for the catchment of rivers of North India.

Envelope Curves

For developing envelope curves, the data regarding the maximum flood discharges and the respective areas of the drainage basins having similar topographical features and climatic conditions are compiled. The maximum flood discharges are then plotted against the areas of the drainage basins and a curve is drawn to cover or envelope the highest plotted points, which is known as envelope curve. If the plotting is done on a log-log paper then the envelope curve is usually a straight line. By using the envelope curve the maximum flood discharge may be estimated if the area of the drainage basin is known. For Indian rivers enveloping curves from observed data of floods have been developed by Kanwar Sain and Karpov which are shown in Fig. 5.13.

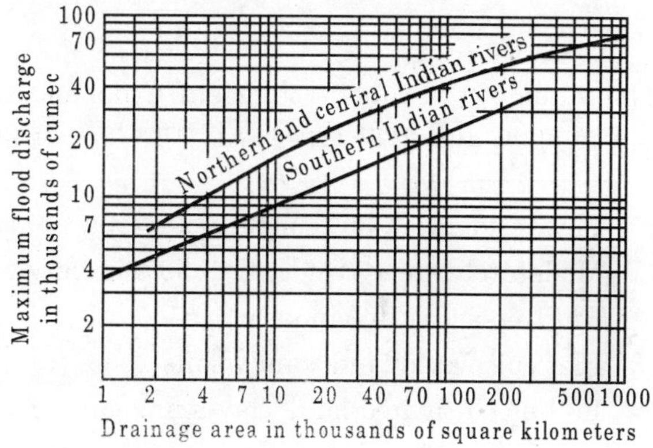

Fig. 5.13 *Envelope curves for Indian rivers*

This method is better than empirical formula in that, the selection of coefficients on the basis of judgement is eliminated. However, this method has the same limitations as in the case of empirical formulae that it also takes into consideration only one drainage basin characteristic i.e., area. The shape and slope of the drainage basin, rainfall, geology etc., are not taken into account.

Another inherent limitation of the envelope curve lies in the fact that it is based on the data of the floods recorded in the past, a still higher flood may occur at some time in future. Therefore as more data is made available the curve needs to be redrawn.

However, these curves are very useful for estimating maximum flood discharges. The curves given by Kanwar Sain and Karpov have been used as a guidance in determining the maximum flood discharge for various projects in our country.

(d) Rational Method. The rational method can also be used to estimate the maximum flood discharge for which the following formula is used

$$Q = C\,I_c\,A \qquad ...(5.26)$$

where Q is the maximum flood discharge in hectare-metre per hour (or cubic metre per hour); A is area of the drainage basin in hectares (or square metres); I_c is the critical intensity (or design intensity) of rainfall in metres per hour corresponding to the time of concentration t_c for the drainage basin ; and C is a runoff coefficient.

The time of concentration is defined as the time which would be required for the surface runoff from the remotest part of the drainage basin to reach the point where the runoff is estimated. This time will vary, generally depending on the slope and character of the drainage basin.

The critical intensity of rainfall I_c can be obtained from the equation

$$I_c = \frac{P}{t_R}\left(\frac{t_R+1}{t_C+1}\right) \quad ...(5.27)$$

where P is maximum precipitation in metres that occurs during a storm period of t_R hours. However, if the time of concentration t_c is not known then

$$I_c \approx \frac{P}{t_R} \quad ...(5.28)$$

The value of the runoff coefficient C depends on the characteristics of the drainage basin and the same for different types of terrain are as given below.

Type of Terrain	*Value of C*
Flat residential, 30% impervious area	0.40
Moderately steep residential, 50% impervious area	0.60
Built up area, 70% impervious	0.80
Flat cultivated land, sandy soils	0.20
Rolling cultivated land, clay-loam soil	0.50
Hilly areas, forests, clay and loamy soils	0.50

(e) Flood Frequency Studies. In this method on the basis of the past records of the maximum flood discharge the predictions for the maximum flood discharges which may occur in future are made by determining the flood frequency and using the probability concept.

Flood frequency means the number of times a flood of given magnitude will be equalled or exceeded in any one year and it is usually expressed in percentage. Thus, a 10% frequency (F) of a flood means that there are 10 out of 100 chances of the flood being equalled or exceeded in any one year. Flood frequency may also be expressed in terms of recurrence interval or return period.

Recurrence interval is the number of years during which a flood of given magnitude will be equalled or exceeded once. It is usually denoted by T and is given by

$$T = \frac{100}{F} \quad ...(5.29)$$

In other words recurrence interval denotes the number of years in which a flood of given magnitude can be expected to occur once. The flood is then designated according to its recurrence interval as 100 year flood, 1000 year flood etc. A 100 year flood is that flood which may be expected once in 100 years.

The recurrence interval for a flood of given magnitude may be determined if record of maximum flood discharge in each year of n number of years is given. The flood discharges are arranged in the descending order of magnitude and assigned serial numbers. Thus, the highest flood is placed at the top and given serial number 1, next highest is given serial number 2 and so on. The lowest flood discharge will be at the bottom with serial number n. If a particular flood discharge has a serial number m then its recurrence interval can be found by any of the following methods.

(i) *California Method*

$$T = \frac{n}{m} \qquad ...(5.30)$$

(ii) *Allen–Hazen Method*

$$T = \frac{2n}{2m-1} \qquad ...(5.31)$$

(iii) *Weibull Method*

$$T = \frac{n+1}{m} \qquad ...(5.32)$$

(iv) *Gumbel Method*

$$T = \frac{n}{m+C-1} \qquad ...(5.33)$$

where C is known as Gumbel's correction the value of which depends on m/n and it can be found from Table 5.2.

Table 5.2 Gumbel's Correction

m/n	1	0.9	0.8	0.7	0.6	0.5	0.4	0.3	0.2	0.1	0.08	0.04
C	1	0.95	0.88	0.845	0.78	0.73	0.66	0.59	0.52	0.4	0.38	0.28

For determining the flood discharge of desired frequency the peak flood data is compiled from the records. There are two methods of compiling flood peak data namely (i) annual flood series and (ii) partial duration series. In *annual flood series* only the highest flood in each year is used ignoring the second or next highest in any year which sometimes may exceed many of the annual maximum in some other years. In *partial duration series* all floods above a selected minimum are taken for analysis, regardless of the time interval, so that in some years there may be a number of floods above the assumed minimum, while in some other years there may not be any such flood at all. However, for flood frequency studies usually the annual flood series is adopted. Some of the commonly used flood frequency analysis methods using annual flood series are as follows.

(i) Weibull's method (or Plotting Positions method)

(ii) Gumbel's method

(iii) Log-Normal Distribution method

(iv) Log-Pearson Type III Distribution method

However, only the first two methods are dealt with here as indicated below.

(i) *Weibull' s method* (*or Plotting Positions method*). In this method the maximum flood data obtained for large number of successive years are arranged in decreasing order of magnitude and the recurrence interval T or frequency F are computed by any of the formulae noted earlier. A plot of annual flood Q v/s T (or F) is made and a best fitting curve is drawn through the plotted points. The plot is usually made on log-log paper. By extrapolation of the curve the magnitude of the flood for any desired recurrence interval or frequency may be obtained.

(ii) *Gumbel's method.* Gumbel's method is based on probability concept. This method is described below.

If Q_1, Q_2, Q_3,......,Q_n are series of n extreme (or maximum) values of flood discharges then according to Gumbel's theory of extreme values for a large value of n (*i.e.*, $n \to \infty$) the probability P of occurrence of a flood equal to or larger than Q is given by

$$P = 1 - e^{-z} \qquad \text{...(5.34)}$$

in which $z = e^{-y}$; e = base of natural logarithms; and y = a reduced variate given by

$$y = \frac{1}{0.78\sigma}\left[Q - \bar{Q} + 0.45\,\sigma\right] \qquad \text{...(5.35)}$$

in which

$$\bar{Q} = \text{Arithmetic mean of all the flood discharges in series;}$$

or

$$\bar{Q} = \frac{Q_1 + Q_2 + Q_3 + \dots\dots\dots\dots\dots + Q_n}{n}$$

$$\sigma = \text{standard deviation of flood series}$$

or

$$\sigma = \sqrt{\frac{\left(Q_1 - \bar{Q}\right)^2 + \left(Q_2 - \bar{Q}\right)^2 + \dots\dots\dots\dots\dots + \left(Q_n - \bar{Q}\right)^2}{(n-1)}}$$

Equation (5.34) may be transposed as

$$y_p = -\log_e\left[-\log_e(1-P)\right] \qquad \text{...(5.36)}$$

in which y_p is the value of y for a given value of P.

Further if T is the recurrence interval for the flood of magnitude Q, then

$$T = \frac{1}{P}; \quad \text{or} \quad P = \frac{1}{T}$$

Introducing the above value of P in Eq. (5.36), we obtain

$$y_T = -\left[\log_e \cdot \log_e\left(\frac{T}{T-1}\right)\right] \qquad \text{...(5.37)}$$

or

$$y_T = -\left[0.834 + 2.303\log_{10} \cdot \log_{10}\left(\frac{T}{T-1}\right)\right] \qquad \text{...(5.37)}$$

where y_T is the value of y for a given value of T.

If a flood of magnitude Q_T has recurrence interval T then from Eq. (5.35), we have

$$y_T = \frac{1}{0.78\,\sigma}\left[Q_T - \bar{Q} + 0.45\sigma\right]$$

From which solving for Q_T, we obtain

$$Q_T = (0.78\sigma)\, y_T + \bar{Q} - 0.45\,\sigma \qquad ...(5.38)$$

or

$$Q_T = \bar{Q} + K_T\,\sigma \qquad ...(5.38\text{ a})$$

where K_T is known as frequency factor and it is given by

$$K_T = (0.78\, y_T \; - 0.45) \qquad ...(5.39)$$

Equations (5.38) and (5.39) constitute the basic Gumbel's equations which are, however, applicable to an infinite sample size (i.e., $n \to \infty$). Since the practical annual data series of extreme events such as floods have finite lengths of record (i.e., n is finite), Eq. (5.39) is modified to account for the finite value of n as explained below.

For finite value of n the value of the frequency factor K is given by the following expression

$$K_T = \frac{y_T - \bar{y}_n}{s_n} \qquad ...(5.40)$$

in which

y_T = reduced variate the value of which is given by Eq. (5.37) or (5.37 a);

$\bar{y}_n$ = reduced mean, a function of sample size n and its value may be obtained from Table 5.3; for $n \to \infty$, $\bar{y}_n \to 0.577$; and

s_n = reduced standard deviation, a function of sample size n and its value may be obtained from Table 5.4; for $n \to \infty$, $s_n = 1.2825$

Table 5.3 Reduced mean in Gumbel's Extreme value distribution

n*	0	1	2	3	4	5	6	7	8	9
10	0.4952	0.4996	0.5035	0.5070	0.5100	0.5128	0.5157	0.5181	0.5202	0.5220
20	0.5236	0.5252	0.5268	0.5283	0.5296	0.5309	0.5320	0.5332	0.5343	0.5353
30	0.5362	0.5371	0.5380	0.5388	0.5396	0.5402	0.5410	0.5418	0.5424	0.5430
40	0.5436	0.5442	0.5448	0.5453	0.5458	0.5463	0.5468	0.5473	0.5477	0.5481
50	0.5485	0.5489	0.5493	0.5497	0.5501	0.5504	0.5508	0.5511	0.5515	0.5518
60	0.5521	0.5524	0.5527	0.5530	0.5533	0.5535	0.5538	0.5540	0.5543	0.5545
70	0.5548	0.5550	0.5552	0.5555	0.5557	0.5559	0.5561	0.5563	0.5565	0.5567
80	0.5569	0.5570	0.5572	0.5574	0.5576	0.5578	0.5580	0.5581	0.5583	0.5585
90	0.5586	0.5587	0.5589	0.5591	0.5592	0.5593	0.5595	0.5596	0.5598	0.5599
100	0.5600									

* n = Sample size

By using these equations the magnitude of the flood corresponding to a given recurrence interval T may be estimated as described below.

1. From the records obtain a series of maximum values of flood discharges for number of successive years (say n years).

2. Calculate $\bar{Q}$ and σ for the series of the noted maximum flood discharges.

3. For the known value of n, determine the values of $\bar{y}_n$ and s_n from Tables 5.3 and 5.4 respectively.

4. For a given value of T calculate y_T by using Eq. (5.37).

5. Determine the value of K_T by using Eq. (5.40).

6. Determine the required flood discharge Q_T by using Eq. (5.38 a).

Table 5.4 Reduced standard Deviation s_n in Gumble's extreme value distribution

n*	0	1	2	3	4	5	6	7	8	9
10	0.9496	0.9676	0.9833	0.9971	1.0095	1.0206	1.0316	1.0411	1.0493	1.0565
20	1.0628	1.0696	1.0754	1.0811	1.0864	1.0915	1.0961	1.1004	1.1047	1.1086
30	1.1124	1.1159	1.1193	1.1226	1.1255	1.1285	1.1313	1.1339	1.1363	1.1388
40	1.1413	1.1436	1.1458	1.1480	1.1499	1.1519	1.1538	1.1557	1.1574	1.1590
50	1.1607	1.1623	1.1638	1.1658	1.1667	1.1681	1.1696	1.1708	1.1721	1.1734
60	1.1747	1.1759	1.1770	1.1782	1.1793	1.1803	1.1814	1.1824	1.1834	1.1844
70	1.1854	1.1863	1.1873	1.1881	1.1890	1.1898	1.1906	1.1915	1.1923	1.1930
80	1.1938	1.1945	1.1953	1.1959	1.1967	1.1973	1.1980	1.1987	1.1994	1.2001
90	1.2007	1.2013	1.2020	1.2026	1.2032	1.2038	1.2044	1.2049	1.2055	1.2060
100	1.2065									

* n = Sample size

In order to verify whether the given data follows the Gumbel's distribution or not it is plotted on Gumbel's probability paper, which consists of specially marked T-scale on its abscissa. To construct the T-scale on the abscissa, first construct an arithmetic scale of y_T values say from – 2 to + 7. Then for slected values of T, say 2, 10, 50, 100, 500 and 1000 determine the values of y_T by using Eq. (5.37) and mark off their positions on the abscissa which forms the T-scale. The ordinate of the Gumbel's probability paper, on which the values of the flood discharge Q are plotted, may have either an arithmetic scale or logarithmic scale. From Eq. (5.38) it is evident that Q varies linearly with y_T and hence if the data follows Gumbel's distribution, it will plot as a straight line on the Gumbel's probability paper. Further by extrapolating this straight line the flood discharge for any desired recurrence interval may be obtained.

The method described above is explained in Illustrative Example 5.8.

5.13 HYDROGRAPH

A hydrograph is a graphical plot of discharge of a natural stream or river versus time. Discharge is usually expressed in cumec, or hectare metre per day (or per month) and time is expressed in hours, days or months. Discharge is plotted on Y-axis and the corresponding time is plotted on

X-axis. Figure 5.14 shows a hydrograph of a typical stream resulting from rainfall of intensity as indicated by a hyetograph plotted at the left top corner. At the beginning there is only base flow, i.e., the ground water contribution to the stream as indicated at *A* in Fig. 5.14. After the commencement of rain when the initial losses like interception and infiltration are met, the surface flow begins and there is progressive increase in the stream flow. The hydrograph therefore gradually rises and reaches a peak value at *B* at time t_p, called time lag or basin lag or time to peak, measured from the centroid of the hydrograph of effective rainfall (defined as that portion of the total rainfall which enters the stream as direct runoff). Thereafter the flow in the stream recedes and hence the hydrograph declines. Thus, a hydrograph consists of rising limb *AB* and recession limb *BC*. Discharge at *B* corresponds to the maximum or peak flood discharge and the area under the curve *ABC* would give the volume of the total runoff during the flood.

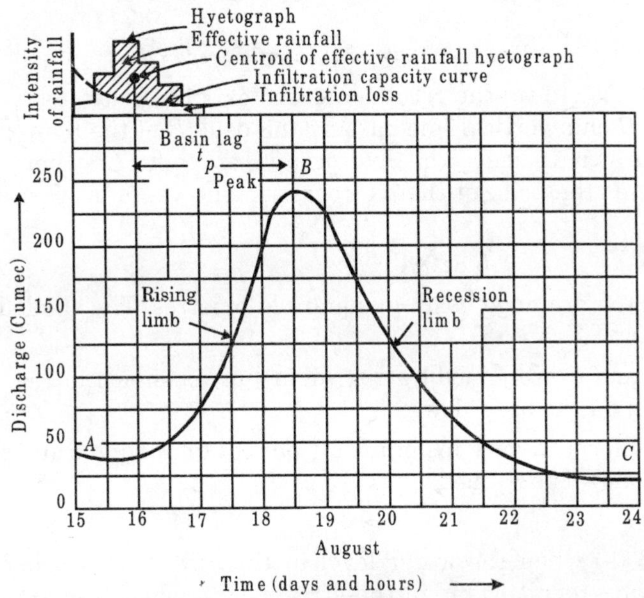

Fig. 5.14 *Typical hydrograph*

A single peaked hydrograph as shown in Fig. 5.14 is developed when there is a single storm causing precipitation of more or less uniform intensity. However, due to complex storms causing precipitations of different intensities with some interval, a multiple peaked hydrograph also known as complex hydrograph is developed. For flood analysis and derivation of unit hydrograph a single peaked hydrograph is required. A complex hydrograph, however, can be resolved into simple single peaked hydrographs by drawing hypothetical recession lines.

Seperation of Base Flow

As described in section 5.9 the total runoff comprises direct runoff and base flow. For the derivation of unit hydrograph, the base flow has to be separated from the total runoff. The following three methods are commonly adopted for the separation of the base flow.

(i) The base flow may be separated by simply drawing a line *AC* tangential to both the limbs of the hydrograph at its lower portion as shown in Fig. 5.15 (a). The shaded portion below line *AC* represents the base flow. This method is very simple but is approximate and can be used only for preliminary estimates.

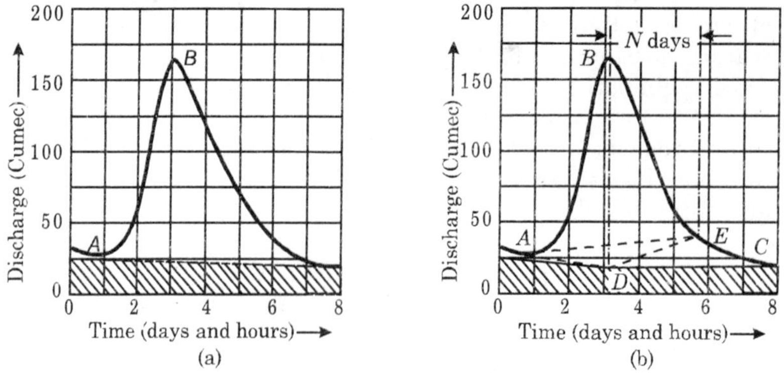

Fig. 5.15 *Methods of base flow separation*

(ii) As shown in Fig. 5.15 (b) the recession curve existing prior to occurrence of the rainfall is extended to a point D on a vertical line passing the peak B of the hydrograph. From point D a straight line DE is drawn to meet the hydrograph at point E, where E is a point on the hydrograph N days after the peak of the hydrograph. The value of N in days is approximately given by

$$N = 0.826\,A^{0.2} \qquad \text{...(5.41)}$$

where A is the area of the drainage basin in square kilometres. The shaded portion below $ADEC$ represents the base flow.

(iii) The points A and E are joined by a straight line as shown in Fig. 5.15 (b) and the area below this line is taken as the base flow.

Since the base flow forms a very insignificant part of total runoff during high floods, often a constant value of base flow is assumed.

Negative Base Flow

As shown in Fig. 5.16 (a) when the water level in the stream is low before the occurrence of flood, there is flow from ground water to the stream. But when the water level in the stream rises due to flood, there is flow of water from the stream to the ground water and this is termed as negative base flow. Again as the flood recedes the water level in the stream is lowered, the ground water starts contributing to the stream. As such the variation in base flow is as shown by dotted line in Fig. 5.16 (b) in which the shaded portion represents the negative base flow. However, the dotted line represents the extreme variation in the base flow and the actual variation is usually much less. Moreover, as stated earlier the contribution of the ground water

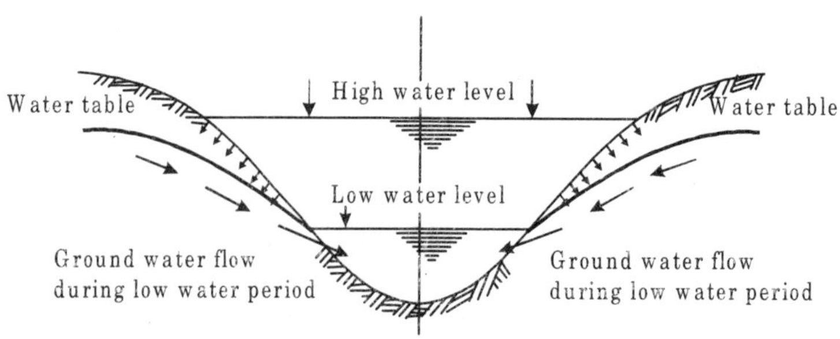

(a)

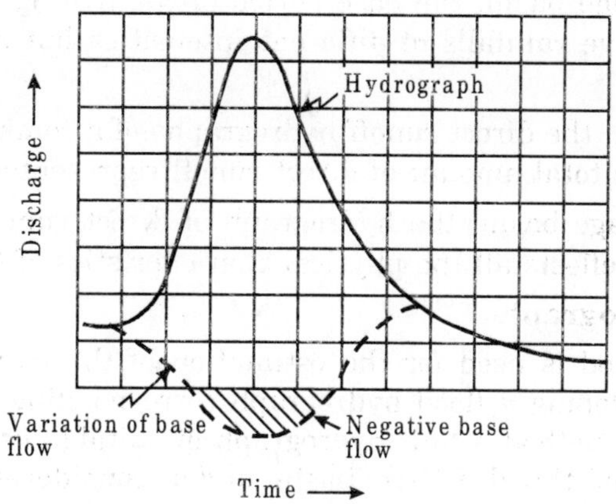

Fig. 5.16 *Variation of base flow with the variation of stream flow*

to the total runoff during high floods is insignificant, it is justified to consider the base flow as given by any one of the methods of base flow separation and as shown by the shaded portion in Fig. 5.15.

5.14 UNIT HYDROGRAPH

A *unit hydrograph* may be defined as a hydrograph of direct surface runoff resulting from one centimetre of effective rainfall falling uniformly over the drainage basin or catchment in space as well as in time for a specified or unit duration. The effective rainfall also called rainfall excess, is that part of the total rainfall that enters the stream as direct runoff. The specified or unit duration is that period within which the effective rainfall is considered to be uniformly distributed. A unit hydrograph is designated according to the unit duration. For example, if a unit hydrograph results from a rainfall of 3 hours unit duration it is known as a 3 hour unit hydrograph, meaning thereby a hydrograph produced by surface runoff from a storm lasting for 3 hours and yielding a rainfall excess of 1 centimetre.

Since the area under a hydrograph represents the total runoff, it is evident from the definition of unit hydrograph that the area under the unit hydrograph represents a direct runoff of 1 cm.

The concept of unit hydrograph was first given by L.K. Sherman in 1932. It is based on this assumption that if two identical storms occur over a drainage basin with identical conditions prior to the rain, the hydrographs of runoff from the two storms would be expected to be the same. This basic premise implies that the rainfall excess of n centimetres resulting from a storm of certain unit duration will give a runoff hydrograph having ordinates n times the ordinates of unit hydrograph of the same unit duration.

Assumptions of Unit Hydrograph Theory

The theory of Unit hydrograph is based on the following assumptions :

(i) The effective rainfall is uniformly distributed within its duration or specified period of time.

(ii) The effective rainfall is uniformly distributed over the entire area of the drainage basin.

(iii) For a given drainage basin, the base period of the hydrographs of direct runoff corresponding to effective rainfalls of different intensities but same unit duration, is constant.

(iv) The ordinates of all the direct runoff hydrographs of a common base period are directly proportional to the total amount of direct runoff represented by each hydrograph.

(v) For a given drainage basin, the hydrograph of direct runoff corresponding to a given period of rainfall reflects all the physical characteristics of the basin.

Derivation of Unit Hydrograph

The unit hydrograph method is used for the estimation of the maximum flood discharge of a stream as well as for developing a flood hydrograph corresponding to any anticipated rainfall. For the application of this method a unit hydrograph of suitable unit duration is derived from an observed hydrograph of the drainage basin under consideration. Then using the unit hydrograph so derived the flood hydrograph corresponding to any rainfall of the same unit duration may be obtained for the drainage basin on the basis of the above noted assumptions. In order to derive the unit hydrograph the following steps are adopted.

(i) Select a hydrograph resulting from an isolated, intense, short-duration rainfall of nearly uniform distribution over the entire drainage basin as well as within its duration.

(ii) Separate the base flow from the total runoff represented by the hydrograph in order to obtain the direct runoff. For base flow separation any of the methods as indicated earlier may be adopted.

(iii) Draw a series of ordinates at some uniform time interval for the entire hydrograph.

(iv) From these ordinates of the hydrograph by deducting the corresponding ordinates of base flow obtain the ordinates of direct runoff.

(v) Compute the volume of the direct runoff which is equal to

$$(\Sigma O \times t \times 3600) \text{ m}^3$$

where ΣO is the sum of the direct runoff ordinates in cumec and t is the time interval in hours between the successive ordinates.

(vi) Divide the volume of direct runoff by the area of the drainage basin to obtain the equivalent depth of direct runoff which is given as

$$\text{Direct runoff depth} = \frac{(\Sigma O \times t \times 3600) \times 100}{A \times 10^6} \text{ cm}$$

$$= \frac{0.36 \times \Sigma O \times t}{A} \text{ cm} \qquad \ldots(5.42)$$

where A is the area of the drainage basin in square kilometres.

(vii) By dividing each of the ordinates of direct runoff by the depth of direct runoff obtain the ordinates of the unit hydrograph. By plotting these ordinates at their respective times and drawing a smooth curve passing through these, unit hydrograph is obtained. The unit duration of the unit hydrograph so derived is same as the duration of effective rainfall corresponding to which the resulting hydrograph has been used for deriving the unit hydrograph.

Use of Unit Hydrograph for Deriving a Flood Hydrograph Resulting from Rainfall of Unit Duration

From the above noted assumptions of the unit hydrograph theory it may be inferred that an effective rainfall of intensity R centimetres occurring uniformly over the entire drainage basin in a unit duration of T hours will produce a direct runoff hydrograph of the same base period as that of a unit hydrograph of T hours unit duration but having its ordinates R times those of the unit hydrograph. The principle involved in this inference is known as the principle of linearity. By applying the principle of linearity the flood hydrograph that would result from a rainfall of known intensity and its duration same as the unit duration of the unit hydrograph may be obtained as follows:

(i) Deduct the infiltration losses from the rainfall intensity to obtain the intensity of effective rainfall.

(ii) Multiply the ordinates of the unit hydrograph by the intensity of effective rainfall to obtain the ordinates of the direct runoff hydrograph.

(iii) Add the corresponding base flow ordinates to the direct runoff ordinates to obtain the ordinates of the anticipated flood hydrograph which may be plotted.

The derivation of unit hydrograph and its use for obtaining a flood hydrograph is shown in Illustrative Examples 5.4 and 5.5.

Use of Unit Hydrograph for Deriving a Flood Hydrograph Resulting from a Series of Rainfalls Each of Same Unit Duration

A unit hydrograph of certain unit duration can also be used for deriving a flood hydrograph that would result from a series of rainfalls, each of which may be of different intensity but of duration same as unit duration of the unit hydrograph. For this by applying the principle of linearity for each rainfall separate hydrograph of direct runoff, beginning at the same time as the rainfall, is obtained. Then applying the principle of superposition the ordinates of the individual direct runoff hydrographs at the same times are summed up to obtain the ordinates of the total direct runoff hydrograph that would result from the series of rainfalls. To the total direct runoff ordinates the corresponding base flow ordinates are added to obtain the ordinates of the anticipated flood hydrograph due to the series of rainfalls.

The same method may also be adopted to derive a flood hydrograph that would result from a rainfall of uniform intensity but lasting for a duration longer than the unit duration of the unit hydrograph. In this case the entire rainfall may be divided into a series of rainfalls each of which is of same intensity and duration same as the unit duration of the unit hydrograph. The resulting flood hydrograph may then be obtained by adopting the same procedure as indicated earlier. However, in this case the duration of the entire rainfall must be an integral multiple of the unit duration of the available unit hydrograph.

The method discussed above is explained in Illustrative Example 5.5.

5.15 S-HYDROGRAPH (OR S-CURVE)

S-hydrograph (or S-curve) is a hydrograph of direct surface discharge that would result from successive storms each of which producing effective rainfall of one centimetre in unit duration. Thus, S-hydrograph may be derived by summing up the ordinates of a series of unit hydrographs of same unit duration spaced at intervals of the unit duration, and hence it is also known as summation hydrograph or summation curve. As shown in Fig. 5.17 S-hydrograph is therefore a continuously rising curve in the form of letter S, till a constant value of the discharge is reached. If the base time of unit hydrograph is T hours and the unit duration is t_r,

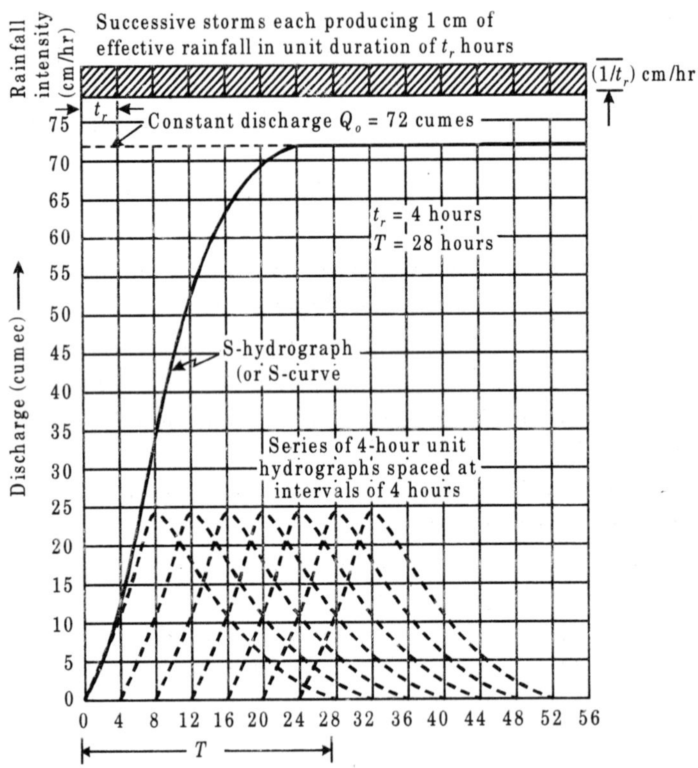

Fig. 5.17. *S-hydrograph*

hours then it is observed that the discharge becomes constant at time equal to the base time of the unit hydrograph minus the unit duration i.e., at $(T—t_r)$ hours. Further since 1 cm of effective rainfall occurs every t_r hours on the drainage basin the corresponding intensity of rainfall is $(1/t_r)$ cm/hour, and the constant discharge developed is given by

$$Q_0 = \frac{(A \times 1000 \times 1000) \times 1}{100 \times t_r \times 3600}$$

$$Q_0 = \frac{2.778A}{t_r} \qquad \ldots(5.43)$$

where Q_0 is the constant discharge in cumec, and A is the area of the drainage basin in square kilometre.

However, if the area A of the drainage basin is in hectares then the constant discharge is given by

$$Q_0 = \frac{(A \times 100 \times 100) \times 1}{100 \times t_r \times 3600}$$

or $$Q_0 = \frac{A}{36t_r} \qquad \text{...(5.43a)}$$

Table 5.5 Derivation of S-hydrograph from a given 4 hour unit hydrograph

Time (hours)	*Ordinates of 4-hr Unit Hydrograph (cumec)*	*Ordinates of Successive 4-hr Unit Hydrographs each lagging from the preceding one by 4 hrs. (cumec)*						*Ordinates of S-hydrograph (cumec)*
(1)	(2)	(3)	(4)	(5)	(6)	(7)	(8)	(9)
0	0	—	—	—	—	—	—	0
2	4.5	—	—	—	—	—	—	4.5
4	10.0	0	—	—	—	—	—	10.0
6	17.5	4.5	—	—	—	—	—	22.0
8	24.5	10.0	0	—	—	—	—	34.5
10	21.5	17.5	4.5	—	—	—	—	43.5
12	17.5	24.5	10.0	0	—	—	—	52.0
14	14.0	21.5	17.5	4.5	—	—	—	57.5
16	11.0	17.5	24.5	10.0	0	—	—	63.0
18	8.5	14.0	21.5	17.5	4.5	—	—	66.0
20	6.0	11.0	17.5	24.5	10.0	0	—	69.0
22	4.5	8.5	14.0	21.5	17.5	4.5	—	70.5
24	3.0	6.0	11.0	17.5	24.5	10.0	0	72.0
26	1.5	4.5	8.5	14.0	21.5	17.5	4.5	72.0
28	0	3.0	6.0	11.0	17.5	24.5	10.0	72.0

The derivation or S-hydrograph is illustrated by an example in which S-hydrograph has been derived from a 4-hour unit hydrograph, the computations for which are shown in Table 5.5. The derived S-hydrograph is shown in Fig. 5.17, which also shows a series of 4-hour unit hydrographs spaced at intervals of 4 hours. Further the 4-hour unit hydrograph is for a drainage basin having area 103.7 sq. km, and hence the constant discharge

$$Q_0 = \frac{2.778 \times 103.7}{4}$$

$$= 72.02 \text{ cumec}$$

which is reached at (28 – 4) = 24 hours as shown in Table 5.5.

As described in section 5.16 the S-hydrograph derived from a given unit hydrograph may be used to derive a unit hydrograph of shorter or longer unit duration than that of the given unit hydrograph.

5.16 DERIVATION OF UNIT HYDROGRAPH OF DIFFERENT UNIT DURATION FROM A UNIT HYDROGRAPH OF GIVEN UNIT DURATION

When a unit hydrograph of desired unit duration is not available it may be derived by conversion of an available unit hydrograph of known unit duration. If a unit hydrograph of unit duration longer than that of the available unit hydrogrph is required to be derived and the unit duration of the required unit hydrograph is an integral multiple of the unit duration of the available unit hydrograph then by the principle of superposition the unit hydrograph of desired unit duration may be derived. However, a more general method involving the use of S-hydrograph may be adopted to derive a unit hydrograph of unit duration either longer or shorter than that of the available unit hydrograph. Further in this method the unit duration of the required unit hydrograph need not be an integral multiple of the unit duration of the available unit hydrograph. Both these methods are described below.

Table 5.6 Derivation of 6-hour unit hydrograph from a given 2-hour unit hydrograph

Time (*hours*)	*Ordinates of 2-hr Unit Hydrograph* (*cumec*)	*Ordinates of 2-hr Unit Hydrograph* (*cumec*) *lagged by 2 hrs*	*Ordinates of 2-hr Unit Hydrograph* (*cumec*) *lagged by 4 hrs*	*Total* (2) + (3) +(4) (*cumec*)	*Ordinates of 6-hr Unit Hydrograph* (*cumec*) (5) ÷ 3
(*1*)	(2)	(3)	(4)	(5)	(6)
0	0	—	—	0	0
1	8	—	—	8	2.7
2	20	0	—	20	6.7
3	40	8	—	48	16.0
4	49	20	0	69	23.0
5	46	40	8	94	31.3
6	39	49	20	108	36.0
7	32	46	40	118	39.3
8	27	39	49	115	38.3
9	22	32	46	100	33.3
10	19	27	39	85	28.3
11	16	22	32	70	23.3
12	13	19	27	59	19.7
13	11	16	22	49	16.3
14	8	13	19	40	13.3
15	6	11	16	33	11.0
16	4	8	13	25	8.3
17	2	6	11	19	6.3

Contd.

Contd.

(1)	(2)	(3)	(4)	(5)	(6)
18	0	4	8	12	4.0
19	—	2	6	8	2.7
20	—	0	4	4	1.3
21	—	—	2	2	0.7
22	—	—	0	0	0

(a) Derivation of Unit Hydrograph of Unit Duration t_r' from a Unit Hydrograph of Unit Duration t_r such that $t_r' > t_r$ and t_r' is an Integral Multiple of t_r.

Let a unit hydrograph of unit duration t_r' hours be required to be derived from a unit hydrograph of unit duration t_r hours such that $t_r' > t_r$ and $t_r' = nt^r$, where n is a positive integer.

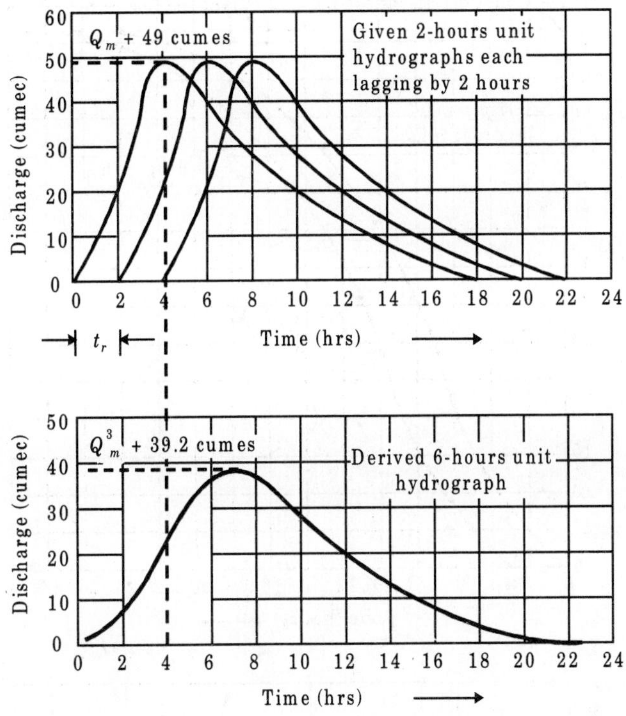

Fig. 5.18 *Derivation of 6-hour unit hydrograph from a given 2-hour unit hydrograph*

For this by taking the sum of the ordinates of n unit hydrographs of t_r unit duration, each lagging from the other by t_r hours, and dividing the sum by n, the ordinates of the required unit hydrograph of unit duration t_r' are obtained. This method is illustrated by an example in which a 6-hour unit hydrograph has been derived from a given 2-hour unit hydrograph, the computations for which are shown in Table 5.6. The derived 6-hour unit hydrograph is plotted in Fig. 5.18 which also shows three 2-hour unit hydrographs each lagging by 2-hours. It may however be observed that the peak discharge Q_m' of the derived unit hydrograph is less than the peak discharge Q_m of the given unit hydrograph, the peak of the derived unit hydrograph occurs later than that of the given unit hydrograph and the derived unit hydrograph has a longer base time by $(n-1)\ t_r$ hours than that of the given unit hydrograph. This is so because increase in the unit duration results in a lower intensity of the rainfall for a longer time which in turn causes the discharge to be extended over a longer time.

(b) Derivation of a Unit Hydrograph of Longer or Shorter Unit Duration than that of a given Unit Hydrograph by the S-Hydrograph Method

Let a unit hydrograph of unit duration t_r' hours be required to be derived from a unit hydrograph of unit duration t_r hours such that $t_r' \gtrless t_r$. Further t_r' may not be an integral multiple of t_r. For this the S-hydrograph method as described below may be used.

From the given unit hydrograph of unit duration t_r derive S-hydrograph as indicated in section 5.15. Figure 5.19 shows the S-hydrograph so obtained which represents a constant effective rainfall of intensity $(1/t_r)$ cm/hour. A lagged S-hydrograph is then drawn by shifting

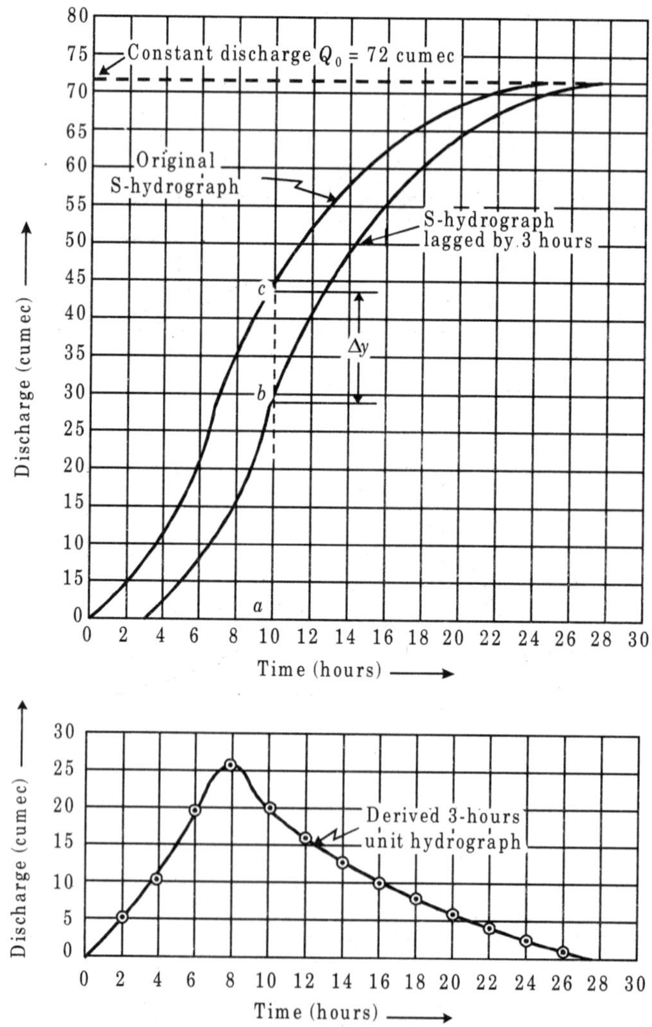

Fig. 5.19 *Derivation of 3-hour unit hydrograph from a given 4-hour unit hydrograph*

the original S-hydrograph by t_r' hours (equal to the unit duration of the desired unit hydrograph) as shown in Fig. 5.19. The area between the original and the lagged S-hydrographs represents the total runoff due to a rainfall of intensity ($1/t_r$) cm/hour in a duration of t_r' hours, i.e., it represents a total runoff of (t_r'/t_r) cm in t_r' hours. Further the difference between the ordinates of the original and the lagged S-hydrographs Dy, represents the rate of runoff at any instant of time. In other words the values of Δy at different instants of time represent the ordinates of a hydrograph having a unit duration of t_r' hours and a total runoff of (t_r'/t_r) cm. As such the ordinates of a hydrograph of unit duration t_r', hours and having a total discharge of 1 cm, i.e., the ordinates of a t_r' -hour unit hydrograph may be obtained by the principle of linearity as

$$\frac{\Delta y}{(t_r'/t_r)} = \frac{(\Delta y)t_r}{t_r'}$$

This method is illustrated by an example in which a 3-hour unit hydrograph has been derived from a 4-hour unit hydrograph. The S-hydrograph derived from a 4-hour unit hydrograph as shown in Table 5.5 is adopted to derive a 3-hour unit hydrograph, the computations for which are shown in Table 5.7. The ordinates of the derived 3-hour unit hydrograph, [Column (5) of Table 5.7] are plotted and a smooth curve is drawn as shown in Fig. 5.19. In general all the ordinates of the unit hydrograph so obtained may not fall on a smooth curve and hence some smoothening has to be done. The values of the ordinates along the smoothened curve are given in Column (6) of Table 5.7. However, as a check the sum of the ordinates in columns (5) and (6) must be the same.

Table 5.7 Derivation of 3-hour unit hydrograph from 4-hour unit hydrograph by the S-hydrograph method

Time (hours)	*Ordinates of S-Hydrograph (cumec) from Table (5.5)*	*Ordinates of lagged S-hydrograph (cumec)*	*Δy (cumec)*	*Ordinates of 3-hr Unit Hydrograph (cumec) = Δy (4/3)*	*Ordinates of 3-hr Unit Hydrograph (smoothened) (cumec)*
(1)	(2)	(3)	(4)	(5)	(6)
0	0	—	0	0	0
2	4.5	—	4.5	6.0	5.0
4	10.0	2.0	8.0	10.7	11.3
6	22.0	7.0	15.0	20.0	20.0
8	34.5	15.0	19.5	26.0	26.0
10	43.5	29.0	14.5	19.3	20.0
12	52.0	39.5	12.5	16.7	16.0
14	57.5	48.0	9.5	12.7	12.7
16	63.0	55.0	8.0	10.7	10.3
18	66.0	60.0	6.0	8.0	8.2
20	69.0	64.0	4.5	6.0	6.2
22	70.5	67.5	3.0	4.0	4.5
24	72.0	70.0	2.0	2.7	2.7
26	72.0	71.2	0.8	1.1	1.0
27	72.0	72.0	0	0	0
28	72.0	72.0	0	—	—
			Σ0	143.9	143.9

5.17 DERIVATION OF UNIT HYDROGRAPH FROM COMPLEX STORMS

In nature a single storm developing a single peaked hydrograph as shown in Fig. 5.14 seldom occurs and in most of the cases a complex storm occurs which comprises successive storms of different intensities resulting in a multiple peaked hydrograph. As such a unit hydrograph may be required to be derived from hydrographs of complex storms. If the storms are sufficiently separated so that the individual storm results in a well defined peak and it is possible to separate the hydrographs resulting from each of the storms, then the unit hydrographs may be derived from each of these hydrographs by using the method discussed earlier for isolated storms. However, if it is not possible to separate the hydrograph resulting from a complex storm then the following method may be adopted to derive the unit hydrograph.

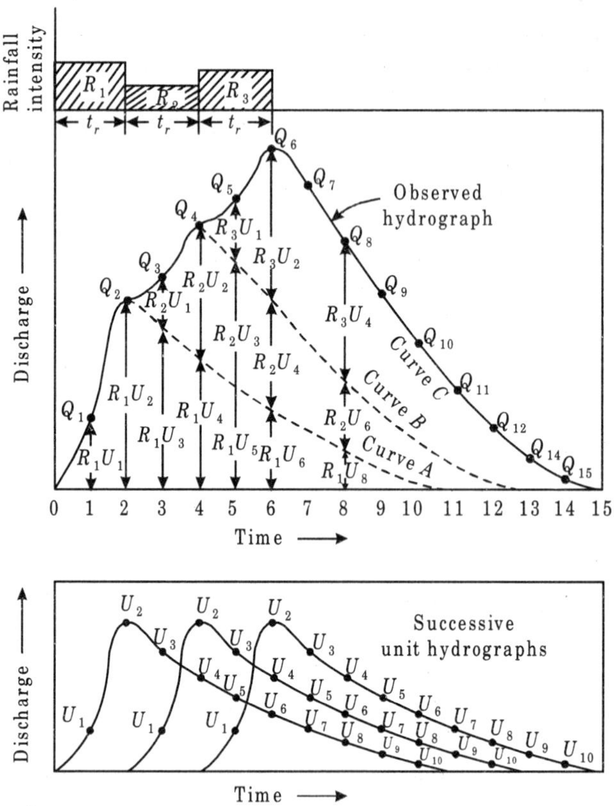

Fig. 5.20 *Derivation of unit hydrograph from a complex storm hydrograph*

As an illustration a complex storm consisting of three successive storms is considered. Let R_1, R_2 and R_3 be the respective rainfalls of the storms each of t_r hours duration.

Let the ordinates of the direct runoff hydrograph due to complex sotrm be represented by $Q_1, Q_2, Q_3, \ldots\ldots\ldots Q_n$ at successive times.

In this case corresponding to three storms there will be three unit hydrographs each operating at a time lag equal to the duration of the storm, i.e., t_r hours. Let the ordinates of the unit hydrograph be designated as $U_1, U_2, U_3, \ldots\ldots U_m$ at successive times.

By considering the complex storm as successive storms of different intensities the runoff hydrograph due to complex storm may be considered to be developed as a result of superposition of the successive storm hydrographs. In other words it means that at any instant of time the direct runoff ordinate of the hydrograph due to complex storm is equal to the sum of the ordinates of hydrographs due to each of the storms. Further by the principle of linearity the ordinates of each storm hydrograph may be obtained as the effective rainfall of the storm times corresponding ordinate of the unit hydrograph.

Thus, for the given complex storm hydrograph shown in Fig. 5.20.

First ordinate $Q_1 = R_1 U_1$

Since Q_1 and R_1 are known, U_1 can be computed

Second ordinate $Q_2 = R_2 U_2$

Since Q_2 and R_2 are known, U_2 can be computed

Third ordinate $Q_3 = R_1 U_3 + R_2 U_1$

Since all the values except U_3 are known the same can be computed

Fourth ordinate $Q_4 = R_1 U_4 + R_2 U_2$

Since all the values except U_4 are known the same can be computed

Fifth ordinate $Q_5 = R_1 U_5 + R_2 U_3 + R_3 U_1$

Since all the values except U_5 are known the same can be computed

... ...

Nth ordinate $Q_N = R_1 U_N + R_2 U_{N-2} + R_3 U_{N-4}$ (where N is less than m)

Thus, by repeating the same procedure the ordinates U_1, U_2, U_3, ...U_m of the unit hydrograph can be determined.

For deriving the above noted ordinates of the complex storm hydrograph the following guideline may be observed.

As shown in Fig. 5.20 the curves *A*, *B* and *C* represent the respective recession limbs of the three storm hydrographs. The ordinates of each of these curves may be determined as follows.

The ordinates of curve *A* are equal to R_1 × the successive ordinates of the First Unit Hydrograph.

The ordinates of curve *B* are equal to (successive ordinates of curve *A*) + (R_2 × successive ordinates of Second Unit Hydrograph).

The ordinates of curve *C* are equal to (successive ordinates of curves *A* + *B*) + (R_3 × successive ordinates of Third Unit Hydrograph).

The derivation of a unit hydrograph from a complex storm hydrograph by using the above described method is shown in Illustrative Example 5.6.

5.18 SYNTHETIC UNIT HYDROGRAPH

There are several drainage basins for which the streamflow records are not available but for such basins unit hydrograph may be required. In such cases synthetic unit hydrographs may be developed. Several methods for developing a synthetic unit hydrograph have been given out of which the following two methods are discussed.

1. *Snyder's Method*

F.F. Snyder (1938) analysed a large number of hydrographs from drainage basins ranging in area from 25 sq. km to 25000 sq. km and selected three elements for the development of unit hydrograph, namely, base time (T), peak discharge (Q_p) and basin lag (lag time or time to peak, t_p) and proposed the following empirical formulae for the three elements:

Basin lag (in hours)

$$t_p = C_t\,(LL_c\,)^{0.3} \qquad \text{...(5.44)}$$

Peak discharge (in cumec)

$$Q_p = C_p\,\frac{A}{t_p} \qquad \text{...(5.45)}$$

Base time (in days)

$$T = 3 + 3\left(\frac{t_p}{24}\right) \qquad \text{...(5.46)}$$

where C_t = coefficient of value ranging from 1.0 to 2.2, lower values being associated with basins of steeper slopes

L = length of the main stream in km, from the outlet or the gaging station to the upstream boundary limit of the drainage basin, Fig. 5.21 (a)

L_c = distance measured along the main stream from the outlet or the gaging station to a point on the stream nearest to the centroid of the drainage basin, Fig. 5.21 (a)

C_p = coefficient of value ranging from 4 to 5

A = area of the drainage basin in sq. km.

From Eq. (5.38) the peak discharge per unit area of the drainage basin q_p may be expressed as

$$q_p = \frac{C_p}{t_p} \qquad \text{...(5.47)}$$

Further for the unit duration of the effective rainfall, which would result in developing a unit hydrograph with its elements given by the above noted equations, Snyder proposed the following expression

$$t_r = \frac{t_p}{5.5} \qquad \text{...(5.48)}$$

in which t_r is the unit duration of effective rainfall in hours.

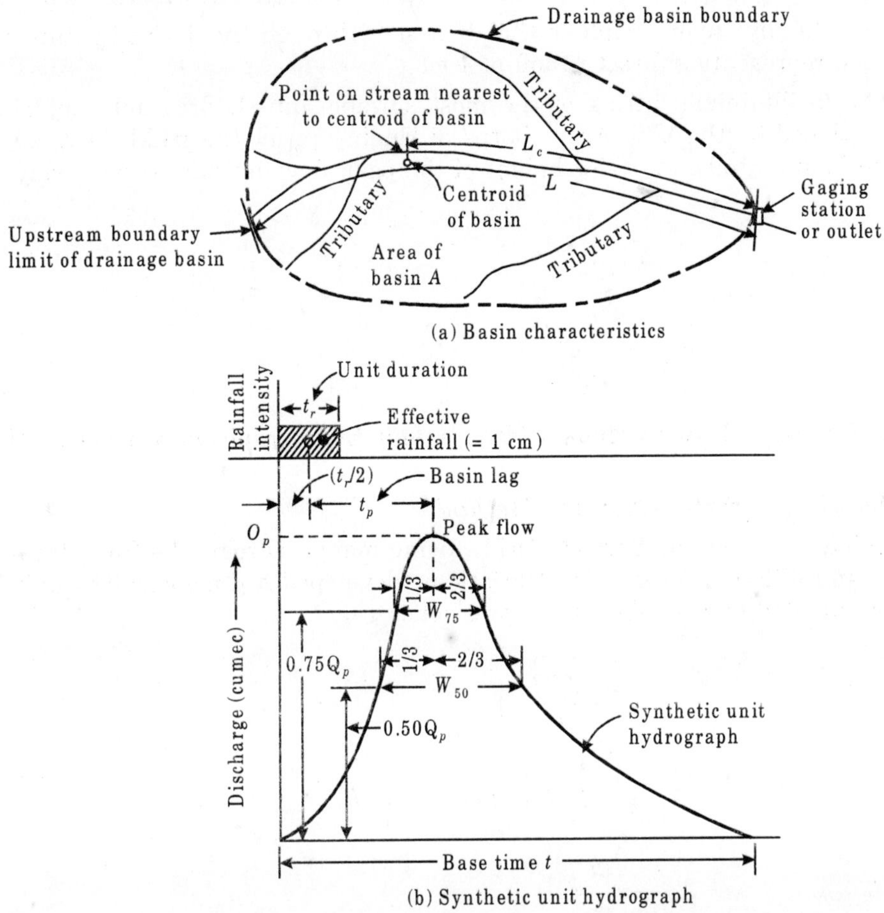

Fig. 5.21

However, if the actual duration of the storm is not equal to t_r given by Eq. (5.48) but is t_r', then in order to allow for some variation in the basin lag with the variation in the unit duration of the effective rainfall, Snyder proposed the following expression

$$t_{pr} = t_p + \frac{t_r' - t_r}{4} \qquad ...(5.49)$$

in which t_{pr} = basin lag for a storm duration of t_r'. This modified basin lag t_{pr} is used in place of t_p in Eqs (5.45), (5.46) and (5.47).

For deriving a synthetic unit hydrograph for a basin for which the streamflow records are not available the following procedure may be adopted.

(1) Collect the data for the basin such as A, L and L_c.

(2) Obtain the values of the coefficients C_p and C_t from adjacent basins for which streamflow records are available and which are hydrologically and meteorologically homogeneous.

(3) Compute basin lag, peak discharge and base time from the Snyder's equations.

(4) Sketch the unit hydrograph so that it passes through the derived points and the area under the curve represents a direct runoff of 1 cm.

In order to facilitate the fitting of the most suitable unit hydrograph empirical formulae have been developed by the U.S. Army Corps of Engineers for the widths W_{50} and W_{75} of the hydrograph in hours at 50% and 75% height of the peak flow ordinate respectively as follows

$$W_{50} = \frac{5.6}{q_p^{1.08}} \qquad \text{...(5.50)}$$

and

$$W_{75} = \frac{3.21}{q_p^{1.08}} \qquad \text{...(5.51)}$$

As shown in Fig. 5.21 (b) with these widths a unit hydrograph with a better shape can be sketched.

2. *U.S. Soil Conservation Service Method*

The U.S. Soil Conservation Service (1971), using many hydrographs from drainage areas of varying sizes and different geographical locations, developed a dimensionless unit hydrograph which is approximated by the following expressions.

(a) $$T_c = \left[\frac{0.87L^3}{H}\right]^{0.385} \qquad \text{...(5.52)}$$

(b) $$T_p = \sqrt{T_c} + 0.6T_c \qquad \text{...(5.53)}$$

(c) $$q_p = \left(\frac{C_p}{T_p}\right) \qquad \text{...(5.54)}$$

where

T_c = time of concentration in hours, defined as travel time for water from hydraulically most distant point in the drainage basin to the outlet or the gaging station

L = length of the main stream in km, from the outlet or the gaging station to the upstream boundary limit of the drainage basin

H = elevation differene in m, between the river bed at the upstream boundary limit of the drainage basin and that at the outlet or the gaging station

T_p = time in hours from start of rise of hydrograph to peak discharge rate

q_p = peak discharge rate in cumec per sq. km area of drainage basin, and

C_p = coefficient of value ranging from 4.0 to 5.0

The peak discharge Q_p of the unit hydrograph is given by

$$Q_p = q_p \times A$$

A curvilinear hydrograph can be constructed from the values of Q_p and T_p by using the ratios of (t/T_p) and (Q/Q_p) which may be obtained from Table 5.8 given by U.S. Soil Conservation Service.

Table 5.8 Values of ratio's (t/T_p) and (Q/Q_p)

(t/Tp)	(Q/Q_p)	(t/T_p)	(Q/Q_p)	(t/T_p)	(Q/Q_p)
0	0	1.0	1.000	2.4	0.180
0.1	0.015	1.1	0.980	2.6	0.130
0.2	0.075	1.2	0.920	2.8	0.098
0.3	0.160	1.3	0.840	3.0	0.075
0.4	0.280	1.4	0.750	3.5	0.036
0.5	0.430	1.5	0.660	4.0	0.018
0.6	0.600	1.6	0.560	4.5	0.009
0.7	0.770	1.8	0.420	5.0	0.004
0.8	0.890	2.0	0.320	∞	0
0.9	0.970	2.2	0.240		

SOLVED ILLUSTRATIVE EXAMPLES

Illustrative Example 5.1 *For the Cauvery basin shown in Figs. 5.8 and 5.8(a) with rainfall observations in mm marked at various rain gage stations. Compute the average rain fall by (i) Arithmetic mean method, (2) Thiessen polygon method, and (3) Isohyetal method.*

Solution

1. Arithmetic Mean Method. Since the stations having gage readings as 530, 810 and 630 mm are located outside the basin, the same are not included in taking the arithmetic mean.

∴ Arithmetic mean

$$= \frac{1}{9}(580+710+690+860+840+550+690+610+790)$$

$$= 702.22 \text{ mm}$$

∴ Average rainfall

$$= 702.22 \text{ mm}$$

2. Thiessen Polygon Method. The polygons are drawn as shown in Fig. 5.8. Area of each polygon is determined by planimeter. The calculations are shown in the accompanying table.

Computations for Thiessen Polygon Method

Area of each polygon A (sq.mm)	*Observed precipitation P (mm)*	*Product P × A*
308	580	178 640
38	630	23 940
153	710	108 630
194	690	133 860

Contd.

Contd.

235	860	202 100
369	840	309 960
484	550	266 200
40	530	21 200
374	690	258 060
194	610	118 340
417	790	329 430
$\Sigma A = 2806$		$\Sigma(PA) = 1950\ 360$

∴ Average rainfall

$$= \frac{\Sigma(PA)}{\Sigma A} = \frac{1950360}{2806} = 695.07 \text{ mm}$$

Note. Since the polygon enclosing the station having gage reading 81 cm lies outside the basin, the same is not included.

3. Isohyetal Method. The isohyets are drawn as shown in Fig. 5.8 (a). Area between the adjacent isohyets are determined by planimeter. The calculations are shown in the following Table.

Isohyet (*mm*)	*Area enclosed between adjacent isohyets* *A (sq. mm)*	*Average precipitation* $\frac{1}{2}(P_1 + P_2)(mm)$	*Product* $A(P_1 + P_2)/2$
860			
850	65	855	55 575
800	318	825	262 350
750	415	775	321 625
700	415	725	300 875
650	490	675	330 750
600	385	625	240 625
550	430	575	247 250
500	288	525	151 200
	$\Sigma A = 2806$		$\Sigma A\ (P_1+P_2)/2 = 1910\ 250$

Average rainfall

$$= \frac{\Sigma A(P_1 + P_2)/2}{\Sigma A} = \frac{1910250}{2806} = 680.77 \text{ mm}$$

Illustrative Example 5.2 *A precipitation station X was inoperative for some time during which a storm occurred. At three stations A, B and C surrounding X the total precipitation recorded during this storm are 75, 58 and 47 mm respectively. The normal annual precipitation amounts at stations X, A, B, and C are respectively 757, 826, 618 and 482 mm. Estimate the storm precipitation for station X.*

Solution

If N_x, N_A, N_B and N_c are the average annual precipitation amounts at *X*, *A*, *B* and *C* and if P_A, P_b and P_c are the precipitations recorded during the storm at the stations *A*, *B* and *C* respectively, then the storm precipitation at station *X* is given by

$$P_x = \frac{1}{3}\left[\frac{N_x}{N_A}\times P_A + \frac{N_x}{N_B}\times P_B + \frac{N_x}{N_C}\times P_C\right]$$

$$= \frac{1}{3}\left[\frac{757}{826}\times 75 + \frac{757}{618}\times 58 + \frac{757}{482}\times 47\right]$$

$$= 71.2 \text{ mm}$$

Illustrative Example 5.3 *Hourly rainfalls of 25, 60 and 35 mm occur over a 20 ha area consisting 4 ha of ϕ = 50 mm/hr, 10 ha of ϕ = 30 mm/hr and 6 ha of ϕ = 10 mm/hr. Derive hourly values of rainfall excess (or net rain).*

Solution

First hour: (P = 25 mm)

$$P_{\text{excess}} = \frac{4(0)+10(0)+6(25-10)}{20} = 4.5 \text{ mm}$$

Second hour: (P = 60 mm)

$$P_{\text{excess}} = \frac{4(60-50)+10(60-30)+6(60-10)}{20} = 32 \text{ mm}$$

Third hour: (P = 35 mm)

$$P_{\text{excess}} = \frac{4(0)+10(0)+6(35-10)}{20} = 7.5 \text{ mm}$$

∴ Total rainfall excess for the 3 hours storm = 44 mm.

Illustrative Example 5.4 *The normal annual rainfall depths recorded at five rain gage stations A, B,C,D, and E are 910, 1070, 1410, 810 and 590 mm respectively. Determine the optimum number of rain gage stations to be established in the drainage basin if it is desired to limit the error in the mean value of rainfall to 10%. What is the percentage accuracy of the existing network in the estimation of the average depth of rainfall over the basin? If the location of the five rain gage stations and the areas of the different zones caused by the isohyets are as given in the following table, indicate the distribution of the additional rain gage stations required if any.*

Zone	*I*	*II*	*III*	*IV*	*V*	*VI*	*Total*
Area (km²)	*415*	*905*	*2855*	*1755*	*725*	*555*	*7210*
Rain gage stations	*1*	*1*	*1*	*1*	*1*	*–*	*5*

Solution

From Eq. (5.1) the optimum number of rain gage stations N to be established in a given drainage basin is given as

$$N = \left(\frac{C_v}{p}\right)^2$$

where C_v = coefficient of variation of the rainfall of the existing rain gage stations, and
p = desired degree of percentage error in the mean value of rainfall.

The value of C_v is calculated as imentioned below.

Station	*Normal Annual Rain x mm*	*Difference* $(x - \bar{x})$	$(x - \bar{x})^2$	*Staistical Parameters* $\bar{x}, \sigma, C_v$
A	910	– 48	2304	$\bar{x} = \frac{4790}{5} = 958$ mm
B	1070	112	12544	
C	1410	452	204304	$\sigma = \sqrt{\frac{\Sigma(x-\bar{x})^2}{n-1}}$
D	810	–148	21904	$= \sqrt{\frac{376480}{(5-1)}} = 307$
E	590	– 368	135424	$C_v = \frac{\sigma}{\bar{x}} = \frac{307}{958} \times 100 = 32\%$
$n = 5$	$\Sigma x = 4790$		$\Sigma(x-\bar{x})^2 = 376480$	

$\bar{x}$ = Arithmetic mean and σ = standard deviation. Given p = 10%

By substituting the values in the above expression, we get

$$N = \left(\frac{C_v}{p}\right)^2 = \left(\frac{32}{10}\right)^2 = 10.24 \simeq 11$$

∴ Additional rain gage stations to be established

$$= (N - n) = (11 - 5) = 6$$

The percentage error p in the estimation of average depth of rainfall in the existing network of raingage stations is given by Eq. (5.1) as

$$p = \frac{C_v}{\sqrt{N}}$$

Puttting $N = n = 5$, we get

$$p = \frac{32}{\sqrt{5}} = 14.3\%$$

∴ Percentage accuracy = (100 – 14.3) = 85.7%

The additional six rain gage stations have to be distributed in proportion to the areas between the isohyets as shown in the following Table:

Zone	*I*	*II*	*III*	*IV*	*V*	*VI*	*Total*
Area km^2	415	905	2855	1755	725	555	7210
Area as decimal	0.06	0.13	0.40	0.24	0.10	0.08	1.01
$N \times$ area in decimal ($N = 11$)	0.66	1.43	4.4	2.64	1.10	0.88	
Rounded as	1	1	4	3	1	1	11
Existing rain gages	1	1	1	1	1	—	5
Additional rain gages	—	—	3	2	—	1	6

Illustrative Example 5.5 *The annual rainfall at rain gage station X and the average annual rainfall at 15 surrounding rain gage stations are given below. Check the consistency of the record at the rain gage station X and determine the year in which a change in the rainfall regime has occurred. Also adjust the records for the change in regime and determine the average annual rainfall (a.a.r) for the period 1960 – 1978 for the changed regime.*

Year	*Annual rainfall mm*	
	Station X	*15 -Station average*
1960	*315*	*232*
1961	*379*	*350*
1962	*435*	*301*
1963	*323*	*276*
1964	*273*	*255*
1965	*321*	*280*
1966	*495*	*363*
1967	*282*	*185*
1968	*247*	*250*
1969	*219*	*241*
1970	*283*	*340*
1971	*175*	*236*
1972	*221*	*362*
1973	*285*	*310*
1974	*243*	*237*
1975	*271*	*234*
1976	*217*	*237*
1977	*296*	*333*
1978	*287*	*271*

Solution

The cumulative values of rainfall at station X and the cumulative values of average rainfall for a group of 15 stations are computed in the following Table.

Year	*Annual rainfall mm*	
	Station X	*15 -Station average*
1960	315	232
1961	694	582
1962	1129	883
1963	1452	1159
1964	1725	1414
1965	2046	1694
1966	2541	2057
1967	2823	2242
1968	3070	2492
1969	3289	2733
1970	3572	3073
1971	3747	3309
1972	3968	3671
1973	4253	3981
1974	4496	4218
1975	4767	4452
1976	4984	4689
1977	5280	5022
1978	5567	5293

These values of cumulative rainfalls are plotted on the graph sheet. It can be seen that there is a distinct change in slope in the year 1967. To make the records prior to 1967 comparable with those after a change in regime has occurred, the earlier records have to be adjusted by multiplying by the ratio of slopes (M_c / M_a). As obtained from the figure M_c = 0.9 and M_a = 1.25.

$$\therefore \qquad \frac{M_c}{M_a} = \frac{0.9}{1.25}$$

Cumulative rainfall from 1967 to 1978

$$= (5567 - 2541) = 3026 \text{ mm}$$

Cumulative rainfall from 1960 to 1966 adjusted for the changed regime

$$= 2541 \times \frac{0.9}{1.25} = 1829.52 \text{ mm}$$

Cumulative rainfall from 1960 to 1978 for the current regime

$$= (3026 + 1829.52)$$

$$= 4855.52 \text{ mm}$$

Average annual rainfall adjusted for the current regime

$$= \frac{4855.52}{19} = 255.55 \text{ mm}$$

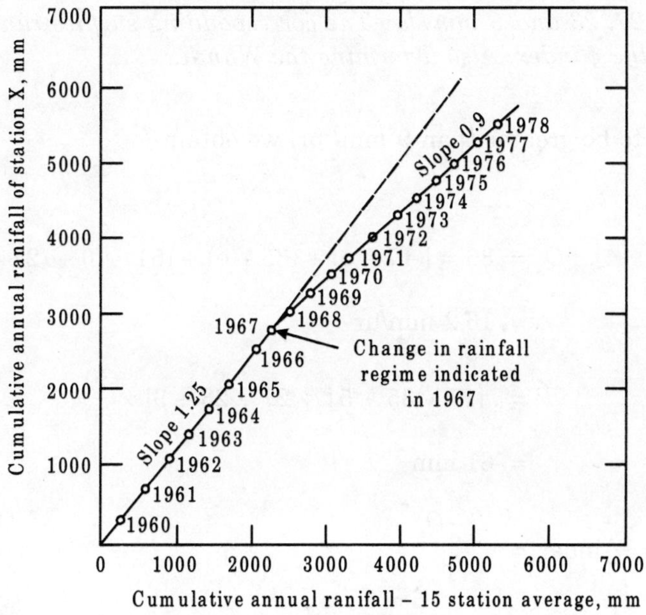

Fig. Ex. 5.5

Illustrative Example 5.6 *The rates of rainfall for successive 30 min. period of 210 min. storm are: 35, 40, 120, 85, 45, 45, and 30 mm/hr. Assuming ϕ-index of 35 mm/hr, find the net rainfall in mm, the total rainfall and the volume of W-index.*

Solution

The net rainfall P_{net} or surface runoff Q_s is given by

$$P_{net} = Q_s = (\Sigma\, (i–Q)\, t$$

$$= [(40 - 35) + (120 - 35) + (85 - 35) + (45 - 35) + (45 - 35)] \times \frac{30}{60}$$

$$= 80 \text{ mm}$$

Total rainfall $$P = [35 + 40 + 120 + 85 + 45 + 45 + 30] \times \frac{30}{60}$$

$$= 200 \text{ mm}$$

W-index $$= \frac{P - Q_s}{t_f}$$

$$= \frac{(200-80)}{\left(\frac{210}{60}\right)} = 34.3 \text{ mm/hour.}$$

Illustrative Example 5.7 *The rate of rainfall for the successive 30 min period of a 3 hour storm are: 17, 35, 51, 27, 23 and 9 mm/hr. The corresponding surface runoff is estimated to be 36 mm/hr. Establish the ϕ-index. Also determine the W-index.*

Solution

Assuming ϕ-index to be greater than 9 mm/ hr, we obtain

Surface runoff

$$Q_s = 36 = [(17-\phi) + (35-\phi) + (51-\phi) + (27-\phi) + (23-\phi)] \times \frac{30}{60}$$

$$\therefore \quad \phi = 16.2 \text{ mm/hr}$$

Total rainfall

$$P = [17 + 35 + 51 + 27 + 23 + 9] \times \frac{30}{60}$$

$$= 81 \text{ mm}$$

$$\text{W-index} = \frac{P - Q_s}{t_f}$$

$$= \frac{(81-36)}{3} = 15 \text{ mm/hr}$$

Illustriative Example 5.8 *The peak values of the flood observed on a river for each of the 15 years from 1951 to 1965 are as tabulated below:*

Year	*Peak flood flow in cumec*	*Year*	*Peak flood flow in cumec*
1951	*2750*	*1959*	*6700*
1952	*4600*	*1960*	*5400*
1953	*6000*	*1961*	*4300*
1954	*3300*	*1962*	*3600*
1955	*2300*	*1963*	*4200*
1956	*4900*	*1964*	*8000*
1957	*3900*	*1965*	*4000*
1958	*3000*		

Estimate the magnitude of flood having a recurrence interval of 100 years by Gumbel's method. The values of frequency factor K_T for Gumbel's method are as follows:

Sample size	*Value of K_T (Frequency factor) for T years*		
	75	*100*	*1000*
15	*3.721*	*4.005*	*6.265*
20	*3.563*	*3.836*	*6.006*

Also verify whether the Gumbel extreme value distribution fit the recorded values.

Solution

The given data is arranged in descending order and the values of T, $\bar{Q}$, $(Q - \bar{Q})$ and $(Q - \bar{Q})^2$ are calculated as shown in the following Table.

Year	*Flood peak*	*Rank*	$T = \frac{n+1}{m}$	$(Q - \bar{Q})$	$(Q - \bar{Q})^2$
1964	8000	1	16	3536.7	125.082×10^5
1959	6700	2	8	2236.7	50.028×10^5
1953	6000	3	5.33	1536.7	23.615×10^5
1960	5400	4	4	936.7	8.774×10^5
1956	4900	5	3.2	436.7	1.907×10
1952	4600	6	2.67	136.7	0.187×10^5
1961	4300	7	2.28	–163.3	0.267×10^5
1963	4200	8	2	–263.3	0.693×10^5
1965	4000	9	1.78	–463.3	2.146×10^5
1957	3900	10	1.6	–563.3	3.173×10^5
1962	3600	11	1.45	–863.3	7.453×10^5
1954	3300	12	1.33	–1163.3	13.533×10^5
1958	3000	13	1.23	–1463.3	21.413×10^5
1951	2750	14	1.143	–1713.3	29.354×10^5
1955	2300	15	1.067	–2163.3	46.799×10^5
	$\Sigma Q = 66950$				$\Sigma (Q - \bar{Q})^2 = 334.424 \times 10^5$
	$\bar{Q} = 4463.3$				

$$\therefore \quad \sigma = \sqrt{\frac{\Sigma(Q - \bar{Q})^2}{(n-1)}}$$

$$= \sqrt{\frac{334.424 \times 10^5}{(15-1)}} = 15.456 \times 10^2$$

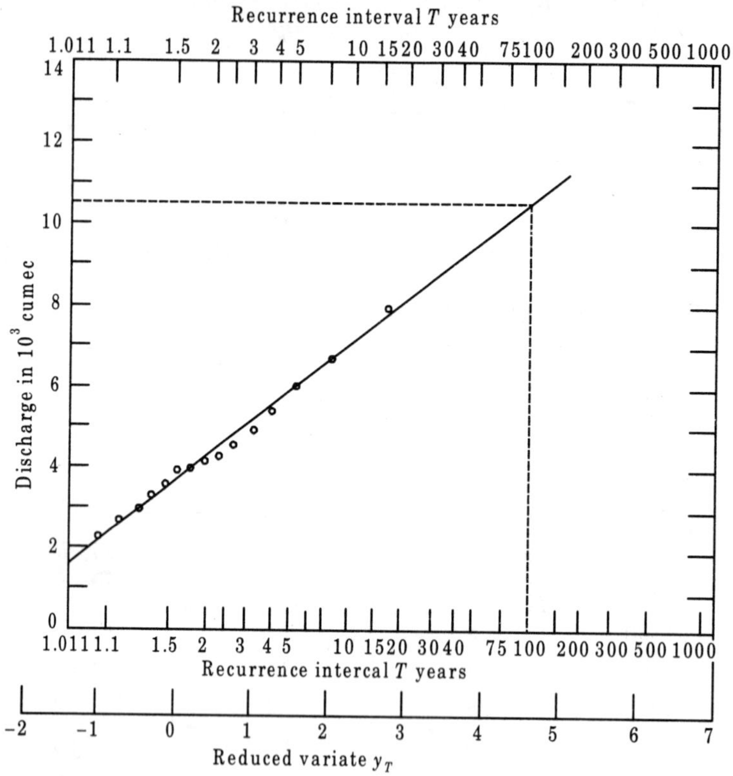

Fig. Ex. 5.8

Thus, by substituting the values in Eq. (5.38 a), for T = 100 years, we get

$$Q_{100} = \bar{Q} + K_T\,\sigma$$
$$= 4463.3 + (4.005 \times 15.456 \times 10^2)$$
$$= 10{,}653.43 \text{ cumec say } 10{,}650 \text{ cumec}$$

Method for computing the value of K_T

The value of K_T as given may be calculated as described below.

From Eq. (5.37), we have

$$y_T = -\left[\log_e . \log_e\left(\frac{T}{T-1}\right)\right]$$

For T = 100 years

$$y_T = -\left[\log_e . \log_e\left(\frac{100}{99}\right)\right] = 4.60015$$

For n = 15, from Tables 5.3 and 5.4, we obtain

$$\bar{y}_n = 0.5128 \text{ and } s_n = 1.0206$$

Thus, by substituting the values in Eq. (5.40), we get

$$K_T = \frac{4.60015 - 0.5128}{1.0206} = 4.005$$

To verify whether the given data follows the Gumbel's distribution or not it is plotted on the Gambel's probability paper as shown in the Fig. Ex. 5.8. It is seen that the given data follows the Gumbel's distribution and hence by extrapolation of the straight line drawn through the plotted points gives the magnitude of flood having a recurrence interval of 100 years as Q_{100} = 10,500 cumec.

Illustrative Example 5.9 *A large sample of peak floods data was available for a river. Flood frequency computations using Gumbel's method, yield the following results:*

Return Period	*Peak Flood*
T (year)	*(m^3/s)*
50	*30 800*
100	*36 300*

Estimate the magnitude of a flood for this river with a return period of 200 years.

Solution

From Eq. (5.38 a), we have

$$Q_T = \bar{Q} + K\sigma$$

Thus, by substitution, we get

$$Q_{100} = 36300 = \bar{Q} + K_{100}\sigma \quad \text{...(i)}$$

$$Q_{50} = 30800 = \bar{Q} + K_{50}\sigma \quad \text{...(ii)}$$

Subtracting Eq. (ii) from Eq. (i), we get

$$(K_{100} - K_{50})\,\sigma = 5500$$

But from Eq. (5.40), we have

$$K_T = \frac{y_T - \bar{y}_n}{s_n}$$

where $\bar{y}_n$ and s_n are constants for the given data series.

$$\therefore \quad (y_{100} - y_{50})\frac{\sigma}{s_n} = 5500$$

Again from Eq. (5.37), we have

$$y_T = -\left[\log_e . \log_e\left(\frac{T}{T-1}\right)\right]$$

Thus, $$y_{100} = -\left[\log_e . \log_e\left(\frac{100}{99}\right)\right] = 4.60015$$

and $$y_{50} = -\left[\log_e . \log_e\left(\frac{50}{49}\right)\right] = 3.90194$$

$$\therefore \quad \frac{\sigma}{s_n} = \frac{5500}{(4.60015 - 3.90194)} = 7877$$

For $T = 200$ years, from Eq. (5.37), we have

$$y_{200} = -\left[\log_e . \log_e\left(\frac{200}{199}\right)\right] = 5.29582$$

Thus,

$$(y_{200} - y_{100})\frac{\sigma}{s_n} = Q_{200} - Q_{100}$$

or $$(5.29582 - 4.60015) \times 7877 = Q_{200} - 36300$$

$$\therefore \quad Q_{200} = 41779.8 \text{ m}^3/\text{s} \simeq 41800 \text{ m}^3\mu$$

Illustrative Example 5.10 *For a data of maximum recorded flood of a river, the mean and standard deviation are 4200 m³/s and 1705 m³/s respectively. Using Gumbel's extreme value distribution, estimate the return period of a design flood of 9550 m³/s. Assume an infinite sample size.*

Solution

From Eq. (5.38 a), we have

$$Q_T = \bar{Q} + K_T\sigma$$

$$Q_T = 9550 \text{ m}^3/\text{s}; \; Q = 4200 \text{ m}^3/\text{s}; \text{ and } \sigma = 1705 \text{ m}^3/\text{s}$$

Thus, by substitution, we get

$$9550 = 4200 + K_T \times 1705$$

$$\therefore \quad K_T = 3.1378$$

Again from Eq. (5.40), we have

$$K_T = \frac{y_T - \bar{y}_n}{s_n}$$

For infinite sample size (i.e., $n \to \infty$)

$$\bar{y}_n = 0.577 \text{ and } s_n = 1.2825$$

Thus, by substitution, we get

$$3.1378 = \frac{y_T - 0.577}{1.2825}$$

$$\therefore \quad y_T = 4.6012$$

From Eq. (5.37 a), we have

$$y_T = -\left[0.834 + 2.303 \log_{10} \log_{10}\left(\frac{T}{T-1}\right)\right]$$

or
$$4.6012 = -\left[0.834 + 2.303 \log_{10} \log_{10}\left(\frac{T}{T-1}\right)\right]$$

or
$$\frac{T}{T-1} = 1.01$$

$\therefore$
$$T = 101 \text{ years}$$

Illustrative Example 5.11 *The observed flows from a storm of 4 hours duration on a stream are as given below:*

Date	*Time*	*Flow (cumec)*	*Date*	*Time*	*Flow (cumec)*
August 15	*0.0 A.M*	*15.0*	*August 17*	*0.0 A.M*	*52.3*
	6.0 A.M.	*117.8*		*6.0 A.M.*	*40.4*
	12.0 Noon	*260.7*		*12.0 Noon*	*30.3*
	6.0 P.M.	*195.3*		*6.0 P.M.*	*22.7*
August 16	*0.0 A.M.*	*152.0*	*August 18*	*0.0 A.M.*	*19.2*
	6.0 A.M	*115.7*		*6.0 A.M.*	*18.6*
	12.0 Noon	*90.3*		*12.0 Noon*	*15.0*
	6.0 P.M.	*70.5*			

Assuming a constant base flow of 15 cumec derive 4 hour unit hydrograph. The area of the drainage basin of the stream is 425 sq. km.

Solution

The solution is given in the following table.

Date and Time	*Ordinates of storm hydrograph (cumec)*	*Base flow (cumec)*	*Ordinates of direct runoff (cumec) (2) – (3)*	*Ordinates of unit hydrograph (4) ÷ 5.04*
(1)	*(2)*	*(3)*	*(4)*	*(5)*
August 15, 0.0 A.M	15.0	15.0	0	0
6.0 A. M.	117.8	15.0	102.8	20.40
12.0 Noon	260.7	15.0	245.7	48.75
6.0 P.M	195.3	15.0	180.3	35.77
August 16, 0.0 A.M.	152.0	15.0	137.0	27.18
6.0 A.M.	115.7	15.0	100.7	19.98
12.0 Noon	90.3	15.0	75.3	14.94
6.0 P.M.	70.5	15.0	55.5	11.01

Contd.

Contd.

August 17, 0.0 A.M	52.3	15.0	37.3	7.40
6.0 A.M.	40.4	15.0	25.4	5.04
12.0 Noon	30.3	15.0	15.3	3.04
6.0 P.M.	22.7	15.0	7.7	1.53
August 18, 0.0 A.M.	19.2	15.0	4.2	0.83
6.0 A.M.	18.6	15.0	3.6	0.71
12.0 Noon	15.0	15.0	0	0

Sum of direct runoff ordinate

$$\Sigma O = 990.8$$

From Eq. (5.35), direct runoff depth

$$= \frac{0.36 \times \Sigma O \times t}{A}$$

$$= \frac{0.36 \times 990.8 \times 6}{425}$$

$$= 5.04 \text{ cm.}$$

Illustrative Example 5.12 *Given below are the monthly rainfall, P, and the corresponding runoff, R, values, for a period of 10 months for a catchment. Develop a correlation between R and P :*

Month No.	*P, cm*	*R, cm*
1	*4*	*0.2*
2	*22*	*7.1*
3	*28*	*10.9*
4	*15*	*4.0*
5	*12*	*3.0*
6	*8*	*1.3*
7	*4*	*0.4*
8	*15*	*2.0*
10	*5*	*0.3*

Solution For the given data

$$\Sigma P = 123$$

$$\Sigma R = 33.3$$

$$N = 10$$

$$\Sigma P^2 = 2083$$

$$(\Sigma P^2) = 15129$$

$$\Sigma PR = 653.2$$

Let the correlation between R and P be

$$R = aP + b$$

in which

$$a = \frac{N(\Sigma PR) - (\Sigma P)(\Sigma R)}{N(\Sigma P^2) - (\Sigma P)^2}$$

and

$$b = \frac{\Sigma R - a\Sigma P}{N}$$

Thus, by substitution, we get

$$a = \frac{10 \times 653.2 - 123 \times 333}{10 \times 2083 - 15129} = 0.43$$

$$b = \frac{33.3 - 0.43 \times 123}{10} = -1.96$$

$$\therefore \quad R = 0.43P - 1.96$$

Illustrative Example 5.13 *In order to compute the flood discharge in a stream by the slope-area method, the following data have been obtained :*

	u/s section	*Middle section*	*d/s section*
Area	*108.6 m²*	*1031. m²*	*99.8 m²*
Wetted perimeter	*65.3 m*	*60.7 m*	*59.4 m*
Gauge reading	*316.8 m*	—	*316.55 m*

Determine the flood discharge assuming Manning's n = 0.029 and length between d/s and u/s sections as 250 m.

Solution Let A_1 and A_2 be the areas at the u/s and d/s sections respectively and A_3 be the areas at the middle section, then

$$\text{Average area } A = \frac{A_1 + 2A_3 + A_2}{4}$$

$$= \frac{108.6 + 2 \times 103.1 + 99.8}{4}$$

$$= 103.65 \text{ m}^2$$

Let P_1 and P_2 be the wetted perimeters at the u/s and d/s sections respectively and P_3 be the wetted perimeter at the middle section, then

Average wetted perimeter

$$P = \frac{P_1 + 2P_2 + P_2}{4}$$

$$= \frac{65.3 + 2 \times 60.7 + 59.4}{4}$$

$$= 61.53 \text{ m}$$

Hydraulic radius

$$R = \frac{A}{P}$$

$$= \frac{103.65}{61.53}$$

$$= 1.685 \text{ m}$$

According to Manning's formula

$$\text{Discharge } Q = \frac{1}{n} AR^{2/3} S^{1/2}$$

Thus, by substitution, we get

$$Q = \frac{1}{0.029} \times 103.65 \times (1.685)^{2/3} S^{1/2}$$

or

$$Q = 5060.98 \sqrt{S}$$

Neglecting the velocity head,

$$\text{Slope of TEL,S} = \frac{h_1 - h_2}{l}$$

$$= \frac{316.8 - 316.55}{250}$$

$$= \frac{1}{1000}$$

Thus, approximate value of the flood discharge

$$Q = 5060.98 \times \sqrt{\frac{1}{1000}} = 160.04 \text{ m}^3/\text{s}$$

For obtaining the correct value of the flood discharge the following trails are made.

Trial 1 :

$$\text{Taking } Q = 160.04 \text{ m}^3/\text{s}$$

$$V_1 = \frac{Q}{A_1} = \frac{160.04}{108.6} = 1.474 \text{ m/s}$$

$$V_2 = \frac{Q}{A_2} = \frac{160.04}{99.8} = 1.604 \text{ m/s}$$

Velocity head at the u/s section

$$\frac{V_1^2}{2g} = \frac{(1.474)^2}{2\times 9.81} = 0.1107 \text{ m}$$

Velocity head at the d/s section

$$\frac{V_2^2}{2g} = \frac{(1.604)^2}{2\times 9.81} = 0.1311 \text{ m}$$

Considering the velocity head,

$$\text{Slope of TEL's} = \frac{(h_1 - h_2) + \left(\frac{V_1^2}{2g} - \frac{V_2^2}{2g}\right)}{l}$$

$$= \frac{(316.8 - 316.55) + (0.1107 - 0.1311)}{250}$$

$$= 0.0009184$$

$$\therefore \quad Q = 5060.98 \times \sqrt{0.0009184}$$

$$= 153.37 \text{ m}^3\text{/s}$$

Trial 2 :

$$\text{Taking } Q = 153.37 \text{ m}^3\text{/s}$$

$$V_1 = \frac{Q}{A_1} = \frac{153.37}{108.6} = 1.412 \text{ m/s}$$

$$V_2 = \frac{Q}{A_2} = \frac{153.37}{99.8} = 1.537 \text{ m/s}$$

$$\frac{V_1^2}{2g} = \frac{(1.412)^2}{2\times 9.81} = 0.1016 \text{ m}$$

$$\frac{V_2^2}{2g} = \frac{(1.537)^2}{2\times 9.81} = 0.1204 \text{ m}$$

$$S = \frac{(316.8 - 316.55) + (0.1016 - 0.1204)}{250}$$

$$= 0.000925$$

$$\therefore \quad Q = 5060.98 \times \sqrt{0.000925}$$

$$= 153.92 \text{ m}^3\text{/s}$$

Since the discharge obtained in Trial 2 is neraly the same as obtained in Trial 1, the flood discharge may be taken as 153.92 m³/s.

Illustrative Example 5.14 *The regression analysis of a 30 year flood data at a point on a river yielded sample mean* $\bar{x}$ = *1200 m³/s and standard deviation* σ = *650 m³/s. For what discharge would you design the structure at this point to provide 95% assurance that the structure would not fail in the next 50 years ? Use Gumbel's method. The values of mean and standard deviation of the reduced variate for n = 30 are 0.53622 and 1.11238 respectively.*

Solution

$$\text{Assurance} = 95\%$$

$$\therefore \quad \text{Risk } R = 100 - 95 = 5\% \text{ or } 0.05$$

$$\text{Life period } N = 50 \text{ years}$$

$$\text{Risk } R = 1 - \left(1 - \frac{1}{T}\right)^{50}$$

where T is the recurrence interval for the flood due to which the structure would not fail in the next 50 years.

Thus, by substitution, we get

$$0.05 = 1 - \left(1 - \frac{1}{T}\right)^{50}$$

or

$$T = 375 \text{ years}$$

From Eq. (5.37), we have

$$y_T = -\left[0.834 + 2.303 \log_{10} \cdot \log_{10}\left(\frac{T}{T-1}\right)\right]$$

or

$$y_T = -\left[0.834 + 2.303 \log_{10} \cdot \log\left(\frac{375}{374}\right)\right]$$

or

$$y_T = 5.9797$$

From Eq. (5.40), we have

$$\text{Frequency factor } K_T = \frac{y_T - \bar{y}_n}{S_n}$$

$$\bar{y}_n = 0.53622 \text{ and } S_n = 1.11238$$

Thus, by substitution, we get

$$K_T = \frac{5.9797 - 0.53622}{1.11238} = 4.8935$$

From Eq. (5.38 a) the design flood

$$Q_T = \bar{Q} + K_T \sigma$$

$$\bar{Q} = 1200 \text{ m}^3/\text{s and } \sigma = 650 \text{ m}^3/\text{s}$$

Thus, by substitution, we get

$$Q_T = 1200 + (4.8935 \times 650)$$
$$= 4380.775 \text{ m}^3/\text{s}.$$

Illustrative Example 5.15. *Find the ordinates of a flood hydrograph resulting from a storm with rainfalls of 2.5, 6.85 and 3.75 cm each during successive 3 hours. The ordinates of a 3-hours unit hydrograph are given below.*

Time (hrs)	*03*	*06*	*09*	*12*	*15*	*18*	*21*	*24*	*03*	*06*	*09*	*12*	*15*	*18*	*21*	*24*
Ordinates of unit hydrograph (cumec)	*0*	*115*	*370*	*510*	*395*	*315*	*252*	*231*	*172*	*127*	*96*	*64*	*43*	*25*	*12*	*0*

Assume an initial loss of 5 mm, infiltration index ϕ = 2.5 mm per hour and base flow of 12 cumec.

Solution

(i) Rainfall excess during the first three hours

$$= 2.5 - (0.25 \times 3) - 0.5 = 1.25 \text{ cm}$$

(ii) Rainfall excess during the second three hours

$$= 6.85 - (0.25 \times 3) = 6.10 \text{ cm}$$

(iii) Rainfall excess during the third three hours

$$= 3.75 - (0.25 \times 3) = 3.00 \text{ cm}$$

The runoff due to rainfall excess of 1.25 cm, 6.10 cm and 3.00 cm will commence at 03, 06 and 09 hours respectively. The computations are shown in the following Table.

Time (hours)	*Ordinate of 3-hour unit hydro-graph*	*Rainfall excess (cm)*	*Runoff from rainfall excess during successive durations of 3 hours (cumec)*			*Base flow (cumec)*	*Ordinate of flood hydro graph (cumec)*
			1.25	*6.10*	*3.00*		
(1)	*(2)*	*(3)*	*(4)*	*(5)*	*(6)*	*(7)*	*(8)*
03	0	1.25	0			12	12.00
06	115	6.10	143.75	0		12	155.75
09	370	3.00	462.50	701.50	0	12	1176.00
12	510		637.50	2257.00	345	12	3251.50
15	395		493.75	3111.00	1110	12	4726.75
18	315		393.75	2409.50	1530	12	4345.25
21	252		315.00	1921.50	1185	12	3433.50

Contd.

Contd.

24	231		288.75	1537.20	945	12	2782.95
03	172		215.00	1409.10	756	12	2392.10
06	127		158.75	1049.20	693	12	1912.95
09	96		120.00	774.70	516	12	1422.70
12	64		80.00	585.60	381	12	1058.80
15	43		53.75	390.40	288	12	744.15
18	25		31.25	262.30	192	12	497.55
21	12		15.00	152.50	129	12	308.50
24	0		0	73.20	75	12	160.20
				0	36	12	48.00
					0	12	12.00

Illustrative Example 5.16 *The stream flows due to three successive storms of 2.9, 4.9 and 3.9 cm of 6 hours duration each on a drainage basin are given below. The area of the drainage basin is 118.8 sq. km. Assuming a constant base flow of 20 cumec, derive a 6 hours unit hydrograph for the drainage basin. An average storm loss of 0.15 cm/hr may be assumed.*

Time (hr)	*0*	*3*	*6*	*9*	*12*	*15*	*18*	*21*	*24*	*27*	*30*	*33*
Flow (cumec)	*20*	*50*	*92*	*140*	*199*	*202*	*204*	*144*	*84.5*	*45.5*	*29*	*20*

Solution

Let the 6-hour unit hydrograph ordinates be u_0,u_1 u_2, u_3, u_4, u_5, u_6, and u_7 at 0, 3, 6, 9, 12, 15, 18 and 21 hours respectively.

The direct runoff ordinates (D.R.O.) due to the three successive storms (of 6 hours duration each are obtained by deducting the base flow of 20 cumec from the stream flows at the corresponding time interval and the same are as given below.

Time (hr).	0	3	6	9	12	15	18	21	24	27	30	33
Flow (cumec)	0	30	72	120	179	182	184	124	64.5	25.5	9	0

The net storm rains are obtained by deducting the average storm loss as

0 — 6 hr : $R_1 = 2.9 - 0.15 \times 6 = 2$ cm

6 — 12 hr : $R_2 = 4.9 - 0.15 \times 6 = 4$ cm

12 — 18 hr : $R_3 = 3.9 - 0.15 \times 6 = 3$ cm

As mentioned in seciton 5.17 the following equations may be obtained.

$$u_0 = 0 \quad \text{or} \quad u_0 = 0$$
$$2u_1 = 30 \quad \text{or} \quad u_1 = 15$$
$$2u_2 + 4u_0 = 72 \quad \text{or} \quad u_2 = 36$$
$$2u_3 + 4u_1 = 120 \quad \text{or} \quad u_3 = 30$$
$$2u_4 + 4u_2 + 3u_0 = 179 \quad \text{or} \quad u_4 = 17.5$$
$$2u_5 + 4u_3 + 3u_1 = 182 \quad \text{or} \quad u_5 = 8.5$$
$$2u_6 + 4u_4 + 3u_2 = 184 \quad \text{or} \quad u_6 = 3$$
$$2u_7 + 4u_5 + 3u_3 = 124 \quad \text{or} \quad u_7 = 0$$
$$\Sigma u = 110$$

As a check the following equations may also be considered

$$4u_6 + 3u_4 = 64.5 \qquad \text{or} \quad u_6 = 3$$
$$4u_7 + 3u_5 = 25.5 \qquad \text{or} \quad u_5 = 8.5$$
$$3u_6 = 9 \qquad \text{or} \quad u_6 = 3$$
$$3u_7 = 0 \qquad \text{or} \quad u_7 = 0$$

Therefore ordinates of the 6 hour unit hydrograph are as follows:

Time (hr)	0	3	6	9	12	15	18	21
Flow (cumec)	0	15	36	30	17.5	8.5	3	0

Another check for the ordinates of the unit hydrograph derived is that the area under the unit hydrograph should give a direct runoff of 1 cm.

$$\text{Direct runoff} = \frac{(\Sigma u)\times t}{A} = \frac{110\times 3\times 60\times 60}{118.8\times 10^6}$$
$$= 0.01 \text{ m} = 1 \text{ cm}$$

Illustrative Example 5.17 *The surface water from airport road side is drained to the longitudinal side drain from across one half of a bituminous pavement surface of total width 7.0 m, shoulder and adjoining land of width 8.0 m on one side of the drain. On the other side of the drain, water flows across from reserve land with average turf and 2% cross slope towards the side drain, the width of this strip of land being 25 m. The inlet time may be assumed to be 10 minutes for these conditions. The runoff coefficients of the pavement, shoulder and reserve land with turf are 0.8, 0.25 and 0.35 respectively. The length of the stretch of land parallel to the road from where the water is expected to flow to the side drain is 400 m. Estimate the quantity of runoff flowing in the drain assuming 10 year frequency. The side drain will pass through clayey soil with allowable velocity of flow as 1.33 m/s. Intensity-duration chart for 10 year frequency is:*

Duration (minutes)	*5*	*10*	*15*	*20*	*30*
Intensity (mm/hour)	*160*	*150*	*125*	*110*	*95*

Solution

Time of concentration is given by Eq. (3.12) as

$$t_c = t_0 + t_f$$

Inlet time $t_0 = 10$ minutes

Time of flow in the sewer

$$t_f = \frac{400}{1.33\times 60} = 5 \text{ minutes}$$

$\therefore$ $t_c = (10 + 5) = 15$ minutes

From the given intensity-duration chart for duration of rainfall equal to time of concentration i.e., 15 minutes the intensity of rainfall is obtained as 125 mm/hour.

Using rational formula (Eq. 5.21) we have

$$Q = \frac{CIA}{360}$$

$$CA = C_1A_1 + C_2A_2 + C_3A_3$$

$$C_1 A_1 = \left(0.8 \times \frac{7}{2} \times 400\right) \times \frac{1}{10000} = 0.112 \text{ ha}$$

$$C_2 A_2 = \left(0.25 \times \frac{8}{2} \times 400\right) \times \frac{1}{10000} = 0.040 \text{ ha}$$

$$C_3 A_3 = (0.35 \times 25 \times 400) \times \frac{1}{10000} = 0.350 \text{ ha}$$

$\therefore$ $$CA = (0.112 + 0.040 + 0.350) = 0.502 \text{ ha}$$

$$I = 125 \text{ mm/hour}$$

Thus, by substitution, we get

$$Q = \frac{0.502 \times 125}{360} = 0.1743 \text{ m}^3\text{/s}$$

$$= 174.3 \; l\text{/s.}$$

REVIEW QUESTIONS

5.1 Describe with the help of neat sketch the hydrologic cycle.

5.2 What do you understand by precipitation ? Write a brief note on mechanism of precipitation.

5.3 Write a note on different types of precipitation.

5.4 List different types of self-recording rain gages. Explain the working of any two of them with the help of neat sketches.

5.5 How will you determine the areal mean rainfall over a basin by (i) Arithmetic mean method, (ii) Thiessen polygon method, (iii) Isohyetal method. Support your answer with illustrative sketches.

5.6 The isohyets for annual rainfall over a catchment were drawn and the areas of strips between isohyets are obtained as below. Determine the average depth of annual precipitation over the area.

Isohyets (mm)	450–550	550–650	650–750	750–950	950–1150	1150–1250
Area (km^2)	1200	3000	2800	1000	900	600

[*Ans.* 723.68 mm]

5.7 What is runoff? Describe the various factors that affect runoff?

5.8 The following are the rates of rainfall for successive 20 minutes storm in mm/hour:

22 22 95 72 12.5 12.5 50

Taking the value of ϕ-index of 31 mm/hour, find out the net runoff in mm and the total rainfall. [*Ans.* 41.33 mm; 95.33 mm]

5.9 Explain how the following parameters affect run-off:

(i) Shape of basin, (ii) Slope of basin, (*iii*) Meteorologic condition, (iv) Storage condition, (v) Vegetative condition.

5.10 List the various methods of determining runoff from a catchment. Give the various empirical formulae for determining the runoff indicating the area for which each of these formulae is applicable.

5.11 Describe the Strange's method for computing runoff from monsoon rainfall over a catchment. Bring out its limitations stating the factors which it ignores and which have influence on the runoff.

5.12 Give the various empirical formulae used for the estimation of peak flood discharges.

5.13 Compute the infiltration index ϕ for the following data:

Total runoff during storm $= 76.25 \times 10^6$ m^3

Estimated groundwater contribution in total runoff $= 1.25 \times 10^6$ m^3

Area of the basin $= 250$ km^2

Rainfall distribution

Hours	0–2	2–4	4–6	6–8	8–10	10–12	12–14	14–16
mm/hour	25	50	50	35	20	20	15	12.5

[*Ans.* 9.688 mm]

5.14 What is Unit Hydrograph? Explain clearly the basic postulates of Unit Hydrograph theory. Describe how can you obtain the Unit Hydrograph from a flood hydrograph resulting from a storm of certain duration.

5.15 Determine the Unit Hydrograph of an effective duration of 3 hours. The flood hydrograph due to an isolated storm of 3 hours duration has 6-hourly ordinates as under in cumec:

2.0 ; 4.0; 33.0; 116.0; 86.0; 64.0; 46.0; 32.0; 18.0; 15.0; 9.0; 7.0; 5.0; and 3.0

Assume a base flow of 2 cumec. The catchment area of the basin is 200 square kilometres.

[*Ans.* 0; 0.45; 6.97; 25.62; 18.88 13.93; 9.89; 6.74; 3.60; 2.92; 1.57; 1.12; 0.67; 0.22]

5.16 Compute and plot the ordinates of a storm hydrograph resulting from a 4 hour storm with rainfall of 40 mm. The ordinates of 4 hour unit hydrograph are given below:

Hours	03	06	09	12	15	18	21	24	03	06	09	12	15	18	21	24
Ordinates of Unit Hydrograph (Cumec)	0	115	370	505	395	315	255	240	180	135	100	70	45	25	15	0

Assume an initial loss of 5 mm and infiltration index 2.5 mm/hour. Base flow is 15 cumec.

[*Ans.* 15; 302.5; 941; 1277.5; 1002.5; 802.5; 652.5; 615; 465; 352.5; 265; 190; 127.5; 77.5; 52.5; 15]

5.17 Explain the procedure for derivation of Snyder's synthetic unit hydrograph for an ungaged catchment.

5.18 Ordinates of a 2 hour unit hydrograph are given as below. Using this derive the ordinates for a 6 hour unit hydrograph for the same catchment.

Time (h)	0	2	4	6	8	10	12	14	16	18	20	22
Discharge (cumec)	0	25	100	160	200	170	110	70	30	20	8	0

[*Ans.* 0; 12.5; 62.5; 130; 180; 185; 140; 90; 50; 25; 14; 4; 0]

5.19 Define (i) ϕ-index; (ii) W-index; (iii) W_{min}-index and (*iv*) f_{ave}-index.

5.20 The information available from a isohyetal map of 1100 sq.km basin is as follows:

Zone	*Area km²*	*Rain Gage Station*	*Normal Annual Rainfall, cm*
I	85	A	120
II	290	B	95
	C	96	
	D	60	
III	395	E	65
	F	70	
IV	230	G	45
V	65	H	21
VI	35	—	—

How many additional rain gage stations will be required if the desired limit of error in the mean value of rainfall is not to exceed 10 cm? Suggest how you propose to distribute these stations.

What factors will you consider in locating the additional rain gage stations between different isohyetals?

[*Ans.* Additional 3 rain gage stations will be required with one each in zones II, III and IV]

5.21 The following is the set of observed data for successive 15 minutes period of 105 minutes storm in a catchment.

Duration (minutes)	Rainfall (cm/hour)
15	2.0
30	2.0
45	8.0
60	7.0
75	1.25
90	1.25
105	4.5

If the value of ϕ-index is 3 cm/hr, estimate the net runoff, total rainfall and the value of W-index. [*Ans.* 2.625 cm/hr; 6.5 cm; 2.214 cm/hr]

5.22 Compute and draw the storm hyetograph and the intensity duration curve for the following storm (of a given frequency) on a drainage basin.

Duration (minutes)	Accumulated precipitation (cm)
0	—
30	5.0
60	7.5
90	8.5
120	9.0

5.23 The average storm rainfall values over a catchment in three successive 2 hour interval are 3, 5 and 2 cm, respectively. The ϕ-index for the catchment is taken as 0.2 cm/hour. The 2 hour unit hydrograph ordinates are given in the Table. Base flow can be taken as 7 m^3/sec at the beginning of the storm, linearly increasing to 9 m^3/sec at 2 hours after the direct runoff peak discharge and later linearly decreasing to 8 m^3/sec at 4 hours after the end of the direct runoff. Compute the resulting flood hydrograph.

Time (h)	0	2	4	6	8	10	12	14	16
Discharge, m^3/sec	0	10	40	70 (Peak)	50	20	10	6	0

CHAPTER

6 Growundwater Hydrology

6.1 INTRODUCTION

Groundwater hydrology may be defined as the science that deals with the occurrence, distribution and movement of water that has infiltrated and passed downward into the earth and occurs beneath the water table in soils and geologic formations that are fully saturated. This water is known as groundwater which constitutes one of the main parts of the hydrologic cycle and it is an important source of water supply throughout the world for irrigation, industries and municipal water supply.

Practically all groundwater originates from surface water. The process in which surface water infiltrates deep into the ground and it is stored as groundwater is known as groundwater recharge. Depending upon the sources of surface water the recharge of groundwater may be classified as *natural recharge* and *artificial recharge*. The main sources of natural recharge of groundwater include precipitation, water flowing in streams or rivers, lakes and other natural water bodies. The artificial recharge of groundwater occurs from excess irrigation, seepage from canals, leakage from reservoirs or tanks or water purposely applied on the ground surface to augment the groundwater storage. The water from any of these sources infiltrates into the ground and moves downward under the action of gravity until it is checked by an impermeable strata to move further down and is stored as groundwater. The groundwater reservoir consists of water held in voids within geologic formation. Water-bearing formations of the earth's crust act as conduits for transmission and as reservoirs for storage of groundwater. The groundwater held by these formations is not static but is slowly moving in a direction determined by the surrounding hydraulic condition.

The discharge of groundwater occurs when water emerges from underground and it may be either in a natural way or in an artificial way. The natural discharge of groundwater occurs as flow into streams or rivers and other surface water bodies such as lakes reservoirs and oceans; and flow to the surface as a spring. Groundwater emerging into surface streams helps in sustaining streamflow when surface runoff is low or non-existent. As indicated in Chapter 5, this flow of a stream is known as base flow. Groundwater near the surface may however, return directly to the atmosphere by evaporation from within the soil and by transpiration from vegetation. The artificial discharge of groundwater occurs through pumping of water from wells.

6.2 AQUIFER, AQUICLUDE, AQUIFUGE AND AQUITARD

Groundwater occurs in many types of geologic formations, out of which the most important ones are those known as aquifers.

An *aquifer* may be defined as a geologic formation that contains sufficient permeable material which permits storage as well as transmission of water through it under ordinary field conditions. Thus, an aquifer contains saturated material which will yield significant quantities of water to wells and springs. Unconsolidated sands and gravels are the typical examples of the formations which serve as aquifers. The other terms commonly used to represent an aquifer are *groundwater reservoir* and *water bearing formation*. In general aquifers have large area extension and these may be overlain or underlain by a confining bed of relatively impervious material.

An *aquiclude* may be defined as a geologic formation of relatively impermeable material which permits storage of water but it is not capable of transmitting water in sufficient quantity. Thus, an aquiclude contains saturated material which does not yield appreciable quantities of water to wells. Clay is an example of such a formation.

An *aquifuge* may be defined as a geologic formation of relatively impermeable material which neither contains nor transmits water. Solid granite belongs to this category of formation.

An *aquitard* may be defined as a geologic formation of poorly permeable material (or semipervious material) which permits storage of water but obstructs groundwater movement and does not yield water freely to wells. However, it may transmit appreciable quantity of water to or from adjacent aquifers. Further where this formation is sufficiently thick it may constitute an important groundwater storage zone. A formation of sandy clay belongs to this category.

6.3 POROSITY, SPECIFIC YIELD AND SPECIFIC RETENTION

Those portions of a rock or soil which are not occupied by solid mineral matter can be occupied by groundwater. These spaces are known as voids, interstices, pores or pore space. Since interstices serve as water conduits they are of fundamental importance to the study of groundwater. The interstices are characterized by their size, shape, irregularity and distribution. With respect to size, interstices may be classified as capillary, supercapillary and subcapillary. Capillary interstices are sufficiently small so that water is held in them by surface tension forces; supercapillary interstices are those larger than capillary ones ; and subcapillary interstices are so small that water is held primarily by adhesive forces. The interstices contained in a rock or soil are measured in terms of porosity which is defined below.

Porosity of a rock or soil is defined as the ratio of the volume of interstices to the total volume of the material and is usually expressed as percentage. Thus, if n is the porosity, then

$$n = \frac{V_i}{V} \times 100 \qquad \ldots(6.1)$$

where V_i is the volume of interstices and V is the total volume of the material. Porosity increases with the fineness of the material and with the sharpness of the soil particles, and is greatest when all the particles are of uniform size. On the other hand when the particles are well graded in size, then since the smaller ones fill the spaces between the larger ones, the porosity is considerably reduced. The porosity of soils varies widely, ranging from as low as 20 percent for graded sand to as high as 65 percent for very heavy clays.

The porosity provides a direct measure of the capacity of a formation to contain water. However, a high porosity of a formation does not indicate that it will yield large quantity of water. The quantity of water which may be obtained from a formation is expressed in terms of specific yield which is defined below.

Specific yield S_y of a soil or rock is the ratio of the volume of water that, after saturation, can be drained by gravity to its own volume. It is usually expressed as percentage. Thus,

$$S_y = \frac{w_y}{V} \times 100 \qquad ...(6.2)$$

where w_y is the volume of water drained and V is the bulk volume of the soil or rock.

The entire quantity of water contained by a saturated soil or rock cannot be drained by gravity because a portion of water is held against gravity by molecular and surface tension forces. The quantity of water retained by a soil or rock against gravity is expressed in terms of specific retention which is defined below.

Specific retention S_r of a soil or rock is the ratio of the volume of water it will retain after saturation against the force of gravity to its own volume. It is also usually expressed as percentage. Thus,

$$S_r = \frac{w_r}{V} \times 100 \qquad ...(6.3)$$

where w_r is the volume of the retained water and V is bulk volume of the soil or rock.

Since w_y and w_r constitute the total volume of water in a saturated material, it is apparent that porosity is equal to the sum of the specific yield and the specific retention. That is

$$n = S_y + S_r \qquad ...(6.4)$$

The value of specific yield depends on grain size, shape and distribution of interstices and compaction of stratum. It may however be noted that fine grained materials yield little water, whereas coarse grained materials permit a substantial release of water and hence the later serve as aquifers.

6.4 DIVISIONS OF SUBSURFACE WATER

As shown in Fig. 6.1 the subsurface water may be divided into the following two zones.

(a) zone of aeration

(b) zone of saturation.

(a) Zone of Aeration. The zone of aeration overlies the zone of saturation and extends upward to the ground surface as shown in Fig. 6.1. In the zone of aeration the interstices are partially occupied by water and partially by air. The water that occurs in the zone of aeration is usually termed as vadose water.* The zone of aeration is further subdivided into the following three zones.

* Vadose is derived from the Latin word vadosus meaning "shallow".

(i) Soil-water zone

(ii) Intermediate zone

(iii) Capillary zone.

(i) *Soil-water zone*. The soil-water zone extends from the ground surface down upto the major root zone. The soil in this zone becomes saturated either during irrigation or when rainfall occurs. However, the soil in this zone remains saturated only for a short duration after irrigation or rainfall, because the excess water drains through the soil under the influence of gravity. This excess water which cannot be retained by the soil is known as *gravitational water*. After the gravitational water drains out the remaining water is held by surface tension forces and is known as *capillary water*. The water of the soil-water zone is gradually depleted by evaporation from within the soil and by transpiration from vegetal growth on the ground surface and if it is not replenished the water content may be reduced to such an extent that only thin film of moisture known as *hygroscopic water* remains absorbed on the surface of the soil particles. The various soil moisture contents of soil-water zone as indicated above are shown in Fig. 3.2 (Chapter 3) where a detailed description of the soil-water is given.

(ii) *Intermediate zone*. The intermediate zone extends from the lower edge of soil-water zone to the upper limit of the capillary zone (Fig. 6.1). This zone usually contains non-moving vadose water (or pellicular water) which is held by molecular and surface tension forces in the form of hygroscopic and capillary water. Temporarily, this zone may also contain some excess water which however, moves downward as gravitational water. The thickness of this zone may vary from zero when the water table is high and is approaching the ground surface, to more than 100 m under deep water table conditions.

(iii) *Capillary zone*. The capillary zone extends from the water table upto the limit of capillary rises of water. The pore space may be considered to represent a capillary tube and hence just above the water table almost all pores contain capillary water which constitute this zone.

(b) Zone of Saturation. In the zone of saturation all the interstices are filled with water under hydrostatic pressure. The zone of saturation is bounded at the top by either a limiting surface of saturation called water table or an overlying impermeable strata and extends down to underlying impermeable strata (or bedrock). At the water table there is atmospheric pressure. Further if a well penetrates the zone of saturation with water table forming its upper surface then the static water level in the well stands at the same elevation as the water table. The saturation actually extends slightly above the water table due to capillary action, which constitutes the capillary zone of the zone of aeration. However, when the zone of saturation is bounded at the top by an impermeable strata, the water is in contact with the bottom of the impermeable strata and is under pressure and consequently there is no water table. When a well penetrates a zone of saturation of this type, the water will rise above the bottom of the confining strata.

The water occurring in the zone of saturation is known as groundwater. The thickness of the zone of saturation may vary from a few metres to hundreds of metres. The formation with the zone of saturation from which sufficient quantity of groundwater can be obtained may be termed as aquifer.

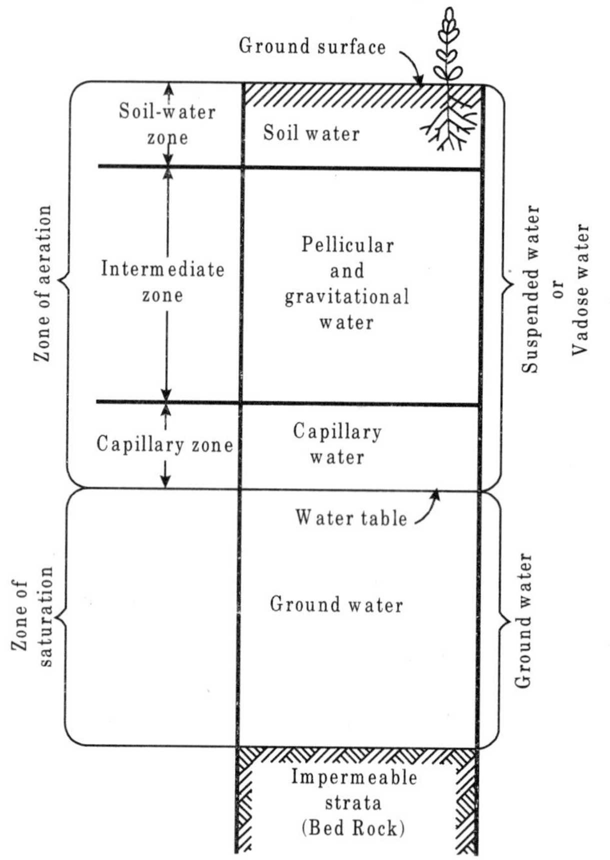

Fig. 6.1 *Divisions of subsurface water*

6.5 TYPES OF AQUIFERS

Aquifers are mainly of two types as indicated below.

(1) Unconfined aquifer

(2) Confined aquifer.

1. Unconfined Aquifer. An unconfined aquifer is the one in which water table forms the upper surface of the zone of saturation. It is also known as water-table aquifer; or phreatic aquifer; or non-artesian aquifer. The upper aquifer shown in Fig. 6.2 is an unconfined aquifer. The water table is not a stationary surface but it rises when more water enters the aquifer from natural or artificial recharge; and it drops when recharge is less and the previously stored water flows out towards springs, streams, wells and other points of groundwater discharge.

2. Confined Aquifer. A confined aquifer is the one in which groundwater is confined under pressure greater than atmospheric pressure by overlying relatively impermeable strata. It is also known as artesian aquifer or pressure aquifer. Water enters a confined aquifer in an area where the confining bed rises to the surface as shown in Fig. 6.2. However, where the confining bed ends underground, the aquifer becomes unconfined. The region supplying water to a confined aquifer is known as recharge area.

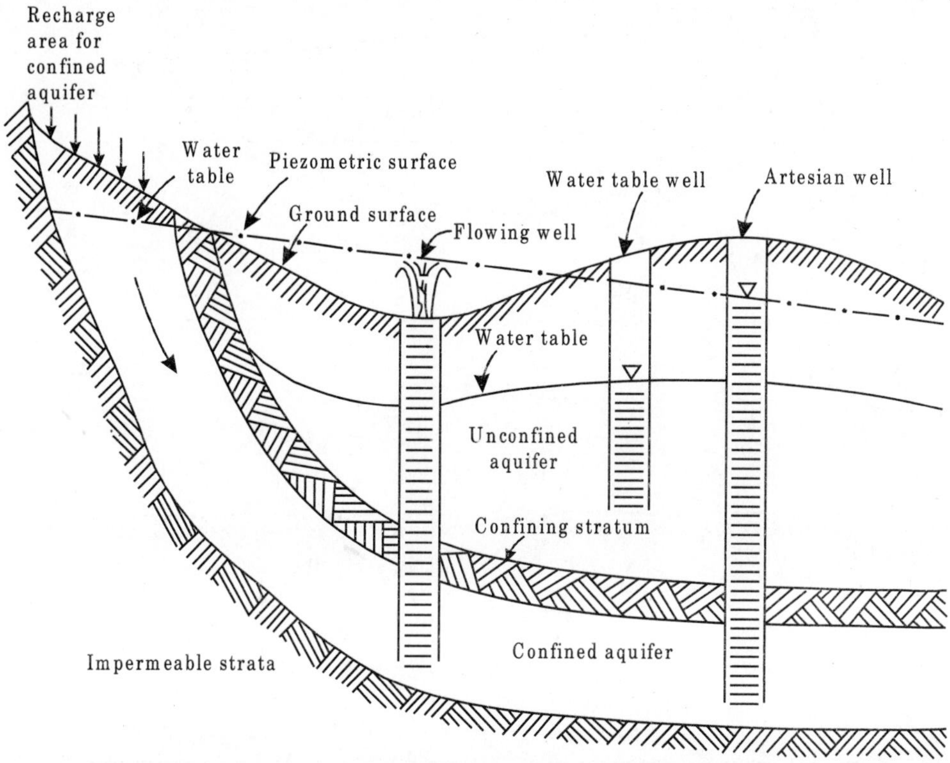

Fig. 6.2. *Unconfined and confined aquifers*

A confined aquifer is analogous to a pipeline. The static pressure at a point within a confined aquifer is equivalent to the elevation of the water table in the recharge area minus the loss of head through the aquifer to the point under consideration. In a well penetrating such an aquifer, the water will rise above the bottom of the confining strata upto the level of the local hydrostatic pressure head or artesian head, as shown by the artesian and flowing wells in Fig. 6.2. An imaginary surface representing the hydrostatic pressure head or artesian head in a confined aquifer is called the *piezometric surface*. A flowing well is developed when piezometric surface lies above the ground surface. The piezometric surface is analogous to water table in an unconfined aquifer. However, a confined aquifer becomes an unconfined aquifer when the piezometric surface falls below the bottom of the upper confining strata.

The rise and fall of water level in a well penetrating a confined aquifer result mainly from changes in pressure rather than changes in storage volumes. Hence confined aquifers show only small changes in storage volumes.

In addition to the above noted two main types of aquifers, there are some special types of aquifers which are as discussed below.

Perched Aquifer

Perched aquifer is a special type of an unconfined aquifer which is shown in Fig. 6.3(a). This occurs wherever a groundwater body is separated from the main groundwater body by a

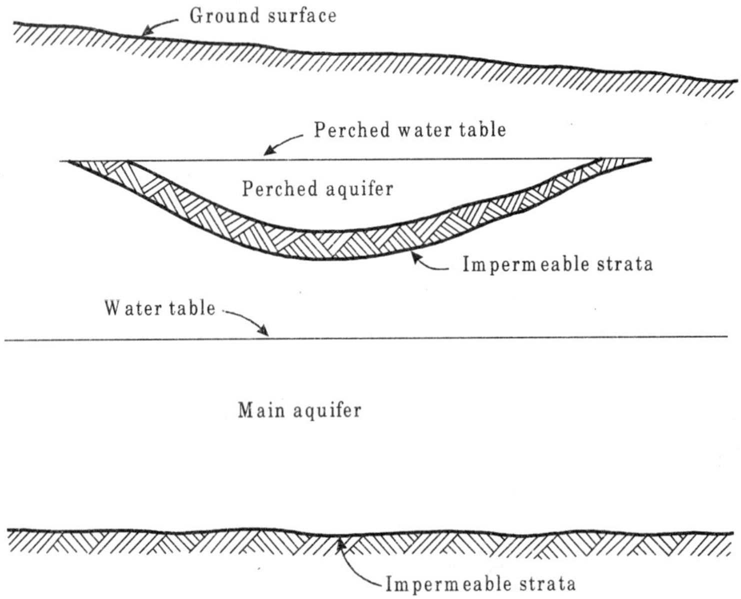

(a) Perched aquifer

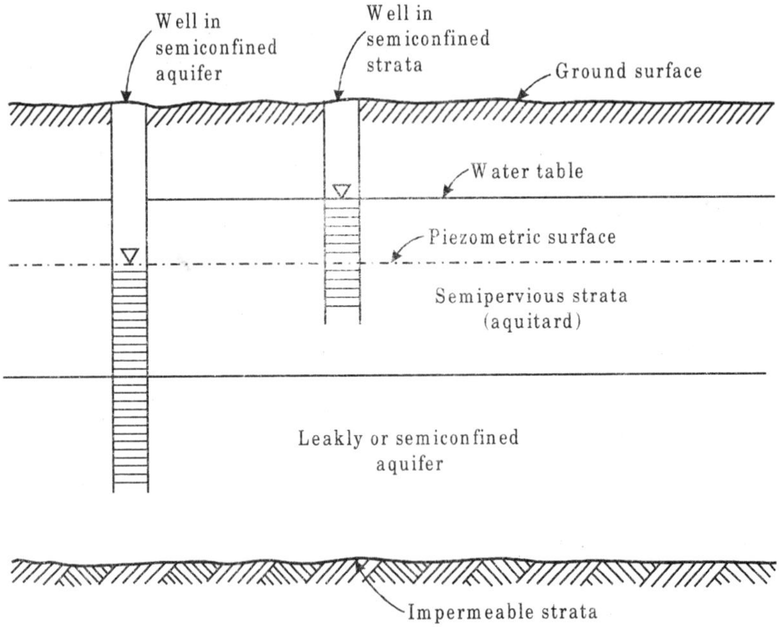

(b) Leaky or semiconfined aquifer

Fig. 6.3

relatively impermeable stratum of small areal extent present in the zone of aeration above the main body of groundwater. The water table of a perched aquifer is termed as perched water table. Wells penetrating perched aquifers yield only small quantities of water and that too only temporarily because usually these aquifers are not recharged.

Leaky Aquifer or Semiconfined Aquifer

Leaky or semiconfined aquifer is the one which is bounded at the top by a semipervious strata (or semiconfining layer) (Fig. 6.3(b)). The water level in a well penetrating a semiconfined aquifer will rise above the bottom of the confining strata and it represents the elevation of the piezometric surface at that point. Further in this case water is also stored in the top semipervious strata which forms an unconfined aquifer lying above the main semiconfined aquifer. The top semipervious strata may be designated as an aquitard. When water is pumped from a well penetrating a semiconfined aquifer the piezometric head is lowered which results in generating a vertical flow of water from the top semipervious strata into the semiconfined aquifer. Thus, in this case as the water is removed by pumping from the semiconfined aquifer it is replenished from the overlying semipervious strata or aquitard.

6.6 GROUNDWATER MOVEMENT—DARCY'S LAW

As stated earlier groundwater is not static but it is slowly moving through aquifers. The rate of flow of groundwater through aquifers, most of which are natural porous media, can be expressed by Darcy's law.

Henry Darcy a French hydraulic engineer investigated the flow of water through the horizontal beds of sand to be used for water filtration. On the basis of his experiments Darcy in 1856 indicated that for laminar flow conditions the velocity of flow of water through saturated porous media is proportional to the hydraulic gradient. It is universally known as Darcy's law which may be expressed as

$$V = ki \qquad \ldots(6.5)$$

where V = velocity of flow

k = coefficient of permeability

i = hydraulic gradient

$= \frac{\Delta h}{L}$; Δh is head loss in a length L of flow path

Further since

$$V = \frac{Q}{A}$$

$$Q = kiA \qquad \ldots(6.6)$$

where Q = rate of flow (or discharge)

A = total cross-sectional area of the porous medium perpendicular to the direction of flow.

The velocity V given by Eq. (6.5) is known as *Darcy velocity* or *specific discharge* or *discharge velocity*. It is an apparent velocity, representing the velocity at which water would move through an aquifer if the aquifer were an open channel. However, the cross-sectional area of flow through a porous medium is actually much smaller than the dimensions of the flow passage, because the flow is actually limited to the pore space only. The cross-sectional area of flow is equal to the product of the effective porosity of the porous material and the total cross-sectional area of the flow passage. The effective porosity is that portion of the pore space through which the flow occurs when it is saturated. The velocity at which the water is actually moving through a porous media is given by

$$V_a = \frac{Q}{nA} = \frac{ki}{n} = \frac{V}{n} \qquad \ldots(6.5\text{ a})$$

in which n is the effective porosity.

V_a is termed as *seepage velocity* or *average linear velocity* or *average interstitial velocity.* This indicates that with a porosity of say 33%, $V_a = 3V$.

Range of Validity of Darcy's Law

Darcy's law is valid only for laminar flow. Since Reynolds number serves as a criterion to distinguish between laminar and turbulent flow, the same may be employed to establish the limit upto which Darcy's law holds good. Reynolds number for this case is expressed as

$$\text{Re} = \frac{\rho V d}{\mu}$$

where ρ is the mass density of fluid;
V is the discharge velocity;
d is the mean diameter of the soil particles; and
μ is the dynamic viscosity of the fluid.

Experiments show that Darcy's law is valid for Re < 1 and does not depart much upto Re = 10, which thus represents an upper limit to the validity of Darcy's law.

Most of the natural groundwater flow occurs with Re < 1 and hence Darcy's law is applicable. However, Darcy's law is not applicable in aquifers containing large diameter solution openings, coarse gravels, rockfills and also in the immediate vicinity of wells, where the flow is no longer laminar due to steep hydraulic gradients and exhibit non-linear relationship between the velocity of flow and hydraulic gradient.

Determination of Direction of Groundwater Movement

With only three groundwater elevations known from wells, estimates of local ground water contours and direction of flow of groundwater through aquifer can be made. As shown in Fig. (6.4) let there be three wells with their water table elevations marked thereon. By linear interpolation the groundwater contours may be drawn as shown in Fig. (6.4). The direction of flow of the groundwater is perpendicular to the contours.

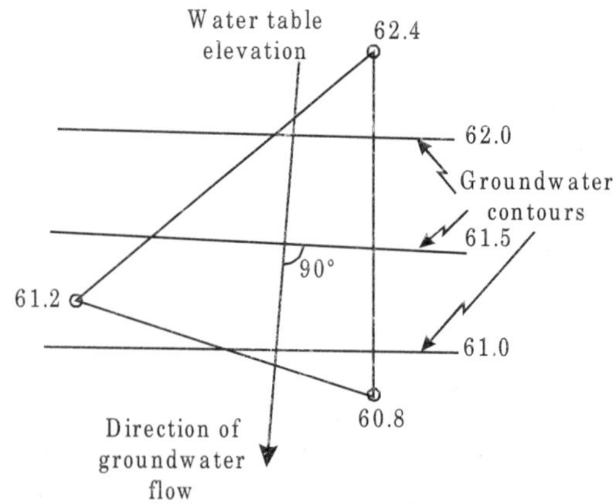

Fig. 6.4 *Groundwater contours and direction of groundwater flow*

Contour maps of groundwater levels together with flow lines are useful data for locating new wells. Convex contours indicate regions of groundwater recharge, while concave contours are associated with groundwater discharge.

Water recharged to or discharged from an aquifer represents a change in the storage volume within the aquifer. For unconfined aquifers this is simply expressed by the product of the volume of aquifer lying between the water table at the beginning and at the end of a period of time and average specific yield of the formation. In confined aquifers, however, water is under hydrostatic pressure which partially supports the weight of the overburden while the solid structure of the aquifer provides the remaining support. When water is pumped from a well penetrating the aquifer, the hydrostatic pressure is reduced and hence the aquifer load increases. A compression of the aquifer results which forces some water from it. In addition, lowering of the pressure causes a small expansion and subsequent release of water. The water yielding capacity of an aquifer can be expressed in terms of storage coefficient or storativity which is defined in the section 6.7.

6.7 STORAGE COEFFICIENT OR STORATIVITY, COEFFICIENT OF PERMEABILITY OR HYDRAULIC CONDUCTIVITY AND COEFFICIENT OF TRANSMISSIBILITY OR TRANSMISSIVITY

Storage Coefficient or Storativity

Storage coefficient or storativity is defined as the volume of water that an aquifer releases from or takes into storage per unit surface area of aquifer per unit change in the component of head normal to that surface. The storage coefficient is a dimensionless quantity because it represents the volume of water per unit volume of aquifer.

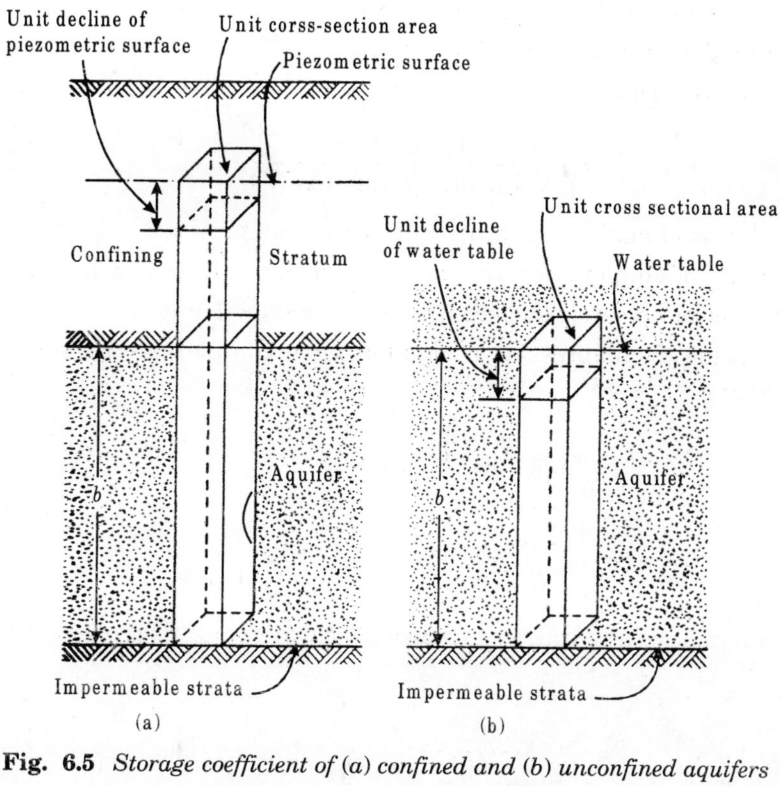

Fig. 6.5 *Storage coefficient of (a) confined and (b) unconfined aquifers*

In a confined aquifer if a vertical column of unit cross-sectional area (say 1 m × 1 m) is considered which is extending upto the piezometric surface as shown in Fig. 6.5 (a) then the storage coefficient S is the volume of water released from the aquifer when the piezometric surface declines by unit distance (say 1 m). For most of the confined aquifers the value of the storage coefficient lies in the range 0.00005 to 0.005. For confined aquifers the value of S can be determined from pumping tests of wells as indicated in section 6.16.

For confined aquifers the value of S can also be determined analytically as described below.

In the case of a confined aquifer when the piezometric surface is lowered by pumping, water is released from storage by the compression of the water bearing material (i.e., aquifer) and by expansion of the water. Thus, the storage coefficient S is a function of the elasticity of both water and the aquifer and it is given by the following expression

$$S = wb\left[\frac{n}{K_w} + \frac{1}{K_{aq}}\right] \qquad \ldots(6.7)$$

In which

w is unit weight of water (9810 N/m^3)

n is porosity of aquifer (fraction)

b is thickness of aquifer

K_w is bulk modulus of elasticity of water; and

K_{aq} is bulk modulus of elasticity of aquifer.

The above expression may be derived as follows.

Bulk modulus of elasticity

$$= \frac{\text{Pressure } p}{\text{Volumetric strain } (\Delta V/V)}$$

If b is the thickness of aquifer, then per unit area, volume $V = b \times 1 = b$.

If n is the porosity of aquifer then volume of water per unit area = $(b \times n)$

For unit decline of piezometric surface pressure $p = (w \times 1) = w$.

∴ Change ΔV_1 due to compression of aquifer when the piezometric surface is lowered by 1 m may be obtained as

$$K_{aq} = \frac{w}{(\Delta V_1 / b)}$$

or

$$\Delta V_1 = \frac{wb}{K_{aq}}$$

Also change in volume ΔV_2 due to expansion of water due to release of pressure may be obtained as

$$K_w = \frac{w}{(\Delta V_2 / bn)}$$

or $$\Delta V_2 = \frac{wbn}{K_w}$$

Since $$S = \Delta V_1 + \Delta V_2$$

$$\therefore \quad S = wb\left[\frac{1}{K_{aq}} + \frac{n}{K_w}\right]$$

For water $K_w = 2.06 \times 10^9$ N/m^2 (or 2.1×108 kg(f)/m^2)

The bulk modulus of elasticity of some of the formation materials are as given in Table 6.1.

TABLE 6.1 Bulk modulus of elasticity of formation material

Material	*Bulk Modulus of Elasticity K*	
	(N/cm²)	*(kg f/cm²)*
Plastic clay	49 – 392	5 – 40
Stiff clay	392 – 784	40 – 80
Medium hard clay	784 – 1470	80 – 150
Loose sand	980 – 1960	100 – 200
Dense sand	4900 – 7840	500 – 800
Dense sandy gravel	9800 – 19600	1000 – 2000
Rock - fissured, jointed	14700 – 294000	1500 – 30000
Rock sand	> 294000	> 30000

In an unconfined aquifer if a vertical column of unit cross-sectional area is considered which is extending up to the water table as shown in Fig. 6.5 (b) then the storage coefficient is the volume of water released from the aquifer when the water table is lowered by a unit distance. The water so released represents, for practical purposes, the gravity drainage from this portion of the column of the aquifer. As such for an unconfined aquifer the storage coefficient is equal to the specific yield. For unconfined aquifers the value of the storage coefficient lies in the range 0.05 to 0.30.

Coefficient of Permeability or Hydraulic Conductivity

The coefficient of permeability or hydraulic conductivity is defined as the rate of flow of water per unit cross-sectional area of aquifer under a unit hydraulic gradient. Thus, coefficient of permeability has the dimensions of velocity and it is usually expressed in mm/s or cm/s and denoted by *k*. The representative values of the coefficient of permeability for some of the soils are given in Table 6.2.

TABLE 6.2 Coefficient of permeability for some soils

Type of soil	*Coefficient of permeability*	
	(mm/s)	*(cm/s)*
Clean gravel	10 and greater	1.0 and greater
Clean sand (coarse)	10 – 0.1	1.0 – 0.01
Sand (mixture)	0.1 – 0.05	0.01 – 0.005
Fine sand	0.5 – 0.01	0.05 – 0.001
Silty sand	0.02 – 0.001	0.002 – 0.0001
Silt	0.005 – 0.0001	0.0005 – 0.00001
Clay	0.00001 and smaller	0.000001 and smaller

Coefficient of Transmissibility or Transmissivity

The coefficient of transmissibility or transmissivity is defined as the rate of flow of water through a vertical strip of aquifer of unit width and extending for the full saturated height under unit hydraulic gradient. Thus, the coefficient of transmissibility T is equal to the coefficient of permeability k multiplied by the aquifer thickness b, i.e.,

$$T = bk \qquad \text{...(6.8)}$$

The value of T for a confined aquifer can be determined by pumping test method or recovery test method as indicated in section 6.16.

6.8 WELL HYDRAULICS

The groundwater is mostly extracted by wells penetrating into the aquifers. Before pumping, the water level in the well stands at the same elevation as the water table or piezometric surface depending on the type of aquifer. When pumping starts, the water is removed from the aquifer surrounding the well, and in and around the well the water table or piezometric surface is lowered and assumes the shape of an inverted cone which is known as *cone of depression*. The area of the base of this cone is known as the *area of influence*, because it is this area which gets affected by the pumping of the well. The boundary of the area of influence is known as the *circle of influence*. The radius of the circle of influence is known as the *radius of influence*. Further at any point the difference in elevation of the water table or piezometric surface before and after pumping is known as *drawdown*. The maximum drawdown occurs at the well and it decreases with increase in the distance from the well. The variation in drawdown with distance from the well is shown by a *drawdown curve* (see Figs. 6.6 and 6.7).

The analysis of radial flow of groundwater towards a well was first proposed by Dupuit (1863) and later modified by Thiem (1906). The Dupuit–Thiem theory is based on the following assumptions.

(i) The aquifer is homogeneous isotropic, of uniform thickness and of infinite areal extent.

(ii) The well penetrates and receives water from the entire thickness of the aquifer.

(iii) The pumping has been continued for a sufficiently long time at a uniform rate so that an equilibrium stage or a steady flow condition has been reached.

(iv) The coefficient of transmissibility is constant at all places and at all times.

(v) The flow lines are radial and the flow of groundwater is horizontal.

(vi) Flow is laminar and Darcy's law is applicable. However the hydraulic gradient may be represented by tan θ instead of sin θ where θ is the angle between the hydraulic gradient line and the horizontal.

(vii) The well is infinitely small with negligible storage and all the pumped water comes from the aquifer.

On the basis of these assumptions the radial flow equations which relate the well discharge to drawdown for steady flow condition have been derived for wells completely penetrating an unconfined aquifer and a confined aquifer as indicated below.

(a) Steady State Flow to Wells in Unconfined Aquifer. Figure 6.6 shows a well of radius r completely penetrating an unconfined aquifer. Let H be the thickness of the aquifer measured from the impermeable strata to the initial level of the water table.

When the well is pumped at a constant rate Q for a long time so that the water level in the well has been stabilised, i.e., an equilibrium stage or a steady flow condition has been reached, then the drawdown curve as shown in Fig. 6.6 is developed. At this stage let h be the depth of water in the well measured above the impermeable strata. Further let R be the radius of influence (or the radius of inappreciable or zero drawdown) measured from the centre of the well to a point where the drawdown is inappreciable.

Considering the origin at a point O at the centre of the well at its bottom, let the co-ordinates of any point P on the drawdown curve be (x, y), if a vertical cylindrical surface passing through point P and surrounding the well located at its centre is considered then the area of the portion of cylindrical surface which is lying within the aquifer below point P is equal to $(2\pi xy)$. Further if (dy/dx) is the hydraulic gradient at P then from Darcy's law the rate of flow of water (or discharge) through the cylindrical surface is equal to $[k(dy/dx)\, 2\pi xy]$. By continuity this rate of flow of water is equal to the well discharge, and hence

$$Q = k \frac{dy}{dx}(2\pi xy)$$

or

$$Q \frac{dx}{x} = 2\pi k y dy \qquad \ldots(i)$$

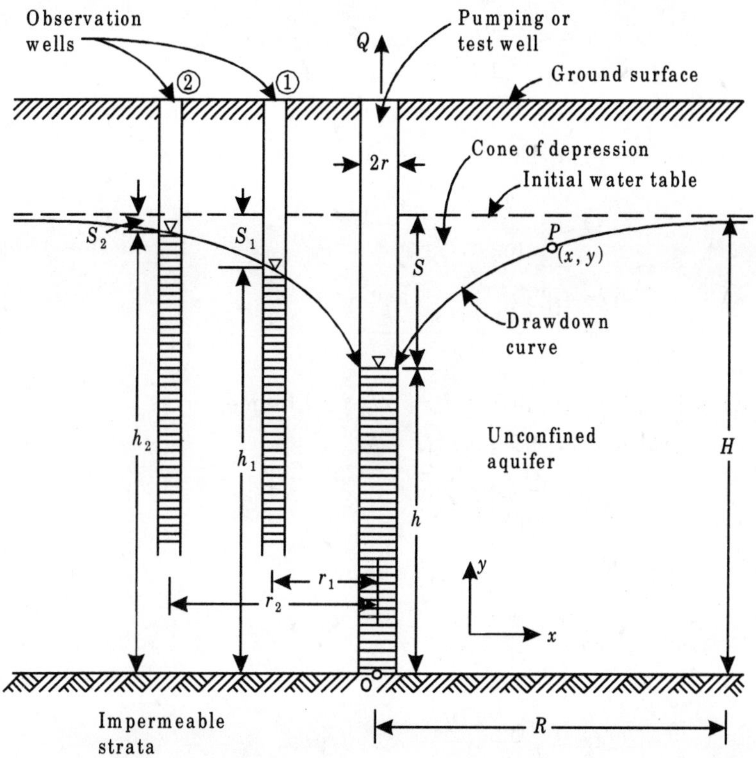

Fig. 6.6 *Well penetrating an unconfined aquifer*

Integrating both sides of Eq. (i) between the limits, at $x = r$, $y = h$ at the well and at $x = R$, $y = H$ at the extemity of the area of influence, we get

$$Q\int_r^R \frac{dx}{x} = 2\pi k \int_h^H y\,dy$$

or
$$Q = \frac{\pi k(H^2 - h^2)}{\log_e(R/r)} \quad ...(6.9)$$

or
$$Q = \frac{\pi k(H^2 - h^2)}{2.303\log_{10}(R/r)} \quad ...(6.9a)$$

or
$$Q = \frac{1.36k(H^2 - h^2)}{\log_{10}(R/r)} \quad ...(6.9b)$$

Equation 6.9 may also be expressed in a different form as indicated below.

If s is the drawdown measured at the well then

$$s = H - h$$

or
$$H = s + h$$

and
$$H + h = s + 2h$$

∴
$$H^2 - h^2 = (H - h)(H + h)$$
$$= s(s + 2h)$$

Introducing this expression in Eq. (6.9) we get

$$Q = \frac{\pi ks(s + 2h)}{\log_e(R/r)} \quad ...(6.10)$$

or
$$Q = \frac{1.36\,ks(s + 2h)}{\log_{10}(R/r)} \quad ...(6.10a)$$

If the drawdown s is small then $(H + h) \approx 2H$ and hence

$$H^2 - h^2 = (H + h)(H - h)$$
$$= 2H(H - h)$$
$$= 2Hs$$

Thus, from Eq. (6.9) we get

$$Q = \frac{2\pi kHs}{\log_e(R/r)} \quad ...(6.11)$$

or
$$Q = \frac{2.72\,kHs}{\log_{10}(R/r)} \quad ...(6.11a)$$

Since for an unconfined aquifer the coefficient of transmissibility $T = kH$, Eq. (6.11) becomes

$$Q = \frac{2\pi Ts}{\log_e (R/r)} \quad \text{...(6.12)}$$

$$Q = \frac{2.72Ts}{\log_{10} (R/r)} \quad \text{...(6.12a)}$$

Equation (6.12) is similar to the one derived later for a confined aquifer.

These equations can however be used only if the radius of influence R is known. In practice, the selection of the radius of influence R is approximate and arbitrary, but the variation in Q is small for a wide range of R. The values of R in general fall in the range of 150 to 300 metres.

Alternatively, R may be computed from the following approximate expression given by Sichardt:

$$R = 3000\ s\ \sqrt{k} \quad \text{...(6.13)}$$

in which

R = Radius of influence in metres

s = Drawdown at the well in metres

k = Coefficient of permeability in metres per second

However, as indicated below the use of R can be avoided if the observation wells are available.

As shown in Fig. 6.6 let there be two observation wells at radial distances r_1 and r_2 and the depths of water in them be h_1 and h_2 respectively. Integrating both sides of Eq. (i) between the limits at $x = r_1$, $y = h_1$ at the observation well No. 1 and at $x = r_2$, $y = h_2$ at the observation well No. 2, the following equation may be obtained which does not involve R.

$$Q = \frac{\pi k(h_2^2 - h_1^2)}{\log_e (r_2 / r_1)} \quad \text{...(6.14)}$$

or

$$Q = \frac{1.36k(h_2^2 - h_1^2)}{\log_{10} (r_2 / r_1)} \quad \text{...(6.14a)}$$

Further if s_1 and s_2 are the respective drawdown in the observation wells, then

$$h_2 = (H - s_2)$$

and

$$h_1 = (H - s_1)$$

Introducing these expressions in Eq. (6.14), we get

$$Q = \frac{\pi k[(H - s_2)^2 - (H - s_1)^2]}{\log_e (r_2 / r_1)}$$

or

$$Q = \frac{2\pi kH(s_1' - s_2')}{\log_e (r_2 / r_1)} \quad \text{...(6.15)}$$

or $$Q = \frac{2.72kH(s_1' - s_2')}{\log_{10}(r_2 / r_1)} \quad \text{...(6.15a)}$$

in which

$$s_1' = \left(s_1 - \frac{s_1^2}{2H}\right)$$

and $$s_2' = \left(s_2 - \frac{s_2^2}{2H}\right)$$

s_1' and s_2' are termed as the corrected or modified steady-state drawdowns at the observation wells.

Further for an unconfined aquifer since $T = kH$, Eq. (6.15) may be expressed as

$$Q = \frac{2\pi T(s_1' - s_2')}{\log_e(r_2 / r_1)} \quad \text{...(6.16)}$$

or $$Q = \frac{2.72T(s_1' - s_2')}{\log_{10}(r_2 / r_1)} \quad \text{...(6.16a)}$$

(b) Steady State Flow to Wells in Confined Aquifer. Figure 6.7 shows a well of radius r fully penetrating a confined aquifer. Let b be the thickness of the aquifer measured between the top and bottom impervious strata, and H be the height of the initial piezometric surface measured above the impervious strata at the bottom.

When the well is pumped at a constant rate Q for a long time so that the water level in the well has been stabilised then the drawdown curve as shown in Fig. 6.7 is developed. At this stage let h be the depth of water in the well measured above the impermeable strata at the bottom. Further let R be the radius of influence.

Let (x, y) be the coordinates of any point P on the drawdown curve with respect to origin O at the centre of the well at its bottom. If a vertical cylindrical surface passing through point P and surrounding the well located at its centre is considered then the area of the portion of the cylindrical surface which is lying within the aquifer is equal to $(2\pi\, xb)$. Further if (dy/dx) is the hydraulic gradient at P, then from Darcy's law the rate of flow of water through this portion of the cylindrical surface is equal to $[k\,(dy/dx)\, 2\pi\, xb]$ which by continuity is also equal to the well discharge, and hence

$$Q = k\frac{dy}{dx}(2\pi xb)$$

or $$Q\,\frac{dx}{x} = 2\pi\, kb\, dy \quad \text{...(ii)}$$

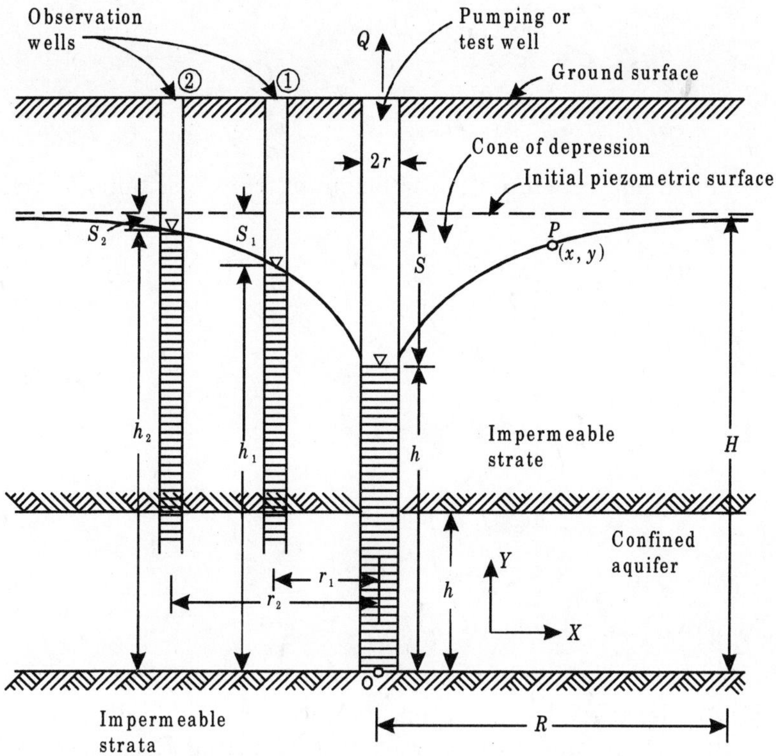

Fig. 6.7 *Well penetrating a confined aquifer*

Integrating both sides of Eq. (ii) between the limits, at $x = r$, $y = h$ at the well and at $x = R$, $y = H$ at the extremity of the area of influence, we get

$$Q\int_r^R \frac{dx}{x} = 2\pi\, kb\int_h^H dy$$

or $$Q = \frac{2\pi kb(H - h)}{\log_e(R/r)} \quad \text{...(6.17)}$$

or $$Q = \frac{2\pi kb(H - h)}{2.303\log_{10}(R/r)} \quad \text{...(6.17a)}$$

or $$Q = \frac{2.72kb(H - h)}{\log_{10}(R/r)} \quad \text{...(6.17b)}$$

Equation (6.17) is known as equilibrium equation or Thiem equation.

If s is the drawdown at the well then since $s = (H - h)$, Eq. (6.17) may be expressed as

$$Q = \frac{2\pi kbs}{\log_e(R/r)} \quad ...(6.18)$$

or

$$Q = \frac{2.72kbs}{\log_{10}(R/r)} \quad ...(6.18a)$$

Further for a confined aquifer since the coefficient of transmissibility $T = kb$, Eqs (6.17) and (6.18) become

$$Q = \frac{2\pi T(H-h)}{\log_e(R/r)} \quad ...(6.19)$$

or

$$Q = \frac{2.72T(H-h)}{\log_{10}(R/r)} \quad ...(6.19a)$$

$$Q = \frac{2\pi Ts}{\log_e(R/r)} \quad ...(6.20)$$

or

$$Q = \frac{2.72Ts}{\log_{10}(R/r)} \quad ...(6.20a)$$

Again as indicated below the use of R can be avoided if the observation wells are available.

As shown in Fig. 6.7 let there be two observation wells at radial distances r_1 and r_2 and the depths of water in them be h_1 and h_2 respectively. Integrating Eq. (ii) between the limits, at $x = r_1$, $y = h_1$ at the well No. 1 and at $x = r_2$, $y = h_2$ at the observation well No. 2, the following equation may be obtained which does not involve R.

$$Q = \frac{2\pi kb(h_2 - h_1)}{\log_e(r_2/r_1)} \quad ...(6.21)$$

or

$$Q = \frac{2.72kb(h_2 - h_1)}{\log_{10}(r_2/r_1)} \quad ...(6.21a)$$

Further if s_1 and s_2 are the respective drawdowns at the two observation wells, then

$$h_2 = H - s_2$$

and

$$h_1 = H - s_1$$

Introducing these expressions in Eq. (6.21) we get

$$Q = \frac{2\pi kb(s_1 - s_2)}{\log_e(r_2/r_1)} \quad ...(6.22)$$

or

$$Q = \frac{2.72kb(s_1 - s_2)}{\log_{10}(r_2/r_1)} \quad ...(6.22a)$$

Further since $T = kb$ Eq. (6.22) may be expressed as

$$Q = \frac{2\pi T(s_1 - s_2)}{\log_e(r_2/r_1)} \quad ...(6.23)$$

or
$$Q = \frac{2.72T(s_1 - s_2)}{\log_{10}(r_2/r_1)} \quad ...(6.23a)$$

6.9 DETERMINATION OF HYDRAULIC PROPERTIES OF AQUIFER

The hydraulic properties of unconfined as well as confined aquifers viz., T and k for steady state of flow may be determined by pumping tests of wells for which in addition to pumping or a test well a number of observation wells are required. The test well is pumped at a constant rate Q for a long time so that an equilibrium stage or a steady flow condition is attained. The steady flow condition is the one in which the water level in the well has been stabilised and hence the change in drawdown with time is negligible. At this stage the drawdowns are recorded at the various observation wells. Further the radial distances of the centres of each of the observation wells from the centre of the pumped or test well are also noted. The values of T and k for the unconfined and the confined aquifers may then be determined as indicated below.

(a) Unconfined Aquifer. For an unconfined aquifer from Eq. (6.16a)

$$Q = \frac{2.72T(s_1' - s_2')}{\log_{10}(r_2/r_1)} \quad ...(6.16a)$$

which shows that for constant values of Q and T, the corrected or modified drawdown s' at any observation well varies linearly with the logarithm of its radial distance from the pumped well. Thus, from the observed steady-state drawdowns the corrected or modified drawdowns are computed as given by the following expression

$$s' = s - \frac{s^2}{2H}$$

in which s' = Corrected or modified drawdown

s = Observed steady-state drawdown

and H = Height of the initial water table measured above the impermeable strata.

On a semi-logarithmic graph paper the corrected drawdown s' of each observation well is plotted against the radial distance r between the observation well and the pumped well. s' is plotted on the vertical axis on arithmetic scale and r is plotted on the horizontal axis on a logarithmic scale. The best fitting straight line is drawn through the plotted points which is the corrected drawdown vs distance curve as shown in Fig. 6.8 (a).

If $\Delta s'$ is the difference between the values of s' over one log cycle of distance r i.e., considering $(r_2/r_1) = 10$, so that log10 $(r_2/r_1) = 1$, then from Eq. (6.16 a), we get

$$T = \frac{Q}{2.72\,\Delta s'} \quad ...(6.24)$$

The value of $\Delta s'$ over one log cycle of distance r may be obtained from the distance v/s corrected drawdown curve (Fig. 6.8 a), and by substituting the values of Q and $\Delta s'$ in Eq. (6.24) the value of T for the unconfined aquifer can be determined.

Further in this case since $T = kH$, by knowing T and H, the value of k for the unconfined aquifer can be determined.

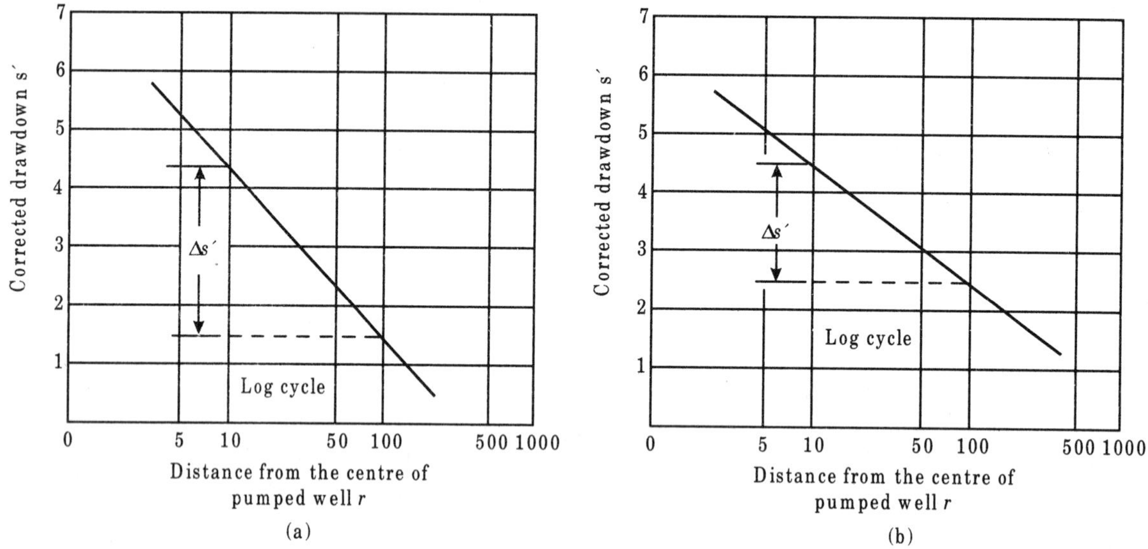

Fig. 6.8 *Plots of (a) corrected drawdown vs distance and (b) drawdown vs distance*

(b) Confined Aquifer. For a confined aquifer from Eq. 6.23 (a)

$$Q = \frac{2.72T(s_1 - s_2)}{\log_{10}(r_2 / r_1)} \quad \text{...(6.23a)}$$

which shows that for constant values of Q and T, the drawdown s at any observation well varies linearily with the logarithm of its radial distance from the pumped well. Thus, on a semi-logarithmic paper the observed steady-state drawdown s of each observation well is plotted against the radial distance r between the pumped well and the observation well. The drawdown s is plotted on the vertical axis on arithmetic scale and the distance r is plotted on the horizontal axis on a logarithmic scale. The best fitting straight line is drawn through the plotted points which is a drawdown vs distance curve as shown in Fig. (6.8b).

If Δs is the difference between the values of s over one log cycle of distance r i.e., considering $(r_2/r_1) = 10$, so that $\log_{10}(r_2/r_1) = 1$, then from Eq. (6.23a), we get

$$T = \frac{Q}{2.72\,\Delta s} \quad \text{...(6.25)}$$

The value of Δs over one log cycle of distance r may be obtained from the distance vs drawdown curve (Fig. 6.8 b), and by substituting the values of Q and Δs in Eq. (6.25) the value of T for the confined aquifer can be determined.

Further in this case since $T = kb$, by knowing T and b, the value of k for the confined aquifer can be determined.

6.10 WELL LOSSES, SPECIFIC CAPACITY OF WELL AND WELL EFFICIENCY

1. Well Losses. When water is pumped from a well the total drawdown caused at the well face is made up of the following.

(i) The drawdown resulting from the head loss due to the resistance to the flow of water in the aquifer which is known as aquifer loss or formation loss.

(ii) The drawdown due to the well losses resulting from the flow through the well screen and the axial flow within the well to the pump intake.

Due to aquifer loss the logarithmic drawdown curve is developed and the resulting drawdown at the well is therefore equal to the steady-state drawdown given by Eqs (6.12) and (6.20) for the wells in unconfined and confined aquifers respectively.

The drawdown due to well loss will however be equal to the total well loss expressed in terms of the head of water. Since near the well face generally turbulent flow occurs, the well loss may be taken to be proportional to the nth power of discharge, i.e., Q^n, where n is a constant greater than one. The well loss and the consequent drawdown is therefore equal to CQ^n, where C is a constant which is known as coefficient of well loss. The value of C depends on the radius, construction and condition of the well.

Thus, taking account of the well loss, the total drawdown s at the well in an unconfined or a confined aquifer may be obtained by adding the drawdown due to well loss to the steady-state drawdown given by Eq. (6.12) or (6.20), and hence

$$s = \frac{Q}{2\pi T} \log_e (R/r) + CQ^n \qquad \text{...(6.26)}$$

Let
$$B = \frac{1}{2\pi T} \log_e (R/r),$$

where B is known as coefficient of aquifer loss, then

$$s = BQ + CQ^n \qquad \text{...(6.27)}$$

In Eq. (6.27), BQ represents the drawdown due to aquifer loss, which is however same as the aquifer loss expressed in terms of the head of water. Figure (6.9a) shows the total drawdown at a well in a confined aquifer as indicated by Eq. (6.27).

If the value of n is taken as 2, then Eq. 6.27 may be expressed as

$$\frac{s}{Q} = B + CQ \qquad \text{...(6.28)}$$

(s/Q) is usually known as *specific drawdown* and is defined as the drawdown at a well per unit discharge of the well. The specific drawdown is reciprocal of specific capacity (Q/s) defined later. By plotting (s/Q) versus Q and fitting a straight line through the points, the slope of the line gives the coefficient of well loss C and the intercept on the (s/Q) axis at $Q = 0$ gives the aquifer loss coefficient B.

Figure (6.9b) shows the variation of total drawdown and well loss with the well discharge. It is apparent from Fig. (6.9b) that for relatively low pumping rates the well loss may be neglected, but for high pumping rates the well loss can be a substantial fraction of total drawdown. However, the well loss can be reduced by increasing the radius of the well, and also by using such well screen which have slots or openings of size compatible with the surrounding porous media and which do not deteriorate or get clogged.

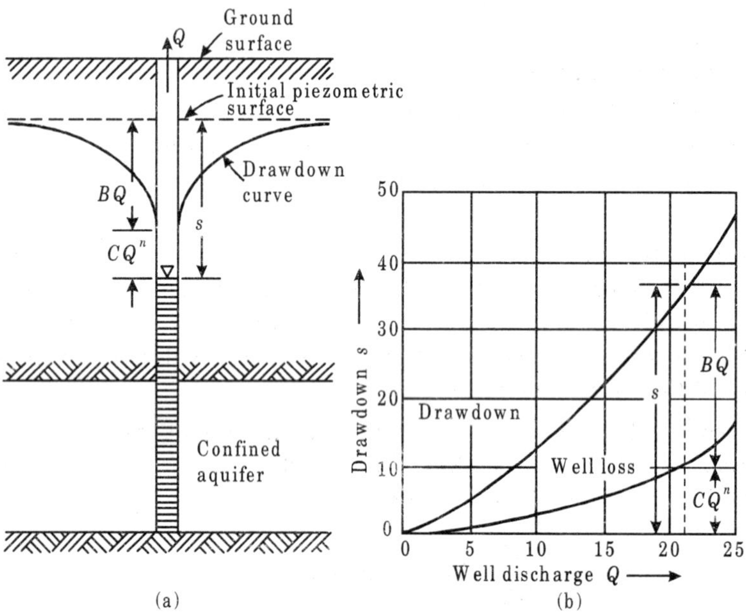

Fig. 6.9 *Relation of well loss to drawdown for a well penetrating a confined aquifer*

(b) Specific Capacity. The specific capacity of a well is defined as the discharge per unit drawdown in a pumping well. In other words the specific capacity of a well is obtained by dividing the discharge by drawdown in a pumping well. Thus, if s is drawdown and Q is well discharge then

$$\text{Specific capacity} = \frac{Q}{s} \qquad ...(6.29)$$

The specific capacity is therefore reciprocal of specific drawdown as defined earlier.

From Eq. (6.27), we have

$$s = BQ + CQ^n$$

$\therefore$ Specific capacity

$$\frac{Q}{s} = \frac{Q}{BQ + CQ^n} \qquad ...(6.30)$$

or

$$\frac{Q}{s} = \frac{1}{B + CQ^{n-1}} \qquad ...(6.30a)$$

Equation (6.30a) indicates that the specific capacity of a well is not costant, but it decreases as the well discharge increases. However, if it is assumed that well loss $CQ^n = 0$, then drawdown $s = BQ$ and specific capacity

$$\frac{Q}{s} = \frac{Q}{BQ} = \frac{1}{B} \qquad \text{...(6.31)}$$

Equation (6.31) indicates that specific capacity of a well is independent of the well discharge, but it gives only theoretical specific capacity because in actual practice well loss is not equal to zero (except at very low pumping rates when the well loss may be negligible).

The specific capacity of a well is a measure of its productivity, and hence larger is the specific capacity, better is the well. Further as indicated by Eq. (6.30a) for a given discharge a well may be assumed to have a constant specific capacity, but in actual practice it is not correct because the specific capacity of a well is found to slightly vary with time.

A significant reduction in the specific capacity of a well may be due to an increase in the well loss resulting from clogging or deterioration of the well screen. Moreover, in the case of an unconfined aquifer due to lowering of the groundwater level the transmissibility may be reduced which in turn may also result in reducing the specific capacity of the well.

(c) Well Efficiency. The efficiency of a well E_w is defined as the ratio of the actual specific capacity (Q/s) of a well measured in the field and the theoretical specific capacity (Q/BQ), and is expressed as percentage. Thus,

$$E_w = \frac{(Q/s)}{(Q/BQ)} \times 100$$

or

$$E_w = \frac{BQ}{s} \times 100 \qquad \text{...(6.32)}$$

Introducing Eq. (6.27) in Eq. (6.32), we have

$$E_w = \frac{BQ}{BQ + CQ^n} \times 100 \qquad \text{...(6.33)}$$

Equation (6.33) indicates that when well loss is large the well is less efficient.

6.11 INTERFERENCE AMONG WELLS

If two or more wells are located so close to each other that when they are discharging their drawdown curves intersect then one well is said to interfere with the other. Observations indicate that due to the interference of wells the discharge of each individual well is decreased, though the total discharge from the wells is increased.

Figure 6.10 shows interference between two wells penetrating a confined aquifer. From the principle of superposition, the drawdown at any point in the area of influence caused by the discharge of several wells is equal to the sum of the drawdown caused by each well individually.

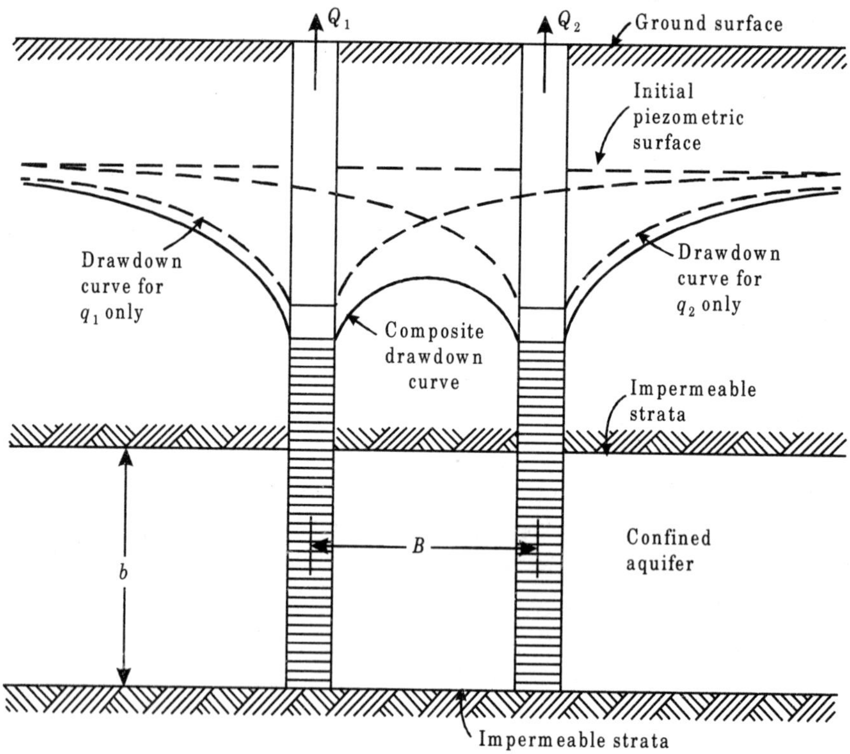

Fig. 6.10 *Interference between two wells*

Thus, for the two wells having same discharge, i.e., $Q_1 = Q_2$, the composite drawdown curve may be obtained from the individual drawdown curves as shown in Fig. 6.10. Further if the two wells are spaced a distance B apart and have same radius r and drawdown $(H - h)$ and discharge over the same period of time then with the help of the method of complex variables it can be shown that the discharge of each of the two wells is given by

$$Q_1 = Q_2 = \frac{2\pi kb(H-h)}{\log_e(R^2/rB)} \qquad \text{...(6.34)}$$

where R is the radius of influence of each well ($R > B$), and the other terms are same as defined earlier.

However, if there was only one well, then under the same drawdown the discharge would be given by Eq. 6.17 as

$$Q = \frac{2\pi kb(H-h)}{\log_e(R^2/r)}$$

Comparing the discharge of one well of the two well system with the discharge of a single well we get

$$\frac{Q_2}{Q} = \frac{\log_e(R/r)}{\log_e(R^2/rB)} \quad ...(6.35)$$

Since $R > B$; $(R^2/rB) > (R/r)$ and hence $Q > Q_1$, i.e., the discharge of each well of a two well system is less than that of a single well.

It can however be shown that $2Q_1 > Q$, thereby indicating that the total discharge of both the wells of a two well system is more than that of the single well.

If instead of confined aquifer, the two wells are penetrating an unconfined aquifer then under similar conditions the discharge of each of the two wells is given by

$$Q_1 = Q_2 = \frac{\pi k(H^2 - h^2)}{\log_e(R^2/rB)} \quad ...(6.36)$$

Again by comparing the discharge of one well of the two well system as given by Eq. (6.36) with the discharge of a single well as given by Eq. (6.9), it can be shown for this case also that the discharge of each well of a two well system is less than that of a single well.

Similar expressions have also been derived for discharge of each of the wells of three, four and five well systems in confined as well as unconfined aquifers and having interference among wells.

6.12 ARTESIAN GRAVITY WELL

Sometimes in an artesian well (i.e., a well in a confined aquifer) due to high rate of pumping the water level in and around the well may fall below the top of the confined aquifer as shown in Fig. 6.11. In such a case, the flow pattern close to the well is similar to that for a gravity well

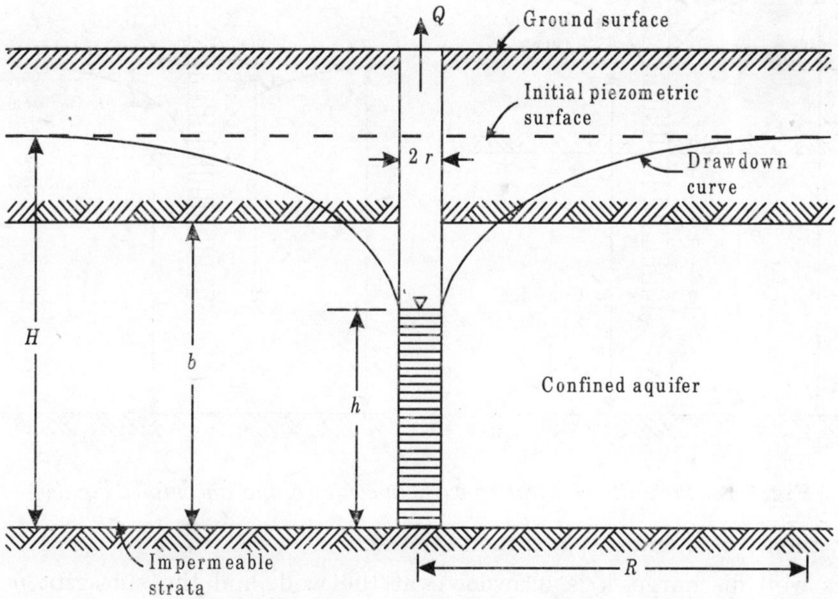

Fig. 6.11. *Artesian gravity well*

(i.e., a well in an unconfined aquifer) whereas at distance farther from the well the flow is similar to that for an artesian well. This type of well is therefore known as *artesian–gravity well*. The discharge for such a well can be determined from the following expression developed by Muskat

$$Q = \frac{\pi k(2bH - b^2 - h^2)}{\log_e (R/r)} \quad \text{...(6.37)}$$

in which the various terms are as indicated in Fig. 6.11.

6.13 PARTIALLY PENETRATING WELLS

A partially penetrating well is the one which extends only partially through the aquifer. Thus, for a partially penetrating well the length of water entry is less than the thickness of the aquifer which it penetrates. In practice such wells are often encountered. Figure 6.12 shows partially penetrating well in confined and unconfined aquifers. The flow pattern in aquifer in the vicinity of a partially penetrating well differs from that for a fully penetrating well. In the case of a partially penetrating well there is a vertical convergence of streamlines near the well resulting in increased entrance velocity and hence a greater resistance to flow is encountered. This results in the following relationships between two similar wells, one partially and one fully penetrating the same aquifer.

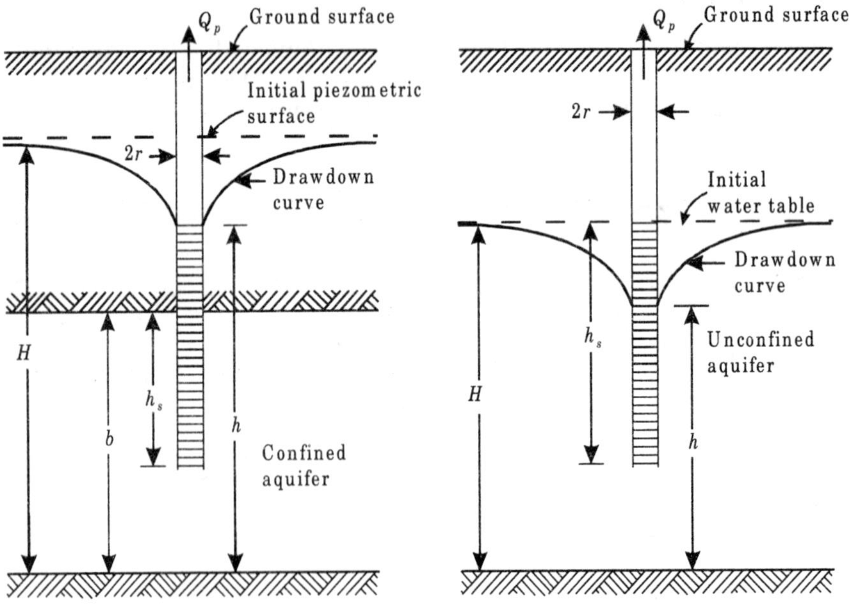

Fig. 6.12. *Partially penetrating wells in confined and unconfined aquifers*

If $Q_p = Q$, then $s_p > s$; and if $s_p = s$, then $Q_p < Q$.

Here Q is well discharge, s is drawdown at the well; and the subscript p refers to the partially penetrating well.

However, beyond a radial distance equal to 2 times the saturated thickness of the aquifer the effect of partial penetration of well on the flow pattern and the drawdown is negligible.

For a well penetrating only the upper portion of a confined aquifer the total drawdown at the well is given as

$$(H - h) = \frac{Q_p}{2\pi k}\left[\frac{1}{h_s}\log_e\left(\frac{\pi h_s}{2r}\right) + \frac{0.10}{b} + \frac{1}{b}\log_e\left(\frac{R}{2b}\right)\right] \quad \text{...(6.38)}$$

where Q_p is discharge in partially penetrating well; k is coefficient of permeability; and the other terms are as shown in Fig. 6.12.

Dividing the above expression by Eq. (6.17) for a fully penetrating well gives

$$\frac{Q_p}{Q} = \frac{\log_e(R/r)}{[(b/h_s)\log_e(\pi h_s/2r) + 0.10 + \log_e(R/2b)]} \quad \text{...(6.39)}$$

where Q is the discharge of a similar fully penetrating well with the same total drawdown.

From Eq. (6.39) the effect of partial penetration on the yield of a well can be determined. Thus, for example if a 300 mm well penetrates only 6 m in a 15 m thick confined aquifer having $R = 150$ m then from Eq. (6.39), $(Q_p/Q) = 0.57$, i.e., in this case a partially penetrating well yield 57 per cent of what a similar fully penetrating well would yield for the same drawdown.

For partially penetrating wells in unconfined aquifers if the drawdown is small in relation to the saturated thickness H, then as a good approximation the value of the ratio (Q_p/Q) is given by the following expression

$$\frac{Q_p}{Q} = \frac{\log_e(R/r)}{[(H/h_s)\log_e(\pi h_s/2r) + 0.10 + \log_e(R/2H)]} \quad \text{...(6.40)}$$

Another expression developed by Kozeny for discharge Q_p of a well partially penetrating a confined aquifer is as follows

$$Q_p = Q\left[\frac{h_s}{b}\left(1 + 7\sqrt{\frac{r}{2h_s}}\cos\frac{\pi h_s}{2b}\right)\right] \quad \text{...(6.41)}$$

where all the terms are same as defined earlier.

The corresponding expression for discharge Q_p of a well partially penetrating an unconfined aquifer is as follows

$$Q_p = Q\left[\frac{h_s}{H}\left(1 + 7\sqrt{\frac{r}{2h_s}}\cos\frac{\pi h_s}{2H}\right)\right] \quad \text{...(6.42)}$$

6.14 SPHERICAL FLOW IN A WELL

A particular case of a partially penetrating well is the one in which a well just penetrates the top surface of a semi-infinite porous media. In this case the flow towards the well is spherical as shown in Fig. 6.13. As such the earlier derived equations which are based on the radial flow towards the well do not apply for this case. The discharge Q_s for such a well is given by

$$Q_s = 2\pi(r')^2 k \frac{\partial h'}{\partial r'} \quad \text{...(i)}$$

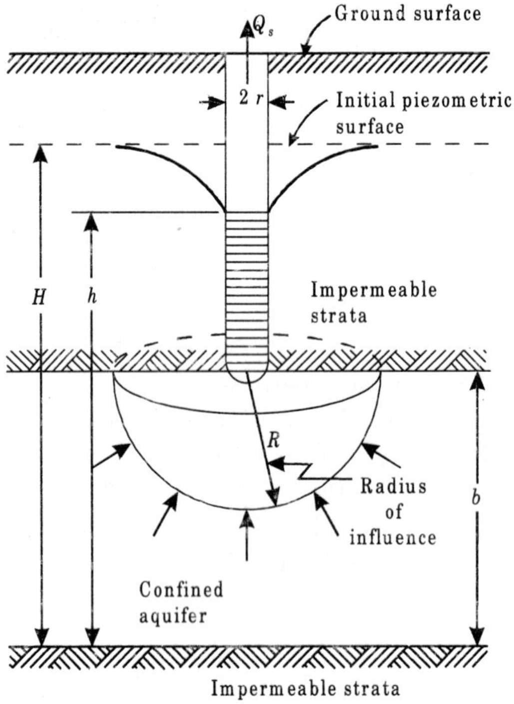

Fig. 6.13. *Spherical flow in a well*

in which $\left(\frac{\partial h}{\partial s}\right)$ is the slope of the hydraulic gradient at a radial distance r' and k is coefficient of permeability.

At $r' = r, h' = h$ and at $r' = R, h' = H$.

The integration of Eq. (i) yields

$$(H - h) = \frac{Q_s}{2\pi k}\left(\frac{1}{r} - \frac{1}{R}\right)$$

Assuming $R >> r$, we obtain

$$(H - h) = \frac{Q_s}{2\pi kr}$$

or $$Q_s = 2\pi kr(H - h) \quad \text{...(6.43)}$$

Also $$Q_s = 2\pi krs \quad \text{...(6.43a)}$$

where $s = (H - h)$ = drawdown at the well

For the case of radial flow towards a well fully penetrating a confined aquifer the discharge Q is given by Eq. (6.17) as

$$Q = \frac{2\pi kb(H-h)}{\log_e (R/r)}$$

$$\therefore \quad \frac{Q_s}{Q} = \frac{r}{b}\log_e\left(\frac{R}{r}\right) \qquad ...(6.44)$$

Since in general the radius of a well is very much smaller than the thickness of the aquifer, it is evident that the spherical flow well is highly inefficient. For example if r = 80 mm, (R/r) = 1000 and b = 18 m then

$$\frac{Q_s}{Q} = \frac{80}{18 \times 1000} \times 2.30 \log_{10}(1000)$$

$$\therefore \quad \frac{Q_s}{Q} = \frac{1}{33}$$

i.e., in this case the discharge of a spherical flow well is only about 3% of that of a radial flow well, and hence a spherical flow well is highly inefficient as compared to a radial flow well.

6.15 UNSTEADY FLOW TOWARDS WELLS

When a well penetrating an extensive aquifer is pumped at a constant rate then if the cone of depression does not vary with time, it may be assumed that steady-state flow exists within the aquifer. The analysis of flow towards wells discussed earlier is based on the assumption that steady-state flow exists within the aquifer. However, actually the steady-state flow does not exist because as water is pumped from the well at a constant rate then since the storage within the aquifer is reduced, the head continues to decline as long as the aquifer is effectively infinite. Although the rate of decline of head decreases continuously as the area of influence expands, this results in the variation of the cone of depression with time and therefore unsteady or transient flow exists within the aquifer. The differential equation which may be applied to analyse the unsteady flow towards wells is derived as follows.

Consider an annular cylinder of radius r and thickness dr surrounding the well as shown in Fig. 6.14. Assuming the flow to be radial, from the principle of continuity the difference in the rate of flow through the inner and outer faces of the cylinder is equal to the rate of change of volume of water within the annular space of the cylinder.

Thus, if Q_1 is the rate of volume of water entering the annular space through the outer face; Q_2 is the rate of volume of water leaving the annular space through the inner face; and $\left(\frac{\partial V}{\partial t}\right)$ is the rate of change of volume of water within the annular space of the cylinder, then

$$Q_1 - Q_2 = \frac{\partial V}{\partial t} \qquad ...(i)$$

As shown in Fig. 6.14 if $\left(\frac{\partial h}{\partial r}\right)$ is the slope of the hydraulic gradient at the inner face of the cylinder then the slope of the hydraulic gradient at the outer face of the cylinder is

$$\left[\left(\frac{\partial h}{\partial r}\right)+\left(\frac{\partial^2 h}{\partial r^2}\right)dr\right]$$

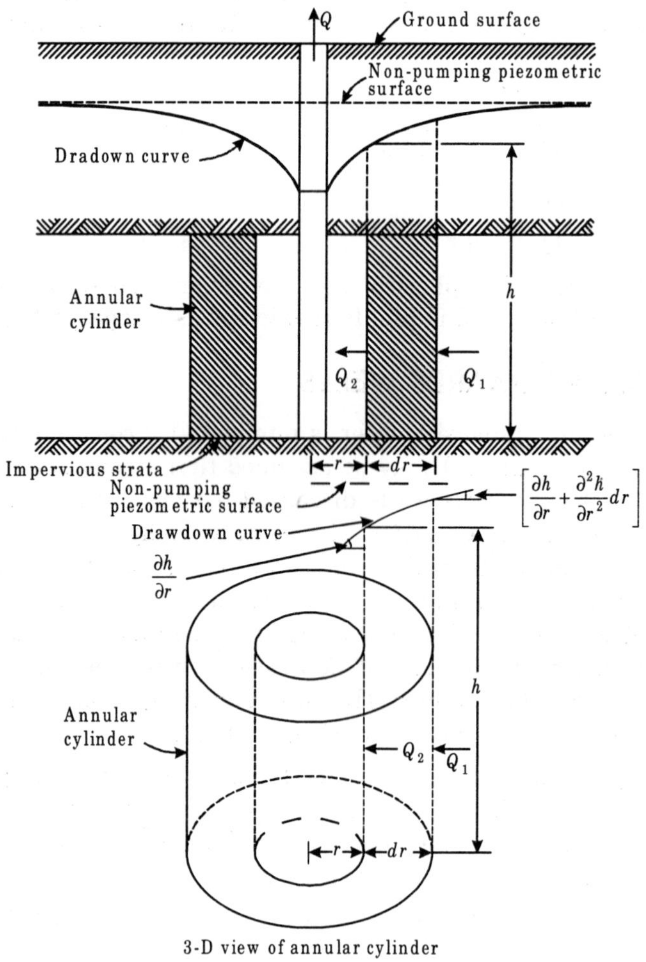

Fig. 6.14. *Unsteady flow towards a well*

Thus, from Darcy's law

$$Q_1 = kb\left(\frac{\partial h}{\partial r}+\frac{\partial^2 h}{\partial r^2}dr\right)2\pi(r+dr)$$

or $$Q_1 = T\left(\frac{\partial h}{\partial r}+\frac{\partial^2 h}{\partial r^2}dr\right)2\pi(r+dr) \qquad \text{...(ii)}$$

and $$Q_2 = kb\left(\frac{\partial h}{\partial r}\right)2\pi r$$

or $$Q_2 = T\left(\frac{\partial h}{\partial r}\right)2\pi r \qquad \text{...(iii)}$$

where k is coefficient of permeability; and T is coefficient of transmissibility.

Further if S is storage coefficient then from the definition of the same given is section 6.6, we have,

$$\partial V = (2\pi r)\, dr\, S\, \partial h$$

or $$\frac{\partial V}{\partial t} = (2\pi r)dr\, S\, \frac{\partial h}{\partial t} \qquad \text{...(iv)}$$

where t is the time since beginning of pumping.

Substituting the values of Q_1, Q_2 and $(\partial V / \partial t)$ in Eq. (i), we get

$$T = \left(\frac{\partial h}{\partial r}+\frac{\partial^2 h}{\partial r^2}dr\right)2\pi(r+dr) - T\frac{\partial h}{\partial r}(2\pi r)$$

$$= (2\pi r)\, dr\, S\, \frac{\partial h}{\partial t}$$

Dividing both sides by $(2\pi r\, T\, dr)$ and neglecting the differentials higher than first order, we have

$$\frac{\partial^2 h}{\partial r^2}+\frac{1}{r}\frac{\partial h}{\partial r} = \frac{S}{T}\frac{\partial h}{\partial t} \qquad \text{...(6.45)}$$

Equation (6.45) is the differential equation applicable to the unsteady flow towards the wells. Theis (1935) obtained a solution of this equation based on an analogy between groundwater flow towards a well and flow of heat towards a sink or point where heat is removed at a uniform rate just as water is removed from the well at a constant rate. Thus, by assuming that the well is replaced by a mathematical sink of constant strength and considering the boundary conditions $h = H$ for $t = 0$, i.e., before pumping; and $h \to H$ as $r \to \infty$ the following solution is obtained.

$$(H - h) = s = \frac{Q}{4\pi T}\int_u^\infty \frac{e^{-u}}{u}du \qquad \text{...(6.46)}$$

where s is drawdown at a radial distance r from the pumping well; Q is the constant well discharge; and

$$u = \frac{r^2 S}{4Tt} \qquad \text{...(6.47)}$$

TABLE 6.3 Values of $W(u)$ for values of u

u	1.0	2.0	3.0	4.0	5.0	6.0	7.0	8.0	9.0
× 1	0.219	0.049	0.013	0.0038	0.0011	0.00036	0.00012	0.000038	0.000012
$\times 10^{-1}$	1.82	1.22	0.91	0.70	0.56	0.45	0.37	0.31	0.26
$\times 10^{-2}$	4.04	3.35	2.96	2.68	2.47	2.30	2.15	2.03	1.92
$\times 10^{-3}$	6.33	5.64	5.23	4.95	4.73	4.54	4.39	4.26	4.14
$\times 10^{-4}$	8.63	7.94	7.53	7.25	7.02	6.84	6.69	6.55	6.44
$\times 10^{-5}$	10.94	10.24	9.84	9.55	9.33	9.14	8.99	8.86	8.74
$\times 10^{-6}$	13.24	12.55	12.14	11.85	11.63	11.45	11.29	11.16	11.04
$\times 10^{-7}$	15.54	14.85	14.44	14.15	13.93	13.75	13.60	13.46	13.34
$\times 10^{-8}$	17.84	17.15	16.74	16.46	16.23	16.05	15.90	15.76	15.65
$\times 10^{-9}$	20.15	19.45	19.05	18.76	18.54	18.35	18.20	18.07	17.95
$\times 10^{-10}$	22.45	21.76	21.35	21.06	20.84	20.66	20.50	20.37	20.25
$\times 10^{-11}$	24.75	24.06	23.65	23.36	23.14	22.96	22.81	22.67	22.55
$\times 10^{-12}$	27.05	26.38	25.96	25.67	25.44	25.26	25.11	24.97	24.86
$\times 10^{-13}$	29.36	28.66	28.26	27.97	27.75	27.56	27.41	27.28	27.16
$\times 10^{-14}$	31.66	30.97	30.56	30.27	30.05	29.87	29.71	29.58	29.46
$\times 10^{-15}$	33.96	33.27	32.86	32.58	32.35	32.17	32.02	31.88	31.76

Equation (6.46) is known as non-equilibrium equation or Theis equation. The integral in Eq. (6.46) is a function of the lower limit u and is known as exponential integral. It can be expressed as a convergent series so that Eq. (6.46) becomes

$$s = \frac{Q}{4\pi T}\left[-0.5772 - \log_e u + u - \frac{u^2}{2\underline{|2}} + \frac{u^3}{3\underline{|3}} - \frac{u^4}{4\underline{|4}} + ...\right] \quad ...(6.48)$$

Further the exponential integral is usually written symbolically as $W(u)$, i.e.,

$$W(u) = \int_u^\infty \frac{e^{-u}}{u} du \quad ...(6.49)$$

where $W(u)$ is known as *well function*. Equation (6.46) may then be expressed as

$$s = \frac{Q}{4\pi T} W(u) \quad ...(6.50)$$

Wenzel tabulated the values of $W(u)$ for a wide range of u from 10^{-15} to 9.9. Table 6.3 gives the values of $W(u)$ for u ranging from 10^{-15} to 9.

Using Eqs (6.46) and (6.47), the aquifer (or formation) constants S and T can be determined by measuring the change in drawdown with time in one or more observation wells when the test well is pumped at a constant rate Q which is also measured. However, on account of mathematical difficulties it is impossible to obtain an exact analytical solution. As such several investigators have developed simpler approximate methods of solution which can be readily used to determine S and T for field purposes. Some of these methods are described in section 6.16.

6.16 PUMPING TEST METHODS AND RECOVERY TEST METHOD

The various methods which are adopted to determine the aquifer constants S and T may be classified as:

(a) Pumping test methods.

(b) Recovery test method.

(a) Pumping Test Methods. The following three pumping test methods are commonly used :

(1) Theis method.

(2) Jacob method.

(3) Chow method.

In all the above noted methods besides a test or pumping well, one observation well is required. The test well is pumped at a constant rate Q which is measured. During the pumping of the test well the drawdown s is measured in the observation well at different instants of time t. The radial distance r of the observation well from the test well is also measured. Thus, knowing the values of Q, s, t and r the aquifer constants S and T can be determined by each of the above noted methods as described below.

1. **Theis Method.** From Eq. (6.50), we have

$$s = \left(\frac{Q}{4\pi T}\right)W(u) \qquad \text{...(6.50)}$$

Rearranging Eq. (6.47) gives us

$$\frac{r^2}{t} = \left(\frac{4T}{S}\right)u \qquad \text{...(6.51)}$$

From Eqs (6.50) and (6.51) it can be seen that the relation between $W(u)$ and u must be similar to that between s and (r^2/t) because the terms in brackets in the two equations are constants. Based on this similarity Theis suggested a graphical method of superposition which provides an approximate solution of the equation for determining S and T as described below.

(i) From the values of $W(u)$ and u given in Table 6.3, $W(u)$ versus u curve is prepared on logarithmic scales as shown in Fig. 6.15 (a). This curve is known as *type curve.*

(ii) Values of drawdown s measured in the observation well are plotted against the values of (r^2/t) on a separate sheet of paper on logarithmic scales of the same size as that for the type curve shown in Fig. 6.15 (b). This plotting is however done on a tracing paper.

(iii) Keeping the coordinate axes of the two plots parallel, the s versus (r^2/t) plot is superimposed on the type curve and so adjusted that most of the points of the s versus (r^2/t) plot fall on a segment of the type curve as shown in Fig. 6.15 (c).

(iv) With both the plots in the best match position an arbitrary point P called matching point is selected on the matching portion of the two plots (see Fig. 6.15 c) and its coordinates are noted for both the upper and the lower plots. In other words a set of values of parameters $W(u)$, u, s and (r^2/t) is determined.

(v) The aquifer constants S and T may then be determined from the expressions obtained by rearranging Eqs (6.49) and (6.50) as follows.

$$T = \frac{Q}{4\pi s} W(u) \qquad \text{...(6.52)}$$

and

$$S = \frac{4T}{(r^2/t)} u \qquad \text{...(6.53)}$$

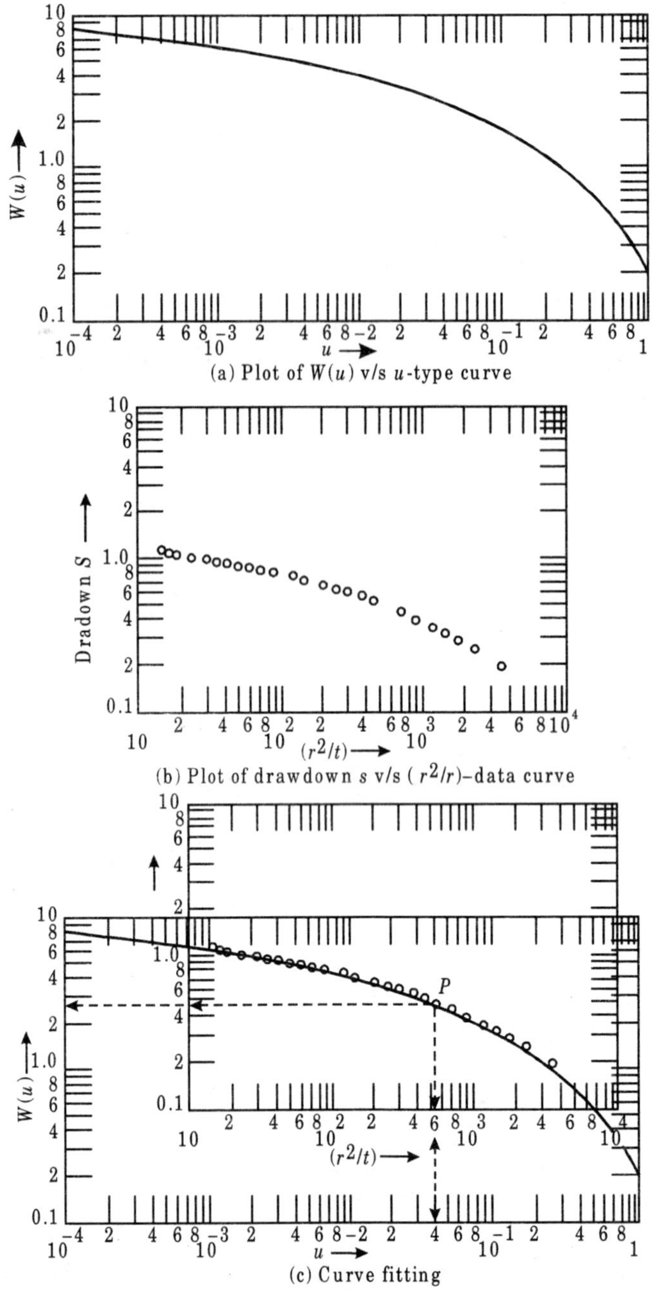

Fig. 6.15. *Theis method for solution of the non-equilibrium equation*

2. Jacob Method. Jacob indicated that for small values of r and large values of t, u is small and hence after the first two terms all other terms of the series in Eq. (6.48) may be neglected. As such the drawdown s can be expressed by the following equation.

$$s = \frac{Q}{4\pi T}\left[-0.5772 - \log_e u\right] \qquad \text{...(6.54)}$$

or

$$s = \frac{Q}{4\pi T}\left[-0.5772 - \log_e\left(\frac{r^2 S}{4Tt}\right)\right] \qquad \text{...(6.54a)}$$

Changing the logarithms to base 10 and grouping together the terms in the brackets, Eq. (6.54a) reduces to

$$s = \frac{2.30Q}{4\pi T}\log_{10}\left(\frac{2.25Tt}{r^2 S}\right) \qquad \text{...(6.55)}$$

Equation (6.55) shows that s varies linearly with $\log t$ since all other terms are constants. Thus, observed values of drawdown s are plotted against $\log t$ and a straight line passing through most of the plotted points is drawn as shown in Fig. 6.16.

Further in the observation well if s_1 and s_2 are the respective drawdowns at times t_1 and t_2 since pumping was started, then from Eq. 6.55, we have

$$(s_1 - s_2) = \Delta s = \frac{2.30Q}{4\pi T}\log_{10}\left(\frac{t_2}{t_1}\right) \qquad \text{...(6.56)}$$

where Δs is the drawdown difference.

If $(t_2/t_1) = 10$, then the drawdown difference Δs per log cycle of t is obtained from Eq. (6.56) as

$$\Delta s = \frac{2.30Q}{4\pi T} \qquad \text{...(6.57)}$$

From Fig. 6.16 value of Δs may be obtained for one log cycle of t and introducing the same in Eq. 6.57 the value of T may be obtained as follows.

$$T = \frac{2.30Q}{4\pi\Delta s} \qquad \text{...(6.58)}$$

Extrapolating the straight line of s versus $\log t$ plot to intersect the zero-drawdown axis permits the determination of S as indicated below.

Let $t = t_0$ when $s = 0$ as shown in Fig. 6.16, i.e., t_0 is the time since pumping started upto which the drawdown in the observation well is equal to zero. Then from Eq. (6.55) we have

$$s = 0 = \frac{2.30Q}{4\pi T}\log_{10}\left(\frac{2.25Tt_0}{r^2 S}\right)$$

from which it follows that

$$\frac{2.25\,T t_0}{r^2 S} = 1$$

or $$S = \frac{2.25T t_0}{r^2} \qquad \text{...(6.59)}$$

Introducing the value of T obtained from Eq. (6.58) in Eq. (6.59), the value of S may be determined.

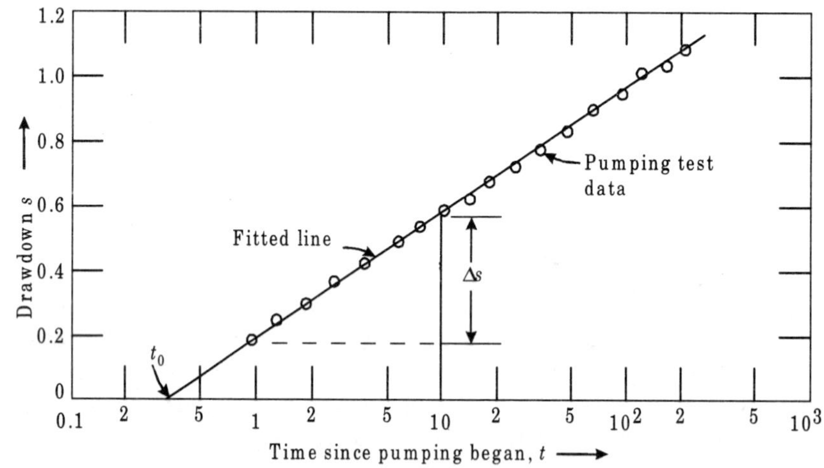

Fig. 6.16. *Jacob method for solution of nonequilibrium equation*

However, the straight line approximation for this method should be restricted to small values of u, i.e., $u < 0.01$ to avoid large errors.

Alternatively if a number of observation wells are available then the values of the aquifer constants S and T may also be obtained by *distance-drawdown relation* as indicated below.

Applying Jacob's Eq. (6.55) at any instant of time t to two observation wells located at radial distances r_1 and r_2 from the pumping or test well, we get

$$s_1 = \frac{2.30Q}{4\pi T}\log_{10}\left(\frac{2.25T_t}{r_1^2 S}\right) \qquad \text{...(i)}$$

$$s_2 = \frac{2.30Q}{4\pi T}\log_{10}\left(\frac{2.25T_t}{r_2^2 S}\right) \qquad \text{...(ii)}$$

Subtracting Eq. (ii) from Eq. (i), we get

$$(s_1 - s_2) = \Delta s = \frac{2.30Q}{4\pi T}\log_{10}\left(r_2^2 / r_1^2\right)$$

or $$(s_1 - s_2) = \Delta s = \frac{2.30Q}{2\pi T}\log_{10}\left(r_2 / r_1\right) \qquad \text{...(iii)}$$

where Δs is the drawdown difference.

If $(r_2/r_1) = 10$, then the drawdown difference Δs per log cycle of distance r measured from the pumped well is obtained from Eq. (iii) as

$$\Delta s = \frac{2.30Q}{2\pi T} \qquad \text{...(iv)}$$

At any instant of time t the drawdown s is noted in each of the observation wells. Equation (6.55) shows that s varies linearly with log r. Thus, a plot of s versus log r is made from which the value of Δs may be obtained for one log cycle of r. Introducing this value of Δs in Eq. (iv) the value of T may be obtained as

$$T = \frac{2.30Q}{2\pi \Delta s} \qquad \text{...(v)}$$

Further extrapolating the straight line of s versus log r plot to intersect the zero-drawdown axis permits the determination of S as indicated below.

If at a radial distance r_o the drawdown $s = 0$, then from Eq. (6.55), we have

$$\frac{2.25Tt}{r_0^2 S} = 1$$

or

$$S = \frac{2.25Tt}{r_0^2} \qquad \text{...(vi)}$$

Introducing the value of T obtained from Eq. (v) in Eq. (vi) the value of S may be determined.

3. Chow Method. Chow (1952) developed a method of solution of This equation which avoids the curve matching technique as in the Theis method and also it is not restricted to large values of t and small values of r as in the Jacob method.

Chow introduced a function $F(u)$ which is related to $W(u)$ and u as follows.

$$F(u) = \frac{W(u)e^u}{2.30} \qquad \text{...(6.60)}$$

The relation between $F(u)$, $W(u)$ and u is shown in Fig. 6.17. The function $F(u)$ is introduced to relate $W(u)$ and u to a certain combination of s and t. The value of $F(u)$ may be determined as follows.

(i) The observed drawdown s is plotted against $\log_{10} t$ and a smooth curve is drawn through the plotted points as shown in Fig. 6.18.

(ii) On the plotted curve an arbitrary point P is chosen and its coordinates s and t are noted.

(iii) A tangent to the curve is drawn at the chosen point P and the drawdown difference Δs per log cycle of time t is determined.

(iv) The value of $F(u)$ is then calculated from the following expression

$$F(u) = \frac{s}{\Delta s} \qquad \text{...(6.61)}$$

Corresponding to the value of $F(u)$ so obtained the values of $W(u)$ and u are determined from Fig. 6.17. The values of the aquifer constants T and S may then be computed from Eqs (6.52) and (6.53).

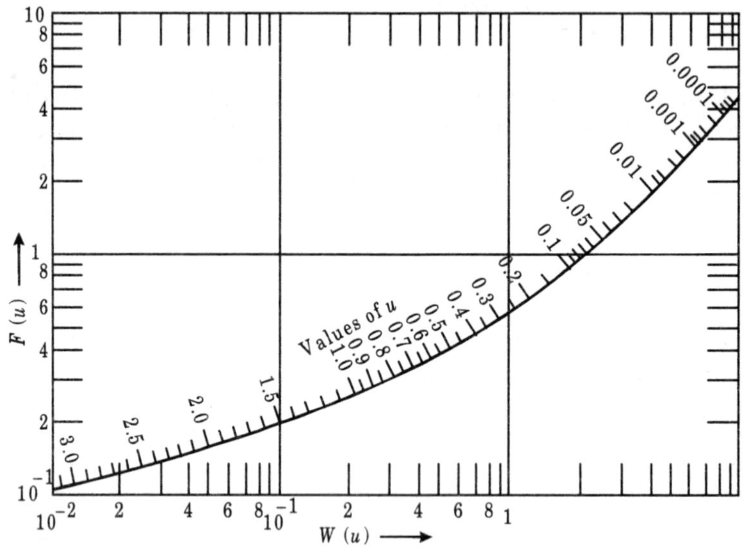

Fig. 6.17. *Relation between F(u), W(u) and u*

Alternatively the values of $W(u)$ and u may also be determined without involving the use of Fig. 6.17 as indicated below.

For small values of u as indicated in Jacob method from Eq. (6.57), we have

$$\Delta s = \frac{2.30Q}{4\pi T}$$

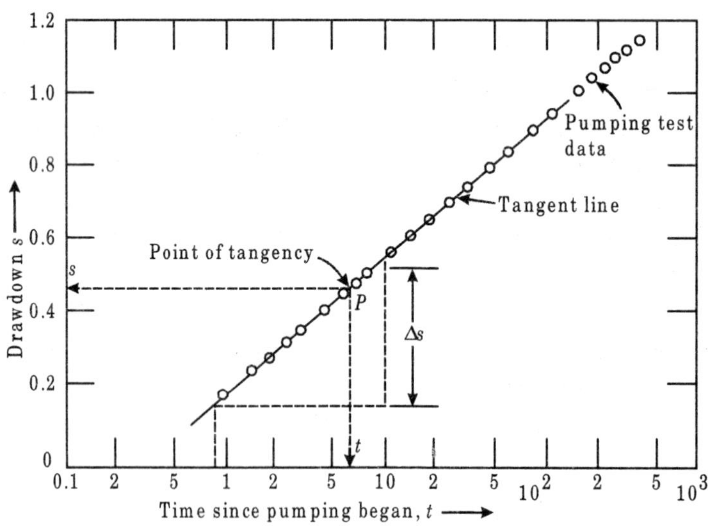

Fig. 6.18. *Chow method for solution of the nonequilibrium equation*

Also from Eq. (6.50), we have

$$s = \frac{Q}{4\pi T} W(u)$$

Substituting the values of s and Δs in Eq. (6.61), we obtain

$$F(u) = \frac{W(u)}{2.30} \quad \text{...(6.62)}$$

or

$$W(u) = 2.30\ F(u) \quad \text{...(6.62a)}$$

Thus, for the value of $F(u)$ obtained as indicated above, the value of $W(u)$ may be determined by using Eq. 6.62 (a). Corresponding to the value of $W(u)$ so determined the value of u may be obtained from Table 6.3. However, this method of determining $W(u)$ and u may be used only for $F(u) > 2$.

All the above noted methods are indicated in the Illustrative Examples 6.12, 6.13 and 6.14.

(b) Recovery Test Method. By the recovery test method only the coefficient of transmissibility T can be determined. In this method the observation well is not required because the required data may be collected on the test or the pumping well. The test well is pumped for a known period of time and then pump is stopped. When the pumping is stopped, the water level in the pumping well begins to rise. This is termed as recovery of the groundwater level. At any instant of time during the recovery period the depth of the water level in the well below the original static level prior to pumping is known as *residual drawdown.* It is assumed that after the pumping is stopped, the drawdown will be identically the same as if the pumping has been continued and a hypothetical recharge well with the same flow rate were superimposed on the pumping well at the instant the pumping is stopped. Based on this assumption Theis showed that the residual drawdown s' can be given as

$$s' = \frac{Q}{4\pi T}[W(u) - W(u')] \quad \text{...(6.63)}$$

where $u = \dfrac{r^2 S}{4T\,t}$ and $u' = \dfrac{r^2 S}{4T\,t'}$ ...(6.64)

and t is the time elapsed since the pumping of the test well was started and t' is the time elapsed since the pumping was stopped.

Further for small values of r and large values of t', the well function can be approximated by the first two terms of the series in Eq. (6.48) so that Eq. (6.63) can be written as

$$s' = \frac{2.30Q}{4\pi T}\log_{10}\left(\frac{t}{t'}\right) \quad \text{...(6.65)}$$

Equation (6.65) shows that s' varies linearly with $\log_{10}\ (t/t')$ since all other terms are constants.

During the recovery period the residual drawdown s' is measured in the test well at different instants of time t'. The observed values of s' are plotted against $\log_{10}\ (t/t')$ and a straight line passing through most of the plotted points is drawn. From this plot the difference of residual drawdown per log cycle of (t/t') is noted. Thus, if $\Delta s'$ is the difference of residual drawdown per log cycle of (t/t') then from Eq. (6.65)

$$\Delta s' = \frac{2.30Q}{4\pi T} \quad \text{...(6.66)}$$

and the value of T may be obtained as

$$T = \frac{2.30Q}{4\pi \Delta s'} \quad \text{...(6.67)}$$

The recovery test method provides an independent check on the value of T determined by pumping test method. For this at the end of the pumping test, when the pump is stopped, the residual drawdowns are measured at different instants of time and from the data so obtained the value of T is determined as indicated above. Further in the pumping test method the test well is required to be pumped at a constant rate which is however difficult to control accurately in the field. On the other hand in the recovery test method the pumping of the well is not involved but during recovery period the rate of recharge Q to the well is assumed constant and equal to the mean pumping rate during the pumping test. Moreover as compared to pumping test method the recovery test method is less costly because no observation well is required.

This method is indicated in the Illustrative Example 6.15.

6.17 TYPES OF WATER WELLS

A water well is a hole or shaft, usually vertical, excavated in the earth for bringing groundwater to the surface. Water wells may be classified as

(1) Open wells or Dug wells

(2) Tube wells.

1. Open Wells or Dug Wells. Open wells are the wells which have comparatively large diameters but low yields (or discharges) and are not very deep. The diameters of the open wells usually vary from 1 m to 10 m. The yield of such wells in most of the cases is about 20 m^3/hour or less. However, a properly constructed open well penetrating a permeable aquifer can yield 100 to 300 m^3/hour. The depths of open wells may generally range from 2 m to 20 m. Since these wells are usually constructed by digging, these are also known as dug wells. The walls of an open well may be built of brick or stone masonry or precast concrete rings. The thickness generally varies from 0.5 m to 0.75 m depending on the depth of the well.

Open wells may be further classified as

(a) Shallow open wells

(b) Deep open wells.

Shallow open wells are those which rest in the top water bearing strata and draw their supplies from the surrounding material as shown in Fig. 6.19. On the other hand deep open wells are those which rest on impervious strata and draw their supplies from the pervious formation lying below the impervious strata through bore holes made in the impervious strata. The impervious strata is generally known as *mota layer* and it is a layer of clay, cemented sand, kankar or other hard materials. The term mota layer is however not applied to layers of hard materials lying above the water table. The main advantage of such a mota layer is that it gives structural support to the open well resting on its surface. Further since the pervious formation below the mota layer generally contain large quantity of water, the yield of deep wells is more

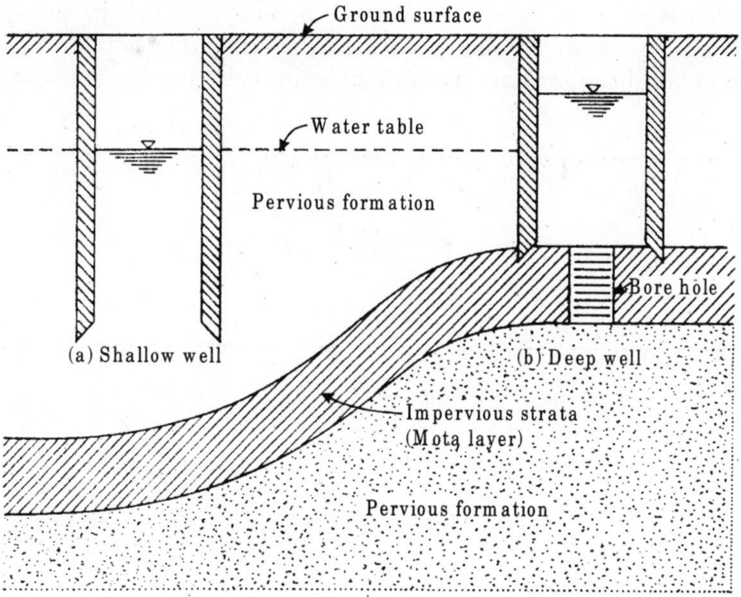

Fig. 6.19. *Shallow and deep open wells*

than that of shallow wells. It may however be mentioned that the nomenclature of shallow and deep open wells is purely technical and it has nothing to do with the actual depth of the well because sometimes a shallow well may have more depth than a deep well.

2. Tube Well. A tube well is a long pipe sunk into the ground intercepting one or more water bearing strata. As compared to open wells the diameter of tube wells are much less and usually range from 80 mm to 600 mm. The tube wells may also be further classified as

(a) Shallow tube wells

(b) Deep tube wells.

Shallow tube wells have their depths limited to about 30 m and may have a maximum yield of about 20 m^3/hour. On the other hand deep tube wells may have maximum depth of about 600 m and may yield more than 800 m^3/hour. As indicated in section 6.20 usually different methods are used for the construction of shallow and deep tube wells.

The tube wells may also be classified as

(a) Strainer type tube well

(b) Cavity type tube well

(c) Slotted type tube well.

A brief description of these three types of tube wells is given below.

(a) *Strainer type tube well*. It is the most common and widely used type of tube well such that the term tube well in general refers only to a strainer type of tube well. In this type of tube well the pipe introduced into the ground is an assembly of strainer pipes (also called strainers) and ordinary pipes (called blind pipes) which are alternately placed. The lengths of the strainer and the blind pipes are so adjusted that the strainer pipes rest against aquifers (or water bearing strata) and the blind pipes rest against aquicludes (or impervious strata) as shown in Fig. 6.20. The various types of strainers that are commonly used for these tube wells are indicated below. In general the strainer consists of a fine wire mesh (or screen) wrapped round a

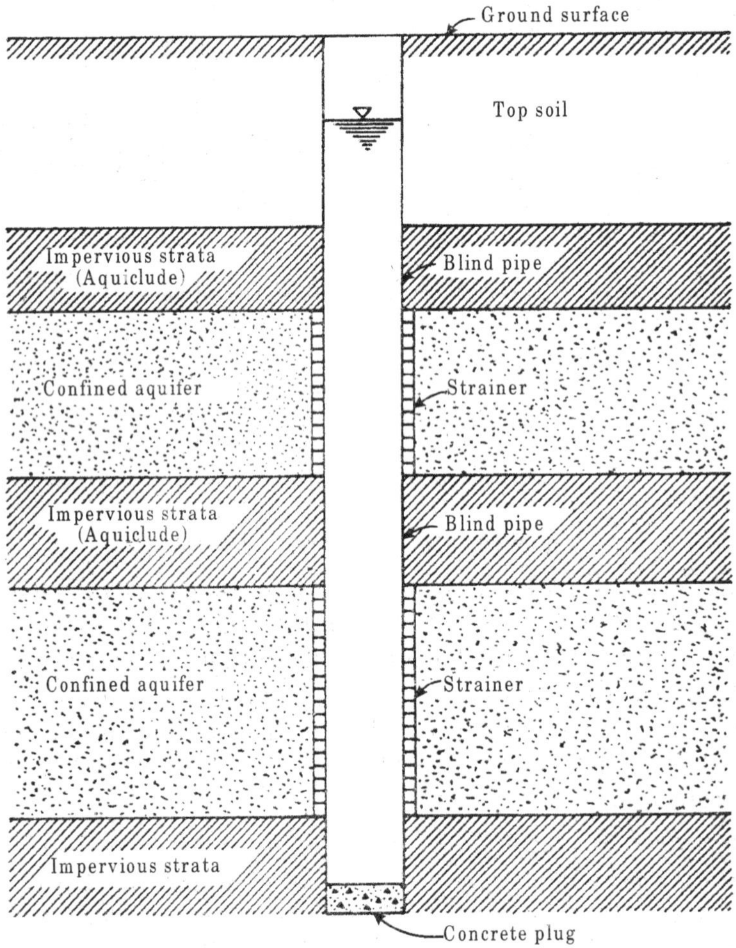

Fig. 6.20. *Strainer type tube well*

slotted or perforated pipe with a small annular space between the two. The total area of the openings of the slotted pipe is kept equal to the area of the openings of the wire mesh so that no change in the velocity of flow occurs between the two. The annular space between the pipe and the wire mesh is required, otherwise the wires of the mesh will cover a large portion of the area of the openings of the pipe. The water enters the well pipe through the wire mesh and hence the sand particles of size larger than the size of the mesh openings are prevented from entering the well pipe. This reduces the danger of sand removal and hence larger velocities of flow can be permitted in this case. The size of the openings of the wire mesh is generally kept equal to D_{60} to D_{70} of the surrounding soil. Some strainers consist of only the slotted pipes without wire mesh wrapped round them, in which case the slots are of relatively very small size so as to prevent the entry of sand particles into the well pipe. The following are the basic requirements of the strainers.

(i) Resistance to corrosion and deterioration.

(ii) Enough structural strength to prevent collapse.

(iii) Suitability to prevent excessive movement of sand into the well.

The strainer type tube well is generally unsuitable for very fine sandy strata, because in that case the size of mesh opening will have to be reduced considerably which may result in choking of the strainer, and if the screen openings are kept bigger, the well will start discharging sandy water.

The diameter of the bore hole or casing pipe for a strainer type tube well is generally kept 50 to 100 mm more than the diameter of the well pipe. Thus, for a 150 mm diameter well, a bore hole of 200 to 250 mm diameter shall be drilled. After the casing pipe is removed the annular space between the well pipe and the surrounding strata is backfilled with gravel.

A strainer type tube well may draw water either from an unconfined aquifer of unlimited extent or from one or more confined aquifers lying one over the other as shown in Fig. 6.20. At the bottom a short blind pipe is provided so as to permit settlement of any sand particles if passed through the strainer. The well is generally plugged at bottom by cement concrete.

Abyssinian tube well is a special type of strainer well in which the diamter of the well pipe is kept equal to 38 mm $\left(1\frac{1}{2}\text{ in.}\right)$ and the strainer is provided only for a length of about 1.2 to 1.5 m (4 to 5 ft).

Type of strainers. Some of the common types of strainers used for the tube wells are as follows.

(i) Cook strainer

(ii) Tej strainer

(iii) Brownline strainer

(iv) Ashford strainer

(v) Leggett strainer

(vi) Phoenix strainer

(vii) Layne and Bowler strainer

(i) *Cook strainer.* It consists of a solid drawn brass tube having wedge-shaped horizontal slots. The slots are made with a slot cutting machine from inside of the tube. The slots are wider at inside and narrower at outside as shown in Fig. 6.21. The gage of the slots depends on the coarseness of the surrounding soil and varies from 0.15 to 0.4 mm. This is a relatively costly strainer and is commonly used in U.S.A. and other western countries.

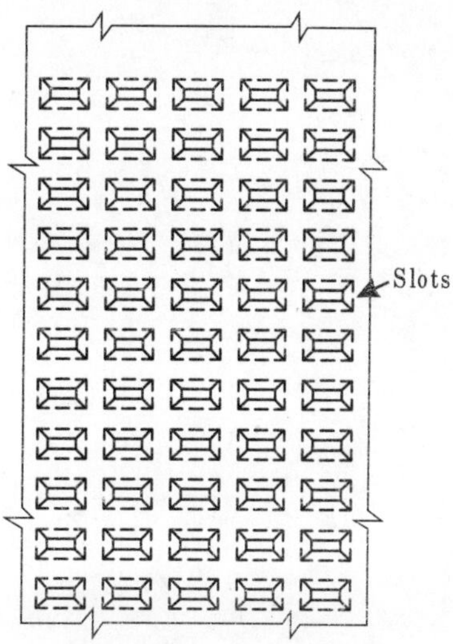

Fig. 6.21 *Cook strainer*

(ii) *Tej strainer.* It is similar to Cook strainer, but it consists of a brass tube made from a brass sheet bent round to form the tube with its vertical joint being brazed. The slots are cut in the sheet before it is bent. This strainer is generally manufactured from 75 mm diameter and upwards, and in 2.5 m length which may be joined together by means of screwed collars of brass. Tej strainers are manufactured in India and are commonly used in our country.

(iii) *Brownlie strainer.* It consists of a polygonal convoluted tube of steel plate having perforations and a copper wire mesh wrapped around it as shown in Fig. 6.22 (a). The mesh consists of heavy parallel copper wires woven with copper ribbons. This is the best type of strainer as it has the advantage of keeping the wire mesh (or straining material) away from the perforated tube. A similar strainer is shown in Fig. 6.22 (b) in which the convolutions are straight sided and the tube is star-shaped.

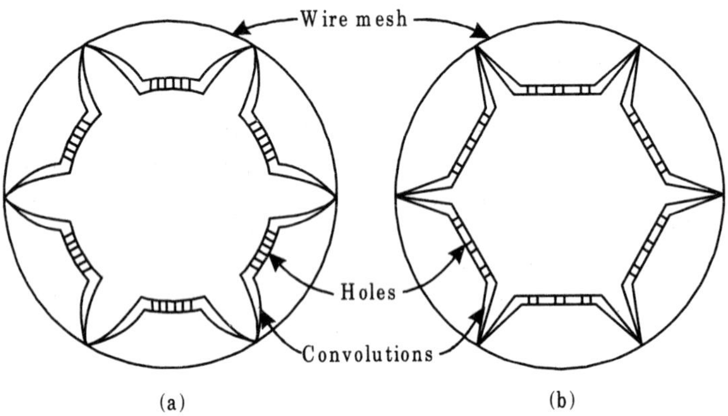

Fig. 6.22 *Brownlie strainer*

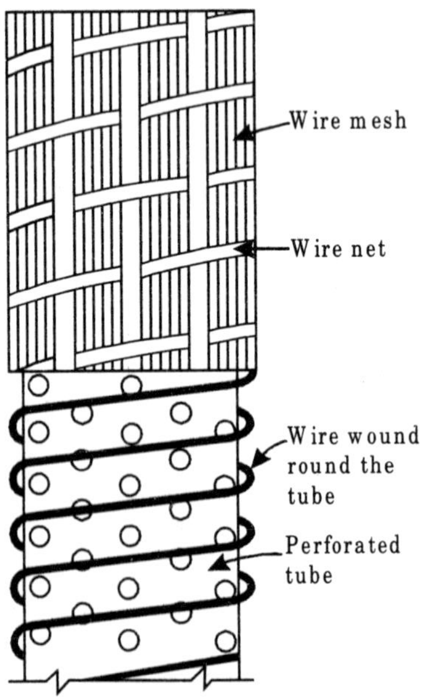

Fig. 6.23. *Ashford strainer*

(iv) *Ashford strainer.* It consists of a perforated tube with a wire wound round it over which a wire mesh is soldered as shown in Fig. 6.23. The wire keeps the mesh away from the tube. The wire mesh is protected and strengthened by a wire net around it as shown in Fig. 6.23. It is a delicate strainer and needs careful handling.

(v) *Legget strainer.* In this strainer a cleaning device is provided which is in the form of cutters. The cutters can be operated from the ground surface. When the cutters are turned in they clean the slits.

(vi) *Phoenix strainer.* It consists of a mild steel tube in which slots are cut from inside by a slot cutting machine. The tube is cadmium plated to keep it free from danger of choking from corrosion caused by chemical action.

(vii) *Layne and Bowler strainer.* It consists of a wedge-shaped steel wire wound to a suitable pitch round a slotted or perforated steel or wrought iron pipe. The strainer pipes are joined together by screwed collars. This strainer is of a very robust type and is generally used in oil wells.

(b) *Cavity type tube well.* As shown in Fig. 6.24 a cavity type tube well consists of a pipe sunk into the ground and resting on the bottom of a strong clay layer. It does not utilise strainers and hence it draws its supplies from the bottom and not from the sides as in the case of a strainer type tube well. In the initial stages of pumping fine sand comes out with water and consequently a cavity is formed at the bottom. The water from the aquifer enters the well pipe through this cavity. As the pumping is continued the spherical area of the cavity increases outwards with the result that the velocity of flow is reduced and consequently the sand particles stop entering the well pipe. Hence in the beginning sandy water is obtained from a cavity type tube well but with the passage of time clear water is obtained.

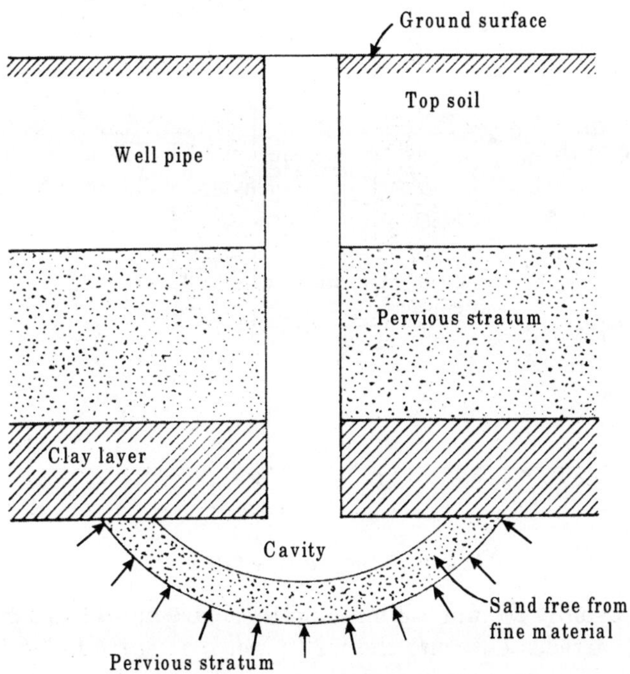

Fig. 6.24 *Cavity type tube well*

Since water is drawn from the bottom, only one particular aquifer can be tapped by the cavity type tube well. Thus, in principle a cavity type tube well is similar to a deep open well, with the difference that whereas a deep open well taps only the first aquifer just below the mota layer, a cavity type tube well may tap any lower strata.

The essential difference in the flow pattern of a strainer type tube well and a cavity type tube well is that due to the provision of strainers the flow in a strainer type tube well is radial, whereas in a cavity type tube well the flow is spherical. Further in a strainer type tube well the area of flow section may be increased by increasing the length of the strainer pipe (only in the case of an unconfined aquifer of large extent) while in the case of a cavity type tube well the area of flow section is increased by enlarging the cavity at the bottom of the well. The cavity formed due to certain discharge gets enlarged if an increased discharge is pumped out from the well.

(c) *Slotted type tube well.* A slotted type tube well consists of a pipe which is slotted for part of its length at one end and for the rest of the length it is plain pipe. The slotted portion of the pipe is usually about 5 m long and it penetrates the confined aquifer as shown in Fig. 6.25.

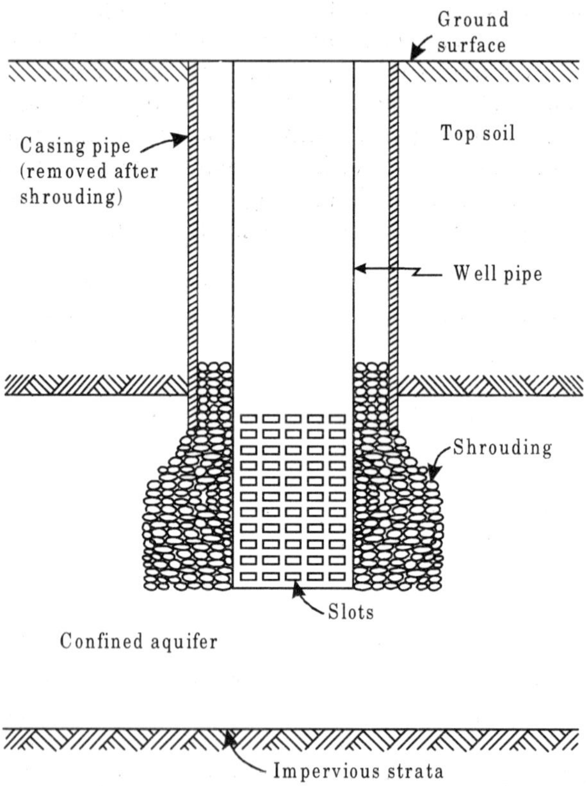

Fig. 6.25 *Slotted type tube well*

The size of the slots is usually 25 mm × 3 mm and these are at 10 to 12 mm spacing. In order to prevent the fine sand particles entering the well pipe, it is surrounded by a mixture of gravel and coarse sand (bajri). The mixture of gravel and coarse sand placed around the well pipe is called *shrouding* and it is poured from the top into the annular space between the well pipe and the casing pipe before withdrawing the casing pipe. The shrouding must extend upto about 3 to 4 m above the top level of the slotted portion of the well pipe. Thus, in this case the diameter of the casing pipe is kept 200 to 250 mm more than the diameter of the well pipe. For example for a well pipe of 150 mm diameter a casing pipe of about 400 mm diameter is required.

A slotted type tube well differs from a strainer type tube well in the following two respects.

(i) In a strainer type tube well the strainer pipes are surrounded by wire mesh to prevent the fine particles from entering the well pipe, while in a slotted type tube well the gravel and coarse sand shrouding serves this purpose.

(ii) A strainer type tube well can tap one or more confined aquifer lying one over the other, whereas a slotted type tube well can tap only one aquifer.

Under certain circumstances as indicated below it may not be possible to provide either a strainer type tube well or a cavity type tube well. For example, if sufficient depth of water bearing stratum is not available even upto a depth of 75 to 100 m then strainer type tube well cannot be provided. Similarly if a suitable strong clay strata is not available then cavity type tube well cannot be provided. In such circumstances a slotted type tube well can be provided if at least one stratum having sufficient quantity of water is available.

6.18 METHODS OF CONSTRUCTION OF OPEN WELLS

From the point of view of construction the open wells may be classified in the following three types.

Type I. *Wells with impervious lining.* These are the most stable and useful type of wells. For constructing such a well first of all a pit is excavated using hand tools such as pick and shovel upto the soft moist soil. A curb is then lowered into the pit. A curb is a circular ring of R.C.C., timber or steel having a cutting edge at the bottom and a flat top wide enough to support the well lining (or steining) of the desired thickness. Masonry or concrete lining is then built up on the curb upto a few metres above the ground surface. The soil in the pit is then excavated and as the excavation proceeds below the curb then the curb along with the lining sinks down. If the lining does not sink under its own weight then it is loaded on the top by sand bags etc. As the lining sinks down it is further extended at the top. To ensure vertical sinking of the well lining plumb bobs are used and if it starts getting tilted it is corrected by adjusting the loads or by removing the soil from below the curb which may be causing the tilt.

After the well has gone upto the water table, further excavation may be done either by continuously removing the water through pumps or the excavation may be carried out from the top by jhams. A jham is a self closing bucket which is tied to a rope and worked up and down over a pulley. When a jham is thrown into the well, its jaws strike the bottom of the well, dislodging some of the soil materials. As the jham is pulled, the soil cuttings get retained but the water oozes out.

In the case of a deep well the sinking is continued till the impervious or mota layer is reached. A bore hole of small diameter is then made through the mota layer in the centre of the well, which is generally protected by a timber lining. In the case of a shallow well since mota layer is not available, the well is sunk as described above upto the required depth and it is partly filled with gravel or stone ballast or brick ballast. This will function as a filter through which sand particles will be prevented from rising up.

In a well lined with an impervious lining on its sides, the flow is not radial. The water enters only from the bottom and the flow becomes spherical when once the cavity has been formed at the bottom.

Type II. *Wells with the pervious lining.* In this type of wells, brick or stone lining without mortar or binding material is used on the sides of the wells. The water thus enters from the sides through the pores in the lining. These wells are generally plugged at the bottom by means of concrete. Therefore the flow in these wells is radial. However, if the bottom is not plugged, then the flow will be a combination of radial and spherical flows. Such wells are generally suitable in gravel or coarse sand strata. When such a well is constructed in finer soil, the pervious lining is surrounded by gravel or stone ballast or brick ballast to prevent the entry of sand into the well along with the percolating water.

Type III. *Kachha wells.* These wells are not provided with lining. As such these are temporary wells of very shallow depths. Such wells can be constructed in hard soils where the wells walls can stand vertically without any support. Further these wells can be constructed only where water table is very near the ground surface. Though these wells are quite cheap and useful, they may collapse after some time and may sometimes prove to be dangerous.

6.19 YIELD OF AN OPEN WELL

The hydraulics of an open well is different from the one discussed in section 6.8 which is applicable to a tube well. The yield of an open well can be found by the following two tests.

(1) Constant level pumping test

(2) Recuperation test.

1. Constant Level Pumping Test. In this test water is pumped from the open well so that the water level in the well is depressed by an amount say h known as *depression head* (or draw-down) as shown in Fig. 6.26. The rate of pumping is then so adjusted that the water level in the well remains constant. In this condition the rate of pumping will be equal to the rate of yield from the well at this depression head. The rate of pumping of water is measured by a suitable discharge measuring device. Thus, if Q is the rate of yield (or discharge) at a depression head h and A is the area of flow section through which water enters the well (which will be equal to the cross-sectional area of the bottom of the well if the lining of the well is impervious) then from Darcy's Law.

$$Q = kAi$$

or

$$Q = kA\frac{h}{L}$$

or

$$Q = CAh \qquad \text{...(6.68)}$$

where $C\ (= k/L)$ is percolation intensity coefficient. This is a constant of the formation around the well. Its value is greater for coarser soils and smaller for finer soils. The value of C may be determined if Q, A and h are known and the same may be used to determine the rate of yield from the well for any other value of the depression head by using Eq. 6.68. Thus, for a constant depression head H, the rate of yield from a well is given by Eq. 6.68 as

$$Q = CAH \qquad \text{...(6.68a)}$$

In case a cavity is formed at the bottom of the well then the area A is taken as (4/3) times the cross-sectional area of the bottom of the well. This is so because if no cavity is formed at the bottom of the well then

$$A = \frac{\pi d^2}{4}$$

On the other hand if a hemispherical cavity is formed at the bottom of the well, then

$$A = \frac{\pi d^2}{2}$$

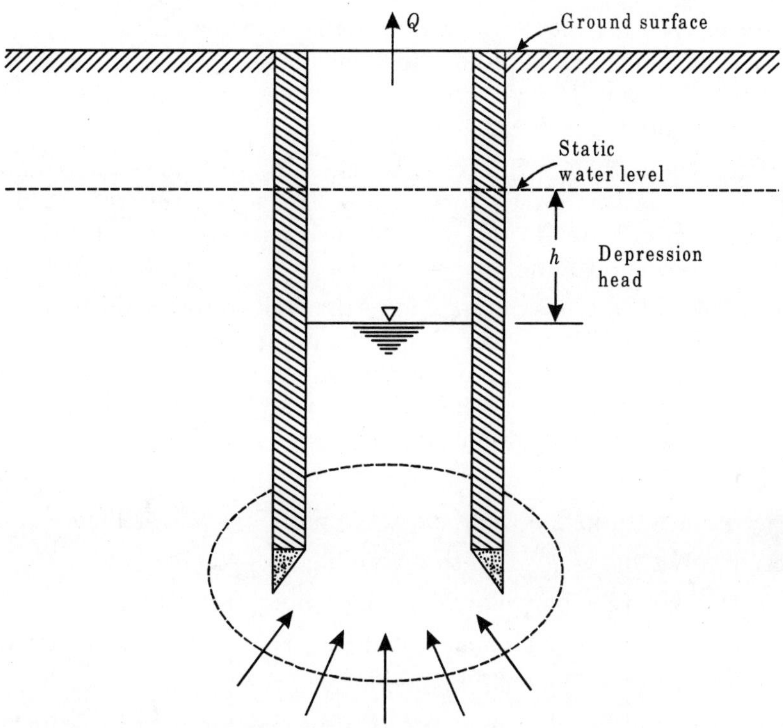

Fig. 6.26 *Constant level pumping test*

Since the area of the actual cavity fromed is in between these two values, the value of A may be taken as the mean of these two values. Thus,

$$A = \frac{1}{2}\left(\frac{\pi d^2}{4} + \frac{\pi d^2}{2}\right)$$

$$= \frac{3}{8}\left(\pi d^2\right)$$

$$= \frac{1}{3}(\pi d^2)$$

$$= \frac{4}{3}\left(\frac{\pi d^2}{4}\right)$$

From Eq. (6.68), it is clear that the discharge increases with the depression head. However, the depression head cannot be increased beyond a certain critical value because otherwise the velocity of water percolating in the well will be excessive and the soil particles will be disturbed and dislodged. The critical value of h at which the percolation velocity is critical is known as *critical depression head.* Normally the depression head is kept equal to 1/3 of the critical depression head and it is known as the working head.

Maximum yield or *critical yield* of a well will therefore be obtained corresponding to the critical depression head. The yield under the working depression head is known as the maximum safe yield of a well. From the constant level pumping test described above the maximum safe yield can be determined.

2. Recuperation Test. Though the constant pumping test gives an accurate value of safe yield of an open well, it is sometimes very difficult to adjust the rate of pumping so as to maintain a constant level of water in the well. In such circumstances, a recuperation test may be adopted which is described below.

In the recuperation test water is pumped from the well so that sufficient depression head is developed. The depression head is measured and the pumping is then stopped. The water level in the well will start rising. The time taken by the water to come back to the initial static level before the pumping was started or any other measured level is then noted. The rate of yield (or discharge) may then be determined as indicated below.

With reference to Fig. 6.27,

Let aa = Static water level in the well before the pumping was started

bb = Water level in the well when the pumping was stopped

h_1 = Depression head in the well when the pumping was stopped

cc = Water level in the well at some noted time t' after the pumping was stopped

h_2 = Depression head in the well at time t' after the pumping was stopped

h = Depression head in the well at time t after the pumping was stopped

dh = Decrease in depression head in time dt.

Thus, in time t after the pumping was stopped the water level in the well recuperated by $(h_1 - h)$. It again recuperates by dh in a time dt after this.

∴ Volume of water entering the well when the head recuperated by dh is

$$dV = Adh \qquad \text{...(i)}$$

where A = cross-sectional area of the well at its bottom.

Again if Q is the rate of recharge into the well at time t under a depression head h, then the volume of water entering the well in a time dt is

$$dV = Qdt$$

But $$Q \propto h$$

or $$Q = Kh \qquad \text{...(ii)}$$

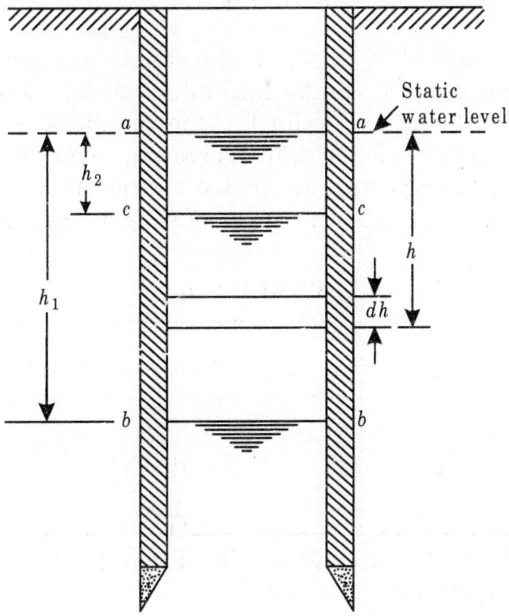

Fig. 6.27 *Recuperation test*

where K is a constant the value of which depends on the type of soil at the base of the well through which water enters the well

$$\therefore \qquad dV = Kh\, dt \qquad \text{...(iii)}$$

Equating the values of dV given by Eqs (i) and (iii), we get

$$Kh\, dt = -Adh \qquad \text{...(iv)}$$

The negative sign is introduced because as time t increases the depression head h decreases.

From Eq. (iv), we have

$$dt = -\frac{A}{K}\frac{dh}{h}$$

Integrating both sides of the above equation between the limits $t = 0$, $h = h_1$; and $t = t'$, $h = h_2$, we get

$$\int_0^{t'} dt = -\frac{A}{K}\int_{h_1}^{h_2} \frac{dh}{h}$$

or $$t' = \frac{A}{K} \log_e (h_2 / h_1)$$

or
$$\frac{K}{A} = \frac{1}{t'}\log_e (h_2 / h_1)$$

or
$$\frac{K}{A} = \frac{2.30}{t'} \log_{10} (h_1/h_2) \qquad ...(6.69)$$

Since h_1, h_2 and $t'$are measured in this test, the same are known and by introducing their values in Eq. (6.69), the value of (K/A) can be calculated. (K/A) is known as the *specific yield* or *specific capacity* of an open well which is defined as the volume of water that percolates into the well per unit time per unit area under a unit depression head. The value of (K/A) is usually expressed in cubic metre per hour per square metre of area under one metre depression head. The approximate values of (K/A) for different types of soils as indicated by Marriot are given in Table 6.4.

TABLE 6.4

Types of soil	(K/A) *(m^3 per hr per m^2 of area under 1 m depression head)*
Clay	0.25
Fine sand	0.50
Coarse sand	1.00

Knowing the value of (K/A) the rate of yield (or discharge) Q from a well under a constant depression head H may be determined as follows:

From Eq. (ii), we have

$$Q = KH \qquad ...(6.70)$$

or
$$Q = \left(\frac{K}{A}\right)AH$$

or
$$Q = \frac{2.30AH}{t} \log_{10} (h_1/h_2) \qquad ...(6.70a)$$

It may however be noted that in Eq. (6.70) if time t' is in hours then Q will be in m³/hr and if t' is in seconds then Q will be in cumec. Further if H is the maximum depression head then the corresponding value of Q obtained from Eq. (6.70) will be the maximum yield of the well. However, if H is the average depression head, Q will be the average yield of the well.

6.20 METHODS OF CONSTRUCTION OF TUBE WELLS

1. Shallow Tube Wells. Shallow tube wells are constructed by boring, driving and jetting. The wells constructed by these methods are designated as bored wells, driven wells and jetted wells. These methods are briefly described below.

(i) *Bored wells.* Where water table exists at a shallow depth in an unconsolidated aquifer, wells may be constructed by boring with hand operated or power driven augers. Hand operated augers are available in different shapes and sizes but in general they have cutting blades at the bottom that bore into the ground as the auger is rotated. When the blades are full of loose earth, the auger is removed from the hole and emptied ; and the operation is repeated until the desired depth is reached. The hand operated augers can bore wells only upto 200 mm in diameter and 15 m in depth. However, the power driven augers can bore wells upto 1 m in diameter and depths exceeding 30 m. A power driven auger consists of a cylindrical steel bucket (Fig. 6.28) with a cutting edge projecting from a slot in the bottom, which cuts into the soil and fill it in the bucket as the auger is rotated. When full, the auger is hoisted to the surface and the excavated material is removed through hinged openings on the side or bottom of the bucket. Reamers, attached to the top of the bucket, can enlarge holes to diameters exceeding the auger size.

A continuous-flight power driven auger has a spiral extending from the bottom of the hole to the surface. In this case cuttings are carried to the surface as on screw conveyor.

Augers work best in formations that do not cave in. However, where loose sand and gravel are encountered in a large-diameter hole, or when the boring reaches the water table a concrete or metal casing pipe is introduced into the hole and the boring is continued inside the casing. After the boring is done upto the desired depth the well pipe together with strainer is introduced centrally in the hole and the casing pipe is removed. The annular space between the well pipe and the surrounding strata is backfilled with gravel.

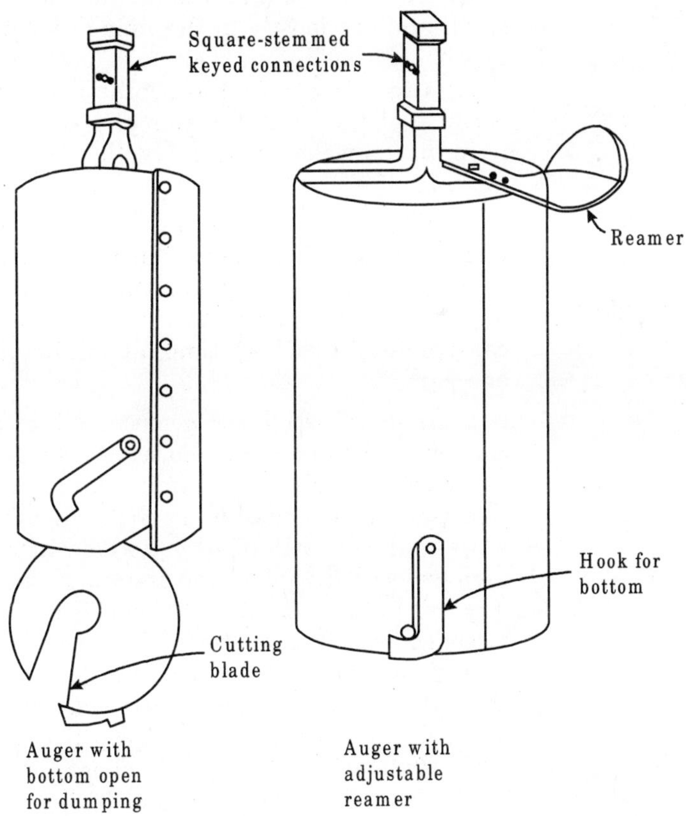

Fig. 6.28 *Augers for boring shallow tube wells*

(ii) *Driven wells.* A driven well is constructed in an unconsolidated formation by the use of a well point. A *well point* consists of a perforated pipe wrapped over by a fine wire mesh (or screen), with a conical steel shoe fixed at the bottom. The conical shoe is known as drive point and it is provided to facilitate the driving of the well point into the soil and also to protect its screened section during driving. The well point is connected to a pipe, and the well point along with the pipe is driven by repeated impacts into the ground. Additional lengths of pipe are connected by threaded couplings and the driving is continued until the desired depth is reached when the entire well point penetrates the aquifer sufficiently below the water table as shown in Fig. 6.29. The impact is provided at the top of the pipe by the falling weight of a drop hammer. A drive cap is provided at the top of the pipe to protect the pipe and its threads against the possible damage due to the impact of the falling weight.

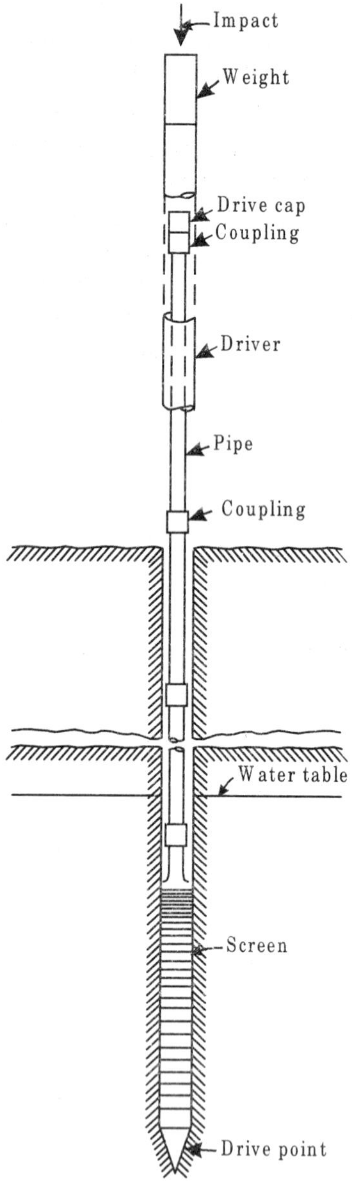

Fig. 6.29 *Well point*

The diameters of driven wells are equal to the diameters of the well points used for their construction. Since from practical considerations the diameters of the well points seldom exceed 100 mm, the diameters of the driven wells are small and are mostly in the range of 30 to 100 mm. Also the depths of the driven wells are generally limited to 15 m though in a few cases the depths may exceed 20 m. As such because of the limitations on the size and depth, the yields from driven wells are small, with discharges of about 4 to 10 m^3/hour.

Driven wells may be constructed only in those unconsolidated formations which do not contain large gravel or rock that may damage the drive point. However, the important advantages of driven wells are that they can be constructed in a short time, at minimum cost, and even by one person.

(iii) *Jetted wells.* A jetted well is constructed by the cutting action of a high velocity jet of water. In this method a casing pipe of diameter slightly larger than that of the well pipe is first introduced in a suitable hole dug at the surface. A jetting pipe (or drill pipe) is lowered in the casing pipe and a high velocity jet of water issuing from the jetting pipe is directed to the bottom of the hole. The soil in the hole is thus loosened and it is carried out of the hole in the form of a slurry by the water returning in the upward direction through the annular space between the casing pipe and the jetting pipe. The hole is thus deepened and the casing pipe is lowered. During the jetting operation the jetting pipe is turned slowly to ensure a straight hole. In penetrating hard clayey soils a jetting drill bit is attached to the end of jetting pipe which is raised and dropped, thus causing the bit to disintegrate the soil. When the casing pipe has penetrated the aquifer to a sufficient depth below the water table, the well pipe with a strainer at the bottom is introduced in the casing pipe. The casing pipe is then pulled out and gravel is inserted in the annular space between the well pipe and the surrounding strata.

By this method generally the wells of small diameters in the range of 30 to 100 mm are constructed but their depths may be more than 15 m. However, the wells of larger diameters upto 300 mm or more may also be constructed by this method. Further the yields from the jetted wells are also small and these wells may also be constructed only in unconsolidated formations consisting of gravel, sand, clay or other soft deposits.

Alternatively a jetted well may also be constructed by using a self jetting well point. In this method casing pipe is not used. Also no separate jetting pipe is used but the well pipe itself is used as a jetting pipe. For this a perforated brass pipe wrapped over by a fine wire mesh (or screen) and ending in a nozzle is screwed to the well pipe. A small hole is dug at the surface and a high velocity jet issuing from the nozzle is directed to the bottom of the hole. As the jetting action progresses, the well pipe goes on sinking. An annular space around the well pipe is automatically created due to the upward motion of water carrying the loosened soil. When the well pipe has been sunk to the desired depth, jetting is stopped, and the annular space is packed with gravel.

2. Deep Tube Wells. Deep tube wells are constructed by drilling. The various methods of drilling which are commonly used for constructing deep tube wells are as follows :

(i) Cable tool method

(ii) Hydraulic rotary method

(iii) Reverse rotary method.

(i) *Cable tool method.* The cable tool method is also known as the percussion method or standard method. This method is capable of drilling wells of 80 to 600 mm diameter through consolidated rock materials to depths of 1500 m. However, in unconsolidated sand and gravel this method is least effective. The method consists of sinking a casing pipe usually sightly larger in diameter than the size of the tube well required with the help of a rig. The function of the casing pipe is to support the hole and prevent it from caving in while the drilling is in progress. The drilling rig for the cable tool method consists of a mast, a multiline hoist, a walking beam, and an engine–all mounted on a truck for ready portability. Drilling is accomplished by regular lifting and dropping of a string of tools. The string of tools consists of swivel socket, a set of jars, a drill stem, and a drilling bit (Fig. 6.30). These tools are made of steel and their total weight may amount to several thousand kilograms. The most important part of the string of tools is the drilling bit which does the actual drilling. The drilling bit is at the lower end of the string of tools and when it is lifted and dropped then with its relatively sharp chisel edge breaks the rock by impact. The drilling bits are manufactured in lengths of 1 to 3 m and weigh upto 1500 kg. The drill stem is a long steel bar that adds weight and length to the drill so that it may cut rapidly and vertically. A set of jars consists of a pair of narrow connecting links and their purpose is only to loosen the tools if they stick in the hole. The swivel socket attaches the drilling cable to the string of tools.

A pit of about 2 to 3 m diameter and 5 to 8 m deep is dug at the site where the well is to be built. A casing pipe with a cutter shoe screwed to its bottom is introduced into the pit. The string of tools is introduced into the casing pipe and the drilling is started by alternately lifting and dropping mechanically the string of tools so that the drilling bit produces a hammering action on the formation resulting in its disintegration. During drilling the tools make 20 to 40 strokes per minute, ranging from 0.4 to 1 m in length. The drilling line is rotated so that the drilling bit forms a round hole. For drilling operations water is necessary and it must be added to the hole until it is encountered in the hole itself. Water forms a paste with the cuttings, thereby reducing friction of the drilling bit as it falls. After the drilling bit has cut 1 to 2 m through the formation the string of tools is taken out and a bailer (Fig. 6.30) is introduced into the hole to remove the cuttings. A bailer consists of a section of pipe with a flap valve at the bottom. When the bailer is introduced into the hole, the valve is automatically opened due to an upward pressure exerted on it by the paste of cuttings and water and the same enters the bailer. The valve, however, prevents the paste from moving out of the bailer during lifting. When the bailer is full, it is lifted to the surface and emptied. The bailers are available in a range of diameters with their lengths varying from 3 to 8 m and capacities upto 0.25 m^3. After the cuttings are taken out from the hole, the string of tools is again introduced into the hole and the drilling is continued until the required depth is reached.

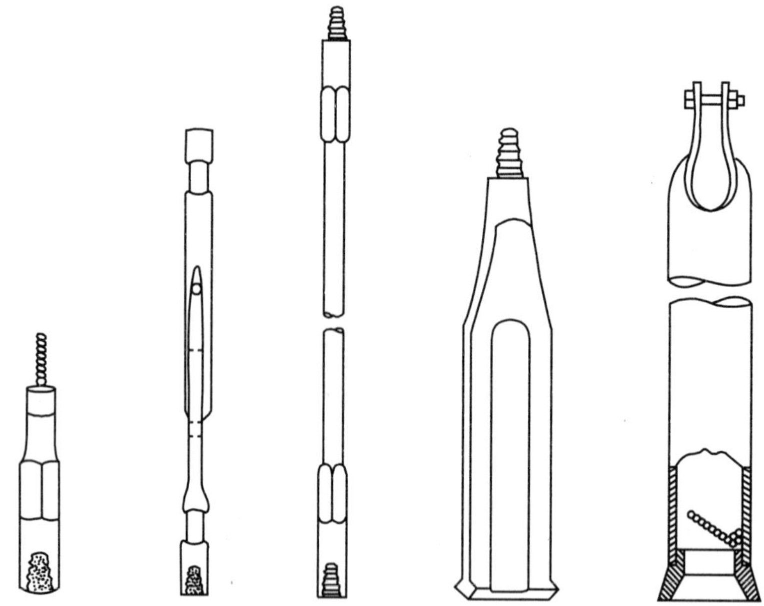

Fig. 6.30 *Basic well drilling tools for the cable tool method*

The casing pipe is required only in unconsolidated formations, in which case as the drilling progresses and the hole is deepened the casing pipe must also be driven down and maintained almost upto the bottom of the hole to avoid caving in of the wall of the hole. The casing pipe is driven down by means of drive clamps fastened to the drill stem ; the up and down motion of the tools striking the top of the casing pipe, protected by a drive head, sinks the casing pipe. The casing pipe is usually in 3 m lengths which are flush jointed by screwing one length on the other. Thus, when the casing pipe has been sunk 3 m another pipe length is screwed on the top.

A record of the material collected by bailer during drilling is kept. A bore log is then plotted to know the depths of various formations. A well pipe assembly comprising strainers and blind pipes is then introduced into the casing pipe in such a way that strainers are placed against the aquifers or water bearing strata which are to be tapped and the blind pipes against aquicludes or the strata which are not to be tapped. The well is then shrouded as indicated in section 6.21 and the casing pipe is taken out.

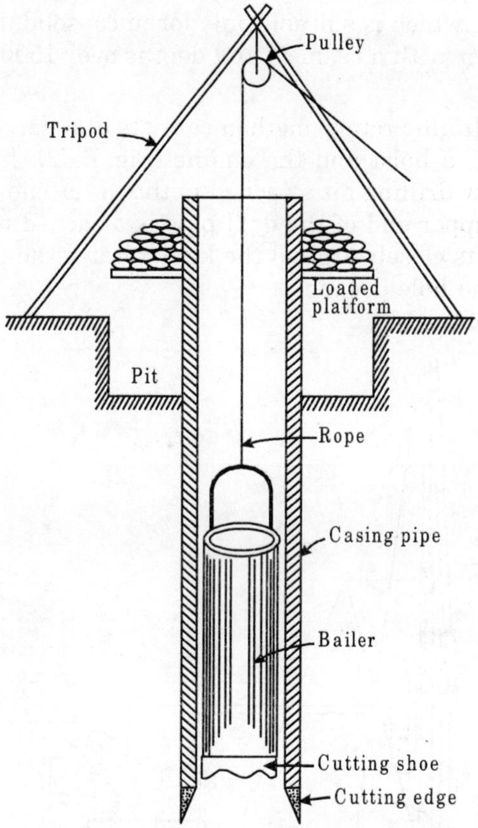

Fig. 6.31 *Boring with the help of a bailer*

As stated earlier the cable tool method is not suitable for drilling tube wells in unconsolidated formations comprising sand and gravel. In the case of such formations the tube well may be drilled only with the help of a *bailer* (also known as *sand pump* or *sledger*). The bailer in this case consists of a steel pipe 2 to 4 m long having a flap valve at the bottom and a cutting shoe of hard steel rivetted to its bottom. A casing pipe is introduced in a pit dug at the well site. Water is added to the casing pipe to facilitate drilling. The bailer is introduced into the casing pipe and it is lifted and dropped manually by means of a rope the other end of which passes over a pulley fixed centrally to a tripod as shown in Fig. 6.31. As the bailer is lifted and dropped it cuts through the formation and a paste of the cuttings and water is formed. The flap valve is automatically opened due to an upward pressure exerted on it by the paste and the same moves into the bailer. When the bailer is full with the paste it is taken out and emptied. The process is repeated till the required depth is reached. As the drilling progresses and the hole is deepened the casing pipe must also be driven to always keep it upto the bottom of the hole to avoid caving in of the wall of the hole. For driving the casing pipe a platform is attached to the upper end of the casing pipe as shown in Fig. 6.31, which when loaded by sand bags, rails etc., drives the casing pipe slowly into the hole. After the drilling is carried out upto the required depth the well pipe is introduced into the casing, the well is shrouded and the casing pipe is taken out.

(ii) *Hydraulic rotary method.* The hydraulic rotary method (or rotary method) is a rapid method of drilling tube wells which is suitable only for unconsolidated strata. Wells upto 0.45 m diameter (upto 1.5 m diameter with a reamer) and depths over 1500 m have been constructed by this method.

The drilling rig for a hydraulic rotary method consists of a derrick or mast, a rotating table, a pump for the drilling mud, a hoist and the engine (Fig. 6.32). In this method the drilling is done with the help of a hollow drilling bit attached to the lower end of a drill pipe which extends to the ground surface. The upper end of the drill pipe is attached to a square section known as kelly. A rotating table that fits closely around the kelly rotates the drill pipe and allows the drill pipe to slide downward as the hole deepens.

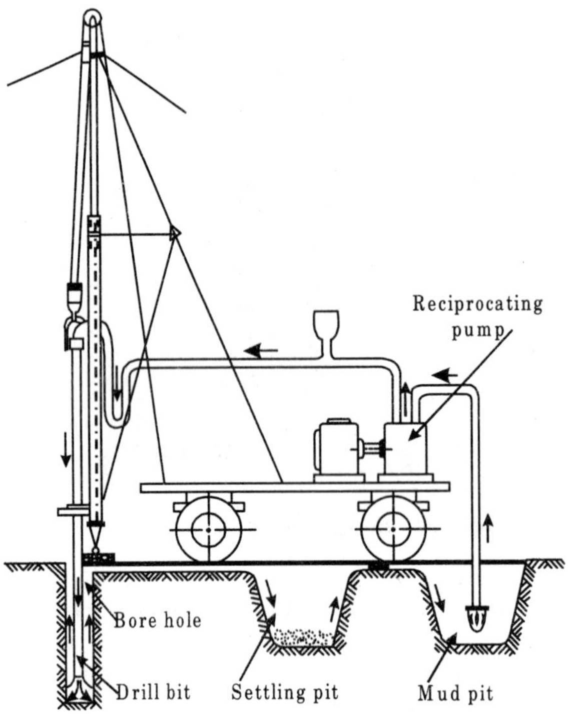

Fig. 6.32 *Hydraulic rotary drilling system*

During drilling a mixture of clay and water, known as *drilling mud,* is continuously pumped into the hole through the drill pipe. The mud leaves the drill pipe through one or more openings of the drilling bit where it cools and lubricates the cutting surface of the drilling bit. The mud flows out of the hole through the annular space between the drill pipe and the walls of the hole. The material loosened by the drilling bit is carried to the surface by the mud flowing out of the hole. In this method no casing pipe is ordinarily required during drilling because the mud forms a clay lining on the wall of the hole and thus prevents caving in of the wall of the hole. The mud flowing out of the hole is taken to a settling basin where the cuttings settle down and the mud is again used for pumping into the hole. To maintain the required consistency of the mud, clay and water are added to the circulation mud from time to time.

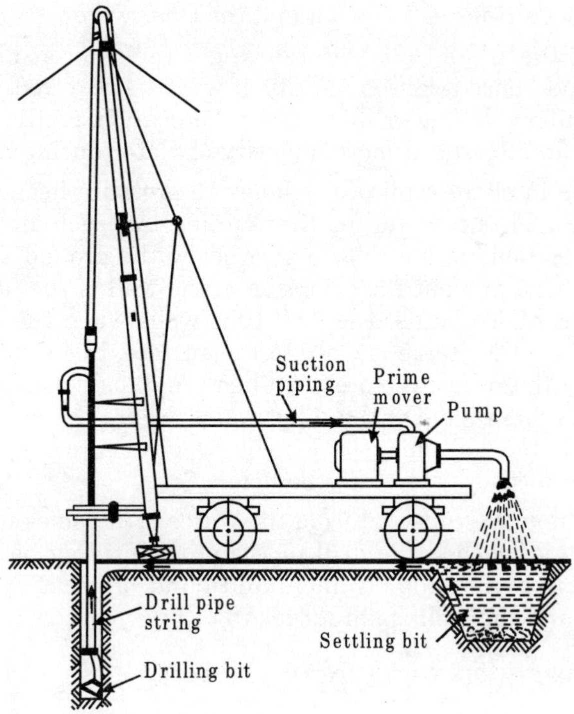

Fig 6.33 *Reverse rotary drilling system*

A complete boring record is maintained to know the type of formations at various depths. When the required depth is reached the drill pipe and the drilling bit are pulled out from the hole and a well pipe containing strainers and blind pipes at appropriate locations is introduced in the hole. Since in this case the wall of the hole is coated with clay, it should be thoroughly washed to obtain more yield from the well. The method of washing is described under well development in section 6.21. The well is then shrouded as indicated in section 6.21.

(iii) *Reverse rotary method.* The reverse rotary method is also suitable for drilling tube wells only in unconsolidated formations. It is probably the most rapid method of drilling and hence it has become increasingly popular. The drilling rig for this method (Fig. 6.33) is similar to that for the hydraulic rotary method. The drilling in this method is also done by a hollow drilling bit attached to the lower end of a drill pipe as in the case of the hydraulic rotary method. However, by this method usually large diameter holes with a minimum diameter of about 0.4 m may be drilled. The drilling bits for this method range in diameter from 0.4 to 1.8 m. Further in this method also no casing pipe is ordinarily used. Instead during drilling the drilled hole is always kept filled with water upto the ground surface, so that the water pressure exerted on the formation prevents the caving in of the wall of the hole. Moreover, the fine-grained soil particles held suspended in the column of water also help in stabilizing the wall of the hole. The material loosened by the drilling bit is carried to the surface by the water which is pumped out of the hole through the drilling bit and the drill pipe. As such in order to maintain the water level in the hole upto the ground surface, water is continuously supplied to the drilled hole from the top in the annular space outside the drill pipe. Moreover, the water pumped out of the hole is taken to a settling basin where the cuttings settle down and the clear water is again supplied to the hole.

In this method due to the use of clear water during drilling, a considerable quantity of water seeps into the aquifers and hence a large quantity of water is required. However, the amount of water taken by the aquifers during drilling is an indication of the yield of the well after completion. Thus, better aquifers need more quantity of water during drilling.

To prevent the caving in of the wall of the hole the pressure head of the water in the hole must be larger than that of the water in the aquifer. As such in general this method is applicable when the water table is about 3 to 4 m below the ground surface so that sufficient pressure head is developed to prevent the caving in of the wall of the hole. However, by using a casing pipe this method can also be used to drill tube wells when the water table is either too high or too low. Where the water table is close to the surface, by using a casing pipe extending above the ground surface the necessary pressure head may be developed by filling water upto the desired level in the casing pipe. On the other hand, where the water table is deep, a large amount of water filled in the hole may be lost by seepage into the aquifers, which may be minimized by introducing a casing pipe into the hole.

Samples of the formation are collected from the water as it comes up through the drill pipe. When the required depth is reached the drill pipe and the drilling bit are pulled out from the hole. A well pipe assembly comprising strainers and blind pipes at appropriate places is then introduced into the hole and the well is shrouded.

6.21 WELL SHROUDING AND WELL DEVELOPMENT

A. Well Shrouding. The shrouding is a layer of coarse material such as gravel and coarse sand interposed in the annular space between the well pipe and the aquifer soil. The shrouding prevents the finer particles of the soil coming in contact with the strainer and choking it, acts as a screen or filter introduced between the strainer and the aquifer soil to allow only water to pass through and not sand, and protects the well pipe from caving in of the surrounding formation. Further the shrouding also increases the effective well diameter. In general a shrouded well has a greater specific capacity than the one of the same diameter but not shrouded. The shrouding is usually not required in rocky and consolidated formations, but it is essential in sandy and unconsolidated formations. It is also essential in a slotted type tube well where wire mesh (or screen) is not used. Such a tube well is usually provided with a thick shrouding and it is also sometimes known as *gravel-packed well.*

The shrouding material should have proper grain size distribution which depends upon the grain size distribution of the aquifer soil and upon the size of the slots or the well screen. When a casing pipe is used during drilling, the shrouding material is introduced before the casing pipe is taken out of the hole. In this case the quantity of the shrouding material per 0.3 m length is determined. In the beginning the shrouding material for 0.6 m length is shovelled from the top in the annular space between the well pipe and the casing pipe. The casing pipe is then taken out by 0.3 m with the help of jacks or crane. Then the quantity of shrouding material for each 0.3 m is added and the casing pipe is taken out by 0.3 m until the strainer is covered.

Sometimes the shrouding material is placed through small pipes. A 50 mm diameter pipe is lowered in the annular space that is to be filled in by the shrouding material. The shrouding material is fed through a hopper provided at the top. The feeding pipe is gradually pulled up as the annular space is filled. Water is fed into the pipe so as to aid in the downward movement of the shrouding material. For very deep wells the reverse circulation system of feeding the shrouding material is more suitable. In this case a pipe is lowered inside the well pipe and it is connected to a pump. A circulation of water is established downwards in the annular space and upwards through the lowered pipe. The shrouding material is fed in the annular space and is carried downwards by the force of the water. However, in an improved method, the shrouding material is pumped down into position by using bladeless pumps developed for this purpose.

B. Well Development. Well development is the process of removing fine material from the formations surrounding the strainers or the screened sections of the well pipe. It is aimed at :

(i) increasing the specific capacity of the well ;

(ii) preventing the sand flowing into the well ;

(iii) obtaining maximum economic well life.

The actual yield of a well can be known only after the well development.

The following methods of well development are commonly used.

(i) By pumping.

(ii) By surging.

(iii) By compressed air.

(iv) By back washing.

(v) By chemicals (or By dry ice).

(i) *By pumping.* In this method the well is pumped in a series of steps from a low discharge to the one exceeding the design capacity. At each step the well is pumped until clear water is obtained, after which the power is shut off. The discharge rate is then increased and the procedure repeated until the maximum capacity of the pump or well is reached. This irregular and noncontinuous pumping agitates the fine material surrounding the well so that it can be carried into the well and pumped out. The coarser fraction entering the well is removed by a bailer or sand pump from the bottom.

(ii) *By surging.* In this method a surging action is imparted to the water in the well by up and down movement of a surge block. A surge block is essentially a plunger which is made by mounting cylindrical wooden blocks with two leather or rubber sandwiched discs and two steel washers at the end of an iron pipe. The diameter of the surge block is such that it can pass easily through the well. The surge block is lowered into the well so that it is 3 to 5 m under water, but above the top of the screened section. The surging acion is transmitted to the screened section by the water column. Surging is started slowly and the speed is gradually increased upto the point the surge block can rise and fall smoothly without jerking. The downward stroke of the surge block causes backwash of water from the well which results in dislodging of the sand particles in the surrounding formation. The upward stroke of the surge block causes sucking action and thus pulls the dislodged sand particles from the surrounding formation into the well. After surging for 10 to 15 minutes the surge block is taken out and a bailer is lowered to the bottom of the well to estimate the quantity of sand pulled in and to bail out the sand. The process is repeated by gradually increasing the period of surging until the sand pulled in and accumulating in the bottom of the well becomes negligible.

(iii) *By compressed air.* In this method an air compressor is used which is connected to an air tank. The air tank has an outlet provided with a quick-opening valve which is connected to an air pipe. Around the air pipe a discharge pipe is fitted. The air pipe and discharge pipe assembly is introduced into the well. The discharge pipe is of diameter smaller than that of the well pipe such that there is a clearance of at least 20 mm all around between the discharge pipe and the well pipe. The air pipe as well as the discharge pipe can be shifted vertically by clamps. Initially these pipes are held with their lower ends near the bottom of the screened section (or strainer) of the well pipe. The lower end of the air pipe is kept about 0.3 m lower than that of the discharge pipe. The valve connecting the air tank and the air pipe is kept closed and the air pressure is allowed to be built up from 690 to 980 kN/m^2 (7 to 10 kg f/cm^2) in the air tank by the compressor. The compressed air is then released suddenly into the well through the air pipe by opening the valve. The inrush of air into the well creates a powerful surge within the well due to which the pressure increases and the water flows outwards through the strainer into the surrounding formation where it loosens the fine material. The air valve is then closed so that the pressure decreases and the water from the surrounding formation enters the well bringing the loosened fine material with it. The water is then pumped out from the well to remove the loosened fine material from the surrounding formation. The pipe assembly may be operated as a conventional air lift (or air jet) pump. For this the air pipe is raised so that its lower end is about 0.3 m higher than that of the discharge pipe and the compressed air is introduced into the well through the air pipe due to which water is pumped out through the discharge pipe. The pumping is continued until water free from sand is obtained. The air valve is then closed and the air pipe is lowered again and the process is repeated until water obtained from the well is free from sand. The same operation is repeated at different levels along the screened section of the well pipe by lifting the pipe assembly every time by a metre or more, until the entire length of the screened section has been fully developed. Finally the pipe assembly is lowered again to the bottom of the well and operated as a pump to clean out any sand that may have accumulated.

(iv) *By backwashing.* In this method the top of the well is fitted with an air tight cover. A compressor and discharge and air pipe are installed in the same manner as in the previous method. However, in this case the lower end of the air pipe is always kept about 0.3 m higher than that of the discharge pipe, because the pipe assembly is operated only as an air lift pump. In addition to these pipes a short air pipe is fitted at the top of the well. A three-way valve is provided to connect the short and the long air pipes with the compressor. In the beginning the compressed air is released into the well through the long air pipe which forces air and water out of the well through the discharge pipe. When clear water starts coming out of the well, the air supply is shut off by closing the valve and the water is allowed to return to its static level. The three-way valve is then turned to admit compressed air into the top of the well through the short air pipe. This backwashes the water from the well through the discharge pipe and at the same time agitates the fine material surrounding the well. Air is supplied through the short air pipe until air begins to escape from the discharge pipe. The valve is then turned so that air is again supplied through the long air pipe so that the pipe assembly works as an air lift pump and the water is pumped out from the well. The operation is repeated till clear water comes and the well is fully developed.

(v) *By chemicals.* In this method use of chemicals is made for the development of a well. In one of the commonly used methods hydrochloric acid and solid carbon dioxide (dry ice) are used for the well development. The hydrochloric acid is first poured into the well. The well is capped at the top and compressed air is introduced into the well which causes the hydrochloric acid to enter the surrounding formation. The cap is then removed from the top of the well and dry ice is poured into the well. The sublimation releases gaseous carbon dioxide and a high pressure of this gas is built up in the well. With the release of this pressure the muddy water is forced up through the well in the form of a jet. Thus, the fine material is automatically forced out with the water.

The addition of any of the polyphosphates to water in the well helps the development process. These compounds act as deflocculants and dispersants of clays and other fine grained materials, thereby enabling the mud cake on the wall of a hole and the clay fractions in an aquifer to be more readily removed during development of well by any of the methods. One such polyphosphate commonly used is sodium-hexametaphosphate (Calgon) which when added to water acts as dispersing agent for clays and other fine grained materials. Calgon is often added to water in the well during development by surging or by backwashing.

6.22 COLLECTOR OR RADIAL WELLS

Collector wells are used when large discharges are required but a thin highly permeable aquifer is available. Such wells are generally constructed near river or lakes where plentiful recharge is available.

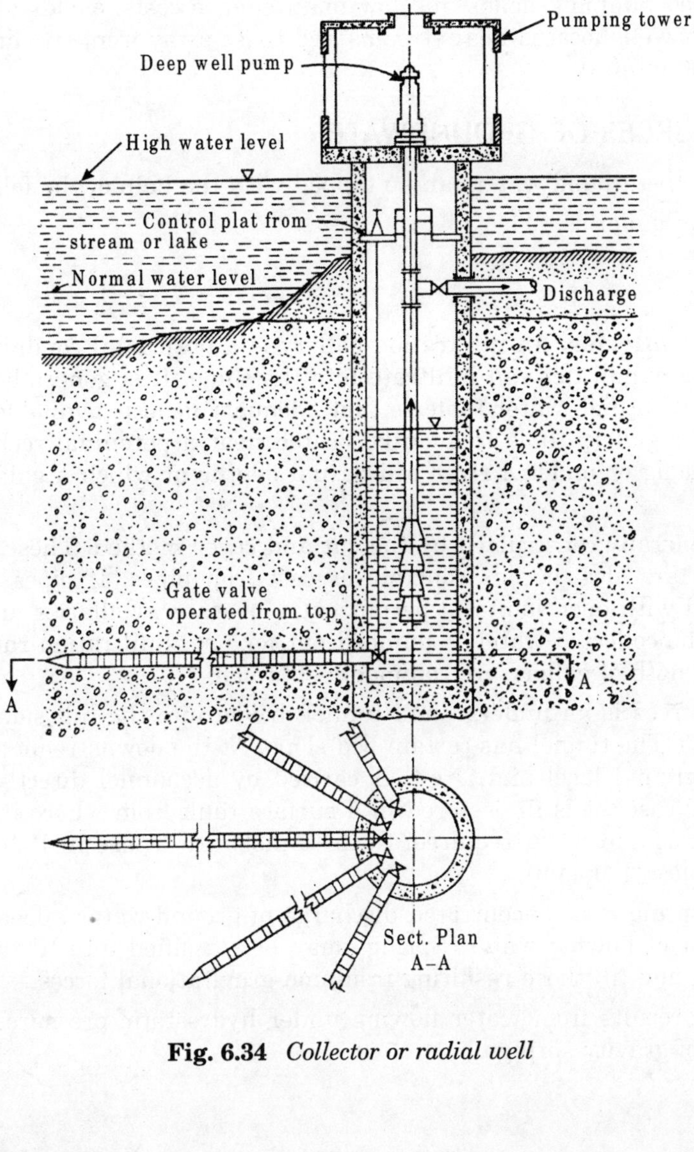

Fig. 6.34 *Collector or radial well*

A collector well (Fig. 6.34) consists of a reinforced concrete caisson or well about 5 m in diameter from which horizontal lateral screened pipes are projected radially near the bottom. The caisson is sunk into the water-bearing stratum by excavating the earth from the inside and its bottom is sealed by a concrete plug. The horizontal screened pipes are heavy steel pipes 150 to 200 mm in diameter; 30 to 150 m in length depending upon the geologic conditions and the discharge required; and are perforated with longitudinal slots. These pipes are jacked hydraulically into the aquifer through precast portholes in the caisson to form a radial pattern of horizontal pipes. In some cases a casing pipe is first introduced through the porthole after which the screened pipe is inserted and the casing pipe is pulled out. Each of these radial collector pipes is then developed either by compressed air or by pumping so that the finer materials of the aquifer are washed out and natural gravel packs are formed around these pipes.

The initial cost of a collector well exceeds that of a vertical well, but the collector wells have larger yields, reduced pumping heads and low maintenance costs. Yields of 27,000 m^3/day are common for collector wells located near streams and those away from streams given an average yield of about 15,000 m^3/day.

6.23 OTHER SOURCES OF GROUNDWATER

In addition to wells the ground water can be tapped with the help of the following source:

(1) Infiltration galleries.

(2) Kareez.

(3) Springs.

1. Infiltration Galleries. An infiltration gallery is a horizontal conduit having permeable boundaries so that groundwater can infiltrate into the same. It is generally provided in highly permeable aquifers with high water table so that adequate head is available for gravity flow of groundwater into the gallery. It is frequently located near a perennial recharge source and is sometimes placed along or under a river bed. Usual depths at which a gallery is placed range from 3 to 6 m.

Galleries are generally constructed by the cut and cover method. These are made of bricks placed without mortar, tiles, porous concrete or cast iron pipes. If pipes are used, they are perforated or placed with open joints to permit entry of water. Galleries are sloped towards a sump from where the collected water is pumped out for use. Infiltration rates of 1500 to 7000 m^3/day per 100 m length of gallery are commonly obtained.

2. Kareez. A kareez is an underground tunnel driven into the hill side to tap water from underground springs. The tunnel has certain bed slope. At the downstream end of kareez water is available above ground level and is either carried by a channel directly to the land to be irrigated or in some cases it is first stored in a surface tank from where it can be taken by a channel to the field as and when required. Such tunnels are in use in Baluchistan and North West Frontier Province (Pakistan).

3. Springs. A spring is a concentrated discharge of ground-water appearing at the ground surface as a current of flowing water. Springs may be classified into (i) those resulting from gravitational forces, and (ii) those resulting from non-gravitational forces.

Gravity springs results from water flowing under hydrostatic pressure. The following are the different types of gravity springs.

(a) *Depression springs.* These are formed where the ground surface intersects the water table.

(b) *Contact springs.* These are created by a permeable water bearing formation overlying a less permeable formation that intersects the ground surface.

(c) *Artesian springs.* These result from release of water under pressure from confined aquifers either at an outcrop of the aquifer or through an opening in the confining bed.

(d) *Impervious rock springs.* These springs occur in tubular channels or fractures of impervious rock.

(e) *Tubular or fracture springs.* These are issuing from rounded channels such as lava tubes or solution channels, or fractures in impermeable rock connecting with groundwater.

Non-gravity springs include volcanic springs and fissure springs. The volcanic springs are associated with volcanic rocks and the fissure springs result from fractures extending to great depths in the earth's crust. These are usually thermal springs. Thermal springs discharge water having a temperature in excess of the normal local groundwater. These are also designated as warm springs and hot springs. Waters of thermal springs are usually highly mineralized and consist for the most part of meteoric water that has been modified in quality by its passage underground.

A *geyser* is a periodic thermal spring resulting from the expansive force of superheated steam within constricted subsurface channels. Water from surface sources and/or shallow aquifers drains downwards into a deep vertical tube where it is heated to above the boiling point. With increasing pressure the steam pushes upward, this releases some water at the surface, which reduces the hydrostatic pressure and causes the deeper superheated water to accelerate upward and to flash into steam. The geyser then surges into full erruption for a short interval until the pressure is dissipated. Thereafter the filling begins again and the cycle is repeated.

SOLVED ILLUSTRATIVE EXAMPLES

Illustrative Example 6.1. *A sample of silty sand has a volume of 7 × 10⁴ mm³. At the natural moisture content it weighs 1.478 N. The sample was then saturated with water and reweighed to a weight of 1.665 N. The sample was drained by gravity until it reached a constant weight of 1.434 N. The sample was then oven dried at 105°C until it reached a constant weight of 1.370 N. Assuming unit weight of water as 9810 N/m³ compute the following: (a) water content under natural conditions; (b) volumetric water content under natural conditions; (c) saturation ratio under natural conditions; (d) porosity; (e) specific yield; (f) specific retention; (g) water content at saturation.*

Solution

(a) Weight of water contained at natural condition

$$= (1.478 - 1.370) = 0.108 \text{ N}$$

∴ Water content under natural condition

$$= \frac{0.108}{1.370} \times 100 = 7.88\%$$

(b) Volume of water contained under natural condition

$$= \frac{0.108}{9810} = 11 \times 10^{-6} \text{ m}^3 = 11 \times 10^3 \text{ mm}^3$$

∴ Volumetric water content under natural condition

$$= \frac{11 \times 10^3}{7 \times 10^4} \times 100 = 15.71\%$$

(c) Weight of water contained by the soil at saturation

$$= (1.665 - 1.370) = 0.295 \text{ N}$$

Volume of water contained by the soil at saturation

$$= \frac{0.295}{9810} = 30 \times 10^{-6} \text{ m}^3 = 30 \times 110^3 \text{ mm}^3$$

∴ Saturation ratio under natural condition

$$= \frac{11 \times 10^3}{30 \times 10^3} \times 100 = 36.67\%$$

(d) $$\text{Porosity} = \frac{30 \times 10^3}{7 \times 10^4} \times 100 = 42.86\%$$

(e) Weight of water drained by gravity

$$= (1.665 - 1.434) = 0.231 \text{ N}$$

Volume of water drained by gravity

$$= \frac{0.231}{9810} = 23.5 \times 10^{-6} \text{ m}^3 = 23.5 \times 10^3 \text{ mm}^3$$

∴ $$\text{Specific yield} = \frac{23.5 \times 10^3}{7 \times 10^4} \times 100 = 33.57\%$$

Weight of water retained by the soil against gravity = (1.434 – 1.370) = 0.064N

Volume of water retained against gravity = $\frac{0.064}{9810}$ = 6.5×10^{-6} m³ = 6.5×10^3 mm³

∴ $$\text{Specific retention} = \frac{6.5 \times 10^3}{7 \times 10^4} \times 100 = 9.29\%$$

(g) Water content at saturation = $\frac{0.295}{1.370} \times 100 = 21.53\%$

Illustrated Example 6.2 *The following observations were made on a 300 mm diameter well penetrating an unconfined aquifer.*

(i) *Rate of pumping = 1800 litres/minute*

(ii) *Drawdown in a test well 30 m away = 1.8 m*

(iii) *Drawdown in a test well 60 m away = 0.6 m*

(iv) *Depth of water in the well before pumping = 50 m.*

Determine (a) the radius of the circle of influence; and (b) the co-efficient of transmissibility of the aquifer.

Solution

From Eq. (6.9b), we have

$$Q = \frac{1.36k(H^2 - h^2)}{\log_{10}(R/r)}$$

By introducing the values of the data obtained at the two observation wells, we get

$$Q = \frac{1.36k(50^2 - 48.2^2)}{\log_{10}(R/30)} = \frac{1.36k(50^2 - 49.4^2)}{\log_{10}(R/60)}$$

Solving the above equation, we get

$$R = 85.39 \text{ m}$$

$$Q = \frac{1800 \times 10^{-3}}{60} = 0.030 \text{ m}^3\text{/sec}$$

By introducing the values of the data obtained at one of the observation wells, we get

$$0.03 = \frac{1.36k(50^2 - 48.2^2)}{\log_{10}(85.39/30)}$$

$$\therefore \quad k = 5.67 \times 10^{-5} \text{ m/s}$$

$$= 5.67 \times 10^{-3} \text{ cm/s}$$

Coefficient of transmissibility

$$T = kH$$

$$= 5.67 \times 10^{-5} \times 50$$

$$= 2.835 \times 10^{-3} \text{ m}^2\text{/sec.}$$

Illustrative Example 6.3 *A tube well of 300 mm diameter penetrates fully a confined aquifer. The length of the strainer is 25 m. Calculate the yield from the well under a drawdown of 4 m. The co-efficient of permeability of aquifer = 50 m/day. Assume radius of circle of influence equal to 200 m.*

Solution

From Eq. (6.18a), we have

$$Q = \frac{2.72kbs}{\log_{10}(R/r)}$$

Here b = Thickness of aquifer

= Length of strainer = 25 m

s = Drawdown = 4 m

k = Coefficient of permeability = 50 m/day

R = Radius of circle of influence = 200 m

r = Radius of well

= 150 mm = 0.15 m

By substituting these values, we get

$$Q = \frac{2.72 \times 50 \times 25 \times 4}{\log_{10}(200/0.15)} \text{ m}^3/\text{day}$$

$$= 4352 \text{ m}^3/\text{day}$$

$$= 50.37 \text{ litres/sec.}$$

Illustrative Example 6.4 *A tube well penetrates fully an unconfined aquifer. Calculate the discharge from the tube well under the following conditions:*

(*i*) *Diameter of the well= 300 mm*

(*ii*) *Drawdown = 3 m*

(*iii*) *Effective length of the strainer under the above drawdown = 10.5 m*

(*iv*) *Coefficient of permeability of aquifer = 0.5 mm/s*

(*v*) *Radius of zero drawdown = 300 m.*

Solution

From Eq. (6.10a), we have

$$Q = \frac{1.36ks(s+2h)}{\log_{10}(R/r)}$$

in which

k = 0.5 mm/s = 5×10^{-4} m/s

s = 3 m

R = 300 m

r = 150 mm = 0.15 m

Since well is fully penetrating the aquifer h will be equal to the effective length of the strainer for the given drawdown and hence

$$h = 10.5 \text{ m}$$

By substituting these values in the above equation, we get

$$Q = \frac{1.36 \times 5 \times 10^{-4} \times 3[3+2(10.5)]}{\log_{10}(300/0.15)}$$

$$= 1.483 \times 10^{-2} \text{ m}^3/\text{s}$$

$$= 14.83 \text{ litres/second}$$

Illustrative Example 6.5 *Design a tube well for the following data.*

(*i*) *Yield required = 0.10 m*^{3}*/s*

(*ii*) *Radius of circle of influence = 200 m*

(*iii*) *Coefficient of permeability = 60 m/day*

(*iv*) *Drawdown* = 6 *m*

(*v*) *Thickness of confined aquifer* = 30 *m.*

Solution

From Eq. (6.18a), we have

$$Q = \frac{2.72kbs}{\log_{10}(R/r)}$$

Here

$$Q = 0.10 \text{ m}^3\text{/s}$$

$$k = 60 \text{ m/day}$$

$$= \frac{60}{24 \times 60 \times 60} = 6.944 \times 10^{-4} \text{ m/s}$$

$$b = 30 \text{ m}$$

$$s = 6 \text{ m}$$

$$R = 200 \text{ m}$$

The value of r the radius of the well is to be determined. Thus, by substituting these values in the above equation, we get

$$0.10 = \frac{2.72 \times 6.944 \times 10^{-4} \times 30 \times 6}{\log_{10}(200/r)}$$

or

$$\log_{10}(200/r) = \frac{2.72 \times 6.944 \times 10^{-4} \times 30 \times 6}{0.10} = 3.3998$$

or

$$\frac{200}{r} = \text{Antilog } 3.3998$$

$$= 2511$$

$\therefore$

$$r = \frac{200}{2511}$$

$$= 0.08 \text{ m} = 80 \text{ mm}$$

Hence provide a tube well of radius 80 mm.

Illustrative Example 6.6 *A tube well fully penetrates a confined aquifer of thickness 30 m and coefficient of permeability 38 m/day. Determine the radius of the well if the yield required is 40 litres/second under a drawdown of 4 m. Use radius of circle of influence as recommended by Sichardt.*

Solution

From Eq. (6.18a), we get

$$Q = \frac{2.72kbs}{\log_{10}(R/r)}$$

Here

$$Q = 40 \ l\text{/s} = 0.04 \text{ m}^3\text{/s}$$

$$k = 38 \text{ m/day}$$

$$= \frac{38}{24 \times 60 \times 60} = 4.398 \times 10^{-4} \text{ m/sec}$$

$$b = 30 \text{ m}$$

$$s = 4 \text{ m}$$

According to Sichardt R is given as

$$R = 3000\, s\sqrt{k}$$

$$= 3000 \times 4 \times \sqrt{4.398 \times 10^{-4}}$$

$$= 251.66 \text{ m}$$

Thus, by substituting the values in the above equation, we get

$$0.04 = \frac{2.72 \times 4.398 \times 10^{-4} \times 30 \times 4}{\log_{10}(251.66/r)}$$

or
$$\log_{10}(251.66/r) = \frac{2.72 \times 4.398 \times 10^{-4} \times 30 \times 4}{0.04}$$

$$= 3.5888$$

or
$$\frac{251.66}{r} = \text{Antilog } 3.5888 = 3880$$

$\therefore$
$$r = \frac{251.66}{3880}$$

$$= 0.065 \text{ m} = 65 \text{ mm}$$

Hence a tube well of radius 65 mm may be provided.

Illustrative Example 6.7 *A well penetrates fully at 10 m thick water bearing stratum of medium sand having coefficient of permeability of 0.004 m/sec. The well radius is 100 mm and is to be worked under a drawdown of 4 m at the well face. Calculate the discharge from the well. What will be the percentage increases in the discharge if the radius of the well is doubled? Take R = 300 m in each case.*

Solution

From Eq. (6.18a), we have

$$Q = \frac{2.72kbs}{\log_{10}(R/r)}$$

Here
$$k = 0.004 \text{ m/sec}$$

$$b = 10 \text{ m}$$

$$s = 4 \text{ m}$$

$$\therefore \qquad Q = \frac{2.72 \times 0.004 \times 10 \times 4}{\log_{10}(300/r)}$$

or
$$Q = \frac{0.4352}{\log_{10}(300/r)}$$

(i) In the first case r = 100 mm = 0.1 m

$$\therefore \qquad Q_1 = \frac{0.4352}{\log_{10}(300/0.1)} \; 0.1252 \text{ m}^3\text{/s}$$

(ii) In the second case r = 200 mm = 0.2 m

$$\therefore \qquad Q_2 = \frac{0.4352}{\log_{10}(300/0.2)} = 0.1370 \text{ m}^3\text{/s}$$

∴ Percentage increase in discharge

$$= \frac{0.1370 - 0.1252}{0.1252} \times 100$$

$$= 9.42\%$$

Thus, by doubling the well diameter the discharge will increase only by 9.42%. It may therefore be noted that the size of the well has little influence on the discharge obtained from the well.

Illustrative Example 6.8 *Pumping test was performed to determine the permeability of an unconfined aquifer. Following observations were recorded.*

(i) Diameter of well = 200 mm

(ii) Discharge from the well = 200 m^3/hour

(iii) R.L. of water level in the well before pumping = 138.5 m

(iv) R.L. of water level in the well at constant pumping = 133.2 m

(v) R.L. of impervious strata at the bottom of the well = 108.3 m

(vi) R.L. of water level in an observation well = 137.6 m

(vii) Radial distance of the observation well from the test well = 40 m.

Calculate the coefficient of permeability k. Also calculate (a) the error in k if observations are not taken in the observation well and the radius of circle of influence is assumed as 300 m; (b) actual radius of circle of influence based on the observations at the observation well.

Solution

From Eq. (6.14a), we have

$$Q = \frac{1.36k(h_2^2 - h_1^2)}{\log_{10}(r_2/r_1)}$$

In this case since there is only one observation well the test well itself may be considered as one of the observation wells. Thus, using suffix 1 for the test well and suffix 2 for the observation well, we have

$$r_1 = 100 \text{ mm} = 0.1 \text{ m}$$
$$r_2 = 40 \text{ m}$$
$$h_1 = 133.2 - 108.3 = 24.9 \text{ m}$$
$$h_2 = 137.6 - 108.3 = 29.3 \text{ m}$$
$$Q = 200 \text{ m}^3\text{/hour}$$

Thus, by substituting the values in the above equation, we get

$$200 = \frac{1.36k(29.3^2 - 24.9^2)}{\log_{10}(40/0.1)}$$

$$\therefore \quad k = 1.605 \text{ m/hour}$$
$$= 38.51 \text{ m/day}$$

(a) For calculating the value of k on the basis of the assumed radius of circle of influence, we have from Eq. (6.9b)

$$Q = \frac{1.36k(H^2 - h^2)}{\log_{10}(R/r)}$$

in which

$$Q = 200 \text{ m}^3\text{/hour}$$
$$H = 138.5 - 108.3 = 30.2 \text{ m}$$
$$h = 133.2 - 108.3 = 24.9 \text{ m}$$
$$R = 300 \text{ m}$$
$$r = 100 \text{ mm} = 0.1 \text{ m}$$

Thus, by substitution, we get

$$200 = \frac{1.36k(30.2^2 - 24.9^2)}{\log_{10}(300/0.1)}$$

$$\therefore \quad k = 1.751 \text{ m/hour}$$
$$= 42.024 \text{ m/day}$$

$\therefore$ Percentage error in the value of k

$$= \frac{42.024 - 38.510}{38.510} \times 100 = 9.12\%$$

(b) For calculating the actual radius of circle of influence, we have from Eq. (6.9b)

$$Q = \frac{1.36k(H^2 - h^2)}{\log_{10}(R/r)}$$

in which

$$Q = 200 \text{ m}^3\text{/hour}$$
$$k = 1.605 \text{ m/hour}$$

$$H = 138.5 - 108.3 = 30.2 \text{ m}$$
$$h = 133.2 - 108.3 = 24.9 \text{ m}$$
$$r = 100 \text{ mm} = 0.1 \text{ m}$$

Thus, by substitution, we get

$$200 = \frac{1.36 \times 1.605(30.2^2 - 24.9^2)}{\log_{10}(R/0.1)}$$

or $$\log_{10}(R/0.1) = \frac{1.36 \times 1.605(30.2^2 - 24.9^2)}{200} = 3.1872$$

or $$(R/0.1) = \text{Antilog } 3.1872$$

$\therefore$ $$R = 153.9 \text{ m}$$

Illustrative Example 6.9 *Design an open well in fine sand to give a discharge of 0.005 cumec when operated under a depression head of 3 m.*

Solution

From Eq. (6.70), we have

$$Q = KH$$

or $$Q = \left(\frac{K}{A}\right)AH$$

$$Q = 0.005 \text{ m}^3/\text{sec} = 0.005 \times 3600 = 18 \text{ m}^3/\text{hour}$$
$$H = 3 \text{ m}$$

For fine sand from Table 6.4

$$(K/A) = 0.5 \text{ m}^3/\text{hour per m}^2 \text{ of area under 1 m depression head}$$

Thus, by substituting the values in the above equation, we have

$$18 = 0.5 \times A \times 3$$

or $$A = \frac{18}{0.5 \times 3} = 12 \text{ m}^2$$

If d is the well diameter, then

$$A = \frac{\pi d^2}{4} = 12$$

$\therefore$ $$d = \sqrt{\frac{4 \times 12}{\pi}} = 3.91 \text{ m}$$

Alternative Solution

The well diameter may also be determined by using the discharge formula in terms of percolation intensity coefficient. Thus, from Eq. (6.68), we have

$$Q = CAh$$

Here
$$Q = 0.005 \text{ m}^3/\text{sec}$$
$$h = H \text{ (constant depression head)}$$
$$= 3 \text{ m}$$

For fine sand the average value of the percolation intensity coefficient C may be taken as 0.075 mm/sec per m.

Thus,
$$C = 0.075 \text{ mm/sec per m}$$
$$= 0.075 \times 10^{-3} \text{ m /sec per m}$$

By substituting these values, we get
$$0.005 = 0.075 \times 10^{-3} \times A \times 3$$
$$\therefore \quad A = 22.22 \text{ m}^2$$

Assuming that a cavity is formed at the bottom of the well, then as indicated in section 6.19
$$A = \frac{4}{3}\left(\frac{\pi d^2}{4}\right)$$

where d is the diameter of the well. Thus,
$$\frac{4}{3}\left(\frac{\pi d^2}{4}\right) = 22.22$$
$$\therefore \quad d = 4.606 \text{ m}$$

However, if the value of C is taken equal to 0.1 mm/sec per m then
$$d = 3.99 \text{ m.}$$

Illustrative Example 6.10 *Two tube wells each of 300 mm diameter are spaced 100 m apart. Both the wells penetrate fully a confined aquifer of thickness 15 m. Calculate the discharge if only one well is discharging under a drawdown of 3 m. What will be the percentage decrease in the discharge of the well if both the wells are discharging under the drawdown of 3 m. Take radius of circle of influence for each well equal to 225 m and coefficient of permeability of aquifer as 50 m/day.*

Solution

When only one well is discharging then from Eq. (6.18a)
$$Q = \frac{2.72kbs}{\log_{10}(R/r)}$$

in which
$$k = 50 \text{ m/day}$$
$$b = 15 \text{ m}$$
$$s = 3 \text{ m}$$
$$R = 225 \text{ m}$$
$$r = 150 \text{ mm} = 0.15 \text{ m}$$

Thus, by substituting these values, we get

$$Q = \frac{2.72 \times 50 \times 15 \times 3}{\log_{10}(225/0.15)}$$

$$= 1926.898 \text{ m}^3/\text{day} = 80.29 \text{ m}^3/\text{hour}$$

When both the wells are discharging then the discharge from each well is given by Eq. (6.34) as

$$Q_1 = Q_2 = \frac{2\pi kb(H-h)}{\log_e(R^2/rB)}$$

$$= \frac{2.72kbs}{\log_{10}(R^2/rB)}$$

$$= \frac{2.72 \times 50 \times 15 \times 3}{\log_{10}\left(\frac{225 \times 225}{0.15 \times 100}\right)}$$

$$= 1734.56 \text{ m}^3/\text{day}$$

$$= \mathbf{72.27\ m^3/hour}$$

∴ Percentage decrease in discharge

$$= \frac{80.29 - 72.27}{80.29} \times 100$$

$$= 9.99\%.$$

Illustrative Example 6.11 *A gravity well has a diameter of 1 m. Before pumping is started the depth of water in the well is 50 m. When pumping is being done at the rate of 2000 litres per minute the drawdown in a well 20 m away is 4 m and in another well 40 m away is 2 m. Determine (a) the radius of the circle of influence; (b) coefficient of permeability; (c) drawdown in the well; (d) specific capacity of the well; and (e) the maximum rate at which water can be pumped from the well.*

Solution

(a) For a well in an unconfined aquifer, we have from Eq. (6.9b)

$$Q = \frac{1.36k(H^2 - h^2)}{\log_{10}(R/r)}$$

Here $H = 50$ m

At $r = 20$ m, $h = (50 - 4) = 46$ m

At $r = 40$ m, $h = (50 - 2) = 48$ m

Applying the above equation at these two locations, we have

$$\frac{50^2 - 46^2}{\log_{10}(R/20)} = \frac{50^2 - 48^2}{\log_{10}(R/40)}$$

or $$\log_{10}(R/40) = 0.5104 \log_{10}(R/20)$$

or $$\log_{10}(R/40) = \log_{10}(R/20)^{0.5104}$$

or $$\frac{R}{40} = \left(\frac{R}{20}\right)^{0.5104}$$

or $$R^{(1-0.5104)} = \frac{40}{(20)^{0.5104}} = 8.6699$$

or $$R = (8.6699)^{\frac{1}{0.4896}}$$

∴ $$R = 82.39 \text{ m}$$

(b) $$Q = 2000 \text{ litres/minute}$$
$$= 2000 \times 10^{-3} \text{ m}^3\text{/minute}$$
$$= 2 \text{ m}^3 \text{ / minute}$$
$$H = 50 \text{ m}$$
$$h = (50 - 4) = 46 \text{ m}$$
$$r = 20 \text{ m}$$
$$R = 82.39 \text{ m}$$

Thus, by substituting values in the above equation, we get

$$2 = \frac{1.36k(50^2 - 46^2)}{\log_{10}(82.39/20)}$$

∴ $$k = 0.00235 \text{ m/minute}$$
$$= 3.3840 \text{ m/day}$$

(c) If h is the depth of water in the well, then we have

$$2 = \frac{1.36 \times 0.00235 \times (50^2 - h^2)}{\log_{10}(82.39/0.5)}$$

or $$(2500 - h^2) = \frac{2}{1.36 \times 0.00235} \times \log_{10}(82.39/0.5)$$
$$= 1387.3$$

or $$h = 33.36 \text{ m}$$

∴ Drawdown $$= (50 - 33.36)$$
$$= 16.64 \text{ m}$$

(d) The specific capacity S_c is defined as the discharge per unit drawdown. The yield from a well may be assumed to be directly proportional to the radius of circle of influence. Thus,

$$Q \propto R$$

or $$Q = CR$$

where C is a constant of proportionality

$$\therefore \qquad C = \frac{Q}{R}$$

Here $$Q = 2 \text{ m}^3/\text{min}$$
$$R = 82.39 \text{ m}$$

$$\therefore \qquad C = \frac{2}{82.39} = 0.0243$$

$\therefore$ In general $$R = \frac{Q}{C} = \frac{Q}{0.0243}$$

Now corresponding to drawdown of 1 m, $Q = S_c$ (by definition)

Hence for this value of Q, the radius of the circle of influence is

$$R = \frac{S_c}{0.0243}$$

Thus, by substituting these values in the above equation, we get

$$Q = S_c = \frac{1.36 \times 0.00235 \times (50^2 - 49^2)}{\log_{10}\left(\dfrac{S_c}{0.0243 \times 0.5}\right)}$$

or $$S_c \log_{10} (82.30\, S_c) = 0.3164$$

Solving by trial and error, we get

$$S_c = 0.2432 \text{ m}^3/\text{minute per unit drawdown.}$$

(e) Maximum discharge Q_m will be obtained when the draw-down at the well is equal to H, i.e., $h = 0$.

For this discharge the value of R is given by

$$R = \frac{Q_m}{0.0243}$$

Thus, by substituting these values in the above equation, we get

$$Q_m = \frac{1.36 \times 0.00235 \times (50^2 - 0^2)}{\log_{10}\left(\dfrac{Q_m}{0.0243 \times 0.5}\right)}$$

or $$Q_m \log_{10} (82.30\, Q_m) = 7.99$$

Solving by trial and error, we get

$$Q_m = 3.2854 \text{ m}^3/\text{minute} = 3285.4 \text{ litres/minute.}$$

Illustratie Example 6.12 *An aquifer of 20 m average thickness is overlain by an impermeable layer of 30 m thickness. A test well of 0.5 m diameter and two observation wells at a distance of 10 m and 60 m from the test well are drilled through the aquifer. After pumping at a rate of 0.1 m^3/s for a long time, the following drawdowns are stabilized in these wells: first observation well, 4 m; second observation well, 3 m. Show the arrangement in a diagram. Determine the coefficient of permeability and the drawdown in the test well.*

Solution

The arrangement of the wells is as shown in Fig. 6.7.

From Eq. (6.22a), we have

$$Q = \frac{2.72kb(s_1 - s_2)}{\log_{10}(r_2 - r_1)}$$

$$Q = 0.1 \text{ m}^3\text{/s};\ b = 20 \text{ m};\ s_1 = 4 \text{ m};\ s_2 = 3 \text{ m};$$

$$r_2 = 60 \text{ m; and } r_1 = 10 \text{ m}$$

Thus, by substitution, we get

$$0.1 = \frac{2.72 \times k \times 20(4-3)}{\log_{10}(60/10)}$$

$$\therefore \quad k = 0.00143 \text{ m/s} = 1.43 \text{ mm/s}$$

Let s be the drawdown in the test well. Thus, applying Eq. 6.22 (a) between the test well and the first observation well, we get,

$$0.1 = \frac{2.72 \times 0.00143 \times 20 \times (s-4)}{\log_{10}(10/0.25)}$$

$$\therefore \quad s = 6.06 \text{ m}$$

Note: Equation (6.22a) may also be applied between the test well and the second observation well, which will also give the same result.

Illustrative Example 6.13 *A well fully penetrating a confined aquifer is pumped at a uniform rate of 2500 m^3/day. During the pumping period drawdowns are measured in an observation well 60 m away and the observations of time t since pumping began and draws are given below. Using Theis method determine the formation constants of the aquifer.*

Time since pumping began t (min)	*Drawdown s (m)*	*Time since pumping began t (min)*	*Drawdown s (m)*
0	0	18	0.67
1.0	0.20	24	0.72
1.5	0.27	30	0.76
2.0	0.30	40	0.81
2.5	0.34	50	0.85
3.0	0.37	60	0.90
4.0	0.41	80	0.93

Contd.

Contd

5.0	*0.45*	*100*	*0.96*
6.0	*0.48*	*120*	*1.00*
8.0	*0.53*	*150*	*1.04*
10.0	*0.57*	*180*	*1.07*
12.0	*0.60*	*210*	*1.10*
14.0	*0.63*	*240*	*1.12*

Solution

Values of (r^2/t) in m²/min are computed as noted in the following Table.

Drawdown *s (m)*	r^2/t *(m² /min)*	*Drawdown* *s (m)*	r^2/t *(m²/min)*
0	∞	0.67	200
0.20	3600	0.72	150
0.27	2400	0.76	120
0.30	1800	0.81	90
0.34	1440	0.85	72
0.37	1200	0.90	60
0.41	900	0.93	45
0.45	720	0.96	36
0.48	600	1.00	30
0.53	450	1.04	24
0.57	360	1.07	20
0.60	300	1.10	17
0.63	257	1.12	15

Values of s and r^2/t are plotted on logarithmic paper as shown in Fig. Ex. (6.13a). Values of $W(u)$ and u from Table 6.2 are plotted on another sheet of logarithmic paper as shown in Fig. Ex. (6.13b) and a curve is drawn through the points.

The two sheets are superimposed and shifted with coordinate axes parallel until the observational points coincide with the curve as shown in Fig. Ex. (16.13c). A match point P as shown in Fig. Ex. (6.13c) is selected and its coordinates are obtained as follows.

$$W(u) = 4.0 \text{ and } u = 1.0 \times 10^{-2}$$

$$s = 0.72 \text{ m and } (r^2/t) = 150 \text{ m}^2/\text{min.}$$

Thus, from Eq. (6.52)

$$T = \frac{Q}{4\pi s} W(u)$$

Here $\quad Q = 2500 \text{ m}^3/\text{day}$

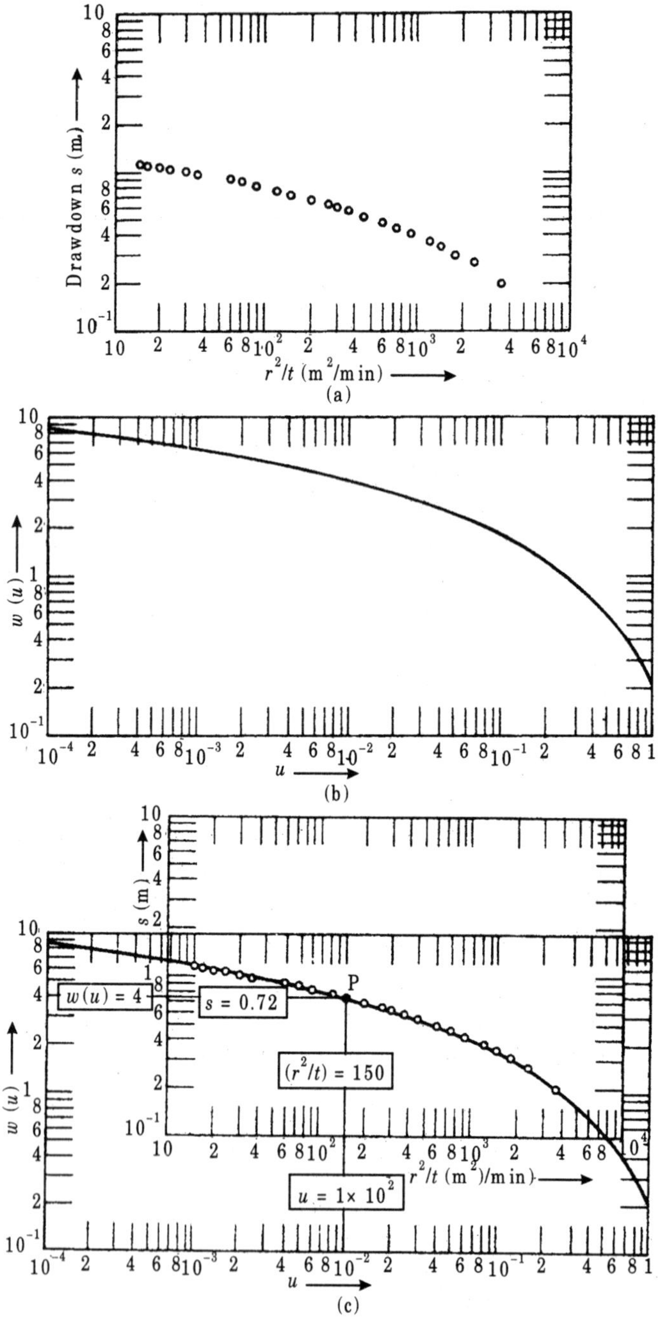
Drawdown s (m)
r^2/t (m^2/min)
(a)
w (u)
u
(b)
s (m)
w (u) = 4
s = 0.72
P
(r^2/t) = 150
u = 1× 10^2
r^2/t (m^2)/min)
w (u)
u
(c)

Fig. Ex. 6.13

$$\therefore \qquad T = \frac{2500 \times 4.0}{4\pi \times 0.72} = 1105 \text{ m}^2/\text{day}$$

$$= \frac{1105}{24 \times 60} = 0.7674 \text{ m}^2/\text{min}$$

Further from Eq. (6.53),

$$S = \frac{4T}{(r^2/t)} u$$

$$= \frac{4 \times 0.7674}{150} \times 1 \times 10^{-2} = 0.000205$$

Illustrative Example 6.14 *Solve example 6.13 by Jacob's method.*

Solution

The observed values of s and t are plotted on a semilogarithmic paper as shown in the Fig. Ex. 6.14. A straight line is fitted through the points. The drawdown difference Δs for one log cycle of time t is obtained from the plot as 0.4. Further from the same plot t_0 = 0.39 min = 2.708 × 10^{-4} day.

Thus from Eq. (6.58)

$$T = \frac{2.30Q}{4\pi(\Delta s)} = \frac{2.3 \times 2500}{4\pi(0.4)} = 1144 \text{ m}^2/\text{day}$$

Further from Eq. 6.59

$$S = \frac{2.25T\, t_0}{r^2} = \frac{2.25 \times 1144(2.708 \times 10^{-4})}{(60)^2} = 0.000194$$

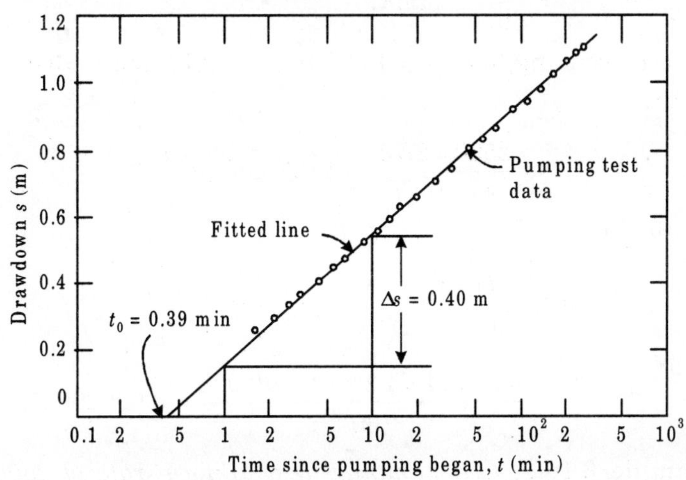

Fig. Ex. 6.14

Illustrative Example 6.15 *Solve example 6.13 by Chow's method.*

Solution The observed values of s and t are plotted on a semilogarithmic paper as shown in the Fig. Ex. (6.15). A point P is selected on the curve where t = 6 min = 4.2×10^{-3} day and s = 0.47 m. A tangent to the curve is drawn at P as shown in Fig. Ex. (6.15) and the drawdown difference per log cycle of time is Δs = 0.38 m.

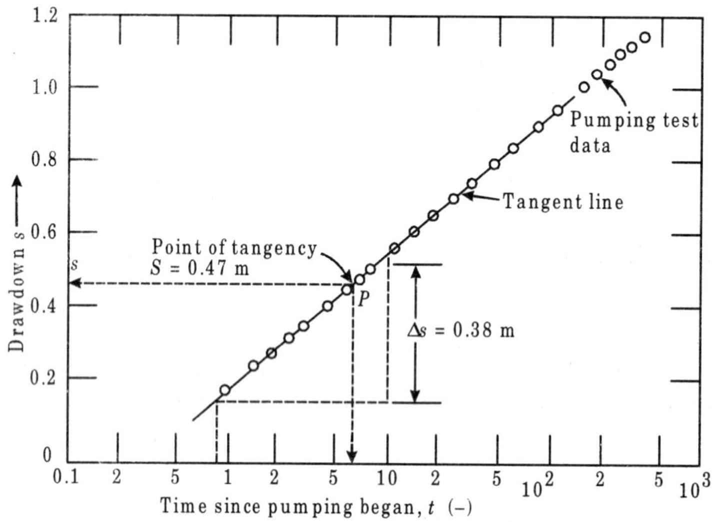

Fig. Ex. 6.15

From Eq. (6.61)

$$F(u) = \frac{s}{\Delta s}$$

$$= \frac{0.47}{0.38} = 1.237$$

From Fig. (6.17) corresponding to $F(u)$ =1.237, $W(u)$ = 2.75 and u =0.038.

Thus, from Eq. (6.52),

$$T = \frac{Q}{4\pi s} W(u) = \frac{2500 \times 2.75}{4\pi \times 0.47} = 1164 \text{ m}^2/\text{day}$$

Further from Eq. (6.53)

$$S = \frac{4Ttu}{r^2}$$

$$= \frac{4 \times 1164 \times 4.2 \times 10^{-3} \times 0.038}{(60)^2} = 0.000206$$

Illustrative Example 6.16 *A well pumping at a uniform rate of 2500 m³/day was shut down after 240 minutes, thereafter the measurements of s′ and t′ were made in an observation well which are as given below. Determine the coefficient of transmissibility T for the aquifer.*

Time since pump shut down t′ (min)	*Time since pumping began t (min)*	*Residual drawdown s′ (min)*
1	241	0.89
2	242	0.81
3	243	0.76
5	245	0.68
7	247	0.64
10	250	0.56
15	255	0.49
20	260	0.45
30	270	0.38
40	280	0.34
60	300	0.28
80	320	0.24
100	340	0.21
140	380	0.17
180	420	0.14

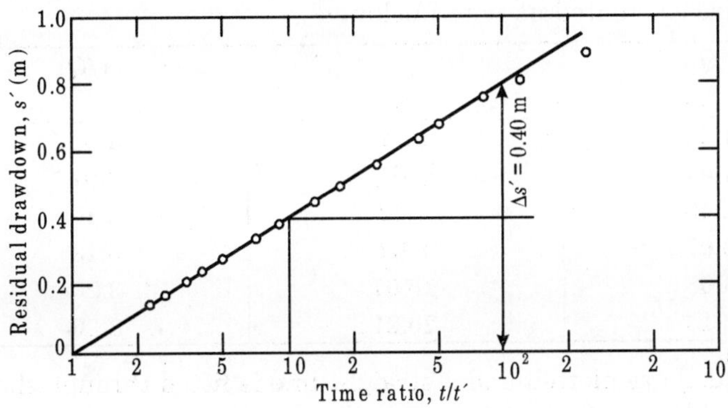

Fig. Ex. 6.16

Solution

Values of (t/t') are computed as noted below.

s'	(t/t')	s'	(t/t')	s'	(t/t')
0.89	241	0.56	25	0.28	5
0.81	121	0.49	17	0.24	4
0.76	81	0.45	13	0.21	3.4
0.68	49	0.38	9	0.17	2.7
0.64	35	0.34	7	0.14	2.3

Values of s' and (t/t') are plotted on semilogarithmic paper as shown in Fig. Ex. (6.16). A straight line is drawn to pass through the points. From this plot the difference of residual drawdown $\Delta s'$ per log cycle of (t/t') is obtained as 0.40.

From Eq. (6.67),

$$T = \frac{2.30Q}{4\pi\Delta s'}$$

$$= \frac{2.30 \times 2500}{4\pi \times 0.40} = 1144 \text{ m}^2/\text{day}$$

Illustrative Example 6.17 *Step drawdown test was carried out in a well constructed to give a yield of 2.3 m³/minute. The data obtained are given below. Determine the well loss and the efficiency of the well.*

Yield in m³/minute	*Drawdown in metres*
1.23	*3.72*
1.84	*5.65*
2.46	*7.62*
3.07	*9.62*
4.08	*13.07*
6.13	*20.21*

Solution

Values of (s/Q) are calculated as noted below.

Q (m^3 /min)	s (m)	(s/Q)
1.23	3.72	3.02
1.84	5.65	3.07
2.46	7.62	3.10
3.07	9.62	3.13
4.08	13.07	3.20
6.13	20.21	3.30

Values of (s/Q) and Q are plotted and a straight line is fitted through the points as shown in Fig. Ex. (6.17). The well loss coefficient C is given by the slope of this line and the formation loss coefficient B is given by the intercept $Q = 0$.

Thus, $C = 5.78 \times 10^{-2}$ and $B = 2.96$

$\therefore$ Well loss $= CQ^2 = 5.78 \times 10^{-2}\, Q^2$

For $Q = 2.3$ m³ /minute

Well loss $= 5.78 \times 10^{-2}\, (2.3)^2$ m $= 0.306$ m

Further

$$s = 2.96Q + 5.78 \times 10^{-2}\, Q^2$$

$$= 2.96 \times 2.3 + 5.78 \times 10^{-2}\, (2.3)^2$$

$$= 7.114 \text{ m}$$

From Eq. 6.31 efficiency of well is given as

$$E_w = \frac{BQ}{s} \times 100$$

$$= \frac{2.96 \times 2.3}{7.114} \times 100 = 95.7\%$$

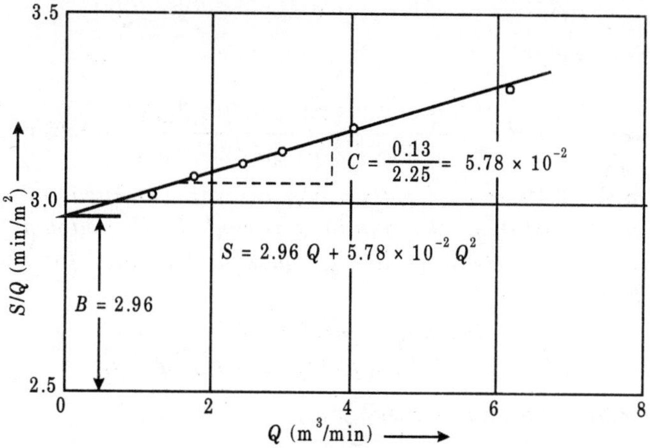

Fig. Ex. 6.17

Illustrative Example 6.18 *A well is located in a 25 m confined aquifer of permeability 30 m/day and storage coefficient 0.005. If the well is being pumped at the rate of 1750 litres per minute, calculate the drawdown at a distance of 100 m from the well after 20 hours of pumping.*

Solution

From Eq. (6.47), we have

$$u = \frac{r^2 S}{4T_t}$$

in which

$$r = 100 \text{ m};\ S = 0.005\ ;\ T = \left(\frac{30 \times 25}{24}\right) \text{m}^2\text{/hour}\ ;\ \text{and } t = 20 \text{ hours}$$

Thus, by substitution, we get

$$u = \frac{(100)^2 \times 0.005 \times 24}{4 \times 30 \times 25 \times 20} = 0.02$$

From Table 6.3 for $u = 0.02$; $W(u) = 3.35$. Alternatively approximate value of $W(u)$ is given by

$$W(u) = [-0.5772 - \log_e u + u]$$

From which for u 0.02

$$W(u) = [-0.5772 - \log_e (0.02) + 0.02] = 3.35$$

From Eq. (6.50) drawdown s is given by

$$s = \frac{Q}{4\pi T} W(u)$$

in which

$$Q = \frac{1750 \times 10^{-3}}{60} \text{ m}^3\text{/s}; \ T = \left(\frac{30 \times 25}{24 \times 3600}\right) \text{m}^2\text{/s}$$

Thus, by substitution, we get

$$s = \frac{1750 \times 10^{-3} \times 24 \times 3600 \times 3.35}{60 \times 4 \times \pi \times 30 \times 25} = 0.896 \text{ m} .$$

Illustrative Example 6.19 *A well fully penetrating a confined aquifer was pumped at a uniform rate of 2500 m³/day. During the pumping period the drawdowns were measured in an observation well 60 m away. The observed data was plotted and for the matching point the following data was obtained.*

$$W(u) = 4.0 \ \text{ and } u = 1.0 \times 10^{-2}$$

$$s = 0.72 \text{ m and } r^2/t = 150 \text{ m}^2/\text{min.}$$

Determine the formation constants of the aquifer.

Solution

From Eq. (6.52), we have

$$T = \frac{Q}{4\pi s} W(u)$$

Thus, by substituting the given values, we get

$$T = \frac{2500}{4 \times \pi \times 0.72} \times 4$$

$$= 1105 \text{ m}^2\text{/day} = 0.767 \text{ m}^2\text{/min}$$

Further from Eq. (6.53), we have

$$S = \frac{4T}{(r^2/t)} u$$

Thus, by substituting the given values, we get

$$S = \frac{4 \times 0.767}{150} \times 1.0 \times 10^{-2} = 0.000205$$

Illustrative Example 6.20 *A 300 mm well penetrates an aquifer of transmissivity 2 × 10⁵ litres per day per metre and storativity 0.005. What pumping rate could be adopted so that the drawdown will not exceed 10 m within the subsequent two years of draught?*

Solution

From Eq. (6.55), we have

$$s = \frac{2.30Q}{4\pi T} \log_{10}\left(\frac{2.25Tt}{r^2 S}\right)$$

By substituting the given values, we get

$$10 = \frac{2.30 \times Q}{4\pi \times 2 \times 10^5} \log_{10}\left(\frac{2.25 \times 2 \times 10^5 \times 2 \times 365}{1000 \times (0.15)^2 \times 0.005}\right)$$

$\therefore$ $Q = 11.544 \times 10^5$ litres per day

$= 802$ litres per minute.

Illustriative Example 6.21 *A well is located in an aquifer with hydraulic conductivity 15 m/day and storativity 0.005. The aquifer is 20 m thick and is pumped at a rate of 2725 m^3/day. What is the drawdown at a distance of 7 m from the well after one day of pumping?*

Solution

From Eq. (6.8), we have

$$T = bk$$

$$= 20 \times 15 = 300 \text{ m}^2\text{/day}$$

From Eq. (6.47), we have

$$u = \frac{r^2 S}{4Tt}$$

$$= \frac{(7)^2 \times 0.005}{4 \times 300 \times 1} = 0.0002$$

From Table 6.3, for $u = 0.0002$, $W(u) = 7.94$

$\therefore$ From Eq. (6.50), we get

$$s = \frac{Q}{4\pi T} W(u)$$

$$= \frac{2725}{4\pi \times 300} \times 7.94 = 5.74 \text{ m}$$

Illustrative Example 6.22 *An artesian well is pumped at a constant rate of 1000 m^3/day from an extensive aquifer of average thickness 35 m. If the specific storage of the aquifer is 3×10^{-4} m^{-1} and hydraulic conductivity is 9.5 m/day find (a) the drawdown at a point 4 m from the well after 12 hour of continuous pumping; (b) the time during which Jacob method cannot be used at the same point in part (a).*

Solution

Specific storage is defined as the volume of water released per unit volume of saturated aquifer per unit decline in head. It is denoted by S_s. Thus,

$$S = S_s \times b$$

$$= (3 \times 10^{-4} \times 35) = 0.0105$$

Also $T = b \times k = (35 \times 9.5) \text{ m}^2\text{/day}$

(a) From Eq. (6.55), we have

$$s = \frac{2.30Q}{4\pi T} \log_{10}\left(\frac{2.25Tt}{r^2 S}\right)$$

By substituting the given values, we get

$$s = \frac{2.30 \times 1000}{4 \times \pi \times 35 \times 9.5} \log_{10} \left(\frac{2.25 \times 35 \times 9.5 \times 12}{24 \times (4)^2 \times 0.0105} \right)$$

$$= 1.84 \text{ m}$$

(b) Considering the limiting value of $u < 0.01$, we have

$$u = \frac{r^2 S}{4Tt} < 0.01$$

or

$$t > \frac{r^2 S}{4T \times 0.01}$$

or

$$t > \frac{(4)^2 \times 0.0105}{4 \times 35 \times 9.5 \times 0.01} \text{ day}$$

or

$$t > 1.26 \times 10^{-2} \text{ day}$$

or

$$t > 18.2 \text{ minutes}$$

That is, the pump test data observed within 18 minutes should not be considered while drawing the straight line graph.

Illustrative Example 6.23 *During a recuperation test, the water in an open well depressed, by pumping, by 2.1 m and it recuperated 1.6 m in 90 minutes. Find the diameter of the well to yield 10 litres per second under a depression head of 2 m.*

Solution

From Eq. (6.70a), we have

$$Q = \frac{2.30AH}{t'} \log_{10}(h_1 / h_2)$$

$$Q = 10 \; l/\text{s} = 0.01 \text{ m}^3/\text{s}; \; H = 2 \text{ m};$$

$$t' = 90 \times 60 = 5400 \text{ s}; \; h_1 = 2.1 \text{ m}; \text{ and}$$

$$h_2 = (2.1 - 1.6) = 0.5 \text{ m}$$

Thus by substitution, we get

$$0.01 = \frac{2.30 \times A \times 2}{90 \times 60} \log_{10}(2.1 / 0.5)$$

or

$$A = 18.84 \text{ m}^2$$

∴ Diameter of the well is

$$D = \sqrt{\frac{18.84 \times 4}{\pi}} = 4.9 \text{ m}$$

REVIEW QUESTIONS

6.1 Briefly explain the role of ground water in water resources development in the country.

6.2 Define the terms (i) Aquifer; (ii) Aquiclude; (iii) Aquifuge; and (iv) Aquitard.

6.3 Define porosity, specific yield and specific retention and obtain a relation between them.

6.4 Describe with neat sketches confined, semi-confined (or leaky), unconfined and perched aquifers.

6.5 Explain the terms "storage coefficient', and 'coefficient of transmissibility'.

6.6 Explain Darcy's law. What are its assumptions? Discuss its validity.

6.7 A 400 mm diameter well fully penetrates a confined aquifer of permeability 40 m/day. The length of the strainer is 20 m. Under steady state of pumping the discharge from the well was 1500 litres per minute. Compute the drawdown at the well if the radius of influence of the well is 400 m. [*Ans.* 3.28 m]

6.8 A 300 mm diameter well penetrates 25 m below the static water-table. After 24 hours of pumping at 6000 litres/minute, the water table in a test well at 90 m is lowered by 0.53 m and in a well 30 m away the drawdown is 1.11 m. What is the transmissibility of the aquifer? [*Ans.* 0.031 m^2/s]

6.9 What do you understand by a partially penetrating well? Support your answer with neat sketches.

6.10 Explain the terms well losses, specific capacity, specific drawdown, well efficiency.

6.11 Derive the following differential equation for unsteady flow towards wells with usual notation

$$\frac{\delta^2 h}{\delta r^2} + \frac{1}{r}\frac{\delta h}{\delta r} = \frac{s}{T}\frac{\delta h}{\delta t}$$

6.12 Explain unsteady flow in a confined aquifer. Hence give main features of Jacob's method for analysis of time drawdown data of pumping tests to determine aquifer parameters S and T.

6.13 During a recuperation test, the water in an open well was depressed by 2.5 m and it recuperated 1.875 m in 1.5 hours. Estimate the yield from the well of 2m diameter under a depression head of (a) 2.5 m; (b) 2.25 m; (c) 2 m.

[*Ans.* (a) 7.25 m3/*h*; (b) 6.52 m^3/h; (c) 5.80 m^3/h]

6.14 Describe recuperation test for an open well.

6.15 List the various methods of drilling used for constructing tube wells. Describe any one of these methods in detail.

6.16 What do you understand by well development? Describe briefly the various methods of well development.

6.17 Describe with neat sketches the different types of strainers used for the tube wells.

6.18 Describe with a neat sketch a collector or radial well.

6.19 An excavation is to be made for a basement 3 m deep and 25 m square. During an investigation of the site it was found out that the water table is 1 m below the ground and the aquifer has conductivity of 25 m/day and a storage coefficient 0.02. The impermeable layer lies 20 m below the ground level. Dewatering is then needed and intended to be carried out using two bore holes, placed at 2 m outside the excavation on either side and it was assumed that the radius of influence is 200 m.

(a) Determine the pumping rate required for full dewatering of the excavation site.

(b) If only single well is used and pumped at the rate of 25 litres per second, how long would it take to fully dewater the excavation.

[*Ans.* (a) 14.37 litres/sec. for each well; (b) 15.20 hours]

6.20 A well located in 20 m thick aquifer is pumped at a rate of 2700 m^3 per day. If after one day of pumping the drawdown observed at a distance of 8 m from the well is 5.49 m determine the storativity of the aquifer. Given that hydraulic conductivity of the aquifer = 15 m/day,

and for $u = 2.0 \times 10^{-4}, W(u) = 7.94$;

for $u = 3.0 \times 10^{-4}$, $W(u) = 7.53$ [*Ans.* 0.005]

6.21 A well in a confined aquifer was pumped at a rate of 900 litres per minute for 8 hours. The aquifer was 6 m thick. Time-drawdown data for an observation well 250 m away was plotted on a semilogarithmic paper and it was found that the drawdown per log cycle was 1.68 m and the time for zero drawdown was 5.2 minutes. Find T, k and S. Also determine the time during which Jacob method cannot be used at the observation well.

[*Ans.* 141.19 m^2/ day; 23.53 m/day; 1.84×10^{-5}; 293 min.]

6.22 Show that the drawdown in the case of a partially penetrating well in a confined aquifer (well just penetrating to the top of an infinite porous medium) is given by an equation

$$s = \frac{Q_s}{2\pi r k}$$

where

Q_s = discharge from well under steady condition;

r = radius of well;

k = coefficient of well; and

s = drawdown at the well

Also find an expression for (Q_s/Q), where Q is discharge from a fully penetrating well in confined aquifer under steady state.

6.23 During a recuperation test, the water in an open well was depressed by pumping by 2.5 m and it recuperated 1.8 m in 80 minutes. Calculate the yield from a well 4 m diameter under a depression head of 3 m. [*Ans.* 10 *l*/s]

CHAPTER

7 Reservior Planning

7.1 INTRODUCTION

The flow rate of a river or a natural stream may vary considerably during different periods of a year. It may carry little or no water during some periods of a year and may carry huge amount of water after heavy rains thus, becomes a raging torrent and a hazard all along its banks. During extremely low flows it may not be possible to meet the demands of the consumers if water is drawn directly from a river. As such it is essential to create a reservoir or an artificial lake by constructing a dam across the river which can retain the excess water from periods of high flows for use during the periods of low flows or droughts. In addition to conserving water for later use, the storage of floodwater may also reduce flood damage on the down stream of the reservoir. In this chapter the general aspects of reservoir planning are discussed.

7.2 TYPES OF RESERVOIRS

The reservoirs are generally classified according to the purpose served by them. The reservoirs are constructed to serve many purposes as indicated below.

(i) Flood control, (ii) Irrigation, (iii) Development of hydroelectric power, (iv) Navigation, (v) Recreation, (vi) Water supply for domestic and industrial use, (vii) Development of fish and wild life, (viii) Soil conservation, and (ix) other miscellaneous purposes.

If a reservoir is constructed to serve only one purpose, then it is known as a *single purpose reservoir*. On the other hand if a reservoir is constructed to serve more than one purpose then it is termed as a *multipurpose reservoir*. For example, if a reservoir is constructed to supply water only for irrigation and no other purpose is served by it then it is known as a single purpose reservoir. However, if a reservoir is used for supplying water for irrigation as well as for the generation of hydroelectric power then it may be designated as a multipurpose reservoir. Moreover, the entire work involved in the construction of these reservoirs as well as for the use of water stored by them is usually designated as a single purpose project and a multipurpose project depending upon the purpose served by the reservoir. Since in most of the cases single purpose reservoirs are not economically feasible, usually multipurpose reservoirs are constructed.

Depending upon the specific purpose served, the reservoirs may also be classified as follows.

1. Storage or conservation reservoirs
2. Flood control reservoirs
3. Distribution reservoirs.

1. Storage or Conservation Reservoirs. A storage reservoir is constructed to store the excess water that flows in the river during the period of high flows for later use for various purposes. Storage reservoirs are primarily used for supply of water for irrigation, development of hydroelectric power and domestic and industrial water supply. However, to some extent these reservoirs also help in moderating the floods and thus provide protection against floods on the downstream side. Depending upon the purpose served, these reservoirs may be designated as single purpose reservoirs and multipurpose reservoirs.

2. Flood Control Reservoirs. A flood control reservoir is constructed to protect the area lying on its downstream side from the damages due to floods and hence it is known as flood protection reservoir or flood mitigation reservoir. Usually these reservoirs serve only one purpose viz., flood control and hence these are designated as single purpose reservoirs.

A flood control reservoir holds some of the flood waters of a river during the rising flood when the flow rate reaches a stage likely to cause damage on the downstream side and releases them gradually at a safe rate when the flood recedes. The effect of holding up of water by a flood control reservoir on a flood is to reduce its peak rate of flow in the river on the downstream of the reservoir by absorbing a part of the flood volume when the flood is rising and releasing the same volume gradually when the flood is receding. Thus, as shown in Fig. 7.1 the firm line *ABC* represents the natural hydrograph of a flood at the dam site and the dotted line *AB′C* represents the hydrograph as moderated by the reservoir. The maximum flood discharge is therefore reduced from Q_1 to Q_2. Further it follows that the reservoir should be able to hold in storage the volume of water represented by the hatched area between the two curves which represents the excess of inflow over the outflow during the rising stage, and the same would then be gradually released during the falling stage as represented by dots. If the reservoir comes back to its original level the total volume under the natural and the moderated hydrographs would be the same and hence the hatched area representing the excess supply held in storage will be equal to the area with dots representing the volume released from storage.

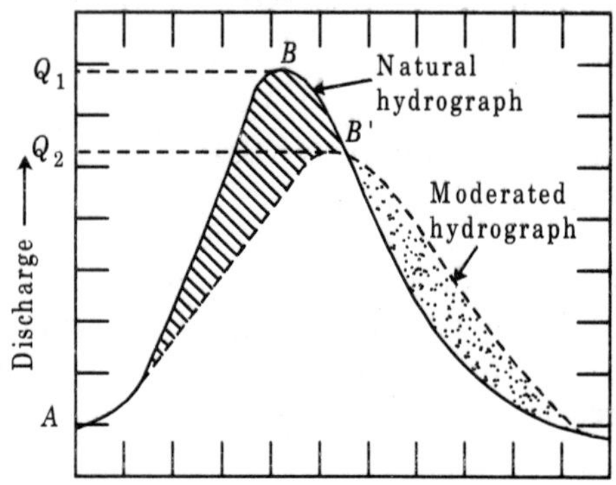

Fig. 7.1. *Hydrograph moderated by flood control reservoir*

Flood control reservoirs are of two types, viz.,

(a) Retarding reservoirs

(b) Detention basins.

(a) *Retarding Reservoirs.* A retarding reservoir is the one which is provided with outlets and spillway not controlled by gates or valves. The discharging capacity of the outlets and spillway is such that the total outflow through them would not exceed the safe capacity of the river downstream during the highest flood. The advantages of a retarding reservoir are:

(1) There are no gates or valves at the outlets and the spillway and hence the possibility of human error or negligence in the operation of a retarding reservoir is eliminated.

(2) The cost of gate or valve installations and their operation is saved.

However, a disadvantage of retarding reservoirs is that when constructed on a number of tributaries joining the same river the automatic control may cause all the modified flood crests from all the tributaries to coincide on the main stream and thus reduce the benefits of the flood control reservoir. In such cases the locations of the reservoirs may be so chosen that the possibility of all the flood waves joining the main stream simultaneously is avoided.

An ideal location for a retarding reservoir is immediately above a city or the area to be protected from flood.

(b) *Detention Basins*. A detention basin is the one which is provided with outlets and spillway controlled by gates or valves. The gate or valve installations provide greater flexibility of operation as in the earlier stages of a flood larger discharge may be released (subject to the safe maximum capacity of the river downstream) and later with increasing heads on the outlets if the discharge tends to rise above the maximum permissible value it can be restricted by gates or valves. Moreover in this case because of the control by the gates or valves larger outlets may be provided through which the reservoir can be emptied more quickly and the reservoir capacity is made available again for moderating a possible subsequent flood after a shorter interval. This enables a better use to be made of the available reservoir capacity. Further in the case of a number of detention basins on the various tributaries of a main river the release of water may be so arranged that the flood crests from the different streams are staggard and thus keep the peak discharge in the main river at a lower value than that in the case of a retarding reservoir.

The disadvantage of detention basins is that they are more costly, initially due to the additional cost of gates or valves and lifting machinery and subsequently due to the maintenance and operation cost as a regular staff has to be employed to maintain and operate the gates or valves. Further the detention basins also involve the possibility of human error or negligence in their operation.

The advantages and disadvantages of retarding reservoirs and detention basins should be considered in any case before a decision is taken. However, a detention basin is generally preferred to a retarding reservoir where the area to be protected from the floods is large and widespread.

3. Distribution Reservoirs. A distribution reservoir is a small storage reservoir used for water supply in a city. Such a reservoir is required to fulfill the varying demands of the consumers at different periods of the day. Water is continuously pumped into the reservoir at a constant rate and is supplied to the consumers from the reservoir. During the periods of no demand or demand less than the pumping rate water is stored in the reservoir but during the periods of demand exceeding the pumping rate water is supplied by the reservoir from its storage. Thus, a distribution reservoir permits pumps to operate at a constant rate and it can supply water at a rate higher than the constant inflow rate of pumping.

7.3 INVESTIGATIONS FOR RESERVOIR PLANNING

For reservoir planning the following three types of investigations are required.

(1) Engineering surveys.

(2) Geological investigations.

(3) Hydrological investigations.

1. Engineering Surveys. The area of the dam site, reservoir and other associated works is surveyed and a contoured plan of the entire area is prepared. From the contoured plan the storage capacity and the water spread area of the reservoir at various elevations are determined as mentioned below.

From the contoured plan of the reservoir site the water spread area of the reservoir at any elevation is directly determined by measuring the area enclosed by the contour corresponding to that elevation with a planimeter. The water spread areas at different elevations are thus determined and the same are plotted against elevation to obtain an *elevation-area curve* as shown in Fig. 7.2.

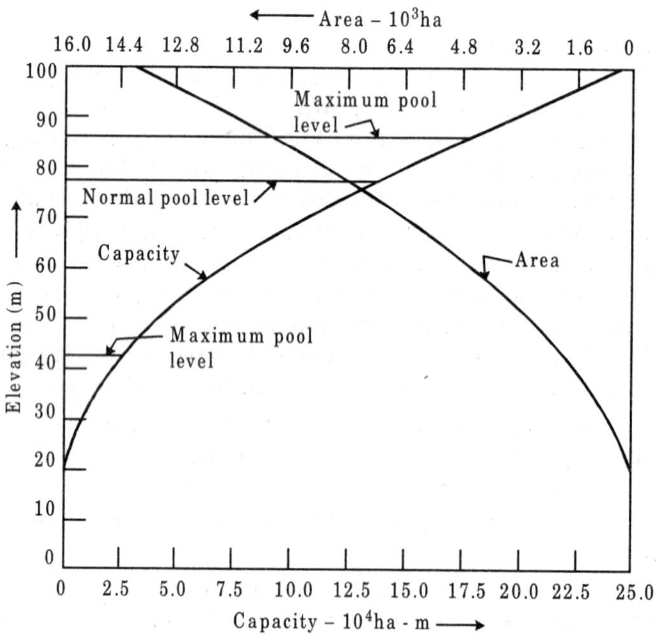

Fig. 7.2. *Elevation-area and elevation-capacity curve*

The storage capacity of the reservoir may be determined by taking contour areas at equal interval and summing up by trapezoidal formula, cone formula or prismoidal formula. Thus, if V is the storage volume and A_1, A_2, A_3,......A_n are the areas enclosed by the successive contours at an interval h, then the various formulae are as follow:

(i) *Trapezoidal formula*

$$V = \sum \frac{h}{2}(A_1 + A_2)$$

$$= h\left[\frac{A_1 + A_2}{2} + A_2 + A_3 + A_4 ... + A_{n-1}\right]$$

(ii) *Cone formula*

$$V = \sum \frac{h}{3}\left(A_1 + A_2 + \sqrt{A_1 A_2}\right)$$

(iii) *Prismoidal formula*

$$V = \sum \frac{h}{3}(A_1 + 4A_2 + A_3)$$

$$= \frac{h}{3}\left[(A_1 + A_n) + 4(A_2 + A_4 + \ldots\ldots) + 2(A_3 + A_5 + \ldots\ldots)\right]$$

The storage volumes upto various elevations are determined by using any of the above noted formulae and the same are plotted against elevation to obtain *elevation-capacity curve* as shown in Fig. 7.2.

The contoured map of the reservoir area also indicates the land and property which would be submerged in the reservoir when filled upto certain elevation and thus enables the estimation of compensation to be paid to the owners and the time schedule according to which they must be moved, as the reservoir is gradually filled during the construction of the dam.

2. Geological Investigations. Geological investigations are required to determine the following items.

(i) Suitability of foundation for the dam.

(ii) Watertightness of the reservoir basin.

(iii) Location of quarry sites for obtaining suitable materials required for the construction and estimation of the quantities likely to be available from them.

Depending upon the requirements of the site and the type and height of the proposed dam, subsurface exploration is to be carried out to determine the depth of overburden to be excavated and removed for laying the foundation, the natural rock formation below and the nature and extent of fault zones, if any, present in the rock. In case of earth dams or low dams to be founded on soil foundations the subsurface exploration provides information about the properties of the foundation soil at various depths. Moreover the information obtained during these explorations will also help in devising a suitable program of foundation treatment by grouting before the construction is taken up.

A geological survey of the basin should be carried out to detect the presence of cavernous (full of cavities or porous) rock formation which may enable the stored water to escape into adjacent valleys. If such areas are of small magnitude, it may be possible to treat them, but if they are widespread the site may have to be abondoned.

For the construction of a dam large quantity of material is required and hence it is necessary of know the locations from where the bulk of the construction material may be obtained. Further the quantities of the material likely to be available from each location should also be known. If the material required for the construction is available in the vicinity of the dam site this will result in considerable reduction in the cost of the project. However, if possible the designs may be adjusted to suit the type of material available at or near the dam site.

3. Hydrological Investigations. The entire planning of a reservoir is based on the estimate of the quantity of water likely to be available in the river from year to year and season to season. As such the hydrological investigations constitute one of the most important aspects of reservoir planning. The hydrological investigations may be divided into two heads :

(i) Study of runoff pattern of the river at the proposed dam site to determine the storage capacity of the reservoir corresponding to a given demand.

(ii) Determination of the hydrograph of the worst flood to determine the spillway capacity and design.

In order to determine the storage capacity of a reservoir it is essential to predict the runoff pattern of the river in future years at the proposed dam site. This prediction is however based on the premise that if the runoff pattern can be determined at the site for a large number of past years, then on the average the same pattern would recur in future years. Thus, if direct records of flow are available from a stream gaging station at or near the dam site the same may be used to determine the storage capacity of the reservoir. However, if such records are not available then the runoff data has to be deduced indirectly by any of the methods discussed in Chapter 5.

For spillway design or for ascertaining the flood moderating capacity of the reservoir the hydrograph for the worst flood is required which may also be obtained by any of the methods discussed in Chapter 5.

7.4 SELECTION OF SITE FOR A RESERVOIR

The selection of site for a reservoir depends on the following factors.

1. Suitable dam site must be available where the reservoir is proposed to be constructed.

2. The river valley at the site should be narrow so that the length of the dam to be constructed is less, but it should open out on the upstream side to provide a large basin for the reservoir.

3. The surrounding hills which constitute the rim of the reservoir should be water tight, so that there is no leakage of water through any part of the rim.

4. The reservoir basin should also be reasonably water tight so that the stored water is not able to escape under the surrounding hills through cavernous rock or other continuous rock or other continuous pervious strata.

5. The site should be such that as far as possible minimum land and property is submerged in the reservoir.

6. The site should be such that it avoids water from those tributaries which carry unusually high content of sediment.

7. The site must be such that adequate reservoir capacity is made available.

8. As far as possible a deep reservoir must be formed so that the land costs per unit of capacity are low, evaporation loss is less and there is less likelihood of weed growth.

9. The reservoir site should be such that there are no objectionable minerals and salts present in the soil and rocks at the site which may get dissolved in water and deteriorate its quality.

10. The quality of water stored in the reservoir must be satisfactory for its intended use.

11. The site should be such that the costs of associated works such as roads, rails, housing colonies for workers and other staff, etc., are not excessive.

7.5 ZONES OF STORAGE IN A RESERVOIR

The entire storage capacity of a reservoir may be divided into a number of zones by certain water surfaces or pool levels in the reservoir as mentioned below.

Normal Pool Level (N.P.L.)

It is the maximum elevation to which the water surface will rise in the reservoir during ordinary operating conditions. As shown in Fig. 7.3 in the case of an ungated spillway the normal pool level is determined by the elevation of the spillway crest. However, if the spillway is gated then the normal pool level is determined by the top of the spillway gates. The normal pool level is also known as *Full Reservoir Level* (F.R.L.) or *Full Tank Level* (F.T.L.).

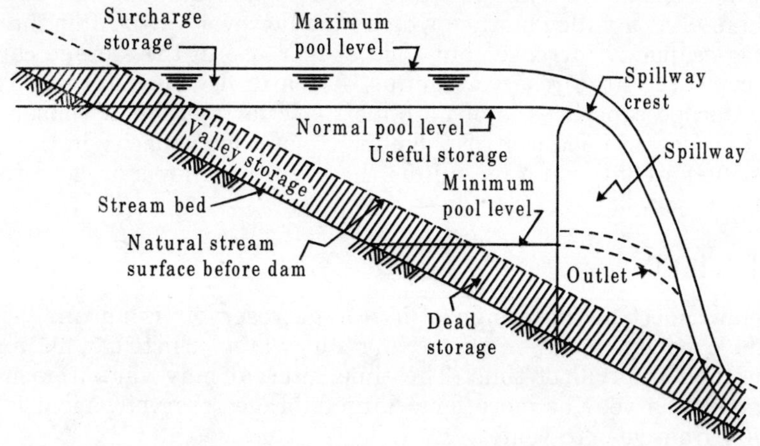

Fig. 7.3. *Zones of storage in a reservoir*

Minimum Pool Level. It is the lowest elevation to which the water is drawn from the reservoir under normal conditions. This level may be fixed by the elevation of the lowest outlet in the dam, or in the case of hydroelectric reservoir, by the minimum head required for efficient functioning of turbines.

Maximum Pool Level. It is the maximum elevation to which the water surface will rise in the reservoir during the design flood (or worst flood). It is also known as *Maximum Water Level* (M.W.L.) or *Pool Level during Design Flood.*

The various zones of storage in a reservoir are as follows.

(a) Useful storage.
(b) Dead storage.
(c) Surcharge storage.
(d) Bank storage.
(e) Valley storage.

(a) Useful Storage. The volume of water stored between the normal pool level and the minimum pool level of a reservoir is called the *useful storage* (Fig. 7.3). In a multipurpose reservoir the useful storage may be subdivided into *conservation storage and flood control storage* in accordance with the adopted plan of operation of the reservoir. The useful storage is also known as *live storage* as it can be used for various purposes required to be served by the reservoir.

(b) Dead Storage. The volume of water held below the minimum pool level of a reservoir is known as *dead storage* (Fig. 7.3). It cannot be used for any purpose under ordinary operating conditions.

(c) Surcharge Storage. The volume of water stored between the normal pool level and the maximum pool level of a reservoir is called *surcharge storage*. The surcharge storage is an uncontrolled storage as it exists only while a flood is occurring and cannot be retained for later use.

(d) Bank Storage. The *bank storage* is the volume of water that is temporarily stored in the permeable banks of a reservoir when the reservoir fills and drains out as the water level in the reservoir is lowered. The bank storage effectively increase the capacity of the reservoir above that indicated by elevation-capacity curve (Fig. 7.2). The amount of bank storage depends on geologic conditions and may amount to several percent of the reservoir volume.

(e) Valley Storage. The volume of water held by a natural stream channel is known as the *valley storage*. Even before a reservoir is constructed certain amount of water is stored in the natural stream channel as a valley storage which may however vary. After the construction of a reservoir the storage capacity increases but the net increase in the storage capacity is equal to the total storage capacity of the reservoir minus the natural valley storage. This distinction of the total and net storage capacities is of no importance for storage or conservation reservoirs, but from the point of view of flood control the effective storage capacity in the reservoir is equal to the useful plus the surcharge storage minus the natural valley storage corresponding to the rate of inflow to the reservoir (Fig. 7.3).

7.6 RESERVIOR YIELD

The most important aspect of the design of a storage reservoir is an analysis of the relation between yield and capacity. Yield is the amount of water which can be supplied from the reservoir in a specified interval of time. The time interval may vary from a day for a small distribution reservoir to a year or more for a large storage reservoir. Yield is dependent upon inflow and will vary from year to year.

For most of the storage reservoirs in addition to yield it is also necessary to know safe or firm yield and secondary yield which are defined below.

Safe or firm yield is the maximum quantity of water which can be supplied during a critical (or worst) dry period. In practice the period of lowest natural flow on record for the stream is usually taken as the critical period. However, there is a possibility that a drier period may occur with a yield even less than the safe yield determined on the basis of the past record of the stream flow.

Secondary yield is the quantity of water available in excess of safe yield during periods of high flows.

Average yield is the arithmetic average of the yield (firm and secondary) over a long period of time.

The yield of a reservoir and its storage capacity are very much dependent on each other. Further the storage capacity of a reservoir also depends on the inflow to a reservoir. The inflow, yield and storage capacity of a reservoir are related by the following storage equation.

$$\text{Inflow} - \text{Yield} = \text{Change in storage}$$

Thus, if inflow to a reservoir is more than the yield of the reservoir then the storage will increase, but the storage will be depleted if the inflow to a reservoir is less than the yield of the reservoir. The value of the yield which may be adopted for the design of a reservoir is known as *design yield* which should be such that the demands of the consumers are met with and at the same time the storage is not unduly depleted. However, the value of the design yield also depends on the urgency of the water requirement or on the amount of risk involved if the actual yield is less than the design yield. For example, a reservoir for domestic water supply must be planned on the basis of firm yield. On the other hand a reservoir for irrigation may be planned with a value of design yield about 20% higher than the firm yield. Similarly power commitments to domestic users must be on a firm basis and should not exceed the power which can be produced with the firm yield unless thermal power (steam or diesel) is available to support the hydroelectric power. However, the hydroelectric power developed from secondary yield may be supplied to large industries on 'as and when available' basis.

7.7 MASS CURVE AND DEMAND CURVE

Mass Curve. A *mass curve* (or mass inflow curve) is a plot of accumulated flow in a stream against time. As indicated below a mass curve can be prepared from the flow hydrograph of a stream for a large number of consecutive previous years. Figure 7.4 (a) shows a typical flow

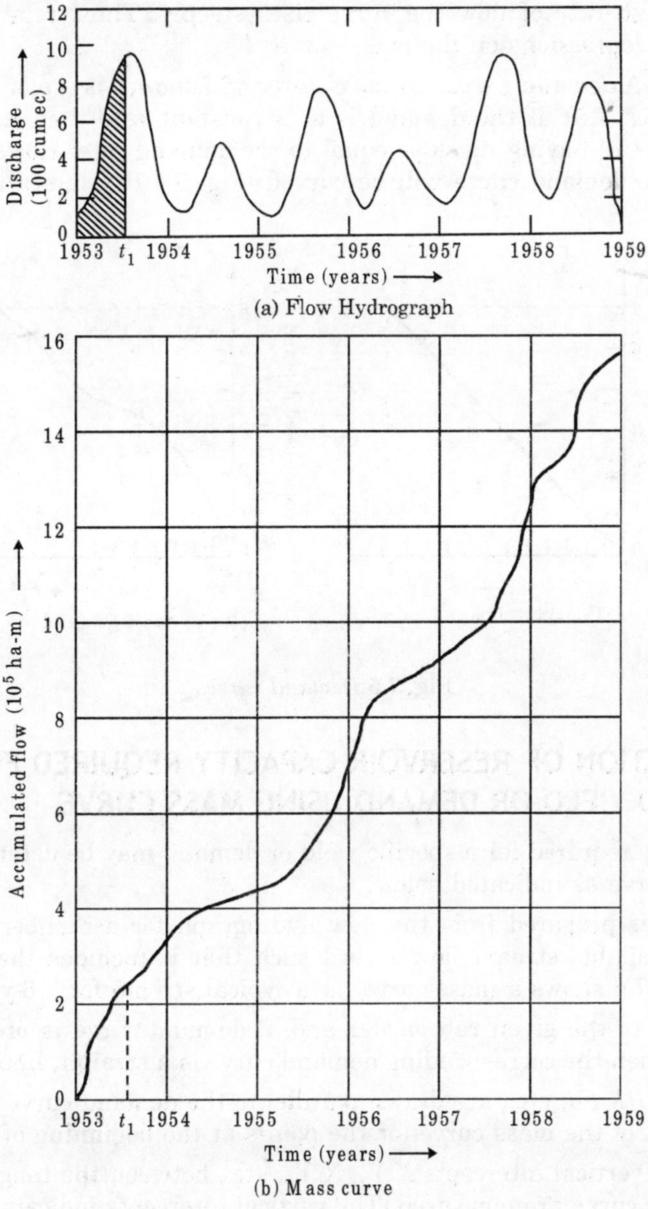

(a) Flow Hydrograph

(b) Mass curve

Fig. 7.4

hydrograph of a stream for six consecutive years. The area under the hydrograph from the starting year (i.e., 1953) upto any time t_1 (shown by hatching in Fig. 7.4 a) represents the total quantity of water that has flown through the stream from 1953 upto time t_1 and hence it is equal to the ordinate of the mass curve at time t_1. The ordinates of the mass curve corresponding to different times are thus determined and plotted at the respective times to obtain the mass curve as shown in Fig. 7.4 (b). A mass curve continuously rises as it shows accumulated flows. The slope of the curve at any point indicates the rate of flow at that particular time. If there is no flow during certain period the curve will be horizontal during that period. If there is high rate of flow the curve rises steeply. Thus, relatively dry periods are indicated as concave depressions on the mass curve.

Demand Curve. A *demand curve* (or *mass curve of demand*) is a plot between accumulated demand and time (Fig. 7.5). If the demand is at a constant rate then the demand curve is a straight line [Fig. 7.5 (a)] having its slope equal to the demand rate. However, if the demand is not constant then the demand curve will be curved [Fig. 7.5 (b)] indicating a variable rate of demand.

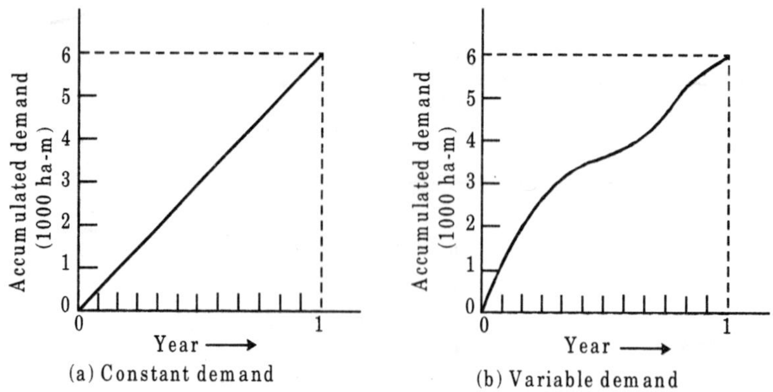

Fig. 7.5 *Demand curves*

7.8 DETERMINATION OF RESERVOIR CAPACITY REQUIRED FOR A SPECIFIED YIELD OR DEMAND USING MASS CURVE

The reservoir capacity required for a specific yield or demand may be determined by using mass curve and demand curve as indicated below.

1. A mass curve is prepared from the flow hydrograph for a number of consecutive years selected from the available stream flow record such that it includes the most critical or the driest period. Figure 7.6 shows a mass curve for a typical stream for a 6-year period.

2. Corresponding to the given rate of demand, a demand curve is prepared. If the rate of demand is constant then the corresponding demand curve is a straight line as shown in Fig. 7.6.

3. Lines such as *GH*, *FJ*, etc., are drawn parallel to the demand curve and tangential to the high points *G*, *F*, etc., of the mass curve (or the points at the beginning of the dry periods).

4. The maximum vertical intercepts X_1Y_1, X_2Y_2, etc., between the tangential lines drawn in step (3) and the mass curve are measured. The vertical intercepts indicate the volume by which

the total flow in the stream falls short of the demand and hence required to be provided from the reservoir storage. For example, assuming the reservoir to be full at G, for a period corresponding to points G and Z_1, there is a total flow in the stream represented by $Y_1 Z_1$ and there is a total demand represented by X_1Z_1, leaving a gap of volume represented by X_1Y_1 which must be met with from the reservoir storage.

5. The largest of the maximum vertical intercepts X_1Y_1, $X_2 Y_2$, etc., determined in step (4) represents the reservoir capacity required to satisfy the given demand. However, the requirement of storage so obtained would be the net storage which must be available for utilization and it must be increased by the amount of water lost by evaporation and seepage.

As shown in Fig. 7.6 the vertical distance between the successive tangential lines such as GH and FJ represents the quantity of water which would spill over from the reservoir through the spillway and go as a waste to the downstream side. This is so because between H and F the reservoir would remain full and all inflow in excess of demand would flow through the spillway to the downstream side.

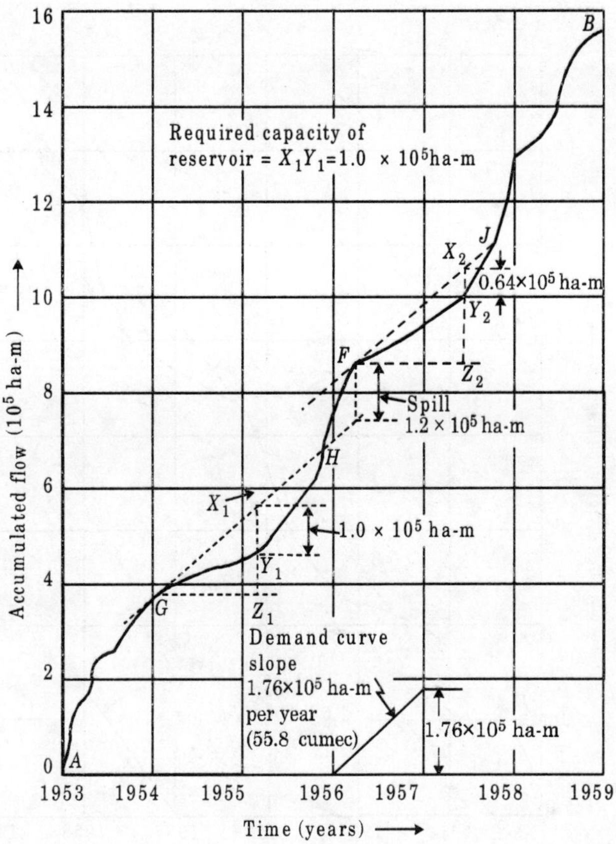

Fig. 7.6 *Use of a mass curve to determine the reservoir capacity required to produce a specified yield*

Further, it may also be noted that the tangential lines drawn parallel to the demand curve when extended forward must intersect the mass curve, such as at H, J, etc., so that the reservoir which was full at G and F will be filled again at H and J. However, if the line does not intersect the mass curve, the reservoir will not be filled again. Moreover, if the reservoir is very large the time interval between the points *G* and *H*, *F* and *J*, etc., may be several years.

For the numerical example indicated in Fig. 7.6, corresponding to a demand of 1.76×10^5 ha-m (or 55.8 cumec) the following results are thus obtained.

(i) The required reservoir capacity is given by $X_1Y_1 = 1.0 \times 10^5$ ha-m.

(ii) Assuming the reservoir to be full at *G*, *it would be empty at* Y_1 and would be full again at *H*.

(iii) Between *H* and *F* the reservoir would remain full and all inflow in excess of the demand would be discharged through the spillway to the downstream side. The spill over from the reservoir would be 1.2×10^5 ha-m.

(iv) Assuming the reservoir to be full at *F*, it would be depleted to $(1.0 \times 10^5 - 0.64\ 10^5) = 0.36 \times 10^5$ ha-m of storage at Y_2 and would be full again at *J*.

In the earlier discussion the rate of demand has been assumed to be constant. However, the rate of demand may not be always constant, in which case the demand curve will be a curve

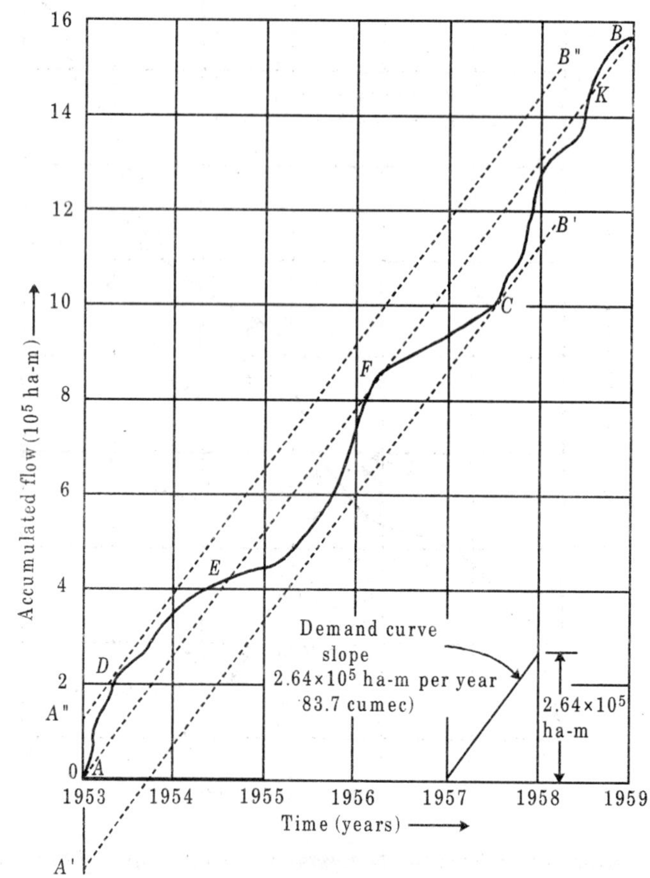

Fig. 7.7 *Use of a mass curve to determine the reservoir capacity required to produce yield equal to the average discharge of the stream*

with its slope varying from point to point in accordance with the variable rate of demand at different times. In this case also the required capacity of the reservoir can be determined in the same way by superimposing the demand curve on the mass curve from the high points (or beginning of the dry period) till the two meet again. The largest vertical intercept between the two curves gives the required reservoir capacity. It is however essential that the demand curve for the variable demand coincide chronologically with the mass curve of streamflow, i.e., June demand must coincide with June inflow and so on. Furthermore, in this case also the storage obtained must be increased to account for the water lost by evaporation and seepage.

As shown in Fig. 7.7 if the end points of the mass curve are joined by a straight line *AB*, then its slope represents the average discharge of the stream over the total period for which the mass curve has been plotted. If a reservoir is to be constructed to permit continuous release of water at this average value of discharge for the entire period then the capacity required for the reservoir is represented by the vertical intercept between the two straight lines *A′B′* and *A″B″* drawn parallel to *AB* and tangent to the mass curve at the lowest tangent point *C* and the highest tangent point *D* respectively. If the reservoir having this capacity is assumed to contain a volume of water equal to *AA′* at the beginning of the period, then the reservoir would be full at *D* and it would be empty at *C*. However, if the reservoir was empty in the very beginning, then it would be empty again at point *E* and also during the period from *F* to *K*. On the other hand if the reservoir was full in the very beginning it would be full again at points *F* and *K*, and between points *A* and *E* there will be spill of water from the reservoir.

7.9 DETERMINATION OF YIELD FROM A RESERVOIR OF GIVEN CAPACITY

Mass curve may also be used to determine the yield which may be obtained from a reservoir of given capacity for which the following procedure may be adopted.

(1) A mass curve as shown in Fig. 7.8 is prepared in the same manner as in the previous case.

(2) Tangents are drawn at the high points *G*, *F*, etc., of the mass curve in such a manner that their maximum departure from the mass curve does not exceed the given capacity of the reservoir.

(3) The slopes of each of these tangents are measured which indicate the yield which can be obtained in each year from the reservoir of given capacity. The slope of the flattest demand curve is the safe or the firm yield.

For the numerical example shown in Fig. 7.8 from a reservoir of capacity 1.2×10^5 ha-m the safe yield of 1.68×10^5 ha-m per year may be obtained. It is given by the slope of the tangent *GH* which is flatter than that of the tangent *FJ*.

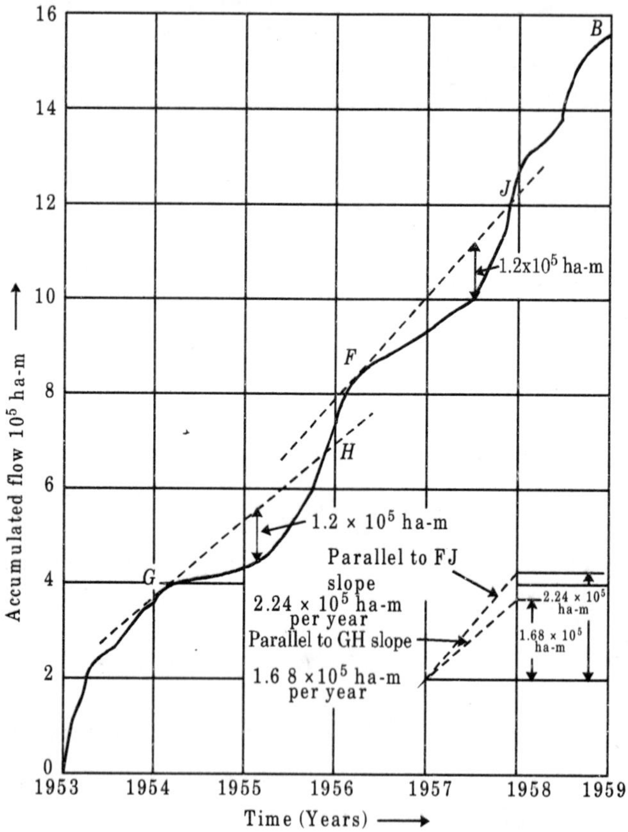

Fig. 7.8 *Use of a mass curve to determine the possible yield from a reservoir of specified capacity*

7.10 DEMAND PATTERNS FOR VARIOUS TYPES OF RESERVOIRS

In the previous sections it has been explained how the capacity required for a reservoir can be determined for a known demand curve with constant or varying demands or *vice versa*. In general the demands are man-made and hence these should be adjusted in such a manner that maximum benefits are obtained with minimum costs. However, the demand pattern mainly depends upon the purpose for which the reservoir has been constructed and hence it is dealt with separately for each main type of reservoir as follows:

(i) *Single purpose conservation reservoir*. A single purpose reservoir is meant for storing water for only one purpose such as domestic and industrial water supply or irrigation or hydroelectric power generation. In all such cases the demand for water can be predicted with fair degree of accuracy for different parts of the year and the storage capacity can be determined as explained earlier. The points to be kept in view are, firstly the probable growth of demand in future and secondly the overall development potential of the river which should not be disturbed by the construction of the reservoir.

(ii) *Single purpose flood control reservoir*. In the case of a single purpose flood control reservoir, there is no specific demand of water except that the release of water from the reservoir should not exceed the safe carrying capacity of the river on the downstream side. As such the capacity of such a reservoir should be sufficient to hold excess water during rising flood and gradually release it when the flood subsides. It is therefore evident that the capacity of such a reservoir does not depend upon any pattern of demand but it mainly depends upon the hydrograph of the worst flood that is likely to occur and also upon the safe carrying capacity of the river on the downstream side. The hydrograph of the worst flood can be obtained from hydrological investigations and the safe peak rate of outflow can be determined from the conditions of river on the downstream side. The capacity of the reservoir required to moderate the inflow peak to a value equal to or less than the safe peak rate of outflow can be found by flood routing which is explained in the next section.

(iii) *Multipurpose reservoirs*. These reservoirs are constructed to serve more than one purpose. As such one of the important aspects of the design of a multipurpose reservoir is the allocation of the storage for various purposes. The two possible extreme cases in allocation of reservoir storage for various purposes are as follows.

1. It is assumed that no storage is jointly used, i.e., a separate allocation of storage is to be made for each of the purposes served by the reservoir.

2. It is assumed that the storage is jointly used for various purposes to be served by the reservoir.

In the first case the storage requirements for all the purposes to be served by the reservoir are pyramided to create a large total storage requirement which will be economically feasible only when unit cost of storage is either constant or decreases as total storage increases. The other extreme case results in maximum economy since the required storage is not greater than that necessary for any one of the several purposes. However, for this case to be practically feasible the water requirements of the several uses must be the same, but such a situation is rare in actual practice. As such a multipurpose project is usually designed to fall somewhere between these two extreme cases and it is by proper operation of the reservoir the available storage is suitably used for various purposes. Therefore, a unique feature of the design of a multipurpose reservoir is to develop an operation plan which would be an effective compromise among the various uses.

In our country for multipurpose reservoirs the most common combination is irrigation, hydroelectric power generation and flood control. The operation plan for such a reservoir may be understood by the following example.

It is assumed that a reservoir is to be located on the tributary of a snow-fed river subject to severe floods during monsoon period from June to September. The reservoir is to be designed for lowering flood stage downstream, providing water for Rabi (winter) irrigation and generating hydroelectric power. The requirements of operation for such a reservoir are as follows:

(i) Reservation of a certain minimum storage at all times for flood control.

(ii) Provide a steady discharge for Kharif (summer) irrigation during April, May and June.

(iii) Provide a larger steady discharge during Rabi (winter) irrigation period from middle of November to middle of February.

(iv) Provide a steady discharge for generating maximum firm power.

On ice-fed North Indian rivers supplies are usually lowest in February and March. This coupled with heavy draw of water for Rabi irrigation during the preceding months may bring the reservoir level at this time down to or near the dead storage level. During the months of

April, May and June the supplies in the river steadily increase due to melting of snow in the head reaches. During these months water is required to be supplied for Kharif irrigation as well as for power generation. Hence, during April, May and June the reservoir level may fall or rise slightly at a slow rate depending on whether the outflow is more or less than the inflow. Thus, just before monsoon a large reservoir capacity would be available for the dual purpose of conservation and flood control. During the monsoon months the demand for irrigation would be small and water will be released mainly for power generation, but due to large inflow the reservoir level would steadily rise and it may be allowed to get filled upto the maximum conservation level by the end of normal monsoon period, say, the middle of October. Even at this time some flood control storage would be available between the conservation level and the maximum water level to take care of unusually late severe floods. The actual date on which the reservoir may be allowed to get filled upto the maximum conservation level depends on the runoff pattern of past years and many a times on the wise discretion of the operating authorities so as to have a full reservoir at the end of the flood season without encroaching upon the storage reserved for flood control. After the monsoon, there would be a release of water for power generation and some demand for irrigation during October and early November after which demand for Rabi irrigation would necessitate heavy draw from the reservoir from November to February. During this period the natural flow in the river will have a steadily downward trend except for some winter freshets in December and January. The reservoir level will therefore go down slowly at first and then move quickly till dead storage level will be approached near about by the end of February.

From the above example it is clear that if the pattern of floods is distinctly seasonal (as in the case of the rivers of Northern India), only a very small proportion of the total reservoir capacity need be specifically reserved for flood control to take care of unusual floods only. This is so because normally before the flood season almost the entire conservation storage will be available for flood control also. However, in regions where severe floods may occur at any time during the year or during more than one season, a larger storage has to be reserved for flood control according to the requirements.

The planning of a multipurpose reservoir should be carried out by trying a large number of tentative schedules prepared for a large number of past years depending on the availability of hydrological data and the frequency of occurrence of high floods. A period of 30 to 40 years is usually considered sufficient for this purpose. The schedule which gives maximum total benefits for the various design purposes without encroaching upon the lower or upper limits of storage is adopted for estimation and for initial operation. This schedule may be tried and if necessary modified on the basis of past experience and future needs.

7.11 APPORTIONMENT OF TOTAL COST OF A MULTI-PURPOSE RESERVOIR

Apportionment of total cost of a multipurpose reservoir means the subdivision of cost for various uses for which the reservoir is planned. This is required because a multipurpose reservoir serves several different groups of beneficiaries and it is often necessary to allocate the cost among the several uses to fix the prices of water or power or to determine the contribution required of flood control beneficiaries.

The total cost of the project may be divided in *separable costs* of each of the single functions (or purposes) of the project and the *joint costs.* The separable costs are those costs which are directly chargeable to a single project function, such as the cost of the power house and its equipments in case of a hydroelectric project, or the cost of navigation lock in case of navigation project, or the cost of canal head works in case of irrigation project etc. The separable costs for a single function of the project is usually determined by computing the cost of the project with that function omitted. The separable cost of the function is then obtained as the total cost of the project minus the cost of the project determined with that function omitted. The separable costs may be directly allocated to the respective single functions of the project.

The joint costs are those costs which are common for all the functions of the project and these are determined by subtracting the sum of the separable costs of all the single functions of the project from the total cost of the project. The real problem in cost allocation is the division of the joint costs among the various functions of the project. Various methods adopted for allocation of joint costs are as follows :

(1) Equal apportionment method

(2) Use of facility method

(3) Alternative justifiable expenditure method

(4) Remaining benefits method.

1. Equal Apportionment Method. In this method the joint costs of the project are equally divided among the various functions of the project irrespective of different benefits derived from them. Though the method is simple, it is not based on any sound reasoning and is used only when the other methods cannot be used due to lack of required data.

2. Use of Facilities Method. In this method the joint costs are divided in the ratio of storage volume provided for each function. This method is also simple but may not give correct allocation of the cost because cost is not directly proportional to storage. As a matter of fact with the increase in the storage the cost per unit volume decreases and hence the unit cost of storage does not remain same.

3. Alternative Justifiable Expenditure Method. In this method the joint costs are distributed in the ratio of the differences between the separable costs and the estimated cost of a single-purpose project which would provide equivalent services and would itself be economically justifiable. However, the difficulty with this method is that of estimating the justifiable cost of the alternate single-purpose project.

4. Remaining Benefits Method. In this method the joint costs are distributed in the ratio of the differences between the separable costs and the estimated benefits of each function or the cost of an alternate single purpose project which would provide equivalent benefits, whichever is lower. These differences are also known as remaining benefits. However, the difficulty with this method is also that of estimating the benefits of each function or the justifiable cost of the alternate single purpose project.

In each of these methods the distributed joint costs are added to the separable cost to obtain the total allocated costs for each of the functions. It may be noted that the separable costs are normally a large part of the total costs.

It may be stated that there is no really satisfactory method of cost allocation which would be equally applicable to all projects and would yield allocations which are unquestionably correct. However, for multipurpose reservoir projects both the alternative justifiable expenditure method and the remaining benefits method give results which are within the probable limits of accuracy obtainable in any allocation. Both these methods are explained in Illustrative Example 7.4.

7.12 FLOOD ROUTING

The process of computing the reservoir storage volumes, outflow rates and water levels in the reservoir corresponding to a particular hydrograph of inflow is known as flood routing. Flood routing is carried out to obtain a hydrograph as moderated or modified by a reservoir when a particular flood passes through it and to determine, therefrom the corresponding maximum rise in the water surface in the reservoir and the maximum discharge in the river on the downstream side. Thus, it is a useful technique for obtaining complete solution of flood control problems.

The relationship governing the computations for flood routing is simple and it states that over any interval of time, the volume of inflow must be equal to the volume of outflow plus the change in storage during this period. This may be expressed mathematically as

$$\bar{I} = \bar{Q} + \Delta s \qquad \text{...(7.1)}$$

where $\bar{I}$ = average inflow during a given time period

$\bar{Q}$ = average outflow during the same period and

Δs = the change in storage during the same period.

If the water level in the reservoir is rising there will be increase in storage and the change in storage will be positive, but if it is falling there will be decrease in storage and the change in storage will be negative.

Although the above relationship is simple its direct evaluation is not possible because of the fact that the relations between time and rate of inflow (i.e., inflow hydrograph), elevation and storage of reservoir (i.e., elevation-capacity curve) and elevation and rate of outflow (i.e., elevation-outflow curve) cannot be expressed by simple algebraic equations. As such a step by step method is usually adopted for flood routing computations. In this method small time intervals are chosen and from the known conditions at the beginning of the interval the conditions at the end of each interval are successively determined. The interval of time should be sufficiently small so that the mean of the outflow rates at the beginning and the end of the period may be taken as the average throughout the period. A number of methods have been developed to carry out step by step, computation out of which the following two methods are commonly adopted and the same are described below:

(1) Graphical method

(2) Trial and error method.

1. Graphical Method (Inflow, Storage, Discharge Curve Method)

For a small interval of time Eq. 7.1 may be expressed as follows

$$\frac{I_1 + I_2}{2}t - \frac{O_1 + O_2}{2}t = s_2 - s_1 \qquad \text{...(7.2)}$$

where I_1 and I_2 are the inflow rates, O_1 and O_2 are the outflow rates, and s_1 and s_2 are the storage volumes at the beginning and end of the interval of time t.

Equation 7.2 can be written as follows:

$$(I_1 + I_2) + \left(\frac{2s_1}{t} - O_1\right) = \left(\frac{2s_2}{t} + O_2\right) \qquad \text{...(7.3)}$$

For carrying out the computations by this method the following three curves are required.

(i) Inflow flood hydrograph (Fig. 7.9 a)

(ii) Elevation-storage curve (Fig. 7.9 b)
(iii) Elevation-outflow curve (Fig. 7.9 c)

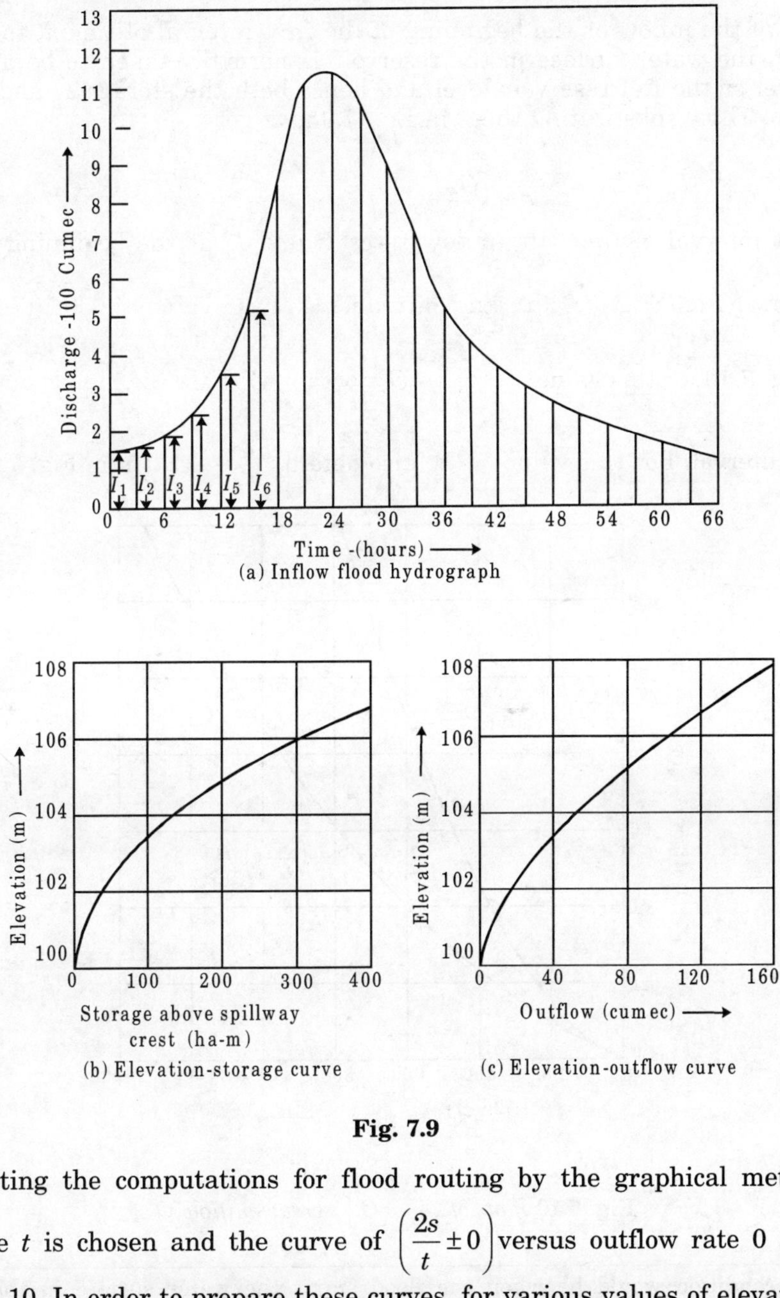

Fig. 7.9

Before starting the computations for flood routing by the graphical method, a suitable interval of time t is chosen and the curve of $\left(\frac{2s}{t} \pm 0\right)$ versus outflow rate 0 are prepared as shown in Fig. 7.10. In order to prepare these curves, for various values of elevations, the values of s and O are obtained from the elevation-storage curve (Fig. 7.9 b) and elevation-outflow curve (Fig. 7.9 c). Knowing the corresponding values of s and O and the chosen interval of time t, values of $\left(\frac{2s}{t} \pm 0\right)$ are computed and the same are plotted against the corresponding values of outflow rate O.

At the start of the inflow or the beginning of the first interval of time t, to account for the *worst condition**, the water surface in the reservoir is normally taken to be at the maximum conservation level or the full reservoir level and hence both the storage s_1 and the outflow O_1 are equal to zero. Thus, substituting these in Eq. (7.3), we get

$$(I_1 + I_2) = \frac{2s_2}{t} + O_2 \qquad ...(7.4)$$

For the first interval of time, the inflow rates I_1 and I_2 at the beginning and end of the interval are known from Fig. 7.9 (a), and introducing these values in Eq. (7.4), $\left(\frac{2s_2}{t} + O_2\right)$ is known. From Fig. 7.10 for this value of $\left(\frac{2s_2}{t} + O_2\right)$ obtain the value of O_2, the outflow rate at the end of the first interval. For this value of O_2, also obtain $\left(\frac{2s_2}{t} - O_2\right)$ from Fig. 7.10.

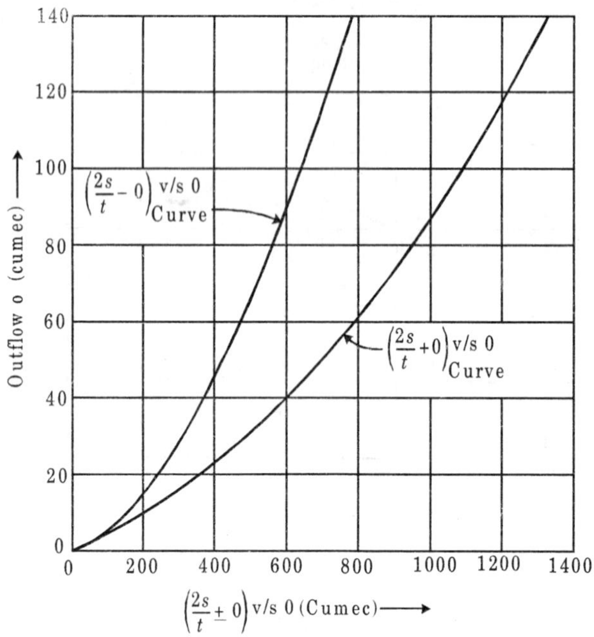

Fig. 7.10 *Plots of* $\left(\frac{2s}{t} \pm O\right)$ *versus outflow O*

* The worst condition would be when the flood starts the water surface in the reservoir is at maximum conservation level so that the flood would be absorbed only by the surcharge storage. For the condition at the start of inflow since the surcharge storage is equal to zero, the storage $s_1 = 0$. Similarly for this condition at the start of inflow there being no flow over the spillway, the overflow rate $O_1 = 0$.

Now applying Eq. (7.3) for the second interval of time, we get

$$(I_2 + I_3) + \left(\frac{2s_2}{t} - O_2\right) = \left(\frac{2s_3}{t} + O_3\right) \quad ...(7.5)$$

In Eq. (7.5), I_2 and I_3 are the inflow rates at the beginning and end of the second interval of time which are known from Fig. (7.9a), and $\left(\frac{2s_2}{t} - O_2\right)$ is known from the previous step as indicated above. Hence $\left(\frac{2s_3}{t} + O_3\right)$ is known. Again from Fig. (7.10) corresponding to this value of $\left(\frac{2s_3}{t} + O_3\right)$ obtain the value of O_3, the outflow rate at the end of the second interval, and corresponding to this value of O_3 obtain the value of $\left(\frac{2s_3}{t} - O_3\right)$.

Similarly applying Eq. 7.3 for the third interval of time, we get

$$(I_3 + I_4) + \left(\frac{2s_3}{t} - O_3\right) = \left(\frac{2s_4}{t} + O_4\right) \quad ...(7.6)$$

from which $\left(\frac{2s_4}{t} + O_4\right)$ is known and the process is repeated for the successive time intervals of the entire time period of the inflow flood hydrograph.

The various steps of graphical method of flood routing may thus be summarized as follows:

(i) From the inflow flood hydrograph Fig. (7.9a) note the inflow rates I_1 and I_2 at the beginning and end of the first interval of time and compute the sum $(I_1 + I_2)$.

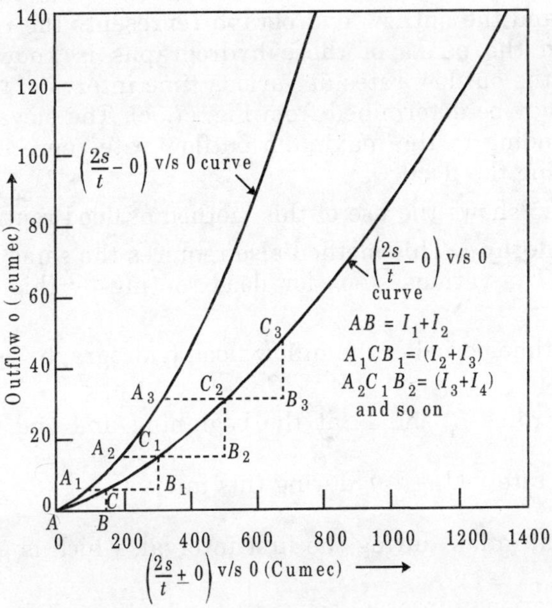

Fig. 7.11 *Graphical method of flood routing*

(ii) Enter this value of $(I_1 + I_2)$ in the plot of $\left(\frac{2s}{t} \pm O\right)$ versus O from A as indicated by AB in Fig. (7.11).

(iii) Draw vertical line from B to meet $\left(\frac{2s}{t} + O\right)$ curve at point C. The ordinate of point C gives the value of O_2, the outflow rate at the end of the first interval of time.

(iv) Draw a horizontal line through C to cut $\left(\frac{2s}{t} - O\right)$ curve at point A_1.

(v) From Fig. 7.9 (a) note the inflow rates, I_2 and I_3 at the beginning and end of the second interval of time and compute the sum $(I_2 + I_3)$.

(vi) Enter this value of $(I_2 + I_3)$ in Fig. 7.11 from A_1 as indicated by A_1CB_1.

(vii) Draw a vertical line from B_1 to meet $\left(\frac{2s}{t} + O\right)$ curve at point C_1. The ordinate of point C_1 gives the value of O_3, the outflow rate at the end of the second interval of time.

(viii) Draw a horizontal line through C_1 to cut $\left(\frac{2s}{t} - O\right)$ curve at point A_2.

The same procedure is repeated for the successive time intervals of the entire time period of the inflow flood hydrograph.

The outflow rates at various time intervals are thus determined as indicated above and the same are plotted against time to obtain the outflow hydrograph or the hydrograph as moderated by the reservoir. The peak of the outflow hydrograph represents the maximum discharge in the river on the downstream side during the flood. The difference in the peak flow rates of the inflow flood hydrograph and the outflow hydrograph represents the reduction in peak and the interval of time between the peaks of these hydrographs is known as peak lag. Further corresponding to each of the outflow rates at various time intervals, the elevation of the water surface in the reservoir may be determined from Fig. (7.9c). The elevation of the water surface in the reservoir corresponding to the maximum outflow rate represents the maximum water level in the reservoir during the flood.

Illusrative Example 7.5 shows the use of this method of flood routing.

2. Trial and Error Method. This method also requires the same three curves as indicated in the previous method. The various steps for flood routing by this method are as indicated below.

(i) Divide the total time period of the inflow flood hydrograph Fig. (7.9a) into a number of small intervals.

(ii) Note the inflow rates I_1 and I_2 at the beginning and end of the first interval and compute the mean inflow rate $\frac{1}{2}(I_1 + I_2)$ during this interval.

(iii) Compute the total inflow during the first interval, which is given by the mean inflow rate multiplied by the duration of interval t, i.e., $\frac{1}{2}(I_1 + I_2)\,t$.

(iv) At the start of the inflow or the beginning of the first interval, the water surface elevation in the reservoir is known because as indicated earlier this is normally taken as the maximum conservation level or full reservoir level to account for the worst condition. Corresponding to this known water surface elevation in the reservoir at the beginning of the first interval obtain the outflow rate from the elevation-outflow curve Fig. (7.9c). It will however be equal to zero as the water surface elevation is taken as maximum conservation level and the outflow is only through a spillway.

(v) Assume a trial value of the water surface elevation in the reservoir at the end of the first interval. Corresponding to this trial water surface elevation at the end of the first interval obtain the outflow rate from the elevation-outflow curve Fig. (7.9c).

(vi) Compute the mean of the outflow rates at the beginning and the end of the first interval, i.e., $\frac{1}{2}(O_1 + O_2)$.

(vii) Compute the total outflow during the first interval, which is given by the mean outflow rate multiplied by the duration of the interval, i.e., $\frac{1}{2}(O_1 + O_2)\ t$.

(viii) The total outflow obtained in step (vii) is subtracted from the total inflow obtained in step (iii) to obtain the storage during the first interval.

(ix) Corresponding to the water surface elevation in the reservoir at the beginning and the end of the first interval find the storage s_1 and s_2 at the beginning and end of the first interval from the elevation-storage curve Fig.(7.9b). Compute the difference $(s_2 - s_1)$ which represents the storage during the first interval.

If the trial value of the water surface elevation in the reservoir at the end of the first interval as indicated in step (v) has been correctly assumed, the value of storage obtained in step (viii) would be equal to that obtained in step (ix). However, if the values of the storage obtained in steps (viii) and (ix) are not equal, another trial value of the water surface elevation in the reservoir at the end of the first interval is assumed and the procedure is repeated until the values of the storage obtained in steps (viii) and (ix) are equal. When this happens, the correct water surface elevation, outflow rate and storage at the end of the first interval are known. The process is then repeated for the successive time intervals of the entire time period of the inflow flood hydrograph.

Illustrative Example 7.6 shows the use of this method of flood routing.

7.13 RESERVOIR LOSSES

Some quantity of water stored in a reservoir is always lost and hence it is necessary to account for the same in the planning and operation of a reservoir. The loss of water from a reservoir may be due to evaporation, absorption and percolation.

The evaporation loss mainly depends on the water surface area of the reservoir. The other factors influencing evaporation loss are temperature, wind velocity and relative humidity. The loss of water due to evaporation is usually expressed in terms of depth of water measured in millimetres which thus represents the volume of water lost per unit area of the water surface of a reservoir. As indicated in Chapter 5 for any reservoir the evaporation loss can be determined by measuring the depth of evaporation in a pan and multiplying the same by a predetermined pan coefficient. Various types of pans have been developed and standardized for the measurement of evaporation. Though the evaporation loss may vary from place to place, the observed values of the same on existing reservoirs can be adopted for the design of a new reservoir after making due allowance for difference in the conditions of the two. Table 7.1 gives the average values of the evaporation loss for North, and South and Central Indian reservoirs on mean exposed water surface area during each month of the year.

TABLE 7.1

	Evaporation loss in mm	
Month	*North India*	*South and Central India*
January	70	100
February	90	100
March	130	180
April	160	230
May	270	250
June	240	180
July	180	150
August	140	150
September	140	150
October	130	130
November	90	100
December	80	100
Total	1720	1820

Various measures taken to reduce the evaporation loss in a reservoir are indicated in the next section.

The loss of water from a reservoir due to absorption will depend on the type of soil forming the reservoir basin. This loss may be quite large in the beginning but will be gradually reduced as the pores get saturated. As such the absorption loss is not significant in reservoir planning.

The percolation or seepage loss is also small in most of the cases. However, it may be significant in some cases where severe leakage may occur under the hills or under the base of the dam through continuous seams of porous strata or cavernous or fissured rock. The reservoir basin should therefore be carefully checked for watertightness during geological investigations and necessary remedial measures such as grouting etc., should be taken to prevent such leakage.

The total loss of water in a reservoir during a given period can be determined by measuring the fall in the reservoir level and knowing the inflow and outflow rates during this period. From the inflow and outflow rates the total inflow and outflow during a given period can be computed and the fall in the reservoir level multiplied by the average water surface area gives the change in storage during this period. The total loss of water in the reservoir during a given period is then given by

$$\text{Total loss} = \text{Inflow} - \text{Outflow} + \text{Change in storage}$$

The values of the total loss of water in existing reservoirs determined as indicated above usually provide valuable guidance for planning of future reservoirs.

7.14 MEASURES TO REDUCE EVAPORATION LOSS IN RESERVOIRS

Various measures taken to reduce loss of water due to evaporation in reservoirs are as follows.

1. By constructing reservoirs of less surface area and more depth, so that total evaporation is less due to less surface area being exposed to atmosphere.

2. By growing tall trees on the windward side of the reservoir which act as wind breakers and hence reduce evaporation.

3. By spraying certain chemicals or fatty acids and thus forming thin film of these substances on the water surface in the reservoir. For this purpose cetyl alchol (Hexadecanol) is commonly used, which when sprayed on the water surface spreads as a monomolecular layer forming a thin film on the water surface which is only about 0.015 micron in thickness. This film allows the precipitation from the top to enter the reservoir but does not allow water molecules to escape through it and hence reduces evaporation. However, this method is effective only when wind velocities are less. Further it is best suited only for small and medium size reservoirs.

Besides cetyl alcohol, certain other alcohols such as alkoxy ethanols are also being tried to reduce evaporation from the water surface in the reservoirs. However, it would be possible to use these alcohols commonly, if these are able to withstand high wind velocities in the field and if these are commercially available.

4. By providing the outlets in such a way that the warmer water is released and hence the temperature of water is reduced and consequently the evaporation is also reduced.

5. By removing the weeds and plants from the periphery of the reservoir.

6. By providing coverings of thin polythene sheets for the water surface. This is possible only for small agricultural ponds and lakes.

7. By developing underground reservoirs, since the evaporation from a ground water table is very much less than the evaporation from a water surface.

8. By growing huge trees and forest around the reservoir so that due to cooler environment the evaporation will be less.

7.15 SEDIMENT TRANSPORT BY RIVERS-RESERVOIR SEDIMENTATION

All rivers carry certain amount of sediment (or sediment load) which is produced due to erosion in their catchment areas. The amount of sediment in a river depends on the extent of erosion in its catchment area which depends on the following factors.

(i) Nature of soil in the catchment area

(ii) Topography of the catchment area

(iii) Vegetal cover

(iv) Intensity of rainfall.

If the soil in the catchment area of a river is loose and easily erodible, the river will bring in large amount of sediment. On the other hand a river will carry less sediment if the soil in the catchment is hard and not easily eroded. Further the catchment areas with steep slopes will give rise to high velocities of flow and hence cause more erosion of the surface soil. As such rivers having catchment areas with steep slopes will bring in more sediment. Similarly the catchment areas having poor or practically no vegetal cover will be more easily eroded and hence rivers having such catchment areas will carry more sediment. On the other hand the catchment areas having thick vegetal cover will produce less sediment. Again higher intensity of rainfall in the catchment area will cause greater runoff and more erosion, thus more sediment will be produced.

The sediment load carried by a river may be divided into the following two parts.

(i) Suspended load

(ii) Bed load

The *suspended load* (or suspended sediment) is that part of the sediment load which does not move in contact with the bed of the river but is held in suspension against gravity by the vertical component of the eddies in the turbulent flow. It consists of relatively finer material which remains dispersed throughout the flow cross-section of the river.

The *bed load* is that part of the sediment load which moves in contact with the bed of the river. It consists of relatively coarser material.

When the sediment laden water of the river approaches the reservoir, the velocity and turbulence are greatly reduced due to which the larger suspended particles and most of the bed load are deposited as a delta at the head of the reservoir as shown in Fig. 7.12. The finer particles remain in suspension for a longer time and are deposited farther down in the reservoir. However, some of the very find particles may be carried with water discharged through the outlets or the spillway. The deposition of the sediment in the reservoir is known as *reservoir sedimentation* or *reservoir-silting*. In order to allow for such deposition of sediment in a reservoir (or silting up of reservoir) certain percentage of the total storage is usually left unutilised which as defined in section 7.5 is called dead storage. The dead storage is usually not more than one fourth of the total capacity of the reservoir. The water surface in the reservoir is never allowed to be dropped below the level of the dead water storage. As such all the outlets drawing water from the reservoir are provided above the level of the dead water storage.

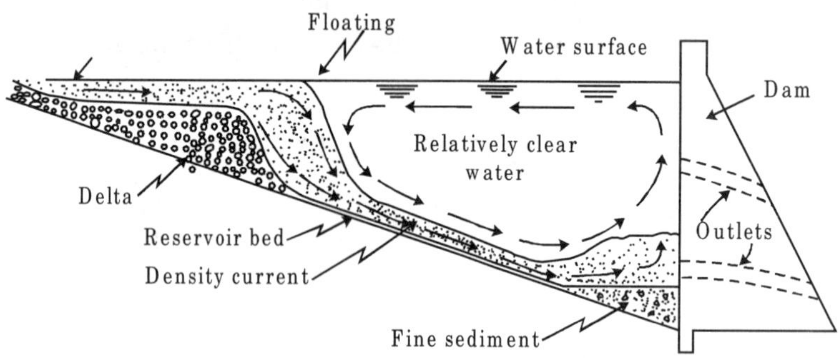

Fig. 7.12. *Sediment deposition in a typical reservoir—Density current*

Measurement of Sediment Load. The suspended sediment load of a stream is measured by taking the samples of water at various depths. The samples are weighed and filtered to remove the sediment. The sediment is then dried and weighed. The sediment load is expressed in parts per million (ppm) which may be obtained by dividing the weight of the sediment by the weight of the water sample (i.e., the weight of the sediment and water in the sample) and multiplying it by 10^6. However, there is no practical device available for the measurement of bed load, which is estimated to be about 15% of the suspended sediment load.

The average annual sediment transported by the stream may be obtained by multiplying the weight of the sediment transported per unit volume of water by the average annual inflow of the stream.

Density Currents. Density current may be defined as a flow of one fluid under another fluid of a slightly different density under the effect of gravity. In many reservoirs density currents are usually developed, especially during floods when rivers carry heavily sediment-laden or muddy water. The two fluids involved in the development of a density current in a reservoir are the sediment-laden water and the water already stored in the reservoir which is relatively clear due to the deposition of the sediment. When a sediment-laden water enters a reservoir it plunges downwards and moves towards the dam along the bed of the reservoir as a density current below the reservoir water as shown in Fig. 7.12. The reservoir water being relatively clear, has a slightly lower density than that of the sediment-laden water. It is because of this density difference, the sediment-laden water of the density current does not mix readily with the reservoir water but maintains its identity for a considerable time. However, in addition to sediment the density difference may also result from dissolved salts or minerals, or from temperature difference which may also develop a density current.

Useful Life of a Reservoir. The deposition of sediment gradually reduces the available storage capacity of the reservoir. As such with more and more deposition of sediment in a reservoir when its useful capacity is reduced so much that it is not able to serve the required purpose, the *useful life* of the reservoir is considered to be over. The sedimentation of a reservoir is measured in terms of trap efficiency. The *trap efficiency* is defined as the percent of the inflowing sediment which is retained in a reservoir. The observations on various reservoirs have indicated that the trap efficiency is a function of the ratio of reservoir capacity to total inflow, i.e.,

$$\text{Trap efficiency} = f\left(\frac{\text{Capacity}}{\text{Inflow}}\right)$$

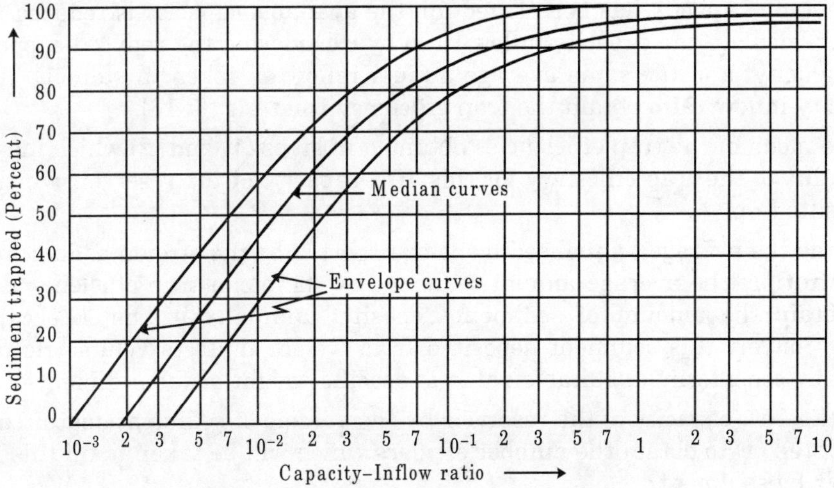

Fig. 7.13. *Plot of reservoir-trap efficiency versus capacity-inflow ratio*

Figure 7.13 shows a plot of trap efficiency versus capacity-inflow ratio on the basis of the data obtained from existing reservoirs. It may be observed from Fig. (7.13) that the trap efficiency is low for reservoirs having small capacity-inflow ratio and it gradually increases as the capacity-inflow ratio increases. Thus if a reservoir of small capacity is constructed on a stream having a large inflow rate, it will have a small capacity-inflow ratio and a low trap efficiency, because most of its inflow will be discharged to the downstream so quickly that the suspended sediment will hardly have any time to settle in the reservoir. On the other hand if a reservoir of large capacity is constructed on the same stream it will have a higher capacity-inflow ratio and a higher trap efficiency because in this case water will be retained for a longer duration and will thus permit most of the suspended sediment to settle in the reservoir. Further the trap efficiency of a reservoir will be more in the beginning but it will decrease as the capacity of the reservoir will be reduced due to sedimentation. Thus, complete silting up of the reservoir may take a very long time. However, in general when about 80% of the useful capacity of the reservoir gets silted up, the useful life of the reservoir is considered to be over.

The trap efficiency may also be determined by the following expression

$$\text{The efficiency} = 100\left[1-\frac{1}{(1+ax)}\right]^n \qquad \ldots(7.7)$$

in which x is capacity-inflow ratio $\left(\text{i.e., } x = \frac{\text{Capacity}}{\text{Inflow}}\right)$; and a and n are constants.

In order to justify the construction of a reservoir it is essential to determine its probable useful life, for which the following method may be used.

Method for Determining Useful Life of a Reservoir

1. Determine the required capacity of the reservoir as indicated in section 7.8. Also determine the average annual inflow of the stream and calculate the capacity-inflow ratio. For this value of capacity-inflow ratio obtain the trap efficiency from Fig. (7.13).

2. Divide the total capacity of the reservoir into suitable interval say 10%. Assuming that 10% of the reservoir capacity has been silted up, the available capacity is reduced to 90% of the total capacity. Calculate the capacity-inflow ratio by considering the reduced capacity (i.e., 90% of the total capacity) and the same average annual inflow as taken in step (1). Again for this value of capacity-inflow ratio obtain the trap efficiency from Fig. (7.13).

3. Find the mean of the trap efficiencies obtained in steps (1) and (2) which may be taken as the average value of the trap efficiency for this 10% capacity of the reservoir which is assumed to have been silted up.

4. Determine the average annual sediment transported by the stream as indicated earlier in this section. Multiply the average annual sediment by the mean trap efficiency determined in step (3) to obtain the amount of sediment deposited annually in the 10% capacity under consideration. Convert this sediment deposited from weight units to volumetric units such as hectare-metre by considering a suitable value of specific weight of the sediment deposited.

5. Divide the 10% capacity of the reservoir by the volume of sediment deposited annually as determined in step (4) to obtain the number of years which will be taken to fill this 10% capacity of the reservoir by sediment.

6. Repeat the same procedure and obtain the number of years which will be taken to fill

each of the next 10% capacities of the reservoir, i.e., by considering 80%, 70%, 60%, 50%, etc., as the available capacities until the available capacity of the reservoir is reduced to 20%. The sum of the years which will be taken to fill each of the 10% capacities of the reservoir will give the probable useful life of the reservoir.

Illustrative Example 7.6 shows the above described method for computing the useful life of a reservoir.

7.16 CONTROL OF SEDIMENTATION OF RESERVOIRS

In order to increase the useful life of a reservoir it is necessary to control the deposition of sediment in the reservoir. The various methods which may be adopted to control the deposition of sediment in a reservoir are as follows.

1. Selection of Reservoir Site. The reservoir sedimentation can be controlled by selecting such a site for the construction of reservoir where the sediment inflow is low. Thus, if alternative sites for the construction of a reservoir are available then a site which excludes runoff from an easily erodible part of the catchment area should be preferred to the one which includes this runoff. Similarly if a tributary of a main river carries more sediment then the site for the construction of a reservoir on the main river should be on the upstream of the tributary in order to exclude its water entering the reservoir.

2. Reservoir Design. As discussed earlier for a reservoir of small capacity on a river having large inflow rate, the rate of sedimentation is less than that for a reservoir of a large capacity on the same river. Thus, the rate of sedimentation of a reservoir may be controlled by designing the reservoir in such a way that instead of providing the entire designed capacity at the first instance the capacity of the reservoir may be increased in stages. For this initially a reservoir of small capacity may be created by constructing the dam upto a certain lower height. Later on when a portion of the reservoir gets silted up its capacity may be increased by raising the dam to the desired height. However, in this case some provisions will have to be made in the design of the dam so that it may be raised in stages.

Further an adequate number of outlets must be provided in the dam at different elevations, so that the flood water which is usually heavily sediment-laden may be discharged to the downstream without much of the sediment being deposited in the reservoir.

3. Control of Sediment Inflow. The inflow of sediment to a reservoir can be controlled by constructing check dams and by providing vegetation screens.

A *check dam* is a small dam usually constructed across a stream to trap most of the coarser sediment transported by the stream. Thus inflow of sediment can be controlled by constructing check dams across those tributaries which might be carrying relatively more sediment. The check dams may also be constructed across gullies to retain some sediment and prevent from entering the streams.

Vegetation screens may be developed by promoting the growth of vegetation in the catchment as well as at the entrance to the reservoir. The vegetation screens would trap a large amount of sediment if flood waters pass through them before entering the reservoir and hence help in reducing the sediment inflow.

4. Control of Sediment Deposition. The deposition of sediment in a reservoir can be controlled to some extent by designing and operating the outlets in a dam in such a manner that the water having higher sediment content is discharged to the downstream through the outlets.

The sediment content of water in the reservoir is relatively high during and immediately after flood and if most of the water can be discharged to the downstream at such times, the amount of sediment trapped in the reservoir will be reduced. Moreover the concentration of sediment is more at some levels and if water can be withdrawn from these levels through the outlets provided there, the deposition of sediment in the reservoir will be reduced.

5. Removal of Sediment Deposits. The sediment already deposited in a reservoir may be removed either by excavation or by scouring through sluices in the dam. The removal of the deposited sediment by excavation is however not feasible in most of the cases. Further as indicated below there are certain limitations in the use of sluices for removing the deposited sediment. For scouring the deposited sediment the sluices are to be provided near the bottom of the dam which involves a lot of structural problem and hence except for some small dams such sluices are usually not provided. Moreover for scouring of the deposited sediment through sluices considerable amount of water is lost. Further it is also found that the sluices are not much effective in the removal of the deposited sediment because the water flowing through the sluice scoops out a narrow channel for itself on the upstream side of the sluice and leaves the bulk of the deposited sediment undisturbed. However, to some extent the scouring action may be made more effective if simultaneously with scouring the deposited sediment is loosened and pushed towards the sluices by mechanical means. The sluices are therefore generally used for allowing the flood waters to escape to the downstream with minimum detention in the reservoir so that much deposition of sediment does not take place.

6. Erosion Control in the Catchment Area. The control of erosion in the catchment area will reduce the inflow of sediment to the stream and consequently there will be less inflow of sediment to a reservoir constructed on the stream. The erosion in the catchment area may be controlled by the use of the methods of soil conservation within the area. The various methods of soil conservation are afforestation, control of deforestation, regrassing and control of grazing, control on terrace cultivation and provision of contour bunds, checking gully formation by providing small embankments, providing check dams in the gullies formed, etc. These methods of soil conservation are quite effective in the control of sedimentation of reservoirs but are costly and would show appreciable results only after considerable time.

SOLVED ILLUSTRATIVE EXAMPLES

Illustrative Example 7.1 *The following table gives the monthly inflows during a critical low water period at a site of a proposed dam, the corresponding monthly evaporation and precipitation at a nearby station, and the estimated monthly demand of water. Prior water rights require a release of full natural flow or 4.5 hectare-metre per month, whichever is less. Assume that 25 percent of the rainfall on the land area to be flooded by the reservoir has reached the stream in the past. Using a net increased pool area of 405 hectares, find the required useful storage. Take pan evaporation coefficient equal to 0.7.*

Month	*Inflow at the proposed dam site (ha-m)*	*Pan-evaporation (mm)*	*Precipitation (mm)*	*Demand (ha-m)*
January	0.0	20	0	5.0
February	0.0	27	0	5.0
March	2.2	30	0	10.0

Contd.

Contd.

April	*5.5*	*57*	*15*	*12.5*
May	*6.2*	*105*	*20*	*15.0*
June	*137.8*	*127*	*85*	*15.0*
July	*209.7*	*112*	*203*	*7.5*
August	*80.3*	*86*	*55*	*7.5*
September	*33.7*	*73*	*40*	*9.0*
October	*8.3*	*54*	*0*	*12.5*
November	*2.5*	*32*	*0*	*15.0*
December	*0.0*	*18*	*0*	*15.0*

Solution

The computations are shown in the Table Ex. (7.1) in which the first five columns are same as given in the problem. Column (6) gives the downstream requirement on account of prior water right and the entries in this column are equal to the river flow or 4.5 hectare-metre whichever is less.

Column (7) gives the quantity of water E which evaporates from the whole of the water surface area of reservoir during each month which is computed by the following expression.

$$E = \frac{\text{Reservoir area} \times \text{Pan evaporation} \times 0.7}{100}$$

$$= \frac{405 \times 0.7 \times \text{Column (3)}}{100}$$

$$= 2.835 \times \text{Column (3) hectare-metre.}$$

Column (8) gives the precipitation P in hectare-metre, falling over the reservoir area. Since 25% of the precipitation is assumed to have reached the stream it is included in the inflow given in Column (2) and hence only 75% of the precipitation falling on the reservoir area is considered which is given by the following expression.

$$P = \frac{\text{Reservoir area} \times \text{Precipitation} \times 0.75}{100}$$

$$= \frac{405 \times \text{Column (4)} \times 0.75}{100}$$

$$= 3.038 \times \text{Column (4) hectare-metre}$$

Table for Example 7.1

Month	*Inflow (ha-m)*	*Pan-Evapor ation (mm)*	*Precipi-tation (mm)*	*Demand (ha-m)*	*Dowinst-ream Require-ment (ha-m)*	*Evapor-ation E (ha-m)*	*Precipit-ation P(ha-m)*	*Adjusted Inflow I(ha-m)*	*Require Storage S (ha-m)*
(1)	*(2)*	*(3)*	*(4)*	*(5)*	*(6)*	*(7)*	*(8)*	*(9)*	*(10)*
January	0.0	20	0	5.0	0.0	5.67	0.00	– 5.67	10.67
February	0.0	27	0	5.0	0.0	7.65	0.00	– 7.65	12.65

Contd.

Contd.

March	2.2	30	0	10.0	2.2	8.51	0.00	– 8.51	18.51
April	5.5	57	15	12.5	4.5	16.16	4.56	– 10.60	23.10
May	6.2	105	20	15.0	4.5	29.77	6.08	– 21.99	36.99
June	137.8	127	85	15.0	4.5	36.00	25.82	+ 123.12	Nil
July	209.7	112	203	7.5	4.5	31.75	61.67	+ 235.12	Nil
August	80.3	86	55	7.5	4.5	24.38	16.71	+ 68.13	Nil
September	33.7	73	40	9.0	4.5	20.70	12.15	+ 20.65	Nil
October	8.3	54	0	12.5	4.5	15.31	0.00	– 11.51	24.01
November	2.5	32	0	15.0	2.5	9.07	0.00	– 9.07	24.07
December	0.0	18	0	15.0	0.0	5.10	0.00	– 5.10	20.10
Total	486.2	741	418	129.0	36.2	210.07	126.99	366.92	170.10

Column (9) gives the adjusted inflow I which is computed as follows.

$$I = \text{Column (2)} - \text{Column (6)} - \text{Column (7)} + \text{Column (8)}.$$

Column (10) gives the quantity of water S required from storage to meet the required demand and it is computed as follows.

$$S = \text{Demand} - \text{Adjusted inflow}$$
$$= \text{Column (5)} - \text{Column (9)}$$

Only the positive values of S are to be considered and entered in column (10) and where negative values are obtained nil should be entered in column (10). This is so because a positive value of S indicates that water is required from the storage since demand is more than the adjusted inflow, whereas a negative value of S indicates that no water is required from storage since demand is less than the adjusted inflow.

The sum of column (10) gives the required useful storage, the value of which in this case is found to be 170.10 hectare-metre.

Illustrative Example 7.2. *The average seasonal discharge of a river for 11 years is as given below. Determine the storage capacity required to supply water at the rate of 570 cumec throughout the year.*

Year	*A* *16th June to 30th September (cumec)*	*B* *Ist October to 31st March (cumec)*	*C* *Ist April to 15th June (cumec)*
1950–51	*3300*	*350*	*80*
1951–52	*1800*	*170*	*100*
1952–53	*1000*	*330*	*55*
1953–54	*1200*	*380*	*90*
1954–55	*4700*	*180*	*70*
1955–56	*650*	*130*	*110*
1956–57	*500*	*180*	*40*
1957–58	*900*	*300*	*120*
1958–59	*2200*	*200*	*105*
1959–60	*4500*	*260*	*70*
1960–61	*2000*	*150*	*95*

Solution

The data for preparing the mass inflow curve are shown in the Table for Example 7.2. The mass curve of inflow is plotted as shown in the Fig. Ex. (7.2).

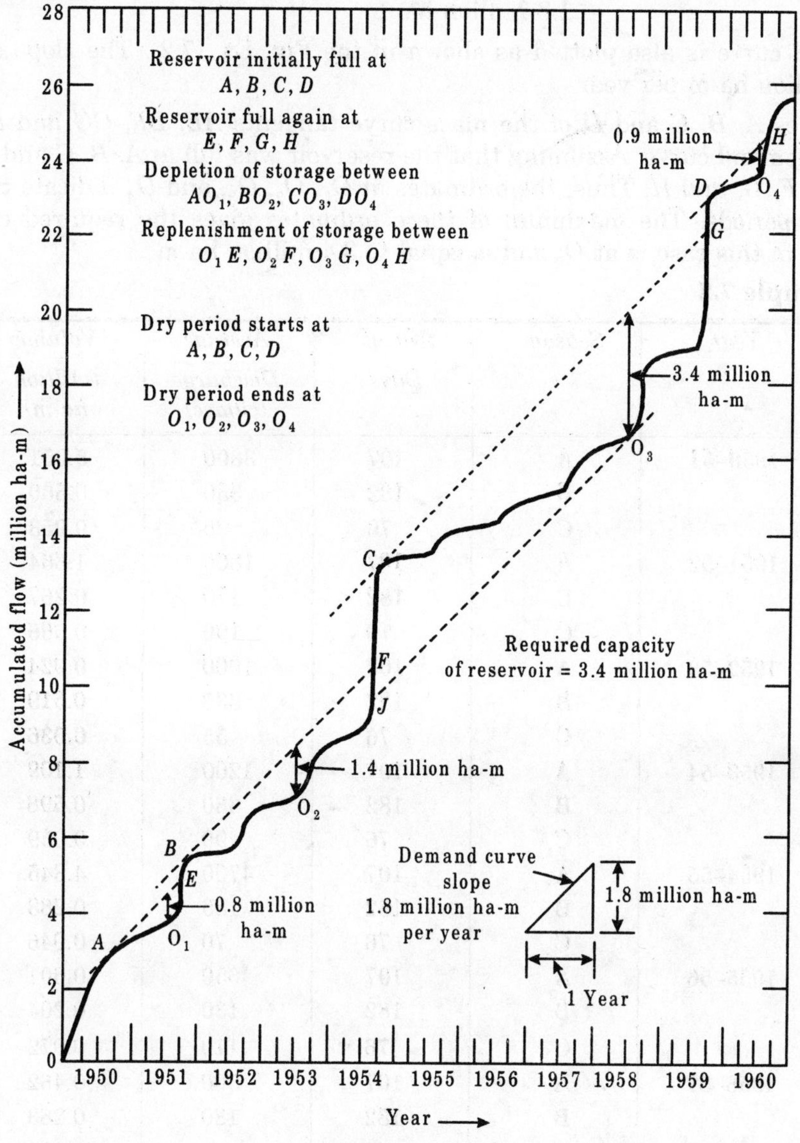

Fig. Ex. 7.2

Total demand of one year

$$= 570 \times 365 \text{ cumec-days}$$

$$= 570 \times 365 \times 24 \times 3600 \text{ m}^3$$

$$= \frac{570 \times 365 \times 24 \times 3600}{10^4} \text{ ha-m}$$

$= 1.7976 \times 106$ ha-m

$\simeq$ 1.8 million ha-m

The demand curve is also plotted as shown in the Fig. Ex. (7.2). The slope of the demand curve is 1.8 million ha-m per year.

At the apexes *A, B, C* and *D* of the mass curve tangents *AE, BF, CG* and *DH* are drawn parallel to the demand curve. Assuming that the reservoir was full at *A, B, C* and *D*, it would be full again at *E, F, G,* and *H*. Thus, the ordinates at O_1, O_2, O_3, and O_4 indicate the deficiencies during the dry periods. The maximum of these ordinates gives the required capacity of the reservoir which in this case is at O_3 and is equal to 3.4 million ha-m.

Table for Example 7.2

S. No.	*Year*	*Season*	*No. of Days*	*Average Discharge (cumec)*	*Volume (million ha-m)*	*Cumulative Volume (million ha-m)*
1	1950–51	A	107	3300	3.051	3.051
		B	182	350	0.550	3.601
		C	76	80	0.053	3.654
2	1951–52	A	107	1800	1.664	5.318
		B	182	170	0.267	5.585
		C	76	100	0.066	5.651
3	1952–53	A	107	1000	0.924	6.575
		B	182	330	0.519	7.094
		C	76	55	0.036	7.130
4	1953–54	A	107	1200	1.109	8.239
		B	182	380	0.598	8.837
		C	76	90	0.059	8.896
5	1954–55	A	107	4700	4.345	13.241
		B	182	180	0.283	13.524
		C	76	70	0.046	13.570
6	1955–56	A	107	650	0.601	14.171
		B	182	130	0.204	14.375
		C	76	110	0.072	14.447
7	1956–57	A	107	500	0.462	14.909
		B	182	180	0.283	15.192
		C	76	40	0.026	15.218
8	1957–58	A	107	900	0.832	16.050
		B	182	300	0.472	16.522
		C	76	120	0.079	16.601

Contd.

Contd.

9	1958–59	A	107	2200	2.034	18.635
		B	182	200	0.314	18.959
		C	76	105	0.069	19.028
10	1959–60	A	107	4500	4.160	23.188
		B	182	260	0.409	23.597
		C	76	70	0.046	23.643
11	1960–61	A	107	2000	1.849	25.492
		B	182	150	0.236	25.728
		C	76	95	0.062	25.790

Further if lines (such as $O_3 J$) are drawn at the points O_1, O_2, O_3, and O_4 parallel to the demand curve and tangential to the mass curve then these must intersect the mass curve at some points (such as J) so that the reservoir becomes full at the start of the dry period. Also between AO_1, BO_2, CO_3, and DO_4 there would be depletion of storage and between O_1E, O_2F, O_3G, and O_4H there would be replenishment of storage.

Illustrative Example 7.3 *The mean monthly discharges of a river in a dry year are as given below.*

Month.	*Mean flow (cumec)*	*Month*	*Mean flow (cumec)*
Jan.	*31.7*	*July*	*43.7*
Feb.	*62.0*	*Aug.*	*70.9*
March	*82.9*	*Sept.*	*96.5*
April	*59.0*	*Oct.*	*74.7*
May	*23.9*	*Nov.*	*46.3*
June	*22.8*	*Dec.*	*24.3*

Determine:

(i) the minimum capacity of a reservoir if the entire annual inflow is to be drawn off at a uniform rate (with no flow going to waste over the spillway) ;

(ii) the amount of water which must be initially stored to maintain the uniform draw off as determined in (i) above ;

(iii) the maximum possible draw off rate and the amount of water wasted over the spillway assuming the same reservoir capacity determined in (i) above, if the amount of water initially stored is 150 M.m^3*;*

(iv) the amount of water wasted over the spillway if the largest reservoir that can be economically constructed is of capacity 150 M.m^3*.*

Solution

The inflow volume in each month and the cumulative inflow are computed and tabulated in the Table for Ex. (7.3).

Table for Example 7.3

Month	*Mean flow (cumec)*	*Inflow volume ($M.m^3$)*	*Cumulative inflow ($M.m^3$)*	*Month*	*Mean flow (cumec)*	*Inflow volume ($M.m^3$)*	*Cumulative inflow ($M.m^3$)*
Jan.	31.7	85	85	July	43.7	117	850
Feb.	62.0	150	235	Aug.	70.9	190	1040
Mar.	82.9	222	457	Sept.	96.5	250	1290
April	59.0	153	610	Oct.	74.7	200	1490
May	23.9	64	674	Nov.	46.3	120	1610
June	22.8	59	733	Dec.	24.3	65	1675

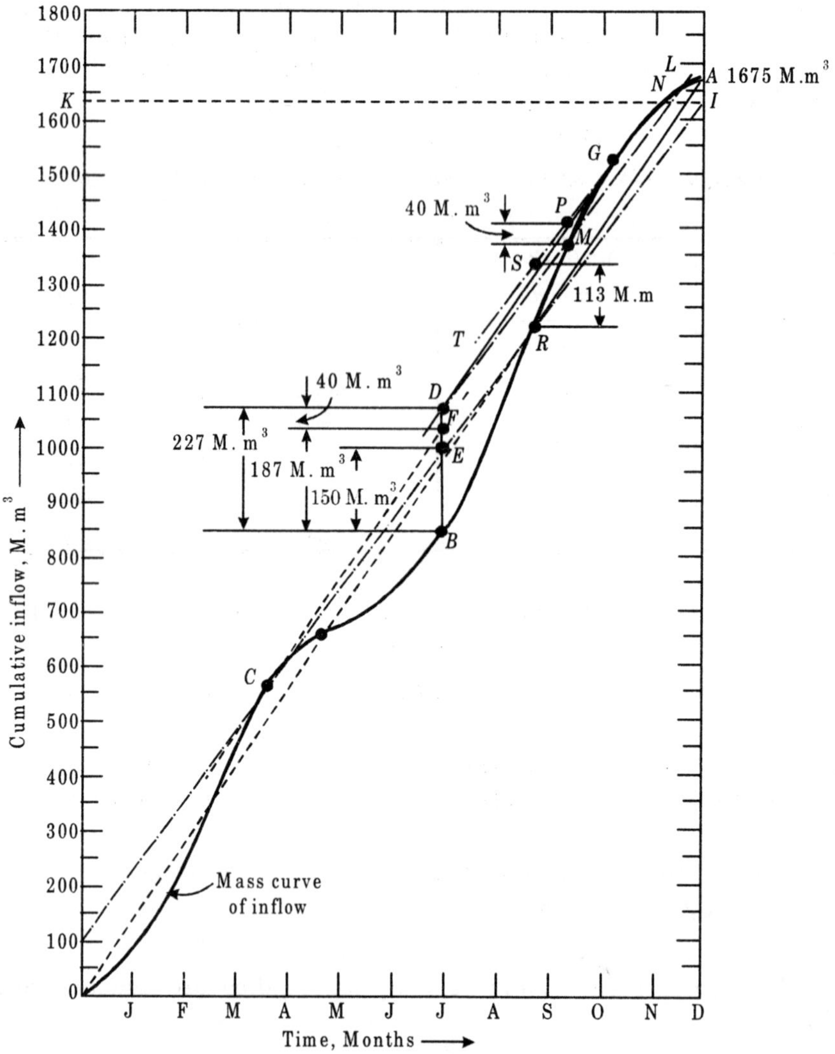

Fig. Ex. 7.3

The mass curve of inflow is plotted as cumulative inflow versus month as shown in the Fig. Ex. (7.3).

(i) The end points *O* and *A* of the mass curve are joined by a straight line *OA*. The slope of the line *OA* i.e., 1675 $M.m^3/yr$ or $(1675 \times 10^6/365 \times 24 \times 3600) = 53.1$ cumec is the uniform draw off rate throughout the year with no spill over the spillway. From the crests *C* and *G* of the mass curve draw lines *CF* and *GD* parallel to *OA*. The minimum capacity of the reservoir

$$= BF + FD$$

$$= (187 + 40) = 227 \text{ M.m}^3$$

(ii) Amount of water to be initially stored for the uniform draw off of

$$53.1 \text{ cumec} = BF = 187 \text{ M.m}^3$$

(iii) If the amount of water initially stored is only 150 $M.m^3$, measure $BE = 150$ $M.m^3$, join *CE* and produce it to intersect the ordinate through *O* (i.e., the cumulative inflow axis) at *H* and the other ordinate at *I*. The slope of the line *HI* gives the maximum possible draw off rate in this case. The vertical intercept $HK = 1533$ $M.m^3$ and the slope of the line $HI = 1533$ $M.m^3/yr = 48.6$ cumec, which is the maximum possible draw off rate.

With the same reservoir capacity of 227 $M.m^3$ being maintained, for determining the amount of water wasted draw a straight line *DL* parallel to *HI* intersecting the mass curve of inflow at *M* and *N*. Draw a straight line *GT* parallel to *HI*. The vertical intercept *PM* gives the amount of water wasted over the spillway (during the time period between points *M* and *N*) which is 40 $M.m^3$.

(iv) If the reservoir capacity is limited to 150 $M.m^3$ from economic considerations the amount of water wasted over the spillway = $RS = 113$ $M.m^3$, where *R* is the point of intersection of the line *HI* and the mass inflow curve and *S* is the point at which the vertical at *R* meets the line *GT* drawn parallel to the line *HI* at the crest *G*.

Illustrative Example 7.4 *Using data given below calculate the allocations to each project purpose by the remaining benefits method and by the alternative justifiable-expenditure method. Assume total cost of project as 20 million rupees.*

(All costs in millions of rupees)

Item	*Flood control*	*Power*	*Irrigation*
Separate costs	2.5	6.7	5.5
Estimated benefits	3.5	12.0	9.7
Alternate single-purpose cost	4.0	8.7	8.5

Solution

Total cost of the project = ₹ 20,000,000

Total separable costs = ₹ 14,700,000

∴ Joint cost = ₹ 5,300,000 = (5.3 million rupees)

The allocation of the cost by each of the two methods is calculated as follows.

(a) *Remaining benefits method*

(All costs in millions of rupees)

Line	*Item*	*Flood control*	*Power*	*Irrigation*	*Total*
1	Benefits limited by alternate cost	3.5	8.7	8.5	20.7
2	Remaining Benefits	1.0	2.0	3.0	6.0
3	Allocated joint costs	0.883	1.767	2.650	5.3
4	Total allocation	3.383	8.467	8.150	20.0
5	Percentage	16.91%	42.34%	40.75%	100%

In the above table in (line 1) the estimated benefits or the alternate single-purpose cost whichever is less for each purpose is entered. The remaining benefits (line 2) are obtained by subtracting the separate cost of each purpose from the corresponding values entered in line (1). The allocated joint costs (line 3) for each purpose are obtained from the following expression

$$\left(\frac{\text{Total joint cost}}{\text{Total remaining benefits}}\right) \text{(Remaining benefits of that purpose)}$$

For example for power the allocated joint cost is obtained as

$$\left(\frac{5.3}{6}\right)\times 2 = 1.767$$

The total allocation (line 4) for each purpose is obtained by adding the separate costs and the allocated joint costs and it may also be expressed in the form of percentage (line 5).

(b) *Alternative justifiable expenditure method*

Line	*Item*	*Flood control*	*Power*	*Irrigation*	*Total*
1	Alternate cost less separate cost	1.5	2.0	3.0	6.5
2	Allocated joint costs	1.223	1.631	2.446	5.3
3	Total allocation	3.723	8.331	7.946	20.0
4	Percentage	18.61%	41.66%	39.73%	100%

In the above table the values in (line 1) are obtained by subtracting the separate cost of each purpose from the corresponding alternate single-purpose cost. The allocated joint cost (line 2) for each purpose are obtained from the following expression

$$\left(\frac{\text{Total joint cost}}{\text{Total alternate cost less separable cost}}\right)\text{(Alternate cost less separate cost of that purpose)}$$

For example, for flood control the allocated joint cost is obtained as

$$\frac{5.3}{6.5}\times 1.5 = 1.223$$

The total allocation (line 3) for each purpose is obtained by adding the separate costs and the allocated joint costs and it may also be expressed in the form of percentage (line 4).

Illustrative Example 7.5 *The data pertaining to an inflow flood hydrograph for a reservoir is given below.*

Time (hr)	0	6	12	18	24	30	36	42
Inflow (cumec)	50	70	160	300	460	540	510	440
Time (hr)	48	54	60	66	72	78	84	90
Inflow (cumec)	330	250	190	150	120	90	80	70

The reservoir has uncontrolled spillway with its crest at elevation 100.0 m. The values of water surface area and outflow discharge at various elevations are given below.

Elevation (m)	100.0	100.5	101.0	101.5	102.0	102.5	103.0	103.5	104.0	104.5	105.0
Area (ha)	400	412	428	448	470	495	516	538	560	585	613
Outflow (cumec)	0	15	42	77	120	160	220	275	336	408	490

Determine the maximum reservoir level and maximum discharge over the spillway. Also draw the inflow hydrograph and outflow hydrograph and indicate the reduction in peak flow and peak lag resulting on account of routing of the flood.

Solution

From the water surface area and the corresponding elevation the storage capacity of the reservoir may be computed by using the cone formula as noted below.

$$V = \sum \frac{h}{3}\left(A_1 + A_2 + \sqrt{A_1 A_2}\right)$$

The values of the storage capacity of the reservoir are as noted in the Table (A) for Ex. (7.5).

TABLE (A) for Example 7.5

Elevation (m) (1)	*Water surface area at elevation noted in Col. (1) (m^2) × 10^4* (2)	*Storage between successive elevations noted in Col. (1) (m^3) × 10^4* (3)	*Total storage capacity upto elevation noted in Col. (1) (m^3) × 10^4* (4)
100.0	400	—	—
100.5	412	203	203
101.0	428	210	413
101.5	448	219	632
102.0	470	229	861
102.5	495	241	1102
103.0	516	253	1355
103.5	538	263	1618
104.0	560	274	1892
104.5	585	286	2178
105.0	613	299	2477

The inflow hydrograph, elevation versus storage and elevation versus outflow curves are plotted as shown in the Table Ex. 7.5 (A) and (B) respectively.

The values of the parameter are computed as given in Table (B) Ex. 7.5. The value of t is assumed to be 6 hours (or 21,600 seconds).

TABLE (B) for Example 7.5

Elevation (*m*)	*Storages* (*m*) × 10^4	$\left(\frac{2s}{t}\right)$ for t = 6 hr $\frac{s}{10800}$	*Outflow O*	$\frac{2s}{t}-O$	$\frac{2s}{t}+O$
100.0	0	0	0	0	0
100.5	203	188	15	173	203
100.5	203	188	15	173	203
101.0	413	382	42	340	424
101.5	632	585	77	508	662
102.0	861	797	170	677	917
102.5	1102	1020	165	855	1185
103.0	1355	1235	220	1035	1475
103.5	1618	1498	275	1223	1773
104.0	1892	1759	336	1423	2095
104.5	2178	2017	408	1609	2425
105.0	2477	2294	490	1804	2784

The plots of outflow O versus $\left(\frac{2s}{t}\pm O\right)$ are prepared as shown in the Fig. Ex. 7.5 (C).

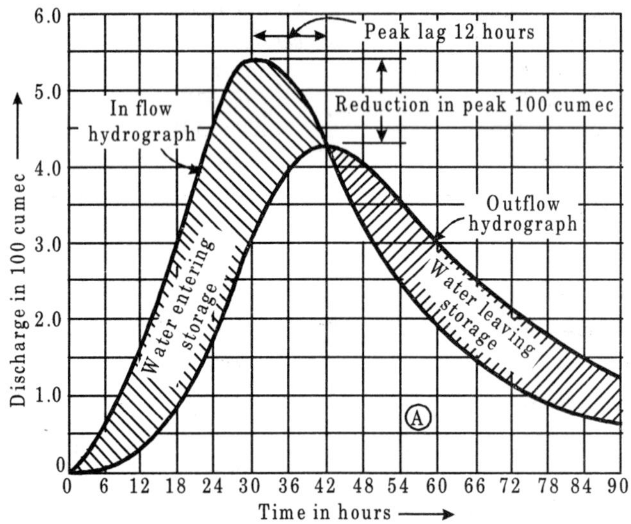

Fig. Ex. 7.5

The computation of flood routing are shown in Table (C) Ex. 7.5. In Cols. (1) and (2) the values of time t and inflow I are entered as given in the problem. However, if the values of t

other than those given are considered then the corresponding values of I may be obtained from the inflow hydrograph (Figure A) Ex. 7.5. In Col. (3) sum of the previous value and the next value of I is entered. In Col. (5) the sum of the values in Cols. (3) and (4) is entered as shown by horizontal arrows. Corresponding to a value of $[(2s/t)+O]$ entered in Col. (5) the value of outflow O is obtained from Fig. Ex. 7.5 (C) and it is entered in Col. (6). Further corresponding to this value of outflow O the value $[(2s/t) - O]$ is obtained from Fig. Ex. 7.5 (C) which is entered in the next line in Col. (4) as indicated by

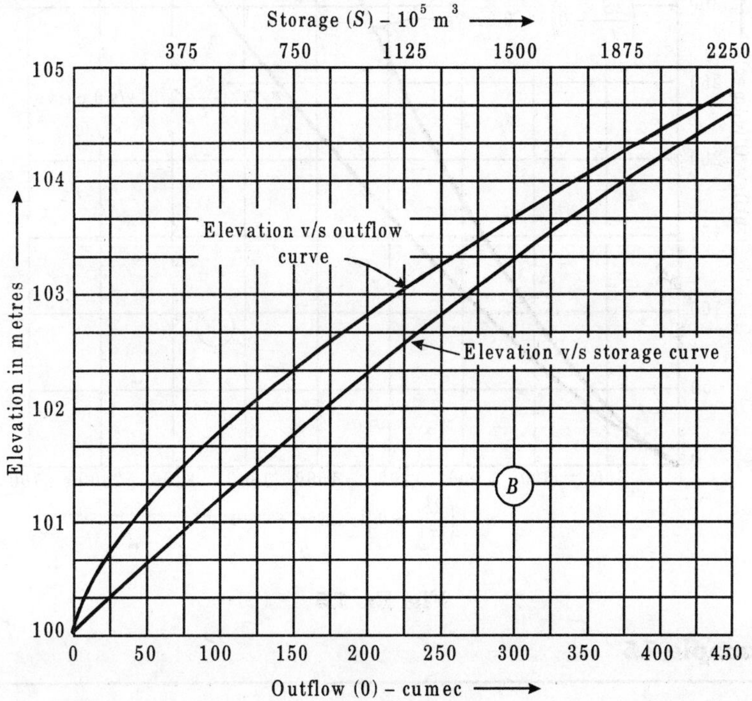

Fig. Ex. 7.5

an inclined arrow. Also for this value of O the reservoir elevation is obtained from Fig. Ex. 7.5 (B) which is entered in Col. (7). The same procedure is repeated till the flood is completely routed through the reservoir and the outflow hydrograph is obtained as shown in Fig. Ex. 7.5 (A). The following results are obtained from the flood routing.

(1) Maximum reservoir level = 104.65 m

(2) Maximum discharge over the spillway = 430 cumec

(3) Reduction in peak discharge

= (Maximum inflow – Maximum outflow)

= (540 – 430) = 110 cumec

(4) Peak lag

= (Time for maximum outflow) – (Time for maximum inflow)

= (42 – 30) = 12 hours.

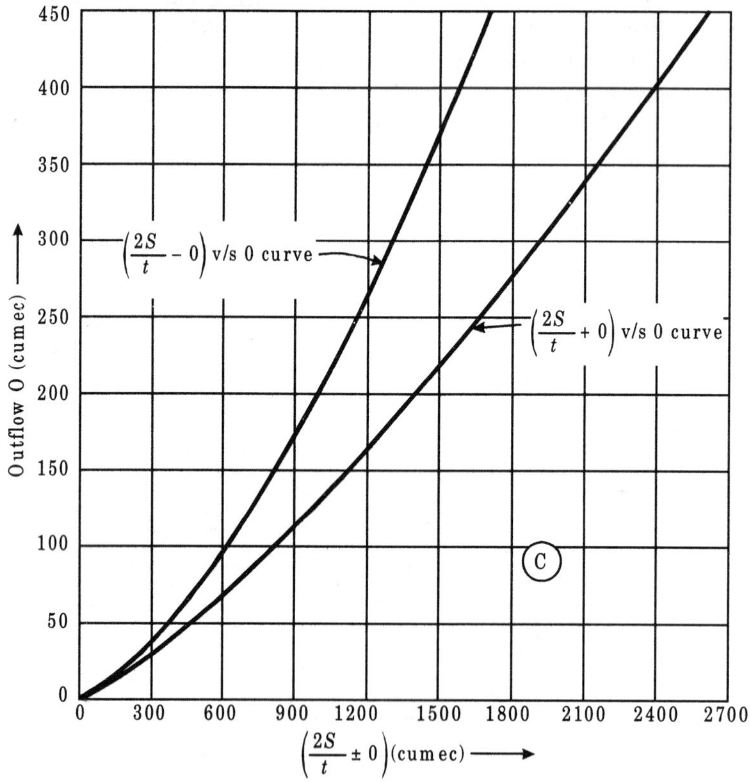

Fig. Ex. 7.5

Table (C) for Exmaple 7.5

Time (hours) (1)	*Inflow* I (2)	$I_1 + I_2$ (cumec) (3)	$\left(\frac{2s}{t} - O\right)$ (cumec) (4)	$\left(\frac{2s}{t} + O\right)$ (cumec) (5)	*Outflow O* (cumec) (6)	*Reservoir Elevation (m)* (7)
0	50	—	—	—	—	—
6	70	120 —	→ 0	→ 120	7.5	100.33
12	160	230 —	→ 87	→ 317	27.5	100.73
18	300	460 —	→ 248	→ 708	85.0	101.58
24	460	760 —	→ 540	→ 1300	187.5	102.70
30	540	1000 —	→ 930	→ 1930	305.0	103.75
36	510	1050	1320	2370	395.0	104.42
42	440	950	1575	2525	430.0	104.65
48	330	770	1657	2427	407.5	104.50
54	250	580	1605	2185	355.0	104.13

Contd.

Contd.

60	190	440	1470	1910	300.0	103.70
66	150	340	1305	1645	252.0	103.30
72	120	270	1155	1425	210.0	102.90
78	90	210	1050	1260	180.0	102.67
84	80	170	900	1070	145.0	102.20
90	70	150	772	922	120.0	101.90

Illustrative Example 7.6 *Solve Example 7.5 by trial and error method.*

Solution

The procedure for solving the problem by trial and error method is explained in section 7.12 and the computations are shown in the given Table for Ex. 7.6. The results obtained are same as in the Illustrative Example 7.5 and are as noted below.

(1) Maximum reservoir level = 104.65 m

(2) Maximum discharge over the spillway = 430 cumec

(3) Reduction in peak discharge = (540 – 430) = 110 cumec

(4) Peak lag = (42 – 30) = 12 hours.

Illustrative Example 7.7 *Find the probable life of a reservoir with an initial capacity of 3700 ha-m if the average annual inflow is 7400 ha-m and the average annual sediment inflow is* 2×10^6 *kN. Assume a specific weight of sediment as 11.2 kN/m*^{3}*. The useful life of the reservoir will terminate when 80 per cent of its initial capacity is filled with sediment. The values of trap efficiency for different values of capacity-inflow ratio as obtained from Fig. 7.12 are as follows.*

Capacity Inflow ratio	*0.1*	*0.2*	*0.3*	*0.4*	*0.5*	*0.6*	*0.7*	*0.8*	*0.9*	*1.0*
Trap Efficiency %	*87*	*93*	*95*	*95.5*	*96*	*96.5*	*97*	*97.3*	*97.4*	*97.5*

Solution

Annual sediment inflow $= 2 \times 10^6$ kN

$$\therefore \text{ Volume of sediment inflow} = \frac{2\times10^6}{11.2} = 1.786 \ \ 10^5 \text{ m}^3 = 17.86 \text{ ha-m}$$

Initial capacity of the reservoir = 3700 ha-m

Average annual inflow = 7400 ha-m

$$\text{Initial capacity-inflow ratio} = \frac{3700}{7400} = 0.5$$

The capacity-inflow ratio goes on reducing as the sedimentation takes place and the capacity reduces.

TABLE for Example 7.6

Time period (hours)	Reservoir Elevation at beginning of period (m)	Inflow volume (ha-m)	Spillway discharge at beginning of period (cumec)	Trial reservoir elevation at end of period (m)	Spillway discharge at end of period (cumec)	Mean outflow rate (cumec)	Outflow volume (ha-m)	Storage capacity between reservoir elevation in beginning and end of period (ha-m)	Outflow volume Col. (8) + Storage capacity Col. 9 (ha-m)	Col. (3) - Col. (10)	Remarks
(1)	(2)	(3)	(4)	(5)	(6)	(7)	(8)	(9)	(10)	(11)	
0–6	100.00	129.60	0.00	100.30	9.00	4.50	9.72	119.88	129.60	0.00	
6–12	100.30	248.40	9.00	100.80	31.10	20.05	43.31	205.11	248.42	0.02	
12–18	100.80	496.80	31.10	101.65	89.75	60.43	130.52	366.26	496.78	0.02	
18–21	101.65	364.50	89.75	102.16	133.75	111.75	120.69	243.78	364.47	0.03	
21–24	102.16	450.90	133.75	102.70	189.75	161.75	174.69	276.22	450.91	0.01	
24–27	102.70	529.20	189.75	103.30	253.25	221.50	239.22	290.00	529.22	0.02	
27–30	103.30	572.40	253.25	103.80	311.38	282.32	304.90	267.50	572.40	0.00	
30–33	103.80	580.50	311.38	104.17	358.58	334.94	361.74	218.75	580.49	0.01	
33–36	104.17	564.30	358.50	104.43	397.20	377.85	408.08	156.25	564.33	0.03	
36–39	104.43	531.90	397.20	104.58	418.83	408.01	440.65	91.25	531.90	0.00	
39–42	104.58	488.70	418.83	104.65*	430.00	424.42	458.37	30.33	488.70	0.00	
42–45	104.65	442.80	430.00	104.60	423.00	426.50	460.62	–17.83	442.79	0.01	
45–48	104.60	383.40	423.00	104.50	407.38	415.19	448.41	–65.00	383.41	0.01	
48–51	104.50	333.45	407.38	104.35	383.75	395.57	427.22	–93.75	333.45	0.00	

*Maximum reservoir elevation.

The computations for determining the life of the reservoir are shown in the following table. The interval of capacity chosen is 740 ha-m, i.e., 20% of the initial capacity.

Capacity %	Capacity (ha-m)	*Capacity Inflow Ratio*	*Trap efficiency %*	*Average Trap efficiency for capacity interval %*	*Volume of sediment trapped (ha-m)*	*Volume of capacity interval (ha-m)*	*Years to fill*
100	3700	0.5	96.0				
80	2960	0.4	95.5	95.7	17.09	740	43.3
60	2220	0.3	95.0	95.2	17.00	740	43.5
40	1480	0.2	93.0	94.0	16.79	740	44.1
20	740	0.1	87.0	90.0	16.07	740	46.0
	Total	...	...	...	...	...	176.9

The useful life of the reservoir is 176.9 years.

REVIEW QUESTIONS

7.1 How are reservoirs classified? Describe briefly the various types of reservoirs.

7.2 Differentiate between a single purpose and a multipurpose reservoir.

7.3 Discuss in brief various investigations required for reservoir planning.

7.4 Explain how the area-elevation curve and elevation-storage curve are prepared. What is the use of these curves in reservoir planning?

7.5 Explain the terms : (i) Dead storage; (ii) Useful storage; (iii) Surcharge storage; (iv) Bank storage; and (v) Valley storage.

7.6 Explain the the terms : (i) safe yield; (ii) secondary yield and (iii) average yield.

7.7 What do you understand by a Mass-inflow curve? How would you find 'safe yield' if the mass inflow curve and the reservoir capacity are given.

7.8 Explain briefly how you will determine long term storage requirement using mass curve. Assume rate of demand as constant.

7.9 Explain with illustrative sketch how would you determine the capacity of a reservoir to avoid drought. How will you find the dead and live storage.

7.10 What is flood routing? Write down basic flood routing equation. Explain in detail any one method of flood routing.

7.11 Discuss the various methods of cost allocation for multipurpose projects.

7.12 What do you understand by life of a reservoir? What measures may be taken to prolong the life of a reservoir?

7.13 Write a note on reservoir sedimentation. How would you determine the useful life of a reservoir?

7.14 Define the terms : (i) Trap efficiency; (ii) Density currents.

7.15 Discuss the various methods of controlling sedimentation of reservoirs.

7.16 Describe briefly the various measures taken to reduce loss of water due to evaporation in reservoirs.

7.17 The inflow into a reservoir in a period of 6 hours is 6.2×10^6 m^3. The reservoir elevations at the beginning and end of this period are 100.00 m and 100.30 m respectively. The crest level of the spillway is 100.00 and its effective length 150 m. If the coefficient C in the equation $Q = CBH^{3/2}$ is 2.0, find the average area of the water spread between the elevations 100.00 and 100.30. [*Ans.* 18.892×10^6 m^2]

7.18 In a reservoir flood comes at 0 hour. The flood hydrograph data are given below:

Hour	0	1	2	3	4	5	6	7
Flood discharge in cumec	0	100	200	300	400	500	400	300

The area-elevation curve is also linear and is given below

Height (m) above spillway crest	0	1	2	3	4
Area (m^2)	7.0×10^5	8.0×10^5	9.0×105^5	10.0×10^5	11.0×10^5

The discharge of the spillway is given by the equation $Q = 300\ H^{3/2}$.

Find the rise in reservoir level at 4.0 hours assuming that the water level is at crest at 0 hour. [*Ans.* 0.9925 m]

7.19 A reservoir has a capacity of 4 million cubic metres and a drainage area of 250 sq. km. The annual inflow is equivalent to 375 mm of runoff from the given drainage area, and annual sediment production is equivalent to a weight of 12.25 million Newton per sq. km. of drainage area. The sediment has an average specific weight of 14715 N/m^3. Assume dead storage as 15 % of initial reservoir capacity, and an average value of trap efficiency 88 %. Determine the number of years it will take for dead storage to be filled with sediment.

[*Ans.* 3.276 years]

7.20 A reservoir is contemplated on a stream which has an annual average runoff 500 million m^3. Measurements indicate that the average sediment inflow is 2×10^9 N/year. Assuming that a cubic metre of settled sediment will dry out to a density of 12000 N/m^3. The original capacity of the reservoir is 24×10^6 m^3. Trap efficiency $= 100 \left[1 - \dfrac{1}{(1+ax)}\right]^n$ where x = capacity-inflow ratio, $a = 65$ and $n = 2.0$. Determine the design life of reservoir, based on sedimentation of 65 % of the original capacity. [*Ans.* 217.49 years]

CHAPTER

8 Dams—General

8.1 DEFINITION

A dam is a barrier constructed across a river or a natural stream to create a reservoir for impounding water, or to facilitate diversion of water from the river, or to retain debris flowing in the river along with water.

8.2 CLASSIFICATION OF DAMS

Dams may be classified in different ways on the basis of their function, hydraulic design, material of construction and structural design.

8.2.1 Classification Based on Function

On the basis of function, dams may be classified in the following categories.

(i) *Storage Dam or Impounding Dam.* A storage dam is constructed to create a reservoir to store water during the period when the flow in the river is in excess of the demand, for utilisation later on during the period when the demand exceeds the flow in the river. The water stored in the reservoir so created may be used for various purposes such as irrigation, hydro-electric power generation, water supply etc.

(ii) *Detention Dam.* A detention dam is primarily constructed to temporarily detain all or part of the flood waters of a river and to gradually release the stored water at controlled rates so that the entire region on the downstream side of the dam may be safeguarded against the possible damage due to floods. A detention dam may also be used as a storage dam. Similarly by suitable operation of the reservoir a storage dam may also be used as a detention dam.

(iii) *Diversion Dam.* A diversion dam is constructed for the purpose of diverting part or all of the water from a river into a conduit or a channel. Mostly for the purpose of diversion of water from a river into an irrigation canal a weir is constructed across a river which is known as *diversion weir*.

(iv) *Coffer Dam*. A coffer dam is a temporary dam constructed to exclude water from a specific area. Such a dam is invariably constructed on the upstream side of the site where actual dam is to be constructed so that the site for the constructional work is rendered dry. The coffer dam constructed in this case also behaves as a diversion dam.

(v) *Debris Dam*. A debris dam is constructed to catch and retain debris such as sand, gravel, silt and drift wood flowing along with water in the river.

8.2.2 Classification based on Hydraulic Design

On the basis of hydraulic design, dams may be classified as follows.

(i) *Overflow Dam or Overfall Dam*. An overflow dam is constructed with a crest to permit the overflow of surplus water which cannot be retained in the reservoir. Usually dams are not designed as overflow dams for their entire length. Only some dams few metres in height have their entire length designed for overflow, which are mainly used for diversion of water and are known as *diversion weirs*.

(ii) *Non-overflow Dam*. A non-overflow dam is that for which water is not allowed to flow over its crest.

In most of the cases a part of the length of dam is designed as an overflow dam, while the rest is designed as non-overflow dam. In such cases the overflow section of a dam is called *spillway*. However, in some cases a non-overflow dam is not combined with an overflow dam, but it is provided with a different type of spillway.

8.2.3 Classification based on Material of Construction

Based on the material used for construction, the dams may be classified in the following categories.

(i) *Rigid Dam*. A rigid dam is that which is constructed with rigid material such as masonry, concrete, steel or timber. Earlier stone masonry was commonly used for the construction of dams, but nowadays it is almost totally replaced by concrete. Bhakra Dam (maximum height 226 m) which is concrete dam and Rana Pratapsagar Dam (maximum height 53.96 m) which is a stone masonry dam, are the example of rigid dams in India. A steel dam is made of steel plates supported on inclined struts and a timber dam is made of wooden planks supported on a wooden framework. The steel and timber dams are constructed only for very small heights and these are rarely constructed.

(ii) *Non-rigid Dam*. A non-rigid dam is that which is constructed with non-rigid material such as earth, tailings, rockfill etc. There are four types of non-rigid dams viz., earth dam, tailings dam, rockfill dam and rockfill composite dam. An *earth dam* (*or earthen dam*) is constructed with gravel, sand, silt and clay. A *tailings dam* is built from the waste or refuse obtained from mines (or mine tailings). A *rockfill dam* consists of fragmental rock material supporting a water tight membrane on the upstream face. A *rockfill composite dam* consists of a rockfill on the downstream side and an earth fill on the upstream side. All these four types of non-rigid dams are normally classified under the category of *embankment dams*.

In most of the cases an earth dam is provided with a concrete or stone masonry overflow or spillway section. Such a dam is known as *composite dam*. Further in some cases a part of the length of a dam is constructed as earth dam and the rest is constructed as stone masonry dam (length of spillway portion being excluded). Such a dam is known as *masonry-cum-earth dam*. Nagarjunasagar dam (maximum height 124.7 m) is an example of masonry-cum-earth dam in India.

8.2.4 Classification based on Structural Behaviour

On the basis of the structural behaviour, dams may be classified in the following categories.

(i) *Gravity Dam*. A gravity dam is a masonry or concrete dam which resists the forces exerted upon it by its own weight. Its cross-section is approximately triangular in shape. If a gravity dam is straight in plan it is known as *straight gravity dam,* while if it is curved in plan it is known as *curved gravity dam*. A curved gravity dam (or arch-gravity dam) however, resists the forces exerted upon it both by gravity action and arch action. Further, a gravity dam is also classified as solid gravity dam and hollow gravity dam. A *solid gravity dam* has its entire body consisting of a solid mass of masonry or concrete. On the other hand a *hollow gravity dam* has hollow spaces left within the body of the dam. Most of the gravity dams constructed in India are straight solid gravity dams and Bhakra Dam is an example of this type of dams.

(ii) *Arch Dam*. An arch dam is a curved masonry or concrete dam, convex upstream, which resists the forces exerted upon it, mainly by arch action. The only example of arch dam in India is Idduki dam which is double curvature (in plan) arch dam in concrete.

(iii) *Buttress Dam*. A buttress dam consists of a water retaining sloping membrane or deck on the upstream which is supported by a series of buttresses which are generally in the form of equally spaced triangular reinforced concrete or masonry walls or counterforts. The sloping membrane is usually reinforced concrete slab. In some cases the upstream slab is replaced by multiple arches supported on buttresses or by flaring the upstream edge of the buttresses to span the distance between the buttresses. The former is known as *multiple arch buttress dam* and the later is known as *bulk head buttress dam* or *massive head buttress dam*. In general the structural behaviour of a buttress dam is similar to that of a gravity dam.

(iv) *Embankment Dam*. An embankment dam is a non-rigid dam, which resists the forces exerted upon it mainly by its shear strength. To some extent an embankment dam also resists the forces exerted upon it by gravity but its structural behaviour is different in many respects from that of a gravity dam and hence it cannot be included in the category of a gravity dam.

8.2.5 Classification based on Size

The dams may be classified according to size by using the hydraulic head (from normal or annual average flood level on the downstream to the maximum water level) and the gross storage behind the dam as given below. The overall size classification for the dam would be greater of that indicated by either of the following two parameters

Classification	*Gross Storage*	*Hydraulic Head*
Small	Between 0.5 and 10 million m^3	Between 7.5 m and 12 m
Intermediate	Between 10 and 60 million m^3	Between 12 m and 30 m
Large	Greater than 60 million m^3	Greater than 30 m

8.3 FACTORS GOVERNING SELECTION OF TYPE OF DAM

It is rare that for any given dam site only one type of dam is suitable. Thus, except in the cases where the selection of the type of dam is obvious, it would be necessary to prepare preliminary designs and estimates for the several types of dams before one can get the best solution from the point of view of direct costs and all other factors. The various factors which affect the choice of the type of dam are described below.

1. Topography—Valley Shape. In general topography dictates the first choice of the type of dam and the most important factor in this respect is the shape of valley or gorge. The shapes of valleys normally found in nature may be broadly classified in three categories and for each of these a particular type of dam is generally suitable as indicated below.

(i) A narrow V-shaped valley with sound rock in abutments has an arch dam as the first choice. However, for economic arch dam it is preferable to have the top width of the valley less than about four times its height.

(ii) A narrow or moderately wide U-shaped valley with sound rock foundation is best suited for gravity dam or buttress dam.

(iii) A wide valley with foundation of soil material to a considerable depth leaves no choice but an embankment dam.

2. Geology and Foundation Conditions. All the forces acting on dam including its weight are transmitted to the foundation. As such the foundation conditions at the dam site need to be thoroughly investigated. The foundation conditions will limit the choice of the type of dam to certain extent, though such limitations can frequently be modified. Foundation conditions depend upon the geological character and thickness of the strata, their inclination, permeability and relation to underlying strata, existing faults and fissures. The common types of foundations generally encountered are discussed below.

(i) *Good Rocky Foundations*. Good rocky foundations have high bearing capacity and offer high resistance to erosion and percolation. Therefore any type of dam can be built on such foundations. However, in this case the choice of the type of dam will be governed by economy of materials or overall cost. Sometimes seams and fractures are present in these rocks which will have to be sealed by grouting. Further, disintegrated rock if any will have to be removed.

(ii) *Gravel and Coarse Sand Foundations*. Gravel and coarse sand foundations are unable to bear the load of high concrete gravity dams. As such these foundations are suitable only for earthfill, rockfill and low concrete gravity dams upto about 15 m height. Since through these foundations water may percolate at high rates, effective water cutoffs or seals will have to be provided.

(iii) *Silt and Fine Sand Foundations*. These foundations are suitable only for earthfill dams and low concrete gravity dams upto about 8 m height, and they are not suitable even for rockfill dams. In these foundations the main problems are settlement, the prevention of piping, excessive percolation of water and protection of the foundation at the downstream toe from erosion.

(iv) *Clay Foundations*. There may be considerable settlement of the dam constructed on clay foundation, if the clay is unconsolidated and the moisture content is high. As such clay foundations are ordinarily not suitable for the construction of concrete gravity dams and totally unsuitable for rockfill dams. However, earthfill dam may be constructed on clay foundations but these may also require special treatment to consolidate the foundation. Moreover, tests of the foundation material in its natural state are usually required to determine the consolidation characteristics of the material and its ability to support the superimposed loads.

(v) *Nonuniform Foundations*. At several dam sites uniform foundations of any of the types described above are not found. As such if the dams are to be built at these sites, the same will have to be constructed on nonuniform foundations of rock and soft material. However, such unsatisfactory conditions of foundation can be overcome by adopting special designs or appropriate foundation treatments.

3. Availability of Construction Materials. The choice of the type of dam also depends on the types of construction materials that may be available in sufficient quantity at or near the dam site. This is so because a dam constructed with locally available materials will be the most economical due to considerable reduction in the transportation costs for the construction materials. Thus, availability of suitable aggregate viz., sand and gravel or crushed stone for concrete is a factor favourable to the constructions of a concrete dam. On the other hand if suitable soils are available, the choice may be for an earthfill dam.

4. Spillway Size and Location. Spillway is a structure provided for a dam to pass surplus or flood water which cannot be contained in the reservoir. The size and type of spillway are mainly decided by the magnitude of the flood to be byepassed and its location depends on the site conditions. The choice of the type of dam is also affected by size, type and location of a spillway. Thus if a large spillway is required to be provided, then generally spillway and dam are combined into one structure, in which case a concrete dam with overflow and non-overflow sections may be adopted. On the other hand if small spillway is required then even in narrow dam sites the choice may be in favour of earthfill or rockfill dams.

An overflow type of spillway in general favours the selection of a concrete or masonry dam while other types of spillways generally have earthfill or rockfill dams as the first choice. However, nowadays overflow concrete or masonry spillways are also provided for embankment dams, though several problems are associated with such designs.

The spillway may be located either away from the dam or within the limits of the dam. When a spillway is located at a site away from the dam, then the entire dam can be non-overflow type and the choice may include rigid as well as non-rigid dams. In some of these cases a separate spillway channel may be required to be excavated, then an earthfill dam may be selected to utilize the excavated material. On the other hand if the spillway is to be located within the limits of the dam then the choice of the type of dam will have be such that it can include an overflow type of spillway. In this case the spillway can be placed so as to occupy only a portion of the main river channel and the remainder could be non-overflow dam of earth, rock or concrete.

5. Environmental Considerations. The environmental considerations can also have a major influence on the choice of the type of dam. The selection of the type of dam, its dimensions and location of spillway and other appurtenances should be such that there are no adverse effects on the environment and as far as possible maximum protection is provided for the environment. For example, a particular river might be having scenic and/or recreational quality for most of its length, which will be spoiled completely if a high dam is constructed and a big reservoir is created in which considerable land will be submerged. A possible alternative in this case may be to construct a low diversion dam at the same site and divert water to an off-stream reservoir created at a suitable site away from the river so that environmental qualities of the river are preserved. Thus, it would be necessary to consider one or more alternative designs and select the one which would provide maximum environmental protection.

6. Earthquake Zone. If a dam is to be constructed in an area that is subject to earthquake shocks then the selection of the type of dam should be such that it is able to resist the earthquake shocks without damage. Although by including the provisions for the added loading due to earthquake in the design any type of dam may be adopted in these areas, but earthfill and concrete gravity dams are the best suited types of dams in this respect.

7. Cost. The overall cost of construction of dam and the cost of subsequent maintenance would often become the deciding factors in the choice of the type of dam. The cost of construction of dam is affected by the availability and price of construction materials and labour, while the cost of subsequent maintenance mainly depends on the nature of the construction materials. Thus, if the choice is between gravity and buttress dams, then though the concrete required for a buttress dam is much less as compared to that in gravity dam, the cost of special type of form work and reinforcement required for a buttress dam should also be taken into consideration to arrive at the final decision. Similarly the initial cost of construction of an earthfill dam may be less but the cost of subsequent maintenance would be more than that of a concrete dam and hence this factor would require consideration in the final selection of the type of dam.

8. General Considerations. The choice of the type of dam also depends on various other considerations such as problem of diverting the stream flow during construction, availability of labour and equipment, inaccessibility of sites, limitation of outlet works and cost of protection from spillway discharges.

Unless a dam is so small that it can be constructed in a single dry season there must be some means of diverting the stream flow during construction usually through a tunnel in the abutment that can be closed after construction. However, at any dam site if diversion of stream flow is not possible then earthfill dam cannot be constructed and the choice will be in favour of masonry or concrete dam. This is so because it is possible to pass flood flows through a partially constructed masonry or concrete dam during construction without causing much damage to the dam section.

The type of the dam may also be selected to utilize the locally available labour. Nagarjunasagar dam in India was built as a stone masonry dam mainly due to the consideration of providing employment for local labour.

8.4 SELECTION OF SITE FOR A DAM

The selection of a suitable site for the construction of a dam depends on various factors which are briefly described below.

1. Suitable foundations as described in section 8.3 should be available at the dam site. It is, however possible to improve the foundation conditions by adopting appropriate foundation treatments.

2. For economy it is necessary that the length of the dam should be as small as possible and for a given height it should store large volume of water. It, therefore, follows that the river valley at the dam site should be as narrow as possible and it should open out upstream to create a reservoir with as far as possible large storage capacity as shown in Fig. 8.1 (a). Often the dam is located on the downstream of the confluence of two rivers, so that advantage of both the valleys to provide larger storage capacity is available as shown in Fig. 8.1 (b).

3. As far as possible the dam should be located on high ground as compared to the river basin. This will reduce the cost and facilitate drainage of the dam section.

4. A suitable site for the spillway should be available in the vicinity of the dam if the spillway is to be located separately from the dam. However, if the spillway is to be an integral part of the dam then there is no special site requirement for the spillway.

5. From the standpoint of economy the bulk of the materials required for the construction of dam should be available at or near the dam site.

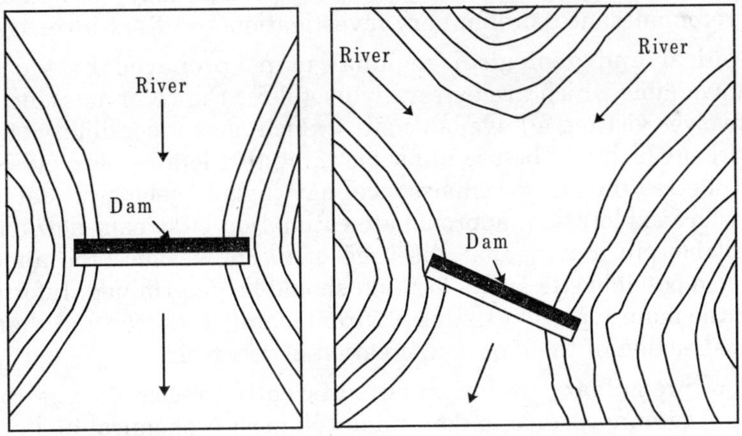

(a) Narrow valley (b) Confluence

Fig. 8.1. *Suitable site for a dam*

6. Immediately on the upstream of the dam site there should be a watertight rim for the reservoir formed by the surrounding hills upto the proposed elevation of the dam. However, if there are some gaps in the surrounding hills, the same will have to be plugged by suitable types of subsidiary dams to prevent escape of water from the reservoir. In some cases if a low saddle is available in the surrounding hills, it may be utilized for providing a separate spillway. Further the reservoir basin should also be reasonably watertight and the stored water should not be able to escape under the surrounding hills through cavernous rock or other continuous previous strata.

7. The value of the property and land submerged in the reservoir created by the proposed dam should be as low as possible.

8. The dam site should be such that the reservoir would not silt up soon. For this if any of the tributaries of the river is transporting relatively large quantity of sediment, then the dam site may be selected on the upstream of the confluence of this tributary with the river.

9. It is preferable to select a dam site which is already connected or can be conveniently connected to a nearby rail head by road or rail, so that the dam site is easily accessible. This would facilitate transportation of men, material, machinery and various other essential items to the dam site.

10. In the near vicinity of the dam site ample space with healthy environment must be available for establishing colonies for labour and other staff members associated with the construction of dam.

11. The dam site should be such that it involves minimum overall cost of construction as well as minimum cost of subsequent maintenance.

8.5 PRELIMINARY AND FINAL INVESTIGATIONS OF DAM SITES

A thorough investigation is required to be carried out to determine the most suitable site for a dam. Such an investigation will include surveys, topographic mapping, geologic studies, subsurface exploration and hydrologic studies. These investigations may be broadly classified under three heads viz., reconnaissance, preliminary investigation and final investigation.

The topographical and geological regional maps prepared by various Government organisations are available which are extremely useful for the reconnaissance work. The reconnaissance work involves visiting all available sites which have a possibility of being utilized and gathering information which will be useful for planning the detaiied surveys and investigations. The information collected during reconnaissance may include geological data obtained without any kind of subsurface exploration, approximate estimation of stream flow data, storage capacity and head available etc. Further, at the time of reconnaissance the approximate spillway requirement and also possible site for its location should be determined. Information should also be collected about the main types of existing environmental resources and changes likely to take place after the construction of the dam and creation of reservoir.

In the course of preliminary investigation sufficiently precise data is collected at several sites selected during reconnaissance to determine the most economical and suitable site among these. The preliminary investigation usually requires the following items.

(i) A not too precise site survey with the resulting topographic site map.

(ii) Some investigation of the overburden.

(iii) A few borings, say from 6 to 50, according to the size of the dam.

(iv) A preliminary geologic investigation and corresponding report.

(v) Investigation of available construction materials, such as earth and gravel and concrete aggregates.

(vi) Determination of public utilities such as roads, railroads, telephone and telegraph lines, pipe lines and power plants which might be affected by the construction of dam.

(vii) Hydrologic studies.

(viii) Determination of the quantity of silt carried by the stream.

(ix) Checking of high-water marks for their use in determining spillway capacity requirements.

After the preliminary investigations at the several possible dam sites have been made and office studies and estimates for each of them completed, one of the several sites is selected for final, precise investigation. The final investigation involves the following principal items.

(i) A sufficiently precise site survey and preparation of topographical maps to serve all the purposes of design and construction of dam.

(ii) Accomplishment of necessary borings, test pits, sub-surface explorations, geologic studies and tests on the materials in foundation and in the proposed borrow lands.

(iii) Determination of the type of dam to be used.

(iv) Planning the foundation treatment on the basis of sub-surface investigation.

(v) Demarcation of lands for the sites of structures and for other necessary purposes.

(vi) Determination of the extent of land, villages etc., which would be submerged in the reservoir and the arrangements for rehabilitation for the residents of that area.

(vii) Determination of the extent and character of relocation of railroads and public highways necessary on account of the creation of reservoir.

(viii) Obtaining sufficient information for an accurate estimate of cost.

(ix) Determination of the final location of the dam, construction equipment, colonies for labour and other staff members, coffer dams, construction highways and railroads, as well as the probable source of materials of construction and all other information needful to the construction engineer.

(x) Obtaining all necessary information affecting the design of the dam.

It may however be mentioned that there is no distinct line of demarcation between the preliminary and the final investigations of dam site and often one blends into the other.

8.6 SALIENT FEATURES OF IMPORTANT DAMS OF INDIA

The salient features of some of the important dams of India are indicated below.

Bhakra Dam

Location : Across river Sutlej at the foot of Shivalik Hills in Himachal Pradesh

Type : Straight gravity, concrete dam

Length : 518 m

Max. Height : 226 m

Reservoir : Gross Storage : 9867.8 M.cu.m
Live storage : 7770.9 M.cu.m

Benefit : Irrigation: 1.48 M.ha (In Haryana, Punjab and Rajasthan)
Power : Left bank 5 units of 90 MW each, Right bank 5 units of 120 MW each

Nagarjunasagar Dam

Location : Across river Krishna near Nandikonda village in Nalgonda district about 144 km from Hyderabad in Andhra Pradesh

Type : Stone masonry and earthen dam

Length : Stone masonry : 1450 m, Earthen : 3414.6 m

Max. Height : 124.7 m

Reservoir : Gross storage : 11538.7 M.cu.m., Live storage: 6797 M.cu.m

Benefit : Irrigation : 0.83 M.ha

Rana Pratap Sagar Dam

Location : Across river Chambal 51.5 km upstream of Kota Barrage and 56.33 km downstream of Gandhi Sagar Dam

Type : Stone masonry gravity dam

Length : 1143.29 m

Max. Height : 53.96 m

Reservoir : Gross storage : 2900 M.cu.m., Live storage : 1567 M.cu.m

Benefit : Irrigation : 0.1214 M.ha (In Madhya Pradesh and Rajasthan from Kota Barrage), Power: 4 units of 43000 kW each

Gandhi Sagar Dam

Location : Across river Chambal about 8 km downstream of Chaurasigarh fort in Mandsaur district near Jhalawar Road Railway station in Madhya Pradesh

Type : Straight gravity stone masonry dam
Length : 514 m
Max. Height : 63.7 m
Reservoir : Gross storage: 7746 M.cu.m., Live storage: 6920 M.cu.m
Benefit : Irrigation: 222 585 ha. (In Madhya Pradesh from Kota Barrage), Power : 5 units of 23000 kW each

Tawa Dam

Location : Across river Tawa, a tributary of Narmada river, 822.96 m downstream of Tawa and Denwa confluence in Hoshangabad district of Madhya Pradesh
Type : Earth-cum-masonry
Length : Stone masonry (including spillway) : 1630.22 m + 2 dykes (earthen) : 183 m each
Max. height : 57.95 m
Reservoir : Gross storage : 2297.99 M.cu.m., Live storage: 2074.51 M.cu.m
Benefit : Irrigation : 32800 ha

Hirakud Dam

Location : Across river Mahanadi 9.7 km from Sambalpur in Orissa
Type : Earth, Masonry and Concrete dam
Length : Main dam : 4801.2 m., Earthen dykes : 20661 m
Max. Height : 61 m
Reservoir : Gross storage : 8100 M.cu.m., Live storage : 5822 M.cu. m
Benefit : Irrigation : 496567 ha., Power : 4 units of 37500 kW each, 5 units of 24000 kW each

Ramganga Dam

Location : Across river Ramganga 3.2 km upstream near Kalagarh in District Garhwal of Uttar Pradesh
Type : Masonry, Earth and Boulder fill dam
Length : 625.8 m
Max. Height : 125.6 m
Reservoir : Gross storage : 2196 M.cu.m., Live storage : 1941.5 M.cu.m
Benefit : Irrigation : 659000 ha., Power : Installed capacity 198000 kW

Mettur Dam

Location : Across river Cauvery about 51.5 km from Salem Town in Tamil Nadu State
Type : Concrete gravity dam
Length : 1615.85 m
Max.Height : 65.24 m
Reservoir : Gross storage : 2650 M.cu.m
Benefit : Irrigation: 18210 ha., Power : 4 units of 50 MW each

Beas Dam I

Location : Across river Beas at Pandoh at a distance of 76.8 km from Joginder Nagar and 20.8 km upstream of Mandi Town in Himachal Pradesh

Type : Rockfill dam

Length : 254.6 m

Max.Height : 60.98 m

Reservoir : Gross storage : 17.76 M.cu.m., Live storage : 10.36 M.cu.m

Benefit : Irrigation : 0.53 M.ha. (In Punjab and Haryana), Power : 6 units of 165 MW each

Beas Dam II

Location : Across river Beas at Pong 38.62 km from Mukerian, a station on the Delhi-Pathankot Railway line in Himachal Pradesh

Type : Earth core-cum-gravel dam

Length : 1950 m

Max. Height : 115.8 m

Reservoir : Gross storage : 8140.94 M.cu.m., Live storage: 6907.49 M.cu.m

Benefit : Irrigation : 0.65 M.ha. (In Punjab and Haryana), Power : 6 units of 60000 kW each

Iddiki Dam

Location : Across river Periyar in Kottayam district of Kerala

Type : Double curvature arch dam in concrete

Length : 365.76 m

Max. Height : 170.68 m

Reservoir : Gross storage : 2106.8 M.cu.m., Live storage : 1544.5 M.cu.m

Benefit : Power : 3 units of 130 MW each

Maithon Dam

Location : Across river Barakar 25.6 km from Asansol railway station in Bihar

Type : Earthen dam with concrete spillway

Length : 994 m

Max. Height : 49.4 m above river bed

Reservoir : Gross storage : 1357 M.cu.m

Benefit : Designed primarily for flood control. Power : 3 units of 20000 kW each

Koyna dam

Location : Across river Koyna at Deshmukhwadi near Helwak in Satara district of Maharashtra. Dam is about 240 km south of Mumbai

Type : Rubble Concrete dam

Length : 853.44 m

Max. Height : 85.3 m above river bed

Reservoir	: Gross storage : 2775.33 M.cu.m., Live storage : 2676.6 M.cu.m
Benefit	: Power : 4 units of 60 MW each, 4 units of 75 MW each, 4 units of 80 MW each

Rihand Dam

Location	: Across river Rihand near village Pipri in Mirzapur District 46 km south of its confluence with river Sone
Type	: Straight gravity concrete dam
Length	: 934 m
Max. Height	: 91.4 m
Reservoir	: Gross storage : 10600 M.cu.m., Live storage : 8970 M.cu.m
Benefit	: Power : 6 units of 50000 kW each

REVIEW QUESTIONS

8.1 How are dams classified? Discuss in detail.

8.2 Discuss the various factors which govern the selection of type of dam.

8.3 What factors affect the selection of site for a dam? Discuss them briefly.

8.4 Discuss the various investigations required to be carried out to determine the most suitable site for a dam.

8.5 Explain the terms

(i) Coffer dam

(ii) Detention dam

(iii) Rockfill dam

(iv) Tailings dam

(v) Composite dam.

CHAPTER

9 Design and Construction of Gravity Dams

9.1 INTRODUCTION

A *gravity dam* is a solid masonry or concrete structure with its cross-section approximately triangular in shape, so proportioned that the external forces exerted on it are resisted by its own weight (see Fig. 9.1). It is also called a *solid gravity dam* in order to distinguish it from a *hollow gravity dam* or *hollow dam*. A gravity dam is mostly straight in plan and is thus known as *straight gravity dam*. However, in some cases it may be slightly curved in plan. Since all the forces acting on a gravity dam including its self weight are transmitted to the foundation, a sound rock foundation is essentially required for the construction of a gravity dam. The other requirements of site for a gravity dam are indicated in the previous chapter.

Most of the gravity dams are provided with an overflow type of spillway which is constructed as an integral part of the dam. The dam then consists of two sections viz., non-overflow section and overflow (or spillway) section as shown in Fig. 9.1.

For the purpose of layout of dam the axis of dam is taken as the reference line which is defined separately in the plan and in the cross-section of the dam. In plan the *axis of dam* is defined as the horizontal trace of the upstream edge of the top of the dam and it is also called the *base line of dam*. The axis or base line of dam may be straight, slightly curved upstream or a combination of end curves and central straight line to take the best advantage of topographic conditions at the site. In the cross-section of the dam the vertical line passing through the upstream edge of the top of the dam is considered as the axis of the dam.

The *length of dam* is the length measured along the axis of the dam at the top of the dam from one abutment to the other abutment.

The *maximum base width* of the dam is the horizontal distance between the outer points of the heel and toe of the cross-section of the dam.

The *maximum height of dam* or *structural height of dam* is the vertical distance between the lowest point in the foundation and the top of the dam.

Some of these parameters of dam are shown in Fig. 9.1.

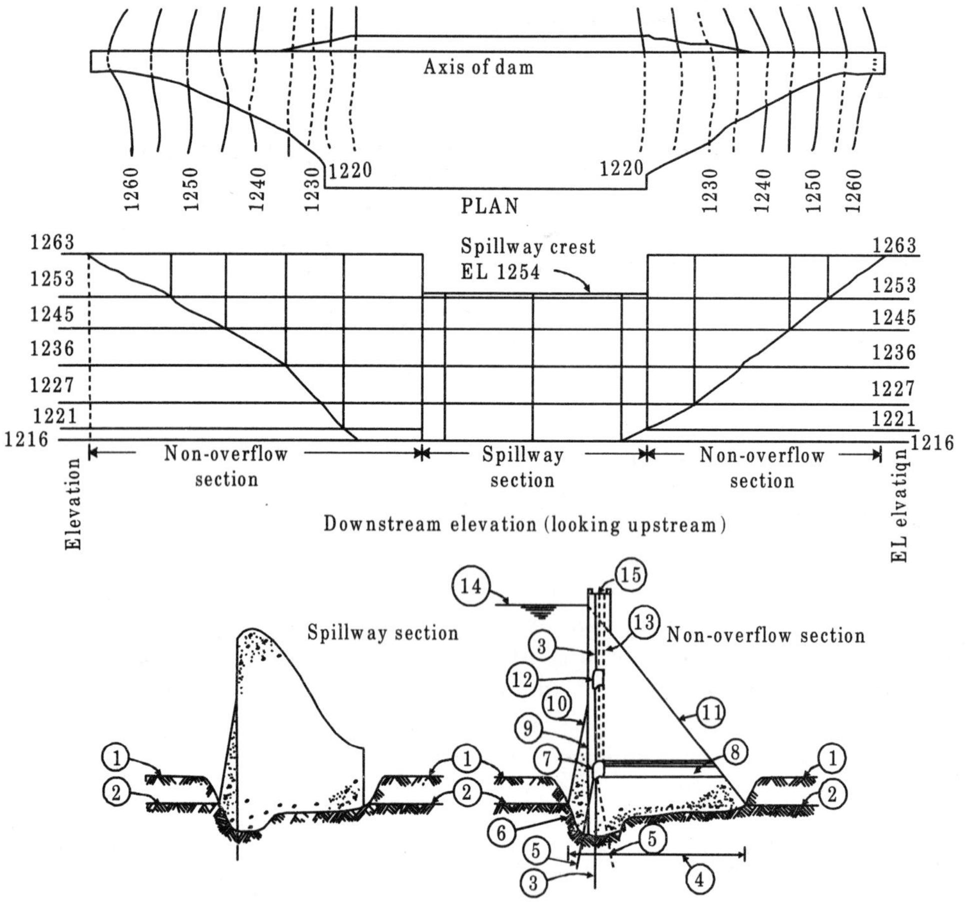

① River bed-level ② Foundation rock level ③ Vertical drain ④ Base width ⑤ Grout hole
⑥ Cutoff trench ⑦ Drainage gallery ⑧ Entrance gallery ⑨ Axis of dam ⑩ Upstream face
⑪ Downstream face ⑫ Inspection gallery ⑬ Inspection shaft ⑭ Maximum water level ⑮ Crest

Fig. 9.1 *Typical layout for a gravity dam*

Grand Dixence dam in Switzerland is the world's highest concrete gravity dam which has the maximum height of 285 m. Bhakra dam in India is the world's second highest concrete gravity dam which has a maximum height of 226 m.

9.2 FORCES ACTING ON A GRAVITY DAM

The various forces acting on a gravity dam are as follows.

(1) Water pressure
(2) Weight of the dam
(3) Uplift pressure
(4) Earthquake (or Seismic) forces

(5) Silt pressure (6) Wave pressure

(7) Ice pressure (8) Wind pressure

Each of these forces are described below.

1. Water Pressure. Water pressure is the force exerted by the water stored in the reservoir on the upstream face of the dam. It can be calculated by the law of hydrostatics according to which in a static mass of liquid the pressure intensity varies linearly with the depth of liquid and it acts normal to the surface in contact with the liquid. For the non-overflow and the overflow (or spillway) sections of the dam water pressure may be calculated as indicated below.

Non-overflow section. For a dam with upstream, face vertical the water pressure is equal to the total pressure on the plane AB as shown in Fig. 9.2 (a). For plane AB, triangle 1 2 3 represents the pressure distribution diagram, with pressure intensity equal to zero at the water

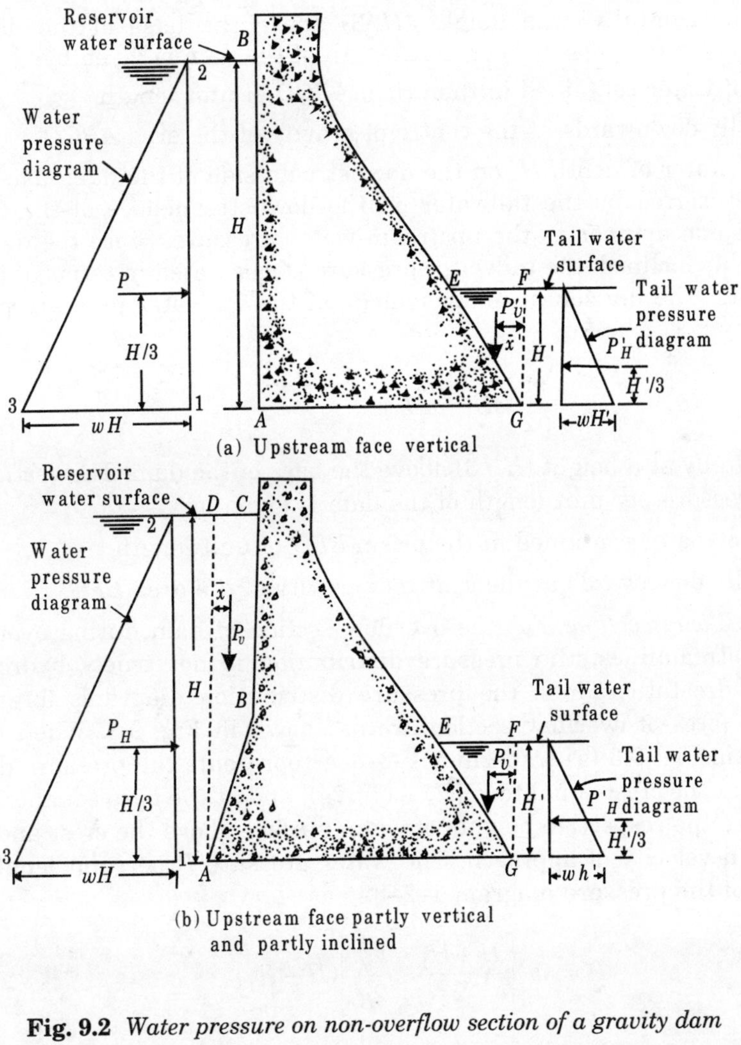

Fig. 9.2 *Water pressure on non-overflow section of a gravity dam*

surface and equal to wH at the base, where w is the specific weight of water and H is the depth of water. The water pressure P, per unit length of the dam is equal to the area of triangle 1–2–3 and is thus given by

$$P = \frac{1}{2} wH^2 \qquad \text{...(9.1)}$$

which acts horizontally at a height $(H/3)$ above the base of the dam.

When upstream face of a dam is either inclined or is partly vertical and partly inclined, as shown in Fig. (9.2b), the water pressure can be conveniently calculated in terms of its horizontal and vertical components. The horizontal component P_H of the water pressure per unit length of the dam is equal to the total pressure on the plane AD and is thus given by Eq. (9.1) as

$$P_H = \frac{1}{2} wH^2$$

which also acts horizontally at a height $(H/3)$ above the base of the dam. The vertical component P_V of the water pressure per unit length of the dam is given by

P_V = Weight of water contained in the prism $ABCD$ of unit length

which acts vertically downwards at the centre of gravity of the area $ABCD$.

If there is tail water of depth H' on the downstream side of the dam, as shown in Fig. 9.2, then the pressure exerted by the tailwater on the downstream face of the dam may also be calculated in the same manner as the upstream water pressure. Since the downstream face of the dam is invariably inclined, the tailwater pressure is calculated in terms of its horizontal and vertical components. The horizontal component P_H' of the tailwater pressure per unit length of the dam is given by

$$P_H' = \frac{1}{2} wH'^2 \, wH'^2$$

which acts horizontally at a height $(H'/3)$ above the base of the dam. The vertical component P_V' of the tailwater pressure per unit length of the dam is given by

P_V' = Weight of water contained in the prism EFG of unit length

which acts vertically downwards at the centre of gravity of the area EFG.

Overflow or spillway section. For the overflow section of dam during overflow, due to the curvature of the streamlines, the pressure distribution is not truely hydrostatic, but it is assumed to be hydrostatic. Thus, the pressure distribution diagrams for the vertical and inclined upstream faces of overflow section are as shown in Fig. 9.3. When upstream face is vertical, as shown in Fig. 9.3 (a) trapezium 1–2–3–4 represents the pressure distribution, with pressure intensity at the crest equal to $w\,[H_1 + (V_a^2/2g)]$ and at the base equal to $w\,[H + (V_a^2/2g)]$, where H_1 and H are the respective heights of water surface above the crest and above the base, and V_a is the mean velocity of approach. The water pressure P per unit length of the dam is given by the area of the pressure diagram 1–2–3–4 as

$$P = w\left[\frac{H + H_1}{2} + \frac{V_a^2}{2g}\right](H - H_1) \qquad \text{...(9.2)}$$

which acts horizontally at a height $\bar{z}$ above the base given by

$$\bar{z} = \frac{1}{3}(H - H_1)\left[\frac{H + 2H_1 + 3h_a}{H + H_1 + 2h_a}\right] \quad ...(9.3)$$

However, if the velocity of approach is negligible then the above noted expressions for P and $\bar{z}$ may be simplified as follows.

$$P = \frac{1}{2} w\,(H^2 - H_1^2) \quad ...(9.2\text{ a})$$

$$\bar{z} = \frac{1}{3}(H - H_1)\left[\frac{H + 2H_1}{H + H_1}\right] \quad ...(9.3\text{ a})$$

If upstream face is inclined as shown in Fig. 9.3 (b), then the water pressure exerted on the upstream face may be resolved into horizontal and vertical components. The horizontal component

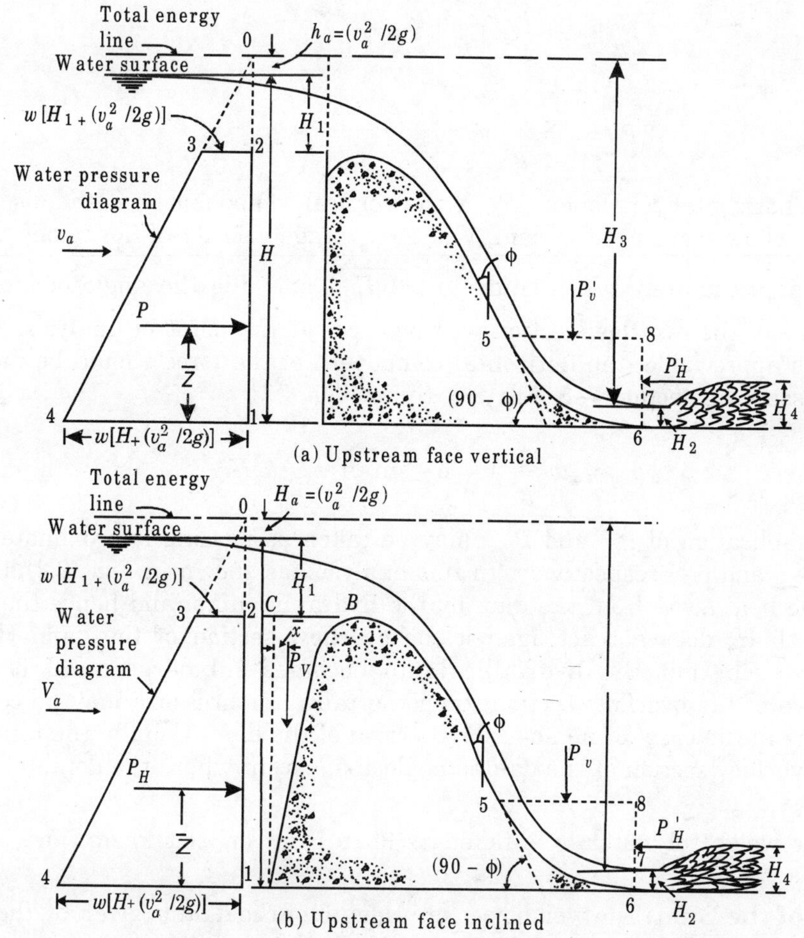

Fig. 9.3 *Water pressure on overflow section of a gravity dam*

P_H of the water pressure per unit length of the dam is equal to the total pressure on the plane AC which is equal to P given by Eq. (9.2), and acting at a height above the base given by Eq. (9.3). The vertical component P_V of the water pressure per unit length of the dam is equal to the weight of water contained in the prism ABC of unit length and acts at its centre of gravity. It may be noted that for computing P_V the weight of water flowing over the crest is not considered in the analysis because the water usually approaches spouting velocity and exterts little pressure on the crest.

Mostly gates are provided on the crest of the overflow section in which case when the gates are closed the water pressure on the overflow section and its point of application can be determined in the same manner as for non-overflow section.

The downstream face of an overflow section is usually curved at the base as shown in Fig. (9.3), to reduce the erosive effect of the falling water. In going around such a curve, the water exerts a force on the over flow section of the dam. For the computation of the horizontal and the vertical components of this force per unit length of the dam, the following approximate equations have been obtained on the basis of momentum principle.

$$P'_H = w \frac{q}{g} V (1 - \sin \phi) \qquad ...(9.4)$$

$$P'_V = w \frac{q}{g} V \cos \phi \qquad ...(9.5)$$

where P'_H is the horizontal component, P'_V is the vertical component, q is the discharge per unit length of the overflow section of the dam, V is the average velocity of flow along the curved face which may be approximately taken equal to $\sqrt{2gH_3}$, and ϕ is the angle of inclination of the downstream face of the overflow section with vertical at the point of analysis. If the depth of tailwater, H_2, is appreciable, the horizontal component of the force should be increased by the hydrostatic pressure on section 7–8, Fig. 9.3, thus,

$$P'_H = w \frac{q}{g} V (1 - \sin \phi) + \frac{1}{2} wH_2^2 \qquad ...(9.6)$$

The points of application of P'_H and P'_V may be taken as passing approximately through the mid-points of 6–8 and 5–8 respectively. In this analysis, as shown in Fig. (9.3), the tailwater is considered to be held away from the dam by the hydraulic jump, and hence the back pressure due to the depth H_4 does not act against the overflow section of the dam. However, if the conditions are such that the hydraulic jump becomes submerged and it forms on the downstream face of the overflow section, then the tailwater pressure may be computed in the same manner as in the case of non-overflow section of the dam. Usually the tailwater pressure acting on the overflow section of the dam is neglected, except where the depth of discharge over the crest is very large.

The specific weight of water w is taken as 9810 N/m^3 or 9.81 kN/m^3, or [1000 kg f/m^3 or 1 t/m^3].

2. Weight of the Dam. The weight per unit length of the dam is given by the product of the area of cross-section of the dam and the specific weight of the construction material viz., concrete or stone masonry and it acts vertically downwards at the centre of gravity of the section. For the sake of convenience the entire cross-section of the dam may be divided into smaller sections of simple geometrical shapes such as triangles, rectangles etc., and the weight of each of these acting at its centre of gravity may be considered.

For a gravity dam the weight of the dam is the main stabilising force, and hence the construction material should be as heavy as possible. The specific weight of concrete and masonry depends considerably on the various materials that go to make them. Thus, in order to obtain heavier concrete/masonry, the coarse aggregate/stones should have greater specific gravity. Further coarse aggregate/stones of larger size should be used to reduce voids and thus to obtain dense concrete/masonry. For preliminary designs, the specific weight of concrete and stone masonry may be taken as 23 544 N/m^3 or 23.544 kN/m^3 or [2400 kg f/m^3] and 22 563 N/m^3 or 22.563 kN/m^3 or [2300 kg f/m^3] respectively. For final designs the specific weights shall be based on actual test data. It is essential to ensure that the actual specific weight obtained for the construction material is more than or at least equal to that assumed in the design.

3. Uplift Pressure. Uplift pressure is the force exerted by water penetrating through the pores, cracks and seams within the body of the dam, at the contact between the dam and its foundation, and within the foundation. It acts vertically upwards at any horizontal section of the dam as well as its foundation and hence it causes a reduction in the effective weight of the portion of the structure lying above this section.

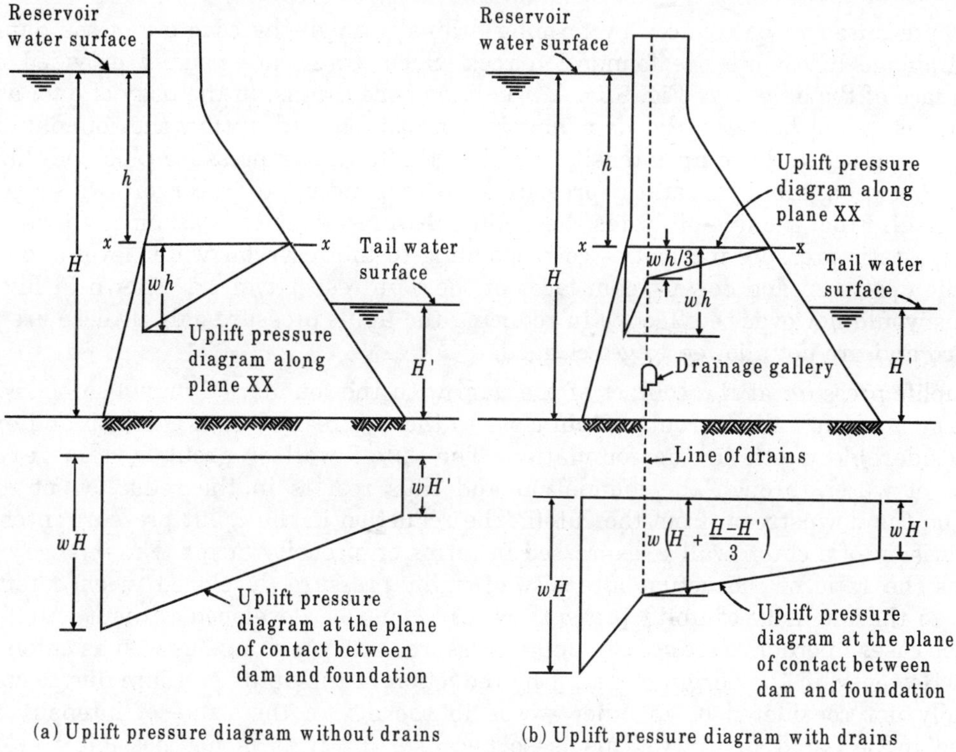

Fig. 9.4 *Uplift pressure diagram without drains and with drains*

The computation of uplift pressure involves the consideration of two constituent elements viz., (i) the area on which the uplift pressure acts and (ii) the intensity of the uplift pressure at various points. Both these elements are discussed below.

The percentage of the area on which the uplift pressure acts is defined as the *area factor*. The value of area factor for concrete has been determined experimentally by several investigators. However, for the foundation rock the value of area factor is not determinable experimentally and hence the same has been estimated on the basis of theoretical considerations. Some of the earlier investigators recommended, for both concrete and rock, a value of area factor ranging from 1/3 to 2/3, i.e., only one-third to two-thirds of the area may be considered as effective area over which the uplift pressure acts. However, Harza, Terzaghi and Leliavasky have indicated that, for both concrete and rock, the value of area factor is nearly equal to unity. As such the present practice followed in the design of dams is that the uplift pressure is assumed to act over 100 percent of the area within the body of the dam as well as its foundation.

It is evident that for the plane of contact between the dam and the foundation, the intensity of uplift pressure at the upstream face of the dam is equal to the hydrostatic pressure corresponding to full reservoir water level and at the downstream face of the dam it is equal to the hydrostatic pressure corresponding to tailwater level, or equal to zero if there is no tailwater. Along this plane the uplift pressure intensities are assumed to vary linearly between these two extreme values at the upstream and downstream faces of the dam, as shown in Fig. (9.4 a). The uplift pressure distribution diagram at any other plane within the body of the dam may also be obtained in the same manner as shown in Fig. (9.4 a).

Uplift pressure can be reduced by forming drains through the concrete of the dam and by drilling drainage holes into the foundation rock. Such drains are usually provided near the upstream face of the dam, (see Fig. 9.1). At the line of the drains, in the body of dam as well as at the contact plane between the dam and its foundation and within the foundation, uplift pressure is assumed to have an intensity equal to the tailwater pressure plus one-third of the difference between the hydrostatic pressures corresponding to reservoir water level and tailwater level. From the line of drains the uplift pressure intensity is assumed to vary linearly to the hydrostatic pressure intensities corresponding to the reservoir water level and tailwater level at the upstream and downstream faces of the dam respectively, as shown in Fig. (9.4 b). The drains would however be effective in reducing the uplift pressure only if these are properly maintained and are not allowed to get clogged.

The uplift pressure at the contact of the dam with the foundation may also be reduced by constructing a cutoff wall or grout curtain close to the upstream face of the dam and extending it for considerable depth in the foundation. The cutoff wall or grout curtain reduces the underflow of water through the foundation and thus results in the reduction of the uplift pressure on the downstream from the cutoff. The reduction in the uplift pressure intensity due to the provision of a cutoff wall is expressed in terms of intensity factor. The *intensity factor* is defined as the ratio of the actual intensity of uplift pressure developed when cutoff wall is provided, to the intensity of uplift pressure which would be developed without cutoff wall. In most of the cases in order to reduce the uplift pressure, both the drains as well as cutoff wall are provided. However, in the design of dams the reduction in the uplift pressure due to cutoff wall is generally not considered or in other words in the design the value of intensity factor is considered to be equal to unity. This is so because, firstly it is not possible to determine quantitatively the effectiveness of the cutoff wall in the reduction of the uplift pressure and secondly a cutoff wall is considered to be an additional factor of safety.

The uplift pressure distribution diagram along any horizontal section of the dam needs to be modified if at the section under consideration the reservoir water pressure exceeds the vertical normal stress (computed without uplift) at the upstream face of the dam. In this case it is assumed that along the section a horizontal crack exists which extends from the upstream face towards the downstream face of the dam upto the point where vertical normal stress (computed without uplift) is equal to the reservoir water pressure at that elevation. The uplift pressure is then assumed to be equal to the reservoir water pressure from the upstream face upto the end of the crack and from there it is assumed to vary linearly to the tailwater pressure at the downstream face or to zero if there is no tailwater, as shown in Fig. (9.5).

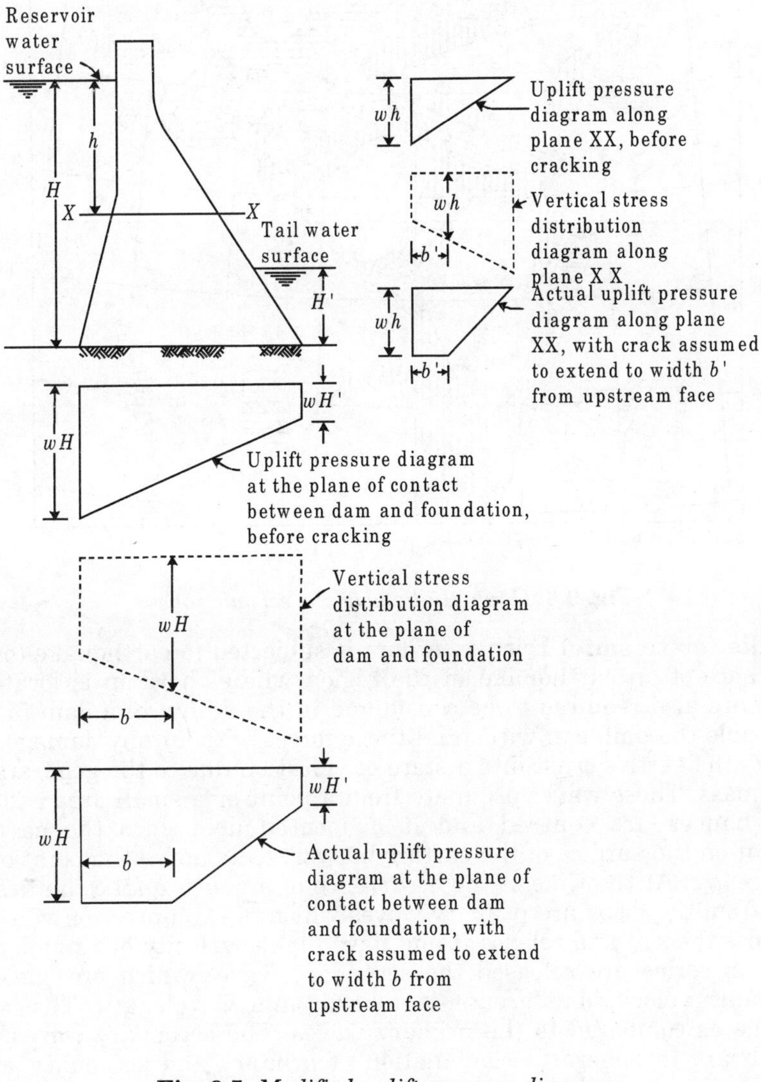

Fig. 9.5 *Modified uplift pressure diagram*

The total uplift pressure U per unit length of the dam at any horizontal section is equal to the area of the uplift pressure distribution diagram at the section obtained as shown in Fig. (9.5) and it acts vertically upwards at the centroid of the uplift pressure distribution diagram.

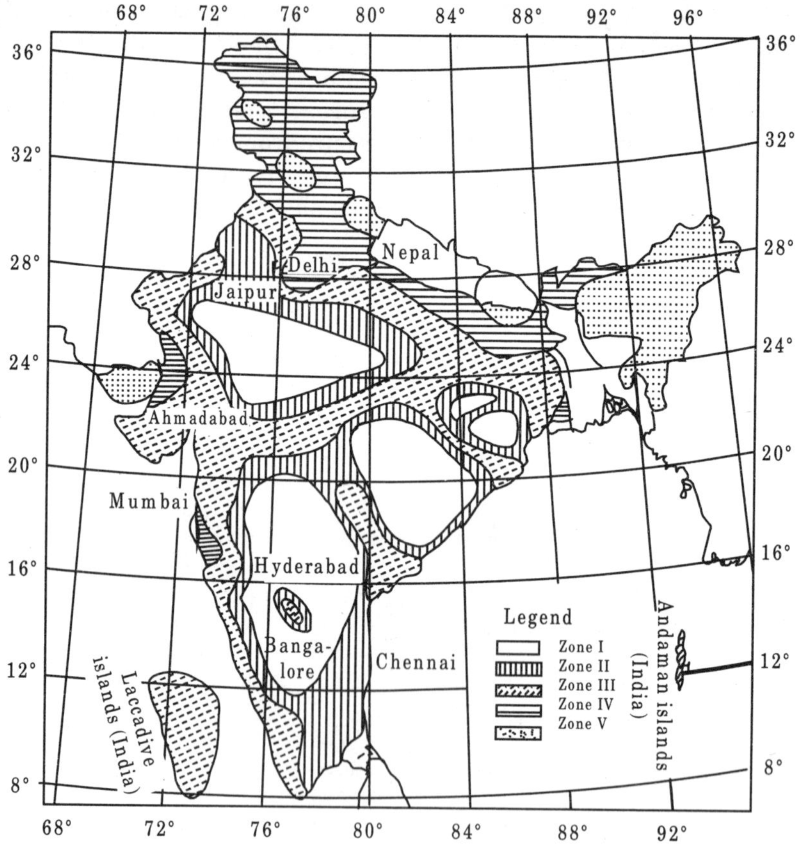

Fig. 9.6 *Map of India showing seismic zones*

4. Earthquake (or seismic) Forces. A dam is subjected to earthquake (or seismic) forces during the occurrence of an earthquake in the region where the dam is located. As such the earthquake forces are also required to be considered in the design of a dam in order to ensure that as far as possible the dam can withstand these forces without any damage being caused to it. Earthquake sets the earth's crust into a state of vibration due to the generation of a series of waves in the soil mass. These waves originate from a point or a small area called *focus* around which the disturbances are centred and it is located deep into the earth's crust. The corresponding point on the surface of the earth lying vertically above the point of disturbance or focus is called *epicentre*. At the time of the occurrence of an earthquake the first type of waves that are released from the focus are primary waves which are compression waves. These waves alternately compress the soil and release it and have higher velocity but smaller amplitude and wavelength. Next in series are released the secondary waves which are shear or transverse waves having smaller velocity but greater amplitude and wavelength. The secondary waves cause more damage as compared to the primary waves. The secondary waves are followed by other waves which are in general a combination of primary and secondary waves. All these waves impart a momentary acceleration to the earth's crust and thus cause its movement in the direction in which the wave is travelling at that instant. Since the dam acts as one with the foundation, the earthquake acceleration imparted to the foundation is also transmitted to the dam. The earthquake forces acting on a dam obviously depend on the earthquake acceleration imparted to it which may be determined as discussed below.

The earthquake acceleration is usually designated as a fraction of g the acceleration due to gravity and it is expressed as αg, where α is known as *seismic coefficient*. The seismic coefficient depends on various factors such as intensity of the earthquake, the part or zone of the country in which the dam is located, the elasticity of the material of the dam and its foundation etc. For the purpose of determining the value of the seismic coefficient which may be adopted in the design of a dam, India has been divided into five seismic zones depending upon the severity of the earthquakes which may occur in the zones. A map of India showing the seismic zones is reproduced from IS : 1893—1984 in Fig. (9.6), which shows that the intensity of earthquake and hence the value of the seismic coefficient gradually increases with a minimum value for zone I to a maximum value for zone *V*. Further as indicated in IS: 1893–1984 the value of the seismic coefficient a may be determined by one of the following two methods.

(a) Seismic coefficient method
(b) Response spectrum method
(a) In *seismic coefficient method*

$$\alpha = \beta I \alpha_0 \qquad \text{...(9.7)}$$

(b) In *response spectrum method*

$$\alpha = \beta I F_0 \left(\frac{S_a}{g}\right) \qquad \text{...(9.8)}$$

where β = Soil-foundation system factor, the value of which for dams is taken as 1.0
I = Importance factor, the value of which for dams is taken as 2.0
α_0 = Basic seismic coefficient, the value of which for each of the five seismic zones is given in Table 9.1
F_0 = Seismic zone factor for average acceleration spectra, the value of which for each of the five seismic zones is given in Table 9.1
$\frac{S_a}{g}$ = Average acceleration coefficient as read from Fig. 9.7 for appropriate natural period of vibration and damping of the structure.

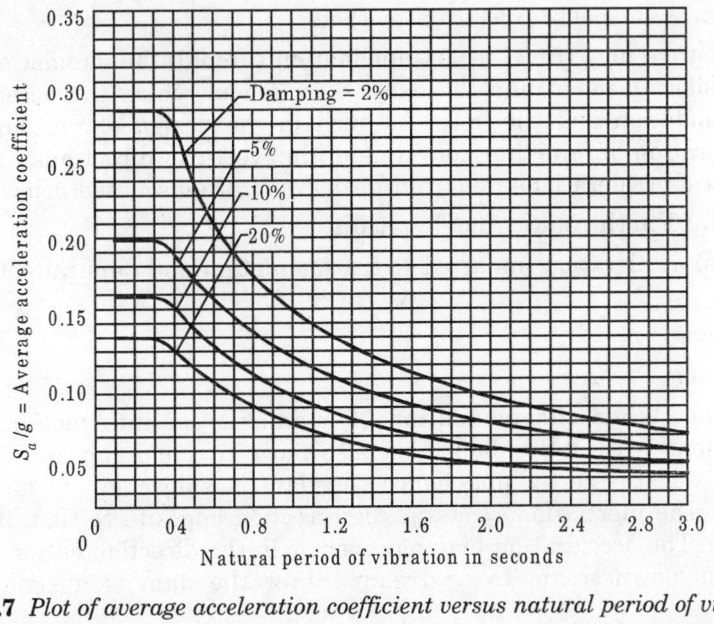

Fig. 9.7 *Plot of average acceleration coefficient versus natural period of vibration*

TABLE 9.1 *Values of basic seismic coefficient and seismic zone factors in different seismic zones*

S. No.	Zone No.	*Method*	
		Seismic coefficient method *Basic seismic coefficient,* α_0	*Response Spectrum method Seismic zone factor for average acceleration spectra,* F_0
1.	I	0.01	0.05
2.	II	0.02	0.10
3.	III	0.04	0.20
4.	IV	0.05	0.25
5.	V	0.08	0.40

The natural (or fundamental) period of vibration of a gravity dam may be determined by the following expression.

$$T = 5.55 \frac{H^2}{B} \sqrt{\frac{w_m}{gE_s}} \qquad \text{...(9.9)}$$

in which

T = Natural period of vibration of the dam in s

H = Height of the dam in m

B = Base width of the dam in m

w_m = Specific weight of the material of dam in N/m^3 [or kg(f)/m^3]

g = Acceleration due to gravity in m/s^2, and

E_s = Modulus of elasticity of the material in N/m^2 [or kg(f)/m^2].

Thus, using the value of T given by Eq. 9.9 and assuming a damping of 5 per cent (as indicated in IS : 1893—1984) the value of (S_a/g) may be obtained which may be introduced in Eq. 9.8 to obtain the value of α by Response Spectrum method.

The earthquake wave may travel at any inclination through the foundation of the dam and hence it is not possible to determine the exact direction in which the acceleration would be imparted to the foundation and the dam. As such in the design of the dams it is usual to consider the acceleration in the horizontal and the vertical directions. The effect of the horizontal and vertical accelerations imparted to the dam are discussed below.

Effect of Horizontal Earthquake Acceleration

Due to the horizontal acceleration imparted to the foundation and dam the following two forces act on the dam.

1. Inertia Force
2. Hydrodynamic Pressure

1. **Inertia Force.** When the foundation of a dam is set into motion by a horizontal acceleration, then along with it the dam must also move to avoid rupture. The dam however opposes the motion due to inertia and hence the dam is subjected to inertia force in the horizontal direction. The inertia force acts in the direction opposite to that of the acceleration imparted to the dam. The acceleration may however be in the direction either from upstream to downstream or from downstream to upstream. Since the dam is designed for the worst

combination of forces, the acceleration is assumed to be in the direction which would produce the worst combination of the forces. Thus, for the case of reservoir full, the worst combination would be produced when the acceleration is in the direction from downstream to upstream so that the inertia force acts on the dam in the direction from upstream to downstream. However, for the case of reservoir empty the acceleration in the direction from upstream to downstream would produce the worst combination. In general the inertia force is equal to the product of the mass of the dam and the acceleration which is thus equal to the product of the weight of the dam and the seismic coefficient, as Inertia Force = Mass × αg = Weight × α. However, the magnitude of the inertia force may be computed by seismic coefficient method for dams upto 100 m height and by Response spectrum method for dams greater than 100 m height as indicated below.

(a) *Seismic coefficient method (for dams upto 100 m height).* In this method a diagram showing the distribution of the horizontal seismic coefficient αh for the entire height of the dam is drawn by taking 1.5 times the seismic coefficient α given by Eq. (9.7) at the top of the dam and reducing linearly to zero at the base (see Fig. 9.8). The inertia force for a part of the dam lying above any horizontal section as well as for the entire dam above its base may be computed as indicated below.

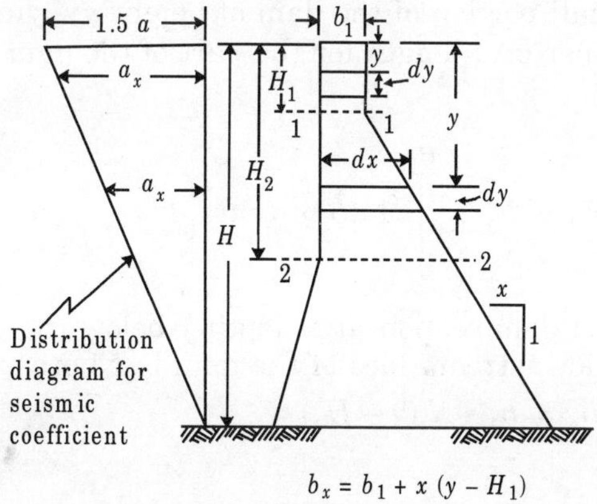

Fig. 9.8 *Determination of horizontal seismic coefficient*

Considering a small portion of the dam of height dy lying above the section 1–1, at a depth y below the top of the dam, as shown in Fig. (9.8), the inertia force dF_1 for this portion of the dam per unit length may be obtained as

$$dF_1 = (1 \times b_1 \times dy)\, w_c \times \alpha_x$$

in which w_c is the specific weight of the construction material of the dam and α_x is the seismic coefficient, the value of which may be obtained from the distribution diagram as

$$\alpha_x = \left[\left(1 - \frac{y}{H}\right) 1.5\alpha\right]$$

By integrating the above expression the inertia force per unit length for the part of the dam lying above the section 1–1 may be obtained as

$$F_1 = \int_0^{H_1} (1 \times b_1 \times dy) w_c \left(1 - \frac{y}{H}\right) 1.5\alpha$$

or $$F_1 = 1.5\alpha\, w_c b_1 H_1 \left(1 - \frac{H_1}{2H}\right) \quad \text{...(9.10 a)}$$

The moment of the inertia force F_1 about any point in the section 1–1 may be obtained as

$$M_1 = \int_0^{H_1} (1 \times b_1 \times dy) w_c \left(1 - \frac{y}{H}\right) 1.5\alpha (H_1 - y)$$

or $$M_1 = 1.5\alpha\, w_c b_1 \frac{H_1^2}{2} \left(1 - \frac{H_1}{3H}\right) \quad \text{...(9.10 b)}$$

Similarly by considering a small portion of the dam of height dy lying between the sections 1–1 and 2–2, the inertia force F_2 per unit length for the part of the dam lying above the section 2–2 may be obtained as

$$F_2 = F_1 + \int_{H_1}^{H_2} (1 \times b_1 \times dy) w_c \left(1 - \frac{y}{H}\right) 1.5\alpha$$

in which b_x is the width of the dam section at a depth y below the top of the dam as shown in Fig. (9.7). If the slope of the downstream face of the dam is x horizontal to 1 vertical then

$$b_x = b_1 + x\,(y - H_1)$$

and hence

$$F_2 = F_1 + 1.5\alpha\, w_c\,(H_2 - H_1) = \left[b_1\left(\frac{2H - H_2 - H_1}{2H}\right) + x\left(\frac{H_2 - H_1}{2} - \frac{2H_2^2 - H_1H_2 - H_1^2}{6H}\right)\right] \quad \text{...(9.10 c)}$$

The moment of the inertia force F_2 about any point in the section 2–2 may be obtained as

$$M_2 = M_1 + F_1\,(H_2 - H_1) + \int_{H_1}^{H} (1 \times b_x \times dy) w_c \left(1 - \frac{y}{H}\right) 1.5\alpha (H_2 - y) \quad \text{...(9.10 d)}$$

in which F_1 and M_1 are given by Eqs (9.10 a) and (9.10 b) respectively.

By adopting the same procedure the inertia force for the entire dam section as well as its moment about any point in the base of the dam may be obtained, which is indicated in Illustrative Example 9.4.

(b) *Response spectrum method (for dams greater than 100 m height).* In this method the value of horizontal seismic coefficient α_h is assumed to be equal to the value of the seismic coefficient α obtained by Eq. (9.8). The inertia force F per unit length of the dam and its moment M about any point in the base of the dam may be obtained by the following expressions:

$$F = 0.6\ W\alpha_h \qquad ...(9.11)$$

$$M = 0.9\ W\bar{h}\ \alpha_h \qquad ...(9.12)$$

where W = Weight of the dam per unit length in N/m [or kg f/m]

α_h = Seismic coefficient obtained by Eq. (9.8), and

$\bar{h}$ = Height of the centre of gravity of the dam above the base in m.

Further for a part of the dam lying above any horizontal section X–X at a depth y below the top of the dam the inertia force F_y per unit length of the dam and its moment M_y about any point in the section X–X may be obtained as follows:

$$F_y = C_F F = C_F\ (0.6\ W\alpha_h) \qquad ...(9.13)$$

$$M_y = C_M M = C_M\ (0.9 W\bar{h}\ \alpha_h\ ) \qquad ...(9.14)$$

where C_F and C_M are the coefficients, the values of which may be obtained from Fig. (9.9).

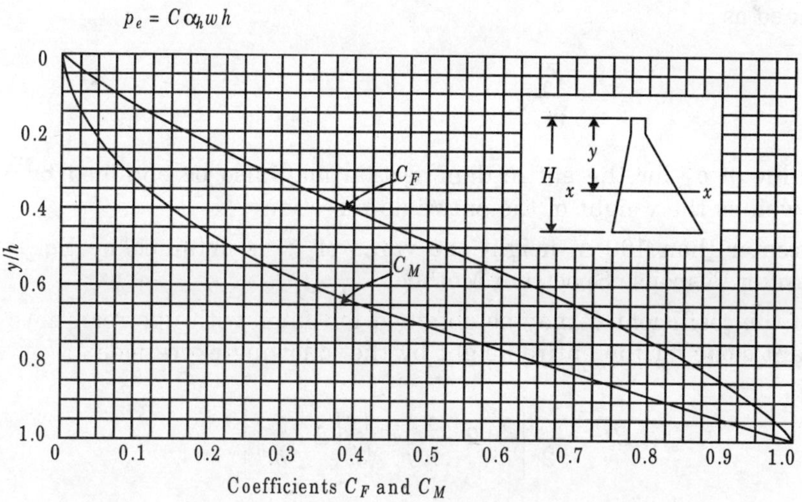

Fig. 9.9 *Coefficients C_F and C_M for different values of (y/H)*

2. Hydrodynamic Pressure. Due to horizontal acceleration the foundation and dam are accelerated either towards the reservoir or away from it depending on the direction of the acceleration being from downstream to upstream or vice versa. However, this movement of the dam is resisted by the water in the reservoir owing to its inertia, on account of which there is an instantaneous hydrodynamic pressure (or suction) exerted on the dam. This hydrodynamic pressure is in addition to the hydrostatic pressure acting on the dam. The direction of the hydrodynamic pressure is opposite to the direction of the earthquake acceleration. For computing the magnitude of the hydrodynamic pressure exerted on the dam, Zanger in 1952 derived certain formulae by electrical analogy, based on the assumption that water is incompressible. These formulae are given in IS : 1893–1975, and the same are indicated below.

The intensity of hydrodynamic pressure at depth y below the water surface in the reservoir is given by the following equation

$$p_e = C\alpha_h wh \qquad ...(9.15)$$

in which

p_e = Intensity of hydrodynamic pressure in N/m^2 [or kg(f)/m^2] at depth y below the water surface in the reservoir

α_h = Horizontal seismic coefficient the value of which may be determined as indicated below

w = Specific weight of water in N/m^3 (or kg(f)/m^3)

h = Total depth of reservoir in m, and

C = A dimensionless coefficient which depends on the shape of the dam and the depth of the reservoir and its value may be determined as indicated below.

For dams upto 100 m height the value of α_h is determined by dividing the inertia force acting on the dam by the weight of the dam. The inertia force is computed by the seismic coefficient method as indicated earlier. Thus, if F_y is the inertia force per unit length for the part of the dam lying above a horizontal section X–X at a depth y below the water surface in the reservoir and W_y is the weight of this part of the dam per unit length then the value of α_h for this part of the dam is obtained as

$$\alpha_h = \frac{F_y}{W_y} \qquad ...(9.16)$$

Similarly the value of α_h for the entire dam section may also be obtained by considering the inertia force as well as the weight of the entire dam section.

For dams greater than 100 m height the value of α_h is obtained by Eq. 9.8 as indicated earlier in the case of Response Spectrum Method.

The value of the coefficient C may be obtained for dams with upstream face either vertical or having constant slope for the entire height by the following expression:

$$C = \frac{C_m}{2}\left[\frac{y}{h}\left(2-\frac{y}{h}\right)+\sqrt{\frac{y}{h}\left(2-\frac{y}{h}\right)}\right] \qquad ...(9.17)$$

where,

C_m = Maximum value of C for a given constant slope of the face of the dam

y = Depth of the horizontal section under consideration below the water surface in the reservoir in m, and

h = Total depth of reservoir in m.

The value of C_m may be obtained from Fig. (9.10) which is a plot of C_m v/s ϕ, where ϕ is the angle in degrees that the face of the dam makes with the vertical. However, an approximate value of C_m is given as

$$C_m = 0.735\left(1-\frac{\phi}{90^\circ}\right) \qquad ...(9.18)$$

Further the value of C as given by Eq. 9.17 may also be obtained from Fig. 9.11 in which the coefficient C is plotted against the ratio (y/h) for different values of ϕ.

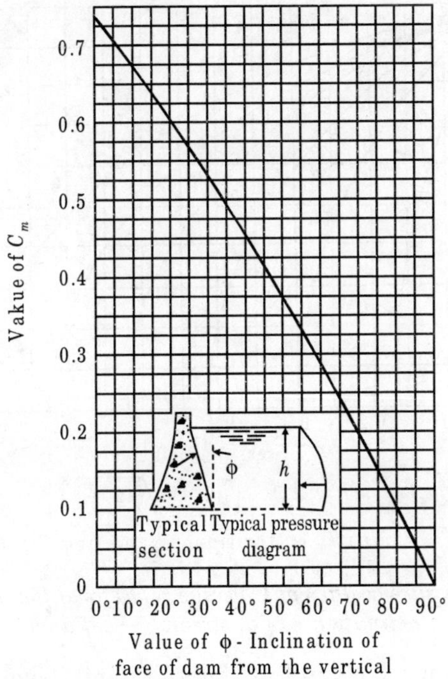

Fig. 9.10 *Plot of maximum pressure coefficient C_m versus angle of inclination ϕ of upstream face of dam*

For a dam with upstream face partly vertical and partly sloping the value of C may be obtained as mentioned below.

(a) If the height of the vertical portion of the upstream face of the dam is equal to or greater than half the total height of the dam then the entire face is considered as vertical.

(b) If the height of the vertical portion of the upstream face of the dam is less than half the total height of the dam, then the slope of the upstream face is considered to be equal to the slope of a line joining the point of intersection of the upstream face of the dam and the water surface in the reservoir; and the point of intersection of the upstream face of the dam and the foundation (i.e., the heel of the dam).

The total hydrodynamic pressure per unit length of the dam acting horizontally for a part of the dam lying above a horizontal section X–X at a depth y below the water surface in the reservoir and its moment about any point in the section X–X are given by the following expression.

$$P_e = 0.726\, p_e\, y \qquad \ldots(9.19)$$

$$M_e = 0.299\, p_e\, y^2 \qquad \ldots(9.20)$$

where P_e = Total hydrodynamic pressure per unit length of the dam in N/m [or kg(f)/m],

M_e = Moment of the total hydrodynamic pressure P_e about any point in the section X–X at depth y in N.m/m [or kg(f)-m/m], and

p_e = Intensity of hydrodynamic pressure in N/m² [or kg(f)/m²] at depth y given by Eq. 9.15.

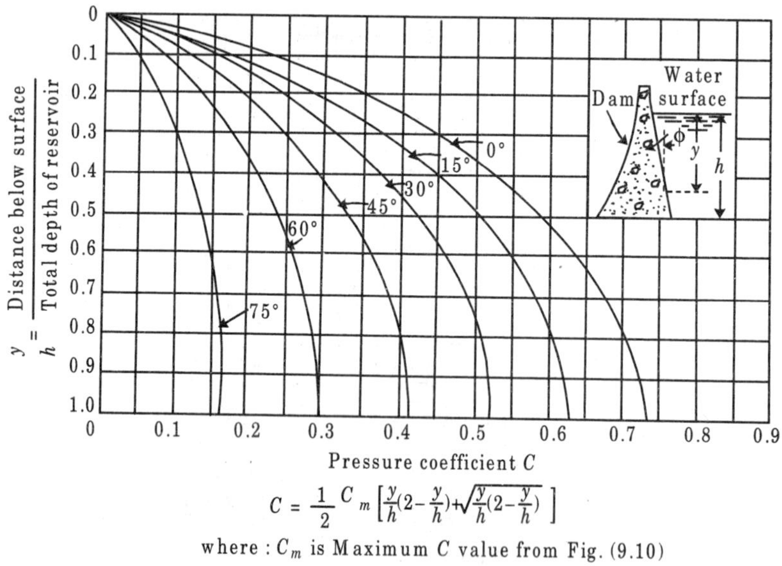

Fig. 9.11 *Plot of pressure coefficient C versus (y / h) for different values of angle of Inclination ϕ of upstream face of dam*

The corresponding values of p_e, P_e and M_e for the dam above its base may be obtained by substituting h for y in the these equations and the same may be expressed as follows.

$$p_{e_h} = C_m \; \alpha_h \, wh \qquad \text{...(9.15 a)}$$

$$P_{e_h} = 0.726 \, p_{e_h} \, h \qquad \text{...(9.19 a)}$$

$$M_{e_h} = 0.299 \, p_{e_h} \, h^2 \qquad \text{...(9.20 a)}$$

If there is tail water on the downstream side of the dam then due to horizontal acceleration hydrodynamic pressure (or suction) will also be exerted on the downstream face of the dam, which may also be computed by these formulae.

The hydrodynamic pressure (or suction) acts normal to the face of the dam. As such if the face of the dam on which this force is acting is sloping then it will have horizontal as well as vertical components. The magnitude of the horizontal component of the hydrodynamic pressure (or suction) is given by Eq. 9.19. However, the magnitude of the vertical component of the hydrodynamic pressure (or suction) for a part of the dam lying above a horizontal section X–X at a depth y below the water surface in the reservoir or the tail water is given as follows.

$$V_e = (P_{e_2} - P_{e_1}) \tan \phi \qquad \text{...(9.21)}$$

where, V_e = Vertical component of the total hydrodynamic pressure (or suction) per unit length of the dam in N/m [or kg(f)/m] for a part of dam upto a horizontal section being considered at depth y below the water surface in the reservoir or the tail water

P_{e_2} = Horizontal component of the total hydrodynamic pressure (or suction) per unit length of the dam in N/m [or kg(f)/m] for a part of the dam upto a horizontal section being considered at depth y below the water surface in the reservoir or the tail water

P_{e_1} = Horizontal component of the total hydrodynamic pressure (or suction) per unit length of the dam in N/m [or kg(f)/m] for a part of the dam upto a horizontal section at which the slope of the face of the dam commences, and

ϕ = Angle between the face of the dam and the vertical.

The moment of V_e about any point in the section X–X may be obtained by determining the location of the point of application of the total hydrodynamic pressure from the hydrodynamic pressure distribution diagram.

Effect of Vertical Earthquake Acceleration

Due to vertical acceleration the dam as well as the water in the reservoir are accelerated vertically upwards or downwards depending upon the acceleration being in the upward or downward direction. However, this movement is resisted by the dam and the water owing to their inertia and hence an inertia force is exerted on the dam as well as on the water in the reservoir in the direction opposite to that of acceleration. The magnitude of this inertia force would be equal to the product of the weight of the dam or water and the vertical seismic coefficient. Thus, if W is the weight per unit length of the dam and α_v is the vertical seismic coefficient then the inertia force exerted per unit length of the dam would be equal to $W\alpha_v$. For an upward acceleration the inertia force $W\alpha_v$ would be acting downwards and hence it would result in an increase in the weight of the dam from W to $(W + W\alpha_v)$ or $W(1+ \alpha_v)$. On the other hand if the acceleration is acting downwards the inertia force $W\alpha_v$ would be acting upwards which would result in the reduction of the weight of the dam from W to $(W - W\alpha_v)$ or $W(1 - \alpha_v)$. Similar expressions may also be obtained for the water in the reservoir. It may therefore be inferred that the effect of the vertical acceleration due to earthquake is to change the weight of the dam and the water in the reservoir and the modified weights may be obtained by multiplying the weight of the dam and the water in the reservoir by a factor $(1 + \alpha_v)$ or $(1 - \alpha_v)$ depending on the acceleration being upward or downward. Since the vertical acceleration does not alter the volumes of the concrete (or masonry) in the dam and the water in the reservoir, the modified weights of the dam and the water may be obtained by considering the modified specific weights of the concrete (or masonry) and water expressed as $wc\,(1 \pm \alpha_v)$ and $w\,(1 \pm \alpha_v)$, where w_c and w are the specific weights of concrete (or masonry) and water respectively. However, the increase or decrease in the weight of concrete (or masonry) and water may be obtained by siesmic coefficient method for dams upto 100 m height and by Response Spectrum Method for dams greater than 100 m height as mentioned below.

(a) *Seismic Coefficient Method.* In this method a diagram showing the distribution of the vertical seismic coefficient α_v for the entire height of the dam is drawn by taking 0.75 times the seismic coefficient a given by Eq. (9.7) at the top of the dam and reducing linearly to zero at the base.

(b) *Response Spectrum Method.* In this method the distribution diagram of α_v for entire height of the dam is drawn by taking 0.75 times the seismic coefficient α given by Eq. (9.8) at the top of the dam and reducing linearly to zero at the base.

From the distribution diagrams for α_v obtained in both the above noted methods, the increase or decrease in the weight of concrete (or masonry) and water may be obtained by adopting the same procedure as indicated for the computation of inertia force due to horizontal earthquake acceleration.

It may however be noted that in the computation of the forces exerted on a dam due to horizontal earthquake acceleration the effect of the vertical acceleration on the specific weights of the dam material and water is not considered.

Effect of Earthquake Acceleration on Uplift Pressure

The uplift pressure acting at any horizontal plane in a dam is assumed to be unaffected by the earthquake. This is so because the duration of the earthquake is too short to permit the building up of pore pressure in the concrete (or masonry) and rock foundation.

Earthquake Forces for Overflow (or Spillway) Section

The earthquake forces acting on an overflow (or spillway) section of a dam may also be obtained in the same manner as in the case of the non-overflow section indicated earlier. However, in the case of the overflow section of a dam the height of the dam is taken from the base of the dam to the top of the bridge provided above its crest for computing the forces and their moments.

5. Silt Pressure. Practically all streams carry silt along with water which during floods may be in enormous quantity. When a dam is constructed on a stream the silt carried by it will get deposited against the upstream face of the dam and it will exert a pressure on the dam. Silt is treated as a saturated cohesionless soil having full uplift and whose value of the angle of internal friction is not materially changed on account of submergence. Further the experiments have indicated that in a submerged fill the silt pressure exerted on a dam is in addition to the water pressure and that the silt pressure on the dam is reduced due to the reduction in the weight of the silt by submergence. For determining the silt pressure Rankine's formula may be used according to which the total silt pressure P_{s_h} acting per unit length of the dam with its upstream face vertical is given as follows :

$$P_{s_h} = \frac{w_s h_1^2}{2}\left(\frac{1-\sin\phi}{1+\sin\phi}\right) \quad \ldots(9.22)$$

where, w_s = Submerged specific weight of silt in N/m³ [or kg(f)/m²]

h_1 = The depth of silt deposited in m, and

ϕ = Angle of internal friction of silt.

The force P_{s_h} is assumed to act horizontally at a distance of $(2h_1/3)$ below the surface of the silt deposited. If the upstream face of the dam is inclined or partly inclined and partly vertical as shown in Fig. (9.12), then the total silt pressure has a horizontal component P_{s_h} as given by Eq. (9.22) and a vertical component P_{s_v} which is equal to the submerged weight of the silt within the area 1–2–3–4 acting vertically downwards at the centroid of this area.

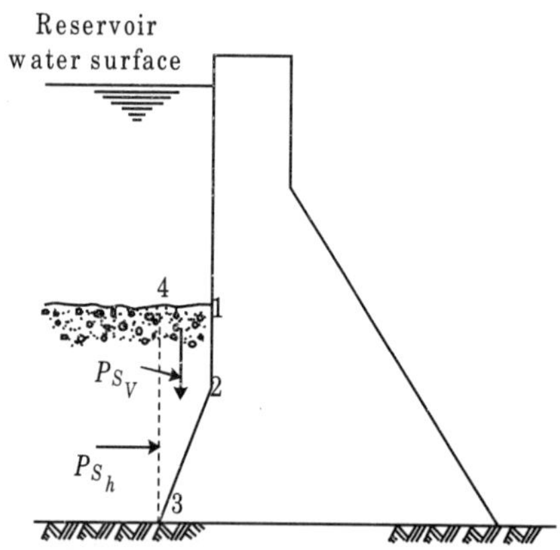

Fig. 9.12 *Silt pressure on a gravity dam*

In Rankine's formula for the computation of silt pressure acting on a dam, submerged weight and angle of internal friction of silt may have to be assumed in the absence of reliable test data, due to which correct value of silt pressure may not be obtained. As such for the determination of the silt pressure acting on a dam another method has been recommended in which the silt pressure acting on the dam is assumed to be equivalent to a pressure exerted on the dam by a liquid comprising a mixture of *silt and water* standing against the dam for a depth upto which the silt would be deposited. The horizontal and vertical components of the combined pressure due to silt and water acting per unit length of the dam may be obtained as follows.

Horizontal component of 'silt and water pressure' acting per unit length of the dam is

$$P'_{s_h} = \frac{1}{2} w'_s h_1^2 \quad \ldots(9.23)$$

which is acting horizontally at a distance of $(2h_1/3)$ below the surface of silt deposited

where w'_s = Specific weight of 'silt and water' in N/m^3 [or kg(f)/m^3], and

h_1 = Depth of silt deposited in m.

Vertical component of the 'silt and water pressure' acting per unit length of dam is

$$P'_{s_v} = \text{Weight of 'silt and water' within the area 1–2–3–4}$$

which is acting vertically downwards at the centroid of this area.

For computing the horizontal and the vertical components of the 'silt and water pressure' different specific weights have been recommended as indicated below.

(a) For horizontal component 'silt and water' is assumed to have specific weight as 13.342 kN/m^3 (1360 kg(f)/m^3).

(b) For vertical component 'silt and water' is assumed to have specific weight of 18.884 kN/m^3 (1925 kg(f)/m^3).

As the specific weight of water is 9.81 kN/m^3[1000 kg(f)/m^3], the specific weights of silt may be obtained in this case as 3.532 kN/m^3 [360 kg(f)/m^3] and 9.074 kN/m^3 [925 kg(f)/m^3] in the horizontal and vertical directions respectively and by using these values of the specific weights in the above expressions the horizontal and vertical components of the pressure due to silt alone acting per unit length of the dam may be obtained.

A higher value of the specific weight of silt in the vertical direction is considered in this case because as more and more silt is deposited it gets consolidated and results in a higher density of the material in the vertical direction.

The depth of silt deposit in the reservoir varies with time and hence its value is suitably assumed for the computation of the silt pressure acting on the dam. The usual practice in this regard is to assume the minimum depth of silt deposit equal to the depth of the dead water storage in the reservoir.

It may however be noted that the increase in the silt pressure due to earthquake is not considered separately but it is assumed to have been taken into account by considering the hydrodynamic pressure of water right upto the base of the dam ignoring the depth of the silt deposit.

In the design of several gravity dams the silt pressure has not been considered. This is so because initially the silt pressure is not present and by the time it might become significant the silt would get consolidated to some extent and therefore it would act less like a liquid. Furthermore, silt deposited in the reservoir would be somewhat impervious and it would help to minimize the uplift pressure under the dam.

6. Wave Pressure. Wind blowing over the water surface in the reservoir exerts a drag on the surface due to which ripples and waves are formed. The impact of these waves produces a pressure on the upper portion of the dam. The magnitude of the wave pressure mainly depends on the dimensions of the waves which in turn depend on the extent and configuration of the water surface, the velocity of wind and the depth of water in the reservoir. The most significant dimension of wave is the height of wave in terms of which the wave pressure exerted on the dam can be expressed.

Earlier as recommended in IS: 6512–1972, for determining the wave height Moliter's formula was used which is as follows:

$$h_w = 0.0322\sqrt{FV} + 0.763 - 0.271\sqrt[4]{F} \quad ...(9.24)$$

where h_w = Height of wave in m

F = Fetch or straight length of water expanse normal to the dam axis in km; and

V = Velocity of wind on the water surface in km/hour.

For $F > 32$ km, Eq. (9.24) may be simplified as

$$h_w = 0.0322\sqrt{FV} \quad ...(9.25)$$

Moliter's formula is, however, particularly suited to the design of structures such as breakwaters and seawalls constructed in oceans in which case the depth as well as water spread area are quite large. On the other hand the artificial reservoirs created by the construction of dams are characteristically deepest near the dam site and gradually decreasing in depth in the upstream direction, the water spread areas are irregular in configuration, open areas are comparatively long and narrow and adjacent terrain is often rugged and covered by forests of varying degrees. Therefore conditions affecting the generation of waves in reservoirs are somewhat different from those prevailing in ocean waters. As such for determining the height of waves produced in reservoirs IS: 6512–1984 has recommended the use of an empirical method given by T. Saville which is based on the research studies on specific cases. This method takes into account the effect of the shape of reservoir and also wind velocity over water surface rather than on land by applying necessary correction. Further in this method the values of different wave heights and the percentage of waves exceeding these height are given so that design wave height for required exceedance can be selected. The details of the procedure to be followed for determining the wave height by the method given by *T*. Saville are given below and illustrated in Illustrative Example 9.7.

Step 1. Select design full reservoir level (FRL).

Step 2. Using reservoir water spread or submergence contour plan showing dam axis and FRL contour work out maximum fetch in km.

Step 3. Select a line *AB,* with *A* on dam axis and *B* on FRL contour so as to cover the maximum reservoir water spread area within 42° angle on either side of line *AB*. Draw 7 rays at 6° interval on either side of *AB* and compute effective fetch. If necessary, do two or three trials, so that the computed effective fetch is maximum.

Step 4. Find out maximum wind velocity on land at dam site from IS: 875 (Pt. 3) – 1987 'Code of Practice for Design Loads for Buildings and Structures'. The maximum wind velocity is assumed to act along *BA*.

Step 5. Convert and velocity on land U_L to wind velocity on water U_w using Table 9.2.

TABLE 9.2 *Value of ratio* (U_w / U_L) *for different values of effective fetch*

Effective fetch (km)	1	2	4	6	8	10	12
(U_W/U_L)	1.10	1.16	1.24	1.27	1.30	1.31	1.31

Step. 6. Using Fig. 9.13 compute significant wave height H_s for the computed wind velocity on water and effective fetch.

Step. 7. Using Table 9.3 obtain design wave height h_w as that specific wave height H which is exceeded by only 4 percent waves. Thus $h_w = 1.27 H_s$.

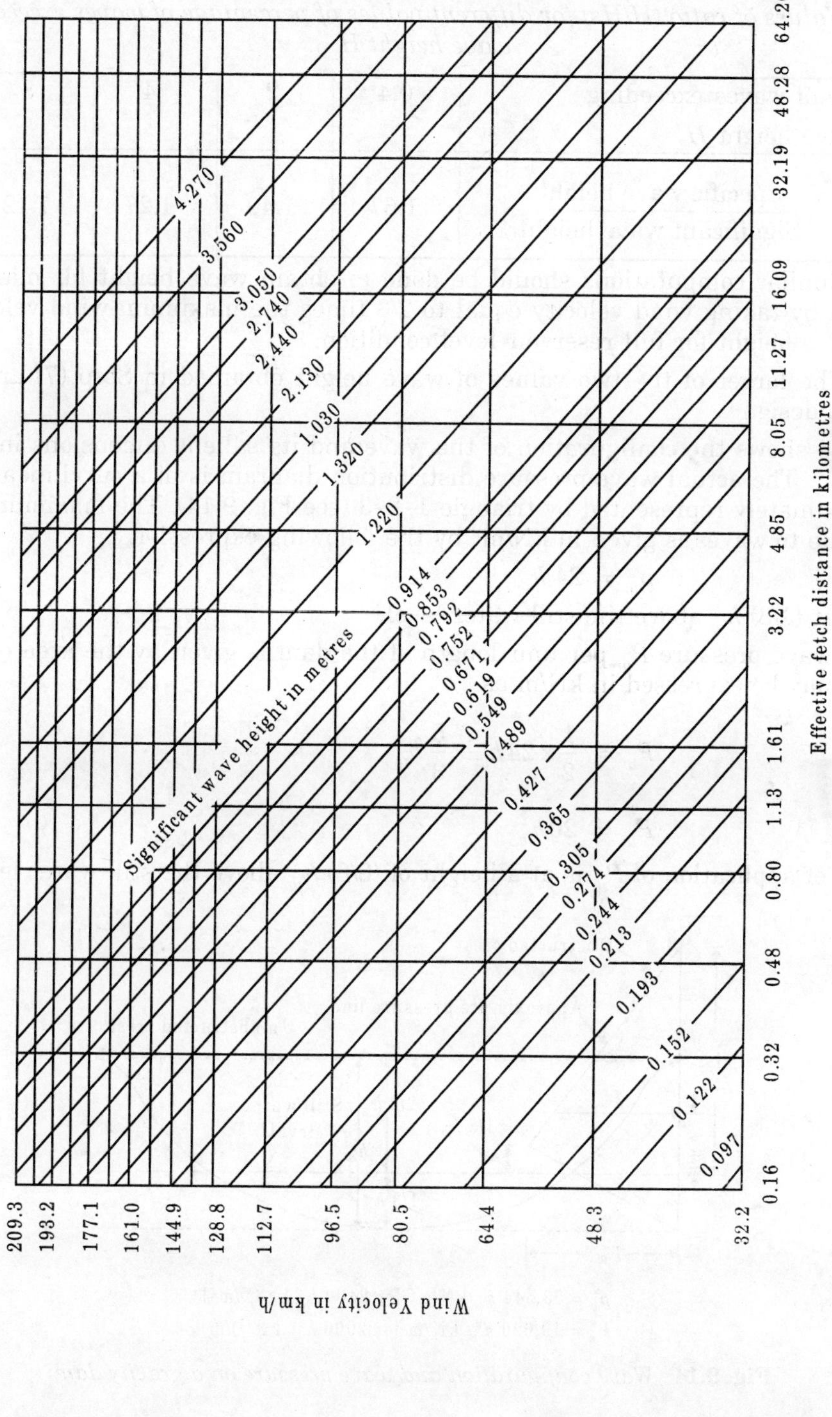

Fig. 9.13 *Correlation of significant wave heights (Hs) with wind velocity and effective fetch*

Higher percentage of wave exceeding H may be considered for small dams not likely to cause damage to life and property in case of breach. Lower percentage may be considered in case of large dams.

TABLE 9.3 *Values of ratio (H/Hs) for different values of percentage of waves exceeding specific wave height H*

Percentage of waves exceeding specific wave height H	0.4	2	4	8	13
Ratio $\frac{H}{H_s}\left(=\frac{\text{Specific wave height}}{\text{Significant wave height}}\right)$	1.67	1.40	1.27	1.12	1.00

Step. 8. Similar computations should be done to obtain wave height for maximum water level condition by taking wind velocity equal to 2/3 times the maximum wind velocity used for calculating wave height for full reservoir level condition.

Step. 9. The larger of the two values of wave height obtained in Step (7) and Step (8) is adopted in the design.

Figure 9.14 shows the configuration of the wave and its salient dimensions in terms of the wave height h_w. The actual wave-pressure distribution diagram is of a curvilinear form, but it may be approximately represented by triangle 1–2–3 (see Fig. 9.14). The maximum intensity of pressure p_w due to waves is given in kN/m² by the following expression

$$p_w = 24\, h_w \qquad \text{...(9.26)}$$

and it occurs at (1/8) h_w above the still water level.

The total wave pressure P_w per unit length of the dam is given by the area of the triangle 1–2–3 and it may be expressed in kN/m as

$$P_w = \frac{1}{2} \times 24 h_w \times \frac{5}{3} h_w$$

or

$$P_w = 20 h_w^2 \qquad \text{...(9.27)}$$

The point of application of P_w is at a height of (3/8) h_w above the still water level.

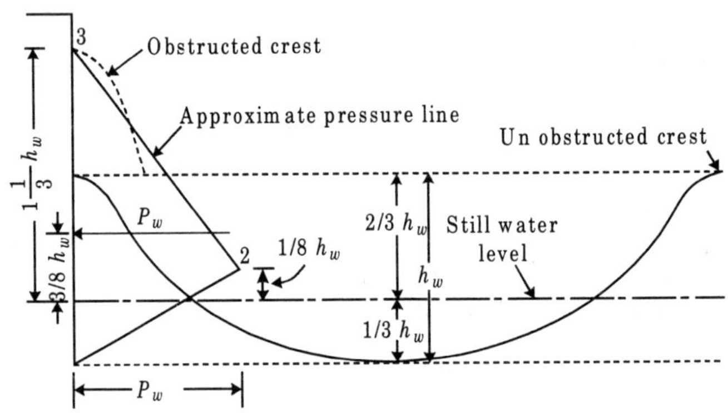

p_w = 23.544 h_w kN/m² [or 2400 h_w kg (f)/m²]

P_w = 19.620 h_w^2 kN/m [or 2000 h_w^2 kg (f)/m]

Fig. 9.14 *Wave configuration and wave pressure on a gravity dam*

7. Ice Pressure. Ice pressure is exerted on a dam by a sheet of ice formed on the entire water surface of the reservoir when it is subjected to contraction and expansion with changes in temperature. In a reservoir having a sheet of ice formed on its entire water surface, if the temperature drops further, the ice sheet may crack due to contraction. These cracks subsequently fill with water and freezed solid. When the next rise in temperature occurs, the ice expands and, if restrained, it exerts pressure on the dam. Ice pressure may also be exerted on a dam due to variation in the water level in the reservoir having ice sheet formed on its entire surface. Further the wind drag acting on the surface of the ice sheet also contributes to the pressure exerted by the ice on a dam.

The magnitude of ice pressure exerted on a dam depends on several variable factors and the existing information on ice pressure is inadequate and somewhat approximate. As such it has been mentioned in IS: 6512–1984 that till specific reliable procedures become available for the assessment of ice pressure it may be provided for at the rate of 250 kN/m^2 applied to the face of dam over the anticipated area of contact of ice with the face of dam. However, in the design of dams in India ice pressure is generally not required to be considered except in a few cases when dams might be located at high altitudes.

8. Wind Pressure. Wind pressure is not a significant force and therefore it is not considered in the design of dams. However, the superstructure of dams carrying very large sluice gates may be subjected to an appreciable amount of wind pressure. As such in the design of a superstructure of a dam wind pressure may be considered at the rate of 1 to 1.5 kN/m^2 over the area exposed to the wind.

9.3 LOAD COMBINATIONS FOR DESIGN

The design of a gravity dam is based on the most adverse combination of the loads (or forces) acting on it, which includes only those loads having a reasonable probability of simultaneous occurrence. The combinations of transient loads such as those due to maximum flood and earthquake are not considered because the probability of occurrence of each of these phenomena is quite low and hence the probability of their simultaneous occurrence is almost negligible. Thus, for the design of gravity dams IS: 6512–1984 specifies the load combination *A, B, C, D, E, F, G* as described below.

(a) *Load Combination A* (*Construction Condition or Empty Reservoir Condition*) – Dam completed but no water in reservoir and no tailwater.

(b) *Load Combination B* (*Normal Operating Condition*) – Full reservoir elevation (or top of gates at crest), normal dry weather tailwater, normal uplift, ice and silt (if applicable).

(c) *Load Combination C* (*Flood Discharge Condition*) – Reservoir at maximum flood pool elevation, all gates open, tailwater at flood elevation, normal uplift, and silt (if applicable).

(d) *Load Combination D* – Combination *A,* with earthquake.

(e) *Load Combination E* – Combination *B,* with earthquake but no ice.

(f) *Load Combination F* – Combination *C,* but with extreme uplift (drains inoperative).

(g) *Load Combination G* – Combination *E,* but with extreme uplift (drains inoperative).

9.4 REACTION OF FOUNDATION AND DISTRIBUTION OF VERTICAL (OR NORMAL STRESS AT THE BASE OF DAM-MIDDLE THIRD RULE

As shown in Fig. 9.15 let ΣV be the sum of all the vertical forces and ΣH be the sum of all the horizontal forces acting on a dam above the base and R be the resultant of all the forces. For the

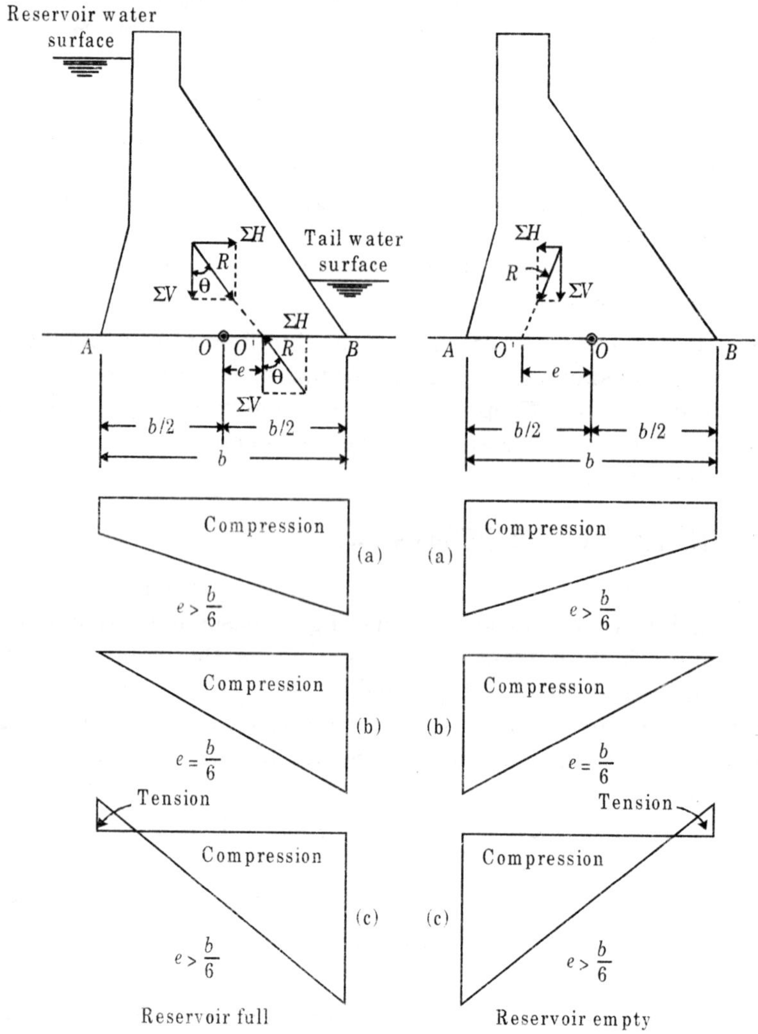

9.15 *Vertical (or normal) stress distribution on the base of a gravity dam*

dam to be in static equilibrium the resultant R, must be balanced by an equal and opposite reaction of the foundation, consisting of the total vertical reaction equal to ΣV and the total horizontal shear or friction equal to ΣH. Further let O be the centroid of the area of the base of the dam and O' be the point of intersection of the resultant with the base. The base of the dam is thus subjected to eccentric loading and hence the vertical (or normal) stress at any point on the base may be obtained as a combination of direct stress and bending stress by the following expression

$$\sigma_y = \frac{\Sigma V}{A} \pm \frac{(\Sigma V)e}{I}x \qquad ...(9.28)$$

where, σ_y = Vertical (or normal) stress at the base

ΣV = Sum of all the vertical forces, including uplift but excluding foundation reaction

A = Area of the base

I = Moment of inertia of the area of the base about an axis passing through its centroid and parallel to the length of the dam

x = Distance from the centroid of the area of the base to the point at which the stress is determined and

e = Distance between the centroid of the area of the base and the point of intersection of the resultant with the base, or eccentricity of loading.

The direct stress at every point on the base is compressive, but the nature of the bending stress depends on the location of the point O'. Thus, for the reservoir full condition the point O' lies between O and B, corresponding to which the bending stress is compressive for any point between O and B and is tensile for any point between O and A. As such for the reservoir full condition in Eq. 9.28 positive sign is to be used for all the points between O and B and negative sign is to be used for all the points between O and A. On the other hand for the reservoir empty condition the point O' lies between O and A and hence the bending stress is compressive for any point between O and A and is tensile for any point between O and B. Accordingly for the reservoir empty condition in Eq. 9.28 positive sign is to be used for all the points between O and A and negative sign is to be used for all the points between O and B.

Equation 9.28 may also be used to obtain the distribution of vertical (or normal) stress at any horizontal section of the dam, in which case the values of the various terms in the equation are considered for the part of the dam lying above the section.

For a straight gravity dam of base width b, if unit length is considered then the base is a rectangle of length b and width 1 m. The centroid of the base is at the mid-point between the heel and the toe of the dam. Also $A = b$, and $I = (b^3/12)$. Substituting these values in Eq. 9.28 gives

$$\sigma_y = \frac{\Sigma V}{b}\left(1 \pm \frac{12e}{b^2}x\right) \qquad ...(9.29)$$

In Eq. (9.29) ΣV thus, represents the sum of all the vertical forces acting per unit length of the dam. Further in this case for the toe and heel of the dam since $x = b/2$, the vertical (or normal) stress at these points may be obtained for reservoir full and empty conditions as indicated below.

For reservoir full

At toe, $$\sigma_{y_D} = \frac{\Sigma V}{b}\left(1 + \frac{6e}{b}\right) \qquad ...(9.29\text{ a})$$

At heel, $$\sigma_{y_U} = \frac{\Sigma V}{b}\left(1 - \frac{6e}{b}\right) \quad \text{...(9.29 b)}$$

For reservoir empty

At toe, $$\sigma_{y_D} = \frac{\Sigma V}{b}\left(1 - \frac{6e}{b}\right) \quad \text{...(9.29 c)}$$

At heel, $$\sigma_{y_U} = \frac{\Sigma V}{b}\left(1 + \frac{6e}{b}\right) \quad \text{...(9.29 d)}$$

Figure (9.15) shows the vertical (or normal) stress distribution at the base as represented by Eq. (9.29) for both the reservoir full and empty conditions. If the value of e is less than $b/6$, then at every point on the base the stress is positive, i.e., compressive for both the cases as shown in Fig. (9.15a). If e is equal to $b/6$, then for the reservoir full condition at heel $\sigma_y = 0$ and at toe $\sigma_y = 2(\Sigma V/b)$; and for the reservoir empty condition at toe $\sigma_y = 0$ and at heel $\sigma_y = 2(\Sigma V/b)$. The corresponding stress distribution is as shown in Fig. (9.15b) which shows that in this case also for both the reservoir full and empty conditions the stress at every point on the base is compressive. Further if the value of e is greater than $b/6$, then at the heel for the reservoir full condition and at the toe for the reservoir empty condition the stress is negative, i.e., tensile which becomes zero at some point on the base and is compressive at the toe for the reservoir full condition and at the heel for the reservoir empty condition as shown in Fig. (9.15c). It is however evident that the magnitudes of both the compressive as well as the tensile stresses increase with the increase in the value of e.

It, thus, follows that if no tension is to be produced at any point on the base of the dam the distance between the centre of the base and the point at which the line of action of the resultant cuts the base should not exceed one-sixth of the base length, i.e., $e \leq (b/6)$. This leads to the well known *middle third rule* which states that for no tension at any point on the base of the dam the resultant for all conditions of loading must cut the base within the middle third of the base.

9.5 PRINCIPAL AND SHEAR STRESS

1. Principal Stresses. A dam section with full reservoir and tail water levels is considered as shown in Fig. (9.16). Both the upstream and the downstream faces of the dam are subjected to water pressure which acts in the direction normal to the face. Since there are no shear stresses acting on both the faces of the dam, these may be considered as the principal planes. Further at any point in a structure the principal planes are mutually perpendicular to each other, and hence the other principal plane at each face of the dam will be at right angles to the face.

Consider a small triangular element *ABC* at the toe of the dam such that the plane *BC* is at right angles to the face *AB*. Thus, *AB* and *BC* are the principal planes which have only normal stresses on them and *AC* is a part of the base of the dam which has both the normal as well as shear stresses on it. The element is so small that the stress intensities on the planes *AB*, *BC* and *CA* may be assumed to be uniform and also in the limit it is reduced to a mere point. Let ds, dr and dx be the lengths of *AB*, *BC* and *CA* respectively, and ϕ_D be the angle of inclination of the downstream face of the dam with the vertical. Further let p' be the intensity of water pressure on the face *AB*; σ_{1_D} be the normal stress on the plane *BC*, which is the principal stress; and

σ_{y_D} and $(\tau_{yx})_D$ be the normal and the shear stresses respectively on the plane *CA*. The value of σ_{y_D} is obtained by Eq. (9.29).

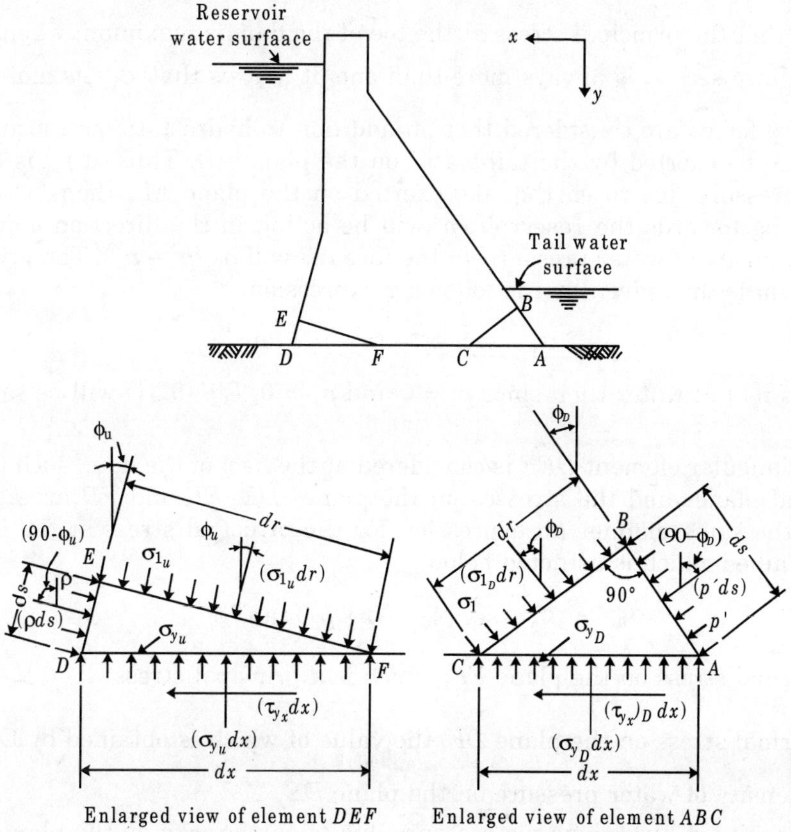

Fig. 9.16 *Principal stress at the heel and toe of a gravity dam*

If unit length of the dam is considered then the normal forces on the planes *AB, BC* and *CA* are $p'ds$, $\sigma_{1_D}\ dr$ and $\sigma_{y_D}\ dx$ respectively. However, on the plane *AC* in addition to normal force there is a shear (or tangential) force $(\tau_{yx})_D\, dx$. Since the element is in equilibrium under these forces, resolving all the forces in the vertical direction and equating their algebraic sum to zero, we get

$$\sigma_{1_D}\ dr\ \cos\phi_D + p'ds\ \sin\phi_D = \sigma_{y_D}\ dx$$

But $$dr = dx\ \cos\phi_D \text{ and } ds = dx\ \sin\phi_D$$

$\therefore$ $$\sigma_{1_D} = \sigma_{y_D}\ \sec^2\phi_D - p'\tan^2\phi_D \qquad ...(9.30)$$

Equation (9.30) gives the principal stress at the toe of the dam. In most of the cases either there is no tail water or if it is there its depth is quite less, so that $p' = 0$ or p' is much less than

σ_{1_D}. As such at the toe of the dam σ_{1_D} is the major principal stress and p' is the minor principal stress. If there is no tail water then $p' = 0$ and Eq. 9.30 becomes

$$\sigma_{1_D} = \sigma_{y_D} \sec^2 \phi_D \qquad ...(9.30\ a)$$

which indicates that the principal stress at the toe of the dam is maximum when there is no tail water. Further since $\sec^2 \phi_D$ is always more than one, it follows that σ_{1_D} is more than σ_{y_D}.

If earthquake forces are considered then in addition to hydrostatic pressure, hydrodynamic pressure will also be exerted by the tail water on the plane *AB*. Thus, if p_e' is the intensity of hydrodynamic pressure due to earthquake exerted on the plane *AB*, then for the earthquake acceleration acting towards the reservoir, p_e' will be acting in the direction opposite to that of p' and the net intensity of water pressure on the face *AB* will be $(p' - p_e')$. The principal stress at the toe of the dam is then given by the following expression.

$$\sigma_{1_D} = \sigma_{y_D} \sec^2 \phi_D - (p' - p_e') \tan^2 \phi_D \qquad ...(9.31)$$

Again if there is no tail water then since $p' = 0$ and $p_e' = 0$, Eq. (9.31) will be same as equation (9.30 a).

If a small triangular element *DEF* is considered at the heel of the dam, such that *DE* and *EF* are the principal planes and the stresses on the planes *DE, EF* and *FD* are as shown in Fig. (9.16), then in the same manner an expression for the principal stress at the heel of the dam may also be obtained which is as noted below.

$$\sigma_{1_U} = \sigma_{y_U} \sec^2 \phi_U - (p + p_e) \tan^2 \phi_U \qquad ...(9.32)$$

where σ_{1_U} = Normal stress on the plane *EF*, which is a principal stress

σ_{y_U} = Normal stress on the plane *DF*, the value of which is obtained by Eq. (9.29)

p = Intensity of water pressure on the plane *DE*

p_e = Intensity of hydrodynamic pressure due to earthquake on the plane *AB*, and

ϕ_U = Angle of inclination of the upstream face of the dam with the vertical.

For the reservoir empty condition since $p = 0$ and $p_e = 0$ Eq. (9.32) becomes

$$\sigma_{1_U} = \sigma_{y_U} \sec^2 \phi_U \qquad (9.32\ a)$$

At the heel of the dam except for the reservoir empty condition the intensity of water pressure p is in general more than the normal stress σ_{1_U}. As such at the heel of the dam p is the major principal stress and σ_{1_U} is the minor principal stress.

If the upstream face of the dam is vertical, i.e., $\phi_U = 0$, then from Eq. 9.32, $\sigma_{1_U} = \sigma_{y_U}$ and the principal plane *EF* coincides with the plane *FD*, thereby indicating that in this case at the heel of the dam the shear stress is equal to zero.

2. Shear Stress. The shear stresses $(\tau_{yx})_D$ and $(\tau_{yx})_U$ at the toe and heel of the dam may be determined by considering the condition for the equilibrium of the elements *ABC* and *DEF* in the horizontal direction. Thus, resolving all the forces acting on the element *ABC* in the horizontal direction and equating their algebraic sum to zero, we get

$$(\tau_{yx})_D\ dx = \sigma_{1_D}\ dr \sin\phi_D - p'ds \cos\phi_D$$

But $\qquad dr = dx \cos\phi_D \quad \text{and} \quad ds = dx \sin\phi_D$

$$\therefore \qquad (\tau_{yx})_D = (\sigma_{1_D} - p')\cos\phi_D \sin\phi_D$$

Substituting the value of σ_{I_D} from Eq. 9.30, we get

$$(\tau_{yx})_D = (\sigma_{y_D} \sec^2\phi_D - p' \tan^2\phi_D - p')\cos\phi_D\ \sin\phi_D$$

or

$$(\tau_{yx})_D = (\sigma_{y_D} - p')\tan\phi_D \qquad \text{...(9.33)}$$

However, if the intensity of hydrodynamic pressure p' due to earthquake is considered then Eq. (9.33) becomes

$$(\tau_{yx})_D = [\sigma_{y_D} - (p' - p'_e)]\tan\phi_D \qquad \text{...(9.34)}$$

Again if there is no tail water then $p' = 0$ and $p'_e = 0$ and hence Eqs 9.33 and 9.34 become same as follows

$$(\tau_{yx})_D = \sigma_{y_D}\ \tan\phi_D \qquad \text{...(9.34 a)}$$

In the same manner an expression for $(\tau_{yx})_U$ may also be obtained which is as noted below

$$(\tau_{yx})_U = [\sigma_{y_U} - (p + p_e)]\tan\phi_U \qquad \text{...(9.35)}$$

Equation 9.35 also indicates that for vertical upstream face since $\phi_U = 0$, $(\tau_{yx})_U = 0$.

Although all these expressions are derived for the toe and heel of the dam, the same may also be used to obtain the principal and shear stresses at the downstream and upstream faces of the dam at any horizontal section, by considering the values of the various terms for the part of the dam lying above the section.

9.6 MODES OF FAILURE OF DAM —STABILITY REQUIREMENTS

A gravity dam must be designed to resist its failure in all the possible ways with ample factor of safety. The various ways in which a gravity dam may fail are as follows :

(1) Overturning.

(2) Sliding.

(3) Overstressing.

1. Overturning. A dam would overturn about its toe, or about the downstream edge of any horizontal section, if the resultant of all the forces acting on the entire dam or on the part of the dam lying above the horizontal section, passes outside the base or the horizontal section of the dam (Fig. 9.17 a). However, such a condition does not arise in actual practice, but a dam may overturn if the toe or the downstream edge of a horizontal section of a dam fails by crushing due to excessive compressive stresses, so that the effective width of the base or that of the horizontal section would be reduced to cause the resultant to pass outside the dam section. It, thus, follows that the failure of a dam by overturning would be preceded by the failure due to excessive compression, which in turn would result if the resultant passes appreciably outside the middle third of the base or that of any horizontal section of the dam. Moreover, if the resultant passes outside the middle third, tensile stress would occur at the heel or at the upstream edge of the dam at any horizontal section. Due to tension a horizontal crack may occur which would result in increasing the uplift pressure as shown in Fig. 9.17 (b) and thus in reducing the stabilising force for the dam. As such a gravity dam may be considered safe against overturning if the criterion of no tension at any point in the dam is satisfied. Furthermore another safety criterion usually adopted is to compute a factor of safety against overturning which is defined as the ratio of the stabilizing moments to the overturning moments about the toe of the dam for the reservoir full condition and about the heel of the dam for the reservoir empty condition, i.e.,

$$\text{F.S.} = \frac{\sum \text{Stabilizing Moments}}{\sum \text{Overturning Moments}} \qquad \text{...(9.36)}$$

The factor of safety against overturning should not be less than 1.5. However, if the dam is safe against compression and sliding then usually a factor of safety between 1.5 and 2.5 is available against overturning.

2. Sliding. The horizontal forces acting on dam tend to slide the entire dam at its base or a part of the dam lying above any horizontal section over the lower part. This tendency to slide is resisted by the sliding resistance developed at the surface of sliding or plane of contact. The resistance to sliding is due to friction at the surface of sliding and cohesion inherent in the materials at contact viz., concrete (or masonry) and foundation rock. For the stability of a dam against sliding the minimum total available resistance to sliding must exceed the total magnitude of the forces tending to cause sliding along any horizontal section of the dam by a safe margin. Thus, for a dam a factor of safety against sliding may be considered which is defined as the ratio of the forces resisting sliding to the forces tending to cause sliding and it may be computed in either of the following two ways.

If the sliding is resisted by friction only then the force resisting the sliding is equal to the product of the coefficient of friction and the vertical reaction which is equal to the sum of all the vertical forces including uplift. The factor of safety against sliding at any horizontal section of the dam is then given by the equation

$$F_1 = \frac{\mu \Sigma V}{\Sigma H} \qquad \text{...(9.37)}$$

where F_1 = Factor of safety against sliding

μ = Coefficient of friction of the materials above and below the section under consideration,

ΣV = Sum of all the vertical forces including uplift acting on the part of the dam lying above the section under consideration, and

ΣH = Sum of all the horizontal forces acting on the part of the dam lying above the section under consideration.

As shown in Fig. (9.14) if θ is the angle between the vertical and the resultant of all the forces, then $\tan\theta$ is known as *sliding factor,* and since $\tan\theta = \left(\frac{\Sigma H}{\Sigma V}\right)$, Eq. (9.37) may also be expressed as

$$F_1 = \frac{\mu}{\tan\theta} \quad ...(9.37\ a)$$

Value of μ for concrete on concrete (or masonry on masonry) and for concrete or masonry on good rock foundation varies from 0.6 to 0.75.

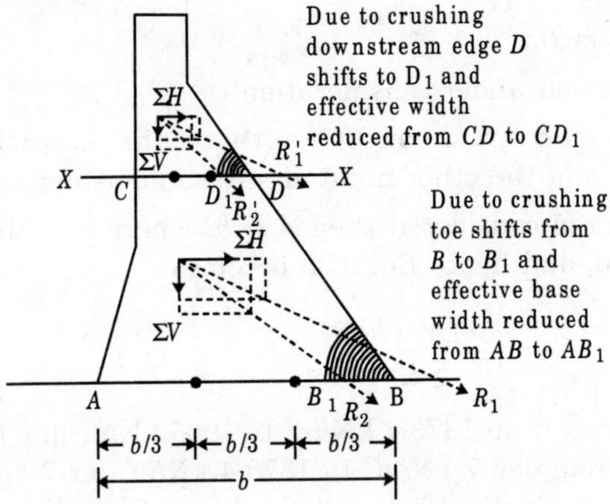

(b) Effect of tension crack on uplift pressure

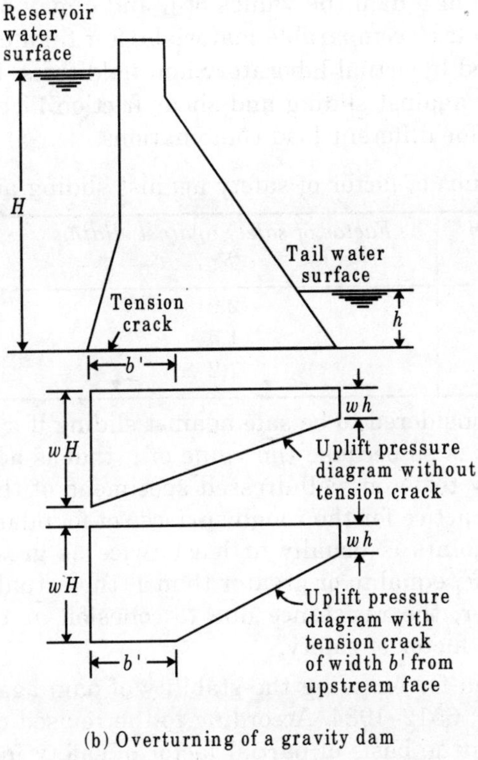

(b) Overturning of a gravity dam

Fig. 9.17

If the sliding is resisted by friction and cohesion or shear strength of the concrete (or masonry) or foundation rock, then the factor of safety against sliding is known as *shear friction factor*. The shear friction factor for any horizontal section of the dam is given by the equation

$$F_2 = \frac{\mu \Sigma V + Ac}{\Sigma H} \quad \text{...(9.38)}$$

where F_2 = Shear friction factor

A = Area of the section under consideration

C = Average cohesion or shear strength of the material at the section under consideration, and the other notation are same as defined under Eq. 9.37.

If a unit length of the dam is considered then $A = b$, where b is the width of the dam at the section under consideration, and hence Eq. 9.38 becomes

$$F_2 = \frac{\mu \Sigma V + bc}{\Sigma H} \quad \text{...(9.38 a)}$$

Value of c for good rock varies from 1373.4 kN/m^2 to 4905 kN/m^2 [or 14 kg(f)/cm^2 to 50 kg(f)/cm^2] and for concrete it varies from 686.7 kN/m^2 to 1373.4 kN/m^2 [or 7 kg(f)/cm^2 to 14 kg(f)/cm^2]. As such for computing the shear friction factor at the base of the dam, out of the cohesion of the foundation rock and the concrete (or masonry) the one which is smaller is to be used in Eq. 9.38.

For the preliminary design of a dam the values of μ and c may be suitably assumed on the basis of available data on similar or comparable materials. For final design, however, the values of μ and c should be determined by actual laboratory and field tests. Further as indicated in IS: 6512–1972 the factor of safety against sliding and shear friction factor should not be less than the values given in Table 9.4, for different load combinations.

TABLE 9.4 Minimum values of factor of safety against sliding and shear friction factor

Sl. No.	*Load combination*	*Factor of safety against sliding* F_1	*Shear friction factor* F_2
1.	*A, B, C*	2.0	4.0
2.	*D, E*	1.5	3.0
3.	*F, G*	1.2	1.5

Generally a dam may be considered to be safe against sliding if a value of F_1 greater than 1 is obtained in the design. This is so because the value of μ that is adopted for computing F_1 is the one obtained by laboratory tests on well-dressed specimens of the same materials but the value of μ obtained in actual practice for the rough surfaces of foundation rock at the base of the dam and at the construction joints is usually at least twice as great as the one obtained by laboratory tests. Therefore for F_1 equal to or greater than 1, the actual factor of safety of at least two will be provided. Moreover, the resistance due to cohesion at the foundation and at the joints will further increase the factor of safety.

The above discussed criteria for ensuring the stability of dam against sliding has, however, been revised as indicated in IS: 6512–1984. According to the revised criteria the factor of safety against sliding is calculated on the basis of partial factor of safety in respect of friction F_ϕ and partial factor of safety in respect of cohesion F_c, the values of which are given in Table 9.5.

TABLE 9.5 Partial factors of safety against sliding

Sl. No.	*Load Combination*	F_ϕ	F_c		
			For Dams and the contact plane with foundation	*For foundation*	
				Thoroughly investigated	*Others*
1.	*A, B, C*	1.5	3.6	4.0	4.5
2.	*D, E*	1.2	2.4	2.7	3.0
3.	*F, G*	1.0	1.2	1.35	1.5

The factor of safety against sliding is then given by the following equation

$$F = \frac{\dfrac{\mu \Sigma V}{F_\phi} + \dfrac{cA}{F_c}}{\Sigma H} \qquad \text{...(9.38 b)}$$

where F = factor of safety against sliding, and the other notation are same as defined earlier.

The factor of safety F against sliding obtained by Eq. 9.38 (b) should not be less than 1.0.

3. Overstressing. A dam may fail if any of its part is overstressed and hence the stresses in any part of the dam must not exceed the allowable working stress for the concrete or masonry. Further in order to ensure the safety of a dam against its failure due to overstressing the strength of the concrete or masonry must exceed the stresses anticipated in any part of the dam by a safe margin which according to IS : 6512–1984 is as indicated below.

(a) Compressive Strength. The compressive strength of concrete and masonry is determined as indicated in section 9.7 and it shall conform to the following requirements:

(i) *Concrete.* The strength of concrete should satisfy early load and construction requirements and at the age of one year the compressive strength should be four times the maximum computed stress in the dam or 14000 kN/m^2, whichever is more. Further the allowable working stress in any part of the dam shall also not exceed 7000 kN/m^2.

(ii) *Masonry.* The strength of masonry should satisfy early load and construction requirements and at the age of one year the compressive strength should be five times the maximum computed stress in the dam or 12500 kN/m^2, whichever is more.

If the stresses at the heel and toe are excessive they can be brought within permissible limits by providing fillets with slopes one horizontal to two vertical at the heel and two horizontal to one vertical at the toe. The height of these fillets above heel and toe are given by the following expressions.

For heel

$$h_f = 6H - 0.658\, H^2 - 1.07$$

For toe

$$h_f = 6.5\, H - 1.1\, H^2 - 0.9$$

where h_f = Height of fillet in metres; and

H = Height of dam in 100 metres.

(b) Tensile Strength. Although concrete and masonry are known to possess some tensile strength, its quantitative evaluation is uncertain and difficult. Moreover, construction joints can be planes of weakness against tensile strength. Further due to tension, cracks may be developed which have significant adverse effects on the structural behaviour of the dam. As such no tensile stress is permitted at the upstream face of the dam for load combination *B* (normal operating condition). However, nominal tensile stresses may be permitted in the case of other load combinations but their values should not exceed the permissible values given in Table 9.6.

TABLE 9.6 Values of permissible tensile stress in concrete and masonry

Load Combination	*Permissible tensile stress*	
	Concrete	*Masonry*
C	$0.01\ f_c$	$0.005\ f_c$
E	$0.02\ f_c$	$0.01\ f_c$
F	$0.02\ f_c$	$0.01\ f_c$
G	$0.04\ f_c$	$0.02\ f_c$

where f_c is the cube compressive strength of concrete/mortar for masonry.

Small values of tension on the downstream face of the dam may be permitted since it is very improbable that a fully constructed dam is kept empty and downstream cracks which are not extensive and for limited depths from the surface may not be detrimental to the safety of the dam.

9.7 QUALITY AND STRENGTH OF CONCRETE AND MASONRY

The strength of concrete and masonry varies with age, the kind of cement and other ingredients as well as their proportions in the work and it can be determined only by experiment. Since different concrete and masonry gain strength at different rates, measurements must be made from specimens of sufficient age to permit proper evaluation. Mix proportions are determined from the results of laboratory tests made with the materials that will be used in the structure. The proportions are selected to produce concrete or masonry of sufficient strength to meet the design requirement multiplied by an appropriate safety factor. Tests are made on concrete or masonry as well as mortar at specified intervals during the progress of the work and the mixes modified as necessary in order to consistently secure the required strength, workability, density and impermeability together with maximum possible economy.

The compressive strength of concrete is determined by compressing to failure 450 m × 900 mm cylinders of full mass mix cured at temperatures approximating to those expected in the structure. However, since it is not practicable to make and test 450 m × 900 mm cylinders for control purposes at the site, representative concrete is wet-screened over a 40 mm screen from which 150 mm × 300 mm cylinder are fabricated and tested. The strength of the 150 mm × 300 mm control cylinders is correlated with the strength of the 450 mm × 900 mm cylinders made in the laboratory on which the selection of the original mix proportions is based.

The compressive strength of masonry is determined by compressing to failure 750 mm cubes of the masonry fabricated and cured at temperatures approximating to those expected in the structure (or 450 mm × 900 mm cylinders cored out of the structure or out of blocks made for the purpose). Further the control of the quality of masonry at the site is exercised by tests on the mortar as the work progresses and proportions adjusted at take care of the changes in materials etc. The mortar strength is determined by compressing to failure 150 mm cubes of the mix which have been cured at temperatures approximating to those expected in the structure. The strength of the control cubes is correlated with the strength of the cubes made in the laboratory on which the selection of the mix proportion is based.

9.8 ELEMENTARY PROFILE OF A GRAVITY DAM

If a gravity dam is subjected to only water pressure, uplift pressure and its self weight, then the stability conditions can be satisfied by a right angled triangular section having zero width at the water surface in the reservoir where water pressure is zero and a base width b at the bottom where the water pressure is maximum. The right angled triangular section of a dam will also provide maximum possible stabilising force against overturning without causing tension at any point in the base for the reservoir empty condition. This is so because when the reservoir is empty the only force acting on the dam is its self weight which acts at a distance of $b/3$ from the upstream face of the dam, and hence satisfies the middle third rule. The rightangled triangular section of a dam is known as the elementary profile of gravity dam. The base width of the elementary profile of a dam for no tension at any point in the base of the dam as well as for no sliding of the dam may however be obtained as indicated below.

Consider an elementary profile of a gravity dam of height H and base width b as shown in Fig. 9.18. The various forces acting on the elementary profile per unit length of the dam are as follows.

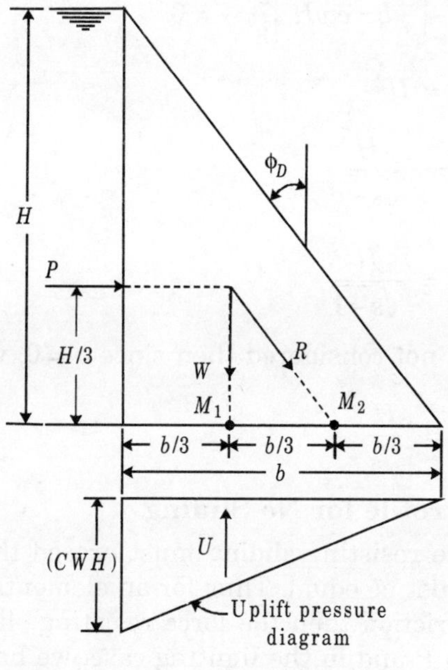

Fig. 9.18 *Elementary profile of a gravity dam*

1. Weight of the dam (W)

$$W = 12\ bH \times sw$$

Acting vertically downwards at $b/3$ from the heel,

where s = Specific gravity of dam material

w = Specific weight of water

2. Water pressure (P)

$$P = \frac{1}{2}wH^2, \text{ acting horizontally at } H/3 \text{ from the base,}$$

3. Uplift pressure (U)

$$U = \frac{1}{2}b\,(cwH), \text{ acting vertically upwards at } b/3 \text{ from the heel,}$$

where c = Uplift pressure intensity factor.

Base Width of Elementary Profile for No Tension

If R is the resultant of all the forces acting on the dam, then for no tension at any point in the base of the dam the resultant R must pass through the middle third of the base. Thus, for the reservoir full condition the outer middle third point M_2 is the limiting position of the point at which the resultant R may meet the base for no tension at any point in the base of the dam. As such assuming the resultant to pass through the outer middle third point M_2, by taking moments of all the forces about M_2 and equating the sum of the moments to zero (since the moment of R about M_2 is zero), we get

$$\left(\frac{1}{2}bH \times sw\right)\frac{b}{3} - \left(\frac{1}{2}wH^2\right)\frac{H}{3} - \left(\frac{1}{2}b \times cwH\right)\frac{b}{3} = 0$$

or $$b^2(s - c) = H^2$$

or $$b = \frac{H}{\sqrt{s-c}} \qquad \text{...(9.39)}$$

If, $$c = 1,\ b = \frac{H}{\sqrt{s-1}} \qquad \text{...(9.39 a)}$$

Further if uplift pressure is not considered then since $c = 0$, we get

$$b = \frac{H}{\sqrt{s}} \qquad \text{...(9.39 b)}$$

Base Width of Elementary Profile for No Sliding

For no sliding to occur the force resisting sliding must exceed the force causing sliding and in the limiting case these forces must be equal. Thus for an elementary profile if it is assumed that the sliding is resisted only by friction then the force resisting sliding is equal to $\mu(W - U)$; the force causing sliding is equal to P and in the limiting case, we have

$$\mu(W - U) = P$$

or $$\mu\left[\left(\frac{1}{2}bH \times sw\right) - \frac{1}{2}b(cwH)\right] = \frac{1}{2}wH^2$$

or $$b = \frac{H}{\mu(s-c)} \qquad \text{...(9.40)}$$

If $$c = 1,\ b = \frac{H}{\mu(s-1)} \qquad \text{...(9.40 a)}$$

Further if uplift pressure is not considered then since $c = 0$, we get

$$b = \frac{H}{\mu s} \qquad \text{...(9.40 b)}$$

The minimum base width to be provided for the elementary profile of a gravity dam should be the greater of the base widths given by Eqs (9.39) and (9.40).

Stresses Developed in the Elementary Profile

The normal stress at the toe and heel of a dam is given by Eq. (9.29) as

$$\sigma_y = \frac{\Sigma V}{b}\left(1 \pm \frac{6e}{b}\right)$$

In the case of elementary profile

$$\Sigma V = (W - U)$$

$$= \frac{1}{2} \quad bwH\,(s - c)$$

$$\sigma_y = \frac{1}{2} \quad wH\,(s - c)\left(1 \pm \frac{6e}{b}\right)$$

Further in the limiting case of no tension at any point in the base of the dam $e = (b/6)$ and hence

At toe, $$\sigma_y = \sigma_{y_D} = wH\,(s - c) \qquad \text{...(9.41)}$$

At heel, $$\sigma_y = \sigma_{y_U} = 0 \qquad \text{...(9.41 a)}$$

The principal stress at the toe of the dam is given by Eq. (9.30 a) as

$$\sigma_{1_D} = \sigma_{y_D} \sec^2 \phi_D$$

Substituting the value of σ_{y_D} for elementary profile from Eq. (9.41), we get

$$\sigma_{1_D} = wH\,(s - c) \sec^2 \phi_D$$

Further for elementary profile

$$\sec^2 \phi_D = \left[1 + \left(\frac{b}{H}\right)^2\right]$$

and for this case from Eq. (9.39)

$$\frac{b}{H} = \frac{1}{\sqrt{s-c}}$$

By introducing these values in the above expression for the principal stress at the toe of the elementary profile of a dam, we get

$$\sigma_{1_D} = wH\,(s - c + 1) \qquad \text{...(9.42)}$$

The shear stress at the toe of the dam is given by Eq. (9.34a) as

$$(\tau_{yx})_D = \sigma_{y_D} \tan \phi_D.$$

Substituting the value of σ_{y_D} from Eq. (9.41), we get

$$(\tau_{yx})_D = wH (s - c) \tan \phi_D$$

Further in this case

$$\tan \phi_D = \frac{b}{H} \text{ and } \frac{b}{H} = \frac{1}{\sqrt{s-c}}$$

and hence

$$(\tau_{yx})_D = wH \sqrt{s-c} \qquad \text{...(9.43)}$$

Since at the heel of the dam as indicated by Eq. (9.41a) the normal stress $\sigma_{y_U} = 0$, the principal and shear stresses are equal to zero.

All these expressions are derived for the elementary profile of a dam for the reservoir full condition. However, when the reservoir is empty the only force acting on the elementary profile is its weight acting vertically downwards through the inner middle third point M_1 and hence

At heel, $$\sigma_y = \sigma_{y_U} = \frac{2W}{b} = wsH \qquad \text{...(9.44)}$$

At toe, $$\sigma_y = \sigma_{y_D} = 0 \qquad \text{...(9.44 a)}$$

Limiting Height of Elementary Profile of Gravity Dam—High and Low Gravity Dams

For a dam having elementary profile the principal stress at the toe is given by Eq. (9.42) as

$$\sigma_{1_D} = wH (s - c + 1)$$

which shows that the value of σ_{1_D} varies only with H, as all the other terms are constant for any dam. In order to avoid failure of the dam due to crushing the value of σ_{1_D} should not exceed the allowable working stress f for the dam material and in the limiting case

$$f = \sigma_{1_D} = wH (s - c + 1)$$

from which the height H is given by

$$H = \frac{f}{w(s-c+1)} \qquad \text{...(9.45)}$$

Thus, Eq. 9.45 gives the maximum height which may be provided for a dam having elementary profile without exceeding the allowable working stress for the dam material.

However, a lower value of H will be obtained when $c = 0$, i.e., when uplift pressure is not considered. Hence to be on the safer side for determining the limiting height of a dam having elementary profile the uplift pressure is neglected. The limiting height of a dam having elementary profile is thus given by

$$H = \frac{f}{w(s+1)} \qquad ...(9.45 \text{ a})$$

A distinction is usually made between a low and a high gravity dam in terms of the limiting height of a dam given by Eq. (9.45 a). Thus, a *low gravity dam* is the one whose height is less than or equal to that given by Eq. (9.45 a). On the other hand if the height of a dam is more than that given by Eq. (9.45 a) then it is known as a *high gravity dam.* As an example taking w = 9810 N/m³ [or 1000 kg f/m³], s = 2.4 and f = 2943 kN/m² [or 3 × 10⁵ kg f/m²] the limiting height of a low gravity dam is obtained by Eq. (9.45 a) as

$$H = \frac{2943 \times 10^3}{9810(2.4+1)} \approx 88 \text{ m}$$

9.9 PRACRICAL PROFILE OF A GRAVITY DAM

The elementary profile of a dam is only a theoretical profile which cannot be adopted in actual practice. This is so because unlike the elementary profile having zero top width and no freeboard, a dam must have substantial top width and also sufficient freeboard. Moreover, in almost all the cases due to stability requirements, beyond certain depth the upstream face of a dam is also required to be made sloping. As such a practical profile of a dam differs considerably from the elementary profile. The practical considerations regarding the provision of the top width and the freeboard for a dam are indicated below.

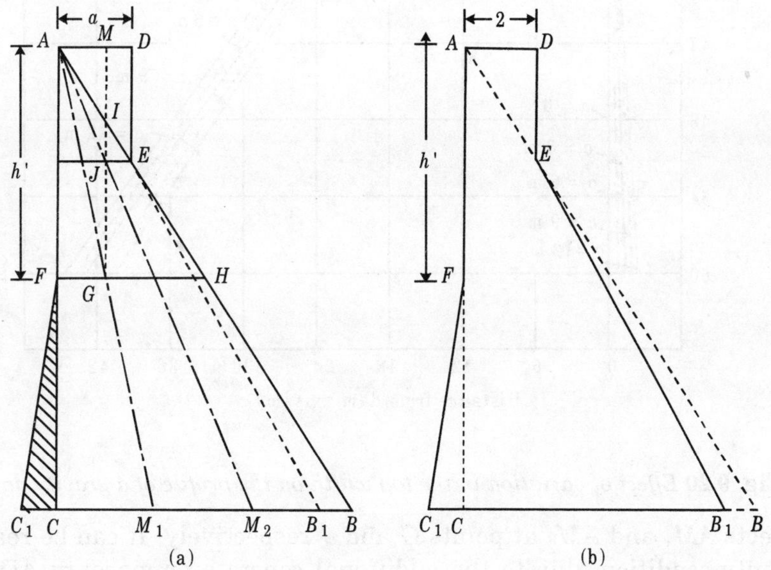

Fig. 9.19 *Effect of top width on the profile of a gravity dam*

Top Width. A dam must have sufficient top width to provide the required roadway and to resist the shock of floating objects. However, the concrete or masonry added for providing the top width affects the cross section of the dam which is illustrated in Figs. 9.19 and 9.20. Figure 9.19 (a) shows an elementary dam section having AM_1 and AM_2 inner and outer third-point lines respectively. Let *ADE* be concrete or masonry added to provide a top width *AD* equal to *a* and *MI* be the line passing through the centroid of the added triangle *ADE,* which when

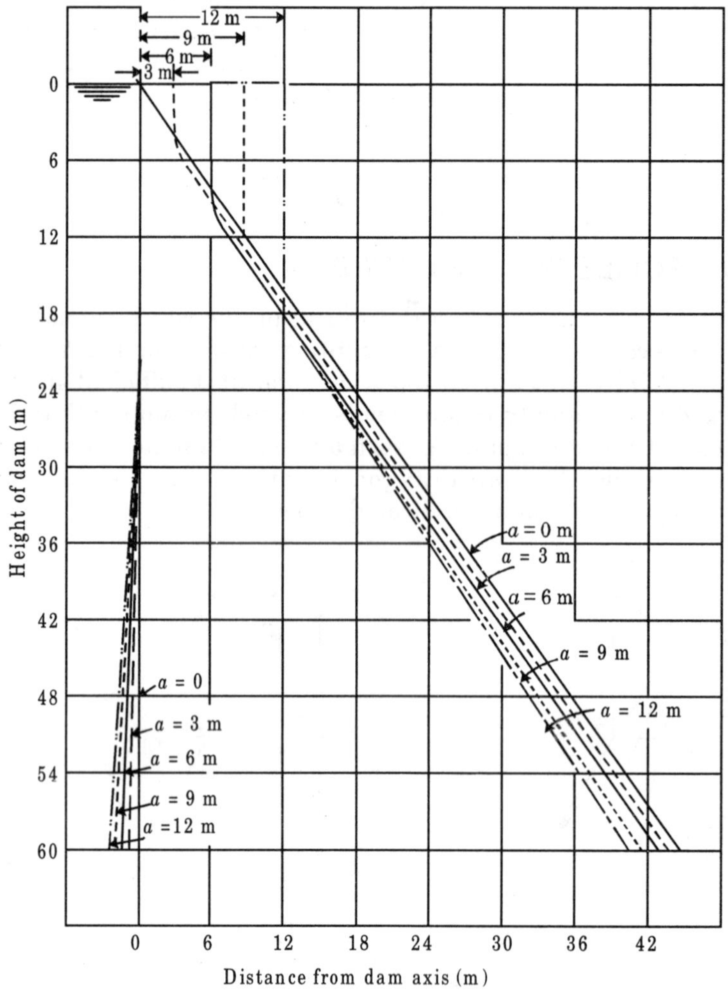

Fig. 9.20 *Effect of variation in the top width on the profile of a gravity dam*

produced intersects AM_1 and AM_2 at points *G* and *J* respectively. It can be readily shown that for reservoir empty condition, due to the additional concrete or masonry *ADE,* the resultant force will intersect at a point on the right of AM_1 within the middle third for all the sections above the plane *FGH* and on the left of AM_1 outside the middle third for all the sections below the plane *FGH.* Hence tension will be developed at the downstream face of the dam below the plane *FGH* and in order to avoid tension, concrete or masonry will have to be added at the lower levels on the upstream side of the dam by providing upstream batter FC_1 as shown in

Fig. (9.19 a). This would result in increasing the total volume of concrete or masonry in the dam. However, in most of the cases the reservoir empty condition is not developed.

For reservoir full condition due to the additional concrete or masonry *ADE,* for all the sections below the plane *JE,* the resultant force will intersect at a point on the left of AM_2 within the middle third. However, for the sake of economy the resultant force must intersect at the outer third point line at all sections. As such in order to make the resultant force to intersect at the outer third point at all the sections the downstream face of the dam may be shifted from *EB* to EB_1, resulting in the reduction of concrete or masonry. Thus, due to the provision of some top width the modified dam section will be $ADEB_1C_1F$ as shown in Fig. (9.19 b).

Figure (9.20) shows the effect of the variation in the top width on a dam section. It may be observed that as the top width is increased the concrete or masonry to be added at the upstream face is increased while there is reduction in the concrete or masonry at the downstream face. It has been further observed that within limits the concrete or masonry added for providing the top width decreases rather than increases the total concrete or masonry volume in the dam. This is so because the increased volume of concrete or masonry in the upper portion of the dam is compensated by the reduction at the lower levels. Hence the most economical top width is a function of the height of the dam. According to Creager (1916) for non-overflow sections of gravity dams of fairly uniform moderate heights, and designed without considering the earthquake forces, the most economical top width is about 14 percent of the height of the dam. However, for low dams the top width is usually greater than that provided on the basis of economy, because a minimum roadway is often required to be provided at the top and also sufficient top width may be required to withstand the shock of floating objects.

In order to obtain the adequate top width without differing much from the elementary profile two arrangements are commonly used as shown in Fig. (9.21). The arrangement shown in Fig. (9.21 a) is applicable to low dams where the truly economic top width is too small to meet

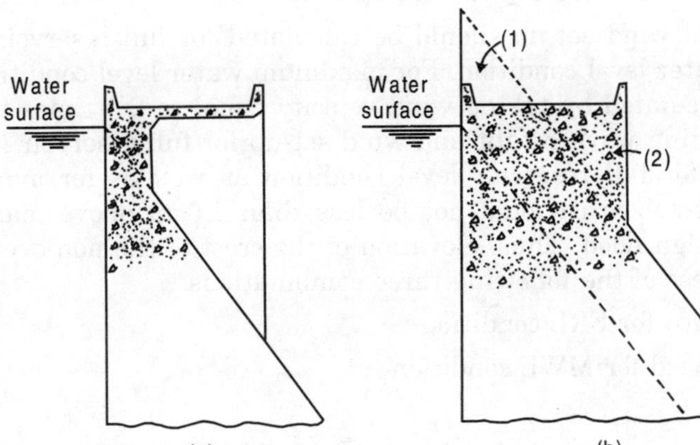

Fig. 9.21 *Alternative arrangements for top width of a gravity dam*

the practical purposes. The arrangement shown in Fig. (9.21 b) is in general suitable for high dams, which has the advantage of reducing the angle ϕ_D for a given volume of concrete or masonry so that the principal stress at the toe of the dam does not become excessive. Further the arrangement of Fig. (9.21 b) has the effect of increasing the height to which the vertical upstream face of the dam may be raised.

Freeboard. Freeboard is the difference in the elevation of the crest of the non-overflow section of the dam and the still water level in the reservoir. It is provided to avoid the possibility of water spilling over the non-overflow section of the dam due to waves formed on the water surface in the reservoir.

Earlier as indicated in IS: 6512–1972 it was recommended that the freeboard should be equal to (3/2) h_w, where h_w is the wave height for the full reservoir level (FRL) condition or for the maximum water level (MWL) condition, whichever gives the higher crest elevation for the non-overflow section of the dam. For calculation of wave height Moliter's formula Eq. (9.24) or Eq. (9.25) is used in which the wind velocity of 120 km/h over water in the case of full reservoir level condition and of 80 km/h over water in the case of maximum water level condition are generally assumed. Further the freeboard above maximum water level should not be less than 0.9 m.

However, the earlier recommendation for freeboard has been revised as indicated in IS: 6512–1984, according to which the freeboard should be equal to (4/3) h_w plus wind set-up, where h_w is the wave height. The wave height and wind set-up should be calculated according to the method given by *T.* Saville in which wind set-up is given by the following equation.

$$\text{Wind set-up} = \frac{V^2 F}{62000 D} \qquad \text{...(9.46)}$$

where

V = Maximum wind velocity over water in km/h;

F = Maximum fetch in km; and

D = Average depth of reservoir in m along maximum fetch.

Equation (9.46) gives the value of wind set-up in m.

The wave height and wind set-up should be calculated for full reservoir level condition as well as for maximum water level condition. For maximum water level condition the wave height and wind set-up are calculated by taking wind velocity equal to 2/3 times the maximum wind velocity used for calculating wave height and wind set-up for full reservoir level condition. The freeboard is calculated for full reservoir level condition as well as for maximum water level condition. Further the freeboard should not be less than 1.0 m above maximum water level corresponding to the design flood. Thus, elevation of the crest of the non-overflow section of the dam should be the highest of the following three combinations

(i) FRL + Freeboard for FRL condition

(ii) MWL + Freeboard for MWL condition

(iii) MWL + 1.0 m

If the design flood is not same as probable maximum flood (PMF) then the crest of the non-overflow section of the dam should not be lower than maximum water level corresponding to probable maximum flood. Further out of the freeboard computed, 1.0 m high solid parapet may be provided on the upstream side above the crest of the non-overflow section of the dam in all cases.

The details of the procedure to be followed for computation of freeboard by the method given by T. Saville are illustrated in Illustrative Example 9.7.

9.10 METHODS OF DESIGN OF GRAVITY DAMS

The various methods used for the design of gravity dams are as follows.

(1) Stability Analysis Method

(2) Zone Method of Determination of Profile of a Dam.

1. Stability Analysis Method. In this method a trial section of the dam is first assumed on the basis of the previous designs, experience, configuration of valley, etc. The stability of the assumed section is then checked at the foundation level as well as at other levels. The various methods used for stability analysis are: (i) Gravity method; (ii) Trial load twist method; (iii) Experimental method; (iv) Slab analogy method; (v) Lattice analogy method; (vi) Finite element method. These methods are briefly described below.

(i) *Gravity method.* In this method the dam is considered to be composed of a series of vertical cantilevers independent of each other and the load acting on the dam is transferred to the foundation through cantilever action. For the sake of convenience a cantilever of unit length contained between two vertical planes normal to the axis of the dam is considered above the deepest foundation level and its stability is checked against all the possible modes of failure described in section 9.6, at the base of the dam as well as at various horizontal sections above the base. If necessary the assumed dam section is modified and when a section satisfying the stability requirements is obtained the same is adopted for the entire dam. The various steps involved in this method are as indicated below.

(a) Calculate all the forces acting per unit length of the dam considered above the deepest foundation level.

(b) Find the horizontal and vertical components of all the forces.

(c) Find ΣH the algebraic sum of all the horizontal forces and ΣV the algebraic sum of all the vertical forces.

(d) Find the moments of all the forces about the toe of the dam and compute ΣM_0 the sum of the overturning moments and ΣM_R the sum of the righting moments. Also find the algebraic sum of all the moments

$$\Sigma M = (\Sigma M_R - \Sigma M_o)$$

(e) Find the distance $\bar{x}$ of the point of intersection of the resultant R with the base from the toe of the dam by using the relation

$$\bar{x} = \left(\frac{\Sigma M}{\Sigma V}\right)$$

(f) Find the eccentricity e, which is the distance between the centroid of the area of the base and the point of intersection of the resultant with the base, as given by the expression

$$e = \frac{b}{2} - \bar{x}$$

where b is the base width of the section.

(g) Find the normal stress at the toe as well as heel of the dam given by Eq. (9.29) as

$$\sigma_y = \frac{\Sigma V}{b}\left(1 \pm \frac{6e}{b}\right)$$

(h) Find the principal and shear stresses at the toe and the heel of the dam given by Eqs (9.31), (9.32), (9.34) and (9.35) respectively.

(i) Find the factor of safety against overturning by the expression

$$\text{F.S.} = \frac{\Sigma M_R}{\Sigma M_{oh}}$$

(j) Find the factors of safety against sliding given by Eqs (9.37) and (9.38).

Most of the dams are designed by this method which has been illustrated in Example 9.4.

A concrete (or masonry) dam is actually constructed in the form of a series of vertical blocks with transverse joints between the adjacent blocks extending from the upstream face to the downstream face of the dam. These joints are provided with water stops (or water seals) as described in section 9.11 and may be either left open or provided with only shear keys and not grouted or provided with shear keys and grouted. The gravity method of design involves two dimensional analysis and hence it may be adopted only if the transverse joints between the adjacent blocks are neither keyed nor keyed and grouted but are provided with water seals and left open. However, if the transverse joints between the adjacent blocks are keyed or keyed and grouted then the structural behaviour of the dam will be altogether different and hence the gravity method will lead to erroneous design. In this case the trial load twist method may be adopted which involves the three dimensional analysis.

(ii) *Trial load twist method.* If a dam is considered to be made up of a series of vertical cantilevers, then each cantilever from the centre of the dam towards the abutment will be shorter than the preceding one. Consequently due to water pressure each cantilever will be deflected less than the preceding one and more than the succeeding one. If the transverse joints between the adjacent cantilever blocks are keyed and either grouted or not grouted the movement of each cantilever will be restrained by the adjacent ones. The longer cantilever will tend to pull the adjacent shorter cantilever forward and the shorter cantilever will tend to hold it back. This interaction between the adjacent cantilever elements causes torsional moment or twist, which materially affects the manner in which the load due to water pressure is distributed between the cantilever elements in the dam. This changes the stress distribution from that found by the ordinary gravity analysis in which these effects are neglected.

In the trial load twist method, the dam is assumed to be divided into vertical cantilever elements and horizontal beam elements each of which occupies the entire volume of the dam. As the structural behaviour of a dam with joints keyed but not grouted is different from the one with joints keyed and grouted, each of these cases is dealt with separately.

(a) Joints keyed but not grouted. If the transverse joints between the adjacent cantilever blocks are only keyed but not grouted then the horizontal beams will behave as shear beams (i.e., beams capable of taking only shear and not the bending moment) and the horizontal elements will be subjected to twisting in the horizontal as well as vertical planes. Thus, in this case the entire dam may be assumed to be made up of the following two structural systems.

(i) Cantilever structure

(ii) Twisted structure.

The load acting on the dam due to water pressure is divided between the cantilever structure and the twisted structure in such a manner that the computed deflection for any point in the dam, considered as a point in the cantilever structure will be identical with its computed deflection, considered as a point in the twisted structure. The division of loading which will cause identical deflections at any point in the dam, considered in each of the two structural systems is found by successive trials. The process is repeated for all the points in the dam. The stresses are then computed from the forces and moments thus known.

(b) Joints keyed and grouted. If the transverse joints between the adjacent cantilevers are keyed and grouted then the entire dam will become a monolithic structure. Thus in this case the horizontal beams will be capable of taking bending moment in addition to shear, and hence an additional structure designated as beam structure is introduced for resistance to bending in the horizontal elements. The entire dam in this case may therefore be assumed to be made up of the following three structural systems.

(i) Cantilever structure

(ii) Twisted structure

(iii) Beam structure.

The load acting on the dam due to water pressure is now divided between these three structural systems in such a manner that the computed deflections are identical for any point in the dam considered in each of the three structural systems. The division of loading which will cause identical deflections at any point in the dam considered in each of the three structural systems is found by successive trials. The process is repeated for all the points in the dam. The stresses are then computed from the forces and moments thus known.

(iii) *Experimental methods.* The experimental methods may be classified as (a) Direct method and (b) Indirect method which are described below.

(a) Direct Method. In this method either two dimensional or three dimensional geometrically similar models, which are the exact replica of the prototype dam, are made of elastic materials. In a two dimensional model a portion of the dam between two vertical planes normal to the axis is considered, while in a three dimensional model complete valley is considered. The model is located in similar manner as the prototype. The model is then acted upon by various pressures proportional to those of the prototype. In order to simulate the high water pressure acing on the prototype, a liquid of high specific gravity such as mercury may be used or the effect may be obtained by the jetting action of water. The deflections occurring at various points of the model are measured and from the correlation developed between the model and the prototype for these stress and deflection, the stresses for the prototype can be obtained. This method has been found to be quite useful for the analysis of high dams.

(b) *Indirect Method.* Under the indirect methods photoelastic model analysis and the methods of magnetic and electrical analogy are usually considered. In the case of photoelastic model analysis, a model of a dam is made of elastic, transparent, isotropic material which is free from initial or residual stresses. Some of the materials which have been successfully used for such models are bakelite, celluloid, gelatin and glass. The model is then proportionately loaded and the stresses developed in it are determined by means of the visible optical effects which are produced by passing polarized light through the model. The effects of stress in a photoelastic model are made visible by means of an optical instrument known as photoelastic polariscope. From the stresses determined in the model the stresses in the prototype can be predicted. However, the studies employing photoelastic models are usually limited to conditions of plain stress or strain and for determining the regions of stress concentration.

(iv) *Slab analogy method.* The slab analogy method is based on the mathematical analogy between the distribution of stresses in a dam section and the deflections of a slab of similar shape by forces and moments around its edges. The slab has the same shape as a cantilever section including a large block of the foundation. The edges of the slab are bent into a form which corresponds to the stresses at the surface of the dam structure. The analysis is made by dividing the analogous slab into horizontal and vertical beams which are brought into slope and deflection agreement by trial loads. The curvatures in the slab are then proportional to the shears in the dam structure and consequently the moments in the horizontal and vertical beams are proportional to the stresses in the vertical and horizontal directions respectively. It is however a very laborious method.

(v) *Lattice analogy method.* In the lattice analogy method the dam is considered as though it were a slice or plate of a unit thickness which in turn is simulated in size and shape by a lattice network composed of interconnected elemental square frames, each diagonally connected at the corners. The solution in this case involves the adjustment of the lattice by a systematic relaxation of restraints at the joints. When the adjustment of the lattice to remove restraint has been completed, the strains are deduced from relative displacements between successive joints and from these the stresses may be computed.

(vi) *Finite element method.* The finite element method can be used to analyse the structural behaviour of high gravity dams. In this method the dam and its foundation may be divided into elementary, contiguous triangles and the elastic properties of each may be analysed and linear simultaneous equations formulated for each nodal point. The solution of these equations would give the stresses and deflections due to a given system of loads.

2. Zone Method of Determination of Profile of a Dam. The zone method of design of dam is also known as multiple step method or step by step method. In this method the dam section is considered to be divided into a number of zones as shown in Fig. 9.22 and starting

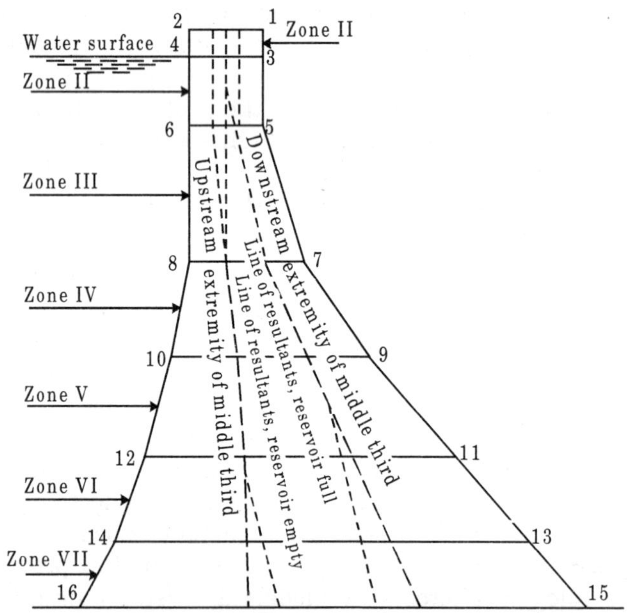

Fig. 9.22 *Zones for non-overflow section of a gravity dam*

from the top of the dam each zone is designed in accordance with stability requirements as indicated below.

Zone I. It is that portion of the dam which begins from the top of the dam and extends upto the maximum water surface, or if there is ice, upto the bottom of the ice sheet. Both upstream and downstream faces of this zone are vertical. If there is no ice, the height of Zone I is fixed by the desired freeboard requirements and the width is determined by practical consideration such as minimum roadway required or economy for the entire dam section. However, when ice pressure occurs, the quantity of concrete or masonry provided in Zone I must have sufficient weight to prevent the portion 1–2–3–4 from sliding.

Zone II. For some distance below the bottom of Zone I the resultants for reservoir full and empty conditions lie well within the middle third, as the width of the top is always greater than necessary to conform to middle third rule. Hence the width same as at the top of the dam may be continued until at some plane 5–6 the resultant for reservoir full condition intersects this plane just at the downstream extremity of the middle third. The portion of the dam between the bottom of Zone I and the plane 5–6 constitutes Zone II. Therefore both upstream and downstream faces of Zone II are also vertical.

Zone III. Below the bottom of Zone II the downstream face of the dam will have to be made inclined in order to conform to middle third rule for reservoir full condition. However, for reservoir empty condition the resultant still lies within the middle third and hence the upstream face may remain vertical until at some plane 7–8 the resultant for reservoir empty condition intersects this plane just at the upstream extremity of the middle third. The portion of the dam between planes 5–6 and 7–8, in which the dimensions are determined by middle third rule for reservoir full condition, constitutes Zone III.

Zone IV. Below the plane 7–8 the upstream face of the dam will also be required to be made inclined in order to conform to middle third rule for reservoir empty condition. For some distance below the plane 7–8 the inclinations of the upstream and the downstream faces of the dam may be so adjusted that the resultants for the reservoir full and empty conditions intersect each plane just at the extremities of the middle third. However, at some plane 9-10, for reservoir full condition the principal stress at the downstream face may reach the maximum allowable limit, which thus fixes the lower limit of Zone IV. Hence the portion of the dam between the planes 7–8 and 9–10 constitutes Zone IV.

Zone V. Below the plane 9–10 for some distance the inclination of the downstream face of the dam will have to be so adjusted that for the reservoir full condition the principal stress at this face does not exceed the maximum allowable limit. In this case the resultant for reservoir full condition intersects well within the middle third. On the other hand the inclination of the upstream face is so adjusted that for reservoir empty condition the resultant continues to intersect just at the upstream extremity of the middle third. However, at some plane 11–12 for reservoir empty condition the principal stress at the upstream face may reach the maximum allowable limit, which thus fixes the limit of Zone V. The portion of the dam between planes 9-10 and 11–12 thus constitutes Zone V.

Zone VI. Below the bottom of Zone V for some distance the inclinations of both upstream and downstream faces of the dam are so adjusted that the principal stress at the upstream face for reservoir empty condition and at the downstream face for reservoir full condition does not exceed the maximum allowable limit. The portion of the dam between planes 11–12 and 13–14, which may be designed according to these rules thus constitutes Zone VI.

Zone VII. As the height of the dam increases, the inclinations of both upstream and downstream faces increase. Consequently, at some plane the value of $\sec^2 \phi_D$ in Eq. (9.30 a) may become so great that the principal stress at the downstream face may exceed the maximum allowable limit. The portion of the dam in which this condition prevails constitutes Zone VII. However, this zone should be eliminated by revision of the entire design.

Comparison between Zone Method and Single Step Method of Design of Dam

In the case of a high dam designed by zone method sometimes it is found that beyond Zone IV the upstream and downstream faces of the dam have unusual shapes. Figure (9.23) shows a typical section of a high dam obtained by zone method of design. The upstream face of the dam

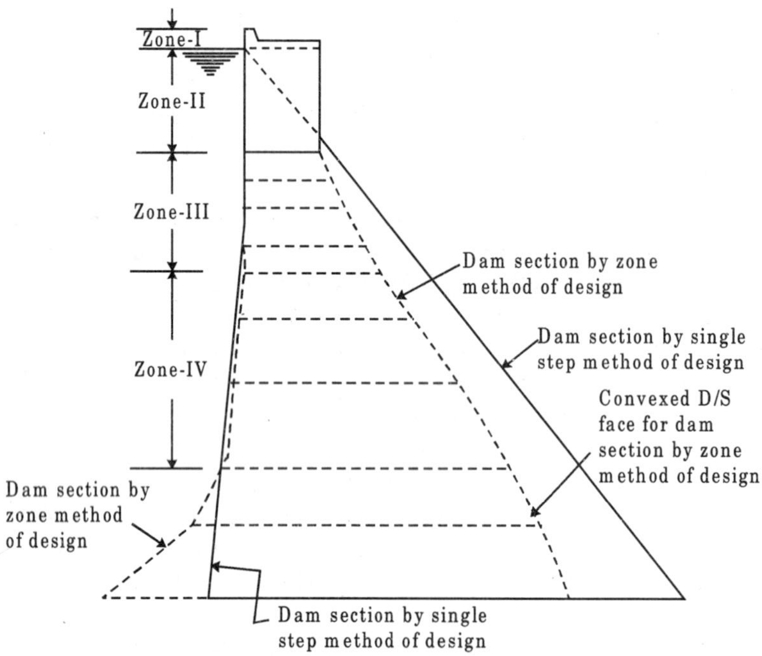

Fig. 9.23 *Comparative dam profiles obtained by zone method and single step method of design of a gravity dam*

has a relatively flat slope while the downstream face has outward convex shape. A convexed face under compression, whether smoothly curved or polygonal, may be subject to tensile stress on surface parallel to the face. Such a shape for the downstream face of the dam is not desirable because the outer layer of such a section tends to buckle outward and unless the buckling force is cancelled by the weight component normal to the face, tension results. The dam section will have to be then redesigned so that such a curvature for the downstream face of the dam is avoided. For this by increasing the concrete (or masonry) strength the dam may be completely designed by dividing it into only four zones thus keeping out of Zones V and VI. Alternatively the dam section may be designed by single step method.

In the single step method of design a section of the dam is suitably assumed and the entire dam section is considered as a single zone. In the assumed section of the dam the upstream face is kept vertical for some depth to be determined by trial. As a first trial the vertical portion of the upstream face of the dam may be assumed to be extended upto the plane *FGH* shown in Fig. (9.19 a), shown in page 419 which it is given some slope. The slope given to the upstream and the downstream faces of the dam are so adjusted that the principal stresses at the upstream face for the reservoir empty condition and at the downstream face for the reservoir full condition reach their maximum values simultaneously. This may be accomplished only by trial and error. The dam section is then checked for all the stability requirements for reservoir empty and reservoir full conditions, which will normally be satisfied at all the points above the base of the dam when the slopes for the upstream and the downstream faces are adjusted as indicated above. Figure (9.23) also shows a typical section of a high dam obtained by single step method of design.

For comparison the dam sections obtained for a high dam designed by zone method and single step method are plotted together in Fig. (9.23). It may be observed that the zone method of design gives a dam section which is more economical at the upper portion of the dam. Further the dam section obtained by the single step method of design is in general under stressed at all points except at the base. Thus, from the foregoing discussion the following conclusions may be drawn:

(i) Dams of smaller height can be designed economically by zone method of design.

(ii) High dams beyond Zone IV are designed by single step method so that convex curvature of downstream face and excessive flat slope of upstream face are avoided.

(iii) It may be economical to increase the concrete (or masonry) strength through the use of more expensive materials, so that even a high dam may be designed by dividing it into only four zones, thus keeping out of Zones V and VI.

9.11 JOINTS, KEYS AND WATER STOPS

Joints. Joints are required to be provided in a dam to permit systematic, convenient and economical construction and to prevent the development of cracks. In general the various joints provided in a dam are known as construction joints and these are classified as

(a) Contraction joints and
(b) Horizontal joints.

(a) Contraction Joints. Contraction joints are the vertical joints provided in a dam to prevent the development of cracks. The cracks may be developed in a dam due to tensile stresses being produced when the volumetric changes of concrete (or masonry) are restrained. The volumetric changes may result from the shrinkage of concrete (or masonry) as it dries and from the temperature variations. The major volumetric change in concrete (or masonry) however results from the change in temperature. Due to high daily difference in temperature between the surface and near surface areas of the dam, surface cracks may be developed. Surface cracks once started may progress into the interior of the dam. However, deep cracks may be developed in a dam by high interior temperature (due to evolution of heat of hydration of cement) which creates steep temperature gradient between the interior and the surface for considerable time even after the completion of the dam. Cracks cause stress concentration and destroy the monolithic nature of the structure there by preventing it from carrying the designed loads. Moreover, cracks adversely affect water-tightness, durability, appearance and internal stresses of concrete (or masonry) and are therefore undesirable.

The contraction joints are of two types viz., (i) transverse joints and (ii) longitudinal joints.

(i) *Transverse joints.* Transverse joints are the joints provided normal to the axis of the dam. These joints extend vertically from the foundation to the top of the dam and are continuous from the upstream face to the downstream face, thus divide the dam into separate blocks. These joints allow the contraction of concrete (or masonry) on the two sides and thus prevent the development of transverse cracks in the dam. As such these joints are also known as *transverse contraction joints.*

The spacing of transverse joints is governed by physical features of the dam site, details of the structures associated with the dam such as spillway gates, outlets, penstocks, spillway bridge piers etc., results of temperature studies and the probable plant capacity. For concrete dams these joints are usually spaced 15 to 20 m apart, but for masonry dams a larger spacing may be adopted.

The transverse joints may be grouted or may not be grouted but in both the cases adjacent to the upstream face of the dam water stops (or water seals) are provided across these joints. When the transverse joints are grouted the water stops prevent the flow of grout to the upstream side and these are then known as grout stops. However, if the transverse joints are not grouted then the water stops prevent the leakage of water from the reservoir through these joints. Moreover, vertical keys are usually provided over the entire surfaces of the adjacent blocks in the transverse joints when these joints are grouted. However, even if the transverse joints are not grouted the vertical keys may be provided to minimize the leakage of water through the joints and to provide shearing resistance at the joints.

(ii) *Longitudinal joints.* Longitudinal joints are the joints provided parallel to the axis of the dam. These joints extend vertically from the foundation and as they approach the downstream or the upstream face of the dam either the direction of the joints is changed from the vertical to effect a perpendicular intersection with the face of the dam with an offset of 1 to 1.5 m, or the joints are terminated at the top when they are within 5 to 6 m of the face. The later arrangement is usually adopted in the case of the joints which may extend upto the upstream face of the dam. The longitudinal joints run between the two adjacent transverse joints and are not continuous all along the length of the dam. These joints are stagged by a minimum distance of 8 m at the transverse joints in the adjacent blocks. A spacing of 15 to 30 m is generally adopted for these joints which is however largely governed by the convenience of construction and the foundation conditions.

The longitudinal joints sub-divide each block formed by transverse joints into several smaller blocks, but since each block must be a monolithic block these joints are invariably provided with horizontal keys over the entire surface of the joint and are grouted. These joints are usually provided only in the case of high dams because as the height of the dam increases the base width also increases and it approaches a limit beyond which vertical cracks parallel to the axis of the dam may be developed if the longitudinal joints are not provided. Thus, these joints serve the same purpose in each block of the dam as the transverse joints serve in the dam as a whole.

It has however been recognized by the designers that the provision of longitudinal joints is basically unsound unless a high degree of perfection may be achieved in grouting these joints and it may be ensured that each block is monolithic. As such the present practice is to avoid the longitudinal joints altogether, even in the case of high dams and a better alternative to this is to achieve necessary temperature control by precooling of concrete supplemented, where necessary, by post cooling. Further in the case of dams built of rubble masonry longitudinal joints are not considered necessary.

Figure 9.24 shows the positions of transverse and longitudinal joints in a dam.

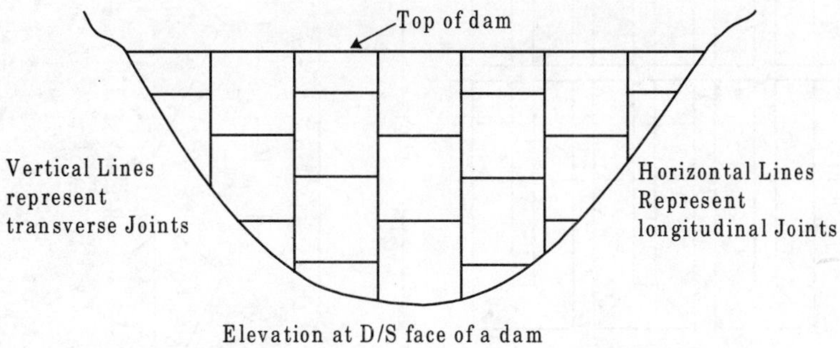

Fig. 9.24 *Transverse and longitudinal joints in a gravity dam*

(b) Horizontal Joints. Horizontal joints are the joints introduced between the successive lifts of concrete (or masonry). A lift is defined as the height by which each block is raised in one continuous operation of pouring concrete or laying masonry. The concrete (or masonry) of the next lift is placed after sufficient time is allowed for the previously placed concrete (or masonry) to cool and attain its initial set and become hard. As such horizontal joints are provided to allow sufficient cooling between the successive lifts of concrete (or masonry). For solid gravity dams, lifts of about 1.5 m are usually adopted and for the layer immediately on the rock a lift equal to half of this, i.e., about 0.75 m is adopted. However, for buttress as well as thin arch dams depending upon thickness higher lifts of about 3 to 4.5 m may be adopted.

Prior to the placement of the concrete (or masonry) of the next lift, the surface of the previously placed concrete or (masonry) is thoroughly cleaned by the use of high velocity jet of water and air as well as wet-sand blasting process. Further immediately before the concrete placing of the next lift begins, a 12.5 mm thick layer of mortar should be applied to permit a proper bond between the previously placed concrete and the new concrete. Such treatment of the surface and the use of good concrete create tight horizontal joints for which no provisions to ensure water tightness, such as water stops, keys etc., are necessary.

Keys. Keys are the regular projections provided in the contraction joints of a dam for transfer of load from one block to other through shear. In transverse joints vertical keys are provided which are designed to transfer horizontal shear. The keys in transverse joints also assist in minimizing the leakage of water through the joints. Figure (9.25 a) shows the vertical keys which are standardized by U.S.B.R. and are commonly adopted.

In longitudinal joints horizontal keys are provided which are designed to take vertical shear. Fig. (9.25 b) shows typical details of the horizontal keys. The faces of these keys are aligned to make them conform approximately with the lines of principal stresses for full water load so that there are no shear stresses acting on the faces of the keys. However, the direction of principal stresses varies from the upstream face to downstream face and from the foundation to the crest of the dam, and hence an unlimited number of key shapes with resulting high cost of form work would be required if close conformity between the key faces and the principal stresses was to be maintained. As such a single key shape determined by the general direction of the lines of principal stresses in the lower downstream portion of the dam where the vertical shear is maximum, is usually adopted for all such keys which would simplify the form work for these keys.

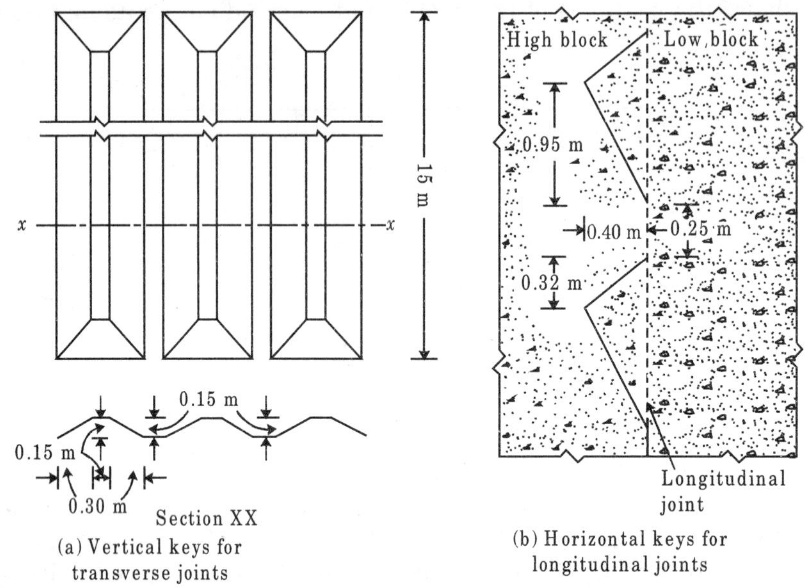

Fig. 9.25 *Types of keys provided for transverse and longitudinal joints of a gravity dam*

Waterstops (or Waterseals). Waterstops are provided across transverse joints, adjacent to the upstream face of the dam, to prevent the leakage of water from the reservoir through the transverse joints if the transverse joints are not grouted. However, if the transverse joints are grouted then also the waterstops are required to be provided to prevent the flow of grout to the upstream side and these are then known as groutstops. In general following three types of waterstops are used:

(i) Metal waterstops.

(ii) Asphalt waterstops.

(iii) Rubber/PVC waterstops.

(i) *Metal waterstops.* The metal waterstops are the most common types of waterstops used in dams. These waterstops are made out of strips of annealed soft copper, or monel metal (Nickle-Copper Alloy), or stainless steel of minimum thickness 1.5 mm. As shown in Fig. 9.26 (a) and (b) two shapes of metal waterstops viz., Z-shape and M-shape are used. For Z-shape waterstop a 600 mm wide strip is used. About 250 mm of the strip from each side is embedded in each of the two adjacent blocks when these are being cast and remaining about 100 mm of the strip is left between the blocks parallel to the joint, thus forming the shape of letter Z. For M-shape waterstop a 700 mm wide strip is used and after embedding about 250 mm of the strip from each side in each of the two adjacent block, the remaining about 200 mm of the strip is folded to form the shape of letter M.

As shown in Fig. (9.26 d) the metal waterstops are generally located at a distance of about 600 mm from the upstream face of the dam.

The metal waterstops are erected in place with the help of anchor rods as shown in Fig. (9.26 d). In the case of masonry dams for erecting the waterstop a concrete blockout 1850 mm × 800 mm (shown by dotted lines in Fig. (9.26 d) is provided in each block of the dam on either side of the transverse joint. The surface adjacent to the blockouts should be irregular and the joints in the masonry should be raked out when mortar is green, with some stones protruding beyond the dotted lines regularly in both directions. Further in order to prevent

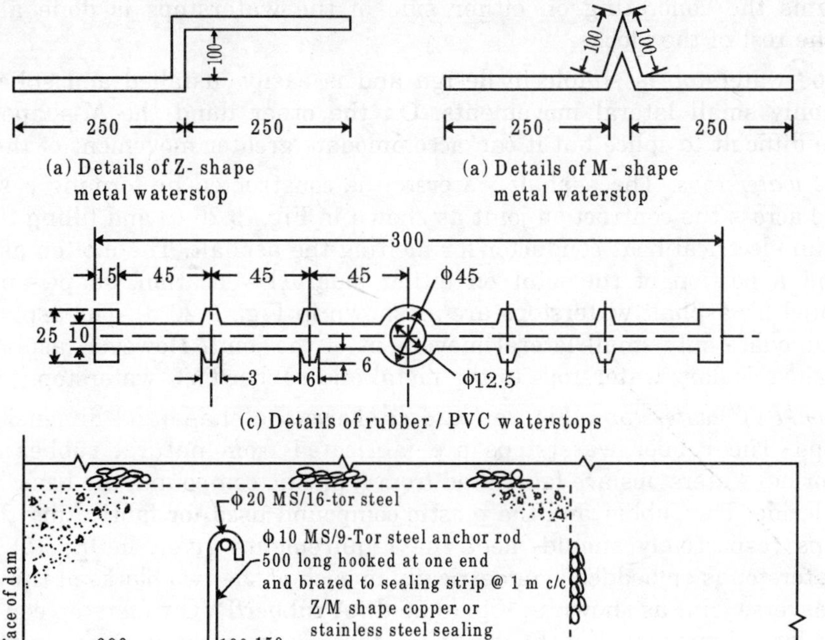

Fig. 9.26 *Types of waterstops (or waterseals)*

shrinkage cracks at the concrete/masonry interface of the blockout, 1500 mm long, 25 mm diameter dowel bars (not shown in Fig. 9.26 (d) are provided at 500 mm c/c in both directions at the interface. The 500 mm length of the dowel bars is embedded in the concrete of the blockout and the remaining 1000 mm length of the dowel bars is embedded in the masonry of the dam.

The blockout of one block of the dam is concreted first and the joint face is given a coat of coaltar black paint and then the blockout of the adjacent block is concreted so as to have a clear contraction joint. However, no such blockouts are provided in concrete dams because in the case of concrete dams the concreting on either side of the waterstops is done along with the concreting of the rest of the blocks.

The Z-shape waterstop is simple in design and is easily installed and spliced but it can accommodate only small lateral movements. On the other hand the M-shape waterstop is relatively more difficult to splice but it can accommodate greater movement of the joint.

(ii) *Asphalt waterstops.* The asphalt waterstop is constructed by forming a square shaped well (or groove) across the contraction joint as shown in Fig. (9.26 d) and filling the well with a steam pipe or an electrical heat conductor for melting the asphalt. The molten asphalt fills the entire well and a portion of the joint on either side. The location, shape and dimensions generally adopted for asphalt waterstops are as shown in Fig. (9.26 d). The asphalt waterstops can also accommodate only small lateral movement of the joint. However, asphalt waterstops are used only as secondary waterstops to the metal and rubber/PVC waterstops.

(iii) *Rubber/PVC waterstops.* Figure (9.26 c) shows the shape and dimensions of rubber/PVC waterstops. The rubber waterstops are fabricated from natural rubber and the PVC (Polyvinyl Chloride) waterstops are fabricated from a plastic compound, the basic resin of which is polyvinyl chloride. The rubber and the plastic compound used for fabricating the rubber and PVC waterstops respectively should meet the requirements given in IS: 12200–1987. The rubber/PVC waterstop is embedded symmetrically in each of the two blocks of the dam on either side of the transverse joint as shown in Fig. (9.26 d). A rubber/PVC waterstop can accommodate a much greater lateral movement of the joint than that can be accommodated by the other types of the waterstops.

All the different types of waterstops described here are as per IS: 12200–1987 [Indian Standard Code of Practice for Provision of Waterstops at Transverse Contraction Joints in Masonry and Concrete Dams]. Further in most of the cases all the three types of waterstops are used simultaneously at each transverse joint as shown in Fig. (9.26 d) to ensure that the leakage of water from the reservoir through the transverse joints is completely stopped.

9.12 OPENINGS IN DAMS

The various openings provided in a dam may be classified on the basis of the purpose for which they are provided in the following two categories.

(i) Waterways

(ii) Gallery systems.

(i) Waterways are the openings provided for obtaining water from the reservoir for various purposes such as irrigation, water supply, power generation etc., as well as for allowing the excess water which cannot be stored in the reservoir during floods to overflow to the downstream side. These include outlets, penstocks, spillways etc.

(ii) Gallery systems comprise openings provided for obtaining access to the interior of the dam for various purposes. These include galleries, adits, vaults, shafts etc.

A detailed description of the openings included in the gallery systems is given in the next sections.

9.13 GALLERIES

A gallery is an opening provided in a dam for obtaining access to the interior of the dam. It is usually rectangular in shape with its top and bottom either flat or semicircular as shown in Fig. (9.27). For a gallery with flat top and bottom all the corners should be rounded to reduce the stress concentration [Fig. (9.27 a)]. The semicircular bottom of the gallery is however filled with an unbonded floor slab to provide a walkway as shown in Fig. (9.27 b). A gallery should be large enough to accommodate a drilling rig for drilling holes for grouting and to provide adequate working space and access for equipment for normal maintenance. As such the size of a gallery in general varies from 1.5 to 1.8 m in width and 2.2 to 2.4 m in height. A gallery may run in longitudinal or transverse direction and it may have horizontal or sloping floor. The various purposes for which a gallery is provided in a dam are as follows.

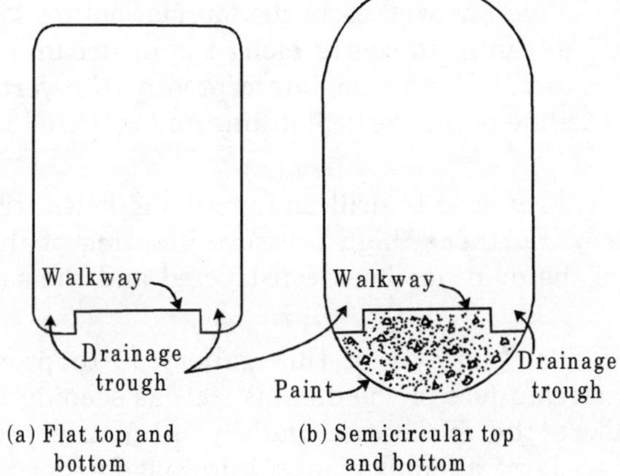

Fig. 9.27. *Typical galleries*

(i) To provide drainage for water percolating from the upstream face of the dam as well as seeping through the foundation of the dam.

(ii) To provide facilities for drilling and grouting operations for foundation of the dam.

(iii) To provide space for the piping system and equipment used for grouting of the contraction joints as well as for the post cooling of the concrete blocks.

(iv) To provide access to the interior of the dam for inspection and also for installing various instruments to study the structural behaviour of the dam after completion.

(v) To provide access to, and room for the mechanical and electrical equipment required for the operation of gates for outlet conduits, power penstocks and spillway crests.

(vi) To provide access through the dam for control cables and/or power cables.

(vii) To provide access routes for visitors.

Classification of Galleries

The galleries are classified on the basis of their location in the dam as well as the purpose for which these are provided as indicated below.

(i) *Foundation Gallery.* It is provided at the foundation level of the dam and it follows the configuration of the valley. This gallery generally extends over the entire length of the dam and it normally runs parallel to the axis of the dam. It is always provided very close to the upstream face of the dam but for the sake of structural safety the minimum thickness of concrete (or masonry) between the upstream face of the dam and the gallery should be either 10% of the maximum head of water on the upstream side or 6 m which ever is more. Further there should be a minimum of 1.5 m of concrete between the floor of the gallery and the foundation rock.

The main purpose of the foundation gallery is to provide drainage for the water percolating through the upstream face of the dam as well as seeping through the foundation of the dam and hence it is also known as *drainage gallery.* The water percolating through the dam is intercepted by vertical drains formed in the concrete (or masonry) and that seeping through the foundation is intercepted by vertical drain holes drilled in the foundation from this gallery. The vertical drains provided in the dam as well as in the foundation are 125 to 300 mm in diameter and spaced approximately 3 m centre to centre along the upstream or the downstream face of the gallery as shown in Fig. (9.27). The water intercepted by the vertical drains flows through open drain provided along the line of the vertical drains and collected in a sump from where it is pumped out.

The foundation gallery is also used to drill and grout the holes for the grout curtain on the upstream edge of the gallery. Further a high pressure grouting of the foundation of a dam is done only after a portion of the dam has been constructed and the same is also done from the foundation gallery.

(ii) *Drainage Gallery.* The purpose of this gallery is to provide drainage for water percolating through the upstream face of the dam as well as seeping through the foundation of the dam. As indicated above the foundation gallery is mainly provided for this purpose. However, in the case of high dams a supplementary drainage gallery is sometimes provided at the foundation level at about two-thirds the base width from the upstream face of the dam for draining the downstream portion of the foundation through the vertical drain holes drilled in the foundation from this gallery. This gallery usually extends only through the deepest portion of the dam.

(iii) *Gate Gallery.* A gate gallery in a dam provides access to, and room for, the mechanical and electrical equipment required for the operation of gates for outlet conduits, power penstocks and spillway crests. The size of this gallery will depend on the size of the gates to be served and that of the equipment to be housed.

(iv) *Grouting Gallery.* This gallery is meant to provide facilities for grouting of contraction joints within the dam. The supply, return and vent headers of the grout piping system are located in this gallery. These galleries are spaced in the vertical direction according to the grout lift which is generally 15 m. These galleries are also used for post-cooling of the concrete blocks and hence the piping system for the same terminates in these galleries.

(v) *Inspection Gallery.* This gallery provides access to the interior of the dam in order to inspect the structure and take observations of the instruments installed to study the structural behaviour of the dam after completion. However, the foundation, drainage, gate and grouting galleries also serve as inspection galleries.

9.14 ADITS, VAULTS AND SHAFTS

Adit. An adit is a passage-way or a gallery which connects the gallery system in the dam with the downstream face of the dam, or the features outside the dam, such as power house, gate house, pump chamber etc. It normally runs in the transverse direction. It is also known as *access gallery* or *entrance gallery.*

Vault. When a gallery is enlarged for installation of some equipment, it is called a vault or a chamber. For example, a portion of a gate gallery may have to be enlarged, to install equipment needed for hoisting of gates, which is then known as *hoisting chamber*.

Shaft. It is a vertical opening provided in the dam for various purposes such as for location of headers of post-cooling system, accommodation of measuring devices, for connecting galleries at different elevations by stair case or lift arrangement, for elevator movement and for hoisting equipment.

Plumb-Line Shaft or Collimator Shaft. It is a shaft located near the maximum section of a dam in order to make observations of the movements of the dam with respect to the base. A plumb-bob held at the top of the dam is suspended through this shaft. As the dam deflects relative to the base, the bob also gets deflected by the same amount which may be measured on a scale provided at the lower end near the bob.

Stilling Well Shaft. It is a special shaft provided sometimes in a dam. It opens to the reservoir below the minimum reservoir level and contains a floating mechanism to record reservoir fluctuations.

9.15 TEMPERATURE CONTROL IN CONCRETE DAMS

As stated earlier the major volumetric changes and the accompanying crack-producing tendencies are developed in a concrete dam due to temperature changes or steep temperature gradients caused by the heat of hydration evolved as the concrete sets. In order to reduce the crack-producing tendencies in a concrete dam contraction joints are provided. However, even with the sub-division of the dam into separate blocks by the contraction joints, the size of each block between the joints is quite large for which if no measures are adopted to control the rise in temperature due to the heat of hydration the cracks may be developed. The various measures commonly adopted to control the rise in temperature of the concrete due to the heat of hydration include precooling, postcooling, or a combination of the both and by the use of a concrete mix designed to limit the heat of hydration, each of which are described below.

Precooling of concrete. This is one of the most effective temperature control measures in which the temperature of the concrete is reduced by cooling each of the ingradients of the concrete before mixing to a minimum possible temperature. The water to be mixed to prepare the concrete can be cooled to a temperature ranging from 0° to 4.5° C. Crushed ice may also be added to the mix for cooling it. The coarse aggregate may be cooled to about 1.7° to 4.5° C by either chilling the aggregate in large tanks of refrigerated water, or by forcing refrigerated air through the aggregate while it is draining in stockpiles, while it is on a conveyor belt, and while it is in the bins of the batching plant. The sand may be cooled by passing it through vertical tubular heat exchangers. Cold air jets directed on the sand as it is transported on conveyor belts can also be used. However, the methods for cooling of cement have not been very successful.

By the use of the above methods for cooling the ingradients of the concrete it has been possible to obtain the concrete having temperatures as low as about 10°C. The low temperature of the concrete counteracts the heat of hydration of the cement, and thus controls the rise in temperature of the concrete.

Postcooling of concrete. The postcooling of concrete is done by circulating cold water through a pipe system embedded in the concrete. The system consists of pipe or tubing placed in grid like coils over the top surface of each lift of concrete after the concrete has hardened. The coils are formed by joining together lengths of thin-wall metal pipe or tubing. The pipe or tubing of 25 mm diameter are commonly used. The number of coils in a block depends upon the size of the block and the horizontal spacing which usually varies from 0.5 m to 2 m depending upon the extent of the cooling required. Each coil is individually connected through manifolds to the supply and return headers which are placed either on the downstream face of the dam or in the galleries or in the cooling shafts. The cooling water is usually pumped through the coils, but sometimes it may be made to flow under gravity. The minimum velocity of flow of the cooling water through the embedded coils is about 0.6 m/s, or the minimum flow rate is about 18 litres/minute in a 25 mm diameter pipe or tubing. The circulation of the cold water is started as soon as the coils are covered in the concrete of the next lift and continued till the temperature of the concrete has been reduced to mean annual temperature for that locality. In order to ascertain the temperature of the concrete, electrical resistance type thermometers are embedded in the concrete at mid-lift.

The postcooling reduces the peak temperatures of the concrete due to the heat of hydration. Moreover the postcooling is necessary if the contraction joints are to be grouted. This is so because the contraction joints can be grouted only after the concrete temperature has been reduced to a desired value at which the contraction joints between the blocks will open sufficiently to be grouted and this may be accomplished in a relatively shorter duration by using postcooling.

Design of concrete mix to limit the heat of hydration. The use of a concrete mix designed to limit the heat of hydration can significantly reduce the peak temperatures. Since the heat generated within the concrete is directly proportional to the amount of cement used per unit volume of concrete, the mix should be such that it would provide the required strength and durability with the lowest cement content. The quantity of cement in concrete can be reduced by replacing a part of it by pozzolans. The more common pozzolans used in mass concrete include calcined clays, diatomaceous earth, volcanic tuffs and pumicites, and fly ash. By using pozzolans a lower peak temperature is obtained. Further pozzolans develop heat of hydration at a much lower rate than do cements and most of the heat is lost to the surface. Pozzolans can also be used as a replacement for part of the cement to improve workability, effect economy and obtain a better quality concrete.

Other temperature control measures include,

1. use of cements which would develop low heat of hydration.

2. use of low placement lifts so that a greater percentage of the heat generated is lost to the surface.

3. use of water curing which reduces the surface temperature to about that of water which is less than the prevailing air temperature. Moreover as the water evaporates the surface temperature is further lowered.

4. use of retarding agents which retard the rate of heat generation allowing more heat to be lost to the surface.

9.16 FOUNDATION TREATMENT

The entire load acting on a gravity dam including the self weight is transmitted to the foundation and hence the foundation must be strong enough to withstand the entire load. However, it is very rare that the rocks in the foundations at the dam sites are ideal from the geological point of view. In most of the cases the rocks have numerous structural and lithologic conditions which may be undesirable in dam foundations. As such it is essential that a thorough geological investigation of the dam site is carried out to determine its feasibility for the construction of the dam as well as the defects in the foundation rock and the remedial treatments to be carried out before the construction is commenced.

Defects in Foundation Rock. The various geological conditions which constitute defects in dam foundations are primary features such as bedding planes, unconformities, shale beds etc., which are developed during the formation of the rocks; and secondary features such as faults, joints, cracks, cavities, seams etc., which result from various stresses, deformations and weathering. Some of these defects are common to all rocks, others are restricted to rocks of certain types. Thus, bedding planes are found in the sedimentary rocks which may also have unconformities. Similarly faults, joints and miscellaneous cracks are found in all consolidated rocks ; solution cavities are found in limestones and related carbonate rocks ; seams and zones of decayed material are found in sillicious, argillaceous and other non-calcareous rocks and flow contracts and flow tunnel in lavas.

Methods of Treatment. The first step in the foundation treatment is that of the surface preparation. This consists of removing all loose overburden of weathered rock as well as any decayed or weak rock till sound bed rock is exposed. For this purpose the excavation should be carried out in such a way that the underlying sound rock is not damaged. Further the final rock surface should be excavated to develop stepped or benched foundation to increase the sliding resistance of the dam. The surface should be thoroughly cleaned with wet sand blasting and washing before the concrete is laid.

Moreover in order to rectify the deficiencies in the deep foundation as well as to improve the water tightness of the foundation, the following methods of treatment are generally used.

(i) Excavation of seams of decayed or weak rock by tunneling and back filling the tunnels with concrete.

(ii) Excavation of weak rock zones by mining methods from shafts sunk to the zone and back filling the entire excavated region with concrete.

(iii) Excavating for and making a subterranean concrete cut-off walls across leakage channels in the dam foundation where the water channels are too large or too wet for mining and grouting.

(iv) Grouting the foundation to increase its strength and to render it impervious. The different types of groutings done for the dam foundation are described below.

Foundation Grouting. This consists of injecting under pressure a mixture of cement and water (called grout) into holes drilled in the foundation rock for the purpose of sealing seams, cracks and fissures or filling voids present in the rock. The foundation grouting may be classified on the basis of the pressure used for grouting as (i) low pressure grouting and (ii) high pressure grouting ; and on the basis of the purpose served by the grouting as (i) consolidation grouting and (ii) curtain grouting. The consolidation grouting is a low pressure grouting and the curtain grouting is a high pressure grouting. A brief description of consolidation grouting and curtain grouting is given below.

Consolidation Grouting. The consolidation grouting is done to consolidate the entire foundation rock and thus increase its strength. It is a low pressure grouting for which shallow holes (called *B* holes) are drilled through the foundation rock on a grid pattern. These holes are drilled to depths ranging from 3 to 15 m at spacings of 6 to 30 m. Prior to the commencement of the grouting operation the holes are thoroughly washed with alternate use of water and compressed air to remove all loose material and drill cuttings. The grout holes are then tested with water under pressure to obtain an idea of the tightness of the hole which is necessary to decide the consistency of the grout to be used and to locate seams or other openings in the rock which are to be plugged. The grout is then injected into these holes at a relatively low pressure which is usually less than about 392.4 kN/m^2 [or 4 kg f/cm^2]. Since this is a low pressure grouting it is accomplished before any concrete for the dam is laid. This grouting results in the consolidation of the foundation into more or less a monolithic mass by bonding together the jointed or shattered rock. Moreover this grouting serves as cutoff against the leakage of the grout of the high pressure grouting which may be required to be done later.

Curtain Grouting. The curtain grouting is done to create a deep curtain or a barrier in the foundation rock at the heel of the dam which will prevent leakage from the reservoir and thus reduce the uplift pressure on the dam. It is a high pressure grouting for which relatively deeper holes (called *A* holes) are drilled which may be either vertical or inclined. The inclined holes are drilled to intercept the principal joints in the rock to develop an effective curtain. In this case a primary series of holes with a spacing of 4.5 to 7.5 m depending upon the nature of the rock is drilled, washed, tested with water under pressure and grouted. A second series comprising an equal number of intermediate holes, is then drilled, washed, tested with water under pressure and grouted. This is then repeated for a third series which thus reduces the spacing of the holes to quarter of that of the first series. As such in this case the spacing of the holes may vary from 1.2 to 1.5 m. The depth of the holes depends upon the nature of the rock in the foundation and in general it may range from 30 to 40% of the head of water on dense foundations to 70% of the head for poorer foundations.

In this case since the grouting pressures are usually very high, this grouting is accomplished only after some concreting has been done so that the foundation strata is not lifted up or dislocated due to higher grouting pressures. Further generally only one line of grout holes parallel to the axis of the dam is sufficient. However, in exceptional cases, where very high pressures are required to be employed, initially one set of holes is drilled and grouted at comparatively lower pressures and after this another set of holes is drilled and grouted at the required higher pressures. Further for a dam of more than 60 m height or for the one constructed on inferior rock the cutoff curtain usually consists of a multiple line arrangement of grouted holes.

SOLVED ILLUSTRATIVE EXAMPLES

Illustrative Example 9.1 *Show that for an elementary profile of a dam Maurice Levy's criterion gives the same requirement of base width as the no tension requirement with unity intensity factor.*

Solution

For no tension at any point in the base of a dam of elementary profile of height *H*, the required base width *b* is given by Eq. (9.39 a) as

$$b = \frac{H}{\sqrt{s-1}}$$

where *s* is the specific gravity of concrete (or masonry).

According to Maurice Levy the normal (or vertical) stress at each point along the upstream face of the dam (computed without considering the uplift pressure) should be compressive and it should be more than or equal to the hydrostatic pressure at that point.

The compressive stress at the upstream face of the dam is given by Eq. (9.29 b) as

$$\sigma_v = \frac{\Sigma V}{b}\left(1 - \frac{6e}{b}\right)$$

in which ΣV is the sum of all the vertical forces acting on the dam; and e is the eccentricity.

Thus, according to Maurice Levy's criterion

$$\frac{\Sigma V}{b}\left(1-\frac{6e}{b}\right) = wH \qquad \text{...(i)}$$

where w is specific weight of water.

Since uplift pressure is not considered in accordance with Maurice Levy's criterion, the other forces acting on the dam in this case are the water pressure and the weight of the dam. The magnitudes of each of these two forces acting per unit length of the dam are as indicated below.

Water pressure $\qquad P = \dfrac{wH^2}{2}$

Weight of the dam $\qquad W = \dfrac{1}{2}\, bHsw.$

Thus, in this case $\Sigma V = W = \dfrac{1}{2}\, bHsw$ and introducing this value in Eq. (i), we get

$$\frac{W}{b}\left(1-\frac{6e}{b}\right) = wH$$

or

$$\frac{\frac{1}{2}bHsw}{b}\left(1-\frac{6e}{b}\right) = wH$$

or

$$e = \frac{b}{6}-\frac{b}{3s} \qquad \text{...(ii)}$$

Further if O' is the point of intersection of the resultant R of these two forces with the base of the dam then by taking the moments of all the forces about O', we get

$$\frac{wH^2}{2}\times\frac{H}{3} = \frac{1}{2}\, bHsw\left(e+\frac{b}{a}\right)$$

or

$$e = \left(\frac{H^2}{3sb}-\frac{b}{6}\right) \qquad \text{...(iii)}$$

Equating the values of e given by Eqs (ii) and (iii), we get

$$\frac{b}{6}-\frac{b}{3s} = \frac{H^2}{3sb}-\frac{b}{6}$$

or

$$b = \frac{H}{\sqrt{s-1}}$$

which is same as the base width required for no tension with unity intensity factor.

Illustrative Example 9.2 *Following data were obtained from the stability analysis of a concrete gravity dam.*

(i) *Total overturning moment about toe = 1 × 10^6 kN-m*

(ii) *Total resisting moment about toe = 2 × 10^6 kN-m*

(iii) *Total vertical force above base = 50 000 kN.*

(iv) *Base width of the dam = 50 m*

(v) *Slope of the d/s face = 0.8 (H) : 1 (V)*

Calculate the maximum and minimum vertical stress to which the foundation will be subjected to. What is the maximum principal stress at toe? Assume there is no tail water.

Solution

Let $\bar{x}$ be the distance of the point of intersection of the resultant with the base of the dam from the toe. Then

$$\bar{x} = \frac{\Sigma M}{\Sigma V}$$

$$= \frac{(2\times10^6)-(1\times10^6)}{50\,000} = 20 \text{ m}$$

Eccentricity $$e = \frac{b}{2} - \bar{x} = \left(\frac{50}{2} - 20\right) = 5 \text{ m}$$

The maximum vertical stress is at the toe of the dam which is given by Eq. (9.29 a) as

$$\sigma_{y_D} = \frac{\Sigma V}{b}\left(1+\frac{6e}{b}\right) = \frac{50\,000}{50}\left(1+\frac{6\times5}{50}\right)$$

$$= 1600 \text{ kN/m}^2 \text{ (compressive)}$$

Similarly the minimum vertical stress is at the heel of the dam which is given by Eq. (9.29 b) as

$$\sigma_{y_U} = \frac{\Sigma V}{b}\left(1-\frac{6e}{b}\right) = \frac{50\,000}{50}\left(1-\frac{6\times5}{50}\right)$$

$$= 400 \text{ kN/m}^2 \text{ (compressive)}$$

The maximum principal stress at the toe of the dam is given by Eq. (9.30 a) as

$$\sigma_{1_D} = \sigma_{y_D} \sec^2 \phi_D$$

Since $$\tan \phi_D = 0.8, \sec^2 \phi_D = 1.64$$

∴ $$\sigma_{1_D} = 1600 \times 1.64 = 2624 \text{ kN/m}^2$$

Illustrative Example 9.3 *Following data were obtained from the stability analysis of a concrete gravity dam:*

(i) Total overturning moment about toe = 1 × 10^5 t-m

(ii) Total resisting moment about toe = 2 × 10^5 t-m

(iii) Total vertical force above base = 5000 t

(iv) Base width of the dam = 50 m

(v) Slope of the d/s face = 0.8(H) : 1(V)

Calculate the maximum and minimum vertical stress to which the foundation will be subjected to. What is the maximum principal stress at toe? Assume there is no tail water.

Solution

Let be the distance of the point of inetersection of the resultant with the base of the dam from the toe. Then

$$\bar{x} = \frac{\Sigma M}{\Sigma V} = \frac{(2\times10^5)-(1\times10^5)}{5000} = 20 \text{ m}$$

Eccentricity
$$e = \frac{b}{2} - \bar{x} = \left(\frac{50}{2} - 20\right) = 5 \text{ m}$$

The maximum vertical stress is at the toe of the dam which is given by Eq. (9.29 a) as

$$\sigma_{y_D} = \frac{\Sigma V}{b}\left(1+\frac{6e}{b}\right)$$

$$= \frac{5000}{50}\left(1+\frac{6\times5}{50}\right)$$

$$= 160 \ t/\text{m}^2 \text{ (compressive)}$$

Similarly the minimum vertical stress is at the heel of the dam which is given by Eq. (9.29 b) as

$$\sigma_{y_U} = \frac{\Sigma V}{b}\left(1-\frac{6e}{b}\right) = \frac{5000}{50}\left(1-\frac{6\times5}{50}\right)$$

$$= 40 \ t/\text{m}^2 \text{ (compressive)}$$

The maximum principal stress at the toe of the dam is given by Eq. (9.30 a) as

$$\sigma_{I_D} = \sigma_{y_D} \sec^2 \phi_D$$

Since
$$\tan \phi_D = 0.8, \ \sec^2 \phi_D = 1.64$$

$$\therefore \quad \sigma_{I_D} = 160 \times 1.64 = 262.4 \ t/\text{m}^2$$

Illustrative Example 9.4 *A 100 m high concrete gravity dam as shown in Fig. Ex. 9.4 is located in seismic zone V. Specific weight of concrete is 24.525 kN/m^3 (2.5 t/m^3) and its modulus of elasticity is 20.601 × 10^6 kN/m^2 (2.1 × 10^6 t/m^2). Calculate the earthquake forces by seismic coefficient method as well as by response spectrum method per unit length of the dam.*

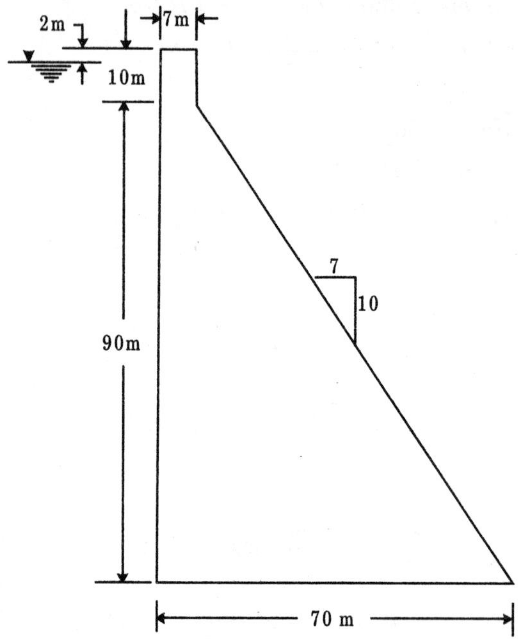

Fig. Ex. 9.4

Solution

(A) Inertia Force

(a) *Seismic Coefficient Method.* Horizontal seismic coefficient α_h at the top of the dam is given by

$$\alpha_{h_T} = 1.5\ \alpha = 1.5\ \beta I \alpha_0$$

in which $\beta = 1.0$; $I = 2.0$; and for zone V $\alpha_0 = 0.08$

$$\alpha_{h_T} = (1.5 \times 1 \times 2 \times 0.08) = 0.24$$

α_h varies linearly from a value of 0.24 at the top to zero at the base.

SI units

For the portion of the dam upto 10 m below the top of the dam the inertia force per unit length of the dam is given by

$$F_{10} = \int_0^{10} (1 \times 7 \times dy)\ 24.525 \times (1 - 0.01y)\ 0.24$$

$$= 391.419 \text{ kN}$$

The moment of the inertia force F_{10} about any point in the horizontal section at 10 m below the top of the dam is given by

$$M_{10} = \int_0^{10} (1 \times 7 \times dy) 24.525 \times (1 - 0.01y)\ 0.24\ (10 - y)$$

$= 1991.43$ kN-m

For the portion of the dam upto 50 m below the top of the dam the inertia force per unit length of the dam is given by

$$M_{50} = F_{10} + \int_{10}^{50} 1[7+0.7(y-10)]\,dy \times 24.525 \times (1-0.01y)\,0.24$$

$$= 391.419 + 3241.224$$

$$= 3632.643 \text{ kN}$$

The moment of the inertia force F_{50} about any point in the horizontal section at 50 m below the top of the dam is given by

$$M_{50} = M_{10} + F_{10}\,(50 - 10)$$

$$+ \int_{10}^{50} 1[7+0.7(y-10)]\,dy \times 24.525 \times (1-0.01y)\,0.24(50-y)$$

$$= 1991.43 + 15\,656.76 + 56\,034.72$$

$$= 73\,682.91 \text{ kN-m}$$

For the entire dam above the base of the dam the inertia force per unit length of the dam is given by

$$F_{100} = F_{10} + \int_{10}^{100} 1[7+0.7(y-10)]dy \times 24.525 \times (1-0.01y)0.24$$

$$= 391.419 + 6674.724$$

$$= 7066.143 \text{ kN}$$

The moment of the inertia force F_{100} about any point in the base of the dam is given by

$$M_{100} = M_{10} + F_{10}\,(100 - 10)$$

$$+ \int_{10}^{100} 1[7+0.7(y-10)]dy \times 24.525 \times (1-0.01y)0.24(100-y)$$

$$= 1991.43 + 35\,227.71 + 325\,392.79$$

$$= 362\,611.93 \text{ kN-m}$$

Metric Units

For the portion of the dam upto 10 m below the top of the dam the inertia force per unit length of the dam is given by

$$F_{10} = \int_{0}^{10} (1 \times 7 \times dy)2.5 \times (1-0.01y)0.24$$

$$= 39.9\ t$$

The moment of the inertia force F_{10} about any point in the horizontal section at 10 m below the top of the dam is given by

$$M_{10} = \int_0^{10} (1 \times 7 \times dy) 2.5 \times (1 - 0.01y) 0.24 (10 - y)$$

$$= 203.0 \; t\text{-m}$$

For the portion of the dam upto 50 m below the top of the dam the inertia force per unit length of the dam is given by

$$M_{50} = M_{10} + \int_{10}^{50} 1[7 + 0.7(y - 10)] dy \times 2.5 \times (1 - 0.01y) 0.24$$

$$= 39.9 + 330.4 = 370.3 \; t$$

The moment of the inertia force F_{50} about any point in the horizontal section at 50 m below the top of the dam is given by

$$M_{50} = M_{10} + F_{10}(50 - 10) + \int_{10}^{50} 1[7 + 0.7(y - 10)]\, dy \times 2.5 \times (1 - 0.01y) 0.24 (50 - y)$$

$$= 203.0 + 1596.0 + 5712.0$$

$$= 7511 \; t\text{-m}$$

For the entire dam above the base of the dam the inertia force per unit length of the dam is given by

$$F_{100} = \text{F}_{10} + \int_{10}^{100} 1[7 + 0.7(y - 10)]\, dy \times 2.5 \times (1 - 0.01y)\, 0.24$$

$$= 39.9 + 680.4 = 720.3 \; t$$

The moment of the inertia force F_{100} about any point in the base of the dam is given by

$$F_{100} = M_{10} + F_{10}(100 - 10) + \int_{10}^{100} 1[7 + 0.7(y - 10)]\, dy \times 2.5 \times (1 - 0.01y)\, 0.24 (100 - y)$$

$$= 203.0 + 3591.0 + 33\,169.5$$

$$= 36\,963.5 \; t\text{-m}$$

(b) *Response Spectrum Method.* The fundamental period of vibration of the dam is given by Eq. 9.9 as

$$T = 5.55 \frac{H^2}{B} \sqrt{\frac{w_m}{gE_s}}$$

SI Units

$$T = 5.55 \times \frac{(100)^2}{70} \times \sqrt{\frac{24.525}{9.81 \times 20.601 \times 106}}$$

$= 0.276\ s$

Horizontal seismic coefficient α_h is equal to α given by Eq. (9.8), and hence

$$\alpha_h = \beta I F_0 \frac{S_\alpha}{g}$$

Values of β and I are same as taken in previous case.

For zone V, $F_0 = 0.4$ and from Fig. 9.6 for $T = 0.276$ s and 5% damping, $(S_\alpha/g) = 0.19$

$$\therefore \quad \alpha_h = 1.0 \times 2.0 \times 0.4 \times 0.19 = 0.152$$

The inertia force per unit length for the entire dam is given by Eq. (9.11) as

$$F_{100} = 0.6\ W\ \alpha_h$$

$$W = 1 \times \left(7 \times 100 + \frac{63 \times 90}{2}\right) \times 24.525$$

$$= 86\ 695.875 \text{ kN}$$

$$\therefore \quad F_{100} = 0.6 \times 86\ 695.875 \times 0.152 = 7906.664 \text{ kN}$$

The moment of F_{100} about any point in the base of the dam is given by Eq. (9.12) as

$$M_{100} = 0.9\ W\ \bar{h}\ \alpha_h$$

$$\bar{h} = \frac{\left(7 \times 100 \times \frac{100}{2} + \frac{1}{2} \times 63 \times 90 \times \frac{90}{3}\right)}{\left(7 \times 100 + \frac{1}{2} \times 63 \times 90\right)}$$

$$= 33.96 \text{ m}$$

$$\therefore \quad M_{100} = 0.9 \times 86\ 695.875 \times 33.96 \times 0.152$$

$$= 402\ 765.45 \text{ kN-m}$$

For the portion of the dam upto 10 m below the top of the dam the inertia force per unit length of the dam is given by Eq. (9.13) as

$$F_{10} = C_F \times F_{100}$$

From Fig. (9.8), for $(y/H) = 0.1$, $C_F = 0.075$

$$\therefore \quad F_{10} = 0.075 \times 7906.664 = 593 \text{ kN}$$

The moment of F_{10} about any point in the horizontal section at 10 m below the top of the dam is given by Eq. (9.14) as

$$M_{10} = C_M \times M_{100}$$

From Fig. (9.8), for $(y/H) = 0.1$, $C_M = 0.015$

$$\therefore \quad M_{10} = 0.015 \times 402\ 765.45 = 6041.482 \text{ kN-m}$$

For the portion of the dam upto 50 m below the top of the dam the inertia force per unit length of the dam is given by Eq. (9.13) as

$$F_{50} = C_F \times F_{100}$$

From Fig. (9.8), for $(y/H) = 0.5$, $C_F = 0.52$

$$\therefore \quad F_{50} = 0.52 \times 7906.664 = 4111.465 \text{ kN}$$

The moment of F_{50} about any point in the horizontal section at 50 m below the top of the dam is given by Eq. (9.14) as

$$M_{50} = C_M \times M_{100}$$

From Fig. (9.8), for $(y/H) = 0.5, C_M = 0.235$

$$\therefore \quad M_{50} = 0.235 \times 402\,765.45 = 94\,649.88 \text{ kN-m}$$

Metric Units

$$T = 5.55 \times \frac{(100)^2}{70} \times \sqrt{\frac{2.5}{9.81 \times 2.1 \times 10^6}}$$

$$= 0.276 \text{ s}$$

Horizontal seismic coefficient α_h is equal to α given by Eq. (9.8), and hence

$$\alpha_h = \beta I F_0 \frac{S_a}{g}$$

Values of β and I are same as taken in previous case.

For Zone = V $F_0 = 0.4$ and from Fig. (9.6) for $T = 0.276\ s$ and 5% damping, $(S_a/g) = 0.19$

$$\therefore \quad \alpha_h = 1.0 \times 2.0 \times 0.4 \times 0.19 = 0.152$$

The inertia force per unit length for the entire dam is given by Eq. (9.11) as

$$F_{100} = 0.6\ W \alpha_h$$

$$W = 1 \times \left(7 \times 100 + \frac{63 \times 90}{2}\right) \times 2.5 = 8837.5\ t$$

$$\therefore \quad F_{100} = 0.6 \times 8837.5 \times 0.152 = 805.98\ t$$

The moment of F_{100} about any point in the base of the dam is given by Eq. (9.12) as

$$M_{100} = 0.9\ W \bar{h}\ \alpha_h$$

$$\bar{h} = \frac{\left(7 \times 100 \times \frac{100}{2} + \frac{1}{2} \times 63 \times 90 \times \frac{90}{3}\right)}{\left(7 \times 100 + \frac{1}{2} \times 63 \times 90\right)} = 33.96 \text{ m}$$

$$\therefore \quad M_{100} = 0.9 \times 8837.5 \times 33.96 \times 0.152 = 41\,056.62\ t\text{-m}$$

For the portion of the dam upto 10 m below the top of the dam the inertia force per unit length of the dam is given by Eq. (9.13) as

$$F_{10} = C_F \times F_{100}$$

From Fig. (9.8), for $(y/H) = 0.1, C_F = 0.075$

$$\therefore \quad F_{10} = 0.075 \times 805.98 = 60.45\ t$$

The moment of F_{10} about any point in the horizontal section at 10 m below the top of the dam is given by Eq. (9.14) as

$$M_{10} = C_M \times M_{100}$$

From Fig. (9.8), for $(y/H) = 0.1, C_M = 0.015$

$$\therefore \quad M_{10} = 0.015 \times 41056.62 = 615.85 \; t\text{-m}$$

For the portion of the dam upto 50 m below the top of the dam the inertia force per unit length of the dam is given by Eq. (9.13) as

$$F_{50} = C_F \times F_{100}$$

From Fig. (9.8), for $(y/H) = 0.5, C_F = 0.52$

$$\therefore \quad F_{50} = 0.52 \times 805\ 98 = 419.11 \; t$$

The moment of F_{50} about any point in the horizontal section at 50 m below the top of the dam is given by Eq. (9.14) as

$$M_{50} = C_M \times M_{100}$$

From Fig. (9.8), for $(y/H) = 0.5, C_M = 0.235$

$$\therefore \quad M_{50} = 0.235 \times 41056.62 = 9648.31 \text{ t-m}$$

(B) Hydrodynamic Pressure (or Suction)

The hydrodynamic pressure at depth y below the reservoir surface is given by Eq. (9.15) as

$$P_e = C\alpha_h wh$$

From Eq. (9.17)

$$C = \frac{C_m}{2}\left[\frac{y}{h}\left(2-\frac{y}{h}\right)+\sqrt{\frac{y}{h}\left(2-\frac{y}{h}\right)}\right]$$

and from Eq. (9.18)

$$C_m = 0.735 \text{ (for } \phi = 0°)$$

Thus,

$$C = \frac{0.735}{2}\left[\frac{y}{98}\left(2-\frac{y}{98}\right)+\sqrt{\frac{y}{98}\left(2-\frac{y}{98}\right)}\right]$$

SI Units

The value of p_e in kN/m² is given by

$$p_e = \frac{0.735}{2}\left[\frac{y}{98}\left(2-\frac{y}{98}\right)+\sqrt{\frac{y}{98}\left(2-\frac{y}{98}\right)}\right]\times \alpha_h \times 9.81 \times 98$$

or

$$p_e = 353.307\left[\frac{y}{98}\left(2-\frac{y}{98}\right)+\sqrt{\frac{y}{98}\left(2-\frac{y}{98}\right)}\right]\alpha_h$$

At depth 8 m from reservoir level or 10 m from the top of the dam

$$P_{e_{10}} = 195.142\ \alpha_h$$

At depth 48 m from reservoir level or 50 m from the top of the dam

$$P_{e_{50}} = 565.201\ \alpha_h$$

At depth 98 m from reservoir level or 100 m from the top of the dam

$$P_{e_{100}} = 706.614\ \alpha_h$$

(a) *Seismic Coefficient Method.* From Eq. (9.16)

$$\alpha_h = \frac{F_y}{W_y}$$

At depth 8 m from reservoir level or 10 m from the top of the dam

$$\alpha_h = \frac{391.419}{1 \times 7 \times 10 \times 24.525} = 0.228$$

$$\therefore \quad P_{e_{100}} = 195.142 \times 0.228 = 44.492 \text{ kN/m}^2$$

At depth 48 m from reservoir level or 50 m from the top of the dam

$$\alpha_h = \frac{3632.643}{\left(7 \times 50 + \dfrac{28 \times 40}{2}\right) \times 1 \times 24.525} = 0.163$$

$$\therefore \quad P_{e_{50}} = 565.201 \times 0.163$$

$$= 92.128 \text{ kN/m}^2$$

At depth 98 m from reservoir level or 100 m from the top of the dam

$$\alpha_h = \frac{7066.143}{\left(7 \times 100 + \dfrac{63 \times 90}{2}\right) \times 1 \times 24.525} = 0.082$$

$$\therefore \quad P_{e_{100}} = 706.614 \times 0.082 = 57.942 \text{ kN/m}^2$$

Total hydrodynamic pressure per unit length of the dam acting horizontally for a part of the dam upto depth y from reservoir level and its moment about any point in the horizontal section at this depth are given by Eqs (9.19) and (9.20) respectively as

$$P_e = 0.726\ p_e y$$

and

$$M_e = 0.299\ p_e y^2$$

At depth 8 m from reservoir level or 10 m from the top of the dam

$$P_{e_{10}} = 0.726 \times 44.492 \times 8$$

$$= 258.410 \text{ kN/m}$$

$$M_{e_{10}} = 0.299 \times 44.492 \times (8)^2$$

$$= 851.399 \text{ kN-m/m}$$

At depth 48 m from reservoir level or 50 m from the top of the dam

$$P_{e_{50}} = 0.726 \times 92.128 \times 48 = 3210.477 \text{ kN/m}$$

$$M_{e_{50}} = 0.299 \times 92.128 \times (48)^2 = 63\,466.608 \text{ kN-m/m}$$

At depth 98 m from reservoir level or 100 m from the top of the dam

$$P_{e_{100}} = 0.726 \times 57.942 \times 98 = 4122.457 \text{ kN/m}$$

$$M_{e_{100}} = 0.299 \times 57.942 \times (98)^2 = 166\,386.010 \text{ kN-m/m}$$

(b) *Response Spectrum Method.* As computed earlier the value of

$$\alpha_h = 0.152$$

and hence at depth 8 m from reservoir level or 10 m from the top of the dam

$$p_{e_{10}} = 195.142 \times 0.152$$

$$= 29.662 \text{ kN/m}^2$$

$$P_{e_{10}} = 0.726 \times 29.662 \times 8 = 172.277 \text{ kN/m}$$

$$M_{e_{10}} = 0.299 \times 29.662 \times (8)^2 = 567.612 \text{ kN-m/m}$$

at depth 48 m from reservoir level or 50 m from the top of the dam

$$p_{e_{50}} = 565.201 \times 0.152 = 85.911 \text{ kN/m}^2$$

$$P_{e_{50}} = 0.726 \times 85.911 \times 48 = 2993.827 \text{ kN/m}$$

$$M_{e_{50}} = 0.299 \times 85.911 \times (48)^2 = 59\,183.740 \text{ kN-m/m}$$

at depth 98 m below reservoir level or 100 m from the top of the dam

$$p_{e_{100}} = 706.614 \times 0.152 = 107.405 \text{ kN/m}^2$$

$$p_{e_{100}} = 0.726 \times 107.405 \times 98 = 7641.651 \text{ kN/m}$$

$$M_{e_{100}} = 0.299 \times 107.405 \times (98)^2 = 308\,423.760 \text{ kN-m/m}$$

Metric Units

The value of p_e in t/m^2 is given by

$$p_e = \frac{0.735}{2}\left[\frac{y}{98}\left(2-\frac{y}{98}\right)+\sqrt{\frac{y}{98}\left(2-\frac{y}{98}\right)}\right]\times \alpha_h \times 1 \times 98$$

$$= 36.015\left[\frac{y}{98}\left(2-\frac{y}{98}\right)+\sqrt{\frac{y}{98}\left(2-\frac{y}{98}\right)}\right]\times \alpha_h$$

At depth 8 m from reservoir level or 10 m from the top of the dam

$$p_{e_{10}} = 19.892\ \alpha_h$$

At depth 48 m from reservoir level or 50 m from the top of the dam

$$p_{e_{50}} = 57.615\ \alpha_h$$

At depth 98 m from reservoir level or 100 m from the top of the dam

$$p_{e_{100}} = 72.03\ \alpha_h$$

(a) *Seismic Coefficient Method.* From Eq. (9.16)

$$\alpha_h = \frac{F_y}{W_y}$$

At depth 8 m from reservoir level or 10 m from the top of the dam

$$\therefore \qquad \alpha_h = \frac{39.9}{1 \times 7 \times 10 \times 2.5} = 0.228$$

$$\therefore \qquad p_{e_{10}} = 19.892 \times 0.228 = 4.535\ t/\text{m}^2$$

At depth 48 m from reservoir level of 50 m from the top of the dam

$$\alpha_h = \frac{370.3}{\left(7 \times 50 + \dfrac{28 \times 40}{2}\right) \times 1 \times 2.5} = 0.163$$

$$\therefore \qquad p_{e_{50}} = 57.615 \times 0.163 = 9.391\ t/\text{m}^2$$

At depth 98 m from reservoir level or 100 m from the top of the dam

$$\alpha_h = \frac{720.3}{\left(7 \times 100 + \dfrac{63 \times 90}{2}\right) \times 1 \times 2.5} = 0.082$$

$$\therefore \qquad p_{e_{100}} = 72.03 \times 0.082 = 5.906\ t/\text{m}^2$$

Total hydrodynamic pressure per unit length of the dam acting horizontally for a part of the dam upto depth y from reservoir level and its moment about any point in the horizontal section at this depth are given by Eqs (9.19) and (9.20) respectively as

$$P_e = 0.726\ p_e y$$

and

$$M_e = 0.299\ p_e y^2$$

At depth 8 m from reservoir level or 10 m from the top of the dam

$$P_{e_{10}} = 0.726 \times 4.535 \times 8 = 26.339\ t/\text{m}$$

$$M_{e_{10}} = 0.299 \times 4.535 \times (8)^2 = 86.782\ t\text{-m/m}$$

At depth 48 m from reservoir level or 50 m from the top of the dam

$$P_{e_{50}} = 0.726 \times 9.391 \times 48 = 327.258\ t\text{/m}$$

$$M_{e_{50}} = 0.299 \times 9.391 \times (48)^2 = 6469.422\ t\text{-m/m}$$

At depth 98 m from reservoir level or 100 m from the top of the dam

$$p_{e_{100}} = 0.726 \times 5.906 \times 98 = 420.200\ t\text{/m}$$

$$M_{e_{100}} = 0.299 \times 5.906 \times (98)^2 = 16959.645\ t\text{-m/m}$$

(b) *Response Spectrum Method.* As computed earlier the value of

$$\alpha_h = 0.152$$

and hence at depth 8 m from reservoir level or 10 m from the top of the dam

$$p_{e_{10}} = 19.892 \times 0.152 = 3.024\ t\text{/m}^2$$

$$P_{e_{10}} = 0.726 \times 3.024 \times 8 = 17.563\ t\text{/m}$$

$$M_{e_{10}} = 0.299 \times 3.024 \times (8)^2 = 57.867\ t\text{-m/m}$$

at depth 48 m from reservoir level or 50 m from the top of the dam

$$P_{e_{50}} = 57.615 \times 0.152 = 8.757\ t\text{/m}^2$$

$$P_{e_{50}} = 0.726 \times 8.757 \times 48 = 305.164\ t\text{/m}$$

$$M_{e_{50}} = 0.299 \times 8.757 \times (48)^2 = 6032.662\ t\text{-m/m}$$

at depth 98 m below reservoir level or 100 m from the top of the dam

$$p_{e_{100}} = 72.03 \times 0.152 = 10.949\ t\text{/m}^2$$

$$p_{e_{100}} = 0.726 \times 10.949 \times 98 = 778.999\ t\text{/m}$$

$$M_{e_{100}} = 0.299 \times 10.949 \times (98)^2 = 31441.103\ t\text{-m/m}$$

Illustrative Example 9.5 (*a*) *Check the stability of the gravity dam section shown in Fig. Ex. 9.5 for reservoir empty and reservoir full conditions. Neglect seismic forces. Also find principal stress and shear at the toe and the heel of the dam.*

(*b*) *Check the stability of the dam section by considering seismic forces also assuming a constant value of the seismic coefficient a = 0.1 for the entire dam both in the horizontal as well as vertical directions.*

Take specific weight of concrete = 2.4 t/m³ (23.544 kN/m³); coefficient of friction m = 0.7; and shear strength of concrete c = 14 kg f/cm² (1373.4 kN/m²).

Solution

(a) **Stability analysis**

Case I. *Reservoir empty condition.* When the reservoir is empty, the only force acting on the

TABLE Ex. 9.5 (A) Forces (other than seismic) and their moments

S.No.	Item	Description and Dimension	Forces (t/m)		Lever Arm from toe of dam (m)	Moment about toe of dam (t-m/m)		Remarks
			Vertical	Horizontal		Anticlockwise (+ve)	Clockwise (–ve)	
1.		*Weight of the dam*						
	W_1	$\frac{1}{2}\times 4.5\times 45\times 1\times 2.4$	243.00		52.75	12 818.25		For reservoir empty condition the only force acting is the weight of the dam
	W_2	$7.5 \times 75 \times 1 \times 2.4$	1 350.00		47.50	64 125.00		
	W_3	$\frac{1}{2}\times 43.75\times 62.5\times 1\times 2.4$	3 281.25		29.17	95 714.06		
		Sum	ΣV_1 = 4 874.25		Sum	ΣM_1 = 172 657.31		
2.		*Water pressure*						
	P_H	$\frac{1}{2}\times 1\times 72\times 72\times 1$		2 592.00	24.00		62 208.25	
	P_{v_1}	$4.5 \times 27 \times 1 \times 1$	121.50		53.50	6 500.25		
	P_{v_2}	$\frac{1}{2}\times 4.5\times 45\times 1$	101.25		54.25	5 492.81		
3.		*Uplift pressure*						
	U_1	$\frac{1}{3}(72)\times 7\times 1$	– 168.00		52.25		8 778.00	
	U_2	$\frac{1}{2}\times\frac{2}{3}(72)\times 7\times 1$	– 168.00		53.42		8 974.56	
	U_3	$\frac{1}{2}\times\frac{1}{3}(72)\times 48.75\times 1$	– 585.00		32.50		19 012.50	
		Sum	ΣV_2 = 4 176.00	ΣH_1 = 2 592.00	Sum	184 650.37	98 973.06	
						ΣM_2 = 85 677.31		

dam is the weight of the dam. The resulting sum of the vertical forces ΣV_1 and the sum of the moments of the forces ΣM_1 for this case are worked out in the Table Ex. 9.5 1.

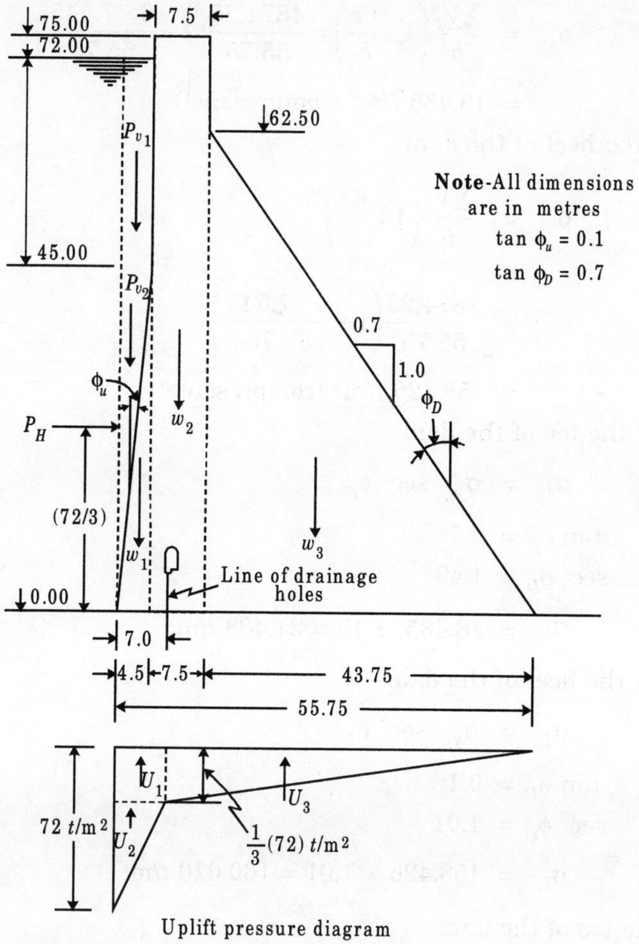

Fig. Ex. 9.5

If $\bar{x}$ is the distance of the point of intersection O' of the resultant force with the base from the toe of the dam, then

$$\bar{x} = \frac{\Sigma M_1}{\Sigma V_1} = \frac{1{,}72{,}657.31}{4{,}874.25} = 35.42 \text{ m}$$

The distance of point O' from the mid-point of the base O, i.e., eccentricity is

$$e = \frac{b}{2} - \bar{x}$$

$$= \frac{1}{2}(55.75) - 35.42 = -7.545 \text{ m}$$

The negative sign indicates that point O' lies on the left of the mid-point O of the base.

Vertical stress at the toe of the dam

$$\sigma_{y_D} = \frac{\Sigma V_1}{b}\left(1-\frac{6e}{b}\right) = \frac{4874.25}{55.75}\left(1-\frac{6\times 7.545}{55.75}\right)$$

$$= 16.435 \ t/m^2 \text{ (compressive)}$$

Vertical stress at the heel of the dam

$$\sigma_{y_U} = \frac{\Sigma V_1}{b}\left(1+\frac{6e}{b}\right)$$

$$= \frac{4874.25}{55.75}\left(1+\frac{6\times 7.545}{55.75}\right)$$

$$= 158.426 \ t/m^2 \text{ (compressive)}$$

Principal stress at the toe of the dam

$$\sigma_{I_D} = \sigma_{y_D} \sec^2 \phi_D$$

$$\tan \phi_D = 0.7$$

$$\therefore \quad \sec^2 \phi_D = 1.49$$

$$\therefore \quad \sigma_{I_D} = 16.435 \ \ 1.49 = 24.488 \ t/m^2$$

Principal stress at the heel of the dam

$$\sigma_{I_U} = \sigma_{y_D} \sec^2 \phi_U$$

$$\tan \phi_U = 0.1$$

$$\therefore \quad \sec^2 \phi_U = 1.01$$

$$\therefore \quad \sigma_{I_U} = 158.426 \times 1.01 = 160.010 \ t/m^2$$

Shear stress at the toe of the dam

$$(\tau_{yx})_D = \sigma_{y_D} \tan \phi_D$$

$$= 16.435 \times 0.7 = 11.505 \ t/m^2$$

Shear stress at the heel of the dam

$$(\tau_{yx})_U = \sigma_{y_D} \tan \phi_U = 158.426 \times 0.1 = 15.8426 \ t/m^2$$

For reservoir empty condition there cannot be any sliding or overturning.

Case II. *Reservoir full condition.* In this case the various forces acting on the dam are the weight of the dam, the water pressure and the uplift pressure. The resulting sum of the vertical forces ΣV_2, sum of the horizontal forces ΣH and the sum of the moments of the forces ΣM_2 for this case are given in the Table Ex. 9.5 (A).

If $\bar{x}$ is the distance of the point of intersection O' of the resultant force with the base from the toe of the dam, then

$$\bar{x} = \frac{\Sigma M_2}{\Sigma V_2} = \frac{85{,}677.31}{4{,}176.00} = 20.517 \text{ m}$$

The distance of point O' from the mid-point of the base O, i.e., eccentricity is

$$e = \frac{b}{2} - \bar{x} = \frac{1}{2}(5.75) - 20.517 = 7.358 \text{ m}$$

Vertical stress at the toe of the dam

$$\sigma_{y_D} = \frac{\Sigma V_2}{b}\left(1 + \frac{6e}{b}\right) = \frac{4176}{55.75}\left(1 + \frac{6 \times 7.358}{55.75}\right)$$

$$= 134.223 \ t/\text{m}^2 \text{ (compressive)}$$

Vertical stress at the heel of the dam

$$\sigma_{y_U} = \frac{\Sigma V_2}{b}\left(1 - \frac{6e}{b}\right) = \frac{4176}{55.75}\left(1 - \frac{6 \times 7.358}{55.75}\right)$$

$$= 15.588 \ t/\text{m}^2 \text{ (compressive)}$$

Principal stress at the toe of the dam

$$\sigma_{I_D} = \sigma_{y_D} \sec^2 \phi_D = 134.223 \times 1.49 = 199.922 \ t/\text{m}^2$$

Principal stress at the heel of the dam

$$\sigma_{I_U} = \sigma_{y_D} \sec^2 \phi_U - p \tan^2 \phi_U$$

$$p = wH = 1 \times 72 = 72 \text{ t/m}^2$$

$$\therefore \quad \sigma_{I_U} = 15.588 \times 1.01 - 72 \times 0.01$$

$$= 15.024 \ t/\text{m}^2$$

Shear stress at the toe of the dam

$$(\tau_{yx})_D = \sigma_{y_D} \tan \phi_D$$

$$= 134.223 \times 0.7 = 93.956 \ t/\text{m}^2$$

Shear stress at the heel of the dam

$$(\tau_{yx})_U = -(\sigma_{y_U} - p) \tan \phi_U$$

$$= -(15.588 - 72) \times 0.1$$

$$= 5.641 \ t/\text{m}^2$$

Factor of safety against overturning

$$\text{F.S.} = \frac{\Sigma M(+\text{ve})}{\Sigma M(-\text{ve})} = \frac{184650.37}{98973.06} = 1.866$$

Since F.S. > 1.5, the dam section is safe against overturning.

Factor of safety against sliding

$$F_1 = \frac{\mu \Sigma V_2}{\Sigma H} = \frac{0.7 \times 4176.0}{2592.0} = 1.128$$

Since $F_1 > 1$, the dam section is safe against sliding.

Shear friction factor

$$F_2 = \frac{\mu \Sigma V_2 + bc}{\Sigma H}$$

$$c = 14 \text{ kg(f)/cm}^2 = \frac{14 \times 10^4}{1000} = 140 \; t/\text{m}^2$$

$$\therefore \quad F_2 = \frac{(0.7 \times 4176.0) + (55.75 \times 140)}{2592.0} = 4.139$$

Since $F_2 > 4$, the dam section is safe against sliding.

(b) Stability analysis by considering seismic forces. If seismic forces are also considered, then additional forces acting on the dam and their moments are as indicated in the accompanying Table Ex. 9.5 (B). The critical condition of loading would occur when due to earthquake the horizontal acceleration acts upstream and the vertical acceleration acts downwards.

Hydrodynamic pressure caused by horizontal acceleration due to earthquake can be found from Zanger's formula.

Since the height of the vertical portion of the upstream face of the dam is less than half the total height of the dam, the slope of the upstream face of the dam is given by

$$\tan \phi = \frac{4.5}{72} = 0.0625$$

$$\therefore \quad \phi = 3° \, 34' = 3.567°$$

Hence
$$C_m = 0.735 \left(1 - \frac{3.567°}{90°}\right) = 0.706$$

At base
$$C = C_m = 0.706$$

The intensity of hydrodynamic pressure at the base

$$p_e = C \alpha w h$$
$$= 0.706 \times 0.1 \times 1 \times 72 = 5.083 \; t/\text{m}^2$$

The horizontal component of the total hydrodynamic pressure acting per unit length of the dam

$$P_e = 0.726 \, p_e \, h = 0.726 \times 5.083 \times 72$$
$$= 265.699 \; t/\text{m}$$

The moment of P_e about the toe of the dam

$$M_e = 0.299 \, p_e h^2$$

$$= 0.299 \times 5.083 \times (72)^2$$
$$= 7\,878.73 \; t\text{-m/m}$$

The sum of all the vertical and the horizontal forces acting on the dam and the sum of their moments about the toe of the dam are indicated in the accompanying Table Ex. 9.5 (B).

If $\bar{x}$ is the distance of the point of intersection O' of the resultant force with the base from the toe of the dam, then

$$\bar{x} = \frac{\Sigma M_3}{\Sigma V_3}$$

$$= \frac{48272.15}{3\,688.575} = 13.087 \text{ m}$$

The distance of O' from the mid-point of the base O, i.e., eccentricity is

$$e = \frac{b}{2} - \bar{x} = \frac{1}{2}(5.75) - 13.087 = 14.788 \text{ m}$$

Vertical stress at the toe of the dam

$$\sigma_{y_D} = \frac{\Sigma V_3}{b}\left(1 + \frac{6e}{b}\right)$$

$$= \frac{3{,}688.575}{55.75}\left(1 + \frac{6 \times 14.788}{55.75}\right)$$

$$= 171.463 \; t/\text{m}^2 \text{ (compressive)}$$

Vertical stress at the heel of the dam

$$\sigma_{y_U} = \frac{\Sigma V_3}{b}\left(1 + \frac{6e}{b}\right)$$

$$= \frac{3{,}688.575}{55.75}\left(1 - \frac{6 \times 14.788}{55.75}\right)$$

$$= -\,39.138 \; t/\text{m}^2 \text{ (tensile)}$$

Since the resultant does not fall within the middle third of the base and the value of the tensile stress is substantial the dam section is unsafe under seismic loading.

Principal stress at the toe of the dam

$$\sigma_{I_D} = \sigma_{y_D} \sec^2 \phi_D = 171.463 \times 1.49$$
$$= 255.48 \; t/\text{m}^2 \text{ (compressive)}$$

Principal stress at the heel of the dam

$$\sigma_{I_U} = \sigma_{y_U} \sec^2 \phi_U - (p + p_e) \tan^2 \phi_U$$
$$= -\,39.138 \times 1.01 - (72 + 5.083) \times 0.01$$
$$= -\,40.30 \; t/\text{m}^2 \text{ (tensile)}$$

Hence the dam section is unsafe.

TABLE Ex. 9.5 (B) Seismic forcess and their moments

S.No.	Item	Description and Dimension	Forces (t/m)		Lever Arm from toe of dam (m)	Moment about toe of dam (t-m/m)		Remarks
			Vertical	Horizontal		Anticlockwise (+ve)	Clockwise (−ve)	
1.		Inertia forced due to earthquake						Inertia force due to vertical accelerarion is considered to act at the C.G. of the entire dam section and that due to horizontal acceleration is considered to act at the C.G. of each subsection
		(a) vertical acceleration						
	$\alpha_v = \Sigma V_1$	0.1 × 4874.25	− 487.425		35.42		17 264.59	
		(b) Horizontal acceleration						
	$\alpha_h W_1$	0.1×243.00		24.300	15.00		364.50	
	$\alpha_h W_2$	0.1×1 350.00		135.00	37.50		5 062.50	
	$\alpha_h W_3$	0.1×3 281.25		328.125	20.83		6 834.84	
2.	P	Hydrodynamic Pressure		265.699			7 878.73	The vertical component of the hydrodynamic pressure has not been considered
		Sum	$\Sigma V = -487.425$	$\Sigma H = 753.124$			$\Sigma M = 37\ 405.16$	
		[Sum of forces and moments from the Table Ex. 9.5 (A)]	$\Sigma V_2 = 4\ 176.000$	$\Sigma H_1 = 2\ 592.000$		184 650.37	98 973.06	
					Sum	184 650.37	136 378.22	
			$\Sigma V_3 = 3\ 688.575$	$\Sigma H_2 = 3\ 345.124$			$\Sigma M_3 = 48\ 272.15$	

Shear stress at the toe of the dam

$$(\tau_{yx})_D = \sigma_{y_D} \tan \phi_D = 171.463 \times 0.7$$
$$= 120.024 \ t/m^2$$

Shear stress at the heel of the dam

$$(\tau_{yx})_U = -[\sigma_{y_U} - (p + p_e)] \tan \phi_U$$
$$= -[-39.138 - (72 + 5.083)] \times 0.1$$
$$= 11.622 \ t/m^2$$

Factor of safety against overturning

$$\text{F.S.} = \frac{\Sigma M(+\text{ve})}{\Sigma M(-\text{ve})} = \frac{184650.37}{136378.22} = 1.354$$

Since F.S. < 1.5, the dam section is unsafe against overturning.

Factor of safety against sliding

$$F_1 = \frac{\mu \Sigma V_3}{\Sigma H_2} = \frac{0.7 \times 3688.575}{3345.124} = 0.772$$

Since $F_1 < 1$, the dam section is unsafe against sliding.

Shear friction factor

$$F_2 = \frac{\mu \Sigma V_3 + bc}{\Sigma H_2}$$
$$= \frac{0.7(3.688.575) + (55.75 \times 140)}{3,345.124}$$
$$= 3.105$$

Since $F_2 < 4$, the dam section is unsafe against sliding.

The given dam section is unsafe under the seismic forces, and hence it will have to be modified.

[**Note** : The problem may be solved in the same manner by using the values given in SI units.]

Illustrative Example 9.6 *For the dam section of Illustrative Example 9.5 determine the factor of safety against sliding as per revised recommendation for reservoir full condition (a) without seismic forces, (b) with seismic forces.*

Solution

The factor of safety against sliding is given by Eq. (9.38 b) as

$$F = \frac{\dfrac{\mu \Sigma V}{F_\phi} + \dfrac{cA}{F_c}}{\Sigma H}$$

(*a*) *For reservoir full condition without seismic forces*

$$\Sigma V = 4176.0 \ t/m; \ \text{m} = 0.7; \ c = 140 \ t/m^2;$$
$$A = b \times 1 = 55.75 \ m^2; \ \Sigma H = 2592.0 \ t/m;$$
$$F_\phi = 1.5 \text{ and } F_c = 3.6 \text{ (From Table 9.5)}$$

Thus, by substitution, we get

$$F = \frac{\dfrac{0.7 \times 4176.0}{1.5} + \dfrac{140 \times 55.75}{3.6}}{2592.0}$$

$$= 1.59 > 1.0$$

Hence safe against sliding.

(b) **For reservoir full condition with seisic forces**

$$\Sigma V = 3688.575 \text{ t/m}; \ \mu = 0.7; \ c = 140 \text{ t/m}^2$$
$$A = b \times 1 = 55.75 \text{ m}^2; \ \Sigma H = 3345.124 \text{t/m}$$
$$F_\phi = 1.2 \text{and } F_c = 2.4 \text{ (From Table 9.5)}$$

Thus by subsitution, we get

$$F = \frac{\dfrac{0.7 \times 3688.575}{1.2} + \dfrac{140 \times 55.75}{2.4}}{3345.124}$$
$$= 1.62 > 1.0$$

Hence safe against sliding.

It may be noted that as per earlier criteria the dam section was found to be unsafe against sliding under seismic forces, but as per the revised criteria it is found to be safe.

Illustrative Example 9.7. *Determine the wave height and freeboard for FRL condition for a dam having reservoir water spread or submergence contour plan as shown in Fig. Ex. 9.7. The full reservoir level (FRL) is 146.3 m, maximum fetch is 11.0 km and maximum wind velocity over land at dam site is 140 km/h.*

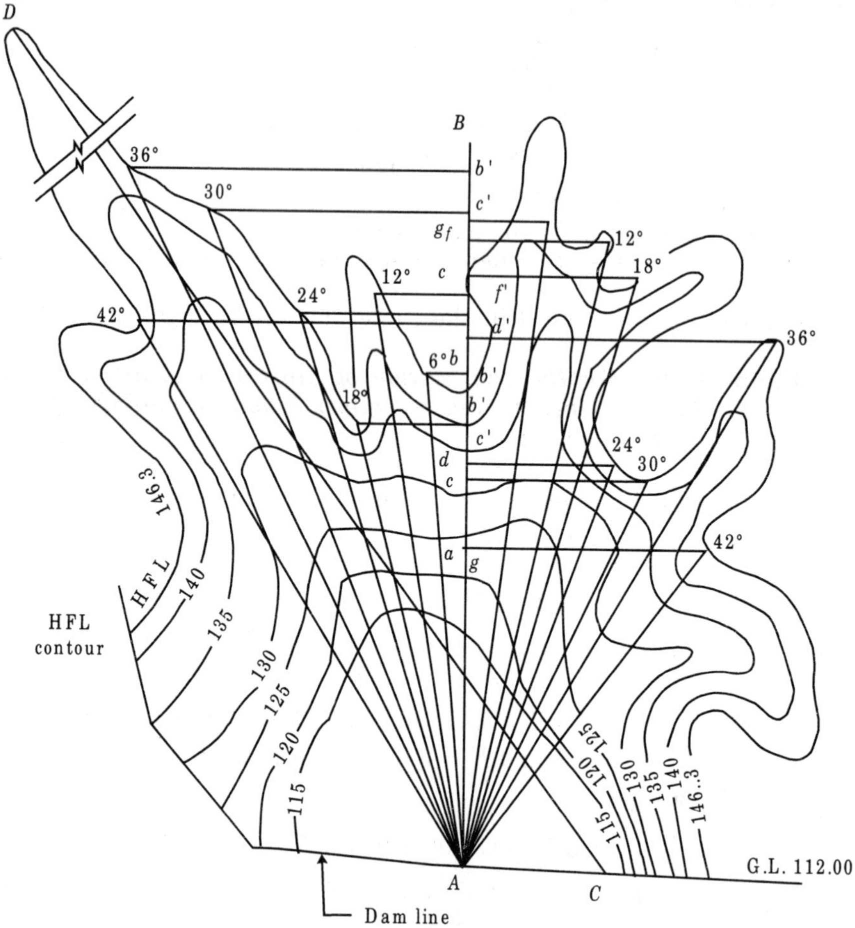

Fig. Ex. 9.7

Solution

Step 1. Full reservoir level (FRL) = 146.3 m

Step 2. Reservoir maximum fetch *CD* = 11.0 km

Step 3. Reservoir effective fetch is calculated as given below in tabular form.

$\alpha°$	$\cos\alpha$	x_1 in km	$x_1 \cos\alpha$
42°	0.7431	*Aa* = 3.00	2.2293
36°	0.8090	*Ab* = 4.85	3.9237
30°	0.8660	*Ac* = 3.65	3.1609
24°	0.9135	*Ad* = 4.80	4.3848
18°	0.9511	*Ae* = 5.50	5.2311
12°	0.9781	*Af* = 5.90	5.7708
6°	0.9945	*Ag* = 6.00	5.9670
0	1.0000	*AB* = 4.50	4.5000
6°	0.9945	*Ag′* = 4.70	4.6742
12°	0.9781	*Af′* = 5.30	5.1839
18°	0.9511	*Ae′* = 4.20	3.9946
24°	0.9135	*Ad′* = 5.25	4.7959
30°	0.8660	*Ac′* = 6.20	5.3692
36°	0.8090	*Ab′* = 6.55	5.2990
42°	0.7431	*Aa′* = 5.10	3.7898
	$\Sigma \cos\alpha = 13.5106$		$\Sigma x_1 \cos\alpha = 68.2742$

$$\text{Effective fetch} = \frac{\Sigma x_1 \cos\alpha}{\Sigma \cos\alpha}$$

$$= \frac{68.2742}{13.5106}$$

$$= 5.053 \text{ km}$$

Step 4. Maximum wind velocity over land = 140 km/h

Step 5. Wind velocity over water

$$= \text{Wind velocity over land} \times \frac{U_W}{U_L}$$

From Table 9.2, for effective fetch = 5.053 km

$$\frac{U_W}{U_L} = 1.256$$

$\therefore$ Wind velocity over water = 140 × 1.256

$$= 175.84 \text{ km/h}$$

$$\simeq 176 \text{ km/h}$$

Step 6. From Fig. 9.13 for wind velocity of 176 km/h and effective fetch 5.053 km, significant wave height H_s = 3.00 m.

Step 7. Design wave height $h_w = 1.27 \times H_s$

$$= 1.27 \times 3.00 = 3.81 \text{ m}$$

Step 8. Wind set-up $= \dfrac{V^2F}{62000\,D}$

V = 176 km/h; F = 11 km; and

D = Average depth of reservoir along maximum fetch which is calculated as given below in tabular form.

Sl. No.	*FRL* (*m*)	*G.L* (*m*)	*Depth* (*m*)
1	146.3	146.3	0
2	146.3	140.0	6.3
3	146.3	135.0	11.3
4	146.3	130.0	16.3
5	146.3	125.0	21.3
6	146.3	120.0	26.3
7	146.3	115.0	31.3
8	At the dam	112.0	34.3
			Total 147.1

$$\text{Average depth } D = \frac{147.1}{8} = 18.39 \text{ m}$$

Thus, by substitution, we get

$$\text{Wind set-up} = \frac{(176)^2 \times 11}{62000 \times 18.39}$$

$$= 0.299 \text{ m} \simeq 0.3 \text{ m}$$

Step 9. Free board $= \dfrac{4}{3}h_w + \text{wind set-up}$

$$= \left(\frac{4}{3} \times 3.81\right) + 0.3 = 5.38 \text{ m}$$

Thus, for FRL condition wave height is 3.81 m and free board is 5.38 m.

REVIEW QUESTIONS

9.1 What is a gravity dam? Enumerate the various forces acting on a gravity dam.

9.2 Explain with sketch how you fill and find the uplift pressure on a gravity dam provided with drainage gallery.

9.3 Prove that the base width b for an elementary (or theoretical) profile of a low gravity dam is given by

$$b = \frac{h}{\sqrt{s-1}}$$

in which

h = Height of the dam, and
s = Specific gravity of the dam material.

Assume that the uplift pressure acts on the full area of the base.
Also calculate the maximum compressive stress developed in the dam.

9.4 Derive an expression for the limiting height of a low gravity dam. Differentiate between low and high gravity dam.

9.5 What are the various modes of failure of a gravity dam? Discuss each of these briefly.

9.6 A 20 m high concrete dam having trapezoidal section has top width 2 m and bottom width 16 m. The face of the dam exposed to water has a batter of 1 : 10. On the reservoir side water stands up to the top. Assuming weight of concrete = 23.544 kN/m^3 [2400 kg(f)/m^3], coefficient of friction = 0.75 and allowable shear stress = 490.5 kN/m^2 [5 kg(f)/cm^2], and taking into account only the weight of the dam, water pressure and uplift pressure, calculate the factor of safety against overturning, against sliding and shear friction factor.

[*Ans.* 1.511; 1.095; 5.095]

9.7 A 100 m high concrete gravity dam trapezoidal in cross-section has upstream face vertical, crest width 6 m, base width 75 m and free board equal to 4 m. Calculate the maximum principal stress at the toe when the reservoir is full. Take unit weight of concrete as 23.544 kN/m^3 [2.4 t/m^3]. Neglect all other forces except hydrostatic water pressure, uplift pressure and self weight. There is no drainage gallery and no tail water.

[*Ans.* 3421.90 kN/m^2 (348.82 t/m^2)]

9.8 Following data were obtained from the stability analysis of a concrete gravity dam :

(i) Total resisting moment about toe

= 14.715 × 10^5 kN-m (1.5 × 10^5 t-m)

(ii) Total overturning moment about toe

= 9.81 × 10^5 kN-m (1.0 × 10^5 t-m)

(iii) Total vertical force above the base

= 49.05 × 10^3 kN (5000 t)

(iv) Base width of the dam

= 50 m

(v) Slope of the downstream face of the dam

= 0.7 (H) : 1(V)

Calculate the maximum and minimum vertical stress to which the foundation will be subjected to. What is the maximum principal stress at toe? Assume there is no tail water.

[*Ans.* 27.468 × 10^2 kN/m^2 (2.8 × 102 t/m^2) compressive; 7.848 × 10^2 kN/m^2 (0.8 × 102 t/m^2) tensile; 40.927 × 10^2 kN/m^2 (4.172 × 10^2 t/m^2)]

9.9 Find the hydrodynamic pressure and moment due to earthquake on a dam with vertical u/s face and depth of water equal to 100 m when intensity of earthquake is 0.2 g.

[*Ans.* (10.469 × 10^3 kN/m (1.067 × 10^3 *t*/m); 43.118 × 103 kN-m/m) (4.395 × 10^3 *t*-m/m)]

9.10 The following are the details of a straight gravity concrete dam (sp. gr. 2.4):

Foundation level, R.L. 100.65

Height of dam 250 m

Maximum u/s water surface elevation, R.L. 347.65

Top width 8 m

Upstream face is vertical and downstream face vertical upto elevation R.L. 340.65 and has a slope of 0.8 horizontal to 1 vertical below this elevation. Assume no tailwater and no drainage gallery to compute the following :

(i) Normal stress at heel and toe

(ii) The principal and shear stresses at the toe

(iii) The sliding factor and shear friction factor,

[*Ans.* μ = 0.7, q = 735.75 kN/m^2 (75 *t*/m^2)];

[(i) 214.84 kN/m^2 (21.9 *t*/m^2) tensile; 3687.58 kN/m^2) (375.9 *t*/m^2) compressive;

(ii) 6047.63 kN/m^2 (616.48 *t*/m^2); 2950.06 kN/m^2 (300.72 *t*/m^2) (iii) 0.862; 1.304)]

9.11 A concrete gravity dam has top width 8 m and height 40 m above the base. Freeboard is 4 m and the width at the base is 36 m. Upstream face of the dam is vertical but the downstream face is sloping uniformly upto the base from a height 10 m below the top of the dam. Determine:

(i) Factor of safety against overturning.

(ii) Factor of safety against sliding (μ = 0.75).

(iii) Maximum principal stress at the toe.

Consider only the forces due to self weight, uplift (full) and hydrostatic thrust. Take unit weight of concrete as 21.582 kN/m^3 (2.2 *t*/m^3).

[*Ans.* (i) 1.0; (ii) 1.134; (iii) 810.42 kN/m^2 (82.612 *t*/m^2)]

9.12 A masonry dam 80 m high is trapezoidal in section with top width 2 m. The face exposed to water has a slope of 1 horizontal to 10 vertical and the downstream face has a slope of 0.75 horizontal to 1 vertical. Test the stability of the dam at its base. Also find the principal stresses at the toe and heel of the dam and shear friction factor:

(i) Assume tail water of depth 5 m.

(ii) Assume a line of drain at 6 m from the heel and a free board of 5 m.

(iii) Take μ = 0.65; specific gravity of masonry as 2.4 and allowable shear stress as 490.5 kN/m^2 (50 *t*/m^2).

(iv) Horizontal acceleration due to earthquake may be taken as 0.1 g.

[*Ans.* 2.06 × 10^3 kN/m^2; 0.45 × 10^3 kN/m^2; 1.932]

9.13 What do you understand by longitudinal and transverse joints in a concrete dam? What considerations govern their provision and spacing?

9.14 Discuss the zone method of design of a gravity dam.

9.15 What purpose is served by consolidation and curtain grouting? Where are these types of grouting done?

9.16 Discuss the various purposes for which galleries are provided in dams.

9.17 What are the various temperature control measures adopted in a massive concrete dam? What factors are to be taken into account in deciding the type of such measures to be adopted?

9.18 Describe with neat sketches the various types of water stops (or water seals) commonly adopted in dams.

9.19 What type of keys are provided at the transverse and longitudinal joints in concrete dams?

9.20 The base width of a concrete gravity dam is 75 m. For the full reservoir condition the resultant passes through a point 12 m from the centre. The slope at the downstream face is 0.7 horizontal to 1 vertical. If the total vertical load is 6850 × 10^4 N; find the normal stress, shear stress and the principal stress at the toe.

[*Ans.* 179.013 × 10^4 N/m^2; 125.309 × 10^4 N/m^2; 266.729 N/m^2]

9.21 Define the terms: (i) Sliding factor; (ii) Shear friction factor; (iii) Load combinations for design of a gravity dam; (iv) Precooling and Postcooling of concrete.

CHAPTER

10 Buttress and Arch Dams

10.1 INTRODUCTION

In addition to gravity dams the other two types of rigid dams which are also commonly used are the buttress dams and the arch dams. Both these types of dams are made of concrete. A brief description of each of these two types of dams is explained below.

10.2 BUTTRESS DAMS

A buttress dam consists of a water retaining sloping membrane or deck on the upstream which is supported by a series of buttresses at right angles to the axis of the dam. The upstream water retaining sloping membrane may be in the form of a reinforced-concrete slab, or a series of arches or thickened buttress heads. At the upstream end a cutoff is provided which extends upto the sound rock in the foundation to prevent or reduce the seepage of water. The buttresses are thin walls of triangular profile with a sloping upstream face. These are usually spaced at equal intervals along the length of the dam and are supported either on a continuous mat foundation for all the buttresses or on a separate spread footing for each buttress. The upstream end of the buttress is usually made wide by providing a haunch or corbel. A number of bracings or struts are usually provided between the adjacent buttresses at right angles to them to reduce the unsupported length and thus provide lateral stiffness and resistance to buckling of the buttresses. The width of the downstream end of the buttress may also be increased by providing flanges or pilasters. In high dams additional flanges or pilasters known as stiffeners may be added along the width of the buttresses either in place of or in addition to the struts. However, the struts and pilasters (or stiffeners) may be eliminated by using double-wall or hollow buttresses. The various components of a buttress dam indicated above are shown in Fig. 10.1.

10.3 TYPES OF BUTTRESS DAMS

The buttress dams may be classified in the following two ways.

1. On the basis of the type of the sloping membrane.

2. On the basis of joint between the sloping membrane and buttresses.

Classification based on the Type of Sloping Membrane

On the basis of the type of the sloping membrane the buttress dams may be classified in the following categories.

(i) Deck slab or Flat slab or Ambursen type buttress dams.

(ii) Multiple-arch type buttress dams

(iii) Multiple-dome type buttress dams

(iv) Massive head type buttress dams or Bulk head type buttress dams.

1. Deck slab or Flat slab or Ambursen type buttress dams. The deck slab or flat slab type buttress dams are also known as Ambursen type buttress dams in honour of Nils F. Ambursen who in 1903 built the first flat slab type of buttress dam. In this type of buttress dam

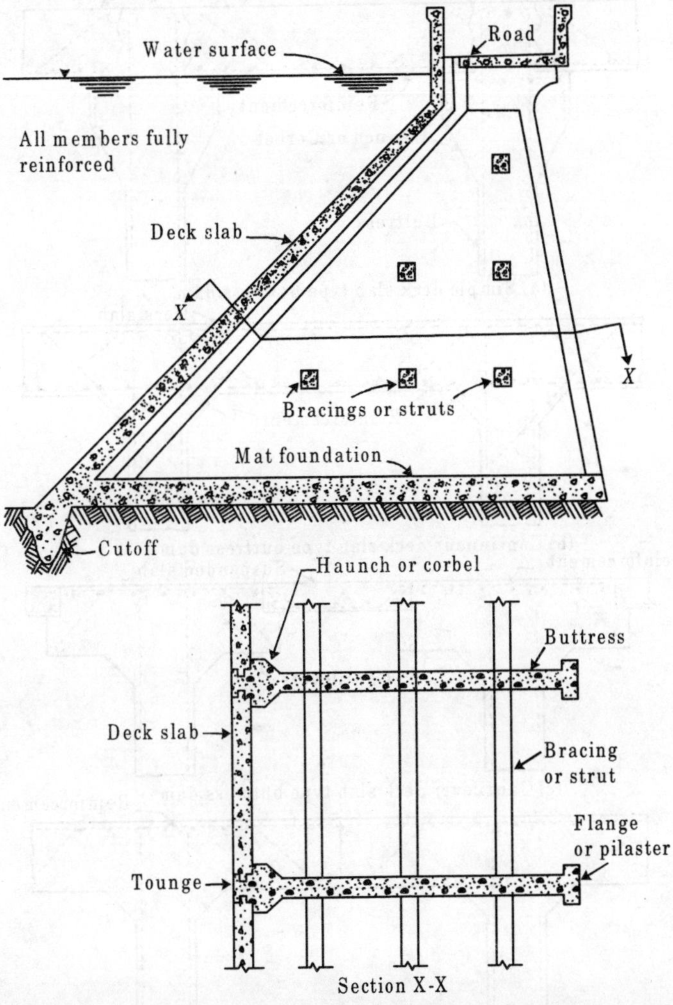

Fig. 10.1 *Deck slab or Flat slab or Ambursen type buttress dam*

the sloping membrane or deck consists of a reinforced-concrete slab supported by a series of buttresses (Fig. 10.1). The inclination of the deck slab is kept between 35° to 45° with the horizontal. In order to provide a wide base for the slab supported by the buttress the upstream end of the buttress where it joins the slab is usually made wide by providing haunch or corbel as shown in Fig. (10.1). Since there may be tensile stresses developed in the haunch or corbel it is always provided with reinforcement, but for the remaining portion of the buttress the reinforcement may be provided or may not be provided. However, the modern tendency is to provide adequate reinforcement for the entire buttress to prevent cracking. A number of bracings or struts of reinforced-concrete are provided between the adjacent buttress as shown in Fig. (10.1).

The deck slab type buttress dams may be further subdivided into the following three types.

(a) *Simple deck slab type buttress dams.* In this type of buttress dam the deck slab is not rigidly attached to the buttresses but it is in the form of a simply supported slab which remains in contact with the buttress tongue as shown in Fig. (10.2 a). The joint between the slab and

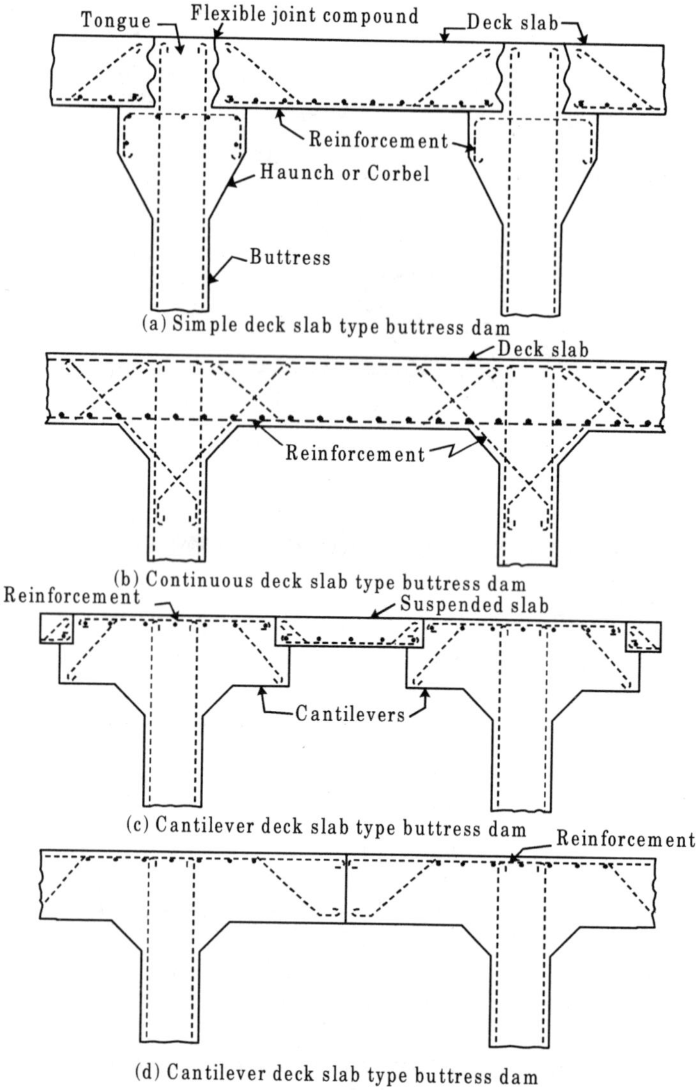

(a) Simple deck slab type buttress dam

(b) Continuous deck slab type buttress dam

(c) Cantilever deck slab type buttress dam

(d) Cantilever deck slab type buttress dam

Fig. 10.2 *Different types of deck slab type buttress dams*

buttress tongue is filled with bituminous mastic (or bitumastic) or asphaltic putty or some flexible joint compound. This permits each slab to act independently and allows free expansion of slab in the direction parallel to the base line of the dam as well as along the slope. However, the movement of the slab along haunch has to be restricted and it is also desirable that the full weight of the deck slab is carried by the buttress. For this keys are provided in the contact surface between the deck slab and the buttress tongue (Fig. 10.2 a). Thus, the most typical feature of this type of dam is its articulation or flexibility due to which minor settlement of the foundation will not seriously harm the structure. Hence for this type of dam the influence of settlement of foundation is not significant. Further the influence of temperature variation and shrinkage of concrete is also not significant for this type of dams.

For a simple deck slab type buttress dam since the deck slab is designed as a simply supported slab the reinforcement is placed only at the downstream face and hence there is no danger of the reinforcement being corroded. However, this type of dam requires a relatively thicker slab. The simple deck slab type buttress dams are particularly suitable for wide valleys where a long dam is required and foundation materials are of inferior strength.

(b) *Fixed or continuous deck slab type buttress dams.* In this type of buttress dam the deck slab is cast monolithic with the buttress and it is in the form of a continuous slab (Fig. 10.2 b). Thus, in this case the reinforcement is provided both on upstream and downstream faces of the slab because of reversing bending moments and there is a possibility of upstream reinforcement being corroded. Further in this type of dam the stresses due to temperature variation, shrinkage of concrete and foundation settlement are significant. However, in this type of dam the deck slab is relatively thin.

(c) *Cantilever deck slab type buttress dams.* In this type of buttress dam the deck slab is cast monolithic with the buttress in such a manner that the deck slab overhangs on either side of each buttress as cantilever. In one of the arrangements as shown in Fig. (10.2 c) a part of the

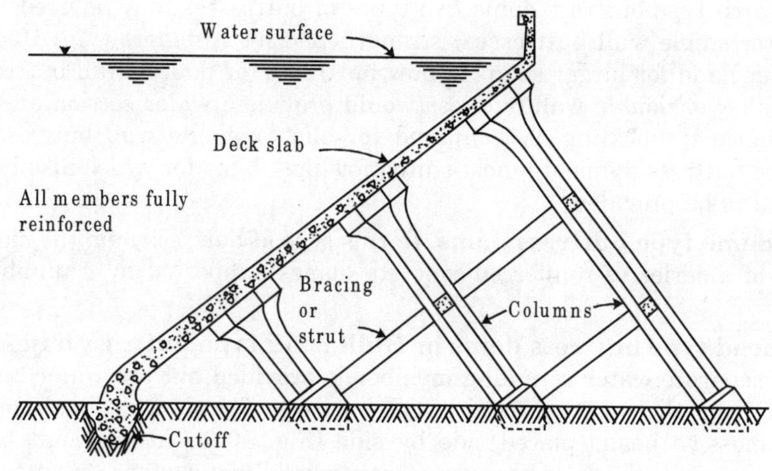

Fig. 10.3 *Columnar type buttress dam*

half of the deck slab overhangs on either side and in the gap between the two adjacent cantilevers a slab known as suspended slab is provided which is supported by the cantilevers with proper joints at both ends. In another arrangement the deck slab extends from either side of each buttress right upto the middle of the span between the adjacent buttresses with a proper joint in the centre as shown in Fig. 10.2 (d). In this type of dam because of cantilever action the reinforcement has to be provided on the upstream face of the dam and hence there is a possibility of the reinforcement being corroded. The cantilever deck slab type buttress dams are subjected to high stresses due to temperature variation, shrinkage of concrete and foundation settlement.

In addition to the above noted deck slab type buttress dams there are two more types of these dams as noted below.

(a) Columnar type buttress dams

(b) Truss type buttress dams.

(a) *Columnar type buttress dams.* This type of buttress dam consists of a continuous flat slab of reinforced concrete supported on a number of inclined columns (Fig. 10.3) instead of buttresses. The columns are also reinforced and are stiffened by a two-directional system of struts.

(b) *Truss type buttress dams.* In this type of buttress dam the reinforced-concrete deck slab is supported on a number of reinforced-concrete trusses instead of buttresses.

2. Multiple-arch type buttress dams. In this type of buttress dam the sloping membrane or deck consists of a series of reinforced-concrete arches supported by a number of buttresses (Fig. 10.4). The upstream face of the dam is usually inclined at 45°. The arches are cast monolithic with the buttresses.

As such a multiple-arch type buttress dam is more rigid than a flat slab type and hence requires a better foundation. For short spans (10 to 15 m) circular arches of uniform thickness are economical while for longer spans greater economy may be achieved by using an arch of variable thickness. The central angle of the arches is usually between 150° to 180°. For arches having central angle 180° (i.e., semicircular arches) the horizontal thrust between adjoining arches is eliminated and buttress design is simplified.

For multiple-arch type buttress dams two types of buttresses may be used. For short spans solid buttresses (or single wall buttresses) strengthened by stiffeners (Fig. 10.4 a) are usually used. On the other hand for larger spans hollow buttreses (or double wall buttresses) (Fig. 10.4 b) are used. A hollow or double wall buttress would provide a wider section and hence it would offer more resistance to buckling as compared to solid or single wall buttress. Moreover the multiple arch type buttress dams are most suited for high dams for which also hollow buttresses would be required to be provided.

3. Multiple-dome type buttress dams. In this type of buttress dam the sloping membrane or deck consists of a series of reinforced-concrete domes supported by a number of buttresses (Fig. 10.5).

4. Massive head type buttress dams or Bulk head type buttress dams. In this type of buttress dam no separate water retaining member is provided but the water retaining member is formed by enlarging the upstream end of the buttress. The dam is thus made of a series of buttresses with massive heads placed side by side (Fig. 10.6). At the joint between the two adjacent massive heads of the buttresses two copper strip water stops (or water seals) with an asphalt water stop (or water seal) in between are provided (Fig. 10.6). Further the entire joint between the two adjacent buttress heads is filled with a bitumastic filling material. Thus, in this case each buttress behaves as an independent structural unit which permits slight movement of one buttress without affecting the other and hence provides certain degree of articulation. In some cases at each joint between the adjacent buttress heads a key is provided as shown in Fig. 10.6 (d) in which case on the upstream side of the key only one copper strip water stop is usually provided and the entire joint is filled with a bitumastic filling material. Further in some cases cross-walls or stiffeners are provided between the adjacent buttresses so that monolithic buttress units each consisting of two buttresses are developed as shown in Fig. 10.7. However, due to the provision of keys as well as cross-walls the degree of articulation would be reduced.

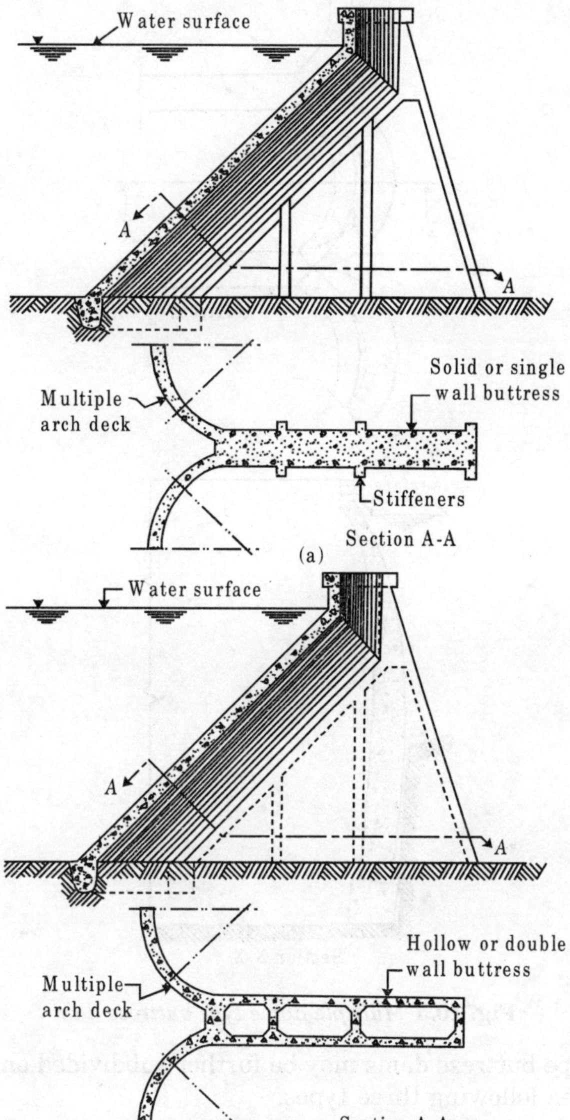

Fig. 10.4 *Multiple-arch type buttress dams*

The upstream face of this type of buttress dam is usually vertical for some portion from the top and for the remaining portion it is sloping with an inclination of about 60° with the horizontal. Further the buttress heads are so shaped that the tensile stresses are not developed in any portion of the buttress heads and hence the buttress heads are not reinforced.

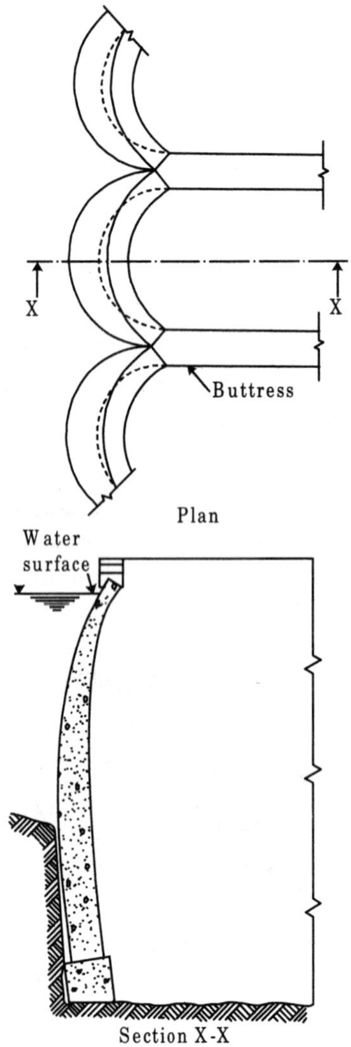

Fig. 10.5 *Multiple-dome type buttress dam*

The massive head type buttress dams may be further subdivided on the basis of the shape of the buttress head into the following three types.

(a) Round head buttress dams

(b) Diamond head buttress dams

(c) Tee head buttress dams.

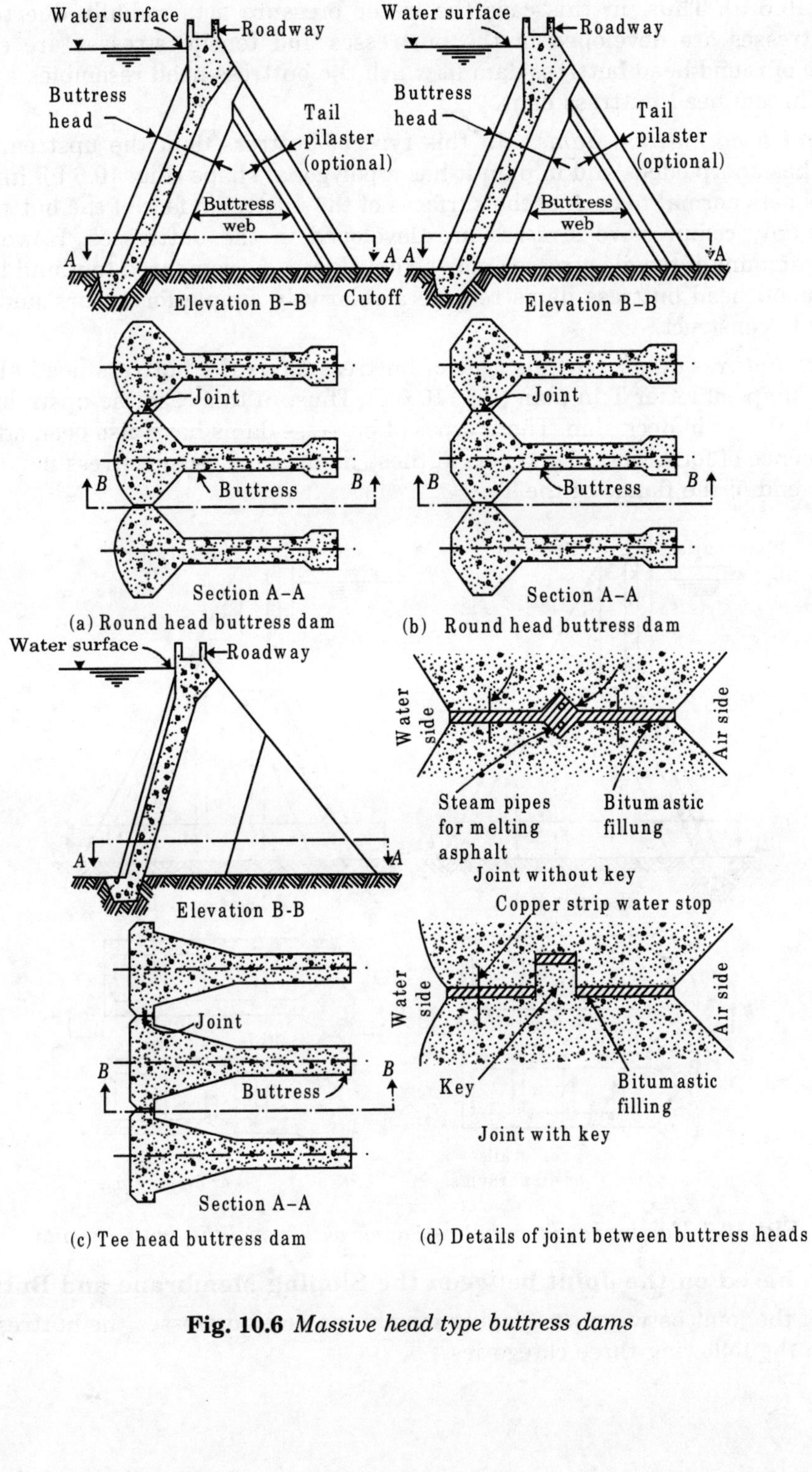

Fig. 10.6 *Massive head type buttress dams*

(a) *Round head buttress dams.* The round head buttress dam was first proposed by F.A. Noetzli in 1926. In this type of buttress dam the upstream face of each of the buttress heads is circular (Fig. 10.6 a). Thus, in this case the water pressure acts radially due to which only compressive stresses are developed in the buttresses and tensile stresses are eliminated. A particular type of round head buttress dam in which the buttress head resembles a mushroom is known as mushroom head buttress dam.

(b) *Diamond head buttress dams.* In this type of buttress dam the upstream face of the buttress head has sharp edges and in plan it has a polygonal shape (Fig. 10.6 b). In this case the water pressure acts normal to each of the surfaces of the upstream face of the buttress head and hence usually only compressive stresses are developed in the buttresses. However, in some portions small amount of tensile stresses may be developed. As compared to round head buttress dams the diamond head buttress dams require a relatively simple form work and hence these dams are easy to construct.

(c) *Tee head buttress dams.* In this type of buttress dams the buttress head along with the buttress has a shape of letter T in plan (Fig. 10.6 c). Thus, in this case the upstream face of the buttress head is flat as in deck slab. These types of buttress dams have also been adopted for the sake of convenience of form work. However, in these dams some tensile stress may be developed near the outer end of the flange of the Tee.

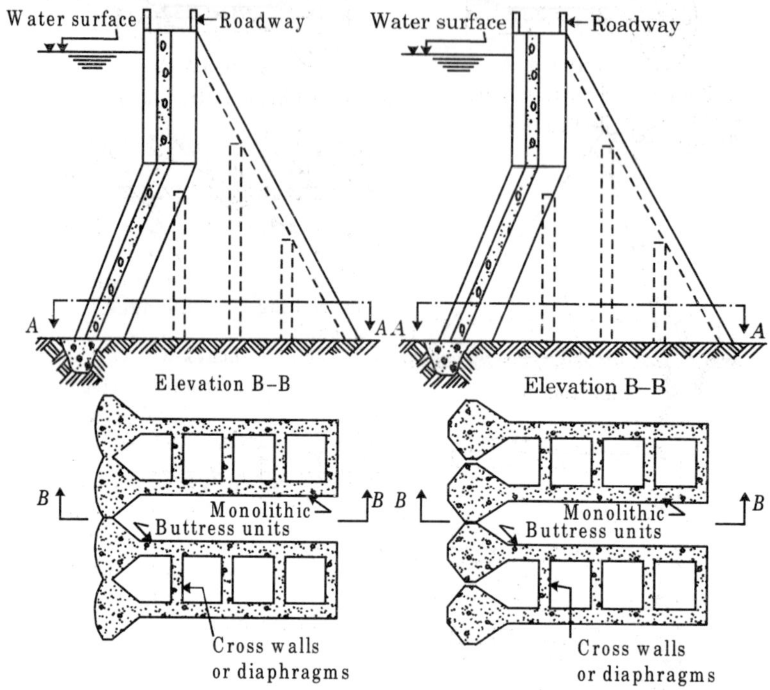

Fig. 10.7 *Massive head type buttress dams with monolithic buttress units*

Classification based on the Joint between the Sloping Membrane and Buttresses

On the basis of the joint between the sloping membrane and buttresses the buttress dams may be classified in the following three categories.

(i) Rigid buttress dams

(ii) Articulated buttress dams

(iii) Intermediate or semi-rigid buttress dams

(i) *Rigid buttress dams.* Rigid buttress dams are those in which the sloping membrane or deck is constructed monolithic with the buttresses. As such for these dams the influence of foundation settlement, temperature variation and shrinkage of concrete are significant and provisions have to be made for these in the design. The continuous deck slab type, multiple arch type and multiple dome type buttress dams are the examples of the rigid buttress dams.

(ii) *Articulated buttress dams.* Articulated buttress dams are those in which the sloping membrane or deck is not constructed monolithic with the buttresses. As such these dams are flexible and hence the influence of settlement of foundation, temperature variation and shrinkage of concrete are not significant for these dams. The simple deck slab type buttress dam is an example of articulated buttress dams.

(ii) *Intermediate or semi-rigid buttress dams.* The intermediate or semi-rigid buttress dams are in between the rigid and articulated buttress dams. Thus, these dams eliminate the undesirable rigidity but also do not possess high degree of articulation or flexibility. The massive head type buttress dams are the examples of this type of buttress dams.

10.4 FORCES ON BUTTRESS DAMS

The various forces acting on a buttress dam are same as those acting on a gravity dam. However, because of the slope of the upstream face the downward component of water pressure is greater and ice pressure is usually not important as ice sheet tends to slide up the dam. The uplift pressure acting on a buttress dam would depend on the type of dam and also on the type of foundation provided for the buttresses. Generally for a massive head type buttress dam the uplift pressure would be more than that for a deck slab type or multiple arch type buttress dam. Thus, in massive buttress heads vertical drains are provided to intercept seepage and reduce uplift pressure. Further if separate spread footing is provided for each buttress then the uplift pressure is relieved by the gaps between the buttresses. As such in this case with rock foundations the uplift pressure due to headwater (or reservoir water) is usuallly neglected and only the uplift pressure due to tailwater (if any) is considered for the entire footing. However, in the same case if the foundation rock is such that uplift pressure due to headwater cannot be neglected then under the deck and haunch section full uplift pressure due to headwater is considered and for the remaining portion of the footing full uplift due to tailwater is considered. Further if for all the buttresses a continuous mat foundation is provided then the uplift pressure may be computed in the same manner as in the case of a gravity dam. Vertical drains may be provided in the foundation to intercept seepage and reduce uplift.

The buttress dams are essentially gravity dams and hence their design should fulfill the same basic principles of stability as for gravity dams. Further in the case of a buttress dam the total force acting per buttress unit must be considered instead of force per unit length of dam as in the case of a gravity dam.

10.5 DESIGN OF FLAT-SLAB TYPE BUTTRESS DAMS

The design of a flat slab type buttress dam involves the following three steps.

(i) Determination of economical buttress spacing and upstream slope.

(ii) Design of deck slab and other details.

(iii) Preliminary design of buttress and check for overall stability.

1. Economical Buttress Spacing and Upstream Slope

(a) *Economical buttress spacing.* Any change in the buttress spacing produces variation in the quantities of material required for the dam as well as the cost of the necessary form work. For wider spacings of buttresses the quantity of concrete and the necessary reinforcing steel for the deck slab are considerably increased but the cost of the form work is reduced. On the other hand smaller spacings of buttresses reduce the amount of concrete and reinforcement in the deck slab but increase the cost of form work. Thus, the most economical spacing of buttress is that which will give the most economical balance of concrete, reinforcing steel and form work and will satisfy the design requirements. The most economical and satisfactory buttress spacing is mainly influenced by (a) the mean height of the dam; (b) the presence of spread footings or continuous mat foundation; (c) the presence of a spillway over the dam; (d) the unusual foundation or sidehill conditions; and (e) the slope of the upstream face of the dam.

The economical spacing of buttress varies considerably with the height of the dam. For high dams greater buttress spacings are economical and satisfactory, whereas for low dams the buttress spacing will be proportionally less. For sound rock foundation the most economical buttress spacings for deck slab type buttress dams are as follows.

Height of dam (*m*)	*Economic spacing centre to centre of buttresses* (*m*)
15 to 30	5 to 6
30 to 45	9 to 12
Above 45	12 to 15

The quantities of material required for spread footing or continuous mat foundation and the apron slab of spillway increase with the increase in the buttress spacing and hence it would be economical to have smaller spacings for buttresses.

Foundation conditions may influence the buttress spacing by the presence of fault zones or steep, unstable sidehills which frequently require an arbitrary selection of buttress spacing to suit the particular condition. For relatively narrow fault zones wide buttress spacings may be used. On the other hand wider fault zones can be bridged with a special arch or girder structure which transmits the buttress loads to either side of the fault and hence in this case smaller spacings may be desirable so as to spread the buttress loads more uniformly. Likewise for steep or unstable sidehill conditions the buttress spacings may have to be increased to avoid the difficult areas or decreased to distribute the loading.

The slope of the upstream face of the dam does not influence the buttress spacing significantly. However, the buttress spacing as well as the slope of the upstream face affect the economy significantly and hence economical buttress spacings for different values of the upstream slopes may be obtained as indicated below.

(*b*) *Upsteam slope.* The slope of the upstream face of a flat-slab type buttress dam usually varies from 35° to 45° and it is governed by the requirements of sliding factor, i.e., $\Sigma H/\Sigma V$. Since the self weight of a buttress dam is relatively small it has to depend on the vertical component of the water pressure acting on the deck slab to develop resistance against sliding. The vertical component of the water pressure varies with the upstream slope. Further for a constant buttress spacing a small variation in the upstream slope results in an appreciable change in the quantity of concrete and hence it affects economy. Likewise for a constant upstream slope the variation in the buttress spacing also results in change in the quantity of concrete and hence affects economy. Thus, for each value of upstream slope there would be an economical buttress spacing which may be determined from a master curve which may be prepared as indicated below. Further from the same curve a combination of the most economical buttress spacing and the upstream slope may also be determined.

Consider a line diagram for a buttress dam as shown in Fig. (10.8). Let

h be the height of dam in metres

ξ be the angle made by the upstream face of dam with the horizontal

P_V be the vertical component of water pressure per metre length of dam

P_H be the horizontal component of water pressure per metre length of dam

W_c be the weight per metre length of dam, i.e., weight of deck slab and buttress computed per metre length of dam

C_q be the quantity of concrete in cubic metres per metre length of dam

w be the specific weight of water

w_c be the specific weight of concrete

f be the sliding factor

Thus,

$$P_H = \frac{wh^2}{2}$$

$$P_V = \frac{wh^2}{2}\cot\xi$$

$$W_c = w_c\, C_q$$

$$f = \frac{\Sigma H}{\Sigma V}$$

or

$$f = \frac{P_H}{P_V + W_C}$$

or

$$f = \frac{\dfrac{wh^2}{2}}{\left(\dfrac{wh^2}{2}\cot\xi + w_c C_q\right)}$$

Taking w = 9.81 kN/m³ [or 1000 kg(f)/m³] and w_c = 23.544 kN/m³ [or 2400 kg(f)/m³] and solving for C_q, we get

$$C_q = 0.02083\, h^2\left(\frac{1}{f} - \cot\xi\right) \qquad \text{...(10.1)}$$

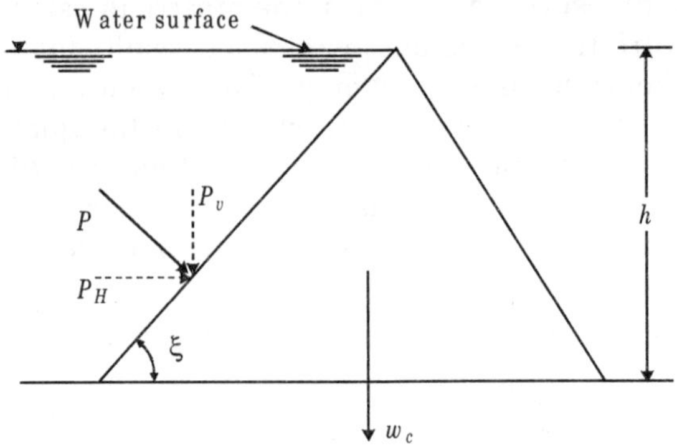

Fig. 10.8 *Forces acting on a buttress*

Equation (10.1) shows that for a given height h and a given value of sliding factor f the quantity of concrete C_q depends on cot x, and for given values of h, f and ξ the quantity of concrete is constant. Thus, using Eq. (10.1) for given values of h and f the quantity of concrete C_q may be obtained for each value of ξ. The values of C_q so obtained are independent of the buttress spacing and hence these will plot as vertical lines on a plot of buttress spacing versus quantity of concrete as shown in Fig. (10.9). Next choosing certain value of ξ the dam is

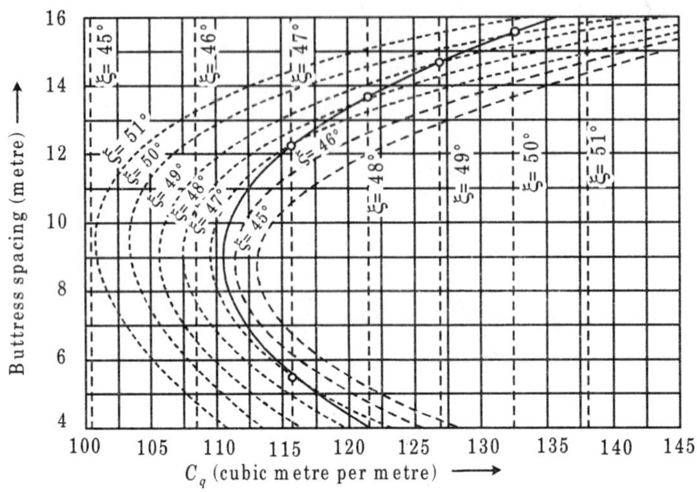

Fig. 10.9 *Master curve for determining economical buttress spacing*

designed by considering a number of buttress spacings and the quantity of concrete required per unit length of the dam for each buttress spacing is determined. The values of the quantity of concrete so obtained are plotted against the corresponding buttress spacing on the same graph sheet to obtain a curve for the chosen value of ξ. By repeating the same procedure for different values of ξ a series of such curves are obtained as shown by dotted lines in Fig. (10.9). A master curve is then drawn by joining the points of intersection of the curves and the vertical lines corresponding to each value of ξ which is shown by a firm line in Fig. (10.9). The master curve gives the value of economical buttress spacing for each value of ξ for the given height of dam.

Further the apex of the master curve gives the value of the most economical buttress spacing as well as the corresponding value of ξ and the quantity of concrete per metre length of dam.

(ii) Design of deck slab and other details

(a) *Deck slab*. The deck slab is designed as a simply supported slab by using the standard methods of reinforced concrete design. The design should however comply with an appropriate Code of practice for water retaining structures. The slab thickness is generally controlled by bending moments at higher levels and by shear at lower levels. As the weight of the slab contributes to overall stability, a thick slab with relatively less reinforcement is preferred to a

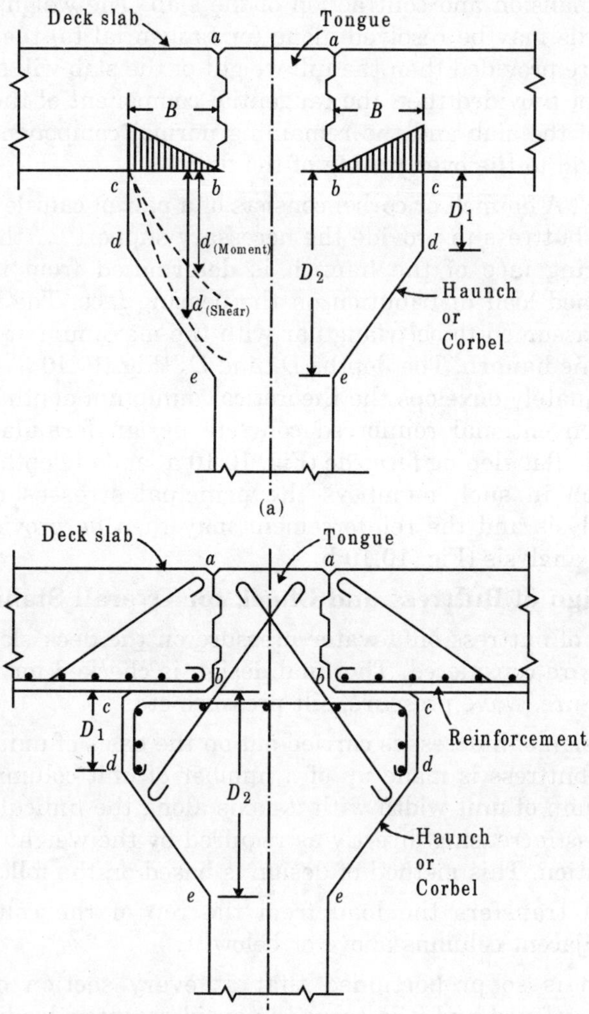

Fig. 10.10 *Details of connection between deck slab and buttress*

thin heavily reinforced slab. Moreover a thick slab has a higher factor of safety against surface deterioration, spalling etc. Shear stresses should be kept low enough to avoid the necessity of shear reinforcement, but some bent up rods are necessary at the supports. Allowance should be made for the direct tensions induced by shrinkage and fall in temperature. Further transverse (vertical) reinforcement should be generously provided to prevent cracks of any kind which is of utmost importance in this case. Concrete cover for reinforcement is generally not less than 75 mm.

(b) *Keys*. The concrete surface between the deck slab and buttress tongue must be provided with special keys (Fig. 10.10). These must be designed to ensure that the full weight of the slab is actually carried by the buttress (and thus contributes to the stability of the dam) and at the same time to permit expansion and contraction of the slab. The weight of the deck slab which acts vertically downwards may be resolved along (or tangential to) the slab and normal to the slab. If adequate keys are provided then the full weight of the slab will act as a stabilising force. However, if keys, are not provided then the tangential component of the weight of the slab will be lost down the axis of the slab and the remaining normal component will have a horizontal component which will add to the overturning of the dam.

(c) *Haunch or Corbel*. A haunch or corbel consists of a pair of cantilever brackets constructed monolithically with the buttress to provide the necessary support for the deck slab (Fig. 10.10). The width of the bearing face of the haunch is determined from the permissible bearing pressure and the assumed load distribution on the bearing face. The load distribution on the bearing face is usually assumed to be triangular with the maximum at the outer edge and zero at the inside corner of the haunch. The depths D_1 and D_2 (Fig. 10.10 a) are so proportioned that the bracket profile adequately envelops the theoretical minimum depth profiles for moment and shear. However, the conventional reinforced concrete design formulae are not applicable to haunches with extremely flat sloping faces de (Fig. 10.10 a) and of depth materially greater than the projection. As such in such members the principal stresses may be determined by photoelastic model analysis and the reinforcement may then be provided to carry the tensile stress indicated by this analysis (Fig. 10.10 b).

(iii) Preliminary Design of Buttress and Check for Overall Stability

For preliminary design of buttress only water pressure on the deck slab and the weight of the deck slab and buttress are considered. The final design is checked for the other forces such as earthquake, uplift pressure, wave pressure, silt pressure etc.

The preliminary design of buttress is carried out on the basis of unit column theory in which it is assumed that the buttress is made up of a number of unit columns (Fig. 10.11 a). A unit column is a curved column of unit width with its axis along the funicular polygon of the load it carries and the thickness increasing directly as required by the weight of the column above the section under consideration. This method of design is based on the following assumptions.

(1) Each column transfers the load from the top of the column to the foundation independently of the adjacent columns above or below it.

(2) Each column is so proportioned that at every section of the column uniform compressive stress is developed and it is curved to avoid eccentric loading.

These conditions may be accomplished if the axis of the column or the line of first principal stress lies along the funicular polygon formed by combining the initial load P_0 with the weight components of the buttress. The magnitude of the initial load P_0 depends on the type of joint between the deck slab and the buttress. Thus, if the joint between the deck slab and the buttress is frictionless then P_0 is equal to the sum of the total water pressure on the deck slab P_w and the component of the weight of the deck slab normal to the buttress face W_{d_n} (Fig. 10.11 b). On the other hand if the joint between the deck slab and buttress is monolithic then P_0 is equal to the resultant of the total water pressure on the deck slab P_w and the weight of the deck slab W_d (Fig. 10.11 c).

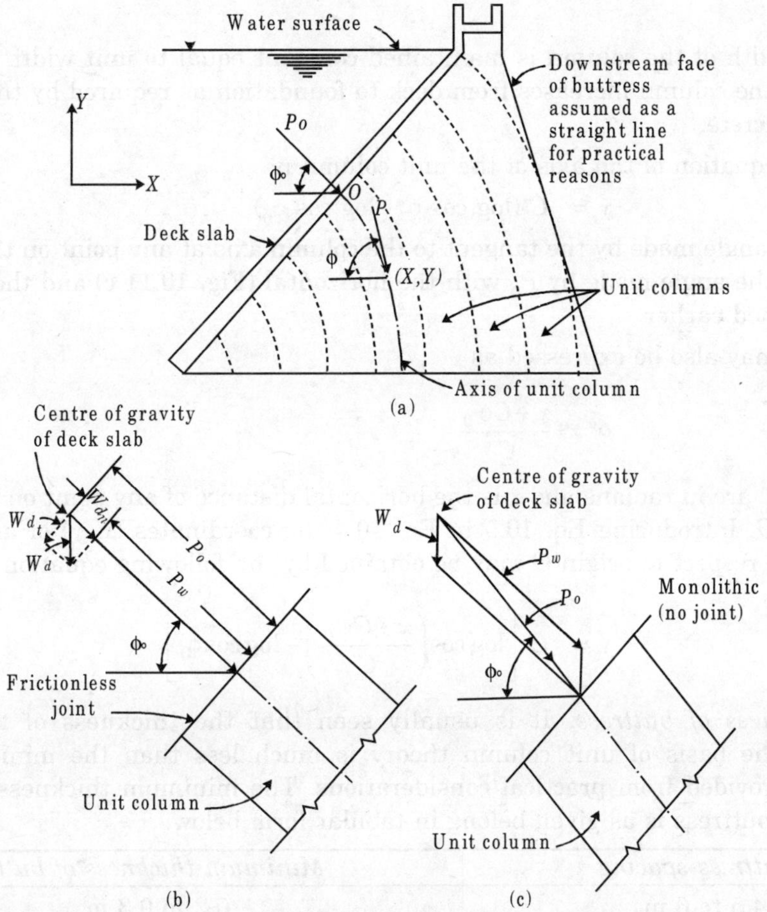

Fig. 10.11 *Unit columns and resultant forces for frictionless joint and monolithic connection between deck slab and buttress*

In this method of buttress design the coordinates of the axis and the thickness of a suitable number of unit columns are determined. In general five columns distributed over the entire buttress will be adequate. The surface profile of the buttress may then be obtained by drawing the lines of same thickness.

For each unit column the value of the initial thickness t_0 is obtained by the relation

$$t_0 = \frac{P_0}{f_c} \qquad \text{...(10.2)}$$

where f_c is permissible compressive stress in concrete.

The thickness t at any point in the unit column at a vertical distance y from the origin O (Fig. 10.11a) is given by the relation

$$t = t_0 \; e^{-y/C} \qquad \text{...(10.3)}$$

in which y is negative downward, and C is a constant equal to (f_c/w_c) where w_c is specific weight of concrete.

Since the width of the column is maintained constant equal to unit width (or 1 m width), the thickness of the column increases from deck to foundation as required by the increments of the weight of concrete.

Further the equation of the axis of the unit column is

$$y = C\,(\log\cos\phi - \log\cos\phi_0) \qquad \text{...(10.4)}$$

in which ϕ is the angle made by the tangent to the column axis at any point on the axis with the horizontal; ϕ_0 is the angle made by P_0 with the horizontal (Fig. 10.11 c) and the other notation are same as defined earlier.

The angle ϕ may also be expressed as

$$\phi = \frac{x + C\phi_0}{C} \qquad \text{...(10.5)}$$

in which ϕ and ϕ_0 are in radians and x is the horizontal distance of any point on the column axis from the origin O. Introducing Eq. 10.5 in Eq. 10.4 the coordinates (x, y) of any point on the column axis with respect to origin O may be obtained by the following equation

$$y = C\left[\log\cos\left(\frac{x + C\phi_0}{C}\right) - \log\cos\phi_0\right] \qquad \text{...(10.6)}$$

Minimum thickness of buttress. It is usually seen that the thickness of the buttress as determined on the basis of unit column theory is much less than the minimum thickness required to be provided from practical considerations. The minimum thickness required to be provided for the buttress is as given belong in tabular form below.

Buttress spacing	*Minimum thickness of buttress*
4.5 to 6 m	0.2 to 0.3 m
> 6 m	0.45 to 1.2 m

The design of the unit column should therefore be carried out with the actual thickness provided as per the requirements of the minimum thickness.

Downstream slope of buttress. The unit column theory provides a curved downstream face for the buttress. However, a buttress with a curved downstream face is difficult to construct and hence from practical considerations a buttress with a plane sloping downstream face is usually provided. The slope of the downstream face of the buttress should be such that the horizontal length of the buttress at all elevations is adequate from the considerations of overall stability. Usually the base length of the buttress varies from 1.2 to 1.5 times the depth below the maximum water level in the reservoir.

Modification of the buttress profile. The theoretical profile of the buttress obtained on the basis of unit column theory needs to be modified on account of practical considerations as discussed below.

The design of buttress on the basis of unit column theory results in a buttress with warped surfaces because the lines of same thickness are not straight but are curved. Since the construction of a buttress with warped surfaces is difficult, the buttress with plane surfaces is developed. For this the curved lines of same thickness are replaced by straight lines such that there is least possible departure of the straight lines from the curved lines. These straight lines will provide a buttress with plane surfaces.

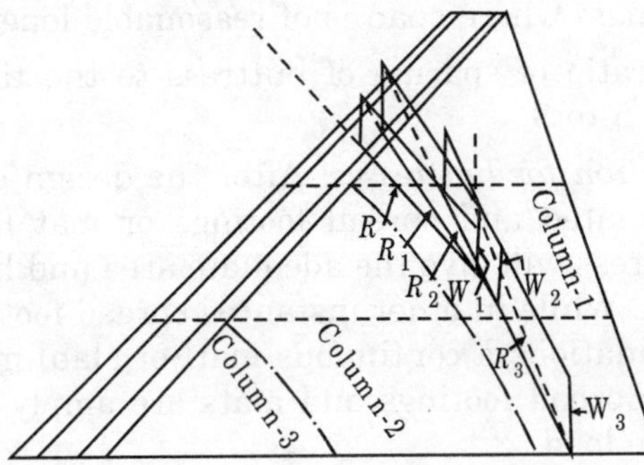

Fig. 10.12 *Modification of the buttress profile*

Further some of the unit columns close to the downstream face of the buttress will be discontinuous, i.e., not resting on the ground because from practical considerations the buttress is provided with a plane sloping downstream face. As such it is necessary to investigate the rationality of the design and also to determine whether the design needs to be modified. For this the buttress is divided into a suitable number of larger columns as shown in Fig. (10.12). These larger columns are continuous upto the foundation and are aligned to follow the principal stress lines so that the joints between the adjacent columns may be assumed to be frictionless. The top column (i.e., column 1) is assigned somewhat more length along the upstream face of the buttress than the other columns. This excess length gives greater latitude in the development of the line of principal stress for this column consistent in direction with other columns because the weight in the upper portion of the buttress is greater than is actually needed for stability and stress in the upper elevations of the dam. For each of these columns the line of principal stress is obtained by funicular polygon method in which the resultant external forces are combined with increments of concrete weight and their resultant indicates the direction of the first (or major) principal stress. If there is deviation of the funicular polygon from the centre of gravity of the corresponding column, it indicates the existence of second principal stress. The second principal stress will produce axial bending in the columns and it must be evaluated to determine the possibility of tension being developed parallel to the upstream face and the reinforcement to be provided. However, an excessive second principal stress will necessitate the modification of the design of the buttress.

Check for overall stability of buttress. After the dimensions of the buttress have been determined by the foregoing methods it should be checked for safety against overturning and sliding to ensure its overall stability. For this usual gravity analysis is adopted and the direct stress, sliding factor and shear-friction factor are computed by using the same formulae as idescribed in Chapter 9 for gravity dams.

Further the buttress which has been found adequate for the axial loads should also be investigated for column ratio (or slenderness ratio) and massiveness factor which are as defined below.

Column ratio is the ratio of unsupported length of buttress measured along the lines of principal stress to the thickness of the buttress. Low buttresses having no lateral supports may have a column ratio in the range of 12 to 15, but for high buttress which require lateral supports the column ratio should be limited to the range of 7 to 10. Provisions for lateral support will consist of curtain walls in double wall (or hollow) buttresses, counterforts in widely spaced buttresses and struts (or bracings) where span is of reasonable length.

Massiveness factor is the ratio of spacing of buttress to the thickness of the buttress. In general its value varies from 2.5 to 3.

Spread footings or Mat foundation for buttresses. After the design of deck slab and buttresses, foundation pressures are calculated and spread footings or mat foundation are designed. In some cases the base of the buttress will give the adequate area and hence spread footings or mat foundation will not be required. While in other instances spread footings or mat foundation may be required. On very poor foundations a continuous mat (or slab) may be provided underneath the entire length of the dam. Spread footings and mats are amply reinforced to improve their effectiveness in distributing the load.

10.6 ADVANTAGES AND DISADVANRAGES OF BUTTRESS DAMS

(A) Advantages

1. A buttress dam is less massive than a comparable gravity dam. As such a buttress dam may be constructed even on a relatively weak foundation on which the gravity dam may not be constructed.

2. The spacing of the buttresses may be adjusted to utilize zones of good foundations.

3. Depending on the degree of articulation or structural isolation provided, buttress dams may be designed to accommodate moderate amounts of foundation movement without serious damages. Hence these dams may be constructed on foundations where differential settlements are anticipated.

4. The amount of concrete required for a buttress dam is about $\frac{1}{3}$ to $\frac{1}{2}$ of the concrete required for a gravity dam of the same height. However, the cost of construction of a buttress dam is not reduced in the same proportion because of increased cost of form work and reinforcement required for a buttress dam.

5. In the case of buttress dams the stress distribution is much more uniform than that in a gravity dam. As such more efficient use of concrete is made in a buttress dam which leads to economy in the quantity of concrete required. However, the financial economy will not necessarily be directly proportional to the quantity of concrete saved.

6. The reduction in the concrete volume and increase in the surface area to volume ratio provide for better heat dissipation during construction and possibly increased speed of construction. This is so because the larger exposed area and thinner sections do not produce problem for cooling.

7. Access is possible to the back of the upstream face of the dam and to foundations between buttresses for periodic inspection and for subsequent grouting and drilling pressure relief holes, if required.

8. The uplift pressure acting on a buttress dam is considerably less which leads to economy in concrete and overall stability of the dam. However, if continuous mat foundation is provided for all the buttresses then there may not be much reduction in the uplift pressure unless vertical drains are provided to intercept seepage.

9. The upstream face of a buttress dam being considerably inclined a large vertical component of water pressure is developed which contributes to overall stability of the dam and saves concrete otherwise required for self weight.

10. The power house, switch yard etc., can be located between the buttresses thus saving some cost of construction.

(B) Disadvantages

1. Skilled labour requirements and the shuttering to concrete ratio are greater for a buttress dam than for a gravity dam. This may lead to higher unit rate of concrete and hence may offset some of the saving due to reduction in the quantity of concrete.

2. Deterioration of upstream surface of relatively thin reinforced-concrete deck slab or arch of a buttress dam has more serious effects on the stability of the dam. If the concrete at the upstream surface deteriorates the leakage of water may occur which would damage the reinforcement and cause ultimate failure of the dam.

3. The number of water seals to be provided and maintained for a buttress dam is usually more than for other dams.

4. The foundation stresses being higher, either sound rock foundation is necessary or spread footings or continuous mat may be required to be provided. The provision of spread footings or continuous mat will however lead to increased cost of construction. Moreover if a continuous mat is provided for all the buttresses the advantage of reduction in the uplift pressure would be lost.

5. The saving due to reduction in the quantity of concrete will also be to some extent offset by the requirement of heavy reinforcement for relatively thin members of a buttress dam.

6. The buttress dams are more susceptible to willful damage (by sabotage or enemy action) than solid gravity dams. The degree of vulnerability will however depend on the thickness of the upstream deck and the facility for access from the downstream side.

10.7 ARCH DAMS

An arch dam is a curved dam that carries a major part of the water load horizontally to the abutments by arch action. The part of the water load so carried depends primarily on the amount of curvature. The balance of the water load is transferred to the foundation by cantilever action. The thrust developed by the water load carried by arch action essentially requires strong abutments or side walls of the valley capable of resisting the arch forces. Further for the construction of an arch dam it is also desirable to have a V-shaped valley. This is so because for such valleys with the increase in the depth the span of the arch ring would decrease and since for a given rise of arch ring greater arch action would be developed for smaller spans of the arch ring, even for higher water load at greater depths the arch ring need not be very thick. Moreover from economical considerations it is essential to have definite ratio of crest length L and dam height H. Earlier it was commonly thought that the ratio L/H should be less than 1.5, but at the present time, ratios of 4 or even 6 are considered practicable. However, it has now been recognised that U-shaped valleys also suit arch dams fairly well.

Many early arch dams were built of rubble or ashlar stone masonry. However, practically all arch dams constructed during recent years have been built of concrete. Several arch dams have been constructed in the various countries. However, in India there is only one arch dam, known as Idduki dam which has recently been constructed on river Periyar in Kerala state. The main dimensions of this dam are as follows : Height 170.68 m; Top width 7.62 m ; Base width 19.81 m and Length at the crest 365.76 m.

10.8 TYPES OF ARCH DAMS

Arch dams may be divided into two types viz., massive arch dams and multiple arch dams. A *massive arch dam* consists of a single curved wall usually vertical or nearly so which spans the full width between abutments. A *multiple arch dam* consists of a number of smaller arches, usually inclined, supported on piers or buttresses. The multiple arch dams are usually considered as a type of buttress dams and the same are described earlier. The massive arch dams may be further divided into the following types.

(1) Constant radius arch dams

(2) Constant angle arch dams

(3) Variable radius arch dams

(4) Double curvature or cupola arch dams

(5) Arch gravity dams.

1. Constant radius arch dams. A constant radius arch dam is the one in which the radius of the extrados (or outside curved surface of the arch) is constant at all elevations from top of the dam to its base (Fig. 10.13). However, the intrados (or inside curved surface of the

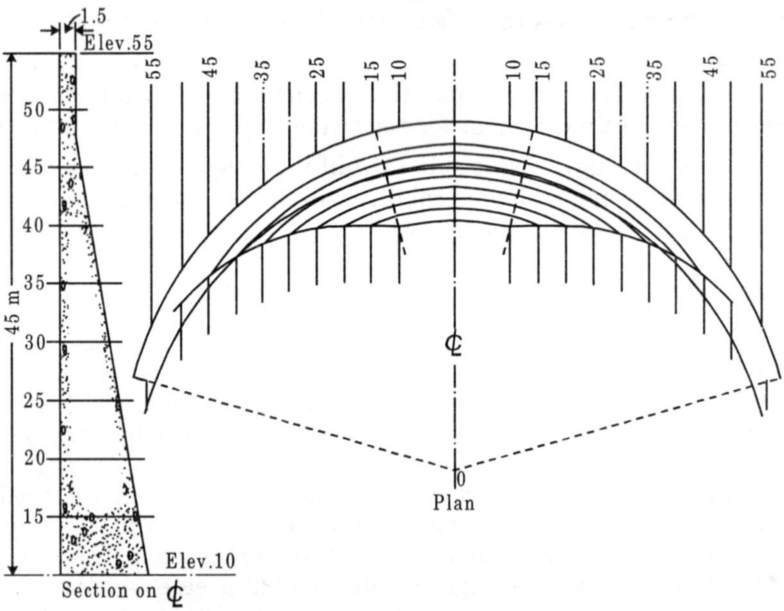

Fig. 10.13 *Constant radius arch dam*

arch) has gradually decreasing radius as the depth below the crest increases, to provide the increased thickness needed for the higher reservoir water pressure. Thus, in this case the upstream face of the dam is vertical and the dam is triangular in cross-section with increased width at the crest. Further in a constant radius arch dam the centres for extrados, intrados and the centre lines of the horizontal arch rings at various elevations lie on a straight vertical line that passes through the centre of the upstream face of the horizontal arch ring at the crest. Hence this type of dam is also sometimes known as *constant centre arch dam*. However, in some cases the intrados may not be concentric with reference to extrados.

In this type of arch dam on account of radius being constant the central angle and the central rise of the arch rings at various elevations decreases as the depth below the crest increases. Due to reduction in the central angle and the central rise of the arch rings at the lower elevations near the base of the dam, considerably low arch action will be developed for these arch rings. Hence relatively large proportions of the water load at the lower elevations will be carried by cantilever action for which the arch rings at the lower elevations will have to be made considerably thick and this will result in an uneconomical design. The reduction in the central angle and the central rise of the arch rings at the lower elevations would be considerable in the case of narrow V-shaped valleys and therefore for such valleys constant radius arch dams would generally be uneconomical. However, for U-shaped valleys the constant radius arch dams are usually found suitable because there is relatively less reduction in the central angle and the central rise of the arch rings at the lower elevations.

2. Constant angle arch dam. A constant angle arch dam is the one in which the central angle of the arch rings at all elevations from top of the dam to its base has the same magnitude (Fig. 10.14). As mentioned in section 10.10, according to thin cylinder theory of design of arch

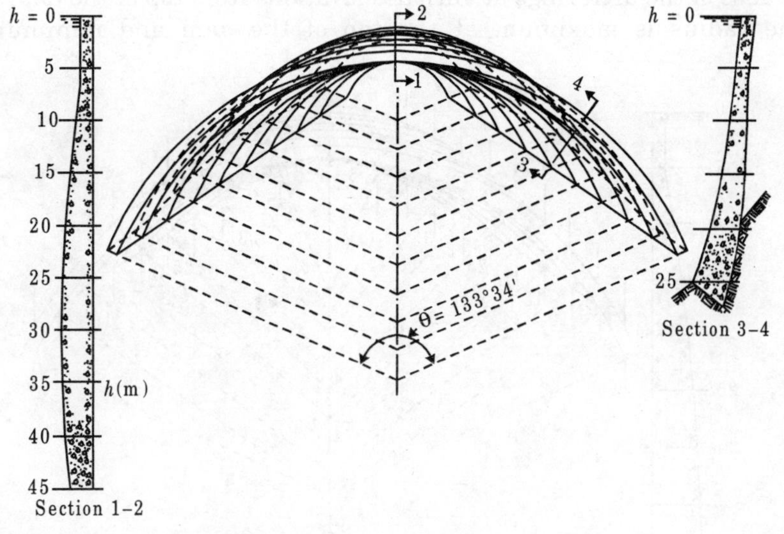

Fig. 10.14 *Constant angle arch dam*

dams the volume of concrete required for an arch dam is minimum if the central angle of the arch rings is 133° 34'. As such theoretically for economical design of constant angle arch dams central angle of arch rings equal to 133° 34' may be adopted. However, in actual practice for the economical design of constant angle arch dams the central angle may be in the range of 100° to 150° depending upon the valley shape. This is so because the cost of shuttering which depends on the valley shape is not included in the calculation of minimum volume of concrete by thin cylinder theory.

In the case of constant angle arch dam because of larger central angle greater arch action will be developed for the arch rings at all elevations. As such in this case large proportions of water load at all elevations will be carried by arch action. Hence in general the thickness of a constant angle arch dam (Fig. 10.14) is smaller than that of a constant radius arch dam (Fig. 10.13). A constant angle arch dam is therefore economical since under same conditions it requires only about 70% concrete as compared to constant radius arch dam.

As shown in Fig. 10.14 the radii of the arch rings of a constant angle arch dam decrease as the depth below the crest increases since the central angle at all elevations has the same magnitude. Moreover, in this case the curvature of each arch ring will be different and the various arch rings may be arranged to have the crowns of either extrados or intrados of all the arch rings flushing. However, any of these arrangements will result in an overhang of the upstream face near the abutments and that of the downstream face near the crown of the arch, on account of which the fromwork needed for a constant angle arch dam will be usually more complicated and costly than that for a constant radius arch dam.

For narrow V-shaped valleys constant angle arch dams are best suited because with larger central angle fairly large arch action will be developed even for arch rings having smaller spans at lower elevations and hence the design will be economical. Likewise these dams are also suitable for U-shaped valleys.

3. Variable radius arch dams. A variable radius arch dam is the one in which the radii of extrados and intrados of the arch rings at various elevations from top of the dam to its base vary (Fig. 10.15). The radius is maximum at the top of the dam and minimum at its base.

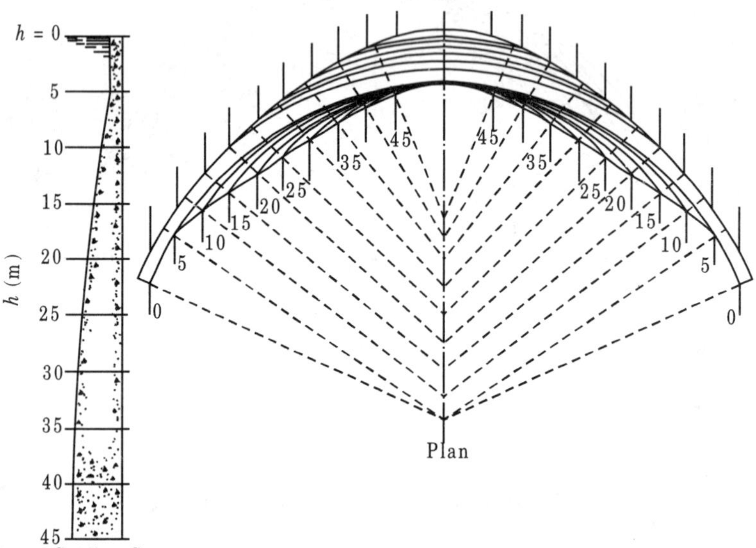

Fig. 10.15 *Variable radius arch dam*

Further in this case the central angle of the arch rings at various elevations from top of the dam to its base also vary, being maximum at the top of the dam and minimum at its base. The central angles of the arch rings usually range from 80° to 150° and these are so adjusted that as far as possible larger arch action is developed even for arch rings at lower elevations. Similarly the radii of the arch rings at various elevations are also so adjusted that as far as possible the overhangs are avoided or minimised. Thus, in a variable radius arch dam neither the radius nor the central angle is constant. Further a variable radius arch dam is also known as variable centre arch dam because the centres of the arch rings at various elevations do not lie on the same vertical line.

The variable radius arch dams are suitable for U-shaped valleys as well as for narrow V-shaped valleys because by adjusting the radii as well as central angles fairly large arch action may be developed even for arch rings having smaller spans at lower elevations.

A constant angle arch dam described earlier may be considered as a special type of variable radius arch dam in which only the radii of the arch rings at various elevations vary but the central angle of the arch rings at various elevations has the same magnitude. However, as compared to a constant angle arch dam a variable radius arch dam is usually less economical since under same conditions a variable radius arch dam requires about 17% more concrete than a constant angle arch dam. On the other hand a variable radius arch dam is in general more economical than a constant radius arch dam since under same conditions the concrete required for a variable radius arch dam is only about 82% of that required for a constant radius arch dam. Hence a variable radius arch dam is a compromise between constant radius and constant angle arch dams.

4. Double curvature or cupola arch dams. A double curvature or cupola arch dam is the one which has curvature in horizontal as well as vertical planes. In other words it is a dam which is curved in plan as well as in sections. These dams are also known as shell arch dams because these are usually designed as shell structures. The double curvature or cupola arch dams were originated and developed in Italy. The famous Vajont dam in Italy is a double curvature or cupola arch dam which is 261.6 m high and it is the second highest arch dam in the world. Moreover, the world's highest arch dam viz., Ingurskaya dam in U.S.S.R. is also this type of dam which is 301 m high. In general these dams have relatively small thickness and hence these are quite economical. For example, the Vajont dam is only 22 m thick at the base. However, only a few dams of this type have been so far built in some of the European countries and the same have not become popular in other countries.

5. Arch gravity dams. A thick arch dam is known as arch gravity dam because structurally it behaves as both gravity and arch dam. The famous Hoover dam in U.S.A. is an arch gravity type dam. It is 221.6 m high, 391 m long at the crest, 13.7 m thick at the crest and 201.3 m thick at the base. For an arch gravity dam generally in the lower part of the valley the extrados of the arch rings have gradually increasing radii so that near the base of the dam the upstream face has a batter and hence the base width of the dam is considerably increased.

10.9 FORCES ON AN ARCH DAM

The various forces acting on an arch dam are same as those which act on a gravity dam. However, the relative importance of the forces is different in arch dams. Because of the narrow base width of an arch dam the uplift pressure is less important than for a gravity dam. On the other hand the stresses caused by ice pressure, temperature changes and yielding of supports (abutments) are very important in an arch dam. The ice pressure causes a continuous concentrated load along the arch element at the elevation of the ice. The temperature changes cause internal forces that move the dam upstream during the summer and downstream during the winter. Thus, the winter condition is more important in the stress analysis since it causes force with acts with the reservoir water pressure. The slight yielding of the abutments may also cause high internal stresses in the arch.

10.10 DESIGN OF ARCH DAMS

The following are some of the methods commonly used for the design of arch dams.

(1) Thin cylinder theory.

(2) Elastic arch theory.

(3) Trial load analysis.

1. Thin cylinder theory. In thin cylinder theory the horizontal section of the arch dam is assumed to behave as a part of a thin cylinder ring at the corresponding elevation. It is further assumed that the horizontal water load is carried entirely by arch action; the pressure on the arch ring is equal to the hydrostatic pressure at the corresponding elevation ; and the stress distribution across the thickness of the arch ring is uniform.

Figure 10.16 shows the section of a thin arch ring having mean radius *r*, angle subtended at the centre of the arch ring equal to θ and thickness *t*. The various forces acting on the arch ring in the horizontal plane are also shown in Fig. 10.16.

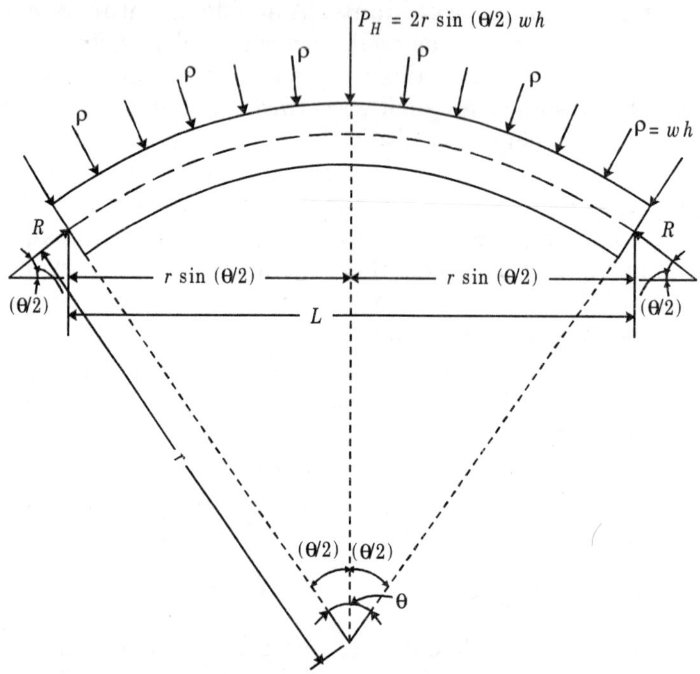

Fig. 10.16 *Thin cylinder theory—definition sketch*

If the arch ring is at a depth h below the water surface in the reservoir, then intensity of hydrostatic pressure p acting on the arch ring radially in the downstream direction is given by

$$p = wh \qquad ...(i)$$

where w is specific weight of water.

The total hydrostatic pressure force P_H acting on the entire arch ring of unit height is given by

$$P_H = (\text{Intensity of pressure}) \times (\text{Projected area})$$

or

$$P_H = wh \times 2r \sin\left(\frac{\theta}{2}\right)$$

$$= 2whr \sin\left(\frac{\theta}{2}\right) \qquad ...(ii)$$

As shown in Fig. (10.16), P_H acts in the downstream direction along the axis of symmetry of the arch ring in the horizontal plane.

The force P_H is balanced by the components of the reactions at the abutments in the upstream direction (i.e., in the direction opposite to that of P_H).

Thus, if R is the reaction at each abutment, its component in the upstream direction is $R \sin(\theta/2)$ and sum of the components at both the abutments is $2R \sin(\theta/2)$. Equating this to P_H for equilibrium we get

$$2R \sin\left(\frac{\theta}{2}\right) = 2whr \sin\left(\frac{\theta}{2}\right)$$

or

$$R = whr \qquad ...(iii)$$

If the thickness t of the arch ring is small as compared with radius r, it may be assumed that uniform compressive stress is developed in the arch ring. Thus, if σ is the compressive stress developed in the arch ring at the abutments then

$$\sigma = \frac{R}{t \times 1} = \frac{whr}{t} \qquad ...(iv)$$

since the arch ring is of unit height.

For safe design of the arch ring the compressive stress σ should not exceed the allowable working stress f_c in compression for the material of arch ring. Thus, in the limiting case $\sigma = f_c$ and hence from Eq. (iv) the required, thickness of the arch ring is given by

$$t = \frac{whr}{f_c} \qquad ...(10.7)$$

Equation (10.7) indicates that the thickness of the arch ring should increase linearly with the depth below the water surface and that for a given pressure the required thickness of the arch ring is proportional to its radius.

Central angle for minimum volume of concrete required for an arch dam

Considering the arch ring as shown in Fig. (10.16), the volume V of concrete per unit height of the arch ring is given by

$$V = r\,\theta\,A$$

where A is the cross-sectional area of the arch ring.

Since $$A = (t \times 1)$$

$$V = r\ \theta\ t \qquad ...(i)$$

Substituting the value of t from Eq. (10.7), we get

$$V = \frac{wh}{f_c} r^2\,\theta \qquad ...(ii)$$

If L is span of the arch ring then

$$L = 2r \sin\left(\frac{\theta}{2}\right)$$

or $$r = \frac{L}{2\sin\ (\theta/2)}$$

$$V = \frac{wh}{f_c}\frac{L^2}{4}\frac{\theta}{\sin^2(\theta/2)} \qquad ...(iii)$$

For minimum volume of concrete

$$\frac{dV}{d\theta} = 0$$

Thus, by differentiating both sides of Eq. (iii) with respect to q and equating it to zero, we get

$$\frac{dV}{d\theta} = \left(\frac{wh}{f_c}\frac{L^2}{4}\right)\left[\frac{\sin^2(\theta/2) - \theta\sin\ (\theta/2)\ \cos\ (\theta/2)}{\sin^4(\theta/2)}\right] = 0$$

or $$\tan\left(\frac{\theta}{2}\right) = \theta$$

which gives $$\theta = 133°\ 34'$$

Thus, on the basis of thin cylinder theory the volume of concrete required for an arch dam would be minimum if the central angle is 133° 34'.

Limitations of thin cylinder theory

The thin cylinder theory of design of arch dams has several limitations which are as given below.

(i) The design of an arch dam by the thin cylinder theory does not take into account the rib shortening effect due to which additional stresses are produced in the dam as indicated below.

A complete circular ring under a uniform external radial load is uniformly shortened and the shape of every portion of the ring remains constant. On the other hand an arch ring of an arch dam is only a part of a circular ring supported at its both the ends and under a uniform external radial load its curved length tends to be shortened which is known as *rib-shortening*.

However, if the end supports are rigid the span of the arch ring does not change and the shortening of the curved length of the arch ring is restrained. Due to such restraint the arch ring is deformed and in addition to normal arch loads, moments and shears are introduced which produce additional stresses in the arch ring. The stresses produced in this manner are called *rib-shortening stresses* which need to be considered in the design of arch dams.

(ii) In thin clinder theory of design of arch dams it is assumed that the stress distribution along the thickness of the arch ring is uniform, but actually the stress distribution along the thickness of the arch ring is not uniform.

(iii) The design of an arch dam by the thin cylinder theory does not take into account the effect of abutment yielding which also produces additional stresses in the dams as indicated below.

It is assumed that the end supports for the arch dams are rigid so that when the dam is subjected to external loading the abutment span does not change. However, actually the abutments being elastic they yield and slightly spread apart by the thrust of the arch. Such spreading of the abutments results in introducing moments and shears in the arch dam thus producing stresses which will be in addition to that due to rib-shortening effect. Further if there are moments at the ends of the arch, the abutment faces will rotate slightly, which will introduce additional moments and shears and hence produce stresses in the arch dam. These stresses are also required to be considered in the design of arch dams.

(iv) In the design of arch dams by the thin cylinder theory the stresses produced due to temperature changes and shrinkage of concrete are also not considered. These stresses are produced as explained below.

A drop in temperature causes shortening of the curved length of an arch ring. Similarly shrinkage resulting from drying out of concrete also causes shortening of the curved length of an arch ring. Both these effects produce moments and corresponding stresses in the arch ring which are in addition to those caused by elastic deformations and the same are required to be taken into account in the design of arch dams. However, a rise in temperature has an opposite effect and hence it is not considered.

(v) The thin cylinder theory can be applied only if the valley is symmetrical about its centre line so that each arch ring has symmetrical loading. However, if the valley is unsymmetrical about its centre line then each arch ring will have different lengths on either side of the centre line and there will be unsymmetrical loading for each of the arch rings. Since the effect of unsymmetrical loading on the arch rings cannot be taken into account in the thin cylinder theory the same cannot be applied in the case of unsymmetrical valleys. Moreover, the thin cylinder theory cannot be applied where inclined or irregularly shaped arches are required to be provided.

(vi) The thin cylinder theory can be applied only to a simple concentric circular ring subject to a uniform radial loading. However, in the design of an arch dam, except water pressure which acts radially, there are various other forces such as earthquake forces, silt pressure etc., which are not radial but need to be considered. Hence the thin cylinder theory cannot be applied when all these forces are required to be considered.

On account of the various limitations of the thin cylinder theory as indicated above it is usually adopted only for the preliminary design of an arch dam and the final design is carried out by the other methods as described below.

2. Elastic arch theory. In the elastic arch theory of design of an arch dam it is assumed that the entire external load acting on the dam is transferred horizontally to the abutments through arch action only. It is further assumed that the entire dam consists of a number of horizontal arch rings placed one above the other, their ends are fixed and each ring is independent of the other. Thus, this method of design of arch dams involves the analysis of an elastic arch with fixed ends which is subject to uniform external loading. This method of design takes into account the additional stresses produced in the arch ring because of the following effects :

(i) Rib-shortening

(ii) Temperature changes

(iii) Shrinkage

(iv) Yielding of abutments

which have been explained earlier and are not considered in the design of arch dams by the thin cylinder theory. Moreover the elastic arch theory can take into account the variable loads along the length of the arch.

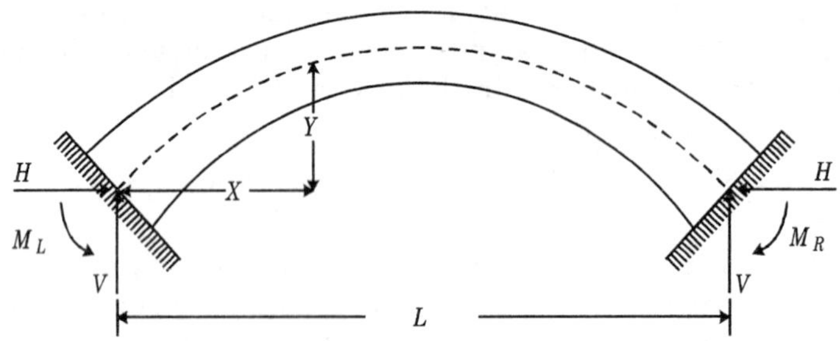

Fig. 10.17 *Elastic arch theory—definition sketch*

Consider an arch as shown in Fig. 10.17 having fixed ends to that the foundation movements are ignored. The following equations may be used for such elastic arches with fixed ends.

$$M = M_0 + M_L + (M_R - M_L) \times \frac{x}{L} + H \cdot y \qquad ...(10.8)$$

$$\int \frac{My}{EI} ds = 0 = \int \frac{M_0 y}{EI} ds + M_L \int \frac{yds}{EI} + \frac{(M_R - M_L)}{L} \int \frac{xy}{EI} ds + H \int \frac{y^2 ds}{EI} \qquad ...(10.9)$$

$$\int \frac{Mx}{EI} ds = 0 = \int \frac{M_0 x}{EI} ds + M_L \int \frac{xds}{EI} + \frac{(M_R - M_L)}{L} \int \frac{x^2}{EI} ds + \int H \frac{xy}{EI} ds \qquad ...(10.10)$$

and

$$\int \frac{Mds}{EI} ds = 0 = \int \frac{M_0}{EI} ds + M_L \int \frac{ds}{EI} + \frac{(M_R - M_L)}{L} \int \frac{x}{EI} ds + H \int \frac{y}{EI} ds \qquad ...(10.11)$$

where M = bending moment at any point on the arch

ML and MR = fixed end bending moments at both the ends

M_0 = bending moment on a straight horizontal freely supported beam

H = horizontal thrust at the abutments

x and y = horizontal and vertical coordinates with respect to origin at the left abutment

L = span of the arch

For symmetrical arch $M_L = M_R$ and Eq. (10.10) vanishes.

Thus, Eqs (10.9) and (10.11) become

$$\int \frac{M_0 y}{EI} ds + M_L \int \frac{y}{EI} ds + H \int \frac{y^2}{EI} ds = 0 \qquad ...(10.12)$$

$$\int \frac{M_0}{EI} ds + M_L \int \frac{ds}{EI} + H \int \frac{y}{EI} ds = 0 \qquad ...(10.13)$$

The effects of temperature change, shrinkage, rib-shortening and abutment yielding can be accounted for as indicated below.

Effect of temperature change

A temperature change causes a change in length of the arch ring which is prevented by the abutments and hence Eq. (10.9) will be modified as indicated below.

$$\alpha TL - \int \frac{My}{EI} ds = 0 \qquad ...(10.9\ a)$$

where α = coefficient of linear expansion

T = temperature variation

However, Eqs (10.10) and (10.11) will remain unchanged.

Effect of shrinkage

Shrinkage of saturated concrete, except for temperature change, is small, and a loaded dam is wet. As such the effect of the change in volume due to shrinkage of concrete is usually not considered separately but it is included in the effect of temperature change.

Effect of rib shortening

This can also be accounted for along with the other loads by adding the expression representing this effect to the left side of Eq. (10.9).

Effect of abutment yielding

This effect may be considered in terms of the various expressions developed by F. Vogt for the deformation caused by thrust, shear and moment introduced in the arch ring due to yielding of the abutments.

The above noted equations are modified in respect of these effects and by the simultaneous solution of these equations the values of *ML, MR* and *H* are found.

3. Trial Load Analysis. In the thin cylinder theory as well as in the elastic arch theory it is assumed that an arch dam consists of a series of horizontal arch rings lying one over the other and each arch ring is independent of the one above and below it. However, this is not true because an arch dam is a monolithic curved mass of concrete supported on the abutments. Thus, neither there can be separate arch rings nor there can be separate cantilevers (as considered in the case of a gravity dam) and the solution obtained on the basis of such assumptions will be quite approximate. Further on account of complexities involved an exact mathematical analysis of an arch dam is not practicable. Hence a method known as trial load method based on trial load analysis has been developed by U.S.B.R. which gives reasonably satisfactory approximations. In this method the dam is assumed to be made up of two systems of elements: (a) horizontal arches and (b) vertical cantilevers. Each system of elements occupies the whole dam. It is further assumed that part of the load acting on the dam is taken by the arch system and the remaining part is taken by the cantilever system. The load is divided between the two systems of elements in such a manner that the computed deflection for any point in the dam, considered as a point in the arch system, will be identical with its computed deflection, considered as a point in the cantilever system. Such division of load which will cause identical deflections at any point is found by successive trials for a number of points on the arch dam.

The trial load analysis can be applied to an arch dam of predetermined dimensions which may be obtained either on the basis of some existing designs of similar type or by thin cylinder theory or by elastic arch theory. The dam profile is divided by horizontal planes, a unit distance apart, into a continuous series of arches. A few representative arches at different elevations are then selected for this analysis. Likewise the dam profile is divided into a continuous series of vertical elements and representative units are chosen for analysis. By this analysis it is determined whether the trial dimensions of the dam are correct and if not the same are modified to obtain the final design.

REVIEW QUESTIONS

10.1 What are the different types of buttress dams? Discuss their suitability for different sites, conditions and project requirements.

10.2 How are the economical buttress spacing and inclination of the upstream face determined?

10.3 Briefly describe the unit column theory of design of buttress dams and derive the necessary equations. What are the limitations of designing a buttress dam on the basis of this theory?

10.4 Give the design procedure for a flat slab type buttress dam.

10.5 What are the advantages and disadvantages of buttress dams?

10.6 A buttress dam is superior to a gravity dam in the utilization of construction materials. Justify the statement by arguments in its favour, indicating how a better utilization is attained.

10.7 For a buttress face inclined at 45° to the horizontal, plot the axis of a 'unit column' graphically. The column is 24 m deep vertically. Permissible stress in concrete is 2943 kN/m^2 (300 t/m^3) and its unit weight is 23.544 kN/m^3 (2.4 t/m^3).

10.8 What are the various types of arch dams? Which type is the cheapest in cost? Explain, giving reasons.

10.9 Discuss the thin cylinder theory for the design of an arch dam. What are the limitations of this theory?

10.10 Design a constant angle arch dam by the cylinder theory for the following conditions: (i) Valley 30 m wide at base and sides rising at 60° to the horizontal on both sides. (ii) Height of the dam 150 m.

(iii) Safe stress 2061 kN/m^2 (210 t/m^2).

10.11 Describe the various methods commonly used for the design of arch dams.

10.12 Briefly discuss the principles of trial load analysis of an arch dam.

CHAPTER

11 Embankment Dams

11.1 INTRODUCTION

The embankment dams are classified as non-rigid type of dams. For the construction of these dams the entire range of soils from clays to boulders or quarried stone are used. These dams usually provide the most economical and most satisfactory solution for sites at which suitable foundation at reasonable depth may not be available for a dam of concrete or masonry.

Two main forms of embankment dams which have been developed and widely used are : the *earth dams* (or earth-fill dams) made predominantly of earth or soil and *rock-fill dams* made predominantly of quarried rock. However, recently composite earth and rock-fill type of embankment dams are also being widely adopted.

Earth dams have been built since the early days of civilization, but until recently these dams were designed by empirical methods and their construction was based mostly on experience and precedent. However, the recent advances in the knowledge of the properties and behavior of soils have provided a rational basis for the design and construction of earth dams which has enabled these dams to be designed with more certainty and confidence. Moreover, with these advancements it has become possible now to construct earth dams of a size which would have been considered neither safe nor practicable a few decades ago. For example, the Nurek dam across river Vakhsh in U.S.S.R. which is the second highest dam in the world, is an earth dam and it is 300 m high. However, the Rogun dam also across river Vakhsh in U.S.S.R. is the highest dam in the world which is a composite earth and rock-fill type embankment dam and it is 335 m high. In India Tehri dam across river Bhagirathi is the highest dam which is also a composite earth and rock-fill type embankment dam and it is 261 m high.

Rock-fill dams had their origin about hundred years ago and since then several successful rock-fill dams have been built in various countries. Moreover, the advances in the art of quarrying and in the manufacture of loading and transportation equipment have made it possible to build fairly high rock-fill dams.

11.2 TYPES OF EARTH DAMS

The earth dams may be classified on the basis of the method of construction into the following three categories.

(1) Rolled fill dam.

(2) Hydraulic fill dam.

(3) Semi-hydraulic fill dam.

11.2.1 Rolled Fill Dam

A rolled fill dam is the one which is constructed in successive, mechanically compacted layers. The material (sand, clay, gravel etc.) is transported from the borrow pits to the dam site by trucks or scrapers. It is then spread within the dam section by bulldozers to form layers of 15 to 45 cm thickness. Each layer is then thoroughly compacted and bonded with the preceding layer by means of power-operated rollers of proper design and weight. Usually sheepsfoot rollers and heavy pneumatic-tired rollers are used either singly or in combination for compacting the fill. Further for proper compaction the moisture content of the material should be near that for optimum density, for which the required quantity of water is sprinkled on each layer during compaction.

11.2.2 Hydraulic Fill Dam

In the case of hydraulic fill dam the transportation of the material from the borrow pits to the dam site as well as its placement in final position in the dam is done by water. Thus in this case at the borrow pits the material is mixed with water to form slush which is transported through flumes or pipes and deposited near the faces of the dam. The coarser materials of the slush stay near the faces of the dam while the finer ones move towards the centre and get deposited there. This would provide a dam section with shoulders of the coarse free-draining particles and an impervious central core of fine grained material such as clay and silt. With proper control a satisfactory dam section can be achieved by this method in which as is evident no compaction is done.

11.3.3 Semi-hydraulic Fill Dam

In the case of a semi-hydraulic fill dam the material is transported from the borrow pits to the dam site and dumped within the section of the dam in the same manner as in the case of a rolled fill dam without the use of water. Some of the dumped material is then moved to its final position in the dam by the action of water. For this jets of water are directed on the dumped fill which causes the finer material from the fill near the faces of the dam to be washed away. The finer material moves towards the central portion of the dam and is deposited there thus forming an impervious central core while the coarser material stays near the faces of the dam. However, in the absence of proper jetting action the dumped fill at the faces of the dam may be more dense and impervious than the material immediately below it on the inside of the dam which may result in the failure of the dam. In this case also no compaction is done.

Out of these three types of dams rolled fill dams are most commonly adopted. This is so because in the case of the other two types of dams lack of control in placing the material may result in the failure of the dams.

Rolled fill dams can be further subdivided into the following three types.

1. Homogeneous type dams
2. Zoned type dams
3. Diaphragm type dams.

1. Homogeneous Type Dams. A purely homogeneous type earth dam is composed of a single kind of material (exclusive of the slope protection). These dams are usually built of impervious or semi-pervious soils, but several successful dams have been built of relatively pervious sands and sand-gravel mixture. Homogeneous dams have been built since the earliest times and are used even today whenever only one type of material is economically available. However, these dams are used only for low to moderate heights.

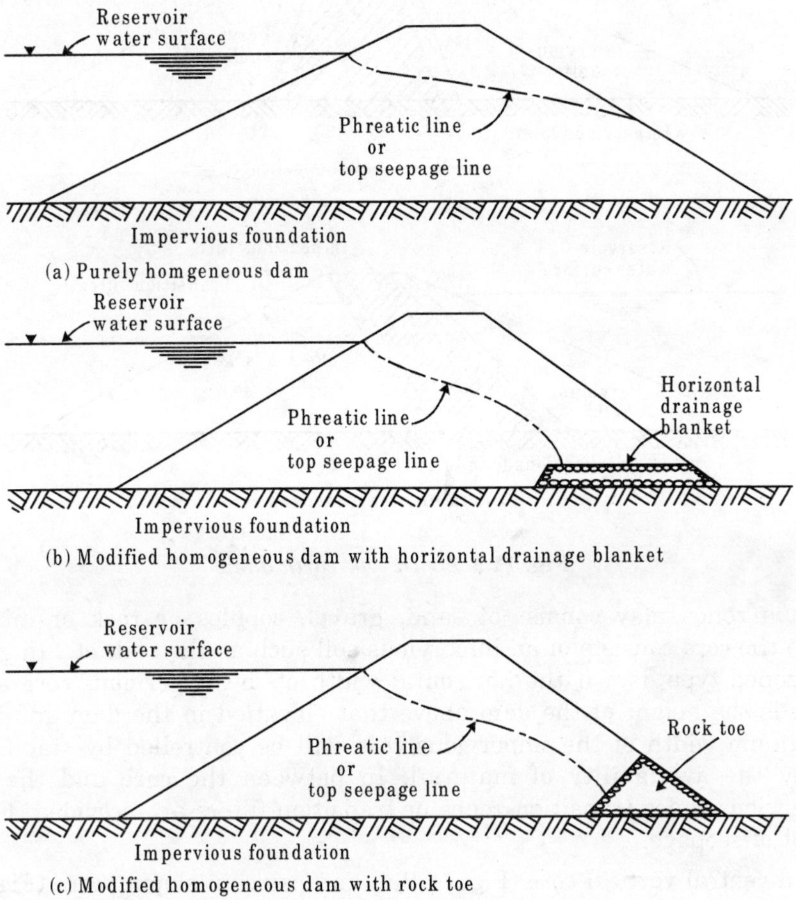

Fig. 11.1 *Purely homogeneous and modified homogeneous earth dams*

A purely homogeneous section has been replaced by a *modified homogeneous section* in which internal drainage system in the form of a horizontal drainage blanket or a rock toe or a combination of both is provided. By providing drainage system, dams with much steeper slopes may be designed and also the phreatic line (or top seepage line) is kept well within the body of the dam (Fig. 11.1).

2. Zoned Type Dams. A zoned type earth dam is composed of more than one kind of material. This is the most common type of a rolled fill dam in which a central impervious core is flanked by zones of materials considerably more pervious (Fig. 11.2). The pervious zones enclose, support and protect the impervious core. The upstream pervious zone provides free drainage during sudden drawdown and the downstream pervious zone acts as a drain to control the phreatic line. The impervious core checks the seepage.

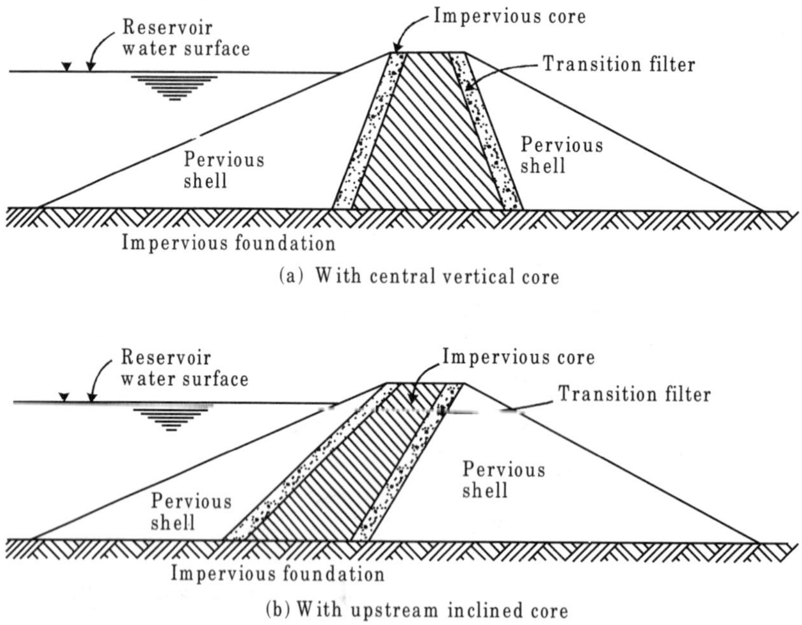

Fig. 11.2 *Zoned type earth dams*

The pervious zones may consist of sand, gravel, cobbles or rock or mixtures of these materials while the core consists of an impervious soil such as clay, silt etc. In general a dam is considered as zoned type dam if the horizontal width of the impervious core at any elevation equals or exceeds the height of the dam above that elevation in the dam and is not less than 3 m. The maximum width of the impervious core will be controlled by stability and seepage criteria and by the availability of material. In between the core and the upstream and downstream pervious zones transition zones or transition filters are provided (Fig. 11.2) for the effective control of seepage.

In place of a central vertical core (Fig. 11.2 a) an upstream sloping core (Fig. 11.2 b) is also provided in a zoned type earth dam. The main advantage of the upstream sloping core is that the downstream portion of the dam can be constructed first and the core placed later. However, in general a central vertical core is preferred to a sloping core.

3. Diaphragm Type Dams. In a diaphragm type earth dam the bulk of the dam is constructed of pervious material (sand, gravel or rock) and a thin core usually known as diaphragm of impermeable material is provided to check seepage. As such these dams are also sometimes known as thin core dams. The position of the diaphragm may vary from a central vertical core (Fig. 11.3 a) to a blanket directly on the upstream face of the dam (Fig. 11.3 b) and in between these extreme positions there may be several intermediate positions of the diaphragm in which an inclined diaphragm sloping upstream may be provided. The inclined diaphragm is also sometimes known as burried blanket. If the diaphragm is provided as an impervious blanket on the upstream face of the dam it needs to be protected against shallow sloughs and slides during drawdown and from erosion by wave action. For this the diaphragm is held burried below a thin layer of pervious material over which the upstream slope protection is provided. The diaphragm may consist of impervious soil, cement concrete, bituminous concrete

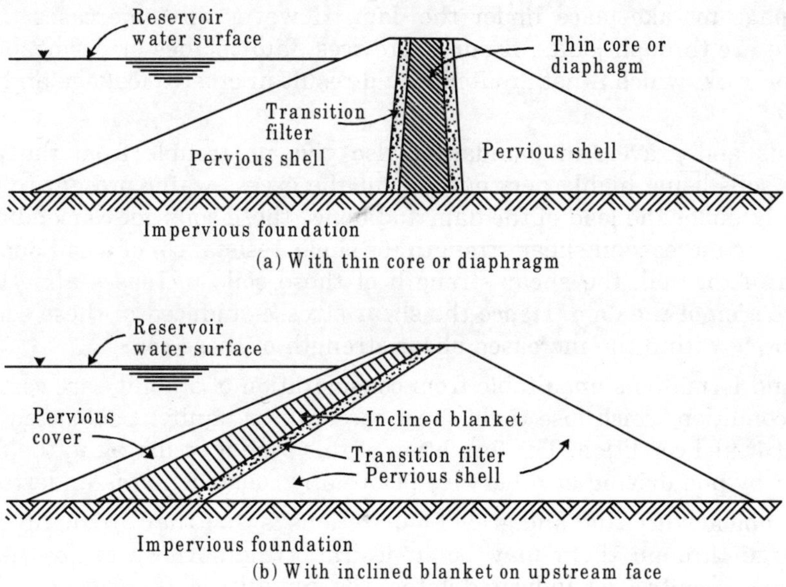

Fig. 11.3 *Diaphragm type earth dams*

or any other suitable material. If the core or blanket is of impervious soil it is considered to be a 'diaphragm' if its horizontal thickness at any elevation is less than 3 m or less than the height of the dam above the corresponding elevation in the dam. However, if the core or blanket equals or exceeds these thicknesses the dam is considered as zoned type earth dam as indicated earlier.

A diaphragm type earth dam is usually constructed where there is an ample supply of pervious material but there is only a limited supply of soil suitable for an impervious core. However, the construction of an internal diaphragm of impervious soil with the necessary filters requires a high degree of precision and control. Moreover, internal diaphragms of rigid materials such as concrete also have the disadvantage of not being readily available for inspection or emergency repair if they are ruptured due to settlement of the dam or its foundation. As such the construction of diaphragm type earth dams is usually restricted to small dams only.

11.3 FOUNDATIONS FOR EARTH DAMS

It is generally possible to construct a safe earth dam on almost any foundation which is available provided that the foundation has been thoroughly explored and tested and the design adopted to suit the conditions thus revealed. The different types of foundations can be considered under three categories viz., (i) Rock, (ii) Sands and gravels and (iii) Clayey material. This is however a broad classification because within each category the properties of materials may vary widely and also the materials of more than one category may be found interspersed in the foundation under the proposed dam. Therefore the design of any one dam may involve a variety of foundation problems. The main problems of each of the above noted three types of foundations are as describe below.

Rock foundations in general do not present any problem. The shear stresses induced in these foundations by earth dams will usually be well within their capacity and they would allow only small seepage to take place under the dam. However, in some cases they may require grouting if there are through joints, fissures, crevices, fault planes or permeable strata present in the foundation rock, which if not grouted would result in erosive leakage and excessive loss of water.

Coarse sands and gravels in foundation also give no trouble from the point of view of stability. These soils being highly pervious permit the excess water present in their voids to be drained out easily under the load of the dam and hence these soils get consolidated very rapidly. Moreover, since the increase in shear strength (or shear resistance) of a soil depends on the rate of consolidation of the soil, the shear strength of these soils increases almost simultaneously with the construction of the dam. Hence the shear stresses induced in these soils by the load of the dam are mostly within the increased shear strength of these soils.

The only sand formations unsuitable from consideration of stability are certain fine uniform sands in loose condition which lose their shear strength or liquify if subjected to sudden shear stresses in a saturated condition. The formation of this type if occurring in the foundations need to be compacted by pile driving or otherwise before any construction is undertaken on them.

The main problem with sand and gravel foundations is that they are highly pervious and the seepage discharge through them may be undesirably high. However, by using the various methods of seepage control as indicated later, the quantity of seepage can be reduced to a considerable extent.

Clay foundations require greatest attention since they are more susceptible to shear failures. Clays being highly impervious do not permit the excess water present in their voids to be drained out easily under the load of the dam and hence their consolidation takes a long time which may even extend to number of years in many cases. As such when a dam is constructed on clay foundations due to lower rate of consolidation the shear strength of the soil increases at a slow rate. Hence the shear stresses induced in the soil by the load of the dam may exceed the shear strength of the soil which may result in the failure of the dam. For clay foundations therefore, it is necessary to ensure that the foundation shear stresses are less than the shear strength of the soil. For this it may sometimes be necessary to flatten the slopes of the dam to spread its load on a larger area though steeper slopes could have been otherwise provided so far as the stability of the dam alone is concerned.

11.4 DESIGN OF EARTH DAMS

The design of an earth dam essentially consists of determining such a cross section of the dam which when constructed with the available materials will fulfil its required function with adequate safety. Thus, there are two aspects of the design of an earth dam viz., (i) to determine the cross-section of the dam; and (ii) to analyse the stability of the proposed cross-section. In this section the various considerations to be made for determining the cross-section of an earth dam are discussed. The stability analysis of the proposed cross-section of an earth dam is described in section 11.12.

For determining the cross-section of an earth dam there are no mathematical analyses or formulae as in the case of rigid dams. The cross-section of an earth dam however depends on the following factors.

(i) Foundation conditions;

(ii) Availability of the materials of construction within reasonable range;

(iii) Physical properties of the materials available for construction;

(iv) Methods of construction to be adopted and the degree of construction control that is anticipated;

(v) Types of earth moving machinery;

(vi) Diversion considerations and construction schedule;

(vii) Climatic conditions in relation to placement moisture content control, subsequent moisture content changes, etc.

(viii) Safety with respect to stability and seepage.

On the basis of these factors the cross-section of an earth dam may be determined which involves the consideration of the following parameters.

(i) Crest width

(ii) Side slopes

(iii) Free board

(iv) Settlement allowance

(v) Seepage control measures.

The first four parameters are described below while the last one is described in section 11.19.

(i) Crest width. The crest width of an earth dam is determined only by the requirement of roadway at the top of the dam. In general the crest width varies from 6 to 12 m, with the wider dimensions being adopted for the higher and more important dams. However, a crest width less than 6 m may also be provided but it should not be less than 3 m because this is the minimum needed for an access road to permit maintenance work.

(ii) Side slopes. No specific rules can be given for determining the side slopes of an earth dam. The general procedure is that on the basis of experience with similar dams side slopes are considered and the same are modified if necessary after the stability analysis is carried out. It is obvious that the most economical dam is obtained when its volume is minimum and therefore when the side slopes have maximum steepness consistent with stability. However, in general the side slopes of a dam depend on the physical properties of the material of construction, the strength of the foundation and the type of the dam, i.e., homogeneous, zoned or diaphragm. The stability of side slopes depends mainly on the physical properties of the materials for construction of the dam. Further if the foundation is weak flatter side slopes will be required but if the foundation is strong steeper side slopes may be provided. Similarly a zoned dam permits the use of steeper side slopes because in this case the stronger materials are placed where they are most effective in resisting shear stresses and the drainage systems control and reduce pore water pressures.

The side slopes of earth dams usually vary in the range between 2 : 1, i.e., 2 horizontal to 1 vertical and 4 : 1, i.e., 4 horizontal to 1 vertical. However, where the foundation is weak side slopes flatter than 4 : 1 may be required, and where the foundation is strong the side slopes may be made steeper than 2 : 1.

It has also been observed that an earth dam with minimum volume is usually obtained when the side slopes are made steeper at the upper elevations and flatter near the bottom. Variable side slopes should be considered for all earth dams higher than about 30 m. The higher the dam and the softer the foundation, the greater is the economic advantage gained from varying the side slopes.

(iii) Freeboard. Freeboard is the difference in the elevation of the crest of the dam and the still water level in the reservoir. Sufficient freeboard must be provided so that there is no possibility whatsoever of the dam being overtopped. The freeboard may be classified as normal freeboard and minimum freeboard. The freeboard above the full reservoir level (FRL) is known as *normal freeboard,* and the freeboard above the maximum water level (MWL) is known as *minimum freeboard.*

Earlier it was recommended that the freeboard should be 1.5 times the wave height (for runup on riprapped slopes) plus a safety factor. The height of wave may be computed by Moliter's formula as indicated in Chapter 9. The safety factors which generally varies from 0.6 to 3 m, is selected by considering the size of the reservoir, the height of the dam, the reliability of the data from which the flood computations are made and the usual practice.

However, the earlier recommendation for freeboard has been revised as indicated in IS: 10635–1983, according to which the freeboard should be equal to the sum of wave run-up and wind set-up. The details of the procedure to be followed for calculating the wave run-up and wind set-up are given in IS: 10635–1983. The wave run-up and wind set-up should be calculated for full reservoir level condition as well as for maximum water level condition, and both normal freeboard as well as minimum freeboard should be calculated. The normal freeboard as well as minimum freeboard should not be less than 2 m. Thus, elevation of the crest of the dam should be the higher of the following two combinations.

(i) FRL + (normal freeboard or 2 m whichever is more)

(ii) MLW + (minimum freeboard or 2 m whichever is more)

(iv) Settlement allowance. Settlement of an earth dam may be caused by consolidation of the soil mass in the dam and in the foundation. It depends on the character of the soil mass in the dam and the foundation and the method of construction used. In order to allow for the settlement it is usual to construct earth dams to a somewhat greater height then the designed height so that there may be no reduction in the designed freeboard above the maximum water level in the reservoir. This extra height of the dam is provided in the form of a longitudinal camber over the designed top level, varying from zero at the abutments to a maximum value at the centre of the valley where the dam will be the highest and the settlement of the dam will be the most. Generally a provision for settlement allowance of 1 to 2 per cent of the height of the dam above the designed top level may be made to account for the settlement of earth dams.

11.5 CAUSES OF FAILURES OF EARTH DAMS

Various causes of failure of earth dams may be grouped into the following three broad categories.

(a) Hydraulic failures

(b) Seepage failures

(c) Structural failures.

On the basis of the investigation reports of a large number of failures of earth dams in the past it has been found that about 35% may be classified as hydraulic failures, about 38% may be classified as seepage failures and about 20% may be classified as structural failures. The remaining about 7% of the failures may be attributed to some miscellaneous causes.

(a) Hydraulic failures. The hydraulic failures include the failure of earth dams on account of the following causes.

(i) Overtopping

(ii) Erosion of upstream slope by waves

(iii) Erosion of downstream slope by wind and rain

(iv) Erosion of downstream toe

(i) *Overtopping*. Overtopping is the most common cause of failure of earth dams as about 30% of the failures of the earth dams in the past have been caused by overtopping. The earth dams may get overtopped if the design flood is underestimated or if the spillway is of insufficient capacity. Sometimes faulty operation of spillway gates may also lead to overtopping. Further overtopping may also be caused due to insufficient freeboard or settlement of foundation and dam. Since earth dams cannot withstand the erosive action of water over the crest during overtopping, either complete or partial failure of the dam is inevitable.

(ii) *Erosion of upstream slope by waves*. The upstream slopes of most of the earth dams are provided with protective covering to safeguard against wave erosion. The protective covering is laid on a filter bed to prevent the soil mass of the dam to be washed out. However, if the protective covering is inadequate and also the filter bed is not properly graded, the covering as well as the filter may be damaged by the waves produced on the water surface of the reservoir. The wave water moving in and out of the protective covering may gradually wash out the filter material lying below the covering if the filter material is too fine. The resulting undermining causes the protective covering to settle and often to be displaced thus exposing the upstream slope of the dam to wave erosion. Moreover, if the protective covering is not heavy enough to resist the hydraulic forces generated by the waves it may be washed out. With the displacement of the protective covering the upstream slope may be eroded which may lead to the failure of the dam. The erosion of upstream slope by waves usually causes partial failure of the dam but in some cases it may also cause complete failure of the dam. Only about 5% of the failures of the earth dams in the past have been caused by the erosion of the upstream slope by waves and hence it is not a common cause of failure of earth dams.

(iii) *Erosion of downstream slope by wind and rain*. In the absence of proper slope protection the downstream slope of an earth dam may get eroded due to wind and rain. The rain water flowing down the slope may result in the formation of gullies on the downstream slope thus damaging the dam which may generally lead to partial failure of the dam and also to its complete failure in some cases.

(iv) *Erosion of the downstream toe*. If the downstream toe of an earth dam is not properly protected it may be eroded due to two reasons:

(a) presence of tail water, and

(b) the cross currents produced from the water discharging from the spillway.

The damage caused due to erosion of the downstream toe of an earth dam may lead to partial failure of the dam.

(b) Seepage failures. The seepage failures include the failure of earth dams on account of the following causes.

(i) Piping through dam and its foundation

(ii) Conduit leakage

(iii) Sloughing.

(i) *Piping through dam and its foundation*. Piping is the progressive backward erosion which may be caused through an earth dam or its foundation by the water seeping through the dam or its foundation. It begins at a point where the water seeping through the dam or its foundation emerges on the downstream side of the dam. As water seeps through the compacted soil of an earth dam or the natural soil of its foundation the pressure head is dissipated in overcoming the resistance to flow through the small soil pores. However, if there is not much dissipation of pressure head seepage water emerges on the downstream side of the dam under sufficient pressure and with sufficient velocity so that the soil particles are washed away. The erosion thus starts at the point where the seepage water emerges on the downstream side of the dam and it progresses backwards towards the upstream side because the seeping water generates erosive forces which tend to pull the soil particles with it as it flows through the dam or its foundation. The erosion of the soil particles is however opposed by resisting forces which depend on the cohesion, the interlocking effect and the weight of the soil particles, as well as on the action of the downstream drain, if any. If the forces resisting erosion are less than those which tend to cause it, the soil particles are washed away thus resulting in the formation of a small tunnel through the dam or its foundation. The tunnel so developed gradually enlarges as more and more soil particles are carried away by the seepage water until in the final stage complete failure of the dam occurs.

The piping mostly results from leaks which may be developed through an earth dam or its foundation. The common causes of the development of leaks through the dam and its foundation which may lead to piping failure are as follows.

1. Poor construction control, which can result in inadequately compacted or pervious layers in the dam, inferior compaction adjacent to concrete outlet pipes or other structures or poor compaction and bond between the dam and the foundation or abutments.

2. Differential settlement of the dam as well as its foundation which may result in developing cracks in the dam.

3. Cracking in outlet pipes which may be caused by foundation settlement, spreading of the base of the dam, or deterioration of the pipe itself.

4. Animal burrows.

5. Surface drying cracks as well as shrinkage cracks.

6. Presence of roots, pockets of gravel or boulders in the dam.

7. Presence of erratic and less dense soils in the foundation of the dam.

Besides the horizontal piping which may occur as indicated above, sometimes vertical piping may occur at the downstream toe of an earth dam. The vertical piping may occur where there is a layer of soil in the foundation which is much more pervious than the strata at the surface of the foundation. In such cases as the head increases the hydrostatic pressure in the pervious layer increases until the water lifts the overlying strata forming 'sand boils' on the surface of the ground. In the sand boils so formed the soil particles are rendered in a state of floatation or suspension and are carried away along with the flow. This is a condition of incipient vertical piping. Generally the sand boils are not disastrous but if unattended they may lead to complete failure of the dam.

Piping is also one of the common causes of failure of earth dams which has resulted in about 23% of the failures of the earth dams in the past.

(ii) *Conduit leakage*. Conduits passing through earth dams have been a source of trouble. The main problem with the conduits passing through earth dams is that of conduit leakage which has caused about 33% of the seepage failure and about 13% of all the failures of earth dams. Conduit leakage may cause the failure of an earth dam in the following two ways :

1. seepage along the conduit wall ; and
2. seepage through cracks developed in the conduit.

Seepage along the conduit wall may be caused if there is a zone of poorly compacted soil around the conduit or if there is a small gap between the conduit and the surrounding soil mass of the dam. Such a gap may sometimes be produced if due to settlement the soil mass shrinks away from the underside of the pipe. The seepage caused along the conduit wall soon develops into piping.

The cracking of a conduit passing through an earth dam may be caused by unequal settlement of the dam or by overloading from the dam. Further the cracking of a conduit may also be caused when the soil mass lying below it settles and the conduit is not sufficiently strong to support the soil mass lying above it. Seepage through cracks in conduits may also develop into piping.

(iii) *Sloughing*. Sloughing may be defined as the falling away of the soil mass of an earth dam. The sloughing may take place when under the full reservoir condition most of the downstream portion of the dam becomes saturated and continuously remains in the same state due to which the soil mass in the downstream portion of the dam gets softened. The extensive saturation of the downstream portion of the dam may be either due to choking of filter toe drain or due to the presence of highly pervious layer in the body of the dam. The sloughing begins when a small amount of softened soil mass at the downstream toe is eroded and produces a small slump or miniature slide. It leaves a relatively steep face which being saturated slumps again, forming a slightly higher and more unstable face. This process is continued until the remaining portion of the dam is too thin to withstand the water pressure and the complete failure of the dam occurs. Sloughing has caused about 2% of the failures of earth dams in the past.

(c) Structural failures. The structural failure includes the failure of earth dams due to the following causes.

(i) Sliding of upstream and downstream slopes.

(ii) Flow slides due to spontaneous liquefaction (or liquefaction slides).

(iii) Damage caused by burrowing animals.

(iv) Damage caused by water soluble materials.

(v) Damage caused by earthquake.

(i) *Sliding of upstream and downstream slopes.* Sliding of upstream and downstream slopes is one of the frequent causes of failure which occurs in earth dams when along any potential sliding surface the forces tending to cause sliding of the soil mass become greater than the forces resisting sliding. The sliding of slopes has caused about 15% of the failures of earth dams in the past. The critical conditions during which the sliding of upstream and downstream slopes may be caused are as follows :

(a) Sliding of upstream of slope during sudden drawdown.

(b) Sliding of downstream slope during steady seepage.

(c) Sliding of upstream or downstream slope (or both) during construction.

These have been discussed in section 11.12.

(ii) *Flow slides due to spontaneous liquefaction or liquefaction slides.* In 'flow' or 'liquefaction' slides the saturated soil mass moves under gravity as if it was a heavy viscous fluid. These slides may occur in the foundation of an earth dam if the foundation consists of fine sand and silt or some types of clays. Similar slides may also occur in the lower portion of an earth dam consisting of loosely compacted granular soil mass. These slides are caused when a large part of the weight of the overlying soil mass is carried momentarily by the pore water of the saturated soil and very little intergranular pressure exists within the soil mass so that the shear strength of the soil mass is reduced almost to that of the liquid and it flows downward. Hence these slides are termed as flow slides or liquefaction slides. The flow or liquefaction slides in the foundation as well as in the lower portion of earth dams has caused about 3% of the failure of earth dams in the past.

(iii) *Damage caused by burrowing animals.* Some of the burrowing animals which may cause damage to earth dams are muskrats, ground squirrels, snakes, beavers, eels, crawfish, mice, woodchucks and rats, out of which muskrats and ground squirrels apparently cause more damage than any of the others. The burrowing animals dig holes in earth dams to make their homes through which free passage of water from the upstream to the downstream face of the dam may be developed which may ultimately lead to the failure of the dam due to piping. However, burrowing animals may cause failure of only small earth dams. In the case of major earth dams there is usually very little possibility of serious damage from the burrowing animals because the holes dug by them may not penetrate to a great depth.

(iv) *Damage caused by water soluble materials.* The leaching of natural deposits of water soluble materials such as gypsum, iron oxide etc., from abutments and foundations may result in creating cavities and flow channels which may cause excessive settlement of the dam. Further the soluble material leached from the natural soil may get deposited in the toe filters and the drains and thus tend to plug the same. Appropriate allowance should therefore be made to account for the settlement of the dam due to this reason.

The soil mass of the dam may also contain water soluble materials. However, no serious damage may be caused to the dam if only small quantities of soluble materials are present in the soil mass. This is so because in the construction of the rolled earth dams the water which is added to the soil mass to bring the water content up to optimum value usually dissolves much of the soluble material.

(v) *Damage caused due to earthquake.* Some of the damages which may be caused to earth dams due to earthquake are as follows.

(a) Longitudinal cracks at the top of the dam.

(b) Liquefaction of loose and saturated soil mass in the lower portion of the dam.

(c) Settlement of the crest of the dam thus reducing the freeboard and increasing the possibility of overtopping.

(d) Cracking of central core wall leading to leakage and piping failure.

(e) Generating large waves on the water surface in the reservoir thus causing overtopping.

(f) Shear failure at the base of the dam.

(g) Rock slide from the side hills into the reservoir causing overtopping.

11.6 CRITERIA FOR SAFE DESIGN OF EARTH DAMS

The following criteria have to be satisfied for safe design of an earth dam.

1. The dam must be safe against overtopping during occurrence of flood and also by wave action.
2. The seepage line should be well within the downstream face.
3. The upstream and downstream slopes should be stable under critical or adverse conditions.
4. The foundation shear stresses should be within safe limits.
5. There should be no opportunity for the free passage of water from the upstream to the downstream face.
6. The dam and foundation should be safe against piping.
7. The upstream slope must be protected against erosion by wave action and the crest and downstream slope must be protected against erosion due to wind and rain.

All these criteria are discussed in the following sections.

11.7 SAFETY AGAINST OVERTOPPING

The safety against overtopping can be ensured by adopting the following two measures.

(i) *Providing ample spillway capacity.* In order to avoid any possibility of overtopping during floods ample spillway capacity should be provided. For designing the spillway capacity the maximum probable flood that is not likely to be exceeded should be considered and the spillway must have sufficient capacity to permit safe disposal of this flood without causing excessive rise of water level above the spillway crest. The maximum rise of water level above the spillway crest can be determined by the process of flood routing as indicated in Chapter 5.

(ii) *Provision of sufficient freeboard.* The minimum freeboard as indicated in section 11.4 should be provided to avoid the possibility of water spilling over the dam due to waves formed on the water surface in the reservoir.

11.8 DETERMINATION OF SEEPAGE LINE

Seepage occurs in all earth dams which will saturate the soil in the lower portion of the dam while the soil in the upper portion of the dam remains relatively dry or moist. The *seepage line* is defined as the line within a dam section below which the soil is saturated and there are positive hydrostatic pressures in the dam. It is also known as *saturation line* or *phreatic line* or *top flow line*. On this line itself the hydrostatic pressure is equal to atmospheric pressure or zero. Above this line there will be a capillary zone in which the hydrostatic pressures are negative. Since the flow through the capillary zone is insignificant it is usually neglected and hence seepage line is taken as the dividing line between the saturated soil below and dry or moist soil above in a dam section. However, the capillary action leads to increased intergranular pressures and greater shear strength within the soil but this effect is usually neglected thus slightly increasing the margin of safety.

It is necessary to predict the position of the seepage line in a proposed dam section to :

(i) ensure that it does not intersect the downstream face much above the toe of the dam and thereby produce softening or sloughing of the toe leading to ultimate failure of the dam.

(ii) obtain the dividing line between dry or moist soil above the seepage line and the saturated soil subjected to seepage forces below the seepage line and carry out the stability analysis accordingly ; and

(iii) obtain the top boundary line for drawing the flow net for determining the quantity of seepage.

Kozeny has indicated theoretically that for flow through uniform isotropic soil lying above a horizontal impervious floor which at a certain point becomes permeable, the flow lines and equipotential lines consist of a family of confocal parabolas with their focus located at the point where the floor becomes permeable. This condition is not exactly fulfilled by any practical earth dam section. However, A. Casagrande has demonstrated that in general for earth dam sections the seepage line for most of its length coincides with Kozeny's theoretical parabola called the *base parabola*, but it deviates from the base parabola for a short distance at the upstream or the entrance face of the dam in all the cases and at the downstream or the exit face of the dam in some cases. The amount and character of the deviations depend on local conditions at these portions of the dam. Thus, for the different types of earth dams the seepage line may be determined as indicated below.

Case I. Homogeneous earth dam with a horizontal drainage blanket

Figure 11.4 shows a homogeneous earth dam with a horizontal drainage blanket FJ. It has been shown by Casagrande that in this case the seepage line for most of its length coincides with the 'base parabola' *AEG* (Fig. 11.4) which has its focus located at *F* the starting point of the drainage blanket. This parabola when extended on the upstream side intersects the water surface at a point *A*. Casagrande has shown that for the dams with reasonably flat upstream slopes the distance *BA* is 0.3*L* where *B* is the point of intersection of the water surface with the upstream face and *L* is the horizontal projection of the wetted portion of the upstream face. The point *A* is known as *corrected entrance point*. From the known positions of the focus *F* and the point *A* the base parabola *AEG* may be drawn by graphical method as indicated below.

(i) The directrix of the parabola may be located by utilizing the basic property of a parabola that any point on the parabola is equidistant from the focus as well as directrix. Thus, with point *A* as the centre and *AF* as the radius, draw an arc to cut the line *AB* produced at *D*. Draw a vertical tangent *DH* to the curve *FD* at *D*. Then since *AF* = *AD*, the vertical line *DH* is the directrix of the parabola.

(ii) The apex (or vertex) *G* of the parabola will lie midway between *F* and *H*.

(iii) A number of intermediate points on the parabola may be obtained by using the basic property of a parabola as indicated above.

(iv) Join all the points by a smooth curve to obtain the base parabola *AEG*.

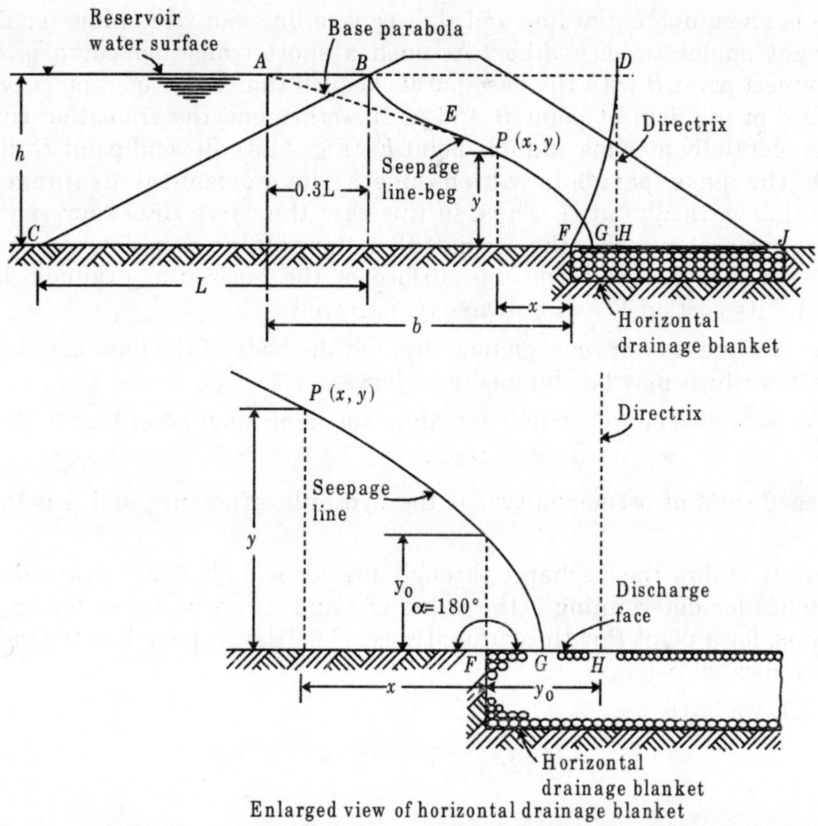

Fig. 11.4 *Seepage line for a homogeneous earth dam with a horizontal drainage blanket*

The base parabola may also be drawn from its equation which may be obtained as indicated below.

Taking the focus *F* as origin if (*x*, *y*) are the coordinated of any point on the parabola then using the basic property of a parabola, the equation of the base parabola may be obtained as

$$\sqrt{x^2 + y^2} = x + y_0$$

or

$$y^2 = y_0 (2x + y_0) \qquad ...(11.1)$$

where y_0 is the focal distance *FH* (or the distance from the focus to the directrix).

The value of y_0 may be obtained as follows.

As shown in Fig. 11.4 the coordinates of the point *A* on the parabola are (*b*, *h*). Substituting these values in Eq. 11.1 and solving for y_0 we get

$$y_0 = \sqrt{b^2 + h^2} - b \qquad ...(11.2)$$

From the base parabola the seepage line may be obtained after necessary modification being made at the upstream face of the dam as indicated below.

The seepage line should actually start from point B which is the point of intersection of the water surface in the reservoir with the upstream face of the dam. Further since the upstream face of the dam is an equipotential line and the seepage line is a streamline (or flow line), they should be at right angles to each other. As such a short transition curve is drawn by eye judgement to connect point B with the base parabola such that the transition curve is normal to the upstream face of the dam at point B and at the other end the transition curve meets the base parabola tangentially at some suitable point E (Fig. 11.4). Beyond point E the seepage line coincides with the base parabola which meets the horizontal drainage blanket FJ perpendicularly (i.e. vertically) at G. Thus, in this case the curve BEG represents the seepage line. Further the surface on which the seepage flow through the dam emerges is known as the *discharge face*. Thus in this case the top surface of the horizontal drainage blanket is the discharge face and its portion FG will always remain wet.

For this case the quantity of seepage flow through the body of the dam may be calculated by a simple expression which may be obtained as follows.

If q is the seepage discharge per unit length of the dam, then according to Darcy's law

$$q = ki\,A \qquad \ldots(i)$$

where k is the coefficient of permeability; i is the hydraulic gradient ; and A is the area of flow section.

For steady state of flow the discharge through any vertical plane across the dam section will be the same. Hence for determining q the value of i and a can be taken for any point on the seepage line. Thus, for a point P with coordinates (x, y) on the seepage line the values of i and a are obtained as indicated below.

From Eq. 11.1, we have

$$y = \sqrt{y_0(2x + y_0)}$$

$$\therefore \qquad i = \frac{dy}{dx} = \frac{y_0}{\sqrt{y_0(2x + y_0)}} = \frac{y_0}{y}$$

and

$$A = y \times 1 = y$$

(i.e., A = the depth of the saturated section of the dam below the seepage line as the length of the dam is taken as unity)

Substituting these values in Eq. (i), we get

$$q = ky_0 \qquad \ldots(11.3)$$

Also substituting the value of y_0 from Eq. (11.2) we get

$$q = k\left(\sqrt{b^2 + h^2} - b\right) \qquad (11.3\text{ a})$$

Case II. Homogeneous earth dam without any arrangement for drainage

Figure 11.5 shows a homogeneous earth dam without any arrangement for drainage. It has been shown by Casagrande that in this case also except for very small angles of downstream slope the seepage line for most of its length coincides with the base parabola AEG (Fig. 11.5) similar to the one given by Kozeny for the Case I. However, in this case the focus F of the base parabola is located at the downstream toe of the dam. Further the base parabola when extended on the upstream side intersects the water surface at a point A. According to Casagrande in this case also for the dams with reasonably flat upstream slopes the point A is locted at a distance of 0.3 L from B, where B is the point of intersection of the water surface with the upstream face and L is the horizontal projection of the wetted portion of the upstream face. From the known positions of the focus F and the point A (which in this case is also known as the corrected entrance point) the base parabola AEG may be drawn by graphical method as indicated in the Case I.

From the base parabola the seepage line may be obtained for this case after necessary modifications being made at the upstream and the downstream faces of the dam as indicated below.

At the upstream face of the dam the modification to be made for this case is exactly same as for the case I. Thus, in this case also a short transition curve is drawn by the eye judgement to connect point B with the base parabola such that the transition curve is normal to the upstream face of the dam at point B and at the other end the transition curve meets the base parabola

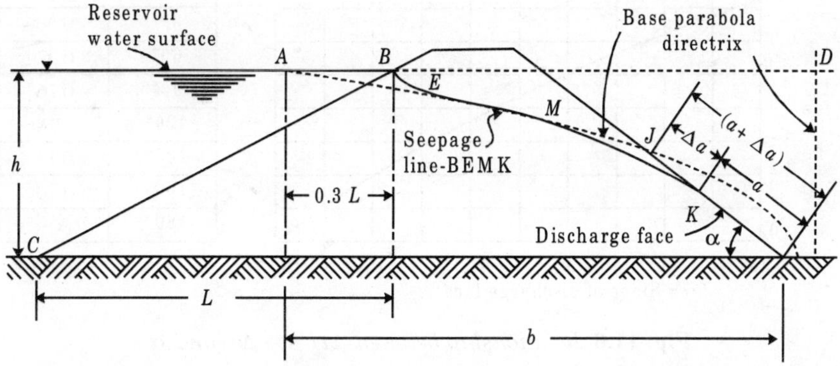

Fig. 11.5 *Seepage line for a homogeneous dam without any arrangement for drainage*

tangentially at some suitable point E (Fig. 11.5). Beyond point E the seepage line coincides with the base parabola for some length until it approaches the downstream face of the dam where again a modification is to be made as indicated below.

In the absence of any drain in this case the seepage flow will emerge at the downstream face of the dam and hence the downstream face of the dam will be the discharge face. The base parabola intersects the downstream face of the dam at point J (Fig. 11.5) at a distance $(a + \Delta a)$ along the downstream face from the toe of the dam. However, the seepage line will actually meet the downstream face tangentially at point K at a distance a along the downstream face from the toe of the dam. The distance between the points J and K is thus Δa which should be known for obtaining the seepage line in this case. Casagrande has shown that the distance Δa varies with the slope angle a of the discharge face and on the basis of graphical studies by means of flow nets he has determined the values of the ratio $[\Delta a/(a + \Delta a)]$ for various values of a which are given in α table as well as a plot of $[\Delta a/(a + \Delta a)]$ versus a in Fig. 11.6. Thus, for a known value of a the value of the ratio $[\Delta a/(a + \Delta a)]$ may be obtained from the table or the plot of Fig. 11.6. The value of $(a + \Delta a)$ may be obtained by measuring the distance FJ after the base parabola is drawn. Alternatively the value of $(a + \Delta a)$ may also be obtained by the following equation (which may be readily derived by using the basic property of a parabola)

$$(a + \Delta a) = \frac{y_0}{(1-\cos\alpha)} \qquad ...(11.4)$$

where y_0, is intercept of parabola on vertical line through focus and hence

$$y_0 = \sqrt{b^2+h^2}-b \qquad ...(11.5)$$

Knowing the values of $[\Delta a/(a + \Delta a)]$ and $(a + \Delta a)$ the value of Δa is obtained. The point K is then marked on the downstream face of the dam at a distance Δa from the point J. The seepage line is then completed by drawing a transition curve from the point K to the base parabola by eye judgement such that the transition curve meets tangentially the downstream face of the dam at point K as well as the base parabola at some suitable point M (Fig. 11.5). The curve *BEMK* thus represents the seepage line in this case and the portion *KF* of the downstream face of the dam will always remain wet.

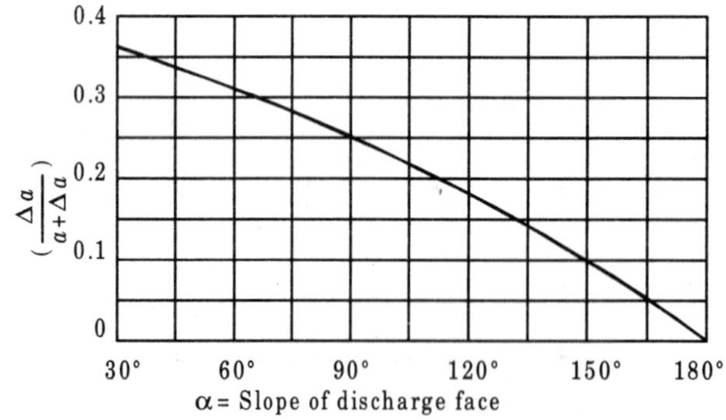

α	Δ a (a+Δ a)
30°	0.36
60°	0.32
90°	0.26
120°	0.18
135°	0.14
150°	0.10
180°	0

Fig. 11.6 *Relationship between* $\Delta a/(a + \Delta a)$ *and* α

For this case also the quantity of seepage flow through the body of the dam may be calculated by Eq. (11.3) derived for the Case I.

The slope angle α of the discharge face is measured clockwise from the horizontal. Thus, in this case $\alpha < 90°$. However, for $\alpha < 30°$, since the value of the ratio $[\Delta a/(a + \Delta a)]$ is not known, an approximate analytical solution as indicated below has been obtained to determine the value of a the distance of point K from the focus F. Further for the Case I, $\alpha = 180°$ for which $\Delta a = 0$ i.e., no modification of the base parabola is needed at the downstream end for obtaining the seepage line. The discharge face with slope angle equal to 90° or greater than 90° is obtained when a rock toe is provided as indicated later in the Case III.

Approximate analytical solution for determining the value of a for $\alpha < 30°$

Assuming that the hydraulic gradient at any point is (dy/dx), the quantity of seepage flow q per unit length of the dam through a vertical section is given by

$$q = ky\frac{dy}{dx} \qquad ...(i)$$

where k is coefficient of permeability and y is the depth of the saturated section of the dam below the seepage line.

Equation (i) is the differential equation for the seepage line. By integration the solution to Eq. (i) is given by

$$qx = \frac{ky^2}{2} + C \qquad \text{...(ii)}$$

where C is a constant.

Equation (ii) represents a parabola.

For $x = b,\ y = h$

$$qb = \frac{kh^2}{2} + C$$

or

$$C = qb - \frac{kh^2}{2} \qquad \text{...(iii)}$$

Substituting this value of C in Eq. (ii), we get

$$q(x - b) = \frac{k}{2}(y^2 - h^2) \qquad \text{...(iv)}$$

Since length of discharge face KF is a, for point K, we have

$x = a \cos \alpha,\ y = a \sin \alpha$ and = tan α

Substituting these values in Eq. (i) the following equation may be obtained for the seepage flow q

$$q = ka \sin \alpha \tan \alpha \qquad \text{...(11.5)}$$

For small values of α since tan ≃ sin α, Eq. 11.5 may also be expressed as

$$q = ka \sin^2 \alpha \qquad \text{(11.5 a)}$$

Further substituting these values in Eq. (iv), we have

$$ka \sin \alpha \tan \alpha\, (a \cos \alpha - b) = \frac{k}{2}(a^2 \sin^2 \alpha - h^2) \qquad \text{...(v)}$$

which reduces to

$$a^2 \sin^2 \alpha - 2ab \sin \alpha \tan \alpha + h^2 = 0 \qquad \text{...(vi)}$$

Solving Eq. (vi) for a, we have

$$a = \frac{b}{\cos \alpha} - \sqrt{\frac{b^2}{cks^2 \alpha} - \frac{h^2}{\sin^2 \alpha}} \qquad \text{...(11.6)}$$

Thus, the value of a may be calculated from Eq. 11.6 and by knowing a, discharge q may be calculated by Eq. 11.5.

Approximate analytical solution for determining the value of a for 30° < a < 60°

It has been observed that the previous solution gives satisfactory results for a < 30°. For steeper slopes the deviation from the correct values increases rapidly beyond tolerable limits. Thus, for this case Casagrande suggested the use of sin α instead of tan α (i.e., dy/ds instead of dy/dx) and hence the following differential equation of the seepage line is obtained

$$q = ky \frac{dy}{ds} \qquad \text{...(i)}$$

in which s is the distance measured along the seepage line.

By integration the solution to Eq. (i) is given by

$$qs = \frac{ky^2}{2} + C \qquad \text{...(ii)}$$

For point *K*, we have

$$s = a, \; y = a \sin \alpha$$

and
$$\frac{dy}{ds} = \sin \alpha$$

Substituting these values in Eq. (i) the following equation may be obtained for the seepage flow q.

$$q = ka \sin^2 \alpha \qquad (11.7)$$

Further substituting these values in Eq. (ii), we get

$$ka^2 \sin^2 \alpha = \frac{ka^2 \sin^2 \alpha}{2} + C$$

or
$$C = \frac{ka^2 \sin^2 \alpha}{2} \qquad \text{...(iii)}$$

Equation (ii) thus becomes

$$qs = \frac{ky^2}{2} + \frac{ka^2 \sin^2 \alpha}{2} \qquad \text{...(iv)}$$

If the total length of the seepage line from *F* to *A* is S_0, then for point *A*, we have

$$s = S_0 \text{ and } y = h$$

Thus, substituting these values as well as the value of q from Eq. 11.7 in Eq. (iv), we have

$$ka \sin^2 \alpha \, S_0 = \frac{kh^2}{2} + \frac{ka^2 \sin^2 \alpha}{2}$$

or $\quad a^2 - 2a \; S_0 + h^2 \operatorname{cosec}^2 \alpha = 0 \qquad \text{...}(v)$

Solving Eq. (v) for a, we have

$$a = S_0 - \sqrt{S_0^2 - h^2 \operatorname{cosec}^2 \alpha} \qquad \text{...(11.8)}$$

The value of a may be obtained from Eq. 11.8 only if the length S_0 is known, which would necessitate a trial construction of the seepage line. However, for slope angle α upto 60° it is tolerable to replace the length S_0 by the straight distance $FA = \sqrt{h^2 + b^2}$, thus eliminating the trial construction. Equation 11.8 then becomes

$$a = \sqrt{h^2 + b^2} - \sqrt{b^2 + h^2 - h^2 \operatorname{cosec}^2 \alpha}$$

or
$$a = \sqrt{h^2 + b^2} - \sqrt{b^2 - h^2 \cot^2 \alpha} \qquad \text{...(11.9)}$$

For α = 90° Eq. (11.9) becomes

$$a = \sqrt{h^2 + b^2} - b \qquad \text{...(11.10)}$$

which however provides a value of a upto 25% deviations from the actual value.

Case III. Homogeneous earth dam with rock toe

Figure 11.7 shows a homogeneous earth dam with rock toe. The upstream face of the rock toe may be either vertical (Fig. 11.7a) for which $\alpha = 90°$; or inclined (Fig. 11.7b) for which $\alpha > 90°$. In

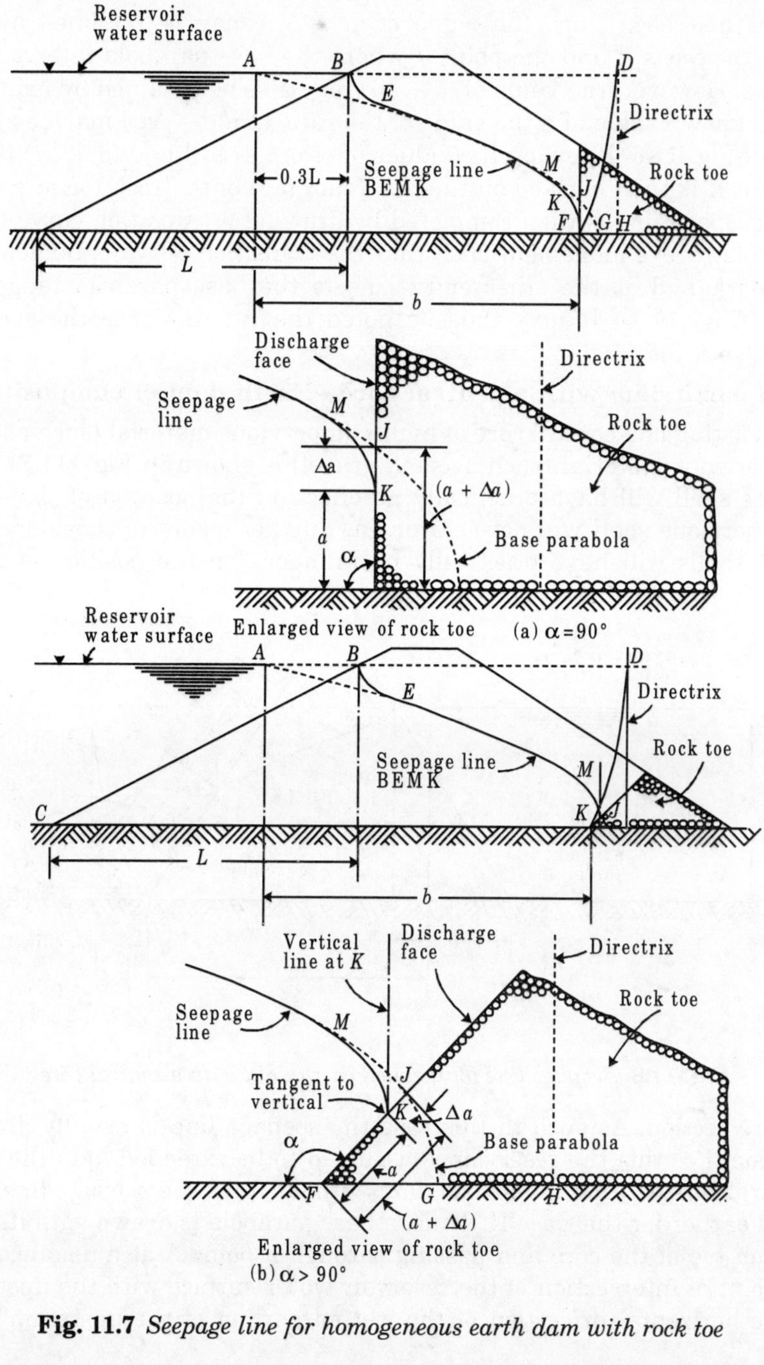

Fig. 11.7 *Seepage line for homogeneous earth dam with rock toe*

this case also the seepage line for most of its length coincides with the base parabola having its focus F located at the upstream end of the rock toe and passing through the corrected entrance point A which is located at a distance of 0.3 L from B as in the previous cases. Thus, at the upstream face of the dam for this case also the seepage line is obtained in exactly the same manner as in the previous cases. Further the upstream face of the rock toe constitutes the discharge face in this case. Thus, the value of $(a + \Delta a)$ may be obtained by measuring the distance between the focus F and the point J where the base parabola intersects the upstream face of the rock toe. However, the value of $(a + \Delta a)$ may also be obtained by using Eqs (11.4) and (11.5). Again for a known value of a the value of the ratio $(\Delta a/(a + \Delta a)]$ may be obtained from the table or the plot of Fig. 11.6. Knowing the values of $[\Delta a/(a + \Delta a)]$ and $(a + \Delta a)$ the value of Δa is obtained. The point K is then marked on the upstream face of the rock toe at a distance "a from the point J. The seepage line is then completed by drawing a transition curve from the point K to the base parabola by eye judgement such that the transition curve is tangential to a vertical line through point K and at the other end it meets the base parabola tangentially at some suitable point M (Fig. 11.7). It may thus be noted that in this case the seepage line drops vertically into the rock toe.

Case IV. Zoned earth dam with a central core—Earth dam of composite cross-section

An earth dam consisting of a central core of highly impervious material (such as a silty clay) and shells of highly pervious material (such as sand gravel) is shown in Fig. (11.8). In this case the upstream pervious shell will have practically no effect on the position of the seepage line and the downstream pervious shell will act as a drain. Further owing to considerable difference in permeability, the shells will have practically no influence on the position of the seepage line

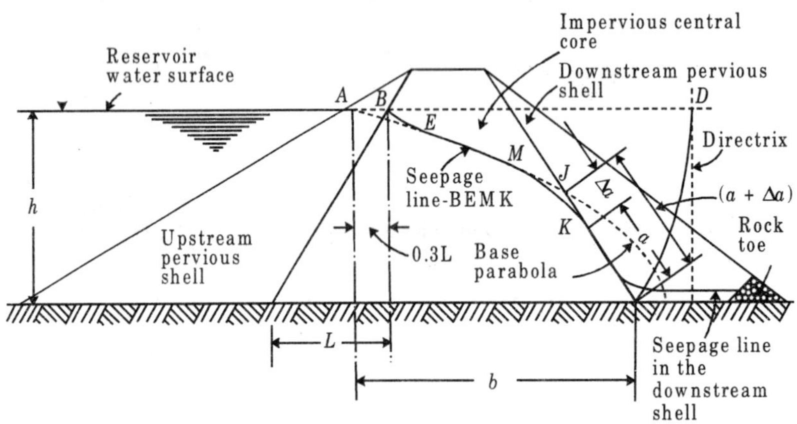

Fig. 11.8 *Seepage line for a zoned earth dam with a central core*

in the central core section. As such in this case the seepage line is usually drawn only for the central core section. For this the reservoir is assumed to be extended upto the central core and treating the central core as a homogeneous dam without filter the seepage line is drawn by the method indicated earlier for the Case II. Thus, a base parabola is drawn with its focus F located at the downstream toe of the core and passing through a point A at a distance of $0.3L$ from B, where B is the point of intersection of the reservoir water surface with the upstream face of the core and L is the horizontal projection of the wetted portion of the upstream face of the core.

From the base parabola the seepage line *BEMK* (Fig. 11.8) is obtained after exactly the same modifications as in the Case II being made at the upstream and downstream face of the core.

In each of these cases it has been assumed that the dam is located on a foundation of impervious material. However, instead of the impervious foundation if a considerable layer of relatively pervious material overlies the rock or an impervious layer, then also the location of the seepage line may be obtained by the methods already described by assuming that the foundation is still a boundary but below this boundary seepage of water takes place through the pervious stratum down to the impervious stratum in the foundation.

11.9 CHARACTERISTICS OF SEEPAFGE LINE

On the basis of the discussion of the previous section the various characteristics of the seepage line may be summarised as follows.

1. At the starting point the seepage line must be normal to the upstream face which is an equipotential line.

2. Since the pressure all along the seepage line is atmospheric, the only change in head along this line is that due to drop in the elevation. Hence the successive equipotential lines will meet the seepage line at equal vertical intervals.

3. The seepage line drops vertically into a horizontal filter drain or a rock toe filter.

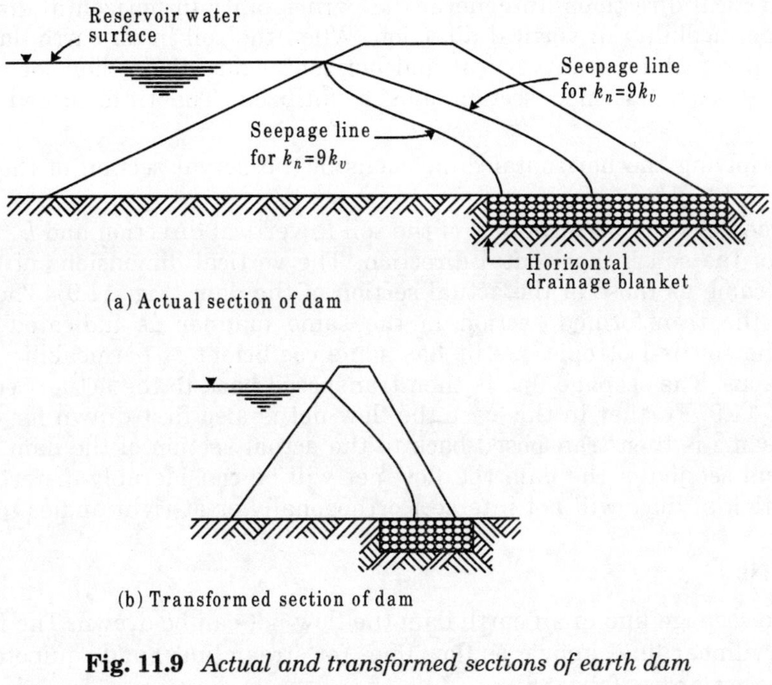

Fig. 11.9 *Actual and transformed sections of earth dam*

4. For an earth dam composed of homogeneous material without filter at downstream toe the seepage line will meet the downstream face at a point above the base of the dam. The location of the seepage line for a homogeneous section and the point where the seepage line meets the downstream face depend only on the geometry of the cross-section of the dam. The

position of the seepage line as well as that of the point where it meets the downstream face are not influenced by the permeability of the material composing the dam so long as the material is homogeneous.

5. For most of the length the seepage line coincides with the base parabola. The focus of the base parabola lies at the break out point of the bottom flow line where the flow emerges out from relatively impervious medium to a highly pervious medium.

6. The presence of pervious stratum in the foundation below the dam does not influence the position of the seepage line.

7. For a zoned earth dam with an impervious core in the centre the effect of the outer shells can be neglected altogether. Thus, in this case the seepage line may be drawn only for the core. The focus of the base parabola is located at the downstream toe of the core. The seepage line is assumed to be practically horizontal through the upstream shell, the downstream shell acts as a drain and is saturated to a depth just enough to carry the seepage flow passing through the core.

11.10 SEEPAGE LINE WHERE VERTICAL AND HORIZONTAL PERMEABILITY DIFFER

Soils which have same permeability in all directions are known as isotropic soils and those which have different permeabilities in different directions are known as anisotropic soils. Soils placed in earth dams may be anisotropic because these may have different permeabilities in horizontal and vertical directions. In general the permeability in horizontal direction may be 4 to 20 times the permeability in vertical direction. When the soil in an earth dam is anisotropic having different permeabilities in vertical and horizontal directions then for drawing seepage line in such cases a transformed section may be utilized. The transformed section may be obtained by multiplying the horizontal dimensions of the actual section of the dam by $\sqrt{\frac{k_v}{k_h}}$, where k_v is the coefficient of permeability of the soil in vertical direction and k_h is the coefficient of permeability of the soil in horizontal direction. The vertical dimensions of the transformed section are kept same as those of the actual section of the dam (Fig. 11.9). The seepage line is then drawn for the transformed section in the same manner as indicated in section 11.8 assuming that the soil is isotropic, i.e., it has same coefficient of permeability in vertical and horizontal directions. The seepage line is then transposed back to the actual section of the dam as shown in Fig. 11.9. Further in this case the flow net is also first drawn for the transformed section and the same is then transposed back to the actual section of the dam. As such in this case for the actual section of the dam the flow net will be considerably distorted and the flow lines and equipotential lines will not intersect orthogonally or at right angles to each other.

11.11 FLOW NET

After locating the seepage line in an earth dam the flow net can be drawn. The flow net consists of two sets of curvilinear lines known as flow lines (or streamlines) and equipotential lines. The properties of flow net are as follows.

(i) The flow lines and equipotential lines intersect at right angles to each other.

(ii) The seepage line is also a flow line of the flow net and accordingly the equipotential lines will intersect it at right angles and also at equal vertical intervals.

(iii) The intersections of the flow lines and equipotential lines form approximately squares so that a circle can be drawn touching all the four sides of the square.

(iv) The quantity flowing between any two adjacent flow lines is the same. Similarly the potential drop between any two successive equipotential lines is also same.

(v) In a homogeneous soil the transitions between the straight and curved sections of flow lines or equipotential lines are smooth of elliptical or parabolic shape.

The flow net may thus be drawn by trial and error keeping in view the properties as indicated above. However, the flow net may also be drawn by the method of electrical analogy which is based on the fact that the Darcy's law governing the flow of water through soil is analogous to the Ohm's law governing the flow of electric current through conductor.

Computation of Rate of Seepage from Flow Net

The rate of seepage through an earth dam may be obtained from the flow net as indicated below.

Figure (11.10) shows a portion of a flow net. Let b and l be the width and length of a portion of the flow net enclosed between two successive flow lines and successive equipotential lines shown hatched in Fig. (11.10). Further let Δq be the discharge passing through each of the flow channels between the adjacent flow lines and Δh be the head drop between the adjacent equipotential lines.

Then considering unit length in the direction perpendicular to the plane of the paper and using Darcy's law, we have

$$\Delta q = k \times \frac{\Delta h}{l}(b \times 1) \qquad \text{...(i)}$$

where k is the coefficient of permeability of the soil.

Further if h is the total head causing the flow (i.e., h is the difference between the upstream and downstream heads) and N_d is the total number of potential drops in the complete flow net, then

$$\Delta h = \frac{h}{N_d}$$

$$\therefore \qquad \Delta q = k\frac{h}{N_d}\left(\frac{b}{l}\right) \qquad \text{...(ii)}$$

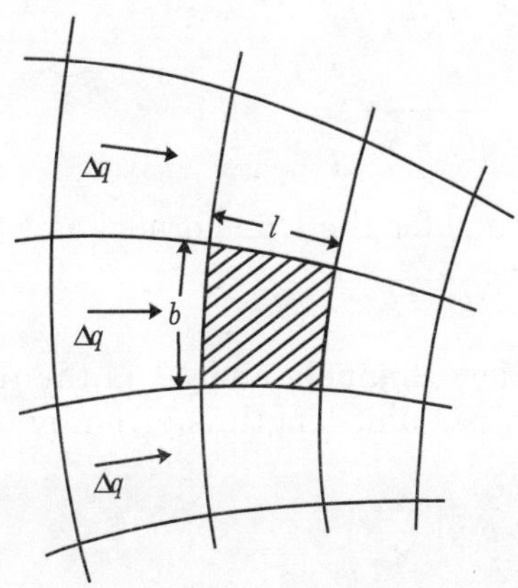

Fig. 11.10 *Portion of a flow net*

Further if N_f is the total number of flow channels in the flow net then the total discharge q per unit length is given by

$$q = \Sigma\Delta q = k\frac{h}{N_d}\left(\frac{b}{l}\right)N_f \qquad \text{...(iii)}$$

Since the portion of the flow net between the two successive flow lines and the successive equipotential lines is a square, $b = l$ and hence Eq. (*iii*) becomes

$$q = kh\left(\frac{N_f}{N_d}\right) \qquad \text{...(11.11)}$$

Thus, Eq (11.11) may be used for computing the rate of seepage through an earth dam after the flow net is drawn and it will give discharge per unit length of the dam. However, Eq. (11.11) is applicable only for isotropic soils for which the coefficient of permeability is same in horizontal and vertical directions. For anistropic soils having different permeability coefficients in horizontal and vertical directions the rate of seepage may be determined as follows.

As indicated earlier for anisotropic soil the flow net is drawn for a transformed section which is obtained by multiplying the horizontal dimensions of the actual section by $\sqrt{k_v/k_h}$, where k_v and k_h are the coefficients of permeability in vertical and horizontal directions respectively. Thus, as shown in Fig. (11.11) a square portion of length of side l of a flow net in the transformed section will be transposed as a rectangle of length $l\sqrt{k_h/k_v}$ and width l in the actual section.

Now if k' is the equivalent coefficient of permeability of soil for the transformed section then for this section, we have

$$\Delta q = k'\frac{\Delta h}{l}(l \times 1) \qquad \text{...(i)}$$

Similarly for the actual section, we have

$$\Delta q = k_h\frac{\Delta h}{l\sqrt{k_h/k_v}}(l \times 1)$$

or

$$\Delta q = \sqrt{k_h \times k_v}\,\frac{\Delta h}{l}(l \times 1) \qquad \text{...(ii)}$$

Since the quantity of flow is same for the transformed as well as actual sections, we have

$$k' = \sqrt{k_h \times k_v} \qquad \text{...(11.12)}$$

Again if N_f is the number of flow channels and N_d is the number of potential drops in the flow net then the total discharge q per unit length is given by

$$q = k'h\left(\frac{N_f}{N_d}\right)$$

$$= \sqrt{k_h \times k_v}\, h \left(\frac{N_f}{N_d} \right) \qquad \text{...(11.13)}$$

Thus, for an earth dam having anisotropic soil the rate of seepage may be computed by Eq. (11.13).

A comparison of Eqs. (11.11) and (11.13) indicates that by considering the equivalent coefficient of permeability as given by Eq. (11.12), Eq. (11.11) may be used for computing the rate of seepage in the case of earth dams having anisotropic soils.

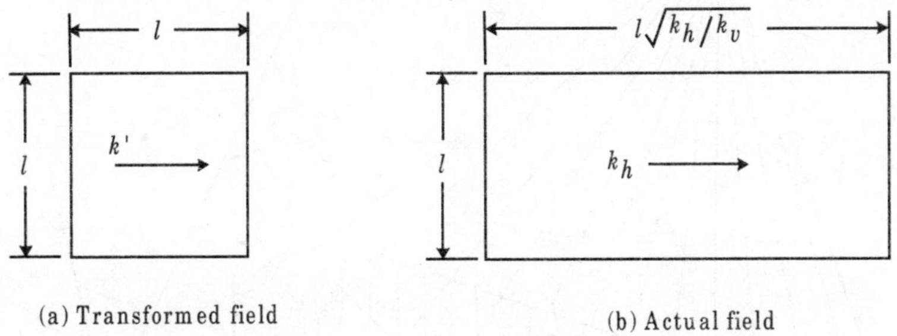

Fig. 11.11 *Anisotropic flow field*

11.12 STABILITY OF SIDE SLOPES OF EARTH DAM—STABILITY ANALYSIS

The side slopes of an earth dam will be stable if the soil mass is not dislodged from the slopes. However, the soil mass in an earth dam is subjected to forces which tend to cause movement or sliding of the soil mass. These forces are known as actuating or driving forces which mainly consist of gravity forces. The movement or sliding of the soil mass in an earth dam is resisted by the resisting or stabilizing forces which are provided mainly by the shear strength of the soil. Thus, if the side slope of an earth dam is to remain stable, the sum of the resisting forces on every possible surface of slippage (or surface of failure) should be more than the sum of the actuating forces, because even if there is any one surface on which this condition is not satisfied, slippage of soil mass can take place on that surface, thus resulting in the failure of the dam. The stability of the side slope of an earth dam is thus analyzed by assuming a surface of slippage within the soil mass and by determining the resisting and the actuating forces acting on this surface and the moments of these forces about the centre of rotation the factor of safety against sliding is calculated. The factor of safety against sliding is defined as the ratio of the moments of the resisting and the actuating forces acting on any surface of slippage. If the factor of safety is more than one, the side slope of the dam is safe on the assumed surface of slippage. On the basis of actual observations made in Sweden on the surface of slippage developed during the failure of a large number of dams due to slipping of the soil mass from the side slopes, it has been established that the surface of slippage is close to circular in most cases. As such for the stability analysis of the side slopes of an earth dam the surface of slippage is assumed as cylinderical or in the cross section as an arc of a circle which is known as slip circle. This method of analysis is therefore known as Swedish method or Slip Circle method. The factor of safety against sliding may be determined in this method as explained below.

Consider a side slope of an earth dam as shown in Fig. (11.12). Let a circular arc be drawn through the toe of the dam with centre at any point O. If this circular arc is assumed to be a possible surface of slippage then the soil mass lying above this surface will slide along this surface when the moment of the actuating forces about the centre O exceeds the moment of the resisting forces about the same point. For determining these forces and their moments about the centre O, the soil mass lying above the circular arc is divided into a number of slices by vertical

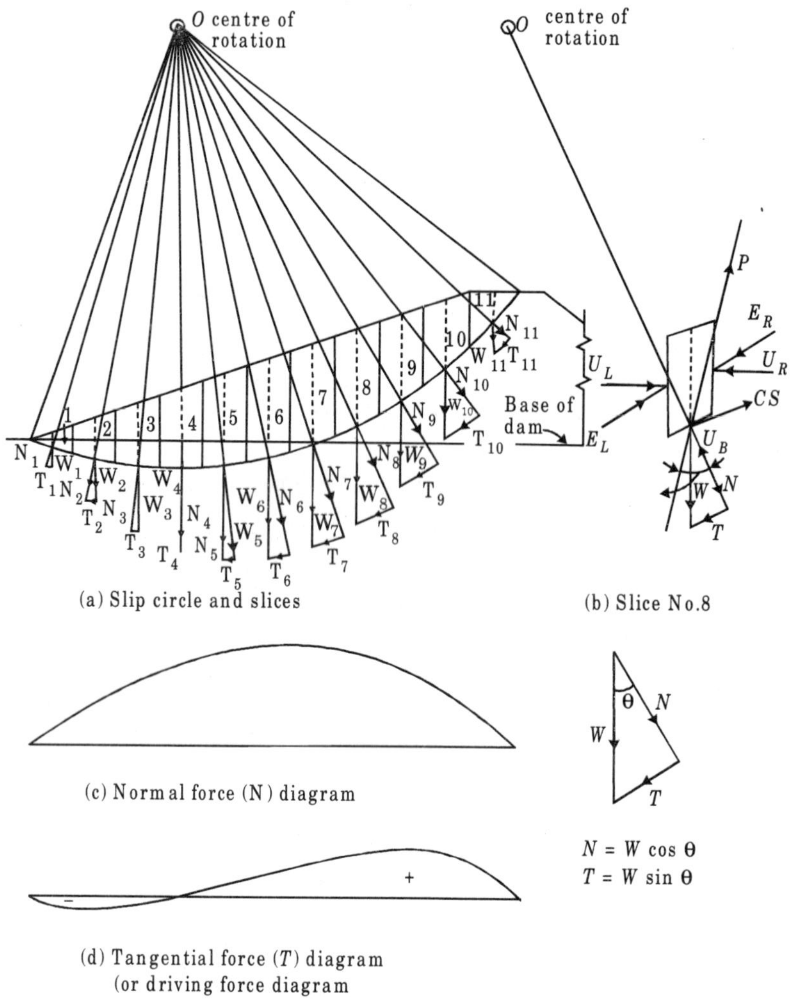

Fig. 11.12 *Slip circle method for stability analysis of side slopes of earth dams (graphical method)*

lines [Fig. 11.12 (a)]. The number of slices may generally be 10 to 15. It is not necessary that the slices should be of equal width but for the sake of convenience it is usual to have the slices of equal width. A typical slice say No. 8 is considered and the forces under which this slice is held in equilibrium are shown in Fig. (11.12 b). The forces acting on this slice are as follows.

1. The self weight W of the slice acting vertically downwards through the centre of gravity.

2. The cohesive forcescs acting tangentially opposite to the direction of probable slippage, where c is the unit cohesive strength of the soil and s is the length of the arc forming the base of the slice.

3. The soil reaction P across the arc. When the soil mass is about to slide the soil reaction P will act at an angle ϕ (the angle of internal friction of the soil) to the normal, i.e., radial direction.

4. The soil reaction on the two vertical sides of the slice, E_R and E_L exercised by the adjacent slices on the right and left respectively.

5. Pore pressures U_L, U_R and U_B as indicated in Fig. 11.12 (b).

Usually it is assumed that E_L and E_R cancel each other and so also U_L and U_R balance each other.

Draw the radial line or the normal at the centre of the slice and resolve the weight W into its normal and tangential components. Let N and T be the normal and the tangential components of the weight W. Now consider the equilibrium of forces in normal and tangential directions. Thus resolving all the forces in the normal (or radial) direction, we get

$$N - U_B = P \cos \phi \qquad \text{...(i)}$$

Also resolving all the forces in the tangential direction, we get

$$T - cs = P \sin \phi \qquad \text{...(ii)}$$

Substituting for P from Eq. (i) in Eq. (ii), we get

$$T = (N - U_B) \tan \phi + cs \qquad \text{...(iii)}$$

The tangential component T is the actuating or driving force and the sum of the frictional resistance $(N - U_B) \tan \phi$ and the cohesional resistance cs constitute the resisting or stabilizing force. Equation (iii) indicates that the actuating force is just balanced by the resisting force. This would be the condition if the soil mass is about to slide on this arc. However, if the actuating force T is more than the resisting force $[(N - U_B) \tan \phi + cs]$ then the soil mass will slide along this arc. On the other hand if the actuating force T is less than the resisting force $[(N - U_B) \tan \phi + cs]$ then the soil mass will remain stable.

If R is the radius of the surface of slippage then the moment of the resisting force about the centre O is $[(N - U_B) \tan \phi + cs]\ R$ and that of the actuating force is $(T \times R)$.

Similar forces are computed for all the slices above the surface of slippage and the same are added to obtain the total resisting and the actuating forces for the entire surface of slippage as follows.

$$\text{Total resisting force} = \Sigma[(N - U_B) \tan \phi + cs] \qquad \text{...(11.14)}$$

$$\text{Total actuating force} = \Sigma T \qquad \text{...(11.15)}$$

Further the moments of the total resisting and the actuating forces for the entire surface of slippage about the centre O are as follows.

Moment of resisting force

$$[\Sigma\{(N - U_B) \tan \phi + cs\}]\ R \qquad \text{...(11.16)}$$

Moment of actuating force

$$= (\Sigma T)\ R \qquad \text{...(11.17)}$$

The factor of safety F_S against sliding is thus given by

$$F_S = \frac{[\Sigma\{(N - U_B)\tan\phi + cs\}]R}{(\Sigma T)R}$$

or

$$F_S = \frac{\Sigma\{(N - U_B)\tan\phi + cs\}}{\Sigma T} \qquad ...(11.18)$$

In Eq. (11.18) ϕ and c are the known soil properties and s is the arc length of each slice which may be actually measured. However, if the width b of the slice is small s may be approximately taken equal to $(b/\cos\theta)$ where θ is the angle made by the radius of the surface of slippage at the base of the slice with the vertical through the centre of gravity of the slice. The weight W of the slice is equal to the area of the slice multiplied by the specific (or unit) weight of the soil as unit length of the dam is considered. However, in a zoned dam the weight of the slice is equal to the areas of the various zones included in the slice multiplied by their respective specific weights. The weight of the slice acts vertically downwards through the centre of gravity of the slice. The normal and tangential components N and T respectively of the weight W of the slice are determined by resolving the weight W in the radial (or normal) and tangential directions and the same may be expressed as follows.

$$N = W \cos\theta \qquad ...(11.19)$$

$$T = W \sin\theta \qquad ...(11.20)$$

The pore pressure U_B acting on the arc of the slice results in an uplift force which reduces the normal component of the weight of the slice. Thus, the net or effective downward force acting on the curved bottom boundary of the slice is the total weight minus the upward force due to pore pressure. The pore pressure may be determined from the flow net. However, the effect of the pore pressure may be accounted for by assuming submerged weight of the soil mass lying below the seepage line. Thus, the normal (or N) component of the weight of the portion of the slice lying below the seepage line is computed by using the submerged unit weight of the soil, but the tangential (or T) component of the weight of the same portion of the slice is computed by using the saturated unit weight of the soil. On the other hand both the normal and the tangential components of the weight of the portion of the slice lying above the seepage line are computed by using the moist or dry unit weight of the soil.

Graphical method for determining ΣN and ΣT

The normal and tangential components of the weight of the slice as well as the sum of these forces for all the slices may also be obtained graphically. A vertical line drawn through the centre of gravity of the slice and intersecting the top and bottom surfaces of the slice may be assumed to represent the weight of the slice infinitely small in width. This may be resolved graphically into normal and tangential components. If the slice is lying partly below and partly above the seepage line then the normal and tangential components are obtained separately for each of these two portions of the slice. Similarly in a zoned dam the normal and tangential components of the various zones included in the slice are obtained separately for each zone. These components for all the slices are plotted separately on two horizontal base lines after projecting the vertical lines on base lines. The extremities of these normal and tangential components of the same zone are joined by smooth curves as shown in Fig. (11.12 c) and (d). The area under these curves represent the sum of the normal and tangential forces acting on the surface of slippage. These areas of the various zones are measured by planimeter and multiplied by the respective specific weights of the soil mass of each zone to obtain ΣN and ΣT. In order to account for the effect of the pore pressure the normal components of the weight of the portions of the slices lying below the seepage line are computed by using submerged unit weights of the soil.

The sum of the normal and tangential components of the weights of the slices may also be determined by a *rectangular plot method.* In this method the end ordinate of each slice is assumed to represent the weight of the slice and it is resolved into normal and tangential components as shown in Fig. (11.13). The values of the normal components N_1, N_2, N_3, etc., and those of the tangential components T_1, T_2, T_3, etc., are plotted to form the base of N-rectangle and T-rectangle as shown in Fig. (11.13 b) and (c), the width of both the rectangles being equal

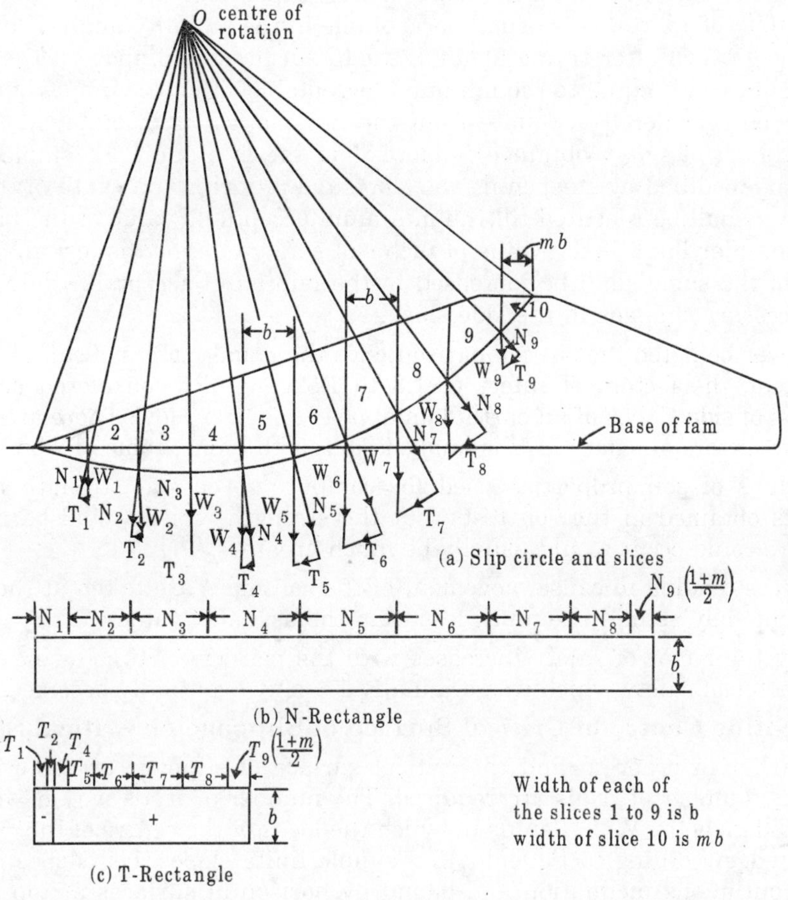

Fig. 11.13 *Slip circle method for stability analysis of side slopes of earth dams (rectangular plot method)*

to the width of the slices. Thus, in this method all the slices should be of the same width. However, if the width of the last slice is not same as that of the other slices, but is less say mb, where m is a multiplying factor and b is the width of each of the other slices, then the last N and T components are reduced by multiplying with the factor $(1 + m)/2$ before being plotted in the rectangles. The area of the N and T rectangles multiplied by the specific weight of the soil gives ΣN and ΣT respectively. However, this method may be used only if the entire slice has the same type of soil mass.

For determining the stability of the proposed side slope of an earth dam it is necessary to find the least factor of safety which may occur on any of the possible surfaces of slippage or slip circles. The surface of slippage or slip circle which yields the least factor of safety is the most critical and hence it is known as *critical surface of slippage or critical slip circle.* For locating the critical surface of slippage it is necessary to try several different surfaces of slippage as one trial arc gives the value of factor of safety for that arc only. Theoretically it is necessary to try an infinite number of possible surfaces of slippage with different centres and radii before the critical surface of slippage may be located. However, in practice a limited number of surfaces of slippage, about 12 to 15, selected on the basis of past experience are considered sufficient. The least factor of safety of 1.3 to 1.5 is usually acceptable in the stability analysis of the side slopes of an earth dam. As such after trying about 12 to 15 surfaces of slippage if the lowest value of factor of safety obtained is equal to the minimum acceptable value the proposed side slope of the earth dam may be considered adequate from the considerations of stability. However, if the lowest value of factor of safety obtained is more than the minimum acceptable value then the side slope shall be modified by steepening the slope and reducing the width of the berms if any. The above process shall be repeated till the minimum acceptable value of the factor of safety is obtained. On the other hand if the value of factor of safety obtained is less than the minimum acceptable value, the same shall be increased to the minimum acceptable value by trials after carrying out necessary changes in the side slope.

It may however be noted that in comparison with the usual factors of safety for structures of concrete and steel, the factors of safety of 1.3 to 1.5 generally considered acceptable in the stability analysis of side slopes of an earth dam appear to be very low. There are several reasons for accepting low factors of safety in this analysis for earth dams, some of which are as follows.

(i) The values of soil properties used for computing the soil strength are usually the minimum values obtained in the soil tests, but the strength which would be mobilized before disastrous failure could occur would usually be much greater.

(ii) The forces tending to cause movement of the soil mass are taken at the upper possible limit but the same may actually be much less than the assumed values.

(iii) Usually the factor of safety increases with the passage of time owing to consolidation etc., so that a factor of safety which was originally 1.3 may eventually become 2.

Method of Locating Centre of Critical Surface of Slippage or Critical Slip Circle

As described earlier in order to find the critical surface of slippage or the critical slip circle ordinarily a large number of trials are required. The number of trials may however be reduced by using a method given by W. Fellenius in which the locus of the centre of the critical surface of slippage is located. According to Fellenius for 'simple finite slopes' i.e., slopes of homogeneous soils with one continuous inclination and bound by horizontal surfaces at top and bottom the centre of the critical surface of slippage lies on a line OP as shown in Fig. (11.14 a) and 11.14 b) for the upstream and the downstream side slopes of an earth dam. The line OP may be determined by locating the points P and O as indicated below.

Point P has its coordinates with respect to the toe of the slope as H vertically downwards and 4.5 H horizontally away from the toe as shown in Fig. (11.14), where H is the height of the dam. On the other hand point O is located at the intersection of the lines AB and CD drawn from the extremities of the side slope. As shown in Fig. (11.14) line AB is drawn from the toe of the slope at an angle α with the side slope and line CD is drawn from the top end of the side slope at an angle β with the horizontal. The angles α and β are known as directional angles and

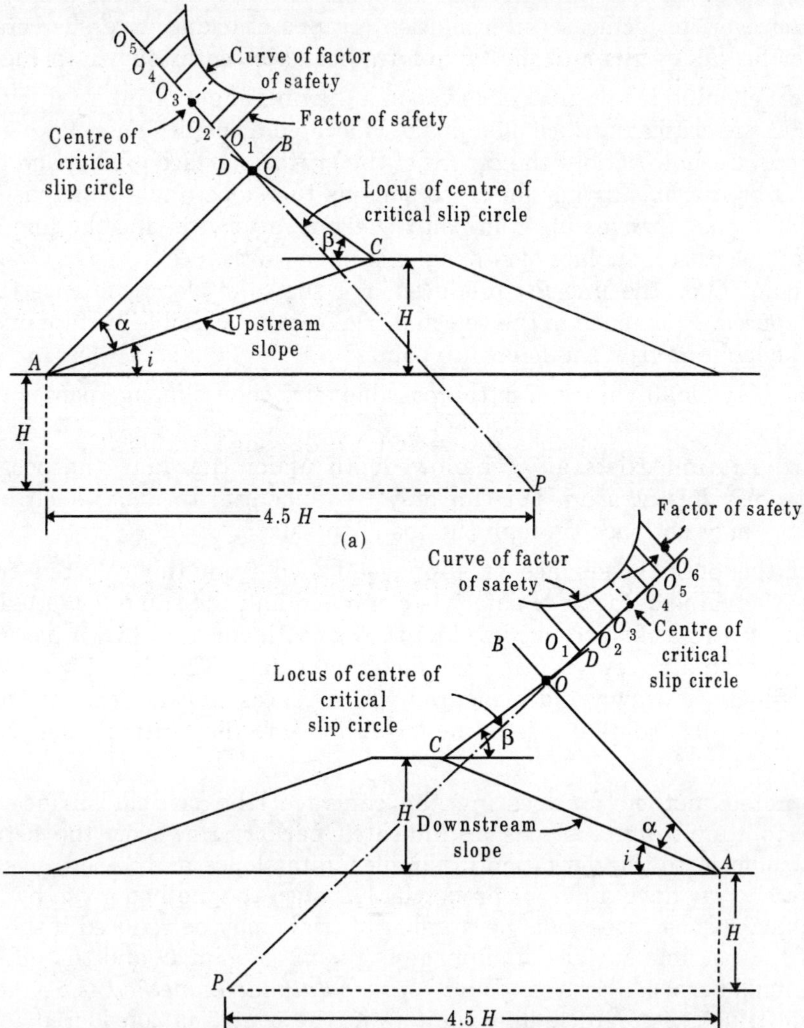

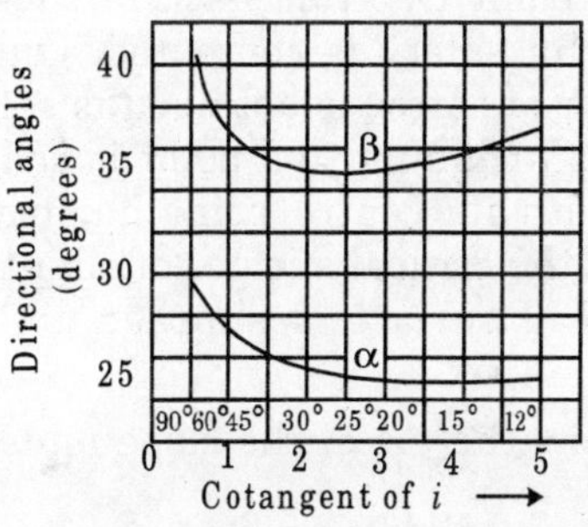

Slope H:V	Slope angle i	Directional Angles α	β
0.58:1	60°	29°	40°
1:1	45°	27°30'	37°
1.5:1	33°41'	26°	35°
2:1	26°34'	25°	35°
3:1	18°26'	25°	35°
4:1	14° 2'	25°	36°
5:1	11°19'	25°	37°

Fig. 11.14 *Fellenius method for locating centre of critical slip circle*

their values depend on the slope angle i (or the angle of inclination of the side slope). For a known value of i the values of α and β may be determined from the table given in Fig. (11.14). For any value of i other than the one given in the table the values of α and β may be obtained by interpolation. However, to facilitate interpolation, graphs of α and β versus cotangent of slope angle i based on the values given in the table have been plotted as shown in Fig. (11.14).

According to Fellenius for purely cohesive soils (having angle of internal friction ϕ equal to zero) the centre of the critical surface of slippage is located at point O and for $c - \phi$ soils (having both cohesion and internal friction) the centre of the critical surface of slippage lies above point O on the line PO produced. Further for $c - \phi$ soils as the slope angle i increases the centre of critical surface of slippage moves up from point O. Thus, after drawing the line PO, in order to find the centre of the critical surface of slippage a number of trial centres O_1, O_2, O_3, O_4, etc., are selected above point O on the line PO produced. For the sake of convenience the trial centres may be equally spaced. From each of the selected trial centres possible surface of slippage or slip circle is drawn keeping in view the following points.

(i) Except for very small values of ϕ, the possible surface of slippage passes through the toe of the slope.

(ii) If a hard stratum exists at a shallow depth under the dam, the possible surface of slippage cannot cross this stratum but can only be tangential to it. As such in this case the surface of slippage may not pass through the toe of the slope.

For each of the possible surface of slippage drawn from the selected trial centres as indicated above, the value of factor of safety is computed and the same is plotted as ordinate at the corresponding centre as shown in Fig. (11.14). A smooth curve is drawn passing through the tops of the plotted ordinates representing the factors of safety. The lowest point on this curve is noted and its ordinate is drawn and measured which represents the least factor of safety and also the point where this ordinate meets line OP represents the centre of the critical surface of slippage.

The above noted method for locating the centre of the critical surface of slippage is applicable only to simple finite slopes as indicated earlier. However, the actual earth dam sections do not conform to the definition of simple finite slopes as the soil mass in the shells, core and foundation may have different properties. As such the method given by Fellenius is not directly applicable for such cases, but the number of trials may be reduced if the initial position of the centre of the possible surface of slippage is taken at point O and the other trial centres are taken in its neighborhood. For this usually *nine point matrix method* is used. In this method a matrix of nine trial centres including point O in the centre is considered in which all the points are equally spaced and are in the vicinity of point O. From each of these trial centres possible surface of slippage is drawn and the factor of safety is computed. Out of these nine points the one corresponding to which minimum factor of safety is obtained is selected and with this point in the centre another matrix of nine points including this point is considered for the next trial. The trials are repeated until a matrix with its centre point having the minimum factor of safety is obtained which is then the centre of the critical surface of slippage. It has been observed that by this method the centre of the critical surface of slippage may be located in about two or three trials each comprising nine point matrix.

The critical surface of slippage may also pass below the toe of the slope under the following conditions.

(i) For small values of ϕ (from O to about 15°) the critical surface of slippage passes below the toe of the slope if the inclination of the side slope is less than 53°.

(ii) If the soils of the dam and foundation are entirely homogeneous then for any practicable side slope the critical surface of slippage may pass below the toe of the dam. Further if the soil in the foundation of the dam has the same or lesser shear strength than that of the dam then also the critical surface of slippage may pass below the toe of the slope.

In such a case according to Fellenius the centre of the critical surface of slippage is likely to be located on a vertical line drawn through the midpoint of the side slope. Further Fellenius found that for this case the critical surface of slippage subtends an angle of about 133½° at the centre. Thus, in order to locate the centre of the critical surface of slippage the following procedure may be adopted.

(i) As shown in Fig. (11.15) a vertical line is drawn at the mid point of the side slope.

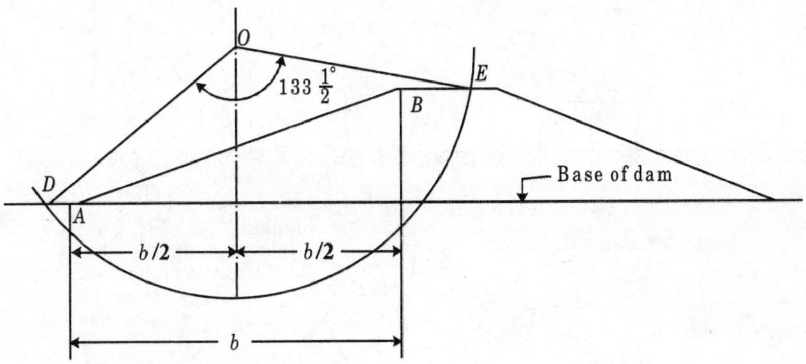

Fig. 11.15 *Method of locating critical slip circle passing below the toe of the slope*

(ii) On this line a trial centre O is taken such that when a possible surface of slippage is drawn with O as centre then there is an angle of 133½° between its two radii at which it intersects the surface of the dam and the foundation.

(iii) For this surface of slippage factor of safety is computed.

(iv) The trial centre is then moved somewhat to the left of the first trial centre O and a new possible surface of slippage of relatively shorter radius is drawn and the corresponding factor of safety is computed.

(v) A number of trial centres and the corresponding possible surfaces of slippage are thus considered and for each of these surfaces the factor of safety is computed to locate the critical surface of slippage.

Critical or Adverse Conditions for Stability of Side Slopes of an Earth Dam

The side slopes of an earth dam must remain stable under all the anticipated critical or adverse conditions. The various critical conditions for which the stability of the side slopes of an earth dam needs to be analyzed are as follows.

(i) Stability of upstream slope during sudden drawdown.

(ii) Stability of downstream slope during steady seepage.

(iii) Stability of both upstream and downstream slopes during and immediately after construction.

(i) *Sudden Drawdown Condition.* For the upstream slope of an earth dam the most critical condition occurs when the reservoir is suddenly emptied without allowing appreciable drainage from the saturated soil mass. This condition is known as sudden drawdown. The rate at which the water withdrawn from the reservoir would develop the sudden drawdown condition mainly depends on the type of soil in the upstream portion of the dam. Thus, if the soil is quite impervious then even if the reservoir is emptied at a relatively slow rate it may lead to the condition of sudden drawndown. On the other hand in the case of pervious soils the condition of sudden drawdown may be developed only if the water is withdrawn from the reservoir at a very fast rate.

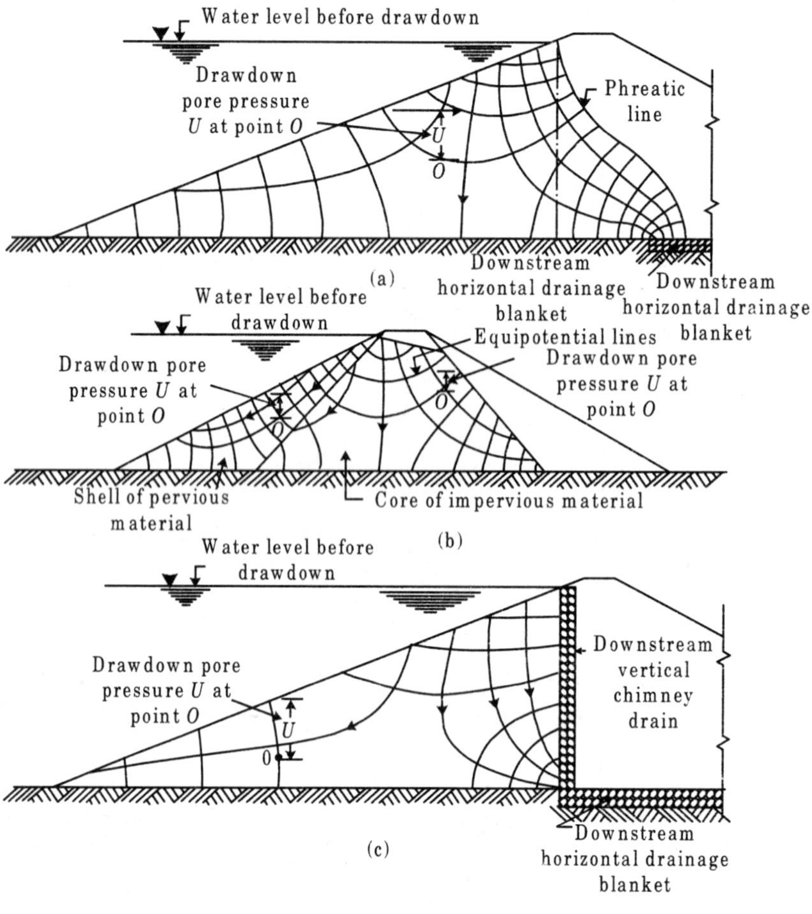

Fig. 11.16 *Sudden drawdown flownet for earth dam with (a) horizontal drainage blanket, (b) central core ; (c) vertical chimney and horizontal drainage blanket*

When sudden drawdown takes place, the water pressure acting on the portion of the upstream slope above the drawdown level is removed but since the drainage is not as rapid as the drawdown the soil mass is subjected to pore pressure. Thus, the resisting forces are considerably reduced on account of the pore pressure while the actuating forces are increased on account of the soil being saturated. To take account of this fact in stability analysis the pore pressures are determined along the portion of the surface of slippage which lies below the water surface for obtaining the net or effective resisting forces and the actuating forces are calculated by taking the saturated weight of the soil mass which lies below the water surface. The pore pressure may be determined by drawing the flownet for this case which is as shown in Fig. (11.16). The pore pressure at any point *O* on the surface of slippage is equal to the vertical distance between the point *O* and the point where the equipotential line passing through the point *O* meets the free surface which in this case is the upstream slope of the dam. The pore pressures determined at various points along the surface of slippage can be plotted to obtain the pore pressure distribution diagram. However, as an approximation the effect of pore pressure may be taken into account without determining the same if the resisting forces are calculated by taking the submerged weight of the soil mass which lies below the water surface. The soil mass below drawdown level is submerged and therefore for this mass both resisting and actuating forces are calculated on the basis of the submerged weight of the soil mass.

(ii) *Steady Seepage Condition.* For downstream slope of an earth dam the critical condition occurs when the reservoir is full and there is steady seepage at its maximum rate. The effect of seepage through the dam is to reduce the stability of the dam by increasing the actuating forces and decreasing the resisting forces. In this case the soil mass below the seepage line being saturated is subjected to pore pressure which must be determined to obtain the net or effective

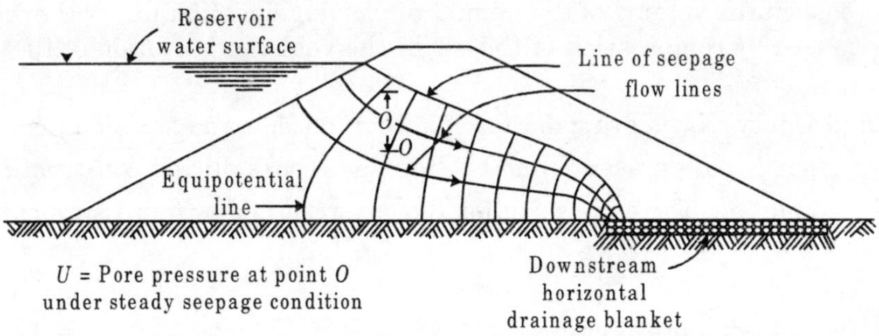

Fig. 11.17 *Steady seepage flow net for homogeneous earth dam with horizontal drainage blanket*

resisting forces. For this case also the pore pressure may be determined by drawing the flownet as shown in Fig. (11.17). The pore pressure at any point *O* on the surface of slippage is equal to the vertical distance between the point *O* and the point where the equipotential line passing through the point *O* meets the free surface which in this case is the seepage line. The pore pressures determined at various points along the surface of slippage can be plotted to obtain the pore pressure diagram. However, as an approximation the pore pressure may be accounted for in the stability analysis (without determining the pore pressure) by taking the submerged weight of the soil mass which lies below the seepage line for computing the resisting forces. The actuating forces are computed by taking the saturated weight of the soil mass which lies below the seepage line.

(iii) *During Construction and End of Construction Condition.* During construction of an earth dam plenty of water is used for compacting the soil mass. Since this water is retained in the pores of the soil mass high pore water pressures may be developed on account of compression due to consolidation of the soil mass under the self weight of the overlying fill. Such pressures may exceed the one which occur later due to seepage from the reservoir and hence they may control the design of the dam from the point of view of stability of side slopes. The construction pore pressures depend on the construction water content, the properties of the soil, the height of the dam, and the rate at which dissipation can occur from drainage. Thus, high construction pore pressures may be developed in homogeneous earth dams constructed of relatively impervious soils or in the impervious sections of zoned earth dams. Further the high pore pressures developed during construction may continue to exist even after the completion of the dam for the first few years of the life of the dam. Since during construction and immediately on the completion of the dam there would be no water load on the upstream slope of the dam, this condition would be critical for both the upstream and the downstream slopes. The construction pore pressure may be estimated by Hilfs' method which is described below.

Hilfs' Method

The method is based on the following assumptions :

(i) Only vertical compression of the dam takes place during construction and there is no lateral bulging.

(ii) The relationship between the compression of the dam and the effective stress is known.

(iii) The pressures in the pore water and the air in the pores are always equal and directly after compaction of the soil mass in the dam they are equal to the atmospheric pressure.

(iv) The decrease in the volume of the dam at any given elevation under the weight of the fill placed above is caused by compression of the air in the voids and solution of the air in the pore water.

(v) Boyle's and Henry's laws are valid for this compression and solution.

(vi) No dissipation of pore pressure due to drainage occurs during construction.

From these assumptions the expression for the construction pore pressure can be derived as below.

Let

u be the induced pore pressure

p_a be the absolute atmospheric pressure (being the pressure of the pore air before start of consolidation)

Δ be the compression of the dam in percent of original total volume of the dam

V_a be the volume of free air in the voids of the soil directly after compaction (or before start of consolidation) in percent of original total volume of the dam

V_w be the volume of pore water in percent of the original volume of the dam, and

$h_c V_w$ be the volume of dissolved air in volume V_w of pore water where h_c is Henry's constant being equal to 0.02 at 20°C.

Applying Boyle's law and equating product of pressure and volume of the pore air before and after the compression of the dam, we have

$$p_a(V_a + h_c V_w) = (p_a + u)(V_a + h_c V_w - \Delta)$$

or
$$u = \frac{p_a \Delta}{V_a + h_c V_w - \Delta} \quad \text{...(11.21)}$$

or
$$u = \frac{p_a \Delta}{V_a + 0.02 V_w - \Delta} \quad \text{...(11.21 a)}$$

The numerical values for the initial air and water volume for use in Eq. (11.21) can be estimated from laboratory compaction tests and from previous experience with dams having average densities. The value of the compression of the dam can be obtained from laboratory consolidation tests, or from the average of field measurements for dams with similar soil mass.

When the compression of the specimen (Δ) becomes equal to the initial volume of air in the voids (V_a) (i.e., $\Delta = V_a$) the entire air goes into solution i.e., the soil becomes saturated. The pore pressure at which this occurs is given by

$$u_{sat} = \frac{p_a V_a}{h_c V_w} \qquad \text{...(11.22)}$$

or

$$u_{sat} = \frac{p_a V_a}{0.02 V_w} \qquad \text{...(11.22 a)}$$

If V_0 denotes the original volume of the soil mass, n_0 its porosity, ΔV the change in volume (being negative for compression) and S_0 is the initial degree of saturation, the volume of voids $= V_0 n_0$ and free air volume $V_a = (1 - S_0)V_0 n_0$ while that of dissolved air

$$h_c V_w = h_c S_0 V_0 n_0$$

Hence

$$V_a + h_c V_w = (1 - S_0 + S_0 h_c) V_0 n_0$$

Substituting this value of $(V_a + h_c V_w)$ in Eq. 11.21 and noting that $\Delta = (\Delta V/V_0)$ the following relationship is obtained which is known as Bishop's relationship

$$u = \frac{-p_a(\Delta V/V_0)}{n_0(1 - S_0 + S_0 h_c) + (\Delta V/V_0)} \qquad \text{...(11.23)}$$

A number of measures may however be taken to reduce the construction pore pressures and thus eliminate the necessity of a more conservative and uneconomical design because of the 'during construction' and 'end of construction' stability conditions. Some of these measures are as indicated below.

(a) The maximum construction pore pressure can be limited by compacting the impervious section of the dam at an average water content a few per cent below the optimum value.

(b) The impervious section can be made thinner, so that high construction pore pressures will have less influence on the stability and will dissipate more rapidly.

(c) Internal drains can be installed within the impervious section to accelerate the pore pressure dissipation.

(d) Longer construction periods may be permitted to allow more time for pore pressure dissipation.

(e) Lower safety factors may be tolerated against the possibility of slope slides during construction than would be permissible when the dam is retaining the reservoir.

11.13 STABILITY OF EARTH DAM AGAINST HORIZONTAL SHEAR DEVELOPED AT THE BASE OF THE DAM

Besides the precise methods discussed in the previous section for the stability analysis of side slopes of earth dams there are certain approximate methods for determining the stability of earth dams from the considerations of horizontal shear developed at the base which are suitable for the preliminary design of the earth dams. These methods are discussed below.

1. Stability of Earth Dam against Headwater Pressure. Due to headwater pressure acting on an earth dam it may tend to slide along the base which is however resisted by the shear resistance developed at the base. Thus, in order to ensure the stability of an earth dam against horizontal shear (or sliding) due to headwater pressure a factor of safety is computed which is defined as the ratio of the force resisting sliding of a dam to the force tending to cause its sliding. The force resisting the sliding of a dam is given by

$$\text{Shear resistance} = W \tan \phi$$

where W is the effective weight of the dam per unit length, and ϕ is the angle of internal friction of the soil mass of which the dam is composed. For computing the effective weight of the dam section seepage line is located and the weighted average of the submerged portion and the moist or dry portion is taken.

The force tending to cause sliding of a dam is the horizontal component of the headwater pressure. The headwater pressure will be perpendicular to the upstream slope of the dam and the horizontal component H of this pressure per unit length of the dam will be

$$H = \frac{wh^2}{2}$$

where w is specific weight of water and h is the maximum depth of headwater.

The factor of safety is then given by

$$Fs = \frac{W \tan \phi}{H} \qquad \ldots(11.24)$$

which should not be less than 2 for the design to be safe.

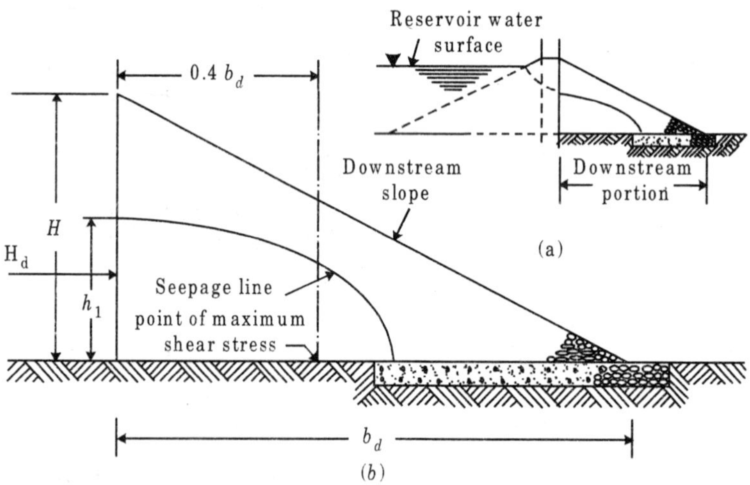

Fig. 11.18 *Shear in downstream portion of dam*

2. Stability of Downstream Portion of an Earth Dam against Horizontal Shear at the Base of the Dam or at any other Horizontal Plane through the Dam. Consider the downstream portion of an earth dam separated from the dam by a vertical plane through the top shoulder of the downstream slope of the dam as shown in Fig. (11.18). A horizontal shear force acting in the downstream direction is exerted on the downstream portion of the dam. As indicated in section 11.11 for the downstream portion of the dam the steady seepage condition is the most critical condition. Thus, for this condition the total horizontal shear force acting on the downstream portion of the dam will be equal to the sum of the earth pressure acting on the vertical section through the top of the slope and the pressure of the water contained by the saturated soil mass lying below the seepage line at this section. The earth pressure may be computed by the use of Rankine's theory and hence the total horizontal shear force H_d acting on the downstream portion per unit length of the dam is given by

$$H_d = \left[\frac{wH^2}{2}\tan^2\left(45° - \frac{\phi}{2}\right) + \frac{w_w h_1^2}{2}\right] \qquad ...(11.25)$$

in which

w is the weighted unit weight of the soil mass in the vertical section through the top shoulder of the downstream slope of the dam

H is the vertical distance from the top of the dam down to the base of the dam or to the horizontal plane under consideration through the downstream portion of the dam

ϕ is the angle of internal friction of the soil mass in the dam

w_w is the unit weight of water, and

h_1 is the vertical distance from the seepage line down to the base of the dam or to the horizontal plane under consideration through the downstream portion of the dam.

The term $\left[w\tan^2\left(45° - \frac{\phi}{2}\right)\right]$ is known as equivalent liquid unit weight which would produce the same pressure as that of the earth.

The weighted unit weight w of the soil mass in the vertical section through the top shoulder of the downstream slope of the dam is given by

$$w = \frac{w_1 h_1 + w_2(H - h_1)}{H} \qquad ...(11.26)$$

where w_1 is the submerged unit weight of the soil mass and w_2 is the moist or dry unit weight of the soil mass.

If b_d is the width of the base of the downstream portion of the dam or that of the horizontal plane under consideration through this portion of the dam then the average shear stress s_d on the base of the downstream portion of the dam or on the horizontal plane under consideration through this portion of the dam is given by

$$s_d = \frac{H_d}{b_d} = \left[\frac{wH^2}{2b_d}\tan^2\left(45° - \frac{\phi}{2}\right) + \frac{w_w h_1^2}{2b_d}\right] \qquad ...(11.27)$$

Further if s_{md} is the maximum shear stress on the base of the downstream portion of the dam or on the horizontal plane under consideration through this portion of the dam, then since the maximum shear stress may be twice the average shear stress, we have

$$s_{md} = 2s_d = \left[\frac{wH^2}{b_d}\tan^2\left(45° - \frac{\phi}{2}\right) + \frac{w_w h_1^2}{b_d}\right] \quad ...(11.28)$$

The assumption that the maximum shear stress is twice the average shear stress is in accordance with the simple elastic theory. However, photoelastic studies indicate that this is a conservative assumption because in such model studies the maximum shear stress is found to be only 1.4 times the average shear stress.

The point where maximum shear stress would occur on the base of the downstream portion of the dam or on the horizontal plane under consideration through this portion of the dam may be taken to be at a horizontal distance of 0.4 b_d from a vertical through the top shoulder of slope (Fig. 11.18), where b_d is the width of the base of the downstream portion of the dam or that of the horizontal plan under consideration through this portion of the dam. The position of the point of maximum shear stress is however empirical but it is found to be in agreement with the photoelastic studies of such models.

A resisting force is however developed against the horizontal shear force acting on the downstream portion of the dam. The resistance against shear is developed due to internal friction and cohesion of the soil mass in the dam. Hence the resisting force R_d developed for the downstream portion per unit length of the dam against the horizontal shear force is given by

$$R_d = W_{ed} \tan \phi + cb_d \quad ...(11.29)$$

in which W_{ed} is the total effective weight of the downstream portion of the dam above the base or above the horizontal plane under consideration through this portion per unit length of the dam; c is cohesion of the soil per unit area ; and the other terms are same as defined earlier. In computing W_{ed} submerged unit weight of the soil should be taken for the soil mass lying below the seepage line and dry or moist unit weight of the soil should be taken for the soil mass lying above the seepage line.

The average factor of safety F_d against shear in the downstream portion of the dam is then given by

$$F_d = \frac{R_d}{H_d} \quad ...(11.30)$$

As mentioned earlier since the horizontal plane is not usually the weakest plane of failure, the factor of safety F_d should be at least 2 for the design to be satisfactory.

The factor of safety at the point of maximum shear stress must also be determined, which is defined as

$$F_{dm} = \frac{\text{Unit shear strength at the point of maximum shear stress}}{\text{Maximum shear stress}} \quad ...(11.31)$$

where F_{dm} is the factor of safety at the point of maximum shear stress for the downstream portion of the dam.

The unit shear strength at the point of maximum shear stress is equal to (wh tan ϕ + c), where w is the weighted unit weight of the soil mass in the section of the dam at the point of maximum shear stress; h is the vertical distance from the downstream slope of the dam down to the point of maximum shear stress ; and the other terms are same as defined earlier.

The factor of safety F_{dm} should be at least 1.5 for the design to be satisfactory.

3. Stability of Upstream Portion of an Earth Dam against Horizontal Shear at the Base of the dam or at any other Horizontal Plane through the Dam. Consider the upstream portion of an earth dam separated from the dam by a vertical plane passing through the top shoulder of the upstream slope of the dam as shown in Fig. 11.19. A horizontal shear force acting in the upstream direction is exerted on the upstream portion of the dam. As indicated in section 11.12 for the upstream portion of the dam the sudden drawdown condition

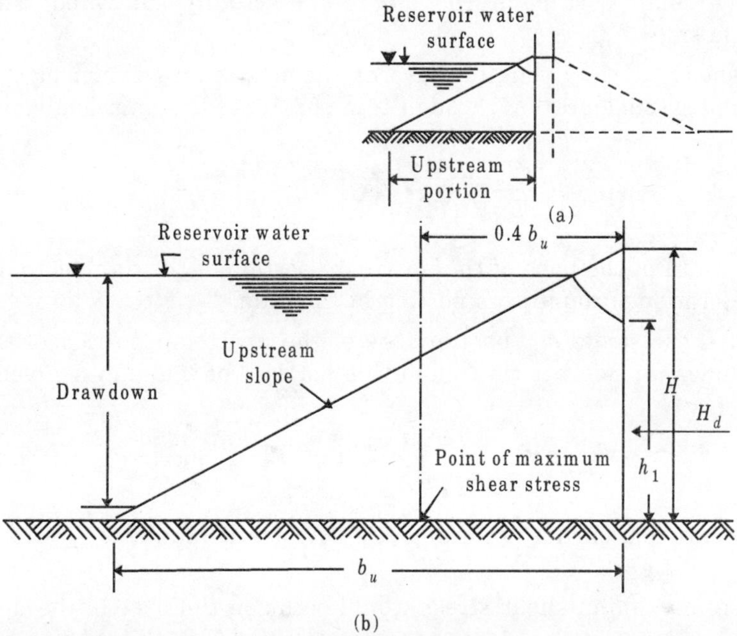

Fig. 11.19. *Shear in upstream portion of earth dam with sudden drawdown*

is the most critical condition. In the case of sudden drawdown the upstream portion of the dam would remain saturated. As such the total horizontal shear force acting on the upstream portion of the dam will be equal to the sum of the earth pressure acting on the vertical section through the top of the slope and the pressure of the water contained in the saturated soil mass lying below the seepage line at this section. For computing the earth pressure saturated unit weight of the soil mass below the seepage line should be used. Further the earth pressure may be computed by using Rankine's theory. Thus, the total horizontal shear force H_u acting on the upstream portion per unit length of the dam is given by

$$H_u = \left[\frac{w_s H^2}{2}\tan^2\left(45° - \frac{\phi}{2}\right) + \frac{w_w h_1^2}{2}\right] \quad ...(11.32)$$

in which

w_s is the saturated unit weight of the soil mass in the upstream portion of the dam

H is the vertical distance from the top of the dam down to the base of the dam or to the horizontal plane under consideration through the upstream portion of the dam

ϕ is the angle of internal friction of the soil mass in the dam

w_w is the unit weight of water, and

h_1 is the vertical distance from the seepage line down to the base of the dam or to the horizontal plane under consideration through the upstream portion of the dam.

It may be noted that in this case the entire upstream portion of the dam is assumed to be saturated though a small portion at the top is not actually saturated. This assumption is however on the safe side.

The average shear stress s_u on the base of the upstream portion of the dam or on the horizontal plane under consideration through this portion of the dam is given by

$$s_u = \frac{H_u}{b_u}\left[\frac{w_s H^2}{2b_u}\tan^2\left(45° - \frac{\phi}{2}\right) + \frac{w_w h_1^2}{2b_u}\right] \qquad ...(11.34)$$

in which b_u is the width of the base of the upstream portion of the dam or that of the horizontal plane under consideration through this portion of the dam.

The maximum shear stress s_{mu} on the base of the upstream portion of the dam or on the horizontal plane under consideration through this portion of the dam is given by

$$s_{mu} = 2s_u = \left[\frac{w_s H^2}{b_u}\tan^2\left(45° - \frac{\phi}{2}\right) + \frac{w_w h_1^2}{b_u}\right] \qquad ...(11.35)$$

However, as mentioned earlier according to photoelastic studies s_{mu} is found to be only 1.4 times s_u.

The point where maximum shear stress would occur on the base of the upstream portion of the dam or on the horizontal plane under consideration through this portion of the dam may be taken to be at a horizontal distance of 0.4 b_u from a vertical through the top shoulder of the upstream slope (Fig. 11.19) where b_u is the width of the base of the upstream portion of the dam or that of the horizontal plane under consideration through this portion of the dam.

A resisting force is however developed against the horizontal shear force acting on the upstream portion of the dam. The resistance to shear is developed due to internal friction and cohesion of the soil mass in the dam. Hence the resisting force R_u developed for the upstream portion per unit length of the dam against the horizontal force is given by

$$R_u = W_{eu} \tan \phi + c b_u \qquad ...(11.36)$$

in which W_{eu} is the total effective weight of the upstream portion of the dam above the base or above the horizontal plane under consideration through this portion per unit length of the dam ; c is cohesion of the soil per unit area ; and the other terms are same as defined earlier. In computing W_{eu} submerged unit weight of the soil should be taken for the soil mass lying below the seepage line and dry or moist unit weight of the soil should be taken for the soil mass lying above the seepage line.

The average factor of safety F_u against shear in the upstream portion of the dam is then given by

$$F_u = \frac{R_u}{H_u} \quad ...(11.37)$$

F_u is also referred to as the factor of safety against sudden drawdown. Further as mentioned earlier since the horizontal plane is not usually the weakest plane of failure, the factor of safety F_u should be at least 2 for the design to be satisfactory.

The factor of safety at the point of maximum shear stress must also be determined, which is defined as

$$F_{um} = \frac{\text{Unit shear strength at the point of maximum shear stress}}{\text{Maximum shear stress}} \quad ...(11.38)$$

where F_{um} is the factor of safety at the point of maximum shear stress for the upstream portion of the dam.

The unit shear strength at the point of maximum shear stress is equal to $(w_1 h \tan \phi + c)$, where w_1 is the submerged unit weight of the soil mass in the section of the dam at the point of maximum shear stress; h is the vertical distance from the upstream slope of the dam down to the point of maximum shear stress ; and the other terms are as defined earlier.

The factor of safety F_{um} should be at least 1.5 for the design to be satisfactory.

11.14 STABILITY OF FOUNDATION OF AN EARTH DAM AGAINST HORIZONTAL SHEAR

Foundation of earth dams consisting largely of coarse sands and gravels or of thoroughly consolidated silts or clays usually have high shear strength. On the other hand foundations consisting of fine, loose, cohesionless soils or of uncoslidated clays and silts may usually be weak in shear and hence require thorough investigation. For determining the stability of the foundation of an earth dam against horizontal shear a factor of safety is determined by a method which is however approximate and is as described below.

Figure 11.20 shows the upstream portion of an earth dam founded on a foundation the safety of which is to be determined. A horizontal shear force acting in the upstream direction is

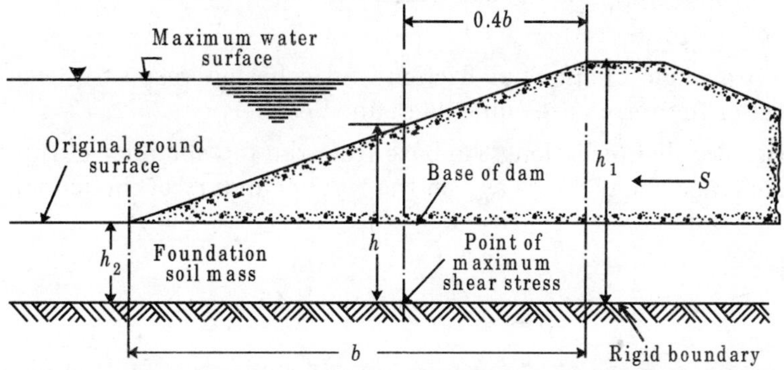

Fig. 11.20. *Shear stress in foundation of earth dam*

exerted on this portion of the dam and the foundation. Again as indicated earlier for the upstream portion of dam sudden drawdown condition is the most critical condition. Thus, for this condition the total horizontal shear force acting on the rigid boundary of the foundation under this portion of the dam is equal to the difference between the lateral thrusts on a vertical through the top shoulder of the slope and a vertical through the toe of the dam. Therefore if S is the total horizontal shear force acting on the rigid boundary of the foundation per unit length of the dam then

$$S = w\left(\frac{h_1^2 - h_2^2}{2}\right)\tan^2\left(45° - \frac{\phi_1}{2}\right) \qquad ...(11.39)$$

where h_1 is the vertical distance from top of dam down to the rigid boundary

h_2 is the vertical distance from base of dam down to the rigid boundary

w is the mean effective unit weight of the soil mass of the dam and foundation which is weighted in proportion to the depth of each, and

ϕ_1 is the equivalent angle of internal friction.

The mean unit weight w is given by

$$w = \frac{w_D(h_1 - h_2) + w_F(h_2)}{h_1} \qquad ...(11.40)$$

in which

w_D is unit weight of soil mass of dam

w_F is unit weight of soil mass of foundation, and rest of the terms are same as defined earlier.

In this case since the soil mass of dam and that of foundation are in a saturated state, in Eq. 11.40 the unit weights w_D and w_F should be taken as saturated unit weights of the soil mass.

The equivalent angle of internal friction ϕ_1 is given by

$$\tan \phi_1 = \frac{wh_1 \tan\phi + c}{wh_1} \qquad ...(11.41)$$

in which ϕ and c are angle of internal friction and cohesion respectively of the soil mass of foundation and rest of the terms are same as defined earlier.

If b is the horizontal distance along the base from the top shoulder of the slope to the toe of the dam then the average shear stress s_{ave} on the rigid boundary of the foundation of the dam is given by

$$s_{ave} = \frac{S}{b} = \frac{w}{2b}(h_1^2 - h_2^2)\tan^2\left(45° - \frac{\phi_1}{2}\right) \qquad ...(11.42)$$

The maximum shear stress s_{max} is given by the following relationship, which has been checked by photoelastic studies

$$s_{max} = 1.4\ s_{ave} \qquad ...(11.43)$$

The point of location of the maximum shear stress on the rigid boundary of the foundation of the dam may be taken to be at a horizontal distance of 0.4b from a vertical through the top shoulder of the slope. This is also in agreement with the photoelastic studies.

For determining the factor of safety against shear in the foundation average unit shear strength (or unit shear resistance) of foundation is determined. The average unit shear strength of foundation is taken to be the mean of the unit shear strengths of foundation below the toe of the dam and below the top shoulder of slope of the dam.

The unit shear strength s_1 below the toe of the dam is given by

$$s_1 = c + w'_F h_2 \tan \phi \qquad ...(11.44)$$

in which w'_F is the submerged unit weight of the soil mass of foundation and rest of the terms are same as defined earlier.

The unit shear strength s_2 below the top shoulder of the slope of the dam is given by

$$s_2 = c + w' h_1 \tan \phi \qquad ...(11.45)$$

in which w' is the mean effective unit weight of the soil mass of the dam and foundation which is given by

$$w' = \frac{w'_D(h_1 - h_2) + w'_F h_2}{h_1} \qquad ...(11.46)$$

where w'_D and w'_F are the submerged unit weights of the soil mass of the dam and the foundation respectively and the rest of the terms are same as defined earlier.

The average unit shear strength s of the foundation is given by

$$s = \frac{s_1 + s_2}{2} \qquad ...(11.47)$$

The overall average factor of safety F against shear in the foundation is given by

$$F = \frac{s}{s_{ave}} \qquad ...(11.48)$$

which should be greater than 1.5 for the design to be satisfactory.

The factor of safety at the point of maximum shear stress must also be determined as described below.

The unit shear strength s' at the point of maximum shear stress is given by

$$s' = c + w' h \tan \phi \qquad ...(11.49)$$

in which

w' is mean effective unit weight of soil mass at vertical section through point of maximum shear stress which is given by

$$w' = \frac{w'_D(h - h_2) + w'_F h_2}{h} \qquad ...(11.50)$$

where h is the vertical distance from the slope of the dam down to the point of maximum shear stress on the rigid boundary of foundation, and rest of the terms are same as defined earlier.

The factor of safety against shear at point of maximum unit shear in the foundation is given by

$$F = \frac{s'}{S_{max}} \quad ...(11.51)$$

which should be greater than unity for the design to be satisfactory.

It may however be stated that at the point of maximum unit shear a factor of safety less than 1.0 may also be accepted. This is so because although in that case there would be local movement at this point, but if the average factor of safety exceeded unity, the resisting forces would be mobilized and failure would not occur.

By adopting the same procedure as indicated for the foundation below the upstream portion of the dam, the stability of the foundation below the downstream portion of the dam may also be investigated. For this case also for the soil mass lying below the seepage line submerged unit weight should be taken for computing the shear strength and saturated unit weight should be taken for computing the shear force and the shear stress.

Further the stability of foundation of an earth dam against horizontal shear should also be investigated for the condition that the dam is just completed but the reservoir is empty. This is so because immediately after the dam is completed and the reservoir is empty then due to lack of consolidation the soil mass will have lower values of angles of internal friction ϕ and cohesion c thus resulting in a lower value of shear strength. For this condition dry or moist unit weight may be taken for the soil mass of dam as well as that of foundation. However, in this case also if it is assumed that the ground water level is at ground surface then since the soil mass of foundation will be in a saturated state for computing the shear force and the shear stress saturated unit weight of the soil mass of foundation should be taken and for computing the shear strength submerged unit weight of the soil mass of foundation should be taken.

11.15 CHECK FOR FREE PASSAGE OF WATER THROUGH EARTH DAMS

Out of the seven criteria for the safe design of earth dams mentioned in section 11.6, the first four criteria are amenable to analysis as discussed in the previous sections. The remaining three criteria are now considered. The fifth criterion states that there should be no opportunity for the free passage of water from the upstream to the downstream face of an earth dam. The free passage of water may be developed through seams or openings in an earth dam which may be caused in a number of ways as indicated below.

(i) By water following the exterior surfaces of pipes or conduits through the dam.

(ii) By burrowing animals such as muskrats.

(iii) By the placing of very pervious material containing large stones in an otherwise impervious dam in such a manner that a blind drain from the upstream to the downstream face is formed.

(iv) By failure to bond and compact the succeeding layers of the dam properly.

(v) By failure to bond the lower layers of an earth dam properly to the foundation.

(vi) By water following the smooth surfaces of concrete abutments or other concrete structures or passing through 'French drains' accidentally created along the sides of such structures by deposition of gravel and stones.

(i) *Pipes or conduits passing through earth dams.* Water from the reservoir may flow along the outer surface of pipes or conduits passing through an earth dam, thus developing a free passage of water through the dam. Further water may also seep through cracks in pipes or conduits passing through an earth dam which may be developed due to various reasons as indicated in section 11.5. This may also result in developing a free passage of water through the dam. As such in order to check the development of free passage of water through an earth dam as far as possible there should be no pipes or conduits passing through the dam but the same may be placed either in the trenches excavated in the original foundation soil mass or in a trench or tunnel excavated through the hills on either side of the dam. However, if the pipes or conduits are required to pass through an earth dam, the same would need special consideration in design and construction. The pipes or conduits passing through an earth dam may usually be placed inside a concrete conduit or tunnel with free space around them.

(ii) *Protection against burrowing animals.* The holes dug by burrowing animals in earth dams may result in developing free passage of water from the upstream to the downstream face of the dam which may cause the failure of the dam due to piping. However, as discussed in section 11.5 burrowing animals may cause failure of only small earth dams and in the case of major earth dams no serious damage may be caused by them. As such in general there is no need to adopt any special measures to safeguard the dam against the burrowing animals. However, in regions which have unusually large number of burrowing animals the earth dam may be provided with a suitable steel plate central core wall running from abutment to abutment for the full height of the dam.

Further all earth dams in regions with large number of muskrats should be carefully inspected at frequent intervals and if necessary the muskrats may be kept under control by trapping and poisoning. It has however been found that creosote floating on the water surface in the reservoir is effective is driving out all muskrats. Moreover, if the outer portions of an earth dam are compacted as a mixture of creosote and soil much less muskrats are likely to burrow in.

(iii) *Formation of a blind drain through an earth dam.* Sometimes the careless placing of coarse material such as stones and coarse gravel may cause a blind drain to be formed from the upstream to the downstream face of an earth dam. Such a condition is however, avoided by preventing coarse material from forming a continuous line from the upstream to the downstream face.

(iv) *Passage of water through uncompacted soil mass.* If succeeding layers of an earth dam are not properly compacted and bonded there is a chance for water to pass through the loose soil mass. In the case of an earth dam if the work has been suspended for some time, the surface may become hard and smooth. On such a surface if additional layers of soil mass are placed without special precautions to secure a bond between the new work and the old, there will be a danger of a leak along the plane of contact with the previous uncompacted layer.

(v) *Bonding earth dams to their foundations.* It is of utmost importance to bond the lower layers of earth dams properly to the foundation soil mass in order to prevent the passage of water along any possible dividing plane between the foundation and the dam. For this all perishable material such as stumps, brush, sod and large roots should be removed from the foundation. Further all top soil containing more than 10 percent of vegetable matter should also be removed from the base of the dam.

(vi) *Cutoff buttresses on spillway and powerhouse walls.* Spillway abutments, powerhouse wall or other concrete walls extending through earth dams in an upstream and downstream direction should be provided with cutoff buttresses projecting well into the dam. The cutoff walls and the rear face of the main wall should have a slight batter (about 1 on 10), so that shrinkage of the dam will tend to bring the soil mass in closer contact with the concrete.

11.16 SAFETY AGAINST PIPING

The dam and foundation should be safe against piping. Piping occurs when seepage water issues from an earth dam or its foundation under sufficient pressure and with sufficient velocity so that the soil particles are carried away. As mentioned in section 11.5 the piping commences at the downstream face of the dam and soon results in the formation of a small tunnel running backwards into the dam or the foundation with the roof continually falling in and being carried away by the piping water. Once piping has started the only remedy is to dump rock grading from fine to coarse right into the downstream face where the piping is occurring, so that an improvised drain and filter will be formed and the piping stopped. Thus safety against piping is obtained by providing a suitable drainage arrangement with properly designed filter at the downstream toe of the dam so that while water can get through freely no soil particles of the dam or foundation are dislocated. The various types of drainage arrangements and the criteria for the design of filter are given in section 11.19.

11.17 PROTECTION OF UPSTREM SLOPE OF AN EARTH DAM

The upstream slope of an earth dam must be protected against erosive action of waves. This is done by providing a surface protection for the upstream slope. Usual types of surface protection for the upstream slope are stone pitching (or stone riprap), precast concrete blocks, monolithic concrete pavement, bituminous paving, soil cement paving, steel facing, etc. Out of the various types of surface protection stone pitching is most commonly adopted and the same is discussed below.

Stone pitching is of the following two types.

(i) Random pitching. (or Dumped riprap)

(ii) Hand-placed pitching.

Random pitching consists of stones dumped in place from cars or trucks or tossed into place by hand. The individual stones must be of sufficient weight to resist displacement by wave action. For determining the minimum average weight of stones in random pitching, the U.S. Army Crops of Engineers have given the following formula

$$W_A = 100 h_w^2 \tan\alpha$$

in which

W_A = average weight of stones in kg

h_w = wave height in m, and

α = upstream slope angle

Further it has been recommended that the stones should have a grading from 1/8 to 4 times the average weight W_A. The thickness of the random pitching should be 1.5 times the linear dimension of the average rock size. Thus, assuming the specific gravity of stones to be 2.65, we have

$$T = 1.5\left(\frac{W_A}{2650}\right)^{1/3}$$

where T is the required thickness in m of random pitching. However, the random pitching should not be less than 1 m thick.

Hand-placed pitching consists of stones carefully laid by hand in a more or less definite pattern with a minimum amount of voids and with the top surface relatively smooth. An effort is made to break joints as much as possible, and the voids are filled with smaller stones. Hand-placed pitching is generally 450 mm thick and the minimum size of individual stones is usually specified as 300 mm by 300 mm by 75 mm. It is also often required that at least 50 percent of the surface shall be of stones which are 450 mm deep.

Both the types of stone pitching are laid on a filter bed consisting of gravel or crushed stone or spalls in order to prevent the underlying soil mass of the dam being washed out through the voids in the pitching by wave action, resulting in undermining of the pitching. The thickness of the filter bed underlying the pitching usually varies from 225 mm to 450 mm according to its character and wave height. The gradation for the filter bed may be determined by the criteria of design of filters given in section 11.19. The filter bed is however not required to be provided under the stone pitching if the upstream portion of the dam consists of gravel.

The upstream slope protection should extend from the crest of the dam to a safe distance below the minimum water level and ordinarily it should terminate on a supporting berm. In the case of reservoirs that are frequently emptied the upstream slope protection should extend upto the upstream toe of the dam.

11.18 PROTECTION OF DOWNSTREAM SLOPE OF AN EARTH DAM

The downstream slope of an earth dam needs to be protected against erosion due to wind and rain which is done by providing a surface protection. If the downstream portion of an earth dam consists of rock or cobble fill, no surface protection for the slope is necessary. On the other hand if the downstream portion of an earth dam consist of fine grained soil and gravel the downstream slope should be protected against erosion by providing a surface protection. For the downstream slope generally stone pitching and turfing (i.e., covering of grass) are used for surface protection. Out of these two types of surface protection, stone pitching is preferred and should be used if the cost is not prohibitive. However, where stone pitching is not practicable, protection from erosion may be obtained by turfing, i.e., by building up a vegetative cover by growing a suitable type of grass on the downstream slope of the dam. If properly developed and maintained turfing also provides adequate protection for the downstream slope of an earth dam against erosion.

In addition to the above noted types of surface protections adopted to protect the downstream slope against erosion, often berms are provided on the downstream slope to minimize surface erosion as described below.

Berms

Berm is a shelf that breaks the continuity of a slope (Fig. 11.21). The berms are provided for the following purposes :

(i) to break the continuity of the slope thereby reducing erosion of the downstream slope from rain

(ii) to provide level surfaces for construction and maintenance operations, and

(iii) to prevent undermining of the lower edge of the stone pitching in case of upstream slope.

For earth dams more than 10 m high, berms from 2 to 6 m wide may be provided on the downstream slope. Usually one berm for every 10 to 15 m difference in elevation is provided. Further a berm is also desirable at elevation of top of the rock toe. The rain water that flows down the slope above the berm is caught at the berm and prevented from flowing further down the slope thus minimizing the erosion of the downstream slope. To prevent rain water from flowing over the outer edge of the berm and down the slope of the dam the surface of the berm should be sloped towards the dam so that the outer edge of the berm is higher than the inner edge. A slope of 1 in 50 is usually provided for this purpose. Further a gutter should be provided at the inner edge of the berm to carry the rain water to the side of the valley where a vertical gutter of drain is provided to carry it to the toe of the dam. However, the berms are generally not necessary for controlling erosion even for high dams if the downstream slope is provided with a suitable type of slope protection. As such the recent practice has been to not to use berms to control surface erosion on the downstream slope.

11.19 MEASURES TO CONTROL SEEPAGE THROUGH EARTH DAMS AND THEIR FOUNDATIONS

The water seeping through the body of an earth dam and through its foundation may have the following adverse effects.

(i) Loss of water.

(ii) Generates erosive forces which may dislodge the soil particles and may cause piping failure of the dam.

(iii) Reduces slope stability which may result in the failure of the dam due to sliding of slopes.

(iv) Causes local sloughing which may result in the failure of the dam.

In order to prevent the adverse effects of the water seeping through the dam and its foundation and thus increase the stability of the dam it is essential to adopt certain measures to control the seepage. The various seepage control measures usually adopted are of the following two types.

(1) Measures adopted for reducing the quantity of seepage.

(2) Measures adopted for safe drainage of the seeping water.

1. Measures adopted for reducing the quantity of seepage. Different measures which are adopted to reduce the quantity of seepage through the dam and through the foundation are as follows.

(a) *In dam.* The only measure adopted to reduce seepage through the dam is the provision of a core of impervious soil, or a core wall of concrete or masonry within the body of the dam. Out of these core of impervious soil (or impervious core) is commonly used which is also known as impervious zone and the same is described below.

Impervious Core

The impervious core forms a barrier within the body of the dam for seeping water. On the basis of location of the cores these may be classified in the following two types viz.

(i) Central vertical cores

(ii) Upstream inclined cores

which are shown in Figs. 11.2 and 11.21. The type of core which may be provided depends mainly on the availability of material, topography of site, foundation condition, etc. The main advantage of central vertical core is that it provides higher pressures at the contact between the core and the foundation thus reducing the possibility of leakage and piping. Another advantage of a vertical core is that for a given quantity of impervious soil the thickness of a vertical core is slightly greater than the thickness of an inclined core. On the other hand the main advantage of the upstream inclined core is that the downstream portion of the dam can be constructed first and the core placed later. Another advantage of the sloping core is that the filter layers between the core and the upstream and downstream pervious zones can be made more thinner with less difficulty in construction than the filter layers for vertical core dams. Further inclined core reduces the pore pressures in the downstream portion of the dam and thereby increases its safety.

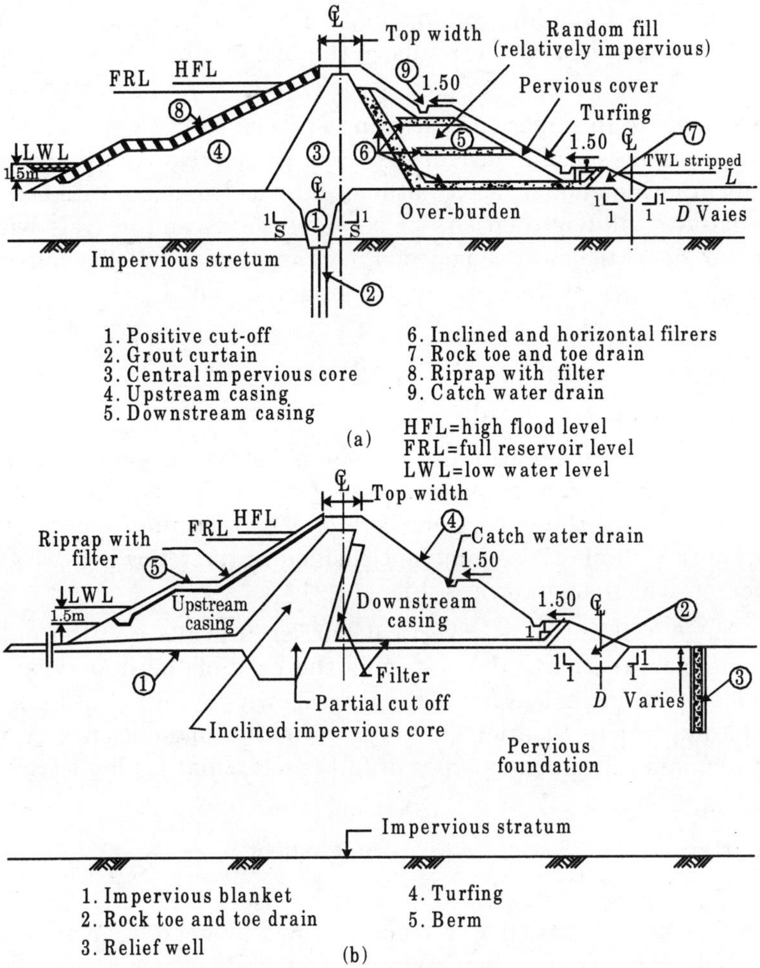

Fig. 11.21 *Earth dam with (a) central core and positive cut-off; (b) inclined core and partial cut-off*

As indicated in section 11.2 zoned type dams have wide cores whereas diaphragm type dams have thin cores. Dams constructed of most fine-grained soils are so impervious that even for very high dams only a thin core would be necessary to reduce the seepage to a negligible value.

Moreover the core material normally has less shear strength than the rest of the dam and hence from the point of view of stability a thinner core is preferred because a wider core will require flatter slopes for the upstream and downstream faces of the dam. However, a wide core has more resistance to piping. The minimum safe core width is usually governed by the following factors: (a) the permissible seepage through the dam ; (b) the minimum width which will permit proper construction ; (c) the type of material available for the core and shells ; (d) the design of proposed filter layers ; (e) precedent on similar projects. However, the width of the core should be such that it will be safe against the failure due to seepage force. Thus, generally cores with a width equal to 30% to 50% of the head of water in the reservoir are found to be satisfactory. However, relatively thin cores of width 15% to 20% of the head of water in the reservoir are also found to be satisfactory if they are adequately designed and constructed and are provided with proper filter layers. Further the minimum top width of the core should be 3 m and the top level of the core should be fixed at least 1 metre above the maximum water level to prevent seepage by capillary syphoning.

(b) *In foundation.* If an impervious stratum such as rock, clay, etc., is available close to the bed of the river then no specific measure is required for reducing the seepage through the foundation and only grouting and surface treatment may be carried out if necessary. However, if the foundation consists of alluvial deposits of pervious sands and gravels with an impervious stratum at a greater depth below the bed of the river then the following measures may be adopted to reduce the seepage of water through the foundation.

1. Cutoff.
2. Upstream impervious blanket.
3. Downstream berm.

1. Cutoff. A cutoff is a vertical impermeable barrier provided within the pervious foundation of an earth dam to reduce the seepage of water through the foundation. It usually extends down from the base of the impervious core provided within the body of the dam. On the basis of the length of the cutoffs these may be classified in the following two types viz., (a) Full cutoff (b) Partial cutoff which are shown in Fig. (11.21.)

(a) Full cutoff. A full cutoff is a vertical barrier provided for the entire depth of the pervious foundation below the earth dam. It joins the base of the impervious core of the dam with the impervious stratum in the foundation as shown in Fig. (11.21 a). Such a cutoff is provided where the impervious stratum is available in the foundation at a reasonable depth below the base of the dam. The various types of full cutoff usually adopted are as follows.

(i) Positive cutoff trench
(ii) Concrete cutoff wall
(iii) Grout curtain
(iv) Steel sheet pile
(v) Slurry trench cutoff.

(i) *Positive cutoff trench* It consists of a trench formed by open excavation in the foundation soil and backfilled by a compacted impervious soil. As such it is also known as open trench and fill back cutoff or rolled earth cutoff. For constructing this type of cutoff the same impervious soil and construction methods are employed as in the impervious core of the dam. These cutoffs are most commonly constructed to depths upto 25 m. However, trenches have been excavated to more than 50 m depth but it becomes expensive and troublesome. The width and side slopes of the trench are generally selected according to the convenience of construction and to ensure stability of excavated slopes. The main difficulty in the construction of these trenches is that of dewatering and holding down the water level till the trench is backfilled. Dewatering of these trenches is however carried out with suitable pumps.

(ii) *Concrete cutoff wall*. It consists of a trench excavated in the foundation soil and backfilled with unreinforced concrete. The trench is of constant width (about 2 m) and its walls are supported by sheets and braces. A new form of concrete cutoff wall is a continuous row of overlapping concrete piers installed with special drilling rigs or other equipment.

(iii) *Grout curtain*. Grout curtain or Grouted cutoff is produced by injection of grout of cement, clay, chemicals or a combination of these materials within the zone assigned to the cutoff. While cement grouts have been used to reduce leakage through fissured rocks under dams, the same has not been found successful to grout alluvial sand and gravel deposits. This is so because cement grout could not be injected uniformly except in coarse materials. Chemical grouts could be injected into sand deposits but these are usually too expensive. However, clay grout has been found to be quite successful for grouting sand and gravel deposits. Usually a grout consisting of a mixture of clay, cement and chemicals such as silicate and phosphate is used.

(iv) *Steel sheet pile*. It consists of steel sheets driven through the pervious foundation. It is however ineffective in controlling seepage because even if it is intact there is considerable leakage through the interlocks between the individual sheets. In addition during driving the pile may be damaged and the interlocks may break because of defect in the steel or when the pile hits some obstacles such as cobbles and boulders. Further if steel sheet piles are driven to hard rock with a very uneven surface, a continuous row of triangular gaps may be present between their lower edges and the rock, or the piles may curl if they are driven too hard. As such a steel sheet pile with a special design has been developed to reduce leakage in the interlocks and to prevent breaking or curling. It consists of a double row of straight sheet piles welded to 450 mm wide flange steel beams. After driving, the space between the beams is excavated and filled with concrete. However, besides the various limitations of steel sheet piles as mentioned above these are also relatively more expensive and hence these are not commonly used.

(*v*) *Slurry trench cutoff*. It also consists of a trench excavated in the pervious deposits of the foundation and backfilled with a suitable impervious soil. However, in this case cutoff is constructed by using slurry trench method. In this method a trench is excavated with a dragline through the pervious deposits down to the impervious stratum of the foundation. The trench is kept continuously filled with bentonite slurry which prevents the trench walls from caving so that no sheeting or shoring is required to be used. After a sufficient length of trench has been excavated the bottom is cleaned with clamshell buckets and air lifts and the backfilling begins. The backfill consists of the excavated material (i.e., a mixture of sand and gravel with bentonite slurry) blended with about 15 to 20% of natural silt brought in from a borrow pit. This mixture is then dumped from one end of the trench displacing the bentonite slurry until the backfilling is complete.

(b) Partial cutoff. In many cases in the foundation of an earth dam the impervious stratum may exist at such a large depth below the bed of the river that it would be impracticable or extremely expensive to provide full cutoff extending upto the impervious stratum. In such cases a partial cutoff may be provided. A partial cutoff is one which extends down from the impervious core of the dam into the underlying pervious strata but does not reach the impervious stratum. A partial cutoff would be effective in reducing the quantity of seepage through the foundation if the horizontal permeability of the foundation soil is much more than the vertical permeability and the permeability decreases with depth i.e., the top strata are more pervious than the lower ones. Thus, partial cutoffs may be effective in alluvial deposits which are usually stratified and hence the horizontal permeability may be 10 to 50 times the vertical permeability. However, at sites where the foundation soil is homogeneous and isotropic a partial cutoff is rather ineffective in reducing the quantity of seepage through the foundation. It has been established experimentally that for homogeneous soil mass in the foundation a cutoff driven to 90% of the depth will reduce the quantity of seepage only by about 64% and that driven to 50% of the depth will reduce the quantity of seepage only by about 25%. As such as far as possible a full cutoff should be provided at a site with homogeneous soil in foundation.

2. Upstream Impervious Blanket. A horizontal blanket of impervious soil may be provided on the river bed on the upstream side to reduce the quantity of seepage through the

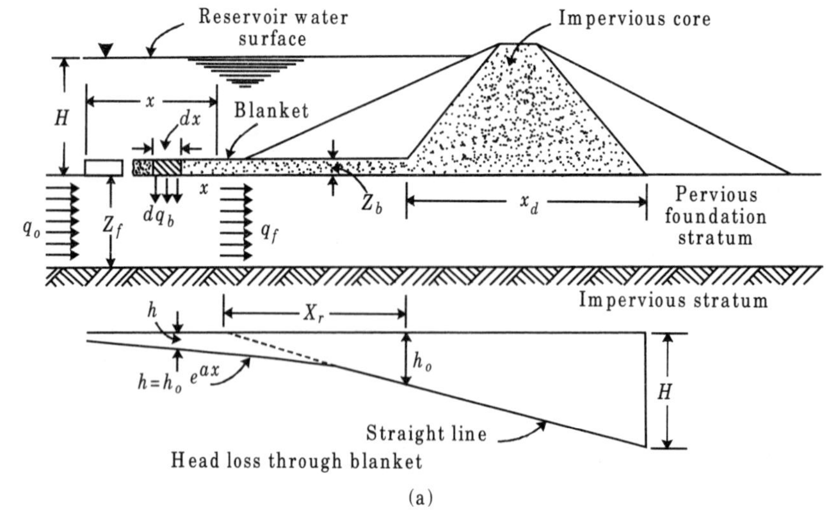

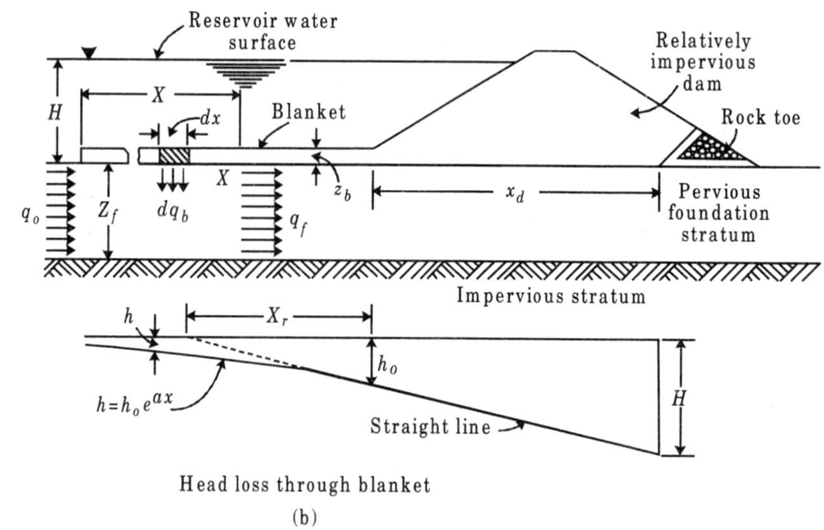

Fig. 11.22. *Upstream impervious blanket-Bennett's solution*

pervious foundation under an earth dam. At some of the dam sites a natural impervious blanket is available if there exists a thin layer of impervious soil of large areal extent at the bed of river overlying a thick layer of previous soil below. The impervious blanket increases the length of the path of seepage under the dam and thus reduce the velocity and quantity of seepage. The impervious blanket should be connected to the impervious core of the dam as shown in Fig. (11.22). However, an impervious blanket may also be provided in the case of a homogeneous dam constructed of relatively impervious soil, in which case the blanket is connected to the upstream toe of the dam. It is however necessary to provide relief wells near the downstream toe of the dam to collect water seeping through the foundation and to reduce the uplift pressure.

The soil used for impervious blanket should have far less permeability than that of the foundation soil. The necessary thickness and length of the blanket depend on the permeability of the soil of the blanket, thickness of the previous foundation, and the maximum depth of water in the reservoir. The blankets of thickness varying from 0.6 to 2 m are most commonly used. The design of an impervious blanket mainly involves the determination of the effectiveness of the blanket of known length and thickness in reducing the quantity of seepage and also the head dissipated through the blanket. For this a mathematical solution developed by P.T. Bennett is used which is as follows.

Bennett's Solution

The Bennett's solution is based on the following assumptions :

(i) the permeability of the foundation soil is equal to or greater than 10 times the permeability of the blanket soil

(ii) the direction of flow of seeping water is vertical through the blanket and horizontal through the foundation

(iii) a vertical line through the upstream end of the blanket is an equipotential line at full reservoir head

(iv) no seepage takes place through the dam or through the core

(v) there is no natural impervious layer over or within the pervious foundation.

Figure 11.22 shows an earth dam with a pervious foundation and an impervious upstream blanket.

At any point X under the blanket the horizontal flow q_f through the foundation is equal to the flow q_b through the blanket upstream from the point X plus the inflow q_o under the upstream end of the blanket. Thus,

$$q_f = q_o + q_b \qquad \text{...(i)}$$

If dq_b is the vertical flow through a small element of the blanket of length dx at a distance x from the upstream end of the blanket, then

$$q_b = \int_0^x dq_b \qquad \text{...(ii)}$$

and hence

$$q_f = q_o + \int_0^x dq_b \qquad \text{...(iii)}$$

If k_b is the coefficient of permeability of the blanket soil; Z_b is the thickness of the blanket ; and h is the loss of head at X (which is the difference of head on the two sides of the blanket under which percolation takes place) then the value of dq_b is obtained by Darcy's law as

$$dq_b = kiA = k_b \frac{h}{Z_b} dx \quad \text{...(iv)}$$

since hydraulic gradient $i = (h/Z_b)$ and area $A = dx$, per unit length of the blanket in the direction perpendicular to the plane of the paper.

Substituting the value of dq_b in Eq. (iii), we have

$$q_f = q_o + \int_0^x k_b \frac{h}{Z_b} dx \quad \text{...(v)}$$

Differentiating both sides of Eq. (v) with respect to x, we get

$$\frac{dq_f}{dx} = \frac{dq_o}{dx} + \frac{d}{dx}\left[\int_0^x k_b \frac{h}{Z_b}\right]$$

or

$$\frac{dq_f}{dx} = \frac{dq_o}{dx} + k_b \frac{h}{Z_b}$$

Since q_o is independent of x

$$\frac{dq_o}{dx} = 0$$

Hence

$$\frac{dq_f}{dx} = k_b \frac{h}{Z_b} \quad \text{...(vi)}$$

The value of q_f is also obtained by Darcy's law as

$$q_f = k_f \frac{dh}{dx} Z_f \quad \text{...(vii)}$$

where k_f is the coefficient of permeability of the foundation soil, Z_f is the thickness of the pervious foundation ; and (dh/dx) is the hydraulic gradient at point X.

Differentiating both sides of Eq. (vii) with respect to x, we get

$$\frac{dq_f}{dx} = k_f Z_f \frac{d^2h}{dx^2} \quad \text{...(viii)}$$

Equating the values of (dqf/dx) given by Eqs (vi) and (viii), we get

$$\frac{d^2h}{dx^2} = \frac{k_b h}{k_f Z_b Z_f} \quad \text{...(11.52)}$$

Equation 11.52 is the differential equation for the pressure head dissipated through the blanket. The solution of this differential equation has been obtained for the following two cases :

(a) Blanket of uniform thickness.

(b) Blanket of variable thickness.

(a) Blanket of uniform thickness. For a blanket of uniform thickness $\left(\dfrac{k_b}{k_f Z_b Z_f}\right)$ is a constant. Let this constant be represented by a^2. Equation 11.52 then reduces to

$$\frac{d^2h}{dx^2} = a^2h \qquad ...(11.53)$$

in which $$a = \sqrt{\frac{k_b}{k_f Z_b Z_f}} = \text{constant}$$

The solution of Eq. 11.53 has been obtained for the following two cases :

(i) Infinite length of the blanket.

(ii) Finite length of the blanket.

(i) *Solution for infinite length of blanket.* A natural impervious blanket of large areal extent if available may be considered as a blanket of infinite length. For a blanket of uniform thickness and infinite length the solution of Eq. (11.53) is

$$h = h_o e^{ax} \qquad ...(11.54)$$

where h is the loss of head through the blanket upto any point at a distance x and h_o is the loss of head through the blanket upto a point $x = 0$. In this case for convenience the point $x = 0$ is taken at the downstream end of the blanket and hence h_o is the total loss of head through the blanket upto the end of the blanket.

As a measure of the efficiency of a blanket of any length x (where x may be infinite or a finite length) a length X_r is considered which is known as equivalent resistance of the foundation and is defined below.

Equivalent resistance X_r of the foundation

It may be defined as the length of a prism of the foundation soil of thickness Z_f and coefficient of permeability k_f which under the loss of head h would carry flow equivalent to the flow which passes the blanket system under the same loss of head. Thus,

$$q_f = k_f \frac{dh}{dx} Z_f = k_f \frac{h}{X_r} Z_f$$

or $$\frac{dh}{dx} = \frac{h}{X_r} \qquad ...(ix)$$

Also from Eq. (11.54), we have

$$\frac{dh}{dx} = h_o a e^{ax} = ah \qquad ...(x)$$

Equating the values of (dh/dx) given by Eq. (ix) and (x), we get

$$ah = \frac{h}{X_r}$$

or $$X_r = \frac{1}{a} = \sqrt{\frac{k_f \; Z_b \; Z_f}{k_b}} \quad ...(11.55)$$

(ii) *Solution for finite length of blanket.* For a blanket of uniform thickness and finite length of blanket the solution of Eq. 11.53 is

$$h = h_n \, (e^{ax} - e^{-ax}) \quad ...(11.56)$$

in which h is the loss of head through the blanket upto any point at a distance x ; and h_n is a constant for computing h. h_n depends on the total head loss of the system of which the blanket is a part and on the ratio of the blanket to the remainder of the system.

From Eq. (11.56) at $x = 0$, $h = 0$, and hence in this case the point $x = 0$ is taken at the upstream end of the blanket.

Differentiating both sides of Eq. 11.56 with respect to x, we get

$$\frac{dh}{dx} = ah_n \, (e^{ax} + e^{-ax})$$

Again if X_r is the equivalent resistance of the foundation, we have from Eq. (ix)

$$\frac{dh}{dx} = \frac{h}{X_r}$$

or $$X_r = \frac{h}{(dh/dx)} = \frac{h_n(e^{ax} - e^{-ax})}{ah_n(e^{ax} + e^{-ax})}$$

or $$X_r = \frac{(e^{ax} - e^{-ax})}{a(e^{ax} + e^{-ax})} \quad ...(11.57)$$

or $$X_r = \frac{(e^{2ax} - 1)}{a(e^{2ax} + 1)} \quad ...(11.57\text{ a})$$

If the length of the blanket is L then by substituting $x = L$ in Eq. (11.57) the value of X_r for the entire blanket is obtained.

Loss of head through the blanket

The loss of head (or the head dissipated) through the blanket of infinite or finite length is given by

$$h_o = \frac{X_r}{X_r + X_d} H \quad ...(11.58)$$

in which h_o is the loss of head upto the end of the blanket ; H is the total reservoir head ; X_d is the base width of the impervious core (or the base width of the dam if it is a homogeneous dam of relatively impervious soil) ; and X_r is same as defined earlier. For determining X_r to be used in Eq. (11.58) the respective equations as derived earlier for the blankets of infinite and finite lengths are used.

With the loss of head h_o upto the end of the blanket being known, if desired, the value of h_n in Eq. 11.56 for a blanket of finite length may be computed by putting $h = h_o$ and $x = L$.

Reduction in quantity of seepage due to provision of blanket

If there was no blanket the quantity of seepage q_f through the foundation is given by

$$q_f = k_f \frac{H}{X_d} Zf$$

in which all the terms are same as defined earlier.

When a blanket of infinite or finite length is provided the quantity of seepage q_f' through the foundation is given by

$$q_f' = k_f \frac{H}{X_r + X_d} Z_f$$

∴ Reduction in the quantity of seepage through the foundation due to the provision of blanket expressed in percentage is

$$\frac{q_f - q_f'}{q_f} \times 100 = \frac{X_r}{X_r + X_d} \times 100 \qquad \text{...(11.59)}$$

It has however been shown that for any value of a, very little benefit is to be gained by extending the blanket beyond that given by the following expression

$$ax = \sqrt{2} \qquad \text{...(11.60)}$$

Hence the length of the blanket

$$x = x_o = \left(\frac{\sqrt{2}}{a}\right)$$

may be considered as the optimum length of the blanket.

Steps for the design of blanket of uniform thickness

(i) Assume suitable thickness Z_b of blanket to be provided of impervious soil of known coefficient of permeability k_b.

(ii) Find a by the expression

$$a = \sqrt{\frac{k_b}{k_f Z_b Z_f}}$$

in which k_f and Z_f are already known.

(iii) Assume various lengths of the blanket including infinite length and find X_r for each of these lengths which is given

(a) for infinite length of blanket by the expression

$$X_r = \frac{h}{(dh/dx)} = \frac{1}{a}$$

(b) for finite length of blanket by the expression

$$X_r = \frac{(e^{2ax} - 1)}{a(e^{2ax} + 1)}$$

(iv) Find the head dissipated through each of the blankets of the assumed lengths by the expression

$$\text{Head dissipated} = \frac{X_r}{X_r + X_d} \times H$$

(v) Find the percentage reduction in the quantity of seepage for each of the blankets of the assumed lengths by the expression

$$\%\text{ reduction in seepage} = \frac{X_r}{X_r + X_d} \times 100$$

(vi) That length of the blanket is recommended to be provided beyond which if the length of the blanket is increased there is no appreciable change in the head dissipated through the blanket and in the percentage reduction of seepage. However, the optimum length of the blanket $x_o = (\sqrt{2}/a)$ may be determined to obtain the maximum length of the blanket to be provided.

(b) Blanket of variable thickness. As the blanket is more effective towards the downstream end, a more efficient use of the blanket would require that it should have an increasing thickness in the downstream direction. In Eq. (11.47) Z_b would then be a variable and it may be written as

$$\frac{d^2h}{dx^2} = \frac{\alpha h}{Z_b} \qquad \text{...(11.61)}$$

where $\alpha = \dfrac{k_b}{k_f Z_f}$ = constant ; and Z_b will vary with x.

For a triangular blanket $Z_b = S_x$, where S is the slope. This however leads to a solution of Eq. (11.56) in terms of a series which is difficult to compute.

However, if it is assumed that the blanket has parabolic profile with its thickness represented by the following equation

$$Z_b = \frac{\alpha x^2}{n(n-1)}$$

then a simple solution of Eq. (11.61) is obtained as noted below

$$h = h_n x^n \qquad \text{...(11.62)}$$

Also for this case

$$X_r = \frac{h}{(dh/dx)} = \frac{x}{n} \qquad ...(11.63)$$

Further the head dissipated through the blanket and the reduction in the quantity of seepage may be obtained for this case also by Eqs (11.58) and (11.59) respectively.

In the case of a blanket of variable thickness at the upstream end the blanket is assumed to have zero thickness, but in actual practice it would be necessary to provide a minimum thickness of 0.4 to 0.5 m at the upstream end. Further the blanket may actually be built to a straight slope approximating the parabolic curve.

Steps for the design of blanket of variable thickness

(i) From the known values of k_b, k_f and Z_f compute the value of α by the expression

$$\alpha = \frac{k_b}{k_f Z_f}$$

(ii) Assume a suitable length of the blanket and taking different values of n such as n = 1.5, 1.8, 2.0, etc., find Z_b at the end of the blanket and X_r by the expressions

$$Z_b = \frac{\alpha x^2}{n(n-1)} \text{ and } X_r = \frac{x}{n}$$

The length of the blanket may however be assumed close to the one given by equation $ax = \sqrt{2}$, for which an average uniform thickness of the blanket is assumed and the value of a is obtained by the expression

$$a = \sqrt{\frac{k_b}{k_f Z_b Z_f}}$$

(iii) The head dissipated through the blanket and the reduction in the quantity of seepage are computed.

(iv) Several other lengths of the blanket are assumed and the procedure is repeated for each of these lengths.

(v) The earth work involved in each case is determined and that length of the blanket is adopted which would cause considerable reduction in the quantity of seepage and would be quite economical.

3. Downstream berm (or landside berm or downstream loading berm)

As shown in Fig. (11.23) a downstream or landside berm extending beyond the downstream toe of the dam is sometimes provided to reduce the quantity of seepage by lengthening the seepage path. It also provides some protection against sloughing of the downstream slope as a result of seepage. Further if at any dam site a relatively thin impervious layer exists at the surface overlying a pervious stratum underneath, then excessive underseepage pressures of the water seeping through the pervious stratum may cause blowing up of the impervious layer at or beyond the downstream toe of the dam. In such cases a downstream berm may be provided so that the weight of the berm plus that of the top impervious layer will be able to resist the excessive underseepage pressures. As such a downstream berm is also known as a downstream loading berm or simply a loading berm.

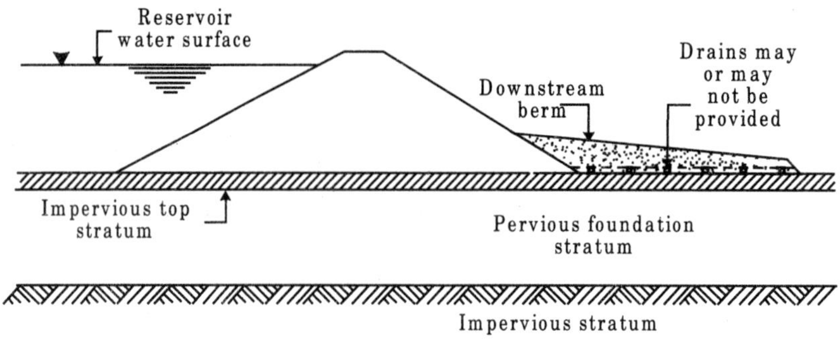

Fig. 11.23 *Earth dam with downstream berm*

For the downstream berms both pervious and impervious soils may be used. However, if the berm consists of impervious soil a drainage layer should be interposed between the berm and the impervious layer. The width of the downstream berm should be sufficient to reduce the underseepage pressure at its toe to a safe allowable value. Usually the maximum width of the downstream berm is limited to about 100 m. Further these berms should have a slope of 1 in 50 or steeper to permit proper drainage.

2. Measures adopted for safe drainage of the seeping water. The various measures that are adopted for safe drainage of water seeping through the dam and through the foundation are as follows.

(*a*) *In dam.* The water seeping through the dam will be drained off if the shells or outer zones on the upstream and downstream sides are highly pervious. However, for homogeneous dams and for zoned dams where the shells are not adequately pervious different types of drainage arrangements may be provided as indicated below.

Drainage of the downstream portion

In the downstream portion of the dam the following different types of drainage arrangements may be provided.

(i) Horizontal drainage blanket.

(ii) Strip drain

(iii) Rock toe

(iv) Chimney drain.

(i) *Horizontal drainage blanket.* As shown in Fig. (11.24 a) a horizontal drainage blanket is provided at the base of the dam which extends from the downstream toe of the dam in the upstream direction. The horizontal drainage blankets are widely used for dams of low to moderate heights. The drainage blanket must be pervious enough to drain off effectively and its design should fulfil the filter criteria as described later. The thickness of a drainage blanket should be sufficient to convey the maximum quantity of seepage estimated to come through the dam. However, it should have a minimum thickness of 1 m. The length of a horizontal drainage blanket usually varies from 25 to 100 percent of the distance from the downstream toe to the centre line of the dam. However, in a zoned dam it would extend only upto the downstream slope of the impervious core.

The main advantage of a horizontal drainage blanket is that in most of the cases it will keep the seepage line farther away from the downstream slope and thereby permit design of a dam section with slightly steeper or more stable downstream slope. From this point of view the drainage blanket should extend farther into the dam, but this may result in somewhat greater quantity of seepage due to shorter seepage path. Another advantage of a horizontal drainage

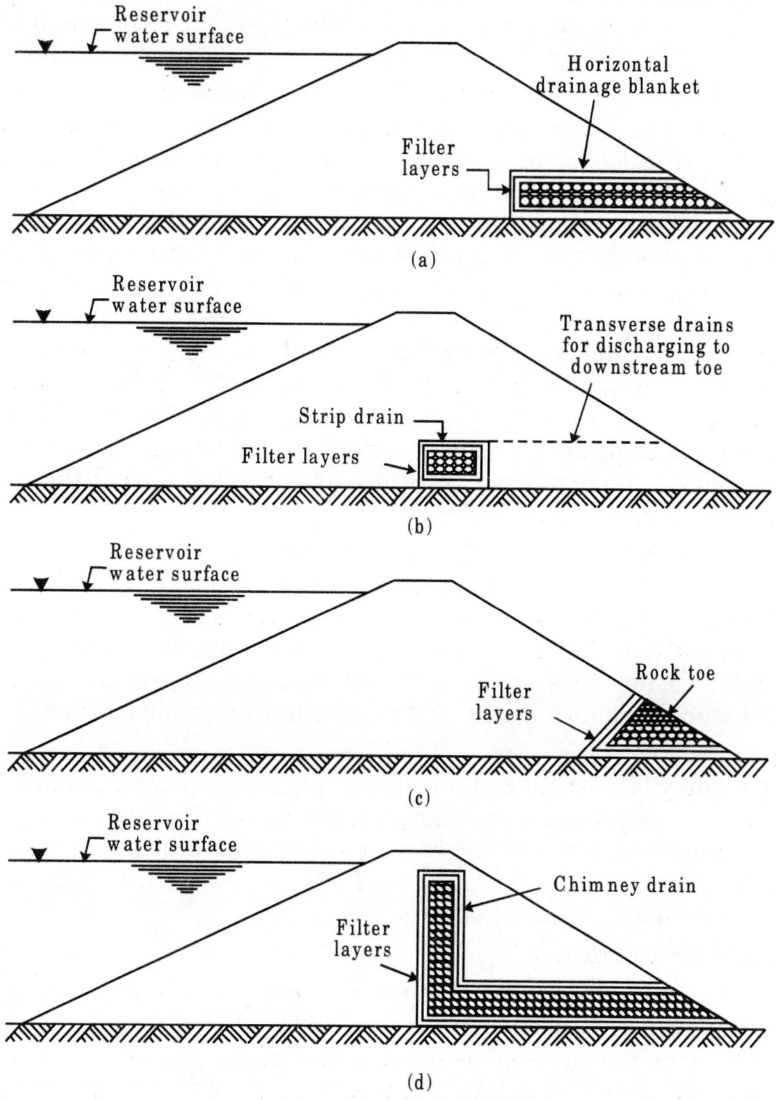

Fig. 11.24 *Drains provided in homogeneous dam (a) horizontal drainage blanket ; (b) strip drain ; (c) rock toe (d) chimney drain*

blanket is that it drains the dam as well as the downstream portion of the foundation. The main disadvantage of a horizontal drainage blanket is that it does not intercept the stratifications in the dam. Thus if there is a relatively pervious layer higher up in the dam either due to careless use of heterogeneous soil mass, or due to negligence in compaction, the water may seep through it and discharge on the face of the downstream slope which may result in surface sloughing.

(ii) *Strip drain.* If there is a paucity of pervious material, instead of a continuous horizontal drainage blanket a strip drain may be provided. As shown in Fig. (11.24 b) a strip drain is a relatively narrow drain provided inside the dam parallel to the axis of the dam. A number of transverse drains are provided for discharging water from the strip drain to another parallel drain provided at the downstream toe of the dam. However, a continuous horizontal drainage blanket is preferred to a strip drain, because in the event of choking of an individual transverse drain a significant length of the dam would become undrained.

(iii) *Rock toe.* As shown in Fig. (11.24 c) a rock toe is provided at the downstream end of an earth dam. The rock toe ccnsists of stones of size usually varying from 150 mm to 200 mm and it may have its upstream face either vertical or inclined. A filter is required to be provided between the rock toe and the soil mass in the dam as well as in the foundation if it is pervious. The rock toe is also suitable for the dams of low to moderate heights. The height of the rock toe is generally kept between 1/3rd to 1/4th of the height of the dam. A rock toe of one third the dam height would require about 10% of the total material required for the whole dam. Besides providing drainage such a rock toe improves the stability of the downstream slope of the dam. The disadvantage of a rock toe is that it does not extend far enough into the dam and the seepage line is usually very close to the downstream slope. However, a rock toe may be combined with a horizontal drainage blanket extending far enough horizontally into the dam and also with a chimney drain extending vertically into the dam so that the advantages of both types of drainage arrangements may be derived.

(iv) *Chimney drain.* As shown in Fig. (11.24 d) a chimney drain extends vertically up into the dam and hence it intercepts all layers of the dam in the seepage zone. Thus, a chimney drain prevents the emergence of seeping water on the downstream slope even if some layers of higher permeability exist in the dam. Besides rendering the downstream portion of the dam free from seeping water and increasing the stability of the downstream slope when the reservoir is full, a chimney drain also helps in reducing pore pressure both during construction and during sudden drawdown.

Chimneys drains may be vertical or they may be slanting upstream to follow the core of the dam or they may be slanting downstream. From the chimney drain the water has to be led out through horizontal drain and hence it is always combined with a horizontal drainage blanket. The chimney drains must also be designed to fulfil the usual filter criteria. However, the chimney drain presents some construction problems.

Drainage of Upstream Portion

If the soil mass in the upstream portion of the dam is not free draining then suitable measures for the drainage of the upstream portion of the dam will also be required to be adopted to relieve this portion of the excessive pore pressure during sudden drawdown and to increase the stability of the upstream slope of the dam. For the drainage of the upstream portion of the dam a horizontal drainage blanket or a chimney drain may be provided. A horizontal drainage blanket provided at the base of the dam, extending from its upstream toe in the downstream direction, is quite effective in reducing the pore pressures during sudden drawdown. Similarly a chimney drain provided in the upstream portion of the dam is also found to reduce drawdown pore pressures effectively. It is however essential to have adequate barrier of soil mass between the upstream and the downstream drainage systems.

Drains for the Control of Construction Pore Pressures

A properly designed drainage system can be used as an effective means of reducing the construction pore pressures. For this purpose the drainage system usually consists of a number of horizontal drainage blankets which are provided in both the upstream and the downstream portions of the dam. These drainage blankets are provided at a vertical interval not exceeding 10 m and extend from the respective slopes into the dam upto such lengths that adequate barrier of soil mass is retained between the drains in the upstream and the downstream portions.

Design of Filters in Earth Dams

As indicated earlier transition filters are required to be provided between the impervious zones of fine grained soils and the pervious zones to prevent the migration of the soil grains from the former to the later due to seeping water. Similarly filters are also required to be provided between the drains of coarse material and the soil mass in the dam as well as in the foundation. The design of these filters is based on the following criteria.

Criteria for the Design of Filters

The two principal requirements for a satisfactory filter are that it must be more pervious than the protected soil in order to act as a drain and that it must be fine enough to prevent particles of the protected soil from washing into its voids. Although several earlier investigators studied this problem, but the first rational approach to filter problems was made by Terzaghi. Later several systematic laboratory investigations were carried out by U.S. Army Corps of Engineers and U.S.B.R. In all of these investigations the specimens of soil of various types protected by filters of varying gradations were subjected to much higher hydraulic gradients than those which normally exist in earth dams. From these results the following rules for the design of filters have been developed which are widely used.

1. The D_{15} size of the filter material (i.e., the particle size which is coarser than the finest 15% of the soil by weight) should be at least 5 times as large as the D_{15} size of the soil to be protected. Further the D_{15} size of the filter material should not be more than 40 times the D_{15} size of the protected soil. That is,

$$\frac{D_{15} \text{ of filter material}}{D_{15} \text{ of protected soil}} = 5 \text{ to } 40$$

2. The D_{15} size of the filter material should not be larger than 5 times the D_{85} size of the protected soil (i.e., the particle size which is coarser than the finest 85% of the soil by weight). That is,

$$\frac{D_{15} \text{ of filter material}}{D_{85} \text{ of protected soil}} = 5 \text{ or less}$$

3. The D_{50} size of the filter material (i.e., the particle size which is coarser than the finest 50% of the soil by weight) should not be larger than 25 times the D_{50} size of the protected soil. That is,

$$\frac{D_{50} \text{ of filter material}}{D_{50} \text{ of protected soil}} = 25 \text{ or less}$$

4. The gradation curve (or grain size distribution curve) of the filter material should have roughly the same shape as the gradation curve of the protected soil.

5. Where the protected soil contains a larger percentage of gravels, the filter should be designed on the basis of the gradation curve of the portion of the material which is finer than the 25 mm sieve size.

6. Filters should not contain more than about 5% of fines passing No. 200 sieve (0.074 mm), and the fines should be cohesionless.

7. In order to minimize segregation and bridging of large particles during placement of filter materials the maximum size of material used in a filter is limited to about 75 mm.

The filter layer designed in accordance with the above criteria would result in one stage of transition from the fine protected material of the dam or foundation to the coarse material of the filter. However, depending on the size of the material to be protected and that of the filter, in order to have more stages of transition from fine to coarse material a multilayered filter having two or three layers of filter may be provided. In a multilayered filter each layer of the filter is designed according to the same criteria as noted above by considering the material of the preceding layer as the protected material and its properties are used to obtain the gradation of the next layer of the filter.

Since horizontal filters are easier to place, the layers of a horizontal filter can be made thinner than those of steeply inclined or vertical filters. Thus, the minimum thickness for the layers of a horizontal filter is about 150 mm for sand and 300 mm for gravel. On the other hand for vertical or inclined filter such as chimney drains or transition filters between the impervious core and the pervious shells, the minimum horizontal width of each layer should be 1.0 to 1.5 m. The filter layers should be compacted to the same density as other noncohesive zones. Every care should be taken in placing the filter materials to avoid segregation and also to prevent contamination from the fines of the core material.

(b) *In foundation.* In general for the safe drainage of water seeping through the foundation, most of the measures adopted for the drainage of water seeping through the dam will serve the foundation also. Besides these the following measures are adopted for the safe drainage of water seeping through the foundation.

(i) Toe drains and drainage trenches.

(ii) Relief wells.

(iii) Vertical sand drains.

(i) *Toe drains and drainage trenches.* Both the toe drains and the drainage trenches are installed along the downstream toes of dams. The toe drains are commonly installed in conjunction with horizontal drainage blankets as shown in Fig. (11.25 a). The purpose of these drains is to collect the seepage from the horizontal drainage blanket and lead it to an outfall pipe which discharges into the spillway stilling basin or into the river channel below the dam. The toe drains may be vitrified clay or perforated asphalt-dipped corrugated metal pipe. The drainpipes are placed in trenches at a sufficient depth below the ground surface to ensure effective interception of the seepage flow. The minimum depth of the trench is about 1.25 m. The drainpipe should be surrounded by a filter to prevent clogging of the drains by inwash of fine material.

Drainage trenches are the open cut trench drains (without drainpipes) backfilled with layers of material in accordance with the filter criteria. The drainage trenches can be used to control the underseepage where a thin impervious top stratum overlies a shallow pervious stratum so that the trench can be built to penetrate the pervious stratum substantially (Fig. 11.25 b). However, a drainage trench will usually be not effective if the underlying pervious stratum is

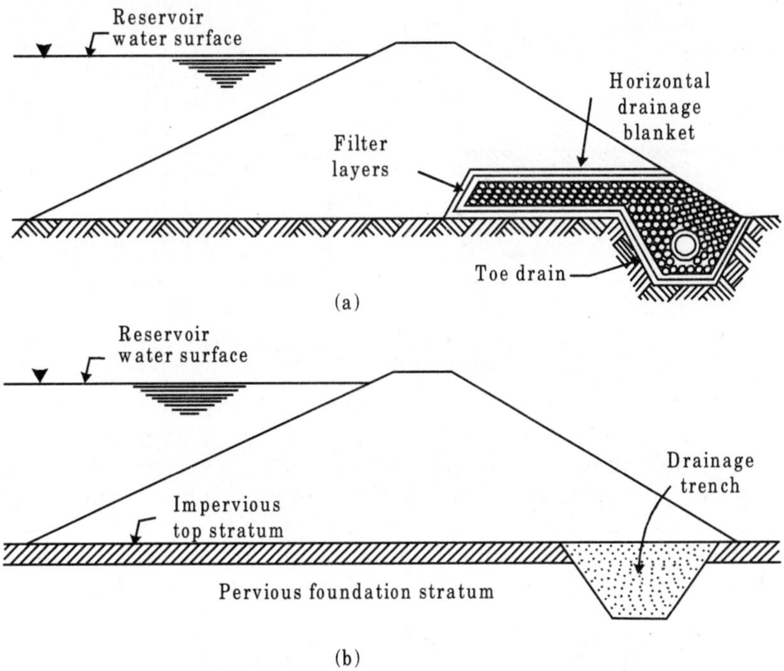

Fig. 11.25 *Earth dam with (a) toe drain ; and (b) drainage trench*

deep and stratified. This is so because a drainage trench of any reasonable depth will attract only a small portion of the underseepage and most of the underseepage will bypass the trench. More effective drainage of deep, stratified, pervious foundations can be accomplished by relief wells.

(*ii*) *Relief wells.* Relief wells are provided at or near the downstream toe of the dam to collect water seeping through the foundation and to reduce the pore pressures in the foundation. They also minimise the risk of piping failure through the foundation. The relief wells are quite effective for deep pervious foundations, particularly where the foundations are stratified and where permeability increases with depth.

As shown in Fig. (11.26) a relief well consists of an interior perforated pipe or well screen with a minimum inside diameter of 150 mm. The well screen is surrounded by a small thickness of gravel graded so as to meet the filter requirements of the surrounding soil. Seepage from the wells is usually discharged on the ground surface at the toe of the dam through a horizontal overflow pipe, from which it is collected in a lined drainage ditch. The relief wells are usually spaced at 15 to 30 m. However, if necessary the spacing may be further reduced.

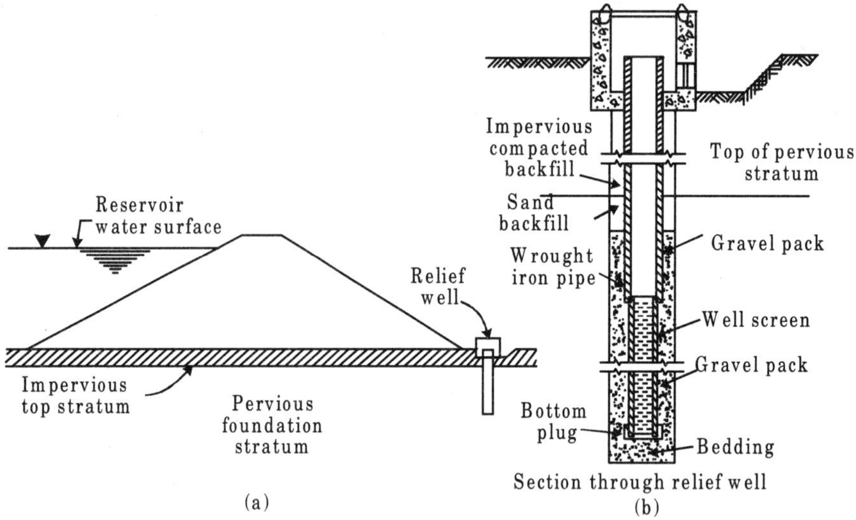

Fig. 11.26 *Relief well*

(iii) *Vertical sand drains.* The vertical sand drains consist of vertical holes drilled in the foundation all along the base of the dam (Fig. 11.27) and filled with clean, coarse sand of high permeability. By providing these drains the path of drainage is considerably reduced which results in keeping the pore pressure low and thereby accelerate the consolidation of the foundation soil. Thus, vertical drains are found to be quite effective for dams on soft clay foundations. In addition to accelerate consolidation of the foundation soil, the sand drains also

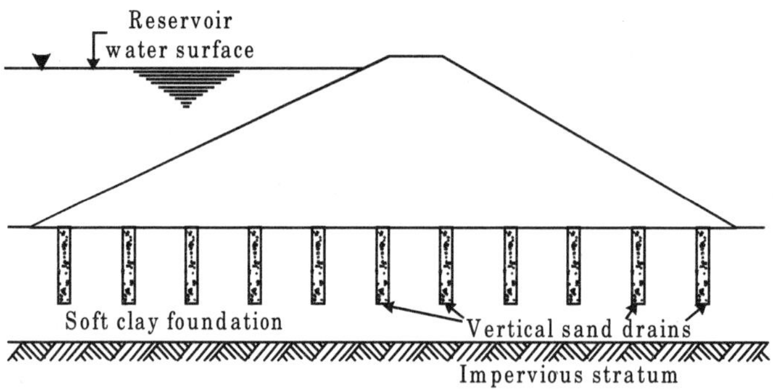

Fig. 11.27 *Vertical sand drains*

act as relief wells and thus help in controlling underseepage. The diameter of the vertical sand drains is usually in the range of 150 to 300 mm. The spacing of these drains however depends on several factors such as characteristics of foundation soil, time for obtaining the required degree of consolidation, etc.

11.20 TYPICAL CROSS-SECTIONS OF EARTH DAMS

The cross-section of an earth dam depends on several factors as mentioned in section 11.4. Out of the various factors the two factors which significantly affect the cross-section of the dam are the foundation conditions and the available materials. In general the following three foundation conditions may exist at the dam sites.

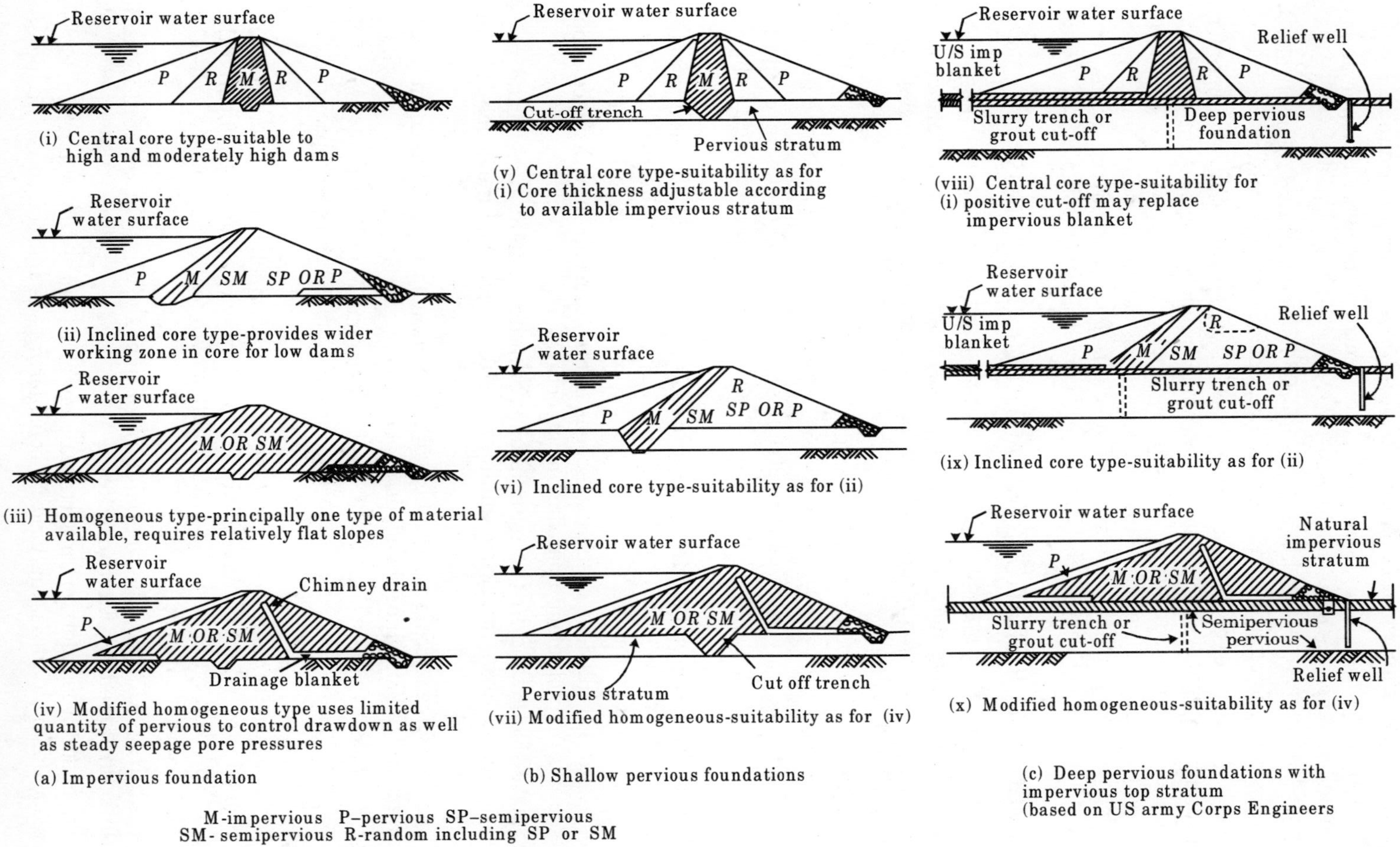

Fig. 11.28 *Typical cross-sections of earth dams for different foundation conditions*

(A) Foundations impervious to a large depth—*impervious foundations.*

(B) Foundations pervious to a moderate depth after which impervious stratum is available—*shallow pervious foundations.*

(C) Foundations pervious to a large depth—*deep pervious foundations.* It also includes the case of deep foundation with a shallow impervious top stratum.

Typical cross-sections of earth dams based on U.S. Army Corps of Engineers for these three foundation conditions are shown in Fig. 11.28. For each foundation condition alternative cross-sections have been given for different kinds of materials. Although the cross-section of an earth dam cannot be standardized and no single type of cross-section is suited for all site conditions, yet the study of a few typical cross-sections of dams for assumed conditions or those of the existing dams will provide very useful guidance for preliminary design.

11.21 DESIGN CONSIDERATIONS IN EARTHQUAKE REGIONS

In general earth dams have notable resistance of earthquake shocks, but the shaking of a dam during a severe earthquake may be quite dangerous. Some of the most serious damages which may be caused to an earth dam due to earthquake and which may even result in the failure of the dam are as indicated below.

1. The shaking of the dam may cause cracks to open through the core of the dam, leakage may develop through these cracks, and in the worst case failure may occur by piping.

2. Shaking of the dam may cause settlement of the crest due to compression of the foundation, or the dam, causing a decrease in the freeboard and leading in the worst case to overtopping.

3. Shaking of the reservoir bottom may cause all the water in the reservoir to oscillate in relatively large, slow waves called seiches which may cause a considerable quantity of water to flow over the top of the dam.

4. Slides of natural hillsides around the edges of the reservoir may displace enough water to overtop the dam or may damage appurtenant structures.

5. Deposits of loose sand in the foundation of the dam might liquefy and move out from underneath the dam causing cracking, sliding or actual horizontal movement of a large part of the dam.

6. Acceleration forces acting on the embankment as it is being shaken may cause a shear slide of an appreciable portion of the slope of the dam.

7. Crustal deformation associated with fault movement may lift the bottom of the reservoir with respect to the dam foundation, reducing the volume of the reservoir and possibly causing over-topping.

8. If a fault passes through the foundation of the dam itself, in addition to all the damages and potential modes of failure envisaged above, the dam may actually be sheared through, with the portions of the dam on either side of the fault moving with respect to each other, creating potentially large concentrated leakage.

Slope Stability Analysis Using Inertial Forces due to Horizontal and Vertical Accelerations Imparted by Earthquake

In order to ensure the safety of an earth dam against failure due to earthquake the stability of the side slopes of the dam is analyzed by considering the inertial forces acting on each slice due to horizontal and vertical accelerations imparted by earthquake. The inertial force may be expressed in terms of the seismic coefficient α, which is given by

$$\alpha = \beta I\, F_o \left(\frac{S_a}{g}\right) \quad \text{...(11.64)}$$

where

β = Soil-foundation system factor, the value of which for dams is taken as 1.0

I = Importance factor, the value of which for dams is taken as 2.0

F_o = Seismic zone factor for average acceleration spectra, the value of which for each of the five seismic zones is given in Table 9.1 (Chapter 9); and

$\left(\frac{S_a}{g}\right)$ = average acceleration coefficient as read from Fig. 9.7 (Chapter 9) for appropriate natural period of vibration and damping of the structure.

The natural (or fundamental) period of vibration of an earth dam may be determined by the following expression

$$T = 2.9\, H \sqrt{\rho/G} \quad \text{...(1.65)}$$

in which

T = Natural period of vibration of the dam in s

H = Height of the dam in m

ρ = Mass density of the shell material in kg/m^3, and

G = Modulus of rigidity of the shell material in N/m^2.

Note: The quantity is known as shear wave velocity through the material of the dam and if instead of ρ and G, shear wave velocity is known it may be used in Eq. 11.65 to calculate T.

For the value of T given by Eq. 11.65 and assuming 10% damping (as indicated in IS: 1893–1984) the value of (S_a/g) may be obtained from average acceleration spectrum curves given in Fig. 9.7 (Chapter 9). By introducing this value of (S_a/g) in Eq. 11.64) the value of seismic coefficient a may be obtained. For each slice the value of seismic coefficient α_y is given by the following expression

$$\alpha_y = \left(2.5 - 1.5\frac{y}{H}\right)\alpha \quad \text{...(11.66)}$$

in which

y = the depth of the centroid of the slice (or the depth of the mid-point of the base of the slice) below the crest of the dam, and

H and α are same as defined earlier.

The inertial forces due to earthquake are considered to be acting on each slice both in horizontal as well as in vertical directions. Thus, if W is the weight of the slice then the horizontal inertial force is taken as α_y W and the vertical inertial force is taken as 0.5 α_y W, (i.e., half of the horizontal inertial force). The horizontal inertial force is considered to be acting in the upstream direction for the upstream slope and in the downstream direction for the downstream slope. The vertical inertial force is, however considered to be acting upwards for both upstream and downstream slopes. The inertial forces will actually be acting at the centroid of the slice, but in order to simplify the computations it is shifted to the base of the slice like other forces as shown in Fig. (11.29).

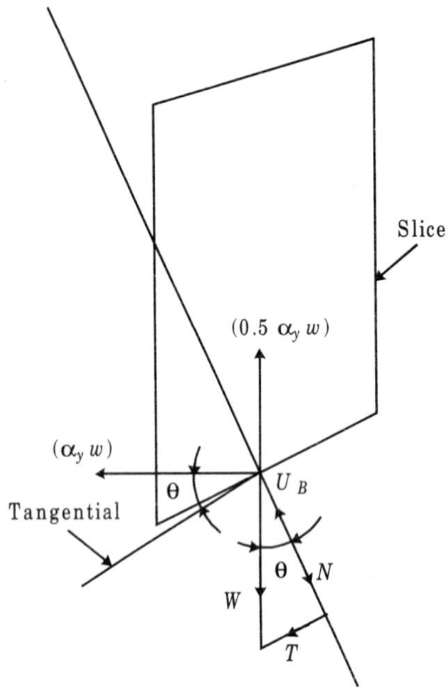

Fig. 11.29 *Inertial forces due to earthquake acting on a slice*

The horizontal inertial force can be resolved along the tangential and normal (or radial) directions as follows:

Component along the tangential direction

$$= \alpha_y W \cos\theta = \alpha_y N$$

since $N = W \cos\theta$ from Eq. (11.19)

Component along the normal direction

$$= \alpha_y W \sin\theta = \alpha_y T$$

since $T = W \sin\theta$ from Eq. (11.20).

Thus, a force equal to $\alpha_y N$ is added to the tangential component T of the weight W of the slice; and a force equal to $\alpha_y T$ is subtracted from the normal component N of the weight W of the slice.

The vertical inertial force can also be resolved along the tangential and normal (or radial) directions as follows

Component along the tangential direction

$$= 0.5\ \alpha_y W \sin\theta = 0.5\ \alpha_y T$$

Component along the normal direction

$$= 0.5\ \alpha_y W \cos\theta = 0.5\ \alpha_y N$$

Thus, a force equal to 0.5 $\alpha_y T$ is subtracted from the tangential component T of the weight W of the slice; and a force equal to 0.5 $\alpha_y N$ is subtracted from the normal component N of the weight W of the slice.

The factor of safety for this case is then given by the following expression

$$F_s = \frac{\Sigma\{(N - U_B - \alpha_y T - 0.5\alpha_y N)\tan\phi + cs\}}{\Sigma(T + \alpha_y N - 0.5\alpha_y T)} \quad ...(11.67)$$

in which the other terms are same as defined in section 11.12.

If only horizontal inertial forces are considered then the factor of safety is given by the following expression

$$F_s = \frac{\Sigma\{(N - U_B - \alpha_y T)\tan\phi + cs\}}{\Sigma(T + \alpha_y N)} \quad ...(11.68)$$

With the consideration of the earthquake forces a factor of safety of unity or slightly more is acceptable.

The earthquake forces are considered only when the stability of upstream and downstream slopes of an earth dam is tested for reservoir full condition. For testing the stability of upstream and downstream slopes of an earth dam for the construction stage or for the reservoir empty condition, the earthquake forces are normally not considered. However, where the construction or operating schedule requires the reservoir empty condition to exist for prolonged periods, earthquake forces may be considered for testing the stability of the upstream and downstream slopes and the same may be taken as 50 per cent of the value considered for the reservoir full condition.

Note: The method described here for considering the earthquake forces for testing the stability of the upstream and downstream slopes of an earth dam is as per IS : 1893–1984.

Apart from the consideration of the earthquake forces as indicated above several special design details as described below may be provided to increase the safety of the dam against the failure due to earthquake.

1. Cohesionless Graded Filters Downstream. If a well graded filter is provided just downstream from the impervious core, it will protect the dam against excessive leakage and progressive piping through the cracks which may be developed in the dam due to earthquake. A crack developed through the impervious core of a dam due to earthquake will not extend into the downstream filter, because a crack will not remain open in a completely cohesionless material. Thus, if the gradation of the filter is so designed that the eroded particles are prevented from entering the filter and the filter is intact, the particles will be stopped, the crack will be sealed, and no piping can develop.

2. Cohesionless Graded Filters Upstream. When the core of the dam is constructed from a very tough clay, the cracks developed in the core due to earthquake will not get sealed of on their own. In such cases a well graded filter with appreciable content of fine sand sizes provided just upstream from the core is found to be quite useful. The fine sand from the filter will tend to wash into the cracks of the core and partially seal them.

3. Very Pervious Downstream Zones. Provision should be made for quick discharge of water flowing through the cracks which may be developed in the core of the dam due to earthquake, so that critically high water pressures may not develop in the downstream portion of the dam. For this the downstream portion of the dam should be highly pervious. If rock is available, the downstream portion may be of large quarried rock which provides the best combination of permeability and stability. Moreover a downstream portion of large rock would provide the greatest security against erosive action of large quantities of water passing over the crest of the dam if it is overtopped.

4. Extra Freeboard. A serious damage due to earth-quake that may be anticipated is overtopping of the dam due to (i) the settlement of the crest of the dam, or (ii) rise in the water level due to landslides into the reservoir basin, or (iii) the upward movement of the reservoir bottom caused by fault movement. In order to ensure safety of the dam against this kind of failure extra freeboard should be provided.

5. Flatter Slopes Near Top of Dam. During an earthquake shock the top of the dam vibrates with considerably greater amplitude than the base and may be subjected to the worst damage. It is therefore desirable to make the top of the dam thicker by either increasing the crest width or by using flatter slopes at the top than would otherwise be required. Moreover with a thicker top longer is the path of seepage through the cracks which may develop and safer will be the dam against the damage due to earthquake.

6. Core Treatment. For the impervious core the soil most resistant to piping should be used. Further a thick core must be provided because it will offer greater resistance to piping and also it will increase the safety of the dam when the earthquake tends to twist the dam alignment. However, concrete or masonry core walls should be avoided in regions of earthquake activity.

7. Foundation Treatment. All the loose or soft soil which might otherwise be left in place must be excavated because a dam on a soft foundation will be more severely shaken during an earthquake than a dam on a hard foundation.

11.22 ROCKFILL DAMS

A rockfill dam is a type of embankment dam which uses variable sizes of rock to provide stability and an impervious membrane to provide watertightness. Many different materials have been used for the membrane which include earth, concrete, steel, asphaltic-concrete and wood. The membrane may be placed either within the rockfill or on the upstream slope. Depending on the location of the membrane the rockfill dams may be classified into two groups as follows: (1) dams with internal membrane, and (2) dams with external or upstream membrane.

The internal membrane is usually constructed of impervious earth materials and it may of two types, viz., central vertical core, and inclined or sloping core. Out of these two types of internal membranes, the central vertical core is preferred because it provides maximum contact pressure with the foundation and requires less strict control during construction than an inclined core. The advantages of the internal membrane include (i) less total area exposed to water, (ii) protection from the effects of weathering and external damage, and (iii) shorter grout curtain lengths. However, the disadvantages of using an internal membrane are (a) inability to construct the membrane after the completion of the rockfill section, (b) inaccessibility of the membrane for damage inspection and inability to rectify the damage if any caused to the membrane, and (c) a smaller section of the dam being available for stability against sliding. Moreover, in an earth core rockfill dam due to considerable gradation difference between impervious core and rockfill, multiple filter layers are required to be provided on both upstream and downstream sides of the core to prevent piping, which would, however, increase the construction cost.

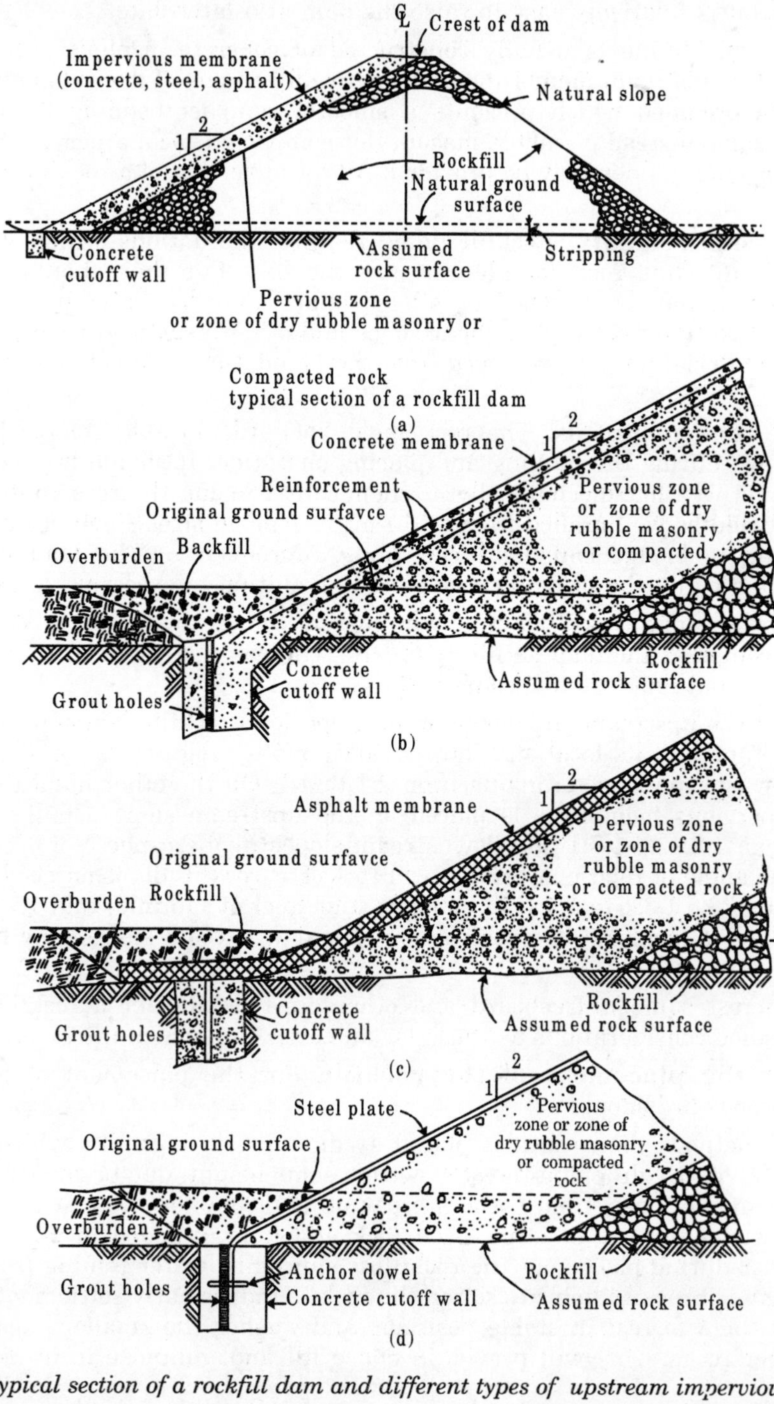

Fig. 11.30 *Typical section of a rockfill dam and different types of upstream impervious membranes*

The upstream membrane has the following advantages: (i) readily available for inspection and repair, (ii) membrane can be constructed after the completion of the rockfill section, (iii) foundation grouting can be done simultaneously with rockfill placement, (iv) a larger portion of the dam is available for stability against sliding, (v) membrane can be used for slope protection, and (vi) it is relatively easy to raise the dam at a later date.

The upstream membrane is usually constructed of concrete, asphaltic-concrete, steel, or wood. In between the upstream membrane and the rockfill a zone of dry rubble masonry or well compacted rock is provided which presents a smooth, compact bedding for the membrane. However, for low dams instead of rubble masonry or compacted rock, a pervious zone of graded sand and gravel or quarry fines may be provided. A typical section of a rockfill dam is shown in Fig. (11.30 a).

Foundation requirements. Rockfill dams require foundations which will result in a minimum settlement. Thus rock foundations are the best but they should consist of hard durable rock which cannot be softened or eroded appreciably by water percolating from the reservoir. It should be free from faults, shear zones and other structural weaknesses. Silt, clay, sand and organic material must be removed from the foundation area before construction of the rockfill dam.

Selection of rock materials. The rock must be hard, durable stone which will resist excessive breakdown during the hauling and placing operation. It should be free from unstable minerals that would weather mechanically or chemically causing the rock to disintegrate. The rock to be used should be well graded from 0.014 m^3 to 1 m^3 in size. Slabby rocks should not be used as they tend to bridge causing large voids. Moreover, as the load is increased by construction of the rockfill, the bridging rocks break, resulting in excessive settlement.

Dam axis. The axis of a rockfill dam may be curved (convex upstream) or straight. A curved axis will tend to compress the dam as filling occurs, whereas a straight axis has the benefit of easy construction layout and less total dam cost.

Dam section. The upstream and downstream slopes of a rockfill dam depend on the type of impervious membrane and its location. The rockfill dams having central or sloping cores have upstream and downstream slopes ranging from 2:1 to 4:1. On the other hand the rockfill dams in which the impervious membrane is placed on the upstream slope usually have upstream slopes ranging from 1.3:1 to 1.7:1 and downstream slopes in the range of 1.3:1 to 1.4:1, which corresponds to the angle of repose of the natural rock. For rockfill dams on good foundation the upstream slopes may be 1.3:1 to 1.4:1 for concrete and steel membranes and 1.7:1 for asphaltic-concrete membranes to facilitate its construction. The downstream slopes may be 1.3:1 to 1.4:1 in all the cases.

Crest width, crest camber, freeboard and other crest details for a rockfill dam are also governed by the same considerations as those for an earth dam which are given in section 11.4.

Methods for the placement of the rockfill. For the placement of the rockfill the following two methods are adopted.

In one of the methods the rockfill is placed by dumping in high lifts which in some cases varied from 20 to 50 m. By dropping the stones from some height, due to impact, the large sized stones break into smaller ones, their corners break off and they tend to rest in stable positions. Thus high lift permits the use of large sized rock and it is intended to minimise the future settlement. Further during placement the rockfill is sluiced by water issuing from high velocity nozzles. The sluicing of rockfill helps to soften the rock to induce early settlement, lubricates the stones and helps them to rest in stable positions and washes the smaller ones into the voids between the larger rocks. This will provide a dense fill and minimise future settlement. The sluicing of rockfill is done by using volume of water equal to two to three times the volume of rock.

In another method well graded rock is placed in relatively thin layers about 1 to 2 m thick and each layer is compacted either with the hauling equipment or with 50 t (or 500 kN) rubber tyre rollers or vibratory steel drum rollers. For compacted rockfill the thickness of each layer is usually 2 m or less. Further for compacted rockfill sluicing may or may not be done. However, it is desirable to sluice each layer if rocks show appreciable loss of strength on wetting, for high dams, and for rockfill material containing higher percentages of fines. For compacted rockfill only a small amount of water is used to saturate the rock pieces and to rearrange the fines to some extent.

Settlement of rockfill dams. The major problem with rockfill dams is that of settlement. In earth core rockfill dams large differential settlement may lead to cracking of the core. Similarly in the case of rockfill dams with upstream membrane there is likelihood of cracking of the membrane due to settlement. The cracking of earth core as well as upstream membrane may result in heavy leakage of water through the dam. As such every effort should be made to reduce the settlement of rockfill dams.

The settlement of a rockfill dam takes place in two stages. The first major settlement occurs during the construction due to the self weight of the rockfill. By proper use of sluicing water this initial settlement may be large and it can be expedited. Further the rapid rise in height of dam should be avoided so that the adjustment of the rockfill may keep pace with loading. The second major settlement occurs after the construction of the dam as the reservoir fills and the thrust due to water load is transmitted to the rockfill.

In order to account for the post-construction settlement of a rockfill dam it is usually constructed to a some what greater height than the designed height so that there is no reduction in the designed freeboard above the maximum water level in the reservoir. This extra height of the dam is provided in the form of a longitudinal camber over the designed crest level, varying from zero at the abutments to a maximum value at the centre of the valley where the dam will be the highest and the settlement of the dam will be the most. For a rockfill dam the maximum value of the post-construction settlement may be assumed as 1 to 2 % of the height of the dam.

Upstream membrane. The upstream membrane is usually placed after the initial settlement has taken place. Between the main rockfill of the dam and the upstream membrane a zone of dry rubble masonry or well compacted rock with a reasonably even outer surface is provided to support the membrane and transmit the load to the main rockfill. This zone may be 1.5 to 3 m thick at the top and increasing at 5% of the height towards the bottom. In some cases this zone with a constant thickness of 4.5 to 8.5 m has also been provided. However, for low dams instead of dry rubble masonry or well compacted rock, a pervious zone of graded sand and gravel or quarry fines may be provided. The pervious zone should have a minimum horizontal width of 4.3 m to facilitate compaction. On the upstream face of this zone the impervious membrane is provided. Fig. 11.30 shows three types of impervious membranes used for the rockfill dams.

Cutoff walls. As shown in Fig. 11.30 a cutoff wall is provided along the contact of the impervious membrane with the foundation and the abutments at the upstream toe of the dam to prevent seepage under the dam. The cutoff wall extends from the upstream toe of the dam into the bedrock. In most of the cases concrete cutoff wall is provided. The width of the cutoff wall is usually governed by construction considerations. The depth of penetration of the cutoff wall into the bedrock depends on the character of the foundation rock. However, even if the rock is sound the minimum depth upto which the cutoff wall should extend into the foundation rock is 1 m. A deeper cutoff wall or special treatment such as grouting is required if the rock is not sound or if open joints or broken rock structure exists. In addition to its function of preventing underseepage, the cutoff wall must provide adequate support for the impervious membrane.

SOLVED ILLUSTRATIVE EXAMPLES

Illustrative Example 11.1 *The upstream slope of a zoned type earth dam is shown in the accompanying figure. The properties of the shell, foundation and core material are as follows :*

(*a*) *Saturated unit weight*

(*i*) *Shell material*	*21.974 kN/m³ [2240 kg(f)/m³]*
(*ii*) *Foundation material*	*21.974 kN/m³ [2240 kg(f)/m³]*
(*iii*) *Core material*	*21.190 kN/m³ [2160 kg(f)/m³]*

(*b*) *Submerged unit weight*

(*i*) *Shell material*	*12.164 kN/m³ [1240 kg(f)/m³]*
(*ii*) *Foundation material*	*12.164 kN/m³ [1240 kg(f)/m³]*
(*iii*) *Core material*	*11.380 kN/m³ [1160 kg(f)/m³]*

(*c*) *Moist unit weight*

(*i*) *Shell material*	*20.405 kN/m³ [2080 kg(f)/m³]*
(*ii*) *Foundation material*	*20.405 kN/m³ [2080 kg(f)/m³]*
(*iii*) *Core material*	*19.620 kN/m³ [2000 kg(f)/m³]*

(*d*) *Angle of internal friction* ϕ

(*i*) *Shell material*	*30°*
(*ii*) *Foundation material*	*30°*
(*iii*) *Core material*	*20°*

(*e*) *Cohesion c*

(*i*) *Shell material*	*7.848 kN/m² [800 kg(f)/m²]*
(*ii*) *Foundation material*	*7.848 kN/m² [800 kg(f)/m²]*
(*iii*) *Core material*	*39.240 kN/m² [4000 kg(f)/m²]*

Check the stability of the upstream side slope of the dam against sudden drawdown along a slip circle as shown in Fig. Ex. 11.1.

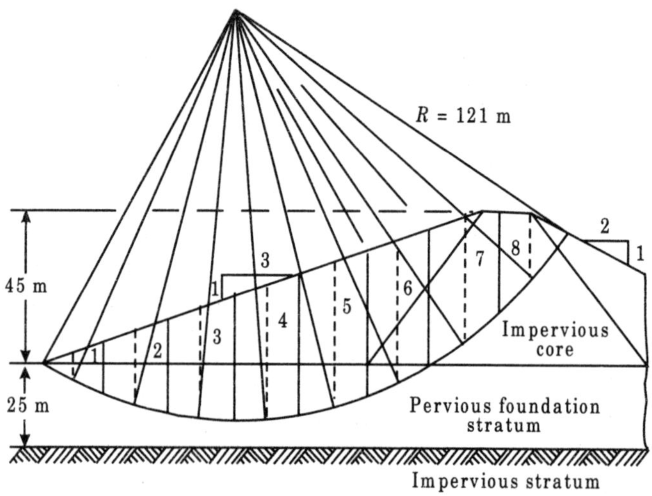

Fig. Ex. 11.1

Table For Example 11.1 (Metric Units)

Slice No.	Area (m^2)		Weight [kg(f)]		θ (Degrees)	cos θ	sin θ	$(N - U_B)$ [kg(f)] Normal Component of Submerged Weight	T [kg(f)] Tangential Component of Saturated Weight	$(N - U_B)$ × tan ϕ [kg(f)]	cs [kg(f)]
	Shell and Foundation	Core	Saturated	Submerged							
1	160	–	358 400	198 400	(–)24.50	0.9100	(–)0.4147	180 544	(–)148 629	104 237	17 308
2	435	–	974 400	539 400	(–)14.50	0.9681	(–)0.2504	522 193	(–)243 990	301 488	16 886
3	635	–	1422 400	787 400	(–)4.75	0.9965	(–)0.0828	784 644	(–)117 775	453 014	16 422
4	770	–	1 724 800	954 800	4.50	0.9969	0.0785	951 840	135 397	549 545	15 662
5	835	–	1 870 400	1 035 400	9.00	0.9877	0.1564	1 022 665	292 531	590 436	16 464
6	551	233	1 234 240 + 503 280	683 240 + 270 280	23.75	0.9153	0.4027	872 757	699 699	503 866	16 886
7	124	629	277 760 + 1 358 640	153 760 + 729 640	35.25	0.8166	0.5771	721 384	944 366	262 562	103 428
8	36	335	80 640 + 723 600	44 640 + 388 600	48.75	0.6593	0.7519	285 635	604 708	103 963	130 868

Tavle For Example 11.1 (SI Units)

Slice No.	Area (m^2)		Weight (kN)		θ (Degrees)	$\cos\theta$	$\sin\theta$	$(N - U_B)$ (kN) Normal Component of Submerged Weight	T (kN) Tangential Component of Saturated Weight	$(N - U_B) \times \tan\phi$ (kN)	cs (kN)
	Shell and Foundation	Core	Saturated	Submerged							
1	160	–	3 515.84	1946.24	(–)24.50	0.9100	(–)0.4147	1 771.08	(–)1 458.02	1 022.53	169.79
2	435	–	9 558.69	5 291.34	(–)14.50	0.9681	(–)0.2504	5 122.55	(–)2 393.50	2 957.50	165.65
3	635	–	13 953.49	7 724.14	(–)4.75	0.9965	(–)0.0828	7 697.11	(–)1 155.35	4 443.93	161.10
4	770	–	16 919.98	9 366.28	4.50	0.9969	0.0785	9 337.24	1 328.22	5 390.86	153.64
5	835	–	18 348.29	10 156.94	9.00	0.9877	0.1564	10 032.01	2 869.67	5 791.98	161.51
6	551	233	12 107.67 + 4 937.27	6 702.36 + 2 651.54	23.75	0.9153	0.4027	8 561.62	6 864.00	4943.05	165.65
7	124	629	2 724.78 + 13 328.51	1 508.34 + 7 158.02	35.25	0.8166	0.5771	7076.95	9 264.35	2 575.80	1 014.63
8	36	335	791.06 + 7 098.65	437.90 + 3 812.30	48.75	0.6593	0.7518	2 802.16	5 931.48	1 019.90	1 283.82

Solution

The soil mass lying above the slip circle is divided into eight slices. The slices have been so taken that except for the upper most slice each slice has its base in one type of material only. To avoid complication the entire soil mass is assumed to be saturated though actually a small portion above the seepage line will not be saturated, this will involve a very slight error on the safer side. The computations for the stability analysis are shown in the accompanying table. For any slice f and c are taken for that material through which the arc of the slip circle is passing.

The factor of safety is given by Eq. (11.18) as

$$F_s = \frac{\Sigma\{(N - U_B)\tan\phi + cs\}}{\Sigma T}$$

SI Units

$$\Sigma\{(N - U_B)\tan\phi = 28\,145.55$$

$$\Sigma cs = 3\,275.79$$

$$\Sigma T = 21\,250.85$$

$$\therefore \quad F_s = \frac{28145.55 + 3275.79}{21250.85}$$

$$= 1.479$$

Since $F_s < 1.5$, this section is not safe.

Metric Units

$$\Sigma\{(N - U_B)\tan\phi + = 2\,869\,111$$

$$\Sigma cs = 333\,924$$

$$\Sigma T = 2\,166\,307$$

$$\therefore \quad F_s = \frac{2\,869\,111 + 333\,924}{2\,166\,307}$$

$$= 1.479$$

Since $F_s < 1.5$, this section is not safe.

Illustrative Example 11.2 *A homogeneous earth dam 20 m high has crest width 6 m, upstream slope 4 : 1 and downstream slope 3 : 1. The properties of the soil mass in the dam are : saturated unit weight = 2160 kg(f)/m³ (21.19 kN/m³) ; submerged unit weight = 1160 kg(f)/m³ (11.38 kN/m³) ; angle of internal friction f = 26 ; and cohesion = 4000 kg(f)/m² (39.24 kN/m²). Check the stability of the upstream portion of the dam against horizontal shear at the base of the dam. At the top shoulder of the upstream slope of the dam the seepage line may be assumed to be 5 m below the crest of the dam.*

Solution

The total horizontal shear force H_u acting on the upstream portion per unit length of the dam is given by Eq. (11.32) as

$$H_u = \frac{w_s H^2}{2}\tan^2\left(45° - \frac{\phi}{2}\right) + \frac{w_w h_1^2}{2}$$

$$H = 20 \text{ m}$$
$$h_1 = (20 - 5) = 15 \text{ m}$$
$$w_s = 2160 \text{ kg(f)/m}^3$$
$$w_w = 1000 \text{ kg(f)/m}^3$$
$$\phi = 26°$$

$$\therefore \quad H_u = \frac{1}{2}\times 2160\times(20)^2\tan^2\left(45° - \frac{26°}{2}\right) + \frac{1}{2}\times 1000\times(15)^2$$
$$= 2.812 \times 10^5 \text{ kg(f)}$$

The resisting force R_u against horizontal shear force is given by Eq. (11.36) as

$$R_u = W_{eu} \tan \phi + cb_u$$
$$b_u = 4 \times 20 = 80 \text{ m}$$
$$W_{eu} = 1160\times\left(\frac{1}{2}\times 20\times 80\right)$$
$$= 9.28 \times 10^5 \text{ kg(f)}$$
$$\therefore \quad R_u = 9.28 \times 10^5 \times \tan 26° + 4000 \times 80$$
$$= 7.726 \times 10^5 \text{ kg(f)}$$

The factor of safety F_u against shear in the upstream portion of the dam at the base of the dam is given by Eq. (11.37) as

$$F_u = \frac{R_u}{H_u}$$
$$= \frac{7.726\times 10^5}{2.812\times 10^5} = 2.748$$

which is satisfactory

Average shear stress s_u is given by Eq. (11.34) as

$$s_u = \frac{H_u}{b_u} = \frac{2.812\times 10^5}{80}$$
$$= 3.515 \times 10^3 \text{ kg(f)/m}^2$$

The maximum shear stress s_{mu} is given by Eq. (11.35) as

$$s_{mu} = 2\, s_u = 7.03 \times 10^3 \text{ kg(f)/m}^2$$

Unit shear strength at the point where maximum shear stress is developed on the base of the upstream portion of the dam is given by

$$s = w_1 h \tan \phi + c$$
$$w_1 = 1160 \text{ kg(f)/m}^3$$
$$\phi = 26°$$
$$h = 0.6 \times 20 = 12 \text{ m}$$

$$c = 4000 \text{ kg(f)/m}^2$$

$$\therefore \quad s = 1160 \times 12 \times \tan 26° + 4000$$

$$= 10.789 \times 10^3 \text{ kg(f)/m}^2$$

The factor of safety F_{um} at the point of maximum shear stress is given by Eq. (11.38) as

$$F_{um} = \frac{s}{s_{mu}}$$

$$= \frac{10.789 \times 10^3}{7.03 \times 10^3} = 1.535$$

which is satisfactory.

Illustrative Example 11.3 *Check the stability of the downstream portion of the dam of Illustrative Example 11.2 against horizontal shear at the base of the dam. At the top shoulder of the downstream slope of the dam the seepage line may be assumed to be 6 m below the crest of the dam. The area of the downstream portion of the dam below the seepage line may be taken as 550 m². The moist unit weight of the soil = 2000 kg(f)/m³ (19.62 kN/m³). The entire section of the dam at the point of maximum shear stress may be assumed to be lying below the seepage line.*

Solution

The total horizontal shear force H_d acting on the downstream portion per unit length of the dam is given by Eq. (11.25) as

$$H_d = \frac{wH^2}{2}\tan^2\left(45° - \frac{\phi}{2}\right) + \frac{w_w h_1^2}{2}$$

$$H = 20 \text{ m}$$

$$h_1 = (20 - 6) = 14 \text{ m}$$

$$\phi = 26° \; ; \; w_w = 1000 \text{ kg(f)/m}^3$$

From Eq. (11.26)

$$w = \frac{w_1 h_1 + w_2(H - h_1)}{H}$$

$$w = 1160 \text{ kg(f)/m}^3$$

$$w_2 = 2000 \text{ kg(f)/m}^3$$

$$\therefore \quad w = \frac{1160 \times 14 + 2000 \times 6}{20}$$

$$= 1412 \text{ kg(f)/m}^3.$$

Thus,

$$H_d = \frac{1}{2} \times 1412 \times (20)^2 \times \tan^2\left(45° - \frac{26°}{2}\right) + \frac{1}{2} \times 1000 \times (14)^2$$

$$= 2.083 \times 10^5 \text{ kg(f)}$$

The resisting force R_d against the horizontal shear force is given by Eq. (11.29) as

$$R_d = W_{ed} \tan \phi + c b_d$$

$$b_d = 3 \times 20 = 60 \text{ m}$$

Total area of the downstream portion of the dam

$$= \frac{1}{2} \times (20 \times 60) = 600 \text{ m}^2$$

The area below the seepage line = 550 m^2 and hence the area above the seepage line = (600 – 550) = 50 m^2.

$\therefore$ $$W_{ed} = 550 \times 1160 + 50 \times 2000$$
$$= 7.38 \times 10^5 \text{ kg(f)}$$

Thus, $$R_d = 7.38 \times 10^5 \tan 26° + 4000 \times 60$$
$$= 5.999 \times 10^5 \text{ kg(f)}$$

The factor of safety F_d against shear in the downstream portion of the dam at the base of the dam is given by Eq. (11.30) as

$$F_d = \frac{R_d}{H_d}$$

$$= \frac{5.999 \times 10^5}{2.083 \times 10^5} = 2.88$$

which is satisfactory.

Average shear stress s_d is given by Eq. (11.27) as

$$s_d = \frac{H_d}{b_d} = \frac{2.083 \times 10^5}{60}$$

$$= 3.472 \times 10^3 \text{ kg(f)/m}^2$$

The maximum shear stress is given by Eq. (11.28) as

$$s_{md} = 2\, s_d = 6.944 \times 103 \text{ kg(f)/m}^2$$

Unit shear strength s at the point where maximum shear stress is developed on the base of the upstream portion of the dam is given by

$$s = wh \tan \phi + c$$
$$w = 1160 \text{ kg/m}^2$$
$$h = 0.6 \times 20 = 12 \text{ m}$$

$\therefore$ $$s = 1160 \times 12 \times \tan 26° + 4000$$
$$= 10.789 \times 10^3 \text{ kg(f)/m}^2$$

The factor of safety F_{dm} at the point of maximum shear stress is given by Eq. (11.31) as

$$F_{dm} = \frac{s}{s_{md}} = \frac{10.789 \times 10^3}{6.944 \times 10^3} = 1.554$$

which is satisfactory.

Illustrative Example 11.4 *A homogeneous earth dam 30 m high has crest width 6 m upstream slope 4 : 1 and downstream slope 3 : 1. The foundation consists of pervious material upto a depth of 20 m. The properties of the soil mass in the dam and the foundation are: saturated unit weight = 2160 kg(f)/m³ (21.19 kN/m³) ; submerged unit weight = 1160 kg(f)/m³ (11.38 kN/m³) ; angle of internal friction = 24° ; and cohesion c = 1200 kg(f)/m² (11.17 kN/m²). Check the stability of the foundation against shear for the upstream portion of the dam.*

Solution

The total horizontal shear force S acting on the rigid boundary of the foundation per unit length of the dam is given by Eq. (11.39) as

$$S = w\left(\frac{h_1^2 - h_2^2}{2}\right)\tan^2\left(45° - \frac{\phi_1}{2}\right)$$

$$w = 2160 \text{ kg(f)/m}^3$$

$$h_1 = 50 \text{ m} ; h_2 = 20 \text{ m}$$

From Eq. (11.41)

$$\tan \phi_1 = \frac{wh_1 \tan\phi + c}{wh_1}$$

$$= \frac{2160 \times 50 \times \tan 24° + 1200}{2160 \times 50} = 0.4563$$

$\therefore$

$$\phi_1 = 24° \; 32' = 24.53°$$

Thus,

$$S = \frac{1}{2} \times 2160(50^2 - 20^2)\tan^2\left(45° - \frac{24.53°}{2}\right)$$

$$= 9.373 \times 10^5 \text{ kg(f)}$$

The average shear stress s_{ave} on the rigid boundary of the foundation is given by Eq. (11.42) as

$$s_{ave} = \frac{S}{b}$$

$$b = 4 \times 30 = 120 \text{ m}$$

$\therefore$

$$s_{ave} = \frac{9.373 \times 10^5}{120}$$

$$= 7.811 \times 10^3 \text{ kg(f)/m}^2$$

The maximum shear stress *smax* is given by Eq. (11.43) as

$$s_{max} = 1.4 \, s_{ave}$$

$$= 1.4 \times 7.811 \times 103 = 10.935 \times 103 \text{ kg/m}^2$$

The unit shear strength s_1 below the toe of the dam is given by Eq. (11.44) as

$$s_1 = c + w_F' \, h_2 \tan \phi$$

$$w_F' = 1160 \text{ kg(f)/m}^3$$

$$\therefore \quad s_1 = 1200 + 1160 \times 20 \times \tan 24°$$

$$= 11.529 \times 10^3 \text{ kg(f)/m}^2$$

The unit shear strength s_2 below the top shoulder of the slope of the dam is given by

$$s_2 = c + w'h_1 \tan \phi$$

$$w' = 1160 \text{ kg(f)/m}^3$$

$$\therefore \quad s_2 = 1200 + 1160 \times 50 \times \tan 24°$$

$$= 27.023 \times 10^3 \text{ kg(f)/m}^2$$

The average unit shear strength s of the foundation is given by Eq. (11.47) as

$$s = \frac{s_1 + s_2}{2} = \frac{1}{2}(11.529 \times 10^3 + 27.023 \times 10^3)$$

$$= 19.276 \times 10^3 \text{ kg(f)/m}^2$$

The overall average factor of safety F against shear in the foundation is given by Eq. (11.48) as

$$F = \frac{s}{s_{ave}} = \frac{19.276 \times 10^3}{7.811 \times 10^3} = 2.468$$

which is satisfactory.

The unit shear strength s' at the point of maximum shear stress is given by Eq. (11.49) as

$$s' = c + w'h \tan \phi$$

$$w' = 1160 \text{ kg(f)/m}^3$$

$$h = (0.6 \times 30 + 20) = 38 \text{ m}$$

$$\therefore \quad s' = 1200 + 1160 \times 38 \times \tan 24°$$

$$= 20.826 \times 10^3 \text{ kg(f)/m}^2$$

The factor of safety against shear at point of maximum unit shear in the foundation is given by Eq. (11.51) as

$$F = \frac{s'}{s_{max}}$$

$$= \frac{20.826 \times 10^3}{10.935 \times 10^3} = 1.905$$

which is satisfactory.

Illustrative Example 11.5 *Design an upstream blanket for a dam shown in the Fig. Ex. 11.5. The thickness of the blanket may be taken as 1.5 m. The coefficient of permeability of the blanket soil is 0.08 m/day and that of the foundation soil is 67 m/day.*

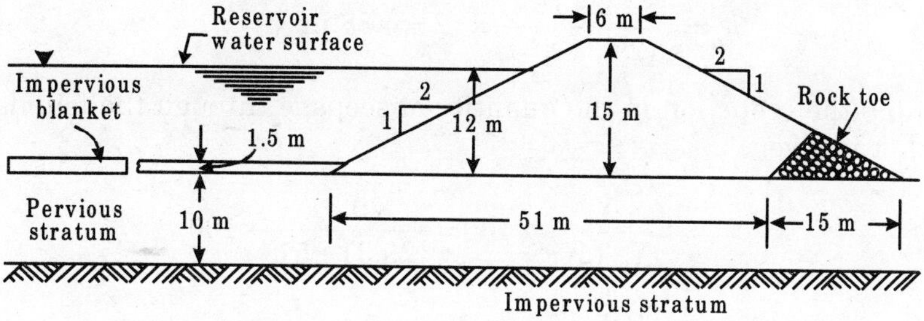

Fig. Ex. 11.5

Solution

The value of the constant a is given by

$$a = \sqrt{\frac{k_b}{k_f Z_b Z_f}}$$

$$k_b = 0.08 \text{ m/day} \; ; \; k_f = 67 \text{ m/day} \; ; \; Z_b = 1.5 \text{ m and } Z_f = 10 \text{ m}$$

$$= \sqrt{\frac{0.08}{67 \times 1.5 \times 10}} = \frac{1}{112.1}$$

From Eq. (11.57 a) the equivalent resistance X_r is given as

$$X_r = \frac{(e^{2ax} - 1)}{a(e^{2ax} + 1)}$$

Assuming the length of the blanket as 30 m we obtain

$$X_r = \frac{\left(e^{\frac{60}{112.1}} - 1\right)}{\frac{1}{112.1}\left(e^{\frac{60}{112.1}} + 1\right)}$$

$$= \frac{112.1(1.708 - 1)}{(1.708 + 1)} = 29.31$$

The head dissipated through the blanket is given by Eq. (11.58) as

$$h_o = \frac{X_r}{X_r + X_d} \times H$$

$$X_d = 51 \text{ m} \; ; \; H = 12 \text{ m}$$

$$\therefore \qquad h_o = \frac{29.31}{(29.31+51)} \times 12 = 4.38 \text{ m}$$

From Eq. (11.59) the reduction in the quantity of seepage through the foundation due to the provision of blanket is given

$$= \frac{X_r}{X_r + X_d} \times 100 = \frac{29.31}{(29.31+51)} \times 100$$

$$= 36.50 \ \%$$

Similarly assuming the lengths of the blanket as 60 m, 90 m, 120 m, 150 m and infinite the above noted quantities are calculated as noted in the following table.

Length of blanket x *(m)*	*Equivalent resistance* X_r *(m)*	*Head dissipated through blanket* h_o *(m)*	*Reduction in seepage* *%*
30	29.31	4.38	36.50
60	54.86	6.22	51.82
90	74.60	7.13	59.39
120	85.03	7.50	62.51
150	104.38	8.06	67.18
∞	112.10	8.25	68.73

From Eq. (11.60) the optimum length of the blanket is obtained as

$$x_o = \left(\frac{\sqrt{2}}{a}\right) = 158.53 \text{ m.}$$

It is observed that in this case by providing a blanket of length more than 150 m there would be only slight increase in the head dissipated through the blanket as well as in the reduction of seepage. Hence it is recommended that a blanket of 150 m length and 1.5 m thickness may be provided.

[**Note**: The various illustrative examples have been solved in metric units. However, these may also be solved in SI units by using the equivalent values given in the parenthesis.]

Illustrative Example 11.6 *Find out the pore pressure in a layer of soil due to continuous loading which results in decrease of volume by 4%. Assume initial porosity 0.35 and saturation ratio 85%. Henry's coefficient may be taken as 0.02 and initial pressures in soil pores as atmospheric.*

Solution

From Eq. (11.23) the pore pressure is given by

$$u = \frac{-p_a(\Delta V / V_0)}{n_0(1 - S_0 + S_0 h_c) + (\Delta V / V_0)}$$

$(\Delta V/V_0) = -0.04$; $n_0 = 0.35$; $S_0 = 0.85$; and $h_c = 0.02$

Thus, by substitution, we get

$$u = \frac{-p_a(0.04)}{0.35(1-0.85+0.85\times0.02)-0.04}$$

or $$u = 2.168\ p_a$$

i.e., pore pressure is equal to 2.168 times the atmospheric pressure.

REVIEW QUESTIONS

11.1 What are the different types of earth dams? Support your answer with neat sketches.

11.2 How earth dams are designed ?

11.3 Discuss the various causes of failure of earth dams.

11.4 What are the criteria for safe design of earth dams ?

11.5 Obtain Kozeny's equation for the phreatic surface in an isotropic and homogeneous earth dam with horizontal under drainage. Also obtain the expression for quantity of seepage flow.

11.6 Describe with neat sketch how top seepage line is drawn in a homogeneous dam without any arrangement for drainage.

11.7 Explain how top seepage line is drawn in the case of an earth dam having different permeabilities in horizontal and vertical directions.

11.8 Write an explanatory note on the procedure for stability analysis of slopes of an earth dam by Swedish slip circle method.

11.9 What do you understand by critical slip circle ? Describe how it can be located for different soils.

11.10 Why is an earth dam zoned ? Stage the characteristics of soil which makes it suitable for a zoned earth dam.

11.11 Describe with neat sketches the various methods adopted for controlling seepage through the body of the dam and through the foundation.

11.12 Discuss the design criteria of filters for earth dams.

11.13 What do you understand by construction pore pressure in earth dams ? Explain how this can be determined.

11.14 Give justification for taking saturated weight for driving forces and submerged weight for resisting forces in determining stability under drawdown conditions for upstream slopes.

11.15 Find the pore pressure in a layer of soil due to continuous loading which results in decrease of volume by 5%. Assume initial porosity 0.40 and saturation ratio 82%. Henry's coefficient may be taken as 0.02 and initial pressure in soil pores as atmospheric. [*Ans.* 1.751 p_a]

11.16 Explain how can you ensure the stability of downstream as well as upstream portions of an earth dam against horizontal shear at the base of the dam.

11.17 Discuss the stability of foundation of an earth dam against horizontal shear stress.

11.18 Explain the terms : (i) Turfing; (ii) Pitching; (iii) Berms; (iv) Cutoff; (v) Relief wells.

11.19 Discuss the Bennet's solution for the provision of an impervious blanket for an earth dam.

11.20 An infinite blanket of 1.5 m thickness overlies a pervious stratum of 12 m thickness. If the impervious base length of the dam is 60 m and total reservoir head is 15 m, determine the head loss through the blanket and seepage through the foundation. Coefficient of permeabilities of the blanket and foundation materials may be taken as 0.3 m/day and 30 m/day respectively. [*Ans.* 6.2/m; 2.2 m^3/h]

11.21 Check the stability of the upstream slope of an earth dam under sudden drawdown condition. For the dam the upstream slope is 3 horizontal to 1 vertical, downstream slope is 212 horizontal to 1 vertical, and crest width is 10 m. The height of the dam is 60 m and maximum depth of water is 57 m. The dam has a central core of impervious soil the side slopes of which are 1 horizontal to 1 vertical. The impervious rock is available at a depth of 20 m below the base of the dam and a sheet pile cutoff is provided at centre of the central core. The probable surface of slippage is cylindrical passing through the toe of the dam and its centre is located 120 m above the base of the dam on a vertical line at a distance of 60 m from the toe of the dam. Assume the following soil properties:

Particulars	*Impervious soil*	*Pervious soil*	*Foundation soil*
(i) Dry weight	17.266 kN/m^3 (1.76 t/m^3)	—	—
(ii) Moist weight	19.620 kN/m^3 (2.00 t/m^3)	20.405 kN/m^3 (2.08 t/m^3)	20.405 kN/m^3 (2.08 t/m^3)
(iii) Submerged weight	10.448 kN/m^3 (1.065 t/m^3)	11.085 kN/m^3 (1.130 t/m^3)	11.085 kN/m^3 (1.130 t/m^3)
(iv) Saturated weight	20.258 kN/m^3 (2.065 t/m^3)	20.895 kN/m^3 (2.130 t/m^3)	20.895 kN/m^3 (2.130 t/m^3)
(v) Angle of internal friction	20°	30°	30°
(vi) Cohesion	42.183 kN/m^2 (4.3 t/m^2)	—	—

11.22 Discuss the various damages which may be caused to an earth dam due to earthquake.

11.23 Describe the various measures which may be taken to prevent the failure of an earth dam due to earthquake.

11.24 What do you understand by piping ? What measures may be taken to ensure safety of an earth dam against the failure due to piping ?

11.25 What is a rockfill dam ? Describe its salient features with a neat sketch.

11.26 Describe with neat sketches the various types of upstream impervious membranes used for rockfill dams.

11.27 What are the different types of settlements in a rockfill dam ? How can they be eliminated ?

CHAPTER

12

Spillways, Energy Dissipators, Spillway Gates and Outlet Works

12.1 INTRODUCTION

A spillway is a waterway provided to dispose of surplus flood waters from a reservoir after it has been filled to its maximum capacity. Spillways are invariably provided for all the dams and these act as safety valves for the dams. A spillway may be located either within the body of the dam or at one end of the dam or entirely away from the dam as an independent structure.

It is essential to provide a spillway of sufficient capacity (or outflow rate through the spillway) so that the surplus flood water is discharged keeping the water level in the reservoir below some predetermined maximum level and no damage is caused to the dam. A spillway of insufficient capacity may however lead to over-topping of the dam resulting in serious damage to the dam or in the failure of the dam. On the other hand an over-estimation of the spillway capacity would result in an uneconomical design.

12.2 ESSENTIAL REQUIREMENTS OF A SPILLWAY

The essential requirements of a spillway are as follows.

(i) The spillway must have sufficient capacity.

(ii) It must be hydraulically and structurally adequate.

(iii) It must be so located that it provides safe disposal of water i.e., spillway discharges will not erode or undermine the downstream toe of the dam.

(iv) The bounding surfaces of the spillway must be erosion resistant to withstand the high scouring velocities created by the drop from the reservoir surface to tail water.

(v) Usually some devices will be required for dissipation of energy on the downstream side of the spillway.

12.3 SPILLWAY CAPACITY

The required capacity of a spillway (i.e., the maximum outflow rate through the spillway) may be determined by flood routing which as described in section 7.12 requires the following data.

(i) Inflow hydrograph (plot of rate of inflow versus time).

(ii) Reservoir capacity curve (plot of reservoir storage versus reservoir water surface elevation).

(iii) Discharge curve (plot of rate of outflow through spillway versus reservoir water surface elevation).

By flood routing, corresponding to a particular inflow hydrograph the maximum outflow rate and maximum rise in the water surface may be determined.

However, the required capacity of a spillway depends on the following factors.

(i) The inflow flood.

(ii) The available storage capacity.

(iii) The discharge capacity of other outlet works.

(iv) Whether the spillway is gated or ungated.

(v) The possible damage if a spillway of adequate capacity is not provided.

Each of these factors are discussed below.

The selection of the inflow flood for the spillway design depends on the degree of protection that ought to be provided to the dam, which, in turn, depends on the type of dam, its location, and consequences of failure of the dam. A high dam storing a large volume of water located upstream of an inhabited area should have a much higher degree of protection than a low dam storing small quantity of water whose downstream reach is uninhabited. Thus, for the safety of the dam the following inflow floods should be taken for the design of spillways for the dams of different sizes.

Size of Dam (as indicated in section 8.2 Chapter 8)	*Inflow Flood for Safety of Dam**
Small	100 year flood
Intermediate	SPF (Standard Project Flood)
Large	PMF (Probable Maximum Flood)

However, floods of larger or smaller magnitude may be used if the hazard involved in the eventuality of a failure of a dam is particularly high or low.

A reservoir with a larger storage capacity will normally require smaller outflow rate through the spillway.

If the outflows through the spillway are supplemented by release through the outlets then the required capacity of the spillway may be correspondingly reduced.

The discharging capacity of a spillway can be modified if it is provided with gates. For a gate controlled spillway more water will be stored and by proper operation of the gates higher heads may be developed so that greater outflow rates through the spillway may be developed.

12.4 COMPONENTS OF SPILLWAY

The various components of a spillway are as described below.

1. Control Structure. The control structure is a major component of a spillway which

*As per IS: 11223–1985–Guideline for Fixing Spillway Capacity.

regulates and controls the outflows from the reservoir. It prevents outflows below fixed reservoir levels and allows the flow of water when the water surface in the reservoir rises above the level. In most of the cases the control structure consists of the weir which maybe sharp (or narrow) crested, ogee shaped or broad crested. Further in order to regulate the flow of water from the reservoir gates may be provided on the crest of the control structure.

2. Discharge Channel. The flow released through the control structure is usually conveyed to the stream bed below the dam in a discharge channel or waterway. The main function of this is to convey the water safely from the reservoir downward to the river. The conveyance structure may be the downstream face of the spillway, an open channel excavated along the ground surface, a closed conduit placed through or under a dam, or a tunnel excavated through an abutment.

3. Terminal Structures-Energy Dissipators. When the water flows from reservoir pool level to downstream river level, the static energy is converted to kinetic energy. This results in high velocities of flow on the downstream side of the spillway which may cause serious erosion or scour of the streambed and banks. The scouring of the streambed usually retrogrades and may cause serious damage to toe of the spillway and the dam and also to the adjacent structures. The high energy of the flow must therefore be dissipated before the flow is returned to the river. For this usually energy dissipators are provided on the downstream side of the spillway.

4. Entrance and Outlet Channels. Entrance channels may be required to be provided to draw water from the reservoir and convey it to the control structure. Similarly outlet channels may be required to be provided to convey the spillway flow from the terminal structure to the river channel below the dam. The entrance and outlet channels are not required where a spillway draws water immediately from the reservoir and delivers it directly back into the river, as in the case of an overflow spillway. However, in the case of spillways placed through abutments or through saddles or ridges, the entrance and outlet channels may be required.

12.5 TYPES OF SPILLWAYS

The spillways may be classified according to different considerations as indicated below.

1. Classification based on the time when the spillways come into operation. On this basis the spillway may be classified as

(a) Main or service spillway

(b) Auxiliary spillway

(c) Emergency spillway.

(a) *Main or service spillway.* A main or service spillway is the one which is the first to come into operation and is designed to pass the entire spillway design flood. However, in some cases where site conditions are favourable it may be economical to provide an auxiliary spillway in conjunction with a smaller main spillway. In such cases the main spillway is designed to pass only small floods which are likely to occur frequently and when these floods are exceeded the main spillway is assisted by the auxiliary spillway in passing the excessive flood waters.

(b) *Auxiliary spillway.* An auxiliary spillway is the one which is provided as a supplement to the main spillway and its crest is so located that it comes into operation only after the floods for which the main spillway is designed are exceeded. Thus, an auxiliary spillway is provided in conjunction with the main spillway. The total spillway capacity is then equal to the sum of the capacities of the main and the auxiliary spillways.

Conditions favourable for the adoption of an auxiliary spillway are the existence of a saddle or depression along the rim of the reservoir which leads into a natural waterway, or a gently sloping abutment where an excavated channel can be carried sufficiently beyond the dam to avoid the possibility of damage to the dam or other structures.

For auxiliary spillways control gates are seldom provided. However, on the crest of an auxiliary spillway instead of control gates a *fuse plug* may be provided. A fuse plug may be a simple earth embankment (called *fuse plug dike*) or flash board or other device which allows the water surface in the reservoir to rise above the crest of the spillway and is so designed that when overtopped it automatically washes out thus releasing the excess flood water.

(c) *Emergency spillway.* An emergency spillway is the one which is provided in addition to the main spillway but it comes into operation only during emergency which may arise at any time and the same might not have been considered in the normal design of the main spillway. Thus, it is provided for additional safety during emergency. Some of the situations which may lead to emergency are

(i) an enforced shutdown of the outlet works

(ii) a malfunctioning of spillway gates

(iii) the necessity for byepassing the regular spillway because of damage or failure of some part of that structure.

Further an emergency may also arise if a recurring flood occurs before a previous flood is evacuated by the main spillway or the outlet works.

Emergency spillways are usually provided in saddles or depressions along the reservoir rim or by excavating channels through ridges or abutments. In the case of an embankment dam emergency spillway is often formed by providing a *breaching section* which is a portion of the dam with its crest lower than the crest of the dam and is so designed that it is washed out when the water in the reservoiur reaches a predetermined elevation. The breaching section is also called *fuse plug spillway.*

Since emergency spillway is not required to function under normal reservoir operation, its crest is placed at or above the designed maximum water surface in the reservoir. Also in considering the design of an emergency spillway usually an encroachment on the freeboard provided for the designed maximum water surface is allowed. Further on the crest of the emergency spillway usually control gates are not provided but fuse plug dike may be provided.

It may, however, be stated that emergency spillways would act as auxiliary spillways if a flood greater than the selected inflow design flood occurred.

2. Classification based on the flow through the spillway being controlled or uncontrolled. On this basis the spillways may be classified as

(a) Controlled or Gated spillway

(b) Uncontrolled or Ungated spillway.

(a) *Controlled or Gated spillway.* A spillway having means to control the outflow from the reservoir is known as controlled or gated spillway.

(b) *Uncontrolled or Ungated spillway.* A spillway, the crest of which permits water to escape automatically, as the water level in the reservoir rises above the crest is known as uncontrolled or ungated spillway.

3. Classification based on the prominent features pertaining to the various components of the spillway. According to the prominent features pertaining to the various components of the spillways such as control structures, discharge channel etc., the spillways may be classified in the following types.

1. Free overfall or straight drop spillway
2. Overflow or Ogee spillway
3. Chute or Open channel or trough spillway
4. Side channel spillway
5. Shaft or Morning glory spillway
6. Conduit or Tunnel spillway
7. Siphon spillway.

A detailed description of each of the above noted types of spillways is given in the following sections.

12.6 FREE OVERFALL OR STRAIGHT DROP SPILLWAY

A free overfall or straight drop spillway is the one for which the control structure is a low height narrow crested weir having its downstream face vertical or nearly vertical (Fig. 12.1). The overflowing water may be discharged as in the case of a sharp crested weir or it may be supported along the narrow section of the crest. However, in either case the water flowing over

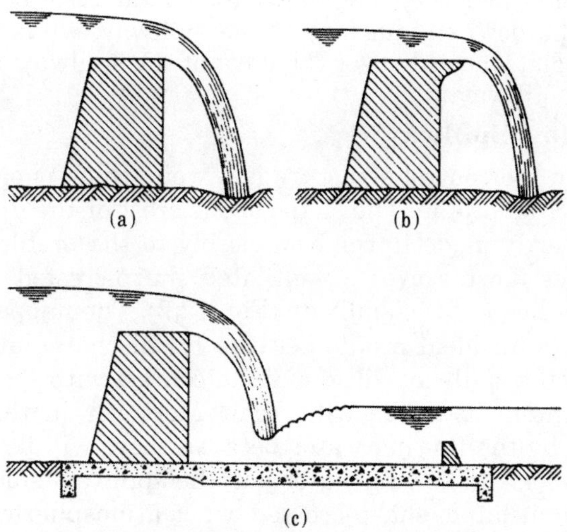

Fig. 12.1. *Straight drop spillway (a) without downstream protection; (b) with overhanging lip ; (c) with downstream protection work*

the crest of this spillway drops as a free jet clearly away from the downstream face of the spillway. As such occasionally the crest of this spillway is extended in the form of an overhanging lip (Fig. 12.1 b) to direct small discharges away from the downstream face of the overfall section. The underside of the nappe is ventilated sufficiently to prevent pulsating or fluctuating jet.

If no artificial protection is provided on the downstream side of the overfall section, the falling jet will usually cause the scouring of the streambed and will form a deep plunge pool. As such in order to protect the stream bed from scouring, an artificial pool may be created by constructing a low auxiliary dam downstream of the main structure or by excavating a basin which is then provided with a concrete apron. However, if tailwater depths are sufficient, a hydraulic jump will form when the jet falls freely from the crest, in which case a sufficiently long flat apron may be provided. Moreover in this case floor blocks and an end sill may be provided as shown in Fig. 12.1 (c) to help in the establishment of the jump and thus reduce the downstream scour.

The free overfall spillway is most commonly used for low earth dams (or earthen bunds). Further this type of spillway is also suitable for thin arch dams and for other dams which have nearly vertical downstream face and would permit free fall of water. However, this type of spillway is not suitable for high drops on yielding foundations, because the apron will be subjected to large impact forces at the point of impingement of the jet. Moreover the impact forces also cause vibrations which may crack or displace the apron and may result in failure by piping or undermining. Thus, ordinarily free overfall spillways are used where the hydraulic drops from head pool to tailwater are not in excess of about 6 m.

12.7 OVERFLOW OR OGEE SPILLWAY

An overflow or ogee spillway is an improvement upon the free overfall spillway and is widely used with gravity, arch and buttress dams. Several earth dams are also provided with this type of spillway. The essential difference between the free overfall spillway and the overflow spillway is that in the case of the former the water flowing over the crest of the spillway drops as a free jet clearly away from the downstream face of the spillway, while in the case of the later the overflowing water is guided smoothly over the crest of the spillway and is made to glide over the downstream face of the spillway.

Crest Shape of Overflow Spillway

For an overflow spillway the control structure is a weir which is ogee or S-shaped (i.e., shaped like an elongated letter S) in profile. The shape of the crest or the upper curve of the ogee profile of this spillway is ordinarily made to conform closely to the profile of the lower surface of the nappe* or sheet of water flowing over a ventilated sharp-crested weir when discharging at a head equal to the design head of the spillway (Fig. 12.2). The nappe-shaped profile for the crest of an overflow spillway is an ideal profile because for discharge at the design head the water flowing over the crest of the spillway will remain in contact with the surface of the spillway as it glides over it and optimum discharge will occur. Moreover in this case no pressure will be exerted on the spillway by the flowing water because there will be atmospheric pressure along the contact surface between the flowing water and the spillway (since along the lower surface of the flow nappe over ventilated sharp-crested weir atmospheric pressure exists). Also for discharge at a head less than the design head the water flowing over the crest of the spillway will remain in contact with the surface of the spillway but in this case positive hydrostatic pressure will be exerted on the spillway by the flowing water because the nappe will tend to be depressed. However, in this case the spillway will support the sheet of flowing water and thus a

* The lower surface of the nappe is also commonly known as lower nappe.

backwater effect will be created and the discharge will be reduced. On the other hand for discharge at a head greater than the design head the overflowing water will tend to break contact with the spillway surface and a zone of separation will be formed (Fig. 12.2 c) in which negative or suction pressure will be produced. The effect of the negative pressure will be to increase the effective head and thereby increase the discharge.

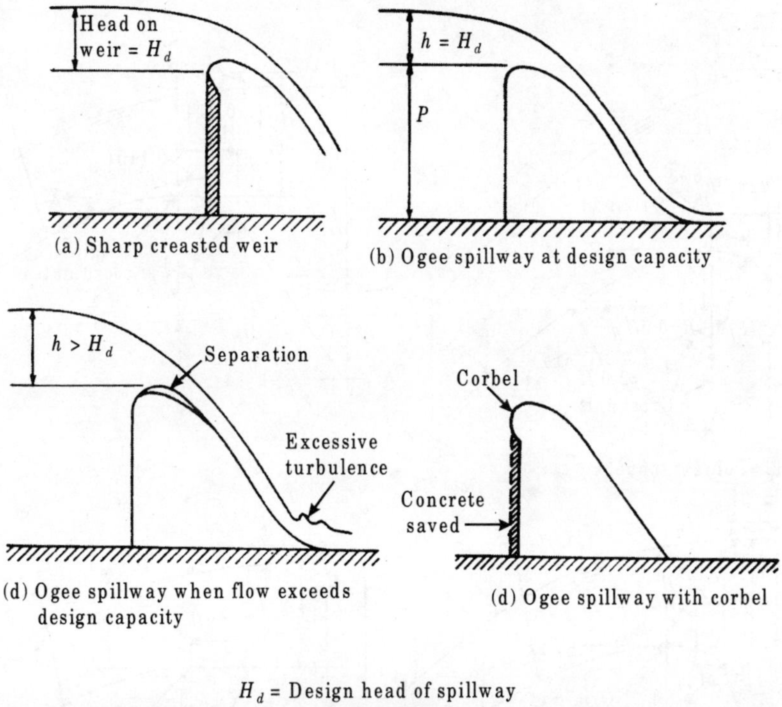

Fig. 12.2 *Ogee spillway*

The shape of the nappe-shaped profile depends upon the head, the inclination of the upstream face of the spillway and the height of the spillway above the stream-bed or the bed of the entrance channel (which influences the velocity of approach to the crest of the spillway). Extensive experiments were conducted by U.S. Bureau of Reclamation (U.S.B.R.) to obtain the nappe-shaped profiles for the crests of the overflow spillways with their upstream face either vertical or inclined. On the basis of the U.S.B.R. data, the U.S. Army Corps of Engineers has developed several standard shapes of the crests of overflow spillways at its Waterways Experiment station at Vicksberg (U.S.W.E.S.). Such shapes (Fig. 12.3) designated as the WES standard spillway shapes, can be represented by the following equation

$$X^n = KH_d^{n-1}Y \qquad \text{...(12.1)}$$

where X and Y are coordinates of the profile of the crest of the spillway with the origin at the highest point of the crest ; H_d is design head excluding head due to velocity of approach ; K and n are constants whose values depend on the slope of the upstream face of the spillway. The values of K and n are given as follows :

Slope of upstream face	K	n
Vertical	2.000	1.850
3 Ver. to 1 Hor.	1.936	1.836
3 Ver. to 2 Hor.	1.939	1.810
3 Ver. to 3 Hor.	1.873	1.776

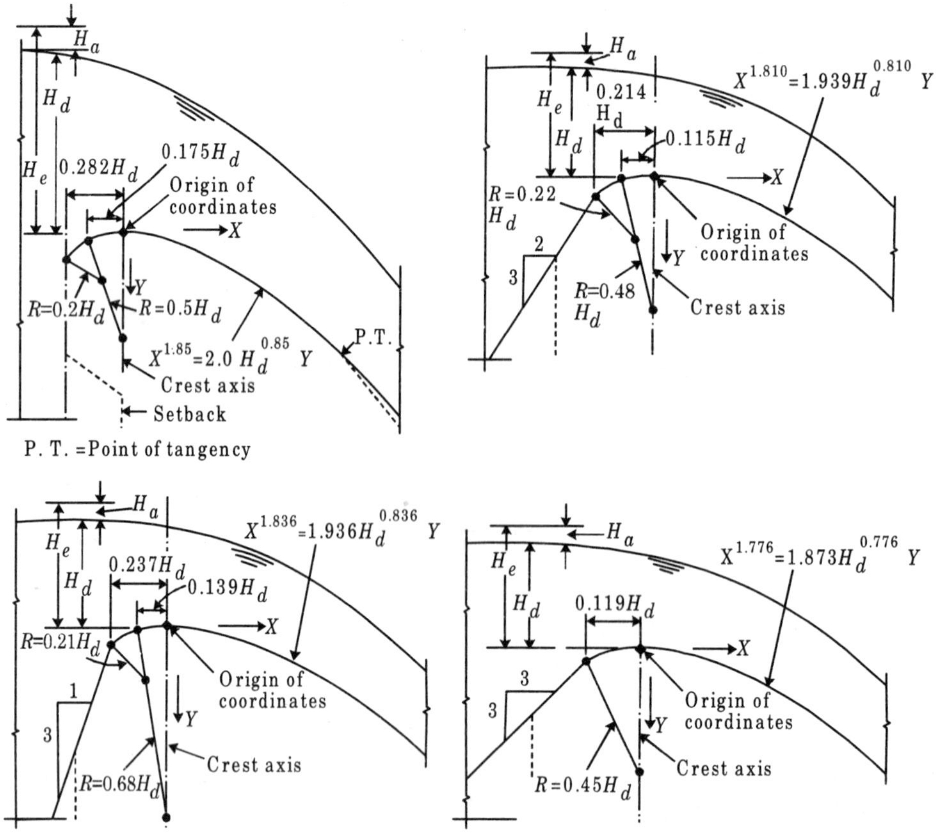

Fig. 12.3 *WES-standard spillway shapes*

For intermediate slopes, approximate values of K and n may be obtained from the plots of the above values against the corresponding slopes as shown in Fig. (12.4).

As shown in Fig. (12.3) X is taken as positive in the downstream direction and Y is taken as positive in the downward direction. Since Eq. (12.1) is applicable only for the positive values of X and Y, it gives the crest shape downstream from the origin of the coordinates.

The curved profile of the crest section is continued tangentially along a straight sloping surface which forms the downstream face of the spillway. The location of the point of tangency however depends on the slope of the straight portion of the downstream face of the spillway which in turn depends on the stability requirements and on the features of the stilling basin at the toe of the spillway. The slope of the straight portion of the downstream face of the spillway is usually in the range of 1 vertical to 0.6 to 0.8 horizontal. At the end of the sloping surface of the spillway a curved bucket is provided (Fig. 12.5) to create a smooth transition of flow from the spillway to the outlet channel or the river on the downstream side and to prevent scouring of the channel or river bed due to the impact of the falling water. The radius R (in m) of the bucket may be obtained approximately by the following empirical formula :

$$R = 10^{(V + 6.4H + 4.88)/3.6H + 19.52} \quad \text{...(12.2)}$$

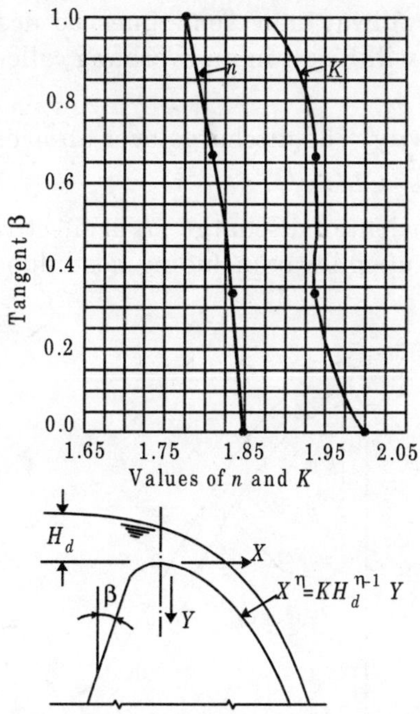

Fig. 12.4 *Plots of n and K versus upstream slope*

where V is the velocity in m/s of the flow at the toe of the spillway ; and H is head in m excluding head due to the velocity of approach, on the spillway crest. Neglecting the energy loss involved in the flow over the spillway the velocity of flow V at the toe of the spillway may be computed by

$$V = \sqrt{2g(Z + H_\alpha - y)} \quad \text{...(12.3)}$$

where Z is fall, or vertical distance from the upstream reservoir level to the floor at the toe ; H_α is head due to velocity of approach on the upstream side ; g is acceleration due to gravity ; and y is depth of flow at the toe.

The crest shape upstream from the origin of the coordinates is usually in the form of a compounded circular curve as shown in Fig. 12.3, or in the form of an elliptical curve. However, on the basis of the recent studies, U.S. Army Crops of Engineers has proposed an equation for the crest shape upstream from the origin of the coordinates which is as follows:

$$Y = \frac{0.724(X + 0.270H_d)^{1.85}}{H_d^{0.85}} + 0.126H_d$$

$$- 0.4315H_d^{0.375}(X + 0.270H_d)^{0.625} \quad \text{...(12.4)}$$

where all the notation are same as defined earlier.

In Eq. (12.4) only negative values of X are to be used. The minimum value of X being $(-0.270H_d)$ corresponding to which the value of Y is equal to $0.126H_d$. Further Eq. (12.4) is applicable only to the spillways which have their upstream face vertical.

The upstream face of the spillway may sometimes be designed to set back as shown by dotted lines in Fig. (12.3) which will result in an overhang called 'corbel' on the upstream side of the spillway.

Discharge of Overflow Spillway. The discharge over an overflow spillway is given by

$$Q = CLH_e^{3/2} \qquad \text{...(12.5)}$$

where Q is discharge ; C is coefficient of discharge ; L is effective length of crest of spillway; and H_e is total head of flow on crest of spillway including head due to velocity of approach.

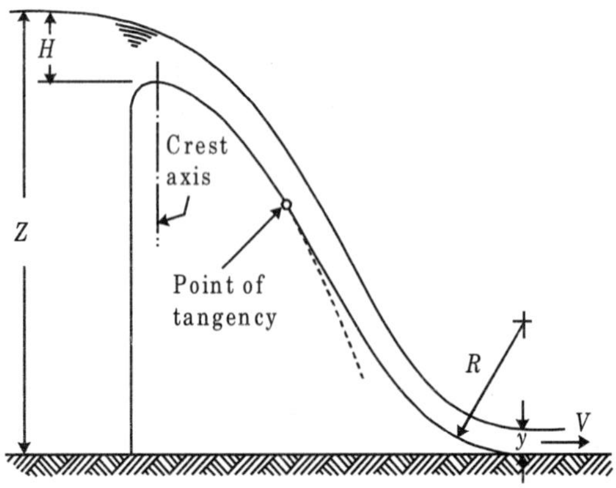

Fig. 12.5 *Profile of an ogee spillway*

Coefficient of Discharge of Overflow Spillway. An overflow spillway has a relatively high coefficient of discharge the maximum value of which may be about 2.2 if no negative or suction pressure is allowed to be developed. However, the value of the coefficient of discharge for an overflow spillway depends on the following factors:

(i) Height of spillway above stream-bed or bed of approach channel.

(ii) Ratio of the actual total head of flow over spillway crest to the design head.

(iii) Slope of upstream face of spillway.

(iv) Extent of downstream submergence of spillway crest.

(i) *Height of spillway above streambed or bed of approach channel.* The height of spillway above stream-bed or bed of approach channel affects the velocity of approach which in turn affects the coefficient of discharge. With increase in the height of spillway the velocity of approach decreases and the coefficient of discharge increases. Model tests of spillways have shown that the effect of the velocity of approach is negligible when the height of spillway above stream-bed is equal to or greater than 1.33 H_d, where H_d is the design head excluding the head due to velocity of approach.

Figure (12.6) shows a plot of coefficient of discharge C versus (P/H_D), where P is the height of the spillway above the streambed or the bed of the approach channel and H_D is the design head including the head due to velocity of approach (i.e., $H_D = H_d + H_a$). The values of C in this plot correspond to $H_e = H_D$. It may be observed from this plot that there is marked increase in the value of C till the height of spillway becomes equal to twice the design head H_D. With further increase in the height of spillway there is not much increase in the value of C.

(ii) *Ratio of the actual total head of flow over spillway crest to the design head.* Figure (12.7) shows a plot of (C/C') versus (H_e/H_D) for a spillway of height P above stream-bed greater than $1.33\ H_d$, where C is coefficient of discharge corresponding to the actual total head of flow H_e and C' is coefficient of discharge corresponding to the design head H_D including the head due to velocity of approach. It may be observed from this plot that with increase in the value of (H_e/H_D) the value of (C/C') increases. In other words with increase in the head H_e the coefficient of discharge increases. However, for $H_e < H_D$, $C < C'$; and for $H_e > H_D$, $C > C'$.

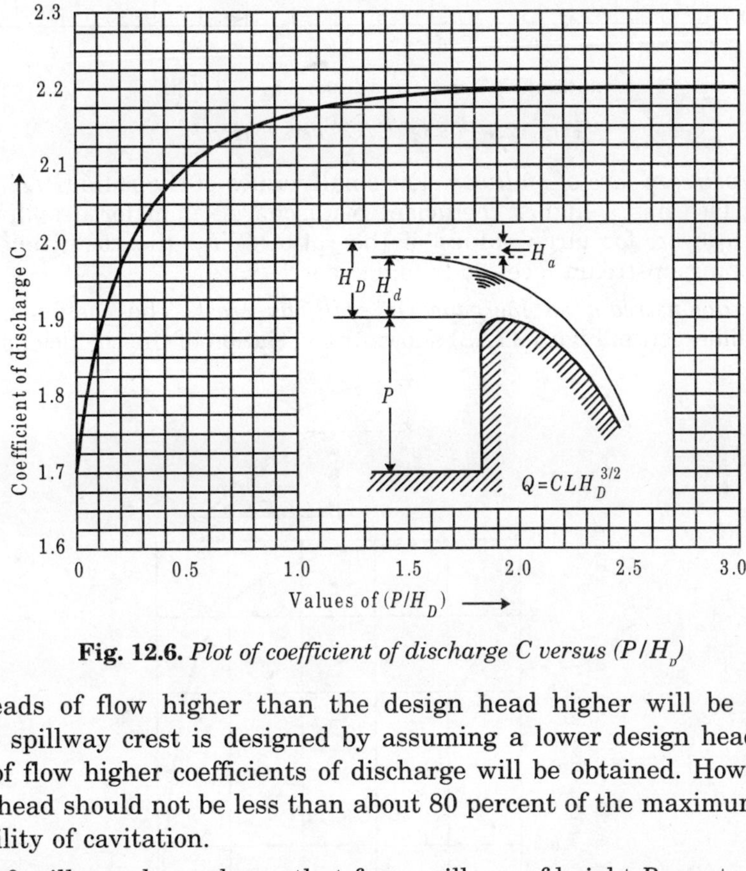

Fig. 12.6. *Plot of coefficient of discharge C versus* (P/H_D)

Since for heads of flow higher than the design head higher will be the coefficient of discharge, if the spillway crest is designed by assuming a lower design head, for most of the range of heads of flow higher coefficients of discharge will be obtained. However, as indicated later the design head should not be less than about 80 percent of the maximum head in order to avoid the possibility of cavitation.

Model tests of spillways have shown that for a spillway of height P greater than $1.33\ H_d$ the head due to velocity of approach is negligible and when the total head of flow is equal to the design head i.e., $H_e = H_D$, the coefficient of discharge is equal to 2.2. For the same spillway the values of the coefficient of discharge corresponding to the heads of flow other than the design head may thus be obtained from Fig. (12.7). Similar curves are also available for spillways of height P less than $1.33\ H_d$.

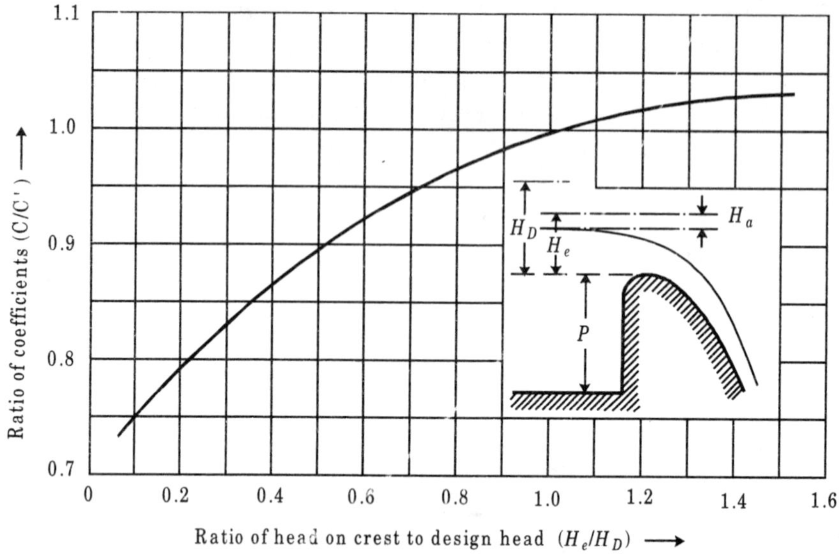

Fig. 12.7 *Plot of (C/C′) versus (H_e/H_D)*

(iii) *Slope of upstream face of spillway.* For small values of the ratio (P/H_D) a spillway with sloping upstream face has a higher coefficient of discharge than for a spillway with vertical upstream face. However, for large values of the ratio (P/H_D) the coefficient of discharge for spillways with sloping upstream face tends to decrease.

(iv) *Extent of downstream submergence of spillway crest.* The coefficient of discharge is reduced due to submergence. Figure (12.8) shows the variation of the coefficient of discharge with

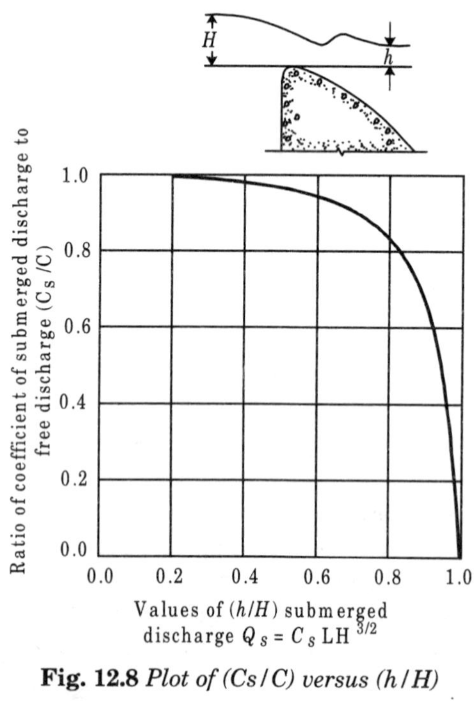

Fig. 12.8 *Plot of (Cs/C) versus (h/H)*

the degree of submergence. It may however be observed that even for a degree of submergence of as much as 60 percent the reduction in the value of the coefficient of discharge is only about 5 percent.

Effective Length of Crest of Overflow Spillway. The effective length of crest of an overflow spillway is given by the following equation :

$$L = L' - 2\,(NK_p + K_a)\,H_e \qquad \text{...(12.6)}$$

in which

L = effective length of crest

L' = net length of crest which is equal to the sum of the clear spans of the gate bays between piers

H_e = total head of flow on crest including head due to velocity of approach

N = number of piers

K_p = pier contraction coefficient, and

K_a = abutment contraction coefficient.

The pier contraction coefficient, K_p depends on several factors such as (i) shape and location of pier nose: (ii) thickness of pier; (iii) approach velocity; and (iv) ratio of actual head on crest to design head. For the flow at the design head the average values of K_p may be assumed as follows.

Pier condition	K_p
1. Square-nosed piers with corners rounded on a radius equal to about 0.1 of pier thickness	0.02
2. Round-nosed piers	0.01
3. Pointed-nose piers	0.00

The abutment contraction coefficient K_a also depends on several factors such as (i) shape of abutment; (ii) angle between upstream approach wall and axis of flow; (iii) approach velocity; and (iv) ratio of actual head on crest to design head. For the flow at the design head the average values of K_a may be assumed as follows.

Abutment condition	K_a
1. Square abutment with headwall at 90° to the direction of flow	0.20
2. Rounded abutment with headwall at 90° to the direction of flow, when $0.5\,H_d \geq r \geq 0.15\,H_d$	0.10
3. Rounded abutments where $r > 0.5\,H_d$ and headwall is placed not more than 45° to the direction of flow	0.00

where r = radius of abutment rounding and

H_d = design head.

For flow at head other than the design head the values of K_p and K_a may be obtained from the graphical plots given in IS: 6934–1973 which deals with the hydraulic design of high ogee or overflow spillways.

Equation (12.3) gives discharge for an ungated overflow spillway or for a gated overflow spillway at full gate opening. The discharge for a gated overflow spillway at partial gate openings will be similar to flow through a low-head orifice and may be computed by the following equation

$$Q = \frac{2}{3} C_d \sqrt{2g} L \left(H_1^{3/2} - H_2^{3/2} \right) \quad \text{...(12.7)}$$

in which Q is discharge ; C_d is coefficient of discharge ; g is acceleration due to gravity ; L is effective length of crest of spillway ; and H_1 and H_2 are total heads (including head due to velocity of approach) above bottom and top of the opening, respectively.

The coefficient of discharge C_d will differ with different gate and crest arrangements and it is also influenced by the approach and downstream conditions, as they affect the jet contractions. Further the effective length L of crest of spillway is equal to the sum of the clear spans of the gate bays between piers.

Pressure on Overflow Spillway. If the profile of the spillway crest is designed according to Eq. (12.1), the pressure on the spillway crest will be very nearly equal to atmospheric pressure when it is operated at the design head. However, in actual practice the spillway will also be operated at heads of flow other than the design head. For heads of flow lower than the design head the pressure on the spillway crest will be above atmospheric pressure. On the other hand for heads of flow higher than the design head the pressure on the spillway crest will be less than the atmospheric pressure and may sometimes become so low as to develop cavitation.

Model tests have shown that when the head of flow exceeds the design head H_d by $33\frac{1}{3}$ percent the observed negative pressure on the crest axis is about 0.35 H_d and on the upstream side of the crest axis at a distance of about 0.2 H_d it is about 0.5 H_d. It has also been shown that the design head may be exceeded by 25 percent, upto which limit no harmful cavitation is seen to develop.

In a gated spillway the flow of water released through partial gate opening occurs as orifice flow. Thus, in this case the profile of the trajectory of the water discharging freely through the partial gate opening will follow the path of a jet issuing from an orifice. The shape of the profile of the spillway downstream from the gate sill must therefore conform to the trajectory profile if negative pressure are to be avoided along the crest of the spillway. Experiments have shown that when gates are operated with small gate openings under high heads, negative pressures will occur along the crest in the region immediately below the gate if the crest profile is thinner than the one conforming the trajectory profile. Tests have also shown that it the crest profile conforms to the ideal nappe profile for any design head H_d as given by Eq. (12.1) and if the gate is operated at small opening then for head of flow exceeding the design head by 33 percent the observed negative pressure on the crest axis at the centre line of the pier bay is about 0.2 H_d and that at the pier is about 0.3 H_d. However, in this case at the pier on the upstream side of the crest axis at a distance of about 0.2 H_d the observed negative pressure is as low as about 0.6 H_d.

The adoption of a trajectory profile rather than a nappe profile downstream from the gate sill will result in a wider spillway crest and reduced discharge efficiency for full gate opening. Thus, where the discharge efficiency is unimportant and where a wider spillway crest is needed for structural stability, the trajectory profile may be adopted to avoid negative pressure zones along the crest. However, where the spillway crest is shaped to the ideal nappe profile as given by Eq. 12.1, the negative pressure zone can be reduced by placing the gate sill downstream from the crest axis. This will provide an orifice which is inclined downstream for small gate openings and thus will result in a steeper trajectory more nearly conforming to the nappe-shaped profile.

12.8 CHUTE OR OPEN CHANNEL OR TROUGH SPILLWAY

In several cases for embankment dams it is not possible to provide overflow spillway. Further for concrete or masonry dams also, the narrowness of the valley, the erodible nature of the stream-bed, or some other reason may make it impossible or undesirable to provide the overflow spillway. In such cases, some sort of spillway must be provided which may be isolated from the main dam. The chute or open channel or through spillway is one such type of spillway.

The chute spillway is provided either in a saddle along the reservoir rim, or along abutment of the dam. It consists of a steep sloped open channel called a chute or trough which carries the

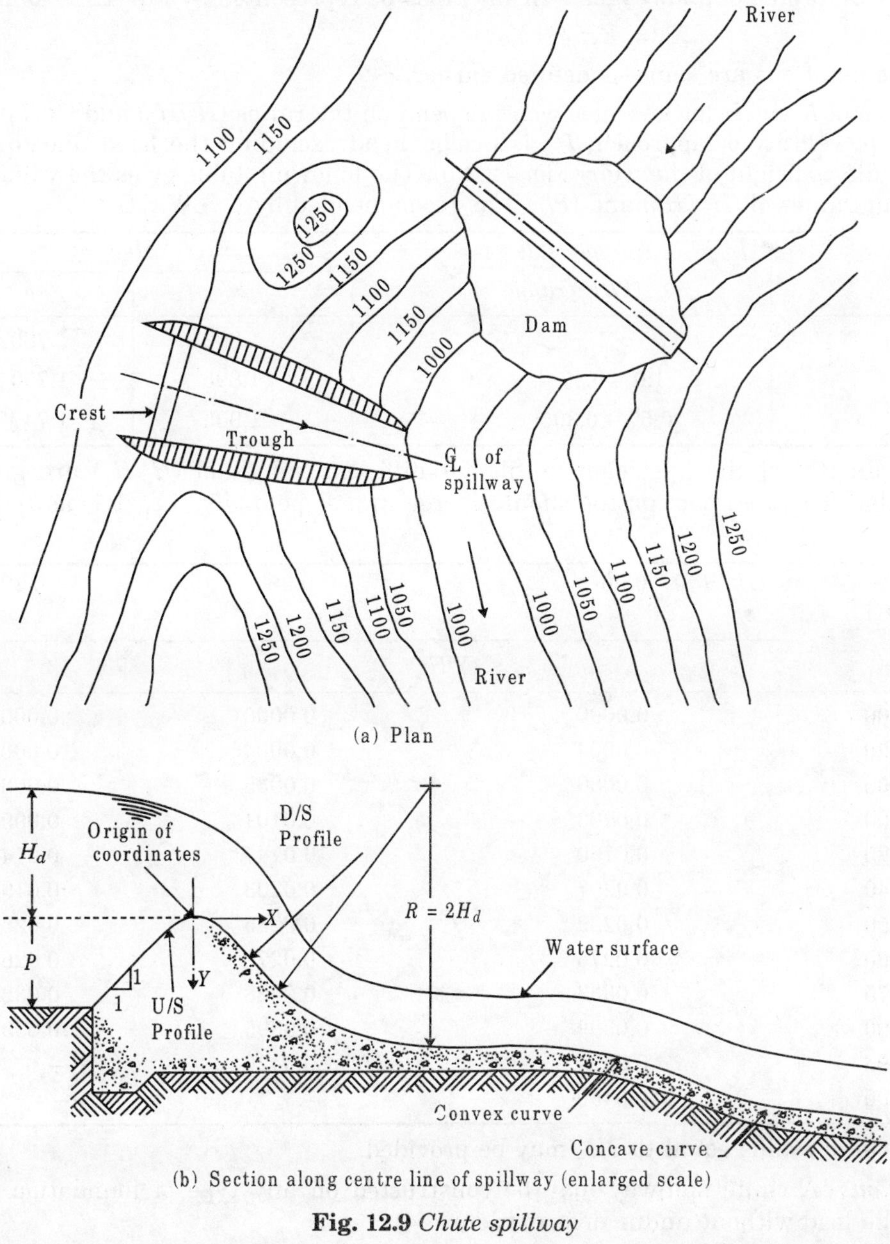

Fig. 12.9 *Chute spillway*

surplus water from the reservoir that flows over the crest of the spillway to the river below the dam (Fig. 12.9). Hence this spillway derives its name from the carrier channel. A control structure may or may not be required to be provided for this spillway depending on the natural levels of the saddle. If the saddle has a higher level than the full reservoir level, it is excavated to the later level and left to serve as flat crested weir. On the other hand if the saddle has a lower level than the full reservoir level then a weir has to be built upto that level which is usually made ogee shaped to obtain a high discharge coefficient.

Design of Low Ogee Weir required as a Control Structure for the Chute Spillway

The downstream profile of a low Ogee weir may also be represented by Eq. (12.1) as noted below

$$X^n = KH_d^{n-1}Y$$

where all the notations are same as defined earlier.

The values of K and n for low ogee weirs depend on the ratios (H_a/H_d) and (P/H_d), where H_a is head due to velocity of approach; H_d is design head excluding the head due to velocity of approach; and P is height of the weir (Fig. 12.9 b). The following table gives the values of K and n for different values of (H_a/H_d) and (P/H_d) as recommended by U.S.W.E.S.

Values of (H_a/H_d)	*Applicable* (P/H_d) *range*	*Values of* K	n
0.00	≥ 1	1.852	1.780
0.08	1.0 – 0.57	1.896	1.750
0.12	0.57 – 0.30	1.905	1.747

The coordinates of the upstream profile for different values of (H_a/H_d) are given in the following Table. The upstream profile should merge in a slope of 45° (i.e., 1:1) as shown in Fig. (12.9 b).

(X/H_d)	(H_a/H_d) 0.00	0.08	0.12
		(Y/H_d)	
– 0.000	0.0000	0.0000	0.0000
– 0.020	0.0004	0.0004	0.0004
– 0.060	0.0036	0.0035	0.0035
– 0.100	0.0103	0.0101	0.0099
– 0.120	0.0150	0.0147	0.0144
– 0.140	0.0207	0.0203	0.0199
– 0.150	0.0239	0.0235	0.0231
– 0.160	0.0275	0.0270	0.0265
– 0.175	0.0333	0.0328	0.0323
– 0.190	0.0399	0.0395	0.0390
– 0.195	0.0424	0.0420	—
– 0.200	0.0450	—	—

The toe curve radius equal to $2H_d$ may be provided.

Foundations. A chute spillway may be constructed on any type of foundation capable of sustaining the load without undue deformation.

General arrangement. The trough is usually widest at the crest and then narrows to a width which is determined by the most economical shape of the trough. At the extreme lower end the trough is again widened to reduce the exit velocity.

Longitudinal slope. The slope of the chute or trough is usually kept just sufficient to meet the requirements of flow for as long a distance from the crest as possible without bringing the trough bed into filling. After that the slope is made as steep as possible without endangering its stability or getting into heavy excavation. This arrangement reduces the excavation to a minimum and keeps high velocities and possibility of scour away from the crest.

Side slopes. The trough bed should be kept in cutting throughout its length and its side slopes designed to be stable with an adequate factor of safety, as any slide blocking the trough at the time of floods may lead to overtopping of the main dam with disastrous results.

Free board. Sufficient freeboard must be provided so that water may no spill over the sides of the trough. The minimum freeboard required to be provided may be computed by the following empirical expression :

$$\text{Freeboard (in metres)} = [0.61 + 0.0371\, V d^{1/3}] \qquad ...(12.8)$$

in which V is mean velocity of flow in m/s; and d is depth of flow in m, in the trough reach under consideration.

Vertical curves. At the sections where the bed slope of the trough changes, the adjacent sloping floors should be interconnected through convex and concave vertical curves. Convex curves are provided where the bed slope changes from mild to steep and concave curves are provided where the bed slope changes from steep to mild.

Convex curves should be flat enough to maintain positive pressures and thus avoid the tendency for separation of the flow from the floor. The convex curve commences tangentially from the end of the upstream sloping floor and it should approximately conform to the shape defined by the following equation

$$-y = x \tan\theta + \frac{x^2}{K[4(d+h_v)\cos^2\theta]} \qquad ...(12.9)$$

in which x and y are coordinates of the convex curve with origin at the end of the upstream sloping floor ; θ is slope angle of the upstream sloping floor ; d and h_v are depth of flow and velocity head respectively at the end of the upstream sloping floor ; and K is a factor. To ensure positive pressure along the entire contact surface of the curve, K should be equal to or greater than 1.5.

Concave curves should have a sufficiently long radius of curvature to minimize the dynamic pressure exerted on the floor by the centrifugal force which results from a change in the direction of flow. The radius of curvature of a concave curve is given by

$$R = \frac{w}{g}\frac{dV^2}{p} = \frac{w}{g}\frac{qV}{p} \qquad ...(12.10)$$

where d is depth of flow ; V is velocity of flow ; q is discharge per unit width which is equal to ($d \times V$) ; p is normal dynamic pressure exerted on the floor ; w is unit weight of water ; and g is acceleration due to gravity. However, the radius of the concave curve should not be less than $10d$. Further the concave curve is so located that it meets both the sloping floors tangentially.

Horizontal curves. In addition to vertical curves, horizontal curves may also be required to be provided if it is not possible to have a straight trough. The horizontal curves should be quite gentle and in order to account for superelevation in the curved portion the trough bed should be provided with a cross slope (i.e., slope along the width of the trough) so that the bed is gradually raised from a lower level at the inner bank to a higher level at the outer bank. However, as far as possible both vertical as well as horizontal curves should be avoided.

Concrete paving. Except in the case of solid rock with no cracks concrete paving of the trough is necessary in all cases. The thickness of the concrete paving may generally range from 300 mm to 375 mm. Light reinforcement of about 0.25 to 1 percent of concrete area is usually provided each way in the top of the paving. To avoid cracking, the concrete paving should be poured in square panels with contraction joints on all sides [Fig. 12.10 (a), (b), (c)]. The panels of size ranging from 9 m to 12 m are usually adopted. The reinforcement should not be continuous through the contraction joints and the surface of the joint should be treated to permit free movement due to contraction and expansion.

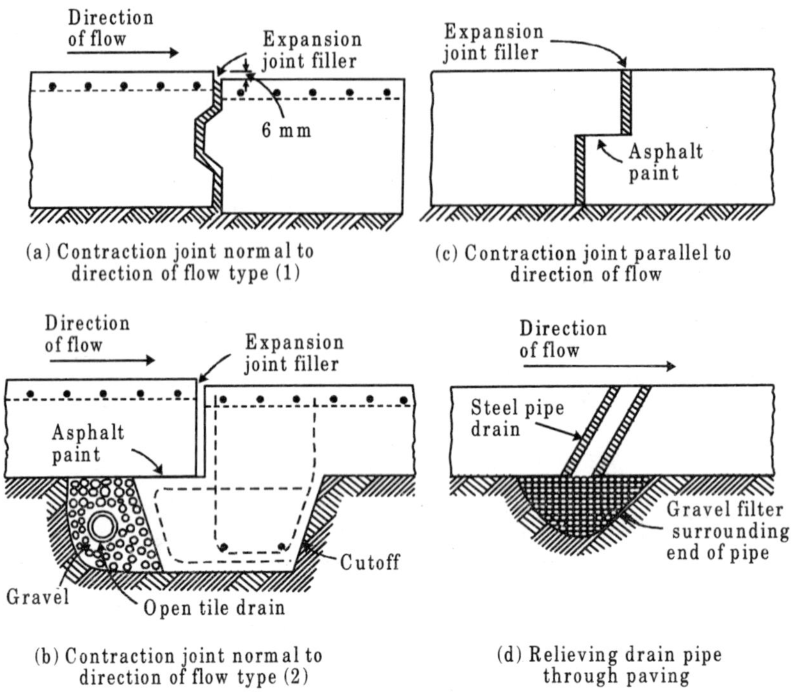

Fig. 12.10 *Contraction joints in paving of a chute spillway ; and drain pipe through paving*

If the paving panel cannot be made heavy enough to resist the uplift pressure it may be provided with hold down piles if the foundation is earth or anchorage rods if the foundation is rock. Anchorages for the paving panels consist of steel rods grouted into holes drilled in the rock and tied to the concrete panel.

Cutoffs. For a chute spillway three types of cutoffs are usually provided. A cutoff at the upper end of the spillway is provided to reduce the uplift pressure on the paving. At the downstream end of the paving a cutoff is provided to prevent undercutting of the paving. Further at the upstream end of each panel a cutoff is provided to prevent creeping of panels resulting from expansion and contraction due to changes in temperature as well as to prevent flow of water from one panel to another along the underside of the paving. A typical cutoff of this type is shown in Fig. (12.10 b).

Drainage. A drainage system is necessary under the concrete paving to prevent uplift from ground water or water that finds its way through the paving during operation of the spillway. If the paving is on a rock foundation, the drainage system generally consists of gravel filled trenches under the paving, with sometimes an open tile drain imbedded in the gravel. The drains are either relieved at intervals through the paving (Fig. 12.10 d), or collected into one or more trunk drains which carry the entire flow to an outlet at the lower end of the trough.

12.9 SIDE CHANNEL SPILLWAY

A side channel spillway (Fig. 12.11) is one in which water after flowing over the crest is carried in a channel parallel to the crest. Thus, the distinctive feature of a side channel spillway which

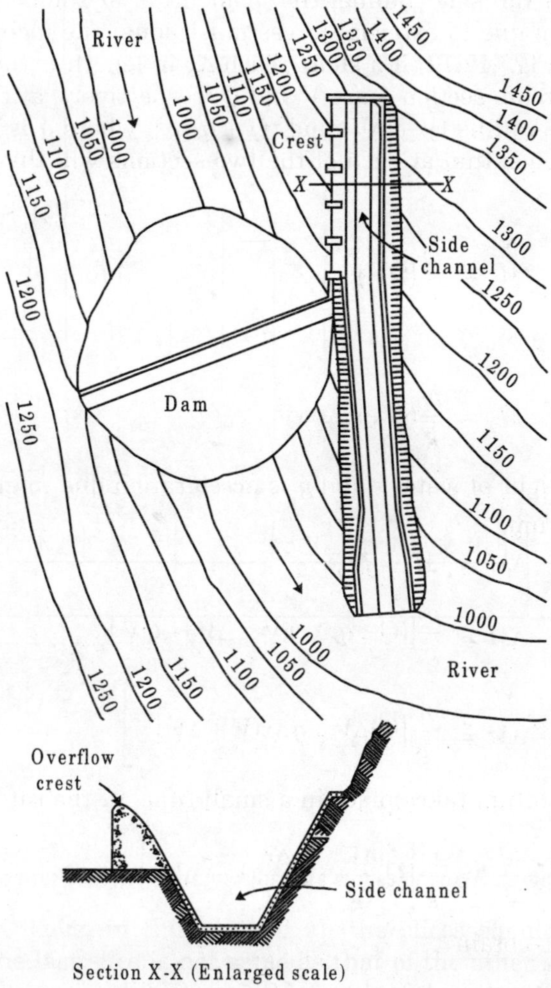

Fig. 12.11 *Side channel spillway*

distinguishes it from a chute spillway is that whereas in a chute spillway water flows at right angles to the crest after passing over it, in a side channel spillway water flows parallel to the crest after passing over it. A side channel spillway is suitable for dams in narrow canyons and for other situations where overflow spillway cannot be provided and where space required for a chutespillway of adequate crest length is not available. The control structure for this spillway is also a weir which is usually ogee shaped.

Hydraulic Theory for the Design of Side Channel Spillway

The theory of flow in a side channel spillway is based principally on the law of conservation of linear momentum with the assumption that the only forces producing motion in the channel result from the fall in the water surface in the direction of the axis of the channel. This premise assumes that the entire energy of the flow over the crest is dissipated through its intermingling with the channel flow and is therefore of no assistance in moving the water along the channel. Axial velocity is produced only after incoming water particles join the channel stream.

For any short reach of the side channel the momentum at the beginning of the reach plus any increase in momentum due to external forces must equal the momentum at the end of the reach. Thus, as shown in Fig. (12.12) if a short reach Δx in length is considered and the velocity and discharge at the upstream section 1 are V and Q, respectively, at the downstream section 2 the velocity and discharge will be $(V + \Delta V)$ and $(Q + q\,\Delta x)$, where q is inflow per unit length of the spillway crest. The momentum at each of the two sections will therefore be :

Momentum at section 1,

$$M_1 = \frac{w}{g}(QV)$$

Momentum at section 2,

$$M_2 = \frac{w}{g}(Q + q\,\Delta x)(V + \Delta V)$$

where w is specific weight of water ; and g is acceleration due to gravity.

∴ Change in momentum

$$\Delta M = (M_1 - M_2)$$

or

$$\Delta M = \frac{w}{g}\left[(Q + q\,\Delta x)(V + \Delta V) - QV\right]$$

or

$$\Delta M = \frac{w}{g}\left[(Q\Delta V + q\Delta x(V + \Delta V)\right]$$

If this change in momentum takes place in a small time Δt, the rate of change of momentum

$$\frac{\Delta M}{\Delta t} = \frac{w}{g}\left[Q\frac{\Delta V}{\Delta t} + q\frac{\Delta x}{\Delta t}(V + \Delta V)\right]$$

As $\Delta t \to 0$, $\Delta V \to 0$, we obtain

$$\frac{dM}{dt} = \frac{w}{g}\left[Q\frac{dV}{dt} + q\frac{dx}{dt}V\right] \qquad \text{...(i)}$$

If (dy/dx) is the slope of water surface, then in accordance with the above noted assumption, we have

$$\frac{dM}{dt} = wQ\frac{dy}{dx} \qquad \text{...(ii)}$$

Thus, equating the values of (dM/dt) given by Eqs (i) and (ii), we get

$$wQ\frac{dy}{dx} = \frac{w}{g}\left[Q\frac{dV}{dt} + q\frac{dx}{dt}V\right]$$

or

$$\frac{dy}{dx} = \frac{1}{g}\left[\frac{dV}{dt} + \frac{q}{Q}\frac{dx}{dt}V\right]$$

or

$$\frac{dy}{dx} = \frac{1}{g}\left[\frac{dV}{dx}\frac{dx}{dt} + \frac{q}{Q}\frac{dx}{dt}V\right]$$

Since $\frac{dx}{dt} = V$, we have

$$\frac{dy}{dx} = \frac{1}{g}\left[V\frac{dV}{dx} + \frac{q}{Q}V^2\right]$$

or

$$dy = \frac{1}{g}\left[V\frac{dV}{dx} + \frac{q}{Q}V^2\right]dx \qquad \text{...(iii)}$$

By integrating both sides of Eq. (iii), we get

$$y = \frac{1}{g}\int_0^x\left[V\frac{dV}{dx} + \frac{q}{Q}V^2\right]dx \qquad \text{...(iv)}$$

This is the equation for water surface profile that will be developed in the side channel when water flowing over the crest of the spillway enters the side channel. The ordinate y of the water surface profile is measured downward from the line AB (Fig. 12.12) which represents the submergence limit.

For solving Eq. (iv) the following assumptions are made. (i) The discharge q per unit length of the spillway crest is constant, thus the total discharge Q at a section x distant from the upstream end of the crest is given by

$$Q = qx \qquad \text{...(v)}$$

(ii) The velocity-distance relation may be expressed by an exponential equation as indicated below

$$V = ax^n \qquad \text{...(vi)}$$

where a and n are arbitrary constants and V and x denote, respectively, average velocity and distance from the upstream end of the crest.

From Eq. (v), we have

$$\frac{dV}{dx} = an\,x^{n-1}$$

Substituting these values for V, Q and (dV/dx) in Eq. (iv) and integrating it, we get

$$y = \frac{a^2}{g}\frac{n+1}{2n}x^{2n} \qquad \text{...(vii)}$$

Since $a^2x^{2n} = V^2$ and $(V^2/2g) = h_v$, Eq. (12.11) may also be expressed as

$$y = \frac{n+1}{n}h_v \qquad \text{...(viii)}$$

The spillway channel may be completely determined by Eqs. (v) to (viii), its cross-sectional shape and the values of a and n. The proper choice of these factors is controlled by economic considerations.

The cross sectional shape of the channel is governed by economy of excavation at site if it will not be lined and by economy of lining if it will be lined.

The water surface profile is controlled by the values of a and n. Figure 12.12 (b) shows the water surface profiles and channel bottoms for three different values of n. If n is less than 0.5,

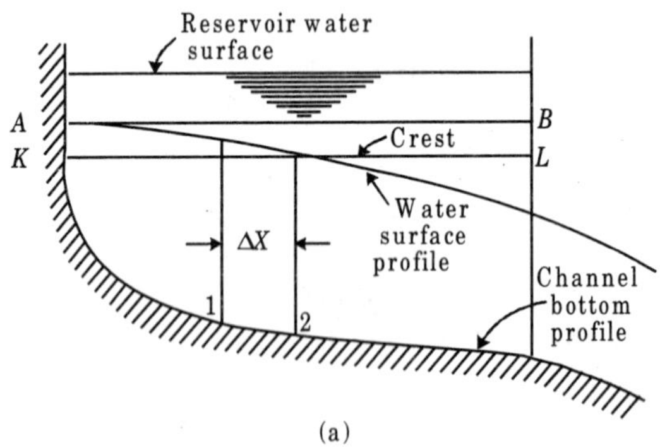

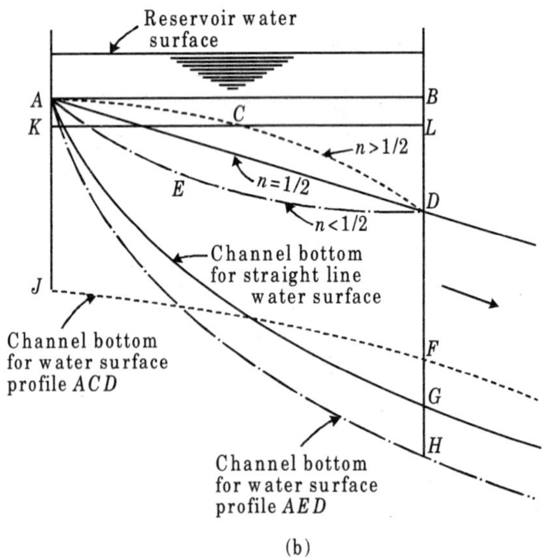

Fig. 12.12 *Profile of water surface and channel bottom for side channel spillway*

the water surface profile will be concaved upward as *AED* and the channel bottom will follow some line such as *AH*. If n is equal to 0.5, the water surface profile is straight and the channel bottom will follow some line such as *AG*. If n is greater than 0.5, the water surface profile will be convexed upward as *ACD*. If n is equal to 1.0, *ACD* will be a parabola and the channel bottom will be a parallel curve such as *JF*. If n lies between 0.5 and 1.0 the bottom line will start at *A*. If n is greater than 1.0, the bottom line theoretically drops to an infinite depth at the upstream end of the crest, then rises rapidly approximately to the line *JF*, which it crosses before reaching *F*. As such from economical considerations the value of n should be between 0.5 and 1.0.

The selection of values of a and n for greatest economy depends on topography and can be made by trial estimates. However, the number of trials required can be reduced by using the condition for minimum excavation at any specified section along the spillway crest which is as derived below.

For minimum excavation at any specified section along the spillway crest the sum of the depth of flow d and the ordinate y should be minimum.

Let $$H = y + d$$

Substituting the value of y from Eq. (viii), we get

$$H = \frac{n+1}{n} h_v + d$$

or $$H = \frac{n+1}{n}\frac{V^2}{2g} + d$$

or $$H = \frac{n+1}{n}\frac{Q^2}{2gA^2} + d$$

Differentiating both sides with respect to d, we get

$$\frac{dH}{dd} = -\frac{n+1}{n}\frac{Q^2}{gA^3}\frac{dA}{dd} + 1$$

Since $(dA/dd) = T$ where T is the width of the channel at the water surface, we have

$$\frac{dH}{dd} = -\frac{n+1}{n}\frac{Q^2}{g}\frac{T}{A^3} + 1$$

For H to be minimum, $(dH/dd) = 0$ and hence, we get

$$\frac{n+1}{n}\frac{Q^2}{2gA^2} = \frac{A}{2T}$$

or $$\frac{n+1}{n} h_v = \frac{A}{2T} \qquad \text{...(ix)}$$

or $$y = \frac{A}{2T} \qquad \text{...(x)}$$

Substituting the value of y from Eq. (vii), we have

$$\frac{a^2}{g}\frac{n+1}{2n}x^{2n} = \frac{A}{2T} \qquad \text{...(xi)}$$

Thus, having chosen a value for either a or n, the corresponding value of the other required to give the greatest economy at a specified section can be found from these equations.

Design Procedure

For designing side channel of the spillway of known discharging capacity Q and crest length L, the following procedure is adopted.

(i) Consider any section and assume a depth of flow d.

(ii) Compute A and T.

(iii) Assume a value of n.

(iv) Find h_v from Eq. (ix)

(v) Compute discharge from the equation

$$Q' = A\sqrt{2gh_v}$$

(vi) If Q' is equal to the required discharge Q, the assumed depth of flow is correct, otherwise assume another depth and repeat the procedure till the computed discharge Q' is equal to the required discharge Q.

(vii) Compute the corresponding value of a.

With a and n being known water surface profile and channel bottom profile may be determined. Since the value of a and n thus determined are based on computations at some selected single section, they are not necessarily the best values for the spillway as a whole. In order to obtain the best values of a and n by taking different combinations of values of a and n a number of channels are designed and their cost estimates are prepared. The best combination of values of a and n will be the one corresponding to which the cost will be minimum.

The limit upto which spillway crest may be submerged at A (Fig. 12.12 b) without seriously reducing the discharge has a significant effect on economy. This is due to the fact that channel bottom profile is obtained by measuring $(d + y)$ downward from line AB which represents the permissible submergence limit. It has been observed that the crest may be submerged at A by $\frac{1}{2}$ to $\frac{2}{3}$ of the head without seriously reducing the total discharge.

12.10 SHAFT OR MORNING GLORY SPILLWAY

A shaft or morning glory spillway consists of a vertical flaring funnel with its top as the crest of the spillway and its lower end connected to a vertical or inclined shaft which in turn is connected at its lower end to a horizontal or near horizontal conduit or tunnel extending through or around the dam (Fig. 12.13). Thus, this spillway has a circular crest. The water flowing over the crest of spillway enters the shaft from which it is delivered by the conduit or tunnel to the river below the dam. There are two types of shaft spillways viz.,(1) standard crested shaft spillway and (2) flat crested shaft spillway. In the case of a standards crested shaft spillway the crest is shaped to conform to the profile of the lower surface of the nappe of water flowing over a circular sharp crested weir of the same diameter discharging at a head equal to the design head of the spillway (Fig. 12.13 a). On the other hand a flat crested shaft spillway has a wide flat crest similar to that of a broad crested weir (Fig. 12.13 b). The coefficient of discharge is higher for a standard crested shaft spillway than that for a flat crested shaft spillway and hence the former requires a smaller crest length or diameter of crest than the later. However, the 'crotch' section where the upper nappe from all around the circumference meets is lower in the case of a standard crested shaft spillway than in the case of a flat crested

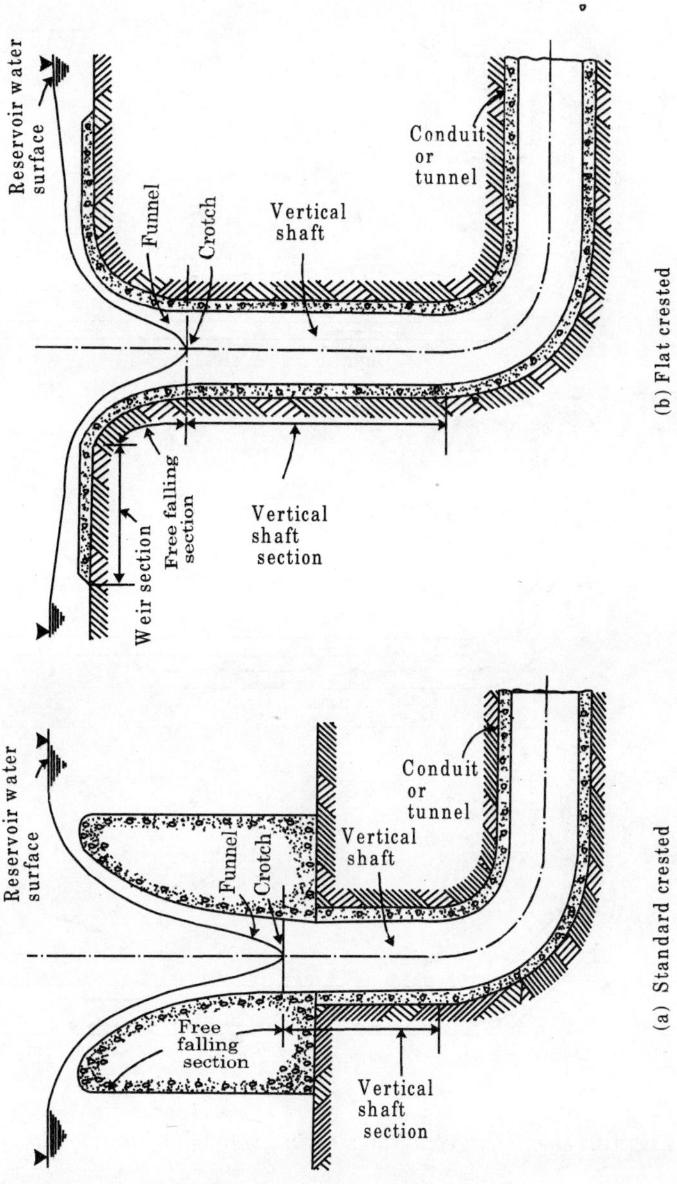

Fig.12.13 *Shaft spillway*

shaft spillway. Thus, a standard crested shaft spillway requires a longer funnel than the one for a flat crested shaft spillway (Fig. 12.13). Moreover, a standard crested shaft spillway has a larger diameter funnel as compared to a flat crested shaft spillway. As such a standard crested shaft spillway is preferable in locations where the spillway has to be constructed in concrete at its top portion as the quantity of concrete required would be smaller, and a broad crested shaft spillway would be preferable where the entire spillway has to be provided in excavation as due to smaller funnel diameter the rock excavation required would be less.

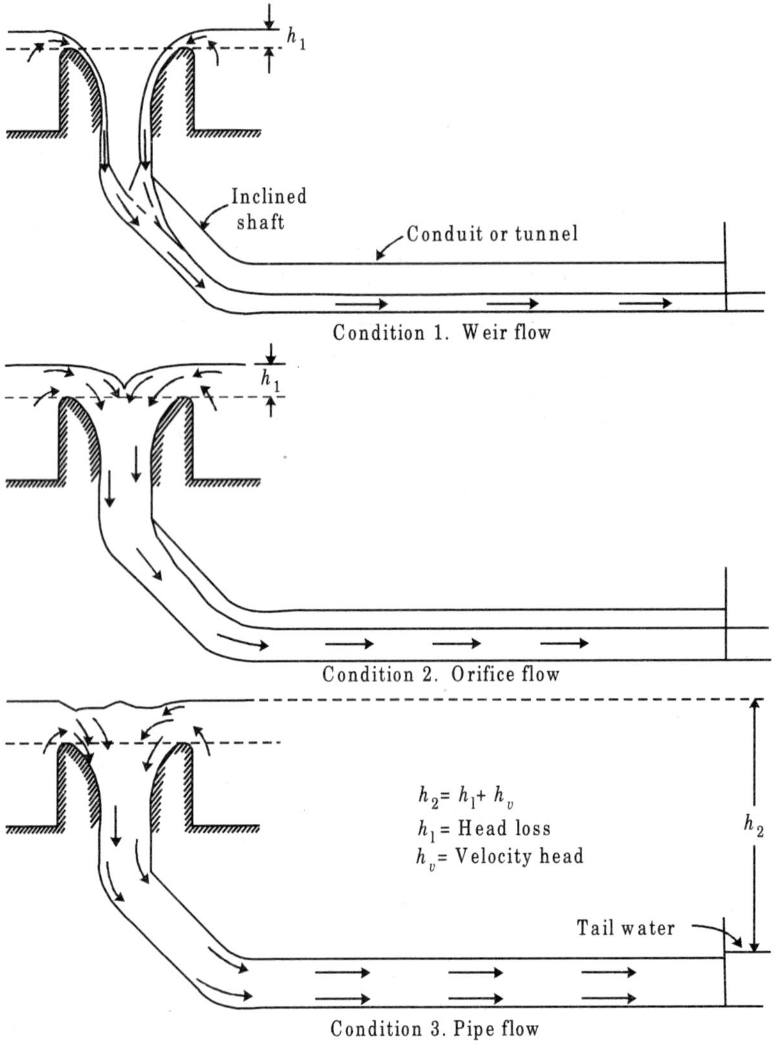

Fig. 12.14 *Different conditions of flow through a shaft spillway*

A shaft spillway may be considered as an alternative to side channel spillway if suitable site is available, and if neither the overflow nor the trough spillways are feasible. A suitable location for a shaft spillway would consists of a rock spur projecting into the reservoir a little distance upstream of the dam. If the top of the spur is lower than the full reservoir level, the spillway has to be constructed above the spur level in concrete ; and if the top of the projecting spur is higher than the full reservoir level, it has to be excavated down to the required crest level. Moreover where diversion tunnels are provided to divert the river during construction, it may be possible to utilise them later for shaft spillway.

On the crest of a shaft spillway radial piers are provided to guide water radially and prevent spiral flow. The same piers also serve to support a bridge around the spillway. As a shaft spillway is surrounded by water on all sides it has to be connected to the dam or the hill side by a bridge.

Depending on the head there are three possible conditions of flow in a shaft spillway (See Fig. 12.14). At low heads the crest of the spillway serves as a weir and hence the discharge of the spillway is proportional to $h^{3/2}_1$ (where h_1, is head over spillway crest) and the outlet conduit flows partly full. As the head is increased water rises in the shaft and the inlet gets submerged, the flow then changes from weir flow to orifice flow, but the outlet conduit still flows only partly full. With further increase in the head the shaft is completely filled with water and the outlet conduit flows full, the flow then becomes a pipe flow and the discharge becomes approximately proportional to $h^{1/2}_2$, where h_2 is the total head on the outlet (Fig. 12.14 c). In this third stage an increase in h_2 results in only a very slight increase in discharge. This in effect, places a limit on the capacity of a shaft spillway.

At the right angle bend between the shaft and the outlet conduit negative pressure may usually be developed. As such particular care has to be taken in designing the bend so that excessive negative pressures are not developed inside the bend which may result in damage due to cavitation. Further an undesirable feature of shaft spillway is that they may get clogged with debris. Thus, trash racks, floating booms, or other types of protection are necessary to prevent debris from entering the inlet.

12.11 CONDUIT OR TUNNEL SPILLWAY

A conduit or tunnel spillway is the one in which a closed channel is used to convey the discharge around or under a dam. The closed channel may be in the form of a vertical or inclined shaft and a horizontal tunnel or conduit. The control structure for this spillway may be in the form of an overflow crest, vertical or inclined orifice entrance and side channel crest. The conduit or tunnel is designed to flow partly full and it is not allowed to flow full because siphonic action may develop due to negative pressure being developed in the conduit. Moreover, in order to ensure free flow in the conduit the ratio of the flow area to the total area of the conduit is often limited to about 75%. Also air vents are provided at critical points along the conduit to ensure an adequate air supply which will avoid unsteady flow through the spillway.

12.12 SIPHON SPILLWAY

A siphon spillway operates on the principle of siphonic action. There are two types of siphon spillways as indicated below.

(i) Saddle siphon spillway

(ii) Volute siphon spillway

(A) Saddle Siphon Spillway

A saddle siphon spillway (or saddle siphon) is shaped like an inverted U tube with unequal legs. The various parts of this spillway are shown in Fig. 12.15. It consists of an air tight reinforced concrete cover called 'hood' or 'cowl' provided over an ogee shaped body wall to form a siphon duct. Hence this spillway is also known as 'hood siphon spillway' or 'hood siphon'. The top of the body wall forms the crest of the siphon spillway and hence it is located at the full reservoir level. The entrance and exit lips of the hood are so shaped that the siphon duct has

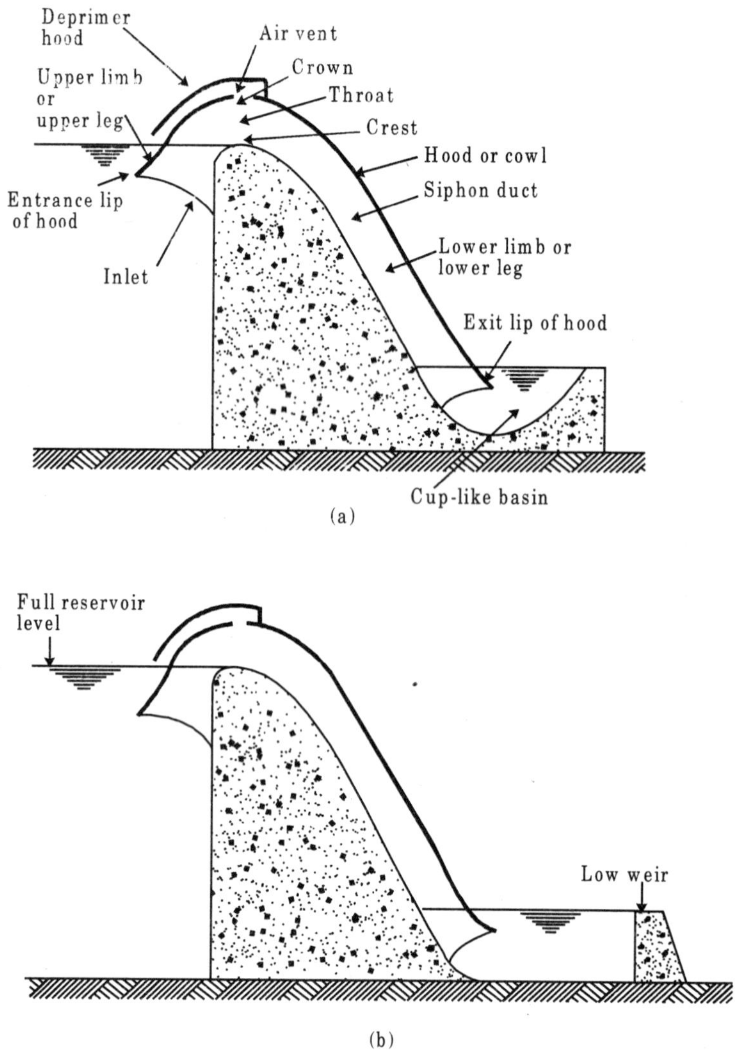

Fig. 12.15 *Saddle siphon spillway*

bell mouth entry and exit. The inlet of the siphon duct is kept submerged well below the full reservoir level so that floating debris etc., does not enter the siphon duct and also the formation of vortices and drawdowns which might break the siphonic action is avoided. The outlet of the siphon duct is kept submerged in a cup-like basin which forms a water seal so that air cannot enter the siphon duct from this end. However, instead of a cup-like basin a cistern may be created by constructing a low weir a little away from the outlet on the downstream of the spillway as shown in Fig. (12.15 b). A small deprimer hood is provided above the main hood and both these hoods are interconnected through an air vent. The inlet of the deprimer hood is kept slightly above the full reservoir level.

At full reservoir level water stands upto the crest of the spillway and hence there is no flow. When the water level rises above the full reservoir level i.e., above the crest of the spillway, water starts flowing over the crest and the inlet of the deprimer hood gets submerged with the result that entry for air into the deprimer hood and the main hood is sealed. Thus, air cannot enter from both the inlet and the outlet of the siphon duct and the air remaining entrapped in

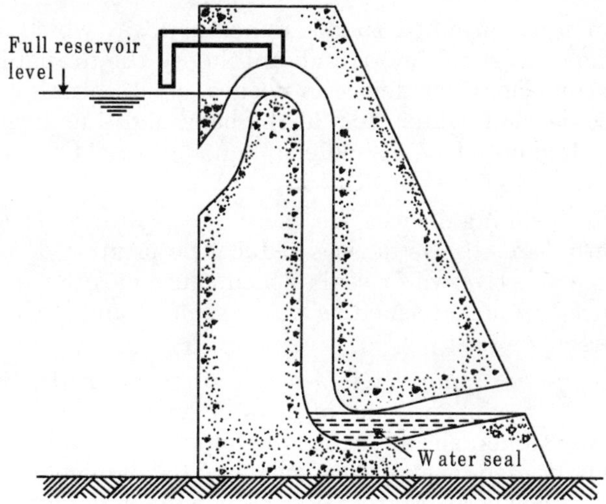

Fig. 12.16 *Saddle siphon spillway with tilted outlet*

the top portion of the siphon duct above the sheet of water flowing over the crest is gradually sucked by the flowing water. As the air is sucked the pressure drops to less than atmospheric in the top portion of the siphon duct which was having atmospheric pressure at the starting of the flow. Thus, a difference of pressure develops between the outside atmosphere and the air inside the siphon duct which creates a suction pull and draws in more water over the crest. This is turn takes more air along with it which further increases the suction pull and draws in more and more water. This sucking action which increases progressively and gradually creates the necessary pull for the commencement of the siphonic action and the siphon duct starts running full. The action of the siphon spillway from the moment the water just begins to flow over the crest to the instant when the siphon duct starts running full in known as *priming,* (or it is said to have 'primed'). It may however be difficult to expel the entrapped air from the siphon duct since it has to be forced through the water in the sealing basin and hence it may not get primed quickly. Moreover relatively large rise in water level above the crest of the spillway may be needed before it gets primed. As such various priming device as described below have been developed to expedite the priming of siphon spillway with relatively less rise in the water level above the crest of the spillway. The depth of water above the crest of a siphon spillway that is required for its priming is known as *priming depth.* The priming depth varies from about $\frac{1}{3}$ to

$\frac{1}{6}$ of the depth of throat depending on the priming device and the rate of rise of water level in the reservoir. It has been observed that priming takes place quickly when the water level rises gradually rather than when there is sudden rise in the water level.

The siphonic action once initiated will continue as long as the water level in the reservoir is above the full reservoir level. The operating head for the spillway is then equal to the difference between the water levels on the upstream and the downstream sides of the spillway. When the water level in the reservoir drops to such a level that the inlet of the deprimer hood gets exposed then air enters the siphon duct and breaks the siphonic action thus stopping the flow. This action is called *depriming* of the siphon spillway.

Figure 12.16 shows another form of a saddle siphon spillway which may be provided within the body of the dam. It has a vertical lower limb and hence the draught of the falling water is enough to cause priming and no other device is needed for its priming. However, its outlet is tilted upwards in order to develop water seal at the bend and thus facilitate early priming. As such it is also known as 'tilted outlet type saddle siphon spillway'. For depriming of the spillway air inlet pipe is provided as shown in Fig. (12.16).

Priming Devices. The priming devices are used to expedite the priming of siphon spillway with relatively small depth. Thus, these devices reduce the priming depth for siphon spillways. The priming devices create effective water seals which cause early priming of siphon spillways. As such the priming devices are also termed as water seals. Some of the priming devices which are commonly used are as follows.

(i) *Baby siphon or auxiliary siphon.* It is a small siphon provided within the main bigger siphon (Fig. 12.17 a). The baby siphon being at a lower level gets primed earlier. The sheet of water issuing from the baby siphon shoots across the lower limb of the main siphon and seals it. The main siphon then gets primed as explained earlier. The arrangement of baby siphon shown in Fig. 12.17 (a) is also known as Maramsilli baby siphon which has been provided for Yaswantsagar Dam, Indore. This arrangement is successful in expediting the priming of the

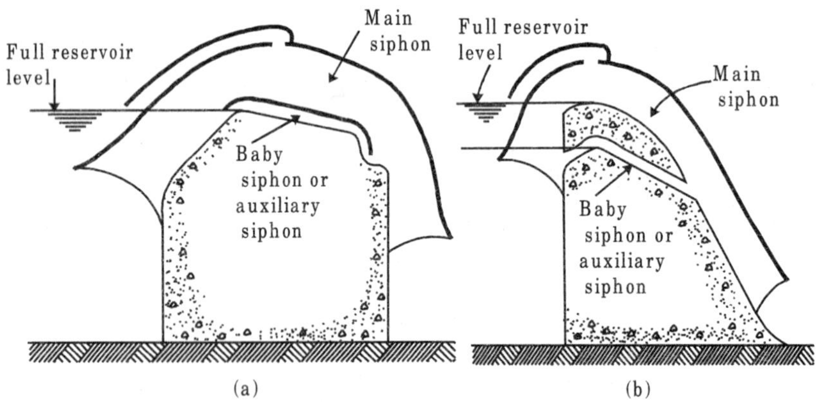

Fig. 12.17 *Baby siphon or auxiliary siphon*

main siphon and also in reducing the priming depth. However, the disadvantage of this arrangement is that the presence of the baby siphon within the main siphon offers obstruction to flow. Further it is also liable to choking by floating debris because its inlet is only slightly lower than the full reservoir level. As such another arrangement as shown in Fig. (12.17 b) may be adopted in which baby siphon is provided within the body wall of the spillway so that both the disadvantages of the earlier arrangement of the baby siphon are eliminated. In this case as soon as the water level rises above the crest of the baby siphon a sheet of water starts flowing through it, thus sealing the lower limb of the main siphon and priming it as in the case of the earlier arrangement. However, in either case as the flow through the main siphon increases, the jet through the baby siphon gets depressed and hence care must be taken that the combined jet of the main and the baby siphons does not strike below the exit lip of the hood. Further the provision of baby siphon is in general expensive and also it is difficult to construct.

(ii) *Joggle or tudel or step.* As shown in Fig. (12.18) a joggle or tudel or step provided on the downstream face of the body wall of a siphon spillway deflects sheet of water flowing over the crest of the spillway to strike against the inner surface of the lower limb of the siphon duct.

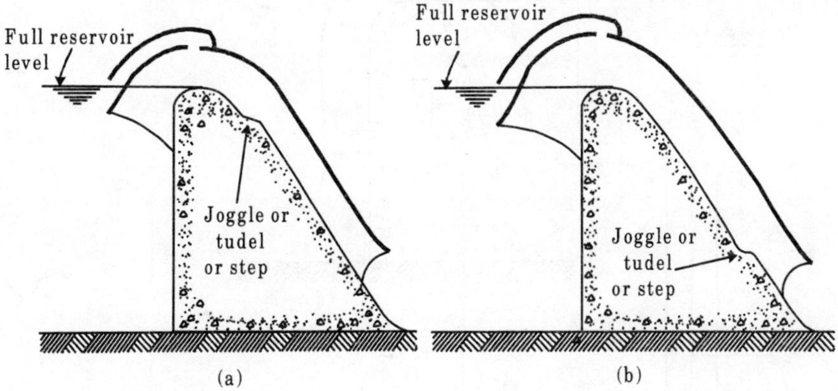

Fig. 12.18 *Saddle siphon spillway with a joggle or tudel or step*

A water seal is thus developed which results in early priming of the spillway. The joggle may be provided either near the crest of the spillway (Fig. 12.18 a) on near the exit lip of the hood (Fig. 12.18 b). In the later case care must be taken to locate the joggle so that the sheet of water deflected by it always strikes the hood above the exit lip. However, the main disadvantage of this device is that the presence of joggle offers resistance to flow when the siphon duct runs full.

(iii) *Hinged steel plate.* A flexible steel plate is provided at the crest of the spillway, which is highed at its top edge and is projecting into the siphon duct as shown in Fig. 12.19. When water starts flowing over the crest of the spillway, the plate deflects the sheet of water to strike

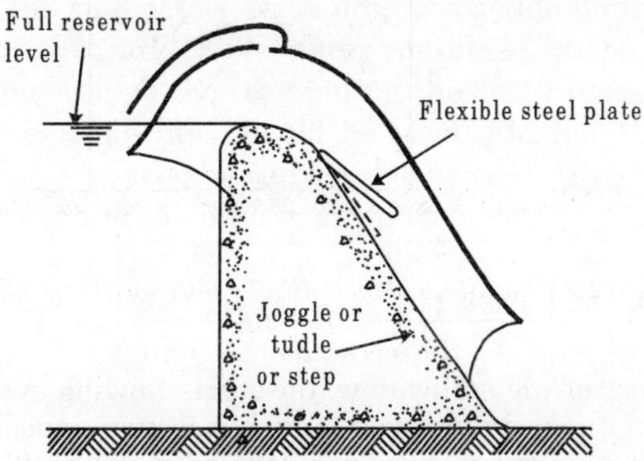

Fig. 12.19 *Saddle siphon spillway with a flexible steel plate*

against the inner surface of the lower limb of the siphon duct. A water seal is thus developed which results in early priming of the spillway. When the siphon duct starts running full the plate is pressed downwards and it remains in contact with the surface of the body wall of the spillway as shown in Fig. (12.19) so that the flow is not obstructed by the plate.

(iv) *Clear overfall.* In this case the downstream face of the body wall of the spillway is vertical or nearly vertical and its top is rounded off (Fig. 12.20) so as to throw the water flowing over the crest of the spillway on to the hood at its exit lip. Thus, a water seal is developed which

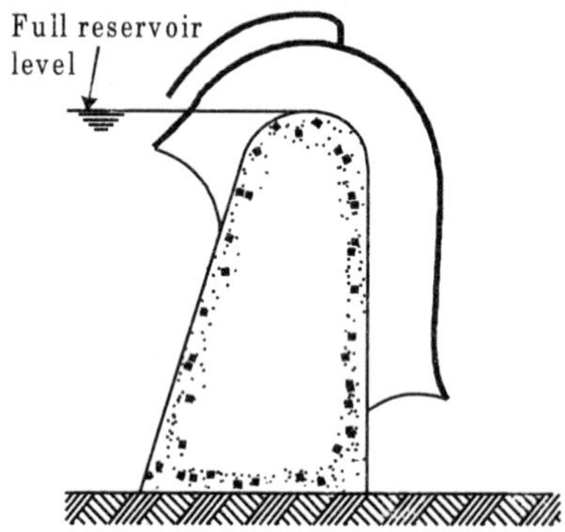

Fig. 12.20 *Saddle siphone spillway with a clear overfall*

results in early priming of the spillway. However, in this case relatively higher priming depth is required to overcome surface tension effects and to induce a higher initial velocity because the distance thrown depends upon the horizontal component of the velocity at the point of throw.

(v) *Priming weir or overfall weir.* In this device a priming weir connected with the main reservoir is provided opposite to the crest of the spillway as shown in Fig. (12.21). The crest of the

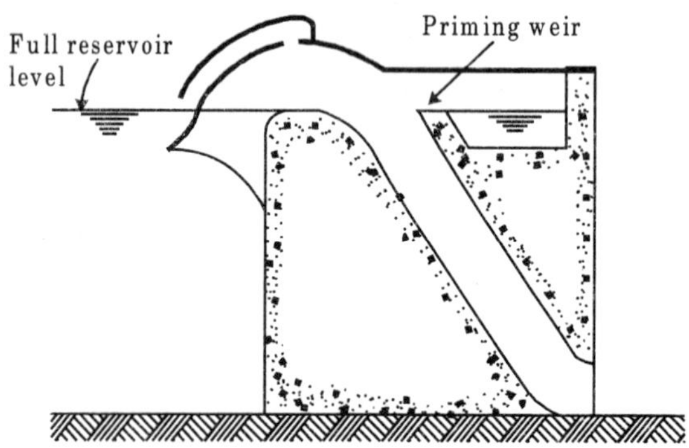

Fig. 12.21 *Saddle siphon spillway with priming weir*

priming weir is at the same level as the crest of the spillway. When the water level in the reservoir rises above the crest of the spillway water starts flowing over the crest of the spillway as well as over the crest of the priming weir and hence two opposite nappes are developed. These nappes intermingle in the siphon duct near the crest of the spillway and thus develop a water seal which results in early priming of the spillway.

Pressure in Siphon Spillway. After the siphon spillway gets primed and is running full, consider point 1 on the water surface in the reservoir, point 2 at the throat section (at the same elevation as point 1) and point 3 at the exit end of the siphon duct if it is discharging freely or on the tailwater surface if the exit end of the siphon duct is held submerged in the tailwater (Fig. 12.22 a).

Let H = operating head for the siphon spillway which is equal to the difference in elevation of points 1 or 2 and 3

V_2 = mean velocity of flow at the throat section of the spillway (i.e., at point 2)

V_3 = mean velocity of flow at the exit end of siphon duct

p_2 = pressure at the throat section of the spillway (i.e., at point 2)

p_a = atmospheric pressure

H_L = head loss in the entire siphon duct (i.e., between points 1 and 3)

h_L = head loss between the throat and the exit end of the siphon duct (i.e., between points 2 and 3), and

h_L' = head loss between the inlet and the throat section (i.e., between points 1 and 2).

Applying Bernoulli's equation between points 1 and 3 both of which being at atmospheric pressure, we get

$$H = H_L + \frac{V_3^2}{2g} \qquad \text{...(i)}$$

or

$$V_3 = \sqrt{2g(H - H_l)} \qquad \text{...(ii)}$$

Similarly applying Bernoulli's equation between points 2 and 3, we get

$$\frac{p_2}{w} + \frac{V_2^2}{2g} + H = \frac{P_a}{w} + \frac{V_3^2}{2g} + h_L \qquad \text{...(iii)}$$

or

$$\frac{p_2}{w} = \frac{P_a}{w} + \left(\frac{V_3^2}{2g} - \frac{V_2^2}{2g}\right) - H + h_L \qquad \text{...(iv)}$$

If the cross-sectional areas of the throat and the exit end of the siphon duct are equal, $V_2 = V_3$ and then from Eq. (iv), we have

$$\frac{p_2}{w} = \frac{P_a}{w} - H + h_L \qquad \text{...(v)}$$

Equation (v) shows that there is negative pressure at the throat section of a siphon spillway as ($h_L < H$). However, on account of the loss of head h_L the negative pressure at the throat section is less than H. Further if the head loss h_L is negligible then Eq. (v) shows that the average absolute pressure at the throat section would fall to zero when the exit end of the siphon duct or the tailwater level were lower by about 10.35 m [as (p_a/w) = 10.35 m of water] below the water surface in the reservoir. This is therefore the limiting value of H which is however only a theoretical value. The practical limit for H is governed by the limit upto which negative or vacuum pressure may be allowed to be developed to avoid cavitation at the throat section. The cavitation may be caused when the pressure drops below the value at which the air dissolved in water starts escaping and hence the pressure p_2 at the throat section may be reduced only upto this value. The limiting value of H may therefore be obtained as follows.

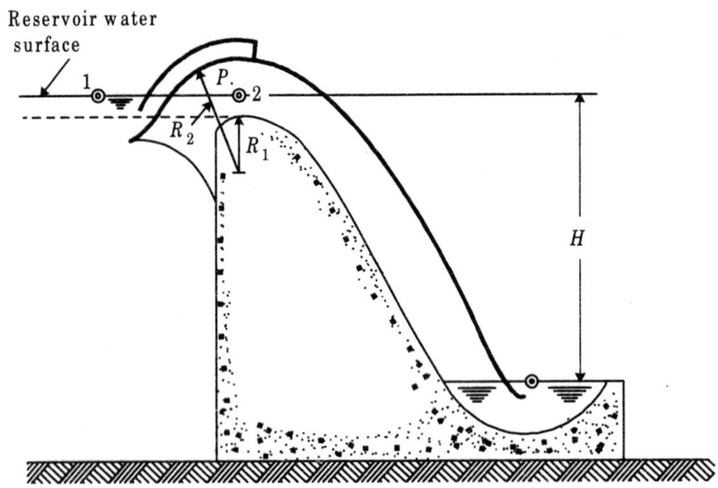

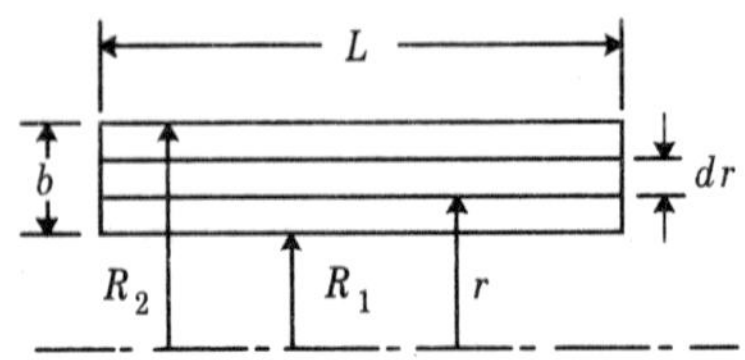

Fig. 12.22 *Analysis of flow through saddle siphon spillway*

From Eq. (v), we have

$$H = \frac{p_a}{w} - \frac{p_2}{w} + h_L \quad \text{...(vi)}$$

If p_v is the absolute pressure at which the dissolved air starts escaping then since $p_2 = p_v$, we have

$$H = \frac{p_a}{w} - \frac{p_v}{w} + h_L \quad \text{...(12.11)}$$

Equation (12.11) gives the limiting value of H and hence in order to avoid cavitation at the throat section

$$H \le \left(\frac{p_a}{w} - \frac{p_v}{w}\right) + h_L \quad \text{...(12.11 a)}$$

For water average value of p_v is equal to about 23.84 kN/m^2 [0.243 kg(f)/cm^2] and hence (p_v/w) = 2.43 m.

Thus, introducing (p_a/w) = 10.35 m and (p_v/w) = 2.43 m in Eq. (12.11), we get

$$H = (10.35 - 2.43) + h_L$$

or
$$H = 7.92 + h_L \qquad ...(12.12)$$

Further in order to avoid cavitation at the throat section

$$H \leq (7.92 + h_L) \qquad ...(12.12\text{ a})$$

Similarly from Eq. (iv) the limiting value of H is obtained as

$$H = \frac{P_a}{w} - \frac{P_v}{w} + \left(\frac{V_3^2}{2g} - \frac{V_2^2}{2g}\right) + h_L \qquad ...(12.13)$$

and in order to avoid cavitation at the throat section

$$H \leq \left(\frac{P_a}{w} - \frac{P_v}{w}\right) + \left(\frac{V_3^2}{2g} - \frac{V_2^2}{2g}\right) + h_L \qquad ...(12.13\text{ a})$$

Further introducing (p_a/w) = 10.35 m and (p_v/w) = 2.43 m in Eq. (12.13), we get

$$H = 7.92 + \left(\frac{V_3^2}{2g} - \frac{V_2^2}{2g}\right) + h_L \qquad ...(12.14)$$

and in order to avoid cavitation at the throat section

$$H \leq 7.92 + \left(\frac{V_3^2}{2g} - \frac{V_2^2}{2g}\right) + h_L \qquad ...(12.14\text{ a})$$

Equation (12.13) indicates that the limiting value of H is increased by losses and may further be increased by reducing the area at the exit end of the siphon duct from that of the throat section so that V_3 will be greater than V_2.

Again, applying Bernoulli's equation between points 1 and 2, we get

$$\frac{P_a}{w} = \frac{p_2}{w} + \frac{V_2^2}{2g} + h_L$$

or
$$\frac{V_2^2}{2g} = \frac{P_a}{w} - \frac{p_2}{w} - h_L'$$

Introducing (p_a/w) = 10.35 m and (p_2/w) = 2.43 m, we get

$$\frac{V_2^2}{2g} = 7.92 - h_L' \qquad ...(12.15)$$

or
$$V_2 = \sqrt{2g(7.92 - h_l')} \qquad ...(12.15\text{ a})$$

Equation (12.15 a) gives the maximum velocity of flow at the throat section.

Discharge through Saddle Siphon Spillway

The discharge through saddle siphon spillway is given by

$$Q = CA\sqrt{2gH} \qquad \text{...(12.16)}$$

where Q = discharge ;

C = coefficient of discharge the value of which may be taken as 0.65

A = area of cross-section of throat = $L \times b$, where L is the length of the spillway and b is the height of the throat

g = acceleration due to gravity, and

H = operating head for the siphon spillway.

However, another expression for discharge through saddle siphon spillway may be obtained on the assumption of free vortex condition for the flow at the throat as indicated below.

As shown in Fig. (12.22 a) let R_1 be the radius of the crest and R_2 be the radius of the crown. Consider any point P on the vertical section at the throat and let its radius be r. The L-section of the throat is shown in Fig. (12.22 b) which is rectangular having length L and height b.

If v_1, v_2 and v are the velocities of flow at crest, crown and point P respectively, then according to the condition of free vortex flow, we have

$$v_1R_1 = v_2R_2 = vr = \text{constant} \qquad \text{...(i)}$$

If h_0 is the maximum negative head at the crest, then

$$v_1 = \sqrt{2gh_0}$$

Hence from Eq. (i), we have

$$v = \frac{v_1R_1}{r} = \sqrt{2gh_0}\,\frac{R_1}{r} \qquad \text{...(ii)}$$

Consider a thin rectangular strip of thickness dr at P and let dQ be the discharge through this strip. Thus,

$$dQ = L\,dr\,v = L\,\frac{v_1R_1}{r} = \sqrt{2gh_0}\,\frac{R_1}{r}\,dr$$

Hence total discharge Q through the throat section of the spillway is given by

$$Q = \int_{R_1}^{R_2} dQ = \int_{R_1}^{R_2} L\sqrt{2gh_0}\,\frac{R_1}{r}\,dr$$

or

$$Q = LR_1\sqrt{2gh_0}\,\log_e\left(\frac{R_2}{R_1}\right) \qquad \text{...(12.17)}$$

As explained earlier if cavitation is to be avoided, the maximum negative head at the crest may be 7.92 m.

Thus, substituting h_0 = 7.92 m in Eq. (12.17), we get

$$Q = LR_1 \log_e \left(\frac{R_2}{R_1}\right)$$

or
$$Q = 12.47\, LR_1 \sqrt{2g \times 7.92} \log_e \left(\frac{R_2}{R_1}\right) \quad ...(12.18)$$

or
$$Q = 28.71 LR_1 \log_{10} \left(\frac{R_2}{R_1}\right) \quad ...(12.18\text{ a})$$

Further an expression for the head loss H_L in the entire siphon duct may be obtained as explained below.

$$\text{Discharge } Q = AV_2$$

where A is area of cross-section of the throat section ; and V_2 is mean velocity of flow at the throat section.

If the cross-sectional areas of the throat and the exit end of the siphon duct are equal

$$V_2 = V_3 = \sqrt{2g(H - H_L)}$$

$$\therefore \quad Q = A\sqrt{2g(H - H_L)}$$

But from Eq. (12.16)

$$Q = CA\sqrt{2gH}$$

Thus, equating both the values of Q, we get

$$A = \sqrt{2g(H - H_L)}\, CA\sqrt{2gH}$$

or
$$(H_A - H_L) = C^2 H$$

$$\therefore \quad H_L = (1 - C^2) H \quad ...(12.19)$$

Classification of Siphon Spillway according to the Operating Head

As mentioned earlier the operating head for a siphon spillway is the difference between the elevations of the water surface in the reservoir and the tailwater surface if the exit end of the siphon duct is held submerged in the tailwater or the exit end of the siphon duct if it is discharging freely. Siphon spillways may be classified according to the operating head into the following two categories.

(i) High head siphon spillway

(ii) Low head siphon spillway.

(i) *High head siphon spillway.* A siphon spillway which operates under a head more than 10.35 m or the barometric height of water column is known as a high head siphon spillway. As indicated earlier the operation of a siphon spillway at such a high head without causing cavitation may be made possible by reducing the area at the exit of the siphon duct from that at the throat section so that the head is increased by an amount equal to $\left(\frac{V_{3^2}}{2g}\frac{V_{2^2}}{2g}\right)$. The operating head for this case is thus given by Eq. (12.13) as

$$H = \left(\frac{p_a}{w} - \frac{p_v}{w}\right) + \left(\frac{V_{3^2}}{2g} - \frac{V_{2^2}}{2g}\right) + h_L$$

It may however be stated that for any siphon spillway even if the operating head $H < 10.35$ m or the barometric height of water column, it is considered as a high head siphon spillway if

$$H > \left(\frac{p_a}{w} - \frac{p_v}{w}\right) + h_L$$

(ii) *Low head siphon spillway.* A siphon spillway which operates under a head less than 10.35 m or the barometric height of water column is known as a low head siphon spillway. In this case the cross-sectional areas of the throat and the exit end of the siphon duct are usually kept equal and hence the operating head is given by the following equation.

$$H = \left(\frac{p_a}{w} - \frac{p_2}{w}\right) + h_L \qquad \text{...(12.20)}$$

However, if in any case very low operating head is available then the negative pressure at the throat section and the discharge of the siphon spillway can be increased by having the siphon duct with its lower limb made divergent so that the area of its exit end is larger than that of the throat section. This will result in making V_3 less than V_2 and in increasing the negative pressure at the throat section.

(B) Volute Siphon Spillway

Volute siphon spillway (or volute siphon) is a special type of siphon spillway which was designed by Ganesh Iyer in India and hence it is also known as Ganesh Iyer siphon. It consists of a vertical pipe or shaft or barrel which opens out in the form of a funnel at the top and at the bottom it is connected to a right angle bend which in turn is connected to a horizontal or nearly horizontal outlet conduit (Fig. 12.23) to lead water away from the toe of the dam. The top or lip of the funnel is kept at the full reservoir level. The inner sloping surface of the funnel is provided with a number of volutes which are the curved vanes or projections (like the blades of a centrifugal pump or turbine) to induce a spiral motion to the water passing along them. A dome supported on a cylindrical drum is provided over the funnel, leaving an annular space around the funnel. The drum is supported on a number of pillars and is open at the bottom. Over the dome small air pipes are provided to serve as deprimers. One end of the air pipe is connected to the interior of the dome at its crown and the other end is open to atmosphere which is kept at the full reservoir level.

When the water level in the reservoir rises above the full reservoir level it spills over the entire circumference of the lip of the funnel and flows along the volutes with a spiral motion forming a vortex in the vertical pipe. The vortex so formed induces a strong suction pull creating a powerful vacuum which sets the siphon in action. When the water level falls to full reservoir level air is let in through the air pipes which stops the siphonic action.

A volute siphon spillway has a lower priming depth and a higher discharge efficiency than a saddle siphon spillway. Further priming of volute siphon spillway takes place on account of vortex flow into the vertical pipe and hence no other priming device is usually necessary. However, sometimes in order to make an effective water seal a baffle or step is provided in the outlet conduit. A volute siphon spillway has an additional advantage that it can be provided with earth dam also because it is quite independent of the dam section, whereas saddle siphon spillway is possible only with a concrete dam.

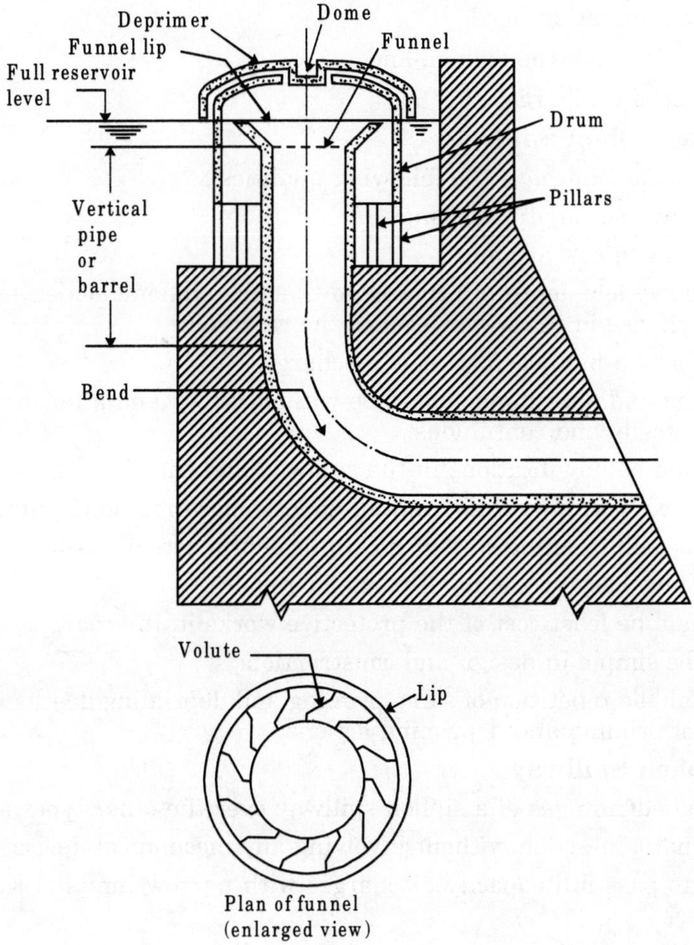

Fig. 12.23 *Volute siphon spillway*

Another type of volute siphon spillway has been developed in which there is no funnel but the volutes are provided on the outer side of the vertical pipe near its top. It is known as external volute siphon spillway in order to distinguish it from the earlier type which is known as internal volute siphon spillway. The idea of providing external volutes is to impart spiral motion to water even before it enters the vertical pipe. However, the external volute siphon spillway has a higher priming depth and its coefficient of discharge is also low.

The discharge through a volute siphon spillway is given by

$$Q = A\sqrt{2g(H - H_L)} = CA\sqrt{2gH} \quad ...(12.21)$$

in which

Q = discharge

A = area of cross-section of pipe

g = acceleration due to gravity

H = maximum operating head

H_L = head loss through the siphon, and

C = coefficient of discharge.

Qualities of a Good Siphon Spillway

A good siphon spillway should have the following qualities :

(1) It should have automatic working.

(2) It should have quick priming.

(3) It should have low depth of priming so that the siphonic action may commence even with a small rise in the water level in the reservoir.

(4) It should have a high coefficient of discharge.

(5) The priming and depriming should be without shock and after it has primed the flow should be steady and continuous.

(6) There should be no vibrations in the spillway and the dam when it is running full.

(7) It should have less initial cost as well as less operation and maintenance cost.

(8) It should be self cleaning so that no silt or debris gets deposited in the siphon duct.

(9) It should have easy access to all parts for repairs.

(10) There should be least cost of the protective works in the rear.

(11) It should be simple in design and construction.

(12) There should be repetition of same priming and depriming depths i.e., there should be certainty of priming and depriming actions.

Advantages of Siphon Spillway

The following are the advantages of a siphon spillway over the other types of spillways.

(1) It is automatic in action without involving any mechanical devices or moving parts.

(2) It is able to pass full capacity discharges with narrow limits of rise of water level in the reservoir.

(3) Its discharge per unit length is more because it has a higher operating head.

(4) There is practically no maintenance cost and it is leak proof.

(5) The cost of acquisition of areas which will get submerged between maximum water level and full reservoir level is minimum in this case.

(6) The height of the dam above the crest of the spillway is also reduced correspondingly.

(7) It is also useful in sucking up sediment from the bed of the reservoir when it is in full action during floods.

(8) Its cost of construction compares favourably well with other types of spillways.

Limitations of Siphon Spillway

The following are the limitations of a siphon spillway.

(1) It is unable to pass ice and debris.

(2) There is a possibility of clogging of the siphon duct and siphon breaker vents with debris or leaves.

(3) As a result of erratic make-and-break action of the siphon sudden surges and stoppages of outflow may occur which may cause considerable fluctuations in the water level in the river on the downstream side.

(4) If a single siphon is provided there is a possibility of outflows from the reservoir being more than the inflows. However, approximate balance between outflow and inflow can be obtained by providing a series of smaller siphons with their siphon breaker vents set at different elevations so that the siphons get primed at different instants as the water level gradually increase in the reservoir.

(5) It requires strong foundation to resist vibrations which are usually quite severe in this case.

(6) Even a minute crack in the cover would interfere with the functioning of a siphon spillway due to sucking up of air.

(7) In the case of volute siphon spillways there may be a problem of negative pressures at the bend.

However, inspite of these limitations siphon spillways may be considered suitable in the case of gorges of limited width and moderate flood discharges in view of their advantages.

12.13 ENERGY DISSPATIONS BELOW SPILLWAY —SCOUR PROTECTION BELOW SPILLWAY

Water flowing over a spillway acquires very high velocity because of the conversion of the potential energy possessed by the water on the upstream side to kinetic energy as it flows down. If the water flowing with such a high velocity is discharged into the river it will scour the river bed and if not checked the scour will extend backwards (or retrograde) and endanger the spillway as well as the dam. As such in order to protect the river bed against scour the higher kinetic energy of the water should be dissipated before it is discharged into the river. For the dissipation of the excessive kinetic energy possessed by the water the following two methods are commonly adopted.

(1) By developing a hydraulic jump.

(2) By using different types of buckets.

1. Hydraulic Jump. Hydraulic jump is defined as a sudden and turbulent passage of water from supercritical to subcritical state. The flow in a hydraulic jump is accompanied by the formation of extremely turbulent rollers and there is considerable dissipation of energy. As such the hydraulic jump is the most suitable and effective means of energy dissipation. However, the amount of energy dissipated in a jump varies with the type of jump as indicated below.

Types of Jump

Different types of hydraulic jumps are developed on a horizontal floor. The jump form and its flow characteristics mainly depend on the Froude number F_1 of the incoming flow which is expressed as $\left(V_1 / \sqrt{gy_1}\right)$, where V_1 is the mean velocity of flow before the jump; g is acceleration due to gravity ; and y_1 is prejump (or initial) depth of flow. Thus, hydraulic jumps may be classified according to the values of F_1 as follows.

For $F_1 = 1$, the flow is critical, and hence no jump can form.

For $F_1 = 1$ to 1.7, the water surface shows undulations, and the jump is called an *undular jump*. The energy dissipation in this case is quite low being only about 5%.

For $F_1 = 1.7$ to 2.5, a series of small rollers develop on the surface of the jump, but the downstream water surface remains smooth. The velocity throughout is uniform. The energy dissipation in this case is also less being only about 20%. This jump is called a *weak jump*.

For $F_1 = 2.5$ to 4.5, the entering jet oscillates back and forth from the bottom to the surface and back again without any periodicity. This jump is thus called an *oscillating jump*. The energy dissipation in this case ranges from 20 to 45%.

For $F_1 = 4.5$ to 9, a stable and well balanced jump is developed. The action and position of this jump are least sensitive to variation in tail water depth. This jump has the best performance. The energy dissipation ranges from 45 to 70%. This jump is called a *steady jump*.

For $F_1 = 9.0$ and more, the jump action is rough which results in a rough water surface with strong surface waves downstream from the jump. The jump action is however effective since the energy dissipation may reach 85%. This jump is called a *strong jump*.

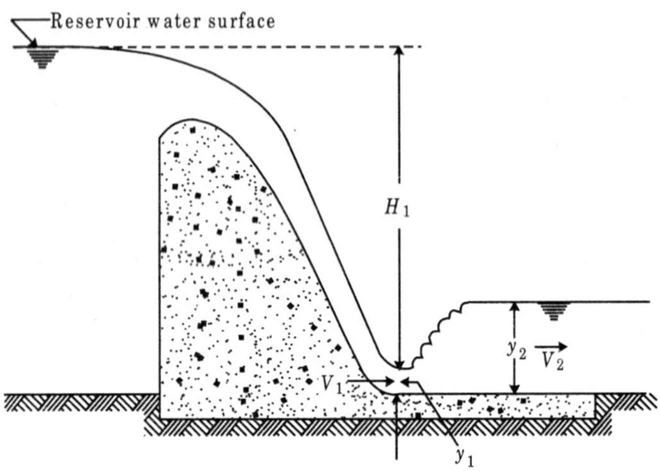

Fig. 12.24 *Hydraulic jump at the toe of a spillway*

Jump Height and Tail Water Rating Curves

For a hydraulic jump to be developed in a rectangular channel the following equation must be satisfied,

$$y_2 = \frac{y_1}{2}\left[\sqrt{1+(8q^2/gy_1^3)}-1\right] \qquad \text{...(12.22)}$$

or $$y_2 = \frac{y_1}{2}\left[\sqrt{1+8F_1^2} - 1\right] \qquad \text{...(12.22 a)}$$

in which y_1 is prejump (or initial) depth; y_2 is post jump (or sequent) depth (Fig. 12.24); g is acceleration due to gravity; q is discharge intensity per unit length ; and F_1 is Froude number of the incoming flow as defined earlier.

For any discharge intensity q, y_1 will be equal to (q/V_1) where V_1 is the mean velocity of incoming flow which will be determined by the drop H_1 as $V_1 = 2gH_1$, if the head loss is neglected. Thus, the values of y_1 are fixed for given values of q, and if a jump is to be developed for all discharge intensities, y_2 must correspond to y_1 in accordance with Eq. (12.22) at all discharge intensities. A number of y_2 values may thus be obtained for different values of q and a plot of y_2 versus q may be made which is known as jump height curve (J.H.C.) or jump rating curves (J.R.C.) (Fig. 12.25 a).

The actual tail water depth y_2' corresponding to any discharge intensity q will however depend on the hydraulic condition of the river channel on the downstream side. The values of y_2' corresponding to different values of q may be obtained by actual observations and a plot of y_2' versus q may also be made which is known as *tail water rating curve* (T.W.R.C.) (Fig. 12.25 b).

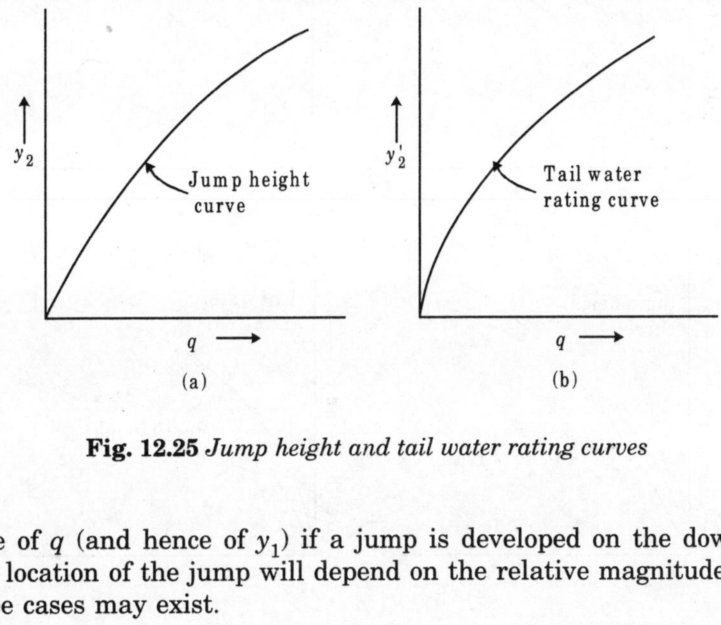

Fig. 12.25 *Jump height and tail water rating curves*

Jump Location

For a given value of q (and hence of y_1) if a jump is developed on the downstream side of a spillway then the location of the jump will depend on the relative magnitudes of y_2' and y_2, and the following three cases may exist.

Case 1. The tail water depth y_2' is equal to the depth of y_2 sequent to initial depth y_1. The jump in this case will be located just at the toe of the spillway. For scour protection purposes, this is an ideal case.

Case 2. The tail water depth y_2' is less than y_2. The jump in this case will be located away from the toe of the spillway on the downstream side. In this case severe erosion may result in the portion of the river between the spillway and the section where jump is located and hence it must be avoided in the design.

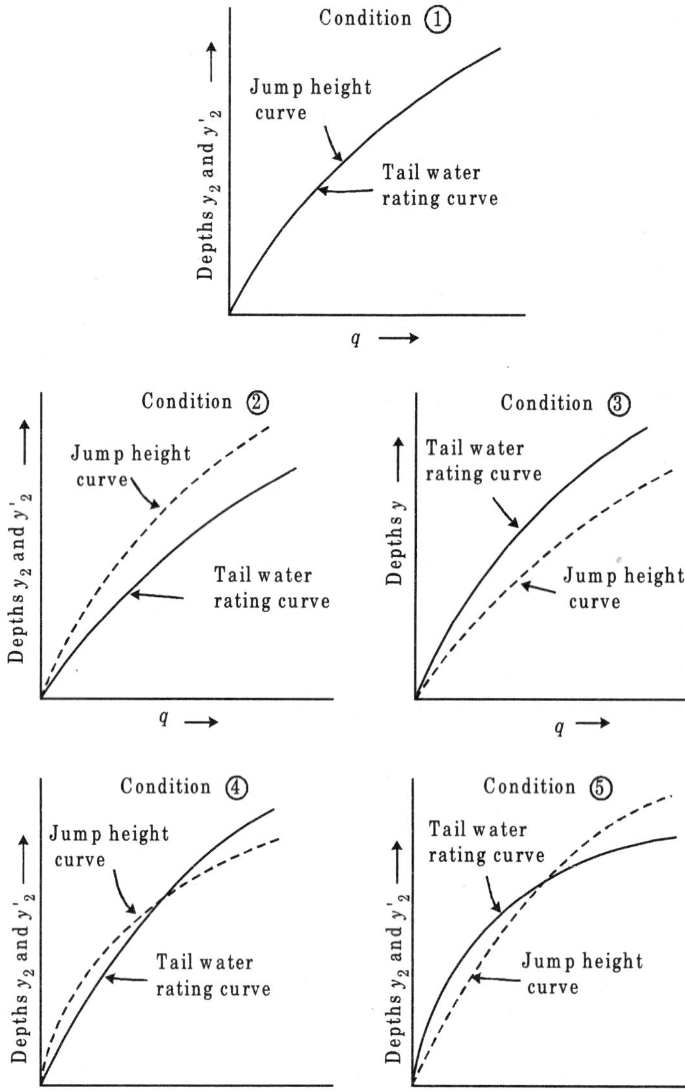

Fig. 12.26 *Tailwater conditions for the design of scour protection works*

Case 3. The tailwater depth y_2' is more than y_2. The jump in this case will be located close to the toe of the spillway but the jump will be drowned or submerged. However, this case must also be avoided in the design because the dissipation of energy in the case of a drowned jump is much less than that in the case of a clear (or undrowned) jump.

The above discussion regarding the location of the jump pertains to a certain value of discharge intensity q corresponding to which there will be a certain value of tailwater depth y_2'. However, as q varies y_2' also varies and hence for determining the location of the jump for different values of q a comparative study of the jump height curve and the tailwater rating curve will have no be made. For this the jump height curve and the tailwater rating curve are plotted on the same paper. Since there may be difference in the relative positions of both these curves any of the five conditions as shown in Fig. 12.26 may exist. For each of these conditions the measures to be taken to develop the jump close to the toe of the spillway are indicated below.

Condition 1. The jump height curve and the tail water rating curve are coinciding with each other for all the discharges. Thus, in this case for all the discharges jump will develop close to the toe of the spillway. As such a horizontal apron may be provided on the river bed downstream from the toe of the spillway. The length of the apron should be equal to the length of the jump corresponding to the maximum discharge over the spillway. This is however, the most ideal condition which would rarely occur in practice.

Condition 2. The jump height curve is above the tail water rating curve for all the discharges. In this case as indicted earlier the jump will develop at a certain section far downstream of the toe of the spillway. Thus, in order to make the jump to develop at a section

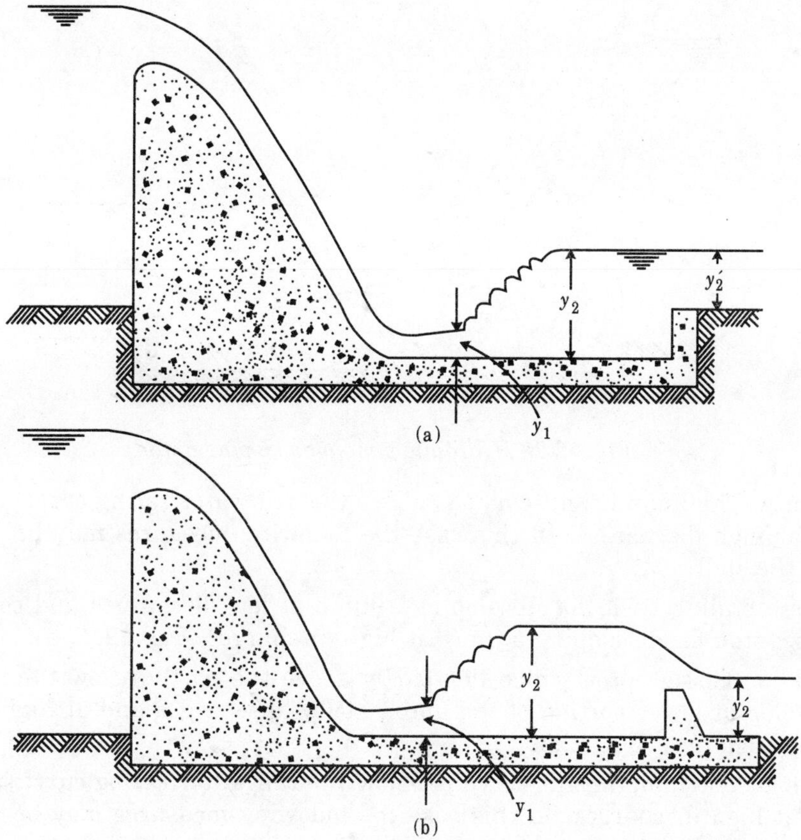

Fig. 12.27 *(a) Depressed horizontal floor ; (b) Stilling basin for hydraulic jump*

close to the toe of the spillway the tail water depth must be increased which may be accomplished in any of the following ways.

(i) Providing a depressed horizontal apron by excavating the river bed on the downstream side from the toe of the spillway (Fig. 12.27 a). The length and depth of the apron should be such that for all the discharges the jump is confined to the apron. Sometimes the depressed apron may be made sloping instead of horizontal.

(ii) Creating a stilling basin by providing a sill or baffle on the river bed at a section downstream of the toe of the spillway (Fig. 12.27 b). Again the length and depth of the stilling basin should be such that for all the discharges the jump is confined to the stilling basin.

Condition 3. The jump height curve is below the tail water rating curve for all the discharges. In this case due to higher tail water depths drowned jump will develop and consequently little energy will be dissipated. As such in order to develop a clear jump the following measures may be taken.

(i) Provide a sloping apron over the river bed extending from downstream surface of the spillway (Fig. 12.28). The slope of the apron may be such that jump will develop on the apron for all the discharges.

(ii) Provide a drop in the river bed so as to lower the tail water level.

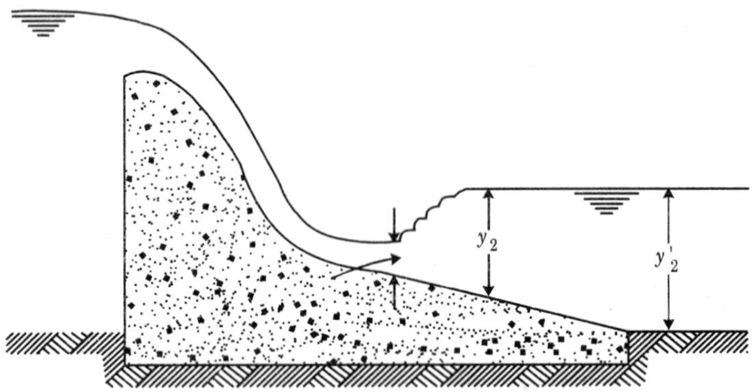

Fig. 12.28 *Hydraulic jump on a sloping apron*

Condition 4. The jump height curve is above the tail water rating curve at low discharges and below it at high discharges. In this case the following measures may be taken to develop jump close to the spillway.

(i) Provide a stilling basin for developing a jump at low discharges and combine the basin with a sloping apron for developing a jump at high discharges (Fig. 12.29 a).

(ii) Provide a sloping apron which lies partly above the partly below the river bed so that jump will develop at lower portion of the apron at low discharges and at higher portion of the apron at high discharges (Fig. 12.29 b).

Condition 5. The jump height curve is below the tail water rating curve at low discharges and above it at high discharges. In this case the following measures may be taken to develop jump close to the spillway.

(i) Provide a sloping apron which lies partly above and partly below the river bed so that jump will develop higher up on the apron at low discharges and at lower portion of the apron at high discharges.

(ii) Create a stilling basin by providing a sill or baffle which will increase the tail water depth and thus develop a jump at high discharges. However, at low discharges the already excessive tail water depth will be further increased by the stilling basin and hence drowned jump will develop which will result in less dissipation of energy. If this is not likely to cause scour the device may be used without any modification for low discharges.

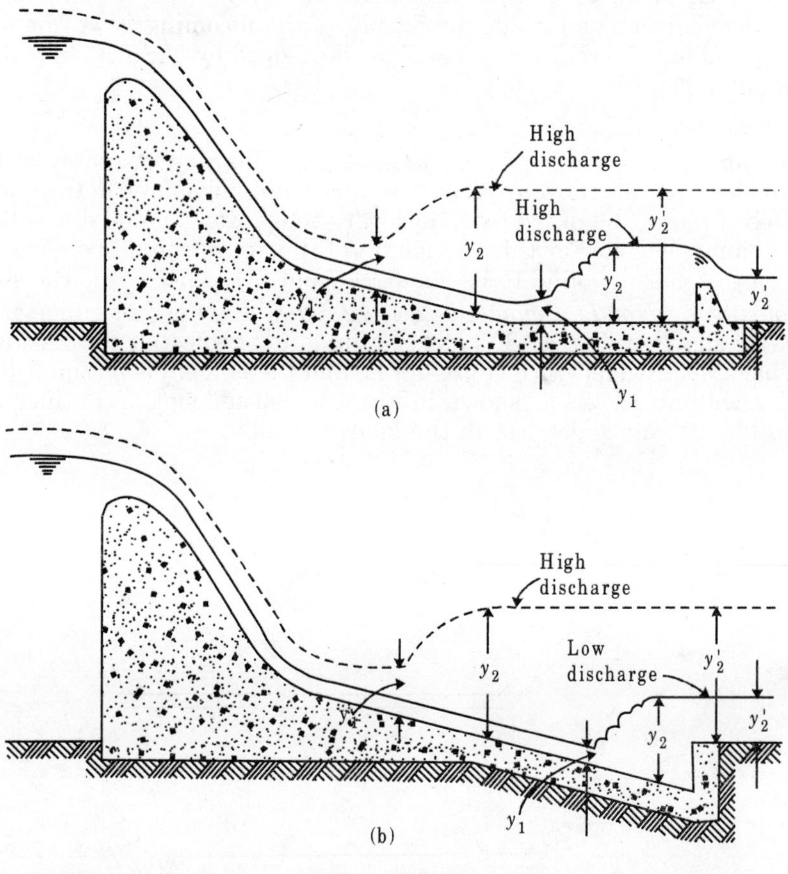

Fig. 12.29 *(a) Sloping apron combined with a stilling basin; (b) Sloping apron partly above and partly below the river bed*

The various devices proposed for developing the jump close to toe of the spillway for the five different conditions would be too expensive if these are designed to confine the entire length of the jump. Moreover these devices are not effective in stabilizing the jump and hence the jump tends to shift to the downstream side thus leaving the apron or the basin. As such certain types of stilling basins have been developed which are relatively more effective in developing stable jumps which would remain confined to the basins. These stilling basins are commonly used with almost all the types of spillways as well as other hydraulic structures such as weirs, barrages, etc. A brief description of these stilling basins is given below.

12.14 STILLING BASINS

A stilling basin is defined as a structure in which a hydraulic jump used for energy dissipation is confined either partly or entirely. Certain accessories such as chute blocks, baffle blocks (or floor blocks), and end sills (or baffles) are usually provided in the stilling basins to reduce the length of the jump and thus to reduce the length and the cost of the stilling basin. Moreover these accessories also improve the dissipation action of the basin and stabilize the jump.

The type of stilling basin to be provided depends on the type of jump which in turn, as indicated earlier, depends on the Froude number F_1 of the incoming flow. The different types of stilling basins suitable for different ranges of F_1 as developed by U.S.B.R. and Indian Standards are described as follows.

U.S.B.R. Stilling Basins

(i) *Stilling basins for Froude number between 1.7 to 2.5.* For this case only a horizontal apron needs to be provided. As the flow in this case does not have much turbulence usually no accessories are required to be provided. However, the apron should be sufficiently long to contain the entire jump over it. For this the length of the apron should be equal to the length of the jump which in this case is found to be equal to about $5y_2$, where y_2 is the sequent depth.

(ii) *Stilling basins for Froude numbers between 2.5 and 4.5.* For this range of Froude number *Type I stilling basin* shown in Fig. 12.30 has been found to be effective for dissipating the energy of the flow. The basin is provided with chute blocks and the end sill is optional. The size, spacing and location of the chute blocks is as shown in Fig. 12.30. The length L of the stilling basin may be obtained for different values of F_1 from the following table :

F_1	2	3	4	5
(L/y_2)	4.3	5.3	5.8	6

where y_2 = sequent depth.

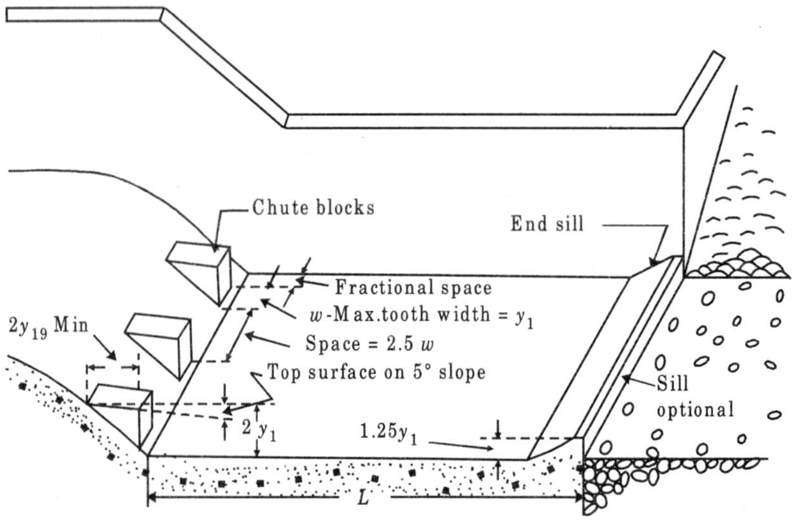

Fig. 12.30 *Type I stilling basin*

In this case due to oscillating jump being developed, wave action is produced which cannot be entirely dampened. However,.in order to suppress the wave action the floor of the basin should be so set that the tail water depth in the basin is 10% greater than the sequent depth y_2 given by Eq. (12.22). This will also help to check the tendency of the jump to sweep out of the basin.

(iii) *Stilling basins for Froude numbers higher than 4.5.* For this case depending upon the velocity of the incoming flow, two types of stilling basins have been developed as indicated below.

(a) When the velocity of the incoming flow is less than 15 m/s. *Type II stilling* basin shown in Fig. (12.31) may be adopted. This basin utilizes chute blocks baffle blocks and an end sill. The size, spacing and location of the chute and baffle blocks are as shown in Fig. (12.31). The length L of the stilling basin and the heights h_3 and h_4 of the baffle blocks and the end sill respectively may be obtained for different values of F_1 from the following table:

F_1	5	6	8	10	12	14	16
(L/y_2)	2.3	2.5	2.6	2.7	2.8	2.8	2.8
(h_3/y_1)	1.5	1.7	2.0	2.3	2.7	3.0	3.3
(h_4/y_1)	1.2	1.3	1.5	1.6	1.7	1.8	1.9

where y_1 = initial depth; and y_2 = sequent depth.

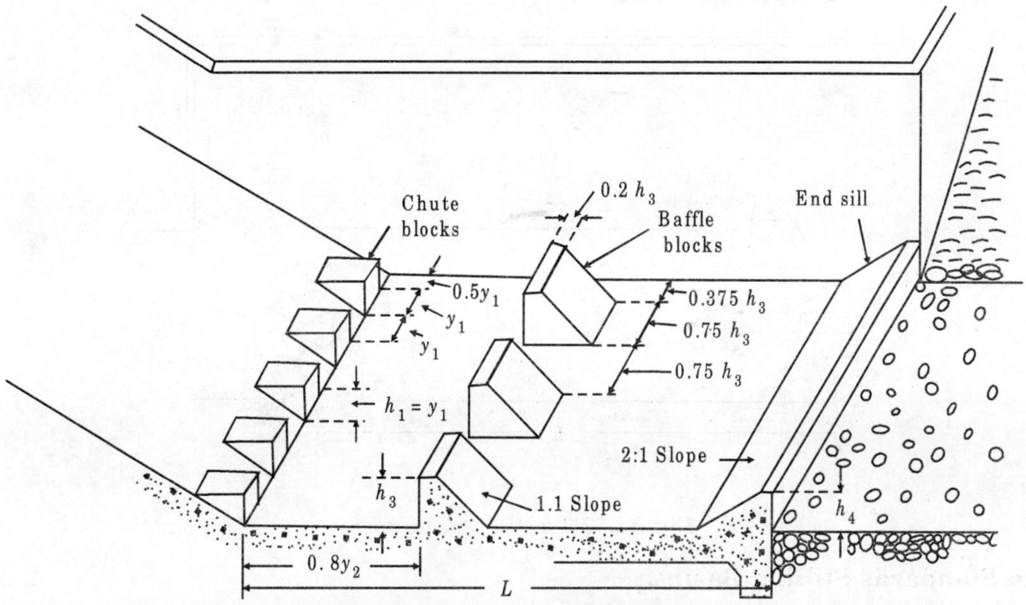

Fig. 12.31 *Type II Stilling basin*

By providing the baffle blocks the length of the stilling basin is considerably reduced because the dissipation of energy is accomplished by the hydraulic jump as well as by the impinging action of the incoming flow against these blocks. However, the baffle blocks will be subjected to large impact forces due to the impingement of the incoming flow. Moreover on the downstream face of the baffle blocks usually suction or negative pressures will be developed which will further increase the force exerted on these blocks. Hence the baffle blocks should be properly anchored at the base. Further the floor of the basin will also be subjected to additional load due to the dynamic force created against the upstream face of the baffle blocks, which should be considered in the design of the floor of the basin.

(b) When the velocity of the incoming flow exceeds 15 m/s. *Type III stilling basin* shown in Fig. (12.32) may be adopted. In this basin only chute blocks are provided and instead of a solid end sill a dentated end sill is provided. In this basin baffle blocks are not provided because (i) due to high velocities of the incoming flows these blocks will be subjected to excessively large impact forces; and (ii) there is a possibility of cavitation along the downstream face of these blocks and the adjacent floor of the basin due to large negative pressure being developed in this region. However, due to baffle blocks being eliminated in this case the dissipation of energy is accomplished primarily by hydraulic jump action and hence the length of the basin will be greater than that indicated for the Type II basin. The length L of the Type III stilling basin may be obtained for different values of F_1 from the following table :

F_1	5	6	8	10	12	14
(L/y_2)	3.85	4.0	4.2	4.3	4.3	4.3

where y_2 = sequent depth.

Again in this case also in order to check the tendency of the jump to sweep out of the basin the floor of the basin is so set that the tail water depth in the basin is 5% greater than the sequent depth y_2 given by in Fig. (12.32).

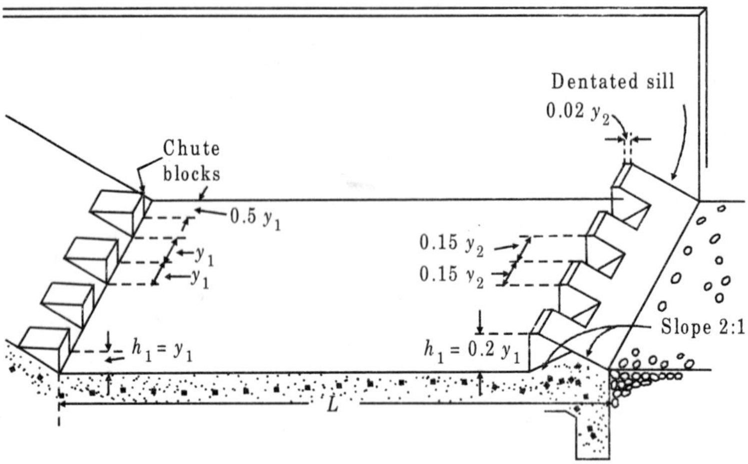

Fig. 12.32 *Type III stilling basin*

Indian Standards Stilling Basins

The stilling basins recommended according to Indian Standards may be classified into the following two types.

(1) Stilling basins with horizontal apron (or floor).

(2) Stilling basins with sloping apron (or floor).

1. Stilling basins with horizontal apron (or floor). The stilling basins with horizontal apron (or floor) may be provided when the jump height curve coincides with the tail water rating curve ; or the former is slightly above or below the later. As such in this case the requisite depth for the development of the jump may be obtained on an apron near or at the ground level. The stilling basins with horizontal apron (or floor) may be further classified into the following two types.

(i) Indian standards stilling basin Type I.

(ii) Indian standards stilling basin Type II.

(i) *Indian standards stilling basin Type I.* It may be provided when the Froude number F_1 of the incoming flow is less than 4.5. This case is generally encountered on weirs and barrages. This stilling basin is provided with chute blocks, basin blocks (or baffle blocks) and a dentated end sill. The size, spacing and location of the chute blocks for this basin are exactly same as those for Type I stilling basin of U.S.B.R. (Fig. 12.30). The height and top thickness of the basin blocks for this stilling basin are same as those for Type II stilling basin of U.S.B.R (Fig. 12.31). However, the width and spacing of the basin blocks for this case is equal to the height of the blocks. Further the size of the dentated end sill for this stilling basin is exactly same as the one for Type III stilling basin of U.S.B.R (Fig. 12.32). The length L of the basin may be obtained for different value of F_1 from the following table :

F_1	2	3	4	4.5
(L/y_2)	3.15	4.0	4.75	5.0

where y_2 = sequent depth

A comparison of the basin lengths of this stilling basin with those of U.S.B.R. Type I stilling basin shows that the former are shorter than the later, which is obviously due to the use of basin blocks in the former.

(ii) *Indian standards stilling basin Type II.* It may be provided when the Froude number F_1 of the incoming flow is greater than 4.5. This case is a general feature for medium and high dams. This stilling basin is also provided with chute blocks, basin blocks (or baffle blocks) and a dentated end sill. However, when the velocity of flow at the location of the basin blocks exceeds 15 m/s, no basin blocks are provided and in that case the floor of the basin should be kept at a depth equal to the sequent depth y_2 below the tail water level. The size, spacing and location of the chute blocks for this stilling basin are exactly same as those for Type II or Type III stilling basins of U.S.B.R. Further the size, spacing and location of the basin blocks if provided for this stilling basin are exactly same as those for Type II stilling basin of U.S.B.R. Also the dentated end sill provided for this stilling basin is exactly same as Type III stilling basin of U.S.B.R. (or Type I stilling basin of Indian Standards). The length L of this stilling basin may be obtained for different values of F_1 from the following table :

F_1	5	6	8	10	12
(L/y_2)	2.9	3.2	3.7	3.9	4.0

where y_2 = sequent depth.

2. Stilling basin with sloping apron (or floor). When the tail water depth is too large as compared to the sequent depth y_2, a drowned jump will develop which is not desirable. In such a case a stilling basin with a sloping apron (or floor) may be provided as it will allow a clear jump to be developed at suitable level on the sloping apron (or floor). The stilling basins with sloping apron (or floor) may be further classified into the following two types.

(i) Indian standards stilling basin Type III.

(ii) Indian standards stilling basin Type IV.

Indian standards stilling basin Type III is recommended for the case where tail water rating curve is above the jump height curve at all discharges. It is usually provided with a sloping apron (or floor) for its entire length.

Indian standards stilling basin Type IV is suitable for the case where the tail water depth at maximum discharge exceeds the sequent y_2 considerably but is equal to or slightly greater than y_2 at lower discharges. It is provided with a partly sloping and partly horizontal apron (or floor).

For both these stilling basins except a solid or dentated end sill no other accessories are provided.

The procedure for the design of all the four types of Indian standards stilling basins is given in I.S.: 4997—1968.

12.15 BUCKET TYPE ENERGY DISSIPATORS

A bucket type energy dissipator consists of an upturned bucket provided at the toe of the spillway to deflect the water flowing over the spillway. The bucket type energy dissipators may be used only for overflow type spillways. This method of energy dissipation becomes more economical than the method of stilling basins when the Froude number F_1 of the incoming flow exceeds 10, because in such cases the difference between the initial and sequent depths being large a long and deep stilling basin would be required. Moreover the bucket type energy dissipators may be used with any tail water condition. However, this method of energy dissipation may be used only when the river bed is composed of stiff rock.

The bucket type energy dissipators are of the following three types.

(i) Solid roller bucket (ii) Slotted roller bucket

(iii) Ski-jump (or flip or trajectory) bucket.

The solid and slotted roller buckets may be used where the tail water depths are too large as compared to the sequent depths required for the formation of hydraulic jump. Both these buckets remain submerged in the tail water and hence these are also termed as submerged bucket energy dissipators. The solid and slotted roller buckets are described below.

(i) *Solid roller bucket*. A solid roller bucket consists of a bucket like apron with a concave circular profile of large radius and a deflector lip as shown in Fig. (12.33 a). When water flows over the bucket the entire sheet of water leaving the bucket is deflected upward by the bucket

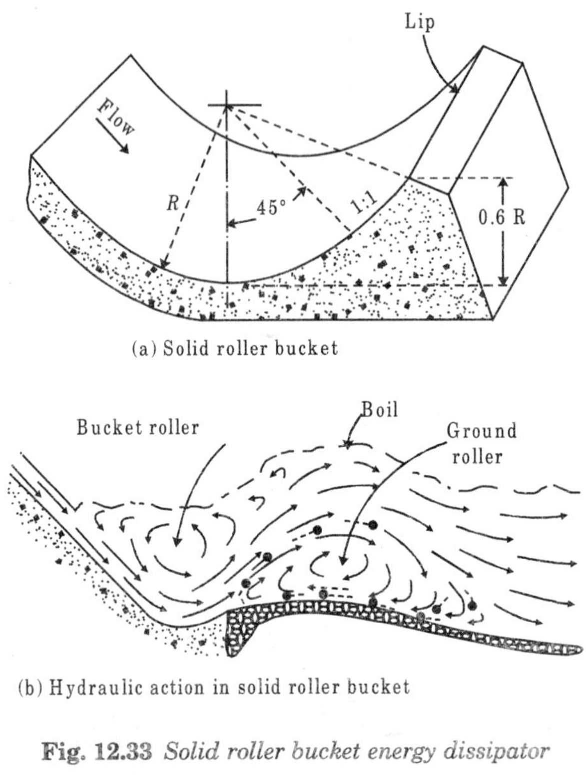

Fig. 12.33 *Solid roller bucket energy dissipator*

lip and two elliptical rollers are developed as shown in Fig. (12.33 b). One of the roller which moves in counter clockwise direction is developed on the surface of the bucket and is contained within the region above the bucket. This is known as bucket roller (or surface roller). The other roller moving in clockwise direction is developed on the ground surface immediately downstream of the bucket, which is known as ground roller. The movement of the rollers, along with the intermingling of the incoming flows causes the dissipation of energy.

The upward deflection of the water by the bucket lip creates a high boil on the water surface and a violent ground roller. This ground roller continuously pulls the loose bed material backwards and deposits the same against the lip of the bucket. Some of this material may move into the bucket, which may get trapped there and damage the surface of the bucket by abrasion. Moreover due to severe surface boil the turbulent flow persists for a considerable distance on downstream side thus resulting in the erosion of the river banks. These drawbacks of the solid roller bucket are remove in the slotted roller bucket.

(ii) *Slotted roller bucket.* A slotted roller bucket also consists of a bucket like apron with a concave circular profile of large radius but it has a slotted (or dentated) deflector lip as shown in Fig. (12.34 a). In general the hydraulic action of a slotted bucket has the same characteristics as those of a solid roller bucket. Thus, in the case of a slotted roller bucket also the same two rollers

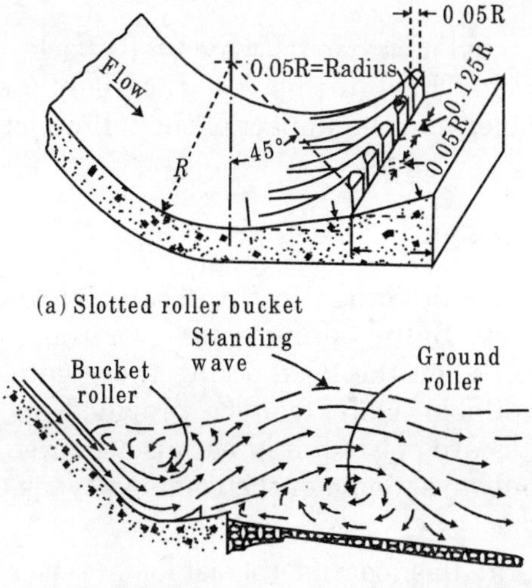

Fig. 12.34 *Slotted roller bucket energy dissipator*

are developed as shown in Fig. (12.34 b). However, in this case water leaves the lip of the bucket at a flatter angle and only a part of it is deflected upwards. Thus, surface boil is considerably reduced and less violent ground roller occurs which results in a smoother flow on the downstream side. Moreover in this case the bed material is neither deposited nor carried away from the bucket lip. Also any debris which might get into the bucket is immediately washed out through the slots. However, the performance of the slotted bucket is considerably affected by the tail water depth. At low tail water depths the bucket roller will be swept out of the bucket and will produce high velocity flow on the downstream side which will scour the river bed. On the other hand when the tail water depth becomes excessively large the jet of water leaving the lip of the bucket will get depressed and dive to the river bed which will scour the bed at the point of impingement. Thus, the slotted roller buckets are suitable only for a limited range of tail water depths.

(iii) *Ski-jump bucket.* A ski-jump bucket may be used where the tail water depth, is less than the sequent depth required for the formation of hydraulic jump and the river bed is composed of stiff rock. As shown in Fig. (12.35) the lip of the bucket is so shaped that the entire sheet of water

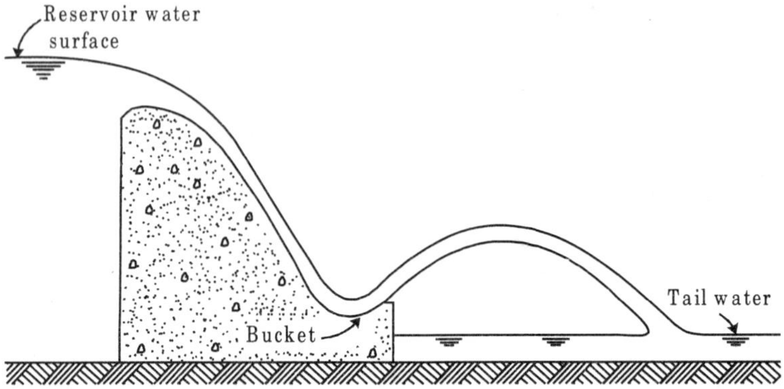

Fig. 12.35 *Ski-jump bucket*

flowing over the bucket is deflected as a free jet which falls back into the river channel at a safe distance away from the spillway. Thus, in this case the energy is dissipated by air resistance, breaking up of the jet into bubbles and the impact of the falling jet against the river bed and tail water.

12.16 SPILLWAY CREST GATES

For an ungated spillway the useful storage in the reservoir can be maintained only upto the level of the crest of the spillway. By installing gates over the crest of the spillway additional storage can be made available. When the flood occurs these gates are removed so that the full spillway capacity is made available. Gates can be provided on all types of spillways except siphon spillways for which gates are not required because the rise of water level during floods is small as compared to other spillways. Some of the common types of gates used for spillways are described below.

1. Flashboards. In general flashboards consist of a series of wooden boards or panels placed on the crest of the spillway. The flashboards may be classified as (i) temporary flashboards ; and (ii) permanent flashboards.

Temporary flashboards consist of a series of wooden boards or panels supported by steel pipes or rods (called pins) inserted loosely into sockets set in the masonry or concrete at the crest of the spillway (Fig. 12.36 a). The steel pipes or rods are designed to bend and release the flashboards when the water surface in the reservoir reaches a certain elevation above the top of the flashboards, thus allowing the excessive floods to pass. Temporary flashboards have been used upto a height of 1.7 m. The operation of the temporary flashboards is automatic, but the disadvantage being that the flashboards are lost each time when the supports fail.

Permanent flashboards are similar in principle to the temporary type, except that they are designed to operate without damage to themselves. Permanent flashboards usually consist of wooden boards or panels which can be raised or lowered from an overhead cableway or bridge. Various types of permanent flashboards have been designed. Figure (12.36 b) shows one of the types of permanent flashboards in which the wooden boards are joined to form shutters. These shutters are placed on the crest of the spillway with their lower edge hinged at the crest and are supported in the raised position by struts or by attaching their upper edge to a bridge or cableway at the top. The shutters drop flat on the crest when the struts are tripped or when they are released from the top. These are therefore not suitable for spillways having curved crests. For raising the shutters either a crane travelling on the rest of the spillway or a trolley on a bridge or an overhead cable if there is no bridge may be used.

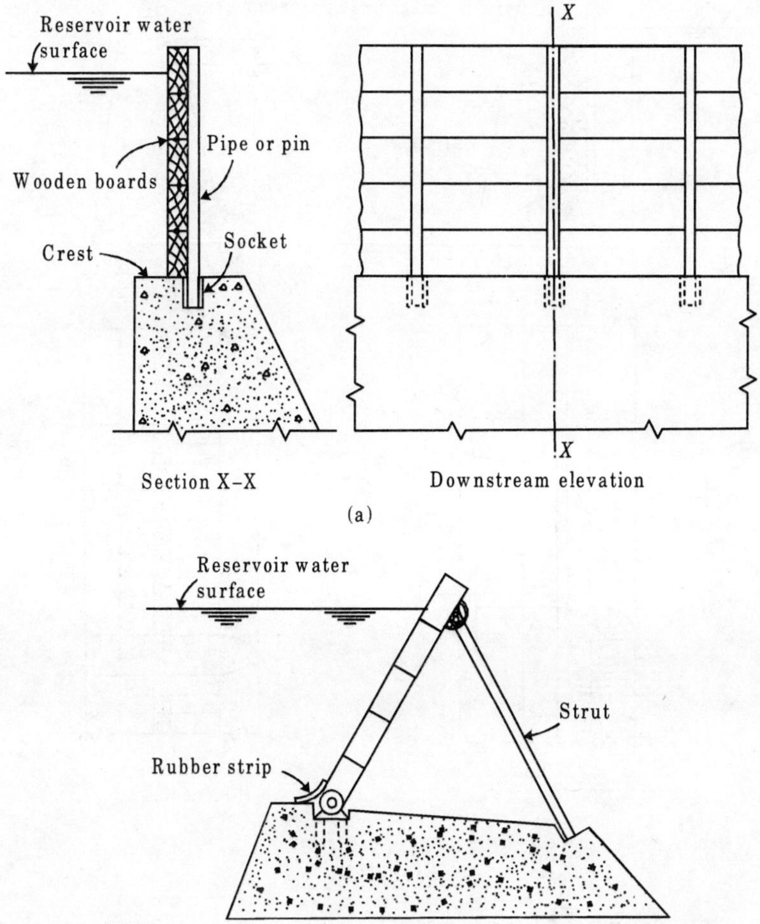

Fig. 12.36 *(a) Temporary flashboards; (b) permanent flashboards*

The flashboards are however used only on low spillways or weirs.

2. Stop logs and Needles. *Stop logs* consist of horizontal wooden logs or beams spanning the space between grooved piers (Fig. 12.37 a). The logs are placed one over the other by pushing them down into the grooves. These may be placed or removed by hand or with a hoist. There is usually much leakage between the logs and also there is difficulty of placing and removing them especially if they become jammed in the grooves due to infrequent operation. Stop logs are ordinarily used for small spillways or weirs.

Needles consist of a row of slightly inclined wooden logs with their ends resting in a keyway on the spillway crest and their upper ends supported by a bridge or beam (Fig. 12.37 b). These are invariably placed and removed by hand and hence should not be too large. Needles are somewhat easier to remove than stop logs but are quite difficult to place in flowing water. These are also used only for small spillways or weirs. However, needles are not frequently used.

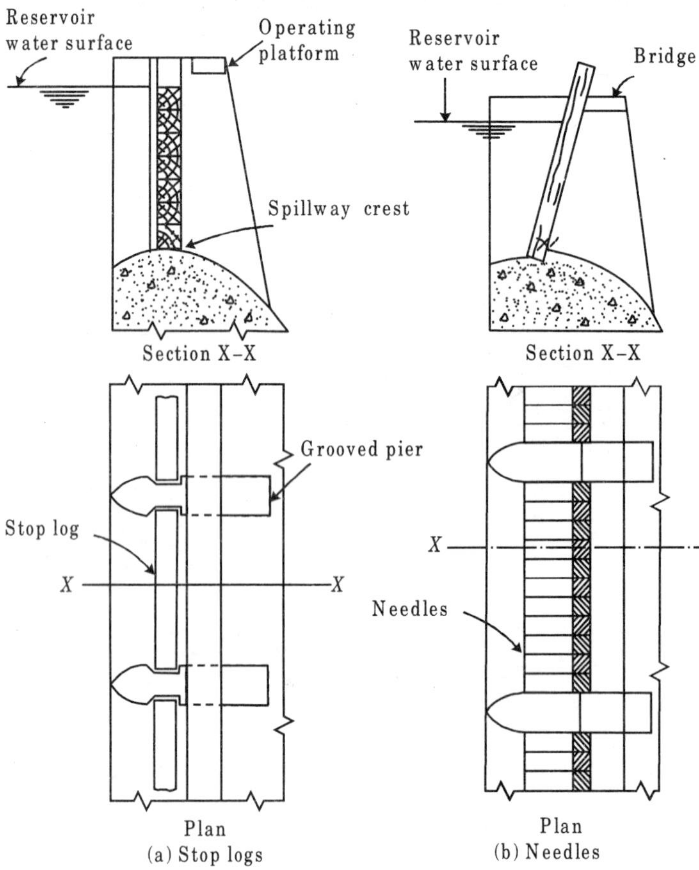

Fig. 12.37

Vertical lift gates. A *vertical lift gate* consists of a framework of steel members to which a skin plate of steel is attached on the upstream side. These gates are rectangular in shape and they move vertically in their own plane in grooves provided in piers at their two ends. The groove is provided with a steel lining in the form of a steel channel section of the required depth in order to provide the necessary bearing strength and also to minimise friction. The gates are raised or lowered by cables attached to them. The different types of vertical lift gates are as follows :

(i) Sliding gates (ii) Fixed-wheel gates

(iii) Stoney gates.

(i) *Sliding gates.* In this type, the frame of the gate bears directly on the downstream surface of the groove and due to water pressure acting on the gate a water tight seal is formed by the contact between the two. When the gate is raised or lowered it slides over the groove surface. As such in this case heavy frictional resistance will have to be counteracted for lifting the gate against water pressure and hence large hoisting capacity will be required. The size of this type of gates is therefore limited by the available hoisting capacity. Moreover on account of these limitations the sliding gates are seldom used.

(ii) *Fixed-wheel gates.* In this type, wheels are attached to the gate along each end and it travels on vertical tracks provided on the downstream surface of the gate grooves. Thus, as compared to the sliding gates the frictional resistance is considerably reduced in the case of fixed-wheel gates. However in the case of these gates both axle and rolling frictions are developed.

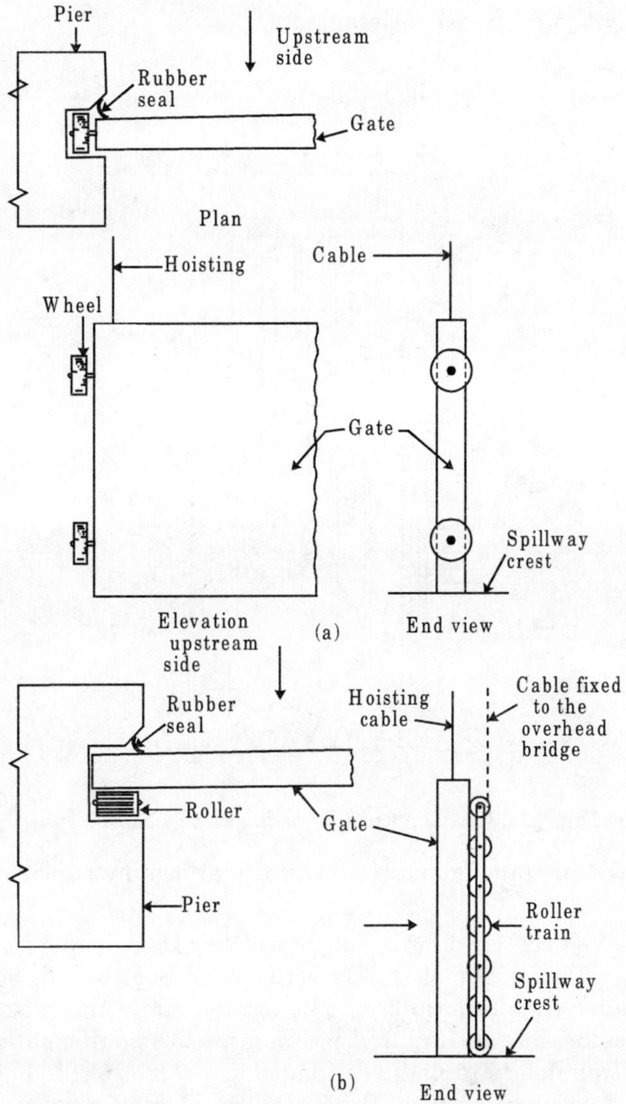

Fig. 12.38 *(a) Fixed wheel gate ; (b) Stoney gate*

(iii) *Stoney gates.* In this type a train of rollers is provided in between the gate and the groove on each side. The roller train is neither attached to the gate nor to the groove but rolls between the two as the gate moves. Thus in this case only roller friction is developed and there is no axle friction because the rollers transmit the water load directly from a bearing strip on the gate to the roller path in the groove and hence there being no pressure on the axle there is no friction at the axle.

Figure (12.38) shows plan and elevation of fixed-wheel and Stoney gates. For both these types of gates rubber sheet or belting is used along the sides on the upstream face of the gate to seal the openings between the gate and sides of the pier.

All vertical lift gates are counterbalanced by a counterweight beam [see Fig. 12.38 (c)] which is loaded to balance almost the entire dead weight of the gate. The lifting force is, therefore, required to balance only the frictional resistance.

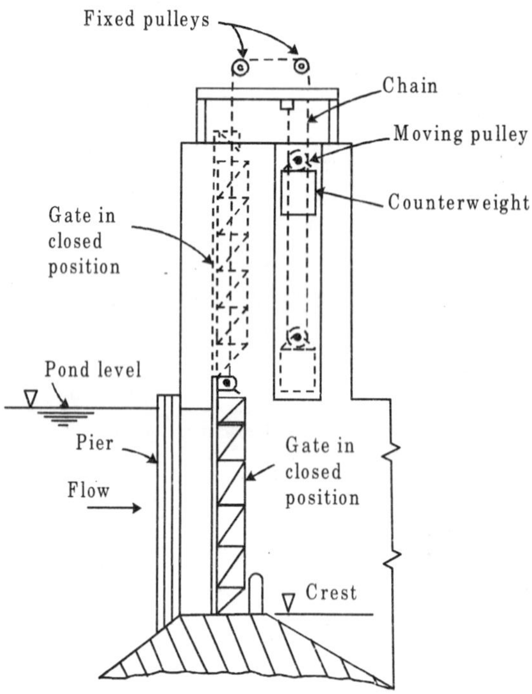

Fig. 12.39 *Counter weight beam for vertical lift gate*

Vertical lift gates are operated from an overhead platform by winches which can be worked manually or by power.

Vertical lift gates have been used for spans as wide as 15 m or more and for heights upto 15 m. With the increase in the height of the gate the headroom or the height of the operating platform increases because the platform has to be located above the raised position of the gate. Moreover the bottom of the gate in its lifted position must be sufficiently above the high flood level to enable the floating debris to pass. To reduce the height of the operating platform, high gates may be built in two or sometimes three horizontal sections, one above the other. The different section of the gate may be placed either in different grooves with each section overlapping the lower one; or in the same groove with different sections flushing with each other. When all the sections are placed in the same groove then for raising the gate fully the upper section of the gate may be lifted and removed from the grooves and the lower section may then be lifted and held in raised position within the grooves. The division of the gate into different sections also reduces the load on the hoisting mechanism.

Tainter gates or Radial gates. A *tainter gate* consists of a skin plate of steel formed to a segment of a cylinder supported on a steel frame-work which is pivoted on trunnions or pins set in the downstream portion of the piers on the spillway (Fig. 12.40). Thus, when the gate is raised or lowered it rotates about a fixed horizontal axis through the trunnions. Rubber seals are provided between the gate and the piers and spillway crest. For this rubber sheet or belting 150 to 200 mm wide is used. One edge of the sheet is attached to the ends and the bottom of the gate around the upstream face of the skin plate, and the other edge is bent back so that a considerable portion of the width is in contact with the pier face and the spillway crest.

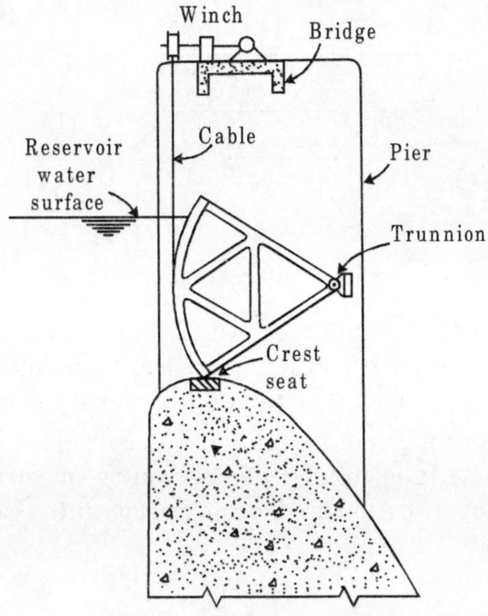

Fig. 12.40 *Tainter gate*

Since the skin plate which supports water is cylindrical the resultant water pressure acting on it passes through its centre. The skin plate is made concentric to the trunnions or the axis of rotation, and hence in the case of these gates the resultant water pressure creates no moment against the lifting of the gate. The hoisting load therefore consists only of the weight of the gate; the friction resistance at the trunnions. Larger tainter gates are also sometimes provided with counterweights to partially balance the weight of the gate, which further reduces the required hoisting capacity. Tainter gates are hoisted by the cables attached to them and extending to the winches on platform above the gates.

Tainter gates are suitable for moderate spans and heights and are often more economical than vertical lift gates of the same size. Moreover these gates require less headroom for the operating platform than vertical lift gates.

Rolling gates or Roller gates. As shown in Fig. (12.40) a *rolling gate* consists of a hollow steel cylinder spanning between the piers. A heavy annular rim having gear teeth on its periphery encircles each end of the cylinder. Each pier has an inclined rack which engages the gear teeth encircling the cylinder. The gate is rolled up the inclined rack by exerting a pull on

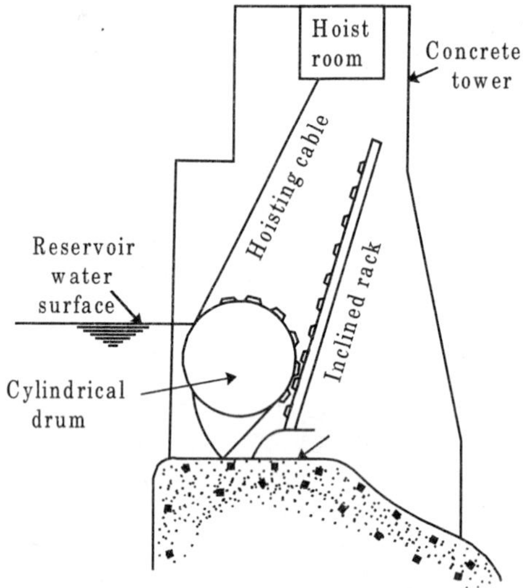

Fig. 12.40 *Rolling gate*

the hoisting cable attached to the cylinder. At the lower portion of the cylinder a cylindrical segment is attached which remains in contact with the spillway crest and increases the gate height. Rolling gates are suitable for long spans and moderate heights.

Drum gates. A *drum gate* consists of a segment of a cylinder formed by skin plates attached to internal bracing (Fig. 12.41). It is hinged at the centre of curvature, which may be either upstream or downstream. In open or lowered position the gate fits in a recesses in the top of

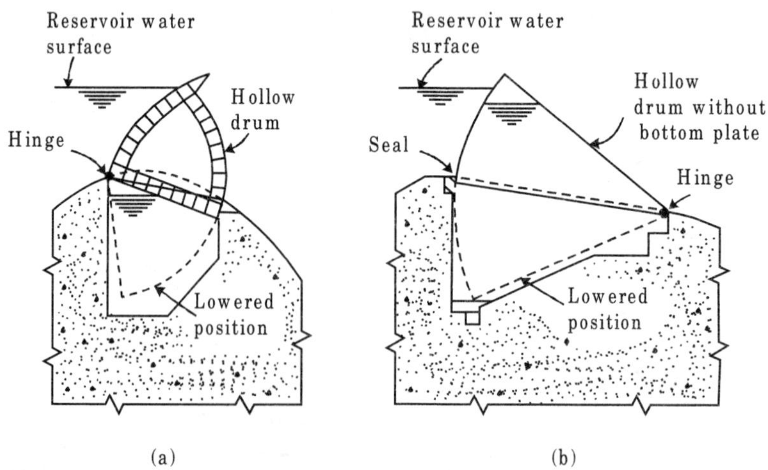

Fig. 12.41 *Drum gates*

spillway. When water is admitted to the recess, the hollow drum gate is forced upward to the closed position. A type of drum gate developed by U.S. Bureau of Reclamation is shown in Fig. (12.41 a) which is a completely enclosed gate, hinged at the upstream edge so that byoyant forces aid in its lifting. When lowered this gate conforms closely to the shape of the ogee crest. This type of gate may be made to operate automatically. The second type of drum gate as shown in Fig. (12.41 b) has no bottom plate and is hinged at the downstream edge. In this case the gate is raised only by the water pressure.

The drum gates are also normally used for long spans and moderate heights. However, on account of large cavity required by drum gates in the lowered position these gates are not suited to low dams.

Bear-trap gates. A *bear-trap gate* consists of two leaves of timber or steel hinged and sealed to the spillway crest (Fig. 12.42). One of the leaves which is on the upstream side is hinged and sealed along its upstream edge and the other which is on the downstream side is hinged and sealed along its downstream edge. When the gate is lowered, the leaves are in horizontal position with one leaf lying on top of the other and the downstream leaf being housed

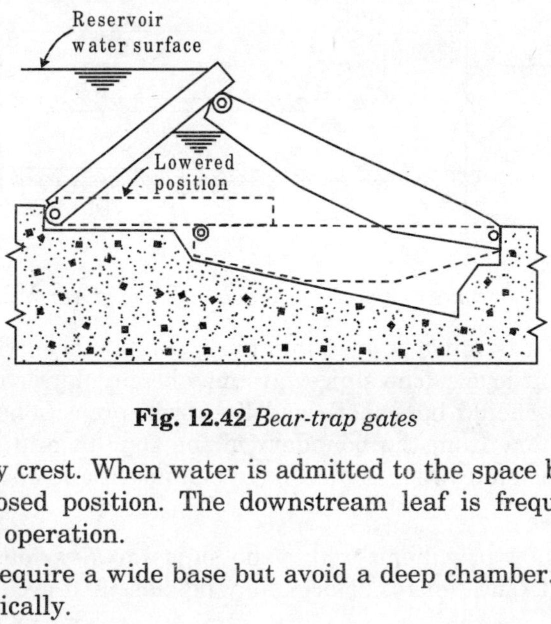

Fig. 12.42 *Bear-trap gates*

in a recess in the spillway crest. When water is admitted to the space below the leaves, they are forced upward to the closed position. The downstream leaf is frequently hollow so that its buoyancy aids the lifting operation.

The bear-trap gates require a wide base but avoid a deep chamber. These gates may also be made to operate automatically.

12.17 OUTLET WORKS

Outlet works are provided to release water impounded by a dam, as and when needed for various purposes. These may be required for several purposes such as feeding water to the turbines for power generation, supplying water to irrigation channels, or for a combination of multipurpose requirements. Sometimes the outlet works of a dam may be used in lieu of a service spillway in conjunction with an auxiliary or secondary spillway. The outlet works may also act as flood regulator, to release waters temporarily stored in flood control storage space or to evacuate storage in anticipation of flood inflows.

The various components of outlet works are as follows.

(i) Sluiceways or waterways
(ii) Control devices
(iii) Intake structures.

Sluiceways or Waterways

A *sluiceway* (also known as outlet) is a pipe or tunnel that passes through a dam or the hillside at either end of the dam. Sluiceways for concrete dams generally pass through the dam, while in the case of earth or rock-fill dams these are preferably placed outside the limits of the dam. However, if a sluiceway must pass through as earth dam, projecting collars should be provided as shown in Fig. (12.43). These collars increase the path of seepage along the outside of the conduit and thus reduce seepage. A common rule is that the collars should increase the length of the seepage path by at least 25 percent. Thus, if there are N collars and x is the projection of each collar then we have

$$2Nx \geq 0.25\, L$$

where L is the length of the sluiceway.

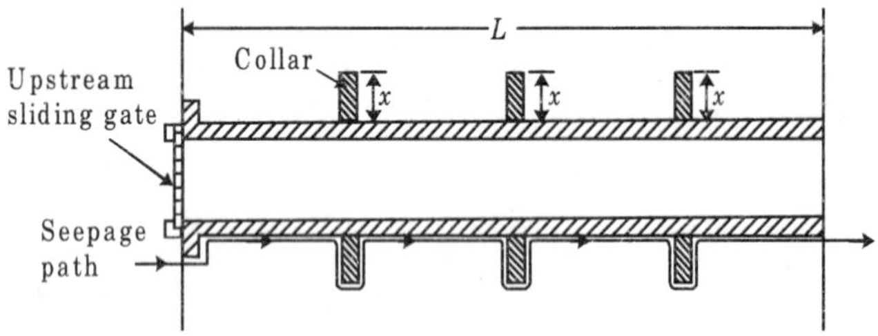

Fig. 12.43 *Sluiceway collars for outlets through earth dams*

The inlet of the sluiceway must be at minimum reservoir level. However, large dams may have sluiceways at different levels. The sluiceways may be circular or rectangular in shape. The interior of the sluiceways should be smooth and free from projections or cavities which might induce separation of the flow from the boundary of the conduit and cause negative pressures and cavitation. Further the sluiceways should have bellmouth entrance.

Control Devices

In order to regulate the flow of water through the sluiceways various control devices such as gates and valves are used. The control devices may be classified according to their location as follows

(i) Entrance gates.
(ii) Interior gate valves.

Most of the sluice ways are provided with some type of gate at their entrance. For low heads upto 15 m sliding gates may be used at the entrance of the sluiceways, but for higher heads roller gates are used. One of the common types of high head entrance gates is a tractor gate (Fig. 12.44). For heads less than 30 m the entrance gates may be used as flow regulators, but with higher heads the partly open gates will be subject to cavitation and vibration ; and hence at heads greater than 30 m the entrance gates are ordinarily used only as emergency gates to permit inspection and repair of the conduit. Further at high head installations an air duct (Fig. 12.44) is provided to admit air to avoid cavitation during the opening and closing of the gate.

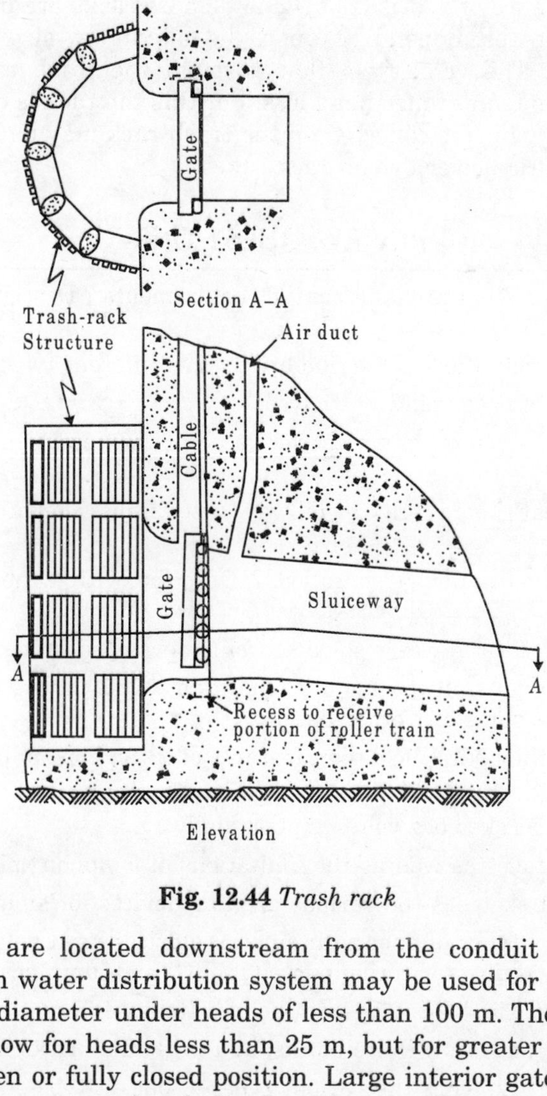

Fig. 12.44 *Trash rack*

Interior gate valves are located downstream from the conduit entrance. Standard gate valves of the type used in water distribution system may be used for interior valves in circular conduits up to 1.25 m in diameter under heads of less than 100 m. The interior gate valves may be used to regulate the flow for heads less than 25 m, but for greater heads they are ordinarily used only in the fully open or fully closed position. Large interior gate valves of different types such as rectangular sliding gate valve, butterfly valve, needle valve etc., have been developed which may be used for conduits of large diameters and under high heads. Some of these valves are described in Appendix V.

Intake Structures. An intake structure forms an entrance into the outlet works and it accommodates the control devices at the entrance. It also supports necessary auxiliary appurtenances such as trash racks, fish screen etc. An intake structure may either be submerged or extended as a tower to some height above the maximum water surface. Submerged intake structure is used where trash cleaning is ordinarily not required and where it serves only as an entrance to the outlet conduit. A tower intake is provided if an operating platform is needed for trash racking, maintaining and cleaning of fish screens, or installing stop logs.

Trash Racks. A *trash rack* is a structure which is provided at the entrance to intakes and sluiceways to prevent entrance of debris (Fig. 12.44). These racks are usually made of steel bars spaced 50 to 150 mm centres, depending upon the maximum size of debris which can be permitted in the conduit. The velocity of flow through the trash rack should be kept low, preferably less than 0.6 m/s to minimize head loss. For this the rack is constructed in the form of a half cylinder. The debris which is collected on the trash rack may be removed either manually or with the help of automatic power driven rack rakes.

REVIEW QUESTIONS

12.1 What is a spillway ? What are its essential requirements ? Describe the various components of a spillway.

12.2 How spillways are classified ? Describe briefly the different types of spillways.

12.3 How is the profile of ogee spillway evolved ?

12.4 Discuss the various factors on which the value of the coefficient of discharge of an Ogee spillway depend.

12.5 Describe with a neat sketch a chute spillway. Also discuss the various design considerations of a chute spillway.

12.6 Why is side channel spillway so called ? Describe with a neat sketch a side channel spillway.

12.7 Derive an equation for the water surface profile developed in a side channel spillway.

12.8 Design a side channel spillway for a discharge of 1200 cumec, crest length being 120 m. Take n = 3/4 and assume any other data.

12.9 Describe in detail the hydraulic design features of (i) Siphon spillway; (ii) Shaft or Morning glory spillway.

12.10 Describe with a neat sketch a volute siphon spillway.

12.11 Discuss the advantages as well as the limitations of a siphon spillway.

12.12 Describe with neat sketches the various priming devices for siphon spillways.

12.13 Discuss the various types of energy dissipation devices used below spillways in relation to the positions of the tailwater rating curve (TWRC) and jump height curve (JHC).

12.14 What is a stilling basin ? Describe with neat sketches different types of stilling basins.

12.15 What are the functions of (i) chute blocks; and (ii) baffle blocks in stilling basins ?

12.16 Describe with neat sketches the various types of bucket type energy dissipators.

12.17 Describe with neat sketches (i) Flash boards; (ii) Stop logs and Needles; (iii) Drum gates; (iv) Roller gates; (v) Tainter gates.

12.18 A spillway 300 m long has a coefficient of discharge of 0.68. If the crest elevation is 100.00 m upto what elevation the water level will rise in the reservoir, when a flow of 1500 m^3/s passes over the spillway. Construct U.S.B.R. spillway profile with this as design discharge. Neglect velocity of approach and assume the upstream face of the spillway as vertical.

12.19 Find the length of spillway from the following data:

Height of spillway crest from river bed	= 100 m
Discharge	= 8500 m^3/s
Permissible level of submergence above F.R.L.	= 10 m

Maximum permissible span (clear) = 12.5 m

Thickness of each pier = 3 m

Piers and abutments are rounded type.

[Provide 12 bays, each with a clear span of 11.55 m, thus total length of spillway = 171.6 m]

12.20 Describe with neat sketches the important types of spillway gates.

12.21 Following data were recorded from an ogee spillway:

(i) Maximum reservoir level = 252.00 m

(ii) Bed level of river = 230.00 m

(iii) Highest flood level downstream = 235.00 m

(iv) Maximum flood discharge = 5000 cumec

(v) Effective length of spillway = 400 m

(vi) Coefficient C (in the equation $Q = CLH^{3/2}$) = 2.1 $m^{1/2}/s$

Determine the R.L. of crest of the spillway. What type of energy dissipator will you recommend ? Assume that there is no loss of head in flow over spillway up to the foot of the spillway.

[*Ans.* 248.715 m; Provide either USBR Type III stilling basin, or Indian Standards Type II stilling basin]

12.22 How is the profile of overflow spillways generally fixed ? Discuss the effect of variation in head on the performance of the spillway under free flow conditions.

12.23 An ogee type spillway has 20 crest gates each having 10 m clear span. Find the maximum flood that can be safely passed by lifting all the gates when the maximum reservoir elevation is 110.00 m and the crest level is 106.00 m. Take coefficient C (in the equation $Q = C\,L_{eff}\,H^{3/2}$) as 2.16 $m^{1/2}/s$. Coefficients of end contractions are 0.05 and 0.10 for the piers and abutments respectively. Neglect velocity of approach. [*Ans.* 31310.85 cumec]

12.24 List the various types of energy dissipators used for protection against scour below dams. Describe ski-jump bucket.

12.25 An Ogee type spillway has 2.5 m head above crest. Depth of flow at the toe of the spillways is 0.6 m. Assume coefficient C (in the equation $q = CH^{3/2}$) as 2.16 $m^{1/2}/s$, compute :

(i) The discharge per metre length of spillway.

(ii) Prejump velocity at the toe.

(iii) Prejump Froude number of the flow.

(iv) Conjugate depth required for hydraulic jump.

If the actual tailwater depth is 4.5 m, what type of energy dissipator would you provide ? Give the approximate length and floor level of the dissipating structure.

[*Ans.* (i) 8.538 $m^3/s/m$; (ii) 14.23 m/s; (iii) 5.865 ; (*iv*) 4.686 m; Provide either USBR Type II stilling basin or Indian Standards Type II stilling basin]

CHAPTER

13 Diversion Headworks

13.1 INTRODUCTION

An irrigation canal takes its supplies from a river or a stream. In order to divert water from the river into the canal it is necessary to construct certain works or structures across the river and at the head of the offtaking canal. These works are known as *canal headworks* or *headworks*.

The canal headworks may be classified into the following two types.

(1) Storage headworks

(2) Diversion headworks.

A *storage headworks* consists of a dam constructed across the river to create a reservoir in which water is stored during the period of excess flow in the river. From the reservoir water is supplied to the canal in required quantity as per the demand. Thus, a storage headworks stores water in addition to its diversion into the canal.

A *diversion headworks* serves to raise the water level in the river and divert the required quantity into the canal. The various purposes served by a diversion headworks are as follows.

(i) It raises the water level in the river so that the commanded area is increased.

(ii) It regulates the supply of water into the canal.

(iii) It controls the entry of silt into the canal.

(iv) It provides some storage of water for a short period.

(v) It reduces the fluctuations in the level of supply in the river.

This chapter deals with the diversion headworks.

13.2 TYPES OF DIVERSION HEADWORKS

The diversion headworks may be classified into the following two types.

(i) Temporary diversion headworks

(ii) Permanent diversion headworks.

A *temporary diversion headworks* consists of a spur or bund constructed across the river to raise the water level in the river and divert it into the canal. These bunds may be required to be constructed every year after the floods, because they may be damaged by the floods.

A *permanent diversion headworks* consists of a permanent structure such as weir or barrage constructed across the river to raise the water level in the river and divert it into the canal. Most of the diversion headworks for important canal system in our country are permanent diversion headworks and the same are described in the following sections.

13.3 LOCATION OF CANAL HEADWORKS

The location of canal headworks depends on the stages of flow (or reaches) of river. Most of the large rivers in our country have the following four stages of flow (or reaches).

(i) *Rocky stage or hilly stage.* In this stage the river is in the hills. The bed slope and velocities are high. The cross-section of the river is made up of rock or very large boulders.

(ii) *Boulder stage.* From the rocky stage the river passes on to the boulder stage. In this stage the bed and banks of the river are composed of boulder and gravel. The river cross-section is usually well defined and confined between non-submersible banks on either side which are close to the main current of the river. In this stage the bed slope and velocity are less than those in the rocky stage. However, there is large subsoil flow in the boulder region because of high permeability of the material.

(iii) *Trough stage or alluvial stage.* From the boulder stage the river passes on to the alluvial plain created by itself. In this stage the cross-section of the river is made up of alluvial sand and silt. The bed slope and the velocity are small. During high floods the river spreads out over a wide area as the banks higher than the high flood level are relatively far away from the main current of the river.

(iv) *Delta stage.* From the trough stage the river passes on to the delta stage as it approaches the ocean. In this stage the bed slope and velocity are reduced so much that it is unable to carry its sediment load. It drops down the sediment and gets divided into channels on either side of the deposit resulting in the formation of a delta.

For the construction of the canal headworks both the rocky and the delta stages are not suitable. In the rocky stage there is no difficulty so far as the construction of the headworks is concerned but the construction of the canal to carry water from the headworks to its commanded area will be too costly. In the delta stage the area available is small and also the irrigation requirements are not significant. The canal headworks may thus be located either in the boulder stage or in the trough stage of the river. The following are the advantages and the disadvantages of locating the canal headworks in the boulder and trough stages.

Advantages of Locating the Canal Headworks in the Boulder Stage

(i) The length of the weir is generally shorter in boulder reaches.

(ii) Since high banks are generally available in boulder reaches, less river training works are required.

(iii) Construction materials like stone, sand etc., are locally available.

(iv) Falls available on the canal taken out in the boulder region may be utilised for the generation of hydroelectric power.

Disdavantages of Locating the Canal Headworks in Boulder Stage

(i) In boulder reach there is excessive loss of water due to subsoil flow.

(ii) In the head reaches the canal passes through similar terrain causing heavy seepage losses.

(iii) A large number of cross-drainage works would normally be required on a canal taking off in the boulder region.

(iv) The demand for irrigation in the head reaches of a canal taking off in the boulder region is usually low due to higher rainfall and smaller areas available for cultivation.

Advantages of Locating the Canal Headworks in Trough Stage

(i) Subsoil flow is comparatively less.

(ii) There is no idle length of canal.

(iii) Relatively larger area is served.

(iv) Less number of cross-drainage works are required on a canal taking off in the trough region.

Disadvantages of Locating the Canal Headworks in Trough Stage

(i) The cost of headworks is usually more due to poor foundation.

(ii) Construction material is usually not available locally.

(iii) Headworks in trough region require extensive river training works.

Selection of Actual Site for Locating Canal Headworks

Having selected the reach of the river, the selection of the actual site for the location of the canal headworks may be made in accordance with the following considerations.

(i) As far as possible a narrow, straight, well defined channel confined between banks not submerged by the highest flood should be selected.

(ii) It should be possible to align the offtaking canal in such a way that the command of its area is obtained without excessive digging.

(iii) The material of construction such as stone, sand etc., should be available in the vicinity of the site.

(iv) The site should be accessible by rail or road. Also suitable site for the location of the colony for workers should be available in the vicinity of the project site.

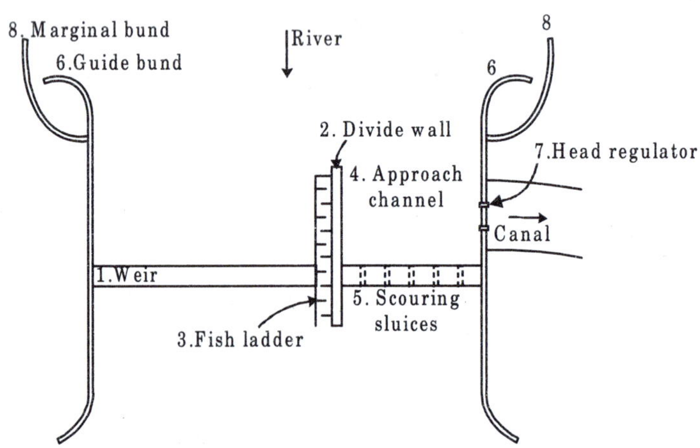

Fig. 13.1 *Typical layout of a diversion headworks*

13.4 COMPONENTS OF DIVERSION HEADWORKS

The various components of a diversion headworks are as follows.

1. Weir or Barrage
2. Divide wall or Divide groyne
3. Fish ladder
4. Pocket or Approach channel
5. Undersluices or Scouring sluices
6. Silt excluder
7. Canal head regulator
8. River training works, such as Marginal bunds and Guide bunds.

A typical layout of a diversion headworks is shown in Fig. 13.1. The description and design details of these components are discussed in the following sections.

13.5 WEIR

A weir is a structure constructed across a river to raise its water level and divert the water into the canal. On the crest of the weirs usually shutters are provided so that part of the raising up of water (or ponding) is carried out by shutters. During floods the shutters may be dropped down to allow water to flow over the crest of the weir. Weirs are usually aligned at right angles to the direction of flow of the river.

Weirs may be classified according to the material of construction and certain design features into the following three types.

1. Masonry weirs with vertical drop or vertical drop weirs
2. Rockfill weirs with sloping aprons
3. Concrete weirs with a downstream glacis.

Masonry Weirs with Vertical Drop or Vertical Drop Weirs. This type of weir consists of an impervious horizontal floor or apron and a masonry weir wall with either both upstream and downstream faces vertical; or both faces inclined; or upstream face vertical and downstream face inclined (Fig. 13.2). At the upstream and the downstream ends of the floor curtain walls and cutoffs or piles are provided. Immediately at the upstream end of the floor a block

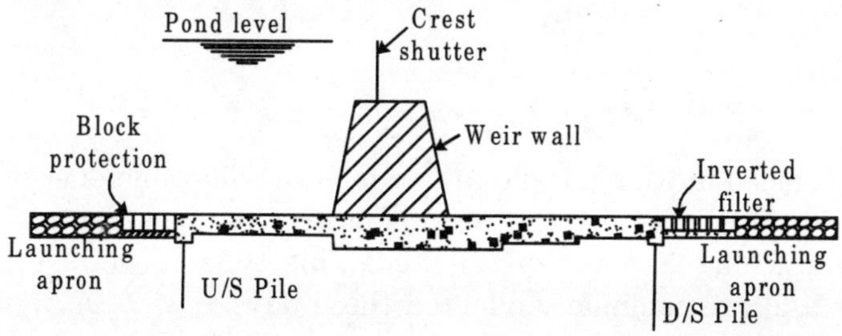

Fig. 13.2 *Vertical drop weir*

protection and at the downstream end a graded inverted filter are provided. After the block protection and the inverted filter launching aprons or pervious aprons (or floors) are provided. This type of weir is suitable for any type of foundation. However, this is an old type of weir for which floor design was usually based on Bligh's theory. One of the important examples of this type of weir is Bhimgoda weir across River Ganga near Haridwar.

Rockfill Weirs with Sloping Aprons. It consists of a masonry weir wall and dry packed boulders laid in the form of glacis or sloping aprons on the upstream and downstream sides of the weir wall with a few intervening core walls (Fig. 13.3). The downstream slope is generally made very flat. It is the simplest type of construction and its stability is not amenable to exact

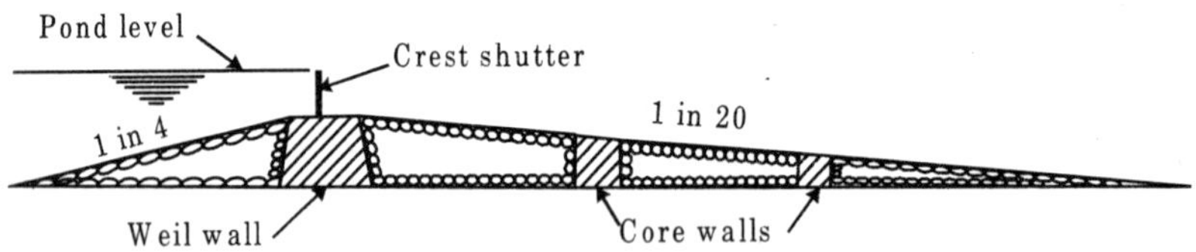

Fig. 13.3 *Rockfill weir with sloping apron*

theoretical treatment. It requires a very large quantity of stone and would be economical only where stone is available in abundance in close proximity. The important example of this type of weir is Okhla weir across River Yamuna near Delhi.

Concrete Weirs with Downstream Galcis. The profile of this type of weir is shown in Fig. 13.4. The weirs of this type are of recent origin and their design is based on modern concept of sub-soil flow, i.e., Khosla's theory. Sheet piles of sufficient depths are provided at the upstream and downstream ends of the floor. Sometimes an intermediate pile is also provided. In this case

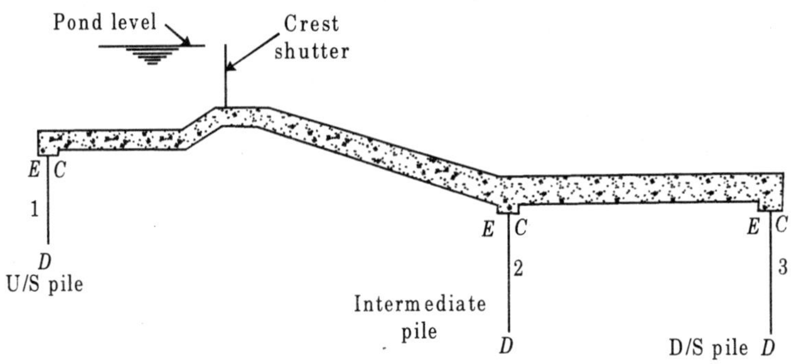

Fig. 13.4 *Concrete weir with downstream glacis*

hydraulic jump is developed on the glacis due to which considerable energy of the flowing water is dissipated. Further in this case also the various protective measures such as block protection, inverted filter and launching aprons are adopted. This type of weirs may be constructed on pervious foundations and are commonly adopted these days.

The design of vertical drop weirs and concrete weirs with downstream glacis is dealt with in detail subsequently.

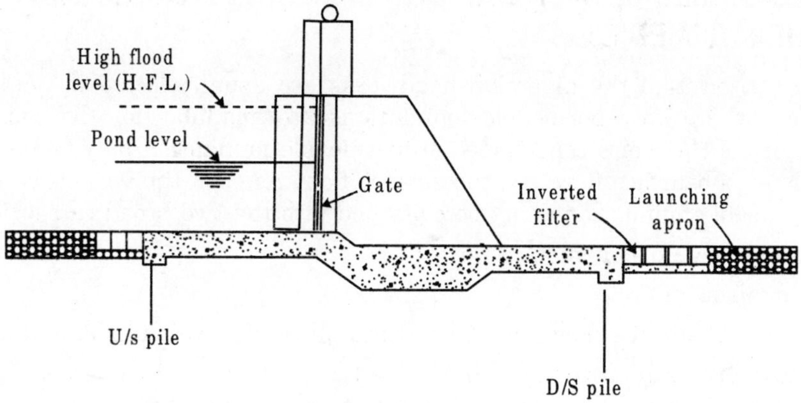

Fig. 13.5 *Barrage*

13.6 BARRAGE

In the case of a barrage the crest is kept at a low level and the raising up of water level (or ponding) is accomplished mainly by means of gates (Fig. 13.5). During floods these gates can be raised clear off the high flood level and thus enable the high flood to pass with minimum of afflux (or heading up of water on the upstream side). A barrage provides better control on the water level in the river but it is comparatively more costly.

The design of a barrage involves the same procedure as a concrete weir.

13.7 EFFECT OF CONSTRUCTION OF A WEIR ON THE RIVER REGIME

The construction of a weir across a river results in creating an obstruction in the path of water flowing in the river, due to which the regime of the river is affected in the following ways.

(i) A weir causes heading up of water which leads to flattening of the water surface slope for some distance on the upstream side of the weir.

(ii) Due to flattening (or reduction) of the water surface slope the sediment carrying capacity of the flowing water is reduced, and part of the sediment load drops and deposits on the river bed, resulting in the formation of irregular shoals on the upstream of the weir.

(iii) As sediment is deposited on the upstream side of the weir, relatively clear water passes over the weir. This water scours the river bed on the downstream side to make up the deficiency in its sediment load and causes a progressive lowering or retrogression of downstream bed level.

(iv) The above effects develop and continue for a number of years. As the shoal formation in the upstream reach increases, the resistance to flow of water in the river is increased. To overcome this resistance increased head is required and the river starts to regain its original slope thus extending the afflux more and more to the upstream. A stage is gradually reached when the upstream reach of the river cannot take up any more sediment and the normal sediment load is passed down the weir which will be deposited in the downstream reach.

(v) Further since the offtaking canal takes comparatively sediment free water, the sediment load going down the weir remains same while the discharge decreases. This will also lead to sediment deposit in the downstream reach and a long range recovery of levels on the downstream side.

13.8 CAUSE OF FAILURE OF WEIRS ON PERMEABLE FOUNDATION AND THEIR REMEDIES

As indicated in section 13.3 the diversion headworks are usually locted in boulder and trough stages of a river, where only permeable foundations are available for the construction of the weirs. The stability of the weirs constructed on permeable foundation may be endangered due to both the seepage or subsurface flow and the surface flow. As such the various causes of failure of the weirs on permeable foundations may be classified into the two broad categories as follows.

1. Due to seepage or subsurface flow
2. Due to surface flow.

The seepage or subsurface flow may cause the failure of a weir in the following two ways.

(i) **By piping or undermining.** If the water percolating through the foundation has sufficient force when it emerges at the downstream end of the impervious floor it may lift up the soil particles at the end of the floor. With the removal of the surface soil, there is further concentration of flow into the resulting depression and more soil is removed. This process of erosion progressively extends backwards towards the upstream side and results in the removal of soil and developing pipe like formation beneath the floor. The floor may subside in the hollows so formed and fail which is known as failure due to piping or undermining.

The following measures may be taken to prevent the failure due to piping or undermining.

(a) Providing sufficient length of the impervious floor so that the path of percolation is increased and exit gradient (i.e., hydraulic gradient at exit) is reduced.

(b) Providing piles at the upstream and the downstream ends of the impervious floor.

(ii) **By uplift pressure**. The water percolating through the foundation exerts an upward pressure on the impervious floor. This pressure is known as uplift pressure. If the uplift pressure is not counterbalanced by the weight of the floor, it may fail by rupture.

The following measure may be taken to prevent the failure of the floor by rupture due to excessive uplift pressure.

(a) Providing sufficient thickness of the impervious floor.

(b) Providing pile at the upstream end of the impervious floor so that the uplift pressure is reduced on the downstream side.

The surface flow may cause the failure of a weir in the following two ways.

1. By Suction due to Standing Wave or Hydraulic Jump. The standing wave or hydraulic jump developed on the downstream side of the weir causes suction or negative pressure which also acts in the direction of uplift pressure. If the floor thickness is insufficient it may fail by rupture due to suction.

The following measures may be taken to prevent the failure of the floor by rupture due to suction or negative pressure.

(a) Providing additional thickness of the impervious floor to counterbalance the suction pressure due to standing wave.

(b) Constructing floor as monolithic concrete mass instead of in different layers of masonry.

2. By Scour on the Upstream and Downstream of the Weir. Both at upstream and downstream ends of the impervious floor the bed of the river may be scoured to considerable depths during floods. If no preventive measures are taken these scours may cause considerable damage to the floor leading to its failure.

The following measures may be taken to prevent the failure of the floor due to such scours.

(a) Providing deep piles both at upstream and downstream ends of the impervious floor. The piles are to be driven upto a depth much below the calculated scour depth.

(b) Providing launching aprons of suitable length and thickness at upstream and downstream ends of the impervious floor.

13.9 CRITERIA FOR THE DESIGN OF WEIRS AND BARRAGES

Alike any other hydraulic structure the design of weirs and barrages also consists of the following two phases.

1. Hydraulic design
2. Structural design

The hydraulic design deals with the evaluation of the hydraulic forces acting on the structure and determination of the configuration of the structure which will be most economical and will have the best functional efficiency. The structural design consists of dimensioning of the various parts of the structure to enable it to resist safely all the forces acting on it. This is done by the accepted norms of structural analysis and design.

The hydraulic design of weirs and barrages on permeable foundation is treated in respect of both subsurface and surface flows which may lead to their failure. The various aspects of design in respect of subsurface flow involves determination of uplift pressures, exit gradient, length of impervious floor, depths of sheet piles or cutoffs at upstream and downstream ends of the impervious floor and protection works. All these aspects of design in respect of subsurface flow are discussed in the next few sections. On the other hand the various aspects of design in respect of surface flow involves determination of pond level, afflux, levels of upstream floor and crest of weir or barrage, shape of weir crest, waterway, effect of regression and measures of energy dissipation. All these aspects of design in respect of surface flow are discussed below.

1. Pond Level. Pond level, in the undersluice pocket, upstream of the canal head regulator may be obtained by adding the working head to the designed full supply level in the canal. The working head should include the head required for passing the design discharge into the canal and the head losses in the regulator. If under certain situations there is a limitation of pond level, the full supply level in the canal is fixed by subtracting the working head from the pond level.

2. Afflux. Afflux may be defined as the rise in water level on the upstream of a weir or barrage over and above that on the downstream caused under free flow conditions as a result of construction of the weir or barrage on the river. The value of afflux corresponding to the design flood is important for the design of the length of weir or barrage, crest levels, downstream cistern, depth of cutoffs, bed protection and river training works. Provision of higher afflux has following advantages and disadvantages.

Advantages

(i) Narrow weir at low cost.

(ii) Makes the weir modular at all stages by fomation of standing wave or hydraulic jump even at maximum flood level.

(iii) Helps operation of silt excluders and silt extractors.

(iv) In general better control on weir.

Disadvantages

(i) Higher cost of gates.

(ii) Greater cost of river training works.

(iii) Greater intensity of discharge.

(iv) Greater depth of probable scour and consequential additional length and thickness of launching aprons upstream and downstream.

(v) Greater depth of cutoffs both upstream and downstream of the weir.

(vi) Greater afflux means greater difference of water level between upstream and downstream of the weir and greater dynamic action as also its effect on the location and depth of standing wave.

(vii) Greater risk of failure due to outflanking.

The maximum permissible value of afflux depends on the river conditions upstream, extent of backwater effect, the area being submerged and its importance. IS: 6966–1989 has recommended an afflux of 1 m in the case of weirs or barrages located on alluvial rivers in upper and middle reaches of the river and 0.3 m in lower reaches of the river having flat gradients. In the case of weirs or barrages located on rivers with shingle/boulder beds and banks afflux more than 1 m is permissible.

3. Levels of Upstream Floor and Crest of Weir or Barrage. The upstream floor level of a weir or a barrage bays (other than undersluices bays) is fixed at the general river bed level at or below the level of the crest of the weir or barrage. Usually the floor level is kept at 0.5 to 1.0 m higher than the upstream floor level of the undersluice bays.

The crest levels of weirs or barrages are fixed as indicated below.

(i) For weirs without shutters the crest level should be at the required pond level.

(ii) For weirs with falling shutters, the crest level should not be lower than 2 m below the pond level as the maximum height of the falling shutters is normally limited to 2 m. It may be suitably raised if possible, from consideration of passing the design flood discharge at the desired afflux and with the waterway provided. If the lowest crest level fixed in the above manner causes too much of afflux, the waterway of weir may be suitably increased.

(iii) For barrages the crest level is determined by the depth required to pass the design flood at the desired afflux. The level of crest in this case should be fixed by adjustment of the waterway. It should in any case be kept higher than the undersluice crest level.

4. Shape of Weir Crest. A vertical drop type weir is usually trapezoidal in cross-section and its dimensions may be obtained on the basis of stability considerations as indicated in section 13.15.

A glacis type weir is provided with a top width of about 2 m, an upstream slope of 2:1 to 3:1 depending on site conditions and downstream slope as required for the glacis of stilling basin. For hydraulic jump type stilling basin IS: 6966–1989 has recommended a slope of 3:1 for the downstream glacis.

5. Waterway. The length of waterway which is equal to the length of weir or barrage is fixed to pass safely the maximum flood discharge. It depends on the following factors.

(i) Maximum flood discharge

(ii) Permissible intensity of discharge

(iii) Afflux and pond level

(iv) Stable river width

(i) *Maximum flood discharge.* Various agencies have recommended the following criteria for fixing the maximum flood to be adopted for the design or the design flood.

Sl. No.	*Agency*	*Criterion*
1.	Central Water Power Commission (CWPC) India	Standard Project Flood (SPF)* or a 100 year flood whichever is higher
2.	Indian Standard Institution, IS: 6966–1989	50 to 100 year frequency flood but of 500 year frequency or Standard Project Flood for freeboard
3.	Khosla Committee of Engineers	Maximum recorded flood for a period greater than 50 years

Discharges smaller than the design flood will result in shoal formations and meandering of the river, causing encroachment on freeboard of river training works. Discharges greater than the design flood will wash off shoals and make available the full waterway. Thus, a higher recurrence interval is preferred for determining the freeboard.

(ii) *Permissible intensity of discharge.* Safe permissible intensity of discharge depends on:

(a) Nature of river bed material; and

(b) Safe allowable afflux depending on topography and economy

In order to prevent scouring of the bed material a limit has to be imposed on the intensity of discharge. Permissible intensity of discharge is lower in alluvial reaches and higher in rocky and boulder reaches.

Higher intensity of discharge reduces the length of the waterway but has the following ill effects:

(a) It increases the velocity of flow, thereby resulting in greater scour. Thus, the cost of protection works increases.

(b) Deepens the downstream floor of the structure. This results in increased cost of dewatering and increase in the thickness of floor.

(iii) *Afflux and pond level.* Higher the afflux, higher will be the pond level and hence the head acting on the weir or barrage will be high. This will reduce the required length of the waterway.

(iv) *Stable river width.* The length of waterway should be equal to the stable river width for the maximum flood discharge so that shoaling upstream is mostly eliminated and a nearly straight and stable approach to the weir or barrage is obtained. In the case of wide rivers, wetted perimeter is practically equal to the width of the river. Further since river under maximum flood may be considered to be in temporary regime, minimum stable river width for the maximum flood discharge may be taken equal to Lacey's regime perimeter. As such the length of the clear waterway to be provided between guide banks or abutments, excluding thickness of piers is usually taken equal to the Lacey's regime perimeter given by

* Standard Project Flood (SPF) is the flood resulting from the most severe combination of hydrological and meteorological factors that are considered reasonably characteristic of the region and is computed by using the Standard Project Storm (SPS).

$$P = 4.83\sqrt{Q}$$

where

P = Lacey's regime perimeter in m, and

Q = design flood discharge in cumec.

The constant in Lacey's equation may vary from 3.6 to 6.15 depending on discharge and nature of bed material. However, the value of 4.83 for the constant in Lacey's regime perimeter equation is the best fit value for regime conditions.

The overall length of waterway or the length of weir or barrage between guide banks or abutments is obtained by adding the thickness of piers to the Lacey's regime perimeter.

Recommended Waterway

Waterway recommended by IS: 6966–1989 is as indicated below.

(i) In deep and confined rivers with stable banks, the overall waterway between abutments including thickness of piers is approximately equal to the actual river width at design flood.

(ii) For shallow and meandering alluvial rivers, the minimum stable width of the river at the design flood may be calculated from Lacey's equation for regime perimeter which is given below

$$P = 4.83\sqrt{Q}$$

where

P = minimum stable width of the river in m; and

Q = design flood discharge in cumec.

Sometimes in order to effect economy clear waterway of length less than the Lacey's regime perimeter may be provided. This is usually so in the case of weirs or barrages constructed on rivers in boulder stage. The restricted waterway has the following advantages and disadvantages.

Advantages

(i) Less length of the weir or barrage

(ii) Less cost due to less length

(iii) Better river control

(iv) Helps in removing shoals formed upstream of the structure.

Disadvantages

(i) Excessive velocities causing deep scour along the guide banks.

(ii) Undesirably high afflux which either submerges a vast area or requires high afflux bund and consequently higher cost of river training works.

(iii) Lower levels of upstream and downstream impervious aprons which entail higher pumping expenses for dewatering before laying the aprons.

For weirs or barrages constructed on rivers flowing in alluvial soils clear waterway of length more than the Lacey's regime perimeter is provided. This is to account for the silt load carried by the river, which is neglected in Lacey's equation. The normal practice in such cases is to provide clear waterway of length 1.1 to 1.25 times Lacey's regime perimeter.

The ratio of the overall length of the weir/barrage provided to the theoretically computed minimum stable width of the river at the design flood obtained by using Lacey's equation is known as *looseness factor*. It is experienced that of weirs constructed on rivers in alluvial reaches with looseness factor 1.5 to 2.0, heavy shoals are invariably formed in the upstream with top upto the crest of the weir and as a consequence it becomes difficult to feed canal due to shifting of the channels caused by shoals. The values of the looseness factor adopted for some of the recently constructed barrages in India on rivers in two important stages are as indicated below.

River stage	*Looseness factor*
Boulder	1.0 to 0.45
Alluvial	1.0 to 1.45

6. Effect of Retrogression. As a result of construction of a weir or barrage progressive retrogression or degradation of the downstream river bed is caused which results in lowering of the downstream river stages (or river water surface levels) and the same has to be suitably provided for in the design of downstream cisterns. The lowering of river water level due to retrogression on the downstream causes increased exit gradients.

Retrogression of river bed and consequent lowering of water levels on the downstream of weir or barrage is more pronounced in alluvial rivers carrying more silt, having finer bed material and having steep slope. As a result of retrogression, low stages of the river are generally affected more compared to the maximum flood levels. A value of 1.25 to 2.25 m may be adopted as retrogression for alluvial rivers at lower river stages depending upon the amount of silt in the river, type of its bed material and its slope. At the design flood in an alluvial river value of retrogression may be considered to vary from 0.3 to 0.5 m depending on whether the river is shallow or confined during floods. Whenever a proposed weir or barrage is situated downstream of a dam, the possibility of heavier retrogression than normal should be considered for the design of downstream floor level and downstream protection works.

7. Discharge and Discharge Intensity. The discharge over the crest of the weir or barrage is given by the following formula

$$Q = CLK^{3/2}$$

where

Q = discharge in cumec

C = coefficient of discharge

L = length of clear waterway of weir or barrage in m, and

K = intercept between the upstream total energy line and the crest, or the static head + the velocity head over the crest, in m.

Thus, K represents the total head causing the flow over the crest of the weir or barrage.

The value of the coefficient of discharge C is usually taken as 1.7.

The discharge intensity q (= Q/L) is thus given by

$$q = CK^{3/2} = 1.7\ K^{3/2}$$

8. Energy Dissipation. For weirs or barrages on alluvial foundations hydraulic jump type stilling basins are normally adopted for dissipation of energy.

13.10 DESIGN OF IMPERVIOUS FLOOR FOR SUBSURFACE FLOW

Upto 19th century the various hydraulic structures constructed in our country on alluvial river beds and plains were mainly designed on experience and intuition as no rational theory of seepage had been developed by then. It was only in 1895 when Khanki weir was damaged, experiments were carried out by Lt. Col. Clibborn, Principal, Thomson College Roorkee, on flow of water through soils to ascertain the criteria of designing weirs on permeable foundations. The results of these experiments indicated that Darcy's law was valid for flow through soils under low heads. Further on the basis of the results of these experiments and certain field observations on Narora weir, it was found that the subsurface flow or foundation seepage may cause the failure of an impervious floor on a permeable foundation in two ways viz., (i) piping ; and (ii) uplift pressure as indicated in section 13.8. Since the beginning of the present century the problem of prevention of the failure due to piping and determination of correct uplift pressures due to subsurface flow has drawn attention of several engineers. Some of the notable contributions for obtaining the solution of this problem are those of Bligh, Lane and Khosla which are discussed in the subsequent sections.

13.11 BLIGH'S CREEP THEORY

In 1910 W.G. Bligh presented a theory for the subsurface flow in his book 'Practical Design of Irrigation Works'. This theory came to be known as *Bligh's Creep Theory*.

Bligh assumed that the percolating water creeps along the base profile of the structure which is in contact with the subsoil. The length of the path thus traversed by the percolating water is called the *length of creep or creep length*. Bligh further assumed that the head loss per unit length of creep which is called the *hydraulic gradient* is constant throughout the percolating passage. From this it follows that the loss of head is proportional to the length of creep. Bligh made no distinction between horizontal and vertical creep.

Consider a barrier impounding water of depth H, which is provided with a horizontal floor of length b with three vertical cutoffs (or sheet piles) as shown in Fig. 13.6. The percolating water will then follow the path indicated by arrows and the creep length L will be given by

$$L = b + 2d_1 + 2d_2 + 2d_3$$

The hydraulic gradient or the loss of head per unit length of creep will therefore be given by

$$\frac{H}{L} = \frac{H}{b + 2d_1 + 2d_2 + 2d_3} \quad \text{...(13.1)}$$

As hydraulic gradient is constant, if L_1 is the creep length upto any point, then head loss upto this point will be $(H/L)\,L_1$ and the residual head at this point will be $[H - (H/L)\,L_1]$. Also there will be losses of head equal to $(H/L)\,2d_1$, $(H/L)\,2d_2$ and $(H/L)\,2d_3$ respectively in the planes of the three vertical cutoffs and the hydraulic gradient line will be drawn as shown in Fig. 13.6.

The reciprocal of hydraulic gradient i.e., (L/H) is known as *Bligh's coefficient of creep C*. Thus,

$$L = CH \quad \text{...(13.2)}$$

According to Bligh, to ensure the safety of the impervious floor against the two possible ways in which failure may be caused by subsurface flow, following criteria are required to be satisfied.

(i) *Safety against piping*. The length of creep should be sufficient to provide a safe hydraulic gradient according to the type of soil.

From Eq. (13.2)

$$L = CH$$

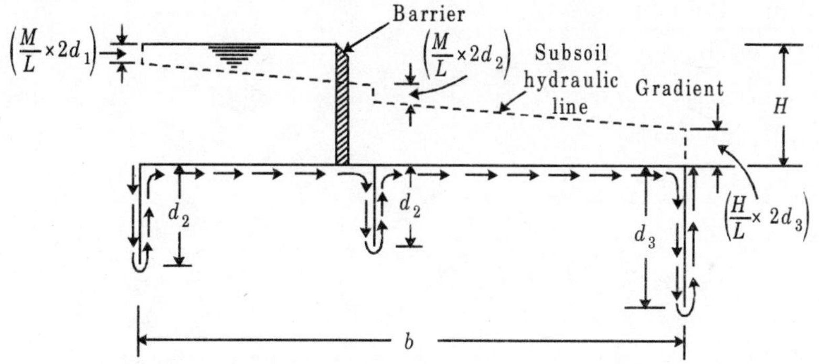

Fig. 13.6 *Bligh's path of creep and sub-soil hydraulic gradient line*

Bligh has recommended certain values of *C* for different soils as given in Table 13.1.

The hydraulic gradient (*H*/*L*) is then equal to (1/*C*) and according to Bligh if the hydraulic gradient ≤ (1/*C*) there will be no danger of piping.

It may be noted that the seepage head *H* is to be measured from the water level upstream to the corresponding lowest water level downstream. The worst condition giving the maximum value of *H* should be selected for design which in most of the cases would occur when water is held up to the highest possible level on the upstream side with no discharge to the downstream side, the downstream water level being taken at the downstream bed level.

TABLE 13.1 Recommended values of Bligh's coefficient of creep C and safe hydraulic gradient

Type of soil	Value of C	Safe Hydraulic Gradient (1/C)
1. Light sand and mud (as in River Nile)	18	1/18
2. Fine micaceous sand (as in North Indian rivers)	15	1/15
3. Coarse grained sand (as in Central and South Indian Rivers)	12	1/12
4. Sand mixed with boulder and gravel ; and for loam soil	5 to 9	1/9 to 1/5

(ii) *Safety against uplift pressure*. The ordinate of the subsoil hydraulic gradient line above the bottom of the floor at any point represents the residual seepage head or the uplift pressure at that point. Thus, if at any point h' is the ordinate of the hydraulic gradient line above the bottom of the floor then at that point the uplift pressure exerted by the percolating water is wh', where w is specific weight of water. If at this point the floor thickness is t and the specific gravity of the material of the floor is G, the downward force per unit area due to the weight of the floor is $(wG)\,t$.

For equilibrium the uplift pressure must be counterbalanced by the weight of the floor. Thus, equating the two, we get

$$wh' = (wG)\,t$$

or $$t = \frac{h'}{G} \qquad ...(13.3)$$

However, the ordinate h' of the hydraulic gradient line above the bottom of the floor can be known only after the floor thickness has been determined. Thus, it is more convenient to express Eq. (13.3) in the following form.

From Eq. (13.3), we have

$$h' = tG$$

Deduct t from both sides

$$\therefore \quad h' - t = t\,G - t = t(G-1)$$

From which

$$t = \frac{h'-t}{G-1} \qquad ...(13.4)$$

or $$t = \frac{h}{G-1} \qquad ...(13.4\text{ a})$$

where $h = (h' - t)$ is the ordinate of the hydraulic gradient line measured above the top of the floor. Dividing this ordinate by $(G - 1)$ i.e., the submerged specific gravity of the floor material, the thickness of the floor may be determined.

The floor thickness given by Eq. 13.4 is usually increased by considering a factor of safety of (4/3) and hence we have

$$t = \frac{4}{3}\frac{h}{G-1} \qquad ...(13.4\text{ b})$$

It may be noted that on the upstream side of the barrier viz., weir wall or the gates of the barrage, which holds up the water and creates the seepage head, the uplift pressures are counterbalanced by the weight of the water standing on the floor. However, when there is no water on the upstream side of the barrier there will be no seepage head and no uplift pressure. As such for the upstream floor only a nominal thickness needs to be provided to resist wear, impact of flowing water or development of cracks. The floor on the downstream side of the barrier must be designed in accordance with Eq. (13.4).

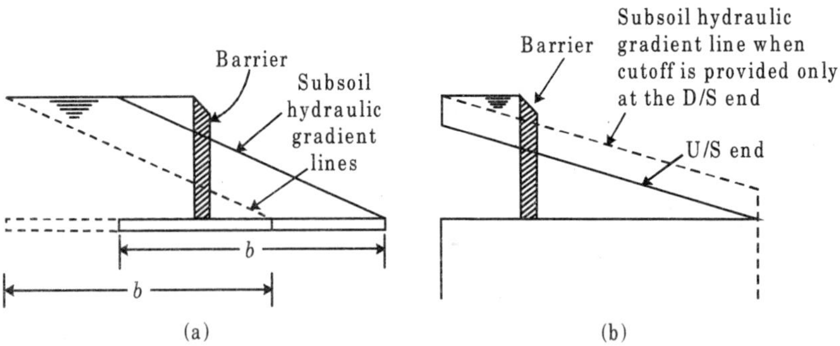

Fig. 13.7 *Effect of shifting of floor relative to the barrier as well as effect of u/s and d/s cutoff on sub-soil hydraulic gradient line*

As the upstream floor needs to be provided with only a nominal thickness as per practical considerations, while the downstream floor has to be designed to resist uplift pressure and is therefore thicker, it would be economical to provide as much of the total required creep length on the upstream of the barrier as possible. However, a minimum floor length is always required to be provided on the downstream side from the consideration of surface flow to resist the action of fast flowing water whenever it is passed to the downstream side of the barrier. Moreover, the provision of maximum creep length on the upstream side of the barrier also reduces uplift pressures on the portion of the floor provided on the downstream side of the barrier (Fig. 13.7 a). This is so because a large portion of the total creep having taken place upto the barrier, the residual heads on the downstream floor are reduced. Further as shown in (Fig. 13.7 b) a vertical cutoff at the upstream end of the floor reduces uplift pressures all over the floor while a vertical cutoff at the downstream end of the floor increases them. Thus, according to Bligh's theory a vertical cutoff at the upstream end of the floor is more useful than the one at the downstream end of the floor.

Limitations of Bligh's Creep Theory

The various limitations of Bligh's creep theory are as follows.

1. Bligh made no distinction between horizontal and vertical creep.

2. Bligh's method holds good so long as the horizontal distance between the cutoffs or pile lines is greater than twice their depth.

3. Bligh made no distinction between the effectiveness of the outer and inner faces of sheet piles and short and long intermediate piles. However, later investigations have indicated that the outer faces of the end sheet piles are much more effective than the inner ones. Also intermediate sheet piles of length shorter than the outer ones are ineffective except for local redistribution of pressure.

4. Bligh did not indicate any significance of exit gradient. However, it has been established that the safety against piping cannot be obtained by simply considering an average hydraulic gradient but the exit gradient should be kept well below the critical value.

5. According to Bligh the loss of head is proportional to the creep length, but it is not true. Also the actual uplift pressure distribution is not linear but it follows a sine curve.

6. Bligh did not specify the absolute necessity of providing a cutoff at the downstream end of the floor, whereas it is absolutely essential to provide a deep vertical cutoff at the downstream end of the floor to prevent undermining.

13.12 LANE'S WEIGHTED CREEP THEORY

Lane approached the problem by making a statistical examination of 290 existing structures on pervious foundations and evolved a theory in 1932 which is known as *Lane's weighted creep theory*. Lane made distinction between horizontal and vertical creep and hence Lane's weighted creep theory in effect may be called 'Bligh's creep theory corrected for vertical contacts'.

Lane indicated that the horizontal creep is less effective in reducing uplift (or in causing loss of head) than the vertical creep. He therefore proposed a weightage factor of (1/3) for the horizontal creep as against 1.0 for the vertical creep. Thus, according to Lane's weighted creep theory the weighted creep length L_w is given by

$$L_w = 13\,N + V \qquad ...(13.5)$$

in which

N is the sum of all the horizontal contacts and all the sloping contacts less than 45° to the horizontal ; and

V is the sum of all the vertical contacts and all the sloping contacts greater than 45° to the horizontal.

Further to ensure safety against piping L_w must not be less than C_1H (i.e., $L_w \geq C_1\,H$) where H is total seepage head i.e., the difference of water levels on the upstream and the downstream; and C_1 is an empirical coefficient depending on the nature of the soil. Values of C_1 or Lane's coefficient of creep for different kinds of soils are given in Table 13.2. Further the hydraulic gradient (H/L_w) is then equal to $(1/C_1)$ and the safety against piping can be ensured if the hydraulic gradient $\leq (1/C_1)$.

Although Lane's theory is an improvement over Bligh's theory, most of the limitations of Bigh's theory exist in Lane's theory also. Moreover Lane's theory is empirical and lacks any rational basis. As such though Bligh's theory is still being used in the design, Lane's theory is not being used and is having only a theoretical importance.

TABLE 13.2 Recommended values of Lane's coefficient of creep C_1 and safe hydraulic gradient

Type of soil	*Value of C_1*	*Safe Hydraulic Gradient $(1/C_1)$*
1. Very fine sand or silt	8.5	1/8.5
2. Fine sand	7.0	1/7
3. Coarse sand	5.0	1/5
4. Gravel and sand	3.5 to 3.0	1/3.5 to 1/3
5. Boulders, gravel and sand	3.0 to 2.5	1/3 to 1/2.5
6. Clayey soil	3.0 to 1.6	1/3 to 1/1.6

13.13 THEORY OF SEEPAGE FLOW

It has been established both theoretically as well as experimentally that if Darcy's law holds good then steady seepage through a homogeneous soil can be represented by the Laplacian equation

$$\frac{\partial^2 \phi}{\partial x^2} + \frac{\partial^2 \phi}{\partial y^2} = 0 \qquad ...(13.6)$$

in which ϕ is flow potential given by $\phi = kh$, where k is coefficient of permeability as defined by Darcy's law and h is the seepage head at any point in the soil. Equation (13.6) represents two sets of curves intersecting each other orthogonally or at right angles (Fig. 13.8). One set of curves is called streamlines or flow lines and the other set of curves is called equipotential lines.

The streamlines indicate the paths adopted by particles of water flowing through the soil. The first streamline follows the outline of the base of the structure and is the same as Bligh's path of creep. Other streamlines follow smooth curves providing a gradual transition from the outline of the base of the structure to a semi-ellipse if the pervious soil medium extends to a very large depth. However, if there is an impervious boundary at a certain depth the last streamline will follow the impervious boundary and the intermediate streamlines will represent a smooth transition from the first streamline to the last.

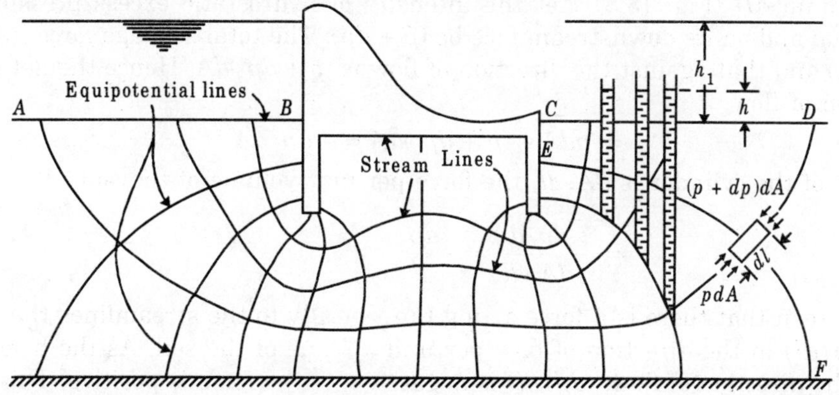

Fig. 13.8 *Flownet for seepage flow through soil below a hydraulic structure*

The water entering the soil at the upstream end has certain head and as it emerges into atmosphere at the downstream end its head is zero. Thus, at the upstream end every streamline has a certain constant head which is entirely lost through the passage of the streamline in the soil and at every intermediate point in its path it has a certain residual head h still to be dissipated in the remaining length to be traversed to the downstream end. If the points having the same value of residual head h on the various streamlines are joined together the curve obtained is called an equipotential line. If h_1 is the total head at the upstream end with downstream bed as datum then since every particle of water enters with the same head h_1 the upstream surface of entry *AB* is the first equipotential line with $h = h_1$. Similarly the downstream surface *CD* is the last equipotential line with $h = 0$ (if no water is standing on it). In between, a number of equipotential lines for different values of h between 0 to h_1 may be drawn. If a number of piezometers are installed on the same equipotential line water will rise in all of them to the same level as the sum of pressure and datum heads is constant all along the equipotential line.

The combined representation of streamlines and equipotential lines is called flow net. The flow pattern for steady seepage through homogeneous soil can be represented by a flow net from which various effects of seepage can be computed.

Seepage Discharge

As described in chapter 11 the seepage discharge q per unit length at right angles to the plane of the figure is given by

$$q = kh_1 \frac{N_f}{N_d}$$

where N_f is the number of spaces between adjacent streamlines (called flow channels), N_d is number of potential drops ; and k and h_1 are same as defined earlier.

Critical Hydraulic Gradient

The seepage water exerts a force called seepage force on the soil in the direction of flow which may be determined as indicated below.

Consider an elementary cylindrical element of soil of cross-sectional area dA and length dl on any streamline AD (Fig. 13.8). Let the intensity of hydrostatic excess pressure on its upstream face be p and on its downstream face be $(p + dp)$. The total seepage force in the direction of flow is (pdA) and that against the direction of flow is $(p + dp)dA$. Hence the net seepage force in the direction of flow

$$= pdA - (p + dp)dA = -dp.dA$$

As the volume of the cylinder is $dA.dl$, the force per unit volume of the soil

$$= \frac{-dp \cdot dA}{dA \cdot dl} = -\frac{dp}{dl}$$

Thus, it is seen that there is a force acting tangentially to the streamlines the magnitude of which is $-(dp/dl)$ in the direction of flow per unit volume of the soil. As the head or pressure decreases in the direction of flow, (dp/dl) is negative and the force is positive. This force has an upward component from the point where the streamline turns upwards. At the downstream or the exit end the streamlines emerge vertically as they must meet the downstream bed, which is an equipotential line, at right angles. At the exit end, therefore, the direction of the force is vertically upwards. If the soil at the exit end is to remain stable and resist undermining, the downward force due to the submerged weight of the soil should be more than the upward force of seepage. When the two forces just balance each other there will be a critical condition.

If G is specific gravity of soil grains and n is porosity of the soil or the proportion of pore spaces to the total volume of the soil, then the submerged weight w_b of a unit volume of soil is given by

$$w_b = w(1-n)(G-1)$$

in which w is specific or unit weight of water.

For critical condition to be reached the upward seepage force $-(dp/dl)$ on unit volume of soil will just balance w_b. Thus,

$$-\frac{dp}{dl} = w(1-n)(G-1)$$

Dividing both sides by w, we get

$$-\frac{dp}{dl} = (1-n)(G-1) \qquad \text{...(13.7)}$$

as $\qquad p = wh$

(dh/dl) represents the rate of loss of head, or the hydraulic gradient. The hydraulic gradient at the exit end is known as *exit gradient.* When exit gradient equals the value given by Eq. (13.7), it is called *critical hydraulic gradient* (or *critical gradient*). If critical hydraulic gradient is reached, failure by undermining is imminent.

For most river sands the value of G is nearly 2.65 and an average value of porosity is 0.4, the value of critical hydraulic gradient is, therefore, $(1 - 0.4)(2.65 - 1) = 0.99$, or nearly 1 on an aveage. However, from the requirements of safety the actual gradient at exit must be much less, say 1/4 to 1/6.

Determination of exit gradient. As indicated later exit gradient may be determined by Eq. (13.9). However, it may also be determined with the help of flownet by using the following expression

$$G_E = \frac{\Delta h}{l} \qquad \text{...(13.7 a)}$$

in which G_E is exit gradient; Δh is the potential (or head) drop between the adjacent length of the last field in the flownet at the exit end.

13.14 KHOSLA'S THEORY

From 1910 onwards Bligh's theory became the accepted basis for designing the structures on permeable foundations. A number of important structures were designed on the basis of this theory, some of which remained stable while others gave trouble or failed. In 1926–27 some siphons on Upper Chenab canal which were designed on the basis of Bligh's theory had undermining trouble. During investigations the actual pressure measurements were made with the help of pipes inserted in the floors of these siphons, which indicated that the actual pressures were quite different from those computed on the basis of Bligh's theory. These investigations were carried out by Dr. A. N. Khosla and his associates which led to the following interim conclusions.

(i) The outer faces of the end sheet piles were much more effective than the inner ones and the horizontal length of the floor.

(ii) The intermediate sheet piles, if smaller in length than the outer ones were ineffective except for local redistribution of pressures.

(iii) Undermining of floors started from the tail end. If the hydraulic gradient at the exit was more than the 'critical gradient' for the particular soil, the soil particles would move with the flow of water thus causing progressive degradation of the subsoil, resulting in cavities and ultimate failure.

(iv) It was absolutely essential to have a reasonably deep vertical cutoff at the downstream end of the floor to prevent undermining.

Khosla and his associates carried out further research to find an ultimate solution to the problem of subsurface flow and provided a complete rational solution of the problem which is known as *Khosla's theory.* The results of this research have been published by the Central Board of Irrigation and Power in the Publication No.12 entitled 'Design of Weirs on Permeable Foundations'. A brief discussion of Khosla's theory is presented as follows.

Khosla and his associates considered the flow pattern below the impervious base of hydraulic structures on permeable foundations. From the flow pattern the distribution of uplift pressures on the base of the structure and the exit gradient can be determined. Various cases were analysed by them, out of which the simplest case is that of a horizontal impervious floor on permeable foundation. This is usually known as *Khosla's First Problem* which is indicated below.

Horizontal Floor

Consider a horizontal impervious floor AB of width b lying on permeable foundation as shown in Fig. 13.9. The flow field in this case may be represented in terms of complex variables by the following expression

$$z = \frac{b}{2} \cosh w \qquad ...(i)$$

where $z = x + iy$; $w = u + iv$; x and y are the coordinates of any point P in the flow field ; and u and v are the stream function and potential (or pressure) function at P.

Substituting these values in Eq. (i), we get

$$(x + iy) = \frac{b}{2} \cosh (u + iv)$$

or

$$(x + iy) = \frac{b}{2} (\cosh u \cos v + i \sinh u \sin v)$$

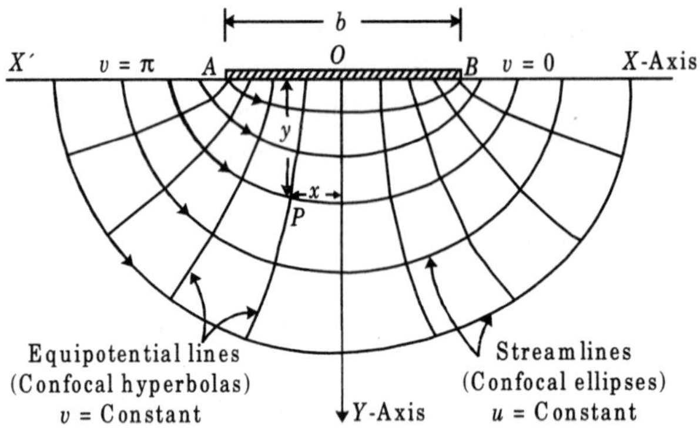

Fig. 13.9 *Horizontal floor on permeable foundation*

Separating the real and imaginary quantities and equating them, we get

$$x = \frac{b}{2} \cosh u \cos v \qquad ...(ii)$$

and

$$y = \frac{b}{2} \sinh u \sin v \qquad ...(iii)$$

From Eqs (ii) and (iii), we have

$$\cos v = \frac{x}{\left(\frac{b}{2}\cosh u\right)} \qquad ...(iv)$$

and

$$\sin v = \frac{y}{\left(\frac{b}{2}\sinh u\right)} \qquad ...(v)$$

Squaring Eqs (iv) and (v) and adding, we get

$$\cos^2 v + \sin^2 v = 1 = \frac{x^2}{\left(\frac{b}{2}\cosh u\right)^2} + \frac{y^2}{\left(\frac{b}{2}\sinh u\right)^2}$$

or
$$\frac{x^2}{\left(\frac{b}{2}\cosh u\right)^2} + \frac{y^2}{\left(\frac{b}{2}\sinh u\right)^2} = 1 \qquad \text{...(vi)}$$

Again from Eqs (ii) and (iii), we have

$$\cosh u = \frac{x}{\left(\frac{b}{2}\cos v\right)} \qquad \text{...(vii)}$$

and
$$\sinh u = \frac{y}{\left(\frac{b}{2}\sin v\right)} \qquad \text{...(viii)}$$

Squaring Eqs (vii) and (viii) and subtracting, we get

$$\cosh^2 u - \sinh^2 u = 1 = \frac{x^2}{\left(\frac{b}{2}\cos v\right)^2} - \frac{y^2}{\left(\frac{b}{2}\sin v\right)^2}$$

or
$$\frac{x^2}{\left(\frac{b}{2}\cos v\right)^2} - \frac{y^2}{\left(\frac{b}{2}\sin v\right)^2} = 1 \qquad \text{...(ix)}$$

Since streamlines are represented by u = constant, the same may be obtained by giving different values to u in Eq. (vi), which is therefore the equation of streamlines for this case. Further for any value of u Eq. (vi) represents an ellipse and hence in this case the streamlines are a set of confocal ellipses with origin at the mid-point O of the floor (Fig. 13.9). One of the streamlines will be AB the base of the floor which is the limiting form of this set of ellipses and is same as Bligh's path of creep.

It may be shown that the streamline along the base of the floor AB is represented by $u = 0$.

Since for $u = 0$, $\cosh u = \cosh 0 = 1$ and $\sinh u = \sinh 0 = 0$, substituting these values in Eq. (vi), we get

$$\frac{x^2}{\left(\frac{b^2}{4}\right)} + \frac{y^2}{0} = 1$$

which is true only when $y = 0$, and this is the equation of the line AB.

The lines of equal potential (or pressure) or equipotential lines are represented by v = constant ; and the same may be obtained by giving different values to v in Eq. (ix), which is therefore the equation of equipotential lines for this case. Further for any value of v Eq. (ix) represents a hyperbola and hence in this case the equipotential lines are a set of confocal hyperbolas with origin at the mid-point O of the floor.

At any point on the base of the floor (where $y = 0$) since $u = 0$, from Eq. (ii), we have

$$x = \frac{b}{2} \cos v \quad \text{(as } \cosh 0 = 1\text{)}$$

or

$$v = \cos^{-1}\left(\frac{2x}{b}\right) \qquad \text{...(x)}$$

At the upstream end of the floor $x = -\left(\frac{b}{2}\right)$ and hence from Eq. (x), we get

$$v = \pi \qquad \text{...(xi)}$$

which gives the values of the potential function at the upstream end of the floor as well as on the upstream bed.

Similarly at the downstream end of the floor $x = \left(\frac{b}{2}\right)$ and hence from Eq. (x), we get

$$v = 0 \qquad \text{...(xii)}$$

which gives the value of the potential function at the downstream end of the floor as well as on the downstream bed.

The value of the potential function at any point on the floor is given by

$$v = \pi\phi \qquad \text{...(xiii)}$$

in which ϕ is the ratio of the residual seepage head or the uplift pressure head P at any point on the floor and the total seepage head H. ϕ is usually expressed in percentage. Thus

$$\phi = PH \qquad \text{...(xiv)}$$

Introducing Eq. (xiii) in Eq. (x), we get

$$\phi = \frac{1}{\pi} \cos^{-1}\left(\frac{2x}{b}\right) \qquad \text{...(xv)}$$

Further from Eq. (xiv), we have

$$\mathrm{P} = \phi H \qquad \text{...(xvi)}$$

Introducing Eq. (xv) in Eq. (xvi), we get

$$P = \frac{H}{\pi} \cos^{-1}\left(\frac{2x}{b}\right) \qquad \text{...(xvii)}$$

Equation (xvii) represents the uplift pressure distribution on the floor. By giving different values to x the uplift pressures at the corresponding points on the floor may be obtained. Figure 13.10 shows the uplift pressure distribution on the floor as given by Eq. (xvii) and also the one given by Bligh's theory for comparison. The following observations may be made from the study of Fig. 13.10.

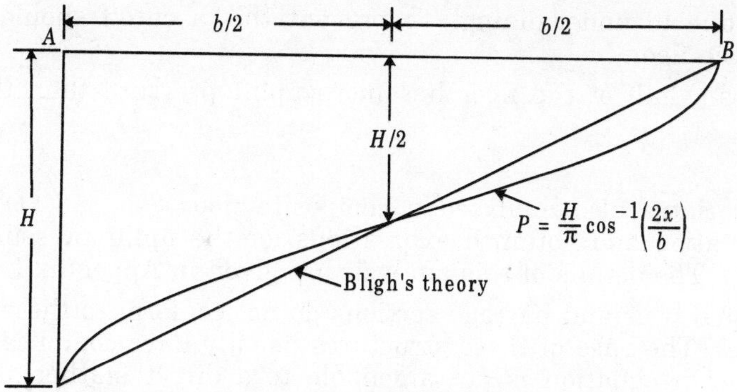

Fig. 13.10 *Uplift pressure distribution on the floor*

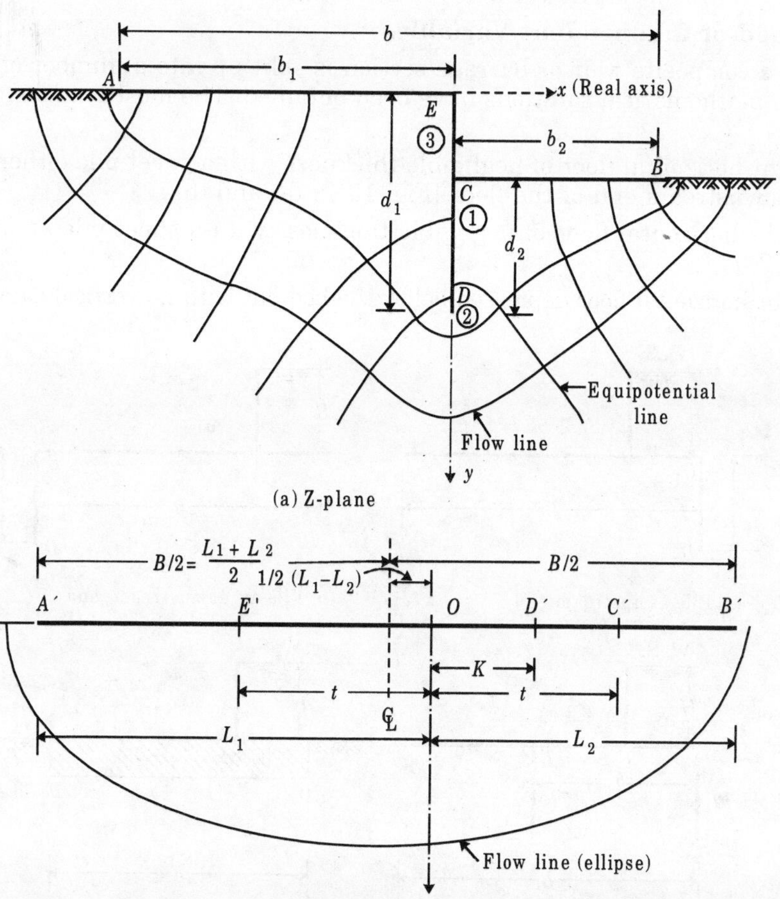

Fig. 13.11 *Composite floor considered by Khosla and his associates*

1. The slope of the uplift pressure diagram at A and B is infinite. Hence theoretically the force at A will be infinite acting downwards and that at B will also be infinite acting upwards. This will cause the dislocation of the soil particles at the end of the floor which may lead to the failure of the floor due to undermining. To prevent this a cutoff should be provided at the downstream end of the floor.

2. The downstream half of the floor has more uplift pressure than that given by Bligh's theory.

Composite Floor

Khosla and his associates also analysed a composite floor (Fig. 13.11) by using Schwarz–Christoffel transformation and obtained expressions for the uplift pressure distribution along the floor and the pile. The details of this analysis are given in Appendix I.

However, the usual weir and barrage sections do not conform to these simple or composite forms of floor profiles. The base of these structures usually have complicated boundary shapes for which the Laplacian equation is not amenable to a direct mathematical integration and hence it is not possible to obtain a mathematical solution of the flownet. To apply the analytical solution to such practical composite profile of a weir or barrage, Khosla and his associates evolved a method of independent variables which is mentioned below.

Khosla's Method of Independent Variables

In this method a composite weir or barrage section is split up into a number of simple standard forms for which mathematical solutions have been obtained. The most useful standard forms are as follows.

(a) A straight horizontal floor of negligible thickness with a sheet pile either at the upstream end or at the downstream end of the floor [Fig. 13.12 (a) and (b)].

(b) A straight horizontal floor of negligible thickness with a sheet pile at some intermediate point [Fig. 13.12 (c)].

(c) A straight horizontal floor depressed below the bed but with no vertical cutoff [Fig. 13.12 (d)].

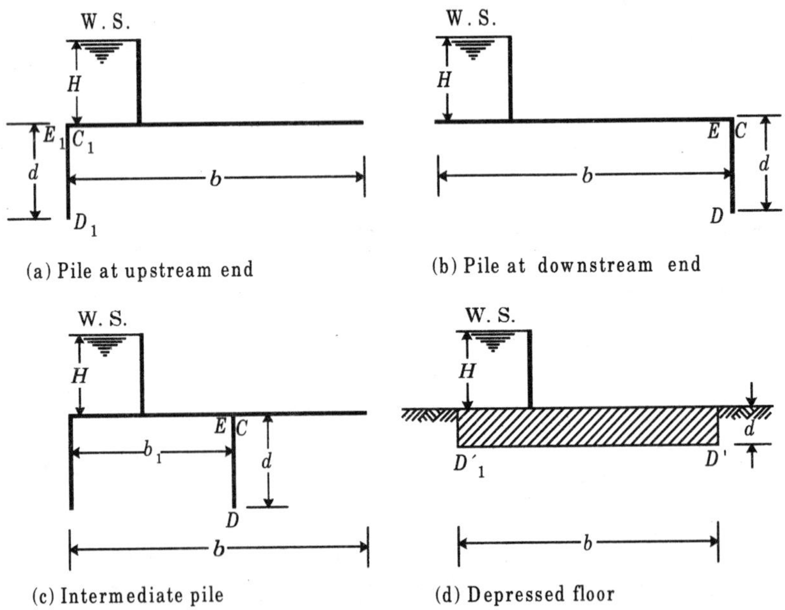

Fig. 13.12 *Standard forms considered in Khosla's method of independent variables*

These cases have been analysed by Khosla and his associates by using Schwarz–Christoffel transformation and as indicated in Appendix I expressions have been derived for determining the residual seepage head or the uplift pressure head at the key points and the exit gradient. The key points are the junction points of pile and floor, the bottom point of pile and the bottom

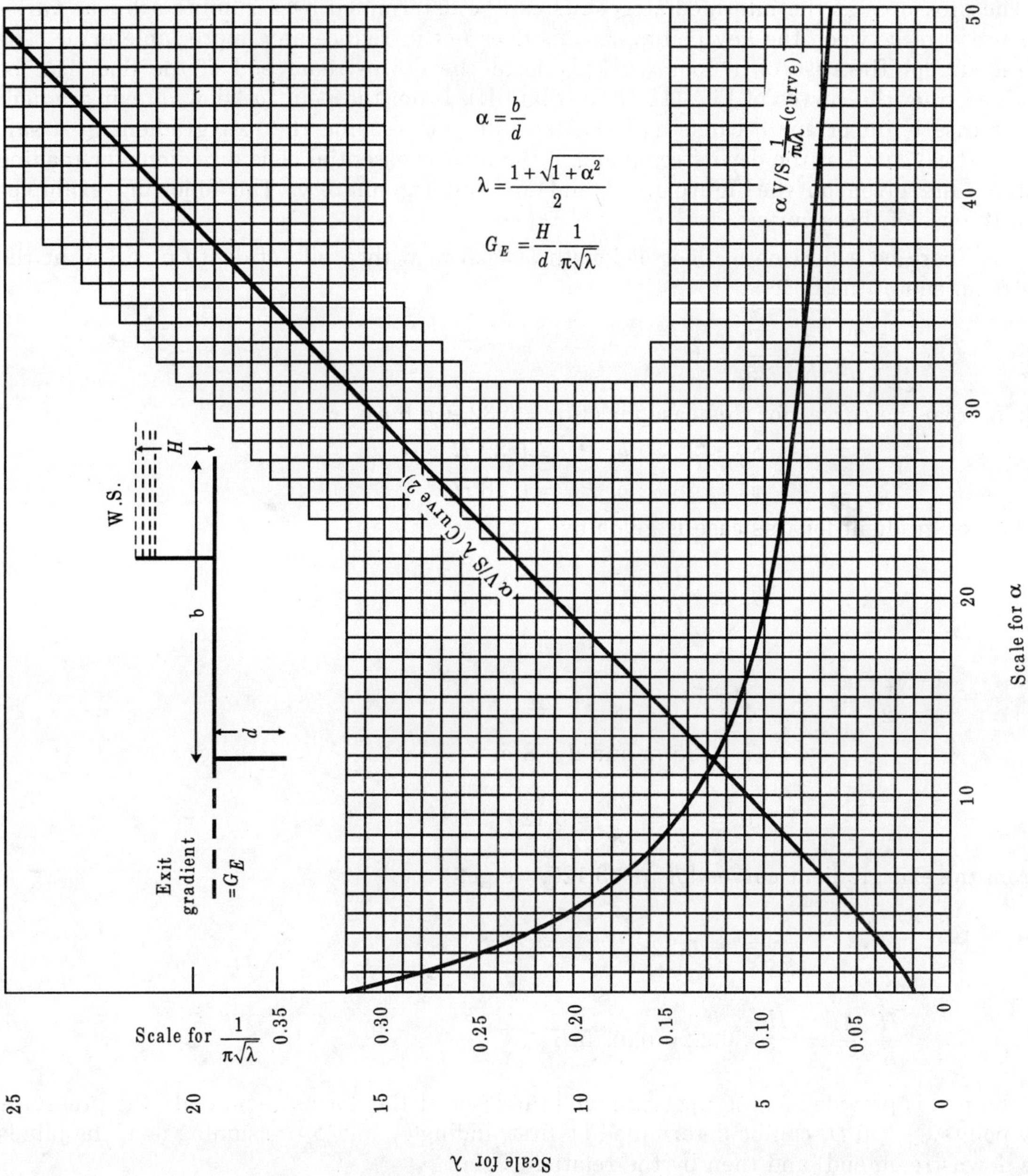

Fig. 13.13 (c) Plate II Khosla's curves

corners of depressed floor. The results of this analysis are presented in the form of curves (Fig. 13.13) from which the values of ϕ, i.e., the ratio of the residual seepage head and the total seepage head (expressed in percentage), can be determined at the key points. Figure (13.13 a) (Plate I) is for the case of a floor with a pile at some intermediate point. Fig. (13.13 b) (Plate I) is for the case of a floor with a pile at either end of the floor as well as for the case of depressed floor. The pressures at the intermediate points can be determined by assuming a linear variation of pressure between the key points, as this does not introduce any appreciable error. The exit gradient for the case of a floor with a pile at the downstream end of the floor can be determined from the curves of Fig. (13.13 c) (Plate II). From the point of view of exit gradient only this case is important because as indicated later for keeping the exit gradient at a safe value a vertical cutoff at the downstream end of the floor is essential. The directions for reading the curves are given on the figure itself and are self explanatory. The following examples explain the use of these curves.

(i) Let there be a horizontal floor of length b = 25 m with a pile of depth d = 8 m at the downstream end of the floor.

Then $$\frac{1}{\alpha} = \frac{d}{b} = \frac{8}{25} = 0.32$$

Thus, from the curves for this case for $(1/\alpha) = 0.32$, we have

$$\phi_D = 32\% \text{ ; or } P_D = 0.32\, H$$

and $$\phi_E = 48\% \text{ ; or } P_E = 0.48\, H$$

The exit gradient for this case is given by

$$G_E = \frac{H}{d}\frac{1}{\pi\sqrt{\lambda}}$$

where $\lambda = \dfrac{1+\sqrt{1+\alpha^2}}{2}$; $\alpha = \dfrac{b}{d}$

$$b = 25 \text{ m and } d = 8 \text{ m}$$

$$\alpha = \frac{b}{d} = \frac{25}{8} = 3.125$$

From the exit gradient curves for $\alpha = 3.125$, we get

$$\frac{1}{\pi\sqrt{\lambda}} = 0.22$$

$$\therefore \quad G_E = \frac{H}{d}\frac{1}{\pi\sqrt{\lambda}} = \frac{H}{8} \times 0.22 = 0.0275\, H$$

If the pile is provided at the upstream end (and not at the downstream end), the pressures at key points C_1 and D_1 can be determined by first finding ϕ_D and ϕ_E (assuming as if the pile is at the downstream end) and then by the relation

$$\phi_{C_1} = 100 - \phi_E$$
$$\phi_{D_1} = 100 - \phi_D$$

These relations are obtained by the consideration of the principle of reversibility of flow, i.e., if the direction of flow is reversed, other conditions remaining the same the flow pattern remains unaltered.

Thus, for the present case since $\phi_D = 32\%$ and $\phi_E = 48\%$, we get

$$\phi_{C_1} = 100 - \phi_E = (100 - 48) = 52\%$$

or $$P_{C_1} = 0.52\ H$$

$$\phi_{D_1} = 100 - \phi_D = (100 - 32) = 68\%$$

or $$P_{D_1} = 0.68\ H$$

(ii) Let there be a horizontal floor of length $b = 20$ m with a pile of depth $d = 5$ m at a distance $b_1 = 8$ m from the upstream end of the floor.

Then $$\alpha = \frac{b}{d} = \frac{20}{5} = 4$$

and $$\frac{b_1}{b} = \frac{8}{20} = 0.4$$

Thus, from the curves for this case for $(b_1/b) = 0.4$ and $\alpha = 4$, we get

$$\phi_C = 41\%\ ;\ \text{or}\ PC = 0.41\ H$$

To determine ϕ_E, obtain ϕ_C for base ratio $[1-(b_1/b)]$ for that value of α and subtract from 100.

Thus, for $(b_1/b) = 0.4$ and $\alpha = 4$

$$\phi_E = 100 - \phi_C \text{ for } (b_1/b) = 0.6 \text{ and } \alpha = 4$$
$$= (100 - 29.1) = 70.9\%\ ;\ \text{or}\ P_E = 0.709\ H$$

To determine ϕ_D for (b_1/b) less than 0.5, obtain ϕ_D for base ratio $[1-(b_1/b)]$ for that value of α and subtract from 100.

Thus for $(b_1/b) = 0.4$ and $\alpha = 4$

$$\phi_D = 100 - \phi_D \text{ for } (b_1/b) = 0.6 \text{ and } \alpha = 4$$
$$= (100 - 44.8) = 55.2\%\ ;$$

or $$P_D = 0.552\ H$$

(iii) Let there be a depressed horizontal floor of width $b = 20$ m and thickness $d = 4.8$ m

Then $$\frac{1}{\alpha} = \frac{d}{b} = \frac{4.8}{20} = 0.24$$

Thus, from the curves for the depressed floor for $(1/\alpha) = 0.24$

$$\phi_D = 20\%\ ;\ \text{or}\ P_D{}' = 0.2\ H$$

From the principle of reversibility of flow

$$\phi_{D_1}{}' = 100 - \phi_D{}'$$
$$= (100 - 20) = 80\%\ ;$$

or $$P_{D_1}{}' = 0.8\ H.$$

The percentage pressures observed from the curves for the simple forms into which the profile has been split may hold for the assembled profile as a whole if the following corrections are applied to the observed values.

1. Correction for thickness of floor.
2. Correction for mutual interference of piles.
3. Correction for the slope of the floor.

1. Correction for Thickness of Floor. In the standard forms with vertical cutoffs the thickness of the floor is assumed to be negligible. Thus, as observed from the curves the pressures at the junction points E and C pertain to the level at the top of the floor whereas the actual junction is with the bottom of the floor. The pressures at the actual junction points E_1 and C_1 are interpolated by assuming a linear variation from the hypothetical point E to D and also from D to C (Fig. 13.14). For different locations of piles the corrections to be applied are as follows.

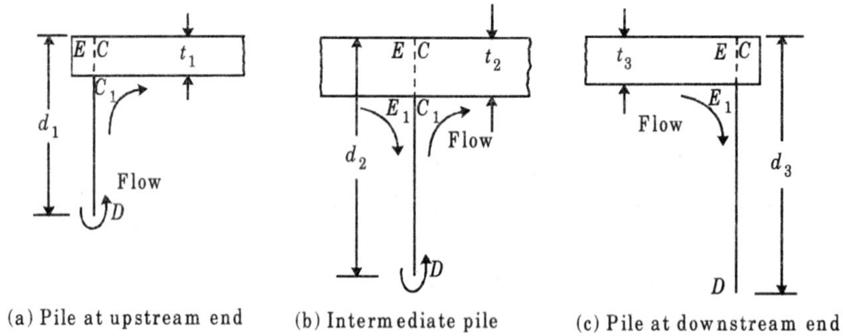

Fig. 13.14 *Correction for floor thickness*

For pile at the *upstream end*

Correction for $C_1 = \dfrac{\phi_D - \phi_C}{d_1} \times t_1$ (Additive)

∴ Pressure at C_1 is

$$\phi C_1 = \phi C + \frac{\phi_D - \phi_C}{d_1} \times t_1$$

where t_1 = Floor thickness

d_1 = Depth of pile.

For *intermediate pile*

Correction for $E_1 = \dfrac{\phi_E - \phi_D}{d_2} \times t_2$ (Subtractive)

∴ Pressure at E_1 is

$$\phi_{E_1} = \phi_E - \frac{\phi_E - \phi_D}{d_2} \times t_2$$

Correction for $C_1 = \dfrac{\phi_D - \phi_C}{d_2} \times t_2$ (Additive)

∴ Pressure at C_1 is

$$\phi C_1 = \phi_C + \frac{\phi_D - \phi_C}{d_2} \times t_2$$

where t_2 = Floor thickness

d_2 = Depth of pile.

For pile at the *downstream end*

Correction for $E_1 = \dfrac{\phi_E - \phi_D}{d_3} \times t_3$ (Subtractive)

∴ Pressure at E_1 is

$$\phi_{E_1} = \phi_E - \frac{\phi_E - \phi_D}{d_3} \times t_3$$

where t_3 = Floor thickness

d_3 = Depth of pile.

2. Correction for Mutual Interference of Piles. The correction for the mutual interference of piles is given by

$$C = 19 \sqrt{\frac{D}{b'}} \left(\frac{d+D}{b} \right) \quad \text{...(13.8)}$$

where C = correction to be applied as percentage of head

b' = distance between the piles

b = total length of floor

D = depth of the pile whose effect is required to be determined on neighbouring pile of depth d, and

d = depth of pile on which the effect of pile of depth D is required to be determined.

Both D and d are to be measured below the level at which the effect of interference is required to be determined as shown in Fig. (13.15).

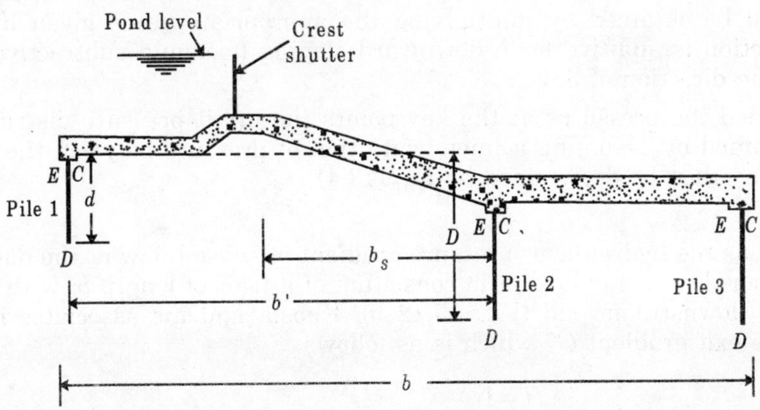

Fig. 13.15 *Correction for mutual interference of piles*

The correction is additive for points in the rear or backwater and subtractive for points forward in the direction of flow. Thus, as shown in Fig. 13.15 when the effect of interference of Pile No. 2 on pile No. 1 is determined the correction will be additive since point C of Pile No. 1 (where correction is to be applied) is in the rear with respect to the direction of flow. This is so because Pile No. 2 causes a blocking effect on the subsoil flow which results in increase in pressure at all points on its upstream side. On the other hand when the effect of interference of pile no. 1 on Pile No. 2 is determined the correction will be subtractive since point E of Pile No. 2 (where correction is to be applied) is *forward* in the direction of flow with respect to Pile No. 1. The above equation, however, does not apply to find the effect of an outer pile on an intermediate pile if the intermediate pile is equal to or smaller than the outer pile and is at a distance less than twice the length of the outer pile. Further the effect of interference of a pile is to be determined only for that face of the adjacent pile which is towards the interfering pile, e.g., Pile No. 2 will interfere only with the downstream face of pile No.1 and upstream face of Pile No. 3.

3. Correction for the Slope of the Floor. A suitable correction is to be applied for a sloping floor. The values of the correction in percent of pressure for different slopes are given in Table 13.3.

TABLE 13.3 Corrections for floor slopes

Slope (Vertical to Horizontal)	*Correction % of pressure*
1 in 1	11.2
1 in 2	6.5
1 in 3	4.5
1 in 4	3.3
1 in 5	2.8
1 in 6	2.5
1 in 7	2.3
1 in 8	2.0

The correction given by this table is to be multiplied by the horizontal length of the sloping floor and divided by the distance between the two pile lines between which the sloping floor is located. The correction is applicable only to those key points of the pile lines fixed at the beginning or the end of the sloping floor which are facing the sloping floor. Thus, as shown in Fig. 13.15 the correction for slope is to be applied only to point E of Pile No. 2 and the correction to be applied will be obtained by multiplying the appropriate value given by Table 13.3 by (b_s/b'). The correction is additive for a downward sloping floor and subtractive for an upward sloping floor in the direction of flow.

Having obtained the pressures at the key points the uplift pressure distribution along the floor may be obtained by assuming a linear variation of pressure between the key points. The floor thickness may then be determined by Eq. (13.4).

Exit Gradient

It may be defined as the hydraulic or pressure gradient of subsoil flow at the downstream or the exit end of the floor. For a standard form consisting of a floor of length b, with a vertical cutoff of depth d at its downstream end (Fig. 13.12 b), Khosla and his associates have derived an expression for the exit gradient G_E which is as follows

$$G_E = \frac{H}{d}\frac{1}{\pi\sqrt{\lambda}} \quad ...(13.9)$$

where H = The total seepage head ;

$$\lambda = \frac{1+\sqrt{1+\alpha^2}}{2}, \text{ and } a = (b/d)$$

From the curves of Fig. 13.13 (c) (Plate II), for any value of α (= b/d), the corresponding value of λ or directly that of the function ($1/\pi\sqrt{\lambda}$) can be obtained. Knowing the values of H and d, the value G_E can be calculated from Eq. (13.9).

It is seen from Eq. (13.9) that if no cutoff is provided at the downstream end of the floor i.e., $d = 0$, the exit gradient G_E is infinite. This is, however, a theoretical value of the exit gradient but it is evident that if there is no cutoff at the downstream end of the floor a higher exit gradient will exist which may lead to the failure of the floor due to piping. It is, therefore, essential that a vertical cutoff should be provided at the downstream end of the floor to reduce the exit gradient. The exit gradient should always be less than the *critical hydraulic gradient* which is defined as that hydraulic or pressure gradient at which the soil particles will be lifted up and which will lead to undermining. As indicated in section 13.13 for alluvial soils the critical hydraulic gradient is found to be approximately equal to 1. However, by adopting a suitable factor of safety permissible or safe hydraulic gradient may be obtained and to safeguard the floor against failure due to piping the exit gradient must not be allowed to exceed the permissible hydraulic gradient. The permissible exit gradients for different soils are given in Table 13.4.

TABLE 13.4 Permissible exit gradients for different types of soils

Type of soil	*Exit gradient*
1. Fine sand	1/6 to 1/7
2. Coarse sand	1/5 to 1/6
3. Shingle	1/4 to 1/5

Depth of Sheet Piles on Upstream and Downstream of Impervious Floor

The sheet pile must be taken upto the level of possible deepest scour below the bed of the river. According to Lacey the depth of scour in alluvial soils is given by

$$R = 1.35\left(\frac{q^2}{f}\right)^{1/3} \qquad ...(13.10)$$

where R = scour depth measured below the high flood level (H.F.L)

q = discharge per unit length, and

f = Lacey's silt factor.

In order to ensure further safety, for the design of sheet piles the scour depth is considered as 1.25 to 2 times R given by Eq. (13.10). Further as indicated earlier an intermediate pile line is not of much use. However, in conservative design an intermediate pile line is provided to serve as an additional factor of safety. The depth of the intermediate pile line if provided should not be less than that of the upstream pile line.

Design of Protection Works at the Upstream and Downstream Ends of the Impervious Floor

In order to further safeguard the impervious floor against the failure due to piping certain protection works are provided at both the upstream and the downstream ends of the impervious floor. These protection works consist of

(i) Inverted filter

(ii) Block protection, and

(iii) Launching apron or pervious apron (or floor) as shown in Fig. (13.16).

(i) *Inverted filter.* An inverted filter consists of layers of materials of increasing permeability from bottom to top. It is of such gradation that while it allows free flow of seepage water the foundation material does not penetrate to clog the filter. The design criteria to satisfy these conditions are discussed in section 11.19 in Chapter 11 on Embankment Dams. The thickness of the inverted filter varies from 0.5 to 1.25 m. To prevent the filter material from dislocation by surface flow they are loaded with large size stone or concrete blocks. The blocks are usually 0.9 to 1.2 m thick and are placed with open joints filled with bajri (i.e., river sand) or filter material.

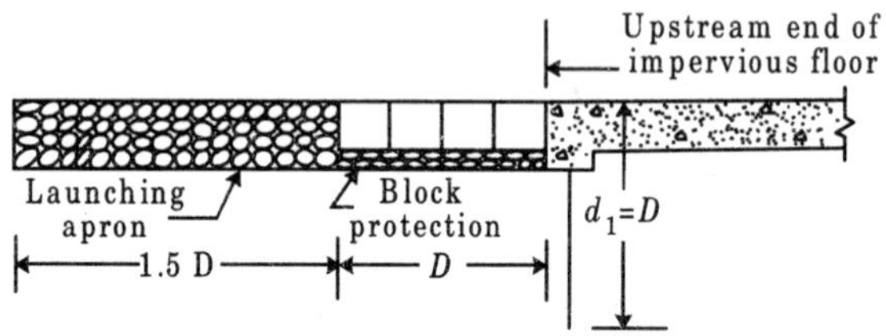

(a) Upstream protection works

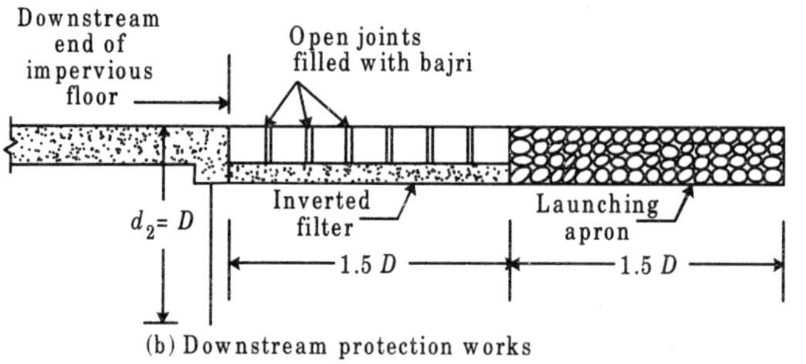

(b) Downstream protection works

Fig. 13.16 *Upstream and downstream protection works*

Immediately at the downstream end of the impervious floor an inverted filter is provided to relieve the uplift pressure. The length of this inverted filter depends on the depth of scour D below the river bed or the impervious floor and it usually varies from 1.5 D to 2 D. The value of D may be obtained from the following equation

$$D = XR - Y \qquad \ldots(13.11)$$

where (XR) is the depth of deepest scour level below high flood level in which X is a multiplying factor which varies from 1.25 to 2 as indicated earlier and R is the scour depth given by Eq. (13.10) ; and Y is the depth of the river bed or the impervious floor below high flood level, i.e.,

$$Y = \text{high flood level} - \text{River bed level (or floor level)}$$

Further since the pile is taken upto the deepest scour level, the depth of scour is also equal to the depth of pile below the river bed or the impervious floor. Hence if d_2 is the depth of pile below the river bed or the impervious floor at the downstream end then the length of the inverted filter may also be considered to vary from 1.5 d_2 to $2d_2$.

(ii) *Block protection.* A block protection consists of 0.6 to 1.0 m thick stone or concrete blocks laid on 0.4 to 0.6 m thick loosely packed stone. A block protection is provided immediately at the upstream end of the impervious floor. The length of the block protection is usually equal to D the depth of scour below the river bed or the impervious floor at the upstream end. Further since the depth of scour D is also equal to the depth of pile d_1 below the river bed or the impervious floor at the upstream end, the length of the block protection is also equal to d_1.

(iii) *Launching apron or pervious apron.* A launching apron is an apron of loosely packed stones. The function of a launching apron is to protect the impervious floor and the pile from the scour holes progressing towards the floor and the pile. The protection is provided by a launching apron by forming a protective covering of stones over a certain slope below the bed of the river at which the apron is originally laid to the bottom of the deepest scour likely to occur. As shown in Fig. 13.17 when scour occurs, the new position attained by this apron is called launched position and the apron is then known as launched apron.

The design of a launching apron depends on the depth of scour, velocity of flow and slope of the launched apron.

The velocity of flow affects the size of the stone because the stone used in the launching apron should be large enough so that it is not washed away during maximum flood. The size of the stone may be determined by the following equation given by U.S.B.R.

$$V_A = 4.915\ d^{1/2} \quad \text{...(13.12)}$$

where V_A = average velocity of flow in m/s ; and

d = mean diameter of stone in m.

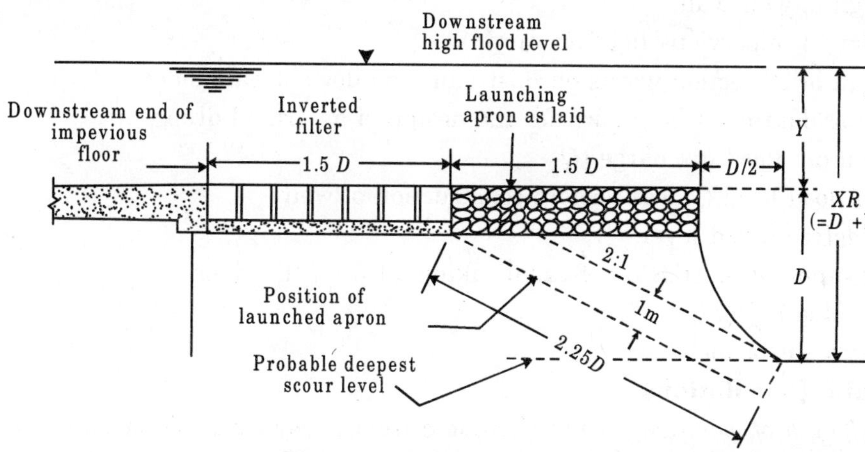

Fig. 13.17 *Launching apron*

However, in actual practice it is possible to procure and handle stones upto 0.4 to 0.5 kN [40 to 50 kg(f)]. When stones of larger size are required, usually cement concrete blocks or small stones in wire crates may be used.

It is generally assumed that the stones launch at a slope of 2:1. Concrete blocks or wire crate boulders launch at a steeper slope, usually 1.5:1.

According to Spring the quantity of stone in a launching apron should be sufficient to provide about 1 m thick cover over a slope of 2 : 1 in the launched position of the apron. Thus, as shown in Fig. (13.17) if D is the depth of scour below the bed of the river (or the top level of the apron in the original laid position), the length of the launched apron would be about $\sqrt{5}\ D = 2.236\ D$. Since the thickness of the launched apron is 1 m, the quantity of stone required is 2.236 D (say 2.25 D) cubic metre per metre length of the apron. The required quantity of stone is provided in the launching apron of uniform thickness and length varying from D to $2D$. Generally the length of the launching apron is kept equal to $1.5D$ in which case the thickness of the launching apron will be 1.5 m.

The launching aprons are provided immediately after the inverted filter on the downstream side and the block protection on the upstream side. Further since D is also equal to the depth of pile below the river bed or the impervious floor the lengths of the launching aprons may also be expressed as follows.

Length of launching apron on the upstream side = $1.5\ d_1$

Length of launching apron on the downstream side = $1.5\ d_2$

where d_1 and d_2 are the depths of piles below the river bed or the impervious floor at the upstream and downstream ends of the floor respectively.

13.15 DESIGN OF VERTICAL DROP WEIR

The design of a vertical drop weir shown in Fig. 13.18 consists of the following parts.

1. Hydraulic calculations for fixing various elevations.
2. Design of weir wall.
3. Design of impervious floor (or apron).
4. Design of protection works on upstream and downstream sides.

The following data must be known for the design of a vertical drop weir.

(a) Maximum flood discharge (Q)

(b) High flood level (H.F.L.) before construction of weir

(c) Downstream bed level

(d) Full supply level (F.S.L.) of canal taking off from the river

(e) Allowable afflux

(f) Lacey's silt factor.

1. Hydraulic Calculations

(i) The *length of waterway* (L) is calculated from Lacey's regime formula

$$L = 4.83\ Q^{1/2} \qquad \text{...(13.13)}$$

where L = Length of waterway in metres

Q = discharge in cumec

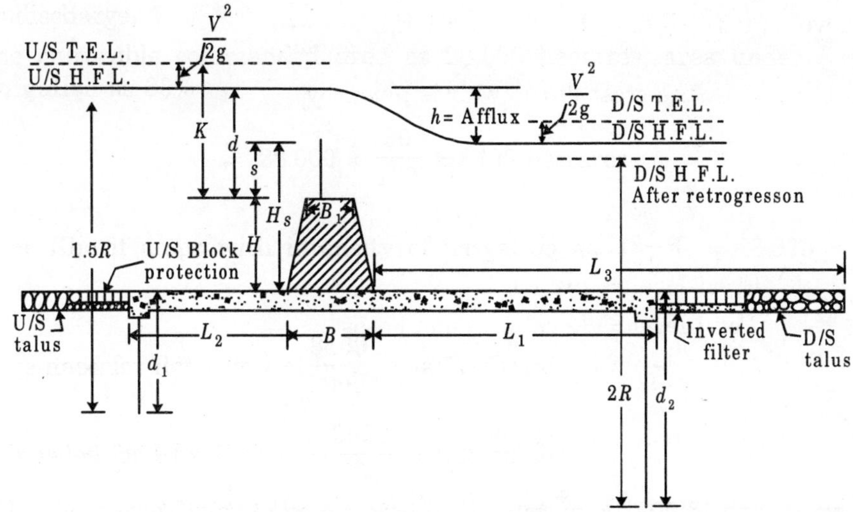

Fig. 13.18 *Vertical drop weir*

(ii) The discharge q per unit length of waterway is calculated from the relation

$$q = \frac{Q}{L} \qquad \text{...(13.14)}$$

where q is in cumec per metre

(iii) The regime scour depth in calculated from Lacey's formula

$$R = 1.35 \left(\frac{q^2}{f}\right)^{1/3} \qquad \text{...(13.15)}$$

where R = scour depth in metres

f = Lacey's silt factor

(iv) The regime velocity and velocity head are calculated from the expressions

$$V = \frac{q}{R} \qquad \text{...(13.16)}$$

and

$$\text{Velocity head} = \frac{V^2}{2g} \qquad \text{...(13.17)}$$

(v) Water levels and total energy line (T.E.L.) on the downstream side (*d/s*) and upstream side (*u/s*) are calculated as follows :

$$\text{Level of } d/s \text{ T.E.L.} = (\text{H.F.L. before construction}) + \frac{V^2}{2g}$$

$$\text{Level of } u/s \text{ T.E.L.} = \text{Level of } d/s \text{ T.E.L.} + \text{Afflux}$$

$$\text{Level of } u/s \text{ H.F.L.} = \text{Level of } u/s \text{ T.E.L.} - \frac{V^2}{2g}$$

(vi) Discharge over the crest of the weir is given by

$$q = 1.70\ K^{3/2} \qquad \text{...(13.18)}$$

or

$$K = \left(\frac{q}{1.70}\right)^{2/3} \qquad \text{...(13.19)}$$

∴ Crest level = u/s T.E.L.– K

(vii) Pond level = Level of top of gates

= F.S.L. of canal + Head loss through regulator

Head loss through the regulator may be taken as $\frac{1}{2}$ to 1 m.

Height of shutters = S= Level of top of gates – Crest level

(viii) Protection against scour :

Level of bottom of u/s pile = u/s H.F.L. – 1.5 R

Level of bottom of d/s pile = H.F.L. after retrogression – 2R.

2. Design of Weir Wall. A weir wall is usually trapezoidal in cross-section with either both upstream and downstream faces inclined or upstream face vertical and downstream face inclined. The design of the weir wall involves the determination of its top and bottom widths such that the section will be stable under the condition of maximum stress. In general the condition of maximum pressure on a hydraulic structure is that in which the head water is at the level of the crest of the weir or at the top of the crest shutters (if any) and no water is flowing over the weir so that there is no water on the downstream side. However, the condition of the maximum stress on the weir may be different in certain cases. Hence it is necessary to check the stability of the weir under the following three states.

State I : When the upstream water or *head water* is at crest level or at the top of the crest shutters (if any) [Fig. 13.19 (a)] and there is no flow.

State II : When water is flowing over the weir crest and the weir is submerged [Fig. 13.19 (b)].

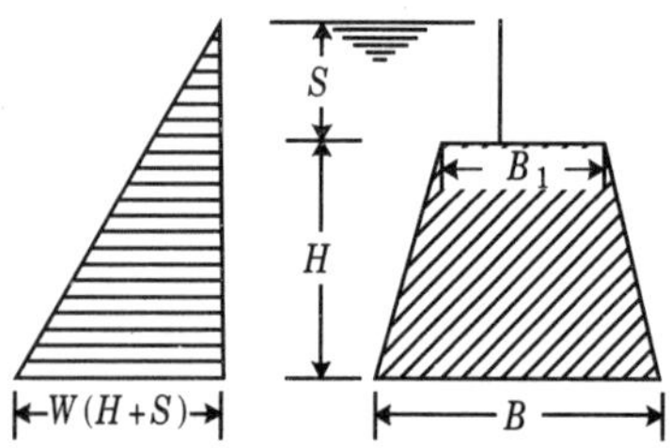

(a) No flow

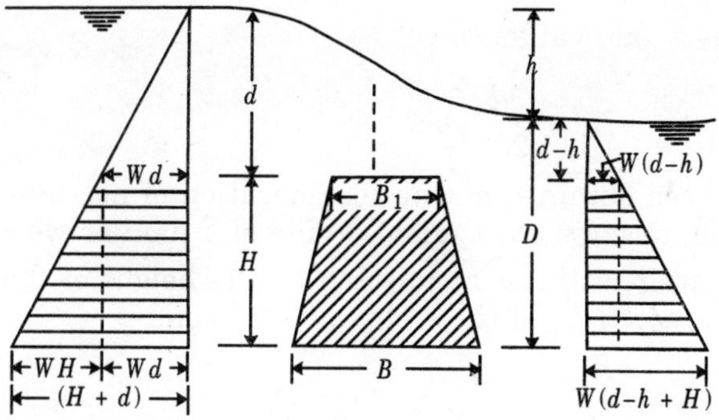

(b) Drowned weir

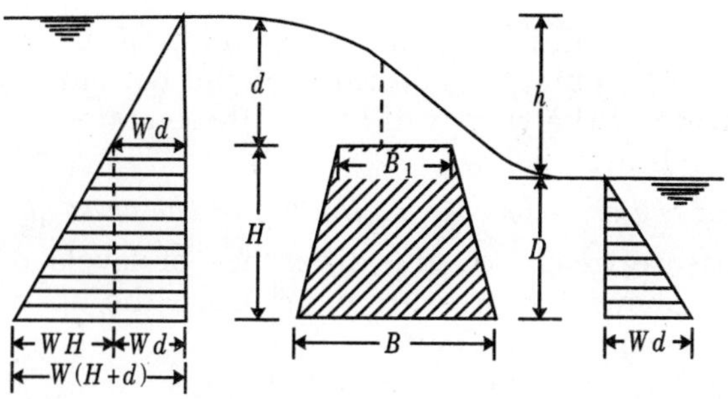

(c) Clear overfall case

Fig. 13.19 *Conditions of flow over vertical drop weir*

State III : When water is flowing over the weir crest and weir is discharging with a clear overfall [Fig. 13.19 (c)].

Top Width of Weir Wall

The top width of weir wall is given by

$$B_1 = \frac{d}{\sqrt{G}} \qquad ...(13.20)$$

where B_1 = top width of the weir wall

d = maximum depth of water above the weir crest, which is equal to *u/s* H.F.L.– crest level ; and

G = specific gravity of the material of the weir.

The above expression has been obtained on the consideration of no tension criterion for determining the width of an elementary (or triangular) profile of a dam or weir.

The top width of the weir is also given by

$$B_1 = \frac{d}{\mu G} \qquad ...(13.21)$$

where μ = coefficient of friction.

Assuming μ = (2/3) as a safe value we get

$$B_1 = \frac{3d}{2G} \qquad ...(13.21\ a)$$

This expression has been obtained on the consideration of no sliding criterion for determining the width of an elementary (or triangular) profile of a dam or weir.

The top width of the weir wall is also affected by the height of the crest shutters and from this consideration the top width B_1 of the weir wall is given by

$$B_1 = S + 1 \text{ (metres)} \qquad ...(13.22)$$

where S = height of crest shutters in metres.

The top width of the weir wall should be greater of the three values of B_1 given by Eqs (13.20), (13.21) and (13.22).

Bottom Width of Weir Wall

The bottom width B of the weir wall is determined by equating the overturning moments to the resisting moments taken at the outer middle third of the bottom width of the weir wall. For determining the bottom width of the weir wall all the three states as indicated above are to be considered. The same are indicated below.

State I : Head water at the crest level or at the top of the crest shutters (if any)

In this state the pressure head i.e., the difference of levels of head and tail water is maximum. Figure (13.19 a) shows the pressure diagram on the upstream face of the weir wall assuming no water on the downstream side.

The overturning moment is given by

$$M_0 = \frac{w(H+S)^3}{6} \qquad ...(13.23)$$

The resisting moment about the outer middle third of the bottom width is given by

$$M_r = \frac{w}{12}\left[\{(G+1.5)H + 2.5S\}B^2 + B_1(GH - H - S) \;\; B - \frac{1}{2}B_1^2(H+3S)\right] \qquad ...(13.24)$$

The above expression is valid when both upstream and downstream faces have same slopes. However, if the upstream face is kept vertical, the moment of resistance is increased and is given by

$$M_r = \frac{wHG}{6}\left(B^2 + BB_1 - B_1^2\right) \qquad ...(13.24\ a)$$

Equating the overturning moment to the moment of resistance the bottom width B can be determined for State I.

State II : Water flowing over the weir crest and the weir is submerged

Figure (13.19 b) shows the pressure distribution diagrams for this case. The overturning moment is given by

$$M_0 = \left\{\frac{wH^3}{6} + \frac{wdH^2}{2}\right\} - \left\{\frac{wH^3}{6} + \frac{w(d-h)H^2}{2}\right\}$$

or $$M_0 = \frac{whH^2}{2} \qquad ...(13.25)$$

For maximum value of M_0, h is taken corresponding to the case when the weir is just submerged.

The moment of resistance about the outer middle third of the bottom width, *with tail water at weir crest level*, considering *full floatation* is given by

$$M_r = \frac{wH(G-1)}{12}(B^2 + B_1B) \qquad ...(13.26)$$

when both faces are having the same slope.

If the upstream face is vertical

$$M_r = \frac{wH(G-1)}{6}\left(B^2 + BB_1 - B_1^2\right) \qquad ...(13.26\ a)$$

Again equating the overturning moment and the moment of resistance B can be determined.

State III : *Water is flowing over the weir crest and the weir is discharging with a clear overfall*

Figure (13.19 c) shows the pressure diagrams, from which the overturning moment is given by

$$M_0 = \frac{wH^3}{6} + \frac{wdH^2}{2} - \frac{wD^3}{6}$$

or $$M_0 = \frac{w}{6}(H^3 + 3dH^2 - D^3) \qquad ...(13.27)$$

In this case there are two unknowns *viz*, d and D and a relation between the two must be known to find the maximum overturning moment. For finding a suitable relation between d and D, river gauging would be necessary for the clear overfall.

When the weir extends over the entire width of the river and the later is of considerable width in comparison to its depth, say width is not less than 30 times the depth, d is roughly proportional to D, i.e.,

$$d = kD \qquad ...(13.28)$$

where k is constant and can be known if the river discharge for any one depth D is known.

Introducing Eq. (13.28) in Eq. (13.27) we get

$$M_0 = \frac{w}{6}\ [H^3 + 3kDH^2 - D^3]$$

For obtaining the condition for the maximum value of M_0

$$\frac{dM_0}{dD} = \frac{w}{6}\ [3kH^2 - 3D^2] = 0$$

or $$D = H\sqrt{k} \qquad ...(13.29)$$

Hence the maximum value of M_0 is given by

$$M_0 = \frac{w}{6}\,[H^3 + 3k^{3/2}H^3 - k^{3/2}H^3]$$

or

$$M_0 = \frac{wH^3}{6}\left[1+2k^{3/2}\right] \quad \text{...(13.30)}$$

The moment of resistance for this case may also be considered to be approximately same as the one given by Eq. (13.26) for the previous case and equating the overturning moment and the moment of resistance B can be determined.

The greater of the three values of B obtained above is adopted.

(3) **Design of impervious floor (or apron).** For underseepage the worst condition would occur when the water on the upstream side is at the level of the weir crest or at the top of the crest shutters (if any), and there is no tail water. The percolation head will then be H_s as shown in Fig. (13.18).

If the floor of the weir is designed on the basis of Bligh's creep theory, the total creep length is given by Eq. (13.2) as

$$L = CH_s$$

which may be computed by taking the value of C from Table 13.1 for the type of soil at the site.

However, if the floor is to be designed on the basis of Khosla's theory, the horizontal length b of the impervious floor is found by the consideration of the permissible exit gradient G_E. From Eq. (13.9)

$$G_E = \frac{H_s}{d}\frac{1}{\pi\sqrt{\lambda}}$$

The value of G_E may be taken from Table 13.4 for the type of soil at the site and the values of H_s and d are known. The value of $\frac{1}{\pi\sqrt{\lambda}}$ may thus by calculated from the above equation. From the exit gradient curve [Fig. 13.13 (c)] for this value of $\frac{1}{\pi\sqrt{\lambda}}$ the corresponding value of α may be determined. Since $\alpha = (b/d)$, knowing α and d, b can be determined.

Out of the total impervious floor length b (or the total creep length L), the lengths L_1 and L_2 on the downstream and upstream of the weir wall respectively as shown in Fig. (13.18) are fixed on the basis of Bligh's recommendations as indicated below.

(i) *Downstream impervious floor or apron* (L_1). For weirs without crest shutters

$$L_1 = 2.21\,C\sqrt{\frac{H_s}{10}} \quad \text{...(13.31)}$$

For weirs with crest shutters

$$L_1 = 2.21\,C\sqrt{\frac{H_s}{13}} \quad \text{...(13.32)}$$

(ii) *Upstream impervious floor or apron* (L_2). According to Bligh's creep theory

$$L_2 = L - L_1 - (B + 2d_1 + 2d_2) \quad ...(13.33)$$

According to Khosla's theory

$$L_2 = b - L_1 - B \quad ...(13.34)$$

4. Design of Protection Works on Upstream and Downstream Sides

(i) *Upstream side.* Immediately at the upstream end of the impervious floor a block protection of length equal to d_1 is provided, where d_1 is the depth of the pile below the river bed or the impervious floor at the upstream end of the floor. Upstream of the block protection a launching apron of length equal to 1.5 d_1 and thickness 1.5 m is provided.

(ii) *Downstream side.* The total length of impervious floor, inverted filter and launching apron (or pervious floor or pervious apron also called 'talus') on the downstream of the weir wall (Fig. 13.18) is also fixed on the basis of Bligh's recommendations as follows.

For weirs without crest shutters

$$L_3 = 18C\sqrt{\frac{H_s}{10} \times \frac{q}{75}} \quad ...(13.35)$$

For weir with crest shutters

$$L_3 = 18C\sqrt{\frac{H_s}{13} \times \frac{q}{75}} \quad ...(13.36)$$

The minimum length of the inverted filter and the launching apron or pervious floor (or apron) is thus equal to $(L_3 - L_1)$.

However, as indicated in section 13.13 an inverted filter of minimum length equal to 1.5 d_2 is to be provided immediately at the downstream end of the impervious floor, where d_2 is the depth of pile below the river bed or the impervious floor at the downstream end of the floor. After the inverted filter a launching apron of length equal to 1.5 d_2 and thickness 1.5 m is provided. If the required length $(L_3 - L_1)$ is more than $3d_2$ the lengths of the inverted filter and the launching apron may be suitably increased.

The design of a vertical drop weir is indicated in Illustrative Example 13.3.

13.16 DESIGN OF GLACIS WEIR

The design of a glacis weir is also based on the same considerations as in the case of a vertical drop weir. However, some of the additional points to be considered in the design of a sloping glacis weir are as follows.

(i) The upstream slope of the glacis may be 1 in 2 while the downstream slope should be 1 in 3 to 1 in 5 for the stability of the hydraulic jump.

(ii) The level of the downstream horizontal floor is determined by the consideration that the hydraulic jump should not form lower than the toe of the glacis.

(iii) The location of the hydraulic jump is made with the help of Blench curves shown in Fig. 13.20 (Plate III). For a known discharge intensity q, the difference between the *u/s* and *d/s* total energy lines is determined which gives the loss of head H_L. Knowing q and H_L , E_{f2} or the specific energy on the downstream side of the jump can be determined from the Blench curves.

Measuring this intercept of E_{f2} below the *d/s* total energy line, the position of the jump on the sloping glacis is obtained.

(iv) The location of the jump should be checked for different discharge intensities because it is usually found that the jump shifts lower for small discharge intensities. The minimum discharge intensity for which a check should be made is that corresponding to a head provided by pond level on the upstream side of the crest.

(v) The length of the horizontal floor should not be less than five times the jump height i.e., $5(D_2 - D_1)$. The maximum value of $(D_2 - D_1)$ which will occur during high flood condition should be used.

(vi) In this case the critical condition of uplift pressure usually occurs when a hydraulic jump develops on the sloping glacis. For determining the uplift pressures at and upstream of the jump section, it would be necessary to plot the water surface profile of the jump. This can be done by means of the 'specific energy' curves prepared by Montague shown in Fig. (13.21) (Plate IV). For plotting the jump profile measure the ordinate of specific energy (say K_1) from the total energy line to the floor at any section and from the specific energy curve for the known discharge intensity read the value of depth corresponding to this specific energy. The unbalanced uplift pressure is obtained by measuring the ordinate between the hydraulic gradient line and the water surface profile at any section. The maximum unbalanced uplift pressure would usually occur at the jump section and the same should be considered for determining the required floor thickness. It may however be stated that if in any case the uplift pressure for no flow condition is greater than the one for the condition of hydraulic jump developing on the floor, then the larger of the two values should be used for determining the required floor thickness.

The unbalanced uplift pressures obtained for the condition of hydraulic jump developing on the floor are reduced because of the following factors.

(a) Due to backward rolling flow of water in the jump developed on the glacis the depth of flow is increased which reduces the unbalanced uplift pressures.

(b) The unbalanced uplift pressure is maximum at the jump section but reduces rapidly on either side. As the floor has some beam action, it may be designed for average uplift pressures rather than for the maximum uplift pressure occurring at any section.

(c) The vertical component of the momentum remains unaffected in the jump which exerts a downward pressure in the vertical direction. The magnitude of this force is quite considerable.

On account of the above factors for the design purposes the unbalanced uplift pressures for the condition of hydraulic jump developing on the floor are taken as 2/3 of the value computed as indicated above.

The design of a glacis weir is indicated in Illustrative Example 13.6.

13.17 UNDERSLUICES OR SCOURING SLUICES

The undersluices are the openings provided in the weir wall with their crest at a low level. These openings are fully controlled by gates. They are located on the same side as the offtaking canal. However, if two canals take off, one on either side of the river, then it would be necessary to provide undersluices on either side.

Functions of Undersluices

The functions of undersluices are as follows.

(i) They preserve a clear and well defined river channel towards the canal head regulator.

(ii) They scour the silt deposited on the river bed in the pocket upstream of the canal head regulator.

(iii) They pass low floods without the necessity of dropping the weir crest shutters.

(iv) They help to lower the high flood level by supplementing the discharge over the weir during high floods.

Discharging Capacity of Undersluices

The discharging capacity of the undersluices is governed by the above noted functions required to be served by them and it is provided as the maximum of the following.

(i) *Two times the maximum discharge of the offtaking canal* (to ensure proper scouring).

(ii) *Maximum winter discharge* (to pass low floods during winter without the necessity of dropping the weir crest shutters).

(iii) *10 to 20% of the maximum flood discharge* (to supplement the discharge over the weir during high floods and thus reduce the length of the weir).

Design Considerations for Undersluices

The design of undersluices is based on the following considerations.

(i) The crest of the undersluices is usually kept at the lowest level in the cross-section of the river at the site of the weir. It is also related to the crest level of the canal head regulator. If no special device to control silt entering into the canal is provided, then for scouring of silt deposit in front of the head regulator the crest of the undersluices should be at least 1.2 m lower than the crest of the head regulator. If a silt excluder or a device designed to reduce entry of silt into the canal is to be provided, then the crest of the undersluices should be about 1.8 to 2 m below the crest of the head regulator.

(ii) The top of the undersluices is placed at the same level as the sill of the canal head regulator if silt excluder is not provided. However, if silt excluder is provided then the top of the undersluices is placed at the same level as the inner surface of the roof of the tunnels of the silt excluder.

(iii) The total gate opening for the undersluices for a given discharge is generally based on an allowable velocity of 2 to 3 m/s. The number and size of gates to be provided depends to a large extent on the judgment of the designer. However, from economic considerations the width of each opening may usually be kept equal to twice the depth of the opening.

(iv) The thickness and the length of the impervious floor for the portion of the undersluices is designed on the same lines as the floor of the weir portion.

(v) Since the discharge intensity through the undersluices is higher, heavier protection should be provided on the downstream side. The length L_2 of the impervious floor on the downstream of the undersluices may be determined according to Bligh's recommendations by the following expression.

$$L_2 = 3.9\, C\sqrt{\frac{H_s}{13}} \qquad \text{...(13.37)}$$

Similarly the total length L_3 of the impervious and the pervious floors (or aprons) on the downstream of the undersluices may be determined according to Bligh's recommendations by the following expression.

$$L_3 = 27\,C\,\sqrt{\frac{H_s}{13} \times \frac{q}{75}} \qquad \text{...(13.38)}$$

A comparison of Eqs (13.37) and (13.38) with the corresponding Eqs (13.32) and (13.36) for the weir portion indicates that for the portion of the undersluices longer floors or aprons are required to be provided.

(vi) From the considerations of retrogression the downstream floor should be depressed upto the likely level of retrogression on the downstream side of the undersluices.

13.18 DIVIDE WALL OR DIVIDE GROYNE

A divide wall is a long masonry or concrete wall or groyne (an embankment protected on all sides by stone or concrete blocks) which is constructed at right angles to the axis of the weir to separate the undersluices from the rest of the weir. If two canals take off, one on either side of the river, then two divide walls are required, one on each side. The top width of divide wall is about 1.5 to 2.5 m. The divide wall extends on the upstream side upto a distance little beyond the beginning of the canal head regulator and on the downstream side upto the end of the loose protection or 'talus' of the undersluices.

Functions of Divide Wall

The functions of a divide wall are as follows.

(i) Since the floor level of the undersluices is generally lower than the floor level of the weir, the two have to be separated from each other and this is done by the divide wall.

(ii) It provides a comparatively quiet pocket in front of the canal head regulator resulting in deposition of silt in the pocket and entry of clear water into the canal.

(iii) It provides a straight approach through the pocket and thus helps to concentrate scouring action of the undersluices for washing out the silt deposited in the pocket.

(iv) It keeps the cross currents, if any, away from the weir. A cross current will develop when the main current in the river tends to approach the bank opposite the canal head regulator and the weir forces the water to flow towards the regulator. The cross current cause formation of vortices and result in deep scour. As such sometimes additional divide walls are provided at equal interval along the weir which will keep this current away from the wier and provide safety against any possible damage.

13.19 FISH LADDER

Large rivers have various types of fish many of which are migratory. They move from one part of the river to another according to season. In our country generally fish move from upstream to downstream in the beginning of winter in search of warmth and return upstream before monsoon for clearer water. Due to the construction of a weir or barrage across the river such migration of the fish will be obstructed and if no arrangement is made in the weir or barrage for this migration, large scale destruction of the fish life may take place in the river. Thus, to enable the fish to migrate *fish ladder* is provided. A typical layout of a fish ladder is shown in Fig. (13.22). Since most fish can travel upstream only if the velocity of flow does not exceed 3 to

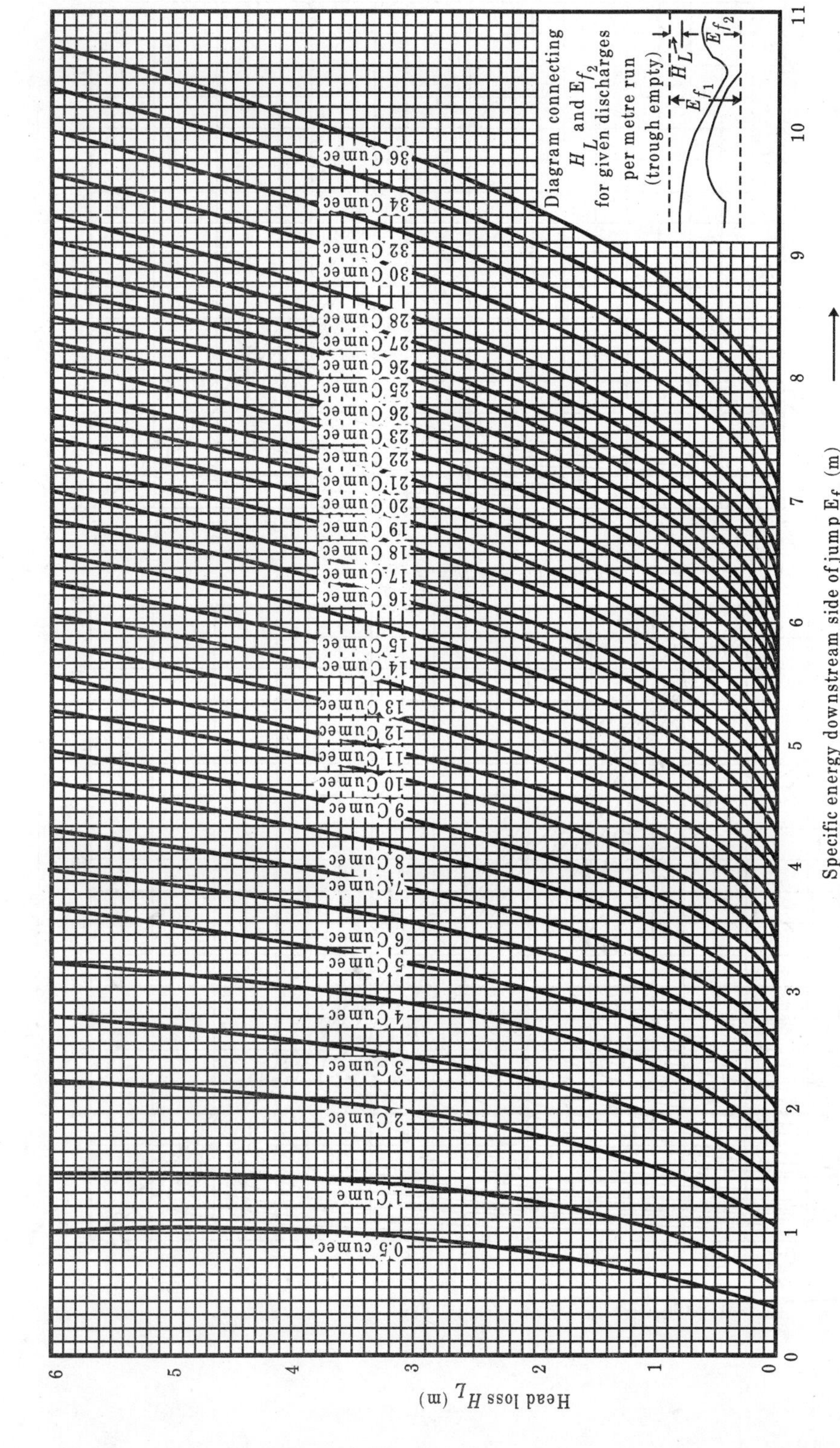

Fig. 13.20 *Plate III Blench curves*

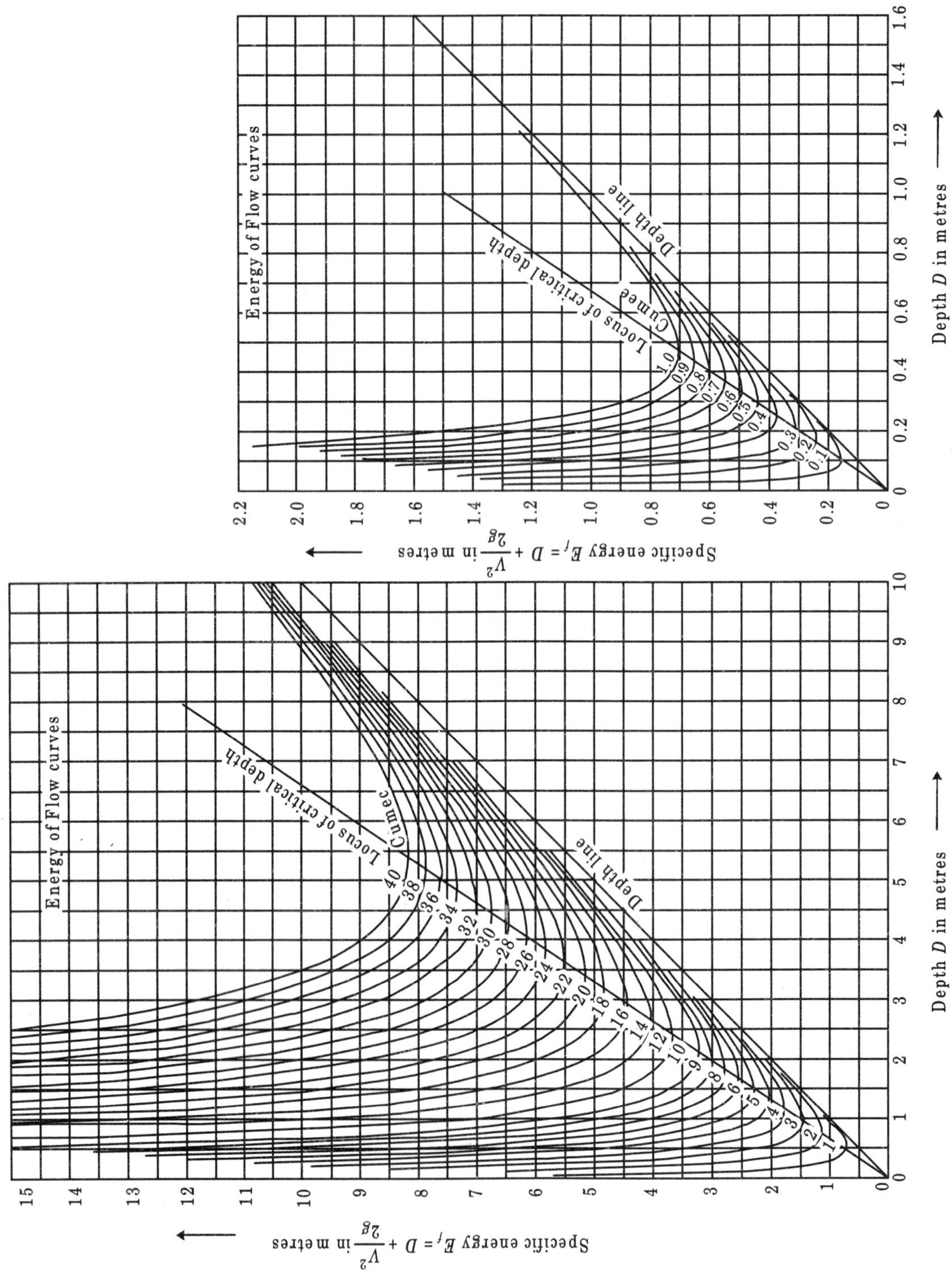

Fig.13.21 Plate IV Energy flow curves–Montague curves

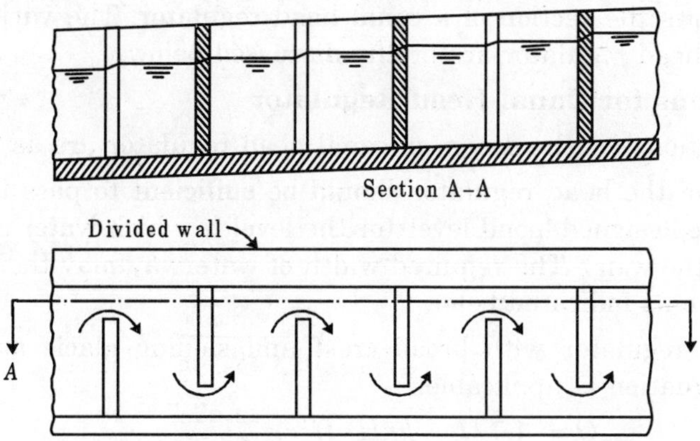

Fig. 13.22 *Fish ladder*

3.5 m/s, the design of the fish ladder should be such that it constantly releases water at a velocity not exceeding this value. To check the velocity of flow in a fish ladder baffles or other staggering devices are provided. Fish ladder is generally located adjacent to divide wall near undersluices because there is always some water in the section below them.

13.20 CANAL HEAD REGULATOR

A canal head regulator is a structure constructed at the head of a canal taking off from the upstream of a weir or a barrage. It consists of a number of spans separated by piers which support the gates provided for regulation of flow into the canal. The spans of 6 to 8 m are commonly used with counterbalanced steel gates which are operated manually by winches. However, larger spans may also be used if necessary and economical.

Functions of Canal Head Regulator

A canal head regulator serves the following functions.

(i) It regulates the supply of water into the canal.

(ii) It controls the entry of silt into the canal.

(iii) It completely excludes the high flood from entering into the canal.

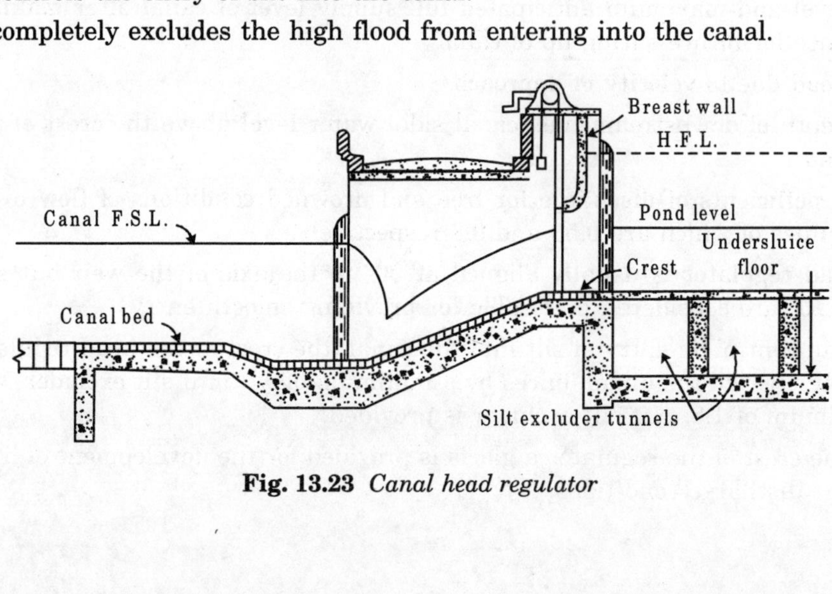

Fig. 13.23 *Canal head regulator*

Figure (13.23) shows the section of a canal head regulator. The various considerations for the design of a canal head regulator are briefly discussed below.

Design Considerations for Canal Head Regulator

The various considerations for the design of canal head regulator are as follows.

1. The waterway of the head regulator should be sufficient to pass the required discharge into the canal with the designed 'pond level' (or the level to which water can be raised up in the river at low stage by the weir). The required width of waterway may therefore be calculated by the discharge formulae as indicated below.

(a) For a head regulator with broad crest and sloping glacis at the downstream the following discharge equation is applicable.

$$Q = 1.7\,(L - knH)\,H^{3/2} \qquad \text{...(13.39)}$$

where Q = total discharge

L = length of waterway

H = head over the crest of the regulator, i.e., difference between pond level and crest level

n = number of end contractions, and

k = a coefficient depending on the shape of the upstream (i.e., river side) cutwater or the nose of the piers and its value varies from 0.01 to 0.03.

(b) If the head regulator is provided with a very wide and shallow waterway it will behave as a drowned weir, and the following discharge equation will be applicable.

$$Q = \frac{2}{3} L_1 C_{d1} \sqrt{2g}\left[(H + h_a)^{3/2}\right] + C_{d2} L_1 d \sqrt{2g\,(H + h_a)} \qquad \text{...(13.40)}$$

where Q = total discharge

L_1 = clear length of waterways

H = difference of upstream and downstream water levels, i.e., difference between pond level and maximum anticipated full supply level of canal after making due allowance for future silting up of canal

h_a = head due to velocity of approach

d = depth of downstream (i.e., canal side) water level above the crest of the regulator, and

C_{d1} and C_{d2}=coefficients of discharge for free and drowned conditions of flow over a weir the values of which are 0.58 and 0.8 respectively.

2. The head regulator is usually aligned at 90° to the axis of the weir but slightly larger angles, upto 110° are considered preferable for providing smooth entry.

3. To obtain control on entry of silt into the canal, the crest of the head regulator should be higher than the crest of the undersluices by a minimum of 1.2 m if silt excluder is not provided and by a minimum of 1.8 m if silt excluder is provided.

4. From the crest of the regulator a glacis is provided for the development of hydraulic jump under different discharge conditions.

5. The calculations for the hydraulic jump developed on the glacis are done in the same way as for a glacis weir.

6. Beyond the glacis a horizontal floor of length five times the maximum height of the jump, i.e., 5 $(D_2 - D_1)$ is provided.

7. The design of the impervious floor should be done on the basis of the most critical condition of uplift pressure in the manner similar to that of glacis weir. The critical condition of uplift pressure occurs when high flood is passing down the weir and there is no flow in the canal so that maximum static head acts. The floor should be able to resist uplift pressures under this condition. These pressures are usually very high and hence if the floor thickness becomes excessive, a reinforced concrete floor may be provided which resist the uplift pressures by its bending strength in addition to its weight.

However, in the jump trough maximum unbalanced uplift pressure may occur when some discharge is passing into the canal. The glacis should therefore be checked for this condition for different discharge intensities upto the maximum.

8. At the end of the horizontal floor a cutoff or sheet pile should be provided to keep the exit gradient well within the safe limit.

9. On the downstream of the impervious floor of the head regulator protection works should be provided for which same principles of design as applicable to the weir are used.

10. The piers separating the various spans of the regulator should be extended beyond the glacis so that they may provide support to the upward bending horizontal floor.

11. In order to prevent spilling of water over the gates into the canal during high floods, a reinforced concrete breast wall is provided from the pond level to well above the high flood level. This breast wall spans from pier to pier and is designed for its self-weight and the water pressure acting on it from the upstream side. When the gate is fully open, it rises up clear off the pond level so that full supplies can be passed into the canal. When the gate is fully closed, it shuts the opening from the crest to the bottom of the breast wall. Thus, above the top of the gate the opening is permanently closed by the breast wall. If breast wall is not provided, very high gates right upto the high flood level would be required to be provided. It would be very uneconomical to provide such high gates due to higher cost of the gates, heavier machinery to operate them on account of greater weights and water pressures and a high operating platform to provide the necessary clearance for lifting the gates.

12. For the operation of the gates a bridge and a working platform are provided across the head regulator.

Types of Regulation

The supplies entering a canal which takes off from the upstream of a weir or a barrage can be regulated in the following two ways.

(i) Still pond regulation

(ii) Semi-open flow regulation.

(i) *Still pond regulation.* In this method of regulation when the canal is running all the gates of the undersluices are kept closed. Thus, in the undersluice pocket only as much discharge is drawn as is required for the canal. The velocity of flow in the undersluice pocket, therefore, gets reduced which causes the silt to deposit in the pocket and relatively clear water enters the canal. The silt is allowed to accumulate in the pocket till it reaches to a level about 0.5 m below the crest of the regulator. The canal is then closed and the gates of the undersluices are opened so that the deposited silt gets scoured and discharged into the river on the downstream side. The scouring operation takes about 24 hours and during this period the supply in the canal is shut off. After the deposited silt is washed out tne gates of the undersluices are closed and the supply in the canal is restored. This method of regulation is very useful in controlling the amount of silt entering the canal. However, the main drawback of this method is that the supply in the canal has to be stopped during the cleaning operation of the undersluice pocket which leads to loss of irrigation during that period.

(ii) *Semi-open flow regulation.* In this method of regulation the gates of the undersluices are always kept partially open so that water in excess of the canal requirement enters the undersluice pocket and the same is allowed to be discharged to the downstream side through the undersluices. The water entering the pocket thus gets divided into two parts viz., the top water (above the crest of the regulator) which is relatively clean enters the canal through the regulator, and the bottom water (below the crest of the regulator) which is silt laden escapes through the undersluices to the downstream side. Due to continuous flow through the undersluices certain velocity is maintained in the pocket which keeps silt in suspension and the same is discharged to the downstream side without being deposited in the pocket. However, this method does not provide proper control on the entry of silt into the canal because the turbulence created in the pocket may sometimes cause even the coarser material to rise up and enter the canal. The advantage of this method of regulation is that there is continuous scouring of silt from the undersluice pocket and the canal need not be closed for this purpose as in the case of still pool regulation method.

13.21 SILT CONTROL DEVICES

Although some control on the entry of silt into the canal taking off from a head works is provided by the two methods of regulation as described in the previous section, but in both the methods of regulation certain amount of silt may enter the canal. As such special devices are required to be provided to control the entry of silt into a canal. These devices are of two types as indicated below.

1. Silt exlcuders
2. Silt extractors or silt ejectors.

The fundamental principle on which both the types of silt control devices operate is that in a stream of water carrying silt in suspension, the concentration of silt in the lower layers is greater than in the upper ones. Hence the device is so designed that the top and bottom layers are separated without any disturbance. The top water which is relatively clear is allowed to flow in the canal while the bottom water which is heavily silt laden is allowed to go as a waste.

1. Silt Excluder. Silt excluders are those silt control devices which exclude silt from water entering the canal. These devices are provided on the river bed in front of the head regulator. A silt excluder consists of a number of rectangular tunnels (Fig. 13.24) resting on the floor of the undersluice pocket. The bottom of the tunnels is formed by the floor of the undersluice pocket. The top level of the roof of the tunnels is kept same as the level of the crest (or sill) of the canal head regulator. The tunnels are of different lengths. The tunnel nearest to the crest of the head regulator is of the same length as the length of the head regulator and the successive tunnels are of decreasing length as shown in Fig. (13.24). The water approaching the head regulator is separated into two parts by the roof slab of the tunnels. The part on the top which is relatively clear enters into the canal through the head regulator, while that at the bottom which carries most of the silt flows through the tunnels and is discharged from the undersluices to the downstream side.

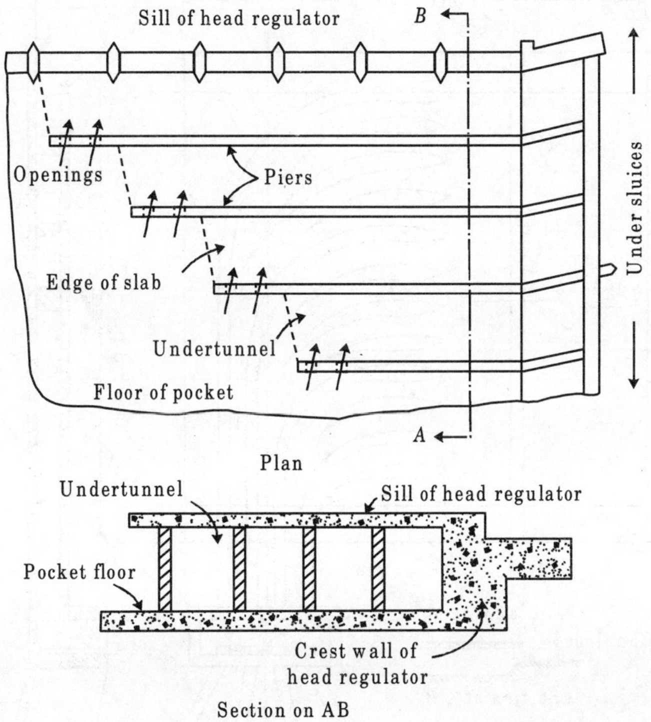

Fig. 13.24 *Silt excluder*

The capacity of the tunnels is kept as about 20% of the canal discharge. Further it is considered that a minimum velocity of 2 to 3 m/s must be maintained through the tunnels to keep them free from silt deposit. Thus, the required cross-sectional area can be obtained by dividing the discharge by the velocity. The height of the opening of the tunnel is equal to the height of the crest of the head regulator above the floor of the undersluice pocket minus the thickness of the roof slab of the tunnel. Knowing the height, the total width of the clear waterway through the tunnels is obtained by dividing the cross-sectional area by the height. This total width of the waterway is suitably divided into a number of tunnels such that the span for the roof slab is not very large and a full number of tunnels including their dividing walls can be accommodated in each undersluice opening. The roof slab of the tunnel must be able to support the maximum water load coming from the top with no water inside the tunnel. This condition may arise as it may be necessary to close the tunnels from upstream for repairs or some other purpose. For closing the tunnels grooves are provided at the entrance to the tunnels in which temporary gates may be inserted.

2. Silt Extractors or Silt Ejectors. Silt extractors are those silt control devices which remove the silt which has already entered the canal from the head. These devices are provided in the canal a little distance downstream from the head of the canal. A silt extractor or ejector consists of a horizontal diaphragm slab a little above the canal bed (Fig. 13.25) which separates out the bottom layers. Under the diaphragm slab there are tunnels to eject the heavily silt laden

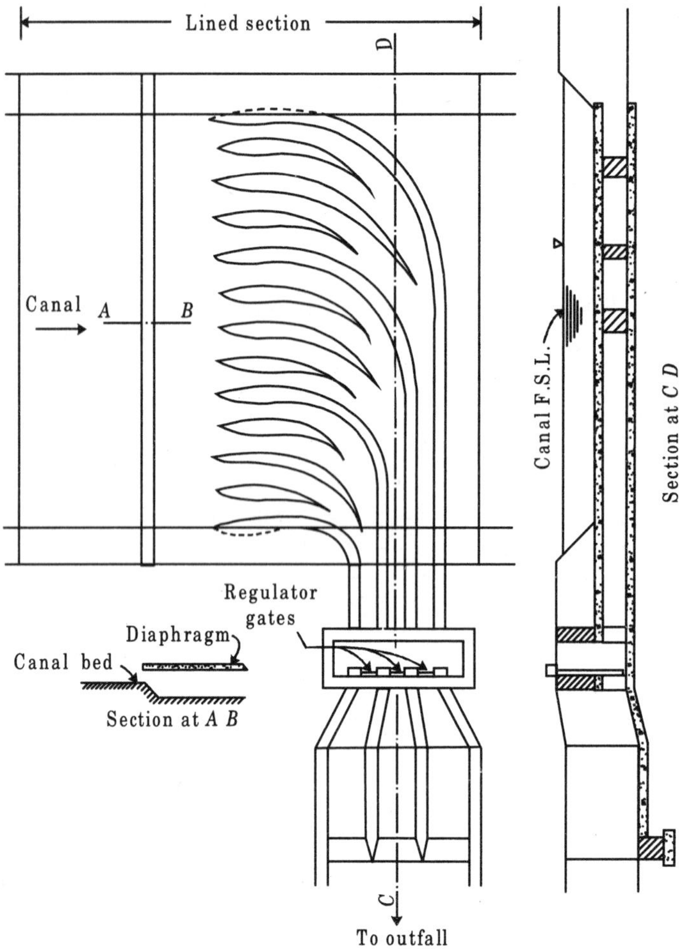

Fig. 13.25 *Silt extractor or silt ejector*

bottom water into an escape channel. The canal bed is slightly depressed under the diaphragm and the height of the tunnels is about 0.5 to 0.6 m. The tunnel entrances are so designed that there is no disturbance at the entry and also the velocity is quickly accelerated to the high value required under the diaphragm to prevent the tunnels from being clogged. The acceleration of velocity is achieved by quickly but steadily reducing the cross-sectional area by streamlined vanes (Fig. 13.25). As the lengths of the tunnels are different, the head loss in each tunnel would be different if all the tunnels take the same discharge. However, as the difference between the water level in the canal and the water level in the escape chamber or the available head is the same for all the tunnels, the shorter tunnel (nearest to the escape channel) would tend to take a higher discharge than the longer one. This would result in a higher velocity of entry in a shorter tunnel than in the others and would lead to disturbance at entry. To counteract this possibility, either the shorter tunnel may serve a larger width of the canal than a longer one or the tunnel cross-sections may be altered, each serving the same width.

The tunnel discharge is regulated by means of gates at the outlet end. The escape channel should have sufficient slope to carry the silt laden water without getting silted up and should lead the water back to the river by the shortest route.

The efficiency of silt excluders and silt extractors may be defined as

$$E = \frac{q_{sa} - q_{se}}{q_{sa}} \quad \text{...(13.41)}$$

where E = Efficiency of the device

q_{sa} = Concentration of silt in the approach channel, and

q_{se} = Concentration of silt in the canal downstream of the device.

13.22 RIVER TRAINING WORKS

At the canal head works river training works are required for providing a smooth non-tortuous approach to the work and prevent the river from outflanking the work. This purpose is usually accomplished by providing guide banks on either side. In addition, marginal bunds are required upstream of the work to prevent additional area from getting submerged due to raised high flood level caused by the afflux created by the weir. The marginal bunds have to be continued till they join high contours above the high flood level. Further 'spurs' or embankments projecting into the stream from the side banks may be required to protect the marginal bunds, or to deflect the current to the opposite bank. The details of the various river training works are given in Chapter 21.

SOLVED ILLUSTRATIVE EXAMPLES

Illustrative Example 13.1 *A river discharges 1000 cumec of water at high flood level of R.L. 103. A weir is constructed for flow diversion with a crest length of 225 m and total length of concrete floor as 40 m. The weir has to sustain the under seepage at a maximum static head of 2.4 m. The silt factor and the safe exit gradient for the bed material are 1.1 and 1/6 respectively. Determine the depth of cut-off required at the downstream end of the concrete floor. Take the level of downstream concrete floor as R.L. 100. Check for exit gradient.*

Solution

Given :

Discharge Q = 1000 m³/s

Length of weir B = 225 m

Length of concrete floor b = 40 m

Silt factor f = 1.1

Thus, discharge per unit width

$$q = \frac{Q}{B} = \frac{1000}{225} = 4.44 \text{ m}^3\text{/s/m}$$

The scour depth R is given as

$$R = 1.35\left(\frac{q^2}{f}\right)^{1/3}$$

Thus, by substitution, we get

$$R = 1.35\left[\frac{(4.44)^2}{1.1}\right]^{1/3} = 3.53 \text{ m}$$

$$\text{Velocity of flow } V = \frac{q}{R} = \frac{4.44}{3.53} = 1.258 \text{ m/s}$$

$$\text{Velocity head} = \frac{V^2}{2g} = \frac{(1.258)^2}{2\times 9.81} = 0.081 \text{ m}$$

$$\text{Downstream TEL} = \text{HFL + Velocity head}$$

$$= 103 + 0.081 = 103.081$$

Depth of cut-off downstream

$$d = 100 - (103.081 - 3.53)$$

$$= 0.449 \text{ m}$$

$$\text{Exit gradient } G_E = \frac{H}{d}\frac{1}{\pi\sqrt{\lambda}}$$

$$H = 2.4 \text{ m}$$

$$\lambda = \frac{1+\sqrt{1+\alpha^2}}{2}$$

$$\alpha = \frac{b}{d} = \frac{40}{0.449} = 89.09$$

$\therefore$

$$\lambda = 45.05$$

and

$$G_E = \frac{2.4}{0.449}\times\frac{1}{\lambda}\frac{1}{\sqrt{45.05}} = 0.253 > \frac{1}{6}.$$

Illustrative Example 13.2 *A flownet for seepage flow through soil below a hydraulic structure gives $N_f = 3$; $N_d = 30$ and total head causing the flow = 10 m. What is the quantity of flow per metre run occurring under the structure? Take $k = 10^{-6}$ m/s.*

If the elementary square of the toe of the structure has dimensions of 0.6 m, what is the exit gradient. Also determine the head when the exit gradient will be equal to 1. What is the implication of this condition to the safety of the structure?

Solution

The quantity of seepage flow per metre run occurring under the structure is given as

$$q = kh\frac{N_f}{N_d}$$

or
$$q = 10^{-6} \times 10 \times \frac{3}{30}$$

$$= 3 \times 10^{-6} \text{ m}^3\text{/s} = 3 \times 10^{-3} \text{ l/s.}$$

From equation 13.7 (*a*) exit gradient

$$G_E = \frac{\Delta h}{l}$$

$$= \frac{10}{30} \times \frac{1}{0.6} = 0.556$$

Let H be the head when the exit gradient will be equal to 1. Thus

$$1 = \frac{H}{30} \times \frac{1}{0.6}$$

$$\therefore \quad H = 18 \text{ m.}$$

When exit gradient becomes equal to 1 the failure of the structure is imminent due to undermining.

Illustrative Example 13.3. *Figure Ex. 13.3 shows the section of a hydraulic structure on permeable foundation. Calculate the average hydraulic gradient according to (i) Bligh's creep theory ; (ii) Lane's weighted creep theory. Also find the uplift pressures at points A, B and C as shown in Fig. Ex. 13.3 and the floor thickness required at these points. Locate the point where the required floor thickness is same from both the theories.*

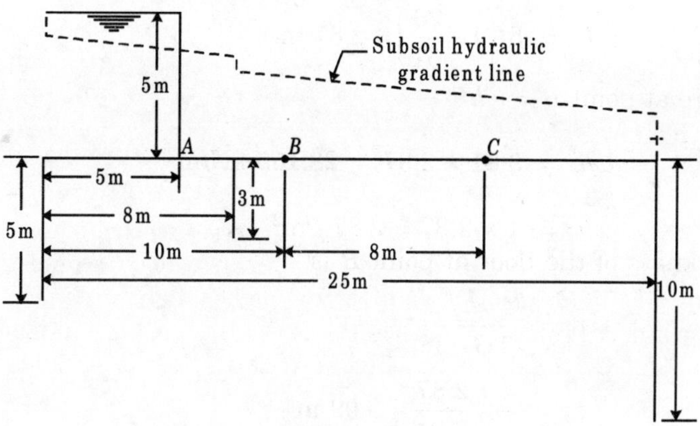

Fig. Ex.13.3

Solution

(i) According to Bligh's creep theory the total creep length is given by

$$L = (2 \times 5) + (2 \times 3) + 25 + (2 \times 10) = 61 \text{ m}$$

$$\therefore \quad \text{Hydraulic gradient} = \frac{5}{61} = \frac{1}{12.2}$$

The structure would be safe on coarse grained sand.

(*a*) *Uplift pressure at point A.* Length of creep upto point A

$$= (2 \times 5) + 5 = 15 \text{ m}$$

∴ Residual seepage head at point A is

$$h_1 = 5\left(1 - \frac{15}{61}\right) = 3.77 \text{ m}$$

∴ Uplift pressure at point A $= wh_1$

In SI Units

Since
$$w = 9810 \text{ N/m}^3 = 9.81 \text{ kN/m}^3$$
$$wh_1 = 9.81 \times 3.77$$
$$= 36.984 \text{ kN/m}^2$$

In metric units

Since
$$w = 1000 \text{ k/m}^3 = 1\ t/\text{m}^3$$
$$wh_1 = 1 \times 3.77$$
$$= 3.77\ t/\text{m}^2$$

From Eq. 13.4 (b) the required thickness of the floor at point A is

$$t_1 = \frac{4}{3}\frac{h_1}{G-1}$$

Taking $G = 2.24$

$$t_1 = \frac{4}{3} \times \frac{3.77}{2.24-1} = 4.05 \text{ m}$$

(*b*) *Uplift pressure at point B.* Length of creep upto point B

$$= (2 \times 5) + (2 \times 3) + 10 = 26 \text{ m}$$

∴ Residual seepage head at point B is

$$h_2 = 5\left(1 - \frac{26}{61}\right) = 2.87 \text{ m}$$

∴ Uplift pressure at point $B = wh^2$

In SI units

$$wh^2 = 9.81 \times 2.87 = 28.155 \text{ kN/m}^2$$

In metric units

$$wh^2 = 1 \times 2.87 = 2.87\ t/\text{m}^2$$

∴ Required thickness of the floor at point B is

$$t_2 = \frac{4}{3}\frac{h_2}{G-1}$$

or
$$t_2 = \frac{4}{3} \times \frac{2.87}{2.24} = 3.09 \text{ m}$$

(c) *Uplift pressure at point C.* Length of creep upto point C

$$= (2 \times 5) + (2 \times 3) + 18 = 34 \text{ m}$$

∴ Residual seepage head at point C is

$$h_3 = 5\left(1 - \frac{34}{61}\right) = 2.21 \text{ m}$$

∴ Uplift pressure at point C $= wh_3$

In SI units

$$wh_3 = 9.81 \times 2.21$$

$$= 21.680 \text{ kN/m}^2$$

In metric units

$$wh_3 = 1 \times 2.21$$
$$= 2.21 \ t/\text{m}^2$$

∴ Required thickness of the floor at point O is

$$t_3 = \frac{4}{3}\frac{h_3}{G-1}$$

or $$t_3 = \frac{4}{3} \times \frac{2.21}{2.24-1} = 2.38 \text{ m}$$

(b) According to Lane's weighted creep theory the total weighted creep length is given by

$$L_w = (2 \times 5) + (2 \times 3) + (2 \times 10) + \frac{1}{3}(25)$$
$$= 44.33 \text{ m}$$

∴ $$\text{Hydraulic gradient} = \frac{5}{44.33} = \frac{1}{8.87}$$

Thus, according to this theory the structure would be safe even on very fine sand or silt.

(*a*) *Uplift pressure at point A.* Weighted length of creep upto point A

$$= (2 \times 5) + \frac{1}{3}(5) = 11.67 \text{ m}$$

∴ Residual seepage head at point A is

$$h_1 = 5\left(1 - \frac{11.67}{44.33}\right) = 3.68 \text{ m}$$

∴ Uplift pressure at point $A = wh_1$

In SI units

$$wh_1 = 9.81 \times 3.68$$
$$= 36.101 \text{ kN/m}^2$$

In metric units

$$wh_1 = 1 \times 3.68$$
$$= 3.68 \ t/\text{m}^2$$

Required thickness of the floor at point A is

$$t_1 = \frac{4}{3} \times \frac{3.68}{2.24-1} = 3.96 \text{ m}$$

(b) *Uplift pressure at point B.* Weighted length of creep upto point B

$$= (2 \times 5) + (2 \times 3) + \frac{1}{3}(25) = 19.33 \text{ m}$$

∴ Residual seepage head at point B is

$$h_2 = 5\left(1-\frac{19.33}{44.33}\right) = 2.82 \text{ m}$$

∴ Uplift pressure at point $B = wh_2$

In SI units

$$wh_2 = 9.81 \times 2.82$$
$$= 27.664 \text{ kN/m}^2$$

In metric units

$$wh_2 = 1 \times 2.82$$
$$= 2.82\ t/\text{m}^2$$

Required thickness of the floor at point B is

$$t_2 = \frac{4}{3}\frac{h_2}{G-1}$$

or

$$t_2 = \frac{4}{3}\frac{2.82}{2.24-1} = 3.03 \text{ m}$$

(c) *Uplift at point C.* Weighted length of creep upto point C

$$= (2 \times 5) + (2 \times 3) + \frac{1}{3}(18) = 22 \text{ m}$$

∴ Residual seepage head at point C is

$$h_3 = 5\left(1-\frac{22}{44.33}\right) = 2.52 \text{ m}$$

∴ Uplift pressure at point $C = wh_3$

In SI units

$$wh_3 = 9.81 \times 2.52$$
$$= 24.721 \text{ kN/m}^2$$

In metric units

$$wh_3 = 1 \times 2.52$$
$$= 2.52\ t/\text{m}^2$$

Required thickness of the floor at point O is

$$t_3 = \frac{4}{3}\frac{h_3}{G-1}$$

or

$$t_3 = \frac{4}{3}\times\frac{2.52}{2.24-1} = 2.71 \text{ m}$$

Let x be the distance of the point from the upstream end of the floor where the required floor thickness is same from both the theories.

According to Bligh's creep theory total creep length upto this point

$$= (2 \times 5) + (2 \times 3) + x = (x + 16)$$

∴ Residual seepage head at this point is

$$h_1 = 5\left(1 - \frac{x+16}{61}\right)$$

According to Lane's weighted creep theory weighted length of creep upto this point

$$= (2 \times 5) + (2 \times 3) + \frac{1}{3}(x) = \left(\frac{x}{3} + 16\right)$$

∴ Residual seepage head at this point is

$$h_2 = 5\left[1 - \frac{(x/3)+16}{44.33}\right]$$

Equating the values of the residual seepage head for both the cases, we get

$$5\left(1 - \frac{x+16}{61}\right) = 5\left[1 - \frac{(x/3)+16}{44.33}\right]$$

Solving for x, we get

$$x = 11.11 \text{ m}$$

Illustrative Example 13.4 *Figure Ex. 13.4 shows the profile of a weir. The various levels shown in the figure are in metres. Determine the uplift pressures at the key points and the exit gradient and find whether the section provided is safe against piping if it is founded on fine sand with permissible exit gradient of 1/6. Also find uplift pressure at point X and check whether it is safe against uplift.*

Solution

(i) Pile line No. 1

$$d = 100 - 91 = 9 \text{ m} \; ; \; b = 65 \text{ m}$$

$$\frac{1}{\alpha} = \frac{d}{b} = \frac{9}{65} = 0.138 \simeq 0.14$$

From the curves

$$\phi_{D_1} = 77\% \; ; \; \phi_{C_1} = 67\%$$

Correction for floor thickness for ϕ_{C_1}

$$= \frac{\phi_{D_1} - \phi_{C_1}}{d} \times t$$

$$t = 100 - 98 = 2 \text{ m}$$

∴ Correction $= \dfrac{77 - 67}{9} \times 2 = 2.2\% \; (+)$

Correction for interference of Pile No. 2 for ϕ_{C_1}

$$C = 19\sqrt{\frac{D}{b'}}\left(\frac{d+D}{b}\right)$$

$$d = 98 - 91 = 7 \text{ m};\ b' = 48 \text{ m}$$

$$D = 98 - 89.5 = 8.5 \text{ m};\ b = 65 \text{ m}$$

$$\therefore \quad C = 19\sqrt{\frac{8.5}{48}}\left(\frac{7+8.5}{65}\right) = 1.9\%\ (+)$$

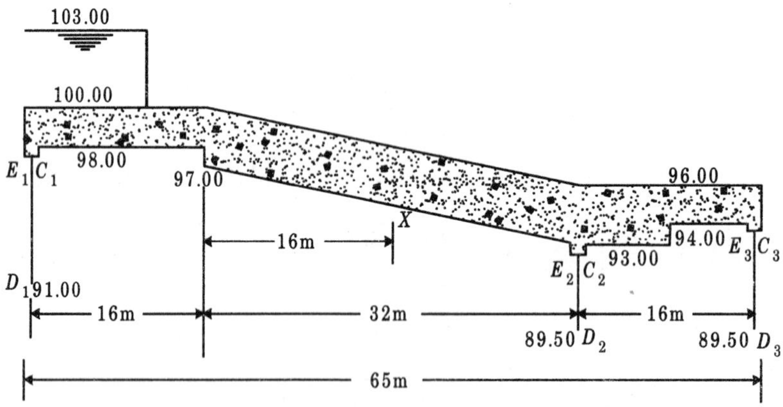

Fig. Ex. 13.4

Hence corrected value of ϕ_{C_1}

$$= 67 + 2.2 + 1.9 = 71.1\%$$

Pile line No. 2

$$d = 96 - 89.5 = 6.5 \text{ m};\ b = 65 \text{ m};\ b_1 = 48.5 \text{ m}$$

$$\alpha = \frac{b}{d} = \frac{65}{6.5} = 10$$

$$\frac{b_1}{b} = \frac{48.5}{65} = 0.746$$

From the curves $\phi_{E_2} = 41\%$; $\phi_{D_2} = 34\%$; $\phi_{C_2} = 26\%$

Correction for floor thickness for ϕ_{E_2}

$$= \frac{\phi_{E_2} - \phi_{C_2}}{d} \times t$$

$$t = 96 - 93 = 3 \text{ m}$$

$$\therefore \quad \text{Correction} = \frac{41-34}{6.5} \times 3 = 3.2\%\ (-)$$

Correction for floor thickness for ϕ_{C_2}

$$= \frac{\phi_{D_2} - \phi_{C_2}}{d} \times t$$

$$t = 96 - 93 = 3 \text{ m}$$

$$\therefore \qquad \text{Correction} = \frac{34 - 26}{6.5} \times 3 = 3.7\% \ (+)$$

Correction for interference of Pile No. 1 for ϕ_{E_2}

$$C = 19 \sqrt{\frac{D}{b'}} \left(\frac{d + D}{b} \right)$$

$$d = 93 - 89.5 = 3.5 \text{ m}; \ b' = 48 \text{ m}$$

$$D = 93 - 91 = 2 \text{ m}; \ b = 65 \text{ m}$$

$$\therefore \qquad C = 19 \sqrt{\frac{2}{48}} \left(\frac{3.5 + 2}{65} \right) = 0.3\% \ (-)$$

Correction due to slope for ϕE_2.

From Table 13.3, for a slope of 1 in 8 the correction % of pressure = 2.0% and hence

$$\text{Correction} = 2.0 \times \frac{b_s}{b'}$$

$$b_s = 32 \text{ m; and } b' = 48 \text{ m}$$

$$\therefore \qquad \text{Correction} = 2.0 \times \frac{32}{48} = 1.3\% \ (+)$$

Correction for interference of Pile No. 3 for ϕ_{C_2}

$$C = 19 \sqrt{\frac{D}{b'}} \left(\frac{d + D}{b} \right)$$

$$d = 93 - 89.5 = 3.5 \text{ m}; \ b' = 16 \text{ m}$$

$$D = 93 - 89.5 = 3.5 \text{ m}; \ b = 65 \text{ m}$$

$$\therefore \qquad C = 19 \sqrt{\frac{3.5}{16}} \left(\frac{3.5 + 3.5}{65} \right) = 1\% \ (+)$$

Hence corrected values of ϕ_{E_2} and ϕ_{C_2} are as follows

$$\phi_{E_2} = 41 - 3.2 - 0.3 + 1.3 = 38.8\%$$

$$\phi_{C_2} = 26 + 3.7 + 1.0 = 30.7 \ \%$$

Pile Line No. 3

$$d = 96 - 89.5 = 6.5 \ ; \ b = 65 \text{ m}$$

$$\frac{1}{\alpha} = \frac{d}{b} = \frac{6.5}{65} = 0.1$$

From the curves

$$\phi_{E3} = 29\% \; ; \; \phi_{D3} = 20\%$$

Correction for thickness of floor for ϕ_{E3}

$$= \frac{\phi_{E_3} - \phi_{D_3}}{d} \times t$$

$$t = 96 - 94 = 2 \text{ m}$$

$\therefore$ Correction $= \dfrac{29 - 20}{6.5} \times 2 = 2.8\% \; (-)$

Correction for interference of Pile No. 2 for ϕ_{E3}

$$C = 19\sqrt{\frac{D}{b'}}\left(\frac{d+D}{b}\right)$$

$$d = 94 - 89.5 = 4.5 \text{ m}; \; b' = 16 \text{ m}$$

$$D = 94 - 89.5 = 4.5 \text{ m}; \; b = 65 \text{ m}$$

$\therefore$ $C = 19\sqrt{\dfrac{4.5}{16}}\left(\dfrac{4.5 + 4.5}{65}\right) = 1.4\% \; (-)$

Hence corrected value of ϕ_{E3}

$$= 29 - 2.8 - 1.4 = 24.8 \; \%$$

Maximum percolation head

$$H = 103 - 96 = 7 \text{ m}$$

The percentage pressures and the residual pressure head at key points are tabulated below.

Point	*% Pressure* (ϕ)	*Residual pressure head* (*P*) *m*
D_1	77.0	5.39
C_1	71.1	4.98
E_2	38.8	2.72
D_2	34.0	2.38
C_2	30.7	2.15
E_3	24.8	1.74
D_3	20.0	1.40

Exit gradient $G_E = \dfrac{H}{d}\dfrac{1}{\pi\sqrt{\lambda}}$

$$H = 103 - 96 = 7 \text{ m}$$

$$d = 96 - 89.5 = 6.5 \text{ m}$$

$$\alpha = \frac{b}{d} = \frac{65}{6.5} = 10$$

From exit gradient curve for

$$\alpha = 10 \; ; \; \frac{1}{\pi\sqrt{\lambda}} = \frac{1}{7.38} = 0.135$$

$$\therefore \qquad G_E = \frac{7}{65} \times \frac{1}{7.38} = \frac{1}{6.85}$$

$$\text{Permissible gradient} = \frac{1}{6}$$

Hence safe.

Uplift pressure at X is

$$P_X = P_{C_1} - \left(\frac{P_{C_1} - P_{E_2}}{48} \times 32\right)$$

or

$$P_X = 4.98 - \left(\frac{4.98 - 2.72}{48} \times 32\right) = 3.47 \text{ m}$$

$\therefore$ Floor thickness required at X

$$= \frac{P_X}{G - 1} = \frac{3.47}{2.24 - 1} = 2.8 \text{ m}$$

Actual thickness provided = 3 m

Hence it is safe against uplift.

Illustrative Example 13.5 *Design a vertical drop weir on the basis of Bligh's theory for the following data obtained at the site.*

(*a*) *Maximum flood discharge = 2585 cumec*

(*b*) *H.F.L. before construction = 255.0 m*

(*c*) *Minimum water level = 248.0 m (d/s bed level)*

(*d*) *F.S.L. of canal = 254.0 m*

(*e*) *Allowable afflux = 1 m*

(*f*) *Coefficient of creep C = 12*

Assume any other data not given.

If permissible exit gradient is 1/6, test the floor of the above weir by Khosla's theory and make changes if found necessary.

Solution

(Refer Fig. 13.18 for various symbols used)

1. *Hydraulic calculations*

Step 1

$$Q = 2585 \text{ cumec}$$
$$L = 4.83 \, Q^{1/2}$$

$$= 4.83\,(2585)^{1/2} = 245.57 \text{ m}$$

$$q = \frac{Q}{L} = \frac{2585}{245.57} = 10.53 \text{ cumec/m}$$

Step 2

Regime scour depth : Take $f = 1$

$$R = 1.35\left(\frac{q^2}{f}\right)^{1/3}$$

$$= 1.35\left[\frac{(10.53)^2}{1}\right]^{1/3} = 6.49 \text{ m}$$

$$\text{Regime velocity } V = \frac{q}{R} = \frac{10.53}{6.49} = 1.62 \text{ m/s}$$

$$\text{Velocity head} = \frac{V^2}{2g} = \frac{(1.62)^2}{2\times 9.81} = 0.13 \text{ m}$$

Step 3

$$\text{Level of } d/s \text{ T.E.L.} = \text{H.F.L. before const.} + \frac{V^2}{2g}$$

$$= 255.00 + 0.13 = 255.13 \text{ m}$$

$$\text{Afflux} = 1 \text{ m}$$

$$\text{Level of } u/s \text{ T.E.L.} = \text{Level of } d/s \text{ T.E.L.} + \text{Afflux}$$

$$= 255.13 + 1 = 256.13 \text{ m}$$

∴ Level of u/s H.F.L.

$$= \text{Level of } u/s \text{ T.E.L.} - \frac{V^2}{2g}$$

$$= 256.13 - 0.13 = 256.00 \text{ m}$$

Actual d/s H.F.L., allowing 0.5 m for retrogression

$$= 255.00 - 0.50 = 254.50 \text{ m}$$

Step 4

$$q = 1.70\,K^{3/2}$$

$$\therefore \quad K = \left(\frac{q}{1.70}\right)^{2/3}$$

$$= \left(\frac{10.53}{1.70}\right)^{2/3} = 3.37 \text{ m}$$

$\therefore$ Crest level = u/s T.E.L. – K

= 256.13 – 3.37 = 252.76 m

Step 5

Pond level = Level of top of gates

= F.S.L. of canal + Head loss through regulator

= 254.00 + 0.50 (say)

= 254.50 m

$\therefore$ Height of shutters S

= Level of top of gates – Crest level

= 254.50 – 252.76 = 1.74 m

Step 6

Level of bottom of *u/s* pile (or cutoff)

= *u/s* H.F.L. – 1.5 R

= 256.00 – (1.5 × 6.49)

= 246.27 m

u/s pile may be taken up to a level of 246.00

$\therefore$ Depth of *u/s* pile = 248.00 – 246.00 = 2.00 m

Hence provide a pile of 2 m depth below the bed of river at the *u/s* end of the floor.

Level of bottom of *d/s* pile

= *d/s* H.F.L. after retrogression – 2 R

= 254.50 – (2 × 6.49)

= 241.52 m

d/s pile may be taken upto a level of 241.00

$\therefore$ Depth of *d/s* pile = 248.00 – 241.00 = 7.00 m

Hence provide a pile of 7 m depth below the bed of the river at the *d/s* end of the floor.

Step 7

Head of water H_s = Level of crest gates – Bed level

= 254.50 – 248.00 = 6.50 m

Height of crest H = Crest level – Bed level

= 252.76 – 248.00 = 4.76 m

(Check : $Hs = H + S$ = 4.76 + 1.74 = 6.50 m)

2. *Design of weir wall*

Step 8

Calculation of top width B_1

d = *u/s* H.F.L. – Crest level

$$= 256.00 - 252.76 = 3.24 \text{ m}$$

From Eq. (13.20)

Top width $$B_1 = \frac{d}{\sqrt{G}} = \frac{3.24}{\sqrt{2.24}}$$

$$= 2.16 \text{ m}$$

From Eq. (13.21 a)

Top width $$B_1 = \frac{3d}{2G} = \frac{3 \times 3.24}{2 \times 2.24}$$

$$= 2.17 \text{ m}$$

From practical considerations

$$B_1 = S + 1 = 1.74 + 1 = 2.74 \text{ m}$$

Hence provide top width

$$B_1 = 3 \text{ m}$$

Step 9

Calculation of bottom width B

Considering state I when the water is upto the top of the crest gates, with no tail water, the overturning moment is given by Eq. (13.23) as

$$M_0 = \frac{w(H+S)^3}{6} = \frac{wH_s^3}{6}$$

In SI units

$$M_0 = \frac{9.81 \times (6.5)^3}{6} = 4\ 49 \text{ kN-m}$$

In metric units

$$M_0 = \frac{1 \times (6.5)^3}{6} = 45.77\ t\text{-m}$$

The moment of resistance is given by Eq. (13.24) as

$$M_r = \frac{w}{12}\left[\{(G+1.5)H + 2.5S\}B^2 + B_1(GH - H - S)B - \frac{1}{2}B_1^2(H+3S)\right]$$

$$H = 4.76 \text{ m} ;\ S = 1.74 \text{ m} ;\ B1 = 3 \text{ m} ;\ G = 2.24.$$

Introducing these values in the above equation and equating it to the overturning moment, we get

$$449 = \frac{9.81}{12}[\{(2.24 + 1.5)\ 4.76 + 2.5 \times 1.74\}\ B^2$$

$$+ 3\ (2.24 \times 4.76 - 4.76 - 1.74)\ B - \frac{1}{2} \times 9\ (4.76 + 3 \times 1.74)]$$

or $$45.77 \times 12 = (22.15\,B^2 + 12.49\,B - 44.91)$$

from which $$B = 4.905 \text{ m}$$

Consider state II when the water is flowing over the weir and the weir is submerged. Considering the tail water just at the crest of weir, the overturning moment is given by Eq. (13.25) as

$$M_0 = \frac{whH^2}{2}$$

When the tail water is at the crest, d and h will be equal. For this case the value of d (i.e., head over the crest) is given by

$$d = \left[\frac{q}{\left(\frac{2}{3}\right)C_d \times \sqrt{2g}}\right]^{2/3}$$

(Neglecting the velocity of approach) where C_d = 0.58, is the coefficient of discharge.

Substituting q = 10.53 cumec/m

$$d = \left[\frac{10.53}{\left(\frac{2}{3}\right)\times 0.58 \times \sqrt{2 \times 9.81}}\right]^{2/3} = 3.36 \text{ m}$$

Since $$d = h = 3.36 \text{ m}$$

$$M_0 = \frac{whH^2}{2}$$

$$= \frac{9.81 \times 3.36 \times (4.76)^2}{2} = 373.42 \text{ kN-m}$$

The moment of resistance is given by Eq. (13.26) as

$$M_r = \frac{wH(G-1)}{12}(B^2 + B_1 B)$$

$$= \frac{9.81 \times 4.76(2.24 - 1)}{12}(B^2 + 3B)$$

$$= 4.825\,(B^2 + 3B)$$

Equating the two, we get

$$373.42 = 4.825\,(B^2 + 3B)$$

From which

$$B = 7.42 \text{ m}$$

Adopting the greater of the two, we get

$$B = 7.42 \text{ m}$$

Provide $B = 7.5$ m

Thus, the weir wall has top width of 3 m and bottom width of 7.5 m with equal slopes of *u/s* and *d/s* faces.

3. *Design of impervious and pervious aprons*

Step 10

Total creep length is

$$L = CH_s$$
$$= 12 \times 6.5 = 78 \text{ m}$$

Step 11

Downstream impervious apron

$$L_1 = 2.21\, C \sqrt{\frac{H_s}{13}}$$

$$= 2.21 \times 12 \sqrt{\frac{6.5}{13}}$$

$$= 18.75 \text{ m}$$

Provide $L_1 = 19$ m

Step 12

Upstream impervious apron (L_2) is given by Eq. (13.33) as

$$L_2 = L - L_1 - (B + 2d_1 + 2d_2)$$
$$d_1 = 2 \text{ m} ;\ d_2 = 7 \text{ m} ;\ \text{and } B = 7.5 \text{ m}$$

$\therefore$

$$L_2 = 78 - 19 - (7.5 + 2 \times 2 + 2 \times 7)$$
$$= 33.5 \text{ m}$$

Step 13

Total length of *d/s* apron

$$L_3 = 18\, C \sqrt{\frac{H_s}{13} \times \frac{q}{75}}$$

$$= 18 \times 12 \sqrt{\frac{6.5}{13} \times \frac{10.53}{75}} = 57.23 \text{ m}$$

Provide $L_3 = 57.5$ m

Step 14

Length of inverted filter + Launching apron on the *d/s* side

$$= L_3 - L_1$$
$$= 57.5 - 19 = 38.5 \text{ m}$$
$$d_2 = 7 \text{ m}$$

Minimum length of inverted filter

$$= 1.5\, d_2$$
$$= 1.5 \times 7 = 10.5 \text{ m}$$

Minimum length of launching apron

$$= 1.5\, d_2$$
$$= 1.5 \times 7 = 10.5 \text{ m}$$

Total minimum length of both

$$= (10.5 + 10.5) = 21 \text{ m}$$

But the total length to be provided for both is 38.5 m.

Hence provide inverted filter of length 17.5 consisting of 1 m thick stone or concrete blocks laid on 0.5 m thick graded filter ; and launching apron 21 m long and 1.5 m thick.

Step 15

Upstream block protection and launching apron

$$d_1 = 2 \text{ m}$$

Hence provide block protection of length 2 m, consisting of 1 m thick stone or concrete blocks laid on 0.5 m thick loosely packed stone.

Minimum length of u/s launching apron

$$= 1.5\, d_1 = 1.5 \times 2 = 3 \text{ m}$$

Hence provide launching apron of length 3 m and thickness 1.5 m.

Step 16

Thickness of impervious floor.

Provide a nominal thickness of 1 m for the portion of the impervious floor upstream of the weir wall, and 1.5 m for the portion below the weir wall.

At point A just at the d/s toe of the weir wall (shown in Fig. Ex. 13.5) the residual seepage head

$$H_r = H_s - \frac{H_s}{L}(2d_1 + L_2 + B)$$

$$= 6.5 - \frac{6.5}{78}(2 \times 2 + 33.5 + 7.5)$$

$$= 6.5 - 3.75 = 2.75 \text{ m}$$

$\therefore$ Required thickness of floor

$$t = \frac{4}{3}\frac{H_r}{G-1}$$

$$= \frac{4}{3} \times \frac{2.75}{2.24-1} = 2.96 \text{ m} \approx 3 \text{ m}$$

Provide a thickness of 3 m from the *d/s* toe of the weir wall upto a point 6 m from it. Residual seepage head at 6 m from the *d/s* toe of the weir wall is

$$H_r = 6.5 - \frac{6.5}{78}(45+6)$$

$$= 6.5 - 4.25 = 2.25 \text{ m}$$

∴ Required thickness of floor

$$t = \frac{4}{3} \times \frac{2.25}{2.24-1}$$

$$= 2.42 \text{ m} \approx 2.5 \text{ m}$$

Provide a thickness of 2.5 m from 6 to 12 m from the *d/s* toe of the weir wall. Residual seepage head at 12 m from the *d/s* toe of the weir wall is

$$H_r = 6.5 - \frac{6.5}{78}(45+12)$$

$$= 6.5 - 4.75 = 1.75 \text{ m}$$

∴ Required thickness of floor

$$t = \frac{4}{3} \times \frac{1.75}{2.24-1}$$

$$= 1.88 \simeq 2 \text{ m}$$

Provide a thickness of 2 m for the remaining 7 m length of the floor.

4. *Check by Khosla's theory*

(*i*) Check for exit gradient

Total length of the impervious floor

$$= 33.5 + 7.5 + 19 = 60 \text{ m}$$

$$\text{Depth of } d/s \text{ pile} = 7 \text{ m}$$

$$\therefore \qquad \alpha = \frac{b}{d} = \frac{60}{7} = 8.57 \simeq 8.6$$

From Khosla's exit gradient curves for

$$\alpha = 8.6, \quad \frac{1}{\pi\sqrt{\lambda}} = 0.145$$

$$\therefore \text{ Exit gradient} \qquad G_E = \frac{H_s}{d}\frac{1}{\pi\sqrt{\lambda}}$$

$$= \frac{6.5}{7} \times 0.145 = \frac{1}{7.4}$$

Permissible exit gradient $= \dfrac{1}{6}$

Hence safe.

(ii) Check for thickness of floor.

Pressures at key points C_1 and D_1 of the u/s pile

$$= \frac{1}{\alpha} = \frac{d_1}{b} = \frac{2}{60} = 0.033$$

From Khosla's curves

$$\phi_{D_1} = 89\% \; ; \; \phi_{C_1} = 82\%$$

Correction for floor thickness for ϕ_{C_1}

$$= \frac{\phi_{D_1} - \phi_{C_1}}{d} \times t$$

$$= \frac{89-82}{2} \times 1 = 3.5\% \; (+)$$

Correction for interference of d/s pile for ϕ_{C_1}

$$C = 19\sqrt{\frac{D}{b'}}\left(\frac{d+D}{b}\right)$$

Here $b = b' = 60 \text{ m} \; ; \; D = 7 \text{ m} \; ; \; d = 2 \text{ m}$

$$\therefore \quad C = 19\sqrt{\frac{7}{60}}\left(\frac{2+7}{60}\right)$$

$$= 0.97 \simeq 1\% \; (+)$$

Hence corrected value of ϕ_{C_1}

$$= 82 + 3.5 + 1 = 86.5\%$$

$\therefore$ Residual pressure head at C_1

$$= H_s \times \phi_{C_1}$$

$$= 6.5 \times 0.865 = 5.62 \text{ m}$$

Pressure at key points E_2 and D_2 of the d/s pile

$$= \frac{1}{\alpha} = \frac{d}{b} = \frac{7}{60} = 0.117$$

From Khosla's curves

$$\phi_{E_2} = 30\% \; ; \; \phi_{D_2} = 21\%$$

Correction for floor thickness for ϕ_{E_2}

$$= \frac{\phi_{E_2} - \phi_{D_2}}{d} \times t$$

Thickness of floor near the pile = 2 m

$$\therefore \quad \text{Correction} = \frac{30-21}{7} \times 2 = 2.6\% \ (-)$$

Correction for interference of *u/s* pile for ϕ_{E2}

$$C = \sqrt{\frac{D}{b'}}\left(\frac{d+D}{b}\right)$$

Here $D = 2$ m ; $d = 7$ m ; $b = b' = 60$ m

$$\therefore \quad C = \sqrt{\frac{2}{60}}\left(\frac{7+2}{60}\right) = 0.5\% \ (-)$$

$$\therefore \quad \text{Corrected value of } \phi_{E2} = 30 - 2.6 - 0.5 = 26.9\%$$

Residual pressure head at E_2

$$= 6.5 \times 0.269 = 1.75 \text{ m}$$

Assuming the linear variation of pressure for intermediate points, the pressures and floor thickness for points *A, B* and *C* are as under.

Point *A* just at the *d/s* toe of the weir wall

$$P_A = 5.62 - \left(\frac{5.62-1.75}{60}\right)(33.5+7.5)$$

$$= 5.62 - 2.64 = 2.98 \text{ m}$$

$$\therefore \quad t_A = \frac{2.98}{2.24-1} = 2.40 \text{ m}$$

Actual thickness provided according to Bligh's theory = 3 m

Hence safe.

Point *B,* 6 m from *d/s* toe of the weir wall

$$P_B = 5.62 - \left(\frac{5.62-1.75}{60}\right)(33.5+7.5+6)$$

$$= 5.62 - 3.03 = 2.59 \text{ m}$$

$$\therefore \quad t_B = \frac{2.59}{2.24-1} = 2.09 \text{ m}$$

Actual thickness provided according to Bligh's theory

$$= 2.5 \text{ m}$$

Hence safe.

Point *C,* 12 m from *d/s* toe of the weir wall

$$P_C = 5.62 - \left(\frac{5.62-1.75}{60}\right)(33.5+7.5+12)$$

$$= 5.62 - 3.42 = 2.2 \text{ m}$$

$$\therefore \qquad t_C = \frac{2.2}{2.24 - 1} = 1.77 \text{ m}$$

Actual thickness provided according to Bligh's theory

$$= 2 \text{ m}$$

Hence safe.

The section of weir is shown in Fig. Ex. (13.5)

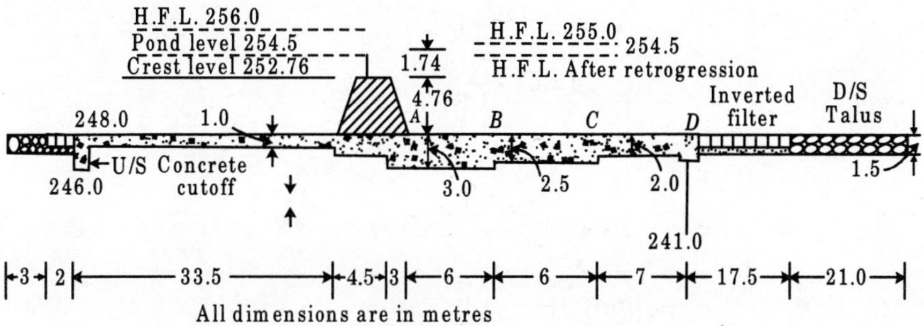

Fig. Ex. 13.5

Illustrative Example 13.6 *Design a glacis weir for the following data obtained at the site.*

(i) *Maximum discharge intensity on weir crest = 12.5 cumec/m*

(ii) *H.F.L. before construction of weir = 225.00 m*

(iii) *River bed level = 218.75 m*

(iv) *Pond level = 224.00 m*

(v) *Height of crest shutters = 1.5 m*

(vi) *Anticipated downstream water level in the river when the weir is discharging with pond level upstream = 221.50 m*

(vii) *Bed retrogression = 0.5 m*

(viii) *Lacey's silt factor = 0.9*

(ix) *Permissible exit gradient = 1/7*

(x) *Permissible afflux = 1 m*

Solution

Step 1 Pond level = 224.00 m

Height of shutters = 1.50 m

$\therefore$ Crest level = 224.00 – 1.50 = 222.50 m

Step 2

$$q = 1.70\, K^{3/2}$$

$$K = \left(\frac{q}{1.70}\right)^{2/3}$$

$$= \left(\frac{12.5}{1.70}\right)^{2/3} = 3.78 \text{ m}$$

∴ Level of *u/s* T.E.L. = Crest level $+K$

$= 222.50 + 3.78 = 226.28$ m

Step 3 Regime scour depth

$$R = 1.35\left(\frac{q^2}{f}\right)^{1/3}$$

$$= 1.35\left[\frac{(12.5)^2}{0.9}\right]^{1/3} = 7.5 \text{ m}$$

∴ Regime velocity $V = \dfrac{q}{R} = \dfrac{12.5}{7.5} = 1.67$ m/sec

Velocity head $\dfrac{V^2}{2g} = \dfrac{(1.67)^2}{2\times 9.81} = 0.14$ m

Step 4 u/s H.F.L. $=$ *u/s* T.E.L. $- \dfrac{V^2}{2g}$

$= 226.28 - 0.14$

$= 226.14$ m

d/s T.E.L. $=$ H.F.L. before construction $+ \dfrac{V^2}{2g}$

$= 225.00 + 0.14$

$= 225.14$ m

∴ Afflux $=$ *u/s* H.F.L. $-$ *d/s* H.F.L.

$= 226.14 - 225.00$

$= 1.14$ m

which is very near to the permissible afflux of 1 m.

The *d/s* H.F.L. after retrogression

$=$ H.F.L. before construction $- 0.50$

$= 225.00 - 0.50$

$= 224.50$ m

d/s T.E.L. after retrogression

$= 225.14 - 0.50$

$= 224.64$ m

∴ Loss of head at high flood is

$H_L =$ *u/s* T.E.L. $-$ *d/s* T.E.L. after retrogression

$= 226.28 - 224.64$

$= 1.64$ m

Step 5 When the weir is discharging with pond level *u/s,* head over the crest

$= 224.00 - 222.50$

$= 1.50$ m

∴ Discharge $q = 1.7\,(K)^{3/2}$

$= 1.7\,(1.5)^{3/2} = 3.12$ cumec/m

Loss of head at this time will be given by

H_L = Pond level – Anticipated d/s water level at this time

$= 224.00 - 221.50$

$= 2.50$ m

Step 6 Hydraulic jump calculations

Hydraulic jump will be developed on the d/s sloping glacis. The hydraulic jump calculations are done in the tabular form as shown in the Table A.

TABLE (A) Hydraulic jump calculations

S.No.	*Items*	*High flood condition*	*Pond level condition*
1.	Discharge intensity q	12.5 cumec/m	3.12 cumec/m
2.	Head loss H_L	1.64 m	2.50 m
3.	Downstream specific energy (E_{f_2}) from Blench curves (Fig. 13.20 Plate III)	4.85 m	2.45 m
4.	Level at which jump will form = (d/s T.E.L. after retrogression $-E_{f_2}$)	224.64 – 4.85 = 219.79	221.50 – 2.45 = 219.05
5.	Upstream specific energy $E_{f_2} = E_{f_2} + H_L$	4.85 + 1.64 = 6.49 m	2.45 + 2.50 = 4.95 m
6.	Prejump depth D_1 corresponding to E_{f_2} from specific energy curves (Fig. 13.21 Plate IV) or calculated from $E_{f_2} = D_1 + \frac{(q/D_1)^2}{2g}$	1.23 m	0.33 m
7.	Post jump depth D_2 corresponding to E_{f_2} from specific energy curves or calculated from $E_{f_2} = D_2 + \frac{(q/D_2)^2}{2g}$	4.45 m	2.36
8.	Height of jump ($D_2 - D_1$)	4.45 – 1.23 = 3.22 m	2.36 – 0.33 = 2.03 m
9.	Length of concrete floor required beyond the jump = 5 ($D_2 - D_1$)	16.10 m Adopt 16.5 m	10.15 m

Step 7 Design of glacis

Let the crest width = 2 m

Upstream slope 2 : 1

Downstream glacis slope 3 : 1

Adopt downstream floor level = 218.85 m, (i.e., about 0.2 m below the lower level at which jump will form).

Horizontal length of floor beyond the toe of the glacis

$$= 16.5 \text{ m}$$

Step 8 Depth of sheet piles

$$R = 7.5 \text{ m}$$

Level of bottom of the possible scour hole at *u/s* end

$$= u/s \text{ H.F.L.} - 1.5\,R$$

$$= 226.14 - 1.5\,(7.5)$$

$$= 214.85 \text{ m}$$

The sheet pile at the *u/s* end may be taken upto a level of 214.85 m.

Level of bottom of the possible scour at *d/s* end

$$= d/s \text{ H.F.L.} - 2R$$

$$= 224.50 - 2(7.5)$$

$$= 209.50 \text{ m}$$

The sheet pile at the *d/s* end may be taken upto a level of 209.50 m.

Provide an intermediate pile at *d/s* toe of the glacis as the second line of defence with its bottom at the same level of 209.50 m.

Step 9 Length of impervious floor

Maximum percolation head

$$= \text{Pond level} - d/s \text{ floor level}$$

$$= 224.00 - 218.85$$

$$= 5.15 \text{ m}$$

$$\text{Depth of } d/s \text{ pile} = 218.85 - 209.50$$

$$= 9.35 \text{ m}$$

Permissible exit gradient

$$G_E = \frac{1}{7}$$

$$G_E = \frac{H}{d}\frac{1}{\pi\sqrt{\lambda}}$$

$$\frac{1}{\pi\sqrt{\lambda}} = \frac{d}{H} \times G_E$$

$$= \frac{9.35}{5.15} \times \frac{1}{7} = 0.259$$

Hence from Khosla's exit gradient curves for

$$\frac{1}{\pi\sqrt{\lambda}} = 0.259$$

$$\alpha = \frac{b}{d} = 1.8$$

∴ Required length of impervious floor

$$b = \alpha d = 1.8 \times 9.35$$
$$= 16.83 \text{ m}$$

Floor length already provided from other considerations

= (horizontal length of *u/s* glacis) + (length of crest) + (horizontal length of *d/s* glacis) + (length of floor beyond the toe of *d/s* glacis)

= {2(222.50 – 218.75) + (2) + 3 (222.50 – 218.85) + (16.5)}

= (7.5 + 2 + 10.95 + 16.5)

= 36.95 m

Thus, the floor length already provided is much more than that is required from the consideration of permissible exit gradient and hence on the upstream side no impervious floor is required. However, a floor of length 1.05 m may be provided on the upstream side so that the total length of impervious floor becomes 38 m.

Step 10 Having known the profile of the glacis and the floor, the pressures at key points are determined by assuming the preliminary thickness of the floor at various points as under.

Floor thickness at *u/s* of crest

= 1.0 m

Floor thickness at *d/s* end of glacis (i.e., at intermediate pile)

= 2.0 m

Floor thickness at *d/s* end of floor

= 1.5 m

The claculations of pressures at various key points have been done in Table (B) with reference to Fig. Ex. 6 (a).

Step 11 Elevation of subsoil hydraulic gradient line (H.G.L.) for different conditions.

Having known the percentage pressures at the various key points, the elevation of subsoil H.G. line for (i) maximum static head, (ii) high flood condition, and (iii) flow at pond level condition can be determined as indicated in Table (C).

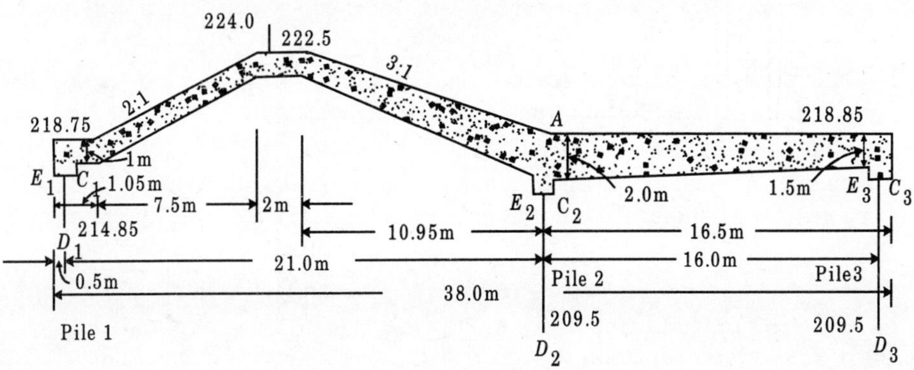

(a)

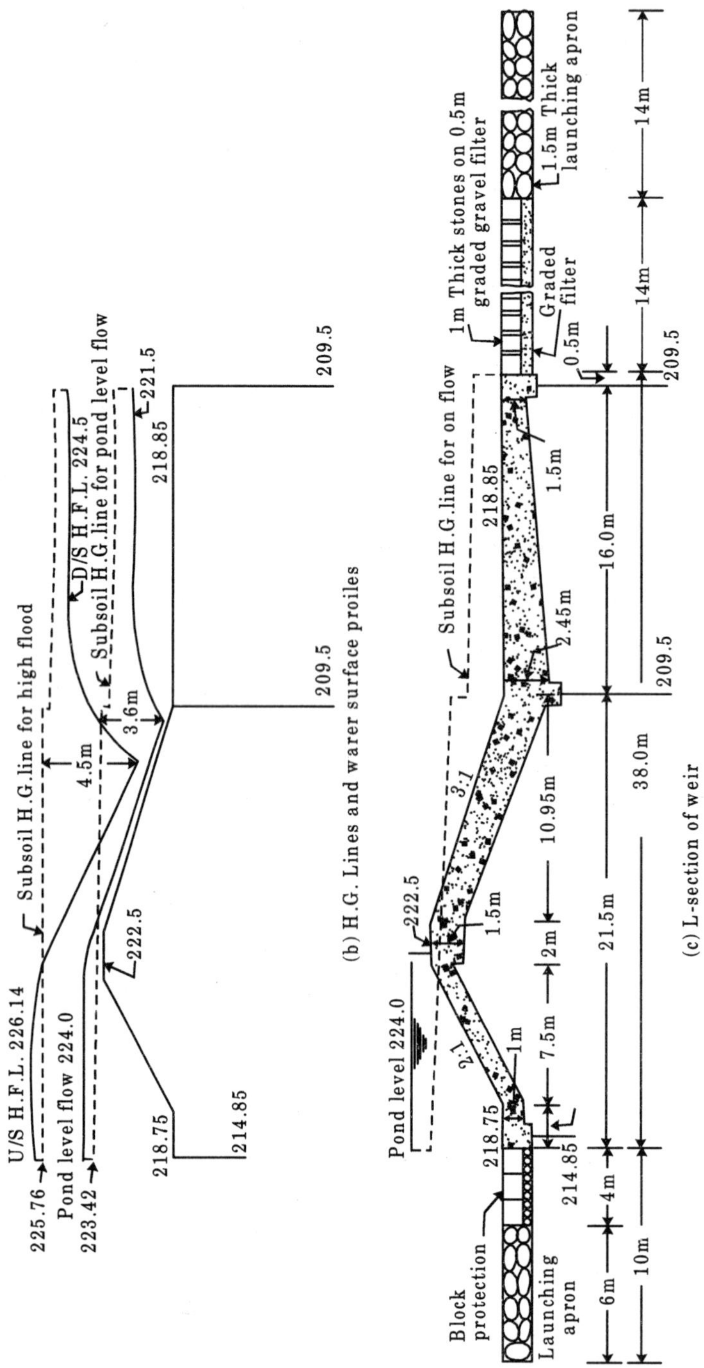

(b) H.G. Lines and warer surface proiles

(c) L-section of weir

Fig. Ex. 13.6

TABLE (B) Calculation for subsoil uplift pressures (b = 38 m; b_s = 10.95 m; b' = 21 m)

Pile line and the Pressure Point	*Depth of pile d (m)*	*1/α = b/d*	*b_1 (m)*	*b_1/d*	*α = b/d*	*Percentage of pressure φ* Before correction	*Correction for thickness*	*Correction for slope*	*Correction for interference*	*Corrected values*
Upstream Pile	3.9	0.103	...	...	...	...	...	...	...	...
Point D_1	...	...	...	...	...	80.0	...	...	...	80.0
Point C_1	...	...	...	...	...	71.0	+2.3	...	+3.5	76.8
Intermediate Pile	9.35	...	21.5	0.566	4.06	...	...	...	...	...
Point E_2	...	...	...	...	...	61.0	–3.0	+2.3	–1.4	58.9
Point D_2	...	...	...	...	...	47.0	...	...	...	47.0
Point C_2	...	...	...	...	...	31.5	+3.3	...	+5.0	39.8
Downstream Pile	9.35	0.246	...	...	...	...	...	...	...	...
Point E_3	...	...	...	...	...	43.0	–2.2	...	–5.5	35.3
Point D_3	...	...	...	...	...	29.0	...	...	...	29.0

Step 12 Plotting the prejump water surface profile for the two flow conditions.

The uplift pressure at any point will be equal to the difference in elevation between the subsoil H.G. line and the elevation of water surface corresponding to the jump profiles for the two flow conditions (or the *d/s* floor for no flow condition). Hence it is essential to plot the water surface profile of the jump.

The water surface profile on the *d/s* glacis before the formation of the jump, can be obtained by finding depth D_1 at various points with the help of curves of Fig. 13.21 (Plate IV). The calculations are shown in Table (D).

Step 13 Calculations for floor thickness

Figures Ec. (13.6) (b) and (c) show the subsoil H.G. lines for the three conditions. The floor is to be designed for the maximum unbalanced head found by subtracting elevation of water surface (or *d/s* bed level as the case may be) from the elevation of the corresponding H.G. Line.

(i) At the point of formation of hydraulic jump for high flood condition.

From Fig. Ex. 13.6 (b) unbalanced head for high flood condition

$$= 4.5$$

$$\frac{2}{3} \text{ of this head } = \frac{2}{3} \times 4.5 = 3.0 \text{ m}$$

Unbalanced static head for no flow condition

$$= 2.21 \text{ m}$$

Since 2/3rd of the unbalanced head for high flood condition is greater than the unbalanced static head for no flow condition, the former will be used for the design. Thus,

$$\text{Design head } = 3.0 \text{ m}$$

TABLE (C) Elevation of H.G. lines for various conditions

Conditions	Datum (d/s) floor level or d/s water level) (m)	Upstream water level (m)		*Height/Elevation of subsoil H.G. line above datum*								
					u/s pile line		*Intermediate pile line*			*d/s pile line*		
				ϕ_{E_1} = 100%	ϕ_{D_1} = 80%	ϕ_{C_1} = 76.8%	ϕ_{E_2} = 58.9%	ϕ_{D_2} = 47%	ϕ_{C_2} = 39.8%	ϕ_{E_3} = 35.3%	ϕ_{D_3} = 29%	ϕ_{C_3} = 0%
No flow condition Head = 224.00 – 218.85 = 5.15 M	218.85	224.00	Height	5.15	4.12	3.96	3.03	2.42	2.05	1.82	1.49	0
			Elevation	224.00	222.97	222.81	221.88	221.27	220.90	220.67	220.34	218.85
High flood condition Head = 226.14 – 224.50 = 1.64 m	224.50	226.14	Height	1.64	1.31	1.26	0.97	0.77	0.65	0.58	0.48	0
			Elevation	226.14	225.81	225.76	225.47	225.27	225.15	225.08	224.98	224.50
Flow at pond level Head = 224.00 – 221.50 = 2.50 m	221.50	224.00	Height	2.50	2.00	1.92	1.47	1.18	1.00	0.88	0.73	0
			Elevation	224.00	223.50	223.42	222.97	222.68	222.50	222.38	222.23	221.50

∴ Required thickness of floor

$$= \frac{3.0}{2.24 - 1} = 2.42 \text{ m}$$

Table (D) Prejump profile (Crest Level 222.50 m)

Distance from d/s end of crest (m)	*Level of glacis (m)*	*High flood condition q = 12.5 cumec/m u/s T.E.L. = 226.28 m*		*Pond level flow q = 3.12 cumec/m u/s H.F.L. = 224.00 m*	
		E_{f_1} = 226.28 – glacis level	*D_1*	*224.00 – glacis level*	*D_1*
3.00	221.50	4.78	1.57	2.50	0.50
6.00	220.50	5.78	1.35	3.50	0.40
8.13	219.79 (Jump location for high flood)	6.49	1.32	—	—
9.00	219.50			4.50	0.35
10.35	219.05 (Jump location for pond level flow)			4.95	0.33

(*ii*) At the point of formation of hydraulic jump for flow at pond level condition.

Unbalanced head for flow at pond level condition

$$= 3.6 \text{ m}$$

$$\frac{2}{3} \text{ of this head} = \frac{2}{3} \times 3.6 = 2.4 \text{ m}$$

Unbalanced static head for no flow condition

$$= 2.86 \text{ m}$$

∴ Design head = 2.86 m

∴ Required thickness of floor

$$= \frac{2.86}{2.24 - 1} = 2.31 \text{ m}$$

Hence provide a thickness of 2.45 m at the *d/s* end of the glacis decreasing it to 1.5 m at the end of the crest. A nominal thickness of 1 m is to be provided for the floor on the upstream of the glacis as shown in Fig. Ex. (13.6 c).

It may be seen from Fig. Ex. (13.6 b) that the design head for the floor on the *d/s* of the glacis will be governed by the unbalanced static head which is greater than 2/3 of the unbalanced head under the two flow conditions.

Pressure head at E_3 at the *d/s* end of the floor

$$= 1.82 \text{ m}$$

∴ Required thickness of floor

$$= \frac{1.82}{2.24 - 1} = 1.47 \text{ m}$$

Hence provide a thickness of 1.5 m at the *d/s* end of the floor.

Thus, the *d/s* floor has a thickness of 2.45 m at the *d/s* of glacis and a thickness of 1.5 m at the *d/s* end. This reduction in thickness may be done linearly as shown in Fig. Ex. (13.6 c).

Step 14 Upstream protection works

At the *u/s* end the scour depth below the river bed is

$$D = 218.75 - 214.85$$
$$= 3.9 \text{ m}$$

Thus, minimum length of block protection at the *u/s* end of the impervious floor

$$= D = 3.9 \text{ m}$$

Hence provide 4 m long block protection consisting of 1 m thick stone or concrete blocks over 0.5 m thick loosely packed stone.

Minimum length of *u/s* launching apron

$$= 1.5\,D = 1.5 \times 3.9$$
$$= 5.85 \text{ m}$$

Hence provide launching apron of length 6 m and thickness 1.5 m immediately at the *u/s* end of the block protection.

Step 15 Downstream protection works

At the downstream end the scour depth below the floor is

$$D = 218.85 - 209.50$$
$$= 9.35 \text{ m}$$

Thus, minimum length of inverted filter at the *d/s* end of the impervious floor

$$= 1.5\,D = 1.5 \times 9.35$$
$$= 14.0 \text{ m}$$

Hence provide 14 m long inverted filter consisting of 1 m thick stone blocks laid on a 0.5 m thick graded filter.

Minimum length of *d/s* launching apron

$$= 1.5\,D = 1.5 \times 9.35$$
$$= 14.0 \text{ m}$$

Hence provide a launching apron of length 14 m and thickness 1.5 m immediately at the *d/s* end of the inverted filter.

The longitudinal section of the weir is shown in Fig. Ex. 13.6 (c).

REVIEW QUESTIONS

13.1 What is a canal headworks? Describe briefly the functions of a canal headworks.

13.2 Sketch the layout of a typical diversion headworks and describe briefly the functions of the various components of diversion headworks.

13.3 Discuss the effects of construction of a weir on the river regime.

13.4 Describe briefly with neat sketches the various types of weirs.

13.5 What is a barrage? How does a barrage differ from a weir?

13.6 Discuss briefly the causes of failure of weirs on permeable foundation and their remedies.

13.7 Discuss Bligh's Creep theory for the design of weir constructed over pervious foundation.

13.8 Write a brief note on Khosla's theory of design of weirs on permeable foundation.

13.9 A weir on a permeable foundation has a level floor of negligible thickness and is 10 m long in the direction of flow. At the two ends of the floor 2.5 m deep piles are provided. Using Khosla's theory calculate the uplift pressure at the mid-length of the floor and at quarter points. The effective head of water can be assumed to be 2 m.

[*Ans.* 1.075 m; 1.000 m; 0.9295 m]

13.10 Outline the steps involved in determining the floor thickness of a weir placed over permeable foundation.

13.11 Explain Khosla's method of independent variables.

13.12 What is the importance of exit gradient? How would you check the exit gradient?

13.13 Discuss briefly the design considerations of a weir on permeable foundation.

13.14 'Khosla's method of design of weirs on permeable foundations is an improvement on Bligh's Creep theory', discuss.

13.15 A weir on a permeable foundation has a level floor of negligible thickness and is 20 m long in the direction of flow. At the u/s end of the floor a 4 m deep pile is provided and at its d/s end a 5 m deep pile is provided. Using Khosla's theory calculate the uplift pressures at the key points *D* and *C* for the upstream pile and *E* and *D* for the downstream pile. Also calculate the exit gradient. The effective head of water may be assumed as 4 m.

[*Ans.* u/s pile : P_D = 2.938 m; P_C = 2.447 m; d/s pile :
P_E = 1.718 m; P_D = 1.165 m; G_E = 1/6.29]

13.16 What is a silt excluder? Describe with neat sketches its working and design considerations.

13.17 Sketch and describe the working of a silt ejector. On what basis is it designed?

13.18 Write short notes on : (i) Launching upron; (ii) Fish ladder (iii) Divide wall; (iv) Under sluices.

13.19 Design a vertical drop weir on the basis of Bligh's theory for the following data:

(i) Maximum flood discharge = 2000 cumec.

(ii) H.F.L. before construction of weir = 250.0 m

(iii) River bed level = 245.0 m

(iv) F.S.L. of canal = 249.0 m

(v) Allowable afflux = 1 m

(vi) Coefficient of creep = 11.

Assume any other data not given. Test the floor of the weir by Khosla's theory.

13.20 What are the considerations for fixing afflux, pond level, waterway and crest levels for a weir on permeable foundation.

13.21 What is the utility of launching aprons? How are these designed?

13.22 Find the crest and bottom widths required for a vertical drop weir with the following data :

Height of weir = m

Pond level = 1 m above crest

u/s H.F.L. = 2 m above crest

d/s H.F.L. = 0.5 below crest

[*Ans.* crest width 2 m; bottom width 4 m if the u/s face is vertical, and 5 m if both faces have same slope]

13.23 A river carries a high flood discharge of 1000 cumec, the average bed level is 200.00 m. A canal carrying 100 cumec is to take off from the head works at 90°. Pond level is 202.00 m and H.F.L. before construction of weir is 203.00 m. If the permissible afflux is 1.0 m propose suitable values for

(a) Length of waterway for the weir and undersluices portion.

(b) Crest levels of weir and undersluices

(c) Depth of upstream and downstream sheet piles for the weir from scour considerations.

Make suitable assumptions whenever necessary.

[*Ans.* (a) 150 m, 83 m; (b) 201.624 m, 200.00 m; (c) 2.75 m, 6.5 m]

13.24 A river carries a high flood discharge of 16000 cumec with its average bed level as 200.0 m. A canal carrying 300 cumec is to take off from the headworks. The FSL of the canal at its head is 203.0 m. Determine the pond level and design the canal head regulator and a silt excluder which is to be provided at the head. Make suitable assumptions wherever necessary.

13.25 A channel carrying 100 cumec has a full supply level of 200.0 m at the head. What pond level is desirable at the headworks? Also if the undersluice floor level is at R.L. 198.5 m, design the crest level and the length to be provided for the canal head regulator. A silt excluder is to be provided at the headworks, determine its tentative dimensions. Make suitable assumptions wherever necessary.

CHAPTER

14 Canal Irrigation (General)

14.1 INTRODUCTION

A canal is an artificial channel constructed to carry water from a river or from a tank or reservoir for various purposes such as irrigation, power generation, navigation, etc. The terms canal and channel are, however, synonymous and hence they mean one and the same. In general the canals have trapezoidal cross-section. The canals are usually designated according to the purpose for which they are constructed as irrigation canals, power canals, navigation canals, etc.

An *irrigation canal* is the one which carries water from the source to the agricultural fields for irrigation. There are several irrigation canals of large discharge capacities in our country such as Ganga canal, Western Yamuna canal, Bhakra canal, Indira Gandhi canal (formerly known as Rajasthan canal), Sirhind canal, etc.

A *power canal* is the one which supplies water to a power house for the generation of hydro-electric power. The power house is usually located on the canal itself at a section where a fall is available on the canal. It is also known as *hydel channel* or *hydel canal.* Generally power generation and irrigation are combined on the same canal so that after the generation of power water is used for irrigation. One of the important examples of power canal in our country is 'Nangal Hydel Canal' which takes off from Nangal barrage on the downstream of Bhakra Dam and on this canal two power houses are located at Ganguwal and Kotla. After the generation of power water is used for irrigation.

A *navigation canal* is a canal which is constructed to provide navigation facilities. The velocity of flowing water in a navigation canal should be small for easy movement of small ships and barges in the direction of flow of water as well as in the opposite direction. Since the requirements of navigation canals are altogether different from those of irrigation canals it is usually not possible to combine navigation canals with irrigation canals. In our country the development of navigation canals is almost negligible.

14.2 CLASSIFICATION OF IRRIGATION CANALS

The irrigation canals can be classified in different ways on the basis of various considerations as follows.

1. Classification based on the Nature of Source of Supply

On the basis of nature of source of supply the irrigation canals may be classified as :

(i) Permanent canals

(ii) Inundation canals.

(i) *Permanent canals.* A permanent canal is the one which is fed by a permanent source of supply. It is a well graded channel and is provided with permanent regulation and distribution works. The permanent canals may be further classified as :

(a) Perennial canals

(b) Non-perennial canals.

Perennial canals are those canals which get continuous supplies from the source throughout the year. The source of supply for these canal is a perennial river. These canals can supply water for irrigation throughout the year during the crop seasons.

Non-perennial canals are those canals which get their supplies only for a part of the year. The source of supply for these canals is evidently a non-perennial river. These canals can therefore supply water for irrigation only during some of the crop seasons.

(ii) *Inundation canals.* An inundation canal is the one which gets its supplies only when the water level in the river, from which it takes off, rises during floods. These canals are not provided with any headworks for diversion of river water to the canal but obtain their supplies through open cuts in the bank of the river, which are called 'heads'. A regulator is, however, provided on the canal 5 to 6 km downstream from the offtake point to control the discharge in the canal. The inundation canals are therefore non-perennial canals and the flow in these canals depend on the periodical rise of water level in the river. As indicated in Chapter 1 the irrigation carried out by inundation canals is known as inundation irrigation.

2. Classification based on the Function of the Canal

Based on the functions performed by the canals these may be classified as

(i) Feeder canals (or Feeders)

(ii) Carrier canals.

(i) *Feeder canals.* A feeder canal is the one which is constructed only to feed another canal. No direct irrigation is carried out from a feeder canal. It is also known as *feeder*. One of the important examples of feeder canal is Indira Gandhi feeder canal (formerly known as Rajasthan feeder canal) which is 216 km long. It takes off from Harike barrage across River Sutlej in Punjab and feed Indira Gandhi canal (formerly known as Rajasthan canal).

(ii) *Carrier canal.* A carrier canal is the one which besides being used for direct irrigation carries water to feed another canal. Thus, it acts as irrigation canal as well as feeder canal. An example of such a canal is Upper Chenab canal in West Punjab (Pakistan).

3. Classification based on Discharge and Relative Importance in a given Network of Canals

The irrigation canals may be classified on the basis of their discharge capacity and their relative importance in the network of canals in the following categories

(i) Main canal
(ii) Branch canal
(iii) Major distributary
(iv) Minor distributary (or Minor)
(v) Water course or Field channel.

(i) *Main canal.* It is the principal canal of a network of irrigation canals. It takes off directly from a river or a reservoir or from the tail end of a feeder canal. It is usually a large capacity canal which supplies water to branch canals and major distributaries. The main canal is generally not used for direct irrigation.

(ii) *Branch canal.* Branch canals (usually called branches) are the irrigation canals which take off from the main canal on either side. The branch canals generally carry a discharge of over 5 cumec. From large branch canals usually little direct irrigation is done but from smaller ones considerable direct irrigation is done. However, the main function of branch canals is to supply water to major and minor distributaries.

(iii) *Major distributary.* Major distributaries are the irrigation canals which take off from the branch canals and sometimes from the main canal. They carry a discharge varying from 0.25 to 5 cumec. The major distributaries are generally used for direct irrigation and hence they supply water through outlets to water courses. They also supply water to the minor distributaries taking off from them. They are also called *Rajbaha.*

(iv) *Minor distributary.* Minor distributaries are the irrigation canals which take off from major distributaries and branch canals. They carry discharge less than 0.25 cumec. The minor distributaries supply water through outlets to water courses for irrigation. They are also called *minors.*

(v) *Water courses or Field channels.* Water courses are small channels which carry water from the outlets of a major or minor distributary or a branch canal to the fields to be irrigated. These are owned, constructed, controlled and maintained by the cultivators. On the other hand the main canal, branch canals, and major and minor distributaries are constructed and maintained by the government. However, under the Command Area Development (C.A.D.) programme nowadays the water courses are also being constructed by the government but they are maintained and regulated by the cultivators.

4. Classification based on Canal Alignment

According to alignment the irrigation canals may be classified as

(i) Ridge canal or Watershed canal
(ii) Contour canal
(iii) Side slope canal.

(i) *Ridge canal or Watershed canal.* A ridge or watershed canal is the one which is aligned along the ridge or the natural watershed line (Fig. 14.1). When the canal runs on a watershed it can irrigate areas on both sides and hence a large area can be brought under cultivation i.e., the canal has a higher command area. Further no drainage can intersect a ridge line or watershed, as all the drainage flows away from the ridge line, a ridge canal does not cross drainage line and hence for this canal cross drainage works are not required to be provided. Thus, ridge canals are quite economical.

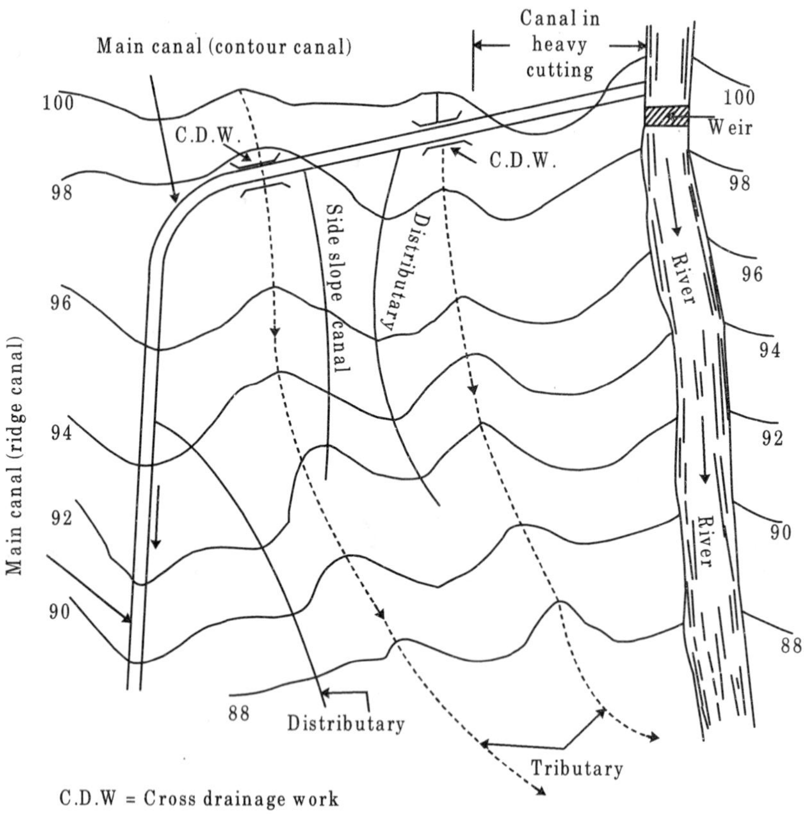

Fig. 14.1 Alignment of canals

(ii) *Contour canal.* A contour canal is the one which is aligned nearly parallel to the contours of the country (Fig. 14.1). It can irrigate areas only one one side. As the ground level on the other side is higher it is not necessary to construct a bank on that side. Such a canal with only one bank is known as *single bank canal.* However, when both the banks are provided it is known as *double bank canal.* A contour canal has to cross drainage and hence for this canal cross-drainage works are required to be provided.

(iii) *Side slope canal.* A side slope canal is the one which is aligned at right angles to the contours of the country (Fig. 14.1). Thus, it is neither on the watershed nor in the valley but it is somewhere in between the two along the slope. As this canal is nearly parallel to the natural drainage of the country it does not intercept drainage and hence cross-drainage works are not required. It can also irrigate areas only on one side as in the case of a contour canal. Further a side slope canal has very steep bed slope, because the direction of the steepest slope of the ground is at right angles to the contours of the country.

5. Classification based on Financial Output

According to financial output the irrigation canals may be classified as

(i) Productive canals

(ii) Protective canals.

(i) *Productive canals.* A productive canal is the one which when fully developed yields enough revenue to cover up its running cost and a net saving at the rate of more than six per cent of the capital invested initially for its construction. Thus, the capital investment for this canal is repaid in a period of about 16 years.

(ii) *Protective canals.* A protective canal is the one which is constructed as a relief work during famine to provide employment to the inhabitants of the famine affected area and to protect that area against famine in future. The revenue from this canal usually does not cover up its running charges and also does not repay its initial capital expenditure.

6. Classification based on the Soil through which Constructed

Based on the soil through which irrigation canals are constructed the same may be classified as follows.

(i) Alluvial canals

(ii) Non-alluvial canals.

(i) *Alluvial canals.* Alluvial canals are those canals which are constructed through alluvial soils. These soils are readily scoured and deposited. Such soils are available in Indo-Gangetic plains in Northern India.

(ii) *Non-alluvial canals.* Non-alluvial canals are those canals which are constructed through hard soils or disintegrated rocks. Such soils are usually found in Central and Southern India.

7. Classification based on the Lining being provided or not

The irrigation canals may also be classified on the basis of lining being provided or not as follows.

(i) Unlined canals

(ii) Lined canals.

(i) *Unlined canals.* An unlined canal is the one which has its bed and banks made up of natural soil through which it is constructed and it is not provided with a lining of impervious material. The velocity of flow in these canals is kept low so that the bed and banks of the canal may not be scoured.

(ii) *Lined canals.* A lined canal is the one which is provided with a lining of impervious material on its bed and banks to prevent the seepage of water. As compared to unlined canals higher velocities of flow can be allowed in the lined canals and hence their cross-sectional areas can be reduced.

14.3 CANAL ALIGNEMENT

For the alignment of an irrigation canal following considerations are made.

1. An irrigation canal should be aligned in such a way that maximum area is irrigated with the least length of channel and its cost including the cost of cross-drainage works is minimum. A shorter length of the canal has less loss of head due to friction and smaller loss of water due to seepage and evaporation so that additional area can be brought under cultivation. As indicated in the previous section a canal aligned along a ridge or watershed can irrigate areas on both sides and hence a large area can be brought under cultivation. Moreover for such a canal no cross-drainage works would be required. As such as far as possible the canal should be aligned along the ridge or watershed.

2. The main canal takes off from the river at some point from where it must be taken to mount the watershed or the ridge in the area. The portion of the canal between the point where it takes off from the river to the point where it mounts the watershed is usually in heavy cutting and also in this portion a number of cross-drainage works will be required to be constructed (Fig. 14.1) which will make this portion of the canal very expensive. As such the alignment of the canal should be such that from the point where it takes off from the river it mounts the watershed in the shortest possible distance. For this if necessary the location of the diversion headworks may also be suitably adjusted.

3. Once a canal is aligned along a watershed it will in general be kept on the watershed. However, if the watershed takes a sharp loop between some points as shown in Fig. 14.2 then considerable length of canal can be saved if it is aligned straight instead of following the watershed. This may necessitate some cross-drainage works. Also in this reach the canal will be able to irrigate areas only on one side of the canal. The watershed may also have to be abandoned in order to byepass towns or villages located on the watershed.

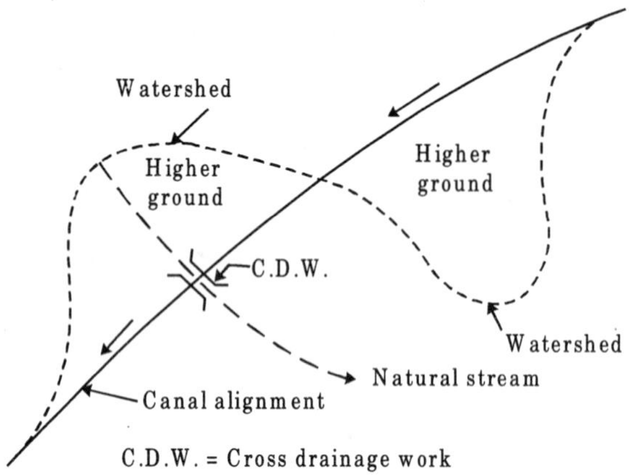

Fig. 14.2 *Alignment of canal for sharp turns in watershed*

4. In hilly areas it may not be possible to align the canal along the watershed or ridge. This is so because in the hills the river flows in the valley below, while the watershed or the ridge may be hundreds of metres above it. It would not be economically feasible to take the canal on top of such a ridge. The canal will thus be aligned as a contour canal.

5. As far as possible curves should be avoided in the alignment of canals because the curves lead to disturbance of flow and a tendency to silt on the inside and to scour on the outside of the curves. However, if curves are to be provided they should be as gentle as possible. According to IS: 5968–1970 the radii of curves should be usually 3 to 7 times the water surface width subject to the minimum as given below.

(A) Unlined Canals

Capacity of canal (*cumec*)	*Minimum radius* (*m*)
80 and above	1500
Less than 80 to 30	1000
Less than 30 to 15	600
Less than 15 to 3	300

Capacity of canal (cumec)	Minimum radius (m)
Less than 3 to 0.3	150
Less than 0.3	90

(B) Lined canals

Capacity of canal (*cumec*)	*Minimum radius* (*m*)
280 and above	900
Less than 280 to 200	750
Less than 200 to 140	600
Less than 140 to 70	450
Less than 70 to 40	300
Less than 40 to 10	200
Less than 10 to 3	150
Less than 3 to 0.3	100
Less than 0.3	50

6. The alignment of the canal should be such that idle length of the canal (i.e., the portion of the canal from which no irrigation is carried out) is minimum.

7. The alignment should be such that as far as possible a balanced depth of cutting and filling is achieved. Moreover it should not involve heavy filling and deep cutting.

8. The alignment should not be made in a rocky, brakish or cracked strata.

9. The alignment should be such that the canal crosses the natural stream where the stream is straight with minimum waterway.

Keeping in view the various considerations to be made for the canal alignment as indicated above, the trial alignments of the different canals viz., main canals, branches and distributaries of a canal system for irrigation are marked on a contoured map of the command of the system. These maps may be obtained from Survey of India and are usually plotted to a smaller scale of about 1 in 15000. The trial alignment of each individual canal of the system is marked on a *shajra* sheet which is a large scale map. This trial alignment of a canal as marked on a *shajra* sheet is then laid out in the field where if necessary it is modified. In the field the alignment of a canal is marked by small masonry pillars at a spacing of about 200 m, with a centre line marked on the top of each pillar to show the exact alignment. In between the pillar the line is marked on the ground by a *dagbel* i.e., a small trench a few centimetres deep and a few centimetres wide excavated in the ground to mark the alignment.

14.4 INUNDATION CANALS

Inundation canals are not provided with headworks but only open cuts called heads are provided in the river bank at the points where these canals take off from the river. The bed level of the cut is same as the bed level of the canal but it is kept much above the bed level of the river (Fig. 14.3). Inundation canals are used for carrying out inundation irrigation. The situations favourable for inundation irrigation are as follows.

(i) When the flood in the river is of such a duration that the water level in the canal remains high and fairly constant for quite a long time.

(ii) When there is an early flood in the river so that kharif crops can be irrigated.

(iii) When there is a late last flood in the river at the end of the rainy season so that rabi crops can be irrigated.

Selection of site for the Offtake point of Inundation Canal

The following factors govern the selection of site for the offtake point or head of inundation canal.

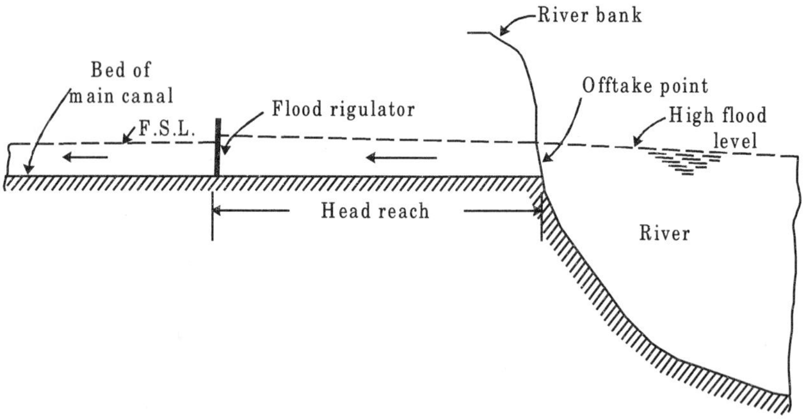

Fig. 14.3 *Inundation canal*

(i) The offtake point should be located as near the area to be irrigated as possible.

(ii) At the site of offtake point the river course should be straight, and the banks should be stable and high.

(iii) If the river is meandering the offtake point should be located on the concave curve or the outer bank to prevent deposition of silt at the head of the canal.

(iv) If shoals are present in the river bed, the site of the offtake point should be in front of the shoal as shown in Fig. 14.4. The shoal will create a pocket in front of the offtake point of the canal. The formation of the pocket will make still pond which will avoid the entry of silt in the canal.

(v) At the offtake point the width of the river and its velocity should be normal so that it is not affected by fluctuations in the water level.

(vi) If a river bye-pass is available, offtake point be located on the bye-pass, because it will cause minimum silting in the canal.

(vii) When one offtake point becomes unserviceable, there should be no difficulty in locating another offtake point.

Considerations in the design of Inundation Canal

The main trouble in an inundation canal is that of deposition of silt and fine sand and formation of bar at its mouth with a sudden fall in the river, which is caused due to lack of control

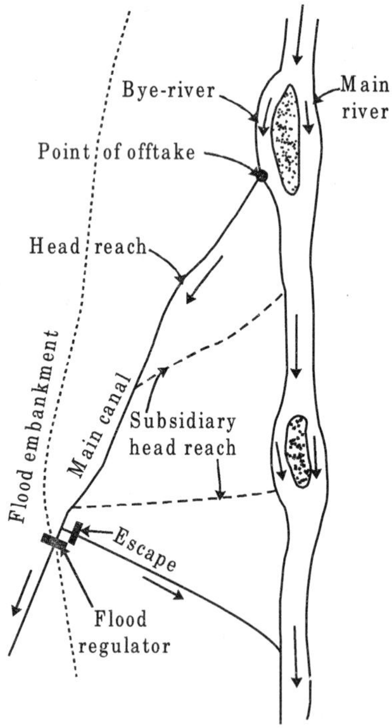

Fig. 14.4 *Head of Inundation main canal*

of water level at the head of the canal. Further during floods, water may find its way through the canal and submerge vast stretch of land on its both sides. Thus, in order to safeguard against the future troubles following considerations should be made in the design of inundation canals.

(i) A flood regulator (Fig. 14.5) is provided on the main canal 5 to 6 km downstream from the offtake point. This is so because if the regulator is provided at the offtake point, there is a danger of the same being washed away during floods. The main function of this regulator is to regulate the flow of water in the main canal and also to prevent heavy river floods from entering the main canal when river is in spate.

(ii) The flood regulator is provided with vertical lift gates in tiers (Fig. 14.5) so as to exclude heavy bed silt from entering the main canal.

(iii) A canal escape is provided just on the upstream side of the flood regulator (Fig. 14.4), through which surplus flood water, during the closure of flood regulator, will be disposed off into the river via., a lead channel.

(iv) Full supply level of main canal is fixed at a *fair-irrigating level,* which is the level at which water in the river is more or less steady for a certain maximum number of days (about 40 to 50 days) during inundation season.

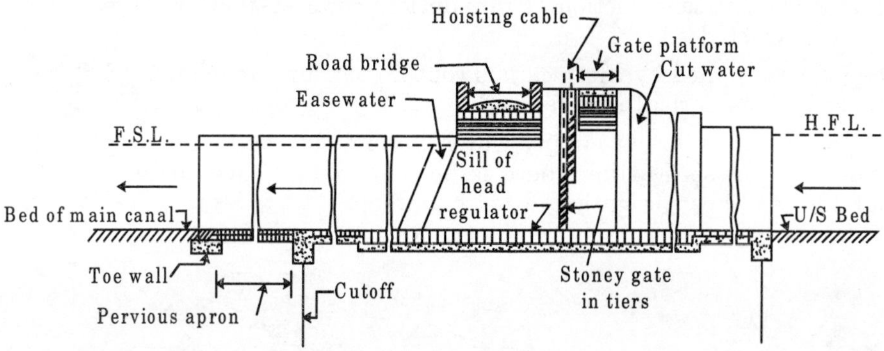

Fig. 14.5 *Flood regulator of inundation main canal*

(v) The canal bed level at the offtake point should be as low as practicable so that canals may get water for a longer period.

(vi) Full supply discharge of the canal is fixed on the basis of the requirements of rabi and kharif crops.

(vii) The cross-section of the canal is kept liberal because the time factor in inundation irrigation is very low and uncertain. Further the inundation main canal is made deeper and narrower (i.e., B/D ratio is small) than other types of irrigation canals.

(viii) The bed slope of the canal is fixed depending upon the slope of the country, keeping in view that excessive velocities causing scouring are not developed. In general the bed slope provided for the canal is in the range of 0.10 to 0.25 m per km.

Maintenance of Inundation Canals

As the silt control devices are not provided for the inundation canals there is an acute silting problem in the main as well as other canals of an inundation irrigation system. In the head reaches of the main canal usually heavy deposition of silt takes place which may vary from 0.3

to 1.75 m in one season. The deposition of silt in the canals reduces their discharge capacity. As such every year after the end of the inundation season, the clearance of silt deposit in irrigation channels is done, which forms the major maintenance work for these canals. Moreover, certain provisions are made during the construction of the inundation canals for their proper maintenance which are as follows.

(i) In addition to the main head or offtake point, subsidiary heads (Fig. 14.4) may be provided which will admit water in the canal, in case the main offtake point gets silted up.

(ii) A flood regulator along with a canal escape may be provided a few kilometres on the downstream of the offtake point. This will regulate the supply of water as well as control the entry of silt into the canal as explained earlier.

(iii) If there are a number of (say 3 or 4) main inundation canals taking off at different points from the same river, a feeder canal taking off from the same river may be constructed as shown in Fig. 14.6, which links up all the main canals and maintains supplies in these canals in case their head reaches get silted up. The provision of a feeder canal avoids the necessity of constructing a number of subsidiary heads and also the other canals receive their supplies under a higher head.

Advantages of Inundation Canals

The various advantages of inundation canals are as follows.

(i) The overall cost of construction of inundation canal system is low because no diversion headworks are constructed.

(ii) As the water supplied by these canals contain silt and fine sand, it helps to improve the fertility of the soil of the agricultural fields.

(iii) Since these canals do not supply water throughout the year, the problem of waterlogging usually does not arise. Further there is no possibility of over irrigation in this case and hence the bad effects of over irrigation are eliminated.

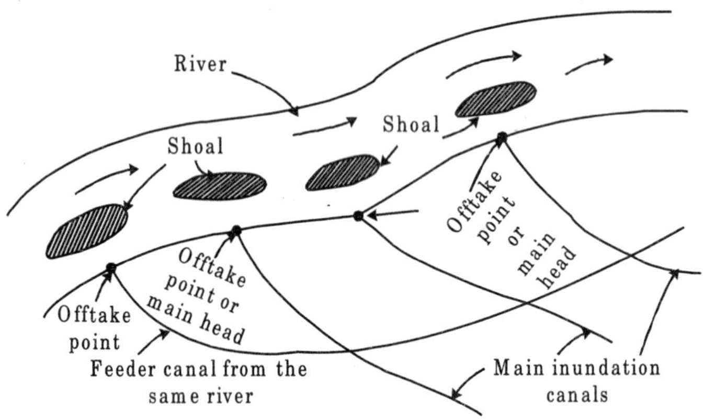

Fig. 14.6 *Feeder canal connecting number of inundation main canals*

Disadvantages of Inundation Canals

The various disadvantages of inundation canals are as follows.

(i) The water supply available from these canals is variable and unreliable.

(ii) As there are no headworks the offtake point or the head of the canal are liable to be damaged during floods.

(iii) Duty of water achieved for the inundation canals is quite low.

(iv) There is a serious silt problem in the inundation canals and also there is a recurring expense on silt-clearance of these canals.

(v) These canals will have to be of larger size to supply the full requirement of water within a limited time.

(vi) The supply in these canals being only during the floods, the supply is not regular which may cause dissatisfaction among the farmers.

(vii) Due to bad and tortuous alignment of these canals they do not function satisfactorily. Moreover, there are plenty of curves in the alignment of these canals where silting and scouring occurs.

14.5 BANDHARA IRRIGATION SCHEME

Bandhara irrigation scheme is essentially a diversion irrigation scheme on small streams which are mostly non-perennial. It is a minor irrigation scheme in which small streams are provided with bandharas. A *bandhara* is a low masonry weir of height 1.2 to 4.5 m, which is constructed

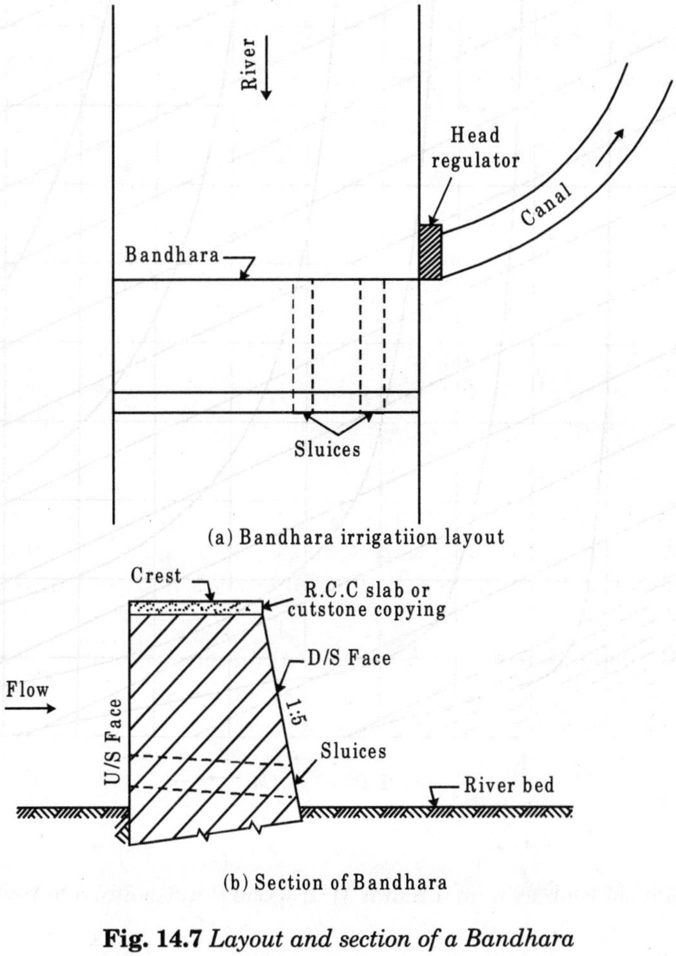

Fig. 14.7 *Layout and section of a Bandhara*

across a small stream for diverting the water of the stream into a small main canal taking off from the upstream side of the Bandhara (Fig. 14.7). The canal usually takes off only from one side and the flow in the canal is controlled by a head regulator. The length of the main canal is usually restricted to about 8 km. In this scheme of irrigation generally direct irrigation is carried out from the main canal. However, in some cases when the main canal is of fairly big size and the irrigated area is wide enough, a distributary may be taken from the main canal.

A series of Bandharas may be constructed one below the other on the same stream. The irrigation capacity of each Bandhara may vary from a few hectares to about 400 hectares. The area irrigated by each Bandhara is usually known as *Thal*. The Bandhara irrigation schemes utilize water from the small streams having isolated catchments, which cannot be economically included in large irrigation schemes. These schemes are successful in places where there are a number of small streams which have sufficient flow in monsoon and upto the end of January so that sufficient water is made available for kharif and rabi crops. However, if the stream is perennial, on some portion of the irrigated area a perennial crop such as sugarcane may also be grown. The Bandhara irrigation scheme is quite economical. The irrigation from such schemes is being done in some parts of Maharashtra in our country from pre-British period.

Selection of Site for Bandhara

The main points to be considered for the selection of site for Bandhara are as follows.

(i) The site should be within a reasonable distance from the area to be irrigated.

(ii) Hard and sound rock foundation must be available at stream bed level or slightly below the bed, at the site of Bandhara.

(iii) Sufficient quantity of water must be available for irrigation.

(iv) At the site the natural banks of stream should be high enough to avoid submergence of land on either side of the banks during floods.

(v) The site should preferably be below the confluence of tributary and main stream so as to have good supply of water and very few cross drainage works.

(vi) The cost of construction of the works should not be high.

Form and Design of Bandhara

1. Bandhara is designed as a low dam with a rectangular section if it is very low and a trapezoidal section if it is fairly high. The trapezoidal section has its upstream face vertical and there is a batter on the downstream face in the range of 1 in 2 to 1 in 5 as required from the stability considerations.

2. The crest width of Bandhara is decided from the stability considerations and it should be wide enough to be used as roadway during fair weather. The minimum crest width of Bandhara is usually 1.2 m.

3. During floods the surplus water flows over its entire length. The discharge formula for a bandhara is as for a broad-crested weir viz.,

$$Q = 1.7\, LH^{3/2}$$

where Q = discharge in cumec

L = length of Bandhara in m

H = height of water above the crest of Bandhara in m.

4. Scouring sluices of size 1 m × 1.2 m should be provided at the bottom at suitable interval to remove the silt.

5. The section of the Bandhara is checked for the following two conditions.

(a) When there is no flow over the Bandhara, the water on the upstream side is assumed to be upto the crest level and there is no water on the downstream side.

(b) When there is flow over the Bandhara, there will be afflux on the upstream and the difference in the water level on the upstream and downstream sides, i.e., differential head, should be considered for checking the stability of the section.

Phad System of Irrigation

The system of irrigation adopted in the areas served by Bandharas is known as phad system of irrigation. In this system, 'thal' or the irrigated area under each Bandhara is divided into 3 or 4 parts called 'phads' or small blocks. Each phad has a number of agricultural plots in it belonging to different cultivators of the same village. In one year, only one kind of crop is grown on all the plots of one phad. On different phads different crops may be grown during the same year. The crop grown in a phad may be changed next year to introduce the rotation of crop for maximum yield.

Advantages of Bandhara Irrigation Schemes

The various advantages of Bandhara irrigation schemes are as follows.

(i) Small quantities of water flowing in the streams which would have otherwise gone as a waste, are utilized for irrigation to the maximum possible extent.

(ii) As the canal system in this case extends over a short length, the seepage and evaporation losses are less.

(iii) The area irrigated being small and close to the source, high duty of water may be achieved and also intensive irrigation may be carried out.

(iv) The initial and maintenance costs of the system are low.

Disadvantages of Bandhara Irrigation Schemes

The various disadvantages of Bandhara irrigation schemes are as follows.

(i) The supply of water in these schemes is unreliable and seasonal in the case of non-perennial streams.

(ii) The area irrigated from each Bandhara is more or less fixed and hence even if more water is available in some year it cannot be utilized.

(iii) Even small streams are provided with Bandharas and their water is used for irrigation, the population on the downstream of the Bandharas do not get enough water for their domestic use during dry periods.

REVIEW QUESTIONS

14.1 What are the different types of canals?

14.2 How irrigation canals are classified?

14.3 What is a contour channel? Where is this constructed?

14.4 Differentiate between a permanent canal and an inundation canal.

14.5 Describe briefly the various considerations made in the alignment of an irrigation canal.

14.6 What do you understand by a productive canal and a protective canal?

14.7 Write a brief note on Bandhara Irrigation?

14.8 Discuss the advantages and disadvantages of Bandhara irrigation scheme.

14.9 What is a side slope canal? Where is this constructed?

14.10 Explain the terms: (i) Feeder canal; (ii) Carrier canal; (iii) Water course; (iv) Branch canal; (v) Distributary; (vi) Phad system of irrigation.

14.11 Distinguish between a ridge canal and a contour canal.

14.12 Write a brief note on Inundation Canals.

CHAPTER

15 Design of Unlined Alluvial Channels by Silt Theories and Sediment Transport

15.1 INTRODUCTION

On account of erosion in the catchment or drainage basin, rivers receive a huge quantity of sediment along with water. The sediment comprises soil particles of various sizes ranging from fine silt to coarse sand which is carried either in suspension or along the bed. A portion of this sediment (mainly silt) is also received by the channels which draw water directly from the rivers. In the case of a rigid boundary (or lined) channel the quantity of soil transported by it remains more or less same as that entering the channel at its offtaking section. Moreover, in such channels relatively higher velocities of flow may be permitted so that the soil is not allowed to get deposited. As such for rigid boundary channels the problem of silt transport does not exist. However, in the case of unlined alluvial channel the quantity of silt transported may vary from section to section due to scouring of bed and sides of the channel as well as due to silting (or deposition of silt) at any section. The bed and sides of unlined alluvial channel may be scoured if velocity of flow is too high and silting may take place if the velocity of flow is too low. Both scouring and silting result in modifying the cross-section of the channel. If the bed and sides of the channel are scoured the cross-section increases and besides other damages because of scour, its full supply depth decreases which will result in reducing its command. On the other hand if silting takes place in the channel its capacity reduces and it will irrigate less area. As such unlined alluvial channels should be designed for such a velocity that neither the bed and sides are scoured nor silt is deposited and a stable channel section is obtained. A velocity of flow which will just keep the silt in suspension and will not scour the bed and sides of the channel is known as non-silting and non-scouring velocity.

Several investigators have worked on the problem of designing a stable, non-silting and non-scouring channel section. The notable contributions are those of R.G. Kennedy, Executive Engineer, Irrigation Department, Punjab and Gerald Lacey, Chief Engineer, U.P. The theories developed by both these investigators are known as silt theories for the design of unlined alluvial channels and are named after them as Kennedy's theory and Lacey's theory. Both these theories are discussed in the subsequent sections.

15.2 FORMULATION OF THE PROBLEM

For the design of a channel the following data are available.

(i) Discharge Q for which the channel is to be designed

(ii) Surface property such a rugosity coefficient N, and

(iii) Soil property such as silt factor f.

The design consists of determining the following :

(i) Area of cross-section A

(ii) Hydraulic mean radius R

(iii) Mean velocity of flow V, and

(iv) Bed slope S.

To commence the design two equations which connect the knowns and unknowns are available as indicated below.

(a) $$Q = A \times V \text{ (Continuity equation)} \quad \ldots(15.1)$$

(b) $$V = \phi\ (N, R, S) \text{ (Flow equation)} \quad \ldots(15.2)$$

Equation (15.2) may be Manning's, Kutter's or any other similar flow equation.

Since there are four unknowns, two more equations must be available for complete and unique solution of the problem. The additional two equations may be obtained from the following criteria.

1. Providing channel of best discharging section which gives another equation between A and R .

2. Adopting an equation which gives non-silting and non-scouring velocity.

3. Adopting a bed slope for the channel depending on the available ground slope.

4. Choosing a suitable B/D ratio on the basis of experience, where B is bed width of channel and D is depth of flow in channel.

Criterion (1) is not suitable for unlined alluvial channels for which criterion (2) is used because as indicated earlier the velocity of flow in these channels must be non-silting and non-scouring. Thus, in Kennedy's method of design, criterion (2) and criterion (3) or (4) are used which provide the additional two equations as required. The suitable value of B/D ratio for this method of design is obtained from Wood's table or from certain empirical formulae. In Lacey's method of design also criterion (2) is used but in this method for the determination of the four unknowns four independent equations are given and the flow equation given by Manning and Kutter are not used.

15.3 KENNEDY'S THEORY

Kennedy carried out a pioneer research work for obtaining a stable, non-silting, non-scouring channel. He made observations on the channels of Upper Bari Doab canal system, one of the oldest in Punjab (Pakistan). On this canal system Kennedy selected a number of sites on various channels which had not required any silt clearance or other maintenance for the last thirty years and were thus considered stable.

On the basis of these observations Kennedy arrived at the following conclusion.

The silt carried by water flowing in a channel is kept in suspension solely by the vertical component of eddies which are generated over the full width of the channel and rise up gently to the surface. These eddies are generated on account of friction of flowing water against the bed and work up (or rise) against the depth of water in the channel. Some eddies may also be generated from the sides of the channel but these are for most of their part horizontal and so do not have any silt supporting power. The silt supporting power of a channel is therefore proportional to the bed width of the channel and not to its total wetted perimeter.

Kennedy also introduced *critical velocity*, denoted by V_0 and defined as the mean velocity which will just keep the channel free from silting or scouring. By plotting the observations of mean velocity and depth of flow Kennedy obtained a relation between the critical velocity and depth of flow which in its converted form from F.P.S. to SI or metric units is as follows.

$$V_0 = 0.55D^{0.64} \qquad \text{...(15.3)}$$

where V_0 = critical velocity in m/s ; and

D = depth of flow in m.

This equation was given by Kennedy in 1895.

Kennedy made his observations on channels of B/D ratio ranging from 3.5 to 12, but the plotted data did not show any effect of the variation of the bed width of the channel and hence Kennedy concluded that the bed width of the channel had no effect on the critical velocity.

Since this equation has been derived on the basis of observations made on channels of only Upper Bari Doab canal system, it is applicable to only those channels which are flowing in sandy silt of the same quality or grade as that of channels in Upper Bari Doab canal system. Kennedy also later recognised that the critical velocity was affected by the grade of the silt and in order to account for the effect of the silt grade he introduced in his equation a factor m called *Critical Velocity Ratio* (C.V.R.). The equation is then written as

$$V = 0.55\, m\, D^{0.64} \qquad \text{...(15.4)}$$

where $m = \dfrac{V}{V_0}$ = C.V.R.

Silt grades finer than that of the channels of Upper Bari Doab canal sytem were assigned a value of m from 0.8 to 0.9 and those coarser than these were assigned a value of m from 1.1 to 1.2.

Thus, Kennedy's V–D relationship given by Eq. (15.4) provides an additional equation for the design of channels.

For determining the mean velocity of flow in the channel Kennedy used Kutter's equation which is as follows.

$$V = \frac{23 + \dfrac{1}{N} + \dfrac{0.00155}{S}}{1 + \left(23 + \dfrac{0.00155}{S}\right)\dfrac{N}{\sqrt{R}}}\sqrt{RS} \qquad \text{...(15.5)}$$

where V = mean velocity of flow in m/s

R = hydraulic mean radius in m

S = bed slope, and

N = rugosity coefficient.

For unlined alluvial channels the value of rugosity coefficient N depends on the nature of the material in the bed and sides of the channel as well as on the condition of the channel. Buckley has recommended the following values of N for earthen channels in Northern India.

Channel condition	*Value of N*
1. Very good	0.025
2. Good	0.0250
3. Indifferent	0.0275
4. Poor	0.0300

Central Board of Irrigation and Power, New Delhi has recommended the following values of N for various discharges.

Discharge (cumec)	*Value of N for unlined channels in ordinary soil*
1. 14 to 140	0.025
2. 140 to 280	0.0225
3. 280 and above	0.02

It may however be mentioned that the limitations of Kutter's equation are incorporated in the Kennedy's method of design of channels.

Kennedy did not give any equation for the bed slope of the channel. The bed slope to be provided for the channel is decided according to the available slope of the ground and it may be determined from the longitudinal section (or L-section) of the alignment the procedure for which is discussed in details in the next chapter. Although the available ground slope limits the range of bed slope for the channel but within the possible range any bed slope may be provided.

Thus, by knowing Q, S, N and m, a channel can be completely designed by using the three equations *viz,* continuity equation (Eq. 15.1), Kennedy's V–D relationship (Eq. 15.4) and Kutter's equation (Eq. 15.5). By varying the bed slope, for the same values of Q, N and m, different channel sections would be obtained. This is indicated by the following example.

Q = 5 cumec ; N = 0.0225 ; m = 1

(side slopes 1 Hor. to 2 Ver.)

Slope	*B*	*D*	*B/D*
1 in 5000	8.59 m	1.00 m	8.59
1 in 4000	4.54 m	1.40 m	3.24
1 in 2000	1.44 m	2.18 m	0.66

All the three channels satisfy Eqs (15.1), (15.4) and (15.5) and hence according to Kennedy's theory they should be non-silting, non-scouring. However, it is evident that in actual practice all of them would not be equally satisfactory, and some may even be highly unsatisfactory from the consideration of silt transportation. Kennedy's theory does not provide any clue as to which of these channels would be the best. Therefore some guidance for fixing a suitable bed-width to depth ratio (i.e., *B/D* ratio) is required. This fact was also recognised by Kennedy who at a later date in 1904 indicated the necessity of fixing suitable *B/D* ratio. These ratios were fixed in various states on the basis of experience. Thus, in Punjab Wood's normal design tables were prepared and used. An extract from Wood's normal design tables duly converted in SI or metric units is given in Table 15.1.

Central Water and Power Commission, New Delhi has recommended *B*/*D* ratio for channels carrying discharges ranging from 0.3 to 300 cumec. These are given in the form of a graph between *B*/*D* ratio and discharge (Fig. 15.1) from which for a given discharge the value of *B*/*D* ratio may be obtained.

If a value of *B*/*D* ratio is known then since a fourth equation is made available a unique design of the channel including its bed slope may be obtained.

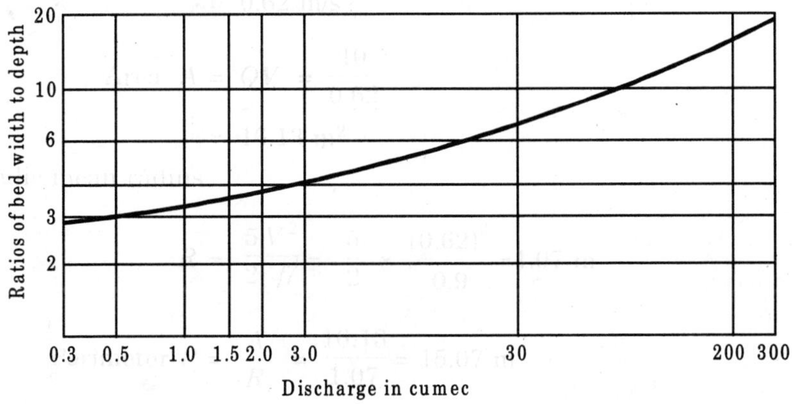

Fig. 15.1. *Bed width to depth ratio for unlined channels recommended by C.W.P.C*

15.4 PROCEDURE FOR DESIGN OF CHANNEL BY KENNEDY'S METHOD

For the design of channel by Kennedy's method the following equations are used.

$$Q = A \times V \qquad \text{...(i)}$$

$$V = \frac{23 + \dfrac{1}{N} + \dfrac{0.00155}{S}}{1 + \left(23 + \dfrac{0.00155}{S}\right)\dfrac{N}{\sqrt{R}}}\sqrt{RS} \qquad \text{...(ii)}$$

$$V = 0.55 \text{ m } D^{0.64} \qquad \text{...(iii)}$$

TABLE 15.1 Wood's normal design table for N = 0.0225

Discharge (cumec)	*B*/*D ratio*	*B (m)*	*D (m)*	*Slope 1 in*	$m = \dfrac{V}{V_0}$	*Means V (m/s)*
0.283	2.9	1.45	0.49	3333	0.92	0.344
0.708	3.4	2.21	0.66	3636	1.01	0.424
1.416	3.7	3.12	0.84	4000	1.00	0.476
2.832	4.2	4.42	1.04	4444	1.00	0.555

Contd.

Contd.

7.079	4.8	6.71	1.43	4444	1.01	0.701
14.158	5.7	9.75	1.72	5000	1.00	0.774
28.317	7.6	15.24	1.98	5000	1.03	0.881
56.634	11.3	25.45	2.26	5714	1.03	0.945
141.584	22.5	56.39	2.50	6666	0.98	0.975
283.168	41.0	{105.00 110.00	{2.59 2.68	{6666 8000	{1.02 0.93	{1.030 0.954
566.336	78.0	212.14	2.72	8000	0.94	0.975

The items which must be known are

(i) Discharge Q

(ii) Rugosity coefficient N

(iii) C.V.R. m, and

(iv) either bed slope S (from L-section) or B/D ratio.

Dependiong upon whether the bed slope S is known or B/D ratio is known the following two procedures may be adopted for the design of channels.

Case (1) Given Q, N, m and S

1. Assume a trial value of D in metres
2. Calculate the velocity V from Eq. (iii),

$$V = 0.55 \text{ m } D^{0.64}$$

3. Cross-sectional area A is obtained from Eq. (i) as

$$A = \frac{Q}{V}$$

4. Assuming the side slopes of the channel as $\frac{1}{2}$ horizontal to 1 (vertical) the bed width B is determined from the following relation

$$A = BD + \frac{D^2}{2}$$

5. Compute perimeter P and hydraulic mean radius R from the following relations

$$P = B + \sqrt{5}\, D$$

$$R = \frac{A}{P} = \frac{BD + \frac{D^2}{2}}{B + D\sqrt{5}}$$

6. Calculate the actual mean velocity of flow from Kutter's equation [Eq. (ii)]. If this value of V is nearly same as that found in step (2), the assumed depth is correct. If not, assume another

trial value of D and repeat the calculations till the two values of the velocities determined in steps (2) and (6) are nearly the same.

The design for this case may also be carried out with the help of Garret's diagrams which provide a graphical solution of Kennedy's and Kutter's equations. The use of these diagrams is explained in the next chapter.

Case (2) Given Q, N, m and B/D ratio

1. Calculate A in terms of D as indicated below.

Let $$\frac{B}{D} = x \text{ (which is given)}$$

$\therefore$ $$B = Dx$$

Since $$A = BD + \frac{D^2}{2}$$

$\therefore$ $$A = xD^2 + \frac{D^2}{2}$$

or $$A = D^2 (x + 0.5)$$

2. Substitute the value of A from step (1) and the value of V from Eq. (iii) i.e., $V = 0.55 \text{ m } D^{0.64}$ in Eq. (i). Thus,

$$Q = A \times V$$

or $$Q = D^2 (x + 0.5) \times 0.55\, m\, D^{0.64}$$

Solving for D, we get

$$D = \left[\frac{Q}{0.55 \quad m(x+0.5)}\right]^{\frac{1}{2.64}}$$

In the above equation Q, m and x are known. Hence D is determined.

3. Knowing D, calculate B and R from the following relations

$$B = xD$$

and $$R = \frac{BD + \frac{D^2}{2}}{B + D\sqrt{5}}$$

4. Calculate the velocity V from Kennedy's equation [Eq. (iii)]

$$V = 0.55\, m\, D^{0.64}$$

5. Knowing V, R and N, determine the bed slope S from Kutter's equation [Eq. (ii)] by trial and error.

15.5 SILT TRANSPORTING CAPACITY OF A CHANNEL ACCORDING TO KENNEDY'S THEORY

According to Kennedy total quantity of silt transported depends directly on the bed width B of the channel and on some power of the critical velocity V_0.

Let Q_t be the quantity of silt transported. Then

$$Q_t \propto BV_0^n$$

or $$Q_0 = aBV_0^n \quad \text{...(i)}$$

where n = an unknown index of $V0$; and

a = a constant.

Further let p be the percentage of silt in water and Q be the discharge of the channel. Then

$$Q_t = pQ \quad \text{...(ii)}$$

Assuming a wide channel

$$Q = BDV_0 \text{ (approximately)} \quad \text{...(iii)}$$

Introducing Eq. (iii) in Eq. (ii), we get

$$Q_t = pBDV_0 \quad \text{...(iv)}$$

Equating the values of Q_t given by Eqs (i) and (iv), we get

$$aBV_0^n = pBDV_0$$

or $$V_0^{n-1} = \frac{1}{a}pD$$

or $$V_0 = \left(\frac{p}{a}\right)^{\frac{1}{n-1}} D^{\frac{1}{n-1}} \quad \text{...(v)}$$

Equation (v) is of the same form as Kennedy's equation

$$V_0 = CD^y = CD^{0.64} \quad \text{...}(vi)$$

Equating the indices of D in the Eqs (v) and (vi), we get

$$\frac{1}{n-1} = 0.64$$

$$\therefore \quad n = 2.56 \simeq \frac{5}{2}$$

Hence the quantity of silt transported is given by

$$Q_t = aBV_0^{5/2} \quad \text{...(15.6)}$$

Kennedy, however, did not determine the value of the constant a.

15.6 DRAWBACKS IN KENNEDY'S THEORY

The vairous drawbacks in Kennedy's theory are as follows.

1. Kutter's equation is used for determining the mean velocity of flow and hence the limitations of Kutter's equation are incorporated in Kennedy's theory.

2. The significance of B/D ratio is not considered in this theory.

3. No equation for the bed slope has been given which may lead to varied designs of the channel with slight variation in the bed slope.

4. Silt charge (or silt concentration) and silt grade are not considered. The complex phenomenon of silt transportation is incorporated in a single factor m called critical velocity ratio.

5. The value of m is decided arbitrarily since there is no method given for determining its value.

6. This theory is aimed to design only an average regime channel.

7. The design of channel by the method based on this theory involves trial and error which is quite cumbersome.

15.7 FURTHER WORK AFTER KENNEDY'S THEORY

Even after the introduction of critical velocity ratio m, the Kennedy's equation could not be accepted universally for the design of channels. However, certain regional formulae of the same general form, $V = CD_y$, were developed on the basis of observations made on stable channels in the canal systems in different regions. In each of these formulae both the coefficient C as well as the power (or index) y of D varied. Some of these formulae developed for different regions are as follows

$V = 0.391\ D^{0.55}$	Godavari Delta
$V = 0.530\ D^{0.52}$	Krishna Western Delta
$V = 0.567\ D^{0.57}$	Lower Chenab Canal
$V = 0.283\ D^{0.73}$	Egyptian Canals

where V = velocity of flow in m/s ; and

D = depth of flow in m.

Lindley's Equations

In view of the various drawbacks of Kennedy's theory and to make the channel design more definite, Lindley of irrigation Department, Punjab, undertook the work of designing a stable channel. Lindley considered bed width of the channel as another variable in addition to depth of flow and on the basis of observations made on Lower Chenab canal system gave two equations one relationg V and D similar to that given by Kennedy and the other relating V and B. Lindley gave these equations in 1919, which in their converted form from FPS. to SI or Metric units are as follows.

$$V = 0.567\ D^{0.57} \qquad ...(15.7)$$

$$V = 0.274\ B^{0.355} \qquad ...(15.8)$$

where V = velocity of flow in m/s

D = depth of flow in m, and

B = bed width in m.

By eliminating V from Eqs (15.7) and (15.8), another equation relating B and D is obtained as follows

$$B = 7.86\ D^{1.61} \qquad ...(15.9)$$

Lindley's Eqs (15.8) and (15.9) if used in conjunction with the continuity equation (Eq. 15.1) and Kutter's equation (Eq. 15.5) provided four equations for the determination of four unknowns B, D, V and S (since A and R are both functions of B and D and can be expressed in terms of B and D). However, Lindley also did not give any equation for bed slope of the channel but recommended the use of Kutter's equation for determining the same.

Lindley was the first to put forward the regime concept which states that the dimensions viz., width, depth and bed slope of a channel constructed in alluvium (or alluvial soil) to carry a given discharge loaded with a given silt charge are all fixed by nature. In other words when an artificial channel is constructed in alluvium to carry silty water, its bed and sides will silt or scour until the width, depth and bed slope attain a state of balance and channel is then said to be in regime. As indicated later Lacey also followed the same idea in the development of his theory.

15.8 LACEY'S THEORY

Gerald Lacey, who retired as Chief Engineer from Irrigation Department, U.P. in 1945, carried out a detailed study of the problem of designing stable channels in alluvium and made significant contribution in the subject. He developed the regime theory and evolved a number of equations on the basis of his own observations and the observations of the earlier investigators. The equations given by Lacey have been widely accepted. In fact the theory developed by Lacey had such an impact in our country that the whole subject became associated with his name and this theory is more popularly known as *Lacey's regime theory*. The salient features of this theory are discussed below.

According to Lacey the width, depth and bed slope of a channel constructed in easily erodible alluvium and carrying a fixed discharge and silt load are uniquely determined by nature. In other words for a channel constructed in alluvium to carry certain discharge, the width, depth and bed slope will undergo modification by silting and scouring till equilibrium is attained. The channel is then termed as regime channel which may be defined as follows.

Regime channel. A regime channel may be defined as a stable channel whose width, depth and bed slope have undergone modification by silting and scouring and are so adjusted that they have attained equilibrium.

Regime conditions. For a channel to be in regime certain conditions called regime conditions, are to be established which are as follows.

(i) The channel should be flowing uniformly in unlimited incoherent alluvium of the same character as that transported by the channel.

(ii) Silt grade and silt charge should be constant.

(iii) Discharge should be constant.

By *incoherent alluvium* is meant loose granular material which can be scoured as easily as it can be deposited. Such a material is however rarely reproduced perfectly in the field.

The regime conditions indicated above are known as *true regime conditions*. If these conditions are met with fully, the channel will be in *true regime*. However, these are the ideal conditions of regime and are seldom fully obtained in actual practice.

Regime on artificial channels. Lacey indicated that in the case of artificial channels there are two stages of regime which are known as initial regime and final regime. These are described below.

Initial regime. It is the first stage of regime attained by a channel after it has been put into service. As explained below the initial regime is attained by a channel only by the variation of bed slope and depth.

If a channel is excavated with somewhat smaller width and flatter bed slope, then as the flow takes place in the channel, the bed slope of the channel is increased due to deposition of silt on the bed of the channel, so that an increased velocity of flow is developed which allows the given discharge to flow through the channel of smaller width. With the increase in the bed slope the depth may also vary. However, the width of the channel remains invariable because the sides of the channel are usually cohesive and hence they resist erosion. The channel thus has a given discharge, silt grade, silt charge and width, and only by varying bed slope and depth it attains stability which is termed as *initial regime.* However, the stability so attained by the channel is only temporary because the width of the channel has not been so far adjusted to suit the requirements of a regime channel.

Final regime. It is the ultimate stage of regime attained by a channel when in addition to bed slope and depth the width of the channel has also been suitably adjusted. For the channel considered earlier due to continuous action of water the resistance of sides is ultimately overcome and a condition is developed when the channel may adjust its width, depth and bed slope so that a stable channel is obtained. The stability so attained by a channel is termed as *final regime* or *true regime.* A channel which has attained final regime is known as regime channel. The various equations developed by Lacey are applicable to a channel which has attained final or true regime.

It may be mentioned that if in the above case the sides of the channel are of stiff clay they may resist erosion almost indefinitely. The channel may then continue to be in initial regime only and may not attain final regime. However, for this channel regime theory is not applicable because it is not in incoherent alluvium. Further if a channel is constructed with a larger width and flatter slope then silting will occur in the channel till final regime is attained. Thus in this case the channel may directly attain the final regime. Similarly if a channel is constructed with either small or larger width and steeper bed slope then scouring and silting will occur till final regime is attained and in these cases the initial regime may or may not exist.

Cross-section of Regime Channel

From the above discussion it is evident that for a true regime channel there is only one cross-section and one bed slope at which the channel will carry a given discharge with a given silt charge of a particular grade. Lacey, however, postulated that a true regime channel has a semi-elliptical section. For a coarser silt the channel is wide and shallow, while for a finer silt the channel is narrow and deep (Fig. 15.2).

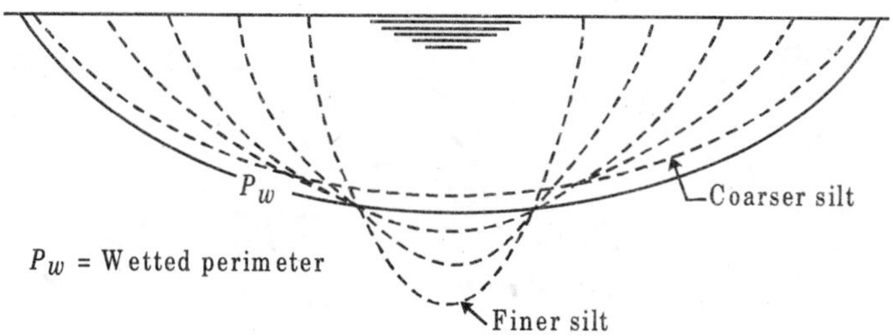

Fig. 15.2. *Cross-section of regime channel*

Lacey's Regime Equations

Lacey also indicated that the silt carried by water flowing in a channel is kept in suspension by the vertical component of eddies but according to him these eddies are generated from bed as well as sides of the channel in the direction normal to the surface of generation. For this reason Lacey considered hydraulic mean radius R as a more pertinent variable rather then depth of flow D considered by Kennedy. Further as a measure of silt grade Lacey introduced a *silt factor* f, with its value equal to unity for the silt of the channels of Upper Bari Doab canal system. Thus, by plotting Kennedy's, Lindley's and Madras data in terms of mean velocity V versus (fR), Lacey obtained a relationship which in its converted form from FPS. to SI or Metric units is as follows

$$V = \sqrt{\frac{2}{5} fR} \qquad ...(15.10)$$

where V = mean velocity of flow in m/s
f = silt factor, and
R = hydraulic mean radius in m.

This was the first regime equation given by Lacey.

For determining the channel dimension second equation would be necessary giving either cross-sectional area A or perimeter P. Lacey on the basis of plotted data thus evolved a relationship between A and V which is as follows

$$Af^2 = 140V^5 \qquad ...(15.11)$$

where A = cross-sectional area in m^2.

1. Silt Factor—Grain Size Relation

According to Lacey the silt factor f is dependent on the mean particle size of the boundary material in the channel. The value of f may be determined by the following relation given by Lacey

$$f = 1.76\sqrt{m_r} \qquad ...(15.12)$$

where m_r = mean particle size in mm
$= d_{50}$ in mm.

2. V-Q-f Relation

A relation between V-Q-f may be obtained from Eq. (15.11) as follows.

Multiplying Eq. (15.11) by V on both sides, we get

$$AVf^2 = 140V^6$$

From continuity equation

$$Q = AV$$

$$\therefore \quad Qf^2 = 140V^6 \qquad ...(15.13)$$

or

$$V = \left[\frac{Qf^2}{140}\right]^{1/6} \qquad ...(15.14)$$

where Q = discharge in cumec.

3. P-Q Relation

By combining Eqs (15.10) and (15.13) a relationship between P and Q can be obtained as follows.

Raising both sides of Eq. (15.10) to the fourth power we get

$$V^4 = \frac{4}{25} f^2 R^2$$

Eliminating f^2 between this equation and Eq. (15.13), we get

$$Q\left[\frac{25V^4}{4R^2}\right] = 140\,V^6$$

or
$$\frac{25}{4}\frac{Q}{R^2} = 140V^2$$

Since
$$R = \frac{A}{P};\; Q = AV, \text{ we get}$$

$$\frac{25Q}{4(A/P)^2} = 140\,V^2$$

or
$$P^2Q = \frac{560}{25}(AV)^2$$

or
$$P^2 = 22.4\,Q$$

or
$$P = 4.75\,Q \qquad \text{...(15.15)}$$

where P = perimeter in m.

Equation (15.15) gives a direct relation between P and Q. It indicates that for a stable channel the perimeter P depends only on discharge Q and involves no other channel characteristic. This equation is commonly used for determining the waterway in the design of several hydraulic structures. The power or index of Q does not vary but the coefficient 4.75 seems to vary between 3.6 and 6.15 depending on the nature of the transported material.

4. Regime Flow Equation

After determining the dimensions of a regime channel the bed slope of the channel is required to be determined. Thus, by plotting a large number of data Lacey obtained the following flow equation

$$V = 10.8\,R^{2/3}S^{1/3} \qquad \text{...(15.16)}$$

where S = bed slope.

Equation (15.16) is called *General regime equation* which does not involve co-efficient of rugosity, as it is considered that rugosity is implicit in R and S. This equation is of considerable practical importance in evaluating flood discharge of rivers. In order to determine the flood discharge of rivers usually Manning's or Kutter's equation is used in which the coefficient of rugosity N is assigned a value on the basis of experience. However, the value of N cannot be specified with a higher degree of accuracy since it depends upon a large number of factors. As such the determination of flood discharge by the use of Manning's or Kutter's equation is unreliable. At the time of floods river is temporarily in regime and hence Eq. (15.16) may be applied which would give quite accurate and reliable results.

From these equations for determining the bed slope of a regime channel three equations may be obtained as indicated below.

5. Regime Slope Equations

(a) **S-f-R Relation**

Cubing both sides of Eq. (15.16), we get

$$V^3 = 1260\,R^2S \qquad \text{...(i)}$$

Also cubing both sides of Eq. (15.10), we get

$$V^3 = (2/5)^{3/2} f^{3/2} R^{3/2} \quad \text{...(ii)}$$

Hence from Eqs (i) and (ii) we have

$$1260\, R^2 S = (2/5)^{3/2} f^{3/2} R^{3/2}$$

or

$$S = \frac{f^{3/2}}{4980 R^{1/2}} \quad \text{...(15.17)}$$

or

$$S = \frac{0.0002 f^{3/2}}{R^{1/2}} \quad \text{...(15.17a)}$$

(b) S-f-Q Relation

Equating the values of V given by Eqs (15.10) and (15.14), we have

$$\sqrt{\frac{2}{5} fR} = \left[\frac{Qf^2}{140}\right]^{1/6}$$

or

$$R^{1/2} = \left[\frac{5}{2f}\right]^{1/2} \left[\frac{Qf^2}{140}\right]^{1/6}$$

or

$$R^{1/2} = \left[\frac{Q}{8.96f}\right]^{1/6} \quad \text{...(iii)}$$

Substituting the value of $R^{1/2}$ from Eq. (iii) in Eq. (15.17), we get

$$S = \frac{f^{3/2}}{4980}\left[\frac{8.96f}{Q}\right]^{1/6}$$

or

$$S = \frac{f^{5/3}}{3340 Q^{1/6}} \quad \text{...(15.18)}$$

(c) S-f-q Relation

Equation (i) can also be written as

$$S = \left(\frac{V^2}{R}\right)^{5/3} \frac{1}{1260 (RV)^{1/3}} \quad \text{...(iv)}$$

Squaring both sides of Eq. (15.10), we get

$$V^2 = \frac{2}{5} fR$$

or $$\left(\frac{V^2}{R}\right) = \frac{2}{5}f \qquad \text{...(v)}$$

Further if q is discharge per unit width of the channel, then assuming it to be a wide channel, we have

$$q = RV \qquad \text{...(vi)}$$

Introducing Eqs (v) and (vi) in Eq. (iv), we get

$$S = \left(\frac{2}{5}f\right)^{5/3} \frac{1}{1260q^{1/3}}$$

or $$S = \frac{0.000178f^{5/3}}{q^{1/3}} \qquad \text{...(15.19)}$$

6. Regime Scour Depth Equation

Squaring both sides of Eq. (15.10), we get

$$V^2 = \frac{2}{5}fR$$

or $$R = \frac{5}{2}\frac{V^2}{f} \qquad \text{...(vii)}$$

Squaring both sides of Eq. (15.14), we get

$$V^2 = \left[\frac{Qf^2}{140}\right]^{1/3} \qquad \text{...(viii)}$$

Substituting for V^2 from Eq. (viii) in Eq. (vii), we get

$$R = \frac{5}{2}\left[\frac{Qf^2}{140}\right]^{1/3}\frac{1}{f}$$

or $$R = 0.47\left[\frac{Q}{f}\right]^{1/3} \qquad \text{...(15.20)}$$

Equation (15.20) gives the hydraulic mean radius (or hydraulic mean depth) of a regime channel. For a wide channel the hydraulic mean depth is almost the same as its depth of flow. Thus, Eq. (15.20) can be used for determining the depth of flow in a river during flood, since at the time of flood a river is temporarily in regime. The depth of flow in a river during flood also represents the scour depth. As such the value of R obtained from Eq. (15.20) corresponding to flood discharge Q gives the scour depth which is called the normal scour depth and is measured below the high flood level to obtain the level of scoured bed of the river in flood.

It is sometimes more convenient to know the scour depth in terms of q, the discharge per unit width rather than the total discharge Q. Thus, an expression for R in terms of q may be obtained as follows.

Since $$q = RV$$

Substituting for R from Eq. (v), we get

$$q = \frac{5}{2}\frac{V^3}{f}$$

Substituting for V from Eq. (15.14), we get

$$q = \frac{5}{2}\left[\frac{Qf^2}{140}\right]^{3/6}\frac{1}{f}$$

or $$q = 0.21\ Q^{1/2} \qquad \text{...(15.21)}$$

or $$Q = \left[\frac{q}{0.21}\right]^2 \qquad \text{...(15.22)}$$

Substituting for Q from Eq. (15.22) in Eq. (15.20), we get

$$R = 0.47\left[\frac{q}{0.21}\right]^{2/3}\frac{1}{f^{1/3}}$$

or $$R = 1.35\left[\frac{q^2}{f}\right]^{1/3} \qquad \text{...(15.23)}$$

Equations (15.20) and (15.23) are commonly used for determining the scour depth which is required to be known for the design of sheet piles or cutoffs for hydraulic structures.

7. Lacey's non-regime Flow Equation

The flow Eq. (15.16) given by Lacey is applicable to only those channels which are in perfect regime and hence it does not contain rugosity coefficient. However, Lacey also gave another flow equation which is a general flow equation applicable to both regime and non-regime channels. This equation may be derived as follows.

Cubing both sides of Eq. (15.16), we get

$$V^3 = 1260\ R^2 S$$

or $$V^2 = 1260\ (R/V)\ RS$$

Taking square root of both sides, we get

or $$V = 35.5\ (R/V)^{1/2}\ \sqrt{RS} \qquad \text{...(15.24)}$$

Equation (15.24) is similar to Chezy's equation in which Chezy's coefficient C is represented by $35.5\ (R/V)^{1/2}$.

From Eq. (15.10) putting

$$V = K\sqrt{fR}$$

Chezy's coefficient C may be expressed as

$$C = \frac{35.5R^{1/2}}{K^{1/2}f^{1/4}R^{1/4}}$$

or $$C = \frac{35.5R^{1/4}}{K^{1/2}f^{1/4}} \qquad \text{...(15.25)}$$

Again putting

$$\frac{35.5}{K^{1/2}f^{1/4}} = \frac{K'}{Na} \qquad \text{...(15.26)}$$

where Na is Lacey's absolute rugosity coefficient, Eq. (15.24) can be written as

$$V = \frac{K'}{Na} \; R^{3/4} S^{1/2} \qquad \text{...(15.27)}$$

Lacey's Eq. (15.27) is comparable to Manning's equation

$$V = \frac{1}{N} R^{2/3} S^{1/2} \qquad \text{...(15.28)}$$

If it is assumed that (i) $Na = N$; and (ii) Eq. (15.27) gives the same velocity as Manning's equation when R = 1 m, then K' =1.

Thus, Lacey's flow equation becomes

$$V = \frac{1}{Na} R^{2/3} S^{1/2} \qquad \text{...(15.29)}$$

Lacey defined Na as an absolute coefficient of rugosity which according to him depends only on grain size of the boundary material of the channel and is independent of the channel condition. Lacey also gave a relation between Na and f which is derived as follows.

8. Na-f Relation

From Eq. (15.26), we have

$$\frac{K'}{Na} = \frac{35.5}{K^{1/2}f^{1/4}}$$

or $$Na = \frac{KK'^{1/2}f^{1/4}}{35.5} \qquad \text{...(15.30)}$$

Since K' and K are numerical coefficients not dependent on the grade of silt, we have from Eq. (15.30)

$$Na \propto f^{1/4}$$

or $$Na = Cf^{1/4} \qquad \text{...(15.31)}$$

where C = a constant

Lacey indicated that for the standard grade silt f = 1, for which the corresponding value of Na = 0.0225. Introducing these values in Eq. (15.31), we get

$$C = 0.0225$$

Equation (15.31) then becomes,

$$Na = 0.0225\, f^{1/4} \qquad \text{...(15.32)}$$

Equation (15.32) may also be derived as follows.

Since $K' = 1$, and from Eq. (15.10) $K = \sqrt{\frac{2}{5}}$, introducing these values in Eq. (15.30), we get

$$Na = \frac{1 \times (2/5)^{1/4} f^{1/4}}{35.5}$$

or

$$Na = 0.0225\, f^{1/4}.$$

Concept of Shock

Lacey indicated that the absolute coefficient of rugosity Na is purely dependent on the grade of boundary material and is independent of the channel condition. It therefore means that Na should be constant for all channels in similar material irrespective of their condition. However, this is not true on account of the following two reasons.

(i) The correct relationship between mean velocity and channel slope, depth and roughness is logarithmic which may, however, be approximated over a wide range by one-sixth power exponential equation. On the other hand Lacey's Eq. (15.29) is of the form

$$\frac{V}{V_*} \propto \left(\frac{R}{K_s}\right)^{1/4}$$

where V^* is shear velocity which is equal to $\sqrt{(\tau_0/\rho)}$ or $\sqrt{gRS}$; and k_s is equivalent sand grain roughness.

Lacey's equation can approximate to the logarithmic equation only over a narrow range of low values of (R/k_s). Beyond that range it will deviate appreciably and lead to considerable error.

(ii) In the case of movable bed channels certain irregularities such as ripples may be formed on the bed. When the flow takes place over these irregularities, the pressure on the front side (*i.e.*, the side facing the flow) of these irregularities being more than the pressure on their rear side, a pressure difference is caused which offers resistance to the flow. The resistance so offered to the flow by the irregularities is called the form resistance and is in addition to the frictional resistance. Thus, in this case the total resistance to the flow is the sum of the frictional resistance and the form resistance. The form resistance depends on size, shape and spacing of the irregularities which themselves depend on the flow conditions. Thus, the form resistance and hence the total resistance offered to the flow in a moveable bed channel cannot be related only to the grain size of the boundary material.

It was thus recognised that the value of rugosity coefficient N may vary even for channels constructed in the same material and hence efforts were made to relate the rugosity coefficient to the channel condition. One such example is that of Buckley who has recommended different values of N depending on the channel condition for earthen channels in Northern India as indicated earlier.

Lacey also recognised that the value of rugosity coefficient Na is affected by the channel condition. However, in order to take into account this effect Lacey introduced the concept of *shock*. Lacey considered that for a channel in very good condition the value of Na depends only on the grain size and attributed any departure from this value (depending on the actual condition of the channel) to the *shock losses* due to irregularities. He further suggested that for all the channels irrespective of their condition the rugosity coefficient should remain constant with its value same as that for a channel in very good condition, but the slope should be divided into two parts—one to overcome friction and the other to meet shock losses. Lacey therefore modified the non-regime flow equation as follows.

$$V = \frac{1}{Na} R^{3/4} (S - s)^{1/2} \qquad \text{...(15.33)}$$

in which s = slope required to meet shock losses.

As indicated above in Eq. (15.33) the value of Na is independent of channel condition so that it remains constant and it is taken as 0.0225 which is the value recommended by Buckley for a channel in very good condition.

Using Eqs (15.29) and (15.33) the value of slope s required to meet shock losses for channels in different conditions may be obtained as follows.

For a channel in 'good' condition Buckley has recommended the value of Na = 0.025. Thus, from Eq. (15.29), we have

$$V = \frac{1}{0.025} R^{3/4} S^{1/2}$$

Again for the same channel from Eq. (15.33), we have

$$V = \frac{1}{0.0225} R^{3/4} (S - s)^{1/2}$$

Thus, equating the values of V given by these equations, we get

$$\frac{1}{0.025} R^{3/4} S^{1/2} = \frac{1}{0.0225} R^{3/4} (S - s)^{1/2}$$

or

$$(S - s)^{1/2} = 0.9\, S^{1/2}$$

or

$$s = 0.19\, S$$

It means that for a channel in good condition 19% of the gross slope is accounted for by shock losses and 81% is available for overcoming friction. Similarly for channels in indifferent and poor conditions 33.06% and 43.75% of the gross slope respectively are accounted for by shock losses.

Lacey's idea of separating form resistance from frictional resistance may be considered as a significant contribution. Lacey, however, indicated that a regime channel is free from shock and hence it has a constant rugosity coefficient for a given size of material in the boundary surface.

It may however be mentioned that there is no reliable method of estimating the form resistance or as Lacey calls it 'shock' and hence Eq. (15.33) cannot be used in practice. As such usually Eq. (15.29) is used by adopting different values of Na even for the channels constructed in the same material depending on their condition.

For ready reference the various equations given by Lacey are listed in Table 15.2.

TABLE 15.2 List of Lacey's equations*

S. No.	*Designation of equation*	*Equation (SI or metric units)*	*Equation No.*
1.	$V-f-R$	$V=\sqrt{\frac{2}{5}fR}$	15.10
2.	$A-f-V$	$Af^2 = 140\ V^5$	15.11
3.	$f-m_r$	$f = 1.76\ \sqrt{m_r}$	15.12
4.	$V-Q-f$	$V=\left[\frac{Qf^2}{140}\right]^{1/6}$	15.14
5.	$P-Q$	$P = 4.75\sqrt{Q}$	15.15
6.	$V-R-S$	$V = 10.8\ R^{2/3}\ S^{1/3}$	15.16
7.	$S-f-R$	$S=\frac{f^{3/2}}{4980R^{1/2}}$	15.17
8.	$S-f-Q$	$S=\frac{f^{5/3}}{3340Q^{1/6}}$	15.18
9.	$S-f-q$	$S=\frac{0.000178f^{5/3}}{q^{1/3}}$	15.19
10.	$R-Q-f$	$R=0.47\left[\frac{Q}{f}\right]^{1/3}$	15.20
11.	$R-q-f$	$R=1.35\left[\frac{q^2}{f}\right]^{1/3}$	15.23
12.	$V-Na-R-S$	$V=\frac{1}{Na}R^{3/4}S^{1/2}$	15.29
13.	$Na-f$	$Na = 0.0225f^{1/4}$	15.32
14.	$V-Na-R-S-s$	$V=\frac{1}{Na}R^{3/4}\ (S-s)^{1/2}$	15.33

* Lacey originally gave his equations in F.P.S., units.

15.9 PROCEDURE FOR DESIGN OF CHANNEL BY LACEY'S METHOD

For the design of channel by Lacey's method the following equations are used.

$$f = 1.76\sqrt{m_r} \quad \text{...(i)}$$

$$V=\left[\frac{Qf^2}{140}\right]^{1/6} \quad \text{...(ii)}$$

$$Q = AV \quad \text{...(iii)}$$

$$P = 4.75\sqrt{Q} \quad \text{...(iv)}$$

$$V = \sqrt{\frac{2}{5} fR} \quad \text{...(v)}$$

$$S = \frac{f^{5/3}}{3340 Q^{1/6}} \quad \text{...(vi)}$$

The items which must be known are as follows

(i) Discharge Q

(ii) Silt factor f or mean particle size m_r.

Procedure of Channel Design

1. Calculate the silt factor f from Eq. (i) if mean particle size (or mean diameter of particles) is given.

2. Calculate the velocity V from Eq. (ii).

3. Calculate area A from Eq. (iii) as

$$A = \frac{Q}{V}$$

4. Calculate perimeter P from Eq. (iv).

5. Knowing A and P and assuming side slopes of the channel as $\frac{1}{2}$ horizontal to 1 (vertical) determine bed width B and depth D from the following relations

$$A = BD + \frac{D^2}{2}$$

$$P = B + \sqrt{5}D$$

6. Calculate hydraulic mean radius R as

$$R = \frac{A}{P}$$

7. Also calculate hydraulic mean radius R from Eq. (v) as

$$R = \frac{5}{2}\frac{V^2}{f}$$

Both the values of R obtained in steps (6) and (7) should be the same, which will provide a check on calculations in steps 1 to 5.

8. Calculate bed slope S from Eq. (vi).

It is, thus, evident that the design of channel by Lacey's method does not involve any trial and error procedure. Further the design of channel by Lacey's method can also be carried out

from his regime diagrams which provide graphical solution of the various equations given by him. The use of Lacey's diagrams is explained in the next chapter.

15.10 COMPARISON BETWEEN KENNEDY'S AND LACEY'S THEORIES

1. Kennedy indicated that silt carried by water flowing in a channel is kept in suspension by the vertical component of eddies generated from the bed of the channel only and hence he gave a relation between V and D. Lacey indicated that the silt carried by water flowing in a channel is kept in suspension by the vertical component of eddies generated from the entire perimeter of the channel and hence he gave a relation between V and R.

2. Kennedy introduced critical velocity ratio m to make his equation applicable to channels flowing in soils of different grades, but he did not give any method to determine the value of m. Lacey introduced silt factor f to make his equation applicable to channels flowing in soils of different grades and also provided a relation between f and mean particle size m_r for determining the value of f.

3. Kennedy recommended the use of Kutter's equation for determining the mean velocity of flow. Lacey gave his own flow equation for determining the mean velocity of flow.

4. Kennedy did not give any equation for the bed slope of the channel, which is decided on the basis of the available slope of the ground or it is determined from Wood's Table. Lacey gave an equation for the bed slope of the channel.

5. Lacey indicated that a true regime channel has a semi-elliptical section. Kennedy did not mention anything about the shape of the section of a stable channel.

6. The design of channel by Kennedy's method involves trial and error procedure. The design of a channel by Lacey's method does not involve any trial and error procedure.

7. Lacey made distinction between two types of resistances offered to the flow in unlined alluvial channels *viz.*, frictional resistance which depends on the nature of the material in the boundary surface; and form resistance or shock which depends on irregularities of the channel. Kennedy did not make any such distinction.

15.11 DRAWBACKS IN LACEY'S THEORY

The various drawbacks in Lacey's theory are as follows.

1. The characteristics of a regime channel are not precisely defined.

2. The true regime conditions defined by Lacey are only theoretical and may not be achieved in actual practice.

3. The derivation of various equations by considering a single factor called silt factor f is not satisfactory. Further the value of f may be different for the bed and the sides and no account of this fact has been made.

4. A considerable increase in the concentration of silt will result due to the loss in absorption which may be 12 to 15% of the total discharge of a channel. In Lacey's theory, however, no consideration has been made regarding the concentration of silt as an independent variable.

5. According to Lacey a regime channel is inherently free from external restrain and shock and has therefore constant *Na* for a given size of the material in the boundary surface. However, a regime channel being a sediment transporting channel and will normally have a changing pattern of bed ripple formations, this statement is unlikely to be correct.

6. Silt charge and silt grade have not been properly defined by Lacey.

7. 'Regime silt charge' was defined by Lacey as 'minimum bed load consistent with 'fully active bed' which is altogether confusing. Moreover, Lacey's equations do not include silt charge.

8. Lacey indicated that a true regime channel has a semi-elliptical section but the same is not supported by any of his equations and also it may not be so in actual practice.

9. The actual dimensions of stable channels are often found to be markedly different from those given by Lacey's equations with *f* based on the size of the bed material only. Moreover, the values of *f* obtained from various equations of Lacey are often quite divergent.

10. Lacey's equations being empirical and based on the data obtained from channels flowing in a particular type of material, before these equations can be applied in general it would be necessary to determine the values of the constants by making observations on existing stable channels flowing in other types of material.

15.12 SEDIMENT TRANSPORT

The cross-section and bed slope of a true regime channel are controlled by the following three independent variables.

1. Discharge carried by the channel.
2. Nature of sediment entering the channel i.e., the grain size distribution, the shape of grains and their specific gravity.
3. Quantity of sediment entering the channel.

The silt theories discussed earlier account for only the first two variables and do not consider the third variable viz., the quantity of sediment entering the channel which is a very important factor and considerably affects the channel design. In fact it has now been recognised that for satisfactory design of unlined alluvial channels a detailed study of sediment transport is quite essential.

Types of Sediment Load

The sediment transported by a stream or channel consists of the following two categories.

(i) Bed load

(ii) Suspended load.

Bed Load

Bed load is defined as that portion of the sediment which moves on or nearer the bed of the stream or channel. The movement of particles of bed load takes place by rolling, sliding or saltation i.e., small leaps.

Suspended Load

Suspended load is that portion of the sediment which moves in suspension in the flowing water. The particles are kept in suspension by the turbulent eddies and they move freely through the flowing water.

Tractive Force Theory for Bed Load Movement

The movement of the bed grains is caused by a force exerted on the grains by the flowing water which is known as tractive force or fluid drag. The tractive force is equal to the tangential component of the weight of water flowing in the channel and hence per unit length of the channel it is given by (wAS), where w is specific weight of water, A is cross-sectional area of the channel and S is its bed slope. The tractive force per unit area will be equal to the boundary shear stress τ_0 and it is obtained by dividing the total tractive force per unit length of the channel by the perimeter P. Thus,

$$\tau_0 = \frac{wAS}{P} = wRS \qquad \text{...(15.34)}$$

where $R = \left(\frac{A}{P}\right)$ is the hydraulic mean radius.

When the tractive exceeds the frictional resistance between the bed grains they start moving. The tractive force required to start general movement of bed grains is called *critical tractive force*. Several investigators have given expressions for the critical tractive force per unit area or critical tractive stress τ_{cr}. Some of these expressions are as given below.

(i) Shield's expression

$$\tau_{cr} = 0.06\, w(G - 1)\, d \qquad \text{...(15.35)}$$

where τ_{cr} = critical tractive stress kN/m^2 or [kg(f)/m^2]

w = specific weight of water in kN/m^3 or [kg (f)/m^3]

G = specific gravity of sediment, and

d = grain diameter in m.

(ii) White's expression

In SI units

$$\tau_{cr} = 0.786\, d \qquad \text{...(15.36 a)}$$

where τ_{cr} = critical tractive stress in kN/m^2, and

d = grain diameter in m.

in Metric units

$$\tau_{cr} = 80.1\, d \qquad \text{...(15.36 b)}$$

where τ_{cr} = critical tractive stress in kg(f)/m^2, and

d = grain diameter in m.

Both the above noted expressions are for turbulent flow. Further the critical tractive stress in sinuous channels may be taken as 90 to 60% of the value in straight reach, the higher the sinuosity, the lower is the value of τ_{cr}.

Estimation of Bed Load

For the estimation of bed load two methods are used viz., stream sampling and analytical. Various types of samplers such as box type sampler, slot type sampler etc., have been used. However, these samplers are not satisfactory. Generally the bed load is taken as a certain percentage of the suspended load. This percentage varies from 3 to 25 per cent of the total suspended load depending upon the nature of the bed material and usually it is taken as 10 per cent.

Analytical Method

For determining the rate of bed load transport various equations have been given by different investigators. The two most widely used equations are those given by Meyer–Peter and

Einstein. The values obtained from these equations agree with each other quite well and the same are given below.

Meyer–Peter's Equation

Meyer-Peter's equation is based on experimental work carried out at Federal Institute of Technology, Zurich. The equation is as given below.

$$\frac{Q_s}{Q}\left(\frac{N'}{N}\right)^{3/2} wSD = 0.047\,(w_s - w)d + 0.25\left(\frac{w}{g}\right)^{1/3}\left(\frac{w_s - w}{w_s}\right)^{2/3}(q_s)^{2/3} \qquad ...(15.37)$$

where Q_s = actual discharge in cumec

Q = discharge in cumec if sides were frictionless

N' = Manning's coefficient for plane bed

N = actual value of Manning's coefficient for rippled bed

w = specific weight of water in kN/m^3 or [kg(f)/m^3]

S = bed slope of channel

D = depth of flow in m

w_s = specific weight of sediment particles in kN/m^3 or [kg(f)/m^3]

d = grain diameter in m

g = acceleration due to gravity, and

q_s = rate of bed load transport per unit width of the channel in kN/m/s or [kg(f)/m/s]

$\left(\frac{Q_s}{Q}\right)$ takes into account correction for sides, and

$$wRS = \tau = wDS \text{ (as } R \simeq D \text{ for a wide channel)}$$

$$\therefore \quad \text{Bed shear } \tau_b = \left(\frac{Q_s}{Q}\right) wDS$$

and $\quad \tau_b\left(\frac{N'}{N}\right)^{3/2}$ = effective stress

Also $\quad \tau_{cr} = 0.047\,(w_s - w)\,d$

Hence the above equation is transformed to

$$q_s = 1500.6\left[\tau_b\left(\frac{N'}{N}\right)^{3/2} - \tau_{cr}\right]^{3/2} \qquad ...(15.38\text{ a})$$

in which

q_s is in N/m/hour ; and τ_b and τ_{cr} are in N/m^2

or

$$q_s = 47453\left[\tau_b\left(\frac{N'}{N}\right)^{3/2} - \tau_{cr}\right]^{3/2} \qquad ...(15.38\text{ b})$$

in which

q_s is in kN/m/hour ; and τ_b and τ_{cr} are in kN/m²

or

$$q_s = 4700\left[\tau_b\left(\frac{N'}{N}\right)^{3/2} - \tau_{cr}\right] \quad ...(15.38\text{ c})$$

in which

q_s is in kg(f)/m/hour, and τ_b and τ_{cr} are in kg(f)/m²

In these equations

$$N' = \frac{(k_s)^{1/6}}{24} \quad \text{and } N = \frac{(k)^{1/6}}{24}$$

where k_s = effective grain diameter in m, and

k = representative size (in m) of roughness of rippled bed.

Einstein's Equation

Einstein's equation is based on the law of probability involving a number of assumptions and experimental coefficients. The equation correlates two dimensionless parameters ϕ_* and ψ_* which are called *bed-load functions* and as given below.

$$\phi_* = \frac{i_B}{i_b}\phi = \frac{i_B}{i_b}\left[\frac{q_s}{Gw}\left(\frac{1}{G-1}\right)^{1/2}\left(\frac{1}{gd^3}\right)^{1/2}\right] \quad ...(15.39)$$

$$\psi_* = \frac{i_B}{i_b}\psi = \frac{i_B}{i_b}\left[\frac{(G-1)d}{R'S}\right] \quad ...(15.40)$$

where q_s = rate of bed load transport per unit width of channel in kN/m/s or [kg(f)/m/s]

d = diameter of sediment particle in m

i_B = fraction of q_s which is of diameter d

i_b = proportion of grains of diameter d in the bed

G = specific gravity of sediment

w = specific weight of water in kN/m³ or [kg(f)/m³]

g = acceleration due to gravity

S = bed slope of channel, and

R' = hydraulic mean radius of the channel if the bed were unrippled.

Since R' represents hydraulic mean radius of a channel having grain roughness only, from Manning's equation, we have

$$V = \frac{1}{N'} R'^{2/3} S^{1/2} \quad ...(\text{i})$$

where N' = Manning's coefficient for plane bed.

Further if R is the actual hydraulic mean radius of channel having rippled bed then

$$V = \frac{1}{N} R^{2/3} S^{1/2} \qquad ...(ii)$$

where N = actual value of Manning's coefficient for rippled bed.

From Eqs. (i) and (ii)

$$R' = R\left(\frac{N'}{N}\right)^{3/2} \qquad ...(15.41)$$

The relationship between ψ_* and ϕ_* has been plotted by Einstein in the form of a graph shown in Fig. (15.3). Thus, if the value of ψ_* is known, ϕ_* can be found from the graph and q_s can be calculated from Eq. (15.39).

For uniform soils

(i) $i_B = i_b$ and hence

(ii) $$\phi_* = \phi = \left[\frac{q_s}{Gw}\left(\frac{1}{G-1}\right)^{1/2}\left(\frac{1}{gd^3}\right)^{1/2}\right] \qquad ...(15.42)$$

(iii) $$\psi_* = \psi = \left[\frac{(G-1)d}{R'S}\right] \qquad ...(15.43)$$

Further the relation between ψ and ϕ for uniform bed material can be expressed as

$$\phi = \frac{1}{0.465} e^{-0.391\psi} \qquad ...(15.44)$$

As indicated in Illustrative Example 15.8 these equations can be used for designing a channel for a given bed load transport. However, these equations do not provide any guidance for fixing the bed width. As such the bed width is usually provided close to Lacey's regime perimeter given by Eq. (15.15). Also the actual value of the Manning's coefficient N for rippled bed will have to be estimated on the basis of experience.

Estimation of Suspended Load

As in the case of bed load, the estimation of suspended load can also be done either by stream sampling or by analytical method. The suspended load can be measured by different types of suspended load samplers.

Analytical Method

The concentration of suspended load C at any height y above the bottom can be determined from the known concentration C_a at a reference point at height a above the bottom by the following equation

$$\frac{C}{C_a} = \left[\frac{D-y}{y}\cdot\frac{a}{D-a}\right]^{\frac{V_s}{kV_*}} \quad ...(15.45)$$

where V_s = settling or fall velocity of sediment grains in still water

D = depth of water

k = Von Karman's universal constant = 0.4

V_* = shear velocity = $\sqrt{(\tau_0/\rho)}$;

τ_0 = boundary shear stress, and

ρ = mass density of water.

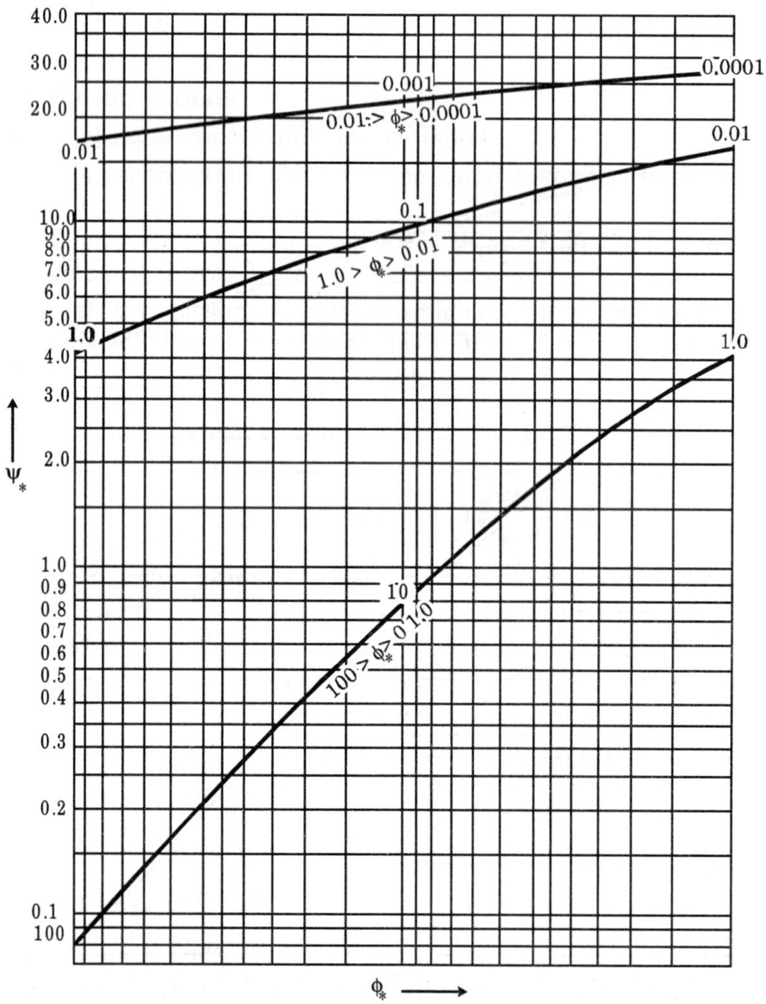

Fig. 15.3 *Einstein's $\psi_* - \phi_*$ curve*

The rate of transport of suspended load may be determined by numerical integration of Eq. (15.45) in conjuction with the Karman–Prandtl velocity distribution equation.

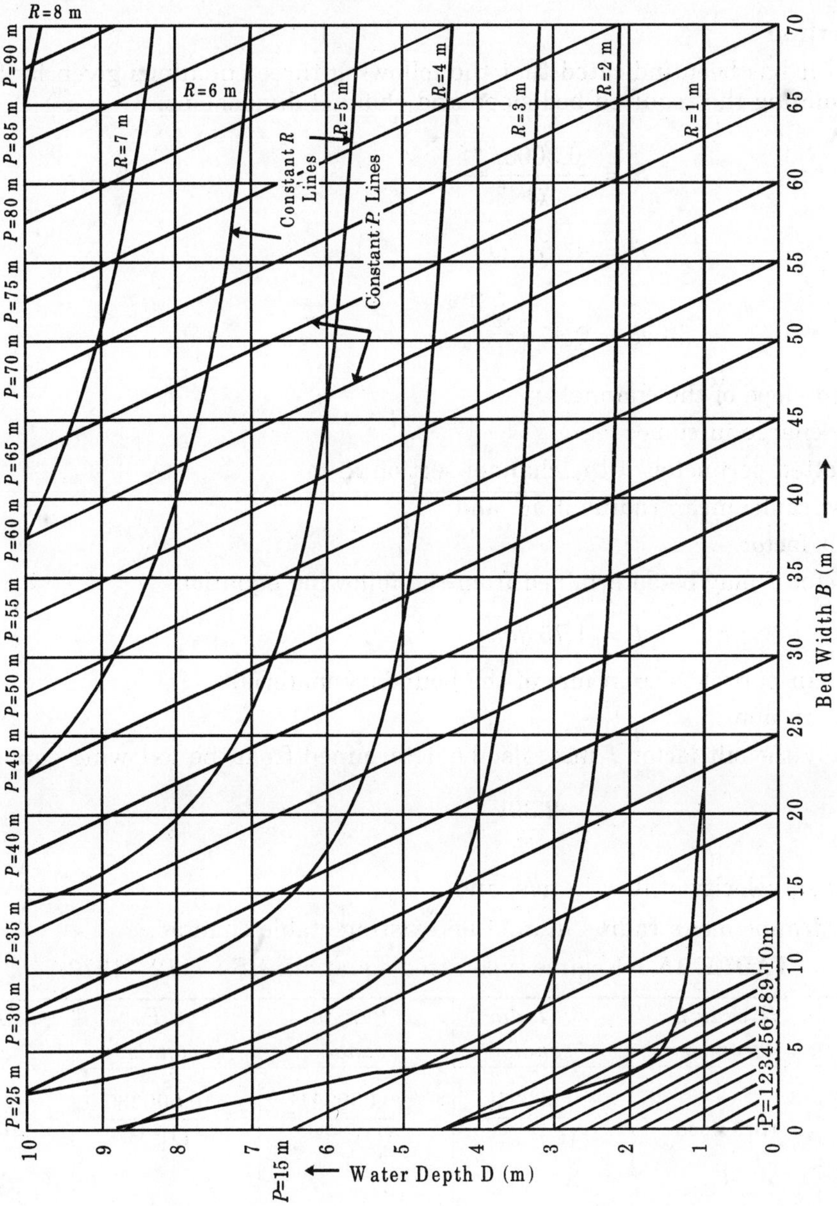

Fig. 15.4 *Hydraulic chart of relationship between B, D, R and P for a channel having side slope (1/2) :1*

15.13 DESIGN OF UNLINED CHANNELS IN ALLUVIAL SOIL ACCORDING TO IS : 7112–1973

For the design of unlined channels in alluvial soil IS: 7112–1973 gives the following two methods.

1. Lacey's Method

In this method it has been indicated that the following three equations given by Lacey may be used for determining the required bed slope and channel dimensions.

$$S = \frac{0.0003 f^{5/3}}{Q^{1/6}} \quad \text{...(i)}$$

$$P = 4.75\sqrt{Q} \quad \text{...(ii)}$$

$$R = 0.47\left(\frac{Q}{f}\right)^{1/3} \quad \text{...(iii)}$$

where S = bed slope of the channel
Q = discharge in cumec
P = wetted perimeter of the channel section in m
R = hydraulic mean radius in m, and
f = silt factor.

The silt factor f may be determined from the following equation

$$f = 1.76\sqrt{m_r} \quad \text{...(iv)}$$

where m_r = mean particle size in mm of the boundary material
= d_{50} in mm.

Alternatively the silt factor f may also be determined from the following equation

$$f = \frac{2.39V^2}{R} \quad \text{...(v)}$$

where V = mean velocity of flow in m/s, and
R = hydraulic mean radius in m of an existing stable channel.

TABLE 15.3 Regime type fitted equations (IS : 7112—1973)

S.No.	*Hydraulic parameter*	*All India canals*	*Punjab canals*	*U.P. canals*	*Bengal canals*
1	S (Bed slope)	$\frac{0.000315}{(Q)^{0.1651}}$	$\frac{0.000251}{(Q)^{0.0961}}$	$\frac{0.000360}{(Q)^{0.1450}}$	$\frac{0.0001346}{(Q)^{0.0375}}$
2	P (Wetted perimeter)	$4.30(Q)^{0.5231}$	$7.00(Q)^{0.6019}$	$3.98(Q)^{0.5020}$	$5.52(Q)^{0.4190}$
3	R (Hydraulic mean radius)	$0.515(Q)^{0.3406}$	$0.466(Q)^{0.3389}$	$0.448(Q)^{0.3649}$	$0.438(Q)^{0.4454}$

Thus, knowing Q and f, the desirable values of S, P and R may be determined from equations (i), (ii) and (iii). Knowing the values of P and R the curves given in Fig. 15.4 may be used for determining the corresponding bed width B and depth D of the channel having side slopes of $\frac{1}{2}$ H : 1 (V) [as it is assumed that the channel attains side slopes of $\frac{1}{2}$ H : 1 (V) when it is in true regime].

Regime Type Fitted Equations for Design of Unlined Channels in Alluvial Soil

The regime type fitted equations evolved on the basis of data collected from various states in our country are given in Table 15.3. In these equations, average boundary condition is taken care of by fitting different equations to data obtained from different states and assuming similar average boundary conditions in a state. Hence silt factor does not appear in these equations.

2. Tractive Force Approach

The unit tractive force (or tractive stress) exerted on the bed of a channel by the flowing water may be calculated from the following equation

$$\tau = wRS \qquad ...(15.46)$$

where τ = unit tractive force in kN/m^2 or [kg(f)/m^2]

w = specific weight of water in kN/m^3 or [kg(f)/m^3] (usually taken as 9.81 kN/m^3 or 1000 kg(f)/m^3

R = hydraulic mean radius in m, and

S = bed slope.

The permissible tractive force (or critical tractive force) may be defined as the maximum tractive force that will not cause serious erosion of the material forming the canal bed on a level surface. The permissible tractive force is a function of sediment concentration and the average particle size (d_{50}) of bed material in case of channels in sandy soils and void ratio in case of channels in clayey soils. For straight channels the values of permissible tractive force per unit area (or critical tractive stress) have been given by some of the investigators on the basis of laboratory experiments, but the same can better be determined by analysis of observed data on existing channels. The value of permissible tractive force for sinuous channels may be reduced by 10% for slightly sinuous ones, by 25% for moderately sinuous ones, and by 40% for very sinuous ones.

Table 15.4 Value of rugosity coefficient N for unlined channels (IS : 7112—1973)

Description of channel	*N*		
	Minimum	*Normal*	*Maximum*
(1) Earth, Straight and Uniform			
(a) Clean, recently completed	0.016	0.018	0.020
(b) Clean, after weathering	0.018	0.022	0.025
(c) Gravel, uniform section, clean	0.022	0.025	0.030
(d) With short grass, few weeds	0.022	0.027	0.033
(2) Earth, winding and sluggish			
(a) No vegetation	0.023	0.025	0.030
(b) Grass, some weeds	0.025	0.030	0.033
(c) Dense weeds or aquatic plants in deep channels	0.030	0.035	0.035

(d) Earth bottom and rubble sides	0.030	0.035	0.040
(e) Stone bottom and weedy banks	0.025	0.035	0.040
(f) Cobble bottom and clean sides	0.030	0.040	0.050
(3) Dragline excavated or dredged			
(a) No vegetation	0.025	0.028	0.033
(b) Light brush on banks	0.035	0.050	0.060
(4) Channels not maintained (weed and brush uncut)			
(a) Dense weeds, high as flow depth	0.050	0.080	0.120
(b) Clean bottom, brush on sides	0.040	0.050	0.080
(c) Same, highest stage of flow	0.045	0.070	0.110
(d) Dense brush, high stage	0.080	0.100	0.140

Note. (1) For normal alluvial soils, it is usual in India to assume a value of $N = 0.02$ for bigger channels ($Q > 15$ cumec) and $N = 0.0225$ for smaller channels ($Q < 15$ cumec).

(2) A suitable value of N should be adopted keeping in view the local conditions and the above values as a guide.

For designing a channel by the tractive force approach, first the sediment concentration X of the flow in the channel and the d_{50} size of the bed material in the case of non-cohesive soils and void ratio of the bed material in the case of cohesive soils are determined and from these corresponding permissible tractive force per unit area (or critical tractive stress) is obtained by use of observed data of existing channels. The value of permissible tractive force per unit area (or critical tractive stress) may also be determined from the expressions given by some of the investigators such as Shield, White etc. A suitable bed slope S of the channel is then selected either with reference to average ground slope along the alignment of the channel or on the basis of experience and the value of R is obtained from Eq. (15.46). Knowing the value of R and assuming a suitable value of N from Table 15.4, the desirable mean velocity of flow V in the channel may be determined by using Manning's equation

$$V = \frac{1}{N} R^{2/3} S^{1/2} \qquad \text{...(15.47)}$$

The area of cross-section A may thus be determined and knowing R and A, the desirable bed width B and depth D of the channel may be determined.

SOLVED ILLUSTRATIVE EXAMPLES

Illustrative Example 15.1 *Design an irrigation channel to carry a discharge of 5 cumec. Assume $N = 0.0225$ and $m = 1$. The channel has a bed slope of 0.2 m per kilometre.*

Solution

1. Assume a trial depth D equal to 1.0 m

2. $$V = 0.55\ m\ D^{0.64} = 0.55 \times 1.0 \times (1.0)^{0.64} = 0.55 \text{ m/s}$$

3. Area $$A = \frac{Q}{V} = \frac{5}{0.55} = 9.09 \text{ m}^2$$

4. $$A = BD + \frac{D^2}{2}$$

or $$9.09 = B \times 1.0 + \frac{(1.0)^2}{2}$$

$\therefore$ $$B = 8.59 \text{ m}$$

5. Perimeter $$P = B + D\sqrt{5} = 8.59 + 1.0\left(\sqrt{5}\right)$$

$$= 10.83 \text{ m}$$

$\therefore$ $$R = AP = \frac{9.09}{10.83} = 0.84 \text{ m}$$

6. $$V = C\sqrt{RS}$$

where $$C = \frac{23 + \frac{1}{N} + \frac{0.00155}{S}}{1 + \left(23 + \frac{0.00155}{S}\right)\frac{N}{\sqrt{R}}}$$

$$R = 0.84 \text{ m} \; ; S = \frac{0.2}{1000} \; ; N = 0.0225$$

$\therefore$ $$C = \frac{23 + \frac{1}{0.0225} + \frac{0.00155}{0.2} \times 1000}{1 + \left(23 + \frac{0.00155}{0.2} \times 1000\right)\frac{0.0225}{\sqrt{0.84}}} = 42.85$$

$\therefore$ $$V = 42.85\sqrt{\frac{0.84 \times 0.2}{1000}} = 0.555 \text{ m/sec}$$

7. Ratio of velocities found in steps (6) and (2)

$$= \frac{0.555}{0.55} = 1.009 \simeq 1.0$$

Hence the assumed D is satisfactory.

Illustrative Example 15.2 *Design an irrigation canal to carry a discharge of 5 cumec. Assume $N = 0.0225$, $m = 1.0$ and $(B/D) = 3.24$.*

Solution

1. $$\frac{B}{D} = 3.24$$

$$A = BD + \frac{D^2}{2}$$

$$= \left[\frac{B}{D}+0.5\right]D^2 = (3.24 + 0.5)\, D^2 = 3.74\, D^2$$

2. $$V = 0.55\, m\, D^{0.64} = 0.55 \times 1.0 \times D^{0.64}$$

$$Q = A \times V$$

or $$Q = (3.74\, D^2)\,(0.55 \times D^{0.64})$$

∴ $$D = \left[\frac{5}{3.74\times 0.55}\right]^{\frac{1}{2.64}} = \left[\frac{5}{3.74\times 0.55}\right]^{\frac{1}{2.64}} = 1.40 \text{ m}$$

3. ∴ $$B = 3.24\, D = 3.24 \times 1.4 = 4.54 \text{ m}$$

$$R = \frac{BD+\dfrac{D^2}{2}}{B+D\sqrt{5}} = \frac{(4.54\times 1.4)+\dfrac{1}{2}(1.4)^2}{4.54+1.4(\sqrt{5})}$$

$$= 0.956 \text{ m}$$

4. $$V = 0.55\, m\, D^{0.64} = 0.55 \times 1.0 \times (1.4)^{0.64} = 0.682 \text{ m/s}$$

5. $$V = \frac{23+\dfrac{1}{N}+\dfrac{0.00155}{S}}{1+\left(23+\dfrac{0.00155}{S}\right)\dfrac{N}{\sqrt{R}}}\sqrt{RS}$$

or $$0.682 = \frac{23+\dfrac{1}{0.0225}+\dfrac{0.00155}{S}}{1+\left(23+\dfrac{0.00155}{S}\right)\dfrac{0.0225}{\sqrt{0.956}}}\sqrt{0.956S}$$

or $$1.04S + 2.42 \times 10^{-6} = 65.96\, S^{3/2} + 1.516 \times 10^{-3}\, S^{1/2}$$

Solving this by trial and error, we get

$$S = \frac{1}{4000}$$

Illustrative Example 15.3 *Determine the dimensions of the irrigation canal for the following data : (B/D) ratio = 3.7 ; N = 0.0225 ; m = 1.0 ; and S = (1/4000). Side slopes of the channel is 1/2 horizontal to 1 vertical. Also determine the discharge which will be flowing in the channel.*

Solution

$$\frac{B}{D} = 3.7$$

∴ $$B = 3.7D$$

For a channel with side slopes 1/2 Hor. 1 Ver.

$$R = \frac{BD+\dfrac{D^2}{2}}{B+D\sqrt{5}}$$

$$= \frac{3.7D^2 + 0.5D^2}{3.7D + D\sqrt{5}} = 0.708\ D$$

From Kennedy's equation

$$V = 0.55\ m\ D^{0.64}$$
$$= 0.55 \times 1.0 \times D^{0.64}$$
$$= 0.55\ D^{0.64}$$

From Kutter's formula

$$V = \frac{23 + \frac{1}{N} + \frac{0.00155}{S}}{1 + \left(23 + \frac{0.00155}{S}\right)\frac{N}{\sqrt{R}}}\sqrt{RS}$$

or
$$V = \frac{23 + \frac{1}{0.0225} + 0.00155 \times 4000}{1 + (23 + 0.00155 \times 4000)\frac{0.0225}{\sqrt{0.708D}}}\sqrt{\frac{0.708D}{4000}}$$

$$= \frac{73.644}{1 + 0.781D^{-1/2}} \times 0.0133\ D^{1/2} = \frac{0.975D^{1/2}}{1 + 0.781D^{-1/2}}$$

Equating the two values of V, we get

$$0.55\ D^{0.64} = \frac{0.9795D^{1/2}}{(1 + 0.781D^{-1/2})}$$

or
$$0.55\ D^{0.64} + 0.4296\ D^{0.14} = 0.9795\ D^{1/2}$$

Solving the above equation by trial and error, we get

$$D = 1.0\ \text{m}$$

$\therefore$
$$B = 3.7 \times 1.0 = 3.7\ \text{m}$$
$$V = 0.55\ \text{m/s}$$

$$A = BD + \frac{D^2}{2} = 3.7 \times 1.0 + \frac{1}{2}(1.0)\ 2 = 4.2\ \text{m}^2$$

$\therefore$
$$Q = A \times V = 4.2 \times 0.55 = 2.31\ \text{cumec}$$

Illustrative Example 15.4 *A channel section is to be designed for the following data :*

Discharge $Q = 5$ *cumec*

Silt factor $f = 1.0$

Side slope $= \frac{1}{2}$ *(H) : 1 (V)*

Also determine the bed slope of the channel.

Solution

1. $$\text{Velocity } V = \left[\frac{Qf^2}{140}\right]^{1/6} = \left[\frac{5\times(1.0)^2}{140}\right]^{1/6} = 0.574 \text{ m/s}$$

2. $$\text{Area } A = QV = \frac{5}{0.574} = 8.71 \text{ m}^2$$

3. $$\text{Perimeter } P = 4.75\sqrt{Q} = 4.75\sqrt{5} = 10.62 \text{ m}$$

4. $$A = BD + \frac{D^2}{2}$$

or $$8.71 = BD + 0.5\,D^2 \qquad \text{...(i)}$$

$$P = B + \sqrt{5}\,D$$

or $$10.62 = B + \sqrt{5}\,D \qquad \text{...(ii)}$$

From Eqs (i) and (ii), we get

$$B = 8.44 \text{ m} \; ; \; D = 0.98 \text{ m}$$

5. $$R = \frac{BD + \dfrac{D^2}{2}}{B + D\sqrt{5}}$$

$$= \frac{(8.44\times 0.98) + 0.5(0.98)^2}{8.44 + 0.98\sqrt{5}}$$

$$= 0.823 \text{ m}$$

6. $$R = \frac{5}{2}\frac{V^2}{f} = \frac{5}{2}\times\frac{(0.574)^2}{1.0} = 0.824 \text{ m}$$ Hence checked.

7. Bed slope $$S = \frac{f^{5/3}}{3340(Q)^{1/6}} = \frac{1}{3340(5)^{1/6}} = \frac{1}{4368} \approx \frac{1}{4400}$$

Hence the channel has bed width B = 8.44 m, depth D = 0.98 m, and bed slope $S = \dfrac{1}{4000}$.

Illustrative Example 15.5 *Design an irrigation channel in alluvial soil according to Lacey's silt theory for the following data.*

Full supply discharge = 10 cumec

Lacey's silt factor = 0.9

Side slopes of channel = $\frac{1}{2}(H):1(V)$.

Solution

1. Velocity $V = \left[\frac{Qf^2}{140}\right]^{1/6}$

$$= \left[\frac{10 \times (0.9)^2}{140}\right]^{1/6}$$

$$= 0.62 \text{ m/s}$$

2. Area $A = QV = \frac{10}{0.62}$

$$= 16.13 \text{ m}^2$$

3. Hydraulic mean radius

$$R = \frac{5}{2}\frac{V^2}{f} = \frac{5}{2} \times \frac{(0.62)^2}{0.9} = 1.07 \text{ m}$$

4. Perimeter $P = \frac{A}{R} = \frac{16.13}{1.07} = 15.07$ m

5. $P = B + D\sqrt{5}$

or $B + D\sqrt{5} = 15.07$ ...(i)

6. $A = BD + \frac{D^2}{2}$

or $BD + \frac{D^2}{2} = 16.13$ ...(ii)

From Eqs (i) and (ii), we get

$$B = 12.27 \text{ m} ; D = 1.25 \text{ m}$$

7. Check $P = 4.75\sqrt{Q} = 4.75\sqrt{10} = 15.02$ m

$$P = B + D\sqrt{5} = 12.27 + 1.25\sqrt{5} = 15.07 \text{ m}$$

The two values of P are nearly equal and hence the computations are correct.

Note. In Illustrative Example 15.4, first A and P were found from which B and D were determined. A check was then made by finding the value of R by Lacey's equation as well as from the computed values of B and D. In this example first A and R have been found to determine B and D. A numerical check has been made in step (7) by finding the values of P from the computed values of B and D and from Lacey's equation. Both these methods are correct and any one can be adopted.

8. Bed slope $S = \frac{f^{5/3}}{3340(Q)^{1/6}} = \frac{(0.9)^{5/3}}{3340(10)^{1/6}}$

$$= \frac{1}{5844} \simeq \frac{1}{5850}$$

Illustrative Example 15.6 *The slope of a channel in alluvium is* $\frac{1}{4000}$, *Lacey's silt factor is 0.9 and side slopes are* $\frac{1}{2}$ *H : 1 (V). Find the channel section and maximum discharge which can be allowed to flow in it.*

Solution

From Eq. (15.18), we have

$$S = \frac{f^{5/3}}{3340Q^{1/6}}$$

or

$$Q = \left[\frac{f^{5/3}}{3340S}\right]^6 = \frac{f^{10}}{(3340S)^6}$$

$$= \left(\frac{4000}{3340}\right)^6 \times (0.9)^{10} = 1.03 \text{ m}^3/\text{sec}$$

From Eq. (15.17), we have

$$S = \frac{f^{3/2}}{4980R^{1/2}}$$

or

$$R = \frac{f^{3}}{(4980S)^2} = \frac{(0.9)^3 \times 4000 \times 4000}{4980 \times 4980} = 0.47 \text{ m}$$

From Eq. (15.15), we have

$$P = 4.75\sqrt{Q} = 4.75\sqrt{1.03} = 4.82 \text{ m}$$

$$A = PR = 4.82 \times 0.47 = 2.27 \text{ m}^2$$

$$A = BD + \frac{D^2}{2} = 2.27 \quad \text{...(i)}$$

$$P = B + D\sqrt{5} = 4.82 \quad \text{...(ii)}$$

Solving the Eqs (i) and (ii), we get

$$B = 3.48 \text{ m}$$

$$D = 0.60 \text{ m.}$$

Illustrative Example 15.7 *A wide channel 2.5 m deep has a bed slope of* $\frac{1}{4500}$. *The bed material consists of uniform grain size of 0.3 mm and having a fall velocity 0.03 m/s in still water. Determine the quantity of bed load transported by the channel according to*

(i) Meyer–Peter's equation

(ii) Einstein's equation

Take representative size of roughness of channel bed, k = 1.5 mm. Also determine the concentration of suspended load 1.0 m above the bed of the channel if the concentration at 0.4 m above the bed is 500 ppm. Take specific gravity of the grains as 2.65.

Solution

(i) According to Meyer–Peter's Eq. (15.38 a) the rate of bed load transported per unit width of channel is given as

$$q_s = 1500.6\left[\tau_b\left(\frac{N'}{N}\right)^{3/2} - \tau_{cr}\right]^{3/2} \text{ N/m/hour}$$

$$N' = \frac{k_s^{1/6}}{24} \text{ and } N = \frac{k^{1/6}}{24}$$

$$k_s = d = 0.3 \times 10^{-3} \text{ m}$$

$$k = 1.5 \times 10^{-3} \text{ m}$$

$$\therefore \quad \left(\frac{N'}{N}\right)^{3/2} = \left(\frac{d}{k}\right)^{1/4}\left(\frac{0.3\times10^{-3}}{1.5\times10^{-3}}\right)^{1/4} = 0.669$$

Since the channel is wide, $(Q_s/Q) = 1$ and $R \simeq D$

$$\therefore \quad \tau_b = \left(\frac{Q_s}{Q}\right)wRS = wDS$$

$$= 9810 \times 2.5 \times \frac{1}{4500} = 5.45 \text{ N/m}^2$$

$$\tau_{cr} = 0.047\,(w_s - w)\,d$$

$$w_s = 2.65 \times 9810 = 25\,997 \text{ N/m}^3$$

$$w = 9810 \text{ N/m}^3$$

$$d = 0.3 \times 10^{-3} \text{ m}$$

$$\therefore \quad \tau_{cr} = 0.047\,(25\,997 - 9810) \times 0.3 \times 10^{-3}$$

$$= 0.228 \text{ N/m}^2$$

$$\therefore \quad q_s = 1500.6\,[5.45 \times 0.669 - 0.228]^{3/2}$$

$$= 9483 \text{ N/m/hour}$$

According to Meyer–Peter's Eq. (15.38 c) the rate of bed load transported per unit width of channel is given as

$$q_s = 4700\left[\tau_b\left(\frac{N'}{N}\right)^{3/2} - \tau_{cr}\right]^{3/2} \text{ kg(f)/m/hour}$$

$$\left(\frac{N'}{N}\right)^{3/2} = 0.669$$

$$\tau_b = wDS = 1000 \times 2.5 \times 14500 = 0.556 \text{ kg(f)/m}^2$$

$$\tau_{cr} = 0.047\,(w_s - w)\,d$$

$$w_s = 2.65 \times 1000 = 2650 \text{ kg(f)/m}^3$$

$$w = 1000 \text{ kg(f)/m}^3$$

$$d = 0.3 \times 10^{-3} \text{ m}$$

$$\therefore \quad \tau_{cr} = 0.047\,(2650 - 1000) \times 0.3 \times 10^{-3}$$

$$= 0.0233 \text{ kg(f)/m}^2$$

$$\therefore \quad q_s = 4700\,[0.556 \times 0.669 - 0.0233]^{3/2}$$

$$= 967.63 \text{ kg(f)/m/hour}$$

(ii) *Einstein's bed-load function method*

For uniform soils

$$\psi_* = \psi = \frac{(G-1)d}{R'S}$$

$$R' = R\left(\frac{N'}{N}\right)^{3/2}$$

For wide channel

$$R = D = 2.5 \text{ m}$$

As derived earlier

$$\left(\frac{N'}{N}\right)^{3/2} = 0.669$$

$$\therefore \quad R' = 2.5 \times 0.669 = 1.673 \text{ m}$$

$$\therefore \quad \psi_* = \frac{(2.65-1)\times 0.3\times 10^{-3}}{1.673\left(\dfrac{1}{4500}\right)}$$

$$= \frac{1.65\times 0.3\times 10^{-3}\times 4500}{1.673} = 1.331$$

From Einstein curves (Fig. 15.3) for

$$\psi_* = 1.331\,; \quad \phi_* = 5.4$$

From Eq. (15.42)

$$\phi_* = \left[\frac{q_s}{Gw}\left(\frac{1}{G-1}\right)^{1/2}\left(\frac{1}{gd^3}\right)^{1/2}\right]$$

or $$q_s = \phi_* Gw(G-1)^{1/2} (gd^3)^{1/2}$$

In SI units

$$q_s = 5.4 \times 2.65 \times 9810 \times (2.65 - 1)^{1/2} \times (9.81)^{1/2} \times (0.3 \times 10^{-3})^{3/2}$$
$$= 2.935 \text{ Nm/s}$$
$$= 2.935 \times 3600 \text{ Nm/hour}$$
$$= 10\ 566 \text{ Nm/hour}$$
$$= 10.566 \text{ kN/m/hour}$$

In metric units

$$q_s = 5.4 \times 2.65 \times 1000 \times (2.65 - 1)^{1/2} \times (9.81)^{1/2} \times (0.3 \times 10^{-3})^{3/2}$$
$$= 0.299 \text{ kg(f)/m/s}$$
$$= 0.299 \times 3600 \text{ kg(f)/m/hour}$$
$$= 1076.4 \text{ kg(f)/m/hour}$$

Concentration of Suspended Load

$$C_a = 500 \text{ ppm at 0.4 m above bed}$$

Fall velocity $$V_s = 0.03 \text{ m/sec}$$

$$V_* = \text{Shear velocity} = \sqrt{\left(\frac{\tau_0}{\rho}\right)} = \sqrt{gR'S}$$

$$R' = 1.673 \text{ m}$$

∴ $$V_* = \sqrt{\frac{9.81 \times 1.673}{4500}} = 0.0604 \text{ m/s}$$

$$k = 0.4$$

∴ $$\frac{V_s}{kV_*} = \frac{0.03}{0.4 \times 0.604}$$
$$= 1.242$$

From Eq. (15.45)

$$\frac{C}{C_a} = \left[\frac{D-y}{y} \cdot \frac{a}{D-a}\right]^{\frac{V_s}{kV_*}}$$

$$\frac{C}{500} = \left[\frac{3-1}{1} \times \frac{0.4}{3-0.4}\right]^{1.242}$$

∴ $$C = 116 \text{ ppm.}$$

Illustrative Example 15.8 *Design an irrigation channel carrying a full supply discharge of 30 cumec with a bed load concentration of 50 parts per million (ppm) by weight. The average grain diameter of the bed material is 0.3 mm.*

Solution

Quantity of bed load to be transported

$$= 50 \text{ ppm}$$
$$= 50 \times 10^{-6} \times 30 \times 9810 \times 3600 \text{ N/hour}$$
$$= 52\,974 \text{ N/hour}$$

The regime perimeter of the channel would be

$$P = 4.75\,Q = 4.75\,30 = 26.02 \text{ m}$$

Assume channel width

$$B = 20 \text{ m}$$

∴ Rate of bed load transported per unit width

$$= \frac{52974}{20} = 2648.7 \text{ Nm/hour}$$

(i) *Design using Meyer–Peter's equation*

From Eq. 15.38 (a)

$$q_s = 1500.6\left[\tau_b\left(\frac{N'}{N}\right)^{3/2} - \tau_{cr}\right]^{3/2} \qquad \text{...(i)}$$

in which

$$N' = \frac{k_s^{1/6}}{24}$$

where k_s is the effective grain diameter.

Since the material is not uniform and the average grain diameter is given as 0.3 mm, let k_s be assumed as 0.4 mm

$$\therefore \qquad N' = \frac{1}{24}\left(\frac{0.4}{1000}\right)^{1/6} = 0.0113$$

From Lacey's Eq. (15.12)

$$f = 1.76\,m_r = 1.76\,\sqrt{0.3} = 0.964 \simeq 1.0$$

Since silt factor f is nearly equal to 1, from Lacey's Eq. (15.32), the value of Manning's N may be taken as 0.0225.

$$\therefore \qquad \frac{N'}{N} = \frac{0.0113}{0.0225} = 0.502$$

$$\tau_b = \frac{Q_s}{Q} wRS$$

Assuming $\dfrac{Q_s}{Q} = 1$, we have

$$\tau_b = wRS$$
$$= 9810\ RS\ \text{N/m}^2$$

Critical tractive stress

$$\tau_{cr} = 0.047\ (w_s - w)\ d$$
$$= 0.047 \times 9810\ (2.65 - 1) \times 0.3 \times 10^{-3}$$
$$= 0.228\ \text{N/m}^2$$

Substituting the values in Eq. (i), we get

$$2648.7 = 1500.6\ [9810 \times R \times S\ (0.502)^{3/2} - 0.228]^{3/2}$$

or $$RS = 484 \times 10^{-6} \qquad \text{...(ii)}$$

From Manning's equation, we have

$$Q = \frac{1}{N} AR^{2/3}\ S^{1/2}$$

$$A = p \times R = 26.02\ R$$

$\therefore$ $$30 = \frac{1}{0.0225}(26.02R)\,R^{2/3}\ S^{1/2}$$

or $$R^{5/3}\ S^{1/2} = 259.4 \times 10^{-4} \qquad \text{...(iii)}$$

from Eq. (ii)

$$S = \frac{484 \times 10^{-6}}{R}$$

or $$S^{1/2} = \frac{22.0 \times 10^{-3}}{R^{1/2}}$$

Substituting the value of $S^{1/2}$ in Eq. (iii), we get

$$R^{5/3} \times \frac{22.0 \times 10^{-3}}{R^{1/2}} = 259.4 \times 10^{-4}$$

or $$R^{7/6} = \frac{259.4 \times 10^{-4}}{22.0 \times 10^{-3}}$$

$\therefore$ $$R = 1.15\ \text{m}$$

$\therefore$ $$S = \frac{484 \times 10^{-6}}{1.15}$$

$$= \frac{1}{2376}$$

If D is the depth of the channel and the side slopes are $\frac{1}{2}$ H : 1 (V) then

$$P = B + D\sqrt{5}$$
$$= 20 + 2.236\,D$$

Also
$$A = BD + \frac{D^2}{2} = 20\,D + \frac{D^2}{2}$$

∴ Hydraulic mean radius

$$R = \frac{A}{P}$$

or
$$1.15 = \frac{20D + \frac{D^2}{2}}{20 + 2.236D}$$

or
$$D = 1.27 \text{ m}$$

Hence the channel has bed width B = 20 m ; depth D = 1.27 m ; and bed slope $S = \frac{1}{2376}$.

(ii) *Design using Einstein's equations*

From Eq. (15.42)

$$\phi_* = \phi = \left[\frac{q_s}{Gw}\left(\frac{1}{G-1}\right)^{1/2}\left(\frac{1}{gd^3}\right)^{1/2}\right]$$

$$= \frac{2648.7}{3600 \times 2.65 \times 9810 \times (2.65-1)^{1/2} \times (9.81)^{1/2} \times (0.3 \times 10^{-3})^{3/2}} = 1.354$$

From Fig. 15.3 for ϕ_* = 1.354, ψ_* = 3.5

From Eq. (15.43)

$$\psi_* = \psi = \left[\frac{(G-1)d}{R'S}\right]$$

or
$$3.5 = \frac{(2.65-1) \times 0.3 \times 10^{-3}}{R'S}$$

or
$$R'S = 141.43 \times 10^{-6} \qquad \text{...(iv)}$$

From Eq. (15.41), we have

$$R' = R\left(\frac{N'}{N}\right)^{3/2} = R\,(0.502)^{3/2} = 0.356\,R$$

Substituting the value of R' in Eq. (iv), we get

$$(0.356\,R)\,S = 141.43 \times 10^{-6}$$

or
$$RS = 397.28 \times 10^{-6} \quad \text{...(v)}$$

Squaring both sides of Eq. (iii), we get

$$R^{10/3}S = 672.88 \times 10^{-6} \quad \text{...(vi)}$$

Dividing Eq. (vi) by Eq. (v), we get

$$R^{7/3} = \frac{672.88 \times 10^{-6}}{397.28 \times 10^{-6}}$$

or
$$R = 1.25 \text{ m}$$

Substituting the value of R in Eq. (v), we get

$$S = \frac{397.28 \times 10^{-6}}{1.25} = \frac{1}{3146}$$

Hydraulic mean radius

$$R = \frac{A}{P}$$

or
$$1.25 = \frac{20D + \dfrac{D^2}{2}}{20 + 2.236D}$$

or
$$D = 1.4 \text{ m}$$

Hence the channel has bed width B = 20 m ; depth D = 1.4 m ; and bed slope $S = \dfrac{1}{3146}$.

REVIEW QUESTIONS

15.1 Describe Kennedy's theory for the design of irrigation channel in alluvial soil.

15.2 Explain the procedure for designing an irrigation channel using Kennedy's theory.

15.3 Discuss the drawbacks of Kennedy's theory.

15.4 Derive an expression for the silt transporting capacity of a channel according to Kennedy's theory.

15.5 For a channel discharge Q, rugosity coefficient N, critical velocity ratio m and bed slope S are given. Explain how would you design the channel using Kennedy's theory ?

15.6 For a channel discharge Q, rugosity coefficient N, critical velocity ratio m and bed width-depth ratio B/D are given. Explain how would you design the channel using Kennedy's theory.

15.7 Describe Lacey's theory for the design of irrigation channel in alluvial soil.

15.8 What do you understand by Lacey's initial and final regime conditions?

15.9 Discuss the drawbacks of Lacey's theory.

15.10 Explain the procedure for designing an irrigation channel using Lacey's theory.

15.11 Compare Kennedy's and Lacey's theories for the design of irrigation channel in alluvial soil.

15.12 Explain the terms: (i) bed load; (ii) suspended load; (iii) critical tractive force.

15.13 Design an irrigation channel to carry a discharge of 10 cumec. Assume $N = 0.0225$ and $m = 1$. The channel has a bed slope of 0.3 m per kilometre. [*Ans.* $B = 4.105$ m; $D = 2.15$ m]

15.14 Design an irrigation canal to carry a discharge of 10 cumec. Assume $N = 0.0225$, m = 1.0 and $(B/D) = 3.5$.

$$\left[Ans.\ B = 6.212\text{m};\ D = 1.775\text{m};\ S = \frac{1}{4200}\right]$$

15.15 A channel section is to be designed for the following data: Discharge $Q = 10$ cumec, silt factor $f = 1.0$, side slope $= \frac{1}{2}$ H : 1(V). Also determine the bed-slope of the channel.

$$\left[Ans. B = 12.34\text{m};\ D = 1.2\text{m};\ S = \frac{1}{4900}\right]$$

15.16 Using Lacey's basic regime equations derive the perimeter-discharge relationship.

15.17 A stable channel is to be designed for a discharge of 40 m^3/s and $f = 1.0$. Calculate the dimensions of the channel using Lacey's regime equations. What would be the bed width of this channel if it is designed on the basis of Kennedy's theory ? Adopt $m = 1.0$ and B/D ratio same as obtained from Lacey's equation.

[*Ans.* Laceys theory: $B = 25.941$ m $D = 1.834$ m,
Kennedy's theory: $B = 25.954$ m $D = 1.835$ m]

CHAPTER

16 Design Procedure for an Irrigation Channel

16.1 INTRODUCTION

The various considerations to be made while fixing the alignment of an irrigation channel and the theories of designing stable channels in alluvium are discussed in some of the earlier chapters. In the present chapter an outline of the actual design procedure for irrigation channels based on the field data such as culturable commanded area, intensity of irrigation, crops to be grown, duty of crops, area under various crops, etc., is discussed.

16.2 LONGITUDINAL SECTION OF A CHANNEL

For preparing the longitudinal section (or L-section) of a channel following procedure is generally adopted.

1. Along the final alignment of the channel marked in the field, ground levels are accurately determined by double levelling, with cross-section at every 200 m or so.

2. The ground levels determined along the alignment of the channel are plotted with reference to convenient datum to obtain the longitudinal section (or L-section) of the ground as shown in Fig. 16.1 (Plate V). For this plotting IS : 5968–1970 recommends a horizontal scale of 1 in 10000 to 1 in 20000 and a vertical scale of 1 in 100. The horizontal scale generally adopted is 1 in 15000. The vertical scale may, however, depend on the magnitude of fall in ground level which is to be plotted.

3. The bed level and full supply level (F.S.L.) of the parent channel are marked just upstream of the head of the offtaking channel for reference purposes.

4. Trial full supply line and bed slope which are parallel to each other are marked for the offtaking channel keeping in view the following points.

(a) The F.S.L. of the offtaking channel at its head should be kept at least 150 mm or 15 cm lower than the water level of the parent channel. This is generally done (i) to provide for the loss of head at the head regulator; (ii) to meet the demand for extra supplies in the channel at a future date ; (iii) to maintain the flow at almost full supply even if the bed silted up to some extent in its head reaches. For main canals taking off from river, F.S.L. is kept about 1 m lower than the pond level of the reservoir.

(b) The full supply line should be above the ground line for most of its length so that most of the commanded area may be irrigated by gravity flow. However, the full supply level needs to be only a few centimetres above the ground level because the channel is running on watershed and hence the ground level on both sides of the channel will be lower and mostly ample cross slopes will be available to provide irrigation for the entire commanded area. Further full supply level need not be taken above isolated high patches of ground. In such reaches the high fields adjacent to the channel may be irrigated by lifting water from the channel. A higher full supply line is undesirable because it results in uneconomical section of a channel, excessive absorption losses and more wasteful use of water by the cultivator due to higher heads being available.

(c) In order to prevent entry of silt into the offtaking channel the bed level of the offtaking channel at its head should be kept higher than the bed level of the parent channel.

(d) The bed slope to be provided for the channel may be chosen close to that obtained by using Lacey's regime slope equation or that given by Wood's normal tables. However, if the available ground slope is less than that obtained by Lacey's equation or Wood's normal table then the maximum slope which is available for the ground may be provided. In case of such channel efforts should be made to reduce their silt factor by preventing the entry of coarse silt at the head of the channel.

(e) If the ground slope exceeds the bed slope to be provided for the channel then the extra fall in ground level should be consumed by providing vertical falls. A fall may be provided at a location where the F.S.L. outstrips the ground level but before the bed of the channel comes into filling. The magnitude of the fall may be such that on downstream of the fall the F.S.L. may be below the ground level for half to three-fourths of a kilometre, as upto this extent the area can be easily irrigated by a watercourse from an outlet at high level upstream of the fall. The other factors which affect the location of falls are discussed in Chapter 19.

(f) As far as possible the channel should be kept in balanced depth of cutting and filling for greatest economy and minimum necessity of borrow pits and spoil banks.

The following information shall be added below the datum line on the longitudinal section (see Fig. 16.1)

(i) Reduced distance
(ii) Natural surface level
(iii) Bed level of the channel
(iv) Full supply level
(v) Full supply depth
(vi) Bed width
(vii) Subsoil water level
(viii) Water surface slope
(ix) Depth of cutting
(x) Height of filling
(xi) Value of N
(xii) Side slopes
(xiii) Full supply discharge
(xiv) Velocity
(xv) Free board
(xvi) Bank width (Left/Right), and
(xvii) Broad details of hydraulic data of outlets, regulators, bridges, drainage crossings, offtaking channels etc.

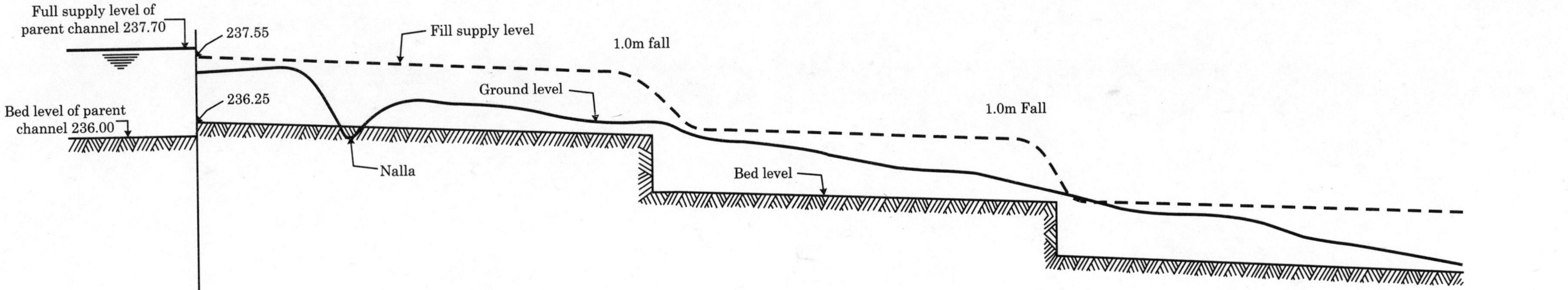

Datum 230.00

	0.0–1.0 km	1.0–2.0 km	2.0–3.0 km	3.0–4.0 km	4.0–5.0 km	
Discharge	5.70 Cumec	5.00 Cumec	4.35 Cumec	3.70 cumec	3.00 cumec	2.75 Cumec
Bed slope	22.5 cm/km	22.5 cm/km	22.5 cm/km	25 cm/km	25 cm/km	25 cm/km
Bed width	5.75 m	5.75 m	5.25 m	4.50 m	4.20 m	4.00 m
Water depth	1.300 m	1.245 m	1.210 m	1.180 m	1.080 m	1.060 m

Outlet

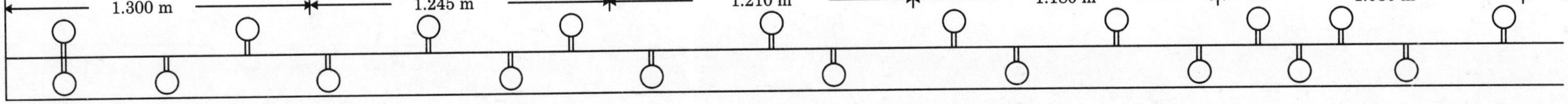

Distance in kilometre	Depth of filling	Depth of digging	R.L. of proposed bed	R.L. of ground
0.0		0.97	236.25	237.22
0.2		1.07	236.21	237.28
0.4		1.15	236.16	237.31
0.6	0.19		236.12	235.93
0.8		0.51	236.07	236.58
1.0		0.51	236.03	236.54
1.2		0.55	235.98	236.53
1.4		0.36	235.94	236.30
1.6		0.24	235.89	236.13
1.8		1.31	234.85	236.16
2.0		1.05	234.80	235.85
2.2		0.94	234.76	235.70
2.4		0.89	234.71	235.60
2.6		0.70	234.67	235.37
2.8		0.57	234.62	235.19
3.0		0.56	234.58	235.14
3.2		0.44	234.53	234.97
3.4		1.24	233.48	234.72
3.6		0.98	233.43	234.41
3.8		0.92	233.38	234.30
4.0		0.88	233.33	234.21
4.2		0.84	233.28	234.12
4.4		0.53	233.23	233.76
4.6		0.42	233.18	233.60
4.8		0.27	233.13	233.40
5.0		0.12	233.08	233.20

Fig. 16.1 Plate -VL - section

16.3 BALANCING DEPTH

Usually a channel section is partly in cutting and partly in filling (Fig. 16.2) so that the material excavated may be used for making the banks. If for a channel section the depth of cutting is such that the quantity of excavation or cutting is equal to the earth filling required for making the banks, then this depth of cutting is known as *balancing depth.* For a channel of given cross-section there is only one balancing depth or depth of cutting at which the earthwork in cutting and filling will exactly balance each other. A channel having balancing depth is quite economical because in this case both excavation and deposition in the bank being at the same place the payment will have to be made for only one of these two operations of earthwork. Moreover, in this case there would be no necessity of borrow pits or spoil banks. The balancing depth may be determined as indicated below.

With reference to Fig. 16.2,

Let h be the height of bank above the channel bed

D be the full supply depth of the channel

B be the bed width of the channel

y be the balancing depth

t be the top width of the bank

$n : 1$ be the side slope of the channel in filling, and

$z : 1$ be the side slope of the channel in cutting.

Thus,

$$\text{Area of cutting} = (B + zy)y$$

$$\text{Area of filling} = 2[t + n(h - y)](h - y)$$

Equating the area of cutting and filling, we get

$$(B + zy)y = 2[t + n\,(h - y)](h - y)$$

or

$$y^2(2n - z) - (B + 4nh + 2t)y + h(2t + 2nh) = 0 \qquad \text{...(16.1)}$$

A channel is usually constructed with a side slope of 1 : 1 in cutting and 1.5 : 1 in filling, i.e., $n = 1.5$ and $z = 1.0$. Introducing these values of n and z in Eq. 16.1, we get

$$2y^2 - (B + 6h + 2t)y + h(2t + 3h) = 0 \qquad \text{...(16.1 a)}$$

From Eq. (16.1) the balancing depth y may be determined.

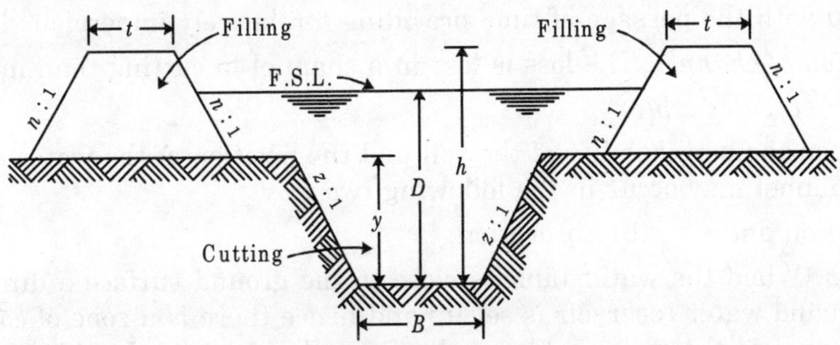

Fig. 16.2. *Balancing depth*

16.4 LOSSES IN CHANNELS

From the headworks where water enters the main canal to the end of the water course where water is supplied to the field there are continuous losses of water which are known as transmission or transit losses. In unlined channels these losses are considerable being 25 to 50 percent of the water that enters the channel at the headworks. However, in lined channels the losses are relatively less. The losses in channels take place on account of the following three causes.

(i) Evaporation

(ii) Transpiration through bank vegetation or weeds

(iii) Percolation

The *loss due to evaporation* is usually a small proportion of the total loss in an unlined channel. For example, on the Upper Ganga Canal evaporation losses have been estimated to be less than 1 percent of the water entering the canal head. The evaporation loss depends on water surface area, relative humidity, wind velocity and temperature. In hot and dry summer months the loss due to evaporation is maximum but it seldom exceeds 10 percent of the total loss. No measure can however be taken to reduce the evaporation loss in channels. Further for design purposes no separate consideration of evaporation loss is made but it is usual to consider the same with percolation loss.

The *loss due to transpiration* through bank vegetation or weeds is also small proportion of the total loss in an unlined channel. This loss can however be reduced by keeping the banks clear of vegetation or weeds.

The *loss due to percolation* also known as *seepage loss* is the main loss as it forms the major portion of the total loss in an unlined channel. The percolation or seepage loss depends on the following factors.

(i) Permeability of the soil in the bed and banks of the channel and the subsoil.

(ii) Depth of water in the channel : greater is the depth, greater is the loss.

(iii) Velocity of flow in the channel : the loss decreases with increase in the velocity of flow in the channel.

(iv) Amount of silt carried in suspension or turbidity of the flowing water. The loss decreases with an increase in the amount of silt carried in suspension.

(v) *Temperature of flowing water.* The loss increases with increase in the temperature of flowing water.

(vi) *Age of channel.* The loss is more in a newly constructed channel and reduces as fine silt gets deposited with the passage of time providing a relatively impervious lining.

(vii) *Condition of channel.* The loss is less in a channel in cutting than in filling.

(viii) *Position of water table.*

Depending upon the permeability of the soil and the position of the water table the seepage from a unlined channel may occur in the following two ways.

(a) Percolation and (b) Absorption.

(a) *Percolation.* When the water table is close to the ground surface a direct flow from the channel to the ground water reservoir is set up and hence there is a zone of continuous saturation from the channel to the water table as shown in Fig. 16.3 (a). In this case almost all the water lost from the channel joins the ground water reservoir. The seepage loss will depend on the total seepage head H, i.e., the difference between the water surface level in the channel and the water table, and will be independent of the water depth in the channel.

(b) *Absorption.* When the water table is at a considerable depth below the ground surface the water seeping from the channel is unable to reach the ground water reservoir, but a small zone of complete saturation is formed round the channel section which is surrounded by a zone of partial saturation as shown in Fig. 16.3 (b). Above the water table below there will be a capillary zone and on the top of that there will be soil of decreasing moisture. Obviously in this case the water seeping from the channel does not augment the ground water storage. Further in this case the seepage loss will depend only on the depth of water in the channel.

Seepage loss by 'percolation' is much more than the seepage loss by 'absorption', the proportion being as high as three times or more.

The losses in a channel are usually measured by a simple method known as *inflow and outflow method.* In this method a long reach of the channel is selected. Discharge observations are taken at the beginning and end of the reach continuously for a number of days. A fairly high level of water should be maintained in the reach. The outlets or offtaking channels if any within the reach should be completely closed during the observation period. The difference between the discharge entering the reach and that leaving the reach is the loss occurring in the reach.

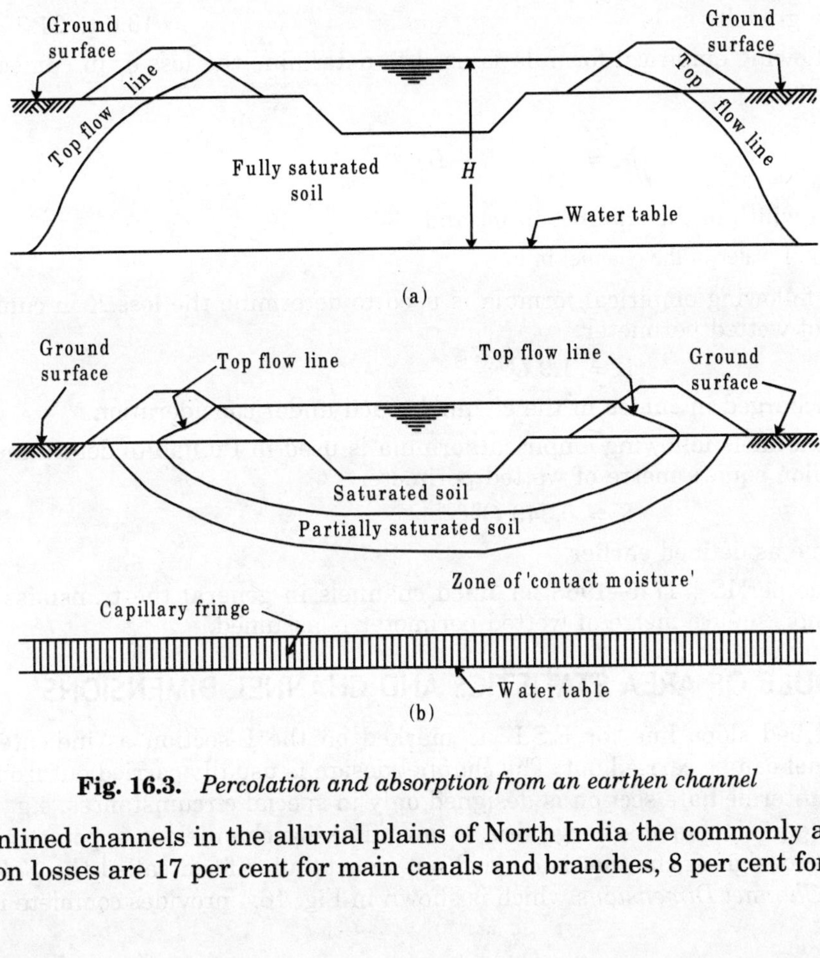

Fig. 16.3. *Percolation and absorption from an earthen channel*

For the unlined channels in the alluvial plains of North India the commonly accepted values of transmission losses are 17 per cent for main canals and branches, 8 per cent for distributaries and 20 percent for water courses which gives a total loss of 45 percent of the water entering the canal head.

In the design of unlined irrigation channels in the alluvial plains of North India the transmission losses are assumed as 2.5 cumec per million square metre of wetted perimeter. However, for channels in various soils following values of transmission losses are assumed :

Type of soil	*Loss in cumec per million square metre of wetted perimeter*
(1) Impervious clay loam	0.9 to 1.2
(2) Medium clay loam with hard pan at a depth not more than 1 m below the bed	1.2 to 1.8
(3) Ordinary clay loam	1.8 to 2.7
(4) Gravelly or sandy clay loam	2.7 to 3.6
(5) Sandy loam	3.6 to 5.2
(6) Loose sandy soils	5.2 to 6.1
(7) Gravelly sandy soils	7.0 to 8.8
(8) Porous gravelly soils	8.8 to 10.6
(9) Very gravelly soils	10.6 to 21.2

In U.P. following empirical formula is used to determine the loss q_L in cumec per kilometre length of channel

$$q_L = \frac{1}{200} (B + D)^{2/3} \quad \ldots(16.2)$$

where B is bed width of the channel in m, and

D is depth of water in the channel in m.

In Punjab following empirical formula is used to determine the loss K in cumec per million square metre of wetted perimeter.

$$K = 1.9\, Q^{0.0625} \quad \ldots(16.3)$$

where Q is discharged in cumec in the channel reach under consideration.

For lined channels following empirical formula is used in Punjab to determine the loss K in cumec per million square metre of wetted perimeter.

$$K = 0.349\, Q^{0.056} \quad \ldots(16.4)$$

where Q is same as defined earlier.

However, as per IS : 4745–1968 for lined channels in general the transmission loss of 0.6 cumec per million square metre of wetted perimeter is assumed.

16.5 SCHEDULE OF AREA STATISTICS AND CHANNEL DIMENSIONS

After the trial bed slope line for F.S.L. is marked on the L-section as indicated earlier, the design of channel can be carried out. The channel design is usually carried out from kilometre to kilometre and intermediate section is designed only in special circumstances, e.g., there may be a substantial reduction in discharge due to an offtaking channel at a point in between two kilometres. The design of channel is carried out in a tabular form called the *Schedule of Area Statistics and Channel Dimensions* which is shown in Fig. 16.4 provides complete information in

Below kilometre	Gross commanded area	Culturable commanded area	Area to be irrigated			Outlet discharge factor	Outlet discharge	Lossee in reach	Total losses	Total discharge	Design discharge	Channel dimensions					Velocity	C.V.R. m=(V/Vo) or silt factor f
			Rabi	Perenial	Kharif							Bed slope	Bed width	Water depth	Height of banks	Width of banks		
	ha	ha	ha	ha	ha	ha per m^3/s	m^3/s	m^3/s	m^3/s	m^3/s	m^3/s		m	m	m	m	m/s	
1	2	3	4	5	6	7	8	9	10	11	12	13	14	15	16	17	18	19
0																		
1																		
2																		
3																		
4																		

ha = Hectare =10,000 m^2

m^3/s = Cubic metre per second = cumec

m = Metre

m/s = Metre per second ha per

ha per
m^3/s = Hectare per cumec

Fig. 16.4. *Schedule of area statistics and channel dimensions*

respect of the area irrigated, outlet discharge factor, discharge, losses, channel dimensions, velocity and silt factor or critical velocity ratio. This table is normally prepared on the same sheet on which L-section is drawn. The working of this table is explained below.

Col. 1. In this column successive channel reaches each of 1 km length commencing from the head of the channel are entered. However, if at any section in between any two kilometres the cross-section of the channel needs to be changed then the location of this section is also entered in this column. The heading of this column is *below kilometre* which indicates that the information given in other columns pertains to the channel reach below the kilometre shown in this column.

Col. 2. In this column gross commanded areas below the successive kilometres are entered. The gross commanded area below a kilometre is the entire area under the command of the channel below the kilometre for which it is being designed.

Col. 3. In this column the culturable commanded areas below the successive kilometres are entered. The culturable commanded area below a kilometre is the area which can be actually cultivated. Thus, it is equal to the gross commanded area entered in Col. 2 minus the areas of the unculturable land included in the gross commanded area.

Cols. 4 to 6. In these columns the areas to be irrigated under Rabi, Perennial (or sugarcane) and Kharif crops below the successive kilometres are entered. These areas are obtained by multiplying the culturable commanded areas in Col. 3 by the respective intensities of irrigation fixed for these crops or seasons.

Col. 7. In this column the outlet discharge factor for the controlling crop is entered. The outlet discharge factor may be obtained as indicated in Chapter 4 and a controlling crop is the one which has a higher demand for water. Usually wheat (Rabi) or rice (Kharif) is the controlling crop and out of these two mostly wheat (Rabi) is the controlling crop. Moreover if the supply is enough for meeting the higher of the demands for these two crops it is usually sufficient for sugarcane as well and if it is so the areas for sugarcane need not be entered.

Col. 8. In this column the outlet discharge required below the successive kilometres is entered. The outlet discharge is obtained by dividing the 'area to be irrigated' of the controlling crop (i.e., one of Cols. 4, 5 or 6) by the outlet discharge factor given in Col. 7.

Col. 7. In this column the losses in each of the channel reaches are entered. By 'losses in reach' is meant the transmission losses occurring in the length of channel lying between the kilometre entered in that horizontal row and the next below it. The transmission losses are generally assumed as 2.5 cumec per million square metre of wetted perimeter for unlined channels and 0.6 cumec per million square metre of wetted perimeter for lined channels. The losses in each reach of the channel are however, calculated on the basis of tentative cross-section of the channel in each reach because the actual cross-section of the channel will be obtained only after its design.

Col. 10. In this column total losses occurring in the entire length of the channel below the successive kilometres are entered. These are obtained by a progressive summation of the previous column from bottom upwards.

Col. 11. In this column total discharge in each of the reaches of the channel below the successive kilometres is entered. The total discharge is obtained by adding the outlet discharge (Col. 8) and the total losses (Col. 10).

Col. 12. In this column the design discharge for each of the reaches of the channel below the successive kilometres is entered. The design discharge is the discharge for which the actual design of the channel is carried out. It is usual practice to design an irrigation channel so that it is capable of taking an overload of 10% above the exact requirements (worked out in Col. 11) in times of keen demand. Thus, the design discharge is obtained by multiplying the total discharge in Col. 11 by 1.1.

Col. 13. In this column the bed slope of each of the channel reaches is entered. The bed slope may be obtained from the L-section.

Cols. 14 to 17. The column pertain to the dimensions of the channel section which are filled up after designing the channel in each reach. The design of channel is carried out by either Kennedy's or Lacey's method. The design of channel by Kennedy's method is usually carried out by Garret's diagrams and that by Lacey's method is carried out by Lacey's regime diagrams. The use of Garret's and Lacey's diagrams for the design of channels is indicated in the next section.

Col. 18. In this column the velocity of flow in each of the channel reaches is entered. This is worked out by dividing the discharge by the cross-sectional area of the channel in each reach.

Col. 19. In this column the value of the critical velocity ratio ($m = V_0/V$) or silt factor f for each of the channel reaches is entered.

All the figures upto Col. 8 can be worked out and filled in as explained above. Beyond this column a step by step procedure is followed from the tail end upwards which is indicated in Illustrative Example 16.4.

16.6 USE OF GARRET'S DIAGRAMS FOR CHANNEL DESIGN

Garret's diagrams provide a graphical method for the design of channels based on Kennedy's theory. The original diagrams were prepared in F.P.S. units, but the same have been converted in metric units. Figure 16.5 [Plate VI (a) and (b)] shows two such diagrams in metric units. In these diagrams discharge is plotted on the abscissa, bed slope is plotted on the left ordinate and the depth and critical velocity are plotted on the right ordinate. The diagrams consist of discharge and bed width curves and in order to distinguish the same the discharge curves are shown by firm line and the bed width curves are shown by dotted lines. The diagrams on different sheets are meant for different ranges of bed slopes and discharges. Further these diagrams have been drawn for $N = 0.0225$, but a nomogram for different values of N varying from 0.018 to 0.03 is provided at the top on each sheet with the help of which these diagrams may be used for any value of N in the range of 0.018 to 0.03.

The procedure for the design of channels by using Garret's diagrams consists of the following steps.

1. Find the discharge for which the channel is to be designed. This may be found from Schedule of Area Statistics.

2. Find the bed slope of the channel from the L-section.

3. Follow the discharge curve corresponding to the design discharge found in step (1) and obtain the point of intersection X of this discharge curve and a horizontal line drawn corresponding to the bed slope obtained in step (2). If necessary interpolation may be done to locate the discharge curve.

4. Draw a vertical line through the point of intersection X obtained in step (3). This line will intersect several bed width curves. A horizontal line drawn through each point of intersection of vertical line and bed width curve gives a depth and critical velocity V_0 corresponding to this depth on the right ordinate.

5. Choose a pair of bed width, depth and V_0 corresponding to a point of intersection obtained in step (4).

6. Calculate the area of channel section corresponding to bed width and depth obtained in step (5).

7. Calculate the velocity V in the channel corresponding to the area A of the channel section obtained in step (6).

8. Calculate the critical velocity ratio (V/V_0). If this is equal to unity or equal to the value given for the channel design, the pair of bed width and depth considered in step (5) may be adopted. If not repeat the procedure with other pairs of bed width and depth till the value of (V/V_0) obtained is equal to unity or equal to the value given for channel design.

In order to simplify the above noted procedure for the design of channels the results may be arranged in a tabular form as indicated in Illustrative Example 16.2. Further it may be noted that Garret's diagrams are applicable only for channel sections having side slopes of 1/2 (H) : $1(V)$.

As stated earlier Garret's diagrams have been drawn for $N = 0.0225$, which is marked by a vertical arrow in the nomogram for N provided at the top on each sheet. The other values of N are shown on left and right of the arrow on the scale. If these diagrams are to be used for any other value of N, then measure the distance between that value of N and the arrow and shift the position of point X obtained in step (3) horizontally by the same distance and in the same

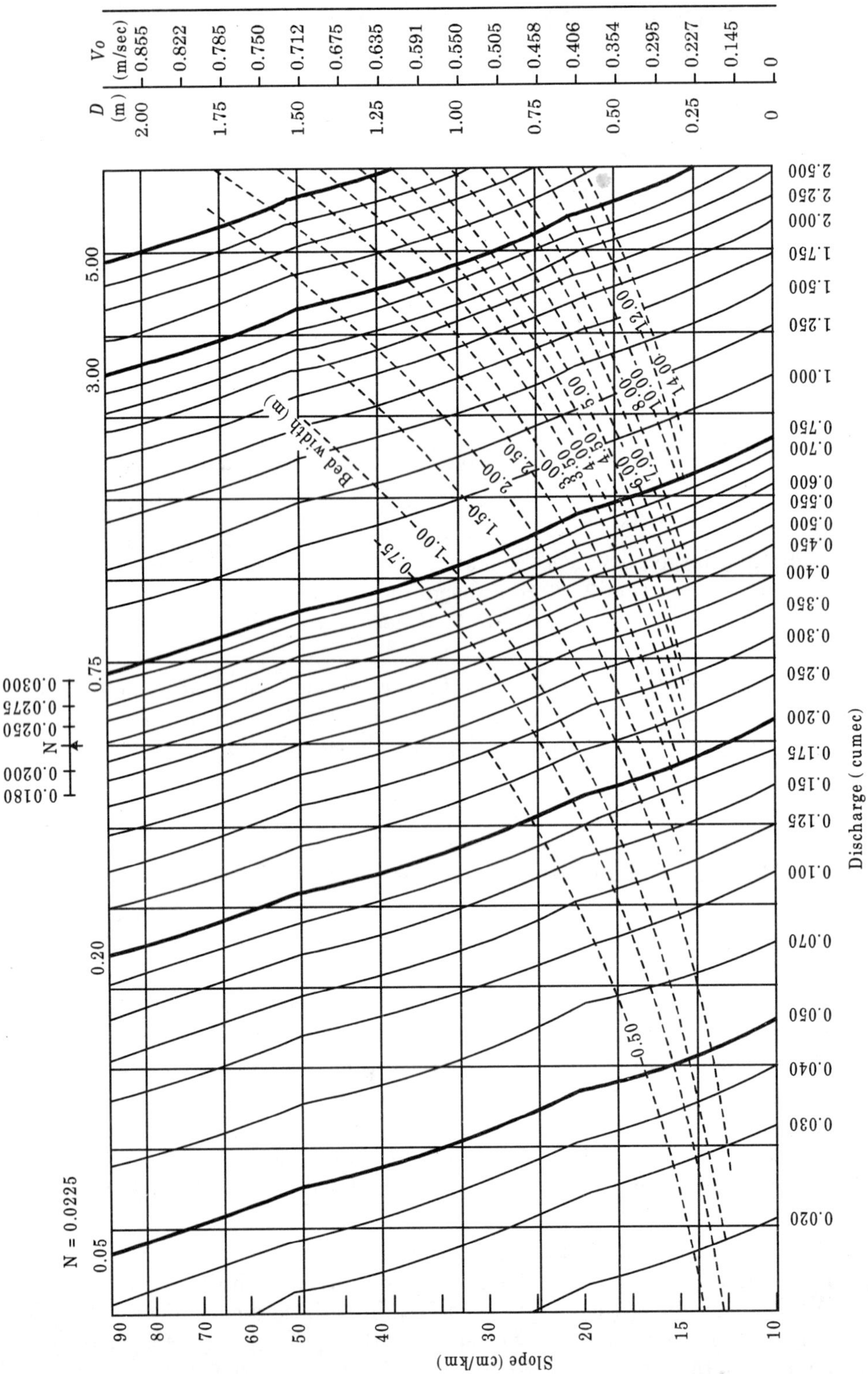

Fig. 16.5 Plate VI (a) Garret's diagram

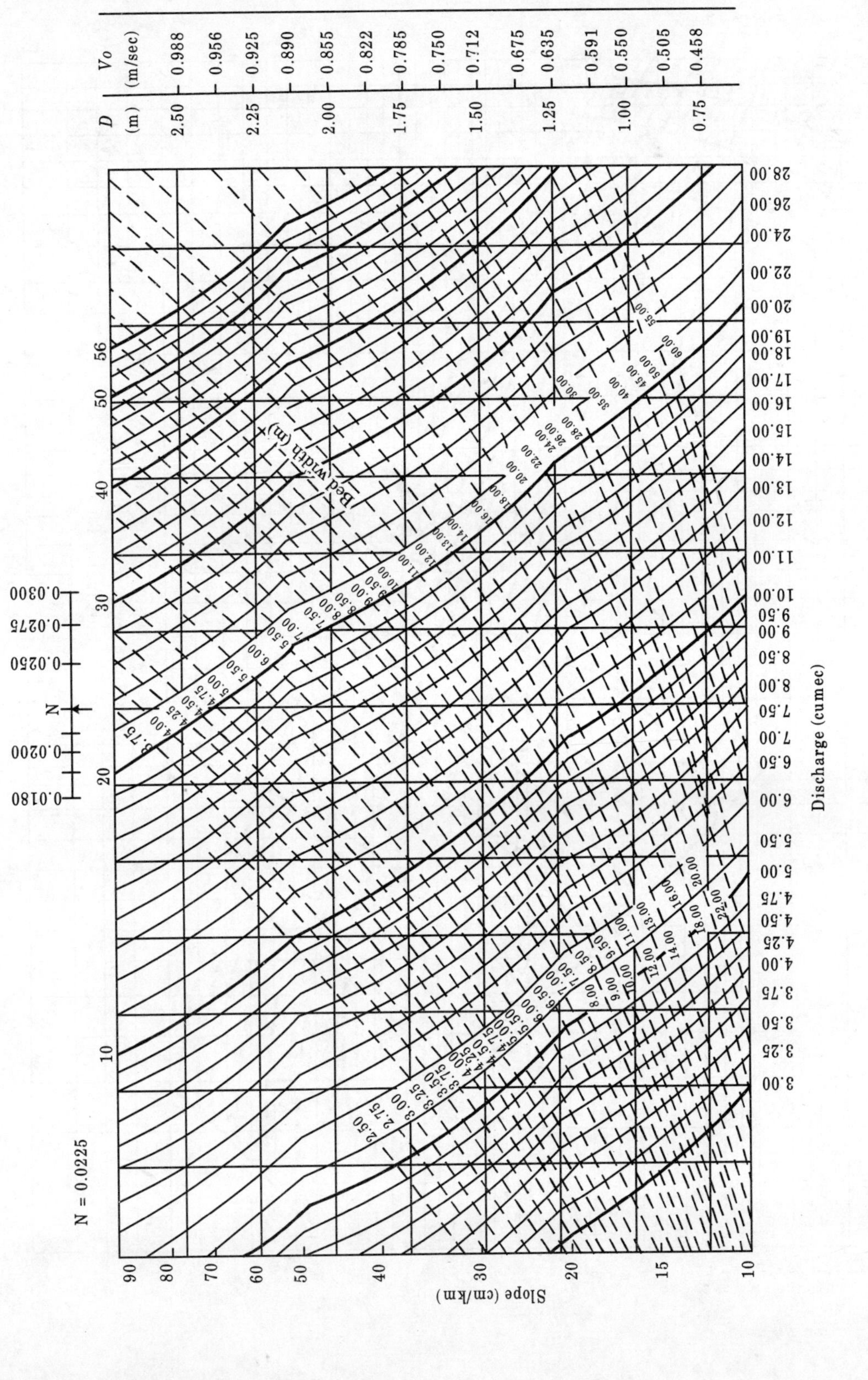

Fig. 16.5 Plate VI (b) Garret's diagram

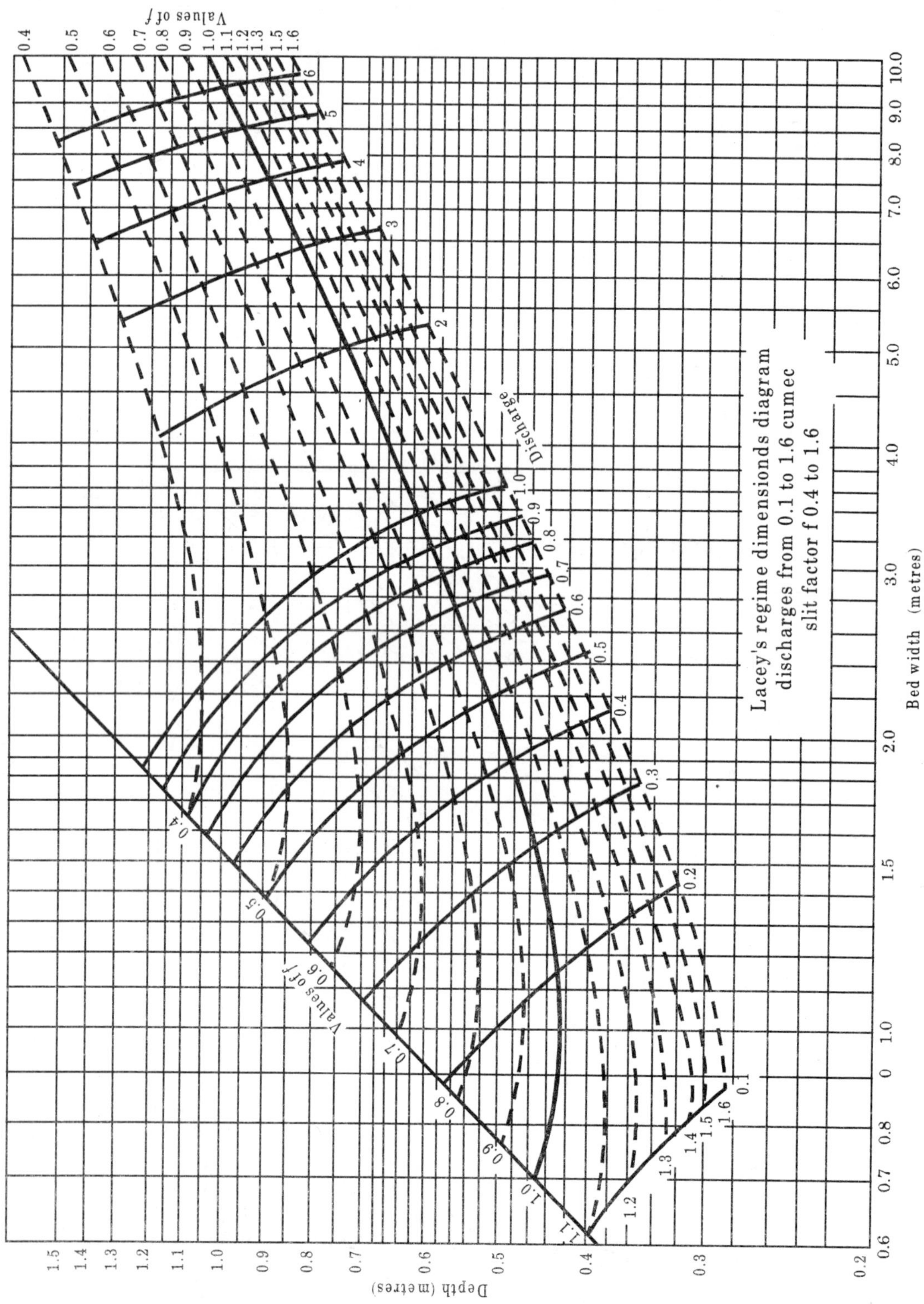

Fig. 16.6 Plate VII (a) Lacey's regime d**iagrams.**

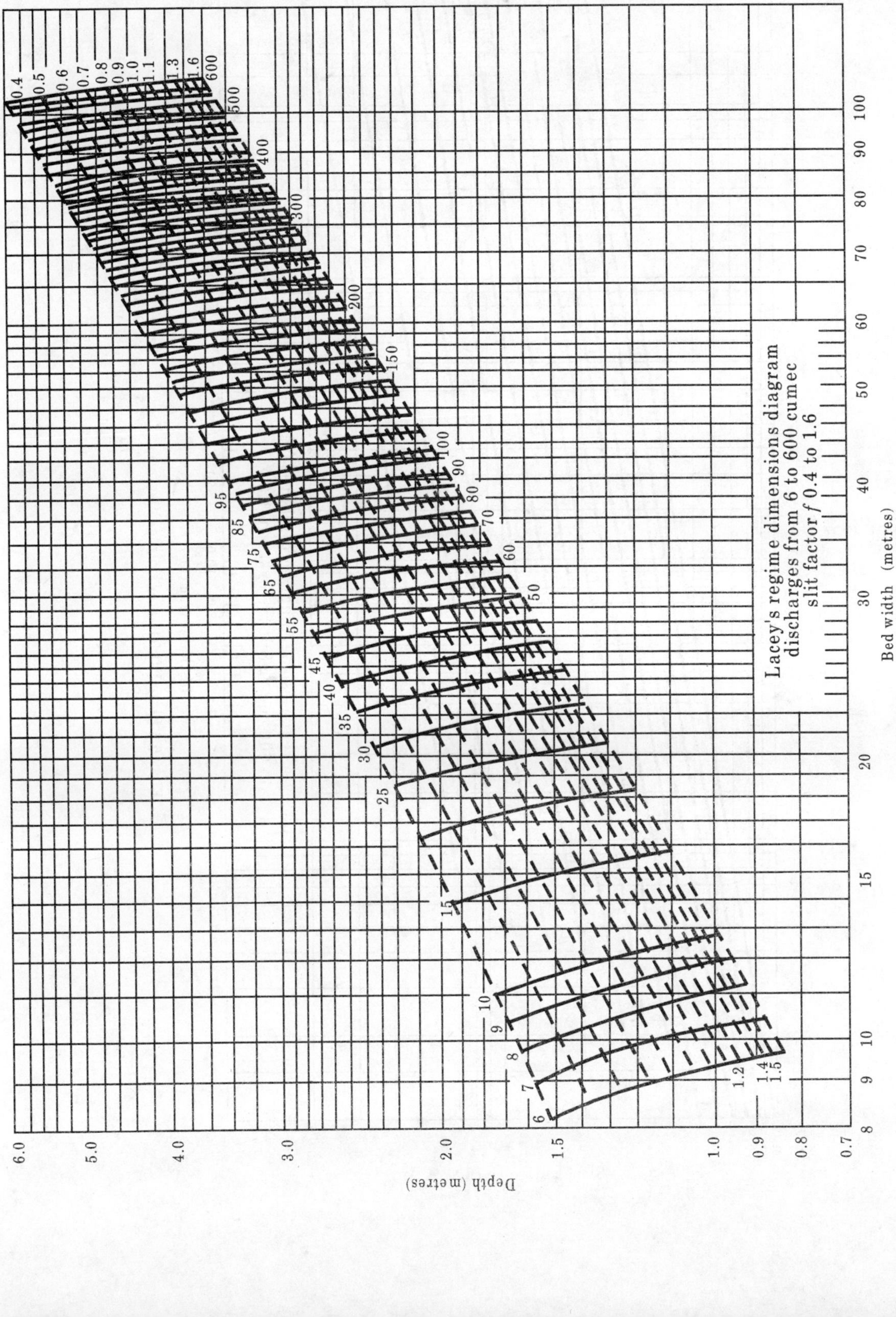

Fig. 16.6 Plate VII (b) Lacey's regime diagrams.

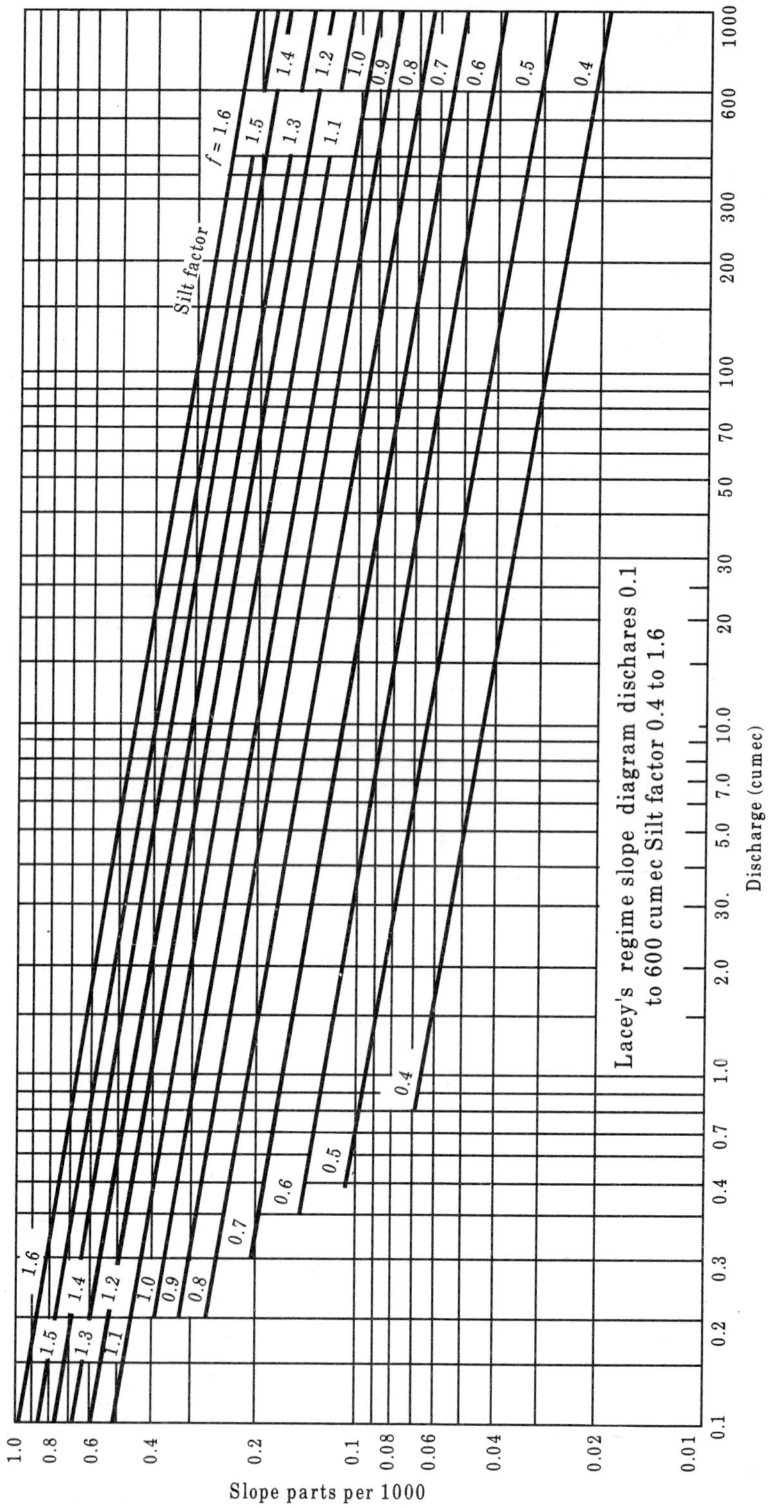

Fig. 16.6 Plate VII (c) Lacey's regime diagrams.

direction i.e., either to the left or the right side in which that value of N lies with respect to the arrow on the nomogram. The vertical line is then drawn through the shifted position of the point and the rest of the procedure for obtaining the bed width and the depth of the channel is same as indicated in steps (5) to (8).

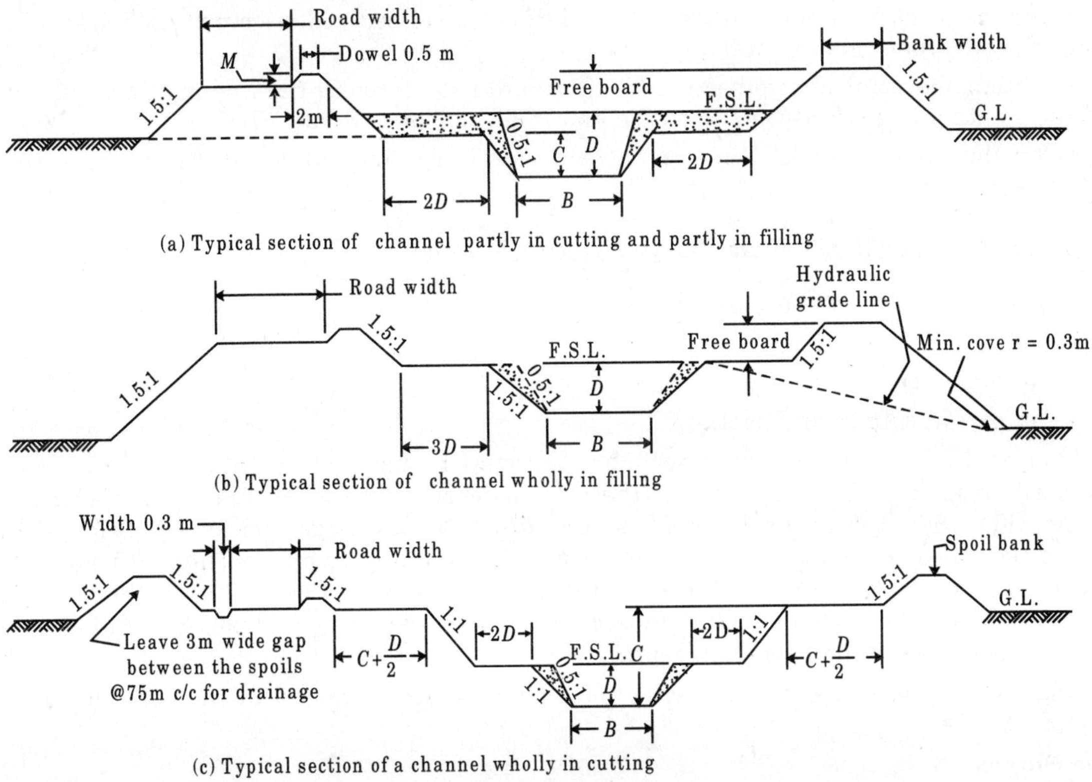

Fig. 16.7 *Typical cross-sections of unlined channels in alluvial soils*

16.7 USE OF LACEY'S REGIME DIAGRAMS FOR CHANNEL DESIGN

The design of channels can also be carried out with the help of regime diagrams prepared by Lacey on the basis of his equations for regime channels. These diagrams were also originally prepared in F.P.S. units but the same have been converted in metric units. Figure 16.6 [Plate VII (a), (b) and (c)] shows Lacey's regime diagrams in metric units. There are two sets of these diagrams. In one of the sets of diagrams depth is plotted on the ordinate and bed width is plotted on the abscissa and these consist of discharge curves shown by firm lines and silt factor curves shown by dotted lines [Plate VII (a) and (b)]. From these diagrams bed width and depth may be determined for the known values of discharge and silt factor. In another set of diagrams bed slope is plotted on the ordinate and discharge is plotted on the abscissa and it consists of a series of slanting lines for different silt factors [Plate VII (c)]. From this diagram regime bed slope may be determined for the known values of discharge and silt factor. The Lacey's regime diagrams have also been plotted for channel sections having side slopes of 1/2(H) : 1(V).

The procedure for the design of channels with the help of Lacey's regime diagrams is as follows.

1. Obtain the point of intersection X of the discharge curve corresponding to the known design discharge and the silt factor curve corresponding to the known silt factor on Plate VII (a) or (b).

2. For the point X obtained in step (1) read the value of bed width from the abscissa and the value of depth from the ordinate.

3. Obtain the point of intersection X, of the ordinate through the known design discharge and the slanting line corresponding to the known silt factor on Plate VII (c).

4. For the point X, obtained in step (3) read the value of its ordinate which gives the regime bed slope.

16.8 CROSS SECTION OF AN IRRIGATION CHANNEL

An irrigation channel section may be either

— in cutting, or

— in filling, or

— in partial cutting and partial filling.

Figure 16.7 shows typical cross-sections of irrigation channels in the above three cases. A channel is said to be in *cutting* when the ground level is above the full supply level in the channel (Fig. 16.7 c). A channel is said to be in *filling* when the ground level is below the bed level of the channel (Fig. 16.7 b). A channel is in *partial cutting and partial filling* when the ground level is in between the bed level and the top of the bank (Fig. 16.7 a). A channel should generally be taken in such a way that its section is partly in cutting and partly in filling, so that it approaches close to the balancing depth.

The various parts of the cross-section of a channel are discussed in the subsequent paragraphs.

Side Slopes

The side slopes depend on the nature of the soil in which the channel is constructed. In average loam soil the side slopes are usually kept 1 : 1 in cutting and $1\frac{1}{2}$: 1 in filling. However, if the soil is comparatively sandy, gentler slopes, say, $1\frac{1}{2}$: 1 in cutting and 2 : 1 in filling may have to be provided.

The minimum side slopes generally adopted for various types of soils are given in Table 16.1.

TABLE 16.1 Side Slopes for channels in different types of soils

Type of soil	*Side slope (Hor. : Ver.)*	
	Cutting	*Filling*
Hard rock	$\frac{1}{8}$: 1 to $\frac{1}{4}$: 1	

Contd.

Contd.

Soft rock	$\frac{1}{4}$: 1 to $\frac{1}{2}$: 1	
Hard clay or gravel	$\frac{3}{4}$: 1 to 1 : 1	$1\frac{1}{2}$: 1
Soft clay, alluvial soils	1 : 1	$1\frac{1}{2}$: 1 to 2 : 1
Sandy loam	$1\frac{1}{2}$: 1	2 : 1
Light sand	2 : 1	2 : 1 to 3 : 1

Berms

A berm is a narrow strip of land left at the ground level between the top edge of cutting and the inner toe of the bank. As the ground level fluctuates irregularly while the bed level of the channel changes very gradually and at a steady rate, the relative depths of cutting d_1 and filling $(d_2 - d_1)$ would fluctuate while the total height from bed to top of bank d_2 remains practically constant (Fig. 16.8). Further since the side slopes in cutting and filling are different, the horizontal distance XY between the bed and bank would also fluctuate from point to point and hence the bed line and bank line will not remain parallel to each other. Thus, in order to keep the bed line and bank line parallel to each other the width of the berm should be so provided that it can absorb all the fluctuations. This would however, result in a berm of variable width which may be determined as follows.

If r_1 : 1 is the slope in cutting and r_2 : 1 is the side slope in filling (i.e., for the bank) then from Fig. 16.8 we have

$$XY = r_1 d_1 + \text{berm width} + r_2(d_2 - d_1) \qquad \text{...(i)}$$

For the bed line and bank line to be parallel to each other, distance XY must be constant which from Fig. 16.8 is obtained as

$$XY = r_2 d_2 \qquad \text{...(ii)}$$

From Eqs (i) and (ii), we get

$$\text{Berm width} = (r_2 - r_1)d_1 \qquad \text{...(16.5)}$$

Equation 16.5 indicates that the berm width depends on the depth of cutting. Since the depth of cutting varies with the variation of the ground level, the berm width also varies.

It has been observed that after the channel runs for some time, silt is deposited on the sides of the channel and its section adopts a shape as shown in Fig. 16.8 with an average side slope of $\frac{1}{2}$: 1. The berm therefore shifts from its initial position at the ground level to the final position at the full supply level (F.S.L.) of the channel and its final width becomes equal to $\left(r_2 - \frac{1}{2}\right)D$, where D is the full supply depth (F.S.D.) of the channel. If the side slope of the bank is $1\frac{1}{2}$: 1, then since $r_2 = 1\frac{1}{2}$, the final width of the berm will be equal to D.

The provision of berm besides maintaining bank line parallel to the bed line of the channel, serves the following useful purposes.

1. The berms reduce bank loads which may cause sloughing of earth into the channel section.

2. The berms developed from fine silt deposition serve as impervious lining and hence help to reduce the seepage losses.

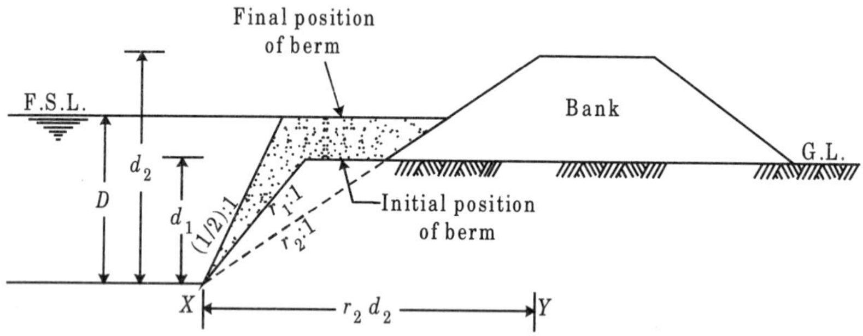

Fig. 16.8 *Initial and final positions of berm*

3. They protect the bank against the eroding action of waves.

4. The berms provide additional strength to the banks and help to protect the banks against breaches.

5. The berms help to break the flow of rain water down the bank slope and thus prevent guttering.

6. They provide scope for future widening of the channel.

7. They sometimes provide an additional inspection path.

8. Borrow pits can be taken on the berms.

In actual practice the berm widths to be adopted have been recommended on the basis of either the capacity of the channel or the relative positions of the full supply level of the channel and the ground level.

The berm widths recommended for the channels of different capacities are as follows :

Channel capacity (cumec)	*Berm width (m)*
Upto 4.25	$0.6 + 0.5\,D$
4.25 to 28	$1.25 + 0.5\,D$
28 and above	$1.25 + \dfrac{\text{F.S. Discharge}}{28} + 0.5\,D$

Minimum berm width = 0.5 m + one fourth of the width of combined side slopes of cutting and filling (or bank).

Maximum berm width = 0.5 m + half of the width of the combined side slopes of cutting and filling (or bank).

IS : 7112–1973 recommends the berm widths to be adopted with the consideration of the relative positions of the full supply level of the channel and the ground level and the same are as follows :

(a) When the full supply level of the channel is above the ground level but the bed of the channel is below the ground level, that is, the channel is partly in cutting and partly in filling, the berm may be kept at the ground level with its width equal to $2D$ (see Fig. 16.7 a) where D is the full supply depth of the channel.

(b) When the full supply level and the bed level of the channel are both above the ground level, that is, the channel is completely in filling, the berm may be kept at the full supply level of the channel with its width equal to $3D$ (see Fig. 16.7 b).

(c) When the full supply level of the channel is below the ground level, that is, the channel is completely in cutting, the berm may be kept at the full supply level with its width equal to $2D$ (see Fig. 16.7 c).

Free Board

Free board is the vertical distance between the full supply level of the channel and the top of the bank. It is provided to ensure that water does not overtop the bank. The magnitude of the free board in a channel is governed by the consideration of the channel size and location, rain water inflow, water surface fluctuation caused by regulators, wind action, soil characteristics, hydraulic gradients, service road requirements and availability of excavated material. IS : 7112–1973 recommends a minimum free board of 0.5 m for channels carrying discharges less than 10 cumec and 0.75 m for channel carrying discharges more than 10 cumec.

Bank Top Width

The banks are required to be provided for the channels which are either completely in filling or are partly in cutting and partly in filling in order to form the channel section. According to IS : 7112–1973 the minimum values recommended for top width of the bank are as given in Table 16.2.

TABLE 16.2 Top width of banks

Discharge (cumec)	*Minimum top width of banks (m)*	
	Inspection bank	*Non-inspection bank*
0.15 to 7.5	5.0	1.5
7.5 to 10.0	5.0	2.5
10.0 to 15.0	6.0	2.5
15.0 to 30.0	7.0	3.5
30.0 and above	8.0	5.0

Hydraulic Gradient Line

When water flows through the channel, seepage occurs through the bank due to which the lower portion of the bank gets saturated while the upper portion remains dry or moist. The line which marks the boundary between the saturated and dry or moist soil within the bank is called saturation line or hydraulic gradient line. The saturation line is actually parabolic in shape but for banks less than 5 m in height it is usually taken as a downward sloping straight line drawn from the point where the full supply level meets the bank. The slope of the saturation line is known as saturation gradient or hydraulic gradient and its value depends on the type of soil of the bank. According to IS : 7112–1973 the following empirical values of the hydraulic gradients (vertical to horizontal) may be used for banks less than 5 m in height.

For silty soils 1 : 4

For silty sands 1 : 5

For sandy soils 1 : 6

Further the hydraulic gradient line shall have a minimum cover of 0.3 m. However, for banks more than 5 m in height true position of the saturation line shall be obtained and the stability of its slopes checked as mentioned in Chapter 11.

Counter Berm or Back Berm

If the section of the bank with normal width and side slopes does not provide the necessary cover over the hydraulic gradient line as indicated earlier then the bank should be strengthened by either flattening the slopes or by providing a counter berm or back berm. A counter berm is a horizontal benching on the outside slope of the bank as shown in Fig. 16.9.

Dowel or Dowla

A dowel or dowla is provided on the side of the service road (or inspection road) between the road and the channel as shown in Fig. 16.7. The top of dowel is kept above the full supply level in the channel by a margin of the free board. The dowels are provided as a measure of safety for vehicles driven on the service road. They act as curbs on the side of the roadway towards the channel. The dowla usually has a top width of 0.5 m, height above road level of 0.5 m and side slopes $1\frac{1}{2}:1$.

Service Road or Inspection Road

For the purpose of inspection, repair and maintenance, roads are provided on the banks of the channels which are usually known as canal roads. In the case of larger channels such as main and branch canals roads are provided on both the banks of the channel while in the case of smaller channels such as smaller branches and distributaries road is provided only on one bank of the channel. Along minors roads are usually not provided. Where roads are provided on both the banks of the channel the one on the left bank (facing in the direction of flow in the channel) is used as inspection road. The road on the other bank is meant for transport of material for repairs and maintenance and other necessary traffic. Where only one road is provided it is located on the left bank of the channel as it is meant for inspection. For main canals and large branches the width of the road may be 4.5 to 6 m and for small branches and distributaries it may be 3.5 to 4.5 m. A suitable cross slope should be given to the roads for proper drainage.

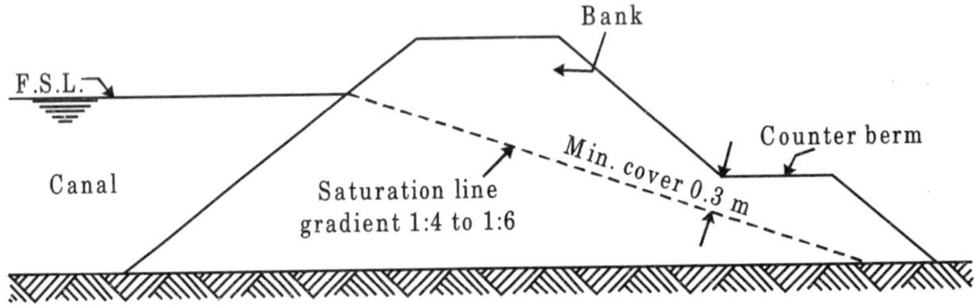

Fig. 16.9 *Counter berm*

As shown in Fig. 16.7 the cross-section of the channel has dowla section on one side and bank section on the other. In the dowla section the road and the bank are combined together, i.e., road is provided on the dowla section which is therefore provided on the left side. If the road is to be provided on the other side also then it may be provided separately by the side of the bank a few centimeters above the ground level to ensure its drainage. However, this arrangement has a drawback that in most of the reaches the channel is not visible while driving on the road. On the other hand the road on the dowla section is always higher than the full supply level and hence an unobstructed view of the channel is obtained from the road. As such a dowla section is preferred to a bank section if the road is to be provided on the other side also. However, provision of dowla section would normally require more earth work. Thus, if enough earth is available from excavation then dowla sections may be provided on both sides, otherwise it should be provided only on the left side and a bank section may be provided on the other side.

The canal roads are usually made up of compacted and dressed earth and are unmetalled and unsurfaced. As such they can take only very light traffic. However, if a canal road is needed for the use of the general public it should be metalled and surfaced before being allowed for general use.

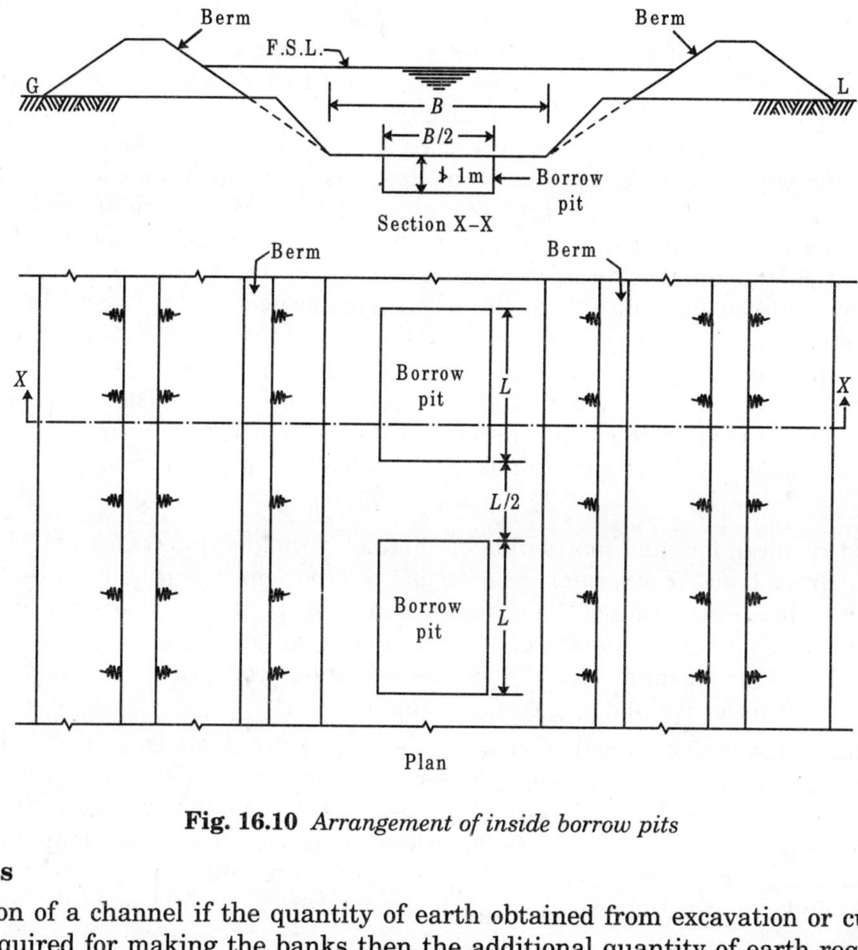

Fig. 16.10 *Arrangement of inside borrow pits*

Borrow Pits

At any section of a channel if the quantity of earth obtained from excavation or cutting is less than that required for making the banks then the additional quantity of earth required for the banks is obtained from borrow pits. The borrow pits may be located either inside the channel or outside the channel. The inside borrow pits may be located on the bed of the channel or on the berms if the later are sufficiently wide. The inside borrow pits are usually preferred because if properly located they get silted up in due course of time after running of the channel and do not spoil the section. In the bed of the channel the borrow pits may be located in the centre of the section with their maximum width equal to half the bed width of the channel and depth not more than 1 m (Fig. 16.10). In between successive borrow pits along the length of the channel a minimum distance equal to half the length of the borrow pit should be left. This will ensure quick silting up of the pits. As far as possible the outside borrow pits should be avoided because these may become breeding places for mosquitoes and may be a source of malaria. However, if inside borrow pits are inadequate to meet the demand the remainder of the earth has to be obtained from the outside borrow pits. The outside borrow pits should be located at a distance of more than 5 m from the toe of the embankment in the case of small channels and 10 m in the case of large channels. These borrow pits should not be deeper than 0.3 m and should be connected by a drain after the measurements of earth work are over, so that they do not hold stagnant water.

Spoil Banks

When the quantity of earth obtained from excavation or cutting of channel section is in excess of the quantity required for forming the banks and service roads then the extra earth is usually deposited in the form of 'spoil banks' on one or both sides of the channel (Fig. 16.7c). Small quantities of extra earth may be utilised in raising or widening the banks beyond the minimum requirements, but when the quantity of earth available is such that if used for banks they would become too high or wide for proper maintenance and they would not be convenient for the inspection of the channel, then the spoil banks are formed. The section of spoil banks depends on the relative cost of land and the labour of lifting earth. Where the adjoining land is cheap and the labour costly, it would be economical to make the spoil banks wide and not very high, but on costly land their height may be increased to minimise the requirements of land. The spoil banks should be provided with a cross slope on either side to ensure proper drainage. A longitudinal drain is provided on the inside (i.e., towards the channel) of the spoil bank. The spoil bank should be discontinuous along its length to enable cross drains to take out the water from the main longitudinal drain. For this usually 3 m wide gap is left between the spoils at 75 m centre to centre.

Land Width

The total land required for the construction of channel is divided into two parts viz, permanent land and temporary land. The *permanent land* is the land which will be required permanently during and after the construction of the channel. Thus, this land is permanently acquired for the construction of the channel. The width of the permanent land is the distance between the outer toes of the banks of the cannel plus a few metres (upto about 1.5 m) on both sides for the construction of a drain or for planting of shade line trees.

During the construction period of a channel some extra land is required for storage of materials and machinery near the site of construction, or for borrow pits or any other similar purpose. This land is known as *temporary land* because it is to be returned to the owner after use with due compensation for its use and deterioration caused if any. As mentioned earlier the borrow pits on temporary land should not be made deeper than 0.3 m, so that the land may be brought back under cultivation without much expenditure after it has been returned to the owner.

16.9 MAINTENANCE OF IRRIGATION CHANNELS

For efficient functioning of irrigation channels their proper and regular maintenance is quite essential. The various operations generally required to be carried out for the maintenance of the channel are as follows.

1. Silt removal
2. Weed control
3. Strengthening of channel banks
4. Maintenance of service roads
5. Plugging of breaches in channel banks.

These are briefly discussed below.

1. Silt Removal. The silt deposited on the bed and sides of a channel reduces the capacity of the channel. If a channel is in regime and takes its full supply, it is not necessary to clear silt to the theoretical cross-section. If a channel is not functioning properly, it may be sufficient merely to clear a portion of silt to get efficient working section or it may be necessary to clear to full theoretical cross-section. Silt deposited at the junction of the bed and slopes may not be removed if the section of the waterway is not unduly affected. Silt is not cleared below falls but if outlets in such places are overdrawing water due to rise in water surface they should be raised. Boning rods are used to see whether the silt has been properly cleared.

In the case of main canals and branches it is not practicable to clear silt from them owing to the large size of the canal and also because the bed is generally covered with water on account of leakage from the head regulator or subsoil contribution. For these canals usually silt excluders and silt ejectors are provided to control the entry of silt into the canal and to remove the silt which has entered the canal. However, even after providing these devices it is not possible to completely check the entry of silt as well as its deposition in the channel. As such the following measures are adopted to remove silt that has already found its way into the channel and impaired its efficient working.

Flushing. Flushing of a channel is carried out by passing a discharge higher than that generally passes through it so that the deposited silt is lifted up and carried away. It is done with water which contains minimum quantity of silt. The flushing is usually over done to cause some scour. This will create room for further silting and will thus reduce the frequency of flushing. It is essential that the flushing water is disposed of to a natural stream unless the silted water is not likely to affect the channels downstream. Thus, the operation of flushing requires huge amount of water.

Silt trap. Scattered pits are dug in the channel bed to arrest the silt flowing with water. When the pits get filled up with silt in course of time, fresh pits may be dug. Thus, in this case silt needs to be cleared only from part of the bed of the channel rather than from the entire bed.

Silt stirring. Various methods are available for stirring and lifting up of deposited silt so that it is held in suspension and removed by the flowing water. In one of the methods water jets from the nozzles are directed towards the bed of the channel to stir and lift up the deposited silt. In another method the deposited silt is stirred and kept in suspension by up and down movement of a number of lower barges fitted with moveable shutters. The lower barges are connected to upper barge by a cable operated by a winch. However, these methods are seldom used in practice.

Dredging. The silt deposited on the bed of a channel may be removed by dredging which is under water excavation of deposited silt. However, for removal of silt from channels dredging is rarely used owing to exhorbitant cost involved.

Excavation. The silt deposited in a channel may be excavated and removed by manual labour. The deposited silt may be dislodged by dragging iron rakes in the channel. Sometimes a bundle of thorny bushes tied together and pressed down by heavy stones are pulled inside the channel by animals to dislodge the deposited silt.

Silt Disposal

The silt removed from the bed of a channel varies in quality from head to tail reaches of the channel. In the head reaches of distributaries coarse silt is found while in the middle reaches medium silt is found and in the tail reaches fine silt is found. The disposal of the silt removed from the channel bed depends on its quality and is recommended to be done according to IS : 4839 (Part I)–1979 as indicated below.

(a) *Course silt.* It contains a large percentage of sand which does not stabilize and is not suitable for use in any repair work. It should therefore be disposed of in filling old outside borrow pits first and then on outer slopes of banks. When silt is thrown along outer slopes of banks care should be taken that the top of the heap is not allowed to come above the bank level and it is disposed of in a regular and uniform manner and not in irregular heaps.

(b) *Medium silt.* It stabilizes after one monsoon passes over it if taken out before or during the monsoon. It may be utilized in (i) closing leaks; (ii) securing outward slopes ; and (iii) widening banks only as supplement to the berm earth if it is insufficient.

(c) *Fine silt.* It contains a good proportion of clay and settles down to a very smooth surface and allows grass to grow freely on it. It is suitable for all repairs and should be used like berm earth.

2. Weed Control. Weeds are unwanted plants which grow in channels under favourable conditions of temperature, light and humidity. Their presence reduces the cross-sectional area of the channel and velocity of flow through it. Thus, the discharging capacity of the channel is reduced and the silting problem is accelerated. The problem of weed growth is more marked in South India where they grow profusely and have reduced the discharge in some cases by as much as 15 percent.

There are mainly three types of water weeds; one growing on channel bed when completely submerged under water, the second growing along the water marks on the banks and the third growing on the water surface. A temperature of 20° to 30°C is favourable for the weed growth. Light accelerates the weed growth. It has also been observed that large growth of weed takes place where silt is deposited. Weed growth is not possible in channels having high velocity of flow but it is generally possible when the velocity of flow is less than 0.6 m/s. Some varieties of weed may even impart harmful qualities to the irrigation water. Thus, in order to permit the channel to function efficiently the nuisance of weed growth has to be controlled. The weed growth can be controlled in the following ways.

(i) By mechanical removal of weed i.e., plucking manually or by some other means.

(ii) By burning the weed when the channel is dry.

(iii) By increasing the velocity of flow in the channel to a value above 0.6 m/s. This will prevent silt from being deposited on channel bed and will keep in suspension and will make water turbid. The turbid water will cut off light rays and prevent weed growth.

(iv) By adopting rush rotation method. In this method the channel is run with full supply discharge for the required period and then the flow is completely stopped and the channel is kept dry for a considerable period. This method helps in controlling weed growth both ways. During full supply discharge the velocity is high and on account of greater depth of water light rays are cut off which prevent weed growth. During closure of the channel the weed gets burnt due to the scorching rays of sun.

3. Strengthening of Channel Banks. In order to prevent breaching of channel banks, they should be properly strengthened. The channel banks are strengthened by following two methods.

(i) Silting

(ii) Berm formation.

(i) *Silting.* In this method the channel banks are strengthened by utilising natural process of silting. This may be done in the following two systems.

(a) *Internal silting system.* In this system the channel banks are set back at a little distance from their normal positions. This results in a section of the channel larger than required and therefore the velocity of flow is low. The section thus gets silted up very quickly. To induce silting and accelerate the process low submersible spurs or groynes projecting from the banks into the channel section are constructed. Instead of fixed spurs or groynes often suspended or hanging spurs or groynes are used to induce silting. The spurs are raised after silting has taken place upto the top of the spur. The silting thus takes place layer by layer till the final level is reached. This is the simplest and most economical system of silting and is applied to a newly constructed channel.

(b) *External silting system.* In this system parallel to the main banks of the channel subsidiary banks are constructed. Within the main bank and the subsidiary bank cross bunds are constructed at a distance of 150 to 1500 m to form compartments of silting or silting tanks. From the channel, water is allowed to get into the compartment through an inlet provided at its upstream end and is held there for some time to allow the deposition of silt before discharging it back to the channel through an outlet provided at its downstream end. The inlet and outlet are provided at 45° to the centre line of the channel. The position of the inlet will have to be changed after some time ; because the maximum deposition of silt will be at the inlet end and it will gradually diminish towards the outlet end and no further silting will take place unless the position of the inlet is moved downstream. It is therefore necessary to close the inlet after working with it for a few weeks, depending upon the silt charge in the water, and then to open another inlet downstream until silting operations are completed.

When the cross bunds are located at a distance of 150 to 300 m, the capacity of the compartment is small and only a portion of full supply discharge is taken into the compartment. This system is then known as *in and out system*. On the other hand when the length of the compartment is large say 1200 to 1500 m, full supply of channel is taken into the compartment and it is then known as *long reach system*. This system is employed only when no water is required downstream of this reach for irrigation purposes.

(ii) *Berm formation.* As already mentioned, berm provides additional strength to the bank, besides keeping the flow away from the bank thereby protecting it against erosion and breaches. As such for new channels berms are usually formed by internal silting. As indicated earlier the formation of berms by internal silting is induced and accelerated by constructing fixed or hanging spurs.

A counter berm or back berm also provides strength to the bank, besides providing necessary cover over the saturation line where it crosses the downstream slope of the bank.

4. Maintenance of Service Roads. The service or inspection roads provided on the banks of the channels are usually unmetalled and, therefore the road surface wears and tears due to the movement of traffic over it. Moreover in monsoon season grass and small bushes may grow on the road surface. As such these roads need proper maintenance to keep them fit for use. The maintenance of these roads generally consists of the following operations.

(i) Removal of grass and small bushes.

(ii) Levelling of road surface.

(iii) Watering and ramming of top surface of road.

Maintenance of these roads is usually carried out after the monsoon season. However, ordinary repairs to these roads should be taken up after first heavy rain falls and should not be postponed till the end of monsoon. Sometimes unauthorised traffic may pass over these roads which would spoil these roads. As such proper control should be exercised to check unauthorised traffic on these roads. For this usually check barriers are constructed at every crossing. The check barrier consists of a small earth mound with a slope of 1 : 4 on upstream and downstream sides with some top width. Jeeps can easily cross over these mounds but for trucks and bullock carts these become real barriers.

5. Plugging of Breaches in Channel Banks. Breach may occur in the bank of a channel which is in filling. During a breach in a channel bank a big gap is formed on the side of the channel through which large portion of water flowing in the channel rushes out thus causing a lot of damage to the adjoining agricultural land and property. Breaches are caused by defective maintenance, bad regulation and sometimes by silt movement besides other causes described below.

Causes of Breaches in Channels

(i) *Weak bank.* One of the main causes of breaches in channel banks is their defective design and construction. For channels in filling if the height of banks is more than 6 m they should be designed as earth dams. Further banks having insufficient width and free board and constructed with material of poor quality are susceptible to breaches.

(ii) *Leakage.* Leakage of water through insect or rat holes in channel bank may ultimately cause breach in the bank if these holes are not repaired in time.

(iii) *Cut.* A wilful cut in the channel bank may cause breach in the bank. A cut may be made in the channel bank for unauthorised irrigation by some cultivators. The cultivators may also sometimes cut the channel bank to drain of their cultivated land which might get flooded due to excessive rainfall.

(iv) *Overflow.* A breach in the channel bank may be caused if water starts overflowing the bank. This may happen when there is excessive supply of water in the channel or when the cultivators stop drawing water from the channel due to local heavy rainfall.

(v) *Piping.* The excessive supply of water in the channel far above the full supply level may result in the failure of bank by piping or sloughing which may cause breach in the bank.

Closing Breaches in Channels

The methods for closing breaches in running channels as recommended by IS : 4839 (Part I)–1979 are as discussed below.

(i) *Closing breach in a small distributary or minor.* Enough earth is collected on both sides of the breach on the existing bank. The earth is however not thrown from the baskets in the water but the closing process is started from both ends by slipping the earth from the heap and protecting the channel side by grassy clods usually available from the berms. Care is taken that the grassy clod is not allowed to be washed down into the breach site. With a rush of earth work at the end, the breach is closed straight away progressing from the bank.

(ii) *Closing breach in a major distributary or small branch.* In this case it is necessary to reduce the flow through the breach otherwise a lot of earth will be washed away before the breach is closed. This is done by driving a double line of stakes as shown in Fig. 16.11 and then putting planks or mattresses against them if available and if not then filling bushes in between the stakes pressing it down with bags filled with sand and by men walking over them. No earthwork is started before the flow through the breach has been checked to some extent in this way. Meanwhile enough earth is piled up on both sides of the breach on the existing bank. The closing is then started from both sides by slipping earth from the heap in the form of a ring bund as shown in Fig. 16.11. All jungle from the site of ring bund is removed before the earthwork is started. The earth is not thrown from the basket in water, but it is slipped from a heap. The last gap of about 3 m is closed when enough earth has been collected on both sides. Straight closure in large canals is usually not possible.

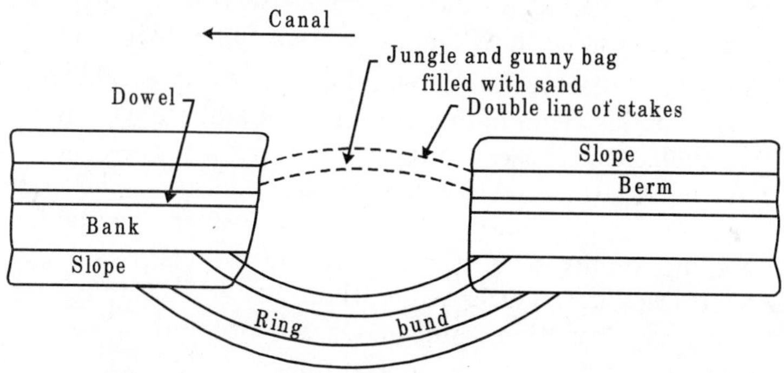

Fig. 16.11. *Method of closing breach in a channel*

(iii) *Closing breach in large channels.* In this case the supply of water in the channel is reduced from the headworks as soon as possible after the breach has occurred. The method of ring bund, described for closing a breach in a major distributary or small branch is used in this

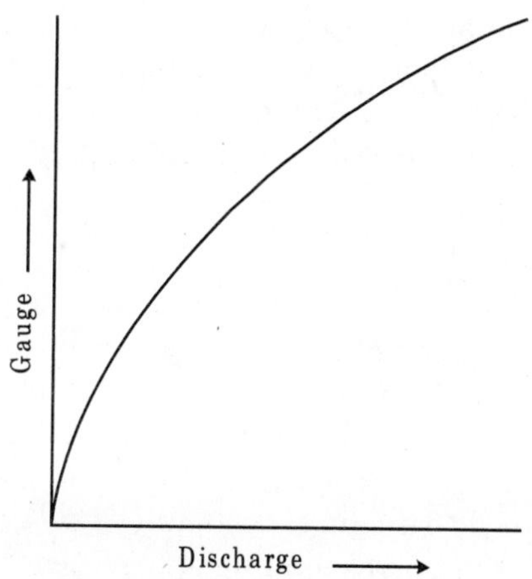

Fig. 16.12 *Gauge-discharge curve*

case also but the bushes or planks would not serve to reduce flow through the breach. The double line of stakes is driven as before if depth permits and a double line of gunny bags filled with sand is put in it. The inter-spaces are plugged with berm earth. A temporary bank of gunny bags is raised in position of stakes and bushing as shown in Fig. 16.11. The closing of the breach is then done by constructing a ring bund.

16.10 REGULATION OF CHANNEL (OR CANAL) SYSTEM

The regulation of channel (or canal) system means distribution of water available for irrigation among the various channels (or canals) of the system. The basic principle involved in regulation is apportioning of the available water supplies in relation to the irrigable area.

An irrigation channel (or canal) is run depending on the demand of the cultivators and the availability of the water supplies. The main canal of an irrigation system usually runs continuously except during annual closure because it is required to serve a large area and there is always a demand somewhere within the channel (or canal) system. The main canal may run to the full designed capacity or with low supplies depending upon the availability of water supplies and the demand. The same applies to larger branches of the main canal. As regards distributaries only those distributaries are run for which there is demand and the rest are closed. The difficulty in the distribution of irrigation water arises when there is a keen demand in the entire system but the available water supplies are not sufficient to meet this demand. The problem may however be solved by adopting either of the two methods of distribution as briefed below. In one of the methods all the distributaries are run with reduced supplies. In the other method the various distributaries are run with full designed capacities in rotation. The former method provides a straight forward but less effective solution of the problem and it has several disadvantages some of which are (i) silting of channels, (ii) increases weed growth, (iii) increased seepage and water logging, (iv) low head on outlets. The later method provides a more effective solution of the problem and it has the advantage that the full closure of some of the channels may facilitate silt clearance, repair of channel section and inspection and repair of masonry works such as falls, regulators etc.

A roster is usually prepared which indicates the supplies allotted to different channels and the names of the channels which will be closed from week to week. As far as possible the roster is adhered to. Sometimes the demand in the area is very fluctuating in which case a flexible regulation from week to week is adopted in accordance with the anticipated demand in the ensuing week. The anticipated demand is furnished by the revenue staff for their respective areas to the engineer incharge who is looking after the regulation of channel (or canal) system.

Gauge-discharge Curve

For effective distribution of irrigation water it is necessary to be able to adjust the discharge in a channel to any known value. For this a gauge indicating the depth of water above the bed of the channel is fixed at a suitable section of the channel. At the gauging station the bed and sides of the channel are pitched for about 50 m to give a straight run of 30 m. The gauge is located in a gauge well which is located centrally in the pitched length. By actual measurement of discharge at different gauge readings a *gauge-discharge curve* is prepared (Fig. 16.12). When a certain discharge is required to be run in the channel the gates or the planks at the head regulator are manipulated till the corresponding gauge reading as indicated by the gauge-discharge curve is obtained in the channel. For preparing the gauge-discharge curve and later on for checking and correcting it, it is necessary to measure the discharge in channels quite frequently. For measurement of discharge in channels the following methods may be employed.

Discharge Measurement in Channels

The various methods used for measurement of discharge in channels are as follows.

(i) Area velocity method

(ii) Chemical method

(iii) Weir method

(iv) Metering flume method.

Out of these the first two methods are described below.

(i) *Area velocity method.* This method is commonly used for measurement of discharge through channels as well as rivers. It consists of measuring the cross-sectional area of channel section and the mean velocity of flow through it. The product of the area and the mean velocity gives the discharge. The measurement of area of channel section presents little difficulty but there is a problem of determining the mean velocity of flow for the entire channel section because of non-uniform velocity distribution in channels. As such for determining the discharge in a channel the cross-section of the channel is divided into a number of compartments each of equal width and the discharge of each compartment is determined. For this the area of each compartment and the mean velocity of flow through it are determined as given below. The product of the area and the mean velocity gives the discharge of each compartment and the sum of the discharges of all the compartments gives the total discharge in the channel.

The area of each compartment is obtained by multiplying its width by the mean depth of water which is measured at the centre line of each compartment by lowering a graduated 'sounding rod' upto the bottom.

The mean velocity of flow through each compartment is assumed to be equal to the mean velocity of flow on a vertical plane in the centre of the compartment which may be measured by any of the following devices.

(a) Surface float

(b) Double float

(c) Velocity rod or Rod float

(d) Current meter

(e) Pitot tube.

(a) *Surface Float.* A surface float is a small piece of a material such as wood, cork etc., which is little lighter than water so that it may float on the water surface. In order to make it clearly visible it is usually painted on the top and sometimes a flag is attached at its top. The float is dropped in water at a certain section of the channel in the centre of one of the compartments and the time taken by it to travel a known distance is noted. The same procedure is repeated for each of the compartments. A number of observations (usually three) are taken for each compartment and the average value of the time is worked out. The distance travelled by the float divided by the average time taken by it gives the average surface velocity at the centre of the compartment. The mean velocity of flow on a vertical plane in the centre of the compartment is obtained by multiplying the surface velocity by a coefficient determined experimentally by direct measurement of mean velocity by other methods.

(b) *Double float.* A double float consists of a surface float to which a hollow metal sphere heavier than water is attached by a cord. The length of the cord is adjustable so that the position of the lower float may be altered. The float is made to travel in the centre of each of the compartments and the velocity is obtained by noting the time taken by the float to travel a known distance. This velocity would be the mean of the surface velocity and the velocity at the depth at which the lower float is carried. By adjusting the lower float at nearly 0.2 times the depth of water above the bottom, the velocity obtained will be approximately the mean velocity on a vertical plane in the centre of the compartment in which the float is made to travel.

(c) *Velocity rod or Rod float.* A velocity rod or rod float consists of a wooden rod or hollow tin tube 25 to 50 mm in diameter and weighted at the bottom so as to float vertically in water wholly immersed except for a small portion at the top. They are made in various lengths to suit different depths. Telescopic rods are also made and their lengths can be adjusted according to the requirement. The procedure for observations is the same as for other floats. In every compartment the longest rod that can be run without touching at the bottom should be used. As the rod floats vertically from surface to very near the bottom it travels with a velocity very nearly equal to the mean velocity on a vertical plane in the centre of the compartment. However, the mean velocity of flow on a vertical plane in the centre of the compartment may be computed by the following empirical formula.

$$V_m = V\left[1.02 - 0.116\sqrt{\frac{d_1}{d}}\right] \qquad \ldots(16.3)$$

in which

V_m = mean velocity of flow

V = observed velocity of the rod float

d = depth of flow at the centre of the compartment, and

d_1 = clearance between lower-end of the rod float and the bottom of the compartment.

(d) *Current meter.* A current meter is a mechanical device which is commonly used for measuring velocity of flow in channels. It consists of a revolving element and a tail made up of flat vanes or fins to keep the instrument aligned in the direction of flow. The revolving element may be either a small wheel formed by a series of conical cups mounted on a spindle, or a small propeller which consists of a shaft or hub with curved blades or vanes mounted around its periphery. For channels of moderate depths the current meter is usually mounted on a rod on which it can be fixed in any required position. The rod is made to rest on the bed of the channel where the velocity is to be measured. For large channels the current meter is suspended in the water flowing in the channel by a cable from a boat or a trolley and to keep the cable vertical a heavy streamlined anchor is attached to the bottom of the current meter. For the measurement of velocity of flow the current meter is held at the required depth such that the wheel or propeller is facing towards the upstream direction. The wheel or propeller is thus rotated by the dynamic thrust exerted on it by the flowing water. The speed of rotation of the wheel or propeller depends on the velocity of flow. Thus, the number of revolutions made per unit time by the wheel or propeller are recorded and from the calibration curve the velocity of flow is determined. In order to facilitate the counting of the number of revolutions made by the wheel or propeller an electrical transmission system with an electric bell or a head phone is provided.

A current meter measures the velocity of flow at the point where it is held in the water flowing in the channel. For determination of mean velocity of flow at the centreline of the compartment either of the following methods may be used.

1. A number of observations of velocity may be taken from surface to bottom and velocity distribution curve plotted from which the mean velocity can be determined by dividing the area of the curve by the depth.

2. The mean of the velocities at depths of 0.2 and 0.8 of the total depth from the surface may be taken. This has been found to be within $2\frac{1}{2}$ % of the mean velocity obtained by (1) which gives an accurate value of the mean velocity.

3. The velocity at 0.6 of the depth from the surface may be taken as the mean velocity. This may involve errors upto 5%.

The first method is normally adopted when the velocity distribution is required for some other purpose. The second method is the most satisfactory for the measurement of discharge in channels.

(e) *Pitot tube*. The pitot tube can also be used for measurement of velocity at any point within a stream. However, it has a fine entrance bore which is likely to get blocked by silt-charged flow. As such pitot tube is not commonly used for discharge measurement in the field though it is quite useful in the laboratory.

(ii) *Chemical method*. The chemical method is a direct method for determination of discharge in channels without determining the mean velocity of flow and cross-sectional area. In this method at a certain section of the channel a concentrated solution of salt of known strength is introduced in the flowing water at a known steady rate from a number of points uniformly distributed over the cross-section of the channel. At a section some distance downstream where the solution may be considered to have properly mixed in the water a number of samples are taken from all over the section and the average concentration of salt is determined.

Let s_0 be the concentration of salt in weight per unit volume in the water flowing in the channel prior to the introduction of the salt solution

s_1 be the concentration of salt in weight per unit volume in the salt solution

s_2 be the concentration of salt in weight per unit volume in the water at the downstream section after the introduction of the salt solution

q be the rate of introduction of the salt solution in volume per unit time (cumec), and

Q be the discharge in the channel in volume per unit time (cumec).

Since the total weight of salt at the two sections of the channel must be the same, we have

$$Qs_0 + qs_1 = (Q + q)s_2$$

from which discharge is given by

$$Q = q\left(\frac{s_1 - s_2}{s_2 - s_0}\right) \quad \text{...(16.6)}$$

The results obtained by this method will be reliable only if the salt solution is thoroughly mixed in the water before it reaches the downstream section where the samples are taken.

SOLVED ILLUSTRATIVE EXAMPLES

Illustrative Example 16.1 *A canal has a bed width of 8 m, full supply depth 2.5 m, bank width 3 m, cutting slope 1 : 1, filling slope 1.5 : 1 and free board 0.5 m. Calculate balancing depth.*

Solution

With reference to Fig. 16.2

$$D = 2.5 \text{ m}$$
$$h = (2.5 + 0.5) = 3.0 \text{ m}$$
$$B = 8 \text{ m}$$
$$t = 3 \text{ m}$$
$$z = 1$$
$$n = 1.5$$

Equating the area of cutting and filling we get

$$(B + zy)y = 2[t + n(h - y)](h - y)$$

Substituting the values we get

$$(8 + y)y = 2[3 + 1.5(3 - y)](3 - y)$$

or
$$8y + y^2 = 18 - 6y + 27 - 18y + 3y^2$$

or
$$2y^2 - 32y + 45 = 0$$

or
$$y = 1.56 \text{ m}$$

Check :

$$\text{Cutting} = (B + zy)y = (8 + 1.56)1.56 = 14.9 \text{ m}^2$$
$$\text{Filling} = 2[t + n(h - y)](h - y)$$
$$= 2[3 + 1.5(3 - 1.56)](3 - 1.56)$$
$$= 14.9 \text{ m}^2.$$

Illustrative Example 16.2 *Design a channel by the use of Garret's diagram with the following data:*

Full supply discharge = 2.5 *cumec*

Bed slope = $\frac{1}{5000}$

Rugosity coefficient $N = 0.0225$

Critical velocity ratio = 0.95

Side slope of channel = $\frac{1}{2}$: 1.

Solution

$$Q = 2.5 \text{ cumec} ; N = 0.0225 ;$$
$$S = 15000 = 20 \text{ cm/km}$$

From Garret's diagram the design of the channel is carried out as shown in the following Table.

Bed width B from diagram (m)	*Depth D from diagram (m)*	*Area A = BD + (D²/2) (m²)*	*Velocity V = (Q/A) (m/s)*	*Critical Velocity V_0 from diagram (m/s)*	*C.V.R. (V/V_0)*	*Remarks*
5.50	0.849	5.030	0.497	0.495	1.004	Large
5.00	0.909	4.958	0.504	0.518	0.973	–do–
4.50	0.972	4.846	0.516	0.540	0.956	–do–
4.30	1.000	4.800	0.521	0.550	0.947	Suitable

Illustrative Example 16.3 *Design the canal of Illustrative Example 16.3 by using Lacey's diagrams. Take silt factor f = 1.*

Solution

From Lacey's diagrams

$$B = 5.56 \text{ m} \; ; \; D = 0.8 \text{ m}$$

Bed slope $S = \dfrac{1}{4000}$

Illustrative Example 16.4 *Design and prepare the longitudinal section and schedule of area statistics and channel dimensions of an irrigation channel from the following data :*

The ground level at every 200 m along the centre line of the channel are as given below :

km	*R.L. in metres*	*km*	*R.L. in metres*
0.0	237.22	2.6	235.37
0.2	237.28	2.8	235.19
0.4	237.31	3.0	235.14
0.6	235.93	3.2	234.17
0.8	236.58	3.4	234.72
1.0	236.54	3.6	234.41
1.2	236.53	3.8	234.30
1.4	236.30	4.0	234.21
1.6	236.13	4.2	234.12
1.8	236.16	4.4	233.76
2.0	235.85	4.6	233.60
2.2	235.70	4.8	233.40
2.4	235.60	5.0	233.20

Assume datum at R.L. 230.00 m

Full supply level of parent channel 237.70 m

Bed level of parent channel 236.00 m

Commanded areas are as given below.

Below km	*Gross commanded area (hectares)*	*Culturable commanded area (hectares)*
0	*40 000*	*28 000*
1	*35 000*	*24 500*

Contd.

Contd.

2	30 000	21 000
3	25 000	17 500
4	20 000	14 000
5	18 000	12 600

Assume channel losses at the rate of 2 cumec per million square metre of wetted perimeter and total losses of 0.4 cumec below km 5.0.

The intensity of irrigation, kor-period and kor-depth are as given below :

Particulars	*Rabi (Wheat)*	*Kharif (Rice)*
Intensity of irrigation	*30%*	$12\frac{1}{2}\%$
Kor-period	*4 weeks*	$2\frac{1}{2}$ *weeks*
Kor-depth	*135 mm*	*190 mm*

Take critical velocity ratio m = 1.0, Kutter's N = 0.0225 and side slope of the channel as $\frac{1}{2}$ *: 1.*

Solution

The longitudinal section of the ground is plotted as shown in Fig. 16.1 (Plate V).

The outlet discharge factor for controlling crops is determined as follows :

(a) Outlet discharge factor for Rabi

$$= \frac{8.64B}{\Delta}$$

$$B = 4 \times 7 = 28 \text{ days}$$

$$\Delta = 135 \text{ mm} = 0.135 \text{ m}$$

∴ Outlet discharge factor

$$= \frac{8.64 \times 28}{0.135} = 1792 \text{ hectare/cumec}$$

$$\simeq 1800 \text{ hectare/cumec}$$

(b) Outlet discharge factor for Kharif $= \dfrac{8.64B}{\Delta}$

$$B = 2.5 \times 7 = 17.5 \text{ days}$$

$$\Delta = 190 \text{ mm} = 0.19 \text{ m}$$

∴ Outlet discharge factor

$$= \frac{8.64 \times 17.5}{0.19} = 795.8 \text{ hectare/cumec}$$

$$\simeq 800 \text{ hectare/cumec}$$

The outlet discharge factor for the controlling crops is the one which gives maximum requirement for discharge.

Taking the culturable commanded area as 28,000 hectares, area under Rabi crop with intensity of irrigation as 30%

$$= 28\,000 \times \frac{30}{100} = 8400 \text{ hectares}$$

and area under Kharif crops with intensity of irrigation as $12\frac{1}{2}\% = 28\,000 \times \frac{12.5}{100} = 3500$ hectares

$$\therefore \text{ Discharge need for Rabi crop} = \frac{8400}{1800} = 4.67 \text{ cumec}$$

$$\text{and discharge needed for Kharif crop} = \frac{3500}{800} = 4.38 \text{ cumec}$$

Hence outlet discharge factor to be adopted in the design = 1800 hectare/cumec.

Design of channel at various sections

The design of channel is started from the tail end where the losses are known.

1. *At km 5.00*

Culturable commanded area

$$= 12\,600 \text{ hectares}$$

Intensity of irrigation for Rabi crop

$$= 30\%$$

Area to be irrigated for Rabi crop

$$= 12\,600 \times \frac{30}{100} = 3780 \text{ hectares}$$

Outlet discharge factor

$$= 1800 \text{ hectare/cumec}$$

∴ Discharge needed for Rabi crop

$$= \frac{3780}{1800} = 2.10 \text{ cumec}$$

Total losses below km 5.0 = 0.4 cumec (given)

∴ Total discharge at km 5.00

$$= 2.10 + 0.4 = 2.50 \text{ cumec}$$

Design discharge Q will be 10% more than the required discharge.

∴ Design discharge = 1.1 × 2.5 = 2.75 cumec

C.V.R. (V/V_0) = 1.0

Kutter's N = 0.0225

Assume bed slope $S = \frac{1}{4000} = 0.25$ m/km

From Garret's diagram Fig. 16.4 (Plate VI a)

Discharge Q (cumec)	*Bed slope S*	*Bed width B (m)*	*Water depth D (m)*	*Area* $A = BD + (D^2/2)$ (m^2)	*Velocity V* $= (Q/A)$ (m/s)	*Critical Velocity* V_0 (m/s)	*C,V,R,* (V/V_0)
		4.50	0.98	4.890	0.562	0.543	1.036
2.75	1/4000	4.30	1.01	4.853	0.567	0.554	1.023
		4.00	1.06	4.802	0.573	0.571	1.004

Adopt bed width $B = 4.0$ m and

Water depth $D = 1.06$ m

2. *At km 4.00*

Culturable commanded area

= 14 000

Intensity of irrigation for Rabi crop

= 30%

Area to be irrigated for Rabi crop

$$= 14\,000 \times \frac{30}{100} = 4200 \text{ hectares}$$

Outlet discharge required

$$= \frac{4200}{1800} = 2.33 \text{ cumec}$$

Losses below km 5.0 = 0.4 cumec.

Losses in Channel Reach between km 5.0 to km 4.0

Although the losses should be calculated on the wetted perimeter of the channel section at km 4.0, but the section cannot be determined before hand. Hence for computation of channel losses in this reach, the perimeter of the section at km 5.0 shall be taken.

Wetted perimeter of the channel section at km 5.0

$$= B + \sqrt{5}\ D$$

$$= 4.0 + \sqrt{5} \times 1.06 = 6.37 \text{ m}$$

Thus losses in the reach

$$= \frac{6.37 \times 1000 \times 2}{10^6}$$

$$= 0.0127 \text{ cumec} = 0.015 \text{ cumec (say)}$$

∴ Total losses below km 4.0

= Losses below km 5.0 + Losses in the reach between km 5.0 to km 4.0

= 0.4 + 0.015 = 0.415 cumec

∴ Total discharge required at km 4.0

$= 2.33 + 0.415 = 2.745$ cumec

Adopt design discharge Q 10% more than the required discharge.

$\therefore$ Design discharge $= 1.1 \times 2.745 = 3.02$ cumec

$= 3.00$ cumec (say)

C.V.R. (V/V_0) $= 1.0$

Kutter's N $= 0.0225$

Assume bed slope $S = \dfrac{1}{4000} = 0.25$ m/km

From Garret's diagram Fig. 16.4 (Plate VI a)

Discharge Q (cumec)	*Bed slope S*	*Bed width B (m)*	*Water depth D (m)*	*Area A = BD + ($D^2/2$) (m^2)*	*Velocity V = (Q/A) (m/s)*	*Critical Velocity V_0 (m/s)*	*C, V, R, (V/V_0)*
		5.00	0.96	5.261	0.570	0.536	1.063
3.00	1/4000	4.50	1.03	5.030	0.596	0.561	1.062
		4.20	1.08	5.119	0.586	0.578	1.014

Adopt bed width $B = 4.2$ m and

Water depth $D = 1.08$ m

3. *At km 3.0*

Culturable commanded area

$= 17\,500$ hectares

Intensity of irrigation for Rabi crop

$= 30\%$

Area to be irrigated for Rabi crop

$$= 17\,500 \times \frac{30}{100} = 5250 \text{ hectares}$$

Outlet discharge required

$$= \frac{5250}{1800} = 2.92 \text{ cumec}$$

Losses below km 4.0 $= 0.415$

Losses in reach between km 4.0 to km 3.0

Wetted perimeter of the channel section at km 4.0

$$= 4.2 + 1.08\sqrt{5} = 6.61 \text{ m}$$

Losses in the reach

$$= \frac{6.61 \times 1000 \times 2}{10^6} = 0.0132 \text{ cumec}$$

= 0.015 cumec (say)

Thus total losses below km 3.0

= 0.415 + 0.015 = 0.430 cumec

Total discharge required at km 3.0

= 2.92 + 0.43 = 3.35 cumec

Adopt design discharge Q 10% more than the required discharge.

∴ Design discharge = 1.1 × 3.35

= 3.685 cumec = 3.70 cumec (say)

C.V.R. (V/V_0) = 1.0

Kutter's N = 0.0225

Assume bed slope $S = \frac{1}{4000}$ = 0.25 m/km

From Garret's diagram Fig. 16.4 (Plate VI b)

Discharge Q (cumec)	*Bed slope S*	*Bed width B (m)*	*Water depth D (m)*	*Area* $A = BD + (D^2/2)$ (m^2)	*Velocity V* $= (Q/A)$ *(m/s)*	*Critical Velocity* V_0 *(m/s)*	*C, V, R,* (V/V_0)
		5.00	1.10	6.105	0.606	0.585	1.036
3.70	1/4000	4.75	1.13	6.006	0.616	0.595	1.035
		4.50	1.18	6.006	0.616	0.611	1.008

Adopt bed width B = 4.5 m and

Water depth D = 1.18 m

4. *At km 2.0*

Culturable commanded area

= 21000 hectares

Intensity of irrigation for Rabi crop

= 30%

Area to be irrigated for Rabi crop

$$= 21\,000 \times \frac{30}{100} = 6300 \text{ hectares}$$

Outlet discharge required

$$= \frac{6300}{1800} = 3.5 \text{ cumec}$$

Losses below km 3.0 = 0.430

Losses in reach between km 3.0 to km 2.0

Wetted perimeter of the channel section at km 3.0

$$= 4.5 + 1.18\sqrt{5} = 7.139 \text{ m}$$

Losses in the reach $= \dfrac{7.139 \times 1000 \times 2}{10^6} = 0.0143$ cumec

= 0.015 cumec (say)

Thus, total losses below km 2.0

= 0.430 + 0.015 = 0.445 cumec

Total discharge required at km 2.0

= 3.5 + 0.445 = 3.945 cumec

Adopt design discharge Q 10% more than the required discharge.

∴ Design discharge = 1.1 × 3.945 = 4.34 cumec

= 4.35 cumec (say)

C.V.R. (V/V_0) = 1.0

Kutter's N = 0.0225

Assume bed slope $S = \dfrac{1}{4444} = 0.225$ m/km

From Garret's diagram Fig. 16.4 (Plate VI b)

Discharge Q (cumec)	*Bed slope S*	*Bed width B (m)*	*Water depth D (m)*	*Area A = BD + ($D^2/2$) (m^2)*	*Velocity V = (Q/A) (m/s)*	*Critical Velocity V_0 (m/s)*	*C, V, R, (V/V_0)*
		4.75	1.29	6.960	0.625	0.647	0.966
4.35	1/4444	5.00	1.25	7.031	0.619	0.635	0.975
		5.25	1.21	7.085	0.614	0.621	0.989

Adopt bed width B = 5.25 m and

Water depth D = 1.21 m

5. *At km 1.0*

Culturable commanded area

= 24 500 hectares

Intensity of irrigation for Rabi crop

= 30%

Area to be irrigated for Rabi crop

$= 24\,500 \times \dfrac{30}{100} = 7350$ hectares

Outlet discharge required

$= \dfrac{7350}{1800} = 4.08$ cumec

Losses below km 2.0 = 0.445 cumec

Losses in the reach between km 2.0 to km 1.0

Wetted perimeter of the channel section at km 2.0

$$= 5.25 + 1.21\sqrt{5} = 7.956 \text{ m}$$

Losses in the reach $= \dfrac{7.956 \times 1000 \times 2}{10^6}$

$= 0.0159$ cumec $= 0.016$ cumec (say)

Thus, total losses below km 1.0

$= 0.445 + 0.016 = 0.461$ cumec

Total discharge required at km 1.0

$= 4.08 + 0.461 = 4.541$ cumec

Adopt design discharge Q 10% more than the required discharge.

∴ Design discharge $= 1.1 \times 4.541 = 4.995$ cumec

$= 5.0$ cumec (say)

C.V.R. (V/V_0) $= 1.0$

Kutter's N $= 0.0225$

Assume bed slope $S = \dfrac{1}{4444} = 0.225$ m/km

From Garret's diagram Fig. 16.4 (Plate VI b)

Discharge Q (cumec)	*Bed slope S*	*Bed width B (m)*	*Water depth D (m)*	*Area $A = BD + (D^2/2)$ (m^2)*	*Velocity V $= (Q/A)$ (m/s)*	*Critical Velocity V_0 (m/s)*	*C,V,R, (V/V_0)*
5.0	1/4444	5.50	1.280	7.859	0.636	0.644	0.988
		5.75	1.245	7.934	0.630	0.633	0.995

Adopt bed width $B = 5.75$ m and

Water depth $D = 1.245$ m

6. *At head i.e., at km 0.0*

Culturable commanded area

$= 28\,000$ hectares

Intensity of irrigation for Rabi crop

$= 30\%$

Area to be irrigated for Rabi crop

$$= 28\,000 \times \frac{30}{100} = 8400 \text{ hectares}$$

Outlet discharge required

$$= \frac{8400}{1800} = 4.67 \text{ cumec}$$

Losses below km 1.0 = 0.461 cumec

Losses in the reach between km 1.0 to km 0.0

Wetted perimeter of the channel section at km 1.0

$$= 5.75 + 1.2455 = 8.534 \text{ m}$$

Losses in the reach $= \dfrac{8.534 \times 1000 \times 2}{10^6} = 0.0171$ cumec

$= 0.018$ cumec (say)

Thus, total losses below km 0.0

$= 0.461 + 0.018$

$= 0.479$ cumec

Total discharge required at km 0.0

$= 4.67 + 0.479$

$= 5.149$ cumec

Adopt design discharge Q 10% more than the required discharge.

∴ Design discharge $= 1.1 \times 5.149$

$= 5.664$ cumec

$= 5.7$ cumec (say)

C.V.R. (V/V_0) = 1.0

Kutter's N = 0.0225

Assume bed slope $S = \dfrac{1}{4444} = 0.225$ m/km

From Garret's diagram Fig. 16.4 (Plate VI b)

Discharge Q (cumec)	*Bed slope S*	*Bed width B (m)*	*Water depth D (m)*	*Area* A = BD + ($D^2/2$) (m^2)	*Velocity V* = (Q/A) (m/s)	*Critical Velocity* V_0 *(m/s)*	*C, V, R,* (V/V_0)
5.7	1/4444	6.00	1.30	8.645	0.659	0.651	1.012

Adopt bed width $B = 6.0$ m and

Water depth $D = 1.3$ m

The schedule of area statistics and channel dimensions are shown in the accompanying Table.

Schedule of Area Statistics and Channel Dimensions

Below kilometre	Culturable commanded area	Culturable commanded area	Area to be irrigated			Outlet discharge factor	Outlet discharge	Losses in reach	Total losses	Total discharge	Design discharge	Channel dimensions					Velocity of flow	Velocity ratio
			Rabi 30%	Perennial	Kharif $12\frac{1}{2}\%$							Bed slope	Bed width	Water depth	Height of bank	Width of bank		
	ha	ha	ha	ha	ha	ha per cumec	cumec	cumec	cumec	cumec	cumec	cm per km	m	m	m	m	m/sec	
0	40 000	28 000	8400	—	500	1800	4.67	0.018	0.479	5.149	5.700	22.5	6.00	1.300	—	2.00	0.659	1.012
1	35 000	24 500	7350	—	3063	1800	4.08	0.016	0.461	4.541	5.000	22.5	5.75	1.245	—	2.00	0.630	0.995
2	30 000	21 000	6300	—	2625	1800	3.50	0.015	0.445	3.945	4.350	22.5	5.25	1.210	—	2.00	0.614	0.989
3	25 000	17 500	5250	—	2188	1800	2.92	0.015	0.430	3.350	3.700	25.0	4.50	1.180	—	2.00	0.616	1.008
4	20 000	14 000	4200	—	1750	1800	2.33	0.015	0.415	2.745	3.000	25.0	4.20	1.080	—	2.00	0.586	1.014
5	18 000	12 600	3780	—	1575	1800	2.10	—	0.400	2.500	2.750	25.0	4.00	1.060	—	2.00	0.573	1.004

REVIEW QUESTIONS

16.1 What is balancing depth in a canal ? Derive an expression for the same.

16.2 Discuss in detail the various causes of losses of water in channels.

16.3 What are the considerations for making a trial water surface on the longitudinal section for channel design ?

16.4 What do you understand by schedule of area statistics ? How is the same prepared ?

16.5 Sketch a typical cross-section of a canal which is partly in cutting and partly in filling.

16.6 What is berm ? List the functions served by a canal berm.

16.7 Sketch a typical cross-section of a canal which is wholly in (i) cutting, (ii) filling.

16.8 Explain the terms: (i) Free board in canals, (ii) Permanent land width, (iii) Temporary land width, (iv) Counter berm, (v) Inspection road, (vi) Dowla. (vii) Borrow pits, (viii) Spoil banks.

16.9 Write a detailed note on the various operations to be carried out for proper maintenance of irrigation channels.

16.10 Describe briefly the various methods of discharge measurement in channels.

16.11 Using Garret's diagrams design an irrigation canal to carry 9.5 cumec at a slope of 1 in 5000. The value of N is 0.025 and critical velocity ratio $m = 0.9$.

[*Ans.* $B = 9.35$ m; $D = 1.48$ m]

16.12 Using Lacey's diagrams design an irrigation channel to carry 25 cumec. Side slopes of the channel are $\frac{1}{2}\ H : 1\ (V)$ and silt factor $f = 1.2$. $\left[Ans.\ B = 20.45\ \text{m};\ D = 1.47\ \text{m};\ S = \frac{1}{4200}\right]$

16.13 A canal has a bed width of 6 m, full supply depth 1.6 m, bank width 2.5 m, cutting slope 1:1, filling slope 1.5:1 and free board 0.4 m. Calculate balancing depth. [*Ans.* 1.34 m]

CHAPTER

17 Waterlogging and its Control-Lining of Channels and Drainage

17.1 INTRODUCTION

An agricultural land is said to be waterlogged when the soil pores within the root zone of the crops are saturated to such an extent that normal circulation of air within the soil pores is totally cut off. The waterlogging affects the productivity or fertility of the land and thus leads to a reduction in the crop yield. The waterlogging is usually caused by a rise of sub-soil water table. The depth of the water table at which it tends to make the land waterlogged and starts harming the crops depends on the height of the capillary fringe and the type of crop. The height of capillary fringe is the height to which water will rise above the water table due to capillary action and is more in fine grained soils and less in coarse grained soils. For most of the agricultural soils the height of the capillary fringe varies from 0.9 to 1.5 m. Thus, for any crop if the yield is adversely affected when soil upto 0.6 m below the surface gets saturated then the land will start deteriorating due to waterlogging when the water table is within 1.5 m (*i.e.*, 0.6 + 0.9) to 2.1 m (i.e., 0.6 + 1.5) below the ground surface. The yield of some of the commonly grown crops is adversely affected when the depth of water table is equal to or less than the one noted against each as follows.

Crops	*Depth of water table*
Wheat	0.9 to 1.2 m
Cotton	1.5 to 1.8 m
Rice	0.6 m
Sugarcane	0.9 m
Fodder crop	1.2 m

In India the problem of waterlogging is of great concern, as large tracts in the Indo-Gangetic plain suffer from waterlogging. The problem of waterlogging has been reported even in several recent projects such as Chambal project in Rajasthan and Madhya Pradesh. As indicated in the Report of Irrigation Commission 1972 the total area of the land affected by waterlogging in different states is estimated as about 3 million hectares with Punjab heading the list where about 1 million hectares of land is estimated to be affected by waterlogging. However, several measures are now being taken under command area development programme to control waterlogging.

17.2 EFFECTS OF WATERLOGGING

Waterlogging causes infertility of land and hence results in reduction of crop yield. A study of the effect of waterlogging on crop production was made at a small farm at Rauni near Patiala. This study showed that the yield per hectare for cotton which was about 7.32 kN or 746 kg(f) in 1951–52 dropped to 0.098 kN or 10 kg(f) in 1959–60 and the fields had become unfit for the cultivation of cotton. The drop in yield of wheat was from 13.18 kN or 1344 kg(f) per hectare in 1951–52 to about 8.79 kN or 896 kg(f) per hectare in 1958–59. The main factors for the infertility of the waterlogged lands are as follows.

(i) *Absence of aeration of soil in the root zone of the plants.* For the growth of plants certain nutrients such as nitrates are quite essential. The supply of these nutrients is maintained in the soil by nitrification of the soil by certain bacteria or microorganisms which break up complex nitrogenous compounds into forms usuable by the plants. These bacteria need oxygen for their life and activity and are unable to live and function under anaerobic conditions which are developed when the land is waterlogged. The cessation of the activity of these bacteria leads to reduced fertility of the soil.

Further when the soil in the root zone becomes saturated the normal circulation of air is cut off, the level of oxygen declines and that of carbon dioxide increases as organic matter decomposes with the saturation of the soil. The depletion of the oxygen supply to plant roots results in wilting and ultimately in the death of the plants.

(ii) *Difficulty in cultivation operations.* It is difficult to carry out cultivation operations such as tilling etc., in a soil which is constantly wet. In an extreme case, when the ground water has risen above the surface, cultivation operation may be impossible.

(iii) *Growth of water weeds and wild aquatic plants.* There are certain plants or weeds which grow very well in marshy lands. These plants find favourable surroundings in waterlogged areas and hence grow in abundance, suppressing the crops growing in these areas. The crops have to therefore compete with these plants which results in reduction of crop yield. However, frequent hoeing may help in proper growth of the crop but it will make the crop uneconomical.

(iv) *Rise of salts in surface layers.* In a waterlogged land continuous upward flow of water is established from the water table to the ground surface. This is so because from the water table water rises into the capillary zone from where it is lost through evaporation and transpiration and more water is taken up from the water table to replenish these losses. As the water rises up it brings with it dissolved salts such as sodium carbonate, sodium chloride and sodium sulphate from the underlying layers to the surface and as water evaporates it leaves behind a deposit of these salts on the surface. The process of salt deposition in this manner is called *salt efflorescene*. Excessive concentration of salt in root zone of plants does not allow the plants to thrive which reults in reduction of crop yield. Moreover with a heavy concentration of salts the soil may become alkaline which is highly unproductive and difficult to reclaim.

(v) *Restricted root growth.* If the water table is high the roots of the plants are confined to the top layers of the soil above the water table and hence their growth is restricted. On the other hand if the water table is low the roots of the plants have more space for their growth. The restricted root growth will result in reduced crop yield.

(vi) *Lower soil temperature.* Well drained soils warm up faster than saturated soils. The low soil temperature which results from excessive moisture in the soil hampers the germination of seed, restricts development of root system and affects the rate of ripening. Moreover at low temperature of soil the activity of the bacteria becomes sluggish and consequently less food is available for the plants. This results in lowering of the crop yield.

(vii) *Plant diseases.* When low soil temperatures are accompanied by poor aeration the root system of plants does not develop properly and becomes vulnerable to attack by pests and diseases. The healthy growth of plants is thus hampered resulting in reduced crop yield.

17.3 CAUSES OF WATERLOGGING

The main factors causing waterlogging are as given below.

(i) *Seepage from canals.* When a canal system with unlined channels is introduced, water percolates from the channels and is added to the ground water reservoir. This causes a general rise in the water table in the region. For example, after the opening of the Ganga Canal, the water table under the Ganga–Yamuna Doab rose from a depth of 12.2 m to about 4.6 m below ground level in 100 years. Similarly in the areas commanded by the Western Yamuna Canal, between 1932 and 1963 the water table rose at an average annual rate of 160 mm or (16 cm). The water table had risen even faster in some areas commanded by the Chambal Canals.

(ii) *Over irrigation of fields.* When irrigation water is liberally applied to the fields the excess water percolates deep into the ground which reults in augmenting the ground water storage and raising up of the water table in the area.

(iii) *Inadequate surface drainage.* Heavy precipitation combined with inadequate surface drainage causes flooding of vast tracts of land. The prolonged flooding or inundation of land results in heavy percolation of water into the ground and a rise in the water table.

(iv) *Obstruction of natural drainage.* If in any area natural drainage is obstructed due to the construction of irrigation channel and road or rail embankment, it will not be able to pass the rain water. There will thus be flooding in the area and consequent waterlogging of land.

(v) *Obstruction of subsoil drainage.* If an impermeable stratum exists at a relatively low depth below the ground surface it will prevent downward movement of water in the subsoil and result in the creation of a high false or perched water table which may cause waterlogging. Similarly masonry structures like causeways with their foundations going into the impermeable stratum may obstruct the movement of water in the subsoil and cause a rise in the water table.

(vi) *Obliteration of natural drainage.* Sometimes cultivators plough up and obliterate (efface or destroy) an existing natural drainage. This results in stoppage of flow of storm water and consequent flooding and waterlogging of land.

(vii) *Nature of soil.* Waterlogging depends upon the nature of soil. A soil having low permeability, such as a black cotton soil is prone to waterlogging due to over irrigation or flooding because it does not allow water to percolate easily through it.

(viii) *Construction of a reservoir.* The seepage from a reservoir augments the ground water storage and consequent rise in water table may cause waterlogging.

(ix) *Incorrect and defective methods of cultivation.* The incorrect and defective methods of cultivation may result in creating pools of stagnant water and consequent waterlogging of land.

17.4 MEASURES FOR PREVENTION OF WATERLOGGING

The following measures can be taken to prevent waterlogging or to relieve area already affected by waterlogging.

(A) Reducing Percolation from Irrigation Channels

For reducing percolation from irrigation channels following methods may be adopted.

(i) *Lining of channels.* By providing a more or less impervious lining for channel bed and sides the percolation of water from the channel can be considerably reduced. The lining of channels is considered to be an effective method to control waterlogging. The different types of linings and various aspects of lining are discussed in sections 17.5 and 17.6.

(ii) *Lowering of full supply level of irrigation channels.* By designing channels with their full supply level as low as possible consistent with flow irrigation for most of their command, the percolation of water from channels may be reduced in the following two ways. Firstly in this case the channels will be mostly in cutting for which the loss of water by percolation is much less than for the channels in embankment or filling. Secondly a low full supply level will result in small difference of head between the F.S.L. and the fields and consequently lesser possibility of wasteful use of water by the cultivators on their fields.

(iii) *Providing intercepting drains along channels.* By providing drains along channels particularly in reaches of high embankment or filling and/or high water table, the water percolating from the channels is intercepted and thus prevented from augmenting the ground water storage and raising the water table. The various aspects of drains are discussed in section 17.8.

(B) Reducing Percolation from Water Courses and Fields

For reducing percolation from water courses and fields the following methods may be adopted.

(i) *Lining of water courses.* This has the same effect as lining of other irrigation channels.

(ii) *Controlling intensity of irrigation.* It means permitting only a fraction of the total culturable commanded area to receive water from irrigation channels in any one year. The usual annual intensity of irrigation is 40% to 50%. In areas where there is a possibility of waterlogging a lower intensity of irrigation should be adopted.

(C) Encouraging Economical use of Water

For this purpose the following methods may be used.

(i) *Educating the cultivators to use water economically.* The cultivators should be educated to use water economically. It should be emphasized upon the cultivators that the wasteful use of water besides being a national loss is a direct loss to them in the form of reduced yields and the long range damage to their land by rise of water table.

(ii) *Changing crop pattern.* In areas susceptible to waterlogging it will be desirable to introduce only those crops which need light irrigation. Crops requiring heavy irrigation should be controlled. Further soil survey data should be made use of in fixing crop patterns.

(iii) *Changing revenue policy.* If water is charged on flat area irrigated basis the cultivators will have a tendency to use more water than the optimum requirement. As such as far as possible volumetric assessment should be introduced to provide an incentive for economical use of water.

(D) Increasing Outflow from the Ground Water Reservoir

The following methods may be adopted to increase the outflow from the ground water reservoir.

(i) *Providing a drainage system.* The provision of a properly designed drainage system is one of the most effective method of controlling the menance of waterlogging. A detailed discussion of the types of drains and the various aspects of design of a drainage system are discussed in section 17.8.

(ii) *Improving the natural drainage of the area.* If the area has a good drainage water due to precipitation would not stand for a long time and thus reduce the percolation considerably. As such the natural drains should be maintained which involves removing obstructions such as weeds, bushes and other vegetation, straightening them or canalising them in shallow wide reaches, and improving their slopes.

(iii) *Pumping ground water.* If ground water is pumped through open or tube wells and is utilised for irrigation it would be a very effective anti-waterlogging measure. In fact a judicious combination of canal and tube-well irrigation (or the conjunctive use of surface and ground waters for irrigation) provides an ideal solution of the problem of waterlogging in many areas which are either waterlogged or are likely to be waterlogged due to the introduction of canal irrigation.

17.5 LINING OF IRRIGATION CHANNELS

Advantages of Lining of Irrigation Channels

The main advantages of lining of irrigation channels are as follows.

1. The lining of channels reduces the seepage loss thus resulting in a saving of water which can be utilised for irrigating additional area. The cost of irrigation per unit area is therefore reduced.

2. The lining of channels is an important anti-waterlogging measure as it reduces the percolation of water to the ground water reservoir and thus prevents rising of water table.

3. The lining provides a smooth surface having a low value of rugosity coefficient. Hence a lined channel has a higher velocity of flow.

4. Due to higher velocity of flow in a lined channel greater discharge can pass through a smaller area of cross-section of the channel.

5. The higher velocity of flow in a lined channel helps in providing a flatter hydraulic gradient or bed slope and hence better command can be obtained. Further this is an important advantage in the case of channels which are utilised for power generation because a smaller slope results in a larger available head at the power house.

6. The higher velocity of flow in a lined channel results in reducing the evaporation loss.

7. The problem of silting is reduced in a lined channel due to higher velocity of flow.

8. Due to smaller cross-section of a lined channel there is saving in cost of earthwork, land width and various structures.

9. Lining of a channel prevents or reduces weed growth.

10. Lining prevents the water flowing in the channel to come in contact with harmful salts which might be present in the soil of the bed and banks of the channel.

11. Lining reduces maintenance costs and possibility of breaching due to increased stability of the channel section.

12. Lining provides a stable channel section which is easier to operate.

Disadvantage of Lining of Irrigation Channels

The lining of irrigation channels has certain disadvantages although the advantages far outweigh the meagre disadvantages. The disadvantages of lining of irrigation channels are as follows.

1. The lining of irrigation channels involve a high initial investment.
2. It is difficult to repair the damaged lining.
3. The lining being permanent, it is difficult to shift the outlets if necessary.
4. A lined channel section is without a berm. As such the additional safety provided by the berms for vehicular and pedestrian traffic is absent in the case of lined channels.

Economics of Channel Lining. The lining of channel will be economically justified if the extra cost of providing the lining is less than or equal to the value of the benefits resulting from it. Thus, the maximum rate of expenditure on lining that is economically justifiable can be worked out as follows.

Let C = cost of lining in rupees per square metre including the additional cost of dressing the banks for lining and accounting for the saving, if any, resulting from the smaller cross sections and hence smaller area of land, quantity of earthwork and structures required for the lined sections. This saving will be available on new canals excavated to lined cross sections from the beginning, but not on lining of existing unlined canals.

s and S = seepage losses in unlined and lined canals respectively in cubic metres per square metre of wetted surface per day of 24 hours.

p and P = wetted perimeter in metres of unlined and lined sections respectively.

T = total perimeter of lining in metres.

d = number of running days of the channel per year.

W = value of water saved in rupees per cubic metre.

L = length of the channel in metres.

Y = life of lining in years.

M = annual saving in rupees in operation and maintenance due to lining.

B = annual estimated value in rupees of other benefits for the length of canal under consideration. These will include prevention of waterlogging, reduced cost of drainage for adjoining lands, reduced risk of breaching etc.

X = percent rate of interest.

The value of water lost by seepage from the unlined section

$$= pLsdW \text{ rupees.}$$

The annual saving in value of water otherwise lost by seepage

$$= (pLsdW - PLSdW) \text{ rupees}$$

$$= \{LdW(ps - PS)\} \text{ rupees.}$$

Total annual benefits resulting from the lining of channel

$$= \{LdW(ps - PS) + B + M\}$$

Let

$$a = \{LdW(ps - PS) + B + M\} \quad \ldots\text{(i)}$$

Additional capital expenditure on the construction of lined channel

$$= TLC \text{ rupees.}$$

This cost must be recovered from the saving during the useful life of the lining, Y, years. If the prevalent rate of interest is X, the *net present worth* (*NPW*) of the total annual benefits a, over the life of the lining (Y years) is determined from the following formula :

$$\text{NPW} = a \frac{(1+X)^Y - 1}{X(1+X)^Y} \quad \text{...(ii)}$$

For the lining to be economically feasible the additional initial cost of the lined canal (TLC) should be equal to or less than the net present worth (NPW) of the annual benefits, i.e.,

$$\text{TLC} \leq \text{NPW}.$$

In the above analysis it may be noted that the actual evaluation of benefits grouped under item B and M is very difficult, particularly on a new project. It can only be approximately estimated on the basis of experience on similar existing projects.

For the purpose of economic analysis, the life expectancy of concrete, brick tile and boulder lining may be assumed to be of the order of 60 years.

Requirements of Channel Lining. The following are the essential requirements of channel lining.

1. *Economy*. The lining should be economical in initial cost as well as in repair and maintenance.

2. *Structural Stability*. The lining should be strong enough to withstand the differential subsoil water pressure from behind the lining due to the sub-grade getting saturated through seepage or rain or due to sudden drawdown of channel. It should also be sufficiently heavy and strong to withstand the effect due to local cavity formation behind the lining.

3. *Durability*. The lining should be able to withstand the effect of velocity of water, rain, sunshine, frost and thawing (where applicable), thermal and moisture changes and chemical action of salts. It should also be able to withstand the damaging effect caused by cattle traffic, rodents and weed growth.

4. *Repairability*. The lining should be such that it can be repaired easily and economically since with lapse of time the lining may get damaged. Brick tile, boulder and precast slab linings are more easily repairable or replaceable than *in-situ* concrete lining.

5. *Impermeability*. The lining should be such that the seepage losses are reduced to a considerable extent.

6. *Hydraulic Efficiency*. The lining should have low coefficient of rugosity so that the channel has a high discharge carrying capacity and the channel section is hydraulically more efficient.

7. *Resistance to erosion*. The lining should be able to withstand abrasion due to sediment transported by the water flowing in the channel. Sometimes water flowing in a channel transports considerable amount of sediment which can damage the lining by abrasion. Cement concrete and boulder linings provide better abrasion resistance as compared to brick tile lining.

17.6 TYPES OF LINING

The various types of linings can be grouped into the following three categories.

(a) *Exposed and hard surface linings* :

1. Cement concrete lining (in-situ)
2. Precast concrete lining
3. Shotcrete lining
4. Cement mortar lining
5. Hydraulic lime concrete lining
6. Brick tile lining or Burnt clay tile lining
7. Stone blocks or undressed stone lining
8. Asphaltic concrete lining

(b) *Buried membrane linings* :

1. Sprayed-in-place asphalt membrane lining
2. Prefabricated asphaltic membrane lining
3. Polyethylene film and synthetic rubber membrane lining
4. Bentonite and clay membrane lining
5. Road oil lining

(c) *Earth lining*

1. Thin compacted earth lining
2. Thick compacted earth lining
3. Loosely placed earth lining
4. Stabilized soil lining
5. Bentonite soil lining
6. Soil-Cement lining

Each of the above noted types of linings are described below.

Cement Concrete Lining (in-situ). The in-situ cement concrete lining consists of a layer of cement concrete of required strength laid on a well prepared and compacted subgrade in the bed and sides of the channel. It is one of the most conventional type of lining which has been successfully used in India and other countries. Cement concrete lining is preferable to any other lining where channel is to carry high velocity flow because of its greater resistance to erosion. Velocities upto 2.7 m/s are generally considered permissible in channels having this type of lining although higher velocities upto 5 m/s have been provided. Cement concrete lining eliminates weed growth and thereby improves flow characteristics. Further burrowing animals cannot penetrate concrete. Also provision of cement concrete lining reduces maintenance charges to a minimum.

A distinct disadvantage of concrete lining is its lack of extensibility which results in frequent cracks due to contraction taking place from temperature changes, drying and shrinkage and settlement of subgrade. It is also likely to be damaged from alkaline water. Further under excessive extrenal water pressure cement concrete lining may be damaged. However, by providing reinforcement in the cement concrete lining it can withstand higher water pressure but it becomes quite costly.

Subgrade preparation. Before the concrete lining is laid the subgrade should be properly prepared as specified in IS : 3837—1978 to form a firm bed for the lining according to the required cross section of the channel. In reaches with expansive soils or where the salt content is over one percent or sodium sulphate is over 0.36 percent and where ground water level is higher or likely to be higher than the water level in the channel then special treatment as recommended in the code should be carried out.

Thickness of concrete lining. The thickness of lining is fixed depending upon the nature of channel requirement, namely, hydel channel or irrigation channel, full supply depth and channel capacity. Hydel channels have a greater thickness than irrigation channels because of draw-down effects and where closure for repairs may not be usual. Deep channels have greater thickness than shallow channels. Minimum thickness of channel lining based on the capacities of the channel are given in Table 17.1.

TABLE 17.1 Thickness of concrete lining

Capacity of channel (*cumec*)	*Depth of water* (*m*)	*Thickness of lining* (*mm*)
0–5	0–1	50 – 60
5–50	1–2.5	60 – 75
50–200	2.5–4.5	75 – 100
200–300	4.5–6.5	90 – 100
300–700	6.5–9.0	120 – 150

A variation of ±10 mm in thickness of the concrete lining is allowed provided average thickness is not less than the specified thickness.

Reinforcement. Normally reinforcement is not provided in cement concrete lining as it does not contribute to the durability of the lining. The use of reinforcement is, therefore, restricted to those lengths of channels where there is an obvious danger of a break-through which would result in destruction of crops and property. Reinforcement upto 0.25 to 0.3 percent of the area of concrete helps to reduce the width of shrinkage cracks and thus minimise seepage. Such reinforcement also gives protection against possible faulting of the cracked slabs where unstable subgrade soils are met with.

The reinforcement adds to the cost of concrete lining and the benefits that accrue are generally not sufficient to justify this increase in cost. Moreover reinforcement renders compaction of concrete difficult and uncertain and also it interferes with the working of mechanical equipment used for the construction of concrete lining.

Joints. Due to temperature variation and shrinkage cracks may be developed in concrete lining which may result in appreciable leakage from the channel. In order to minimise these effects contraction joints should be provided as indicated below.

When lining is laid as a continuous slab, transverse grooves or longitudinal and transverse grooves both in case of concrete lined channels with lined perimeter more than 10 m are formed as shown in Fig. 17.1 and in accordance with Table 17.2.

TABLE 17.2 Contraction joint spacing

t (*mm*)	*b* (*mm*)	*c* (*mm*)	*Approximate groove spacing centre to centre* (*m*)
50	8	17	3
65	8	20	3
75 and 80	11	27	4 to 5
90	11	30	4 to 5
100 and more	11	33	4 to 5

Note. Allowable tolerance on b and c shall be ±1.5 mm.

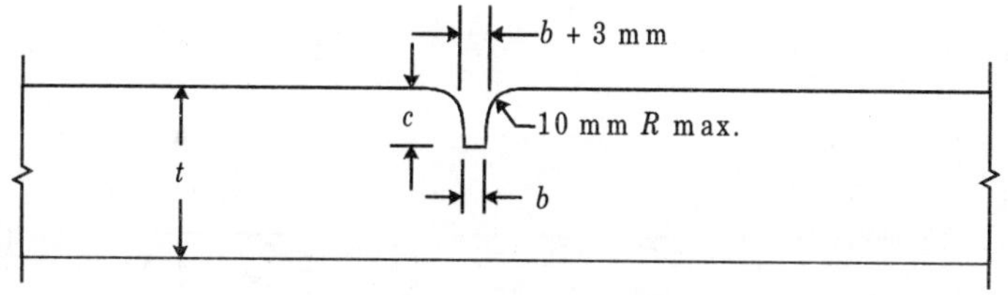

Fig. 17.1 *Shape of groove*

Under the joints, in-situ sleepers in the bed and precast sleepers in the sides are provided. The sleepers are 200 mm wide and 100 mm deep in case of channels with capacity more than 15 cumec and 150 mm wide and 75 mm deep in case of channels with capacity 15 cumec or less. The sleepers are placed centrally below the joint. The concrete used for the sleepers is also of the same grade as for lining. Before laying the cement concrete slabs, the top of the sleepers both in bed and sides are treated with two layers of sealing compound as shown in Fig. 17.2 to reduce the leakage across the joints. Slabs are laid in alternate compartments with an interval of at least one day for setting and contraction. The faces of the previously placed concrete slabs are painted with sealing compounds to ensure that no bonding takes place.

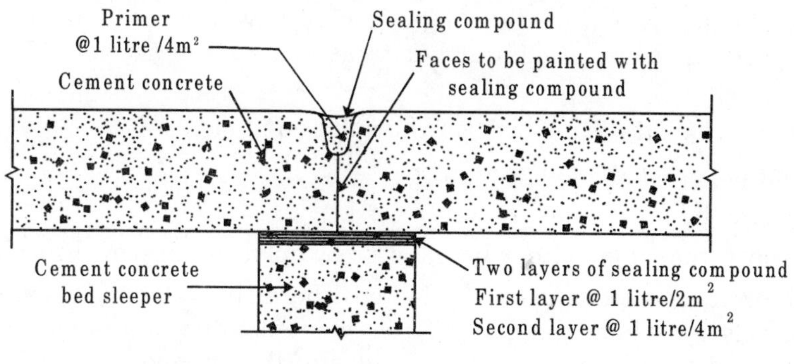

Fig. 17.2 *Details of construction joints when lining cast in panels*

The grooves at the joints in the canal lining are filled with hot-applied sealing compound. The filling of the grooves is taken up after curing period is over.

The joints described above and shown in Figs. 17.1 and 17.2 are as recommended by IS : 3873–1978 for code of practice for laying in-situ cement concrete lining on canals. However, some other types of joints which are also used for this type of lining are shown in Fig. 17.3.

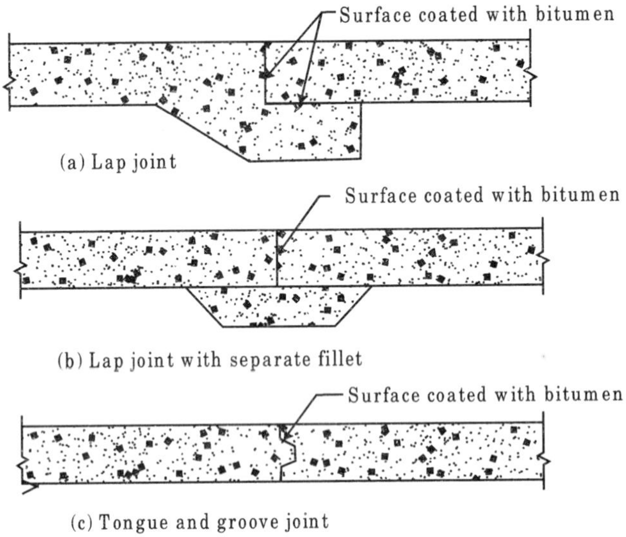

Fig. 17.3 *Other types of joints*

Concrete. For lining generally concrete of grade M10 or M15 conforming to IS : 456–1978 is used. However, in special cases, such as channels subjected to high velocity and/or drawdown, concrete of higher strength may be used or sometimes even two layers of concrete with the lower layer of grade M7.5 and the upper one of grade M10 or M15 may be used.

Laying of concrete. Before laying concrete all the absorptive surfaces of the subgrade are thoroughly saturated to a depth of 300 mm in sandy soil and 150 mm in other soils in order to prevent the subgrade from absorbing moisture from the freshly laid concrete and making it weak and porous. The saturation of the subgrade may however result in muddy and uneven surface of the subgrade and hence it is further treated in any of the following manner before laying concrete.

(i) Spread 1 : 3 cement sand slurry on the subgrade ; or

(ii) Spread oil paper on the subgrade ; or

(iii) Spread crude oil on the subgrade.

The laying of concrete is done either by hand or by specially designed machines.

The *hand laying* of concrete is usually adopted where cheap labour is available and in the case of small channels for which the quantity of concrete to be laid is small. In this method concrete is dumped and spread on the bed and sides of the channel in panels of required size with suitable joints in between. In order to maintain the desired grade and thickness of the lining during its laying, screed guides are laid on the subgrade and the concrete is screeded and finished upto the top surface of the screed guides.

The *mechanical laying* of concrete lining is carried out by means of slip-forms. A slip-form is a steel plate curved at the leading edge of the slip-form machine, extending across the bottom and up the side slopes of the channel to form the finished surface of the lining. The two types of slip-forms used are subgrade guided slip-forms and rail guided slip-forms. The subgrade guided slip-form is supported directly on the subgrade and is used for lining small to moderate size channels. On the other hand the rail guided slip-form is suppored on rails placed along both berms of the channel and is used for lining larger channels of considerable length. Both the types of slip-forms operate longitudinally and in both the cases concrete is spread uniformly and screeded mechanically on the bed of the channel and on the sides from bottom to top. The mechanical laying of concrete lining is quite fast and it results in considerable economy where large lengths of channels are to be lined.

Precast Concrete Lining. The precast concrete lining consists of precast concrete slabs of 1 : 3 : 6 concrete of size 500 mm × 250 mm × 50 mm set in 1 : 6 cement mortar. The slabs have a 10 mm wide and 16 mm deep rebate on all the four sides so that when laid side by side 20 mm wide and 16 mm deep open joints are formed alround each slab. The open joints formed by the rebates are sealed with 1 : 3 cement mortar of a dry consistency and properly packed to eliminate shrinkage. However, in order to render the 1 : 3 cement mortar more impervious and reduce its shrinkage, 5 percent crude oil (B.O.C. furnace oil) is added as an admixture to the mortar. The precast slabs are directly laid on compacted subgrade both in the bed and sides. Where subgrade is weak, reinforcement comprising 6 mm diameter mild steel bars is provided in the joints in both directions at certain intervals depending upon the stability of the compacted soil. Expansion joints are provided at intervals of 3.5 to 5 m which are grouted with a suitable bituminous mastic. In order to obtain a more impervious lining the top surface of the slabs may be provided with a 3 mm monolithic finish of 1 : 2 cement mortar.

The precast concrete lining has following advantages over the in-situ concrete lining laid and compacted manually.

(i) Precast concrete slabs are manufactured under controlled conditions and lining with such slabs provides a better, more impervious and durable lining than that laid with in-situ concrete owing to difficulty of exercising efficient control at the site on the various operations of laying the same.

(ii) As precast concrete slabs of comparatively thinner section and sufficiently strong can be manufactured on mass scale, a lining with such slabs is cheaper than that laid with in-situ concrete.

(iii) Precast concrete lining involves less site operations and therefore takes much shorter time to lay than an in-situ concrete lining. Moreover, in the case of lining an existing channel, precast concrete lining has the advantage that the channel needs to be closed for a shorter period or a longer length can be laid during a certain closure period than that laid with in-situ concrete.

(iv) Precast concrete slabs can be made during non-working season and can be even placed all along the canal for curing purposes thus reducing the cost of curing these slabs.

(v) Precast concrete slabs being of small size, the risk of cracking of lining due to shrinkage of concrete is eliminated.

(vi) In the case of damage to the lining due to settlement of the subgrade, it is easier and cheaper to repair such damage in precast concrete lining than in an in-situ one.

Shotcrete Lining or Gunited Lining. In this type of lining 1 : 4 cement mortar is applied pneumatically (i.e., by compressed air) through a nozzle on the subgrade in the bed and sides of the channel. The pneumatically applied mortar is known as shotcrete and hence the lining is known as shotcrete lining. Further the equipment for pneumatic application of the mortar is generally known as the cement gun and hence the process is known as guniting and the lining is also known as gunited lining. The pneumatic application of mortar is accomplished by the following two processes.

(a) Dry mix process

(b) Wet mix process

In the dry mix process cement and moist sand (having moisture content at least 4 percent but not more than 8 percent) are thoroughly mixed. The dry mixture is introduced into the chamber of the delivery equipment from where it is forced into the delivery hose and carried in suspension by compressed air to a nozzle fitted at the end of the hose. A water hose is connected to the nozzle. As the dry mixed material passes through the nozzle, water is introduced under pressure from the water hose and the resulting mortar is jetted from the nozzle and applied on the subgrade.

In the wet mix process all the ingredients viz., cement, sand and water are thoroughly mixed and mortar is prepared. The mortar is introduced into the chamber of the delivery equipment from where it is forced into the delivery hose and carried by compressed air to a nozzle lifted at the end of the hose. The mortar is jetted from the nozzle and applied on the subgrade.

The sand used for shotcrete lining should be well graded with maximum size 9.5 mm.

The usual thickness of shotcrete lining is 40 mm but it may vary from 25 to 65 mm. Further wire mesh reinforcement is frequnetly used in shotcrete lining but sometimes this lining is provided without any reinforcement. Shotcrete can be laid on irregular surface and hence there is no necessity of fine dressing of subgrade. This is of particular advantage in case of rock cuts. Moreover shotcrete is well suited for resurfacing of badly cracked and leaky but structurally sound old concrete lining. However, due to higher consumption of cement, shotcrete lining is more costly than concrete lining of same thickness and also it is less permanent than the concrete lining of usual thickness.

Cement Mortar Lining. In this type of lining a layer of cement mortar of uniform thickness is laid on the compacted subgrade. The usual tickness of cement mortar lining is 25 mm. The sand used for the mortar should be such that 85 percent or more will pass the No. 4 (6.35 mm) sieve and not more than 12 percent will pass the No. 100 (0.15 mm) sieve. Further the sand should be reasonably well graded from coarse to fine. This type of lining is however not very common.

Hydraulic Lime Concrete Lining. It consists of lime concrete made from hydraulic lime in suitable proportions. The hydraulic lime is usually made from 'kankar' or 'modular limestone'. An outstanding example of this type of lining is Gang Canal in Bikaner for which a lining of lime, grit and 'kankar' in the proportion 1 : 1 : 6 was laid over 50 years ago. In this case since 'kankar' was available throughout the length of the canal it was burnt to form lime. Further 'kankar' was also used as coarse aggregate and since good sand for concrete had to be led from long distances 'kankar' was crushed to form grit and the same was used instead of sand. Thus all the materials used for the lining were derived from 'kankar'. The lining has served satisfactorily for several years after its laying. However, recent investigations have revealed that the lining had been considerably cracked and damaged. In some portions the lining had completely failed in lengths of as much as 50 m.

The success of hydraulic lime concrete lining for Gang Canal has been due to the availability of good 'kankar' in abundance along the alignment of the canal. However, recent laboratory tests have shown that as compared to cement concrete, hydraulic lime concrete would make very poor lining material and hence it is not commonly adopted.

Brick Tile Lining or Burnt Clay Tile Lining. This type of lining consists of either a single or double layer of burnt clay tiles (or brick tiles) laid in cement mortar (Fig. 17.4). The size of the tiles is generally restricted to 300 mm × 150 mm × 50 mm for convenience of handling. Before laying the tiles the subgrade is properly dressed and thoroughly saturated to a depth of 300 mm in sandy soil and 150 mm in other soils to prevent the subgrade from absorbing moisture from freshly laid mortar. According to IS : 3872–1966 the following procedure is adopted for laying the single and double tile lining.

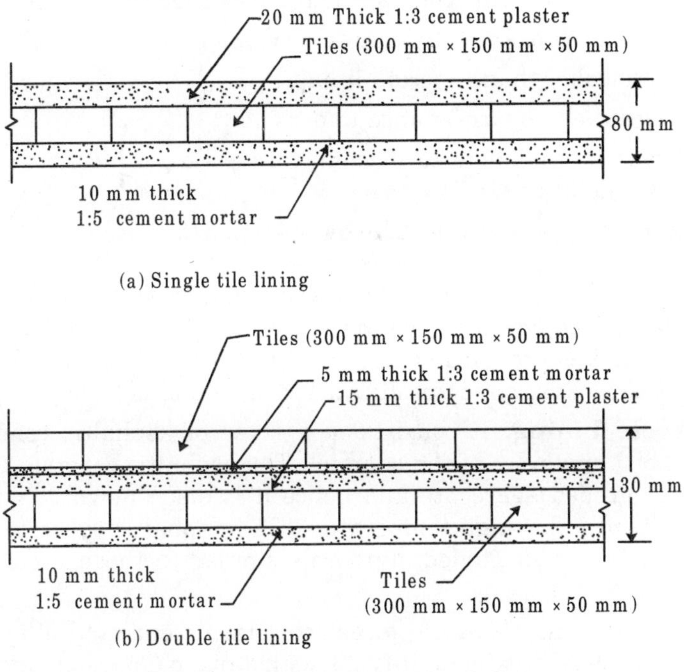

Fig. 17.4 *Single and double tile lining*

In the case of single tile lining a 10 mm thick layer of 1 : 5 cement mortar is laid on the well prepared subgrade. On this layer of mortar the tiles are laid. The tiles are firmly embedded in mortar and the vertical joints are completely filled with mortar. On the top of the tiles a 20 mm thick 1 : 3 cement plaster is laid.

In the case of double tile lining the lower layer of tiles is laid on a 10 mm thick layer of 1 : 5 cement mortar laid on the well prepared subgrade as in the case of single tile lining. On the top of this layer of tiles a 15 mm thick layer of 1 : 3 cement mortar is laid. After this layer of mortar has set it is made rough and on this the top layer of tiles is laid in 5 mm thick layer of 1 : 3 cement mortar. The layer of cement mortar in between the two layers of tiles is usually known as sandwich layer and hence double tile lining is also known as cement-mortar sandwich lining.

In channels having brick tile lining generally velocities upto 1.8 m/s are permissible. This type of lining has been used on Bhakra main canal and Indira Gandhi Canal.

The brick tile lining has certain advantages over cement concrete lining as indicated below :

(i) Brick tile lining is economical in initial cost.

(ii) Highly skilled labour is not required.

(iii) No elaborate equipment is needed.

(iv) Rigid control on work is not required.

(v) Expansion joints are not required as shrinkage is practically eliminated and coefficient of expansion of brick tiles with temperature is half that of concrete.

(vi) Repairs are easy and expeditious.

(vii) Transportation cost is usually less as kilns for burning tiles can be established locally along the channel.

(viii) Thickness of lining remains constant being governed by the thickness of tiles.

The various disadvantages of brick tile lining are as follows :

(i) It is more permeable than concrete lining.

(ii) It has comparatively low resistance to erosion.

(iii) Concrete lining is amenable to mechanical working which may be more economical than hand placed brick tile lining.

Stone Blocks or Undressed Stone Lining. It consists of a layer of undressed stones (or boulders) set in 1 : 6 cement mortar and laid on a well prepared subgrade in the bed and sides of the channel. Since fine dressing of stones is too costly, undressed stones are used but this would result in a higher resistance to flow. As such to reduce the resistance to flow a 20 to 25 mm thick 1 : 4 to 1 : 6 cement plaster is provided on the top of the stones. This type of lining is however suitable only where stones of required specification are available in abundance locally.

Asphaltic Concrete Lining. It consists of a layer of asphaltic concrete laid on a well prepared subgrade in bed and sides of the channel. The asphaltic concrete consists of a mixture of asphalt, cement and graded aggregate and hence it is also known as asphaltic cement concrete. It is mixed and laid under high temperature. The asphaltic concrete lining is laid either by hand ; or by subgrade or rail guided slip-forms similar to those used for cement concrete lining.

Asphaltic concrete lining may be used to replace cement concrete lining where it is cheaper. Asphaltic concrete lining has greater ability to withstand changes in the subgrade. Further asphaltic concrete lining can be used for repairing cement concrete lining by laying a resurfacing layer of asphaltic concrete.

The disadvantages of asphaltic concrete lining are as follows :

(i) Velocities in channels having asphaltic concrete lining are limited to 1.5 m/s.

(ii) It permits certain types of weed growth which results in puncturing of lining.

(iii) It provides insufficient resistance to external hydrostatic pressure.

(iv) There is danger of sliding during hot weather.

Sprayed-in-place Asphalt Membrane Lining. It consists of a layer of 6 mm thick layer of a special high-softening point asphalt sprayed in place at a high temperature of about 150°C on a prepared subgrade to form a water proof barrier. The asphalt layer is usually covered with a 300 mm layer of earth and gravel to protect it from damage and weathering. This type of lining provides an effective and cheap means of seepage control in channels and can be satisfactorily laid in cold and wet weather. This lining however does not decrease rugosity of the channel and retards weed growth for only a limited period. Moreover special equipment and trained workers are required for spraying of hot asphalt.

Prefabricated Asphaltic Membrane Lining. It consists of a prefabricated asphalt membrane which is laid directly on the subgrade and covered with a layer of earth for protection. The membrane consists of either a heavy kraft paper coated with a 4.76 mm thick layer of asphalt or a thin mat of glass fibres coated with a 3.2 mm thick layer of asphalt. The membrane is delivered to the site in rolls. Such a lining has proved durable and can be handled and transported easily.

Polyethylene Film and Synthetic Rubber Membrane Lining. It consists of polyethylene film or synthetic rubber membrane laid on the subgrade and provided with aprotective cover over the film or the membrane. The protective cover may consist of earth or earth and gravel or tiles laid with cement mortar or cement concrete or precast cement concrete slabs laid with cement mortar. This type of lining is not recommended where the problem of weed growth exists because the film or the membrane may be damaged by the weed roots. However, in such cases combination lining may be adopted which consists of lining of sides of the channel with rigid or hard surface lining and the bed of the channel with suitably protected polyethylene film or synthetic rubber membrane.

Bentonite and Clay Membrane Lining. It consists of a layer of bentonite 25 to 50 mm or more in thickness spread as a membrane over the subgrade and provided with a 150 to 300 mm thick protective cover of earth or earth and gravel over the membrane. Bentonite has a peculiar characteristic of becoming impervious on wetting due to swelling and imbibing of water and hence it helps to control seepage from channels. This type of lining may however be used where necessary quantity of such clay is available locally at low cost.

Road Oil Lining. In this type of lining road oil is sprinkled over the subgrade in a thickness of 1.5 mm which is sufficient to secure penetration of the oil to a depth of 50 to 80 mm into the subgrade. The subgrade is then rolled so that oil enters the soil pores. However, this type of lining is not so impermeable but helps to reduce the seepage losses.

Thin Compacted Earth Lining. It consists of 150 to 300 mm layer of cohesive soils thoroughly compacted and often protected by 150 to 300 mm cover of coarse soil or gravel. The thickness of the cover and lining may, however, vary with the type of soil used, velocity of flow and other conditions. In general the following soils are suitable for compacted earth linings.

(i) Gravel with sand-clay binder
(ii) Clayey gravel
(iii) Sand with clay binder
(iv) Clayey sands
(v) Clays

However, on account of the risk of relatively severe damage that may result from the erosion or cleaning operation, thin compacted earth lining has found limited use.

Thick Compacted Earth Lining. It consists of a relatively thick layer of thoroughly compacted suitable impervious soil both at the bottom and sides of the channel. The thickness of lining varies from 0.6 to 0.9 m on sides and from 0.3 to 0.6 m on the bed. The compaction of the soil is carried out in horizontal layers. The performance of this type of lining has generally been found to be relatively more satisfactory than that of other types of earth linings.

Loosely Placed Earth Lining. It consists of uncompacted suitable clayey soils spread over the bed and sides of the channel in layers upto a thickness of 300 mm. Although this type of lining may reduce seepage losses, it is not as effective and durable as compacted earth linings. Moreover it is subject to excessive erosion and scour.

Stabilized Soil Lining. In this type of lining the soil in the bed and sides of the channel is rendered impervious by using specially treated resins, chemicals such as sodium silicate and sodium chloride, commercial resin-cement, lime, asphalt, petrochemicals etc. Although some of these materials are very costly but under special circumstances these may provide a permanent type of earth lining.

Bentonite Soil Lining. In this type of lining the soil in the bed and banks is mixed in place with bentonite and a 50 to 100 mm thick finished lining is formed. The amount of bentonite may vary from 5 to 25 percent depending on the type of soil and is determined on the basis of laboratory tests. Alternatively this type of lining may also be provided by allowing the bentonite slurry to penetrate into the soil in the bed and sides of the channel.

Soil Cement Lining. Soil cement lining consists of a layer of a mixture of cement and soil with certain moisture content laid on the subgrade. There are two types of soil cement linings *viz.*, compacted and plastic. The compacted soil cement lining is also known as standard soil cement lining in which the soil cement mixture is compacted to its maximum density keeping the moisture content of the mix at about optimum value. The plastic soil cement lining is laid with soil cement mixture in plastic form. This type of lining has higher water and cement content and a consistency similar to that of masonary mortar. These properties permit placement of plastic soil cement lining by means of a slip form similar to that used in the placement of cement concrete lining.

The usual thickness of the lining is 75 to 100 mm. Further for compacted soil cement lining the quantity of cement may vary from 5 to 10 percent by weight of the dry soil. On the other hand for plastic soil cement lining the quantity of cement may vary from 10 to 20 percent by weight of the dry soil. The Central Board of Irrigation and Power has specified that for soil cement lining the soil should have as far as possible the following grading.

Fraction	*Percent by weight*
Clay (below 0.002 mm)	8 to 15
Silt (0.002 to 0.02 mm)	12 to 25
Sand (0.02 to 2.0 mm)	60 to 80

17.7 DESIGN OF LINED CHANNEL

In case of channels lined with hard surfaced materials two types of sections are adopted. A triangular section with circular bottom [Fig. 17.5 (a)] is adopted for smaller channels and a trapezoidal section with rounded corners is adopted for large channels. The triangular section with circular bottom is adopted when discharge is less than 85 cumec, assuming the limiting velocity as 1.8 m/s. For discharges greater than 85 cumec trapezodial section with rounded corners is adopted. However, IS : 10430–1982 recommends only trapezodial section with rounded corners for all discharges. Typical cross-sections of the lined channelise in cutting and filling as recommended by IS : 10430–1982 are shown in Fig. 17.6. The various considerations for the design of lined channels in accordance with IS : 10430–1982 are as follows.

Side Slopes. The side slopes should be such that they are nearly equal to the angle of repose of the natural soil in the subgrade so that no earth pressure is exerted over the back of the lining.

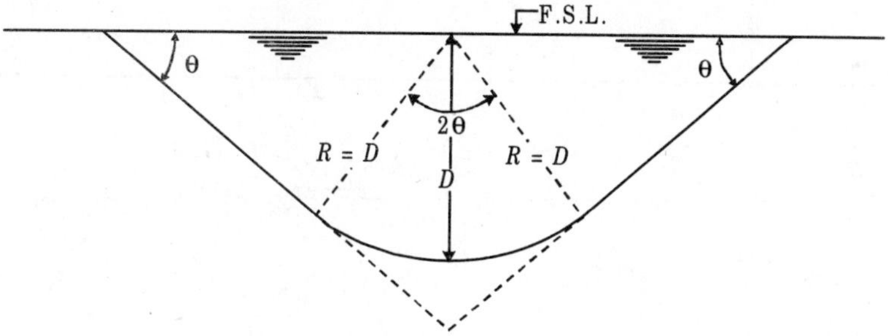

(a) Triangular section with circular bottom

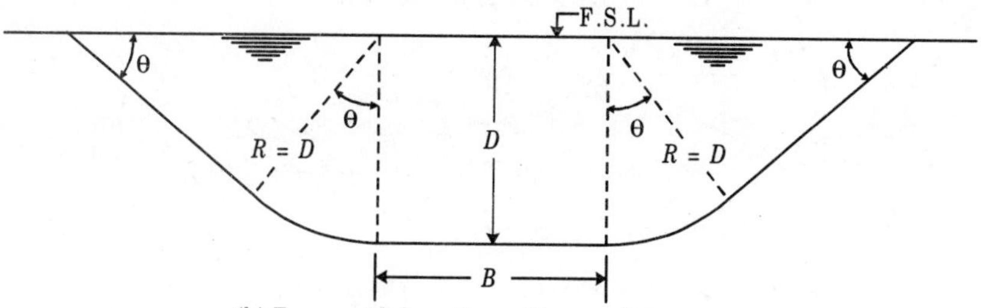

(b) Trapezoidal section with rounded corners

Fig. 17.5 *Cross-section of lined channel*

Freeboard. For channels of different discharge carrying capacities the values recommended for freeboard are as follows.

Type of channel	*Discharge* (*cumec*)	*Freeboard* (*m*)
(i) Main and Branch canals	> 10	0.75
(ii) Branch canals and Major distributaries	5 to 10	0.60
(iii) Major distributaries	1 to 5	0.50
(iv) Minor distributaries	< 1	0.30
(v) Water courses	< 0.06	0.10 to 0.15

The freeboard is measured from the full supply level to the top of the lining.

Berms. In deep cut reaches it is desirable to provide berms of width equal to the full supply depth of the channel to facilitate maintenance, silt clearance, etc., (see Fig. 17.6 c). In such sections the inner sides above the berms may be provided with turfing or pitching.

Bank widths. The minimum values recommended for top width of the bank are as follows.

Discharge (cumec)	*Maximum Bank Inspection Bank (m)*	*Top Width Non-inspection Bank (m)*
0.15 to 7.5	5.0	1.5
7.5 to 10.0	5.0	2.5
10.0 to 15.0	6.0	2.5
15.0 to 30.0	7.0	3.5
30.0 and above	8.0	5.0

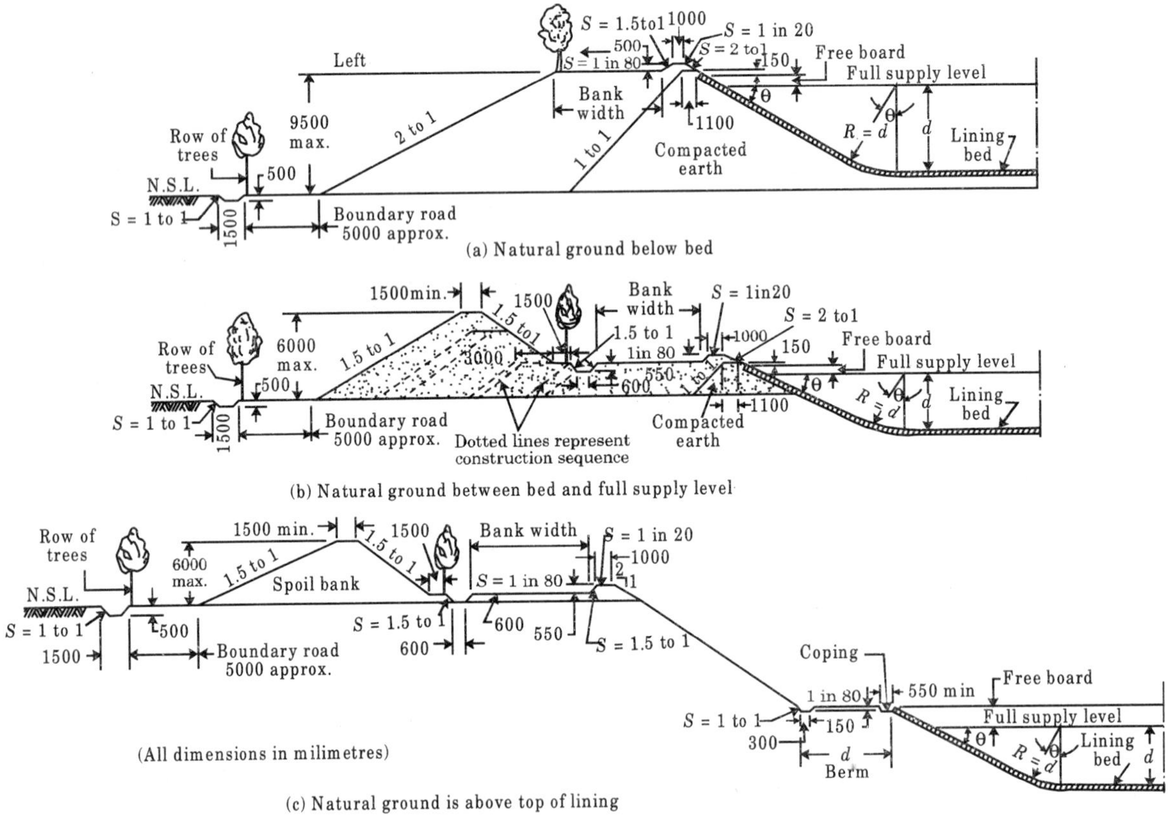

Fig. 17.6 *Typical cross-sections of lined channelis*

Dowla (Dowel). Dowels as shown in Fig. 17.6 shall be provided on the channel side of the service road, on both the banks of the lined channels.

Drainage. Adequate drainage as shown in Fig. 17.6 should be provided so that rain water does not flow towards the channel slope behind the lining.

Limiting Velocities of Flow in Lines Channels

The limiting safe velocities of flow in channels with some types of linings are given below.

(a) Cement concrete lining—2.7 m/s.

(b) Brick tile lining or burnt tile lining—1.8 m/s.

(c) Boulder lining—1.5 m/s.

Values of Rugosity Coefficient for Lined Channels

The values of rugosity coefficient N for various types of linings are given below.

Surface characteristics	*Value of N*
1. Concrete with surface as indicated below :	
(a) Formed, no finish/PCC tiles or slabs	0.018–0.020
(b) Trowel float finish	0.015–0.018
(c) Gunited finish	0.018–0.022
2. Concrete bed trowel finish and sides as indicated below :	
(a) Hammer dressed stone masonry	0.019–0.021
(b) Coursed rubble masonry	0.018–0.020
(c) Random rubble masonry	0.020–0.025
(d) Masonry plastered	0.015–0.017
(e) Dry boulder lining	0.020–0.030
3. Brick tile lining	0.018–0.020

Procedure for Design of Lined Channel Section

For design of lined channel section following data is required.

(i) Discharge Q

(ii) Rugosity coefficient N

(iii) Bed slope S

(iv) Side slopes

(v) Limiting velocity V

The following equations are used in the design.

(i) $Q = A \times V$ (Continuity equation)

(ii) $V = \dfrac{1}{N} R^{2/3} S^{1/2}$ (Manning's equation)

(a) The procedure for design of trapezoidal section with rounded corners is as follows.

1. Knowing limiting velocity V, rugosity coefficient N, and bed slope S, find hydraulic mean radius R using Manning's equation as

$$R = \left(\frac{VN}{S^{1/2}}\right)^{3/2}$$

2. Using continuity equation find area of cross-section

$$A = \frac{Q}{V}$$

3. Find wetted perimeter

$$P = \frac{A}{R}$$

4. For trapezoidal section with rounded corners shown in Fig. 17.5 (b)

$$A = BD + \pi D^2 \frac{\theta}{\pi} + \frac{2D^2 \cot\theta}{2}$$

or $$A = BD + D^2\theta + D^2 \cot \theta$$

or $$A = BD + D^2(\theta + \cot \theta) \qquad ...(17.1)$$

$$P = B + 2D\theta + 2D \cot \theta$$

or $$P = B + 2D(\theta + \cot \theta) \qquad ...(17.2)$$

Since A and P are known from steps (2) and (3), the values of bed width B and depth of flow D can be determined from Eqs (17.1) and (17.2).

Alternatively, if limiting velocity is not given but B/D ratio is given then also the channel section may be designed as indicated in Illustrative Example 17.2.

(b) The procedure for the design of traingular section with circular bottom is as follows.

The area of cross-section A of triangular section with circular bottom shown in Fig. 17.5 (a) is given by

$$A = \pi D^2 \frac{\theta}{\pi} + \frac{2D^2 \cot\theta}{2}$$

or $$A = D^2 (\theta + \cot \theta) \qquad ...(17.3)$$

The wetted perimeter P for this section is given by

$$P = 2\pi D \frac{\theta}{\pi} + 2D \cot \theta$$

or $$P = 2D (\theta + \cot \theta) \qquad ...(17.4)$$

Thus, for this section hydraulic mean radius

$$R = \frac{D^2(\theta + \cot\theta)}{2D(\theta + \cot \theta)}$$

or $$R = \frac{D}{2} \qquad ...(17.5)$$

Equation (17.5) indicates that triangular section with circular bottom is a best discharging section.

From continuity equation mean velocity of flow is given by

$$V = \frac{Q}{A}$$

or $$V = \frac{Q}{D^2(\theta + \cot\theta)} \qquad ...(17.6)$$

From Manning's equation

$$V = \frac{1}{N} R^{2/3} S^{1/2}$$

Substituting the values of R and V from Eqs (17.5) and (17.6), we get

$$\frac{Q}{D^2(\theta + \cot\theta)} = \frac{1}{N}\left(\frac{D}{2}\right)^{2/3} S^{1/2} \quad \text{...(17.7)}$$

From which the value of depth of flow D may be determined.

17.8 LAND DRAINAGE

A properly designed drainage system is an effective means to prevent land from getting waterlogged as well as to relieve the land already waterlogged. The importance of drainage is so well recognised in advanced countries that it is always considered an integral part of any irrigation scheme. In our country also an extensive drainage programme has been undertaken to relieve the land already waterlogged and in most of the new irrigation projects provisions are made for drainage works.

Before undertaking a drainage project, investigations should be carried out which include topographical, geological and soil surveys. The nature of soils and sub-soils from the point of view of permeability should also be studied. A knowledge of the water table and its fluctuations and the quality of groundwater in the area proposed for irrigation is also essential. These studies will also be of assistance in assessing the possibilities of waterlogging. Further data regarding precipitation in the past, and its intensities should also be collected.

Types of Drains

The various types of drains used as anti-waterlogging measure may be classified into two categories as follows :

1. Open drains 2. Closed drains

1. Open Drains. Open drains can be further classified into the following two types.

(i) *Shallow surface drains.* These drains are used to drain away excess irrigation water supplied to fields and also to quickly dispose of storm water. These drains therefore help to reduce percolation of water to groundwater reservoir by reducing the detention of surface water and thus prevent waterlogging of the irrigated land. However, these drains provide very little under-drainage (or drainage of subsoil) and hence they are not useful for relieving the land already waterlogged.

(ii) *Deep open drains.* These drains are used to drain out the subsoil and hence they are useful for prevention of waterlogging of land as well as for relieving the land already waterlogged. These drains are commonly used as outlet drains for a closed drain system for the reclamation of waterlogged land.

Design and Maintenance of Open Drains

A brief discussion on the important factors which govern design and maintenance of open drains is given below.

(i) *Layout.* The drains should be aligned to follow the paths of the natural drainage. Every drain should be provided with an outfall, either into a bigger drain or natural stream. The location of the outfall is adjusted to obtain the required slope in the drain.

(ii) *Capacity*. The drains are provided to carry storm runoffs, seepage water from subsoil and surface flow resulting from excess irrigation of fields. Since storm runoffs are usually far in excess of the other flows, the drains are designed for storm runoffs only. The other small flows viz., seepage flow and surface flow from excess irrigation will be automatically accommodated by the drains designed to take storm runoffs. The storm runoffs can be estimated by hydrological studies of the area. It is, however, not necessary to design the drains for exceptionally intense storms because it will be quite uneconomical. Thus, a judicious estimate of surface runoff should be made to obtain the design capacity of the drains. For example the drainage system on Sarda Canal Project in Uttar Pradesh was designed for a capacity of 0.109 cumec per square kilometre of catchment area. Similarly drainage system in Punjab were designed for a capacity of 0.044 cumec per square kilometre of the catchment area.

(iii) *Section*. The drains have trapezoidal cross-section and are designed on the same general principles as that of irrigation channels. Since deep drains receive both seepage and storm water, a cunnette (a small drain in the bed of larger drain) is usually provided to carry the small seepage discharge. The cunnette is provided in the centre of the drain with the bed of the drain gently sloping towards it from either end. With this arrangement the full section of the drain would be operative only during the storm and for the rest of the time it would be free of water. This would reduce the problem of maintenance because the seepage water which would otherwise have flown at small depth and small velocity over the entire width of the drain leading to weed growth over the entire section, would flow only through the cunnette at a faster velocity and hence only the cunnette section would be required to be maintained to make it free from aquatic weeds.

Depending on the type of soil, the depth and bottom width of deep drains generally varies from 0.6 to 1.5 m and 0.3 to 1.2 m respectively.

(iv) *Maintenance*. Since for most of the time the drains carry low discharges of silt free seepage water, they have profuse weed growth. As such the main problem of maintenance in case of drains is to keep them free of weeds. In our country this work is usually carried out manually. Certain implements have also been evolved for the removal of weed from the drains. The weed growth may however be retarded by having larger velocities of flow in the drains.

Drawbacks of Open Drains

There are certain drawbacks of open drains as explained below.

(i) For the construction of open drains valuable agricultural land is wasted.

(ii) These drains obstruct the farming operations.

(iii) Bridges are required to be constructed to facilitate communication across wide drains.

(iv) Valuable plant nutrients are washed down in the open drains.

2. Closed Drains. Closed drains are useful for prevention of waterlogging of land as well as for relieving the land already waterlogged. These are tile drains, usually of porous earthen ware, which are laid below the ground level butting each other with open joints and covered up by earth. Thus, these drains do not put any area out of cultivation and do not obstruct the farming operations. These drains can thus be more closely spaced where necessary. The usual spacing of these drains is from 15 to 45 m, closer spacing being used for soils of low permeability. The usual diameter of tile drains is 100 mm. The drains are usually located about 0.3 m lower than the desired water table. These drains are laid at a gradient steeper than 1 in 500. As far as possible the closed drains should be placed in a permeable stratum to ensure effective drainage. However, if the drain has to be placed in a stratum of low permeability, it should be surrounded by a filter of coarse sand and 'bajri' (river sand). The closed drains have their outlets in natural or artificial channels.

Closed drains are designed to carry only the seepage water and their capacity is determined by the rate of infiltration. As indicated below the rate of infiltration into a drain can be estimated theoretically if the soil permeabilities are known.

Layout of a Closed Drains System

A closed drain system in general consists of a number of lateral drains (also called laterals or branch drains) which collect water from different areas and discharge into main drain (also called main). The main drain is usually an open drain. Sometimes laterals discharge into a submain which is usually a closed drain and the submain discharges into main drain. The main drain in turn discharge into outfall drain which takes water to natural drain. The layout of a closed drain system is governed by the topography of the area to be drained. The various layout plans for closed drain system are shown in Fig. 17.7 and are discussed below.

(i) *Natural system.* In this system the main/submain and the connecting laterals are provided in natural courses as shown in Fig. 17.7(a). The natural system is preferred in rolling topography.

(ii) *Grid-iron layout.* In this system the laterals are provided only on one side of the main/submain as shown in Fig. 17.7(b). This system is used when the land is practically level or where the land slopes away from the main/submain on one side and when the entire area has to be drained.

(iii) *Herringbone pattern.* In this system laterals join main/submain from each side alternately as shown in Fig. 17.7(c). It is adopted when the main/submain is laid in a depression. In this case the land along the main/submain is double drained which is however, necessary because it is in depression and hence requires more drainage than the land on the adjacent slopes.

(iv) *Double main system.* In this system two separate mains/submains with separate laterals for each are provided as shown in Fig. 17.7(d). This system is provided when the bottom of the depression is wide. This arrangement helps to reduce the length of the laterals and eliminates the break in slope of the lateral at the edge of the depression.

(v) *Intercepting drain system.* In this system there are no laterals but only a main/submain is provided at the toe of the slope as shown in Fig. 17.7(e). This arrangement is adopted when the hilly land is to be drained.

Discharge and Spacing of Closed Drains

The closed drains are placed about 0.3 m below the desired highest level of the water table to ensure effective drainage. The water table should be maintained at a depth such that it would not damage the crops grown. Depending on the type of soil, crop grown and salinity the desired depth of water table may vary from 0.6 to 2.5 m. The water table may be maintained at the desired depth by providing closed drains of adequate capacity at appropriate spacing. An approximate theoretical method to determine discharge and spacing of closed drains is as follows.

With reference to Fig. 17.8, let L be the spacing of the drains and a be their height above the impervious stratum. Let b be the maximum height of the drained water table above the impervious stratum.

At a distance x from the centre of drain, let the drained water table be at a height y above the impervious stratum.

The various assumptions made are as follows.

(i) The hydraulic gradient at a distance x from the centre of the drain is (dy/dx).

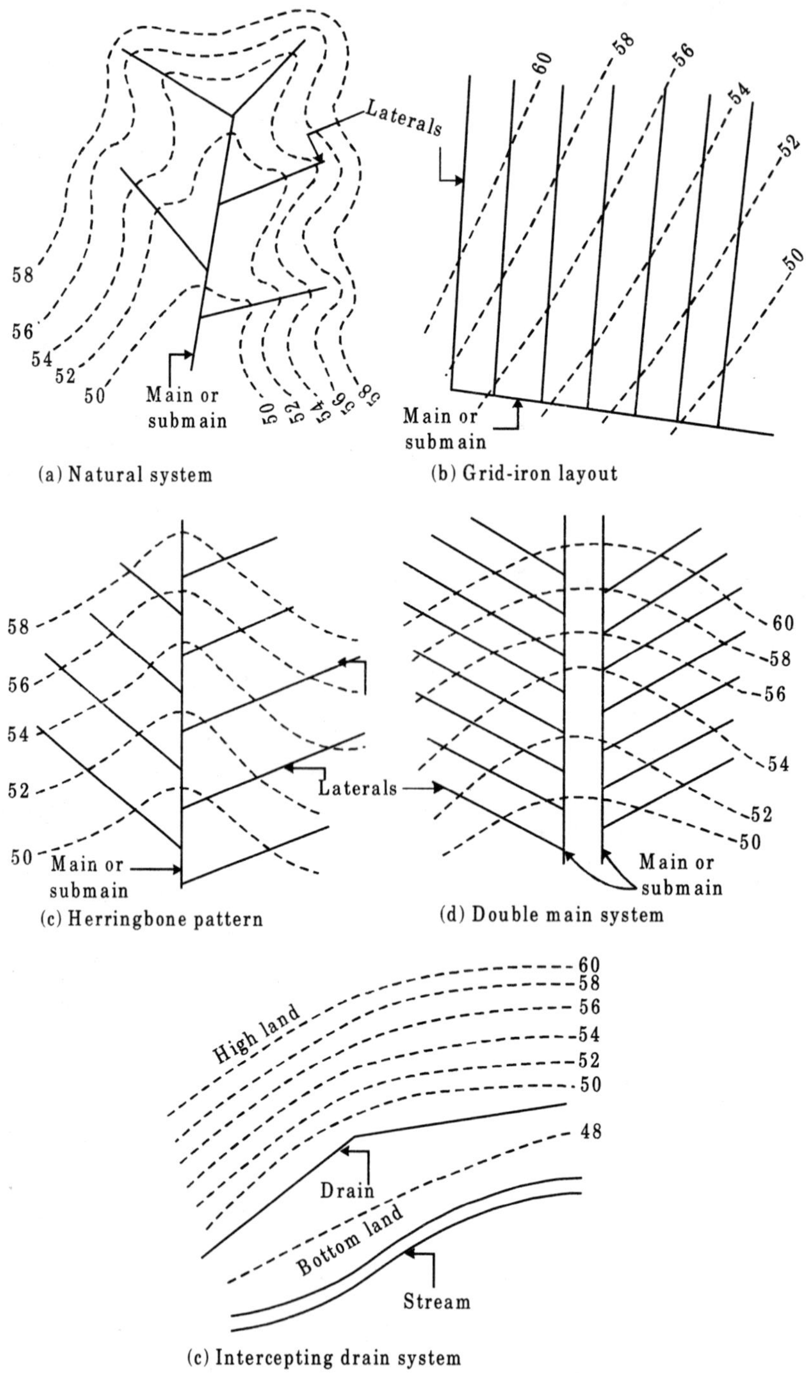

Fig. 17.7 *Layout plans of tile drains*

(ii) Flow lines are parallel and area of flow section at a distance x from the centre of the drain per unit length of the drain is $y \times 1$ $(= y)$.

(iii) Discharge towards the drain is inversely proportional to the distance from the drain.

With the above assumptions according to Darcy's law

$$Q_y = ky\frac{dy}{dx} \quad \text{...(17.8)}$$

where Q_y is the discharge passing through the section at a distance x from the centre of the drain per unit length of the drain.

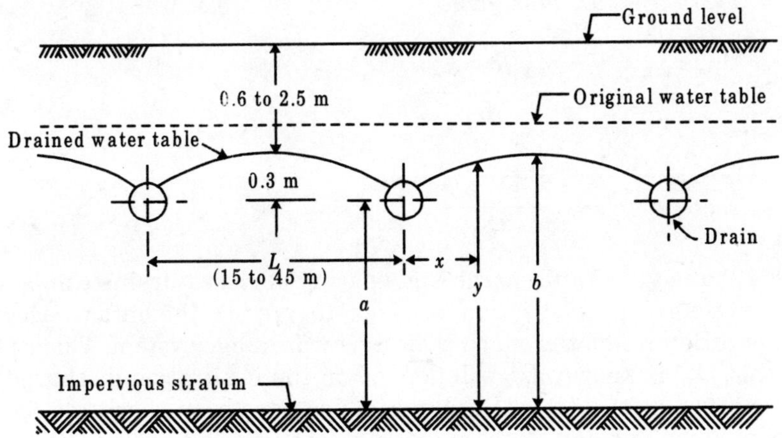

Fig. 17.8 *Spacing of closed drains*

If Q_D is the total discharge per unit length carried by the drain, then $(Q_D/2)$ enters the drain from either side. Since discharge is inversely proportional to the distance from the drain, discharge = 0 when $x = (L/2)$ and discharge = $(Q_D/2)$ when $x = 0$.

Thus,
$$Q_y = \frac{Q_D}{2}\frac{(L/2)-x}{(L/2)}$$

or
$$Q_y = \frac{Q_D}{2L}(L-2x) \quad \text{...(17.9)}$$

Equating the value of Q_y from Eqs (17.8) and (17.9), we have

$$\frac{Q_D}{2L}(L-2x) = ky\frac{dy}{dx}$$

or
$$\frac{Q_D}{2kL}(L-2x)\, dx = y\, dy$$

Integrating, we get

$$\frac{Q_D}{2kL}(Lx - x^2) = \frac{y^2}{2} + C$$

When $x = 0, \quad y = a,$

Hence $C = -\dfrac{a^2}{2}$

Making this substitution and simplifying, we get

$$Q_D = \frac{Lk\left(y^2 - a^2\right)}{\left(Lx - x^2\right)}$$

Also when $x = (L/2)$, $y = b$. This gives

$$Q_D = \frac{4k\left(b^2 - a^2\right)}{L} \qquad \text{...(17.10)}$$

and

$$L = \frac{4k\left(b^2 - a^2\right)}{Q_D} \qquad \text{...(17.11)}$$

From Eqs (17.10) and (17.11) discharge and spacing of closed drains can be determined. For a given drainage system Eq. (17.10) can be used to predict the inflow. However, the more common problem is to determine the spacing for a new drainage system. This can be done by Eq. (17.11) provided that Q_D is known. Q_D will depend on the infiltration discharge into the ground which should be removed by the drain. Different values have been suggested for Q_D out of which the commonly adopted value being 1% of average annual rainfall drained of in 24 hours.

Equation (17.11) shows that spacing of closed drains is independent of drain size. Further it has been observed that an increase in drain diameter produces only slight increase in discharge of drain.

The above method is also applicable to deep open drains which penetrate below the water table.

SOLVED ILLUSTRATIVE EXAMPLES

Illustrative Example 17.1 *Design a trapezoidal shaped concrete lined channel to carry a discharge of 200 cumec at a slope of 30 cm/km. The side slopes of the channel are 1.5 : 1. The value of N may be taken as 0.017. Assume limiting velocity in the channel as 2 m/s.*

Solution

From Manning's equation hydraulic mean radius R is given by

$$R = \left(\frac{VN}{S^{1/2}}\right)^{3/2} \qquad \text{...(i)}$$

$V = 2$ m/s ; $N = 0.017$; and

$$S = \frac{30}{1000 \times 100} = 3 \times 10^{-4}$$

Substituting the above values in Eq. (i), we get

$$R = \left[\frac{2\times0.017}{(3\times10^{-4})^{1/2}}\right]^{3/2}$$

or $$R = 2.75 \text{ m}$$

For side slope 1.5 : 1

$$\cot \theta = 1.5$$

∴ $$\theta = \cot^{-1} (1.5)$$

or $$\theta = 0.588 \text{ radians}$$

Area of cross-section $$A = \frac{Q}{V} = \frac{200}{2} = 100 \text{ m}^2$$

Wetted perimeter $$P = \frac{A}{R} = \frac{100}{2.75} = 36.36 \text{ m}$$

From Eq. (17.1)

$$A = BD + D^2 (\theta + \cot \theta)$$

Substituting the above values, we get

$$100 = BD + D^2 (0.588 + 1.5)$$

or $$100 = BD + 2.088 D^2 \quad \text{...(ii)}$$

From Eq. (17.2)

$$P = B + 2D (\theta + \cot \theta)$$

Substituting the above values, we get

$$36.36 = B + 2D (0.588 + 1.5)$$

or $$36.36 = B + 4.176D \quad \text{...(iii)}$$

From Eq. (iii), we have

$$B = 36.36 - 4.176 D$$

Substituting the value of B in Eq. (ii), we get

$$100 = (36.36 - 4.176 D) D + 2.088D2$$

or $$2.088 D^2 - 36.36 D + 100 = 0$$

or $$D^2 - 17.41 D + 47.89 = 0$$

or $$D = 3.43 \text{ m}$$

Thus, $$B = 36.36 - (4.176 \times 3.43)$$

or $$B = 22.04 \text{ m}$$

Thus required bed width = 22.04 m and depth of flow = 3.43 m.

Illustrative Example 17.2 *Illustrative Example 17.1 if limiting value of the velocity is not given but (B/D) = 5.*

Solution

From Eq. (17.1) and (17.2)

$$A = BD + D^2 (\theta + \cot \theta)$$

and $$P = B + 2D (\theta + \cot \theta)$$

Thus, $$A = BD + 2.088 D^2$$

and $$P = B + 4.176 D$$

Velocity of flow $$V = \frac{Q}{A} = \frac{200}{BD + 2.088 \quad D^2}$$

Hydraulic mean radius

$$R = \frac{A}{P} = \frac{BD + 2.088D^2}{B + 4.176D}$$

From Manning's equation

$$V = \frac{1}{N} R^{2/3} S^{1/2}$$

$$N = 0.017 \text{ and } S = 3 \times 10^{-4}$$

$\therefore$ $$V = \frac{1}{0.017} (R)^{2/3} \left(3 \times 10^{-4}\right)^{1/2}$$

or $$V = \frac{R^{2/3}}{0.981}$$

or $$R^{2/3} = 0.981 V$$

Substituting the values of R and V, we get

$$\left(\frac{BD + 2.088D^2}{B + 4.176D}\right)^{2/3} = 0.981 \left(\frac{200}{BD + 2.088D^2}\right)$$

Given $$B = 5D$$

Thus,

$$\left(\frac{5D^2 + 2.088D^2}{5D + 4.176D}\right)^{2/3} = 0.981 \left(\frac{200}{5D^2 + 2.088D^2}\right)$$

or $$\left(\frac{7.088D}{9.176}\right)^{2/3} = 0.981 \left(\frac{200}{7.088D^2}\right)$$

or $$D^{8/3} = \frac{0.981 \times 28.217}{0.772}$$

or $$D^{8/3} = 35.856$$

$\therefore$ $$D = 3.83 \text{ m}$$

and $$B = 19.15 \text{ m}$$

Thus, required bed width = 19.15 m and depth of flow = 3.83 m.

Illustrative Example 17.3 *Design a concrete lined channel to carry a discharge of 40 cumec at a slope of 1 in 8100. The side slopes of the channel are 1.25 : 1. The value of Manning's N may be taken as 0.016.*

Solution

The channel section is assumed to be triangular with circular bottom as shown in Fig. 17.5 (a).

From Eq. (17.5) for this case hydraulic mean radius

$$R = \frac{D}{2}$$

Given $N = 0.016$; and $S = \dfrac{1}{8100}$

From Manning's equation

$$V = \frac{1}{N} R^{2/3} S^{1/2} \qquad \text{...(i)}$$

Substituting the above values in Eq. (i), we get

$$V = \frac{1}{0.016}\left(\frac{D}{2}\right)^{2/3}\left(\frac{1}{8100}\right)^{1/2}$$

or

$$V = \frac{D^{2/3}}{2.286} \qquad \text{...(ii)}$$

From Eq. (17.3) area of cross section

$$A = D^2 (\theta + \cot\theta)$$

For side slope 1.25 : 1

$$\cot\theta = 1.25$$

$\therefore$ $$\theta = \cot^{-1}(1.25)$$

or $$\theta = 0.675 \text{ radians}$$

$\therefore$ $$A = D^2 (0.675 + 1.25)$$

or $$A = 1.925\, D^2$$

From continuity equation

$$V = \frac{Q}{A}$$

or

$$V = \frac{40}{1.925 D^2} \qquad \text{...(iii)}$$

Equating Eqs (ii) and (iii), we get

$$\frac{D^{2/3}}{2.286} = \frac{40}{1.925 D^2}$$

or $$D^{8/3} = 47.50$$

$\therefore$ $$D = 4.25 \text{ m}$$

Illustrative Example 17.4 *In a system of closed drains the drains are placed with their centres 7.7 m above the impervious stratum and the maximum height of the drained water table above the impervious stratum is 8 m. If the spacing between the drains is 25 m and discharge entering the drain per unit length is 4.5 × 10^{-6} cumec/m, find coefficient of permeability of the soil.*

Solution

From Eq. (17.11), we have

$$L = \frac{4k(b^2 - a^2)}{Q_D}$$

or $$k = \frac{LQ_D}{4(b^2 - a^2)}$$

$$L = 25 \text{ m} ; Q_D = 4.5 \times 10^{-6} \text{ cumec/m} ; b = 8 \text{ m} ;$$

and $$a = 7.7 \text{ m}$$

Substituting these values, we get

$$k = \frac{25 \times 4.5 \times 10^{-6}}{4[(8)^2 - (7.7)^2]}$$

or $$k = 5.97 \times 10^{-6} \text{ m/sec}$$

or $$k = 5.97 \times 10^{-4} \text{ cm/sec.}$$

Illustrative Example 17.5 *In a drainage system closed drains are to be placed with their centres 2 m below the ground level to keep the highest position of the water table 1.7 m below the ground level. The impervious stratum is at a depth of 9.6 m below the ground level. If the average annual rainfall in the area is 850 mm, find the spacing of drains. Assume 1% of the average annual rainfall to be drained in 24 hours and coefficient of permeability k = 1 × 10^{-5} m/s.*

Solution

1% of average annual rainfall over an area A m^2 catered by each drain

$$= \frac{850 \times A \times 1}{1000 \times 100} \text{ m}^3$$

$$= 85 \times 10^{-4} A \text{ m}^3$$

This is to be drained of in 24 hours and hence the volume of water to be drained of per second

$$= \frac{85 \times 10^{-4} A}{24 \times 3600} \text{ cumec}$$

If L is the spacing between the drains then area A catered by each drain per unit length of the drain is

$$A = (L \times 1) = L \text{ m}^2$$

$$\therefore \qquad Q_D = \frac{85 \times 10^{-4} \times L}{24 \times 3600}$$

or $$Q_D = 9.84 \times 10^{-8}\, L \text{ cumec/m}$$

From Eq. (17.11), we have

$$L = \frac{4k(b^2 - a^2)}{Q_D}$$

$$k = 1 \times 10^{-5} \text{ m/sec} ; b = (9.6 - 1.7) = 7.9 \text{ m} ;$$

and $$a = (9.6 - 2) = 7.6 \text{ m}$$

Substituting these values, we get

$$L = \frac{4 \times 1 \times 10^{-5}[(7.9)^2 - (7.6)^2]}{9.84 \times 10^{-8} L}$$

or $$L^2 = 1890.24$$

$$\therefore \qquad L\ 43.48 \text{ m} = 43.5 \text{ m}$$

Hence drains should be provided at a spacing of 43.5 m.

Illustrative Example 17.6 *In a drainage system closed drains are placed with their centres 10.5 m above the impervious stratum and the maximum height of the drained water table above the centre of the drains is 0.3 m. If the spacing of the drains is 36 m and the drains carry 1% of the average annual rainfall in 24 hours, find the average annual rainfall. Take coefficient of permeability* $k = 5 \times 10^{-6}$ *m/s.*

Solution

From Eq. (17.10), we have

$$Q_D = \frac{4k(b^2 - a^2)}{L}$$

$$k = 5 \times 10^{-6} \text{ m/s} ; b = (10.5 + 0.3) = 10.8 \text{ m} ;$$

$$a = 10.5 \text{ m} ; \text{ and } L = 20 \text{ m}$$

Substituting these values, we get

$$Q_D = \frac{4 \times 5 \times 10^{-6}[(10.8)^2 - (10.5)^2]}{36}$$

or $$Q_D = 3.55 \times 10^{-6} \text{ cumec/m.}$$

If the average annual rainfall is x mm then one percent of this rainfall over an area A m^2 catered by each drain

$$= \frac{x \times A \times 1}{1000 \times 100} \text{ m}^3$$

This is to be drained of in 24 hours and hence the volume of water to be drained of per second

$$= \frac{x \times A}{1000 \times 100 \times 24 \times 3600} \text{ cumec}$$

If L is the spacing between the drains then area A catered by each drain per unit length of the drain is

$$A = (L \times 1) = L \text{ m}^2$$

$$\therefore \qquad Q_D = \frac{x \times L}{1000 \times 100 \times 24 \times 3600}$$

Substituting the values of QD and L, we get

$$3.55 \times 10^{-6} = \frac{x \times 36}{1000 \times 100 \times 24 \times 3600}$$

$$\therefore \qquad x = 852 \text{ mm}$$

The average annual rainfall is 852 mm.

Illustrative Example 17.7 *Sugarcane (root zone depth 1.8 m) is grown in a particular area where the ground water table is 2.0 m below ground. If the size of soil pores is 0.08 mm in diameter and surface tension σ = 0.054 N/m, is the field waterlogged. If so, determine the vertical location of closed drains below ground spaced at 15 m. Take drainage coefficient as 0.116 cumec/km^2, coefficient of permeability as 10^{-6} m/s and the impervious stratum to occur at 7.0 m below ground. Assume the capillary rise from ground water table not to interfere with the root zone of plant.*

Solution

$$\text{Capillary rise} = \frac{2\sigma}{wr}$$

$$\sigma = 0.054 \text{ N/m}; \; w = 9810 \text{ N/m}^3; \text{ and}$$

$$r = (0.08) = 0.04 \text{ mm} = 0.04 \times 10^{-3} \text{ m}$$

Thus, by substitution, we get

$$\text{Capillary rise} = \frac{2 \times 0.054}{9810 \times 0.04 \times 10^{-3}} = 0.275 \text{ m}$$

Thus, due to capillary rise the depth of ground water table below the ground will be reduced to (2.0 – 0.275) = 1.725 m, which is less than the root zone depth of 1.8 m, and hence the field will be waterlogged.

From Eq. (17.11), we have

$$L = \frac{4k \; (b^2 - a^2)}{Q_D}$$

$$L = 15 \text{ m}; \; k = 10^{-6}; \; Q_D = 0.116 \times 10^{-6} \times 15 \text{ cumec/m}$$

For the capillary rise to not to interfere with the root zone of plant, we have

$$b = 7.0 - (1.80 + 0.275)$$

$$= 4.925 \text{ m}$$

Thus, by substitution, we get

$$15 = \frac{4 \times 10^{-6} \times \left[(4.925)^2 - a^2 \right]}{0.116 \times 10^{-6} \times 15}$$

or $$a = 4.21 \text{ m}$$

Thus, drains may be provided at a depth of (7.0 – 4.21) = 2.79 m below the ground.

Illustrative Example 17.8. *An unlined canal giving a seepage loss of 4.0 cumec per million square metres of wetted area is proposed to be lined with 12 cm thick cement concrete lining, which costs ₹ 20.00 per square metre. Work out the economics of lining and show if the scheme is justified on the basis of the following data:*

Annual revenue from crops = ₹ 400 000 per cumec of water

Discharge of the channel = 100 cumec

Area of the channel = 50 m^2

Wetted perimeter of the channel = 22.4 m

Wetted perimeter of the lining = 22 m

Annual maintenance cost of unlined channel = 12 paise per m^2

Seepage loss in lined channel = 0.01 cumec per million square metre of wetted perimeter

Saving in annual maintenance as a consequence to lining = 40%

Life of lining = 40 years

Rate of interest = 6 per cent

Assume additional suitable data, if required.

Solution

Let L be the length of the canal.

The value of water lost per year by seepage from the unlined section

$$= (22.4 \times L) \times \frac{4}{10^6} \times 400\,000 = 35.84\ L \text{ rupees}$$

The value of water lost per year by seepage from the lined section

$$= (22 \times L) \times \frac{0.01}{10^6} \times 400\,000 = 0.088\ L \text{ rupees.}$$

∴ The annual saving in value of water otherwise lost by seepage

$$= (35.84 - 0.088)\ L = 35.752\ L \text{ rupees.}$$

Annual saving in rupees in operation and maintenance due to lining

$$= \frac{12}{100} \times \frac{40}{100} \times 22.4 \times L = 1.0752\ L \text{ rupees.}$$

∴ Total annual benefits resulting from the lining of canal = (35.752 L + 1.0752 L)

$$= 36.8272\ L \text{ rupees.}$$

The net present worth (NPW) of the total annual benefits is given by

$$\text{NPW} = (36.8272\ L)\left[\frac{(1+0.06)^{40}-1}{0.06(1+0.06)^{40}}\right] = 554.08\ L \text{ rupees.}$$

Assuming total perimeter T of lining to be about 15% more than the wetted perimeter of the lined canal, $T = 22 \times 1.15 = 25.3$ m.

Thus, additional capital expenditure on the construction of lined channel is

$$\text{TLC} = 25.3 \times L \times 20 = 506\ L \text{ rupees.}$$

Since NPW > TLC, the lining is economically justified.

REVIEW QUESTIONS

17.1 What is waterlogging ? What are the ill-effects of waterlogging ?

17.2 What are the causes of waterlogging ?

17.3 Discuss the various measures which may be taken to prevent waterlogging as well as to relieve the land already waterlogged.

17.4 How can drainage of land be used for reclamation of water logged land? How is the spacing of drains determined ?

17.5 Find the spacing between drains for the following data :

(i) Annual rainfall = 1000 mm

(ii) Height of drains above impervious stratum = 4.5 m

(iii) Maximum height of the drained water table above the impervious stratum = 5.0 m

(iv) Coefficient of permeability $k = 10^{-6}$ m/s. [*Ans.* 12.81 m]

17.6 Find the inflow into the drain for the following data :

(i) Spacing = 13 m

(ii) Height of drains above impervious stratum = 4.5 m

(iii) Maximum height of the drained water table above the impervious stratum = 5.0 m

(iv) Coefficient of permeability $k = 10^{-6}$ m/s [*Ans.* 0.126 m^3/day]

17.7 What is canal lining ? Discuss the advantages and disadvantages of canal lining.

17.8 Write a note on economics of channel lining.

17.9 What are the requirements of channel lining ?

17.10 List the various types of linings. Discuss the salient features of cement concrete lining.

17.11 Describe with neat sketches brick tile lining used for irrigation channels.

17.12 Explain the method of design of a lined canal.

17.13 Design a concrete lined channel to carry a discharge of 20 cumec at a slope of 1 in 10000. The side slopes of the channel are 1.25 : 1 and Manning's N may be taken as 0.014.

[*Ans.* Triangular section with circular bottom and Depth D = 3.25 m]

17.14 Design a concrete lined channel to carry a discharge of 400 cumec at a slope of 1 in 10 000. The side slopes of the channel are 1.25 : 1 and Manning's N may be taken as 0.014. The depth is limited to 5.0 m.

[*Ans.* Trapezoidal section with rounded corners,
Bed width B = 34.11 m and Depth D = 5.0 m]

17.15 What do you understand by salt efflorescene ?

CHAPTER

18 Canal Outlets

18.1 INTRODUCTION

An outlet is a device through which water is released from a distributing channel into a water course or field channel. Thus, an outlet is a sort of head regulator for a water course which supplies water to the fields. The importance of an outlet lies in the fact that it is a connecting link between the government staff responsible for the operation and maintenance of the distributing channel and the cultivators who take water from the water course and are responsible for the maintenance of the water course. The discharge through an outlet is usually not less than 0.03 cumec and not more than 0.085 cumec.

18.2 REQUIREMENTS OF A GOOD OUTLET

The requirements of a good outlet are as follows.

(i) The outlet should be simple in design, construction and working.

(ii) It should not be easily tampered with by the cultivator, but if tampered with it should be easily detected.

(iii) It should be worked efficiently with a small working head.

(iv) The outlet should draw its fare share of silt carried by the distributing channel.

(v) It should be strong, with no moving parts liable to be damaged or requiring periodic attention and maintenance.

(vi) From the point of view of cultivators the outlet should give a fairly constant discharge irrespective of any fluctuations in the distributing channel. However, from the point of view of proper distribution the outlet should draw proportionately more or less discharge with the varying supply in the distributing channel.

(vii) The cost of the outlet should not be high.

18.3 TYPES OF OUTLETS

Outlets may be classified in the following three types.

(1) Non-modular outlets
(2) Semi-modular outlets (or Semi-modules or Flexible modules).
(3) Modular outlets (or Rigid modules or Modules).

1. Non-modular Outlets. Non-modular outlets are the outlets whose discharge depends on the difference in water levels in the distributing channel and the water course. The discharge of such outlets, therefore, varies with the variation of the water levels in the distributing channel and the water course.

2. Semi-modular Outlets. Semi-modular outlets are the outlets whose discharge varies with the variation of the water level in the distributing channel but it is independent of the water level in the water course, so long as the minimum working head required for their working is available.

3. Modular Outlets. Modular outlets are the outlets whose discharge is independent of the water levels in the distributing channel and the water course, within reasonable working limits. In other words modular outlets maintain a constant discharge irrespective of variation of the water levels in the distributing channel and the water course.

18.4 NON-MODULAR OUTLETS

The non-modular outlets consist of a rectangular or circular opening with a pavement or a submerged pipe, out of which submerged pipe outlets are most commonly used. Figure 18.1 shows a submerged pipe outlet. The diameter of the pipes vary from 100 to 250 mm and these

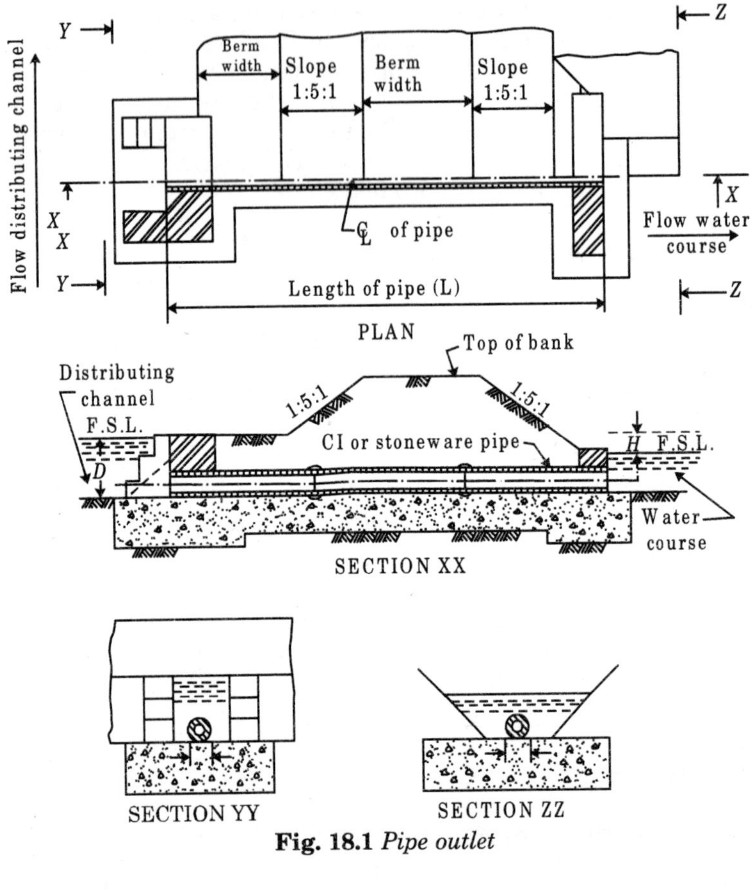

Fig. 18.1 *Pipe outlet*

are usually laid on a light concrete foundation to prevent uneven settlement and consequent leakage. The pipes are placed horizontally and at right angles to the centre line of the distributing channel. The inlet and exit ends of the pipe are fixed in masonry to prevent tampering. In Uttar Pradesh the pipes are usually fixed above the bed of the distributing channel at a depth of 225 mm below the water surface level in the distributing channel. There being considerable disturbance at the entrance to pipe outlets they usually take their due share of silt. However, the silt conduction can be further improved by depressing the upstream end of the pipe and placing the pipe either horizontally or sloping upwards at a slope not steeper than 1 in 12. Since the silt conduction is governed by the position of the upstream end of the pipe, it is a common practice to place the pipe at the bed of the distributing channel (Fig. 18.1) to enable the outlets to draw fair share of silt carried by the distributing channel.

The discharge through a pipe outlet is determined by equating the difference in water levels in the distributing channel and the water course to the sum of the loss of head at entry, loss of head due to friction and residual velocity head at exit. Thus, if H is the difference of water levels in the distributing channel and water course, then we have

$$H = \frac{0.5V^2}{2g} + \frac{fLV^2}{2gd} + \frac{V^2}{2g} \quad \text{...(i)}$$

where V is velocity of flow through pipe

L is length of pipe

d is diameter of pipe

f is coefficient of friction, and

g is acceleration due to gravity.

From Eq. (i), we have

$$V = \sqrt{2gH}\left(\frac{d}{1.5d + fL}\right)^{1/2} \quad \text{...(ii)}$$

The discharge q through the pipe outlet is given by

$$q = AV$$

or

$$q = A\sqrt{2gH}\left(\frac{d}{1.5d + fL}\right)^{1/2} \quad \text{...(18.1)}$$

where A is cross-sectional area of pipe.

Equation (18.1) is usually expressed as

$$q = CA\sqrt{2gH} \quad \text{...(18.2)}$$

where C is coefficient of discharge which depends on friction factor, length and size of pipe outlet.

From Eqs (18.1) and (18.2) the value of C is given by

$$C = \left(\frac{d}{1.5d + fL}\right)^{1/2} \quad \text{...(18.3)}$$

The discharge of the pipe outlet can be increased by deepening the water course and thereby lowering the water surface level in it.

The main advantage of non-modular outlets is that they can work with very small available heads. However, there are several drawbacks in this type of outlets as described below :

(i) The discharge of a non-modular outlet depends on the difference in the water levels in the distributing channel and the water course and hence it varies as the water level in either channel changes.

(ii) As the discharging capacity of the outlet is quite uncertain, proper and equitable distribution of water is very difficult with non-modular outlets.

(iii) A given discharge supplied at the head of the distributing channel may be quite insufficient to feed the tail end when the outlets in the head reaches are drawing high discharges, whereas the same discharge may be more than sufficient and flood the tail end when the outlets in the head reaches are taking lesser discharge.

18.5 CRITERIA FOR JUDGING THE BEHAVIOUR OF SEMI-MODULAR AND MODULAR OUTLETS

In order to judge the behaviour of semi-modular and modular outlets the following terms are used.

1. Flexibility. It is defined as the ratio of the rate of change of discharge of an outlet to the rate of change of discharge of the distributing channel.

Thus,
$$F = \left(\frac{dq/q}{dQ/Q}\right) \quad \text{...(18.4)}$$

where F = flexibility of the outlet

q = discharge through the outlet, and

Q = discharge of the distributing channel.

Now if H is the head acting on the outlet, the discharge through the outlet may be expressed as

$$q = C_1 H^m \quad \text{...(18.5)}$$

where C_1 = constant, and

m = outlet index

the values of which depend on the type of outlet.

From Eq. (18.5), we have

$$dq = mC_1 H^{m-1} dH \quad \text{...(18.6)}$$

Dividing Eq. (18.6) by Eq. (18.5), we get

$$\frac{dq}{q} = m\frac{dH}{H} \quad \text{...(18.7)}$$

Similarly discharge Q of the distributing channel can be related to its depth D by an expression of the form

$$Q = CD^n \quad \text{...(18.8)}$$

where C = constant, and

n = channel index

From Eq. (18.8), we have

$$dQ = nCD^{n-1}dD \qquad ...(18.9)$$

Dividing Eq. (18.9) by Eq. (18.8), we get

$$\frac{dQ}{Q} = n\frac{dD}{D} \qquad ...(18.10)$$

Hence flexibility may be expressed as

$$F = \frac{dq/q}{dQ/Q}$$

or

$$F = \frac{mdH/H}{ndD/D}$$

or

$$F = \frac{m}{n}\frac{D}{H}\frac{dH}{dD} \qquad ...(18.11)$$

Since a change in the water level in the distributing channel would result in equal change in the head acting on the outlet,

$$dD = dH$$

Thus, from Eq. (18.11), we have

$$F = \frac{m}{n}\frac{D}{H} \qquad ...(18.12)$$

A semi-modular outlet or semi-module possesses flexibility and hence it is also known as flexible module. On the other hand for a modular outlet flexibility is equal to zero and hence it is also known as rigid module.

The semi-modular outlets may be further classified according to different values of flexibility in terms of another parameter called proportionality which is described below.

2. Proportionality. If the flexibility is equal to one then the outlet is termed as *proportional outlet*. Thus, for a proportional outlet the rate of change of discharge of the outlet is equal to the rate of change of discharge of the distributing channel. Thus, for proportionality to be attained we have from Eq. (18.12)

$$F = \frac{m}{n}\frac{D}{H} = 1$$

or

$$\frac{H}{D} = \frac{m}{n} \qquad ...(18.13)$$

The ratio (*H*/*D*) is known as *setting* which may be defined as the ratio of the depth of the crest level of the outlet below the full supply level of the distributing channel to the fully supply depth of the distributing channel. For a proportional outlet the setting is therefore equal to (*m*/*n*) or outlet index/channel index.

For a wide trapezoidal channel with side slopes (1/2) : 1 discharge $Q \propto D^{5/3}$ (very nearly) and hence $n = 5/3$. For a pipe or orifice type outlet $q \propto H^{1/2}$ and hence $m = \frac{1}{2}$. Similarly for a weir type outlet $q \propto H^{3/2}$ and hence $m = 3/2$.

The pipe or orifice type outlet or modulus will therefore be proportional when

$$\frac{H}{D} = \frac{1}{2} \times \frac{3}{5} = 0.3$$

i.e., the outlet is set with its crest at 0.3 times the depth below the water surface in the distributing channel.

Similarly the weir type outlet or module will be proportional when

$$\frac{H}{D} = \frac{3}{2} \times \frac{3}{5} = 0.9$$

i.e., the outlet is set with its crest at 0.9 times the depth below the water surface in the distributing channel.

If the flexibility is more than one then the outlet is termed as *hyper-proportional outlet.* Thus, in this case if the discharge in the distributing channel changes by a certain percentage, the discharge of the outlet changes by a larger percentage. Further for a hyper-proportional outlet since

$$F > 1$$

or $$\frac{m}{n}\frac{D}{H} > 1$$

or $$\frac{H}{D} < \frac{m}{n} \qquad \text{...(18.14)}$$

Equation (18.14) shows that for a hyper-proportional outlet the value of setting (H/D) is less than the ratio (m/n), or in other words an outlet is hyper-proportional if the setting is higher than required for a proportional outlet.

Again if the flexibility is less than one then the outlet is termed as *sub-proportional outlet.* Thus, in this case if the discharge in the distributing channel changes by a certain percentage the discharge of the outlet changes by a smaller percentage. Further for a sub-proportional outlet since

$$F < 1$$

or $$\frac{m}{n}\frac{D}{H} < 1$$

or $$\frac{H}{D} > \frac{m}{n} \qquad \text{...(18.15)}$$

Equation (18.15) shows that for a sub-proportional outlet the value of setting $\left(\frac{H}{D}\right)$ is more than the ratio $\left(\frac{m}{n}\right)$, or in other words an outlet is sub-proportional if the setting is lower than required for a proportional outlet.

3. Sensitivity. It is defined as the ratio of the rate of change of discharge of an outlet to the rate of change in the level of the water surface in the distributing channel, referred to the normal depth of the channel.

Let a gauge be provided for the distributing channel which is so set that when the discharge through the outlet is zero the reading on the gauge is zero. Let G be the reading on the gauge when the discharge in the distributing channel is Q.

From the above definition it follows that

$$S = \frac{(dq/q)}{(dG/D)} \quad \text{...(18.16)}$$

where S = sensitivity of the outlet

q = discharge through the outlet

dq = change in the discharge through the outlet

dG = change in the gauge reading, and

D = depth of flow in distributing channel.

Let dD be the change in the depth of flow in the distributing channel, then

$$dG = dD$$

Introducing this value of dG in Eq. (18.14), we get

$$S = \frac{dq/q}{dD/D} \quad \text{...(18.17)}$$

Equation (18.15) indicates that for a given rise or fall in the water surface level of the distributing channel, if greater is the variation in the discharge through the outlet, the larger is its sensitivity. Further for a rigid module since $dq = 0$, the sensitivity is equal to zero.

A relationship between sensitivity and flexibility may be obtained as indicated below.

$$F = \frac{dq/q}{dQ/Q}$$

Also from Eq. (18.10), we have

$$\frac{dQ}{Q} = n\frac{dD}{D}$$

$$\therefore \qquad F = \frac{dq/q}{n(dD/D)} \qquad \text{...(18.18)}$$

Comparing Eqs (18.17) and (18.18), we get

$$S = nF \qquad \text{...(18.19)}$$

4. Efficiency. It is defined as the ratio of the head recovered to the head put in. In the case of a weir type outlet the efficiency is same as the drowning ratio. Thus efficiency is a measure of conservation of head by an outlet.

5. Drowning Ratio. It is the ratio between the depths of water levels over the crest on the downstream and upstream sides of the outlet.

6. Minimum modular Head. It is the minimum head required for the functioning of an outlet. In other words it is the minimum difference between the upstream and downstream water levels which is required to be maintained to enable the outlet to pass its design discharge. It is also termed as minimum modular loss.

7. Modular Limits and Modular Range. The *modular limits* are the upper and lower limits of any one or more factors beyond which a module or semi-module ceases to function as a module or semi-module.

The *modular range* is the range of the modular limits, i.e., it is the range of various factors within which a module or semi-module works as designed.

8. Minimum Modular Head Ratio. It is the ratio between the minimum modular head and the depth of upstream water level on the crest of the outlet.

9. Setting. It is the ratio of the depth below full supply level of the crest of an outlet to the full supply depth of the channel at that point.

10. Adjustability. Adjustability means capacity for modification of the outlet to take into account changed condition such as silting or scouring of the distributing channel, or a necessity to increase or decrease the outlet discharge due to a change in the command.

18.6 TYPES OF SEMI-MODULAR OUTLETS

The different types of semi-modular outlets are as mentioned below.

(1) Pipe outlet

(2) Kennedy's gauge outlet

(3) Open flume outlet

(4) Orifice semi-module.

1. Pipe Outlet. If the pipe outlet is so set that its discharge has free fall into the water course the discharge through it becomes independent of the water level in the water course and hence it acts as a semi-modular, outlet. It is the simplest and the oldest type of semi-modular outlet. In this case the working head H is the difference between the water level in the distributing channel and the centre of the pipe at its exit end. Its discharge is proportional to and hence as indicated earlier it will be proportional if it is set at 0.3 times the depth below the water surface in the distributing channel. However, a pipe outlet is usually set lower than this and is therefore sub-proportional.

The pipe is usually laid with its inlet end above the bed of the distributing channel and sloping downwards through the bank. In order to obtain free flow condition the exit end of the pipe is placed 15 cm above the highest water level in the water course. Sometimes the pipe is laid with its inlet end at the bed level of the distributing channel and sloping upwards through the bank until the exit end of the pipe is located 15 cm above the highest water level in the water course.

The efficiency of the pipe outlet is high and its silt conduction is good.

2. Kennedy's Gauge Outlet. It is one of the earliest types of semi-modules which was evolved by Kennedy in 1906. It consists of a cast iron bell mouthed orifice for entry from the distributing channel. From the orifice the jet of water discharges into a chamber to which a vertical vent pipe is connected which is covered with a cowl pierced with air holes. The vent pipe extends above the water level in the distributing channel and thus allows free circulation of air around the jet. This arrangement enables the discharge through the orifice to take place in exactly the same condition as if it was discharging freely in atmospheric air and hence makes the discharge independent of the downstream conditions. After crossing the air chamber the jet enters an expanding pipe connected to the air chamber and having its entrance diameter slightly larger than that of the orifice. The expanding pipe is usually 3 m long and is made of cast iron or sheet steel. At the end of the expanding pipe a concrete pipe extension is usually added through which water is discharged into the water course.

Since the orifice is discharging freely in atmospheric air, its discharge varies as $\sqrt{H_0}$ where $\sqrt{H_0}$ is the difference between the level of the centre of the orifice and the water surface in the distributing channel. The discharge through the outlet is thus given by

$$q = Ca\sqrt{2gH_0} \qquad \text{...(18.20)}$$

where q = discharge in cumec

a = area of orifice in m^2

H_0 = head as defined above in m

g = acceleration due to gravity

C = coefficient of discharge which may be as high as 0.97.

It has been found by experiments that the losses of head by friction, eddies, etc., are about 0.22 H_0. As such the available working head or the minimum modular head h_1, that is the difference of water levels in the distributing channel and the water course, should not be less than $0.22H_0$. So long as this condition is satisfied the discharge of the outlet is given by Eq. (18.20) whether the outfall end of the outlet is submerged or not and it is independent of the water level in the water course.

This outlet could however be easily tempered with by the cultivators by closing the air vents of the vent pipe. With the closing of the air vents the jet would gradually suck away the entrapped air in the chamber and reduce the pressure in the chamber below atmospheric, thus increasing the discharge. In order to overcome this defect Kennedy in 1915 introduced a 'Modified and Temper Proof Gauge Outlet' which is shown in Fig. 18.2. In the modified outlet the former vertical vent pipe is substituted by a sloping air pipe protected by an angle iron on the outside. An enamelled gauge scale is provided on the outside of this angle iron. The sloping air pipe is connected to a 3 m long horizontal perforated pipe which is also protected on its top by an angle iron and laid on a layer of dry ballast to ensure entry of air. The outer 1 m length of the horizontal pipe is surrounded by expanded metal wound round it and welded. The new design of the outlet made its tampering more difficult but it could also be interfered with by plugging the air pipe through the bell mouth entry. On account of this difficulty and due to its high cost this outlet is no longer in general use.

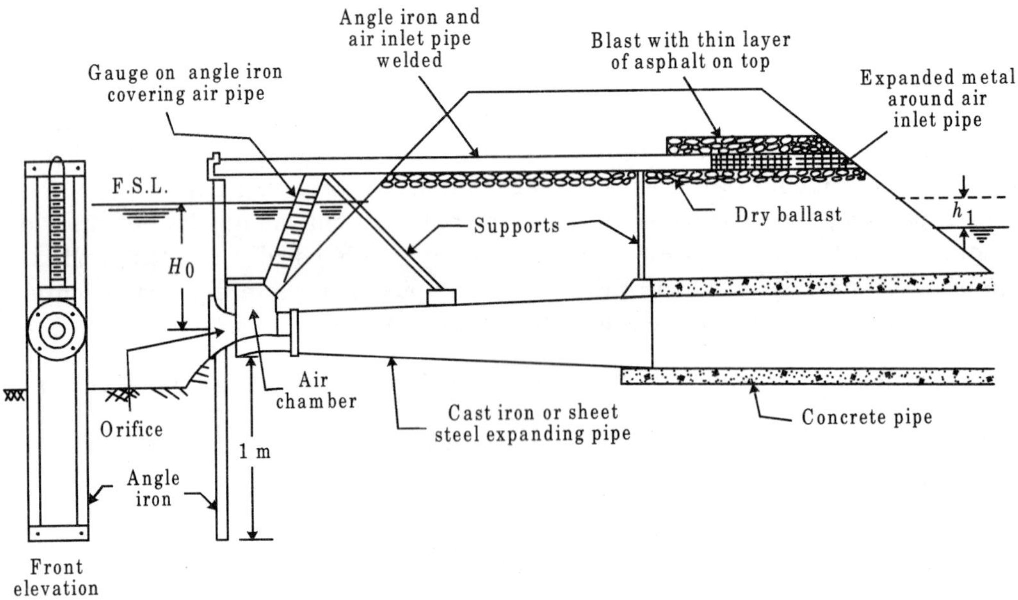

Fig. 18.2. *Kennedy's gauge outlet*

3. Open Fume Outlet. An open flume outlet is a smooth weir with a constricted throat to ensure hyper-critical velocity (i.e., velocity above critical). The length of the throat is such that the controlling section remains within the throat at all discharges. A gradually expanded flume is provided below the throat to ensure the formation of hydraulic jump in this portion and thus obtain maximum recovery of head. The formation of the hydraulic jump makes the discharge through the outlet independent of the water level in the water course. The entire work is built in brick masonry but the controlling section is generally provided with cast iron or steel bed and check plates. Based on this principle various types of open flume outlets have been evolved. Figure 18.3 shows a type of open flume outlet which is commonly used in Punjab and hence it is also known as Punjab open flume outlet. Similarly Fig. 18.4 shows an open flume outlet with roof block.

The discharge through an open flume outlet is given by

$$q = K B_t H^{3/2} \qquad ...(18.21)$$

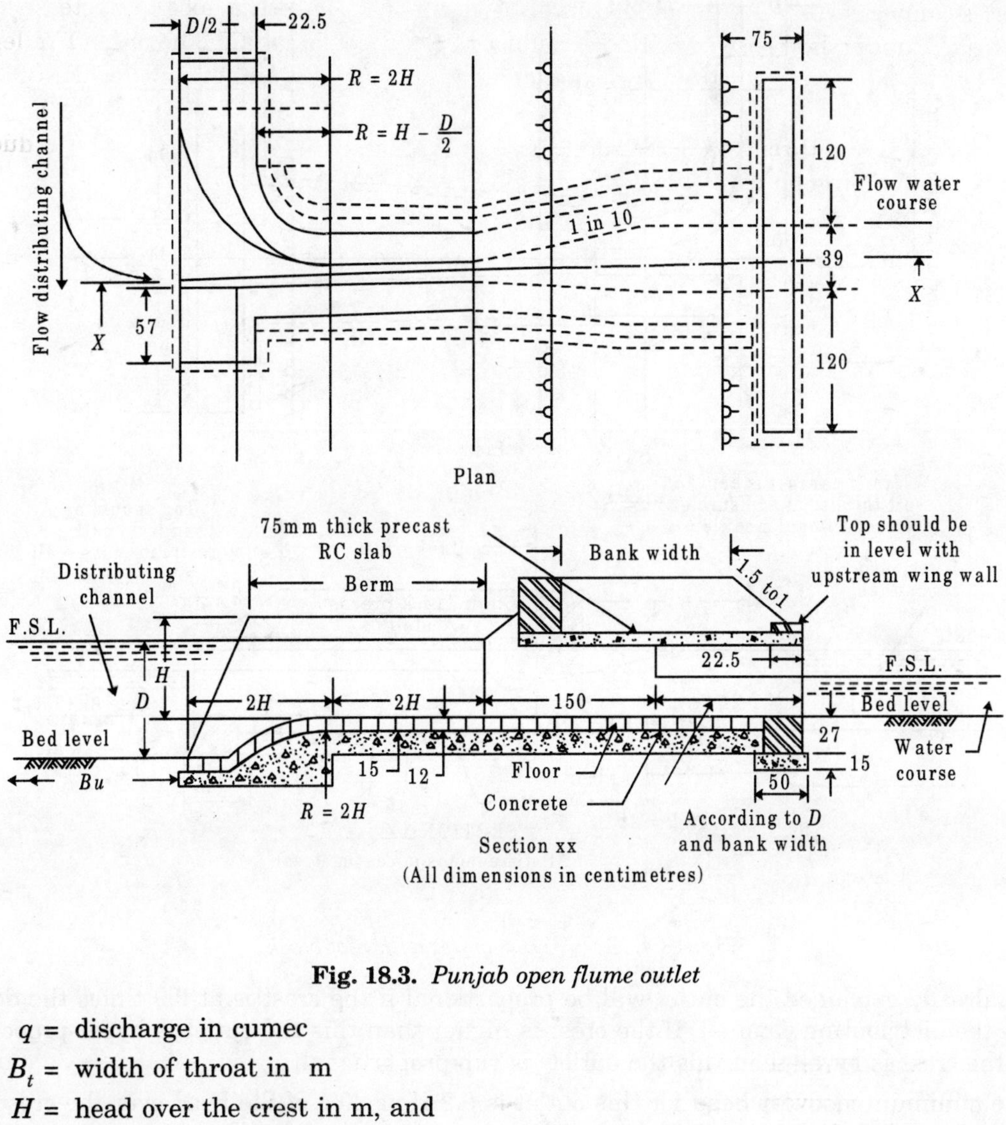

Fig. 18.3. *Punjab open flume outlet*

where q = discharge in cumec

B_t = width of throat in m

H = head over the crest in m, and

K = coefficient of discharge.

The theoretical value of K is 1.71 (*i.e.*, same as for a broad crested weir) but due to losses the actual value of K is less than theoretical value. The following values of K are usually adopted for different throat widths.

Throat width	*Value of K*
0.06 to 0.09 m	1.60
0.09 to 0.12 m	1.64
Over 0.12 m	1.66

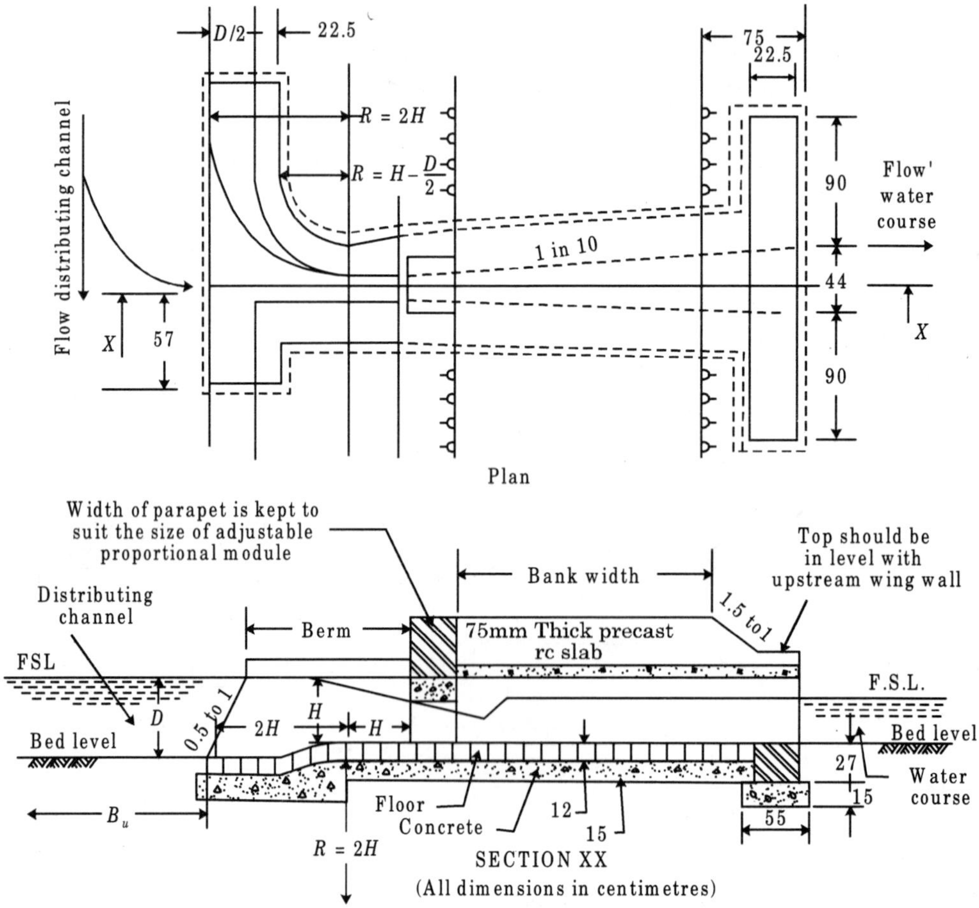

Fig. 18.4. *Open flume outlet with roof block*

As already explained the outlet will be proportional if the crest is at 0.9 times the depth of flow in the distributing channel. If the crest is higher than this the outlet is hyper-proportional and if the crest is lower than this the outlet is sub-proportional.

The minimum modular head for this outlet is 0.2 H or 20% of the head over the crest of the outlet with a minimum value of 6 cm.

The efficiency of the outlet is high being 80 to 90%.

In order to avoid choking of the outlet by floating material the open flume outlets with throat width less than 0.06 m are not constructed. As such for small requirements of discharge a high setting of the weir crest would be necessary but this would result in poor silt conduction of the outlet.

The open flume outlets are not easily adjustable. In order to change the discharge it would be necessary to change crest level or throat width and this could only be done by dismantling and rebuilding one of the side walls. Further the width and depth of the outlet needs to be suitably adjusted because if it is deep and narrow it is easily blocked and if it is shallow and wide it is hyper proportional and does not take its fair share of silt.

4. Orifice Semi-module. An orifice semi-module consists of an orifice provided with a gradually expanding flume on the downstream side. The flow through the orifice is hyper-critical which results in the formation of hydraulic jump in the expanding flume. This makes the discharge through the outlet independent of the water level in the water course. There are various forms of this outlet, the earliest of these is the *Crump's adjustable proportional module* (commonly known as Crump's A.P.M.). The modified form of Crump's adjustable proportional module is commonly used in Punjab which is usually called *Adjustable orifice semi-module* (or A.O.S.M.) and the same is described below.

Figure 18.5 shows a typical layout of adjustable orifice semi-module. It consists of a cast iron roof block, a cast iron base plate (or bed plate) and check plates on either side around which masonry is built. The roof block is shaped to a lemniscate curve with a tilt of 1 in 7.5. This ensures smooth converging flow through the orifice and a relatively steady discharge coefficient under varying heads. The roof block is fixed to the check plates by bolts which can be removed and depth of the orifice adjusted after the masonry is dismantled. The base plates and the roof blocks are manufactured in standard sizes, such as B_t = 6.1, 7.6, 9.9, 12.2, 15.4, 19.5, 24.4 and 30.5 centimetres. These standard sizes with required opening of the orifice are used to obtain desired supply through the outlet. In order to enable the outlet to take a fair share of silt the upstream wing is set back by an amount W (Fig. 18.5) the value of which is given by

$$W = k\frac{q}{Q}\left(B_u + \frac{D}{2}\right) \qquad \text{...(18.22)}$$

where q = discharge of outlet in cumec

Q = discharge of distributing channel upstream of the outlet in cumec

B_u = bed with of the distributing channel upstream of the outlet in m

D = depth of flow in the distributing channel in m, and

k = ratio between the mean velocity for the entire distributing channel and mean velocity in the part of the distributing channel, wherein outlet is to be installed.

The value of k may be obtained for different values of Q as mentioned below:

Q (*cumec*)	k
Below 0.283	1.00
Over 0.283 and upto 1.415	1.25
Over 1.415 and upto 5.660	1.50
Over 5.660	2.00

Below the orifice the side walls expand along circular arcs starting tangentially from the orifice as shown in Fig. 18.5.

The discharge of this outlet is given by the expression

$$q = 4.03 B_t Y \sqrt{H_s} \qquad \text{...(18.23)}$$

where q = discharge in cumec

B_t = width of the orifice in m

Y = height of the orifice at the smallest section in m, and

H_s = head measured from the water surface in the distributing channel to the lowest point of the roof block in m.

The minimum modular head for this outlet is given by

$$H_m = 0.82\, H_s - 0.5\, B_t \qquad ...(18.24)$$

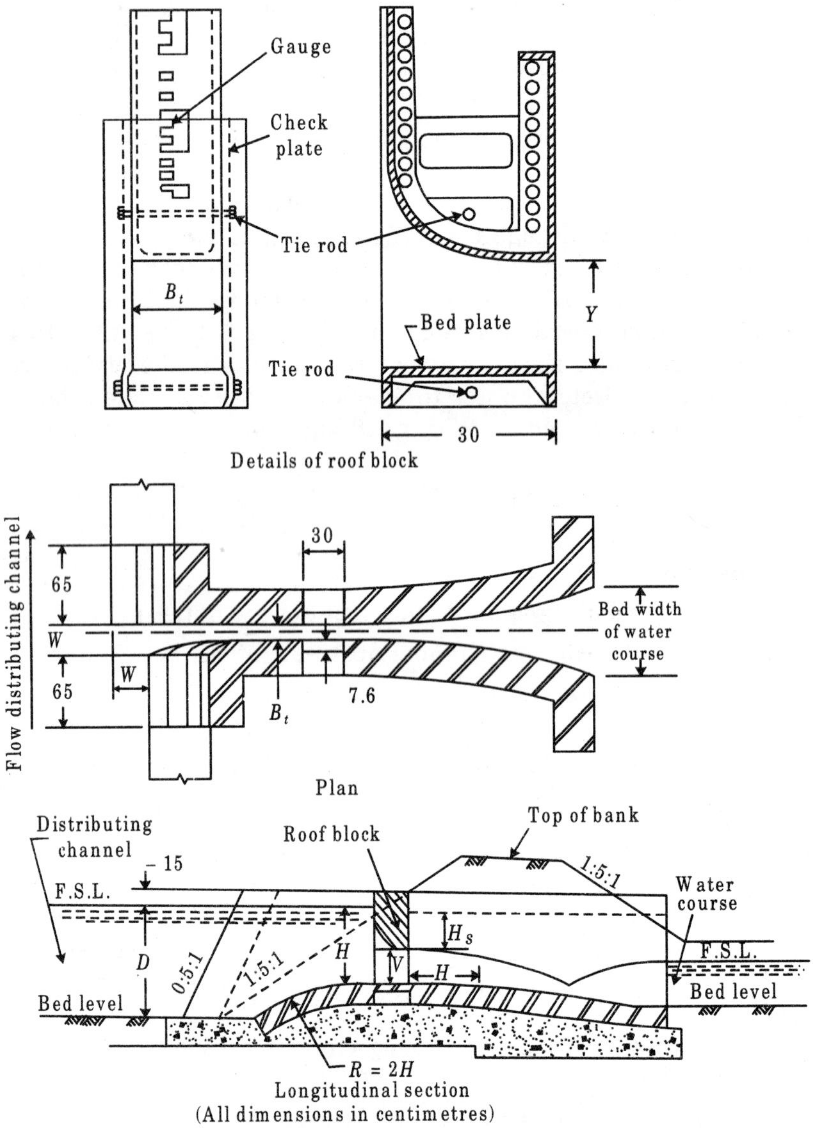

Fig. 18.5. *Adjustable orifice semi-module*

and hence if the available working head is less than this, Eq. (18.23) for the discharge will no longer be applicable.

It is an orifice type outlet and hence the value of (m/n) is 0.3. Further in this case the controlling head is H_s, or the head measured from the water surface in the distributing channel to the lowest point of the roof block. Hence the outlet will be just proportional when H_s is equal to 0.3 times the depth of flow in the distributing channel.

The main advantage of this outlet is that it can be conveniently adjusted at a small cost. For adjustment the masonry is dismantled and by removing the bolts the roof block is raised or lowered as necessary and after refixing the roof block by the bolts the masonry is rebuilt. Moreover any tampering with the roof block by the cultivators is easily detected as it requires the breaking of masonry.

18.7 TYPES OF MODULAR OUTLETS OR RIGID MODULES

A large number of rigid modules have been evolved, some of which have moving parts and some do not have moving parts. The rigid modules with moving parts are usually expensive and are liable to frequent breakdown due to floating debris or silt clogging the moving parts. As such due to higher initial cost and difficulty in subsequent maintenance the rigid modules having moving parts are usually not used and instead of these the rigid modules having no moving parts being more suitable are generally used. The two common types of rigid modules which do not have moving parts are as follows.

(1) Gibb's module

(2) Khanna's module.

Both these modules are described below.

1. Gibb's Module. It is the most commonly used rigid module which was evolved by A.S. Gibb. Figure 18.6 shows plan and section of the module. It consists of a bell mouthed inlet pipe connected to a curved eddy chamber through a 180° rising spiral pipe. Water from the distributing channel enters the inlet pipe and is led into the eddy chamber through the rising spiral pipe in which free vortex flow is developed. The characteristic feature of free vortex flow is that the product of velocity and radius is constant for all filaments (i.e., $V \times r$ = constant). Since water at the outer circumference of the eddy chamber has greater radius it has lower velocity, resulting in rise of the water level there. Thus the water surface in the eddy chamber slopes down towards the inner circumference. In the eddy chamber a number of baffles are inserted. The baffles are suspended from the roof of the eddy chamber with their lower ends sloping against the direction of flow at the required height above the bottom. As the head causing flow through the outlet increases, the water banks up at the outer circumference of the eddy chamber and impinges against the baffles imparting an upward rotational direction of flow to the water, which spins round in the compartment between two successive baffles and finally drops on the oncoming stream of water, thus dissipating excess energy and keeping the discharge constant. According to the requirement of discharge and range of head the spiral turn may vary from half a turn (i.e., semi circle), as shown in Fig. 18.6, to one and a half complete turns. Also the number of baffles coming into action depends on the variations in the head, as more and more baffles come into action with the increase in the head so that the discharge through the outlet remains constant.

Gibb gave the following formula for the discharge through the outlet,

$$q = r_0\sqrt{2g}\,(d_1 + h_0)^{1.5}\left\{\frac{m^2-1}{m^3}\log_e m + \frac{1}{m}(\log_e m) - \frac{m^2-1}{2m^2}\right\} \quad \ldots (18.24)$$

where q = discharge of the outlet

r_0 = radius of the outer circumference of the eddy chamber

r_1 = radius of the inner circumference of the eddy chamber

m = (r_0/r_1) i.e., ratio of outer radius to inner radius

d_1 = depth of water at inner circumference, and

h_0 = head of water at outer circumference.

Gibb's formula given above is based on free vortex flow and holds good only for his standard design in which $m = 2$ and $(h_0/D) = 1/7$, where D is the difference of level measured from the minimum water level in the distributing channel to the floor of the eddy chamber.

The silt conduction of Gibb's module is however not good and also the outlet is comparatively costly.

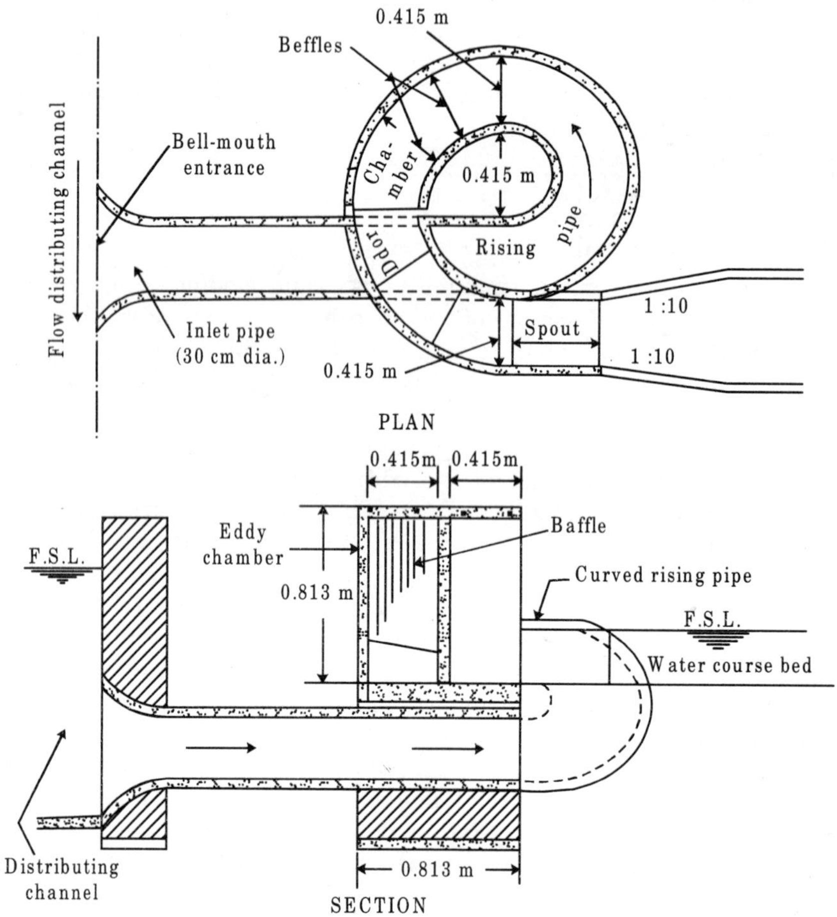

Fig. 18.6 *Gibb's module*

2. Khanna's Module. Figure 18.7 shows Khanna's module. It is essentially an orifice semi-module with the addition of sloping shoots fixed in the roof block to counteract increase in discharge with the rise in the water level in the distributing channel. Water from the distributing channel enters the outlet and the differential chamber *A* through a larger opening *O* and smaller openings *O*′ and *O*″. As long as water surface in the distributing channel is at or below its normal supply level, the outlet acts as an ordinary orifice semi-module. When the water surface in the distributing channel rises above its normal supply level, water rises in the chamber *A* and enters the first sloping shoot. The back flow from the shoot impinges on the flow through the outlet and causes dissipation of additional energy of flow and thus tends to keep the discharge through the outlet constant. Three shoots at a slope of 1 in 2, with their upper ends at the normal supply level in the distributing channel, 150 mm higher than the normal supply level and 300 mm higher than the normal supply level respectively were provided in the original design. However, the number of shoots may be varied to suit different conditions. The shoots being provided with a roof covering their tampering is prevented. Further even if the shoots are choked, the outlet continues to function as orifice semi-module. However, inspite of these advantages these outlets are not being commonly used.

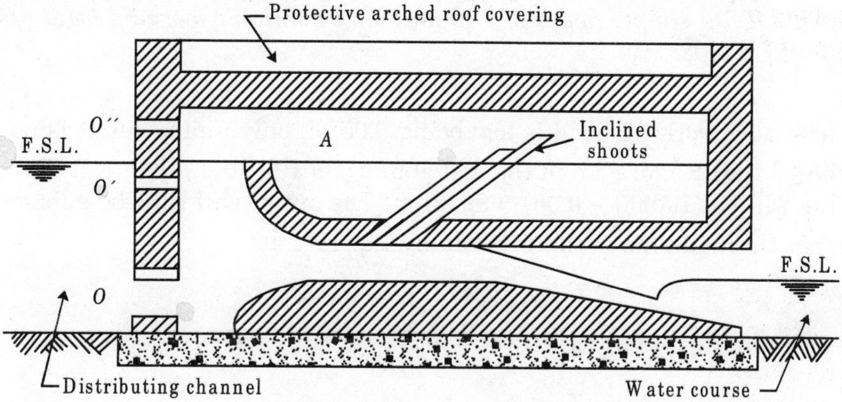

Fig. 18.7 *Khanna's module*

18.8 TAIL CLUSTER

Tail cluster is the work constructed at the tail end of an irrigation canal for the distribution of water there. It usually consists of two or more open flume outlets clustered at the tail end of the irrigation canal. The aggregate discharge of these outlets is over 0.07 cumec and the side offtakes have equal angles with respect to the centreline of the canal. The offtaking angles are either 45°, 60°, 90° or straight inline with the canal alignment. The crests of all the outlets are kept at the same level and proportionality is secured by adjusting their throat width (B_t). The depth of water or head H over the crest of the outlet is kept 30 cm, and the value of K is 1.66 in the discharge formula for the open flume outlet given by Eq. (18.21).

The tail clusters are useful in distributing the water proportionately and in easily absorbing any excess water that reaches the tail end of the canal. The throat width is increased by 20 cm at full supply level to facilitate passing down any excess water reaching the tail end of the canal without endangering safety of the structure and the canal reach upstream.

The low working head of only 6 cm required for the open flume outlets with head H over the crest of each outlet at the tail cluster being kept as 30 cm, avoids undue raising of the full supply level in the canal and thereby resulting in economy in the construction cost of the canal system.

SOLVED ILLUSTRATIVE EXAMPLES

Illustrative Example 18.1 *Design an outlet for a discharge of 0.0185 cume on a distributary having a full supply depth of 1 m and with available working head of 5 cm. Assume length = 12 m and f = 0.02.*

Solution

Since the available working head is low being 0.05 m, only a pipe outlet is possible.

If it is assumed that R.L of FSL in the distributary is 100.00 m, then R.L of water surface in the water course will be (100.00 – 0.05) = 99.95 m. The outlet will thus be submerged.

The discharge through pipe outlet is given by Eq. 18.2 as

$$q = CA\sqrt{2gH}$$

in which C is given by Eq. (18.3) as

$$C = \left(\frac{d}{1.5d + fL}\right)^{1/2}$$

Assume diameter of pipe d = 20 cm = 0.2 m

$$L = 12 \text{ m}; \text{ and } f = 0.02$$

Thus, by substitution, we get

$$C = \left[\frac{0.2}{(1.5 \times 0.2) + (0.02 \times 12)}\right]^{1/2} = 0.609$$

$$A = \frac{\pi}{4}(0.2)^2; \text{ and } H = 0.05$$

Thus, by substitution, we get

$$q = 0.609 \times \frac{\pi}{4} \times (0.2)^2 \times \sqrt{2 \times 9.81 \times 0.05}$$

$$= 0.0189 \text{ cumec}$$

Hence O.K.

Thus, 20 cm diameter submerged pipe outlet may be adopted.

Illustrative Example 18.2 *Design a semi-modular pipe outlet for a discharge of 0.08 cumec on a distributary with full supply depth of 1 m and working head of 0.75 m.*

Solution

The discharge through pipe outlet is given by Eq. (18.2) as

$$q = CA\sqrt{2gH}$$

in which C is given by Eq. (18.3) as

$$C = \left(\frac{d}{1.5d + fL}\right)^{1/2}$$

Assume diameter of pipe d = 25 cm = 0.25 m;
Length of pipe L = 12 m; and f = 0.02.
Thus, by substitution, we get

$$C = \left[\frac{0.25}{(1.5\times 0.25)+(0.02\times 12)}\right]^{1/2} = 0.638$$

$$q = 0.08 \text{ cumec; and } A = \frac{\pi}{4}(0.25)^2$$

Thus, by substitution, we get

$$\therefore \quad H = 0.333 \text{ m}$$

The pipe outlet may be set with its centre 0.333 m below the water surface in the distributary.

Illustrative Example 18.3 *Design an open flume outlet with a discharge of 0.08 cumec, on a channel with full supply depth of 0.9 m. The available working head is 15 cm.*

Solution

The discharge through an open flume outlet is given by Eq. (18.21) as

$$q = K B_t H^{3/2}$$

Assume B_t = 0.08 m for which K = 1.60.

$$q = 0.08 \text{ cumec}$$

Thus by substitution we get

$$0.08 = 1.60 \times 0.08 \times H^{3/2}$$

$$\therefore \quad H = 0.731 \text{ m}$$

The minimum modular head required for the open fume outlet

$$= 0.2\,H$$

$$= 0.2 \times 0.731$$

$$= 0.1462 \text{ m} = 14.62 \text{ cm}$$

which is less than the available working head of 15 cm or 0.15 m, and hence the design is suitable.

The setting of the outlet is

$$\frac{H}{D} = \frac{0.731}{0.9} = 0.812 < 0.9$$

Hence the outlet will behave as hyper-proportional outlet.

REVIEW QUESTIONS

18.1 What is an outlet ? Write the requirements of a good outlet.

18.2 What are the different types of outlets ?

18.3 What are modular, semi-modular and non-modular outlets ?

18.4 Describe with a neat sketch a non-modular type of outlet.

18.5 What do you understand by flexibility of an outlet ? Derive an expression for the same.

18.6 Define proportionality of an outlet. Distinguish between a proportional outlet, a hyper-proportional outlet and a sub-proportional outlet. Obtain the expressions for the setting of these three types of outlets.

18.7 Define sensitivity of an outlet. Find the relation between sensitivity and flexibility of an outlet.

18.8 Explain the terms : (i) Efficiency; (ii) Drowning ratio; (iii) Modular limits; (iv) Minimum modular head; (v) Minimum modular head ratio; (vi) Adjustability as applied to outlets.

18.9 Describe with a neat sketch Kennedy's gauge outlet.

18.10 Describe with neat sketches the working of Crump's adjustable proportional module.

18.11 What is a rigid module ? Describe with neat sketches the working of a Gibb's module.

18.12 Describe with a neat sketch the working of Khanna's rigid module.

18.13 Describe with neat sketches Punjab open flume outlet.

18.14 Define the following terms as used in outlets : Flexibility, Proportionality, Sensitivity and Setting.

18.15 Give the classification of the outlets used and their characteristics.

18.16 Design an open flume outlet for a discharge of 0.075 cumec on a channel with full supply depth of 1 m. The available working head is 20 cm.

18.17 Write a brief note on tail cluster.

18.18 What are the salient features of an open flume outlet? Describe open flume outlets with and without roof block.

CHAPTER

19 Regulation Works

19.1 INTRODUCTION

The structures constructed to regulate the discharge, full supply level or velocity of flow in a channel are known as regulation works. Such structures are necessary for the efficient working and safety of the channel. The various regulation works discussed in this chapter are as indicated below.

1. Falls or Drops
2. Distributary head regulators
3. Cross regulators
4. Escapes

A. FALLS OR DROPS

19.2 DEFINITION

A fall is a structure constructed across a channel to permit lowering down of its water level and dissipate the surplus energy possessed by the falling water which may otherwise scour the bed and banks of the channel.

19.3 NECESSITY AND LOCATION OF FALLS

When the natural slope of the ground over which channel is to be constructed is greater than the designed bed slope of the channel, the difference in the slopes is adjusted by providing vertical falls or drops in the bed of the channel at suitable intervals. The location of a fall is decided according to various considerations as described below.

1. A fall may be provided at a location where the F.S.L. of the channel outstrips the ground level but before the bed of the channel comes into filling.

2. A fall should be so located that as far as possible there is no loss of the commanded area of the channel.

3. A fall should be such that below the fall the F.S.L. of the channel remains below the ground level for $\frac{1}{2}$ to $\frac{3}{4}$ kilometre but not much more. This will, however, not result in any loss of commanded area as upto this extent the area can be irrigated by a watercourse from an outlet at high level upstream of the fall.

4. The location of a fall may also be affected by the possibility of combining it with a regulator or a bridge or some other structure. Such combination often results in economy.

5. For the location of falls relative economy of providing a large number of small falls or small number of large falls should also be considered subject to the condition that the commanded area is not reduced. For a small length of the channel both upstream and downstream of a fall, there is unbalanced earthwork which should be kept at a minimum and almost equal on either side so that extra earth from below may be used to meet the requirements above. For a larger fall the quantity of unbalanced earthwork will be larger. As such the provision of even small number of large falls will result in large quantity of unbalanced earthwork, but the reduction in the number of falls will result in saving in the cost of the falls. This saving has to be balanced against the extra cost of earthwork to determine the relative economy.

19.4 DEVELOPMENT OF DIFFERENT TYPES OF FALLS

In older canals such as Eastern Yamuna Canal initially constructed by Mughal Emperors, falls were avoided by increasing the length of the canal by providing sinuous curves. This method besides being uneconomical did not lead to an efficient command of the proposed area. In our country the falls were first constructed by British engineers in the nineteenth century and since then various types of falls with different shapes, length and height of crest have been constructed for different canal systems. Among the earlier types of falls are: Ogee falls, Rapid falls or Rapids and Stepped falls. Each of these types of falls are briefly described below.

Ogee Fall

The Ogee fall was evolved by Sir Proby Cautley and this type of falls were constructed on the Ganga Canal. This fall was designed with the aim to provide a smooth transition from the upstream to the downstream water level and avoid disturbance and impact as far as possible. Thus, as shown in Fig. 19.1 this fall is provided with gradual convex and concave curves. However, these falls had the following shortcomings.

(i) On account of heavy drawdown on the upstream side of the fall there was erosion of bed and banks of the channel.

(ii) Due to smooth transition the kinetic energy was almost fully preserved which caused erosion of bed and banks of the channel on the downstream side of the fall.

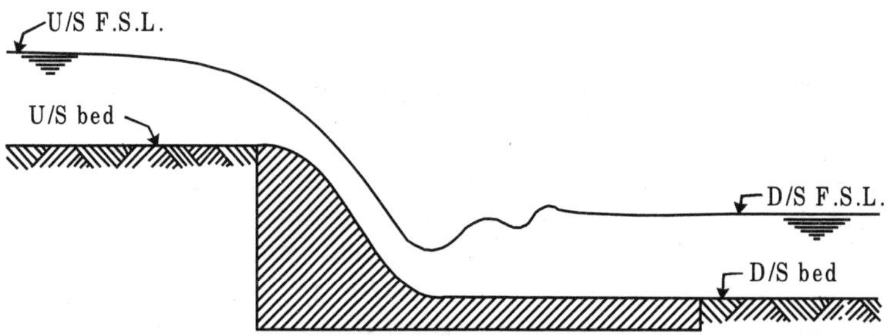

Fig. 19.1 *Ogee fall*

Rapid Fall or Rapid

The rapid fall or rapid was evolved by R.F. Croften and this type of falls were constructed on Western Yamuna Canal. As shown in Fig. 19.2 the rapid falls were constructed with boulder facing and were provided with glacis having gentles slopes in the range of 1 in 10 to 1 in 20. The

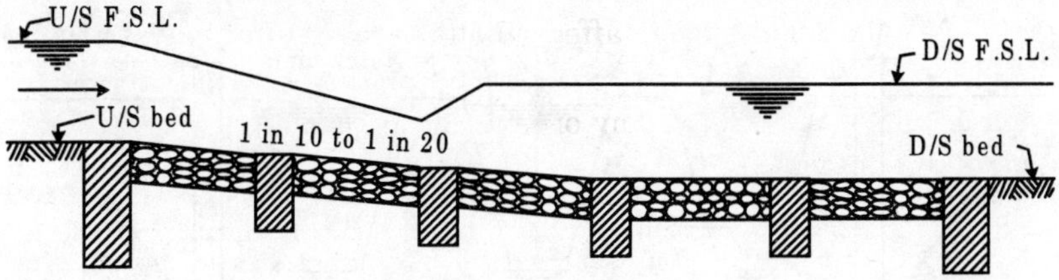

Fig. 19.2 *Rapid fall*

long glacis assured the formation of hydraulic jump and hence these falls worked quite satisfactorily. Further these falls also admitted of timber traffic. However, the cost of construction of these falls was very high.

Stepped Fall

The stepped falls were the modified form of rapid falls in this respect that the long glacis of the rapid falls were replaced by floors in steps in the stepped falls as shown in Fig. 19.3. However, the cost of construction of the stepped falls was also very high.

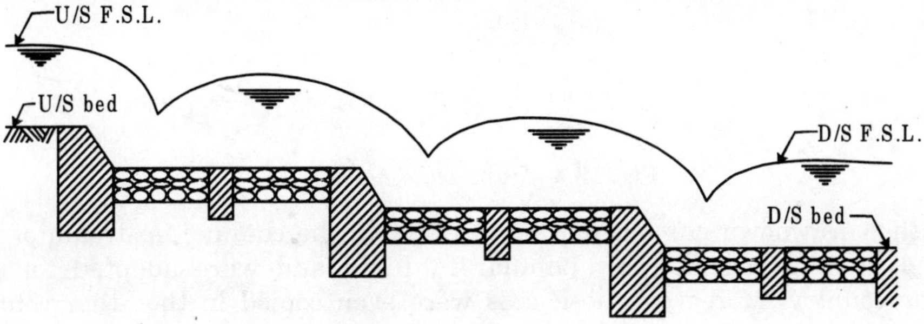

Fig. 19.3 *Stepped fall*

After the development of stepped falls it was recognised that better dissipation of energy could be achieved through vertical impact of falling jet of water on the floor. As such vertical falls with cistern were evolved. However, the earlier types of vertical falls were not well developed and gave trouble. As such these were superseded for some time by Trapezoidal notch falls. After the First World War, major canal projects were taken up in various states in our country for which it was considered necessary to evolve new types of falls which would be suitable for greater discharges and would be less expensive. Thus, experiments were carried out at U.P., Punjab and Poona Research Stations and two types of falls viz., Vertical drop type falls and Glacis type falls were developed. Each of these types of falls are briefly described below.

Trapezoidal Notch Fall

The trapezoidal notch fall was evolved by Reid in 1894. This fall consists of a number of trapezoidal notches in a high breast wall across the channel with a smooth entrance and a flat circular lip projecting downstream from each notch to disperse water (Fig. 19.4). The notches were designed to maintain the normal depth of flow in the upstream channel at any two discharge values, the variation at intermediate values being small. Thus, the depth discharge relationship of the channel remained practically undisturbed by the introduction of the fall i.e.,

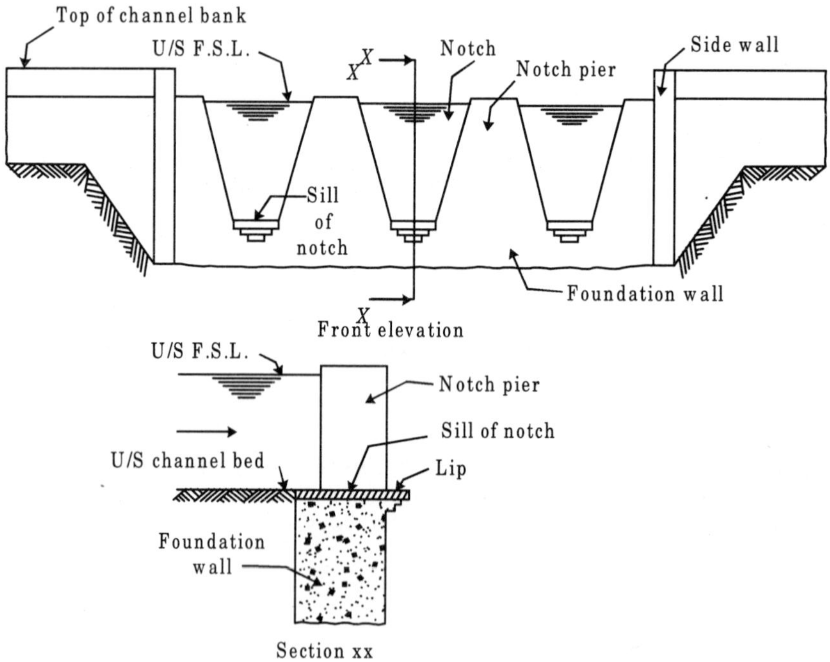

Fig. 19.4 *Trapezoidal notch fall*

there was neither drawdown nor heading up of water in the channel upstream of the fall. The trapezoidal notch falls became very popular in India and were adopted for many years throughout the country. Moreover, these falls were even copied in the other countries of the World also. However, since the development of the other types of falls the trapezoidal notch falls have become obsolete.

Vertical Drop Type Fall

In the vertical drop type fall the clear nappe leaving the crest is made to impinge into a cistern below. The cistern provides a water cushion and helps to dissipate the surplus energy of the falling jet. In the earlier types of vertical drop type falls the dimensions of the cistern were arbitrarily decided on the basis of the past experience of the designers. Further a grid consisting of baulks of timber spaced a few centimetres apart was provided in the cistern to intercept the falling nappe and thus dissipate its surplus energy. These were later abandoned because the timber grid got clogged with floating debris and rotted thus necessitating its frequent replacement.

The Sharda type fall developed on the Sharda Canal Project in U.P. is a recent type of vertical drop type fall which has proved quite successful. In this fall the dissipation of energy is affected by the turbulent diffusion as the high velocity jet enters the deep pool of water in the cistern below.

Glacis Type Fall

This type of fall utilises hydraulic jump (or standing wave) for the dissipation of energy. Earlier a flumed fall with straight glacis was evolved in Punjab. However, there was serious trouble with some of these falls. One of the causes of trouble was that even after the formation of hydraulic jump there was considerable surplus energy in water. Another cause of trouble was due to too rapid expansion after fluming eddies were developed which caused deep scours. Thus, further research was carried out to eliminate these defects and two types of falls viz., Montague and Inglis type falls were evolved. In the Montague type fall the straight glacis has been replaced by a parabolic glacis commonly known as 'Montague Profile'. In the Inglis type fall straight glacis is provided but at a certain distance from the toe of the glacis a baffle wall of certain height is provided. Both these types of falls have been designed with the aim to hold the hydraulic jump stable on or close to the glacis and thus affect maximum dissipation of energy.

Metre or Non-metre Fall

Metre fall is the one which can be used to measure discharge of the channel. On the other hand a Non-metre fall is the one which cannot be used to measure discharge of the channel. If a fall is adopted for measuring discharge of the channel it must have a broad crest so that the discharge coefficient is constant under varying heads. Generally a vertical drop type fall is not suitable as a meter fall due to the formation of partial vacuum under the nappe. On the other hand glacis type falls are quite reliable as meters under different working heads.

Flumed and Full Width Falls

A fall may either be constructed of the full channel width or it may be contracted. The falls of contracted width are known as flumed falls. For a flumed fall smooth upstream transition should be provided to avoid turbulence and to maintain accurate depth discharge relationship.

19.5 CLASSIFICATION OF FALLS

The various types of falls are generally divided with reference to the approach conditions into the following four classes.

Class I. Falls designed to maintain the normal depth-discharge ratio

In this class those falls are included which maintain the normal depth-discharge ratio or the normal depth-discharge curve of the channel with a fair degree of accuracy. In other words for all the discharges more or less normal or uniform flow is maintained in the channel and there is neither drawdown nor heading up of water as the flow in the channel approaches the fall. The falls under this category are:

(i) Trapezoidal notch falls and

(ii) Low crested weir or rectangular notch falls.

The trapezoidal notch falls maintain the normal depth-discharge relationship with accuracy. Further in this case since the notches provide an opening for flow down to the bed level, there is no obstruction to flow of silt in the channel. However, these falls cannot be used as regulators or metres for discharge measurement.

The low crested weir or rectangular notch falls do not maintain the normal depth-discharge relationship with accuracy and also they are liable to cause silting if the flow in the channel is at a lower level than designed for long periods. However, these falls are cheaper to construct and also these can be used as metres for discharge measurement.

Class II. Falls designed to maintain fixed water surface level in the channel upstream of the fall

In this class those falls are included which maintain a nearly fixed water surface level in the channel upstream of the fall. This type of falls are necessary under the following circumstances.

(i) When a subsidiary channel takes off some distance upstream of the fall.

(ii) When a hydro-power station is combined with the fall.

The principal types of falls in this class are

- Siphon falls (or Siphon spillways) ; and
- High crested weir falls.

Siphons are used as spillways discharging surplus waters below the dam. They maintain a nearly constant water level in the reservoir and have a high discharge intensity per unit length but they would be too expensive for ordinary canal falls. They are also not suitable for discharge measurement.

The high crested weir fall would usually maintain fixed water surface level in the channel only when the channel is run at the full supply discharge for which the height of the crest of the weir has been designed. However, if the channel is run for long period at less than the designed full supply it may lead to deposition of silt in a reach of the channel upstream of the fall and there may be heading up of water. These falls are generally constructed without fluming and hence the discharge intensity per unit length of the crest is low and the problem of energy dissipation is simple. Moreover, these falls can also be used as meters for discharge measurement.

Class III. Falls designed to admit the variation of the water surface level in the channel upstream of the fall at the will of the operator.

The falls included in this class are such that they admit the variation of the water surface level in the channel upstream of the fall according to the requirement. Thus, these falls also serve as regulators and are therefore combined falls and regulators. In practice falls of this type are confined to rectangular notches as the trapezoidal notches cannot be used as regulators. The regulation is done with the help of

- sluice gates
- horizontal stop logs and
- vertical needles.

A fall of this type may be required where a subsidiary channel takes off just upstream of the fall.

Class IV. Miscellaneous types of falls

In this class those falls are included which are designed from special considerations without any reference to the condition of approach. The principal types of falls in this class are as follows.

1. Cylinder Falls or Well Falls. As shown in Fig. 19.5 in a cylinder fall water flowing over the crest is dropped into a well from where it emerges near the bottom. The energy is dissipated due to turbulence inside the well. Such falls are suitable and economical for high drops and low discharges and hence these are used on tail escapes of small channels.

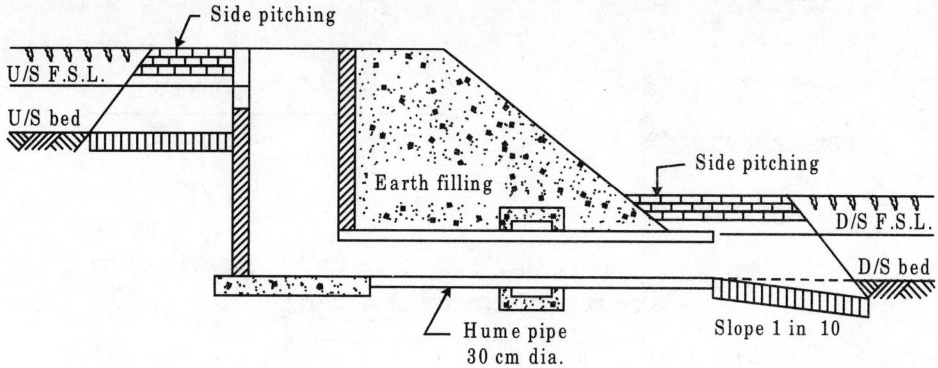

Fig. 19.5 *Cylinder fall or Well fall*

2. Chute or Rapid Falls. In this type of fall a chute or rapid is provided for conducting the channel flow. These falls may be necessary in a channel transporting floating timber.

3. Inclined Pipe Falls. In this type of fall inclined pipes are provided through which water flowing in the channel is carried.

19.6 DESIGN OF CISTERN OR CISTERN ELEMENTS

The *cistern* or *cistern element* is that portion of the fall on the downstream of the crest wall in which the surplus energy of water leaving the crest is dissipated and the subsequent flow stilled before it passes on to the lower level channel. The object of providing a cistern is three-fold : (i) to reduce the intensity of impact of the dropping jet of water against the downstream floor, (ii) to provide water cushion to dissipate the energy of the falling jet, and (iii) to produce reverse flow by providing a suitable end wall to ensure an impact in the cistern. The complete cistern element consists of (i) sloping glacis (if any), (ii) cistern, (iii) roughening devices, and (iv) device for deflecting the residual high velocity jet.

The various types of cistern elements adopted in practice may be divided into the following four classes.

Class I. Cistern element having impact from a stream of water falling freely under gravity

In this case the energy is dissipated by means of impact of a practically vertical stream with the horizontal cistern and due to sudden deflection of velocity from the vertical to the horizontal direction. To protect the floor from the impact of falling water, water cushion is provided by depressing the floor below the downstream bed of the channel as shown in Fig. 19.6. The length and depth of the cistern should be sufficient to cause maximum dissipation of energy and allow relatively calm flow on the downstream side. However, the required length and depth of the cistern are not amenable to theoretical treatment. The best dimensions of the cistern are therefore determined by experience and model experiments. Some of the empirical formulae developed for the length and depth of cistern are given below.

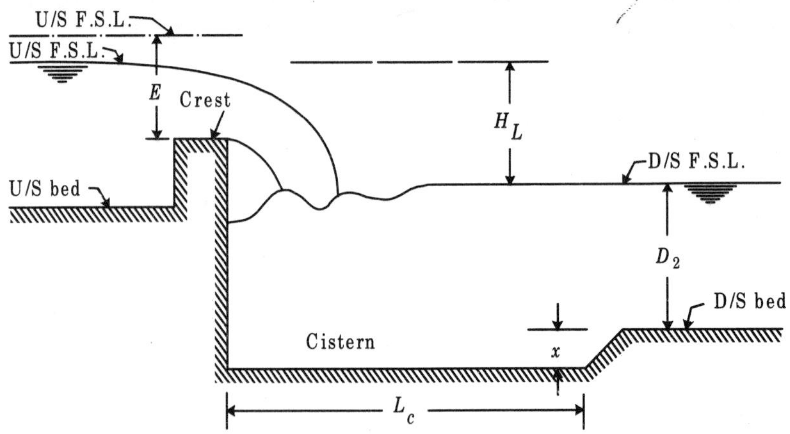

Fig. 19.6 *Cistern or cistern element of vertical drop fall*

(i) *Sir Proby Cautley's recommendations* : Cautley recommended that the depth of the cistern should be equal to the height of the fall.

(ii) *Dyas formula* :

$$x = 0.82\, H_L^{1/2} D_2^{1/3} \qquad \text{...(19.1)}$$

where x = depth of cistern below downstream bed in m

H_L = height of drop in m, and

D_2 = depth of water in the downstream channel in m.

(iii) *Glass formula* :

Glass developed the following formulae for the falls in Bihar and Orissa.

$$x = 1.85 E^{1/2} H_L^{1/3} - D_2 \qquad \text{...(19.2)}$$

$$L_c = (x + D_2) \qquad \text{...(19.3)}$$

where E = depth of crest below upstream total energy line in m, and

L_c = length of cistern.

(iv) *Etchevery formulae* :

These formulae were developed in America

$$L_c = 3\sqrt{H_L \cdot E} \qquad \text{...(19.4)}$$

$$x = \frac{1}{6} L_c \qquad \text{...(19.5)}$$

(v) *Montague formulae* :

These formulae were developed in Punjab

$$x = \frac{1}{2} E_{f_2} \qquad \text{...(19.6)}$$

$$L_c = 4 E_{f_2} \qquad \text{...(19.7)}$$

where E_{f_2} = energy of flow in downstream channel for discharge intensity q and drop H_L.

(*vi*) *Blench formula* :

$$x = 2D_c + \frac{1}{4}(H_L - 3/8D_c) - D_2 \qquad \text{...(19.8)}$$

According to Blench the fall should be designed in such a way that H_L is not less than 3/8 D_c, where D_c is critical depth.

(vii) *U.P. Irrigation Research Institute formulae* :

$$L_c = 5\sqrt{E \cdot H_L} \qquad \text{...(19.9)}$$

$$x = \frac{1}{4}\ (E.H_L)^{2/3} \qquad \text{...(19.10)}$$

These cisterns are generally not provided with roughening devices as considerable energy gets dissipated due to vertical impact. However, in this case aeration of nappe is necessary to prevent it from adhering to the crest wall. This can be provided by aeration pipes embedded in

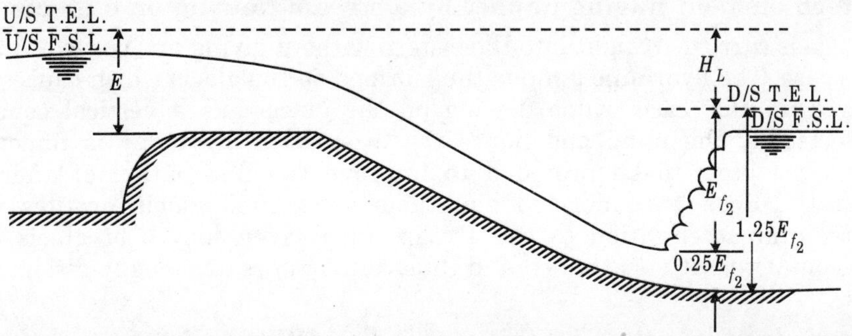

Fig. 19.7 *Cistern element having impact by a horizontal stream*

the wing walls just downstream of the crest wall. Further the exit from the cistern should be smooth so that streamlined flow is established as water enters the channel. If the cistern bed is below the channel bed, a gentle upward slope should be provided to join it to the channel bed.

Case II. Cistern element having impact by a horizontal stream

In this type of cistern the energy is dissipated by the formation of hydraulic jump. The water after flowing over the crest is carried over a glacis which is given a reverse curve at its lower end to turn the hypercritical jet to a horizontal direction before it strikes against the subcritical flow of the lower channel resulting in the formation of a hydraulic jump in the cistern as shown in Fig. 19.7. It is, however, known that for a given drop in the energy line H_L, and discharge intensity q, there will be a definite value of downstream specific energy E_{f_2} , or the downstream depth D_2 required for the formation of hydraulic jump. Thus, theoretically the bed of the cistern should be provided at the lowest level of the jump formation. However, on a horizontal floor, small changes in discharge intensity result in large changes in the jump location. Also if a perfect jump is not formed the high velocity flow persists for long distances causing scour of bed and sides of the channel. As such as an additional safety the depth of the cistern is increased by 25% of E_{f_2} , that is the bed of the cistern is kept at a depth of 1.25 E_{f_2} below the downstream total energy line. However, if the bed level of the downstream channel is lower than the bed level of the cistern determined from the above consideration, then the cistern should be provided at the bed level of the downstream channel.

Alternatively in order to stabilize the hydraulic jump within the cistern, the bed of the cistern is provided with a gentle slope of, say, 1 in 4 from the calculated position of the hydraulic jump for the design discharge intensity, till the depth of the cistern has been increased by 25%. However, when the hydraulic jump shifts down to the sloping portion the impact is no longer horizontal. Moreover, Inglis has adopted a low baffle to hold the jump stable on a horizontal floor.

The length of the cistern should be sufficient to keep the jump confined to the cistern and it is usually kept equal to 5 to 6 times E_{f_2}. Further in this case roughening devices may be required to be provided for efficient dissipation of energy. Also if the cistern bed level is lower than the channel bed level then the cistern bed should be joined to the bed of the lower channel by a gentle upward slope. However, this type of energy dissipators are more commonly used for dam spillways than for canal falls because of the difficulty of holding the jump within specified limits and the comparatively expensive roughening devices required to be provided for the purpose of energy dissipation.

Case III. Cistern element having impact by a stream flowing on a glacis

In this case glacis is carried straight into the cistern without giving any curve at its lower end as in the previous case. The hydraulic jump is thus formed on the glacis which causes dissipation of energy. However, in this case water flowing on the glacis has a vertical component which remains unaffected by the jump and hence the dissipation of energy is imperfect. As such roughening devices have to be provided to improve the dissipation of energy. However, Montague and Dr. Bose have derived equations for curved glacis profiles which impart maximum horizontal acceleration to the stream in a given length of glacis due to which maximum horizontal velocity is attained and thus result in greater energy dissipation than on a straight glacis.

The minimum depth of cistern for this case is also determined as in the case of horizontal impact and the length of the cistern is determined by experience.

Class IV. Cistern element without impact

In this case there is no impact by the flowing stream and hydraulic jump is also not formed. As such for dissipation of energy roughening devices are provided. The depth of the cistern in this case also cannot be determined theoretically but generally the bed of the cistern is depressed below the bed of the lower channel. It increases the volume of the cistern and the upward slope at its exit helps to suppress large scale turbulence.

19.7 ROUGHENING DEVICES

As mentioned in the previous section roughening devices are provided in the cisterns for the dissipation of energy. In the case of cisterns with impact the roughening devices may be used as secondary means for dissipating residual energy of water. However, in the case of cisterns without impact the roughening devices are the only mean available for dissipation of energy. These devices depend on turbulence and boundary friction for this purpose. Figure 19.8 shows the commonly used roughening devices which are briefly described below.

1. Staggered friction blocks. Staggered friction blocks are one of the most useful and simple devices used for dissipation of energy. They consist of rectangular blocks of concrete

(Fig. 19.8 a). The blocks are securely anchored into the floor and projecting upto a height of $\frac{1}{4}$th the depth of water. The spacing between the blocks is about twice the height of the blocks. Two or more staggered rows of these blocks may be provided according to necessity.

2. Arrows. Arrows are specially shaped friction blocks (Fig. 19.8 b) which are provided in the same way and serve the same purpose as rectangular blocks. They are able to withstand the action of the high velocity jet striking against them in a better way.

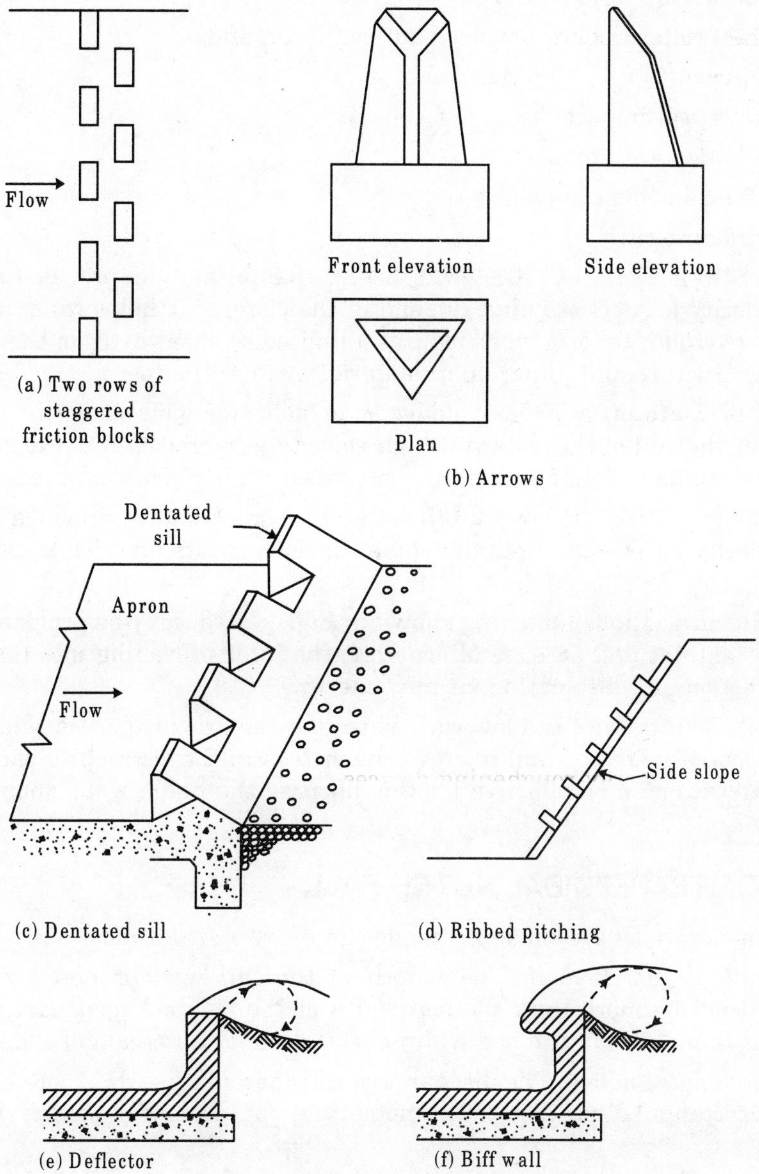

Fig. 19.8 *Roughening devices*

Montague gave the following empirical formula for the determination of the distance upto which roughening of bed is required to be done by the friction blocks or arrows.

$$L = C\frac{D_2^{3/2}H_L^{1/2}}{x} \quad \text{...(19.11)}$$

where L = length of cistern requiring roughening in m

D_2 = depth of water in the downstream channel in m

H_L = height of drop in m

x = depth of cistern below downstream bed in m, and

C = a coefficient

= 1 for vertical impact

= 2 for horizontal impact

= 4 to 6 for inclined impact

= 8 for no impact.

3. Dentated Sill. A dentated sill shown in Fig. 19.8 (c) may be provided at the end of the cistern if high velocity jet persists upto the end of the cistern. It helps to break up the jet and also causes a reverse roller to form which causes a limited scour away from the toe where it does not endanger the structure and piling up of material against the toe.

4. Deflector or Deflector Wall. A deflector or deflector wall shown in Fig. 19.8 (e) may also be provided at the end of the cistern if high velocity jet persists upto the end of the cistern. Its function is also similar to that of dentated sill except that it does not break the jet.

5. Biff Wall. Figure 19.8 (f) shows a biff wall which may also be provided at the end of the cistern if high velocity jet persists upto the end of the cistern. Its function is also similar to that of a deflector.

6. Ribbed Pitching. Ribbed pitching shown in Fig. 19.8 (d) may be provided on the sides. It consists of bricks laid flat and on edge alternately, the later projecting into the stream thereby increasing side friction and dissipating surplus energy.

7. Baffle Wall. A baffle wall is a low wall which may be provided at the end of the cistern. It provides a deep pool of water behind itself in the cistern and thus help to cause dissipation of surplus energy. However, to be effective for this purpose the baffle walls should be sufficiently high.

19.8 DESIGN OF TRAPEZOIDAL NOTCH FALL

The following considerations are made in the design of trapezoidal notch fall.

Length of fall wall or notch wall. The length of the fall wall or notch wall between the abutments should not be more than the bed width of the channel upstream of the fall and it should not be less than 7/8th of the bed width of the channel upstream of the fall.

Discharge of a trapezoidal notch. The discharge of a trapezoidal notch is obtained by adding the discharge of the rectangular portion of the notch and the discharge of the sloping triangular portions.

Let Q = discharge of trapezoidal notch.

L = length of sill of the notch.

n = 2 tan α, where α is the angle made by each of the sides of the notch with the vertical.

C = coefficient of discharge of the notch, which may be taken as 0.7 for notches in channels having depth of flow upto 1.5 m (i.e., for notches in distributary canals) and 0.78 for notches in larger channels having higher velocities (i.e., for notches in main canals).

H = depth of water above sill of the notch measured at some distance upstream of the notch.

h_d = height to which the notch is drowned or depth of tail water over the sill of the notch.

For free overfall conditions

$$Q = \frac{2}{3}C\sqrt{2g}\left(LH^{3/2} + \frac{2}{5}nH^{5/2}\right) \quad ...(19.12)$$

or

$$Q = 2.953\ CH^{3/2}\ (L + 0.4nH) \quad ...(19.12\ a)$$

This equation contains two unknowns L and n. For its solution two values of H with corresponding values of Q must be assumed. It is the general practice to design for the full supply and the half full supply discharges taking values of H equal to the normal depths in the two cases. Thus, if Q_1 and Q_2 represent the two discharges and H_1 and H_2 the corresponding depths then from Eq. (19.12 a) the values of L and n are obtained as

$$L = \frac{Q_1H_2^{5/2} - Q_2H_1^{5/2}}{2.953CH_1^{3/2}H_2^{3/2}(H_2 - H_1)} \quad ...(19.13)$$

$$n = \frac{Q_2H_1^{3/2} - Q_1H_2^{3/2}}{1.181CH_1^{3/2}H_2^{3/2}(H_2 - H_1)} \quad ...(19.14)$$

For drowned conditions

$$Q = C\sqrt{2g}\,(H - h_d)^{1/2}\left(L + \frac{1}{2}h_d n\right)h_d + \frac{3}{2}C\sqrt{2g}\,(H - h_d)^{3/2} \times \{(L \times nh_d) + 0.4n(H - h_d)\} \quad ...(19.15)$$

$$\text{or } Q = 2.953C(H - h_d)^{1/2}\left[\left(\frac{1}{2}h_d + H\right)L + \left\{\frac{3}{4}h_d^2 + (H - h_d)h_d + 0.4(H - h_d)^2\right\}n\right] \quad ...(19.15\ a)$$

For this case also giving two sets of known values to QH and h_d for two stages of the channel, the unknowns L and n can be obtained from the two equations formed by substitution in Eq. (19.15).

Number and sill level of notches. The sills of the notches should be at the bed level of the channel. The number of the notches should be suitable to distribute the discharge fairly throughout the length of the fall wall. Generally the number of the notches is so adjusted that the top width of the notch lies between 3/4th to full water depth above the sill of the notch.

Notch piers. The thickness of the notch piers should not be less than half the full supply depth of the channel upstream of the fall and may be made more if they have to carry a heavy superstructure.

The top of the piers should be at F.S.L., unless they are utilized for carrying a foot bridge or other superstructure.

Details of plan of trapezoidal notch. Figure 19.9 shows the details of a trapezoidal notch fall. The plan of the notch originally used in Punjab is shown in Fig. 19.9 (a). In this fall the plane of the notch profile is set back 0.5 m from the downstream face of the notch wall for the main

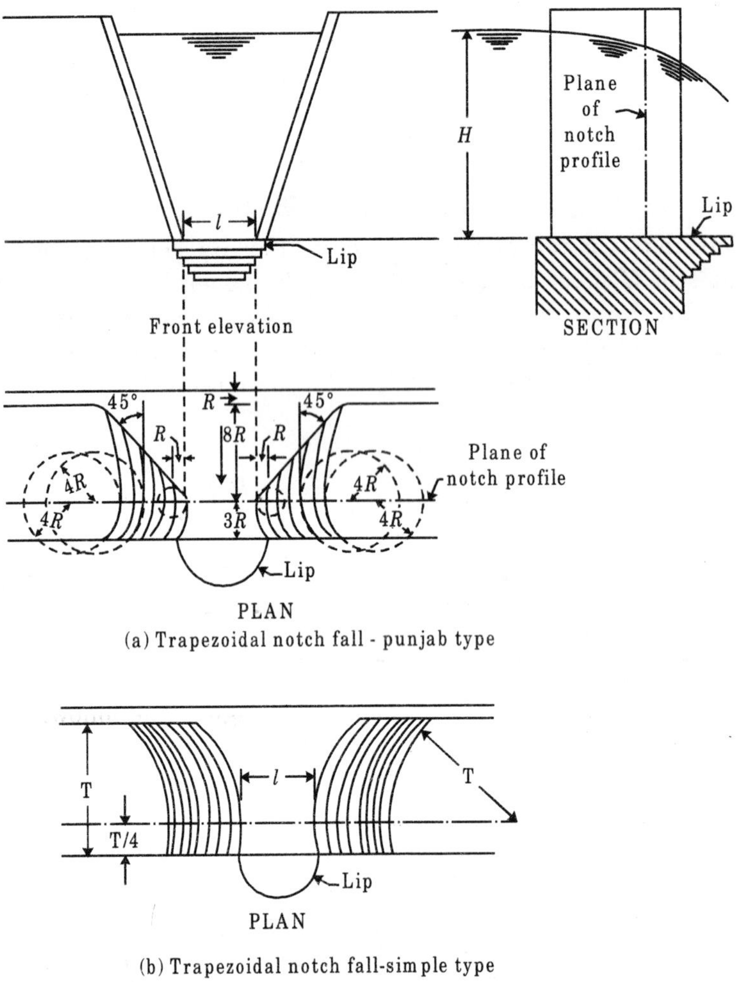

Fig. 19.9 *Details of trapezoidal notch fall, (a) Punjab type ; (b) Simple type*

canal falls and 0.25 m for the distributary falls. The plan of sides of the notch has a series of circular arcs the centres of all of which lie in the plane of the notch profile. Further from the notch section there is a splay of 45° on the upstream side and that of $22\frac{1}{2}°$ on the downstream side. From each notch a circular lip supported by corbelling projects beyond the downstream face of the notch wall. The amount of projection of the lip is 0.6 m for the falls on main canals and 0.3 m for those on distributaries. Also the radius $R = 0.15$ m for main canal falls and 0.10 m for distributary falls.

Since the trapezoidal notch falls used in Punjab involved a complicated construction of curves with a very wide splay on the upstream side, which in turn required wide piers between the notches, simple type of notch fall was evolved the plan of which is shown in Fig. 19.9 (b). In the simplified design of the notch fall if T is the thickness of the notch wall, the plan of sides of the notch are series of circular arcs struck with radius T and centres on a line $\frac{1}{4}\ T$ from the downstream face of the fall wall. Further in this case the projection of the lip is limited to 0.3 m or half the width of the notch sill at its downstream edge whichever is smaller.

The design of floor for these falls may be carried out as already explained in Chapter 13. The layout of wings etc., is explained in detail in connection with Sarda type fall and the same may be adopted for the notch fall also.

19.9 DESIGN OF SARDA TYPE FALL

It is a vertical drop type fall having a raised crest or crest wall and vertical impact of the falling jet. This type of fall was first introduced to replace the notch fall on the Sarda canal system in U.P. owing to its economy and simplicity. In the earlier design of these falls adopted on the Sarda canal system no depressed cisterns were provided and also the downstream wings were not flared or warped. Although the behaviour of these falls was, on the whole, very satisfactory, on some of the falls erosion of banks downstream of the fall was noticed. In order to eliminate this defect extensive model experiments were carried out at Bahadarabad Research Station U.P., and based on these experiments some recommendations were made to modify the design of these falls. The design criteria for the Sarda type fall based on these recommendations are given below.

The complete design of Sarda type fall consists of the design of the following components.

1. Crest wall (or crest)
2. Cistern
3. Impervious floor
4. Downstream protection
5. Upstream approach.

A. Design of Crest Wall (or crest)

1. Length of Crest Wall. The length of the crest wall is kept equal to the bed width of the channel and no fluming is done in this type of fall. However, if future development of irrigation and increase in the discharging capacity of the channel is anticipated, the crest length may be kept more than the bed width of the channel, say it may be kept equal to bed width of the channel plus water depth.

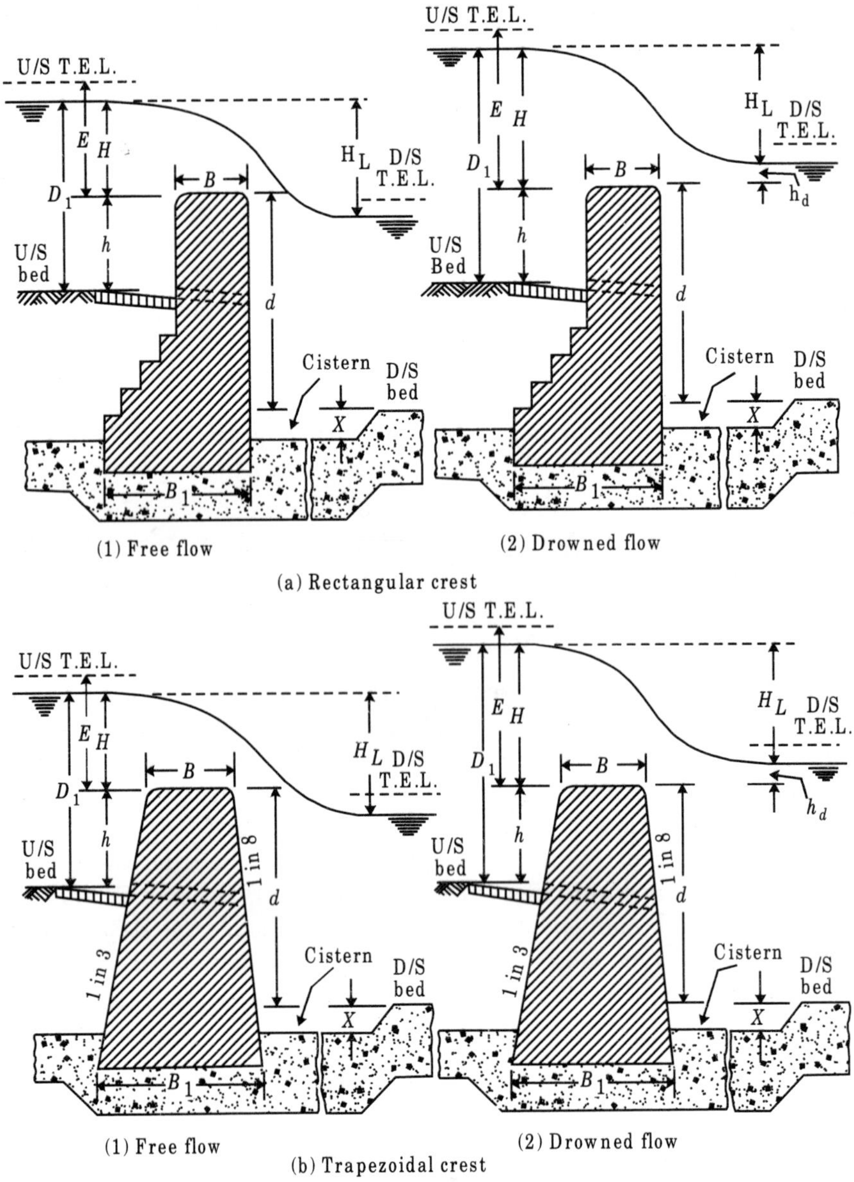

Fig. 19.10 *Crest walls for Sarda type fall*

2. Shape of Crest Walls. Two types of crest walls are used as shown in Fig. 19.10. For discharges upto 14 cumec crest walls rectangular in section with both upstream and downstream faces vertical are used (Fig. 19.10 a). For discharges over 14 cumec crest walls trapezoidal in section with upstream face having a slope of 1 in 3 and downstream face having a slope of 1 in 8 are used (Fig. 19.10 b). The top and base widths of the two types of crest walls are as given below.

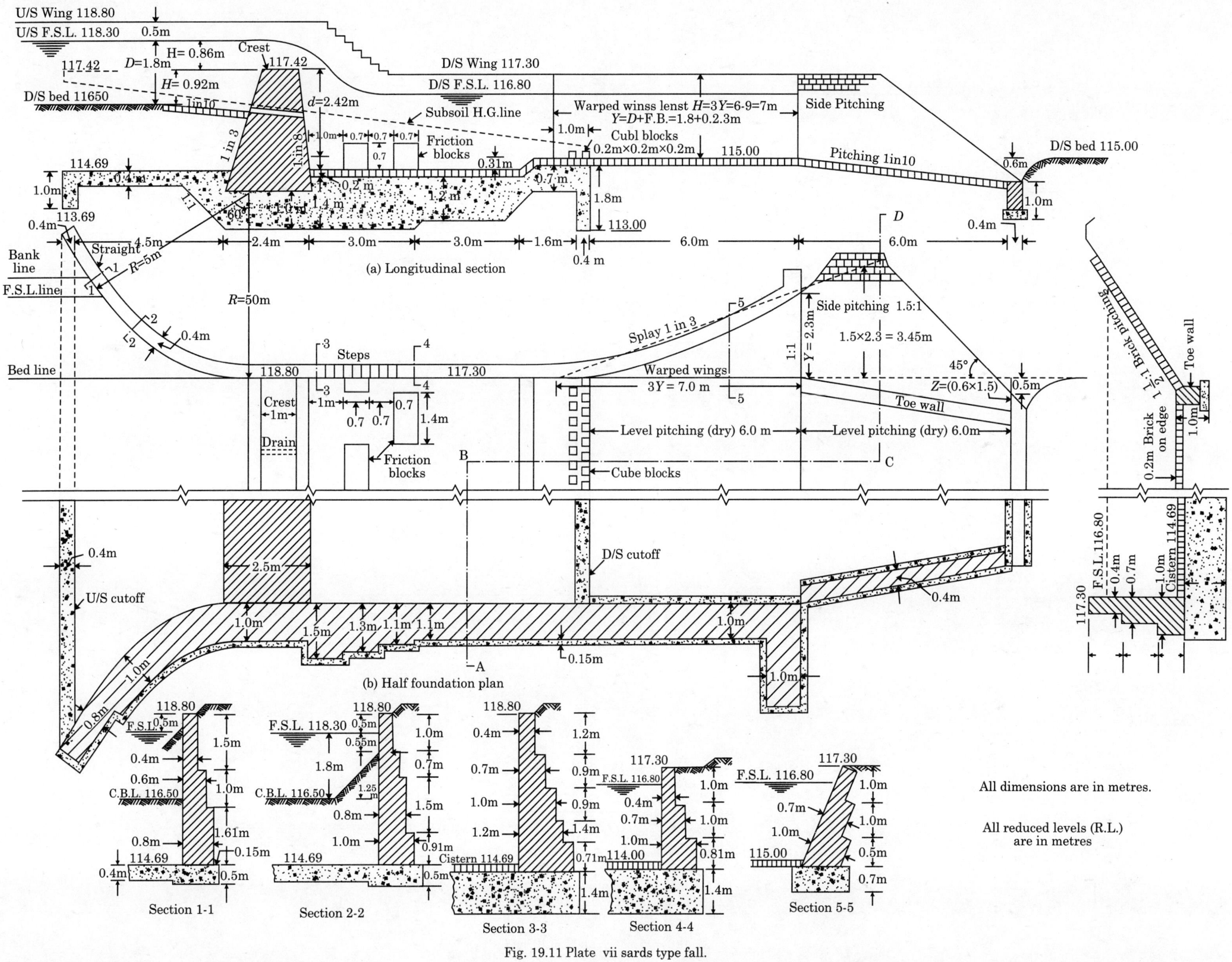

Fig. 19.11 Plate vii sards type fall.

Rectangular crest wall

For a rectangular crest wall top width is given by

$$B = 0.55\sqrt{d} \qquad \text{...(19.16)}$$

where B— top width of the crest wall in m, and

d— height of the crest wall above the downstream bed level in m.

Base width of the crest wall is given by

$$B_1 = \frac{H+d}{G} \qquad \text{...(19.17)}$$

where B_1— base width of the crest wall in m, and

H— height of water surface above the top of the crest wall on the upstream side in m

G— specific gravity of the material of the crest wall, and d is same as defined earlier.

For masonry G may be taken equal to 2.

Trapezoidal crest wall

For a trapezoidal crest wall top width is given by

$$B = 0.55\sqrt{(H+d)} \qquad \text{...(19.18)}$$

where all the terms are same as defined earlier.

Slope of upstream face = 1 in 3

Slope of downstream face = 1 in 8

Thus, base width of the crest wall is determined by the above noted slopes of both the faces.

Further for falls over 1.5 m the stability of the crest wall should be checked by actual analysis.

3. Discharge Formula. With free overfall conditions discharge for the two types of crest walls is given by the following formulae.

For rectangular crest wall

$$Q = 1.835\, LH^{3/2}\, (H/B)^{1/6} \qquad \text{...(19.19)}$$

where Q— discharge in cumec

L— length of the crest wall in m, and

H and B are same as defined earlier.

For trapezoidal crest wall

$$Q = 1.99\, LH^{3/2}\, (H/B)^{1/6} \qquad \text{...(19.20)}$$

where all the terms are same as defined earlier.

For drowned conditions, neglecting velocity of approach and taking an average discharge coefficient C_d for free and drowned portions the discharge for both types of crest walls is given by

$$Q = C_d L\sqrt{2g}\left(\frac{2}{3}H_L^{3/2} + h_d H_L^{1/2}\right) \qquad \text{...(19.21)}$$

where Q = discharge in cumec

L = length of the crest wall in m

H_L = drop in water surface or difference in water surface levels on the upstream and downstream sides of the crest wall in m

h_d = height of the water surface on the downstream side above the top of the crest wall in m, and

g = acceleration due to gravity.

The value of C_d may be taken as 0.65.

4. Crest Level. For free overfall condition

$$\text{Crest level} = u/s\ \text{F.S.L.} - H$$

in which H is same as defined earlier and its value may be obtained from Eq. (19.19) or Eq. (19.20) depending on the type of the crest wall.

Further if D_1 is upstream full supply depth and h is height of the crest wall above the upstream bed, then

$$h = D_1 - H \quad ...(19.22)$$

However, for drowned condition height of the crest wall above the upstream bed, is given by

$$h = D_1 - (H_L + h_d) \quad ...(19.23)$$

in which all the terms are same as defined earlier. The value of h_d is determined from Eq. (19.21).

5. Bed protection upstream of the crest wall. Adjacent to the crest wall on the upstream side, brick pitching is laid on the channel bed for 2 to 4 m length, sloping downwards towards the crest wall at 1 in 10. Also at the level of brick pitching a few drain holes are provided in the crest wall to drain out the upstream bed during the closure of the channel.

B. Design of Cistern

A depressed cistern is provided having length and depth below downstream bed given by Eqs (19.9) and (19.10) as noted below

$$L_c = 5\sqrt{E \cdot H_L}$$

$$x = \frac{1}{4}(E \cdot H_L)^{2/3}$$

Neglecting the small velocity of approach head, E may be replaced by H in these equations.

C. Design of Impervious Floor

The length and thickness of impervious floor may be determined by Bligh's theory for small and medium falls and by Khosla's theory for large falls. The maximum seepage head occurs when there is water on the upstream side upto the top of the crest wall and there is no flow on the downstream side. Thus, as indicated in Fig. 19.10, the maximum seepage head is equal to d.

On the basis of Bligh's theory the length of the impervious floor may be determined by considering the safe hydraulic gradient which may be taken as 1 in 5 to 1 in 8 for mixed or loam soils. Similarly according to Khosla's theory the length of the impervious floor may be determined by considering the safe exit gradient as discussed in Chapter 13. Out of the total length of the impervious floor a minimum length l_d is to be provided on the downstream of the toe of the crest wall which is given by

$$l_d = 2(D_1 + 1.2)\, H_L \qquad \text{...(19.24)}$$

where l_d is in metres and

D_1 and H_L are same as defined earlier and are also in metres.

The balance of the impervious floor length is provided under and upstream of the crest wall.

Having determined the length of the impervious floor and the placing of the floor relative to the crest wall, the hydraulic gradient line may be drawn and the floor thickness determined as explained in Chapter 13. On the upstream of the crest wall the uplift pressures are counterbalanced by the weight of the water standing on the floor. Also the upstream floor is laid below the bed of the channel on the upstream side and is not subjected to flow. As such for the floor on the upstream of the crest wall a minimum thickness of 0.3 m is provided. For the floor on the downstream of the crest wall the thickness of the floor is determined by uplift pressures subject to a minimum of 0.3 to 0.4 m for small falls and 0.4 to 0.6 m for large falls. The cistern and the downstream floor are provided with a top lining of brick on edge in lime or cement mortar and the balance of the required thickness is provided by cement concrete. This facilitates repair of worn out surface.

Cutoffs and Curtain Walls

At the upstream and downstream ends of the impervious floor vertical cutoffs or curtain walls are provided. The minimum depth of the cutoffs or curtain walls below the surface of the floor are as indicated below.

(a) *Upstream cutoff or curtain wall*

$$\text{Minimum depth} = \frac{D_1}{3}$$

where D_1 is upstream full supply depth.

(b) *Downstream cutoff or curtain wall*

$$\text{Minimum depth} = \frac{D_2}{2}$$

where D_2 is downstream full supply depth.

D. Downstream Protection

The downstream protection consists of the following.

(i) Bed protection

(ii) Downstream wings

(iii) Side protection

(iv) Energy dissipators

(i) *Bed protection.* The bed of the channel needs to be protected for some length on the downstream of the impervious floor. The bed protection consists of dry brick pitching (i.e., brick laid dry without mortar) about 200 mm thick (one brick on edge laid over one flat brick) resting on 100 mm ballast. Table 19.1 gives the length of the pitching and the number of curtain walls (or cutoffs) required to hold the pitching. Where only one curtain wall is specified it should be provided at the downstream end of the pitching, where two are specified, the second one should be provided between the horizontal and sloping pitching and where three are specified the third is located in between to sub-divide the longest reach of the pitching.

TABLE 19.1 Details bed Pitching

Head over crest (m)	*Total length of pitching on the downstream (m)*	*Remarks*	*Curtain wall*	
			Number	*Depth (m)*
Up to 0.3	3.0	Sloping at 1 in 10	1	0.30
0.30 to 0.45	$3.0 + 2H_L$		1	0.30
0.45 to 0.60	$4.5 + 2H_L$	Horizontal upto	1	0.45
0.60 to 0.75	$6.0 + 2H_L$	end of masonry	1	0.60
0.75 to 0.90	$9.0 + 2H_L$	wings and then	1	0.75
0.90 to 1.05	$13.5 + 2H_L$	sloping	2	0.90
1.05 to 1.20	$18.0 + 2H_L$	downwards at	2	1.05
1.20 to 1.50	$22.5 + 2H_L$	1 in 10	3	1.35

(ii) *Downstream wings (or wing walls).* After the crest wall the wings are stepped down to the required level of the downstream wings. The downstream wings are kept vertical for a length varying from 5 to 8 times $\sqrt{E \cdot H_L}$ from the crest. The wings are then flared or warped, that is their water face is gradually inclined from vertical to a slope of $1\frac{1}{2}$: 1 or 1 : 1. In the later case warping is continued in the side pitching till a slope of $1\frac{1}{2}$: 1 is attained. The top of the wings is given an average splay of 1 in 2.5 to 1 in 4 for attaining the required slope. The wings follow a circular arc tangential at the beginning of the warp. The sections of the wings shown in Fig. 19.11 (Plate VIII) clearly illustrate the process of warping.

All wing walls are designed as retaining walls to sustain full earth pressure on their back when the channel is closed. In the absence of elaborate stability calculations the width of the wings at any level may be kept equal to $\frac{1}{3}$rd of the height above that level. However, for heavy works the sections of the wing walls are provided as per actual design.

(iii) *Side protection.* After the warped wings, pitching protection is provided on the sides. The side pitching consists of either one brick on edge or $1\frac{1}{2}$ brick on edge (i.e., one brick on edge placed over one flat brick) laid in cement mortar. The side pitching is curtailed at an angle of 45° from the end of the bed pitching in plan. If the warping of the wings is carried out only to a slope of 1 : 1, the side pitching is warped from a slope of 1 : 1 to $1\frac{1}{2}$: 1. However, if the wings are already warped to a slope of $1\frac{1}{2}$: 1 the side pitching is laid on this constant slope. A toe wall is provided between the bed pitching and the side pitching to provide a firm support for the side pitching. The toe wall is usually $1\frac{1}{2}$ brick (i.e., about 0.4 m) thick and of depth equal to half the downstream water depth.

(iv) *Energy dissipators.* Energy dissipators are not provided for small discharges. However, for large discharges, additional energy dissipators are provided. The energy dissipators consist of two rows of friction blocks in the cistern and two rows of cube blocks on the impervious floor at its downstream end. Both the friction blocks and the cube blocks are staggered and the size and position of these blocks are as indicated below.

(A) *Size and position of friction blocks*

(a) Length of block = $2d_c$

(b) Width of block = d_c

(c) Height of block = d_c

(d) Distance of first row of blocks from downstream toe of the crest wall = $1.5\ d_c$

(e) Spacing between the two rows of blocks = d_c

(f) Spacing between blocks in the same row = $2d_c$

where d_c is the critical depth.

(B) *Size and position of cube blocks*

(a) Length of block $= \frac{1}{10} D_2$

(b) Width of block $= \frac{1}{10} D_2$

(c) Height of block $= \frac{1}{10} D_2$

(d) Spacing between the two rows of blocks

$= \frac{1}{10} D_2$

(e) Spacing between blocks in the same row

$= \frac{1}{10} D_2$

where D_2 is downstream full supply depth.

One of the two rows of cube blocks in provided just at the downstream end of the impervious floor and the other one is provided on the upstream side at the above noted spacing.

E. Design of Upstream Approach

The upstream approach consists of wing walls (or wings). For discharges upto 14 cumec, the wing walls may be spayed straight at the angle of 45° from the upstream edge of the crest wall. For greater discharges the wing walls are made segmental (or curved) from the upstream edge of the crest wall, with radius equal to 5 to 6 times H, subtending an angle of 60° at the centre and then carried along straight lines tangential to the segment. The wing walls should be continued inland till they are embedded in the earthen banks of the channel by a minimum of 1 m from the line of F.S.L. Further the wing walls must be extended to the end of the upstream impervious floor. The foundations of the wing walls are laid on the upstream impervious concrete floor (at the level of the downstream bed) which is to be extended on both sides to serve this purpose.

The various steps involved in the design of Sarda type fall are indicated in Illustrative Example 19.1 and the design details are shown in Fig. 19.11 (Plate VIII).

19.10 DESIGN OF STRAIGHT GLACIS FALL

The design of a non-metre straight glacis fall involves the following steps.

1. Design of Crest

(i) The u/s glacis is provided with a slope of $\frac{1}{2}$: 1 and it is joined tangentially to the crest with a radius equal to $\left(\frac{E}{2}\right)$, where E = height of u/s T.E.L. above the crest (Fig. 19.12).

(ii) The width of the crest is kept equal to $\frac{2}{3}E$.

(iii) The d/s glacis is provided with a slope of 2 : 1 and it is joined tangentially to the cistern bed with a radius equal to E.

(iv) Sometimes the glacis fall may be flumed when it is to be combined with a bridge etc. However, the minimum clear length of the crest, expressed in percentage of the bed width of the channel is kept as indicated below :

Drop (m)	*Clear length expressed as percentage of the bed width of the channel*
Below 1.25	65%
1.25 to 3	75%
Above 3	85%

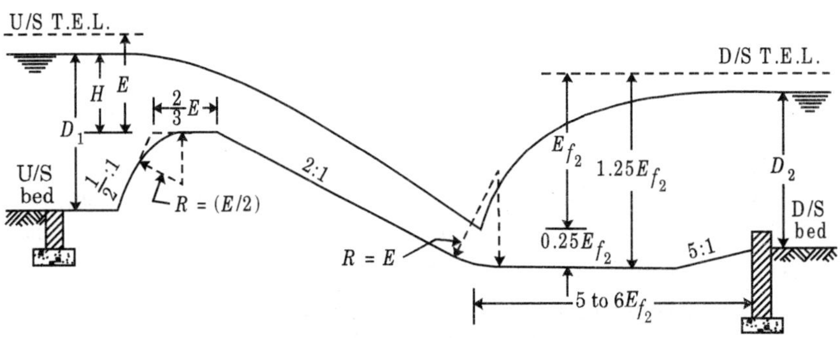

Fig. 19.12. *Straight glacis fall*

(v) The value of E is calculated from the following discharge formula :

$$Q = 1.84\, L_t\, E^{3/2} \qquad ...(19.25)$$

where Q = discharge in cumec, and

L_t = crest length in m.

If there are n number of piers, effective crest length to be substituted in Eq. (19.25) is equal to $(L_t - 0.2nE)$.

(*vi*) Crest Level = *u/s* T.E.L. – E

u/s T.E.L. = *u/s* F.S.L. + velocity head.

2. Design of Cistern

Knowing the discharge intensity q and the drop H_L, the downstream specific energy E_{f_2} (or the energy below the hydraulic jump) may be determined from the Blench curves [Fig. 13.19 (Plate III)]. The depth and length of the cistern may then be obtained as follows :

(i) Bed level of cistern = d/s T.E.L. – 1.25 E_{f_2}

Depth of cistern below d/s bed of channel

= bed level of d/s channel – bed level of cistern

However, minimum depth of cistern below d/s bed of channel

$$= 0.25\, E_{f_2}$$

(ii) Length of cistern = 5 to 6 times E_{f_2}

3. Design of Impervious Floor

The total length of the impervious floor is determined from the considerations of the permissible exit gradient. The total length of the impervious floor may be provided in the following manner.

(i) Length of cistern

(ii) Horizontal length of d/s glacis

(iii) Crest width

(iv) Horizontal length of u/s glacis

(v) Balance to be provided on the u/s side from the toe of u/s glacis.

Cutoffs or curtain walls. At the u/s and d/s ends of the impervious floor vertical cutoffs or curtain walls are provided. The minimum depth of the cutoffs or curtain walls below the surface of the floor are as indicted below :

(a) u/s *cutoff or curtain wall*

$$\text{Minimum depth } d_1 = \frac{D_1}{3}$$

where D_1 is *u/s* full supply depth.

(b) d/s *cutoff or curtain wall*

$$\text{Minimum depth } d_2 = \frac{D_2}{2}$$

where D_2 is d/s full supply depth.

The d/s cutoff or curtain wall is projected above the bed of the channel by $\frac{D_2}{10}$ to act as deflector.

4. Design of d/s Protection

(a) *Bed protection.* Since a deflector wall (or deflector) is provided at the d/s end of the floor (Fig. 19.12), no bed pitching is provided.

(b) *Side protection.* For side protection dry brick pitching about 0.2 m thick (one brick on edge laid over one flat brick) is done at 1 : 1 slope for a length equal to $3D_2$ beyond the impervious floor, where D_2 is full supply depth in the channel on the d/s side. The pitching is supported on a toe wall of width 0.4 m and depth $(D_2/2)$ subject to a minimum of 0.5 m. Further at the end of the pitching a 0.4 m wide profile wall is provided to protect the side pitching.

5. Design of Energy Dissipators

In the case of an unflumed glacis fall only deflector (or deflector wall) is provided. However, in the case of a flumed glacis fall in addition to deflector, friction blocks are provided. Further in both the cases for falls of more than 2 m, glacis blocks are provided. Each of these types of energy dissipators are described below.

(a) *Deflector (or Deflector wall).* A deflector (or deflector wall) of minimum width 0.4 m and height $D_2/10$ above the d/s bed of the channel is provided at the d/s end of the impervious floor or cistern.

(b) *Friction blocks.* Generally four rows of staggered friction blocks are provided. The width and height of the blocks is kept equal to $D_2/8$ and length equal to $3D_2/8$. The spacing between the blocks in the same row is kept equal to $3D_2/8$ and the distance between the rows of the blocks is kept equal to $D_2/8$. The *u/s* edge of the first row of the blocks is kept at $5D_2/8$ from the toe of the *d/s* glacis.

(c) *Glacis blocks.* One row of wedge shaped glacis blocks having same height and spacing as that of friction blocks is provided at the end of the *d/s* glacis.

6. Design of d/s Expansion

In the case of an unflumed glacis fall no expansion is provided on the *d/s* side, but parallel and vertical wing walls (or side walls) are provided upto the end of the *d/s* impervious floor.

In the case of a flumed glacis fall upto the toe of the *d/s* glacis parallel and vertical wing walls (or side walls) are provided, and thereafter on the *d/s* side expansion is provided. A rectangular hyperbolic expansion as shown in Fig. 19.13 is generally provided. For this expansion the bed width B_x at any distance x from the toe of the *d/s* glacis is given by

$$B_x = \frac{B_1 \times B_2 \times L_e}{(L_e \times B_2) - (B_2 - B_1)x} \quad \text{...(19.26)}$$

where L_e is total length in which the expansion from a flumed width B_1 to the normal width B_2 is to be done.

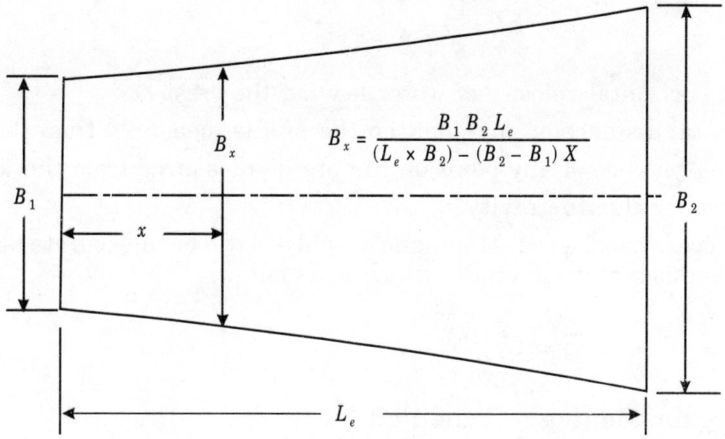

Fig. 19.13 *Rectangular hyperbolic expansion*

In the case of small falls a straight 1 in 3 expansion is provided.

The wing walls in expansion may be flared out from vertical to a slope of 1:1 and extended into the earthen banks of the channel. However, in view of the difficulty and extra cost in construction of walls with varying batter, the wing walls may be splayed straight.

7. Design of u/s Approach

In this case if impervious floor is not required to be provided on the u/s side then the wing walls are splayed straight at an angle of 45° from the u/s edge of the crest wall. However, if impervious floor is required to be provided on the u/s side then parallel wing walls are provided from the crest wall upto the end of the floor and from the u/s end of the floor the wing walls are splayed straight at an angle of 45°. The wing walls are continued inland till they are embedded in the earthen banks of the channel by a minimum of 1 m from the line of F.S.L.

The various steps involved in the design of straight glacis fall are indicated in Illustrative Example 19.2 and the design details are shown in Fig. 19.14 (Plate IX).

19.11 MONTAGUE TYPE FALL

In the case of a hydraulic jump developing on a glacis the vertical component of velocity remains unaffected and hence there is always considerable residual energy with the water below the jump. The problem of excessive residual energy below hydraulic jump is solved in Montague type fall by giving such a shape to the glacis that maximum horizontal acceleration is imparted to the flowing water in a given length of the glacis. By imparting maximum horizontal acceleration, maximum horizontal velocity is attained with the result that considerable energy gets dissipated in the jump. It is obvious that when water is flowing on a horizontal bed in level with the crest no horizontal acceleration will be imparted to it because the reaction being normal to the bed its component in the horizontal direction (which is the accelerating force) is equal to zero. Similarly when water is flowing over a parabolic bed with an initial horizontal velocity U, there will be no horizontal acceleration because the acceleration will be purely vertical under the influence of gravity. In between these two there lies such a parabolic glacis profile which would impart maximum horizontal acceleration to the water flowing over it. Montague derived analytically the equation for such a glacis profile which is as follows :

$$x = U\sqrt{\frac{4y}{g} + y} \qquad ...(19.27)$$

where U = initial horizontal velocity of water leaving the crest

x = horizontal distance of any point on the profile measured from the crest

y = vertical distance of any point on the profile measured from the crest, and

g = acceleration due to gravity.

Dr. Bose, however, considered Montague's analysis to be inaccurate and therefore gave another equation for such a glacis profile which is as follows

$$x = U\sqrt{\frac{2y}{g} + y} \qquad ...(19.28)$$

Design Procedure for Montague Type Fall

A general arrangement of Montague type fall is shown in Fig. 19.15. The various considerations made in the design of this type of fall are as mentioned below :

1. A Montague type fall is always flumed. The extent of fluming may be determined as follows.

Assuming that the jump forms at the downstream bed level, the downstream specific energy E_{f_2} is known. The loss of head H_L is equal to the difference between the energy lines upstream and downstream which is in general same as the height of the fall and hence it is also known. Knowing E_{f_2} and H_L, the discharge intensity q can be determined from Blench curves [Fig. 13.20 (Plate III)]. If Q is total discharge then the required crest length L_t is found by $L_t = (Q/q)$ and hence the extent of fluming is determined.

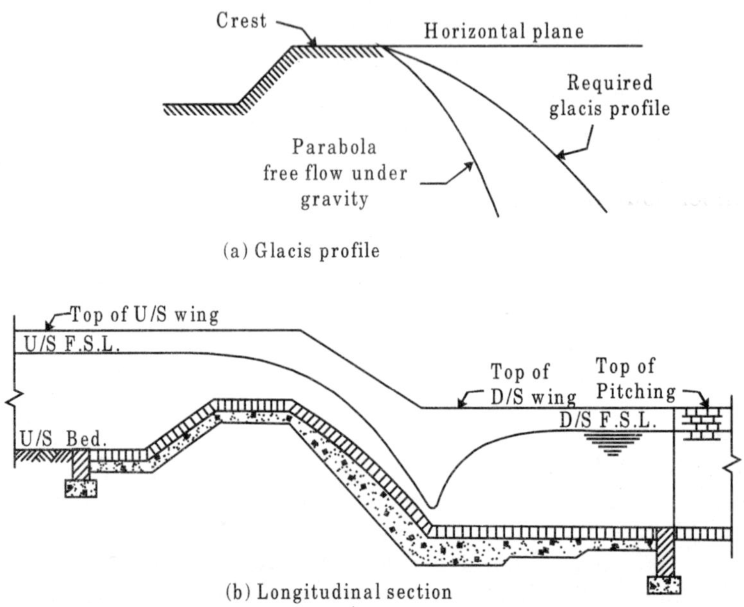

Fig. 19.15. *Montague type fall*

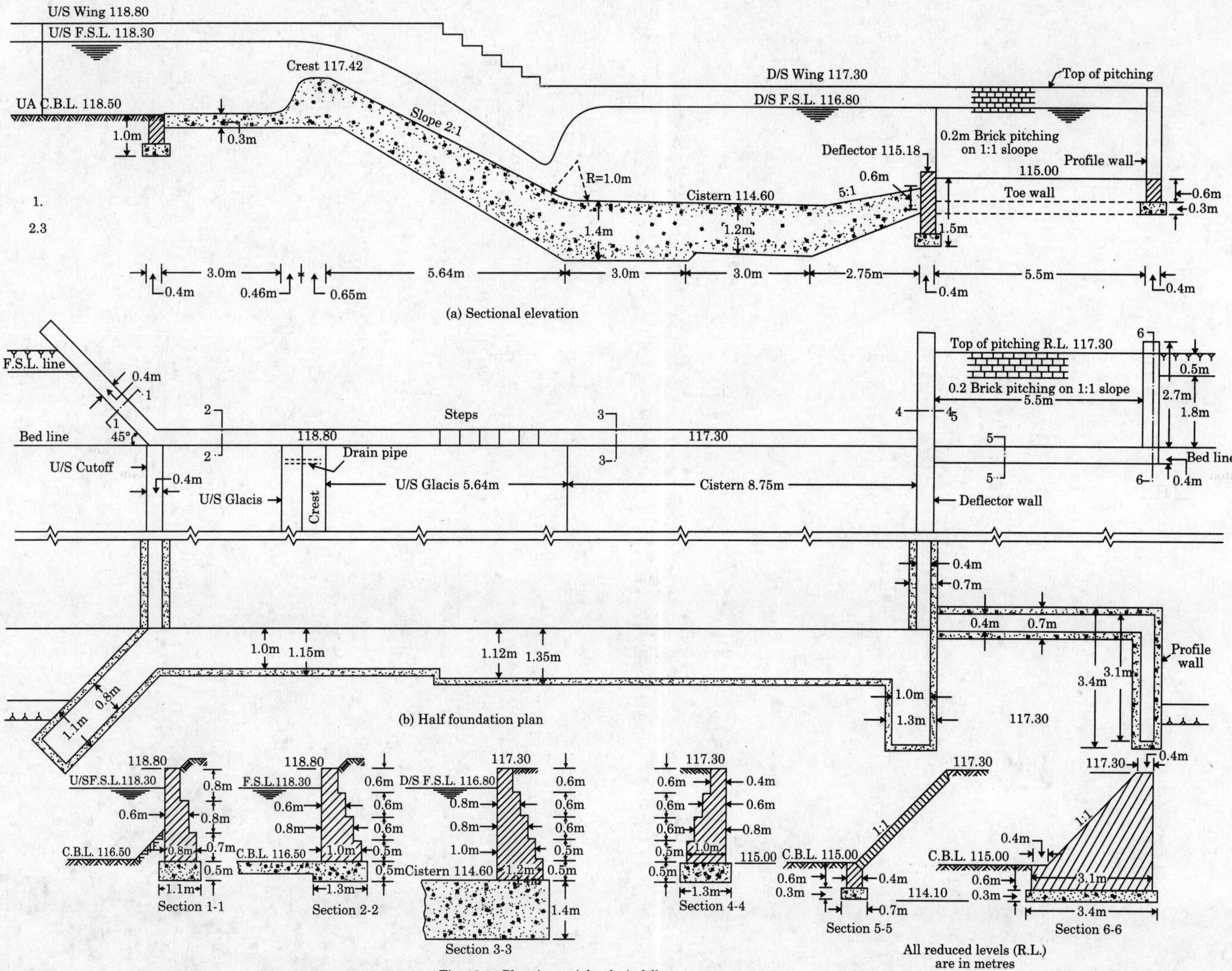

Fig. 19.14 Plate ix straight glacis fall.

2. The width of the crest is kept equal to at least 2 times the head over the crest and hence the discharge formula for a broad crested weir is applicable. The head over the crest E, may thus be obtained from the following formula

$$Q = 1.7\, L_t\, E^{3/2} \qquad ...(19.29)$$

3. The initial velocity U is found from the expression

$$U = \frac{q}{d_c} \qquad ...(19.30)$$

where d_c is critical depth.

Knowing U, the glacis profile may be obtained by using Eq. (19.27) or Eq. (19.28).

4. The depth of cistern below the downstream bed is kept equal to $\frac{1}{2}\, E_{f_2}$ and the length of the cistern is kept equal to four times the depth of the cistern.

5. The impervious floor is designed according to Khosla's theory. The glacis thickness is checked for uplift pressures in the jump trough at different discharge intensities as indicated in Chapter 13. At the upstream and downstream ends of the impervious floor vertical cutoffs or curtain walls are provided. The minimum depth of upstream cutoff is kept equal to $\frac{1}{3}$rd of upstream full supply depth and minimum depth of downstream cutoff is kept equal to $\frac{1}{2}$ of downstream full supply depth. However, the downstream cutoff is taken upto the lowest level of excavation.

6. The upstream contraction is usually 2 : 1 while the downstream expansion has a minimum average splay of 3 : 1. The upstream wings are kept vertical while the downstream wings are flared.

7. Friction blocks are usually provided with this type of fall.

8. After the impervious floor and masonry wings pitching protection of stones or bricks is provided in the bed and sides.

19.12 INGLIS TYPE FALL

This fall was designed by C.C. Inglis. It makes use of a horizontal impact for energy dissipation. As shown in Fig. 19.16 the design consists of a standard long throated weir flume followed by a glacis and a horizontal platform or pavement on which a baffle is fixed to hold the hydraulic jump stable on the platform. On the downstream of the baffle a cistern is provided and at the downstream end of the cistern a deflector is provided. The impervious floor is provided only upto the end of the deflector. On the downstream of the deflector often a second cistern is provided which is surfaced with stone or brick pitching only.

The maximum dissipation of energy by a hydraulic jump occurs when the jump forms at the toe of the glacis. As such the horizontal platform is provided at such a level that for normal discharge conditions the jump forms at the toe of the glacis. For this the position of the hydraulic jump on the glacis is ascertained for the normal discharge conditions and horizontal

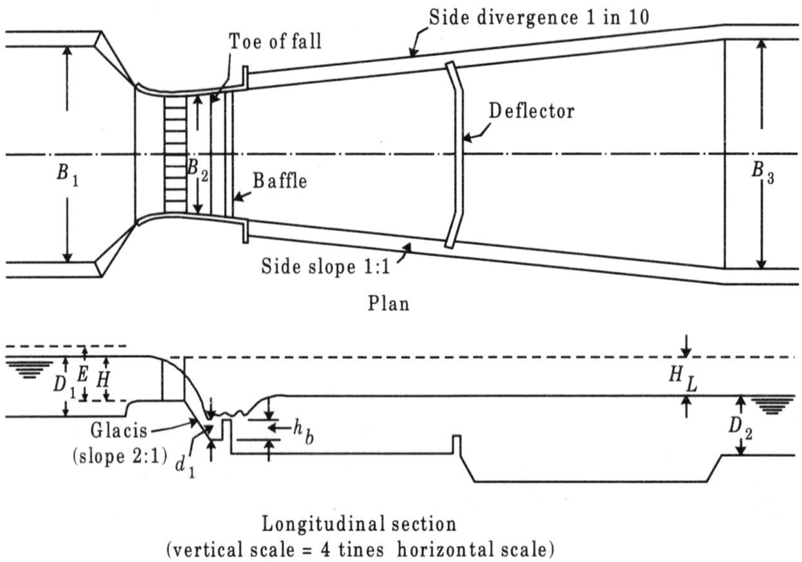

Fig. 19.16. *Inglis type fall*

platform is provided at this level which may however be either at the same level as the downstream bed or may be higher or lower than the downstream bed. The jump would then form at the toe of the glacis on the horizontal platform and the surging or instability of the jump would be prevented by the baffle. The cistern and deflector downstream of the baffle are meant to restore normal distribution of velocities in the channel.

The various considerations made in the design of Inglis type fall are as follows :

1. The crest is raised above the upstream bed by $\frac{1}{8}$th of the downstream water depth D_2. This determines the discharge intensity q and hence the required length of crest. The discharge formula for broad crested weir is applicable since the width of crest is $2\frac{1}{2}$ times the head over the crest E.

2. The glacis is provided with a slope of 2 : 1. If expansion is carried out in the glacis length it should be very gentle. The glacis is connected to the horizontal platform by a circular arc of radius = $2E$.

3. The upstream wings are curved with a radius equal to $2E^{3/2}$ when E is more than 0.3 m and equal to $2E$ when E is less than 0.3 m. The circular wings are continued till they subtend an angle 60° at the centre after which they are produced tangentially for the necessary length.

4. The horizontal platform is provided at the level at which the jump would normally form.

5. The length of the platform = 2.25 h_b where h_b is the height of the baffle. The height of the baffle = $(d_c - d_1)$ above the platform where d_c is critical depth and d_1 is the prejump hypercritical depth.

6. The downstream divergence is provided at a rate varying from 1 in 4 to 1 in 10 depending on bed width depth ratio of the downstream channel.

7. The depth of cistern below downstream bed should be 10% of downstream water depth D_2.

8. The length of cistern is kept equal to $5d_x$, where d_x is the subcritical depth necessary to form the hydraulic jump and is approximately given by the following expression :

$$d_x = 0.982\, q^{0.52}\, H_L^{\;0.21} \qquad ...(19.31)$$

where q— discharge per unit length of crest in cumec/m, and

H_L— height of drop in m.

9. The cistern bed is 'bowed', i.e., deepened in the centre of the cross-section by $\frac{1}{4}$th of hydraulic mean depth.

10. At the end of cistern, a deflector having a height equal to 1/10th of the water depth in the cistern is provided across the bed and a deflector twice that height on the sides.

11. Two rows of blocks are provided, one at the toe of the glacis and the other at $\frac{2}{3}$rd of cistern length from the upstream end of the cistern. The height of the blocks is $0.26d_x$. The spacing of the glacis blocks is equal to twice the height and that of the cistern blocks is equal to the height.

B DISTRIBUTARY HEAD REGULATORS AND CROSS REGULATORS

19.13 OFFTAKE ALIGNMENT

When a distributing channel (or distributary) takes off from another channel usually termed as parent channel, the alignment of the two channels at the offtake (or offtake alignment) should be carefully designed. The best alignment of offtake is when the offtaking channel makes zero angle with the parent channel initially and then separates out along transition curves as shown in Fig. 19.17 (a). The transition curve should be given to the offtaking channel as well as parent channel to avoid accumulation of silt jetty and ensure equitable distribution of silt. If the transition curves are not given, both the parent and the offtaking channels may make an angle with the parent channel upstream of offtake as shown in Fig. 19.17 (b). In case it is essential to keep the parent channel straight both upstream and downstream of the offtake, the edge of the channel rather than the centre line should be considered in deciding the angle of offtake as shown in Fig. 19.17 (c). However, in this case the section should not be narrowed down equally on both sides. Further an unbalanced offtake results in the formation of a silt jetty as shown in Fig. 19.17 (d). The reduction in sectional area caused by the silt jetty would result in scouring of the bed along the line of the deviated current.

19.14 CROSS REGULATORS AND DISTRIBUTARY HEAD REGULATORS

Cross regulator and distributary head regulator are provided to control the supplies passing down the parent channel and the offtaking channel respectively (Fig. 19.17). A cross regulator is provided on the parent channel at the downstream of the offtake to head up the water level and to enable the offtaking channel to draw the required supply. A distributary head regulator is provided at the head of the offtaking channel (or distributary) to control the supplies entering the offtaking channel. Figure 19.18 (Plate X) shows cross regulator and distributary head regulator.

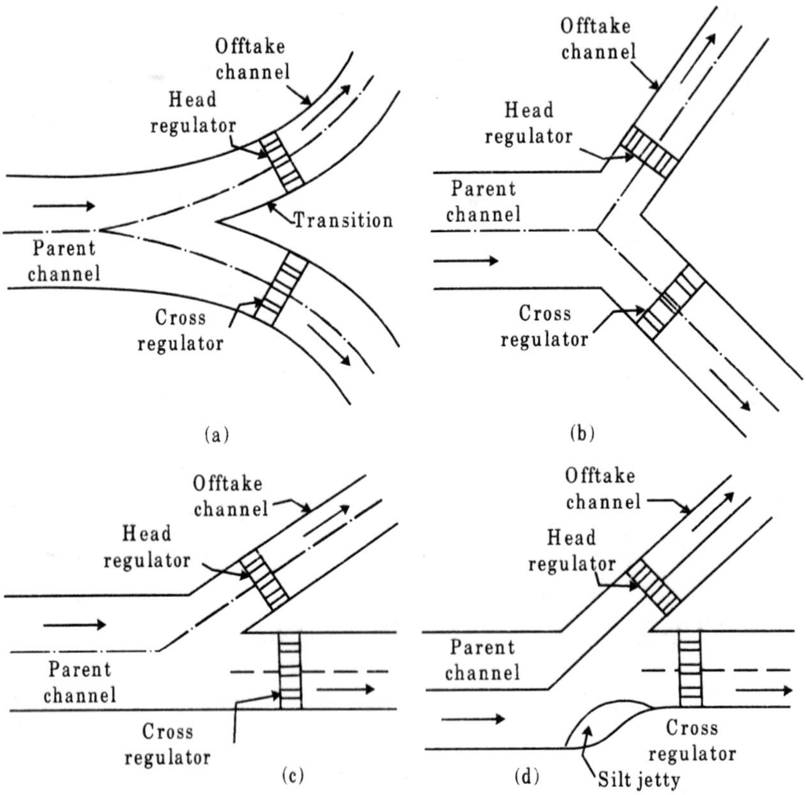

Fig. 19.17. *Offtake alignment*

Functions of Cross Regulators

1. Cross regulator enable effective regulation of the entire canal system.
2. When the water level in the parent channel is low, they help to raise the water level and feed the offtaking channels to their full demand in rotation.
3. They help in closing the supply to the downstream of the parent channel for the purposes of repairs and construction works.
4. In conjunction with escapes they help water to escape from the channels.
5. They facilitate communication, since a road can be taken over them with a little extra cost.
6. They help to absorb fluctuations in the various sections of the canal system, and thus to prevent possibilities of breaches in the tail reaches.
7. They help to control discharge at an outfall of canal into another canal or lake.
8. In conjunction with falls they help to control water surface slope for bringing the canals to regime slope and section.

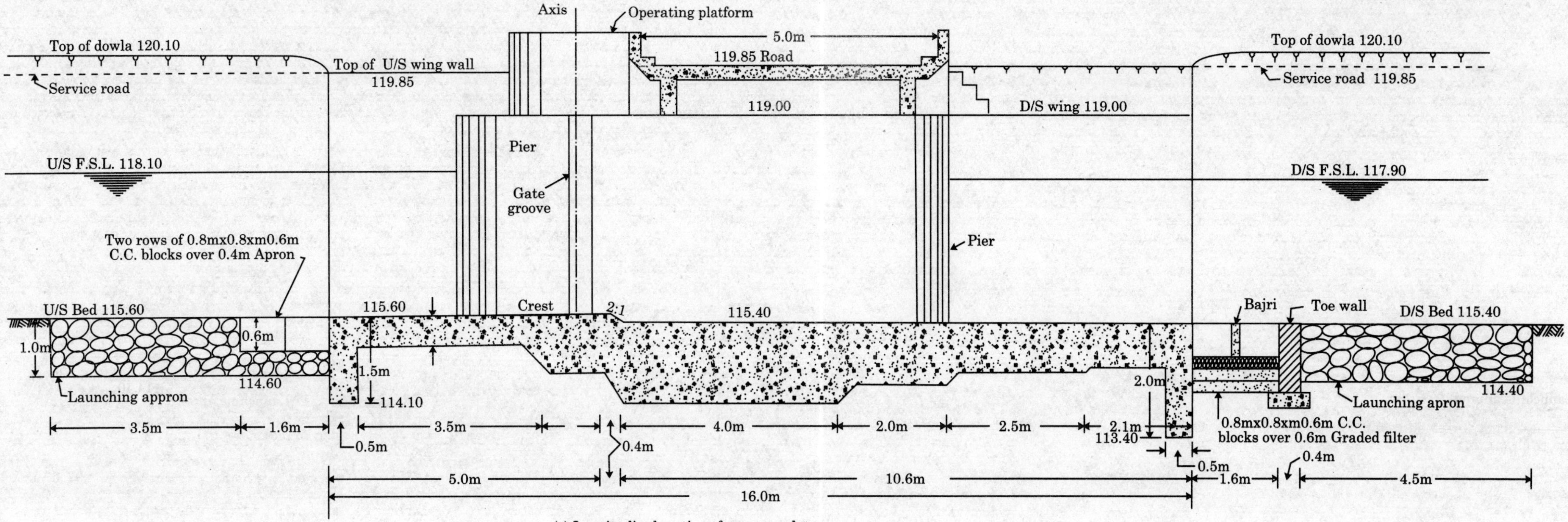

(a) Longitudinal section of cross regulator

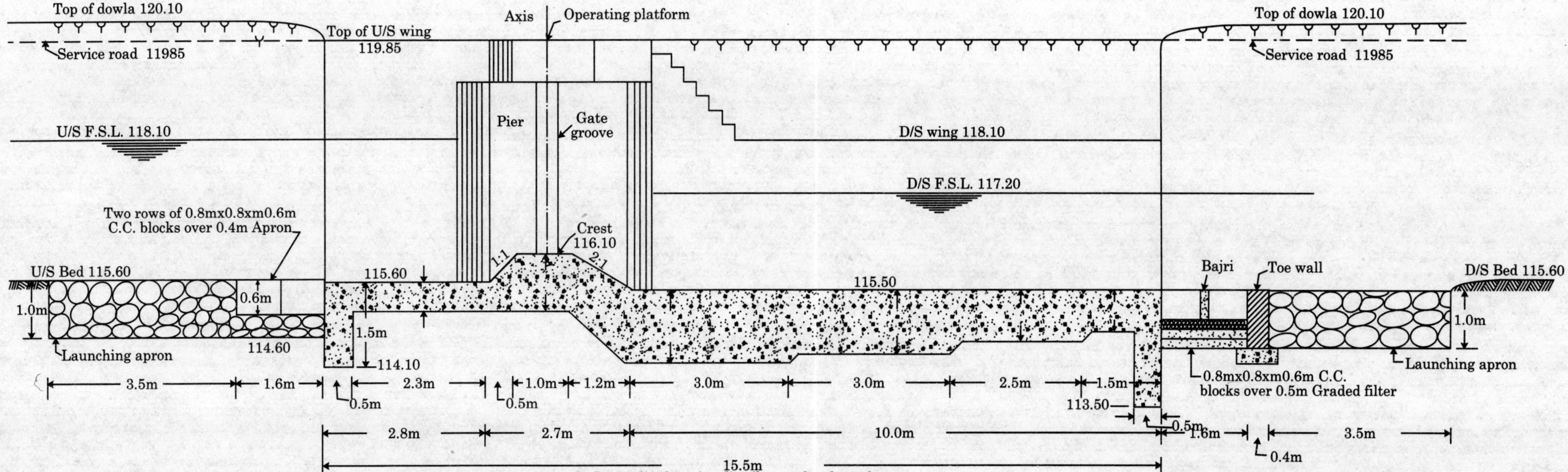

(b) Logitudinal section distributary head regulator

All reduced levels (R.L.) are in metres

Fig. 19.18 Plate X cross regulator and distributary head regulator.

Functions of Distributary Head Regulators

1. They regulate or control the supplies to the offtaking channel from the parent channel.
2. They control the entry of silt in the offtaking channel.
3. They serve as a meter for measurement of discharge entering the offtaking channel.
4. They help in shutting off the supplies when not needed in the offtaking channel or when the offtaking channel is required to be closed for repairs.

19.15 DESIGN OF CROSS REGULATOR AND DISTRIBUTARY HEAD REGULATOR

A. Design of Crest and Waterway

Crest level. The crest of a cross regulator is generally provided at the upstream bed level of the channel. The crest level of distributary head regulator is generally kept 0.3 m to 1.0 m higher than the crest level of the cross regulator.

Length of waterway. The length of waterway is obtained by using the discharge formula for a drowned weir which is as follows

$$Q = \frac{2}{3}C_{d_1}\sqrt{2g}\left[(h+h_a)^{3/2} - h_a^{2/3}\right] + C_{d_2}Ld\sqrt{2g\,(h+h_a)} \quad \text{...(19.32)}$$

where Q— discharge in cumec

L— length of clear waterway in m

h— difference in water levels on u/s and d/s of the crest in m

h_a— head due to velocity of approach

d— depth of d/s water level in the channel above the crest in m

g— acceleration due to gravity in m/s^2

C_{d_1} — coefficient of discharge for freely discharging portion = 0.557, and

C_{d_2} — coefficient of discharge for submerged portion = 0.80.

Generally the head due to velocity of approach h_a being small is neglected.

B. Design of Impervious Floor

Level and length of downstream floor. The level and length of the downstream floor is determined for the following two flowing conditions :

(i) Full supply discharge is passing through both the cross regulator and distributary head regulator with all gates fully open.

(ii) The discharge in the parent channel is insufficient but the offtaking channel is running full, as F.S.L. is maintained by the partial opening of the gates of the cross regulator.

Downstream floor level. For both these conditions, the discharge intensity q and the head loss $H_L (= h)$ are known. The corresponding value of E_{f_2} (i.e., height of d/s T.E.L. above the d/s floor) is thus obtained from Blench curves [Fig. 13.19 (Plate III)]. The downstream floor level is then given by the following relation :

$$\text{d/s floor level} = \text{d/s T.E.L} - E_{f_2}$$

$$\simeq \text{d/s F.S.L.} - E_{f_2}$$

Generally the first condition of flow would govern the design but in some cases, the second condition of flow is more critical. This is so because although the discharge intensity q is small in the second case but H_L is high. Further even if the downstream floor level obtained for the worst condition from the above relation is higher than the downstream bed level of the channel, the floor is provided at the downstream bed level and it should never be provided higher than the downstream bed level.

Length of downstream floor. If E_{f_1} is height of upstream T.E.L. above the downstream floor, then

$$E_{f_1} = D_2 + H_L \qquad \text{...(19.33)}$$

The depth D_1 and D_2 corresponding to E_{f_1} and E_{f_2} respectively are found from specific energy curves [Fig. 13.20 (Plate IV)] for different flow conditions.

$\therefore$ Length of d/s floor = $5\,(D_2 - D_1)$

However, the length of the downstream floor should at least be equal to $\frac{2}{3}$rd of the total length of the impervious floor.

Cutoff

Upstream cutoff. The minimum depth of upstream cutoff below the stream floor level is given by

$$d_1 = \frac{1}{3} u/s \text{ water depth} + 0.6 \text{ m}$$

Downstream cutoff. The minimum depth of downstream cutoff below the downstream floor level is given by

$$d_2 = \frac{1}{2} d/s \text{ water depth} + 0.6 \text{ m}$$

Total Length of Impervious Floor

Having fixed the depth of the downstream cutoff (d_2), the total length of impervious floor (b) is calculated with the consideration of safe exit gradient (G_E). The regulator floor is subjected to maximum static head when the channel is closed and the full supply level is maintained in the upstream for feeding the offtaking channel. Thus,

Maximum static head

$$H_s = u/s \text{ F.S.L.} - \text{d/s floor level}$$

$$G_E = \frac{H_s}{d_2} \frac{1}{\pi\sqrt{\lambda}}$$

or
$$\frac{1}{\pi\sqrt{\lambda}} = \frac{G_E d_2}{H_s}$$

Knowing maximum static head H_s, downstream cutoff depth d_2, and safe exit gradient G_E, the value of $\left(\frac{1}{\pi}\sqrt{\lambda}\right)$ is determined. Then from Khosla's exit gradient curve [Fig. 13.12 c) (Plate II)] for this value of $\left(\frac{1}{\pi}\sqrt{\lambda}\right)$ the value of $\alpha = \left(\frac{b}{d_2}\right)$ is obtained. The total length of impervious floor b is given by

$$b = \alpha d_2$$

After providing the required length of impervious floor on the downstream side the balance length of the impervious floor is provided on the upstream side. The crest and the downstream floor are joined by a glacis having a slope of 2 : 1. Further in the case of distributary head regulator the crest and the upstream floor are joined by a glacis having a slope of 1 : 1.

Thickness of Impervious Floor

The thickness of the impervious floor is found from the considerations of uplift pressure. However, a minimum thickness of 0.3 to 0.5 m is provided from the practical considerations.

C. Design of Upstream and Downstream Protection Works

The upstream scour depth d_1 below the upstream bed level is taken equal to (1/3 u/s water depth + 0.6 m) and the downstream scour depth d_2 below the downstream bed level is taken equal to (1/2 d/s water depth + 0.6 m). The upstream and downstream protection works are designed corresponding to these scour depths as indicated below :

Upstream Protection Works

The upstream protection work consists of a block protection and a launching apron. The cubic content of block protection is kept equal to d_1 cubic metre per metre width of the regulator. The cubic content of the launching apron is kept equal to 2.25 d_1 per metre width of the regulator.

Downstream Protection Works

The downstream protection consists of an inverted filter and a launching apron. The cubic content of an inverted filter is kept equal to d_2 cubic metre per metre width of the regulator. The cubic content of launching apron is kept equal to 2.25 d_2 cubic metre per metre width of the regulator.

The various steps involved in the design of cross regulator and distributary head regulator are shown in Illustrative Example 19.3 and the design details are shown in Fig. 19.18 (Plate X).

D. Devices to Contol Silt Entry into The Offtaking Channel

In order to control entry of silt into the offtaking channel the distributary head regulator is provided with a raised crest. However, the raised crest is not enough by itself to control the silt entry because if there is turbulence at the head, the top water may carry, as much, or even more, silt as the bottom water. Moreover, the silt would heap up against the crest forming a ramp thus making it still more convenient for the silt particles in the bottom layers to climb over into the offtaking channel. As such to reduce silt entry certain devices are used at the head of the offtaking channel. Some of these devices are described below.

E. King's Vanes

As shown in Fig. 19.19 (a) the vanes are vertical thin walls (or diaphragm walls) parallel to each other and curved at a radius of 7.5 m or more. These are of low height and are provided in the parent channel at the head of the offtaking channel such that they deflect the bottom silt laden water away from the offtaking channel at an angle of 30° from the direction of flow and thus help to prevent silt from entering the offtaking channel. The vanes may be made of 75 mm thick R.C.C. or steel plate. The height of the vanes is 1/4th of the depth of water in the parent channel and the spacing of the vanes is $1\frac{1}{2}$ times their height. The vanes should extend from 0.6 m to 1.5 m beyond a line drawn at an inclination of 2 : 1 to the axis of the parent channel from the downstream end of the offtake. They should be provided for a space equal to half the width of the offtaking channel for large channels and this proportion should be more for smaller channels and the minimum should be 0.6 m.

The bed of the parent channel covered by vanes and for a distance of 15 to 30 m upstream of the vanes should be pitched smoothly so that the suspended sediment may fall as low near the bed as possible. To minimise silting tendency, the pitched floor on which the vanes are built should be 0.15 m higher than the normal bed of the parent channel. The flooring should be sloped off at 1 : 10 for a distance of at least 5 m upstream.

The vanes are quite effective in controlling the silt entry into the offtaking channel. However, these vanes would be ineffective if the design of the head is so defective that violent turbulence is caused and silt is drawn out of the vanes.

F. Gibb's Groyne Wall

It is an extension of the downstream wing wall of the offtaking channel into the parent channel in a smooth curve as shown in Fig. 19.19 (b). The groyne wall is usually so provided that it divides the discharge of the parent channel in proportion of the discharge requirement of the offtaking channel and that of the downstream parent channel. The groyne wall is extended upstream into the parent channel to cover 3/4th to full width of the offtake. If the groyne wall divides the discharge proportionately between the offtaking channel and the downstream parent channel, the silt is also divided proportionately. Thus, excessive draw of silt into the offtaking channel is avoided. If it is desired to take in lesser percentage of silt into the offtaking channel than that in the parent channel, the groyne wall should take in more discharge than required and the surplus should be escaped through a hole in the groyne wall near its bottom.

G. Cantilever Skimming Platform

It consists of a slab cantilevered into the parent channel for some distance a little below the crest level as shown in Fig. 19.19 (c). The cantilevered slab cuts apart the top water and the bottom water and prevent their intermixing. It also makes it very difficult for the silt particles at the bottom to make a ramp to climb over into the offtaking channel.

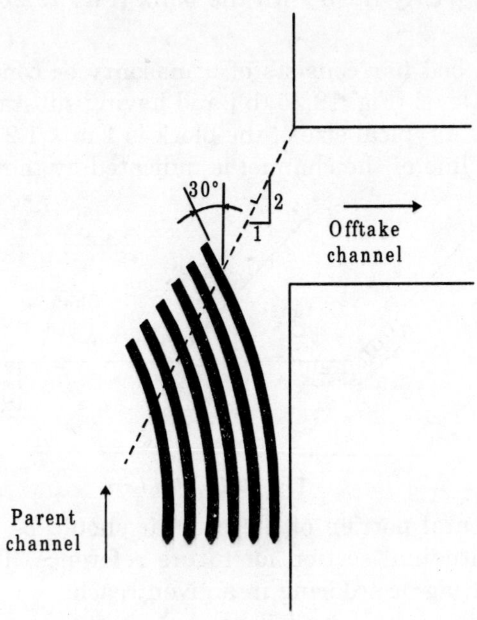

(a) King's vanes

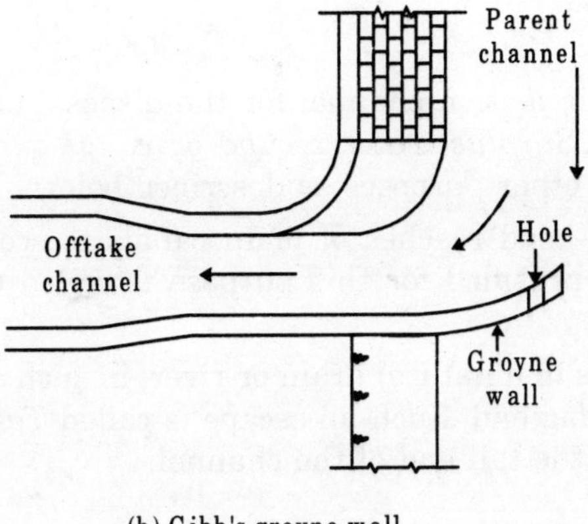

(b) Gibb's groyne wall

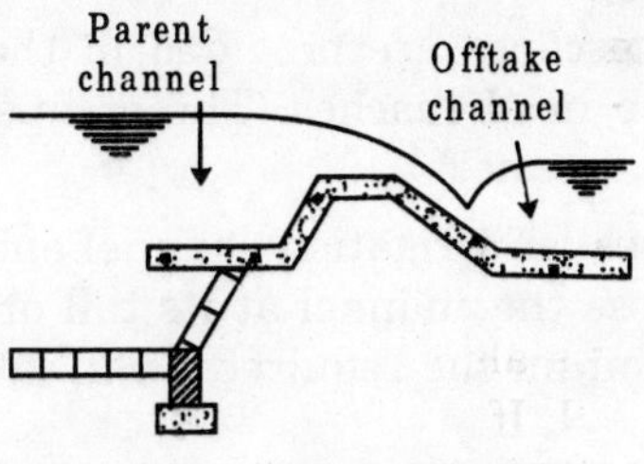

(c) cantilever skimming platform

Fig. 19.19 *Devices to control silt entry into the offtaking channel*

19.17 BED BARS

In the bed of an unlined channel bed bars are constructed at suitable intervals to serve as a permanent mark of reference indicating the correct alignment and theoretical bed level of the channel. The bed bars are constructed at intervals of 200 m in small channels and 500 m in large channels.

For small channels, a bed bar usually consists of a masonry wall partly extending into the bed and flush with it and partly flush with the bank [Fig. 19.20 (a)]. The foundation should be deep and substantial.

For large channels, a bed bar consists of a masonry or concrete block with its upper face flush with theoretical bed level [Fig. 19.20 (b)] and having substantial depth and cross-section to withstand action of water. A typical size of the block is 1 m × 1.2 m × 1.5 m deep. The blocks are so located that the centre line of the channel is indicated by the centre of the blocks.

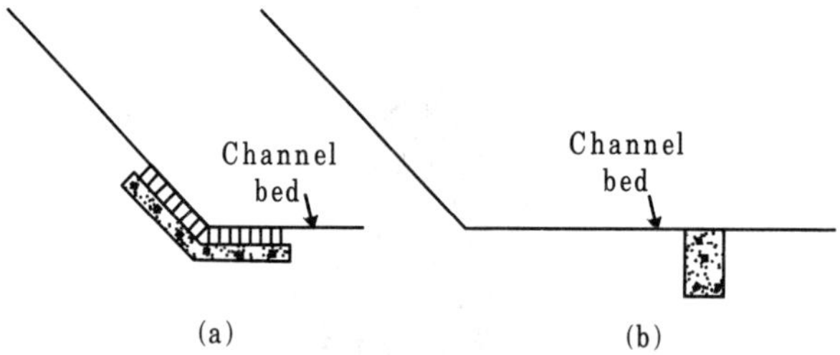

Fig. 19.20 *Bed bars*

The R.L. of the horizontal portion of the bed bar should be taken by accurate double level and recorded on the longitudinal section for future reference. The bed bars enable to indicate whether the channel is silting or scouring in a given reach.

C. ESCAPES

19.18 DEFINITION

An *escape* is a structure constructed on an irrigation channel for the disposal of surplus water from the channel. Hence it is also called *Surplus water escape* or *canal surplus escape* to distinguish it from the escapes provided for other purposes as described below.

Sometimes escapes are provided in the head reaches of main canals to scour out bed silt deposited in the head reaches. The escape provided for this purpose is called *Canal scouring escape.*

In some cases, an irrigation channel ends in a natural drain or river; in such cases an escape is provided across the channel at its tail or fag end. Such an escape is called *Tail escape* and is provided to maintain the required F.S.L. at the tail end of the channel.

19.19 NECESSITY OF SURPLUS WATER ESCAPE

Surplus water may result at any point in an irrigation channel in the following circumstances.

(i) Mistake or difficulty in regulation at the head of a channel.

(ii) Heavy rainfall in upper reaches of a channel.

(iii) Sudden closure of outlets by cultivators due to sudden cessation of demand.

(iv) Sudden closure of any offtaking channel due to breach.

If the surplus water is allowed to go to the lower reaches the water may overflow the banks and damage the same. Although the supplies may be reduced from the head of the channel, but the effect of such reduction would be felt only after a certain time depending on the distance of the affected reach from the head. As such immediate action is necessary to prevent damage and it is done by escapes. This is particularly so in case of a 'breach' or failure of channel bank somewhere. When a breach takes place the channel has to be closed from the head and as soon as intimation reaches the head, this is done. But the channel from the head to the breach site is full of water and if all that water were to rush out of the breach it will widen and deepen the breach, delay closing operations and cause widespread damage to land and property. If escapes of sufficient capacity exist on the upstream of the breach site all the water can be passed out through the escapes, the regulator downstream of the lowest escape closed and the channel dewatered quickly for repairing the breach. The surplus water escapes are therefore the safety valves of an irrigation channel system.

19.20 TYPES OF ESCAPE

The various types of escapes may be classified according to two different considerations as classified below.

Classification based on the purpose for which escape is provided

Depending on the purpose for which escape is provided, there are three types of escapes as noted below.

1. Surplus water escapes or canal surplus escapes.
2. Canal scouring escapes.
3. Tail escapes.

Classification Based on the Structural Design of Escape

On the basis of structural design escapes may be classified in the following two types.

1. Regulator (or Sluice) types escapes.
2. Weir type escapes.

Each of the different types of escapes noted above are described in the following sections.

A. Surplus Water Escapes or Canal Surplus Escapes

The surplus water escapes or Canal surplus escapes may be either Regulator type or Weir type escapes. These escapes are provided in the banks of the channel at intervals depending on the importance of the channel and the vicinity of a suitable natural drain or river for disposal of escaped water. The channel leading surplus water from escape to natural drain is called escape channel or outfall channel. The escape must be so located that the surplus water is led through the shortest possible escape channel to a natural drain which can safely take the maximum discharge of the escape without flooding. The discharging capacity of surplus water escape may be 1/3 to 1/2 of the capacity of the channel at the site of escape. Usually there should be a cross regulator across the channel on downstream side of the location of escape. As such surplus water escapes may be provided at suitable points in the bank of the channel on upstream side of every second or third cross regulator across the channel as per actual requirements.

B. Canal Scouring Escapes

The canal scouring escapes are Regulator type escapes. These escapes are also provided in the banks of the channel but are usually provided only in head reaches of main canals. The discharging capacity of the canal scouring escape should be about 1/2 to 2/3 of the capacity at the head of the main canal. From the escape, water with the scoured silt is led through the shortest possible escape channel to the nearest natural drain. In this case also for efficient working of the escape a cross regulator is provided across the channel just on the downstream side of the location of escape. However, nowadays instead of scouring escapes, silt ejectors are provided in the head reach of main canal as they are found to be more efficient for silt removal.

C. Tail Escapes

The tail escapes are weir type escapes. These escapes are provided across the channels at their tail ends in the case of irrigation channels which end in natural drains. The tail escapes are provided to maintain the required F.S.L. at the tail end of the channel.

1. Regulator (or Sluice) Type Escapes

A regulator (or sluice) type escape is like a small regulator which is provided in the channel bank and has gates supported on pier (Fig. 19.21). The flow through the escape is controlled by the gates which are operated from a raised platform.

As mentioned earlier the regulator type escapes are used as surplus water escapes as well as canal scouring escapes. When a regulator type escape is used as surplus water escape, its sill (or crest) is kept at the bed level of the channel and the bed of the escape channel is kept at or below the bed level of the channel. Also in this case the top of the gates is above the F.S.L. of the channel [Fig. 19.21 (a)]. The discharging capacity of this escape may be $\frac{1}{3}$ to $\frac{1}{2}$ of the capacity of the channel at the site of escape. Water from the escape is led through the escape channel to the nearest natural drain.

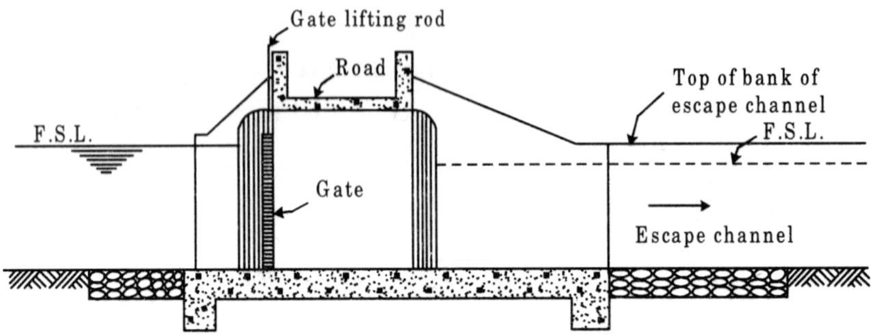

(a) Surplus water escape

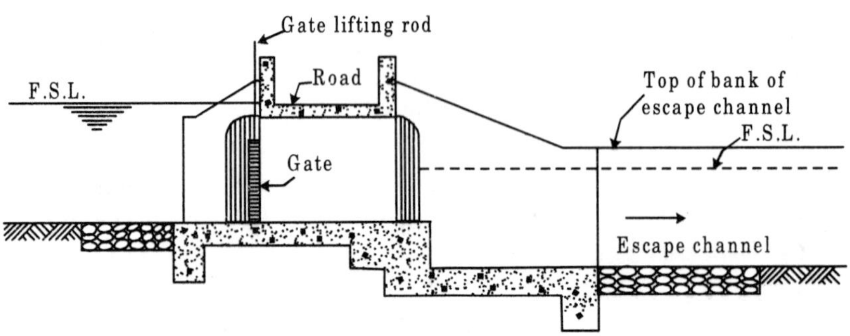

(b) Canal scouring escape

Fig. 19.21 *Regulator (or Sluice) type escapes*

When a regulator type escape is used as a canal scouring escape its sill (or crest) is usually kept about 0.3 m below the bed level of the channel. Also in this case the bed of the escape channel is kept below the bed level of the channel and is provided with such a slope that it is able to carry the silt scoured from the channel bed. Further in this case the top of the gates is usually below the F.S.L. of the channel [Fig. 19.21 (b)]. This escape being provided in the head reach of main canal, its discharging capacity should be about $\frac{1}{2}$ to $\frac{2}{3}$ of the capacity at the head of the main canal. The waterway provided for this escape should be such that high velocity of entry (of the order 4.5 to 6 m/s) is developed through the escape sluices or openings. When silt deposit is to be scoured out, water in excess of the full supply discharge is allowed in the channel from the head regulator and the gates of the escape are raised up to produce scouring velocity which scours the deposited silt. From the escape, water with scoured silt is led through the escape channel to the nearest natural drain.

2. Weir Type Escape

A weir type escape is like a weir with the crest of the weir wall at the full supply level (F.S.L.) of the channel at the site of escape (Fig. 19.22).

As mentioned earlier the weir type escapes are used as surplus water escapes as well as tail escapes. When a weir type escape is used as surplus water escapes the weir wall is provided in the bank of the channel with the crest of the weir wall being kept at the F.S.L. of the channel at the site of escape. As soon as the water level in the channel rises above the F.S.L. the water starts flowing over the crest of the escape and is led through the escape channel to the nearest natural drain.

When a weir type escape is used as a tail escape the weir wall is constructed across the channel at its tail end. The crest of the weir wall is kept at the F.S.L. of the channel at the tail.

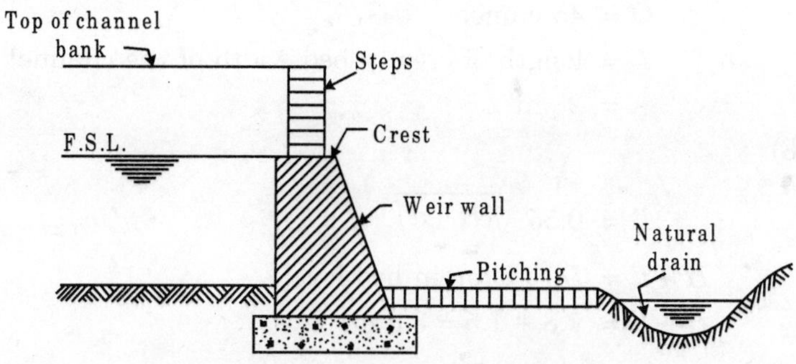

Fig. 19.22. *Weir type channel escape*

A few sluices are provided at the centre of the weir wall with their sills at the bed level of the channel near the tail. These sluices are useful in making the tail reach dry when required and they can also be used for scouring the silt deposit in tail reach of the channel.

SOLVED ILLUSTRATIVE EXAMPLES

Illustrative Example 19.1 *Design a Sarda type fall for the following data :*

(*i*) *Full supply discharge :* $\frac{u/s}{d/s}$ = 45 cumec

(ii) *Full supply level* : $\dfrac{u/s}{d/s} = \dfrac{118.30\ m}{116.80\ m}$

(iii) *Full supply depth* : $\dfrac{u/s}{d/s} = \dfrac{1.8\ m}{1.8\ m}$

(iv) *Bed width* : $\dfrac{u/s}{d/s} = \dfrac{28\ m}{28\ m}$

(v) *Bed level* : $\dfrac{u/s}{d/s} = \dfrac{116.50\ m}{115.00\ m}$

(vi) *Drop* : *1.5 m*

Design the floor on the basis of Bligh's theory taking coefficient of creep = 8. Check the design by Khosla's theory and modify the design if necessary. Safe exit gradient may be taken as 1/5.

Solution

Step 1. *Calculation of H and d*

Since the discharge is more than 14 cumec, trapezoidal crest will be provided for which discharge is given by Eq. (19.20) as

$$Q = 1.99\ LH^{3/2}\left(\frac{H}{B}\right)^{1/6} \quad \text{...(i)}$$

$$Q = 45 \text{ cumec}$$

$$L = \text{length of crest} = \text{bed width of the channel}$$

$$= 28 \text{ m}$$

From Eq. (19.18)

$$B = 0.55\sqrt{(H+d)}$$

But $$H + d = D_1 + \text{drop in bed level}$$

or $$H + d = 1.8 + 1.5 = 3.3 \text{ m} \quad \text{...(ii)}$$

∴ $$B = 0.55\sqrt{3.3} = 1.0 \text{ m} \quad \text{...(iii)}$$

Substituting the values in Eq. (i), we get

$$45 = \frac{1.99 \times 28 \times H^{3/2} \times H^{1/6}}{(1.0)^{1/6}}$$

or $$H^{5/3} = \frac{45 \times 1}{1.99 \times 28}$$

or $$H = (0.0808)^{3/5} = 0.88 \text{ m} \quad \text{...(iv)}$$

From Eq. (ii), we have

$$d = 3.3 - H$$

Substituting the value of H from Eq. (iv), we get

$$d = 3.3 - 0.88 = 2.42 \text{ m} \quad \text{...(v)}$$

Height of crest wall above the upstream bed is given by Eq. 19.29 as

$$h = D_1 - H$$

or

$$h = 1.8 - 0.88 = 0.92 \text{ m.}$$

Step 2. *Design of crest wall*

Top width of the crest wall as calculated in step 1 is

$$B = 1.0 \text{ m}$$

Slope of upstream face = 1 in 3

Slope of downstream face = 1 in 8

Assuming channel side slope 1 : 1, velocity of approach is given by

$$V_a = \frac{45}{(28+1.8)1.8}$$

$$V_a = 0.84 \text{ m/sec}$$

$$\text{Velocity head,} \quad \frac{V_a^2}{2g} = \frac{(9.84)^2}{2\times 9.81} = 0.036 \text{ m}$$

$$\therefore \quad \text{u/s T.E.L.} = \text{u/s F.S.L.} + \frac{V_a^2}{2g} = 118.30 + 0.036$$

$$= 118.336 \text{ m}$$

$$\text{Crest level} = \text{u/s F.S.L.} - H = 118.30 - 0.88$$

$$= 117.42 \text{ m}$$

$$\text{Value of } E = \text{u/s T.E.L.} - \text{Crest level}$$

$$= 118.336 - 117.42 = 0.916 \text{ m}$$

Step 3. *Design of cistern*

From Eq. (19.9) length of cistern

$$L_c = 5(E.H_L)^{1/2}$$

or

$$L_c = 5(0.916 \times 1.5)^{1/2} = 5.86 \text{ m}$$

Thus, provide cistern of length 6 m

From Eq. (19.10) depth of cistern

$$x = \frac{1}{4}(E \cdot H_L)^{2/3}$$

or

$$x = \frac{1}{4}(0.916 \times 1.5)^{2/3} = 0.31 \text{ m}$$

Thus, provide cistern of depth 0.31 m

R.L. of bed of cistern = R.L. of d/s bed $- x$ = 115.00 – 0.31

$= 114.69$ m

Step 4. *Design of impervious floor*

Seepage head $Hs = d = 2.42$ m

Bligh's coefficient $C = 8$

∴ Required creep length $= 8 \times 2.42 = 19.36$ m

Provide u /s cutoff of depth $d_1 = 1.0$ m

and d/s cutoff of depth $d_2 = 1.5$ m

Thus, vertical length of creep

$= 2\ (1.0 + 1.5) = 5.0$ m

∴ Length of horizontal impervious floor

$= 19.36 - 5.0 = 14.36$ m

Provide 15 m length of impervious floor.

Minimum length of impervious floor to be provided on the d/s of the crest wall

$$l_d = 2\ (D_1 + 1.2) + H_L = 2\ (1.8 + 1.2) + 1.5$$
$$= 7.5 \text{ m}$$

Provide $l_d = 8$ m. The balance of the length = 15 – 8 = 7 m is provided under and u/s of the crest wall.

Calculations for Uplift Pressure and Thickness

Total creep length $= 15 + 2\ (1.0 + 1.5) = 20$ m

(i) The uplift pressure under the u/s floor will be counter balanced by the weight of water and hence no thickness is required. However, provide a minimum thickness of 0.4 m.

(ii) For other points the minimum vertical ordinate between the Bligh's H.G. line and the floor level gives the uplift pressure.

Thus, maximum unbalanced head under the d/s toe of the crest wall

$$= 2.42 \left(1 - \frac{7 + 2 \times 1.0}{20}\right) + x$$
$$= 2.42\ (1 - 0.45) + 0.31$$
$$= 1.64 \text{ m}$$

∴ Thickness required $= \dfrac{1.64}{G-1} = \dfrac{1.64}{2.24-1}$ (Since for cement concrete $G = 2.24$)

$= 1.32$ m

Provide 1.4 m thick cement concrete floor over laid by 0.2 m brick pitching.

(**Note** : In working out the thickness the expression has not been multiplied by the safety factor 4/3 since the additional 0.2 m brick pitching will provide the additional safety).

(iii) Provide a minimum thickness of 0.6 m overlaid by 0.2 m thick brick pitching at the *d/s* end of the floor.

(iv) The thickness of the floor at intermediate points may be varied as per requirements of uplift pressure.

Step 5. *Checking of floor thickness by Khosla's theory*

Exit gradient $$G_E = \frac{1}{\pi\sqrt{\lambda}} \frac{H_s}{d_2}$$

$$\frac{1}{5} = \frac{1}{\pi\sqrt{\lambda}} \frac{2.42}{1.5}$$

or $$\frac{1}{\pi\sqrt{\lambda}} = \frac{1}{5} \times \frac{1.5}{2.42} = 0.124$$

From Khosla's exit gradient curve [Fig. 13.12 (c) (Plate II)] for

$$\frac{1}{\pi\sqrt{\lambda}} = 0.124$$

$$\alpha = \frac{b}{d^2} = 12.14$$

$\therefore$ $$b = 12.14 \times 1.5 = 18.21 \text{ m}$$

Length of floor actually provided by Bligh's theory = 15 m which is less than required. Hence either the length of the floor is to be increased by about 3 m or the depth of the d/s cutoff is to be increased.

Let the depth of the downstream cutoff be increased to 1.8 m and the floor length be kept same = 15 m as before.

$\therefore$ $$\alpha = \frac{b}{d^2} = \frac{15}{1.8} = 8.33$$

From exit gradient curve for $\alpha = 8.33$, we have

$$\frac{1}{\pi\sqrt{\lambda}} = 0.147$$

$\therefore$ $$G_E = \frac{1}{\pi\sqrt{\lambda}} \frac{H_s}{d_2} = \frac{0.147 \times 2.42}{1.8} = \frac{1}{5.06}$$

which is less than permissible exit gradient = 1/5 and hence it is safe.

Calculation of uplift pressure

(i) U/S Cutoff

$$d_1 = 1.0 \text{ m}$$
$$b = 15 \text{ m}$$

$\therefore$ $$= \frac{1}{\alpha} = \frac{d_1}{b} = \frac{1}{15} = 0.0667$$

From Khosla's curves Fig. 13.12 (b) (Plate I) for $(1/\alpha)$

$$= 0.0667$$

$$\phi_{C_1} = 100 - 24 = 76\%$$

and $$\phi_{D_1} = 100 - 17 = 83\%$$

Assuming floor thickness = 0.4 m, correction for floor thickness for ϕ_{C_1}

$$= \frac{83-76}{1} \times 0.4 = 2.8\ \%\ (+)$$

Correction for interference of d/s pile for ϕ_{C_1}

$$= 19\sqrt{\frac{D}{b'}}\frac{d+D}{b}$$

$$b = b' = 15 \text{ m} ;\ D = 1.8 \text{ m};\ d = d_1 = 1.0 \text{ m}$$

∴ Correction $$= 19\sqrt{\frac{1.8}{15}}\left(\frac{1+1.8}{15}\right) = 1.2\%\ (+)$$

∴ Corrected $$\phi_{C_1} = 76 + 2.8 + 1.2 = 80\%$$

(*ii*) *d/s Cutoff* $$d_2 = 1.8 \text{ m} ;\ b = 15 \text{ m}$$

∴ $$\frac{1}{\alpha} = \frac{d_2}{b} = \frac{1.8}{15} = 0.12$$

From Khosla's curves Fig. 13.12 (b) (Plate I) for $1/\alpha = 0.12$

$$\phi_{E_2} = 31\% \text{ and } \phi_{D_2} = 21.5\%$$

Assuming floor thickness = 0.6 m, correction for floor thickness for ϕ_{E_2}

$$= \frac{31-21.5}{1.8} \times 0.6 = 3.2\%\ (-)$$

Correction for interference of u/s pile for ϕ_{E_2}

$$= 19\sqrt{\frac{D}{b'}}\frac{d+D}{b}$$

$$b = b' = 15 \text{ m} ;\ D = 1 \text{ m} ;\ d = d_2 = 1.8 \text{ m}$$

∴ Correction $$= 19\sqrt{\frac{1}{15}}\left(\frac{1.8+1}{15}\right) = 0.9\%\ (-)$$

∴ Corrected $$\phi_{E_2} = 31 - 3.2 - 0.9 = 26.9\%$$

Thus, the pressure on the floor varies linearly from 80% at the u/s to 26.9% at the d/s.

Calculation of floor thickness

(i) Provide a minimum thickness of 0.4 m for the u/s floor.

(ii) Thickness of the floor at the d/s toe of the crest wall is determined as follows :

$$\% \text{ pressure} = \left(26.9 + \frac{80 - 26.9}{15} \times 8\right) = 55.2\%$$

$$\therefore \quad \text{Maximum static head} = 0.552\, H_s + x = 0.552 \times 2.42 + 0.31 = 1.65 \text{ m}$$

$$\therefore \text{ Floor thickness required} = \frac{1.65}{G-1} = \frac{1.65}{2.24-1} = 1.33 \text{ m}$$

Floor thickness required by Bligh's theory is 1.4 m.

∴ Provide 1.4 m thick cement concrete floor overlaid by 0.2 m thick brick pitching for a length of 3 m.

(iii) At 3 m from the d/s toe of crest

$$\% \text{ pressure} = \left(26.9 + \frac{80 - 26.9}{15} \times 5\right) = 44.6\%$$

$$\therefore \text{ Maximum static head} = 0.446\, H_s + x = 0.446 \times 2.42 + 0.31 = 1.39 \text{ m}$$

$$\therefore \quad \text{Floor thickness required} = \frac{1.39}{G-1} = \frac{1.39}{2.24-1} = 1.12 \text{ m}$$

Hence provide 1.2 m thick cement concrete floor overlaid by 0.2 m thick brick pitching for a length of 3 m (i.e., upto the end of the cistern).

(iv) At 6 m from the d/s toe of crest

$$\% \text{ pressure} = \left(26.9 + \frac{80 - 26.9}{15} \times 2\right) = 34\%$$

$$\therefore \text{ Maximum static head} = 0.34\, H_s = 0.34 \times 2.42 = 0.82 \text{ m}$$

(**Note.** The value of x has not been added here because the cistern ends at 6 m)

$$\therefore \text{ Floor thickness required} = \frac{0.82}{G-1} = \frac{0.82}{2.24-1} = 0.66 \text{ m}$$

Hence provide 0.7 m thick cement concrete floor overlaid by 0.2 m thick brick pitching for the last 2 m length of the floor.

Summary of Floor Thickness

(i) *u/s* floor : 0.4 m

(ii) *d/s* toe of crest to 3 m : 1.4 m

(iii) 3 m to 6 m : 1.2 m

(iv) Last 2 m : 0.7 m

Step 6. *Design of d/s wings*

Provide d/s wings vertical for a length of

$$6\sqrt{E \cdot H_L} = \sqrt{0.916 \times 1.5} = 7.03 \text{ m} \simeq 7 \text{ m}$$

Then the wings should be warped to 1 : 1 slope at a splay of 1 in 3.

Height of top of d/s wings above bed

= water depth + free board = (1.8 + 0.5)

= 2.3 m

∴ Horizontal projection of this on 1 : 1 slope = 2.3 m. With a splay of 1 in 3, length of warped wings measured along the centre line of the channel

= 2.3 × 3 = 6.9 ≃ 7 m

The sections of the wing wall have been shown in Fig. 19.11 (Plate VIII)

Step 7. *D/S bed protection*

(*a*) *Bed pitching.* Provide about 200 mm thick dry brick pitching consisting of one brick on edge laid over one flat brick resting on 100 mm ballast. From Table 19.1, for $H = 0.88$ m, length of bed pitching = $(9.0 + 2H_L) = (9.0 + 2 \times 1.5) = 12$ m. This should be provided horizontal upto the end of masonry wings and then sloping downwards at 1 in 10. Since the warped wings commence from 1 m u/s of the d/s end of the impervious floor, the length of the horizontal pitching = (7 – 1) = 6 m. The length of the sloping pitching is therefore = (12 – 6) = 6 m.

(b) *Curtain wall*

Thickness of curtain wall $= 1\frac{1}{2}$ brick = 0.4 m (app.)

Depth of curtain wall (from Table 19.1)

= 0.75 m say 1 m

Thus, provide 0.4 m thick and 1 m deep curtain wall at the *d/s* end of the bed pitching.

(c) *Side pitching.* Provide about 200 mm thick side pitching consisting of one brick on edge laid over one flat brick in cement mortar. The side pitching should be warped from a slope of 1 : 1 to $1\frac{1}{2}$: 1 and it should be curtailed at an angle of 45° from the end of the bed pitching in plan.

(*d*) *Toe wall*

Thickness of toe wall $= 1\frac{1}{2}$ brick = 0.4 m (approx.)

$$\text{Depth of toe wall} = \frac{d/s \text{ water depth}}{2}$$

$$= \frac{1.8}{2} = 0.9 \text{ m say 1 m}$$

Thus, provide 0.4 m thick and 1 m deep toe wall between the bed pitching and the side pitching.

(e) *Energy dissipators.* The size and position of the friction blocks and cube blocks are as indicated below.

A. *Size and position of friction blocks*

Critical depth $$d_c = \left(\frac{q^2}{g}\right)^{1/3}$$

$$q = \frac{45}{28} = 1.61 \text{ cumec/m}$$

$$\therefore \quad d_c = \left[\frac{(1.61)^2}{9.81}\right]^{1/3} = 0.64 \text{ m say } 0.7 \text{ m}$$

Thus,

(i) length of block = 2 × 0.7 = 1.4 m

(ii) width of block = 0.7 m

(iii) height of block = 0.7 m

(iv) distance of first row of blocks from d/s toe of the crest wall

= 1.5 × 0.7 = 1.05 say 1 m

(v) spacing between the two rows of blocks

= 0.7 m

(vi) spacing between blocks in the same row

= 2 × 0.7 = 1.4 m

∴ Provide two rows of staggered friction blocks of size 1.4 m × 0.7 m × 0.7 m in the cistern at a distance of 1 m from the d/s toe of the crest wall with a spacing of 0.7 m.

B. *Size and position of cube blocks*

D/S water depth = 1.8 m

Thus,

(i) length of block = $\frac{1}{10} \times 1.8 = 0.18$ m say 0.2 m

(ii) width of block = $\frac{1}{10} \times 1.8 = 0.18$ m say 0.2 m

(iii) height of block = $\frac{1}{10} \times 1.8 = 0.18$ m say 0.2 m

(iv) spacing between two rows of blocks

$$= \frac{1}{10} \times 1.8 = 0.18 \text{ m say } 0.2 \text{ m}$$

(v) spacing between blocks in the same row

$$= \frac{1}{10} \times 1.8 = 0.18 \text{ m say } 0.2 \text{ m}$$

∴ Provide two rows of staggered cube blocks of size 0.2 m × 0.2 m × 0.2 m on the impervious floor at its d/s end with a spacing of 0.2 m.

Step 8. *Design of u/s approach*

Radius of segmental (or curved) portion of the u/s wings

$$= 5 \text{ to } 6 \text{ times } H$$
$$= 5 \text{ to } 6 \text{ times } 0.88$$
$$= 4.4 \text{ to } 5.28 \text{ m}$$

Thus, provide u/s wings having segmental (or curved) portion of radius 5 m and subtending an angle of 60° at the centre from the u/s edge of the crest wall. The wings should then be carried along straight lines tangential to the segment and embedded in the earthen banks of the channel by a minimum of 1 m from the line of F.S.L.

The complete details of the fall are shown in Fig. 19.11 (Plate VIII).

Illustrative Example 19.2 *Design an unflumed straight glacis non-meter fall for the data of Illustrative Example 19.1 noted below :*

(i) Full supply discharge : $\frac{u/s}{d/s} = 45$ *cumec*

(ii) Full supply level : $\frac{u/s}{d/s} = \frac{118.30 \text{ m}}{116.80 \text{ m}}$

(iii) Full supply depth : $\frac{u/s}{d/s} = \frac{1.8 \text{ m}}{1.8 \text{ m}}$

(iv) Bed width : $\frac{u/s}{d/s} = \frac{28 \text{ m}}{28 \text{ m}}$

(v) Bed level : $\frac{u/s}{d/s} = \frac{116.50 \text{ m}}{115.00 \text{ m}}$

(vi) Drop : *1.5 m*

(vii) Permissible exit gradient : *1/5*

Solution

Step 1. Design of crest

From Eq. (19.25), we have

$$Q = 1.84\, L_t E^{3/2} \qquad \ldots(i)$$

$$Q = 45 \text{ cumec}$$

$$L_t = \text{crest length}$$
$$= \text{bed width of the channel}$$

$$= 28 \text{ m}$$

Substituting these values in Eq. (i), we get

$$45 = 1.84 \times 28 \times E^{3/2}$$

or $$E = \left(\frac{45}{1.84 \times 28}\right)^{2/3}$$

or $$E = 0.914 \text{ m}$$

Assuming side slopes of the channel as 1 : 1, velocity of approach is given by

$$V_a = \frac{45}{(28+1.8)1.8}$$

or $$V_a = 0.839 \text{ m/s}$$

Velocity head $$\frac{V_a^2}{2g} = \frac{(0.839)^2}{2 \times 9.81} = 0.036$$

$\therefore$ $$\text{u/s T.E.L.} = \text{u/s F.S.L.} + \frac{V_a^2}{2g}$$

$$= 118.30 + 0.036 = 118.336 \text{ m}$$

Crest level $$= \text{u/s T.E.L.} - E = 118.336 - 0.914$$

$$= 117.422 \simeq 117.42 \text{ m}$$

Width of the crest $$= \frac{2}{3} E = \frac{2}{3} \times 0.914 = 0.61 \text{ m (say 0.65 m)}$$

Provide u/s glacis with a slope of 12 : 1, joined tangentially to the crest with a radius

$$= \frac{E}{2} = \frac{1}{2} \times 0.914 \simeq 0.5 \text{ m.}$$

Provide d/s glacis with a slope of 2 : 1, joined tangentially to the cistern bed with a radius

$$= E = 0.914 \text{ m } 1.0 \text{ m.}$$

Step 2. *Design of cistern*

$$q = \frac{Q}{L_t} = \frac{45}{28} = 1.61 \text{ cumec/m}$$

$$H_L = 1.5 \text{ m}$$

Thus, from Blench curves Fig. 13.19 (Plate III) corresponding to the value of $q = 1.61$ cumec/m and $H_L = 1.5$ m.

$$E_{f_2} = 1.5 \text{ m}$$

$$\text{d/s T.E.L.} = \text{d/s F.S.L.} + \frac{V_a^2}{2g}$$

$= 116.80 + 0.036 = 116.836$ m

Cistern bed level $= $ d/s T.E.L. $- 1.25\ E_{f_2}$

$= 116.836 - (1.25 \times 1.5) = 114.96$ m

Depth of cistern below the d/s bed of the channel

$= 115.00 - 114.96$

$= 0.04$ m

The minimum depth of cistern below the d/s bed of the channel is usually kept

$= 0.25\ E_{f_2}$

$= 0.25 \times 1.5 = 0.375$ m (say 0.4 m)

$\therefore$ Keep cistern bed level $= 115.00 - 0.4 = 114.60$ m

Length of cistern $=$ 5 to 6 times E_{f_2}

$=$ 5 to 6 times 1.5 $=$ 7.5 to 9.0 m

$\therefore$ Provide cistern of length 8.75 m and depth 0.4 m. The cistern shall be joined to the d/s bed of the channel in a slope of 5 : 1.

Step 3. *Design of impervious floor*

Minimum depth of u/s curtain wall or cutoff is given by

$$d_1 = \frac{D_1}{3} = \frac{1.8}{3} = 0.6 \text{ m}$$

$\therefore$ Provide 0.4 m wide and 1 m deep curtain wall at the u/s end of the floor.

Minimum depth of d/s curtain wall is given by

$$d_2 = \frac{D_2}{2} = \frac{1.8}{2} = 0.9 \text{ m}$$

$\therefore$ Provide 0.4 m wide and 1 m deep curtain wall at the d/s end of the floor. This curtain wall shall be projected above the d/s bed by

$$\frac{D_2}{10} = \frac{1.8}{10} = 0.18 \text{ m}$$

so as to act as a deflector

Exit gradient $$G_E = \frac{1}{\pi\sqrt{\lambda}} \frac{H_s}{d_2}$$

$H_s =$ maximum seepage head

$=$ crest level $-$ d/s bed level

$= 117.42 - 115.00 = 2.42$ m

$d_2 =$ depth of d/s curtain wall $= 1$ m

Permissible exit gradient $= \frac{1}{5}$

$$\therefore \quad \frac{1}{5} = \frac{1}{\pi\sqrt{\lambda}}\frac{2.42}{1}$$

or $$\frac{1}{\pi\sqrt{\lambda}} = \frac{1}{5\times 2.42} = 0.083$$

From Khosla's exit gradient curve Fig. 13.12 (c) (Plate II) for

$$\frac{1}{\pi\sqrt{\lambda}} = 0.083$$

$$\alpha = \frac{b}{d_2} = 28.4$$

$$\therefore \quad b = 28.4 \times 1 = 28.4 \text{ m}$$

This length is too excessive, hence increase the depth d_2 of the d/s curtain wall to 1.5 m

$$\therefore \quad \frac{1}{5} = \frac{1}{\pi\sqrt{\lambda}}\frac{2.42}{1.5}$$

or $$\frac{1}{\pi\sqrt{\lambda}} = \frac{1.5}{5\times 2.42} = 0.124$$

From Khosla's exit gradient curve for $\frac{1}{\pi\sqrt{\lambda}} = 0.124$

$$\alpha = \frac{b}{d_2} = 12.1$$

$$\therefore \quad b = 12.14 \times 1.5 = 18.21 \text{ m (say 18.5 m)}$$

This length of 18.5 m shall be provided as follows :

(i) Length of cistern = 8.75 m

(ii) Length of d/s glacis $= \frac{1}{2}(117.42 - 114.60) = 5.64$ m

(iii) Width of crest = 0.65 m

(iv) Length of u/s glacis $= \frac{1}{2}(117.42 - 116.50) = 0.46$ m

Total of (i) to (iv) = 15.5 m

Balance to be provided on the u/s side from the toe of the u/s glacis

= 18.5 – 15.5 = 3.0 m

Calculation of Uplift Pressure

(*i*) *u/s curtain wall*

$$d_1 = 1 \text{ m}$$
$$b = 18.5 \text{ m}$$

$$\therefore \quad \frac{1}{\alpha} = \frac{d_1}{b} = \frac{1}{18.5} = 0.054$$

From Khosla's curves Fig 13.12 (b) (Plate I) for (1/α) = 0.054

$$\phi_{C_1} = 100 - 22 = 78\%$$

and
$$\phi_{D_1} = 100 - 15 = 85\%$$

Assume a minimum thickness of 0.3 m for the floor on the u/s side.

∴ Correction for floor thickness for ϕ_{C_1}

$$= \frac{85-78}{1} \times 0.3 = 2.1\% \ (+)$$

Correction for interference of d/s curtain wall for ϕ_{C_1}

$$= 19\sqrt{\frac{D}{b'}}\frac{d+D}{b}$$

$$b = b' \ 18.5 \text{ m} \ ; D = 1.5 \text{ m} \ ; d = 1.0 \text{ m}$$

$$\therefore \text{ Correction} \quad = 19\sqrt{\frac{1.5}{18.5}}\left(\frac{1+1.5}{18.5}\right) = 0.7\% \ (+)$$

$$\therefore \text{ Corrected} \quad \phi_{C_1} = 78 + 2.1 + 0.7 = 80.8\%$$

(ii) *d/s curtain wall*

$$d_2 = 1.5 \text{ m}$$
$$b = 18.5 \text{ m}$$

$$\therefore \quad \frac{1}{\alpha} = \frac{d_2}{b} = \frac{1.5}{18.5} = 0.081$$

From Khosla's curves Fig. 13.12 (b) (Plate I) for (1/α) = 0.081

$$\phi_{E_2} = 26\%$$

and
$$\phi_{D_2} = 18\%$$

Assuming a minimum thickness of 0.6 m for the floor at the d/s end correction for floor thickness for ϕ_{E_2}

$$= \frac{26-18}{1.5} \times 0.6 = 32\%(-)$$

Correction for interference of u/s pile for ϕ_{E_2}

$$= 19\sqrt{\frac{D}{b'}}\frac{d+D}{b}$$

$$b = b' = 18.5 \text{ m} ; D = 1 \text{ m} ; d = 1.5 \text{ m}$$

$\therefore$ Correction $= 19\sqrt{\frac{1}{18.5}}\left(\frac{1.5+1}{18.5}\right) = 0.6\%(-)$

$\therefore$ Corrected $\phi_{E_2} = 27 - 3.2 - 0.6 = 23.2\ \%$

Thus, the pressure on the floor varies linearly from 80.8% at the u/s to 23.2% at the d/s.

Calculation of floor thickness

(i) Provide a minimum thickness of 0.3 m for the u/s floor.

(ii) Floor thickness at the toe of the d/s glacis :

$$\% \text{ pressure} = \left(23.2 + \frac{80.8-23.2}{18.6} \times 9\right) = 51\%$$

$\therefore$ Maximum static head $= 0.51\ H_s + (115.00 - 114.6)$

$= 0.51 \times 2.42 + 0.4 = 1.63$ m

$\therefore$ Floor thickness required $= \frac{1.63}{G-1} = \frac{1.63}{2.24-1} = 1.32$ m say 1.4 m

Thus, provide 1.4 m thick cement concrete floor for a length of 3 m from the toe of the d/s glacis

(iii) Floor thickness at 3 m from the toe of the d/s glacis :

$$\% \text{ pressure} = \left(23.2 + \frac{80.8-23.2}{18.5} \times 6\right) = 42\%$$

$\therefore$ Maximum static head $= 0.42\ H_s + (115.00 - 114.6)$

$= 0.42 \times 2.42 + 0.4 = 1.42$ m

$\therefore$ Floor thickness required $= \frac{1.42}{G-1} = \frac{1.42}{2.24-1} = 1.15$ m (say 1.2 m)

Thus, provide 1.2 m thick cement concrete floor from 3 m to 6 m from the toe of the d/s glacis.

(iv) Floor thickness at the d/s end :

% pressure at d/s curtain wall = 23.2 %

$\therefore$ Maximum static head $= 0.232 \times 2.42 = 0.56$ m

$$\therefore \quad \text{Floor thickness required} = \frac{0.56}{G-1} = \frac{0.56}{2.24-1} = 0.45 \text{ m}$$

However, a minimum thickness of 0.6 m is provided for the cement concrete floor at its d/s end.

The floor thickness at 6 m from the toe of the d/s glacis is 1.2 m, which is thus reduced linearly to 0.6 m at its d/s end as shown in Fig. 19.14 (Plate IX).

Step 4. *Design of d/s protection*

(i) *Bed protection.* No bed protection is needed since a deflector wall has been provided.

(ii) *Side protection.* Length of side protection

$$= 3D_2 = 3 \times 1.8 = 5.4 \text{ m}$$

Hence provide 200 mm thick brick pitching in 5.5 m length beyond the impervious floor at 1 : 1 slope. The pitching is supported on a toe wall 0.4 m wide and = 0.9 m deep. Provide a 0.4 m wide profile wall at the end of the pitching to protect the side pitching.

(iii) *Friction blocks.* No friction blocks are required since fluming has not been done.

Step 5. *Design of u/s approach*

Provide parallel wing walls from the crest wall upto the end of the impervious floor on the u/s side. From the u/s end of the floor the wing walls shall be splayed straight at an angle of 45° and continued inland till they are embedded upto 1 m into the earthen banks of the channel from the line of F.S.L.

The complete details of the fall are shown in Fig. 19.14 (Plate IX).

Illustrative Example 19.3 *Design a cross regulator and a head regulator for a distributary taking off from a branch canal for the following data :*

Discharge of branch channel = *105 cumec*

Discharge of distributary = *15 cumec*

F.S.L. of branch channel : $\dfrac{u/s}{d/s} = \dfrac{118.10 \text{ m}}{117.90 \text{ m}}$

Bed width of branch channel : $\dfrac{u/s}{d/s} = \dfrac{45 \text{ m}}{41 \text{ m}}$

Depth of water in branch channel : $\dfrac{u/s}{d/s} = \dfrac{2.5 \text{ m}}{2.5 \text{ m}}$

F.S.L. of distributary = *117.20 m*

Bed width of distributary = *15 m*

Depth of water in distributary = *1.6 m*

Permissible exit gradient = *(1/5)*

Solution

(a) **Design of Cross Regulator**

Step 1. *Design of crest and water way*

Crest level = u/s bed level of branch canal

= F.S.L. of branch canal – water depth

= 118.10 – 2.50 = 115.60 m

Neglecting the velocity of approach, the discharge is given by

$$Q = \frac{2}{3} C_{d_1} \sqrt{2g} h^{3/2} + C_{d_2} L d \sqrt{2gh}$$

$$Q = 105 \text{ cumec}$$

$$C_{d_1} = 0.577 \; ; \; C_{d_2} = 0.80$$

h = u/s F.S.L. – d/s F.S.L.

= 118.10 – 117.90 = 0.20 m

d = d/s F.S.L. – crest level

= 117.90 – 115.60 = 2.30 m

Substituting the values in the above expression, we get

$$105 = \frac{2}{3} \times 0.577 \times L \sqrt{2 \times 9.81} (0.20)^{3/2} + 0.8 \times L \times 2.30 \sqrt{2 \times 9.81 \times 0.20}$$

or $$L = 27.65 \text{ m}$$

Provide 4 bays each with a clear water way of 7 m so that total clear waterway = 28 m

Provide 3 piers each of 1.5 m width.

∴ Total length of waterway of cross regulator

= 28 + 4.5 = 32.5 m

Step 2. *Design of impervious floor*

Level and length of downstream floor

$$Q = 105 \text{ cumec}$$

$$q = \frac{105}{28} = 3.75 \text{ cumec/m}$$

H_L = h = u/s F.S.L. – d/s F.S.L.

= 118.10 – 117.90 = 0.20 m

From Blench curves [Fig. 13.19 (Plate III)] for q = 3.75 cumec/m and H_L = 0.2 m, we have

$$E_{f_2} = 1.95 \text{ m}$$

∴ d/s floor level = d/s F.S.L. – E_{f_2}

= 117.90 – 1.95 = 115.95 m

d/s bed level = d/s F.S.L. – d/s water depth

= 117.90 – 2.50 = 115.40 m

Since the d/s floor level obtained above is higher than the d/s bed level, d/s floor level is kept same as d/s bed level, i.e., 115.40 m

$$E_{f_1} = E_{f_2} + H_L = 1.95 + 0.20 = 2.15 \text{ m}$$

From specific energy curves [Fig. 13.20 (Plate IV)] values of D_1 and D_2 corresponding to

$$E_{f_2} = 2.15 \text{ m and } E_{f_2} = 1.95 \text{ m are :}$$

$$D_1 = 0.70 \text{ m}$$

$$D_2 = 1.70 \text{ m}$$

∴ Cistern length (i.e., length of d/s floor)

$$= 5\,(D_2 - D_1) = 5\,(1.70 - 0.70) = 5 \text{ m}$$

Total length of impervious floor as obtained in the next step

$$= 16 \text{ m}$$

$\frac{2}{3}$rd of length of impervious floor

$$= \frac{2}{3} \times 16 = 10.6 \text{ m}$$

Hence provide d/s floor of length

$$= 10.6 \text{ m}$$

Cutoffs

Depth of u/s cutoffs $\quad d_1 = \frac{1}{3}$ u/s water depth + 0.6 m

$$= \frac{1}{3} \times 2.5 + 0.6 = 1.43 \text{ m (say 1.5 m)}$$

Hence provide u/s cutoff of depth 1.5 m

Depth of d/s cutoff $\quad d_2 = \frac{1}{2}$ d/s water depth + 0.6 m

$$= \frac{1}{2} \times 2.5 + 0.6 = 1.85 \text{ m (say 2 m)}$$

Hence provide d/s cutoff of depth 2 m

Width of each of the cutoffs is kept equal to 0.5 m.

Total length of impervious floor

The floor will be subjected to maximum uplift pressure when full supply is maintained in the branch canal on the upstream of the cross regulator for feeding the distributary but no water is flowing down the cross regulator.

∴ Maximum static head $\quad H_s$ = u/s F.S.L. – d/s floor level

$$= 118.10 - (117.90 - 2.50) = 2.7 \text{ m}$$

Depth of d/s cutoff $\quad d_2 = 2.0$ m

$$G_E = \frac{H_s}{d_2}\frac{1}{\pi\sqrt{\lambda}}$$

or
$$\frac{1}{5} = \frac{2.7}{2}\frac{1}{\pi\sqrt{\lambda}}$$

or
$$\frac{1}{\pi\sqrt{\lambda}} = \frac{2}{5\times2.7} = 0.148$$

From Khosla's exit gradient curve [Fig. 13.12 (c) (Plate II)] for $\frac{1}{\pi\sqrt{\lambda}} = 0.148$.

$$\alpha = \frac{b}{d_2} = 8$$

$\therefore$
$$b = \alpha . d_2 = 8 \times 2 = 16 \text{ m}$$

$\frac{2}{3}$ rd of this length $= \frac{2}{3} \times 16 = 10.6$ m

which is more than 5 m obtained earlier. Hence the total floor length of 16 m shall be provided as follows:

(i) d/s floor length = 10.6 m

(ii) d/s glacis length with 2 : 1 slope

$$= 2\,(115.60 - 115.40) = 0.4 \text{ m}$$

(iii) Balance floor length including crest to be provided on u/s side = 5.0 m

Pressure calculations

(i) *Upstream cutoff*

$$d_1 = 1.5 \text{ m} ;\ b = 16 \text{ m}$$

$\therefore$
$$\frac{1}{\alpha} = \frac{d_1}{b} = \frac{1.5}{16} = 0.094$$

From Khosla's curves [Fig. 13.12 (b) (Plate I)] for $(1/\alpha) = 0.094$

$$\phi_{C_1} = 100 - 28 = 72\%$$

and
$$\phi_{D_1} = 100 - 19 = 81\%$$

Assuming u/s floor thickness near cutoff = 0.5 m (minimum thickness) correction for floor thickness for ϕ_{C_1}

$$= \frac{81-72}{1.5}\times 0.5 = 3\%\ (+)$$

Correction for interference of d/s pile for ϕ_{C_1}

$$= 19\sqrt{\frac{D'}{b}}\,\frac{d+D}{b}$$

$$b = b' = 16 \text{ m} ; D = 2.2 \text{ m} ; d = d_1 = 1.5 \text{ m}$$

$$\therefore \quad \text{Correction} = 19\sqrt{\frac{2.2}{16}}\left(\frac{1.5+2.2}{16}\right) = 1.6\% \; (+)$$

$$\therefore \quad \text{Corrected } \phi_{C_1} = 72 + 3 + 1.6 = 76.6\%$$

(ii) *Downstream cutoff*

$$d_2 = 2 \text{ m}; b = 16 \text{ m}$$

$$\therefore \quad \frac{1}{\alpha} = \frac{d_2}{b} = \frac{2}{16} = 0.125$$

From Khosla's curves [Fig. 13.12 (b) (Plate I)] for $(1/\alpha) = 0.125$

$$\phi_{E_2} = 31\%$$

and

$$\phi_{D_1} = 22\%$$

Assuming a minimum floor thickness of 0.6 m at the d/s end, correction for floor thickness for ϕ_{E_2}

$$= \frac{31-22}{2} \times 0.6 = 2.7\%(-)$$

Correction for interference of u/s pile for ϕ_{E_2}

$$= 19\sqrt{\frac{D'}{b}}\,\frac{d+D}{b}$$

$$b = b' = 16 \text{ m} ; D = (1.5 - 0.2) = 1.3 \text{ m} ;$$

$$d = d_2 = 2.0 \text{ m}$$

$$\therefore \quad \text{Correction} = 19\sqrt{\frac{1.3}{16}}\left(\frac{2.0+1.3}{16}\right) = 1.1\% \; (-)$$

$$\therefore \quad \text{Corrected } \phi_{E_2} = 31 - 2.7 - 1.1 = 27.2\%$$

Thus, the pressure on the floor varies linearly from 76.6% at the u/s to 27.2% at the d/s.

Calculation of floor thickness

(i) Provide a minimum thickness of 0.5 m for the u/s floor.

(ii) Thickness of floor at the toe of the glacis :

$$\%\text{ pressure} = 27.2 + \frac{76.6-27.2}{16} \times 10.6 = 60\%$$

$$\therefore \text{ Maximum static head} = 0.60 \times 2.7 = 1.62 \text{ m}$$

$$\therefore \text{ Floor thickness required} = \frac{1.62}{G-1} = \frac{1.62}{2.24-1} = 1.31 \text{ m say } 1.4 \text{ m}$$

Thus, provide 1.4 m thick cement concrete floor for a length of 4 m from the toe of the glacis.

(iii) Thickness of floor at 4 m from the toe of the glacis :

$$\% \text{ pressure} = 27.2 + \frac{76.6-27.2}{16} \times 6.6 = 48\%$$

$\therefore$ Maximum static head= 0.48 × 2.7 = 1.296 m

$$\therefore \text{ Floor thickness required} = \frac{1.296}{G-1} = \frac{1.296}{2.24-1} = 1.05 \text{ m say } 1.1 \text{ m}$$

Thus, provide 1.1 m thick cement concrete floor from 4 m to 6 m from the toe of the glacis.

(iv) Thickness of floor at 6 m from the toe of the glacis :

$$\% \text{ pressure} = 27.2 + \frac{76.6-27.2}{16} \times 4.6 = 41\%$$

$\therefore$ Maximum static head = 0.41 × 2.7 = 1.107 m

$$\therefore \text{ Floor thickness required} = \frac{0.107}{G-1} = \frac{0.107}{2.24-1} = 0.89 \text{ m (say 0.9 m)}$$

Thus provide 0.9 m thick cement concrete floor from 6 m to 8.5 m from the toe of the glacis.

(v) Thickness of floor at 8.5 m from the toe of the glacis :

$$\% \text{ pressure} = 27.2 + \frac{76.6-27.2'}{16} \times 2.1 = 34\%$$

$\therefore$ Maximum static head= 0.34 × 2.7 = 0.918

$$\therefore \text{ Floor thickness required} = \frac{0.918}{G-1} = \frac{0.918}{2.24-1} = 0.74 \text{ m (say 0.8)}$$

Thus, provide 0.8 m thick cement concrete floor for the last 2.1 m length.

Summary of floor thickness

(i) u/s floor : 0.5 m

(ii) Toe of glacis to 4 m : 1.4 m

(iii) 4 m to 6 m from toe of glacis : 1.1 m

(iv) 6 m to 8.5 from toe of glacis : 0.9 m

(v) Last 2.1 m : 0.8 m

(vi) Under the gates for a distance of 1 m floor thickness = 1.0 m

Step 3. *Design of upstream protection*

$$\text{Depth of upstream cutoff } d_1 = \frac{1}{3} \text{ u/s water depth} + 0.6 \text{ m}$$

$$= 13 \times 2.5 + 0.6 = 1.43 \text{ m}$$

(i) *Block protection*

Volume of block protection = d_1 cubic metre/m

$= 1.43$ cubic metre/m

Keep thickness of block protection

$= 1.0$ m

∴ Length of block protection = 1.43 m

Thus, provide 2 rows of 0.8 m × 0.8 m × 0.6 m thick cement concrete blocks over 0.4 m thick apron of loose stones in a length of 1.6 m on the upstream of the impervious floor.

(ii) *Launching apron*

Cubic content of launching apron

$= 2.25\ d_1$ cubic metre/m

$= 2.25 \times 1.43 = 3.22$ cubic metre/m

Keep thickness of launching apron

$= 1.0$ m

∴ Length of launching apron = 3.22 m

Thus, provide 1 m thick and 3.5 long launching apron on the upstream of the block protection.

Step 4. *Design of downstream protection*

Depth of downstream cutoff

$$d_2 = \frac{1}{2} \text{ d/s water depth} + 0.6 \text{ m}$$

$$= \frac{1}{2} \times 2.5 + 0.6 = 1.85 \text{ m}$$

(i) *Inverted filter*

Volume of inverted filter $= d_2$ cubic metre/m

$= 1.85$ cubic metre/m

Thickness of graded filter is kept as 0.6 m which is overlaid by 0.6 m thick cement concrete blocks, making a total thickness of 1.2 m

$$\therefore \text{Length of inverted filter} = \frac{1.85}{1.2} = 1.54 \text{ m}$$

Thus, provide 2 rows of 0.8 m × 0.8 m × 0.6 m thick cement concrete blocks over 0.6 m thick graded filter.

(ii) *Launching apron*

Cubic content of launching apron

$= 2.25\ d_2$ cubic metre/m

$= 2.25 \times 1.85 = 4.16$ cubic metre/m

Keep thickness of launching apron

$= 1.0$ m

∴ Length of launching apron = 4.16 m

Thus, provide 1 m thick launching apron for a length of 4.5 m.

Provide 0.4 m thick and 1.5 m deep masonry toe wall between the inverted filter and the launching apron.

The design details of the cross regulator are shown in Fig. 19.18 (Plate X).

(b) **Design of Distributary Head Regulator**

Step 1. *Design of crest and waterway*

$$\text{Crest level} = \text{u/s bed level} + 0.5 \text{ m}$$
$$= (118.10 - 2.50) + 0.5$$
$$= 116.10 \text{ m}$$

Neglecting the velocity of approach, the discharge is given by

$$Q = \frac{2}{3} C_{d_1} L \sqrt{2g} h^{3/2} + C_{d_2} L d \sqrt{2gh}$$

$$Q = 15 \text{ cumec}$$
$$C_{d_1} = 0.577 \,;\, C_{d_2} = 0.8$$
$$h = u/s \text{ F.S.L.} - d/s \text{ F.S.L.}$$
$$= 118.10 - 117.20 = 0.9 \text{ m}$$
$$d = d/s \text{ F.S.L.} - \text{Crest level}$$
$$= 117.20 - 116.10 = 1.1 \text{ m}$$

Substituting these values in the discharge formula

$$15 = \frac{2}{3} \times 0.577 \times L \sqrt{2 \times 9.8} (0.9)^{3/2} + 0.8 \times L \times 1.1 \sqrt{2 \times 9.81 \times 0.9}$$

or $$L = 2.91 \text{ m}$$

This is too less in comparison to 15 m bed width of the offtaking channel.

Hence provide two bays of 3.5 m each with 1 m thick pier in between.

∴ Total length of waterway of the regulator

$$= 8.0 \text{ m}$$

Step 2. *Design of impervious floor*

Level and length of downstream floor

$$Q = 15 \text{ cumec}$$

$$q = \frac{15}{7} = 2.14 \text{ cumec/m}$$

$$H_L = \text{u/s F.S.L.} - \text{d/s F.S.L.}$$
$$= 118.10 - 117.20 = 0.9 \text{ m}$$

From Blench curves [Fig. 13.19 (Plate III)] for $q = 2.14$ cumec/m and $H_L = 0.9$ m, we have

$$E_{f2} = 1.65 \text{ m}$$

∴ $$\text{d/s floor level} = \text{d/s F.S.L.} - E_{f2}$$
$$= 117.20 - 1.65 = 115.55 \text{ m}$$

$$\text{d/s bed level} = \text{d/s F.S.L.} - \text{d/s water depth}$$
$$= 117.20 - 1.60 = 115.60 \text{ m}$$

Since the required d/s floor level is lower than the d/s bed level, the d/s floor level is kept as 115.50 m

$$E_{f_1} = E_{f_2} + H_L = 1.65 + 0.9 = 2.55 \text{ m}$$

From specific energy curves [Fig. 13.20 (Plate IV)] values of D_1 and D_2 corresponding to E_{f_1} = 2.55 m and E_{f_2} = 1.65 m are :

$$D_1 = 0.32 \text{ m}$$
$$D_2 = 1.56 \text{ m}$$

∴ Cistern length (i.e., length of d/s floor)

$$= 5\,(D_2 - D_1) = 5\,(1.56 - 0.32) = 6.2 \text{ m}$$

Total length of impervious floor as obtained in the next step

$$= 15 \text{ m}$$

$\frac{2}{3}$ rd of length of impervious floor

$$= \frac{2}{3} \times 15 = 10 \text{ m}$$

Hence provide cistern or d/s floor of length

$$= 10 \text{ m}$$

Cutoffs

Depth of u/s cutoffs $\quad d_1 = \frac{1}{3}$ u/s water depth + 0.6 m

$$= \frac{1}{3} \times 2.5 + 0.6 = 1.43 \text{ m (say 1.5 m)}$$

Hence provide u/s cutoff of depth 1.5 m

Depth of d/s cutoff $\quad d_2 = \frac{1}{2}$ d/s water depth + 0.6 m

$$= \frac{1}{2} \times 1.6 + 0.6 = 1.4 \text{ m}$$

In this case maximum static head

$$H_s = \text{u/s F.S.L.} - \text{d/s floor level}$$
$$= 118.10 - 115.50 = 2.6 \text{ m}$$

In the case of the cross regulator H_s was equal to 2.7 m and d_2 was equal to 2 m. Hence in this case also provide d_2 = 2.0 m (as against the calculated value of 1.4 m) otherwise the floor length required will be excessive (i.e., much more than 16 m length of the previous case).

Hence provide d/s cutoff of depth 2.0 m

Width of each of the cutoffs is kept equal to 0.5 m.

Total length of impervious floor

$$G_E = \frac{H_s}{d_2}\frac{1}{\pi\sqrt{\lambda}}$$

$$G_E = 1/5\ ;\ H_s = 2.6 \text{ m}\ ;\ d_2 = 2.0 \text{ m}$$

Substituting these values, we get

$$\frac{1}{5} = \frac{2.6}{2.0} \times \frac{1}{\pi\sqrt{\lambda}}$$

or

$$\frac{1}{\pi\sqrt{\lambda}} = \frac{2.0}{5 \times 2.6} = 0.154$$

From Khosla's exit gradient curve [Fig. 13.12 (c) (Plate II)] for $\frac{1}{\pi\sqrt{\lambda}} = 0.154$

$$\alpha = \frac{b}{d_2} = 7.5$$

$$\therefore \quad b = \alpha.d_2 = 7.5 \times 2 = 15 \text{ m}$$

The total floor length of 15 m shall be provided as follows :

(i) d/s floor length $= \frac{2}{3} \times 15 = 10$ m

(ii) Length of d/s glacis length with 2 : 1 slope

$= 2\,(116.10 - 115.50) = 1.2$ m

(iii) Width of crest $= 1.0$ m

(iv) Length of u/s glacis length with 1 : 1 slope

$= (116.10 - 115.60) = 0.5$ m

(v) Balance floor length to be provided on u/s side

$= 15 - (10 + 1.2 + 1.0 + 0.5) = 2.3$ m

However, the upstream floor length is increased by 0.5 m to make it 2.8 m, thus making the total length of impervious floor = 15.5 m. This increase of 0.5 m length of floor is made to accommodate the piers and u/s wings.

Pressure calculations

(i) *Upstream cutoff*

$$d_1 = 1.5 \text{ m}\ ;\ b = 15.5 \text{ m}$$

$$\therefore \quad \frac{1}{\alpha} = \frac{d_1}{b} = \frac{1.5}{15.5} = 0.097$$

From Khosla's curves [Fig. 13.12 (b) (Plate I)] for $(1/\alpha) = 0.097$

$$\phi_{C_1} = 100 - 28 = 72\%$$

and $\phi_{D_1} = 100 - 19 = 81\%$

Assuming u/s floor thickness = 0.5 m (minimum thickness) correction for floor thickness for ϕ_{C_1}

$$= \frac{81-72}{1.5} \times 0.5 = 3\% \ (+)$$

Correction for interference of d/s pile for ϕ_{C_1}

$$= 19\sqrt{\frac{D}{b'}}\left(\frac{d+D}{b}\right)$$

$$b = b' = 15.5 \text{ m} \ ; D = 2.1 \text{ m} \ ; d = d_1 = 1.5 \text{ m}$$

$$\therefore \quad \text{Correction} = 19 = \sqrt{\frac{2.1}{15.5}}\left(\frac{1.5+2.1}{15.5}\right) 1.6\% \ (+)$$

$$\therefore \quad \text{Corrected } \phi_{C_1} = 72 + 3 + 1.6 = 76.6\%$$

(*ii*) *Downstream cutoff*

$$d_2 = 2 \text{ m} \ ; b = 15.5 \text{ m}$$

$$\therefore \quad \frac{1}{\alpha} = \frac{2}{15.5} = 0.129$$

From Khosla's curves [Fig. 13.12 (b) (Plate I)] for $(1/\alpha) = 0.129$

$$\phi_{E_2} = 32\%$$

$$\phi_{D_2} = 22\%$$

Assuming a minimum floor thickness of 0.6 m at the downstream end, correction for floor thickness for ϕ_{E_2}

$$= \frac{32-22}{2} \times 0.6 = 3\% \ (-)$$

Correction for interference of u/s pile for ϕ_{E_2}

$$= 19\sqrt{\frac{D}{b'}}\left(\frac{d+D}{b}\right)$$

$$b = b' = 15.5 \text{ m} \ ; D = (1.5 - 0.1) = 1.4 \text{ m} \ ; d = d_2 = 2.0 \text{ m}$$

$$\therefore \quad \text{Correction} = 19\sqrt{\frac{1.4}{15.5}}\left(\frac{2.0+1.4}{15.5}\right)$$

$$= 1.3\% \ (-)$$

$$\therefore \quad \text{Corrected } \phi_{E_2} = 32 - 3 - 1.3 = 27.7\%$$

Thus, the pressure on the floor varies linearly from 76.6% at the upstream to 27.7% at the downstream.

Calculation of floor thickness

(i) Provide a minimum thickness of 0.5 m for the upstream floor

(ii) Thickness under the crest = 1 m

(iii) Thickness at the d/s toe of the glacis :

$$\% \text{ pressure} = 27.7 + \frac{76.6 - 27.7}{15.5} \times 10$$

$$= 59\%$$

$\therefore$ Maximum static head $= 0.59 \times 2.6$

$= 1.534$ m

$\therefore$ Floor thickness required

$$= \frac{1.534}{G - 1}$$

$$= \frac{1.534}{2.24 - 1}$$

$= 1.24$ m say 1.3 m

Thus, provide 1.3 m thick cement concrete floor for a length of 3 m from the toe of the glacis.

(iv) Thickness of floor at 3 m from the toe of the glacis :

$$\% \text{ pressure} = 27.7 + \frac{76.6 - 27.7}{15.5} \times 7 = 50\%$$

$\therefore$ Maximum static head $= 0.50 \times 2.6 = 1.30$ m

$\therefore$ Floor thickness required

$$= \frac{1.30}{G - 1} = \frac{1.30}{2.24 - 1} = 1.05 \text{ m say } 1.1 \text{ m}$$

Thus, provide 1.1 m thick cement concrete floor from 3 m to 6 m from the toe of the glacis.

(v) Thickness of floor at 6 m from the toe of the glacis :

$$\% \text{ pressure} = \frac{76.6 - 27.7}{15.5} \times 4 = 40\%$$

$\therefore$ Maximum static head $= 0.40 \times 2.6 = 1.04$ m

$\therefore$ Floor thickness required

$$= \frac{1.04}{G - 1} = \frac{1.04}{2.24 - 1} = 0.84 \text{ m say } 0.9 \text{ m}$$

Thus, provide 0.9 m thick cement concrete floor from 6 m to 8.5 m from the toe of the glacis.

(vi) Thickness of floor at 8.5 m from the toe of the glacis :

$$\% \text{ pressure} = 27.7 + \frac{76.6 - 27.7}{15.5} \times 1.5 = 32\%$$

$\therefore$ Maximum static head $= 0.32 \times 2.6 = 0.832$ m

∴ Floor thickness required

$$= \frac{0.832}{G-1} = \frac{0.832}{2.24-1} = 0.67 \text{ m say } 0.7 \text{ m}$$

Thus, provide 0.7 m thick cement concrete floor for the last 1.5 m length.

Summary of floor thickness

(i) u/s floor : 0.5 m

(ii) Under the crest : 1.0 m

(iii) d/s toe of glacis to 3 m : 1.3 m

(iv) 3 m to 6 m from d/s toe of glacis : 1.1 m

(v) 6 m to 8.5 m from d/s toe of glacis : 0.9 m

(vi) Last 1.5 m = 0.7 m.

Step 3. *Design of upstream protection*

Depth of upstream cutoff

$$d_1 = \frac{1}{3} \text{ u/s water depth} + 0.6 \text{ m}$$

$$= \frac{1}{3} \times 2.5 + 0.6 = 1.43 \text{ m}$$

(i) *Block protection*

Volume of block protection

$= d_1$ cubic metre/m

$= 1.43$ cubic metre/m

Keep thickness of block protection

$= 1.0$ m

∴ Length of block protection

$= 1.43$ m

Thus, provide 2 rows of 0.8 m × 0.8 m × 0.6 m thick cement concrete blocks over 0.4 m thick apron of loose stones in a length of 1.6 m on the upstream of the impervious floor.

(ii) *Launching apron*

Cubic content of launching apron

$= 2.25\ d_1$ cubic metre/m

$= 2.25 \times 1.43 = 3.22$ cubic metre/m

Keep thickness of launching apron

$= 1.0$ m

∴ Length of launching apron

$= 3.22$ m

Thus, provide 1.0 m thick and 3.5 m long launching apron on the upstream of the block protection.

Step 4. *Design of downstream protection*

Depth of downstream cutoff

$$d_2 = \frac{1}{2} \text{ d/s water depth} + 0.6 \text{ m}$$

$$= \frac{1}{2} \times 1.6 + 0.6 = 1.4 \text{ m}$$

(i) *Inverted filter*

Volume of inverted filter

$= d_2$ cubic metre/m

$= 1.4$ cubic metre/m

Thickness of graded filter is kept as 0.5 m which is overlaid by 0.5 m thick cement concrete blocks, making a total thickness of 1.0 m

∴ Length of inverted filter

= 1.4 m

Thus, provide 2 rows of 0.8 m × 0.8 m × 0.5 m thick cement concrete blocks over 0.5 m thick graded filter.

(ii) *Launching apron*

Cubic content of launching apron

$= 2.25\ d_2$ cubic metre/m

$= 2.25 \times 1.4$

$= 3.15$ cubic metre/m

Keep thickness of launching apron

= 1.0 m

∴ Length of launching apron

= 3.15 m

Thus, provide 1 m thick launching apron for a length of 3.5 m.

Provide 0.4 m thick and 1.3 m deep masonry toe wall between the inverted filter and the launching apron.

The design details of the distributary head regulator are shown in Fig. 19.18 (Plate X).

REVIEW QUESTIONS

19.1 What is a fall in a canal ? Why is it necessary to provide a fall in a canal ?

19.2 Discuss the various considerations according to which the location of a fall is decided.

19.3 Describe briefly with neat sketches the different types of falls.

19.4 How falls are classified ?

19.5 Write a note on Notch type fall.

19.6 What is a cistern element in a fall ? Describe briefly the various types of cistern elements.

19.7 Describe with neat sketches the various types of roughening devices provided in cisterns for the dissipation of energy.

19.8 Discuss the procedure for designing a Sarda type fall.

19.9 Describe the procedure for designing a straight glacis fall.

19.10 Write a note on Montague type fall. How does it differ from a straight glacis fall ?

19.11 Describe briefly with a neat sketch an Inglis type fall.

19.12 Design a Sarda type fall for the following data :

(i) Full supply discharge : $\dfrac{u/s}{d/s} = \dfrac{20 \text{ cumec}}{20 \text{ cumec}}$

(ii) Full supply level : $\dfrac{u/s}{d/s} = \dfrac{215.40 \text{ m}}{214.40 \text{ m}}$

(iii) Full supply depth : $\dfrac{u/s}{d/s} = \dfrac{2.0 \text{ m}}{2.0 \text{ m}}$

(iv) Bed width : $\dfrac{u/s}{d/s} = \dfrac{20 \text{ m}}{20 \text{ m}}$

(v) Bed level : $\dfrac{u/s}{d/s} = \dfrac{213.40 \text{ m}}{212.40 \text{ m}}$

(vi) Drop : 1.0 m

Design the floor on the basis of Bligh's theory taking coefficient of creep = 10. Check the design by Khosla's theory and modify the design if necessary. Safe exit gradient may be taken as 1/5. Sketch the longitudinal section of fall.

19.13 Design an unflumed straight glacis non-meter fall for the following data :

(i) Full supply discharge : $\dfrac{u/s}{d/s} = \dfrac{20 \text{ cumec}}{20 \text{ cumec}}$

(ii) Full supply level : $\dfrac{u/s}{d/s} = \dfrac{215.40 \text{ m}}{214.40 \text{ m}}$

(iii) Full supply depth : $\dfrac{u/s}{d/s} = \dfrac{2.0 \text{ m}}{2.0 \text{ m}}$

(iv) Bed width : $\dfrac{u/s}{d/s} = \dfrac{20 \text{ m}}{20 \text{ m}}$

(v) Bed level : $\dfrac{u/s}{d/s} = \dfrac{213.40 \text{ m}}{212.40 \text{ m}}$

(vi) Drop : 1.0 m

(vii) Permissible exit gradient : 1/5

19.14 A Sarda type fall is to be designed for a drop of 1.5 m in a channel carrying 20 cumec with bed width 20 m and depth 1.5 m. Determine (*a*) the crest dimensions, (*b*) the minimum length of floor to be provided on downstream and (*c*) the length and depth of the cistern. Specific gravity of crest material = 2.

19.15 What is a cross regulator ? What are the functions of a cross regulator?

19.16 What is a distributary head regulator ? What are the functions of a distributary head regulator ?

19.17 Design a cross regulator and a head regulator for a distributary taking off from a branch canal for the following data :

(i) Discharge of branch canal	:	100 cumec
(ii) Discharge of distributary	:	25 cumec
(iii) F.S.L. of branch canal	:	$\frac{u.s}{d/s} = \frac{115.20 \text{ m}}{115.00 \text{ m}}$
(iv) Bed width of branch canal	:	$\frac{u.s}{d/s} = \frac{42 \text{ m}}{38 \text{ m}}$
(v) Depth of water in branch canal	:	$\frac{u.s}{d/s} = \frac{2.3 \text{ m}}{2.3 \text{ m}}$
(vi) F.S.L. of distributary	:	114.30 m
(vii) Bed width of distributary	:	12 m
(viii) Depth of water in distributary	:	1.5 m
(ix) Permissible exit gradient	:	1/6

19.18 Describe briefly with neat sketches (i) King's vanes, (ii) Gibb's groyne wall, (iii) Cantilever skimming platform, (iv) Bed bars.

19.19 Describe the procedure for designing a head regulator for a distributary.

19.20 Describe the procedure for designing a cross regulator.

19.21 What is a canal escape ? What are the different types of canal escapes ?

CHAPTER

20 Cross Drainage Works

20.1 INTRODUCTION

A cross drainage work is a structure constructed for carrying a canal across a natural drain or river intercepting the canal. Cross drainage works are unavoidable in any canal system. In order to minimise the number of cross drainage works, canals are generally aligned on a watershed so that they intercept a minimum number of natural drains. However, a canal taking off from a river requires a certain distance before it can mount the watershed. In this reach the

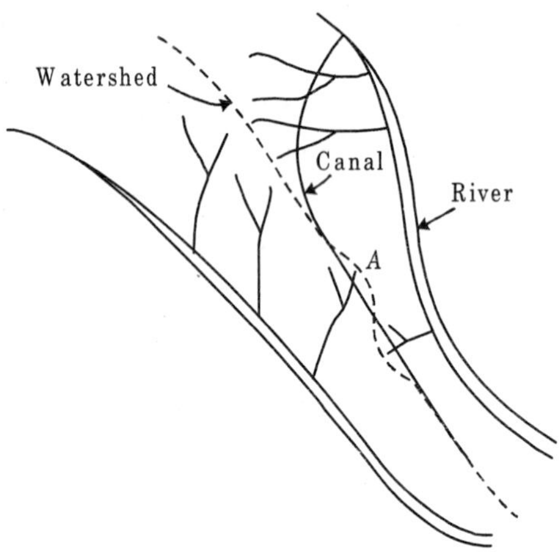

Fig. 20.1 *Canal crossing natural drains*

canal intercepts a number of natural drains from the watershed towards the river and it is necessary to carry the canal across these drains (Fig. 20.1). Hence it is in this reach major cross drainage works are required to be constructed. Once the canal reaches the watershed no cross drainage work would normally be required as all the drains will be away from the watershed towards the two rivers on either side of the watershed. But it may be necessary for the canal to leave the watershed for a short distance to cut across the loop as at *A* in Fig. 20.1. In that case the canal may intercept a drain in the pocket between the watershed and the canal must be passed across the drain for which a cross drainage work will be required. In the case of canals which are aligned as contour canals a number of cross drainage works would be required.

20.2 TYPE OF CROSS DRAINAGE WORKS

Depending upon the relative bed levels and water levels of the canal and the drain and their relative discharge the cross drainage works may be of the following types.

I. Cross drainage works carrying the canal over the natural drain

In this type of cross drainage works the canal is carried over the natural drain. This may be accomplished by the following two types of cross drainage works.

(a) Aqueduct (b) Syphon aqueduct

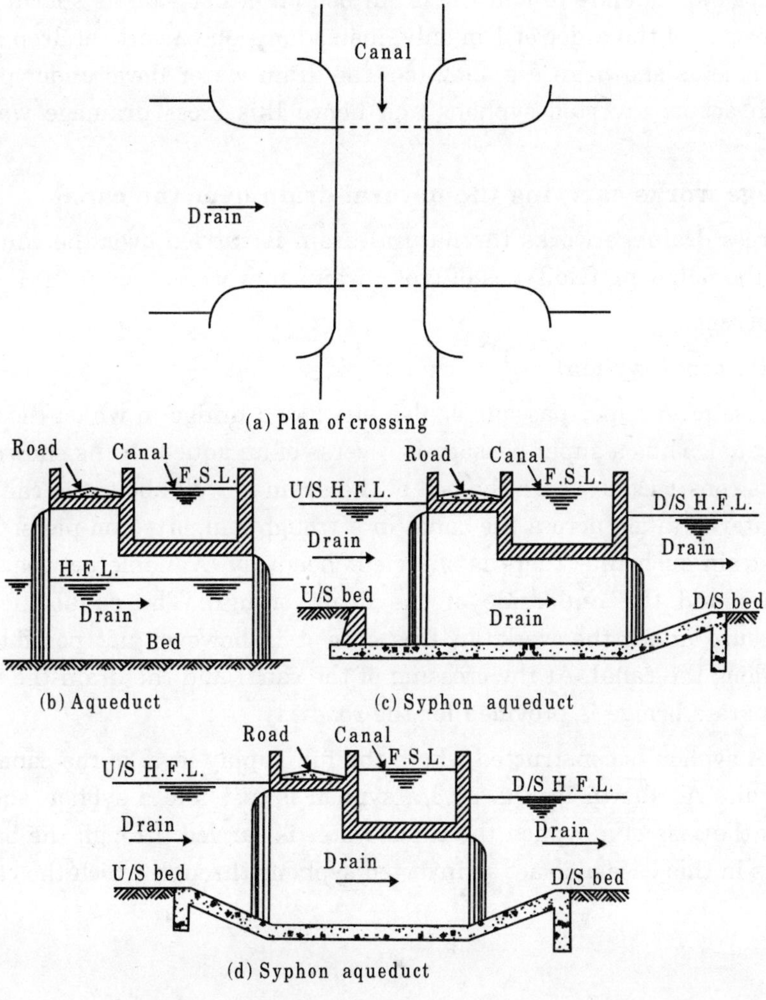

Fig. 20.2 *Aqueduct and Syphon aqueduct*

1. Aqueduct. An aqueduct is just like a bridge in which instead of road or a railway, a canal is carried over a natural drain. As shown in Fig. 20.2 (b) an aqueduct is constructed where the bed of the canal is well above the high flood level (H.F.L.) of the drain. The canal water is taken across the drain in a trough supported on piers. The drain water flows under the canal such that there is sufficient headway available between the H.F.L. of the drain and the underside of the canal trough. The drain therefore flows at atmospheric pressure under the work. Further as shown in Fig. 20.2 (b) generally an inspection road is provided along with the trough.

2. Syphon aqueduct. A syphon aqueduct is constructed where the water surface level of the drain at high flood is higher than the canal bed. As shown in Fig. 20.2 in the case of a syphon aqueduct the bed of the drain is usually depressed and provided with pucca floor thus forming barrels between the piers to pass the drain water. The horizontal floor of the barrels is provided with slopes at its ends to join the drain bed on either side as shown in Fig. 20.2 (d). However, if the drop is of the order of 1 m only, instead of slope a vertical drop may be provided on the upstream side as shown in Fig. 20.2 (c). The drain water flows under pressure through the barrels which act as inverted syphons and hence this cross drainage work is known as syphon aqueduct.

II. Cross drainage works carrying the natural drain over the canal

In this type of cross drainage works the natural drain is carried over the canal. This may be accomplished by the following two types of cross drainage works.

(a) Super passage

(b) Syphon (or canal syphon)

(a) **Super passage**. A super passage is also just like a bridge in which the natural drain is carried over the canal. Thus a super passage is reverse of an aqueduct. As shown in Fig. 20.3 (b) a super passage is constructed where the bed of the drain is well above the canal F.S.L. In this case the drain water is taken across the canal in a trough supported on piers. The canal water flows under the drain such that there is sufficient headway available between the full supply level of the canal and the underside of the drain trough. The canal therefore flows at atmospheric pressure under the work. In this case it is however not possible to provide an inspection road along the canal. At the crossing of the canal and the drain the canal road often detours and a separate bridge is provided for the roadway.

(b) **Syphon**. A syphon is constructed where the full supply level of the canal is higher than the bed of the drain. As shown in Fig. 20.3 a syphon is just like a syphon aqueduct with the difference that in the case of a syphon the canal water is carried through the barrels under the drain. The barrels in this case also act as inverted syphons through which the canal water flows under pressure.

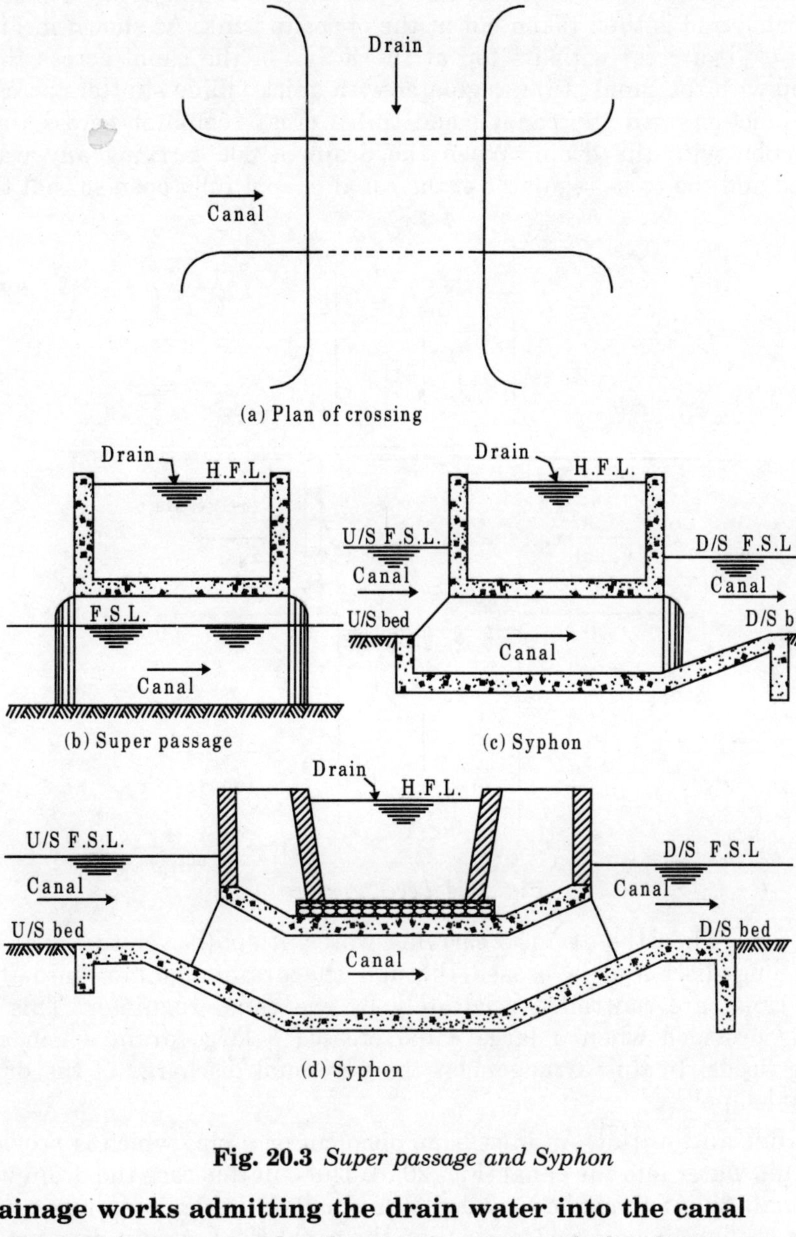

Fig. 20.3 *Super passage and Syphon*

III. Cross drainage works admitting the drain water into the canal

In this type of cross drainage works the canal water and the drain water are allowed to intermingle with each other. This may be achieved by the following two types of the cross drainage works.

(a) Level crossing

(b) Inlet or inlet and outlet

1. Level crossing. A level crossing may be provided when the beds of the canal and the drains are practically at the same level. In this type of cross drainage work the drain water is passed into the canal and is then taken out at the opposite bank. As shown in Fig. 20.4 a level crossing consists of (i) a crest with its top at the F.S.L. of the canal across the drain at its upstream junction with the canal ; (ii) a regulator with quick falling shutters across the drain at its downstream junction with the canal ; and (iii) a cross regulator across the canal at its downstream junction with the drain. When the drain is not carrying any water the drain regulator is closed and the cross regulator of the canal is kept fully open so that the canal flows

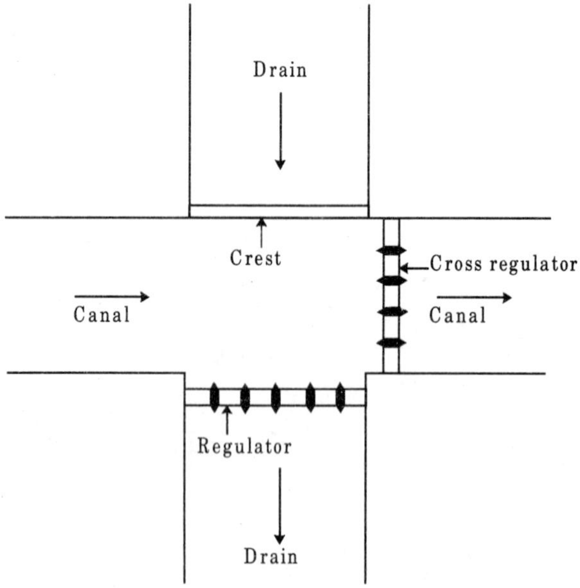

Fig. 20.4 *Level crossing*

without interruption. When the drain is carrying water, it spills over the crest into the canal and a corresponding discharge is passed through the drain regulator into the drain. The supplies in the canal are controlled accurately by the canal regulator. This type of cross drainage work is provided when a large canal crosses a large drain which carries a high discharge during floods. In this arrangement the perennial discharge of the drain is used to augment the canal supplies.

2. Inlet or inlet and outlet. An inlet is an open cut or a pipe which is provided in a canal bank to admit drain water into the canal (Fig. 20.5). Thus, in this case the drain water is simply taken into the canal. An inlet may be provided for a small drain coming across a canal if the bed level of the drain is slightly higher or lower than the canal F.S.L. and it does not carry a heavy load of silt.

If the drain water admitted into the canal through an inlet is discharged through an outlet provided at a suitable site a little downstream along the canal, the arrangement is called inlet and outlet. An outlet is usually combined with some other work where arrangement for escaping is otherwise required to be made.

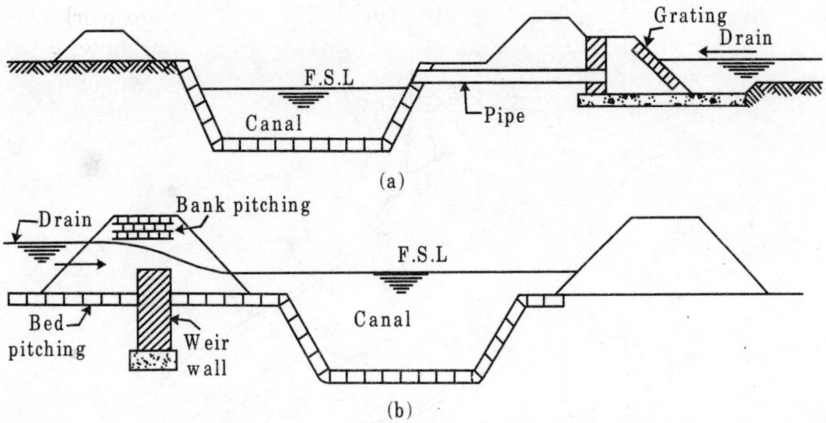

Fig. 20.5 *Inlet (a) pipe type ; (b) open cut type*

20.3 FACTORS AFFECTING SUITABILITY OF AQUEDUCT AND SYPHON AQUEDUCT

The relative size of the canal and the river or natural drain to be crossed is the main factor considered in determining whether for a particular location the cross drainage work should be an aqueduct or a syphon aqueduct. In the case of a large river the lowering of its bed by a drop for providing a syphon aqueduct would be very costly. Moreover, if the amount of bed silt carried by the river is large, there is a risk of choking up of syphon due to silting which would increase with the size of the river. Thus on a large river as far as possible only an aqueduct should be provided. This may however, result in heavily banked approaches of the canal towards the crossing. This is so because in plains the rivers do not flow confined between high banks and the

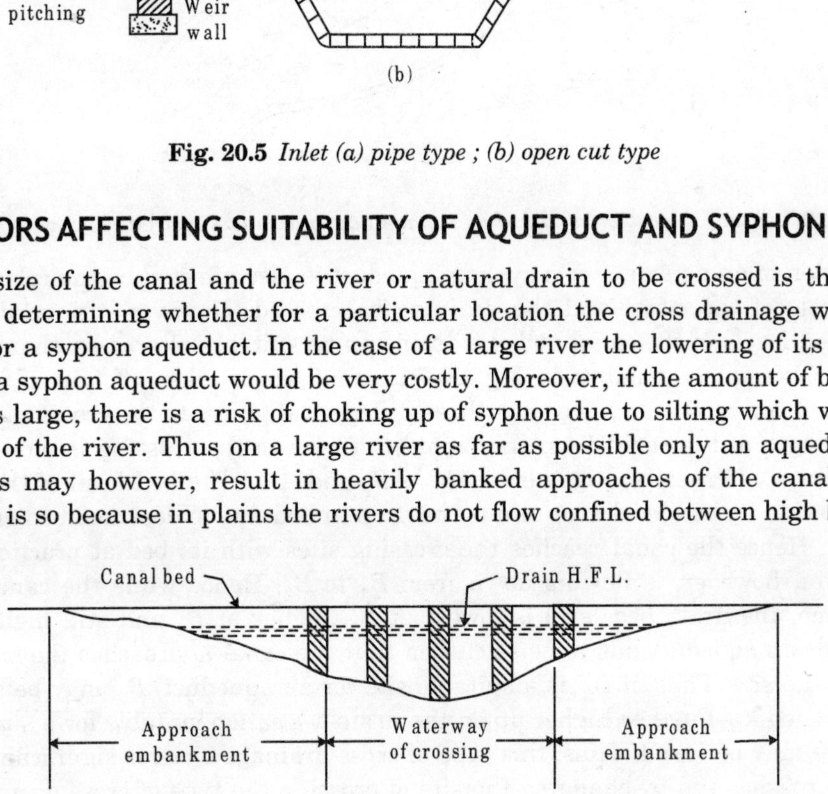

Fig. 20.6 *Approach embankments for reducing the waterway of crossing*

flood cross-section of the rivers is very wide. The crossing is constructed in only small portion of this cross-section. For providing an aqueduct since the canal must be brought to the crossing with its bed level higher than the H.F.L., of the river, it necessarily results in banked approaches on either side as shown in Fig. 20.6. On the other hand in the case of a syphon aqueduct since the canal bed is not clear above the H.F.L., of the river, the relative difference between the canal and the river is much smaller and heavy embankments are practically eliminated. It, therefore, follows that when the river is small as compared to the canal it would be preferable to provide a syphon aqueduct.

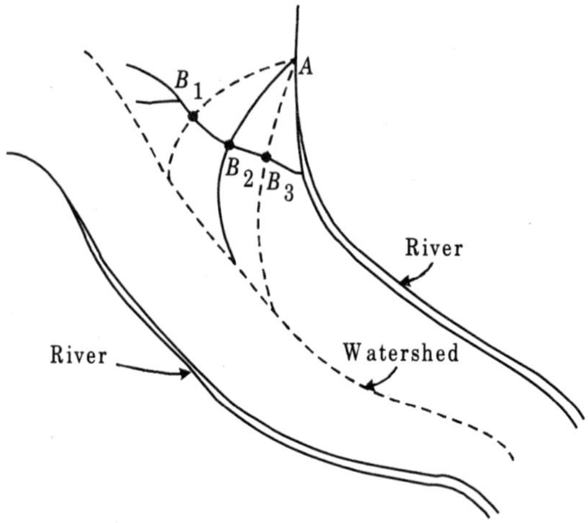

Fig. 20.7 *Sites for location of aqueduct syphon aqueduct and super passage (or syphon)*

The provision of an aqueduct or a syphon aqueduct is also affected by the relative bed levels and water levels of the canal and the drain to be crossed. However, the relative difference between the bed levels of the canal and the drain can be suitably altered by changing the canal alignment. For example, as shown in Fig. 20.7 a canal which takes off from a river at A and comes across a drain at B may have with three different alignments the crossing sites at B_1, B_2 and B_3. The length of the canal from the point of offtake to the crossing is more or less same along all the three alternative alignments AB_1, AB_2 and AB_3. The bed level of the canal at the crossing is equal to its bed level at A minus the drop in bed due to designed bed slope from A to B_1, B_2 and B_3. Hence the canal reaches the crossing sites with its bed at practically the same level. The drain, however, is flowing down from B_1 to B_3. Hence while the canal bed level is nearly the same, the drain bed level is lowest at B_3, higher at B_2 and still higher at B_1. The crossing site for an aqueduct has to be so chosen that the canal approaches the site clear above the H.F.L., at the site. Thus, if B_3 is a suitable site for an aqueduct, B_2 may be a suitable site for a syphon aqueduct. Further higher up on the drain a location suitable for a siphon or even a supper passage may be found. Thus, this type of cross drainage work is significantly influenced by the site of crossing and by changing the site of crossing the type of cross drainage work can be changed.

In addition to the above noted factors the choice between aqueduct and syphon aqueduct is also affected by the following factors.

(i) Suitable canal alignment.

(ii) Nature of foundation.

(iii) Ground water likely to be met with at foundation level and available dewatering equipment.

(iv) Suitability of soil for making embankment.

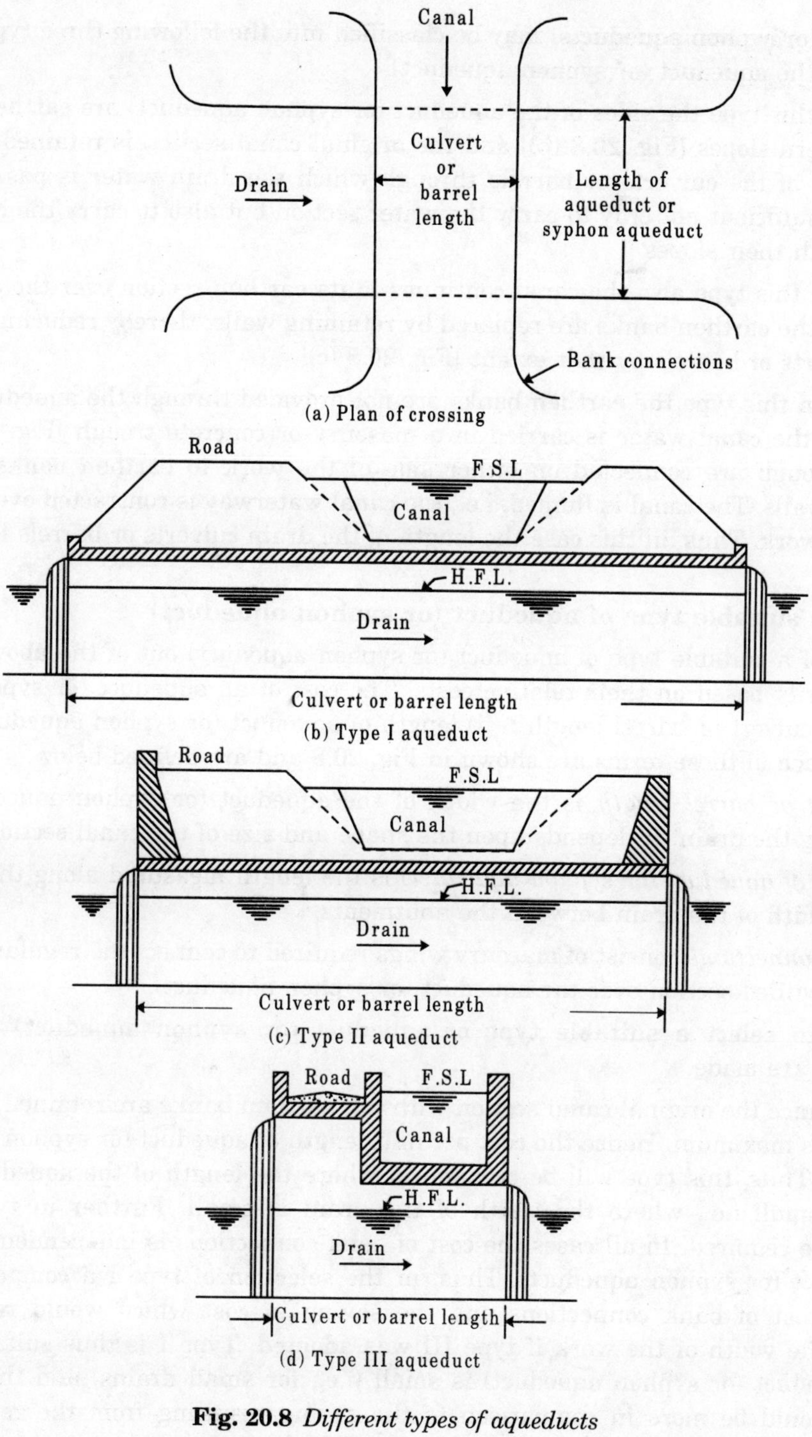

Fig. 20.8 *Different types of aqueducts*

20.4 CLASSIFICATION OF AQUEDUCTS AND SYPHON AQUEDUCTS

The aqueducts (or syphon aqueducts) may be classified into the following three types depending on the sides of the aqueduct (or syphon aqueduct).

Type I. In this type the sides of the aqueduct (or syphon aqueduct) are earthen banks with complete earthern slopes [Fig. 20.8 (b)] and the original canal section is retained. Thus in this case the length of the culverts or barrels through which the drain water is passed under the canal must be sufficient not only to carry the water section but also to carry the earthen banks of the canal with their slopes.

Type II. In this type also the canal continues in its earthen section over the drain, but the outer slopes of the earthen banks are replaced by retaining walls, thereby reducing the length of the drain culverts or barrels by that extent [Fig. 20.8 (c)].

Type III. In this type the earthen banks are not provided through the aqueduct (or syphon aqueduct) and the canal water is carried in a masonry or concrete trough [Fig. 20.8 (d)]. The sides of the trough are connected on either side of the work to earthen banks of the canal through wing walls. The canal is flumed, i.e., the canal waterway is contracted over the crossing in this type of work. Thus, in this case the length of the drain culverts or barrels is considerably reduced.

Selection of a suitable type of aqueduct (or syphon aqueduct)

The selection of a suitable type of aqueduct (or syphon aqueduct) out of the above noted three types is primarily based on their relative costs. The cost of an aqueduct (or syphon aqueduct) depends on (i) culvert or barrel length ; (ii) length of aqueduct (or syphon aqueduct) ; (iii) bank connections. Each of these terms are shown in Fig. 20.8 and are defined below.

The *culvert or barrel length* is the width of the aqueduct (or syphon aqueduct) which is measured along the drain. It depends upon the shape and size of the canal section.

The *length of aqueduct (or syphon aqueduct)* is the length measured along the canal and is equal to the width of the drain between the abutments.

The *bank connections* consist of masonry wings required to connect the regular section of the canal to its modified section over the aqueduct (or syphon aqueduct).

In order to select a suitable type of aqueduct (or syphon aqueduct) the following considerations are made.

In type I since the original canal section with full earthen banks are retained, the culvert or barrel length is maximum. Hence the cost per unit length of aqueduct (or syphon aqueduct) will be maximum. Thus, this type will be economical where the length of the aqueduct (or syphon aqueduct) is small i.e., where the width of the drain is small. Further in type I no bank connections are required. In all cases the cost of bank connections is independent of the length of the aqueduct (or syphon aqueduct). Thus, in the selection of type I a comparison is made between the cost of bank connections and the saving in cost which would result from the reduction in the width of the work if type III was adopted. Type I is thus suitable where the length of aqueduct (or syphon aqueduct) is small (i.e., for small drains) and the cost of bank connections would be more in comparison to the saving resulting from the reduction in the width of the work if type III was adopted.

In type III the culvert or barrel length is minimum and the cost of bank connections is maximum. Hence this type is suitable where the length of aqueduct (or syphon aqueduct) is large (i.e., for large drains) and the cost of bank connections would be less in comparison to the excess cost of long culverts or barrels required to be provided if type I was adopted.

Type II aqueduct (or syphon aqueduct) is however, suitable for intermediate conditions.

From the above discussion it may be concluded that the selection of a suitable type of aqueduct (or syphon aqueduct) depends mainly on the width and magnitude of drain to be crossed in relation to the size of the canal. On a small drain Type I is the most economical while on a drain of considerable size type III is more economical than any other type. Type II is in between these two types. It may however, be stated that the limiting size of the drain for which each type of aqueduct (or syphon aqueduct) may be used varies with local conditions and rates, and can be determined by preparing comparative estimates for any given case.

20.5 FEATURES OF DESIGN OF CROSS DRAINAGE WORKS

The important features of design of cross drainage works are as follows.

A. Hydraulic Design

The hydraulic design involves the following aspects.

1. Determination of maximum flood discharge and high flood level (H.F.L.).
2. Determination of waterway of the drain.
3. Head loss through syphon barrels.
4. Contraction of canal waterway or fluming.
5. Determination of uplift pressure on the underside of the trough (or the barrel roof).
6. Determination of uplift pressure on the floor of the barrel.
7. Design of bank connections.

B. Structural Design

The structural design deals with the following aspects.

1. Design of side walls of trough and roof and floor of barrels.
2. Design of piers and abutments.
3. Design of foundations.

The various features of design of cross drainage works are discussed in the following sections.

20.6 DETERMINATION OF MAXIMUM FLOOD DISCHARGE

The maximum flood discharge for small drains may be determined by using empirical formulae and for large drains by using other reliable methods such as hydrograph analysis, Rational method, etc. The empirical formulae as well as the various methods are described in Chapter 5.

20.7 DETERMINATION OF WATERWAY OF THE DRAIN

The waterway required to be provided for the drain is given by Lacey's equation for regime perimeter as

$$P = 4.75Q \quad ...(20.1)$$

where P— wetted perimeter in m, and

Q— maximum discharge in cumec.

In general, the total waterway between the abutments is made equal to Lacey's regime perimeter. For small drains a contraction upto 20% beyond Lacey's regime perimeter is also permissible. Since the piers will occupy a part of this total waterway, the clear waterway will be less than Lacey's regime perimeter to the extent of total thickness of the piers. This is however, permissible because under the work the regime conditions do not exist. Moreover, the idea of relating the waterway to Lacey's regime perimeter is to develop a stable channel between training banks upstream of the work. In wide shallow rivers no distinction is made between the width of waterway and wetted perimeter but in small drains the width of water may be adjusted to provide the required perimeter.

Size of barrels. After having fixed the waterway the size of barrels is to be fixed. In the case of an aqueduct, the canal trough is carried clear above the H.F.L. and hence the height of openings is automatically fixed. In the case of a syphon aqueduct, the required area of waterway is obtained by dividing the flood discharge by the permissible velocity of flow through the barrels. The velocity of flow through the barrels is normally limited 2 to 3 m/s. Higher velocities may cause quick abrasion of the barrel surfaces by rolling grit and will result in high afflux upstream of the syphon aqueduct thus requiring higher and longer marginal banks to avoid flooding of the country side. The total area divided by the number of spans gives the area required by each span. Knowing the width of span and the shape of the opening, the required depth of the opening can be determined which will fulfill the area requirement of the span. The span to be provided depends entirely on structural and economical considerations. For instance where foundations are costly fewer piers or longer spans are provided.

From the canal bed level, deducting the thickness of the culvert or barrel roof plus the depth of the opening, the level of the floor of the barrel is obtained. Further in the case of syphon aqueduct it is desirable to leave some clear headway between the bed of the drain on the downstream side of the crossing and the underside of the culvert or barrel roof. This headway is desirable to minimise the risk of blocking of culverts or barrels by silt rolling along the bed of the drain and it should be either 1 m or half the height of culvert or barrel, whichever is less.

20.8 HEAD LOSS THROUGH SYPHON BARRELS

In the case of syphon aqueducts and syphons the head loss through the syphon barrels may be calculated by using Unwin's formula which is as follows.

$$h = \left(1 + f_1 + f_2 \frac{L}{R}\right)\frac{V^2}{2g} - \frac{V_a^2}{2g} \qquad \ldots(20.2)$$

where h = head loss through the syphon barrel in m, which is equal to the difference between water levels upstream and downstream of syphon barrels

L = length of the barrel in m

R = hydraulic mean radius of the barrel in m

V = velocity of flow through the barrels in m/s

V_a = velocity of approach in m/s which is generally neglected

f_1 = coefficient for loss of head at entry, which may be taken as 0.505 for an unshaped mouth of the same cross-sectional area as the barrel and 0.08 for bell mouth, and

f_2 = coefficient such that the loss of head in the barrel due to friction is given by $\left(f_2 \frac{L}{R} \frac{V^2}{2g}\right)$.

The value of f_2 is given by

$$f_2 = a\left(1 + \frac{b}{R}\right) \qquad \text{...(20.3)}$$

in which the values of a and b for different materials may be taken from the following table.

Nature of surface of the barrel	*a*	*b*
1. Smooth iron pipe	0.00497	0.025
2. Encrusted pipe	0.00996	0.025
3. Smooth cement plaster	0.00316	0.030
4. Aslar or brickwork	0.00401	0.070
5. Rubble masonry or stone pitching	0.00507	0.250

As stated earlier the velocity of flow through the syphon barrels is limited to 2 to 3 m/s. Thus, knowing the velocity of flow the head required to generate the same can be found from Eq. 20.2. The high flood level of the drain downstream of the work remains unchanged even after the construction of the work, but on the upstream side the high flood level increases by the amount of afflux or the total head loss as given by Eq. (20.2). The top level of guide banks and marginal bunds on the upstream side of the work is governed by this raised high flood level (or the afflux). Further the protection required in the bed of the drain on the downstream side of the work also depends on the afflux. Thus, by permitting higher velocities of flow through the syphon barrels (and thereby increasing afflux) the cross-sectional area of syphon barrels is reduced, but there is corresponding increase in the cost due to higher guide banks etc., and greater length of bed protection downstream of the work. As such there should be an economic balance between the saving in barrel area and the increase in afflux due to increased velocity of flow in the syphon barrels. Further in order to reduce afflux, entry should be bell mouthed and the inside surface of the barrels should be made as smooth as possible.

20.9 CONTRACTION OF CANAL WATERWAY OR FLUMING

As indicated earlier in the case of type III aqueducts (or syphon aqueducts) the canal is flumed, i.e., the canal waterway is contracted, to economise the cost of cross drainage works. The flumed portion or the trough is gradually connected to the normal section with smooth transition wings at both ends. The contraction of canal waterway although reduces the barrel lengths or the width of the works but it involves the provision of extra transition wings. As such the canal waterway should be flumed to such an extent that the cost of the works as a whole is minimum.

The fluming or the contraction of the canal waterway should be done in such a way that the velocity in the trough is not more than 3 m/s and the flow remains subcritical to avoid the possibility of hydraulic jump forming in the trough. The approach or contraction transition should not be steeper than 30° which corresponds to a splay of 2 : 1 and the departure or expansion transition should not be steeper than $22\frac{1}{2}^\circ$ which corresponds to a splay of 3 : 1. The transitions consist of curved and flared wing walls so that there is minimum loss of head and the flow is streamlined.

The transitions can be designed according to two different conditions as indicated below.

1. Design of transitions when the water depth remains constant.
2. Design of transitions when the water depth varies.

1. Design of transitions when the water depth remains constant. In case the water depth in the transition and the trough remains constant the following two methods may be adopted for the design of transitions.

(a) *Mitra's hyperbolic transitions.* For water depth to remain constant in the transition and the trough, A.C. Mitra proposed hyperbolic transitions which may be obtained as follows.

Let B_c and B_f be the bed widths of the normal and flumed sections respectively, and let B_x be the bed width at a distance x from the flumed section (Fig. 20.9). Let V_c, V_f and V_x be the mean velocities at the corresponding sections. Let L_f be the total length of the transition.

The transition is designed on the basis that the rate of change of velocity per unit length of the transition is constant throughout the transition length. Thus,

$$\frac{V_f - V_x}{x} = \frac{V_f - V_c}{L_f} \quad \text{...(i)}$$

Let D be the water depth and Q be the discharge. Since D is assumed to be constant, we have from continuity equation

$$V_f \times B_f \times D = V_x \times B_x \times D = V_c \times B_c \times D = Q$$

or

$$V_f \times B_f = V_x \times B_x = V_c \times B_c = \frac{Q}{D} = k \text{ (say)} \quad \text{...(ii)}$$

where k is a constant.

Hence

$$V_f = \frac{k}{B_f}$$

$$V_x = \frac{k}{B_x}$$

and

$$V_c = \frac{k}{B_c}$$

Substituting these values of velocities in Eq. (i) we get

$$\frac{(k/B_f)-(k/B_x)}{x} = \frac{(k/B_f)-(k/B_c)}{L_f}$$

or $$B_x = \frac{B_c B_f L_f}{L_f B_c - (B_c - B_f)x} \quad \text{...(20.4)}$$

Equation (20.4) represents Mitra's hyperbolic transition.

(b) *Chaturvedi's semi cubical parabolic transition.* R.S. Chaturvedi on the basis of his experiments proposed the following equation for semi-cubical parabolic tansitons when water depth remains constant.

$$x = \frac{L_f B_c^{3/2}}{B_c^{3/2} - B_f^{3/2}} \left\{1 - \left(\frac{B_f}{B_x}\right)^{3/2}\right\} \quad \text{...(20.5)}$$

Choosing various convenient values of B_x, the corresponding distance x can be computed from Eq. (20.5).

2. Design of transitions when the water depth also varies—Hind's method

When the water depths in the trough and the normal section are different, Hind's method given below may be used. It is however, general method which may also be used when the water depth remains constant.

As shown in Fig. 20.9 the contraction transition starts at section 1–1 and ends at section 2–2. The flumed section continues from section 2–2 to section 3–3. The expansion transition starts from section 3–3 and ends at section 4–4. From section 4–4 the channel flows in its normal cross section and the conditions at this section are completely known. Let D and V with appropriate subscripts refer to depths and velocities at the various sections. The design is done in the following steps.

Step 1. Let the bed level and cross section of the canal at section 4–4 be completely known.

Water surface elevation at section 4–4

= Bed level at section 4–4 + D_4

∴ Elevation of T.E.L. at section 4–4

$$= \text{Water surface elevation at section } 4-4 + \frac{V_4^2}{2g}$$

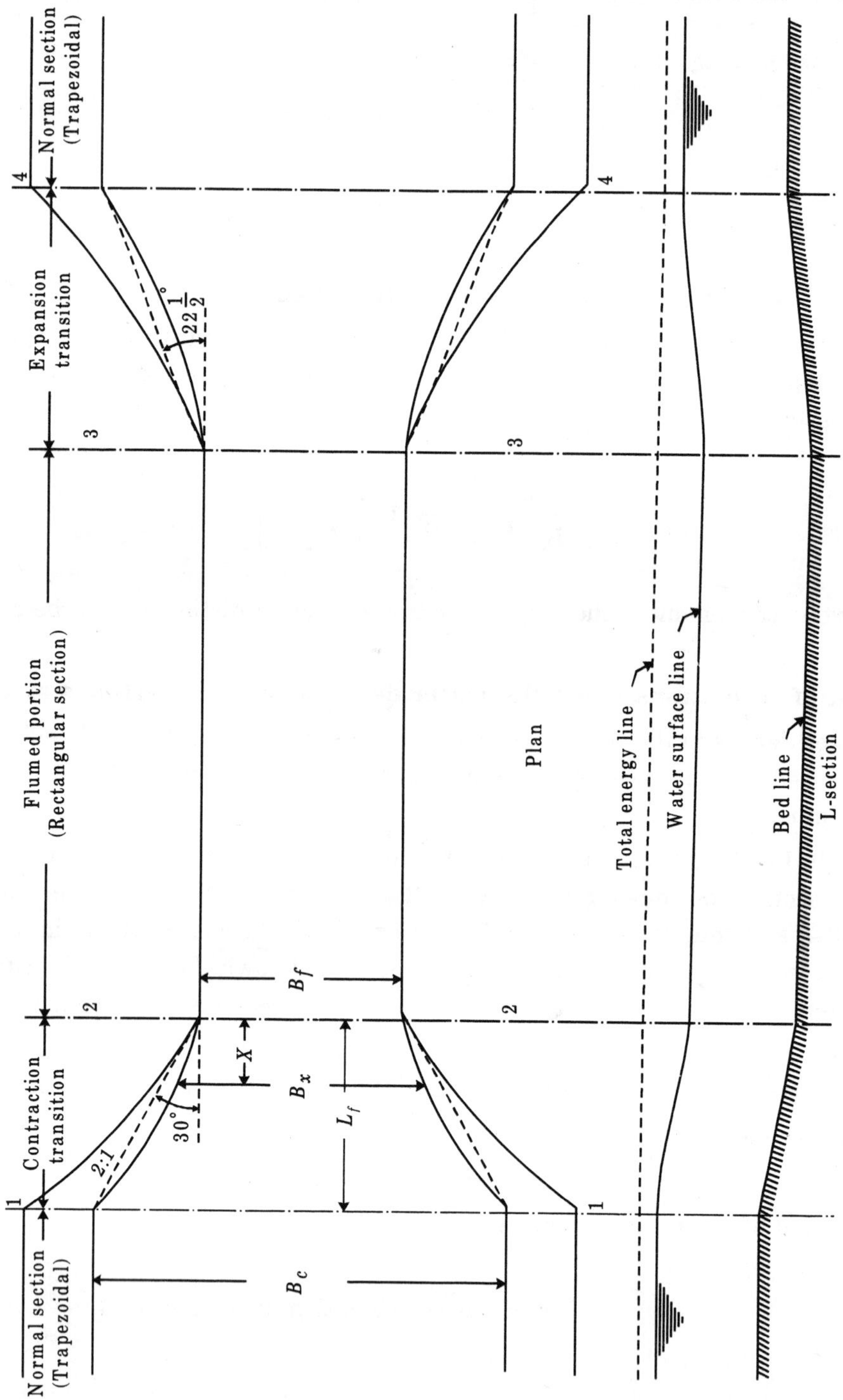

Fig. 20.9 *Contraction and expansion transitions*

Step 2. Between sections 3–3 and 4–4 there is energy loss due to expansion and due to friction. The energy loss due to expansion may be taken as $0.3\left(\frac{V_3^2 - V_4^2}{2g}\right)$. The energy loss due to friction is generally small and may be neglected.

Since trough dimensions at section 3–3 are known, V_3 is known.

∴ Elevation of T.E.L. at section 3–3

$$= \text{Elevation of T.E.L. at section 4–4} + 0.3\left(\frac{V_3^2 - V_4^2}{2g}\right)$$

Water surface elevation at section 3–3

$$= \text{Elevation of T.E.L. at section 3–3} - \frac{V_3^2}{2g}$$

Bed level at section 3–3

$$= \text{Water surface elevation at section 3–3} - D_3.$$

Step 3. Between sections 2–2 and 3–3 the channel flows in a trough of constant cross-section and there is uniform flow. The only loss in the trough is therefore friction loss H_L which can be computed by Manning's equation as indicated below.

$$Q = \frac{1}{N} AR^{2/3} S^{1/2}$$

or

$$Q = \frac{1}{N} AR^{2/3} \left(\frac{H_L}{L}\right)^{1/2}$$

or

$$HL = \frac{Q^2 N^2 L}{A^2 R^{4/3}}$$

∴ Elevation of T.E.L. at section 2–2

$$= \text{Elevation of T.E.L. at section 3–3} + H_L$$

Water surface elevation at section 2–2

$$= \text{Elevation of T.E.L. at section 2–2} - \frac{V_2^2}{2g}$$

Bed level at section 2–2

$$= \text{Water surface elevation at section 2–2} - D_2$$

Step 4. Between sections 1–1 and 2–2 there is energy loss due to contraction as well as due to friction. The energy loss due to contraction may be taken as $0.2\left(\frac{V_2^2 - V_1^2}{2g}\right)$. The energy loss due to friction is generally small and may be neglected.

∴ Elevation of T.E.L. at section 1–1

$$= \text{Elevation of T.E.L. at 2–2} + 0.2\left(\frac{V_2^2 - V_1^2}{2g}\right)$$

Water surface elevation at section 1–1

$$= \text{Elevation of T.E.L. at section 1–1} - \frac{V_1^2}{2g}$$

Bed level at section 1–1

$$= \text{Water surface elevation at section 1–1} - D_1.$$

Step 5. The bed level, water surface elevation and elevation of total energy line at all the four sections are known. The total energy line is drawn assuming it to be a straight line between the adjacent sections. The bed line may also be drawn straight between adjacent sections if the fall or rise of bed is small. The corners should, however, be rounded off in this case. If the change in bed level is considerable, the bed line in the transition section should be drawn as a smooth reverse curve, tangential to the bed lines at ends.

Step 6. In the contraction transition, from section 1–1 to section 2–2 there is a drop in water due to (i) drop in the energy line and (ii) the increased velocity head at section 2–2. This drop in water surface has to be negotiated by a smooth curve tangential at both ends. This can be achieved by two parabolic curves, the first convex and the second concave upwards, and both the curves meeting tangentially at the centre of the transition. As shown in Fig. 20.10 let the length of the contraction transition be $2x_1$ and the drop in water surface be $2y_1$. Thus, at the middle point of the transition, the distance from section 1–1 is x_1 and the drop in water surface is y_1. Taking water surface at section 1–1 as origin the equation of the first parabola may be written as

$$y = cx^2$$

Since at $x = x_1, y = y_1,$

$$c = \frac{y}{x^2} = \frac{y_1}{x_1^2}$$

Hence the equation becomes

$$y = \frac{y_1}{x_1^2}x^2 \qquad \text{...(20.6)}$$

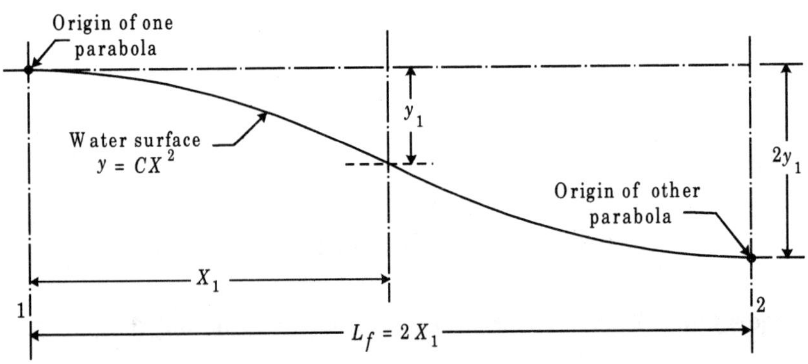

Fig. 20.10 *Water surface profile in contraction transition*

Using Eq. 20.4 the first parabolic curve may be plotted downstream and downwards from the origin at section 1–1. The second parabolic curve is exactly similar, but for it the origin is on the water surface at section 2–2 and the y values are to be measured upstream and upwards from the origin at the water surface at section 2–2.

Step 7. In the expansion transition from section 3–3 to section 4–4 there is rise in water surface. This rise in the water surface has also to be negotiated with a smooth curve tangential at both ends. The water surface is determined in the same way as explained for contraction transition. However, in this case the first half of the curve from section 3–3 is concave upwards and the second half convex upwards.

Step 8. After plotting the water surface profile, the velocity head h_a can be found by measuring the vertical distance between the total energy line and the water surface line at any point. The velocity head can be converted into equivalent velocity by the relationship

$$V = \sqrt{2g\, h_a} \qquad \text{...(20.7)}$$

Thus, the velocity at each section can be known.

Step 9. The cross-sectional area A required to pass the discharge Q is given by

$$A = \frac{Q}{V} \qquad \text{...(20.8)}$$

As the bed line is also drawn before hand the depth D at every section is known.

In a trapezoidal channel of bed width B, if side slopes are $z : 1$, the cross-sectional area is given by

$$A = BD + zD^2 \qquad \text{...(20.9)}$$

In flared wings, the side slopes are gradually brought to vertical from an initial slope and the side slope at any section can be interpolated in proportion to the length of transition undergone. Thus, at any section since A, D and z are known, the bed width B may be obtained from Eq. (20.9). Thus, all the dimensions of the transition are fully determined.

20.10 UPLIFT PRESSURE ON THE UNDERSIDE OF THE TROUGH (OR THE BARREL ROOF)

When water is flowing through the syphon barrel and the water level on the downstream side at the barrel is higher than the underside of the trough (or the barrel roof), uplift pressure is exerted on the underside of the trough (or the barrel roof). At the downstream end of the barrel the uplift pressure on the underside of the trough is equal to the difference between the downstream water level and the level of the underside of the trough. At any other point along the barrel, the uplift pressure is given by the ordinate between the hydraulic gradient line and the underside of the trough. The afflux having been calculated by Unwin's formula the hydraulic gradient line can be drawn as shown in Fig. 20.11. The maximum uplift pressure occurs on the underside of the trough at the upstream end of the barrel. The uplift pressure would be maximum when the highest flood is passing through the barrel and the worst condition would occur when there is no water in the trough at that time. This is therefore one of the critical conditions to be considered for the design of the trough bed (or barrel roof). The other critical condition to be considered for the design of the trough bed (or barrel roof) is when the trough is running full but there is no water flowing through the barrel.

The thickness of the slab forming the trough bed (or the barrel roof) required to counterbalance the uplift (i.e., for the first condition noted above) is generally less than what is required to support the load of trough water (i.e., for the second condition noted above). However, in some cases the thickness of the slab required to counterbalance the uplift may exceed the thickness required to support the water load. In each of these two cases if the slab is designed to balance the uplift or the water load by gravity only (i.e., by the self weight of the slab) then the required thickness of the slab may be quite excessive. The slab of large thickness

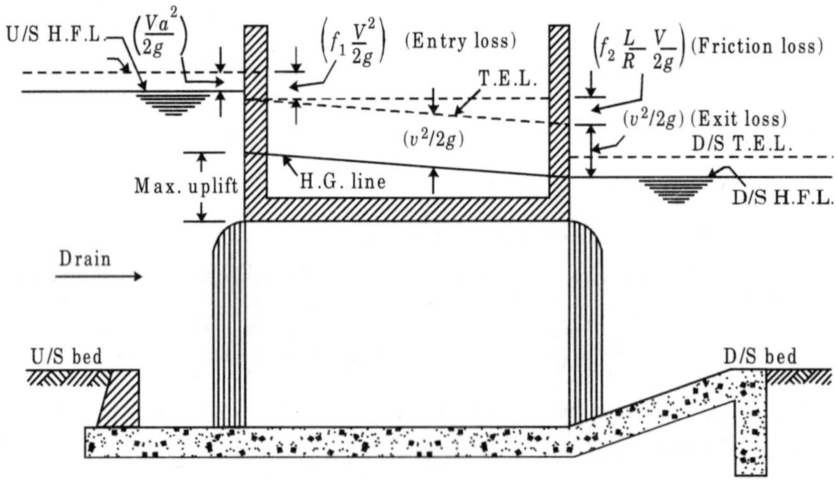

Fig. 20.11 *Uplift pressure on the underside of the trough*

is usually undesirable because it leads to lowering of the levels of both the underside of the trough bed (or the barrel roof) and the barrel floor which in turn results in increasing the uplift on the trough bed as well as on the barrel floor. As such one of the alternatives would be to provide a reinforced concrete slab with reinforcement at the bottom to take the water load of the trough and the reinforcement at the top to resist uplift by bending. It may however be noted that since a part of the uplift is resisted by gravity (i.e., by the self weight of the slab) only the remainder of the uplift after deducting the part counter balanced by gravity is to be resisted by bending. Hence the reinforcement at the top of the slab is to be provided only for the remainder of the uplift. Another alternative may be to provide the thickness of the slab to support the water load in the trough by gravity and to provide reinforcement only at the top of the slab to resist the remainder of the uplift by bending. In both these cases it would, however, be necessary to anchor the slab through the piers to the bottom floor by steel bolts to provide the necessary end reactions for upward bending. A better alternative may be to provide R.C.C. box culvert.

20.11 UPLIFT PRESSURE ON THE FLOOR OF THE CULVERTS OR BARRELS

The floor of the culverts or barrels is subjected to uplift pressure due to the following two causes.

1. *Uplift pressure due to subsoil water in the drain bed.* This uplift pressure acts when the floor of the culverts or barrels is depressed below the bed of the drain. The maximum uplift pressure under the worst condition would occur when there is no water flowing in the drain and

the water table has risen upto the drain bed. There will then be a static uplift pressure on the floor equal to the vertical distance between the drain bed and the bottom of the floor.

2. *Uplift pressure due to seepage of water from the canal to the drain.* On account of difference of head between the canal and the drain, seepage flow will take place from the canal to the drain which causes uplift pressure on the floor of the culverts or barrels. The seepage head will attain a maximum value when the canal is running full and the drain is dry. The seepage flow will start from either side of the impervious floor in the canal trough and will reappear in the drain bed on either side of the impervious floor in the drain bed. Since in this case the seepage flow is three dimensional in character, the uplift pressure due to seepage of water from the canal is difficult to compute. However, an approximate method based on Bligh's creep theory may be used to find the uplift pressure, which for major works needs to be checked by model studies. This method is described below.

As shown in Fig. 20.12 the flow starts from a point A, on the underside of the canal trough at its upstream end and then comes out at point C under the centre of the floor of the first curlvert or barrel at its end. Point B is the point under the centre of the floor of the first culvert.

The length of the seepage path from A to B and from B to C can be known. The total creep length L will then be equal to the sum of the seepage path L_1 from A to B and the seepage path L_2 from B to C i.e., $L = L_1 + L_2$. If H_s = total seepage head = F.S.L. of canal—downstream bed level of the drain, the residual seepage head H_r at the point B is given by Bligh's consideration as

$$H_r = \frac{H_s}{L} \times L_2 \qquad \text{...(20.10)}$$

The residual seepage head given by Eq. (20.10) may be used for the design of the entire floor of the culverts or barrels.

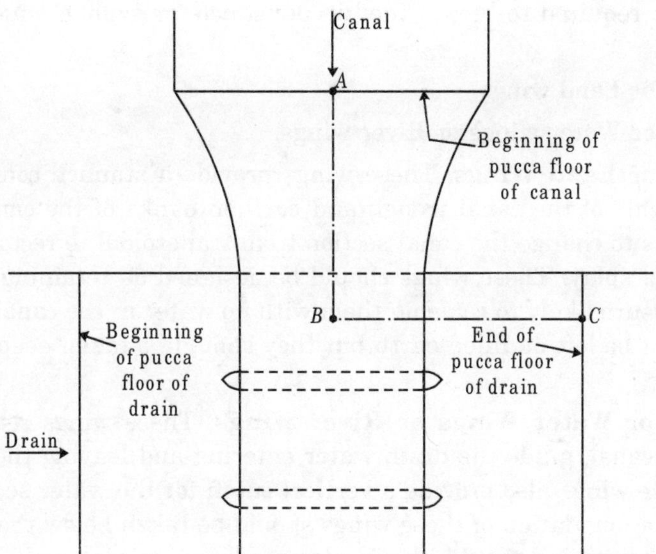

Fig. 20.12 *Uplift pressure on barrel floor*

The floor of the culverts or barrels should be designed for the total uplift pressure which is equal to the sum of the uplift pressure due to static head and the uplift pressure due to seepage head. If this total uplift pressure is to be counterbalanced by the weight of the floor only, the required thickness of the floor will generally be found to be excessive and uneconomical. As such in order to reduce the thickness of the floor reinforcement should be provided on the upper side of the floor. The total uplift pressure will then be partly resisted by the weight of the floor and partly by the bending action.

Methods of Reducing Uplift on the Floor

If the uplift pressure on the floor is unduly high, it may be reduced by adopting the following two methods.

1. By extending the impervious floor of the canal trough on either side. The creep length is thus increased and the uplift pressure is reduced. In place of concrete floor this extension may be provided in clay puddle buried under 0.3 m of soil where economical.

2. By providing openings or holes in the floor. The seepage water from beneath the floor emerges through these holes into the culverts or barrels and hence the uplift pressure is reduced. If the holes are closely spaced there would be practically no uplift pressure acting on the floor. An inverted filter is provided below the floor so that soil particles are prevented from moving out through the holes along with the seepage water. It would, however, be necessary to provide 'flap valves' opening upwards on the top of the holes so that while seepage water from beneath the floor may move out, the silt laden water of the drain may not enter the filter and choke its voids.

20.12 DESIGN OF BANK CONNECTIONS

Two sets of wings are required to be provided in aqueducts or syphon aqueducts. These are as follows:

1. Canal wings or Land wings.
2. Drain wings or Water wings or River wings.

1. Canal Wings or Land Wings. These wings provide a staunch connection between the masonry or concrete side of the canal trough and earthen banks of the canal. These wings are generally warped so as to change the canal section from trapezoidal to rectangular. They should extend upto the end of splay. These wings should be designed as retaining walls for maximum differential earth pressure likely to come on them with no water in the canal. The foundations of these wings should not be left on filled earth but they should be taken deep enough to give safe creep length.

2. Drain Wings or Water Wings or River Wings. These wings retain and protect the earthen slopes of the canal, guide the drain water entering and leaving the work and join it to the guide banks. These wings also provide a vertical cutoff for the water seeping from the canal into the drain bed. The foundation of these wings should be taken below the deepest anticipated scour in the river. These wings should also be designed as retaining walls for maximum earth pressure likely to come on them with no water in the drain.

SOLVED ILLUSTRATIVE EXAMPLES

Illustrative Example 20.1 *Design a syphon aqueduct for the following data :*

(i) Discharge of the canal : 30 cumec

(ii) Bed width of the canal : 20 m

(iii) Depth of water in the canal : 1.6 m

(iv) Bed level of the canal : 260.00 m

(v) High flood discharge of the drain : 450 cumec

(vi) High floor level of the drain : 261.00 m

(vii) Bed level of the drain : 258.00 m

(viii)General ground level : 260.00 m

(ix) Silt factor : 0.9.

Solution

Since the size of the drain is quite large, type III syphon aqueduct will be economical and hence the same is adopted. The canal will be flumed and the canal water will be taken in a concrete trough across the drain.

Step 1. Design of drain water way

Lacey's regime perimeter

$$P = 4.75\sqrt{Q}$$

$$= 4.75\sqrt{450} = 101 \text{ m.}$$

Let clear span of each barrel be kept 7 m, and the pier thickness be kept 1.5 m. Thus, if the clear waterway may be contracted by about 20%, provide 12 spans to give a clear waterway of 12 × 7 = 84 m and an overall waterway of (12 × 7 + 11 × 1.5)

$$= 100.5 \text{ m.}$$

Let the velocity of flow through the barrels be kept

$$= 2.0 \text{ m/s.}$$

∴ Height of barrels required

$$= \frac{450}{84 \times 2.0} = 2.68 \text{ m.}$$

Provide height of barrels as 2.6 m.

∴ Actual velocity of flow through barrels

$$= \frac{450}{84 \times 2.6} = 2.06 \text{ m/s.}$$

Thus, provide rectangular barrels 7 m wide and 2.6 m high.

Step 2. Design of canal waterway

As stated earlier Type III syphon aqueduct is adopted in which the canal is flumed.

Bed width of canal = 20 m

Let the bed width of the canal be reduced to 10 m in the flumed section and provide a 2 : 1 splay in contraction and 3 : 1 splay in expansion.

Length of contraction transition

$$= \frac{20-10}{2} \times 2 = 10 \text{ m.}$$

Length of expansion transition

$$= \frac{20-10}{2} \times 3 = 15 \text{ m}$$

Length of the flumed portion from abutment to abutment

= 100.5 m.

In the transitions the side slopes of the canal will be warped from $1\frac{1}{2}:1$ to vertical.

Step 3. Design of levels at different sections of canal trough

Refer Fig. 20.9.

(a) *At section 4–4*

Area of section with side slope $1\frac{1}{2}:1$

$$= (B + 1.5\ D)D = (20 + 1.5 \times 1.6)1.6$$

$$= 35.84 \text{ m}^2.$$

Velocity $V = \dfrac{Q}{A} = \dfrac{30}{35.84} = 0.84$ m/s.

Velocity head $\dfrac{V^2}{2g} = \dfrac{(0.84)^2}{2 \times 9.81} = 0.036$ m.

R.L. of canal bed = 260.00 (given)

∴ R.L. of water surface at section 4–4

= 260.00 + Water depth = 260.00 + 1.6

= 261.60 m.

R.L. of T.E.L. at section 4–4

= 261.60 + 0.036 = 261.636 m.

(b) *At section 3–3*

Area of trough $= 10 \times 1.6 = 16 \text{ m}^2$

∴ Velocity $V = \dfrac{Q}{A} = \dfrac{30}{16} = 1.88$ m/s

Velocity head

$$\frac{V_2}{2g} = \frac{(1.88)^2}{2\times 9.81} = 0.18 \text{ m}$$

Loss of head in expansion transition from section 3–3 to section 4–4

$$= 0.3\left(\frac{V_3^2 - V_4^2}{2g}\right) = 0.3(0.18 - 0.036)$$

$$= 0.043 \text{ m}$$

∴ R.L. of T.E.L. at section 3–3

= R.L. of T.E.L. at section 4–4 + Head loss

= 261.636 + 0.043 = 261.679 m

R.L. of water surface at section 3–3

= R.L. of T.E.L. at section 3–3–Velocity head

= 261.679 – 0.18 = 261.499 m

∴ R.L. of bed to maintain constant water depth

= R.L. of water surface at section 3–3 – Water depth

= 261.499 – 1.6 = 259.899 m

From section 3–3 to section 2–2, area and velocity are constant.

(c) *At section 2–2*

Hydraulic mean radius of the trough

$$R = \frac{A}{P} = \frac{10\times 1.6}{10+2\times 1.6} = 1.212 \text{ m}$$

By Manning's formula

$$V = \frac{1}{N}R^{2/3}\,S^{1/2}$$

or

$$S = \frac{V^2N^2}{R^{4/3}}$$

Taking Manning's $N = 0.016$, we get

$$S = \frac{(1.88)^2\times(0.016)^2}{(1.212)^{4/3}} = 7.0\times 10^{-4}$$

Length of flumed portion of trough

= 100.5 m

Head loss in trough = $100.5 \times 7.0 \times 10^{-4} = 0.07$ m

∴ R.L. of T.E.L. at section 2–2

= R.L. of T.E.L. at section 3–3 + Head loss in trough

$= 261.679 + 0.07 = 261.749$ m

R.L. of water surface at section 2–2

$= $ R.L. of T.E.L. at section 2–2 – Velocity head

$= 261.749 - 0.18 = 261.569$ m.

R.L. of bed at section 2–2 to maintain constant water depth

$= $ R.L. of water surface at section 2–2 – Water depth

$= 261.569 - 1.6 = 259.969$ m.

(d) *At section 1–1*

Loss of head in contraction transition from section 1–1 to section 2–2

$$= 0.2\left(\frac{V_2^2 - V_1^2}{2g}\right) = 0.2\left[\frac{(1.88)^2 - (0.84)^2}{2 \times 9.81}\right]$$

$$= 0.029 \text{ m.}$$

R.L. of T.E.L. at section 1–1

$= $ R.L. of T.E.L. at section 2–2 + Head loss

$= 261.749 + 0.029 = 261.778$ m

R.L. of water surface at section 1–1

$= $ R.L. of T.E.L. at section 1–1 – Velocity head

$= 261.778 - = 261.742$ m.

R.L. of bed at section 1–1 to maintain constant depth

$= 261.742 - 1.6$

$= 260.142$ m.

Step 4. Design of contraction transition

The transition is designed on the basis of Chaturvedi's formula. Thus from Eq. (20.5) we have

$$x = \frac{L_f B_c^{3/2}}{B_c^{3/2} - B_f^{3/2}}\left\{1 - \left(\frac{B_f}{B_x}\right)^{3/2}\right\}$$

$$B_c = 20 \text{ m} ;\ B_f = 10 \text{ m} ;\ L_f = 10 \text{ m}$$

$$\therefore \quad x = \frac{10 \times (20)^{3/2}}{(20)^{3/2} - (10)^{3/2}}\left\{1 - \left(\frac{10}{B_x}\right)^{3/2}\right\}$$

or

$$x = 15.469\left\{1 - \left(\frac{10}{B_x}\right)^{3/2}\right\}$$

The values of x for various values of B_x are calculated from the above relation and are tabulated below:

B_x	10	11	12	13	14	15	16	17	18	19	20
x	0	2.06	3.70	5.03	6.13	7.05	7.83	8.49	9.06	9.56	10.00

The distance x is measured from the flumed section as shown in Fig. 20.12.

Step 5. Design of expansion transition

The expansion transition is also designed on the basis of Chaturvedi's formula. Thus,

$$x = \frac{L_f B_c^{3/2}}{B_c^{3/2} - B_f^{3/2}} \left\{1 - \left(\frac{B_f}{B_x}\right)^{3/2}\right\}$$

In this case

$$B_c = 20 \text{ m} ; B_f = 10 \text{ m} ; L_f = 15 \text{ m}$$

$$\therefore \qquad x = \frac{15(20)^{3/2}}{(20)^{3/2} - (10)^{3/2}} \left\{1 - \left(\frac{10}{B_x}\right)^{3/2}\right\}$$

or

$$x = 23.204 \left\{1 - \left(\frac{10}{B_x}\right)^{3/2}\right\}$$

The values of x for various values of B_x are calculated from the above relation and are tabulated below:

B_x	10	11	12	13	14	15	16	17	18	19	20
X	0	3.09	5.55	7.55	9.20	10.57	11.74	12.74	13.60	14.34	15.00

Step 6. Design of the trough

The width of the trough carrying canal water is 10 m. The trough will be divided into two compartments each 5 m wide separated by 0.3 m thick partition wall. The road 5 m wide will be carried over one of the compartments. Providing for a freeboard of 0.5 m, the height of the trough will be kept equal to 1.6 m + 0.5 m = 2.1 m. The entire trough section will be constructed in monolithic reinforced concrete and it can be designed by the usual methods of structural design. The outer walls and the bottom slab of the trough may be tentatively kept equal to 0.4 m thick. Thus, the outer width of the trough (or the total length of the syphon barrel) = 10 + 0.3 + (2 × 0.4) = 11.1 m.

Step 7. Head loss through syphon barrels

The loss of head through the syphon barrels can be calculated from the Unwin's formula as noted below :

$$h = \left(1 + f_1 + f_2 \frac{L}{R}\right)\frac{V^2}{2g} \qquad \text{...(i)}$$

$$V = 2.06 \text{ m/s} ; f_1 = 0.505 ; L = 11.1 \text{ m}$$

$$f_2 = a\left(1 + \frac{b}{R}\right)$$

For cement plaster

$$a = 0.00316 \text{ and } b = 0.030$$

$$R = \frac{A}{P} = \frac{7 \times 2.6}{2(7+2.6)} = 0.948 \text{ m}$$

$$\therefore \qquad f_2 = 0.00316\left(1+\frac{0.03}{0.948}\right) = 0.0033$$

Thus, substituting these values in Eq. (i) we get

$$h = \left[\left(1+0.505+\frac{0.0033 \times 11.1}{0.948}\right)\frac{(2.06)^2}{2 \times 9.81}\right]$$

or $$h = 0.334 \text{ m}$$

Downstream H.F.L. = 261.00 (given)

∴ Upstream H.F.L. = Downstream H.F.L. + Loss of head

= 261.00 + 0.334 = 261.334 m

Step 8. Uplift pressure on the underside of the trough or the barrel roof.

R.L. of underside of the trough

= R.L. of canal bed – Slab thickness

= 260.00 – 0.40 = 259.60 m.

Loss of head at the entry $= 0.505\dfrac{V_2}{2g} = 0.505\dfrac{(2.06)^2}{2 \times 9.81} = 0.109$ m.

Velocity head in the barrel $= \dfrac{V^2}{2g} = \dfrac{(2.06)^2}{2 \times 9.81} = 0.216$ m.

∴ Uplift pressure on the underside of the trough at the upstream end of the barrel

= 261.334 – 0.109 – 0.216 – 259.60

= 1.409 m

= 13.822 kN/m^2 [or 1.409 t/m^2]

The bottom slab of the trough is 0.4 m thick which exerts a downward load of

0.4 × 9.81 × 2.4 = 9.418 kN/m^2 [or 0.4 × 2.4 = 0.96 t/m^2]

The balance of the uplift pressure

= (13.822 – 9.418) = 4.404 kN/m^2

or = [(1.409 – 0.96) = 0.449 t/m^2]

has to be resisted by the bending action, for which the necessary reinforcement should be provided at the top of the slab. Since the trough as a whole is heavier than the total upward force no anchorage to the piers is necessary. Further when the water level in the drain is low, the bottom slab of the trough is to be designed for the full water load of the canal and the necessary reinforcement should be provided at the bottom of the slab.

Step 9. Uplift pressure on the floor of the barrels

(a) *Static head*

R.L. of barrel floor = R.L. of under side of the trough – Height of the barrel

= 259.60 – 2.60

= 257.00 m

Let the floor thickness be kept equal to 0.8 m

∴ R.L. of bottom of floor = 257.00 – 0.80

= 256.20

R.L. of drain bed = 258.00 (given)

Assuming the sub-soil water level to be upto the bed level of the drain, the static uplift on the floor

= 258.00 – 256.20 = 1.80 m

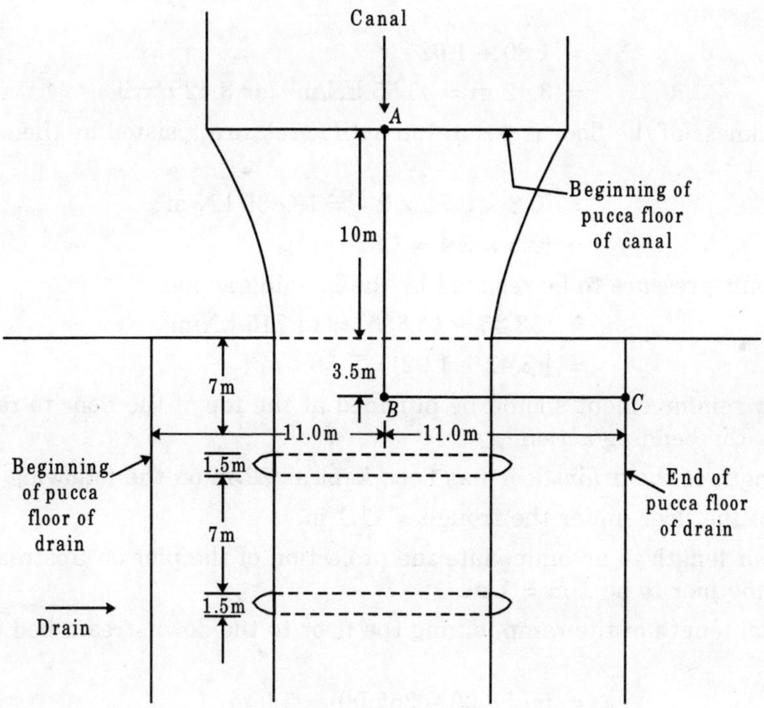

Fig. Ex. 20.1

(b) *Seepage head*

The residual seepage head at the centre of the first barrel (point *B*) can be calculated with reference to Fig. Ex. 20.1. The seepage line according to Bligh's theory will follow the path *ABC*. Neglecting the floor depression or floor thickness the various lengths are as follows :

(i) Length of upstream transition = 10 m

(ii) Half the barrel span = 3.5 m

(iii) End of drain floor from the centre of the barrel, assuming a total length of 22 m = 11.0 m.

∴ Total creep length

$$= 24.5 \text{ m}$$

Creep length upto the centre of the barrel, i.e., upto point B

$$= 13.5 \text{ m}$$

Total seepage head

$$= \text{Canal F.S.L.} - \text{R.L. of drain bed}$$

$$= 261.60 - 258.00 = 3.60 \text{ m}$$

∴ Residual seepage head at point B

$$= 3.6 = \left(1 - \frac{13.5}{24.5}\right) 1.62 \text{ m.}$$

Total uplift pressure

$$= 1.80 + 1.62$$

$$= 3.42 \text{ m} = 33.55 \text{ kN/m}^2 \text{ [or } 3.42 \ t/\text{m}^2]$$

Since the thickness of the floor is 0.8 m the uplift pressure resisted by the self weight of the floor

$$= 0.8 \times 9.81 \times 2.4 = 18.835 \text{ kN/m}^2$$

or $$= [0.8 \times 2.4 = 1.92 \ t/\text{m}^2]$$

∴ Balance uplift pressure to be resisted by the bending action

$$= (33.35 - 18.835) = 14.715 \text{ kN/m}^2$$

or $$= [(3.42 - 1.92) = 1.50 \ t/\text{m}^2]$$

The necessary reinforcement should be provided at the top of the floor to resist the balance uplift pressure by the bending action.

Note. The length of the drain floor has been kept as 22 m on the following considerations :

(i) Length of the floor under the trough = 11.1 m.

(ii) Extra floor length to accommodate the projection of the pier on upstream side, keeping the projection of the pier to be 1 m = 1 m.

(iii) Horizontal length of the ramp joining the floor to the downstream bed of the drain at a slope of 5 : 1

$$= 5(258.00 - 257.00) = 5.0 \text{ m}$$

(iv) Width of downstream cutoff beyond ramp

$$= 0.5 \text{ m}$$

(v) Length of the extra floor on the upstream side

$$= 4.4 \text{ m}$$

Thus, total length of floor = 22 m

Step 10. Design of cutoff and protection works for the drain floor.

The scour depth R

$$= 0.47 \left(\frac{Q}{f}\right)^{1/3}$$

$$= 0.47 \left(\frac{450}{0.9}\right)^{1./3} = 3.73 \text{ m}$$

∴ Depth of upstream cutoff below H.F.L.

$= 1.5\,R = 1.5 \times 3.73$

$= 5.595 \simeq 5.6$ m.

∴ R.L. of bottom of upstream cutoff

= Upstream H.F.L. – 5.60

= 261.334 – 5.60

= 255.734

≃ 255.73 m.

Depth of downstream cutoff below H.F.L.

$= 2.0\,R = 2.0 \times 3.73$

= 7.46 m.

∴ R.L. of bottom of downstream cutoff

= Downstream H.F.L. — 7.46

= 261.00 – 7.46 = 253.54 m.

Length of downstream protection consisting of 0.4 m thick brick pitching

= 2.5 (Depth of downstream cutoff below the bed of the drain)

= 2.5 (R.L. of downstream bed of the drain – R.L. of bottom of downstream cutoff)

= 2.5 (258.00 – 253.54)

= 11.15 m 11.2 m.

The pitching should be supported by a toe wall 0.4 m thick and 1.5 m deep at its downstream end.

Length of upstream protection consisting of 0.4 m thick brick pitching

= 2 (Depth of upstream cutoff below the bed of the drain)

= 2 (R.L. of upstream bed of the drain – R.L. of bottom of upstream cutoff)

= 2 (258.00 – 255.73)

= 4.54 m = 4.6 m.

The pitching should be supported by a toe wall 0.4 m thick and 1 m depth at its upstream end.

The cross-section of the syphon aqueduct is shown in the Fig. Ex. 20.1.

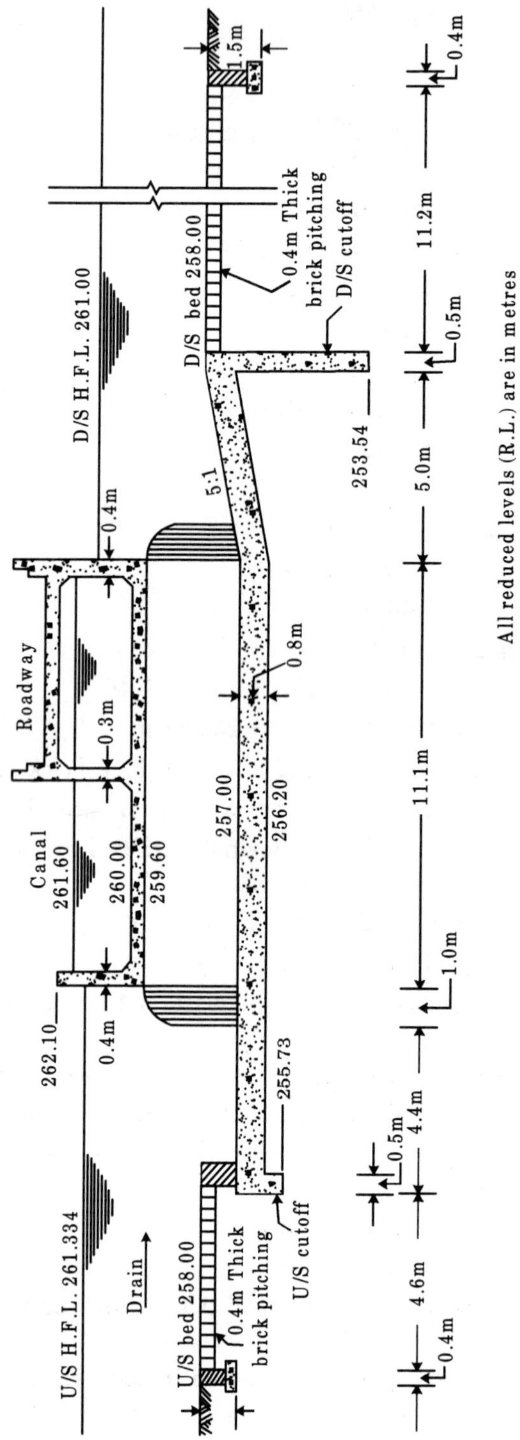
U/S H.F.L. 261.334
262.10
0.4m
Canal
261.60
260.00
259.60
Roadway
0.3m
0.4m
D/S H.F.L. 261.00
Drain
U/S bed 258.00
0.4m Thick
brick pitching
U/S cutoff
255.73
257.00
256.20
0.8m
5:1
D/S bed 258.00
0.4m Thick
brick pitching
D/S cutoff
253.54
1.5m
0.4m
4.6m
0.5m
4.4m
1.0m
11.1m
5.0m
0.5m
11.2m
0.4m
All reduced levels (R.L.) are in metres

Fig. Ex. 20.1

REVIEW QUESTIONS

20.1 Describe with the help of neat sketches the various types of cross drainage works.

20.2 Differentiate between (a) aqueduct and syphon aqueduct, (b) super passage and canal syphon.

20.3 What is level crossing ?

20.4 Discuss the various factors affecting the suitability of aqueduct and syphon aqueduct.

20.5 How aqueducts (or syphon aqueducts) are classified? Indicate the circumstances under which each one is used.

20.6 Explain the method of fixing the waterway of drain in an aqueduct.

20.7 Describe Mitra's method of hyperbolic transition. Derive the expression which represents this transition.

20.8 Discuss Chaturvedi's semi-cubical parabolic transition.

20.9 Describe Hind's method of design of transition.

20.10 Explain the method of determining uplift pressure on the roof of a syphon aqueduct.

20.11 Describe Bligh's approximate method of determining uplift pressure on the floor of the culvert of a cross drainage work.

20.12 What is meant by bank connections ? How are these designed ?

20.13 A syphon aqueduct is to be designed across a stream for the following data :

Canal — Discharge = 56 cumec, Bed width = 32 m,
Water depth = 2.0 m, R.L. of bed = 267.0 m.

Drainage— High flood discharge = 425 cumec, High flood level = 268.20 m, General bed level of low water cross-section = 265.50 m, General ground level = 267.20 m.

Find the number and size of barrels required for the drainage and also design the expanding transition for the canal. Also find the uplift on the barrel roof. Make suitable assumptions where required.

20.14 Using Mitra's method design the canal transition both contraction and expansion for an aqueduct from the following data:

Canal — Discharge = 100 cumec, Side slopes 1 (V) : $\frac{1}{2}H$, Depth of flow = 2.7 m, Bed width of canal = 35.0 m, R.L. of canal bed = 100.0 m at the end of expansion transition. Bed width of the canal in the flumed portion is reduced to 12.5 m and the length of the flumed portion is 60 m. Make suitable assumptions where required.

20.15 A syphon aqueduct is to be designed to pass a maximum flood discharge of 450 m^3/s. R.C.C. rectangular barrels 7.50 m wide and 2.0 m high are proposed to be provided to pass the drainage discharge. Total length of barrel below the canal is 20.0 m. Determine the total number of barrels required and plot the hydraulic gradient line and determine the maximum uplift on the barrel roof. Take a = 0.00316 and b = 0.030 in Unwin's formula.

20.16 A syphon is to be designed for the following data :

Canal— Discharge = 20 cumec

Bed width = 16.50 m

Depth = 1.45 m

Bed level = 249.00 m

Drain —High flood discharge = 200 cumec

Bed level = 249.50 m

High flood level = 251.50 m

Silt factor = 2.0

Design the following :

(a) Waterway for the drainage and the drainage bed at the syphon.

(b) Canal waterway including the number of R.C.C. barrels needed.

(c) Contraction transition for the canal if it is flumed to 4.75 m on the inlet side.

CHAPTER

21 River Engineering

21.1 INTRODUCTION

Rivers are one of the principal sources of water required for satisfying the demand for various purposes such as municipal and industrial water supply, irrigation, power generation, etc. Further rivers also form inland waterways and thus serve as important means of communication. However, the river waters can be utilized only if the rivers are properly trained and controlled. The river engineering in general deals with behaviour, control and training of rivers. Some of these aspects of river engineering are discussed in the following sections.

21.2 CLASSIFICATION OF RIVERS

As indicated in chapter 13 rivers pass through the following four stages :

1. Rocky stage
2. Boulder stage
3. Trough stage
4. Deltaic stage

In the rocky or hilly stage the river cross-section is made up of rock or very large boulders and hence in this stage river training works are generally not required. In the other three stages the river flows through the deposits created by the river itself. In the boulder stage the bed and banks of the river consists of a mixture of boulders, gravels, shingles and alluvial sand deposits. In the trough and the deltaic stage the river flows through alluvial sand deposits. It is in these three lower stages of the river, the river training works are required.

The rivers on alluvial plains may be broadly classified into the following three types :

1. Meandering type
2. Aggrading type
3. Degrading type

1. **Meandering Type.** A 'meandering type' of river flows in consecutive curves of reverse order connected with short straight reaches called 'crossings' (Fig. 21.1). The deepest part of the section is found near the apex of the curve, whereas the shallowest portion is found near the crossing. The geometry of a meander is defined by the following parameters.

Width of meander belt or *Meander width* M_b. It is the transverse distance between the apex point of one curve and the apex point of other curve.

Length of meander M_l. It is the air distance along the river measured from the tangent point of the one curve to the tangent point of the next curve of the same order.

Meander ratio (M_b/M_l). It is the ratio of meander width to meander length.

Degree of sinuosity or Tortuosity. It is the ratio of the curved length of the river to the straight air distance (or meander length).

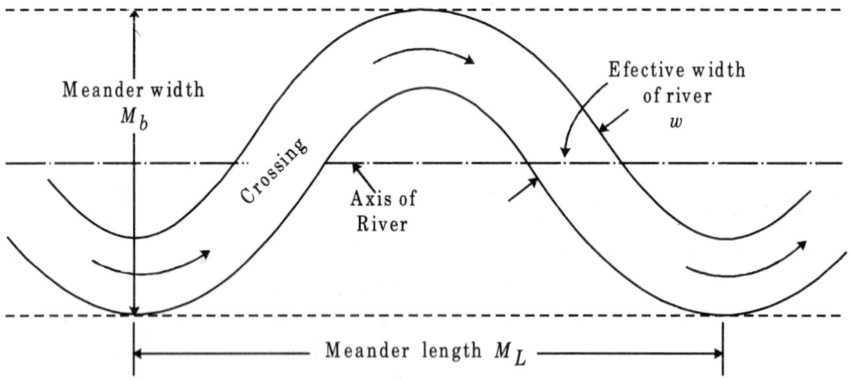

Fig. 21.1 *River meander*

The causes of meandering given by various investigators are as described below.

Causes of Meandering

According the *Inglis* the development of meanders is explained as follows.

When there is a heavy load of bed material in movement during floods, excess turbulent energy is developed due to unevenness of the bed. In such a condition symmetrical axial flow is not maintained and flow tends to concentrate towards one of the banks. The bank towards which the flow concentrates resists the component of velocity causing flow towards itself and converts part of it into pressure energy which results in increasing the water level near this bank as shown in Fig. 21.2 (a). This pressure head causes a cross current of bottom water towards the other bank as shown in Fig. 21.2 (b), leading to the deposition of bed material towards that bank. To replace the bottom water, surface water from the centre and the inner bank develops a curved diving flow towards the outer bank [Fig. 21.2 (a)] accompanied by forced vortex condition. This leads to the undercutting of the bank near its toe and falling down of this material into the river. Such slipping of material occurs to a marked extent when the river stage is falling as the bank is then saturated. The material eroded from the outer bank is carried to the inner bank and gets deposited there as shown in Fig. 21.2 (c), thus leading to embayment on one side and deposition on the other till equilibrium is attained and a more or less stable meandering pattern is developed for the river. The equilibrium conditions represent the total effect of all the varying water discharges and sediment discharges entering the river during various months of the year. So long as this total effect remains substantially unchanged, the river would maintain reasonably stable conditions on a long term basis, though there may be temporary deposition and scour at low and high stages respectively. In general the equilibrium

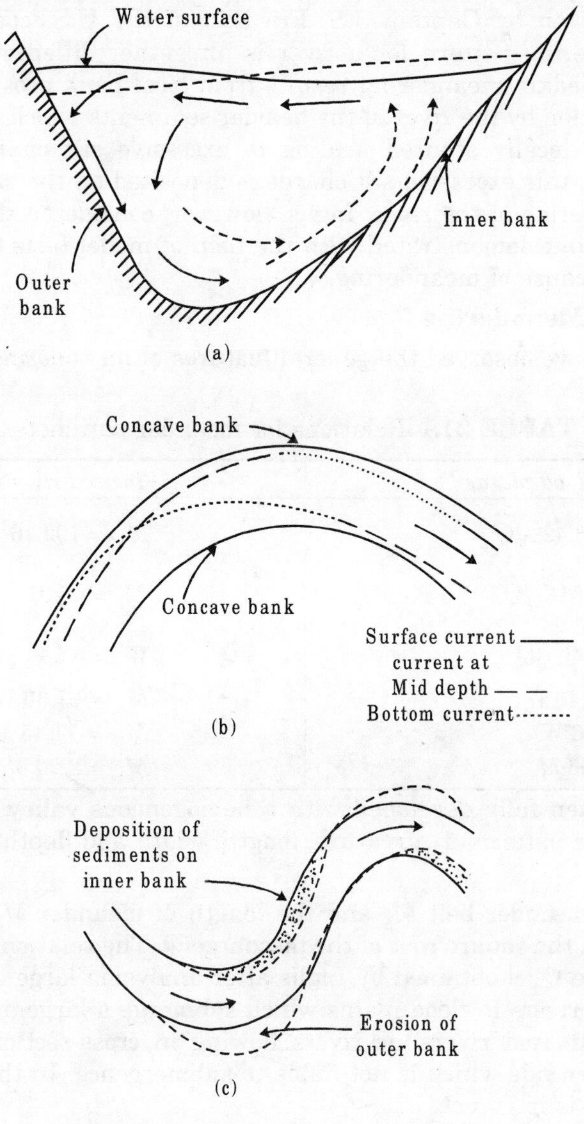

Fig. 21.2 *Flow at a bend in a river*

conditions attained by rivers are only temporary but they may appear to be stable if there is no material change in the conditions from year to year. However, if the conditions change materially from one year to the next, the pattern of meandering will change and try to readjust itself to the changed conditions. The discharge at which the river is temporarily in equilibrium has been termed as *dominant discharge* by Inglis. In many rivers, the dominant discharge agrees closely with the bankful stage. However, in most rivers the dominant discharge varies between $\frac{1}{2}$ and $\frac{2}{3}$ of the maximum discharge.

The explanation given by Captain J.S. Friedkin of the U.S. corps of engineers for the development of meandering pattern for a river is altogether different from the one given by Inglis. According to Friedkin meandering results from local bank erosion and consequent local overloading and deposition by the river of the heavier sediments which move along the bed. This means that a bank is locally scoured leading to excessive silt charge beyond the carrying capacity of the stream ; this excessive silt charge is deposited on the same bank a little further down leading to meandering of the river. Inglis, however, considered this as a possible cause of secondary meandering but demonstrated with the help of model tests the explanation given by him to be the primary cause of meandering.

General Features of Meandering

Various investigators have observed the general features of meandering river and the same are as discussed below :

TABLE 21.1 Relations for meander parameters

Rivers in flood plains	*Incised rivers*
$M_b = 153.42\sqrt{Q}$	$M_b = 102.16\sqrt{Q}$
$M_l = 53.61\sqrt{Q}$	$M_l = 46.01\sqrt{Q}$
$W = 8.84\sqrt{Q}$	$W = 4.53\sqrt{Q}$
$M_b = 17.40\ W$	$M_b = 27.30\ W$
$M_l = 6.06\ W$	$M_l = 11.45\ W$
$M_b = 2.86\ M_l$	$M_b = 2.20\ M_l$

(i) A meander when fully developed with a homogeneous valley material and a constant discharge has a definite pattern of curvature, length, width and depth of channel which can be reproduced in models.

(ii) The width of meander belt M_b and the length of meander M_l as well as the width of river W vary roughly as the square root of the discharge Q. The relation between M_b , M_l , W and the maximum discharge Q as obtained by Inglis after analysing large data on Indian rivers are given in Table 21.1 for rivers in flood plains which submerge a large area on either side during floods as well as for 'incised rivers' or rivers flowing in cross-section cut below the natural ground surface on either side which is not liable to submergence. In these relations M_b, M_l and W are in metres and Q is in cumec.

(iii) The pattern of meander changes whenever there is change in discharge or the material transported by the river.

(iv) Generally the size of bends, their length, width and degree of sinuosity increase with increase in discharge as well as with increase in sediment load.

(v) With time, the meanders progress downstream. However, in most cases the rate of the progress is very slow due to resisting bank material.

From the foregoing discussion it is observed that every phase of meandering depends upon the combined influence of the following three closely related factors :

(i) The discharge and hydraulic properties of the river channel.

(ii) The sediment load of the river.

(*iii*) The relative erodibility of bed and banks of the river.

2. Aggrading Type. An aggrading type of river or an aggrading reach of a river is in the process of building up its bed to a certain slope. This may be due to the following causes :

(i) Excessive sediment entering a river with a sudden reduction of slope on the plain.

(ii) Excessive sediment entering a river with water surface slope flattened due to construction of a weir or a barrage or a dam across a river downstream of a reach.

(iii) The sudden influx of sediment from a tributary.

(iv) Extension of delta at the river mouth.

A river of aggrading type or an aggrading reach of a river is usually straight and wide with shoals in the middle and divided flow.

3. Degrading Type. A degrading type of river or a degrading reach of a river is in the process of losing its bed gradually in the form of sediment load of the river. A degrading reach may develop on a river downstream of a weir or a barrage or a dam due to holding up of the sediment load above and the same being replenished by the water flowing on the downstream side by scouring the river bed. It may also result due to the sudden increase in slope which takes place when a loop of a meandering river is cut off by a straight river channel.

A river has to adjust its alignment, cross-section and slope according to the particular combination of discharge and sediment load prevailing on it. Meandering, aggrading and degrading are the various ways in which a river adjusts itself. The meandering and aggrading results from an excess of sediment while the degrading results from a deficit of sediment. Generally where a river has enough capacity to carry the incoming sediment load without forming large deposits the whole or part of it would be of the meandering type. It may be stated that out of the three types of rivers, the meandering type is the full and final state of development of an alluvial river. The other two are interim phases which are maintained so long as the factors causing them remain operative. Moreover, even the meandering stage also does not represent a permanently stable condition as any change in the balance caused by natural or artificial forces would affect its equilibrium.

21.3 RIVER TRAINING AND ITS OBJECTIVES

The term *river training* implies various measures adopted on a river to stabilize the river channel along a certain alignment with a certain cross-section. These measures are required to be adopted because rivers in alluvial plains frequently alter their courses and cause damage to land and property adjacent to their banks. The main objectives of river training are as follows :

1. To provide a safe passage to flood discharges without overflowing to the banks for protection of cultivated or inhabited area.

2. To prevent outflanking of a work like a bridge, weir or aqueduct constructed across the river and to bring the river on to the work in a straight non-tortuous approach.

3. To protect the banks from erosion and improve the alignment by stabilizing the river channel.

4. To deflect the river away from the bank which it might be attacking.

5. To provide minimum depth of flow and a good course for navigation purposes.
6. To transport efficiently the bed and suspended sediment load.

21.4 CLASSIFICATION OF RIVER TRANING WORKS

On the basis of the purpose required to be served the river training works may be classified into the following types :

1. High water training 2. Low water training 3. Mean water training

1. **High water training.** High water training is undertaken with the purpose of providing expeditious disposal of maximum floods and thus provide protection against damage due to floods. It is mainly concerned with the most suitable alignment and height of marginal embankments for disposal of floods and may also include other measures of channel improvement for the same purpose. Thus, high water training can also be called *Training for discharge*.

2. **Low water training.** Low water training is undertaken with the purpose of providing sufficient depth for navigation during the low water season. This is achieved by contracting the width of the channel at low water and is usually carried out with the help of groynes. Thus, low water training can also be called *Training for depth*.

3. ***Mean water training.*** Mean water training is undertaken to provide efficient disposal of bed and suspended sediment load and thus to preserve the river channel in good shape. Thus, mean water training can also be called *Training for sediment*.

Out of the three types of river training, mean water training is the most important. This is so because a river training work adopted to alter the river cross-section and alignment must obviously be designed in accordance with that stage of the river at which the maximum movement of sediment takes place during any period under consideration. Although there is maximum activity of the bed of the river at the high stage of flow but such a stage is maintained only for a short duration. On the other hand there is little movement of sediment at low stages which persist for a very long duration. In between the two there is a stage at which the combined effect of forces causing sediment movement and the time for which such forces are maintained is maximum. This is therefore the most important stage as it has considerable influence on the configuration of the river. This stage is some where near the mean water. As such mean water training forms the basis on which both high and low water training are planned.

21.5 METHODS FOR RIVER TRAINING

The various methods adopted for river training including bank protection are as follows :

1. Marginal embankments or levees
2. Guide banks or Guide bunds
3. Groynes or spurs
4. Cutoffs
5. Pitching of banks and provision of launching aprons.
6. Pitched islands
7. Miscellaneous methods such as sills, closing dykes, bandalling etc.

Each of the above noted methods are described in the following sections.

21.6 MARGINAL EMBANKMENTS OR LEVEES

Marginal embankments or levees are the earthen embankments which are provided to confine the flood waters of the river within the cross-section available in between the embankments. Thus, spreading of floods beyond the marginal embankments is prevented.

The alignment of marginal embankments should follow the normal pattern of meandering of the river. The retirement (i.e., spacing away from the main channel) of the embankments is governed by technical as well as human considerations, as the land falling on the river side of the embankments remains unprotected. The section of the embankments is governed by the usual considerations for the design of earthen dams as indicated in Chapter 11. If the marginal embankments are likely to come in contact with high velocity flow then the waterside of the embankments should be provided with pitching protection. Moreover, launching apron may also be provided if the embankment is close to the main river channel. However, for retired embankments both these would normally be unnecessary.

Effect of Marginal Embankments on River Flow during Floods

The effect of confining the flood waters of a river between marginal embankments or levees is :

1. to increase the rate at which the flood wave travels down the stream
2. to increase the water surface elevation at flood
3. to increase the maximum discharge at all points downstream
4. to reduce the water surface slope of the stream on the upstream of the leveed portion, and
5. to increase the velocity and the scouring action through the leveed sections.

Each of these points are explained below.

When marginal embankments are constructed, they restrict the valley storage to a greater or smaller extent in accordance with their spacing from the main river. The reduction in the valley storage would lead to quicker passage of the flood wave, higher peak discharge rates and higher flood stages which explains the first three points noted above. The fourth point follows from the fact that the restriction of the flow passage in the leveed reach would lead to formation of backwater and hence flattening of water surface slope on the upstream of that reach. The fifth point viz., increase in the velocity and the scouring action through the leveed section depends on the effect of rising and falling stages of the flood as well as on the spacing of the marginal embankments. During rising stage of the flood since the discharge and velocity keep on rising, the sediment transporting capacity of the stream will be increased and hence scouring of section will usually be in excess of silting or deposition of material in the bed. On the other hand during falling stage (i.e., when the flood recedes) the silting will exceed the scouring of the cross-section. Thus, the total effect of the passage of a flood on the river cross-section will be the net result of scouring and silting during the rising and falling stages of the flood. Further as the marginal embankments increase the flood stage and the velocity, they would normally increase the sediment transporting capacity of the stream and hence increase the scouring action. However, this would depend on the spacing of the marginal embankments. Thus, if the embankments are placed far apart or in other words if 'retired embankments' are provided, there would be relatively more valley storage and small rise in high flood level, the increase in velocity would also be small which would result in a lower sediment transporting capacity and hence reduced scouring action. On the other hand for closely spaced embankments reverse would be the case.

It is thus seen that the effect of marginal embankments on the bed levels of a river depends on the factors affecting the natural tendencies of a river briefed above and the spacing of the embankments relative to the main channel and apart from each other. As such the effect of marginal embankments will not be the same in all the cases.

Merits and Demerits of River Training by Marginal Embankments

The various merits of marginal embankments as a method of river training for flood protection are as follows.

(i) Embankments are the only means to prevent the spreading of flood waters over the large areas of flood plains. The spreading of flood waters would cause considerable hardship to the inhabitants of the villages on the flood plains and would result in damage to their houses and other belongings. Such damages would, however, be avoided by providing the embankments.

(ii) The initial cost of embankments is low, although when raised subsequently, they become expensive.

(iii) The construction of embankments is easy and presents no difficulty, as it can be done by utilising local materials and unskilled labour. Also the maintenance of embankments is simple and cheap.

(iv) The embankments may be constructed in parts, provided that the ends are properly protected.

The various demerits of marginal embankments as a method of river training are as follows.

(i) Embankments cause raising of high flood levels.

(ii) Embankments may fail by piping due to boring by small animals like crabs and worms. As such they need to be supervised closely during flood and protected as soon as they are in danger.

(iii) In the event of a breach there is a sudden and considerable inflow of water which may cause damage to the neighbourhood and may result in the deposition of considerable quantities of sand rendering vast areas unproductive. Moreover, embankment breaches may result in flooding almost the entire areas, protected by the embankment system.

(iv) In a flood plain unprotected by embankments the flood waters spread over the plain during every flood season and leave a deposit of fine silt behind them as they recede. Thus, the land gets benefitted by way of inundation irrigation as well as adding new fertile soil during every flood season. When embankments are provided for flood protection the land in the flood plains would be deprived of these benefits.

21.7 GUIDE BANKS OR GUIDE BUNDS

The guide banks are the earthen embankments provided to confine the flood water of alluvial rivers within reasonable waterway and provide a straight non-tortuous approach towards a work constructed across the river. They also tend to prevent the river from changing its course and outflanking the work. In India the guide banks were first designed by Bell and hence these are often called Bell's bunds. The Bell's design was later modified by Spring and Gales and the same is being adopted these days.

The alluvial rivers in flood plains spread over a large area during floods and hence it would be too expensive to provide bridges or any other work across them if they were to span across the entire natural water spread during the flood. By providing the guide banks an artificial gorge section is created for the river in which it can flow without causing abnormal velocities or scours, and hence the length of the work to be constructed is considerably reduced.

As shown in Fig. 21.3 the main parts of a guide bank are as follows.

(i) Upstream curved head or impregnable head.

(ii) Downstream curved head.

(iii) Shank or straight portion which joins the two curved heads.

(iv) Slope protection and Launching apron.

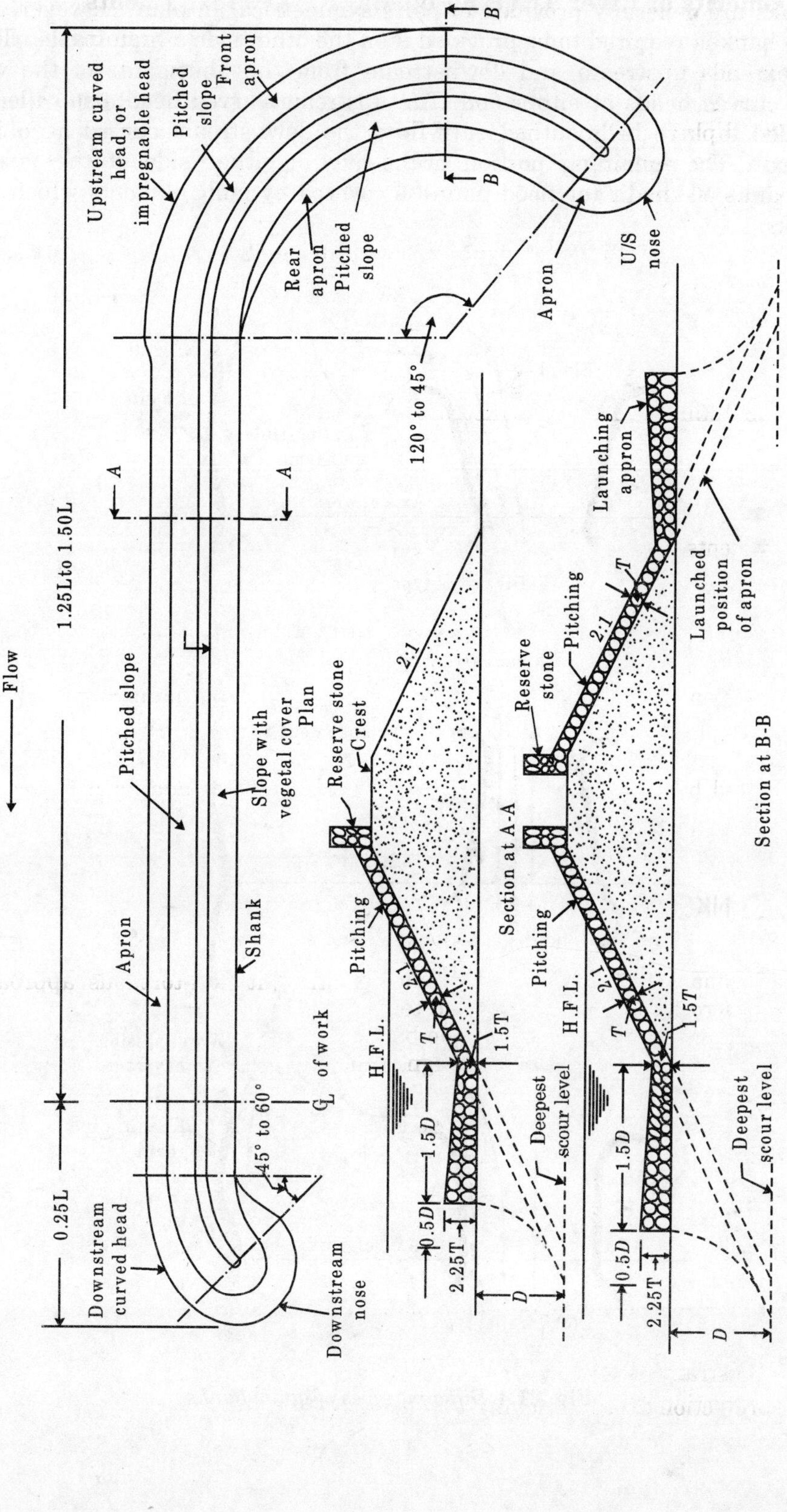

Fig. 21.3 *Guide bank*

The guide banks are generally provided in pairs symmetrical in plan. However, in some cases only one guide bank is required to be provided if on the other side a high stable bank exists. The guide banks extends upstream and downstream from the abutments of the work and are provided with curved heads at either end. The upstream curved head also called *impregnable head* is provided to have bellmouthed entry and the downstream curved head is provided to have smooth exit. The remaining portion of the river on either side of the guide banks upto the edges or banks of khadir (or flood plain) is covered by embankments which are known as approach banks.

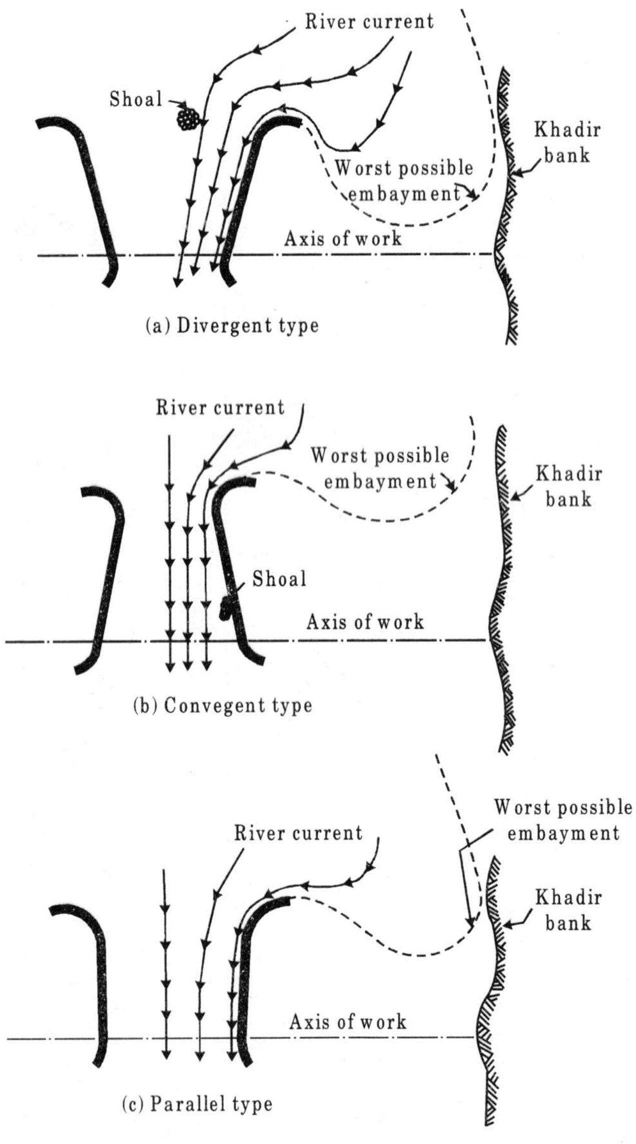

Fig. 21.4 *Different types of guide banks*

As shown in Fig. 21.4 the guide banks may be classified according to their layout upstream of the work as (a) Divergent guide banks, (b) Convergent guide banks, and (c) Parallel guide banks.

Divergent guide banks tend to attract the flow towards them and are suitable where the river current has been oblique to the structure or where the work is located at one edge of khadir. However, as shown in Fig. 21.4 (a) with divergent guide banks the worst possible embayment tends to be very close to the approach bank and hence the same gets relatively less protection. Further these guide banks also have the disadvantage of shoal formation in between them if the splay is excessive. *Convergent guide banks* have the disadvantage of excessive attack and heavy scour at the upstream head and shoaling all along the bank thus rendering the end bays ineffective [Fig. 21.4 (b)]. Due to these drawbacks divergent and convergent guide banks with excessive splay are rarely used. As such in most of the cases *parallel guide banks* [Fig. 21.4 (c)] or slightly convergent guide banks with a splay of 1 in 20 or 1 in 40 (Fig. 21.5) are used unless the local conditions necessitate the other types.

In general the guide banks have proved to be quite successful for the purpose for which they are provided. However, the guide banks sometimes require heavy maintenance and a reserve of stone is usually kept ready on the top of the bank for dumping whenever the bank is endangered.

Design Criteria for Guide Banks

The design of guide bank involves the following considerations.

1. Length of Waterway. As described in section 13.9 of Chapter 13 the length of the clear waterway to be provided between the guide banks or the abutments of the work is taken equal to the Lacey's regime perimeter which is given by the following equation

$$P = 4.83\sqrt{Q} \qquad \text{...(21.1)}$$

where P = Lacey's regime perimeter in m, and

Q = design flood discharge in cumec.

The constant in Lacey's equation may vary from 3.6 to 6.15 depending on discharge and nature of bed material.

The length of the overall waterway L (or the length of the work or the structure) between the guide banks or the abutments of the work is obtained by adding the thickness of piers to the Lacey's regime perimeter. For railway bridges in India, twice the pier thickness is taken as the total obstruction to waterway to allow for end contractions of piers. Further as an approximation the value of L may also be taken as 1.1 to 1.25 times Lacey's regime perimeter.

2. Length of Guide Bank. According to Spring the length of the guide bank on the upstream side from the axis of the work should be equal to 1.1 L, where L is the length of the work between the abutments or the guide banks. The length of the guide bank on the downstream side from the axis of the work should be between 0.1 L and 0.2 L.

However, Gales has recommended various lengths depending on the river discharges. Thus, according to Gales the length of the guide bank on the upstream side from the axis of the work should be 1.25 L for rivers carrying maximum discharges upto 21,000 cumec ; between 1.25 L and 1.5 L for rivers carrying maximum discharges between 21,000 and 42,500 cumec ; and 1.5 L for rivers carrying maximum discharges more than 42,500 cumec. Further a covergence of 1 in 20 has been recommended for the guide banks for rivers carrying maximum discharges upto

21,000 cumec ; between 1 in 20 and 1 in 40 for rivers carrying maximum discharge between 21,000 cumec and 42,500 cumec ; and 1 in 40 for rivers carry maximum discharge more than 42,500 cumec. The length of the guide bank on the downstream side from the axis of the work should be 0.25 L for all sizes of rivers.

3. Radius of Curved Heads. According to Spring, with a velocity of 2.4 to 3.1 m/s which is generally obtained during floods, the upstream curved head of guide banks may have radius varying from 183 to 244 m. Gales, on the other hand, has recommended a radius of 250 m (equivalent to a 7° curve) for rivers carrying maximum discharge upto 21,000 cumec and radius of 582 m (equivalent to a 3° curve) for rivers carrying maximum discharges more than 42,500 cumec. For rivers carrying maximum discharges between 21,000 and 42,500 cumec the value of the radius may be obtained by interpolation. Further according to Gales sharper curves with radius less than 250 m are permissible for rivers carrying maximum discharges less than 7000 cumec.

Besides the above noted values the radius of the upstream curved head of the guide banks may also be taken equal to 0.45 L, where L is length of the work between the abutments or the guide banks.

The upstream curved head should have a sweep angle of 120° to 145°, i.e., the upstream curved head should be extended to subtend an angle of 120° to 150° at its centre.

For the downstream curved head the radius may be half of the radius of the upstream curved head and it should have a sweep angle of 45° to 60°.

Based on these recommendations the general layout plans of guide banks are shown in Fig. 21.5.

4. Cross-section of Guide Bank. The top width of the guide bank should not be less than 3 m and is generally more being as much as 6 to 9 m to allow material trains to pass over the bank. The banks are constructed of locally available material, usually sand. The side slopes of the bank are generally not steeper than 2 : 1. A free board of about 1.25 m to 1.5 m above maximum anticipated flood level is provided.

5. Slope Protection for Guide Bank. The water face of the guide bank is protected by stone pitching. The pitching is usually done with 'one man' stone each weighing 0.4 to 0.5 kN [or 40 to 50 kg(f)], so that the stones are heavy enough and cannot be easily displaced by the current and at the same time each stone can be lifted by one man. Along the shank pitching is done only on the water face and the rear slope is provided with vegetal cover to protect it against rain and wind erosion. At both the curved ends the pitching is done on both the front and the rear faces. The pitching must extend upto 1 m above the maximum flood level. The thickness of pitching as recommended by Inglis is given by the relation

$$T = 0.06\, Q^{1/3} \qquad \ldots(21.2)$$

where T is thickness of stone pitching in m ; and

Q is maximum discharge in cumec.

6. Launching Apron. The slope of the guide bank may be damaged due to scour which may occur at the toe of the bank with consequent undermining and collapse of the stone pitching. In order to protect the slope against such damage a stone cover known as launching apron is laid from the toe of the bank on the horizontal river bed, so that when scour occurs it first undermines the apron starting at its farther end and extending backwards towards the slope. The apron then launches to cover the face of the scour with stones forming a continuous

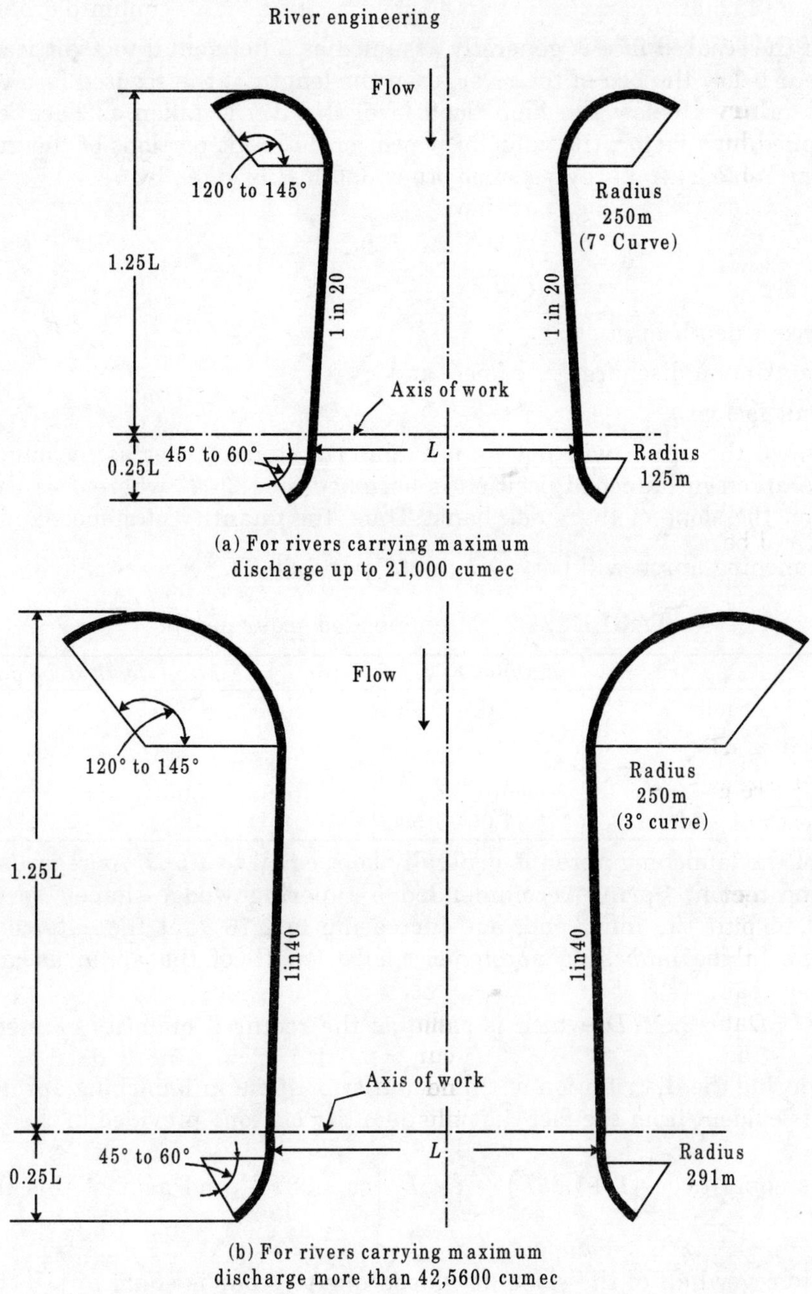

Fig. 21.5 *Guide banks for rivers carrying maximum discharges*

carpet below the slope of the guide bank. In order to ensure complete protection of the whole of the scoured face, adequate quantity of stone should be provided in the launching apron. The quantity of stone required to be provided in the launching apron may be calculated as described below.

The slope of the scoured face is generally assumed as 2 horizontal to 1 vertical. Thus, if D is the depth of scour below the bed of the river, then the length of the scoured face will be equal to $\sqrt{5}\,D$. The depth of scour below the high flood level (H.F.L.) is taken as Lacey's regime scour depth R, multiplied by a factor, the value of which for different portions of the guide bank may be obtained from Table 21.2. Lacey's regime scour depth R is given by

$$R = 0.47\left(\frac{Q}{f}\right)^{1/3} \qquad \text{...(21.3)}$$

where R is scour depth in m

Q is maximum discharge in cumec, and

f is silt factor.

In launching, since the apron will not form a uniform stone carpet as by hand packing, the thickness of the apron in launched position is assumed as 1.25 T, where T is the thickness of stone pitching on the slope of the guide bank. Thus, the quantity of stone required per metre length of the launching apron will be $\left(\sqrt{5}D \times 1.25T\right)$ or 2.80 TD.

TABLE 21.2 Recommended scour depths

Location	*Range of scour depth*	*Mean depth to be adopted in design*
1. Noses of guide bank	2.00—2.50 R	2.25 R
2. Transition from nose to straight portion	1.25—1.75 R	1.50 R
3. Straight reach of guide bank	1.00—1.50 R	1.25 R

The width of the launching apron is generally kept equal to 1.5 D. As regards the thickness of the launching apron, Spring recommended a tapering wedge-shaped section having a thickness equal to T at the inner end, and increasing to 2.76 T at the outer edge. Thus, the quantity of stone in the launching apron per metre length of the apron as laid is equal to $\left[\frac{1}{2}(T + 2.76T) \times 1.5D\right]$ or $2.82TD$, which is same as the required quantity as mentioned above. However, considering the distribution of this quantity of stone in launching apron as laid is not adequate. This is evident from the fact that the quantity of stone provided in the first half of the wedge-section is equal to $\left[\frac{1}{2}(T + 1.88T) \times \frac{1}{2}(1.5D)\right]$ or $1.08TD$, and against this the quantity of stone required to cover half of the slope developed due to scour is equal to $\left[\frac{1}{2}\left(\sqrt{5}D\right) \times 1.25T\right]$ or $1.4TD$.

Thus, when the apron is completely launched its thickness in the upper portion will be less than T.

In order to overcome the above noted deficiency in the distribution of stone in the launching apron, T.S.N. Rao proposed a modified wedge-shaped section having thickness equal to $1.5T$ at the inner end, and increasing to $2.25T$ at the outer edge. Thus, with the same quantity of stone as recommended by Spring, Rao has made the apron design more sound by altering only the distribution.

For the shank or the straight portion of the guide bank the launching apron is provided only on the front side (or the water face side) of the bank. On the other hand for the curved portion of the guide bank the launching apron is provided on both the front and the rear sides of the bank.

A detailed design of guide bank is given in Illustrative Example 21.1.

21.8 GROYNES OR SPURS

Groynes or spurs are the structures constructed transverse to the river flow extending from the bank into the river. Groynes may be aligned either perpendicular to the bank line or at an angle pointing upstream or downstream. These are also known as spur dikes or transverse dikes, and constitute the most widely used river training works.

Functions of Groynes. Groynes serve one or more of the following functions.

(i) Training the river along a desired course by attracting, deflecting or repelling the flow in the river.

(ii) Creating a slack flow with the object of silting up the area in the vicinity.

(iii) Protecting the river bank by keeping the flow away from it.

(iv) Contracting a wide river channel, usually for the improvement of depth for navigation.

Type of Groynes. Groynes may be classified according to different considerations as mentioned below.

1. According to methods and materials of construction.
2. According to height of groyne.
3. According to function served.
4. Special types.

Classification according to methods and materials of construction

According to methods and materials of construction the groynes may be classified as

(i) Impermeable groynes

(ii) Permeable groynes

1. Impermeable groynes. Impermeable or solid groynes do not permit appreciable flow of water through them. These groynes consist of either rockfill, or a core of sand, or sand and gravel, or soil as available in the river bed, protected on the sides and top by a strong armour of stone pitching or concrete blocks. The section or groyne has side slope varying from 2 : 1 to 3 : 1 depending on the material used for its construction and front or head slope varying from 3 : 1 to 5 : 1. Since the head of the groyne is subjected to severe attack by the stream, it needs protection. Hence at the groyne head a launching apron is generally provided and also the thickness of pitching is increased (Fig. 21.6). The impermeable groynes may be designed to attract the flow towards any bank or repel or deflect the flow away from the bank and thus make the river to follow a desired course.

2. Permeable groynes. The permeable groynes permit restricted flow of water through them. These groynes obstruct the flow and slacken it to cause deposition of sediment carried by the river. The permeable groynes are therefore classified as sedimenting groynes and are best suited for rivers which carry considerable amount of sediment in suspension. As the sediment

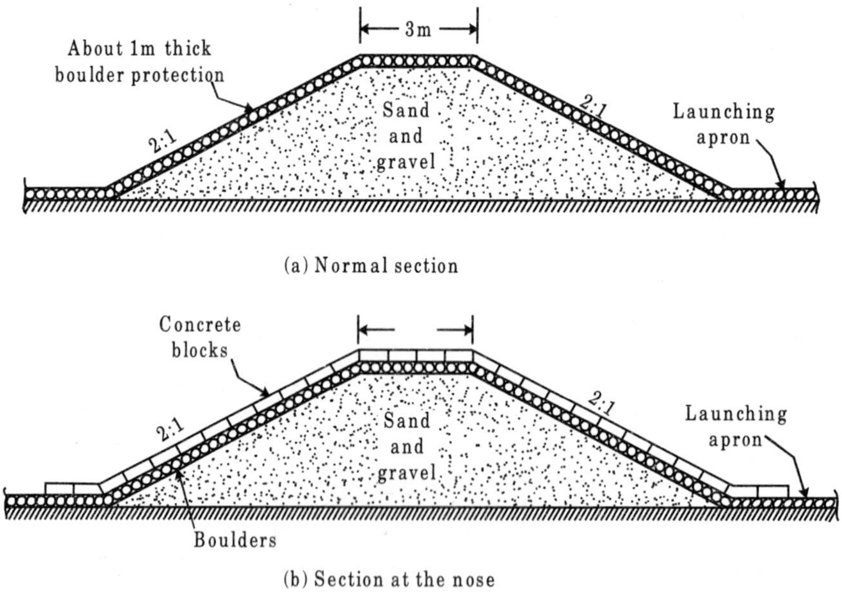

Fig. 21.6 *Cross-sections of impermeable groynes*

accumulates between the groynes, the foreshore becomes more or less permanent so that there is no need to use any other material for its protection. However, in rivers carrying relatively clear water the action of these groynes is to dampen the erosive strength of the current and thus prevent the local bank erosion.

The common types of permeable groynes are

(a) Tree groynes

(b) Pile groynes.

(a) *Tree groynes.* A tree groyne consists of a thick wire rope 25 mm diameter firmly anchored at one end to the bank and tied at the other end to a heavy buoy or a concrete block. The length of the wire rope is kept equal to the length of the groyne with some allowance being given for the sag due to the weight of the trees. Entire leafy trees with abundant branches are then fixed to the wire rope starting from the bank and proceeding outwards to the free end. For fixing the trees a hole is drilled through the stem of each tree, in which an iron ring is inserted and attached either directly or by another piece of wire rope to the main rope. In order to make the groyne more effective the trees should be packed as closely as possible.

(b) *Pile groynes.* This type of groynes may be constructed out of timber or R.C.C. piles or sheet piles. In pile groynes, the piles constitute the main verticals which are driven down 6 to 9 m inside the river bed, 2.4 to 3 m apart and in at least 2 or 3 similar rows. The rows of the piles are not more than 1.2 to 1.8 m apart. Between the main piles there can be two intermediate piles embedded at least 1.2 m below the bed. Each row is closely inter-twined either by brushwood branches going in and around each pile, or by horizontal railings. The upstream row is braced to the downstream row by transverses and diagonals. Every other main pile of the rear row is strutted, the struts being embedded a minimum of 2.4 m below the bed. Between two rows, the space is filled with alternate layers of 1.8 m thick brushwood branches weighted by 0.6 m thick stones or sand bags. The brushwood branches should be closely packed and tamped.

With a properly designed, constructed and functioning permeable groyne, there is generally no tendency for scour around the nose of the groyne or for undetermining of piles. However, when debris collects upstream and the groyne becomes sand bound it may function like an impermeable groyne. To guard against scour occurring under such conditions, it is desirable to protect the bed both upstream and downstream of the groyne and around the nose with a 0.9 m thick stone apron over a width of 3 m along the shanks and 6 m along the nose.

Advantages of Permeable Groynes

The various advantages of permeable groynes are as indicated below :

1. The permeable groynes require only temporary or semi-permanent construction and hence they are relatively cheap.
2. Small quantity of stone is required for the construction of permeable groynes and hence they are specially suitable where stone is scarce.
3. It has been observed that permeable groynes are more effective than solid or impermeable groynes in the regulation of river courses or in the protection of banks especially in a silt-ladden river.
4. The flow through a permeable groyne does not change abruptly like that passing an impermeable groyne and hence there are no serious eddies and scour holes in the case of permeable groynes.
5. In the case of deep and narrow rivers where depths are considerable solid groynes become expensive and may cause undesirable flow conditions. In such cases permeable groynes can be effective in providing necessary bank protection.

Disadvantages of Permeable Groynes

The permeable groynes are not strong enough to resist shocks and pressures from debris, floating ice and logs. They are, therefore, unsuitable for upper reaches of rivers.

Classification according to Height of Groynes. According to height of groynes these may be classified as :

(i) Non-submerged groynes

(ii) Submerged groynes

For training of rivers and protection of banks generally non-submerged groynes are used but in some cases submerged groynes are also used. For example, in the case of deep rivers where depths are considerable submerged groynes may be used. Submerged groynes can also be either solid groynes or permeable groynes. However, submerged permeable groynes are preferable to submerged solid groynes since the former do not create turbulent and eddy conditions as strong as with the later. Further where deep scour has occurred in the bed of a river submerged groynes may be used as a curative measure. These submerged groynes are usually known as 'submerged sills' or 'submerged dykes' and are considered as a separate type of river training works as indicated later.

Classification according to function served. According to function served by the groynes these may be classified as

(i) Attracting groynes
(ii) Repelling groynes
(iii) Deflecting groynes
(iv) Sedimenting groynes

(i) *Attracting groynes*. A groyne pointing downstream tends to attract the river flow towards the bank on which it is provided and hence it is called an attracting groyne [Fig. 21.7 (a)]. Such a groyne causes scour hole to form closer to the bank than the groyne inclined at right angles to the bank or that inclined slightly upstream and therefore it tends to maintain the deep current close to the bank. The attracting groyne usually makes an angle of 60° with the bank. However, the angle of inclination of an attracting groyne with the bank may be in the range of 30° to 60°. Further in the case of an attracting groyne the main attack of the stream is on its upstream face and therefore it needs better protection as compared to the downstream face.

The attracting groynes safeguard the opposite bank against the attack of the current as they attract the current towards the bank adjacent to them. However, the attracting groynes are not useful for bank protection and may sometimes even endanger the adjacent banks since silting between successive groynes is absent. On account of this trouble caused by attracting groynes these are not commonly used.

(ii) *Repelling groynes*. A groyne pointing upstream tends to repel the river flow away from the bank on which it is provided and hence it is called a repelling groyne [Fig. 21.7 (b)]. The angle of inclination of repelling groynes with the bank varies from 60° to 80°, or with a line perpendicular to the bank varies from 10° to 30°. On the upstream side of a repelling groyne a

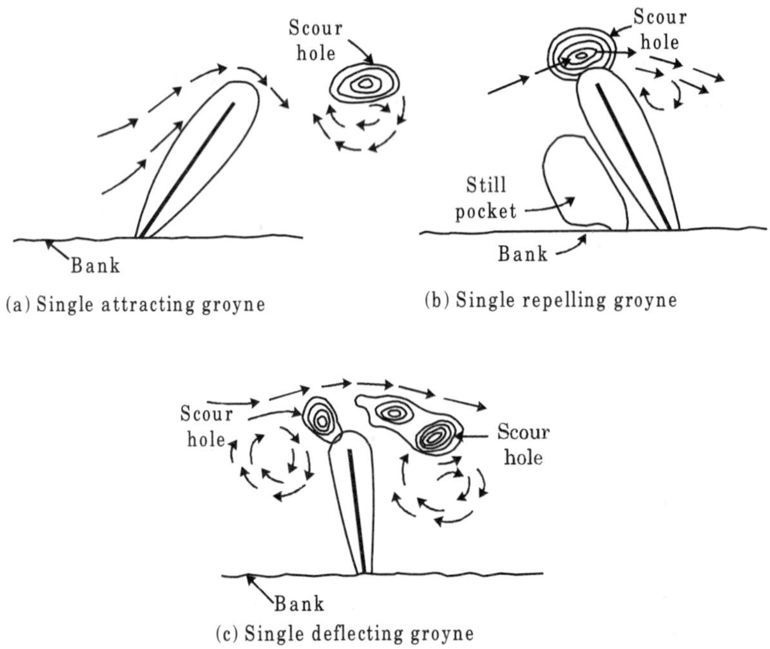

Fig. 21.7 *Attracting, repelling and deflecting groynes*

still water pocket is formed and the suspended sediment carried by the river gets deposited in the pocket. The head of a repelling groyne causes disturbances in the flow at its nose and heavy scour occurs at the nose and slightly downstream of it due to eddy formation. Further the head of a repelling groyne needs a very strong protection because it is subjected to direct attack of a swirling current.

As compared to attracting groynes, repelling groynes are more effective and do not cause any trouble. As such repelling groynes are commonly used for the purpose of river training and bank protection.

(iii) *Deflecting groynes*. A groyne either perpendicular to the bank or pointing slightly upstream and having a relatively short length tends to only deflects the flow without repelling it and hence it is called a deflecting groyne [Fig. 21.7 (c)]. A deflecting groyne, however gives only local protection.

(iv) *Sedimenting groynes*. A groyne which dampens the velocity of flow and thus causes deposition of sediment carried by the river without repelling or deflecting the flow, is called a sedimenting groyne. As indicated earlier permeable groynes are usually classified as sedimenting groynes. These groynes are best suited for rivers carrying considerable amount of sediment in suspension.

Special Types of Groynes

Some of the special types of groynes that have also been used are

(i) Denehy's T-headed groynes

(ii) Hockey head groynes or Hockey groynes

(c) *Denehy's T-headed groynes*. These groynes which are also called Denehy's groynes or T-headed groynes were developed in India by Denehy and were first used by him in 1880 for the purpose of river training and bank protection at the Okhla Headworks, Delhi on the Yamuna River. Later on in 1887 these groynes were also used at the Narora Headworks (U.P.) on the Ganga River. This groyne consists of a main groyne projecting from the bank at the head of which a straight short cross groyne is provided in the direction normal to the main groyne so that the entire groyne has a T-shape [Fig. 21.8 (a)]. The cross groyne protects the main groyne

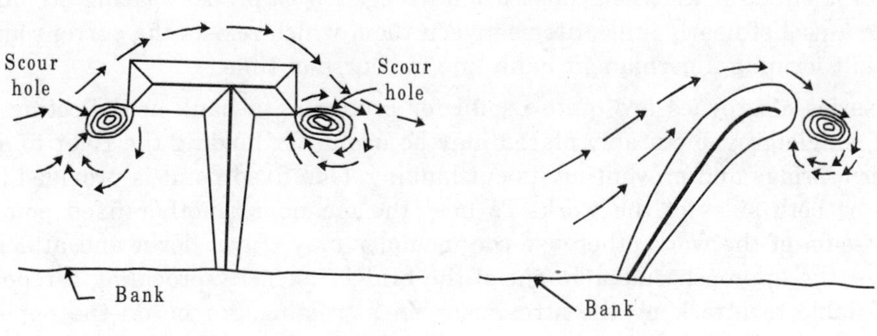

Fig. 21.8 *Denehy's T-headed groyne and Hockey head groyne*

on the same principle as the main groyne protects the bank. A greater length of the cross groyne projects on the upstream side and a smaller length on the downstream side of the main groyne. In the case of a T-headed groyne scour holes are formed on both ends of the cross groyne. These groynes are usually spaced 800 m apart with the T-heads on a regular curved or straight line to train the river channel along a required course.

(ii) *Hockey head groynes or Hockey groynes.* These groynes are provided with a curved head such that at the lower end their shape is like a hockey stick [Fig. 21.8 (b)]. The hockey head groynes behave more like attracting groynes and hence they are not likely to be useful for bank protection or repelling of the current away form the bank.

General Considerations in the Design of Groynes

(i) *Length of groynes.* The length of the groynes depends upon the position of the original bank line and the designed normal line of the trained river channel. Too long groynes constructed out of easily erodible material are liable to damage and failure. As such to start with groynes of shorter length may be provided and they may be extended gradually as silting between them proceeds.

(ii) *Spacing of groynes.* Since the length of the bank which may be protected by each groyne depends on the length of the groyne, the main factor that governs the spacing between adjacent groynes is their length. The spacing of groynes is thus taken as a certain proportion of the length of the groynes. The other factors which affect the spacing of groynes are as follows :

(a) *Width of the river.* For rivers of equal flood discharges, a larger ratio of spacing to length of groynes may be used for a wide river than for a narrow river.

(b) *Location of the groynes.* Larger spacing may be used for convex banks than for concave banks. The spacing for straight reaches or crossings being kept in between the two.

(c) *Type of construction.* The permeable groynes may be spaced farther apart than solid or impermeable groynes.

In general a spacing of 2 to 2.5 times the length of groynes at convex banks, and equal to the length of the groynes at concave banks may be adopted.

(iii) *Number of groynes.* Depending upon the purpose, groynes can be used singly or in series. They can also be used in combination with other training measures. The need for using groynes in series arises if the reach of the river to be protected is long, or if a single groyne is not strong enough to deflect the current nor quite effective for silt deposition upstream and downstream of itself. Groynes when constructed in series with proper spacing are quite effective as they create a pool of nearly still water between them which resists the current and gradually accumulates silt forming a permanent bank line in course of time.

While a series of groynes are quite useful for protecting a bank or deflecting the current away from it, single groyne suitably placed may be useful for holding the river to a fixed point such as under a bridge and preventing its outflanking. One fixed point is provided by the guide bank on one or both sides of the work. To hold the meander, another fixed point is usually required upstream of the work otherwise the meander may travel down and attack one of the guide banks or the approach embankment of the bridge itself. By providing a repelling groyne on the bank liable to attack by the stream; or an attracting groyne on the opposite bank a second fixed point to hold the river permanently to a nearly axial course under the bridge is obtained (Fig. 21.9). According to Inglis if these groynes are placed at a distance of 0.4 times the meander length upstream of the bridge, good second fixed point would be provided to serve this purpose.

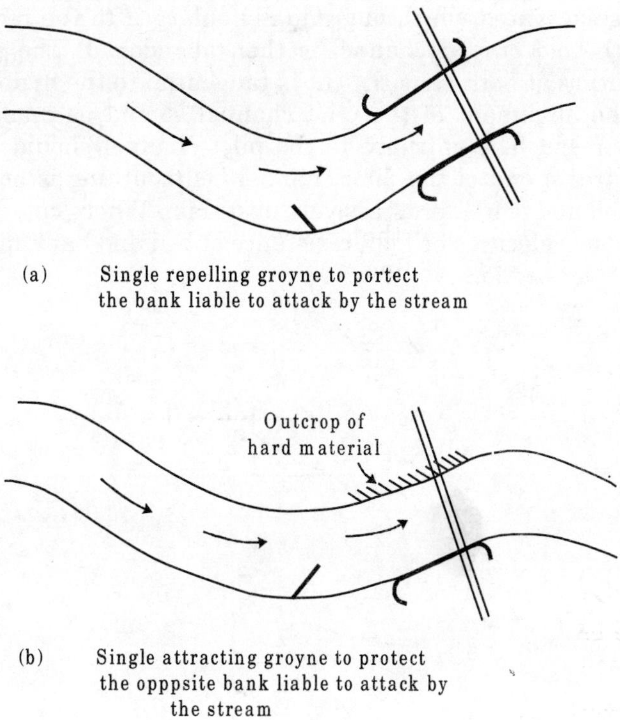

Fig. 21.9 *Use of single groyne for bank protection*

It may, however, be stated that the design and location of the groynes can not be determined on the basis of theoretical considerations, but can best be determined from model tests.

21.9 CUTOFFS

Cutoffs may be defined as a process by which an alluvial river flowing along curves or bends abandons a particular bend and establishes its main flow along a comparatively straight and shorter channel. Moreover the shorter channel so formed is also known as cutoff. A cutoff may occur when the meander of a river reaches extreme conditions and develops into a horse-shoe bend with a narrow neck such that at high flood the river may cut across the neck and create a straight channel for itself. Such a natural cutoff is accompanied by a more or less violent disturbance of the river regime for some distance upstream and downstream of the reach where cutoff occurs. Due to reduction of length there is steepening of slope which results in drawdown and erosion upstream of the cutoff and deposition of some of this excess eroded material downstream of the cutoff. The disturbed conditions continue for some time till the equilibrium is attained.

Artificial cutoffs may be induced for the purpose of river training to divert the river from a curved flow which may be endangering valuable land or properly or to straighten its approach to a work. For flow around a bend the ratio of the length of the bend to that of the chord is called the cutoff ratio. It has been observed that when the cutoff ratio exceeds 1.5 to 2.5 a natural cutoff may occur and hence an artificial cutoff is also induced when cutoff ratio is in this range.

For inducing an artificial cutoff in the case of rivers in easily erodible material, only a pilot channel (or pilot cut) is excavated which is gradually enlarged to the required cross-section by the flood waters and the old curved channel is then abandoned. The alignment of the pilot channel should be such that at both ends the cut is tangential to the main direction of river flow (Fig. 21.10). Further the alignment of the pilot channel should be made on a slight curve to promote speedy development. The entrance to the pilot channel should be made bellmouthed. However, at the downstream end of the pilot channel bellmouthing is not necessary as the cut develops at the lower end and works progressively upstream. Where cuts are unlikely to develop by scour because of too low velocities or high resistance of bed and bank material, they should be excavated to the main river section.

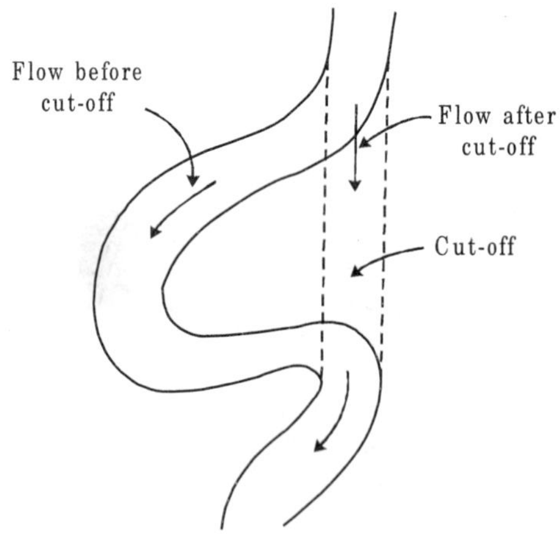

Fig. 21.10 *Cut-off*

Alike natural cutoff an artificial cutoff also causes disturbance of river regime upstream and downstream till readjustment is made.

21.10 BANK PITCHING AND LAUNCHING APRONS

In general bank protection works may be classified as (i) Direct and (ii) Indirect. Direct protection includes works done on the bank itself. Indirect protection includes works constructed not directly on the banks, but in front of them for reducing the erosive action of the current. The pitching of a bank and the provision of a launching apron are the direct methods of bank protection while groynes deflecting the river current and cutoffs altering its course are the indirect methods of the bank protection.

The materials used for the pitching of banks are stone, bricks, concrete blocks, bamboos and brush wood, woven mattresses of fibrous material, naturally growing grasses and willows etc. Out of the various materials stone is the most commonly used material where available locally. Bricks and concrete blocks are usually too costly to be used for bank pitching except in small reaches for protection of a town or an engineering work. On the other hand pitching by brush wood or bamboos or mattresses is usually temporary. The bank to be protected should be cut to a stable slope before pitching is provided. The stable slope vary from 1 : 1 to 2 : 1 depending on the material of the bank and the type of pitching provided.

Launching apron. Even a well protected bank may be destroyed if its toe is not protected from undercutting by the action of the flood. A launching or falling apron of loose stone is provided at the toe of the bank to ensure safety against such damage. The various considerations made in the design of a launching apron are already discussed in section 21.7.

21.11 PITCHED ISLANDS

A pitched island (Fig. 21.11) is an artificially created island in the river bed which generally consists of a sand core protected by revetment or stone pitching on all sides. A launching or falling apron is provided around the island to protect the toe of its side slopes. The pitched island causes redistribution of velocity and tractive force. The tractive force near a pitched island begins to increase rapidly after the construction of the island, with the result that deep scour begins to form round the island and gradually draws the main current towards itself and ultimately holds it permanently. The pitched island may thus be utilised to attract the current towards itself to reduce undue concentration on the river banks or to maintain the approach to some engineering work in the desired direction.

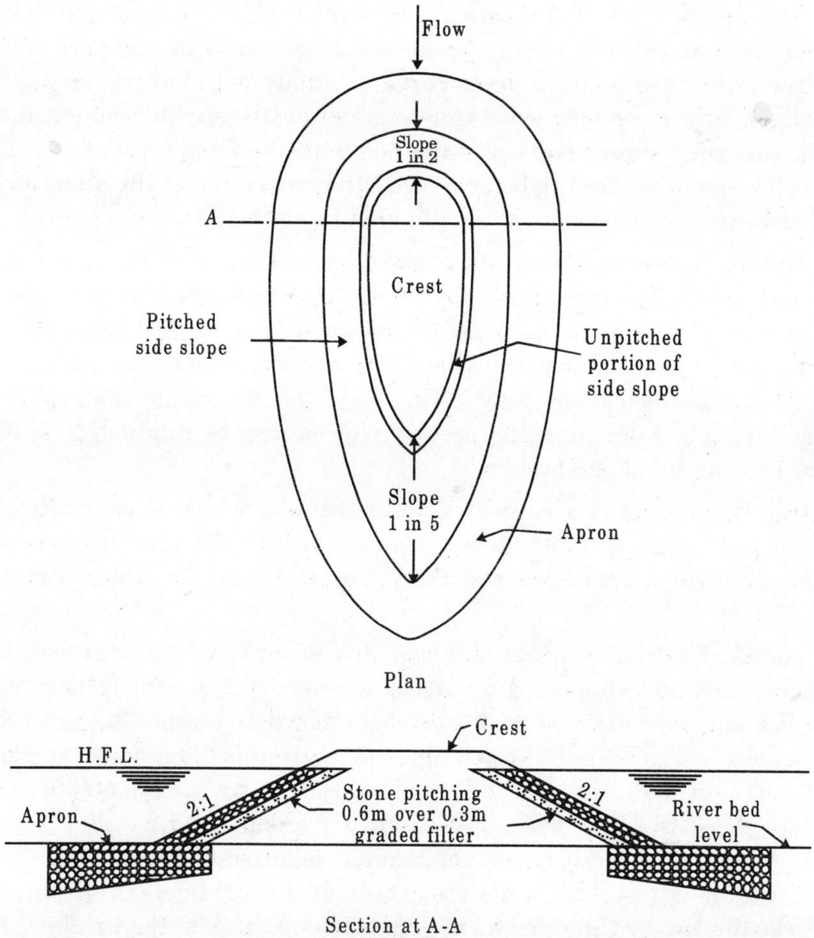

Fig. 21.11 *Pitched island*

As deep scour occurs around the pitched island creating heavy concentration of flow in its vicinity, the stone pitching and launching apron is designed for the maximum scour under worst conditions which is generally taken equivalent to $2R$, where R is Lacey's regime scour depth. The other aspects of the design are, more or less, similar to the design of guide banks. The length and shape of the island can, however, be determined with the help of model tests.

The success of a pitched island depends on the extent of scour around the island. In the case of wide and shallow rivers, where obstruction caused by the island would be very small, scour will not develop around the island readily. Similarly, in flashy rivers, there is not enough time for the development of scour around the island, and even if scour develops during rising stage it may get quickly filled up during the falling stage of the river. In such cases, the pitched island may not be successful in achieving the desired results.

21.12 MISCELLANEOUS METHODS

1. Sills. Sills, which are also sometimes called 'submerged sills' or 'submerged dykes' are used to counteract the tendency to cause excessive scour and large depth in any part of the river cross-section. This situation may arise at a sharp curve or adjacent to natural or artificial length of hard inerodible bank in a river otherwise consisting of easily erodible material. In such cases sills are placed across the scoured portion of the bed with their top level at or a little below the designed bed level, which is desired to be achieved after correction of the abnormal scours. They must be spaced close enough to ensure their functioning properly to stop scour.

2. Closing Dykes. A closing dyke is an obstruction put across a branch of a river which is desired to be closed so that the river flow may be diverted to single channel. The dykes may be either solid or permeable. A solid dyke must be designed to resist the pressure of water on its upstream face and to be safe against the possibility of undermining by seepage. A permeable dyke retards the flow and causes a gradual silting up of the branch above itself. If the velocity of flow in the branch is high, more than one pervious dykes may be required to sufficiently retard the flow to induce silting up of the branch.

3. Bandalling. Bandalling is a method of river training which is generally used to confine the low water flow in a single channel for maintaining required depth for navigation. The work of bandalling usually commences after the flood season as soon as the water levels begin to lower.

A *bandal* consists of a framework of bamboos driven into the river bed, set 0.6 m apart by means of horizontal ties and supported by struts at every 1.2 m (Fig. 21.12). To this bamboo framework, bamboo mattings are tied with coir ropes at water levels. The bamboos used on the framework are generally 3 to 6 m in length and the matting is 0.9 m wide, strengthened at the edges by strips of split bamboo. Bandals are placed at an angle of 30° to 40° inclined downstream. They check the flow and cause sand to be deposited parallel to and behind the bandals. Thus a channel confined between the bandals is formed with sand banks on either side and the whole discharge of the river is directed through this channel. Deepening of the channel is generally achieved in two or three weeks after the construction of the bandals. The depth thus achieved subsists until the next flood season without requiring any maintenance.

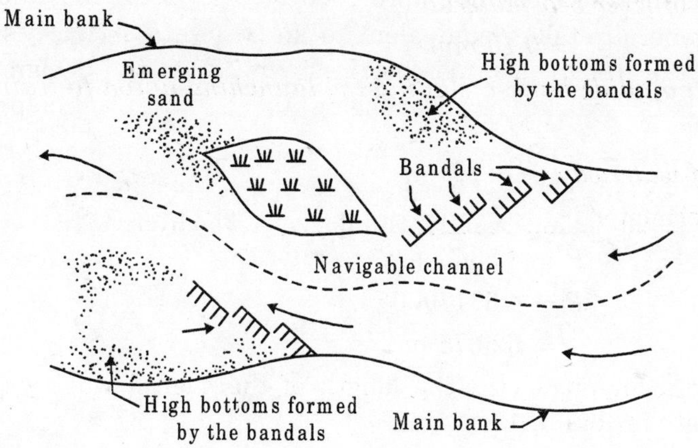

Layout plan of bandals

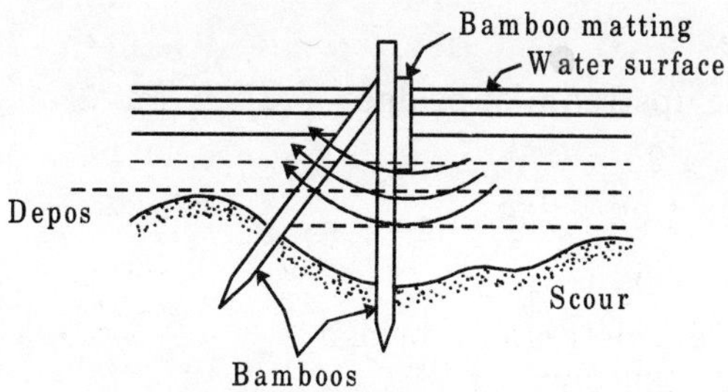

Side view of a bandal

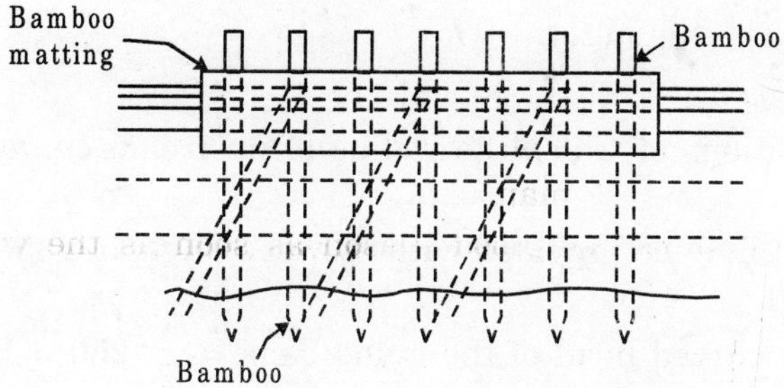

Elevation of a bandal

Fig. 21.12 *Layout plan and other views of bandals*

SOLVED ILLUSTRATIVE EXAMPLES

Illustrative Example 21.1 *The following data pertain to a bridge site of a river.*

Maximum discharge = *17000 cumec*

Highest flood level = *288.00 m*

River bed level = *280.00 m*

Average diameter of river bed material

= *0.10 mm*

Design and sketch a guide bank including the launching apron to train the river.

Solution

Step 1. *Length of waterway*

Lacey's regime perimeter

$$P = 4.83Q$$

or

$$P = 4.8317000$$

$$= 629.75 \text{ m}$$

Allowing 20% extra for piers etc., the length of the overall waterway between two guide banks at the bridge site is obtained as

$$L = 1.2 \times 629.75 \text{ m}$$

$$= 755.7 \text{ m} \simeq 756 \text{ m}$$

Step 2. *Length of guide bank*

The Length of the guide bank upstream of bridge

$$= 1.25 \times L$$

$$= 1.25 \times 756 \text{ m}$$

$$= 945 \text{ m}$$

The length of the guide bank downstream of bridge

$$= 0.25 \times L$$

$$= 0.25 \times 756 \text{ m}$$

$$= 189 \text{ m}$$

Step 3. *Radius of curved head*

Following the recommendations of Gales the radius of upstream curved head of the guide bank = 250 m.

The upstream curved head may be provided with a sweep angle of 135°.

The radius of downstream curved head of the guide bank = $\frac{1}{2}$ 250 = 125 m and it may be provided with a sweep angle of 60°.

Step 4. *Cross section of guide bank*

The top width of the guide bank is assumed as 6 m and the side slopes as 2 : 1.

H.F.L. at the bridge site = 288.00 m

Assume a free board = 1.5 m

∴ Top level of guide bank = 288.00 + 1.50

= 289.50 m

River bed level = 280.00 m

∴ Height of guide bank above the bed of the river

$$= 289.50 - 280.00 = 9.50 \text{ m.}$$

Step 5. *Slope protection*

Thickness of stone pitching on the side slopes of the guide bank is given by

$$T = 0.06\, Q^{1/3}$$

or $$T = 0.06\,(17000)^{1/3}$$

∴ $$T = 1.54 \text{ m say } 1.55 \text{ m}$$

The water face of the shank will be pitched with one man stone weighing 0.4 to 0.5 kN [or 40 to 50 kg(f)]. The rear slope of the shank will be provided with vegetal cover.

Volume of stone required for pitching per m length

$$= \sqrt{5} \times (9.50 \times 1.55)$$

$$= 32.93 \text{ m}^3/\text{m say } 33 \text{ m}^3/\text{m}$$

Step 6. *Launching apron*

Lacey's regime scour depth is given by

$$R = 0.47\left(\frac{Q}{f}\right)^{1/3}$$

$$f = 1.76\sqrt{m_r}$$

$$m_r = 0.10 \text{ mm}$$

∴ $$f = 1.76\sqrt{0.10} = 0.56$$

∴ $$R = 0.47\left(\frac{17000}{0.56}\right)^{1/3} = 14.66 \text{ m}$$

At the nose of the guide bank scour depth below H.F.L.,

$$= 2.25\, R = 2.25 \times 14.66$$

$$= 33 \text{ m}$$

Scour depth below the river bed is

$$D = (33 - 8) = 25 \text{ m}$$

∴ Width of launching apron

$$= 1.5\, D = 1.5 \times 25 = 37.5 \text{ m}$$

Thickness of launching apron at the inner edge

$$= 1.5\, T = 1.5 \times 1.55 = 2.33 \text{ m say } 2.4 \text{ m}$$

Thickness of launching apron at the outer edge

$$= 2.25\, T = 2.25 \times 1.55$$

$$= 3.49 \text{ m say } 3.5 \text{ m}$$

At the transition from nose to straight portion scour depth below H.F.L.,

$$= 1.5R = 1.5 \times 14.66 = 22 \text{ m}$$

Scour depth below the river bed is

$$D = (22 - 8) = 14 \text{ m}$$

∴ Width of launching apron

$$= 1.5\,D = 1.5 \times 14 = 21 \text{ m.}$$

Thickness of launching apron at the inner and outer edges being 2.4 m and 3.5 m respectively as indicated above.

At the shank of the guide bank scour depth below H.F.L.

$$= 1.25R = 1.25 \times 14.66$$

$$= 18.33 \text{ m say } 18.5 \text{ m}$$

Scour depth below the river bed is

$$D = (18.5 - 8.0) = 10.5 \text{ m}$$

∴ Width of launching apron

$$= 1.5\,D = 1.5 \times 10.5 = 15.75 \text{ m say } 16 \text{ m}$$

Again thickness of launching apron at the inner and outer edges being 2.4 m and 3.5 m respectively as mentioned earlier.

Figure 21.3 shows the details of the guide bank and the launching aprons.

REVIEW QUESTIONS

21.1 What do you understand by (i) meandering type, (ii) aggrading type (iii) degrading type rivers?

21.2 What are the causes of meandering of rivers?

21.3 What is meant by river training ? List the various objectives of river training.

21.4 How are river training works classified?

21.5 Mention the methods for the high water river training.

21.6 Describe briefly how a river is controlled with the help of embankments. Discuss the effect of embankments on river regime.

21.7 What are the merits and demerits of river training by embankments?

21.8 Write a note on 'guide banks' used for river training works.

21.9 What is a groyne (or spur)? List the various types of groynes (or spurs) and describe them briefly.

21.10 Explain the terms 'meandering' and 'helicoidal flow' as applied in river engineering. The dominant discharge of a river is 10000 cumec. Estimate the width and length of the meander of the river which may be assumed as a flood-plain river. [*Ans.* 15342 m; 5361m]

21.11 Describe the functions of (i) guide banks, (ii) spurs or groynes as river training measures.

21.12 Discuss the economic aspects of spacing and height of levees.

21.13 Describe the different approaches to the determination of the length of guide bank on the upstream side.

21.14 What is the utility of launching aprons for guide bunds ? How are these designed?

21.15 Distinguish between (i) repelling and attracting spurs, (ii) solid and permeable spurs.

21.16 Describe with neat sketches Denehy's T-headed groyne and Hockey head groyne.

21.17 Discuss the various considerations adopted in the design of groynes.

21.18 What is a cut off? Describe briefly how a cut off may be used as a river training measure?

21.19 Describe with a neat sketch a pitched island.

21.20 What is bandalling? Describe with a neat sketch the use of bandalling as a river training measure.

21.21 A bridge is to be constructed across a river having the following hydraulic data :

Maximum flood discharge : 10 000 m^3/s

Highest flood level : 254.0 m

River bed level : 250.0 m

Average diameter of river sand : 0.25 mm

Design and sketch a guide bank including the launching apron to train the river.

CHAPTER

22 Planning for Water Resources Development

22.1 INTRODUCTION

The development of water resources involves the conception, planning, design, construction, and operation of facilities to control and utilize water. It is basically a function of civil engineers, but the services of specialists from other fields such as economists, political scientists, geologists, electrical and mechanical engineers, chemists, biologists, and other specialists in the natural and social services are also required. Thus, water resources problems are of interdisciplinary nature. Further each water resources development project encounters a unique set of physical conditions to which it must conform, hence standard designs leading to simple solutions applicable to all the projects are rarely available. The special conditions of each project must be met through an integrated application of the fundamental knowledge of the various disciplines.

Planning may be defined as the orderly consideration of a project right from its conception through the evaluation of various alternatives to the final decision on a course of action. It includes all the work associated with the design of a project except the detailed engineering of the structures. It forms the basis for the decision to proceed with (or to abandon) a proposed project and hence it is the most important aspect of the total engineering for the project. As stated earlier since each water resources development project is unique in its physical and economic setting, it is impossible to describe a simple process which will inevitably lead to the best decision. Although there is no substitute for 'engineering judgement' during planning, but each individual step towards the final decision should be supported by quantitative analysis rather than mere judgement. In this chapter some guidelines for the planning of water resources development project and some pitfalls in project planning are presented.

22.2 PURPOSE SERVED BY WATER RESOURCES DEVELOPMENT PROJECTS

The water resources development projects are planned to serve the following purposes :

1. Flood Control. The various objectives to be achieved in this case are prevention or reduction of flood-damage, protection of economic development, conservation storage, river regulation, recharging of groundwater, water supply, development of power, protection of life.

The various works and measures adopted to achieve the above objectives are dams, storage reservoirs, levees, flood walls, channel improvements, floodways, pumping stations, flood plains zoning, flood forecasting.

2. Irrigation. The only objective to be achieved in this case is agricultural production. The various works and measures adopted to achieve the above objective are dams, reservoirs, wells, canals, pumps and pumping plants, weed control and desilting works, distribution systems, drainage facilities, farmland grading.

3. Hydroelectric Power. In this case the only objective to be achieved is generation of electrical power for economic development and improved living standards.

The various works and measures adopted to achieve the above objective are dams, reservoirs, penstocks, power channels, power plants, transmission lines.

4. Navigation. Transportation of goods and passengers is the only objective to be achieved in this case.

The various works and measures adopted to achieve the above objective are dams, reservoirs, canals, locks, channel improvements, harbour improvements.

5. Domestic and Industrial Water Supply. The only objective to be achieved in this case is provision of water for domestic, industrial, commercial, municipal, and other uses.

The various works and measures adopted to achieve the above objectives are dams, reservoirs, wells, conduits, pumping plants, treatment plants, desalination plants, distribution systems.

6. Watershed Management. The various objectives to be achieved in this case are conservation and improvement of soil, sediment abatement (reduction), runoff retardation, forests and grassland improvement, protection of water supply.

The various works and measures adopted to achieve the above objectives are soil-conservation practices, forest and range management practices, headwater-control structures, debris-detention dams, small reservoirs, farm ponds.

7. Recreational Use of Water. In this case increased well-being and health of people is the only objective to be achieved.

The various works and measures adopted to achieve the above objective are reservoirs, facilities for recreational use, works for pollution control, reservation of scenic and wilderness areas.

8. Fish and Wild Life. The various objectives to be achieved in this case are improvement of habitat for fish and wild life, reduction or prevention of fish and wild life losses associated with man's works enhancement of sports opportunities, provision for expansion of commercial fishing.

The various works and measures adopted to achieve the above objectives are wild life refuges, fish hatcheries, fish ladders and screens, reservoir storage, regulation of stream flows, stocking of streams and reservoirs with fish, pollution control, land management.

9. Pollution Abatement. The only objective to be achieved in this case is protection or improvement of water supplies for municipal, domestic, industrial, and agricultural use, and for aquatic life and recreation.

The various works and measures adopted to achieve the above objective are treatment facilities, reservoir storage for augmenting low flows, sewage-collection systems, legal control measures.

10. Insect Control. The various objectives to be achieved in this case are public health, protection of recreational values, protection of forests and crops.

The various works and measures adopted to achieve the above objectives are proper design and operation of reservoirs and associated works, drainage, extermination measures.

11. Drainage. The various objectives to be achieved in this case are agricultural production, urban development, protection of public health.

The various works and measures adopted to achieve the above objectives are ditches, tile drains, levees, pumping stations, soil treatment.

12. Sediment Control. The various objectives to be achieved in this case are reduction or control of silt load in streams, protection of reservoirs.

The various works and measure as adopted to achieve the above objectives are soil conservation, sound forest practices, desilting works, channel and revetment works, bank stabilization, special dam construction and reservoir operations.

13. Salinity Control. The only objective to be achieved in this case is abatement or prevention of salt-water contamination of agricultural, industrial and municipal water supplies.

The various works and measures adopted to achieve the above objective are reservoirs for augmenting low stream flow, barriers, groundwater recharge, coastal jetties.

14. Artificial Precipitation. The control of precipitation within meteorological limits is the only objective to be achieved in this case.

The various works and measures adopted to achieve the above objective are portable cloud-seeding equipment, ground generators.

15. Employment. The only objective to be achieved in this case is stimulation of employment and sources for increased income in depressed areas of unemployment and under-development.

16. Public Works Acceleration. Sometimes the projects are so planned that they may be executed on cost sharing basis among the various agencies such as Central government, State government and other Local bodies. The main objective to be achieved in such cases is the acceleration of the construction of these projects.

17. New Water Resources Policies. In this case the main objective is to develop new policies affecting the economics of project justification as well as project formulation and composition. For our country *National Water Policy* has been formulated by the Ministry of Water Resources Government of India in 1987 which is given in Appendix IV.

22.3 CLASSIFICATION OF WATER RESOURCES DEVELOPMENT PROJECTS

The water resources development projects may be classified according to the purpose served by them in the following two types:

1. Single purpose project
2. Multipurpose project

A *single purpose project* is the one which is designed and operated to serve only one purpose.

A *multipurpose project* (or *multiple-purpose project*) is the one which is designed and operated to serve two or more purposes, It may however, be stated that a project which is designed for single purpose but which produces incidental benefits for other purposes should not be considered a multipurpose project.

Water resources are one of the most important natural resources which are available in abundance. However, with the world population increasing at a considerably faster rate it is imperative that maximum use be made of the water resources and the concept of multiple use is essential to such maximum use. Moreover multiple use of project facilities may increase benefits without a proportional increase in costs and thus enhance the economic justification for the project. As such most of the major water resources development projects are planned as multipurpose projects.

22.4 FUNCTIONAL REQUIREMENTS IN MULTIPURPOSE PROJECTS

The successful joint use of storage space in a multipurpose project depends on the extent to which the various purposes are compatible. It is therefore necessary to review the requirements of the various uses and to consider the ways in which these uses may be coordinated.

Irrigation. Water requirements for irrigation are usually seasonal. The maximum demand for irrigation water in our country is for the rabi crop during the winter months. However, some demand for the kharif crop also exists during the months prior to the monsoon months during which there is little or no demand for irrigation water. Water requirements for irrigation do not vary greatly from year to year, although low rainfall years usually create a greater irrigation demand than high-rainfall years. Further unless the project area is increased, the average annual demand will remain nearly constant. Since irrigation storage is insurance against drought, it is desirable to maintain as much reserve storage as possible consistant with current demands.

Water Supply. Requirements for domestic water supply are more nearly constant throughout the year than are the irrigation requirements, but a seasonal maximum is usually experienced in the summer. The demand normally increases slowly from year to year, and provision must be made in planning for this increase. Maintenance of an adequate reserve to avoid water shortage during droughts is necessary.

Hydroelectric Power. Power demand usually has a marked seasonal variation depending on the type of area served, which therefore affect the water requirement for this purpose. However, most power plants are parts of an interconnected power system, and hence considerable flexibility is possible in coordinating power needs with other uses of water. Moreover, power production is not a consumptive use of water and is therefore more compatible with other uses. Water released from a reservoir for irrigation use downstream may be passed through the turbine to produce power as well. Thus, in an extreme case, power production may be limited to those times when release of water is necessary for other uses or to discharge excess water. This may, however, result in a low load factor for the plant, with a loss in overall efficiency.

Navigation. Water required to be released from a storage reservoir to sustain downstream flows for navigation has a marked seasonal variation, with peak releases being required during the later portion of the dry season.

Flood Control. The basic requirement of flood control is that sufficient empty storage space must be available to permit withholding of flood water during the flood season.

Recreation. It is rare to design a large reservoir for recreational purposes only, and any recreational benefits are usually entirely incidental to the other functions of the project. The ideal recreational reservoir is one which remains nearly full during the recreation season to permit boating, fishing, swimming, and other water sports. Moreover, the reservoir should not be subject to large drawdowns as it may cause several problems.

Fish and Wild Life. The problems of fish and wild life in large reservoirs are mainly those of protection. The construction of a reservoir results in a major change in habitat for existing wild life and may result in a decrease in one species of fish or other wild life and perhaps an increase in other species. Protection of existing fisheries may require the provision of fish ladders or other arrangements to permit migratory fish to travel upstream. Large and rapid fluctuations in water level are harmful to fish, particularly at critical periods such as the spawning period. Further complete stoppage of flow below the dam is also destructive of fish and wild life.

Sanitation. The maintenance of adequate flows downstream is the main sanitary function of a reservoir. Usually this may be easily combined with the release of water for other function. A secondary sanitary requirement in some areas may be that the reservoir be operated to control mosquito growth. This is usually done by causing rapid fluctuations of water level which strand larvae on the shore.

22.5 COMPATIBILITY OF MULTIPURPOSE USES

The successful planning of a multipurpose project involving the joint use of storage space for various purposes depends upon the compatibility of the various purposes (i.e., ability to exist together). For example considemr a multipurpose project proposed to supply water from the reservoir for water supply, irrigation, hydro-power generation and navigation. The estimated release of water for each of these purposes are shown in Fig. 22.1. The water supply and irrigation are the consumptive uses of water while the navigation and power generation are the

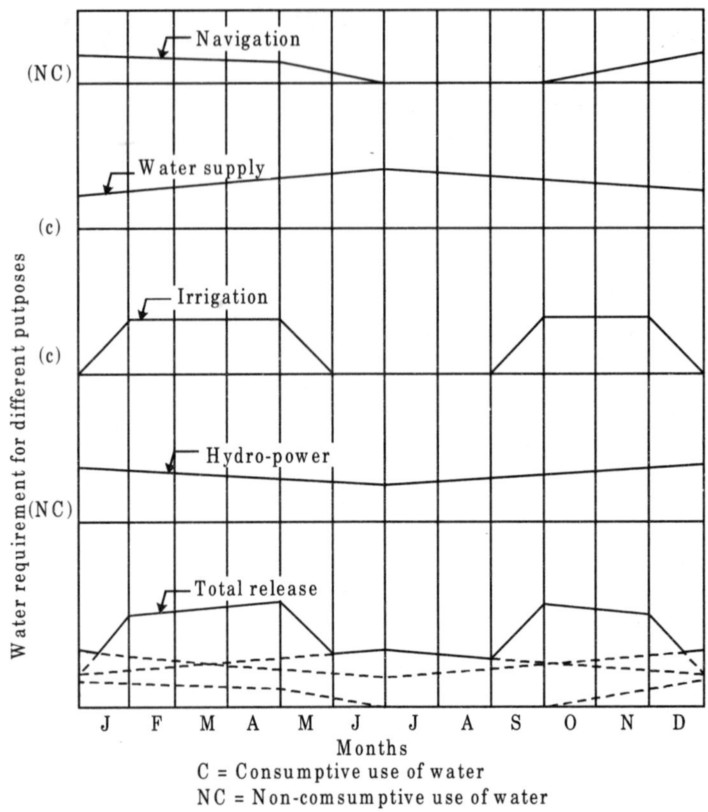

Fig. 22.1 *Water requirement for different purposes and total release of water v/s time curve*

non-consumptive uses of water. Thus, water required to be released for the consumptive uses may be first used for the hydro-power generation. The total release of water v/s time curve is determined by adding the water required to be released for the consumptive uses viz., water supply and irrigation and comparing it with the water required for the non-consumptive uses viz., hydro-power generation and navigation. In this case as shown in Fig. 22.1 the release of water for water supply and irrigation added together will satisfy the demand for hydro-power generation and navigation, in all months except in January and December when some additional release of water is needed. Total release of water integrated over time will represent the total demand which will also represent the useful storage required to be provided for the project.

Flood control with its requirement for empty storage space is the least compatible of all uses. The storage for flood control can be obtained in two ways, either (1) by permanent allocation of space exclusively for flood control or (2) by seasonal allocation of space for flood storage. In the first case, the space is usually allocated at the top of the pool, and water is allowed to encroach on this space only during floods. Often the allocated space is above the crest level of the spillway and is made available when required by closing the spillway crest gates. Any additional space available for flood control is used but it is not dependable. In the second case, the space for flood control is made available by operating the reservoir in such a way that the required space is available for flood storage. Seasonal allocations to flood control can be made where a definite flood season occurs.

Recreational benefits of a reservoir are usually taken as opportunity permits. However, in some cases it may be possible to operate the reservoir in such a way that it is not drawn down excessively during the height of the recreation season. Further picnic grounds, access roads, boating facilities, etc., add to the usefulness of a reservoir for recreation only if the operation is such that pleasant conditions are created.

For fish and wild life protection also the operation of reservoir should be such that there is no excessive drawdown and an adequate flow is maintained in the river on the downstream side. If a minimum flow must be maintained downstream, this may require a separate allocation of storage space unless release of water for other purposes will be more than adequate. Further if the permissible drawdown is limited to some minimum level for fish and wild life protection, the storage below this level is not useful except to provide sediment storage and power head. Thus, some portion of this space may be considered to be allocated for the fish and wild life aspect of the project.

The plan of operation for a multipurpose project combining irrigation, hydroelectric power and flood control is described in Chapter 7 section.

22.6 STEPS INVOLVED IN PLANNING OF A WATER RESOURCES DEVELOPMENT PROJECT

The various steps involved in the planning of water resources development project are as follows :

1. Statement of objective
2. Collection of data
3. Projections for planning
4. Project formulation
5. Project evaluation

Each of these step are discussed below.

1. Statement of Objectives. Before the planning of a project is undertaken, the planner must have a clear statement of objectives of the project. The objectives may, however, depend on the organization planning the project. For example, a national government may have the broad goal of achieving maximum economic and social benefits of development of water resources for the entire nation as a whole, while the government of a state or district which is prone to floods may have a single objective of reducing the flood damage within its boundaries to the extent economically justified in terms of benefits to the residents of that state or district as the case may be. Similarly the corporations or metropolitan councils, or village panchayats needing water supply or irrigation facilities may have only such objectives to the extent justified in terms of benefits to their own residents.

The objectives of a project also depend on the availability of funds. As the central or national government has to allocate funds for serval other items such as defence, education, highways, public health and medical care, etc., unless there are unlimited funds it may not be possible to include all the objectives of a water resources development project. However, in such cases also a comprehensive plan for the project including all the possible objectives may be prepared but it may be implemented in phases depending on the availability of the funds. Moreover it is not necessary that the entire project with comprehensive planning is implemented by a single agency, as certain private groups and various government agencies may carry the development forward in accordance with the comprehensive plan.

2. Collection of Data. For realistic planning of a water resources development project and to arrive at reliable decisions it is absolutely necessary that thorough reliable data are collected. The data required comprise not only the hydrologic and site data but also a substantial geological, economic and social data and data pertaining to other related fields. The services of specialists in many fields are needed to collect and interpret these data. Except hydrologic data, all other data are current and may be collected by a survey at the start of planning. The hydrologic data are essentially historical and the data collected for the past several years are required. Hence intensive hydrologic data-collection programme must be started immediately after the funds for planning of the project are allotted.

The various data required for the planning of a water resources development project are classified as general data and special data and the same are listed below.

(A) General Data. The following general data are required for the planning of a water resources development project.

1. *Physical data* — location, size, physiography, climate, history, population, cultural background both rural and urban, previous investigations.
2. *Hydrologic data* —precipitation, stream flow, evapotranspiration, water quality sediment.
3. *Geological data* —explorations, geological formations, soil survey, erosion, groundwater, minerals.
4. *Cartographic data* — maps.
5. *Ecological data* — vegetation, fish and wild life.
6. *Demographic data*— people and institutions.

7. *Economic data*—industry, transportation, communication, markets, tourism, recreation, land uses, taxes, community needs, national needs.
8. *Legal data*—water rights, pollution control, land zoning, land ownership, administrative pattern, intrastate, interstate and international problems, development of project rights, operation rights.
9. *Public opinion data.*
10. *Data on existing projects.*

(B) Special Data. The following special data are required for the different purposes of the project.

1. *Agricultural use (Irrigation and related drainage)* — land classification, crop water requirements in respect of quantity and quality, climatic studies and limitations.
2. *Municipal and Industrial use* — population and their water needs, industrial water needs in respect of quantity and quality.
3. *Hydroelectric power*—existing capacities, projected needs for future, alternative sources of power.
4. *Flood control*—extent of past flooding and damages, local storm drainage requirements.
5. *Navigation* — present water traffic patterns, alternative means of transportation such as roads, railroads, airways.
6. *Recreational uses*—natural attractions, present recreation pattern in respect of types, place and time.
7. *Fish and wild life* —water requirements, biological data in respect of species and their habits.
8. *Pollution control*— existing waste discharges in respect of location, time and character of waste, water pollution regulations or quality standards.
9. *Drainage*— existing projects, rainfall excess, soil condition, topography.

3. Projections for Planning. Since all projects are for the future, they must be planned not only to meet the present day requirements but also to meet the future needs. For this the projections of future needs will be required to be made. The historical and current data collected for project planning are utilized to serve as the basis for projections of future needs. The projections for future should not be simple extrapolation of past growth curves, because the social, economic and technological developments may cause significant change in the future growth rate as compared to the past growth rate. The planner must anticipate these changes and should take them into account while evaluating the future needs.

The projections for planning should be worked out with utmost care and caution, because unrealistic higher values of 'water needs' may lead to over design and excessive investments, while the lower values may fail to accomplish the full purpose of the project and may lead to shortage of water. Thus for rational planning all attempts should be made to estimate true economic demand rather than a fictitious demand.

4. Project Formulation. Once the basic data and the projections of future needs are obtained, actual formulation of the project can commence. First of all a list of all the possible alternatives is prepared. Such a list should be comprehensive. After the various alternatives are listed, the first step in project formulation is to define the boundary conditions which restrict the project. For example,

(i) One or more of the possible aspects of the development can be eliminated on the basis of physical limitations, e.g., no navigation on torrential mountain stream.

(ii) Certain problem may be fixed in location, e.g., flood mitigation for an existing city.

(iii) The available water may be limited or subject only to minor changes.

(iv) A policy decision may reserve certain lands for specific purposes, e.g., parks and recreation areas.

(v) Certain water use already exists and must continue to be supplied.

(vi) Possible sites for water storage (both surface and underground) can be defined and their limiting capacities evaluated.

The restraints may help to eliminate some of the alternatives from consideration and thus simplify the process of project formulation.

The next step in project formulation is to define alternative land-use plans for each of the possible project units, as the land-use plans influence water requirements. Further the land-use plans can serve as a guide for the selection of the project units which may have some utility. The project units which serve no useful purpose should not be included.

The next step in project formulation is the cataloguing of the possible project units. In the catalogue all the possible project units along with the alternative plans for each project unit should be indicated. For each of the catalogued project units preliminary cost estimates are prepared. Detailed estimates, at this stage need not be prepared, because a few of these alternatives may prove to be too costly and undesirable, and will ultimately be discarded. However, all the preliminary estimates should be based on one particular cost index so that they are comparable.

5. Project Evaluation. A number of alternative proposals and their costs having been worked out in the previous step, the next step is to select the unit or combination of units which is economically most efficient. If this most efficient unit (or units) has benefit-cost ratio more than 1 or any other specified minimum value, the unit or units may be recommended for construction.

If an alternative individual unit is completely independent both physically and economically of all other units, it may be evaluated alone.

A *Physically independent unit* is one for which there are no other units either upstream or downstream which would affect the inflows to the unit or be affected by the outflows from it. In addition, no other unit can contribute to the same end purpose. For example, if there are two reservoirs on two different streams, but both providing flood control for a single downstream point, then they are not physically independent.

An *Economically independent* unit is one which is having no economic interconnection with any other unit. For example, two power plants on two different streams are physically independent as far as stream flow is concerned, but if they are serving the same power system, they are not economically independent, because their power production has to be integrated into the same power system.

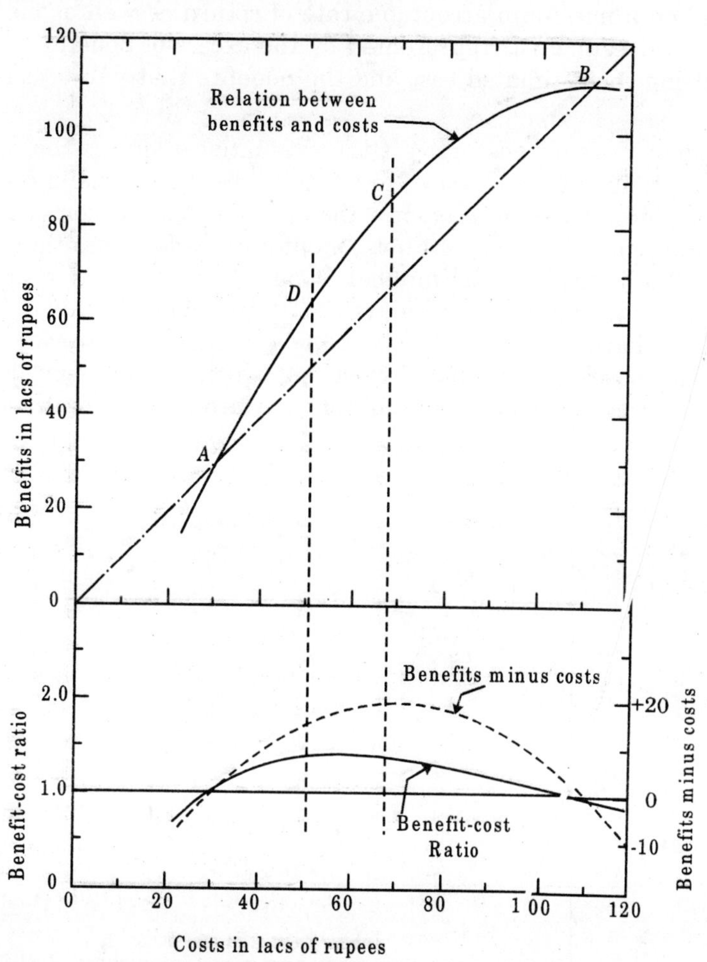

Fig. 22.2 *Relation between costs and benefits for a hypothetical project*

For an independent, single-purpose project the evaluation is relatively simple. In such a case the project may be evaluated in terms of *benefit-cost ratio* as well as *rate of return.* For this the following three curves are plotted (Fig. 22.2).

(i) Benefits v/s costs

(ii) Benefit-cost ratio v/s costs

(iii) Benefits minus costs v/s costs.

Figure 22.2 shows that from point *A* to *B,* benefits exceed costs, i.e., benefit-cost ratio exceeds one. The curve of benefits minus costs (i.e., net benefits) shows a maximum at *C.* Thus beyond point *C*, each rupee of costs returns less than a rupee of benefits that is, (Δ benefits/Δ costs) <1. The maximum benefit-cost ratio occurs at Δ, and this should be the limit of project size if it is desired to obtain a maximum rate of return on the investment. However, the increment from *D* to *C* will also be economical since the rate of return on the increment exceeds the minimum attractive rate of return. It therefore means that the decision to proceed with the project can be based on a minimum acceptable rate of return as well as on a benefit-cost ratio. The rate of return on investment is represented by the excess of benefits over costs and it may be determined by using the estimated cost and the benefits that will accrue over the period of the project life.

For an independent multipurpose project the evaluation requires the preparation of a net-benefit surface (or multidimensional space) to locate the point of maximum net benefits, which determines from the point of view of economy, the optimal combination of the various purposes served by the project. Figure 22.3 shows a net-benefit surface for a multipurpose project serving two purposes. With the increase in the number of purposes served by a project the number of net-benefit surfaces required to be prepared will increase, because a large number of combinations of the various purposes will have to be considered. However, in most of the cases usually two major purposes govern the project, in which case the optimum can be closely approximated with a three dimensional net-benefit surface by holding the other purposes at a constant level.

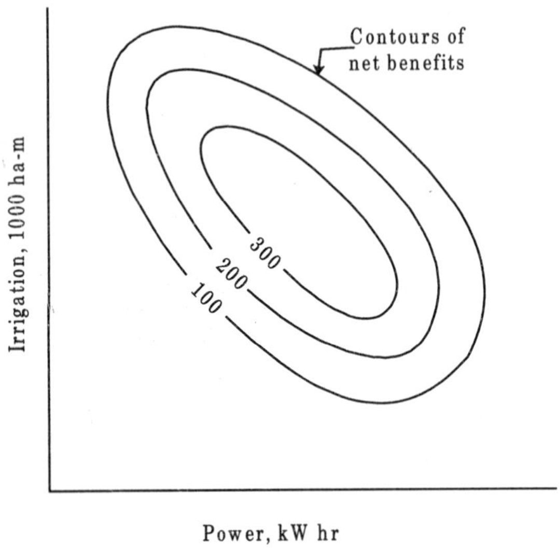

Fig. 22.3 *Typical net-benefit surface for multipurpose project*

In the case of interdependent units the problem becomes much more complex. This problem is that of advanced mathematics and is dealt with by systems analysis. However, the discussion of specific methods of systems analysis as applied to the planning of water resources development projects is beyond the scope of this book.

22.7 FINANCIAL ANALYSIS OF A PROJECT

The financial analysis of a project involves the computation of (i) the present worth of a certain future amount or the benefits which would be derived from a project during certain number of years; (ii) amount to be recovered annually from the beneficiaries of the project so that if needed the project may be replaced by a new one; and (iii) annual instalments for the recovery of the investment made for a project. All these aspects of financial analysis of a project are explained below.

(i) *Present worth of a future amount:*

Consider an investment P made today (i.e., at time O year) which is earning interest @ $i\%$ per annum compounded yearly.

At the end of 1 year the amount would be

$$= (P + iP) = P(1 + i)$$

At the end of 2 years the amount would be

$$= P(1 + i) + iP\,(1 + i) = P(1 + i)^2$$

Thus, at the end of N years the amount grows to

$$F = P(1 + i)^N$$

or

$$P = \frac{1}{(1+i)^N} F \qquad ...(22.1)$$

Equation 22.1 indicates that P is the present worth of an amount F after N years, i.e., an amount F after N years is equivalent to an amount P today.

The term $[1/(1 + i)^N]$ is known as *discount factor* [D.F. (i %, N)], and hence

$$P = [\text{D.F}(i\ \%, \text{N})] \times F \qquad ...(22.1\text{ a})$$

(ii) *Amount to be recovered annually from the beneficiaries of the project to develop a fund to replace the project.*

Let A be the amount recovered from the beneficiaries of the project at the end of every year and it is invested to earn interest @ $i\%$ per annum compounded yearly. It grows to an amount F at the end of N years, (F being the probable cost of the new project to be undertaken to replace the earlier project). Thus,

$$F = A + A(1 + i) + A(1 + i)^2 + ...+ A(1 + i)^{N-2} + A(1 + i)^{N-1} \qquad ...(\text{i})$$

By multiplying Eq. (i) by $(1 + i)$, we get

$$(1 + i)\,F = A(1 + i) + A(1 + i)^2 + A(1 + i)^3 + ...+ A(1 + i)^{N-1} + A(1 + i)^N \qquad ...(\text{ii})$$

By subtracting Eq. (i) from Eq. (ii), we get

$$iF = A[(1 + i)^N - 1]$$

or

$$A = \left[\frac{i}{(1+i)^N - 1}\right] F \qquad ...(22.2)$$

The term $\left[\frac{i}{(1+i)^N - 1}\right]$ is known as *sinking fund factor* [S.F.F ($I\%$, N)], and hence

$$A = [\text{S.F.F.}\ (i\%, N)] \times F \qquad ...(22.2\text{ a})$$

Thus, sinking fund factor indicates the amount A that must be recovered from the beneficiaries of the project at the end of every year and it is invested to earn interest @ $i\%$ per annum compounded yearly to accumulate an amount F at the end of N years.

Alternatively sinking fund factor also indicates the amount A that must be invested uniformly at the end of every year at $i\%$ per annum interest compounded yearly to accumulate an amount F at the end of N years.

(iii) *Annual instalments for the recovery of the investment made for a project*

Let P be the investment made for a project, which if invested for a period of N years at the interest rate $i\%$ per annum compounded yearly, would have grown to an amount F. Thus, from Eq. (22.1), we have

$$F = (1 + i)^N P$$

Introducing this value of F in Eq. (22.2), we get

$$A = \left[\frac{i(1+i)^N}{(1+i)^N - 1}\right] P \qquad ...(22.3)$$

where A is the annual instalment for the recovery of the investment made for a project.

The term $\left[\frac{i(1+i)^N}{(1+i)^N - 1}\right]$ is known as *capital recovery factor* [C.R.F. ($i\%$, N)], and hence

$$A = [\text{C.R.F } (i\%, N)] \times P \qquad ...(22.3\text{ a})$$

Thus, capital recovery factor indicates the amount A that must be recovered every year from the beneficiaries of a project for which amount P has been invested and the interest rate of $i\%$ per annum is considered.

Alternatively capital recovery factor also indicates the amount A that can be withdrawn uniformly at the end of each year for N years, if an amount P is initially invested at $i\%$ per annum interest compounded yearly.

The recoveries may be required to be made from the earning of a project if a loan has been taken for the project and the same is to be repaid; or if a fund is to be created by investing at interest the annual instalments, which may be utilized for undertaking a new project. The fund so created is termed as *sinking fund* and the process of creating this fund is termed as *amortization.*

The above discussed aspects of the financial analysis of a project are indicated in some of the illustrative examples.

22.8 SOME COMMON PITFALLS IN PROJECT PLANNING

It has been observed that there are many pitfalls which may negate an otherwise excellent effort to make a sound project evaluation. Some of the most common pitfalls are discussed in the following paragraphs.

(i) *The preliminary report.* Because of large number of possible projects it is fairly common practice to prepare a preliminary report on a project as the first phase of its analysis. If this report is unfavourable, the project can be dropped without excessive expenditure of funds. If it is favourable, the decision to proceed on more thorough studies is made. The project can be rejected even after a favourable preliminary report, but political considerations may work to avoid any such rejection. As such the preliminary report becomes an important step in the analysis of projects, and it is important that this report is not based on approximations and short-cut procedures.

In particular, the study of hydrology and the factors related to expected water use should be carried to the same level as would be necessary for the final design. Further accurate maps are essential in a preliminary report also. While this may increase the cost of preliminary studies, it would reduce the cost of the final design for projects which reach that phase.

(ii) *Adherence to standards.* Several standards of design have been developed as a result of traditions or administrative rule. However, these may not always be appropriate to the particular problems of a project. For example, a spillway which will flow only once in 50 years might profitably be built of soil cement or other cheaper material rather than conventional material such as reinforced concrete, since repair or replacement on the average of once in 50 years in the former case might be less costly than the more permanent construction in the later. Moreover technological advance may also obsolete a specific standard. As such strict adherence to such standards should not be required when professional judgement suggests otherwise.

(iii) *Early construction.* Construction of a project substantially before it is needed constitutes a waste of funds which might profitably be employed gainfully in the interim period. It may however, be argued that there might be an increase in the cost of construction during the period of postponement of the project, but this need not necessarily represent a loss since increased costs will be at least partially offset by the benefits that would be derived by the investment of these funds elsewhere during the interim period. For example assuming that the economy is functioning fully, a rupee investment postponed for 10 years at 9 per cent interest is equivalent to ₹ 2.37.

Stage construction or investment may be appropriate. For example, the construction of a dam may be planned for stage construction so that with growing water requirements the reservoir size may be increased. Similarly if a project is known to be needed 10 years hence it may be quite desirable to buy right-of-way (or legal right) at once to prevent any other development that would greatly increase cost. Likewise it would be erroneous to build flood-control works on the basis of anticipated future growth of the area to be protected. Also flood control should be provided only if it is justified by existing values.

(iv) *A priori decisions.* Sometimes decisions regarding certain features of a project are made before economic analysis and are never checked. For example in a flood mitigation project decision may be taken to adopt some measures to prevent the flood damage corresponding to certain maximum flow rate, though the flood damage might not begin until the flow is far in excess of the assumed maximum flow. *A priori* decisions of this type can lead to incorporation of uneconomic increments in many projects.

(v) *Failure to consider all alternatives.* The failure to consider all alternatives is probably the most common pitfall. Often certain non-engineering alternatives such as flood plain management may be overlooked by the engineering designer. Moreover, sometimes simple engineering alternatives are also overlooked, especially when they depart from the traditional solutions. For example to reduce the seepage through a small earth dam it may be recommended to provide a blanket on the upstream face and a clay cutoff in the foundation, but the possibility of providing relief wells on the downstream side might not have been investigated though it would have been a useful source of filtered water as well as a seepage control measure.

(vi) *Use of market prices.* Market prices of water do not always reflect its real value. Hence, market prices are not necessarily a satisfactory basis for computing benefits. As far as possible the real benefit should be estimated and utilized in the economic analysis.

(vii) *Use of next best alternative.* In lieu of any other estimate of the project, it is quite common to use the next best alternative as a measure of the benefits of the proposed project. By judicious selection of the next best alternative, the benefits of the proposed project may indeed be made very large. However, the only alternative acceptable as a basis for estimating benefits is the one which would be built if the project under study was dropped.

22.9 AUGMENTATION OF WATER SUPPLIES

Augmentation of water supplies includes methods for providing an entirely new supply of fresh water and techniques for increasing the utility of available supplies of water by conservation or regulation which permits greater use. Techniques for providing completely new supplies of water include *weather modification* which may include many possible plans for altering the atmospheric circulation to provide a more favourable distribution of precipitation or to decrease *evapotranspiration*. For the present the main effort has been devoted to localized precipitation increases by cloud seeding. Recent studies have shown that under favourable conditions on an average about 10 per cent increase in precipitation can be expected by this technique. Further Lumb has demonstrated that the increase in annual runoff ΔR resulting from a small increase in precipitation ΔP is given approximately by

$$\Delta R = \Delta P\left(0.29 + 1.2\frac{R_{avg}}{P_{avg}}\right) \qquad \text{...(22.4)}$$

where R_{avg} and P_{avg} are the mean annual runoff and precipitation respectively. Equation (22.4) is based on the analysis of seven rural basins. It should be used only for preliminary estimates,

where $0.1 < \left(\frac{R_{avg}}{P_{avg}}\right) < 0.5$.

Another feasible but costly method of augmenting fresh water supplies is *desalting of seawater*. Many possible methods are available for this purpose but the cost is still too high for wide spread use.

Conservation of available supplies of water can also be achieved by a number of methods as indicated below.

(i) Sewage disposal by water transport and dilution degrades vast amounts of pure water, and techniques which will use less water in the home and industry could result in substantial savings of water.

(ii) Agricultural water requirements in many areas could be greatly reduced by reducing conveyance losses and wasteful irrigation practices.

(iii) By providing roof over small reservoirs evaporation loss can be reduced. The use of surface agents to reduce evaporation from reservoirs is still under study.

(iv) Large parts of fresh water supply are utilized for evapotranspiration by native vegetation. Substantial amounts of water can be saved by elimination of tamarisk and willow in arid regions. Also several measures may be adopted to reduce evapotranspiration by crops.

(v) Since substantial amounts of water are made unfit for use by the addition of relatively small amounts of pollutants, reclamation and reuse of waste water is another method by which existing water supplies can be accomplished by direct reuse for industrial processes with a minimum of treatment or more elaborate reclamation plants could be constructed to make the water biologically and chemically suitable for higher uses. Cost is a controlling factor, but continuing research should result in lower costs while the value of water will probably increase. At some point of time the two curves should cross.

(vi) Since groundwater is free from evapotranspiration losses, increased use of groundwater storage in conjunction with moderate surface storage will reduce the large evaporation losses occurring from large reservoirs in arid regions.

(vii) Sealing the soil surface with chemicals or asphalt can greatly increase water yield in arid regions. Since nearly all the rain which falls on the protected area is recovered, the yield is easily estimated as the mean annual precipitation over the sealed area. If the runoff water is conveyed to a covered tank, evaporation loss is prevented. The water stored in this manner may be used for various purposes. Cost is dependent on the cost of surface sealing.

SOLVED ILLUSTRATIVE EXAMPLES

Illustrative Example 22.1 *A multipurpose reservoir is planned to serve flood control, water supply, irrigation, power and navigation. Navigable reach lies between dam and power house and a diversion weir is provided for water supply and irrigation. Requirements of water are: power @ 70 cumec for 12 months; water supply @ 48 cumec for 12 months; irrigation @ 86 cumec for 4 months (November to February) and navigation @ 104 cumec for 6 months (April to September). Flood control requires 190 million m^3 and dead storage is estimated as 22% of useful storage. Determine the total storage upto maximum water level. Take each month of 30 days.*

Solution

As shown in Fig. Ex. 22.1 total release of water required for various purposes (or useful storage)

$$= (134 \times 2 + 70 \times 1 + 104 \times 6 + 70 \times 1 + 134 \times 2) \times 30 \times 24 \times 3600$$

$$= 3369.6 \text{ million m}^3$$

$$\text{Dead storage} = 3369.6 \times \frac{22}{100}$$

$$= 741.3 \text{ million m}^3$$

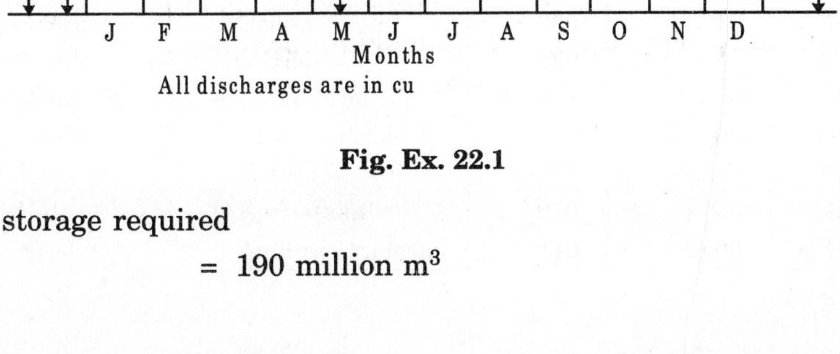

Fig. Ex. 22.1

Flood control storage required

$$= 190 \text{ million m}^3$$

∴ Total storage required

$$= (3369.6 + 741.3 + 190)$$

$$= 4300.9 \approx 4301 \text{ million m}^3$$

Illustrative Example 22.2 *What is the equivalent present worth at 10% interest of three ₹ 6 million investments, one made now, one at the end of third year, and last at the end of tenth year from now ?*

Solution

From Eq. (22.1), we have

$$P = \frac{1}{(1+i)^N} F$$

$$F_1 = F_2 = F_3 = ₹\ 6 \times 10^6$$

$$\text{Rate of interest } i = 10\% = 0.1$$

For the investment made now, since $N = 0$,

$$P_1 = F_1 = ₹\ 6 \times 10^6$$

For the investment made at the end of third year, $N = 3$

$$\therefore \quad P_2 = \frac{1}{(1+0.1)^3} \times (6 \times 10^6)$$

$$= 0.7513 \times 6 \times 10^6$$

$$= ₹\ 4.5078 \times 10^6$$

For the investment made at the end of tenth year, $N = 10$

$$\therefore \quad P_3 = \frac{1}{(1+0.1)^{10}} \times (6 \times 10^6)$$

$$= 0.3855 \times 6 \times 10^6$$

$$= ₹\ 2.3130 \times 10^6$$

The present worth of the three investments is

$$P = (P_1 + P_2 + P_3)$$

$$= (6 \times 106 + 4.5078 \times 10^6 + 2.3130 \times 10^6)$$

$$= ₹\ 12.8208 \times 10^6$$

Illustrative Example 22.3 *What amount should be charged at the end of every year from the beneficiaries of a project so as to replace it with a new one which will cost ₹ 160 million at the end of 20 years. The amount charged may be assumed to earn interest @ 8%.*

Solution

From Eq. (22.2), we have

$$A = \left[\frac{i}{(1+i)^N - 1}\right] F$$

$i = 8\% = 0.08$; $N = 20$ years; and $F = ₹\ 160 \times 10^6$

Thus, by substitution, we get

$$A = \left[\frac{0.08}{(1+0.08)^{20}-1}\right] \times (160 \times 10^6)$$

$$= ₹\ 3.5 \times 10^6$$

Illustrative Example 22.4 *The capital investment for a project is ₹ 300 million. What annual year-end payments are needed at 7% interest to amortize the investment in 10 years ?*

Solution

From Eq. (22.3), we have

$$A = \left[\frac{i(1+i)^N}{(1+i)^N-1}\right] P$$

$i = 7\% = 0.07$; $N = 10$ years; $P = ₹\ 300 \times 10^6$

Thus, by substitution, we get

$$A = \left[\frac{0.07 \times (1+0.07)^{10}}{(1+0.07)^{10}-1}\right] \times (300 \times 10^6)$$

$$= ₹\ 42.71 \times 10^6$$

Illustrative Example 22.5 *The total cost of lining of a certain canal is ₹ 10 million. If the annual benefits resulting from the lining amounts to ₹ 1 million, determine whether the lining would be economically feasible. The rate of interest i = 8% per annum and the life of the lining is estimated to be 20 years. If not, determine the estimated life of the lining which would render it economically feasible.*

Solution

The net present worth (NPW) P of the annual benefits A accruing over the life of the lining is given by Eq. (22.3) as

$$P = A\left[\frac{(1+i)^N-1}{i(1+i)^N}\right]$$

$i = 8\% = 0.08$; $N = 20$ years; and $A = ₹\ 1 \times 10^6$

Thus, by substitution, we get

$$P = 1 \times 10^6 \left[\frac{(1+0.08)^{20}-1}{0.08(1+0.08)^{20}}\right]$$

$$= ₹\ 9.82 \times 10^6$$

Total cost of lining = ₹ 10 × 10^6, which is more than the net present worth of the annual benefits accruing over the life of the lining. Hence in this case the lining would not be economically feasible.

Let N years be the estimated life of the lining which would render it economically feasible.

Thus by substitution, we get

$$P = 1\times10^6\left[\frac{(1+0.08)^N - 1}{0.08(1+0.08)^N}\right]$$

For the lining to be economically feasible the net present worth P should be equal to or greater than the total cost of lining i.e., ₹ 10×10^6.

Assuming N = 21 years in the above equation, we get

$$P = ₹\ 10.02 \times 10^6 > ₹\ 10 \times 10^6$$

Hence if the estimated life of the lining is 21 years then it would become economically feasible.

Illustrative Example 22.6 *Two projects A and B are to be compared in terms of their benefit-cost ratios and net benefit on present worth basis. The applicable discount rate is 10% per annum compounded yearly. The following table shows the benefits and the costs in thousands of Rupees at the end of each year. Which project is more economical ?*

Year	*Construction cost*	*Operation cost*	*Benefits*	*Remarks*
1	100	—	—	
2	—	10	20	
3	—	10	40	Project A
4	—	10	60	
5	—	10	80	
1	100	—	—	
2	100	—	—	
3	—	20	40	
4	—	20	60	Project B
5	—	20	80	
6	—	20	100	
7	—	20	140	
8	—	20	160	

Solution

The benefit-cost ratio and the net benefit for each of the two projects A and B are computed as shown in the following tables.

Project A

Year	Cost	Benefit	(B/C)	Net Benefit	$\frac{1}{(1+i)^N}$	Present worth of Cost	Present worth of Benefit	Present worth of Net Benefit	Net (B/C)
	C	B		(B – C)		C	B	(B – C)	
1	100	—	—	– 100	0.909	90.90	—	– 90.90	—
2	10	20	2.0	10	0.826	8.26	16.52	8.26	2.0
3	10	40	4.0	30	0.751	7.51	30.04	22.53	4.0
4	10	60	6.0	50	0.683	6.83	40.98	34.15	6.0
5	10	80	8.0	70	0.621	6.21	49.68	43.47	8.0
					Total	119.71	137.22	17.51	

Project B

Year	Cost	Benefit	(B/C)	Net Benefit	$\frac{1}{(1+i)^N}$	Present worth of Cost	Present worth of Benefit	Present worth of Net Benefit	Net (B/C)
	C	B		(B – C)		C	B	(B – C)	
1	100	—	—	– 100	0.909	90.90	—	– 90.90	—
2	100	—	—	– 100	0.826	82.60	—	– 82.60	—
3	20	40	2.0	20	0.751	15.02	30.04	15.02	2.0
4	20	60	3.0	40	0.683	13.66	40.98	27.32	3.0
5	20	80	4.0	60	0.621	12.42	49.68	37.26	4.0
6	20	100	5.0	80	0.564	11.28	56.40	45.12	5.0
7	20	140	7.0	120	0.513	10.26	71.82	61.56	7.0
8	20	160	8.0	140	0.467	9.34	74.72	65.38	8.0
					Total	245.48	323.64	78.16	

For Project A

$$\frac{B}{C} = \frac{137.22}{119.71} = 1.146\text{; and}$$

Present worth of net benefits = ₹ 17.51 $\times 10^3$

For Project B

$$\frac{B}{C} = \frac{323.64}{245.48} = 1.318\text{; and}$$

Present worth of net benefits = ₹ 78.16 × 10^3.

As the (*B*/*C*) ratio is greater for project *B* than that for project *A* and also the present worth of net benefits is more for project *B* than that for project *A*, project *B* is more economical.

Illustrative Example 22.7 *A lift canal is proposed to be constructed to deliver 16 cumec of water. The estimated cost of lift canal and its distribution system including water courses is ₹ 21 crores. The cost of pumps is ₹ 2.1 crores. The share cost of the dam, barrage and lined main canal which may be added to this project may be considered @ ₹ 33 lakh per cumec. It is proposed to irrigate 59210 hectares of land and 1.2 cumec of water is used at the tail end of the canal for the domestic and industrial water supply. Compute the benefit-cost (B.C) ratio for the scheme with the following considerations.*

(i) Annual rate of interest on the capital cost of the project @ 10%.

(ii) Annual maintenance and depreciation on canals including water courses @2% of the capital cost excluding the cost of the pumps.

(iii) Annual depreciation charges on pumps @ $3\frac{1}{3}$% of the cost of the pumps.

(iv) Total head for pumping 30 m

(v) Efficiency of the pumps 60%

(vi) Transformer losses 3%

(vii) Capacity factor for canals 0.67.

(viii) Power charges or pumps @ 15 paise per kWh

(ix) Annual maintenance and operation charges for pumps @ ₹ 16000 per cumec.

(x) Annual administrative expenditure @ ₹ 25 per hectare of land irrigated.

(xi) Net value of the agriculture produce after irrigation @ ₹ 1350 per hectare.

(xii) Net value of the agriculture produce before irrigation ₹ 63 lakh.

(xiii) Water charges for the domestic and industrial supply @ 15 paise per 1000 litres.

Solution

(A) Computations for power requirements

Maximum power required for pumping water

$$= \left[\frac{wQH}{1000} \times \frac{1.03}{0.6} \times 24 \times 365\right] \text{kWH}$$

Average power required for pumping water

$$= \left[\frac{wQH}{1000} \times \frac{1.03}{0.6} \times 24 \times 365\right] \times 0.67 \text{ kWH}$$

$$= \left[\frac{9810 \times 16 \times 30}{1000} \times \frac{1.03}{0.6} \times 24 \times 365 \times 0.67\right] \text{kWH}$$

$$= 4.74 \times 10^7 \text{ kWH}$$

(B) Computations for B.C. ratio

1. *Total cost of the project*

	₹
(i) Cost of lift canal and its distribution system including water courses	210 000 000
(ii) Cost of the pumps	21 000 000
(iii) Share cost of the dam, barrage and lined main canal @ ₹ 33 lakh per cumec, (₹ 3 300 000 × 16)	52 800 000
Total cost	**283 800 000**

2. *Annual costs*

(i) Interest on the capital cost of the project @ 10%	28 380 000
(ii) Maintenance and depreciation on canal and water courses @ 2% of the capital cost excluding the cost of pumps	5 256 000
(iii) Power charges for the operation of the pumps @ 15 paise per kWh, $\left(₹\ \frac{15}{100} \times 4.74 \times 10^7\right)$	7 110 000
(iv) Depreciation charges for pumps @ $3\frac{1}{3}$ % of the cost of the pumps, $\left(₹\ \frac{10}{3 \times 100} \times 21000000\right)$	700 000
(v) Maintenance and operation charges for pumps @ ₹ 16 000/= per cumec, (₹ 16 000 × 16)	256 000
(vi) Administrative expenditure @ ₹ 25/= per hectare of land irrigated. (₹ 25 × 59 210)	1,480 250
Total annual cost	**43 182 250**
Deduct revenue on account of water supply for domestic and industrial water supply @ 15 paise per 1000 litres, $\left(₹\ \frac{15}{100} \times \frac{1.2 \times 3600 \times 24 \times 365 \times 10^6}{10^3 \times 10^3}\right)$	5 676 480
Net annual cost for irrigation purposes	37 505 770

3. *Annual benefits*

(i) Net value of the agriculture produce after irrigation @ ₹ 1350/= per hectare (₹ 1350 × 59 210)	79 933 500
(ii) Net value of the agriculture produce before irrigation	6 300 000
Net benefits	**73 633 500**

$$\therefore \quad \text{B.C ratio} = \frac{73\,633\,500}{37\,505\,770} = 1.96$$

Illustrative Example 22.8 *The construction of a hydel project would cost ₹24 400 000. The project has an annual operation and maintenance cost of ₹ 1,40,000 and a 50 year life; a planning discount rate of 3% is to be used. What is the annual cost of the project?*

Solution

The annual cost (AC) of the project is equal to the sum of the annual operation and maintenance cost and C.R.F. Thus

$$AC = 140\,000 + P\left[\frac{i(1+i)^N}{(1+i)^N - 1}\right]$$

$$P = 24\,400\,000;\ i = 3\% = 0.03;\ \text{and } N = 50 \text{ years}$$

By substitution, we get

$$AC = 1{,}40{,}000 + 24{,}400{,}000\left[\frac{0.03(1.03)^{50}}{(1.03)^{50} - 1}\right] \text{ rupees}$$

$$= (1{,}40{,}000 + 9{,}50{,}000) \text{ rupees}$$

$$= 1{,}090{,}000 \text{ rupees}$$

Illustrative Example 22.9 *A multipurpose reservoir project is planned for power and irrigation benefits. Its capital investment is estimated at Rs. 36 million and useful life as 70 years. The annual firm yield with conservation storage provided is estimated as 32 000 hectare-metre, and its release at uniform rate all over the year will generate 2 500 kW power. Somewhere downstream the released water may be used to irrigate sugarcane crop (duty 600 hectare/cumec). Assuming the yearly maintenance cost at ₹0.52 million and rate of interest on capital investment as 4%, fix a suitable revenue rate on power (in paise per kWh) and irrigation (in ₹ Per hectare) such that the total load on the two beneficieries is almost equal and the project is economically justifiable.*

Solution

The total amount of power developed annually

$$= 2500 \times 24 \times 365 \text{ kWh}$$

$$= 21900000 \text{ kWh}$$

Uniform rate of discharge available through out the year

$$= \frac{32000 \times 10^4}{365 \times 24 \times 3600} = 10.147 \text{ cumec}$$

Duty of sugarcane crop = 600 hectare/cumec

∴ Area of sugarcane crop which may be irrigated

$$= (10.147 \times 600) = 6088.2 \text{ hectares}$$

The amount that must be recovered annually from the beneficiaries of the project is given by Eq. (22.2) as

$$A = \left[\frac{i}{(1+i)^N - 1}\right] \times F$$

$i = 4\% = 0.04$; $N = 70$ years; and $F =$ ₹ 36×10^6

Thus, by substitution, we get

$$A = \left[\frac{0.04}{(1+0.04)^{70} - 1}\right] \times 36 \times 10^6$$

$$= ₹\ 98\ 820$$

Yearly maintenance cost = ₹ 520 000

Total yearly cost = ₹ (98 820 + 520 000)

= ₹ 618 820

Considering the total load on the two beneficiaries to be equal, we get

$$\text{Revenue rate on power} = \frac{618820}{2 \times 21{,}900{,}000}$$

= ₹ 0.014 per kWh

= 1.4 paise per kWh

$$\text{Revenue rate of irrigation} = \frac{618\ 820}{2 \times 6088.2}$$

= ₹ 51 per hectare

Illustrative Example 22.10 *A spillway is being designed to pass a flood of 2000 m*^{3} *per second. The annual damages in case the flood exceeds the designed flood are estimated as ₹ 15,000. Which of the following spillway is the most economical ?*

Type of spillway	*Construction cost*	*Life (years)*	*Probability of flood exceeding the designed flood in any one year*
A	20,000	20	0.20
B	30,000	25	0.10
C	55,000	30	0.15

The attractive rate of return is 6% per annum compounded yearly.

Solution

The most economical spillway is the one for which the total annual cost is the least. The total annual cost of a spillway is equal to the sum of the annual construction cost and the annual cost of damage.

The annual construction cost of a spillway may be computed by Eq. (22.3) as noted below

$$A = \left[\frac{i(1+i)^N}{(1+i)^N - 1}\right] P$$

in which

A = annual construction cost of spillway

i = rate of return (or rate of interest)

N = life of spillway in years, and

P = total cost of construction of spillway.

The annual cost of damage is obtained by multiplying the estimated annual cost of damage (which in this case is ₹ 15,000) by the probability of the flood exceeding the designed flood in any one year.

For Type A spillway

N = 20 years; P = ₹ 20,000; and i = 6% = 0.06

Thus, by substitution the annual construction cost of the spillway is obtained as

$$A = \left[\frac{0.06(1+0.06)^{20}}{(1+0.06)^{20}-1}\right] \times 20000$$

$$= ₹\ 1744$$

The annual cost of damage for this spillway

$$= (15000 \times 0.20)$$

$$= ₹\ 3000$$

∴ Total annual cost for Type A spillway

$$= ₹\ (1744 + 3000)$$

$$= ₹\ 4744$$

In the same manner for Type B and Type C spillways annual construction cost, annual cost of damage and total annual cost are computed as indicated in the following table.

Type of spillway	*Construction cost* (₹)	*Life (years)*	*Probability construction cost* (₹)	*Annual of flood exceeding design flood*	*cost of damage* (₹)	*Total annual cost* (₹)
A	20 000	20	1744	0.20	3000	4744
B	30 000	25	2347	0.10	1500	3847
C	55 000	30	3996	0.15	2250	6246

The total annual cost for the Type B spillway is less than that for the other two types of spillways and hence the Type B spillway is the most economical out of the three types of spillways.

REVIEW QUESTIONS

22.1 Give a detailed list of the various purposes for which a multipurpose project is undertaken.

22.2 Write a brief note on compatibility of multipurpose uses in a water resources project.

22.3 List the various data required to be collected for planning a water resources project.

22.4 Discuss the various steps involved in the planning of water resources development project.

22.5 Describe some common pitfalls in planning of a water resources project.

22.6 Write a note on augmentation of water supplies.

22.7 How an independent, single purpose project may be evaluated ?

22.8 Explain the terms (i) physically independent unit and (ii) economically independent unit.

22.9 How an independent, multipurpose project may be evaluated ?

22.10 A multipurpose reservoir is planned to serve flood control, power, water supply, irrigation and navigation. Navigation reach lies between dam and powerhouse and diversion weir for irrigation and water supply. Water requirements are: @ 125 cumec for water supply, @ 140 cumec for power, @ 162 cumec for navigation uniformly for 12 months and @ 86 cumec for two three-monthly periods for irrigation. Flood control requires a space of 25 million m^3, and dead storage is estimated as 18% of conservation storage. Determine the total storage capacity of the dam upto maximum storage reservoir level. Take each month of 30 days.

[*Ans.* 6870 million m^3]

22.11 Define the terms (i) discount factor, (ii) sinking fund factor, (iii) capital recovery factor.

22.12 The total cost of lining of a certain canal is ₹ 10 million. If the annual benefits resulting from the lining amounts to ₹ 1 million, determine the maximum rate of interest at which the loan may be obtained to make the lining economically feasible. The estimated life of the lining is 20 years

$\left[\textit{Ans.}\ 7\frac{3}{4}\%\ \text{per annum}\right]$

22.13 Two alternative plans are suggested for a proposed aqueduct as follows:

Plan A: Uses a tunnel. Cost of tunnel construction ₹ 4,50,000, with estimated life 100 years and annual maintenance cost ₹ 4000.

Plan B: Uses a section of brick lined canal and a section of a concrete flume. Cost of canal construction ₹ 1,20,000, with estimated life 100 years, cost of canal lining ₹ 50,000, with estimated life 20 years, cost of flume construction ₹ 90,000, with estimated life 50 years. Annual maintenance of canal and flume costs ₹ 10,500.

Calculate the total annual costs for each of the two plans of the project and suggest which one would you recommend assuming discount rate as (i) 8%, (ii) 4%, per annum.

All salvage values are assumed to be negligible.

[*Ans.* (i) Plan *A*– ₹ 40,016, Plan *B*– ₹ 32,554, hence Plan *B* recommended;
(ii) Plan *A*– ₹ 22,364, Plan *B*– ₹ 23,266, hence Plan *A* recommended.]

22.14 A power company is considering construction of a 100 MW plant. The costs for two alternative plants have been estimated as follows :

	Thermal	*Hydel*
First cost	₹ 10 million	₹ 24 million
Estimated life	25 years	50 years
Taxes and insurance	4% per annum of first cost	3.5% per annum of first cost
Fuel cost (annual)	₹ 1.5 million	—
Labour and maintenance cost (annual)	₹ 0.35 million	₹ 0.14 million
Extra transmission facilities (first cost)	—	₹ 0.40 million

Assuming discount rate as 10% and no energy loss in transmission, determine

(i) which plant would be better selection;

(ii) annual cost per kW of installed capacity for each plant.

[*Ans.* (i) Thermal, (ii) Thermal; ₹ 33.52, Hydel; ₹ 34.41]

22.15 An equipment used in the construction activity costs ₹ 40,000 at first purchase. The average annual expenditure there on for OMR is estimated to be ₹ 9000 for the next three years for which funds must be procured at the beginning of each year. At the end of the third year it can be scrapped at a then value of ₹ 10,000. If the acceptable rate of return is 15% per annum, what should be the annual income by this machine as reckoned at the end of each year? [*Ans.* ₹ 17186]

CHAPTER

23 Flood Control

23.1 INTRODUCTION

A flood is any relatively high flow that overtops the natural or artificial banks in any reach of a stream. The flood is the result of runoff from rainfall and/or melting snow in quantities too great to be confined in the low water channels of streams. When the banks are overtopped, water spreads over the flood plain and thus cause damage to crops and property within the flood plain of the stream. It is not possible to prevent floods but it is possible to prevent or reduce the damage due to floods by controlling the floods. Thus, flood control or flood management is defined as the prevention or reduction of the flood damage.

In our country about 40 million hectares of area has been identified as flood prone. Out of this, upto March 1982 about 12 million hectare has been provided with reasonable protection by construction of about 12 5000 km of embankments and 25600 km of drainage channels. In addition, about 4 700 villages and 340 towns have been given protection. However, inspite of the various programmes of flood control undertaken in the past and a massive programme currently on, there is still a long way to go. Accordingly Government of India, in July 1976, set up Rashtriya Barh Ayog (National Flood Commission) to study in depth the effectiveness of the works so far carried out and recommend measures to be adopted. The Ayog submitted its report in March 1980 which is a landmark in the evolution of the national policy on floods. A broad outline of the National Policy on floods is given at the end of this chapter.

23.2 CLASSIFICATION OF METHODS FOR FLOOD CONTROL OR FLOOD MANAGEMENT

The various methods of flood control (or flood management) and reduction of flood damage can be classified under different categories, depending on the attempts made to solve this problem. The broad classification as adopted by the Rashtriya Barh Ayog are as under.

(i) Attempt to modify the flood.

(ii) Attempt to modify the susceptibility to flood damage.

(iii) Attempt to modify the loss burden.

(iv) Bearing the loss.

The various methods of flood control under these broad classification are given in Table 23.1 as adopted from the Report of Rashtriya Barh Ayog. Out of this detailed list of the methods of flood control, the commonly adopted methods are as indicated below.

1. Reduction of peak flow by *reservoirs*
2. *Confinement* of the flow within a predetermined channel by *Levees, flood walls,* or a *closed conduit.*
3. Reduction of peak stage by increased velocity resulting from *channel improvement.*
4. Diversion of flood waters through *floodways* or *by passes.*
5. Temporary or permanent *evacuation* of the flood plain.
6. *Flood proofing* of specific properties.
7. Reduction of flood runoff by *land management.*

TABLE 23.1 Methods of flood control or flood management

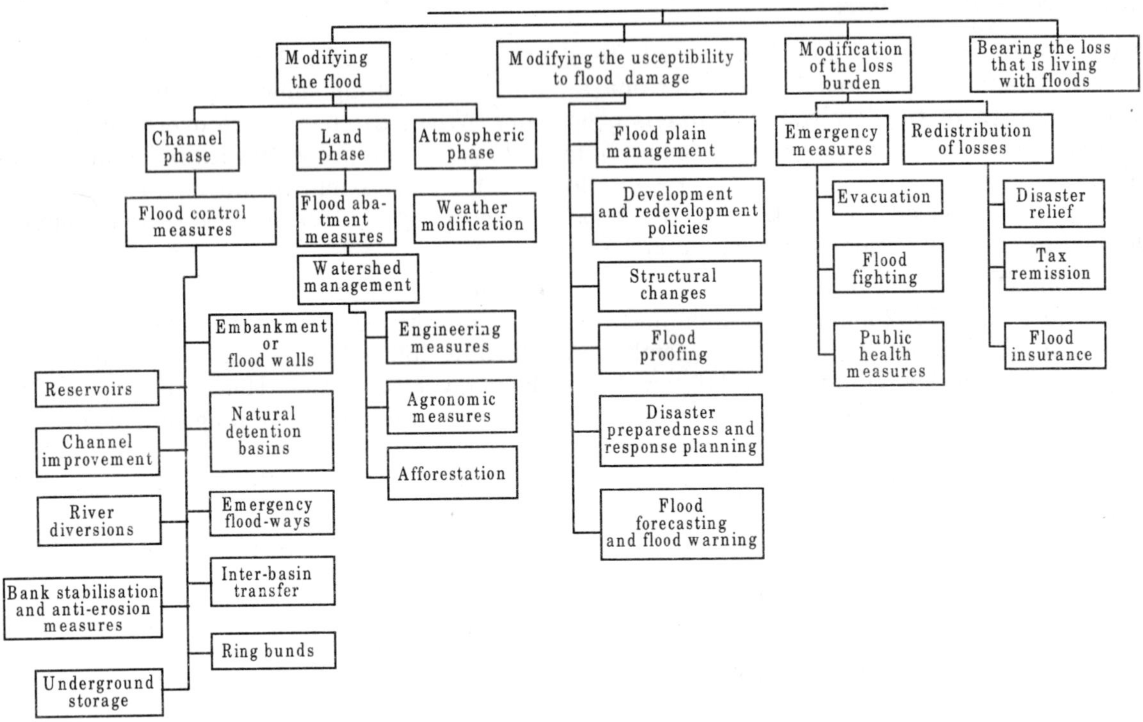

Each of these methods are discussed in the following sections. It may however, be stated that most flood control projects utilize a combination of these methods.

23.3 DESIGN FLOOD AND ITS ESTIMATION

The *design flood* is the flood adopted for the design of a flood control project. It may be either the maximum probable flood or the standard project flood or a flood corresponding to some desired frequency of occurrence depending upon the degree of protection that should be provided by the flood control project.

The *maximum probable flood* is the flood that may be expected from the most severe combination of critical meteorologic and hydrologic conditions that are reasonably possible in the region. It is estimated from the maximum probable storm applying the unit hydrograph principle. The maximum probable storm is an estimate of the physical upper limit to storm rainfall over the basin. It is obtained from the studies of all the storms that have occurred in the region and maximizing them for the most critical atmospheric conditions.

The *standard project flood* is the flood that may be expected from the most severe combination of meteorologic and hydrologic conditions that are reasonably possible in the region. It is computed from the standard project storm rainfall applying the unit hydrograph principle. The standard project storm rainfall may be taken as the largest storm rainfall which has occurred in the region of the basin during the period of weather record. It is however, not maximized for the most critical, atmospheric conditions as in the case of maximum probable flood. The standard project flood is usually about 50 percent of the maximum probable flood for the area.

In the design of a flood control project it would be desirable to provide protection against the maximum probable flood, if this were feasible within acceptable limits of costs. However, it is seldom practicable to provide absolute flood protection by any of the methods. As such some risk must be accepted in the selection of the design flood. The various methods in vogue for estimation of design flood are broadly classified as under.

1. Application of a suitable factor of safety to maximum observed flood or maximum historical flood.
2. Empirical flood formulae.
3. Envelope curves.
4. Frequency analysis.
5. Rational method of derivation of design flood from storm studies and application of unit hydrograph principle.

1. Application of a suitable factor of safety to maximum observed floor or Maximum historical flood. The design flood is obtained by applying a safety factor to the observed or estimated maximum historical flood at the project site or nearby site on the same stream. The safety factor, however, depends on the judgement of the designer. This method is limited by the highly subjective selection of a safety factor and the length of available stream flow record which may give a quite inadequate sample of flood magnitudes likely to occur over a long period of time.

2. Empirical flood formulae. The empirical formulae commonly used in our country are the Dicken's formula, Ryve's formula and Inglis' formula in which the peak flow is given as a function of the catchment area and a coefficient. The values of the coefficient vary within rather wide limits and have to be selected on the basis of judgement. They have limited application, should be used with caution and only when a more accurate method cannot be applied for lack of data.

3. Envelope curves. In the envelope curve method maximum flood is obtained from the envelope curve of all the observed maximum floods for a number of catchments in a homogeneous meteorological region plotted against drainage area. The method, although useful for generalizing the limits of floods actually experienced in the region under consideration, cannot be relied upon for estimating maximum probable floods as an aid to judgement.

4. Frequency analysis. The frequency method involves the statistical analysis of observed data of a fairly long (at least 25 years) period. However, this being a purely statistical approach when applied to derive design flood for long recurrence intervals several times larger than the data has many limitations and hence this method has to be used with caution.

5. Rational method of derivation of design flood from storm studies and application of unit hydrograph principle. This method involved the following steps :

(i) analysis of rainfall v/s runoff data for derivation of loss rates under critical conditions ;

(ii) derivation of unit hydrograph by analysis (or by synthesis, in cases where data are not available) ;

(iii) derivation of the design storm ; and

(iv) derivation of design flood from the design storm by the application of the rainfall excess increments to the unit hydrograph.

Maximization for unit hydrograph peak. It has been observed in practice that the unit hydrograph peak obtained from heavier rainfalls is about 25 to 50 percent higher than those obtained from the smallest rainfalls. Therefore, the unit hydrographs from the observed floods may have to be suitably maximized upto a limit of 50 percent depending on the judgement of the hydrologists. In case the unit hydrograph is derived from very large floods, then the increase may be of very small order, but if it is derived from low floods the increase may have to be substantial.

Limitation of unit hydrograph. The following are the limitations of the unit hydrograph principle.

(a) The unit hydrograph principle is not applicable for drainage basins having an area of more than 5000 km2 where valley storage effects are not reflected and where variation of rainfall in space and time shows a tendency to become too great to be reflected in the unit hydrograph.

(b) Application of the unit hydrograph principle is also not recommended for catchments having an area less than about 25 km^2.

(c) Large number of raingauges suitably located should be available in the entire catchment to reflect the true weighted rainfall of the catchment.

(d) Unit hydrograph principle is not applicable when appreciable proportions of the precipitation occurs in the form of snow or when snow covers a significant part of the catchment.

23.4 FLOOD-CONTROL RESERVOIRS

1. **Purpose of Flood-control Reservoirs**. The function of flood-control reservoir is to store a portion of the flood flow so as to minimize the flood peak at the point to be protected. In an ideal case the reservoir is situated immediately upstream from the protected area and is operated to 'cutoff' the flood peak. This is accomplished by discharging all reservoir inflow until the outflow reaches the safe capacity of the channel downstream. All flow above this rate is stored until inflow drops below the safe channel capacity, and the stored water is released to recover storage

capacity for the next flood. Since the reservoir is situated immediately upstream from the point to be protected, the hydrograph at that point is same as that released at the dam, and the peak has been reduced by an amount *AB* (Fig. 23.1). If there is some distance between the reservoir and the protected area but no local inflow between these points, the reservoir operation will be

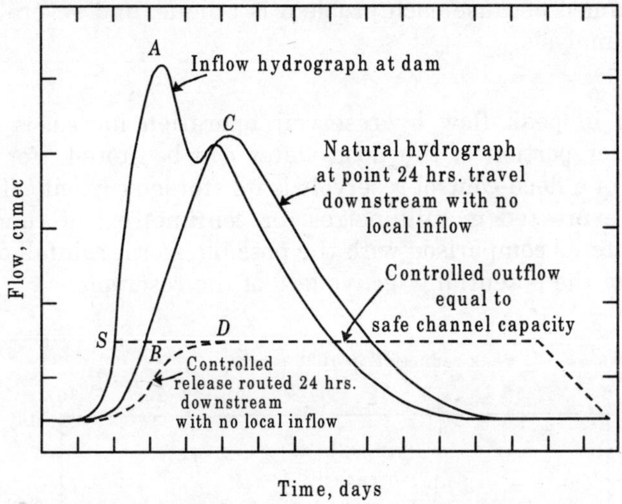

Fig. 23.1 *Idealized flood-control reservoir operation*

quite similar. However, in this case the natural hydrograph would be reduced by storage in the reach downstream from the reservoir so that a reduction *AB* at the dam is decreased to *CD* (Fig. 23.1) at the control point. If there is a substantial local inflow between the dam and the control point, the reservoir must be operated to produce a minimum peak at the protected area rather than a minimum peak at the dam. If, as is the usual case, the local inflow causes the peak to be developed sooner than the inflow from upstream the operation usually requires low releases early in the flood with relatively high releases set to arrive after the peak of the local inflow.

Location of Reservoirs

As indicated earlier the most effective flood control is obtained from an adequate reservoir located immediately upstream from the point (or reach) to be protected. Such a reservoir would often be located in a broad flood plain where a very long dam would be necessary and a large area of valuable land would be submerged in the reservoir. On the other hand the sites farther upstream require smaller dams and less valuable land but are less effective in reducing flood peaks. The loss in effectiveness results from the influence of channel storage (Fig. 23.1) and from the lack of control over the local inflow between the reservoir and the area to be protected. If the local area in between the reservoir and the point to be protected is sufficiently large, it may produce a flood over which the reservoir would have little or no control. Further it should also be noted that a single reservoir cannot give equal protection to a number of points located at different distances downstream. A significant, although largely qualitative, criterion for evaluating a flood control reservoir or a system of reservoirs is the *percent of the total drainage area controlled by the reservoirs.* In general, at least one-third of the total discharge area should be under reservoir control for effective flood reduction.

Economic analysis generally favours the upstream site despite its lesser effectiveness. Moreover it is often preferred to have several small reservoirs instead of a single large reservoir. The use of several small reservoirs offers the possibility of developing initially only those units of the system which would yield the highest economic return and constructing the additional units later on as the development of the area increases the potential benefits. However, no general rules can be framed because each problem is unique, and several alternatives must be evaluated by economic analysis.

Size of Reservoir

The possible reduction in peak flow by reservoir operation increases as reservoir capacity increases, since a greater portion of the flood water can be stored. For this reason a second criterion for evaluation of a flood-control reservoir is its storage capacity. The storage capacity of a reservoir is usually expressed in millimetres (or centimetres) of runoff from its tributary drainage area to facilitate its comparison with the possible storm rainfall over the area and thus to obtain a rough idea of the potential effectiveness of the reservoir.

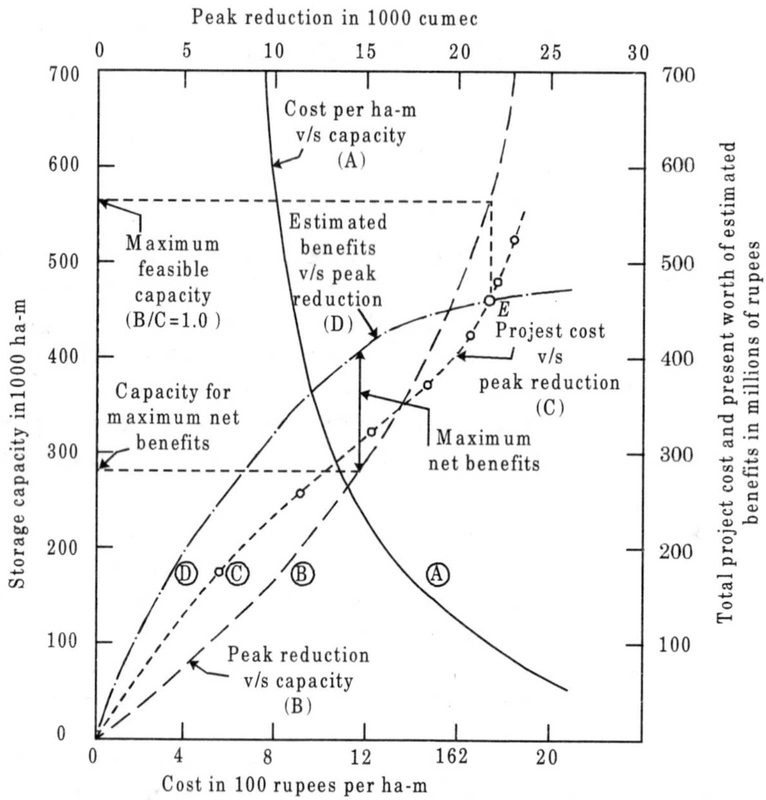

Fig. 23.2 *Cost analysis for a food control reservoir*

However, from the above discussion it must not be presumed that larger is the storage capacity better it is because many factors control the decision. The maximum storage capacity required is equal to the difference between the volume of safe release from the reservoir and the volume of the design flood inflow. As the reservoir size is increased the law of diminishing return may come into play (i.e., the returns may not commensurate with costs). Since the hydrograph is wider at low flows, more water will have to be stored to reduce the peak by a given amount as the total peak reduction is increased. Assuming a hydrograph to be a triangle, the storage capacity required to achieve a given peak reduction varies as the square of the reduction. Moreover, the benefits achieved by a unit of peak reduction are usually less as the reduction is increased, since more marginal area is being protected. These factors are illustrated in Fig. 23.2. Curve A shows the variation in cost of a unit of storage as capacity is varied. Curve B shows the amount of flood peak reduction which can be achieved for various capacities. These curves are used to plot curve C, which shows the total cost of a given peak reduction. Curve D, represents the present worth of the estimated benefits of the different peak reductions. From the curves C and D it may be observed that the greatest net benefits are obtained with a peak reduction of 14 800 cumec, corresponding to which the storage capacity is 280 000 ha-m, which is therefore, the optimum storage capacity in this case. Investment beyond this point results in incremental benefits that are smaller than the incremental costs and hence it would be uneconomical to build storage capacity beyond 280 000 ha-m. Further the curves C and D intersect at point E for which the benefit cost ratio, is equal to one. The storage capacity corresponding to point E is 564 000 ha-m, which is the maximum feasible storage capacity in this case because a storage capacity in excess of this will result in benefit-cost ratio to be less than one and hence it is not justified.

Types of Flood-control Reservoirs

There are two basic types of flood-control reservoirs viz., *detention reservoirs* (or *storage reservoirs,* or *detention basins*) and *retarding reservoirs* (*or retarding basins*), which differ only in the type of outlet works provided for these.

A *detention reservoir* is provided with outlets and spillways controlled by gates and valves which are operated on the basis of the judgement of the project engineer. The detention reservoirs for flood control differ from conservation reservoirs only in this respect that the former need a larger sluiceway capacity to permit rapid drawdown in advance of or after a flood.

A *retarding reservoir* is provided with fixed, ungated outlets which automatically regulate the outflow in accordance with the volume of water in storage. The outlet usually consists of a large ungated spillway or one or more ungated sluiceways. Generally the ungated sluiceway functioning as an orifice is preferable because its discharge equation $\left(Q = C_d A\sqrt{2gh}\right)$ results in relatively greater throttling of flow when the reservoir is nearly full than would a spillway operating as a weir. A simple spillway is normally undesirable because some storage capacity below the crest of the spillway cannot be used. However, a spillway for emergency discharge in case of a flood exceeding the design magnitude is necessary.

The discharge capacity of a retarding basin with full reservoir capacity should equal the maximum flow which the channel downstream can pass without causing serious flood damage. The reservoir capacity must equal the flow volume of the design flood less the volume of water released during the flood. As the flood occurs, the reservoir fills and the discharge increases until the flood has passed and the inflow has become equal to the outflow. After this time, water is automatically withdrawn from the reservoir until the stored water is completely discharged.

Operation Problems. The various problems encountered during the operation of flood-control reservoirs are as indicated below.

1. *Excessive reservoir inflow flood.* The idealized reservoir operation shown in Fig. 23.1 is determined solely by the limiting downstream channel capacity. However, the operations will have to be necessarily different, if the flood volume approached or exceeds the storage capacity of the reservoir. The variations to be made in the reservoir operation may be foreseen only with an accurate forecast of reservoir inflow floods.

2. *Forecast of local inflow between the dam and the control point.* The reservoir operations involving local inflow between the dam and the control point cannot be effectively planned without forecasts of the local inflow. Thus, stream flow forecasts become necessary in planning reservoir operations for flood control. These forecasts are usually made on the basis of reports received by telephone, telegraph, or radio from a network of rainfall and river gauges in the basin. These reports permit the use of rainfall, runoff relations, unit hydrographs, and flood routing. Such forecasts may be quite accurate (±10 percent) under favourable conditions, but if rain occurs after a forecast is made, it may be greatly in error.

3. *Recurring flood.* A flood-control reservoir has its maximum potential for flood reduction when it is empty. After a flood has occurred, a portion of the flood-control storage is occupied by the collected flood-waters and is not available for use until this water is released. Since a second storm may occur before the drawdown is complete, it is often necessary to reserve a portion of the storage capacity as protection against a second flood. In other words the full capacity of the reservoir cannot be assumed to be available for the control of any single flood. If a second flood occurs while the reservoir is full, the effect of the reservoir may be to make this flood worse.* These two effects viz., uncertainty about future inflows during the flood and the need to reserve storage against a possible second flood, mean that a flood-control reservoir cannot be fully effective. As such the potential flood reduction to be expected from a flood-control reservoir should be taken as substantially less than that calculated on the basis of ideal operation, except for small floods which require only a small portion of the reservoir capacity for adequate control.

4. *Synchronization of flood flows.* An operational problem develops when flows in excess of natural flows are released from a reservoir and synchronize at some point downstream with flood flows from a tributary. The resulting flows below this tributary may be greater than the natural flood flows would have been. This situation has occurred in many cases and is one of the hazards of flood-control operation, especially on the large rivers. It can be minimized only by weather forecasts several days or even weeks in advance.

23.5 LEVEES AND FLOOD WALLS

Levees and flood walls are one of the oldest and most widely used methods of protecting land from floodwater. These are essentially longitudinal dams erected roughly parallel to a river, to serve as artificial high banks of the river and thus prevent the river water from spilling over the country side during floods. The flood water is confined between the levees or flood walls and is made to flow down the river without causing any damage to the country side of the levees or flood walls.

A *levee* is an earth dyke or embankment, while a *flood wall* is usually of masonry or concrete. In general, levees and flood walls must satisfy the same criteria of design as regular dams.

* The celerity of a flood wave through a reservoir is approximately equal to $\sqrt{gd}$ where d is the depth of the water in the reservoir. This celerity is usually greater than the celerity in the same wave in the natural channel without the reservoir. Hence, a full reservoir accelerates the flood peak, and this may result in a closer synchronization with peak inflows to the stream below the dam.

Levees are most frequently used for flood control because they can be built at relatively low cost of materials available at the site. Levees are usually built of material excavated from borrow pits parallel to the levee line. The material should be placed in layers and compacted, with the least previous material being placed along the river side of the levee. Usually there is no suitable material for a core, and hence most of the levees are homogeneous embankments.

Levee cross section must be adjusted to fit the site and the available materials. Material is excavated from a borrow ditch parallel to the levee, and an adequate berm is left between the toe of the levee and the ditch to avoid collapse of the ditch bank. Top width of levee is usually determined by the use to which they are to be put, with a minimum width of about 3 m to permit movement of maintenance equipment. Bank slopes are usually very flat because of the relatively poor construction materials. The slopes should be protected against erosion by sodding (or turfing), planting of shrubs and trees or use of rip rap.

Because of the flat side slopes of levees, a levee of any considerable height requires a very large base width. In developed areas it is often difficult to obtain enough land or the land is too valuable to have an earthen embankment. In such cases masonry or concrete flood walls may be used instead of embankments. Flood walls are designed to withstand the hydrostatic pressure as well as uplift pressure exerted by the water when at design flood level. If the wall is backed by an earth fill, it must also serve as a retaining wall against the earth pressures when stages, (or flow depths) are low.

Location of Levees. Levees should be so located that a sufficient channel width is provided to transmit the design flow with a reasonable freeboard against wave action. The channel width between levees (or the spacing of levees) and the height of the levees are closely related. If the flood plain of the river is flat, an increase in channel width will permit lower levees. The cost of a levee system consists of the cost of the land for levees and channel plus the cost of levee construction. Hence it is necessary to determine by trial the channel-width (i.e., spacing of the levees) and levee-height combination which offers the greatest net benefits. Most of the alluvial valleys have natural levees or high ground along the edges of the river channel as a result of deposition of sediment when the stream overflow. It is often cheaper to place the levees along this high ground. Further in any case, full advantage should be taken of ridges, which permit lower levees and often provide better foundation conditions.

When a city or agricultural district is to be protected a *ring levee* which completely encircles the area may be provided. The alternative to a ring levee is to carry the levee line back until it can be terminated in high ground.

Effect of Levees on River Stages. The effect of confining the flood waters of a river between the levees as well as their merits and demerits are discussed in details in section 21.6 of Chapter 21. However, the effect of levees on river stages is briefly discussed in the next paragraph.

Levees restrict the river channel width by preventing the flow on the flood plain, and this results in increased stages in the leveed reach of the river. Channel improvements, which usually accompany levee construction, increase velocity and thus may offset some or all of this increase in stages. With the increase in the stages in the leveed reach, the stages will also be higher upstream from the leveed reach. Downstream from the leveed reach, peak flows will be increased because of lesser channel (or valley) storage as a result of the increase in the velocity of flow. Although the net result of levee construction depends very much on the physical characteristics of the situation, but usually it results in a general increase in flood stages along a river, unless other flood control measures such as reservoirs or extensive channel improvements are used in combination with levees.

The increase in stage following levee construction has sometimes led to severe damaging effects. As such while considering the use of levees as a flood control measure the associated problems and undesirable side effects should be kept in view.

23.6 CHANNEL IMPROVEMENT

A stream can be made to carry higher discharges at lower levels (or stages) by merely improving its hydraulic conditions. Such improvements may be brought about by (a) increase in size of cross-section ; (b) realigning the channel along a shorter route ; (c) increasing velocity of flow in the channel ; and (d) removing deficiency of waterways in drainage crossing.

(a) *Increase in channel section.* The increase in size of cross-section to cater for a high designed flood without spilling over the bank, is practicable from economic consideration, only on channels of small streams carrying manageable high flood discharge. In cases of larger rivers, such a method is generally impracticable on account of excessive cost and also the impracticability of disposing off the excavated channel material over the developed areas.

(b) *Realignment.* The realigning of the channel through a shorter route is practicable only in the case of smaller streams, due to more or less the same reasons as indicated above.

(c) *Increase in velocity.* The velocity of flow in the channel can be increased by deepening, straightening and shortening the channel length by cutoffs, removing barriers in the channel section and lining the channel to improve its coefficient of rugosity. These measures may, however, be adopted keeping in view their practical feasibility as well as cost.

(d) *Removing deficiency of waterways in drainage crossing.* The inadequacy in the waterways of bridges, culverts, canal cross-drainage works etc., may result in aggravating the flood situation. These obstructions of the channel are man-made and can best be avoided by proper evaluation of the ultimate drainage requirement of the basin at the time of any proposed crossing across the drainage line.

All these methods tend to reduce the prevailing water level (or stage) for the corresponding flood discharge with the result that after adopting the measures of channel improvement, the banks of the channel are not overtopped or less frequently overtopped during high flood discharges, thereby providing protection to previously flooded area.

It may however, be stated that although the channel improvement is a well recognised method for flood control, it should be adopted with caution. This is so because the measures for improving channel capacity are essentially local protection measures which may increase flood magnitude at downstream points on account of accelerated flow of runoff from the upper improved reaches of channel. Hence unless well planned and programmed, channel improvement work may merely shift the flood problem from one area to the other.

23.7 FLOODWAYS

Floodways are the vast depressions into which a portion of the flood water diverted from a river through a natural or artificial channel is temporarily stored during the rising flood. After the flood receds in the river, the water stored in these depressions is permitted to flow back to the river. Thus, floodways serve two functions in flood control. First, they create large, shallow reservoirs which store a portion of the flood water and hence decrease the flow in the main channel below the diversion. Second, they provide an additional outlet for water from upstream, thus increasing velocity and decreasing stage for some distance above the point of diversion.

The construction of floodways is limited by the topography of the valley and the availability of low-value land which can be used for the floodway. The floodway is ordinarily used only during major floods, and the land in the floodway may be used for agriculture at other times.

Water may be admitted to floodways in a number of ways. In many cases flow occurs over a low spot in the natural bank or a gap in the levee line. In some cases a *fuseplug levee* is provided. This is a low section of levee which, when once overtopped, washes out rapidly and develops full discharge capacity into the floodway. In other cases a concrete sill or weir, is provided so that overflow occurs when the water in the river reaches a pre-determined stage. This is advantageous as it eliminates the need for replacement of a levee section each time the flow into the floodway occurs. Admission of water to the floodway may also be accomplished by dynamiting a section of levee when the flood situation warrants. Sometimes a spillway section with openings closed by stop logs or needles is provided to obtain more complete control on the flow admitted to floodway.

In our country this method of flood control has been used to protect Srinagar from the floods of River Jhelum. In this case a flood spill channel is provided upstream of Srinagar, which diverts part of the Jhelum flood flow into the Wullar Lake. By this process, the flood situation in the downstream reach of the Jhelum river after the offtake of the spill channel, i.e., in the neighbourhood of Srinagar is considerably improved.

23.8 EVACUATION AND FLLOOD-PLAIN ZONING

1. Emergency Evacuation. The emergency evacuation of the threatened area is one of the most effective means of reducing damage due to floods under certain circumstances. With reliable flood forecasts this technique may be adopted for sparsely populated areas where property values do not justify the use of other methods of flood control and the loss of life can be prevented by prompt evacuation. The success of this method however, depends on the hydrologic characteristics of the stream in question as well as on the availability of a good flood-forecasting service which can provide adequate warning sufficiently in advance to permit orderly and complete evacuation.

2. Flood-Plain Zoning. Flood plain zoning means dividing the entire flood plain area into different zones and to restrict the occupancy of the different zones of the flood plain to uses which will suffer little or no damage during floods. The objectives of zoning of the flood plain area are : (i) avoiding encroachment in that portion of the flood plain which is the usual waterway of the river as, such encroachment, besides resulting in heavy flood-damage, aggravates the flood problem elsewhere ; and (ii) directing activities in other portions of the flood plain in such a manner that they are less susceptible to damages in case of floods. Further the zones which are subject to floods only on rare occasions are also demarcated where the persons are permitted to occupy the land and carry out construction at their own risk. An indirect form of zoning adopted in some cases is the creation of parks along streams where frequent flooding makes other uses impracticable.

Like any other measure of flood control, the flood-plain zoning must also be adopted only if it is economically justified. Thus if the return from an investment in any other form of flood control is greater than that from flood plain zoning, then the zoning plan should not be adopted.

3. **Flood Proofing Individual Units.** In cases where certain isolated units of high value are threatened by flooding, they may sometimes be individually flood proofed. For example an industrial plant comprising building, storage yards, roads, etc., may be protected by a ring levee or flood wall. Similarly individual buildings sufficiently strong to resist the dynamic forces of the flood water are sometimes prmotected by building the lower stories (below the expected high water mark) without windows and providing some means of watertight closure for the doors. Thus, even though the building may be surrounded by water, the property within it is protected from damage and many normal functions can be carried on.

23.9 LAND MANAGEMENT AND FLOOD CONTROL

1. **Vegetation and Floods.** It has been established that vegetal cover removes moisture from the soil by transpiration and it also promotes loose organic soil, which is favourable for the infiltration of rainfall. Further a heavy vegetal cover also results in high interception loss during storms. As such less flood runoff may be expected from a well vegetated area than from an area bare of vegetation. The vegetal cover thus creates a sort of retarding basin which stores a portion of the runoff which might otherwise contribute to floods. However, while this storage may be a substantial quantity in minor storms, it may be quite negligible during a major storm. As such it may be expected that a good vegetal cover on a drainage basin will result in a reduction in frequency and severity of minor floods but it will have relatively little effect on major floods.

2. **Water Conservation and Flood Control.** Water conservation measures help to conserve or store water in the soil as well as in the farm ponds and thus reduce surface run-off and the flood flows. Some of the water conservation measures are as follows.

(i) Use of contour farming and terracing to retard surface runoff and promote infiltration of water into the soil.

(ii) Create farm ponds to retain the flow of small creeks (or small streams) for irrigation and stock water.

(iii) Use of cover crops in fields to avoid bare, fallowed ground during the non-growing season.

Generally the capacity of the soil to store water is not large unless it is initially very dry, and the farm ponds control only a small percentage of the area of the drainage basin. Hence the water conservation methods are useful in reducing flood flows in small streams only and are not very beneficial in the control of unusually large floods. However, since the various measures of water conservation are quite useful for reduction of soil erosion and preservation of soil moisture, they are invariably used and any flood reduction resulting from these may be considered as an incidental benefit.

23.10 FLOOD PLAIN MANAGEMENT

The flood plain of a river is formed by sediment deposition or removal accompanying intermittent overflows of the stream above its low water channel. Since the flood plains are nearly level they are attractive sites for cities, with the result that the occupancy of the flood plain continues to increase. This will however, require large annual expenditures for flood protection to prevent the potential damage from floods and such expenditure will continue to increase unless a systematic approach to flood plain management is adopted.

Flood plain management should minimize the costs of flood plain occupance which consist of (1) the initial cost of development, (2) the cost of flood protection, (3) the residual flood damage, and (4) costs of relief and rehabilitation.

The various tools of flood plain management are as follows :

(i) *Flood hazard surveys*. These indicate the extent of flooding at various probability levels and provide the basic information for defining flood risk.

(ii) *Flood plain zoning*. It is a device through which local jurisdictions can restrict occupancy of the flood plain to uses which will suffer little or no damage during floods. Development of flood plain lands can be prohibited leaving them in their natural state or in agricultural uses. Flood plain land may also be used only for parks and other recreational purposes.

(iii) *Flood proofing*. It may be adopted to protect certain individual properties from damage due to floods.

(iv) *Flood insurance*. A programme of flood insurance may be formulated in which the premium should reflect the risk involved in undertaking the construction in the flood prone areas. This will discourage the unwise use of flood plains.

23.11 ECONOMICS OF FLOOD CONTROL

It is rare that only one type of flood control method is adopted for the entire stream. Moreover economic analysis also indicates that a combination of the different flood control methods is usually most desirable. A project in which several flood control methods are jointly used is known as *combined project*. Often reservoirs are combined with levees and channel improvement works at key points along the stream. The cost analysis for such a project is shown in Fig. 23.3. In this case several possible reservoir sites are available, and with more reservoirs

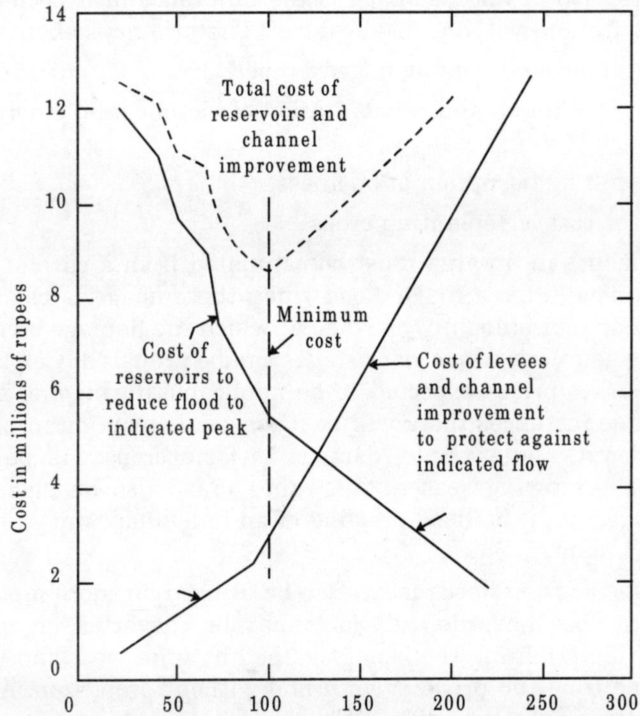

Fig. 23.3 *Cost analysis for a combination of flood control measures*

being constructed, smaller regulated peak flow is obtained (i.e., with increase in cost the regulated peak flow decreases). Curve *A* indicates the cost of reservoirs to reduce a certain design flood to various lesser values. Since reservoirs would not provide complete protection, levees and other channel improvements would be necessary. Curve *B* indicates the costs of these works for protection against various flows. The sum of these curves indicates the total cost of protection against the design flood. The minimum point in the total cost curve represents the least expensive combination, i.e., reservoirs to reduce the peak to about 10 000 cumec and levees and channel improvements to provide protection against this flow.

The construction of a diagram such as Fig. 23.3 requires the design of channel works for protection against several different peak flows to define curve *B* and the determination of the effect of several combinations of reservoirs on peak flow to establish curve *A*. A similar sort of analysis would be possible for other combinations of protective measures.

23.12 ESTIMATING THE BENEFITS OF FLOOD CONTROL

The benefits of flood control are broadly classified as :

(a) tangible benefits ; and

(b) intangible benefits.

Tangible benefits from flood control are of two kinds : (1) those arising from prevention of flood damage ; and (2) those occurring from more intensive use of protected land. The tangible benefits may be further classified as *primary benefits* and *secondary benefits*. The primary benefits from the prevention of flood damage is the difference in expected damage throughout the life of the project with and without flood control. The primary benefits may include :

(i) Cost of replacing or repairing damaged property.

(ii) Cost of evacuation, relief, and rehabilitation of victims, and emergency flood protection measures.

(iii) Losses as a result of disruption of business.

(iv) Loss of crops, or cost of replanting crops.

The probable flood damages in an area must be estimated from a careful survey of the area of flooding. This is often done after a major flood when the damage is clearly evident. The data collected by the survey are classified by the stage at which the damage occurs, so that a curve or table showing the damage from various river stages can be prepared (Col. 2 Table 23.2). In these estimates generally market prices expected to be applicable during the anticipated life of the project are used. In some instances there will be a seasonal variation in potential flood damage which must be considered. For example, damage to farm crops will be maximum for floods during the height of the growing season. Allowance must also be made for any changes in damage potential as a result of the flood proofing of an individual unit or the removal of a large industry from the flood plain.

Land which is protected from floods may often be utilized for more productive purposes than when it is subjected to flood hazards. Idle land may be converted for use by agricultural or industry. A method for determining the benefit gained by improved land use is to estimate the difference in net income from the property with and without protection. An alternative method is to estimate the increase in market value of the land, convert this to an average annual benefit by use of a rate of return appropriate to private enterprise in the type of activity under consideration.

Secondary benefits from flood control may arise in activities which stem from use or processing of products and services directly affected by floods. For example, if a steel mill is closed by a flood and steel reserves are short, factories far removed from the flood area might have to be closed until steel production is again started. Secondary benefits are difficult to assess and are, in part at least, offset by secondary costs usually not included in the cost estimates. Thus, secondary benefits are not normally included in estimates of flood control benefits.

Intangible benefits of flood control include prevention of loss of life and reduction in disease resulting from flood conditions. It is difficult to place a monetary value on the intangible benefits.

Average annual benefits may be computed by multiplying those benefits expected by prevention of flooding at a given stage by the probability of occurrence of that stage in an year ($1/t_p$). The computation is illustrated in Table 23.2 where the increment of prevented loss in 1 m intervals of stage is divided by the expected return period of a flood at the midheight of the interval. The summation of these increments indicates the total benefits with protection against floods of various sizes. It may be noted that even though the total damage for large floods is great, the annual increment of benefits is small because of the long return period for such floods.

The final step in the economic analysis of a project is a comparison of the benefits and costs. This is accomplished by estimating the cost of various degrees of protection. In the example of Table 23.2 the costs are those for a levee providing protection below the given stage. Protection against the rare floods is uneconomical because of the large investment and infrequent flood occurrence. In the example, levees designed to withstand a 10 m stage would be most satisfactory because benefit-cost ratio is maximum for this value of stage.

The preceding analysis ignores possible intangible benefits which may warrant the selection of another design stage. Actually for many flood control works the design criteria cannot be based on a direct economic analysis. However, any project should show economic feasibility before it is constructed. Thus only those portions of a project which are in themselves beneficial should be constructed. For example levees and channel improvement works for the protection of an agricultural district might not be justified by the expected benefits. However, if a proposed reservoir will have some benefit for the agricultural area, these savings should be included in considering the justification of the reservoir. Moreover reservoirs or channel improvements change the flood-frequency curve at the protected point and increase the expected return period for floods of all magnitudes. Benefits from such works are computed by comparing the damage expected before and after the works are constructed.

23.13 BROAD OUTLINE OF THE NATIONAL POLICY ON FLOOD

The national policy on floods and flood management is broadly outlined as under.

1. Sustained and systematic efforts are needed to deal with the problem of flood in a comprehensive basis for which the administration and the people both to share the responsibility and undertake tasks of huge magnitude so that progressively increasing measures of protection from floods could be achieved.

2. Even the best known methods of flood control aim only at providing a reasonable degree of protection and complete protection even if technically feasible would not be economically justifiable in all areas and all communities.

TABLE 23.2 Cost and benefit analysis for a flood control project

(All costs in millions of rupees)

Peak stage (m) (1)	*Total damage below indicated stage* (2)	*Increment of loss* (3)	*Return period (yr)* (4)	*Annual period for protection of increment* (5)	*Total benefits with protection given stage* (6)	*Project cost* (7)	*Ratio of benefit to cost* (8)
6	0	—	—	—	0	0.4	0
		5	10	0.50			
7	5				0.50	0.6	0.83
		6	15	0.40			
8	11				0.90	0.8	1.13
		10	22	0.45			
9	21				1.35	1.1	1.23
		12	30	0.40			
10	33				1.75	1.3	1.35
		13	70	0.19			
11	46				1.94	1.6	1.21
		15	150	0.10			
12	61				2.04	1.8	1.13
		20	300	0.07			
13	81				2.11	2.0	1.06

3. The State Governments are primarily responsible for the execution of the projects for control of floods in their areas, but Central Government has also to take initiative in the matter.

4. Flood control has to be viewed in a broad perspective forming an integrated part of the overall water resources development of the region and also the economic perspective envisaged in the area and also keep in view the ecological and humanitarian consideration

5. The objective of flood management in a particular area may require a package of measures and hence, while preparing a comprehensive plan for a basin, different alternatives with various degree of mix of component measures should be worked out and analysed for optimisation.

6. Reservoirs, to the extent technically and economically feasible, must be considered as an important component in any package of measures for flood management.

7. While conditions permit, the use of natural detention basins for flood moderation may be availed of. However, if reclamation of the detention basin has been carried out, the storage may be made use of only during abnormal floods.

8. While considering the use of embankments, the associated problems and side effects should be kept in view, and minimized to the extent possible.

9. Because of high initial and maintenance costs, channel improvement measures should be considered with due caution.

10. Measures attempting to modify the susceptibility of life and property to flood damage, like flood plain regulation, flood forecasting and warning, flood proofing, disaster preparedness, should be adopted to much more increasing degree than heretofore as engineering measures by themselves cannot provide protection from all floods.

11. Disaster relief, remission of taxes, provision of taccavi loans etc., should be continued on humanitarian considerations and crop insurance and insurance schemes for damage to properties and installations should be taken up.

12. Anti-erosion works should normally be taken up only for protection to towns, industrial areas, group of thickly populated village abadis, for agricultural lands where the benefit cost ratio justifies such works and in the case of railway line and roads, where relocation is not possible on socio-techno-economic grounds.

13. Afforestation and soil conservation measures are useful complements to other measures and should be taken up specially in the watersheds of river with heavy silt charge.

SOLVED ILLUSTRATIVE EXAMPLES

Illustrative Example 23.1 *For a proposed flood control project the design flood is 25000 cumec. The project is to have two components (i) a flood control reservoir and (ii) levees on both banks and channel improvement. A study of these two components gave the following cost figures:*

Reduced peak flow in cumec	*5000*	*10000*	*15000*	*20000*
Reservoir cost in million ₹	*19.7*	*11.7*	*8.5*	*5.3*
Levees and channel improvement cost in million ₹	*3.0*	*6.6*	*13.9*	*21.3*

Determine the size of the two components of the project from economic considerations.

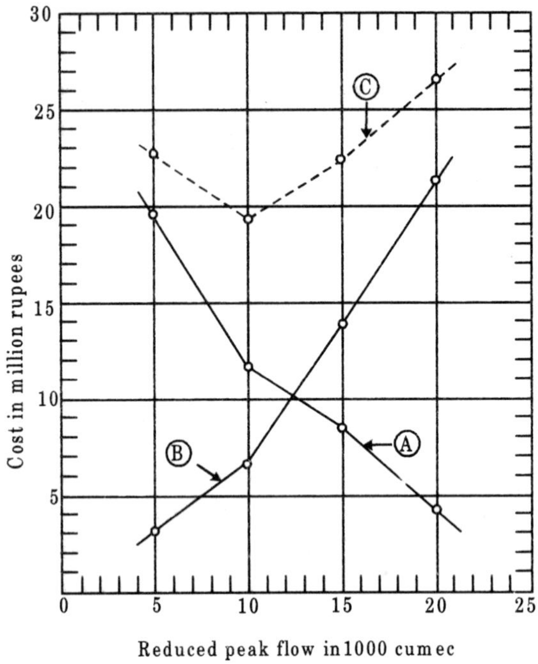

Fig. Ex. 23.1

Solution

The given data are plotted as shown in the accompanying figure in which curve *A* is a plot of cost of reservoir v/s reduced peak flow and curve *B* is a plot of cost of levees and channel improvement v/s reduced peak flow. Curve *C* is a plot of total cost of reservoir, and levees and channel improvement v/s reduced peak flow.

The minimum point in the total cost curve represents the least expensive combination of the two components of the flood control project. Thus, the reservoir should be provided to reduce the peak flow to 10000 cumec and levees and channel improvement should be provided for protection against this flow. The cost of each of the two components would be as follows :

Cost of Reservoir = ₹ 11.7 million

Cost of Levees and Channel improvement = ₹ 6.6 million.

REVIEW QUESTIONS

23.1 How are the methods of flood control classified ?

23.2 Explain the terms :

(i) Design flood, (ii) Maximum probable flood, and (iii) Standard project flood.

23.3 Describe the various methods used for estimating design flood.

23.4 Explain the principles of flood control with the help of reservoirs.

23.5 What are the different types of flood control reservoirs ?

23.6 Discuss the various problems encountered during the operation of flood control reservoirs.

23.7 Describe the use of levees as flood control measure.

23.8 Write a note on channel improvement as a flood control measure.

23.9 Describe briefly floodways.

23.10 What do you understand by evacuation and flood plain zoning ?

23.11 Write a note on flood plain management.

23.12 Discuss briefly the economics of flood control.

23.13 How are the benefits of flood control estimated ?

23.14 Write broad outline of the National policy on floods and flood management.

23.15 An inflow flood with peak flow of 24000 m^3/s comprises of a design flood for a proposed flood control project. The project is to have two components: (i) a storage reservoir to reduce the above peak to a lower value; (ii) levees on both banks and channel improvement to guard against reduced peak. A study of these two components gave following cost figures :

Reduced peak (m^3/s)	5000	10000	15000	20000
Reservoir cost in million rupees	19.2	11.1	7.9	4.8
Channel Imp., Levee cost in million rupees	2.5	6.1	13.4	20.8

From economic considerations determine the size of the project components.

CHAPTER

24 Water Power Engineering

24.1 INTRODUCTION

Water power (or hydro power) is one of the major sources of energy. The other sources of energy being the thermal power developed by the fuels such as coal, oil, natural gas etc., and the nuclear (or atomic) power. These are some of the conventional sources of energy. However, certain non-conventional sources of energy have also been developed such as tidal power, solar energy, geo-thermal energy, wind power etc., but these are not major sources of energy and are still in the developing stage. In this chapter various aspects of water power engineering are briefly described.

24.2 HYDROELECTRIC POWER DEVELOPMENT OF INDIA AND THE WORLD

India has a vast hydropower potential which is estimated to be about 90 million kW at 60% load factor. So far very little percentage of this vast potential has been harnessed. As such in order to develop this vast potential of power, efforts are being made and several projects have been undertaken to generate more and more hydroelectric power.

The major portion of power in India is generated by hydroelectric power plants and thermal power plants, in a ratio of about 2 : 3. A start has, however, been made to develop large scale atomic power in the country with Tarapur Atomic Power Station having installed capacity of 0.44 million kW, Rana Pratap Sagar Atomic Power Station having installed capacity of 0.22 million kW, and a few others. But even with all this the present (upto 1982–83) total installed capacity in India is only about 35.34 million kW. This is made up of 13.04 million kW hydropower, 21.44 million kW thermal power and 0.86 million kW nuclear power. Table 24.1 gives the salient features of some of the major hydropower projects of India.

As regards the world's water power potential considerably varying estimates have been given. According to one assessment the world's water power potential at 100% efficiency and utilisation is estimated at about 5 609 million kW, while according to another assessment the world's technically exploitable water power potential is estimated as only about 2 724 million kW. It will thus be observed that the later figure of the world's water power potential is almost half of the former. The reason given for so much variation in the two assessments being that the former estimate is based on theoretical considerations while the later estimate is based on practical considerations. However, as indicated later such a vast variation in the assessment of the world's water power potential is bound to exist. The world's hydroelectric capacity installed uptodate is about 200 million kW, which is thus about 9% of the later assessment (i.e., 2724 million kW) of the world's total hydropower potential.

The development of the hydropower in the various countries of the world varies considerably. In Europe about 50% of the available water power potential has been developed, and in U.S.A. only about 30% of the available water power potential has been developed. Furthermore, the share of the hydropower in the total energy developed by any country also varies considerably for different countries. For example, in the developed countries the share of the hydropower ranges from about 17% (in U.S.A., U.S.S.R.) to 98% (in Sweden, Norway) of the total energy. However, in developing countries the usage of their hydropower resources has only been started and so far relatively small percentage of it has been developed.

TABLE 24.1 Major hydro power projects of India

Name of power Station	*River*	*State*	*Installed capacity MW*	*Total cost in ₹ Million*	*Year of*	
					Commencement	*Completion*
(1)	*(2)*	*(3)*	*(4)*	*(5)*	*(6)*	*(7)*
Bhakra		Himachal Pradesh	1 050			
Ganguwal : Kolta	Sutlej	Punjab	77	1 383.33	1948	1968
		Haryana	77			
Gandhi Sagar	Chambal	Madhya Pradesh	115	136.03	1954	1960
Rana Pratap Sagar	-do-	Rajasthan	172	278.40	1961	1970
Jawahar Sagar	-do-	-do-	99	184.20	1962	1973
Rihand	Rihand	Uttar Pradesh	300	378.00	1952	1966
Obra	-do-	-do-	99	242.40	1963	1971
Ukai	Tapti	Gujarat	300	1 044.00	1960	1974
Koyna Stage I and II	Koyna	Maharashtra	540	564.39	1954	1968
Koyna Stage III	-do-	-do-	320	380.00	1966	1975
Machkund	Machkund	Andhra Pradesh	114.75	182.79	1946	1961
Upper Sileru	Sileru	-do-	120	185.00	1958	1968
Lower Sileru Stage I	-do-	-do-	460	540.00	1964	—
Lower Sileru Stage II	-do-	-do-	230	—	—	—
Sharavathy	Sharavathy	Karnataka	891	1 067.73	1956	1974
Sabarigiri	Pamba & Kakki	Kerala	300	430.00	1961	1968
Balimela Stage I	Sileru	Orissa	360	447.10	1961	—
Balimela Stage II	-do-	-do-	480	—	—	1977
Ramganga	Ramganga	Uttar Pradesh	198	96.86	1961	1978
Srisailam Stage I	Krishna	Andhra Pradesh	440	747.00	1963	1984
Idukki	Periyar	Kerala	390	682.10	1963	1976
Kalinadi	Kali	Karnataka	910	—	1968	—
Nagarjuna Sagar	Krishna	Andhra Pradesh	440	1 649.00	1956	1974

Contd.

Contd.

Nagarjuna Sagar Pumped Storage	Krishna	Andhrs Pradesh	110	467.59	1970	—
Yamuna Stage I	Tons	Uttar Pradesh	84.75	168.30	1960	1970
Yamuna Stage II	-do-	-do-	240	—	1966	1975
Kundah	Kundah	Tamil Nadu	435	—	1956	1967
Mettur (Tunnel Power House)	Cauvery	-do-	200	118.16	1960	1966
Hirakund	Mahanadi	Orissa	270	827.60	1948	1964
Beas Unit I (Beas Sutlej Link)	Beas	Himachal Pradesh	990	1 468.70	1961	1977
Beas Unit II	-do-	-do-	360	1 676.70	1960	1974
Salal	Chenab	Jammu & Kashmir	540	551.50	1969	1976
Baira-Suil	Baira	Himachal Pradesh	201	204.90	1969	—
Loktak	Manipur	Manipur	105	102.50	1969	—

24.3 COMPARISON OF HYDROELECTRIC AND THERMAL POWER PLANTS

A comparison of hydroelectric and thermal power plants is made as indicated in the following table.

Criterion	*Hydroelectric power Plant*	*Thermal power Plant*
1. Initial cost and life expectancy	(i) Higher initial cost due to costly infrastructure such as dam, spillway, etc., and its location at a place far off from the consumer or load centre.	(i) Lower initial cost due to its proximity to consumer or load centre
	(ii) Higher life expectancy, may be about 50 years or more	(ii) Lower life expectancy.
2. Operation, Maintenance and Repair cost (OMR)	Low OMR costs due to (a) no fuel cost,	High OMR costs due to (i) costly fuel and the cost of fuel transport
	(b) less maintenance and repair costs and staff needed for operation is less.	(ii) high maintenance and repair costs, and more staff needed for operation.
	(c) water is replenished every year and is put to non-consumable use, however supply of water may be erratic year after year, and	(iii) non-replenishable resource
	(iv) low taxes due to its location in remote areas.	(iv) high taxes due to its location near cities.
3. Transmission losses	More due to long transmission lines.	Less due to short transmission lines
4. Pollution	(i) No problem of pollution or ecological unbalance later.	(i) Causes air pollution as well as stream pollution resulting in ecological unbalance unless cooling towers or cooling ponds are provided.

5. Actuating time	(ii) No waste product and hence no problem of its disposal. A few seconds to 3 to 4 minutes depending on the length of conduit. Thus quite suitable to take up peak load.	(ii) Problem of haulage and disposal of waste product viz., smoke, ash. About 30 minutes (1/2 hour) Thus not suitable for use as peak load plants

A hydroelectric power plant is much simple to operate and maintain and as it does not require any fuel, the operating costs of a hydroelectric power plant are much less then those of thermal power plant. Although the improvements in design have brought about a reduction in average coal consumption for a thermal power plant per kWh of energy generated from 1.4 kg in 1920 to 0.4 kg in 1960, yet the cost of power production in a thermal power plant is usually higher. Moreover the cost of coal is largely dependent on the cost of transport of coal from the mines to the power house. In India the ratio of the cost of hydroelectric power generation to that of thermal power generation ranges from about (1/3) to (3/4). However, power system in these days incorporate both hydro (or hydel) as well as thermal power plants. This enables each of these to be used to better advantage.

24.4 ASSESSMENT OF AVAILABLE POWER

A hydropower development is essentially to utilize the hydraulic power (or energy) possessed by the water flowing in a stream and to develop from it electric power (or energy) through hydraulic turbines coupled to electric generators. The total amount of hydroelectric power that would be made available from a stream depends on continuous flow rate of the stream and the head possessed by the flowing water. Thus if Q is the rate of flow of water that may be continuously made available from any stream and it possesses a head H, then the amount of hydropower P in kW that would be available may be expressed as

$$P = \frac{wQ(H - h_f) \times \eta}{1000}$$

or

$$P = \frac{9810Q(H - h_f) \times \eta}{1000}$$

or

$$P = 9.81Q\,(H - h_f) \times \eta \qquad ...(24.1)$$

where $(H - h_f)$ is the net or effective head on the turbine and η is efficiency of the hydropower plant.

From Eq. (24.1) the energy in kilowatt-hours (kWh) may be expressed as

$$E = 9.81\,(H - h_f)\,\eta \times T \qquad ...(24.2)$$

where T is the period of flow in hours.

From the above equation for power (or energy) it is obvious that the correct assessment of the power (or energy) that would be available from a stream would depend on the correct assessment of the discharge Q, of the stream and the net head $(H - h_f)$. Since the flow rate of stream varies considerably, for the assessment of power that would be available from it, what value of discharge Q should be adopted is questionable. Moreover, as Q varies h_f also varies and therefore the net head available will also vary. As such the assessment of the hydropower that would be available from a stream would depend on the value of Q adopted in the above noted expressions and this would result in considerable variation in the assessment of the amount of the hydropower available from a stream. However, if a storage reservoir is developed by constructing a dam across a stream, then the variation of flow can be largely evened out and more or less constant discharge Q can be supplied to the turbines.

24.5 POWER SYSTEMS AND LOAD

In the earlier days most of the power systems were served by a single plant. With such an arrangement, one or more reserve units were needed in the plant. However, these days nearly all power systems are served by several plants, and some systems are interconnected with others to form a grid. In general a power system should have sufficient capacity to supply the expected peak load plus additional capacity to take care of the breakdown and necessary shutdowns for maintenance.

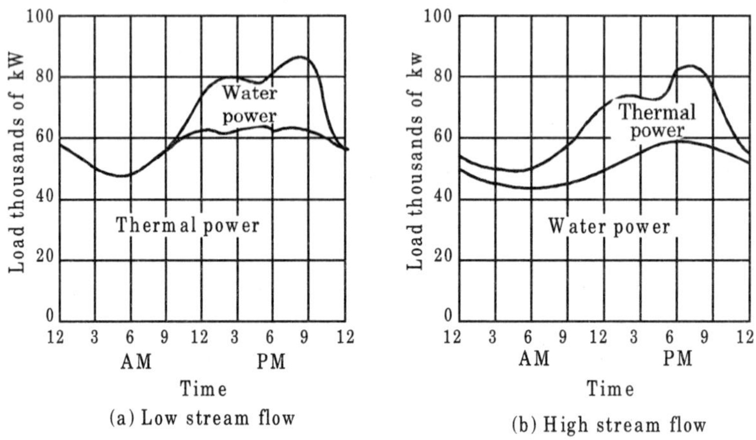

Fig. 24.1 *Load distribution in a combined power system*

Generally by having a combined power system with both hydro and thermal plants, economy and reliability can be achieved. One advantage of having hydroelectric plant in a system is that a hydroelectric plant, when used as standby, can be put into operation in a very short time. The actual time to put a hydroelectric plant into operation may vary from a few seconds to 3 or 4 minutes depending on the length of conduit. Thus, a hydropower plant is suitable for providing power on short notice, which might be required if some other unit of the system might fail suddenly. On the other hand it would take at least 30 minutes for a thermal plant to be put into operation. Moreover, it is much more costly to keep thermal plants on standby. Similarly a power system having only hydroelectric plants may not be able to meet the demands during severe drought if it does not have adequate carry over storage. Also during a flood it may not be possible to generate enough power at low head plants where the net head may be greatly reduced due to high tailwater at the power house.

Figure 24.1 shows typical daily load curves in a combined power system. During periods of low stream flow, thermal plants are operated almost continuously near capacity to carry the base load, while hydroelectric plants with storage are used intermittently to generate power for peak loads. However, during periods of high flow, hydroelectric plants may be operated continuously near capacity to carry the base load while thermal plants may be used intermittently to carry peak loads. This mode of operation results in the fullest utilization of the available water power and minimum consumption of fuel for thermal power. In any combined power system there is a unique division in capacity between hydroelectric and thermal power which will result in greatest overall economy. The optimum point will usually occur when the hydroelectric capacity is somewhere between one-third and two-thirds of the total capacity.

24.6 DEFINITIONS OF HYDROELECTRIC-POWER TERMS

Some of the definitions of hydroelectric power terms are given below.

Gross Head. The gross head for a hydroelectric plant is the total difference in elevation between the water surface in the stream at the diversion (i.e., head race level) and the water surface in the stream at the point where the water is returned after having been used for power generation (i.e., tail race level). In other words the gross head for a hydroelectric plant is the difference between the head race level and the tail race level.

Net or Effective Head. The net or effective head is the head available for energy production after deducting losses in friction, entrance, unrecovered velocity head in the draft tube, etc., from the gross head.

Hydraulic Efficiency. The hydraulic efficiency of a hydroelectric plant is the ratio of net head and gross head.

Overall Efficiency. The overall efficiency of a hydroelectric plant is equal to the hydraulic efficiency multiplied by the efficiency of the turbines and generators. The overall efficiency of hydroelectric plants operating at optimum conditions is usually between 60 and 70 percent.

Capacity or Installed Capacity. The capacity of a hydroelectric plant is the maximum power which can be developed by the generators at normal head with full flow.

24.7 STORAGE AND PONDAGE

As stated earlier the flow rate of a stream varies considerably with time. For instance, during rainy season when the stream is in floods it carries a huge quantity of water as compared to other times of the year when the quantity of water carried by it is considerably less. Although the demand for electric power varies continuously, but it ordinarily does not correspond to such variations of the natural flow. Moreover, since electrical energy cannot be stored, the generation has to closely conform to the demand. As such some arrangement in the form of storage and pondage of water is required to be made to regulate the flow of water so that it may be made available in requisite quantity at the time when the generation of power is required.

Storage may be defined as impounding of a considerable amount of excess runoff during seasons of surplus flow for use during dry seasons. It is achieved by constructing a dam across a stream at suitable site, thus forming a storage reservoir on the upstream side of the dam.

Pondage may be defined as a regulating body of water in the form of a relatively small pond or reservoir provided just at the hydropower plant. It is provided for temporarily storing water when the load on the plant is reduced and supplying the same to meet the instantaneous increased demand when the load is increased. The pondage is usually provided with a capacity necessary to regulate the flow to suit the variation in daily or weekly power (or load) demand. It is required to be provided either when the hydropower plant is located at some distance away from the storage reservoir, or when the storage reservoir is not provided. When the hydropower

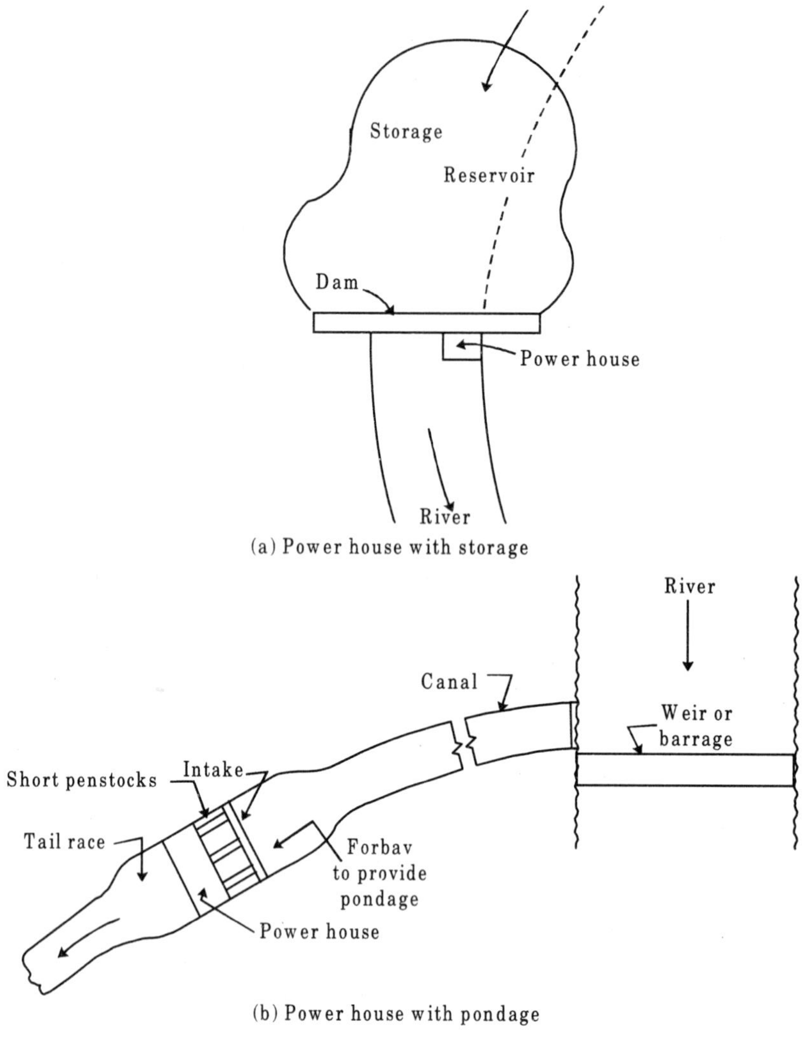

Fig. 24.2 *Hydro-power houses with storage and pondage*

plant is located at some distance away from the storage reservoir and water is supplied to the hydropower plant through a power channel then the regulation of outflow of water from the storage reservoir, to take care of the variation in daily or weekly power demand, may not be possible and a regulating reservoir or pond is needed directly at the hydropower plant. Similarly when the storage reservoir is not provided and the hydropower plant is located directly on the natural stream, then the pondage is necessary to regulate the natural flow of the stream to take care of the variation in daily or weekly power demand. As indicated later a forebay is provided to serve as a regulating reservoir or pond for the purpose of pondage.

Thus, in general the term *storage* is used to indicate the building up of the low natural flow of the stream to a uniform discharge and the term pondage is used to indicate the regulation of the resulting uniform flow, or the natural flow if there is no storage, to suit the variation in daily or weekly load demand. If the hydropower plant is close to the storage reservoir then the storage reservoir itself will also provide the necessary pondage. Fig. 24.2 shows the location of power houses with storage and pondage.

24.8 ESSENTIAL STREAM FLOW DATA FOR WATER POWER STUDIES

The essential stream flow data required for the assessment of the hydropower that would be available from any stream are as follows :

(i) Daily (weekly or monthly) flow over a period of years (the longer is the data better it is) to determine the plant capacity and estimated output, both of which are a function of the average flow of the stream as well as its distribution during the year.

(ii) Minimum or low-water flow to determine the amount of primary or firm or dependable power and also the size of thermal or steam auxiliary plant if used.

(iii) Maximum or flood-flow to enable proper design of spillway of the dam and to provide for the safety of the plant.

The above noted data for any stream may be developed by regularly measuring the stream flow and maintaining the records of its daily flow extending over a long period of time. The detailed past performance of a stream thus determined by measurement is the best index or measure of what may be expected in the future. The stream flow may also be estimated in an indirect manner by adopting some of the methods as indicated in section 5.11 of Chapter 5. However, by these methods only average annual stream flow and maximum or peak value of the flood discharge can be obtained.

From the daily (weekly or monthly) stream flow data a mass curve and a flow duration curve for the site of power plant may be prepared. A mass curve may be used to determine the required storage for a given pattern of demand or load. The method of drawing a mass curve and its uses are discussed in section 7.7 of Chapter 7. A flow duration curve may be used to determine the plant capacity and available power at all times. In the following section the method of drawing a flow duration curve and its uses have been discussed.

24.9 FLOW DURATION CURVE

The *flow duration curve* as ordinarily constructed consists of a plot of values of stream flow (daily, weekly or monthly) in increasing order of magnitude, as ordinate and the percent of time under consideration the flow equalled or exceeded, as abscissa. It may however be noted that instead of discharge, any function of discharge such as power in kW per metre of fall or for a given head, may also be used as ordinate. The curve thus obtained is termed as *power duration curve*. Moreover, in general these curves are also termed as *duration curves*.

Computations involved in the construction of flow duration curve (Fig. 24.3) are shown in Table 24.2. Rates of flow, differing by convenient amounts are set up in Col. 1 in ascending order. In Col. 2 is indicated the number of times the daily flow of the stream equalled a value between the corresponding flow in Col. 1 and the flow next below. For example, during the year, a daily flow between 38 and 40 cumec occurred 13 times. A summation of occurrences is tabulated in Col. 3. It shows that a flow of 38 cumec was equalled or *exceeded* on 303 days during the year. Column 3 is then reduced to percent of time and recorded in Co. 4. Columns 1 and 4 are used to plot the flow duration curve. For a flow duration curve of 1 year period, Col. 2 will total 365 days and for a longer period than a year it will be more than 365 days.

TABLE 24.2 Computations for daily flow duration curve of typical river for the year having the lowest flow of the record

Flow rate cumec (1)	*Number of days* (2)	*Days equalled or exceeded* (3)	*Percent of time* (4)
18.00	1	365	100
20.00	2	364	99.7
22.00	10	362	99.2
24.00	8	352	96.4
26.00	6	344	94.2
28.00	9	338	92.6
30.00	10	329	90.1
32.00	6	319	87.4
34.00	7	313	85.7
36.00	3	306	83.8
38.00	9	303	83.0
40.00	13	294	80.5
45.00	6	281	77.0
50.00	14	275	75.3
55.00	21	261	71.5
60.00	14	240	63.7
65.00	10	226	61.9
70.00	10	216	59.2
75.00	5	206	56.4
80.00	5	201	55.0
85.00	18	196	53.7
90.00	11	178	48.7
95.00	17	167	45.7
100.00	13	150	41.1
105.00	13	137	37.5
110.00	10	124	34.0
115.00	2	114	31.2
120.00	8	112	30.7
130.00	13	104	28.5
140.00	13	91	25.0
150.00	14	78	21.4
160.00	4	64	17.5
170.00	5	60	16.5
180.00	0	55	15.1
190.00	6	55	15.1
200.00	10	49	13.4
210.00	10	39	10.7
220.00	8	29	7.9
230.00	7	21	5.7
240.00	5	14	3.8
250.00	9	9	2.4
		365	

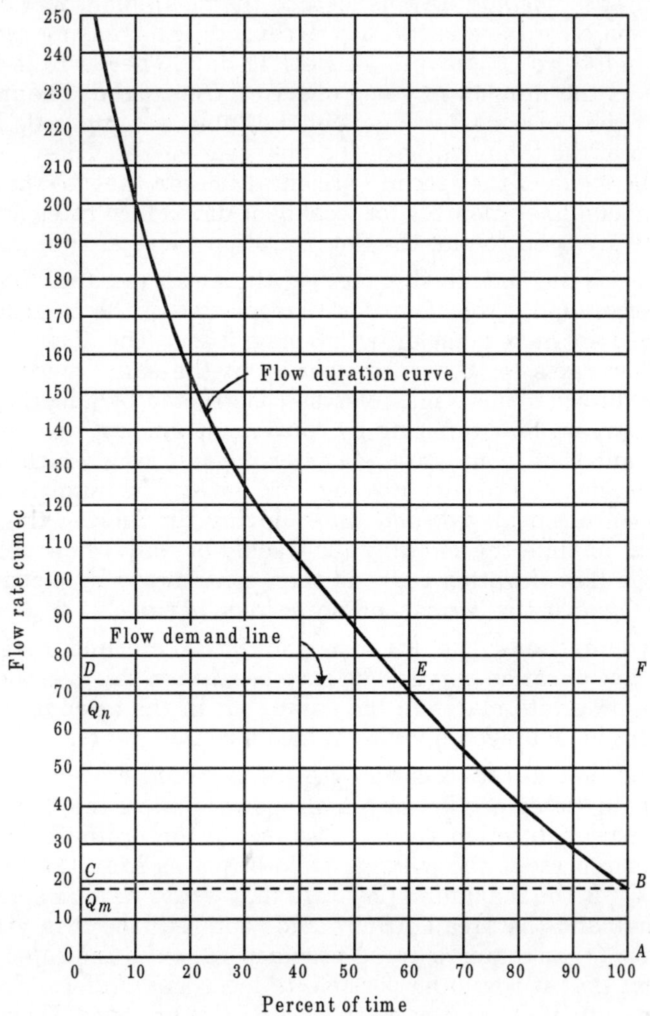

Fig. 24.3 *Daily flow duration curve of typical river for the year having lowest flow of the record*

A flow duration curve may also be constructed in the same manner with the help of mean weekly or monthly flow rates instead of daily flow rates. However, the flow duration curves computed from mean monthly flows are only approximate, because in each month there is a considerable variation in flow not indicated in the use of mean monthly flows for the construction of the duration curves. The error involved in the use of mean monthly flows for the construction of the duration curves usually ranges from 5 to 15% depending on the characteristics of the stream and the extent of utilization of the flow. The difference between the daily and monthly flow duration curves would be negligible for a stream with a steady flow, but much greater for very flashy streams.

The flow duration curve may either be constructed only for that year which is having the lowest flow record, or it may be constructed by considering the entire stream flow record. For the later case two methods are in general use for constructing the flow duration table for the plotting of the curve, which are as mentioned below.

(i) *Calendar-year basis method.* In this method the mean monthly (or weekly or daily) flow rates of each year of the record are separately arranged in increasing order of magnitude and tabulated as first, second, third etc., month (or week or day). There will thus be as many vertical columns in the tables as the number of years of record considered. The mean value of the flow rates noted in each of the horizontal columns of the table is computed. Thus the mean of the first horizontal column of the table will give the flow rate for the driest month (or week or day) of the average year, the mean of the second horizontal column that for the next driest month (or week or day) etc. In this manner monthly (or weekly or daily) flow rates for the average year are obtained and the same are used to plot the flow duration curve as indicated earlier.

(ii) *Total period basis method.* In this method the mean monthly (or weekly or daily) flow rates for the entire period under consideration (irrespective of the calendar year in which they occurred), are arranged in increasing order of magnitude. The flow duration curve is then prepared using the flow rates for the entire period in the same manner as indicated above. Alternatively the flow duration curve may also be constructed by adopting following procedure. Continuous ranges of stream flow differing by convenient amount are considered and for each year separately the number of months (or weeks or days) during which monthly (or weekly or daily) flow of the stream lies in a particular range are noted. The number of months (or weeks or days) for all the years are considered, and those having the flow in the same range are then added, and the percent of time the monthly (or weekly or daily) flow occurred in a particular range is computed. The flow duration curve is then constructed by plotting the lower value of the range of flow rate against the corresponding percent of time.

Out of these two methods the total-period basis method gives a relatively true representation of the flow duration curve. This is so because in this method the actual flow rates of stream appear at appropriate places in the curve, but in the calendar-year basis method the stream flow rates might be averaged up with values for other years.

The area under the flow duration curve (Fig. 24.3) would evidently represent the average yield from the stream. As stated earlier for a hydropower project the flow duration curve can be easily converted into power duration curve by changing the ordinate, in which case the area under the curve would represent the average yield of power from the project. It can be noted that without storage, Q_m is the maximum flow rate that would be available for all the times (i.e., for 100 percent of time), and the area *OABC* would represent the firm yield of water or power. However, if a flow rate Q_n is required to be provided which is indicated by flow demand line *DEF*, then it is evident that it would be possible to meet this uniform demand of flow rate (or power) for all the times only if storage equal to area *BEF* is provided. However, an alternative to storage, for power projects is to instal a thermal power unit of maximum capacity equal to *BF* (on power duration curve ordinate in kW), to work as supplementary to hydropower unit. The curve also shows that natural flow sufficient to meet the flow demand Q_n is available for 53.5% of time or 195 days in this year of the lowest flow of the record.

24.10 TYPES OF HYDROELECTRIC PLANTS

Hydroelectric plants may be classified according to different considerations as indicated below.

Classification based on storage characteristics

Based on the storage characteristics hydroelectric plants may be classified as (i) Run-of-river plants, (ii) Storage or Reservoir plants, (iii) Pumped-Storage plants, and (iv) Tidal plants.

1. Run-of-river Plants. Run-of-river plants are those which utilize the stream flow as it comes, without any storage being provided. As such these plants would be feasible only on such streams which have a minimum dry weather flow of such magnitude which makes the development worthwhile. A weir or a barrage may be constructed across the river close to the power plant to maintain a given water level. These are generally low head plants and often at times of flood, tail water rises to such an extent that the plants are inoperative.

Some run-of-river plants are provided with pondage to permit storing of water during off-peak hours for use during peak hours of the same day as well as to take care of fluctuations in load on the plant throughout the period of a week. In the usual run-of-river plants with pondage, installation is on a high basis in relation to minimum stream flow. In some plants discharge is 10 to 20 times minimum stream flow.

2. Storage or Reservoir Plants. Storage plants are those which are supplied water from large storage reservoirs developed by constructing dams across rivers. Depending upon the storage capacity, the reservoirs can hold over surplus water from the period when stream flow exceeds demand, for utilization during the period when demand exceeds the stream flow. The storage reservoir permits to develop the flow for power generation substantially more than the minimum stream flow. Thus with such plants better utilisation of hydropower potential is achieved.

3. Pumped-storage Plants. Pumped-storage plants are those which pump all or portion of the water used by these plants, back to the head-water pond, to be made available for the

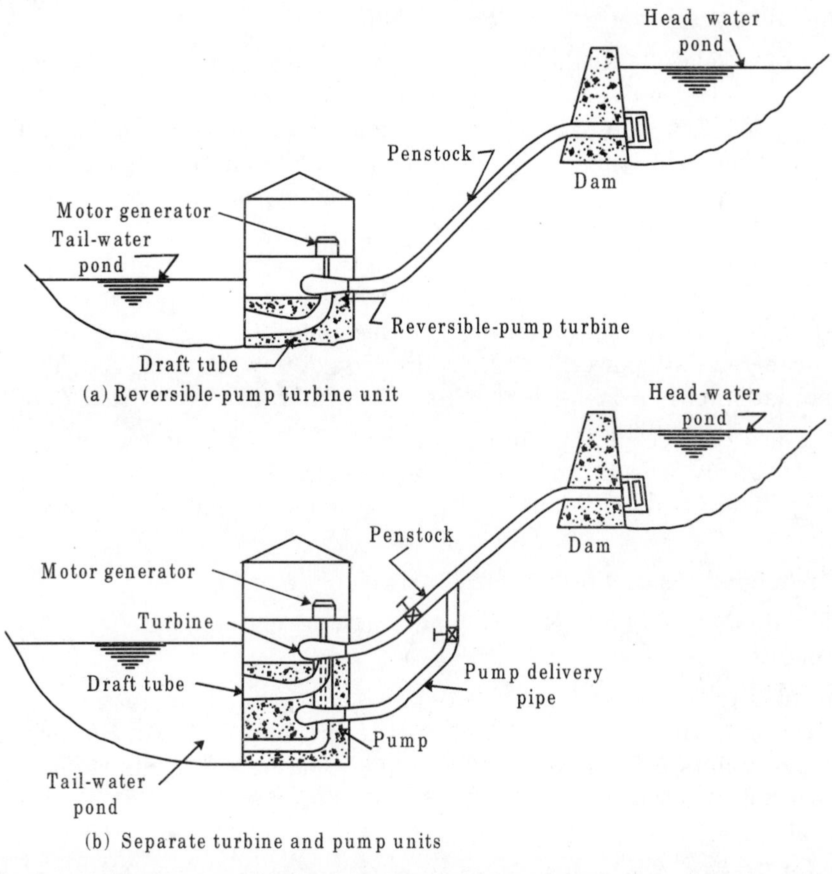

Fig. 24.4. *Pumped-storage plant*

power generation. They essentially consist of a head-water pond and a tail-water pond (Fig. 24.4). During times of peak load, water is drawn from the head-water pond through the penstocks to operate the turbine-generator units. This water after having been used for generation of power is collected in the tail-water pond. During the off peak hours pumps are operated to pump the water back from the tail-water pond to the head-water pond. Power for operating the pumps is provided by some off peak thermal or hydropower plants.

A unique feature of pumped-storage plants is that they utilize the same water again and again and additional supplies are needed only to meet the evaporation and seepage losses.

For heads upto about 120 m special Francis turbines have been developed for the pumped storage plants, the runners of which are so shaped that they can be used both as turbine as well as pump. These turbines are known as *reversible-pump turbines* or *reversible turbines*. In this case the same electrical unit serves as generator and motor by reversing poles. Such a machine may reduce the cost of a pumped-storage project by eliminating the extra pumping equipment and pump house. However, in the absence of reversible pump-turbines or for high heads pumped-storage plants may have separate pumping and power generating units. For high heads usually multistage centrifugal pumps are used for pumping water and high head Francis turbines are installed for power generation.

4. Tidal Plants. Tidal plants do not involve any storage of water. These plants work on the principle that there is a rise in sea water during the high tide period and a fall during the low tide or ebb tide period. Sea water rises and falls twice a day, each full cycle occupying about 12 hours 25 minutes. The tidal range, or the difference between the high tide and the low tide levels is utilized to generate power. This is accomplished by constructing a basin separated from the sea by a wall and installing turbines in openings through this wall. During high tide water passes from the sea to the basin, thus running the turbines and generating power. During low tide water from the basin flows back to the sea which can also be utilized to generate power provided another set of turbines operating in the opposite direction is installed, or special turbines which can generate power for either direction of flow are installed. Thus, depending on the type of installation either one or both directions of flow can be utilized to generate power.

Tidal plants are the recent innovations in the field of hydro-power generation. However, these may be used only where the tidal range is high. Rance power plant in France is an example of this type of plants where the tidal range is about 11 m and it has 9 units of 38 000 kW each.

Classification According to Functional Basis

According to functional basis hydroelectric plants may be classified as (i) Base-load plants, and (ii) Peak-load plants.

1. Base-load Plants. Base-load plants are those which are capable of substantially continuous operation in the base of the load curve throughout the year. Both run-of-river plants as well as storage plants can be used as base-load plants. When run-of-river plants without pondage are used as base load plants, their full plant discharge is seldom more than the minimum flow of river.

2. Peak-load Plants. A peak-load plant is one designed and constructed primarily for taking care of the peak-load of a power system. Pumped-storage plants are peak-load plants. Run-of-river plants with pondage can operate both as peak-load and base-load plants as river flow permits.

Classification on the basis of head

On the basis of head hydroelectric plants may be classified as (i) Low head plants, (ii) Medium head plants, and (iii) High head plants.

1. Low Head Plants. A low head plant is the one which is operating under a head less than about 30 m. Run-of-river plants are usually low head plants. For low head plants generally axial flow turbines such as Kaplan turbines are used.

2. Medium Head Plants. A medium head plant is the one which is operating under a head between 30 and 250 m. The lower ranges of medium head may be made available by utilizing a steep slope or a fall in a river or a channel where run-of-river plants may be provided. However, the higher ranges of medium head may be obtained by constructing dams and hence these are storage plants. Again for lower ranges of heads for these plants axial flow turbines may be used. But for higher ranges of head for these plants usually mixed flow turbines such as Francis turbines are used.

3. High Head Plants. A high plant is the one which is operating under a head more than 250 m. These heads may be obtained by constructing dams of sufficient height and installing the power plant either at the toe of the dam close to it or in a deep depression available at some distance away from the dam. Thus these are invariably storage plants. For these plants usually impulse turbines such as Pelton wheel turbines are used.

It may however be stated that the above noted head ranges for the different types of hydroelectric plants are arbitrary. Moreover, with the advances in the turbine design it has become possible to use axial and mixed flow turbines for higher heads. Consequently the ranges of head indicated above also move up.

Classification on the basis of Plant Capacity

On the basis of plant capacity the hydroelectric plants may be classified as (i) Micro hydel plants, (ii) Medium capacity plants, (iii) High capacity plants, and (iv) Super plants.

(i) ***Micro hydel plants.*** A micro hydel plant is the one which has a capacity less than 5 MW.

(ii) ***Medium capacity plants.*** A medium capacity plant is the one which has a capacity in the range 5 to 100 MW.

(iii) ***High capacity plants.*** A high capacity plant is one which has a capacity in the range of 101 to 1000 MW.

(iv) ***Super plants.*** A super plant is the one which has a capacity more than 1000 MW.

24.11 FIRM (OR PRIMARY)POWER AND SECONDARY (OR SURPLUS) POWER

Firm (or primary) power is the power which a plant can deliver throughout the year or 100 percent of time. *Secondary (or surplus)* power is the power in excess of firm power which a plant can deliver only for a part of the year or for some percentage of time. Thus for a run-of-river plant without any storage the firm power would correspond to the minimum flow of the river which would be available throughout the year. However, by providing the storage the firm power can be considerably increased. Similarly, by having one or more thermal plants which can be utilized to generate extra power when hydroelectric generation is low, a part of the secondary power can be converted into firm power.

As indicated earlier a flow or power duration curve can be used to estimate the firm and secondary power that would be available from a stream. Thus, as shown in Fig. 24.3 the flow rate Qm would correspond to the firm power that would be available from the river without any storage being provided. But if the provision for the storage equal to area *BEF* is made then the firm power for the same river may correspond to a higher flow rate Q_n represented by line *DEF*. However, in the absence of any storage, area *BCDE* would correspond only to the secondary power that would be available from the river.

24.12 LOAD FACTOR, UTILIZATION FACTOR AND CAPACITY FACTOR

Load factor is defined as the ratio of the average load during a certain period to the peak or maximum load during that period. The load factor is thus related to a certain period and therefore these are expressed as daily load factor, weekly load factor, monthly load factor and yearly load factor. For instance, if during a certain week a power plant generates 8 400 000 kW-hr and the peak load during the week is 100 000 kW, then the load factor during the week is

$$\frac{8400000}{24 \times 7 \times 100000} = 0.50 \text{ or } 50\%$$

The load factor of a power plant would vary greatly with the character of the load. The load factor for a power plant serving a highly industrialized area may be as high as 80%, but in residential areas the load factor may be as low as 25 to 30%. The installed capacity of a power plant has to be equal to peak load but the total number of units (kW-hr) generated will be governed by the average load. Thus if the load factor of a power plant is low, a large proportion of the generating capacity of the plant will remain idle for most of the time, and the cost per unit of power generated will be high. Conversely, if the load factor of a power plant is high the generating capacity is better utilized and the cost per unit of power generated will be comparatively less.

Utilization factor (*or Plant-use factor*) is defined as the ratio of the peak load developed during a certain period to the installed capacity of the plant. It thus represents the maximum proportion of the installed capacity utilized during any period. In the case of a hydroelectric plant, with constant head, utilization factor would also be the ratio of water actually utilized for generating maximum power corresponding to peak load to that available in the river, and usually there will be little difference in this factor whether expressed as a ratio of power or water. For a hydroelectric plant, utilization factor commonly varies from about 0.40 to 0.90, depending on plant capacity, load factor, available pondage and storage etc.

Capacity factor (or plant factor) may be defined as the ratio of the energy that the plant actually produces during any period to the energy that it might have produced if operated at full capacity throughout this period. For instance, if during a certain week the peak load on a power plant with a capacity of 1,00,000 kW was 65 000 kW and if the energy produced by the plant was 6,720,000 kW-hr, the capacity factor for the week (168 hr)

$$\frac{6720000}{100000 \times 168} = 0.4 \text{ or } 40\%$$

During the same period the load factor

$$= \frac{6720000}{65000 \times 168} = 0.615 \text{ or } 61.5\%$$

Capacity factor will be identical with load factor when the maximum or peak load just equals the plant capacity. For a hydroelectric plant, capacity factor commonly varies from about 0.25 to 0.70 or more depending on load factor, plant capacity, available pondage and storage etc.

The thermal plants may be operated at any desired capacity factor, whereas the capacity factor at which hydropower plants may operate is usually limited by the variation in the flow of water in the river. Theoretically a thermal power plant might operate at 100% annual capacity factor, but practically because of the necessity for an annual maintenance period the maximum annual capacity factor is much lower and usually it does not exceed 80%. Moreover there is always a decline in the annual capacity factor of thermal power plants with their age.

24.13 GENERAL ARRANGEMENT OF A HYDROELECTRIC PROJECT

A hydroelectric development ordinarily includes a diversion structure (dam, weir or barrage), a conduit (penstock) or a channel (or canal) to carry water to the turbines, turbines and governing mechanism, generators, control and switching apparatus, housing for the equipment, transformers, and transmission lines to the distribution centres. In addition, trash racks at entrance to the conduit, channel and penstock gates, a forebay, surge tank and other appurtenances may be required. A tailrace, or waterway, from the powerhouse back to the river has to be provided if the powerhouse is situated so that the water after flowing through the turbines is not discharged directly into the river. It may however be stated that no two power developments are exactly alike and each will have its own unique problem of design and construction. The type of plant best suited to a given site depends on several factors, including head, available flow and general topography of the area.

In general hydroelectric developments may be classified as

(i) Concentrated-fall development, and

(ii) Divided fall development.

A *concentrated-fall hydroelectric development* is the one in which the power house is located close to the dam on the downstream side. The powerhouse may be located at one or both ends of the dam [Fig. 24.2 (a)].

In a *divided fall hydroelectric development* the powerhouse is located at a considerable distance away from the dam and water is carried to the power house through a canal, tunnel or penstock (Fig. 24.5). This type of development is adopted to utilize a steep fall in the ground surface which might be available at some distance away from the dam. Thus in this case with favourable topography it is possible to achieve a high head even with a low dam. Moreover, with this arrangement, head variations in the reservoir may be small as compared with the total head, and the turbine can operate near optimum head (peak efficiency) at all times.

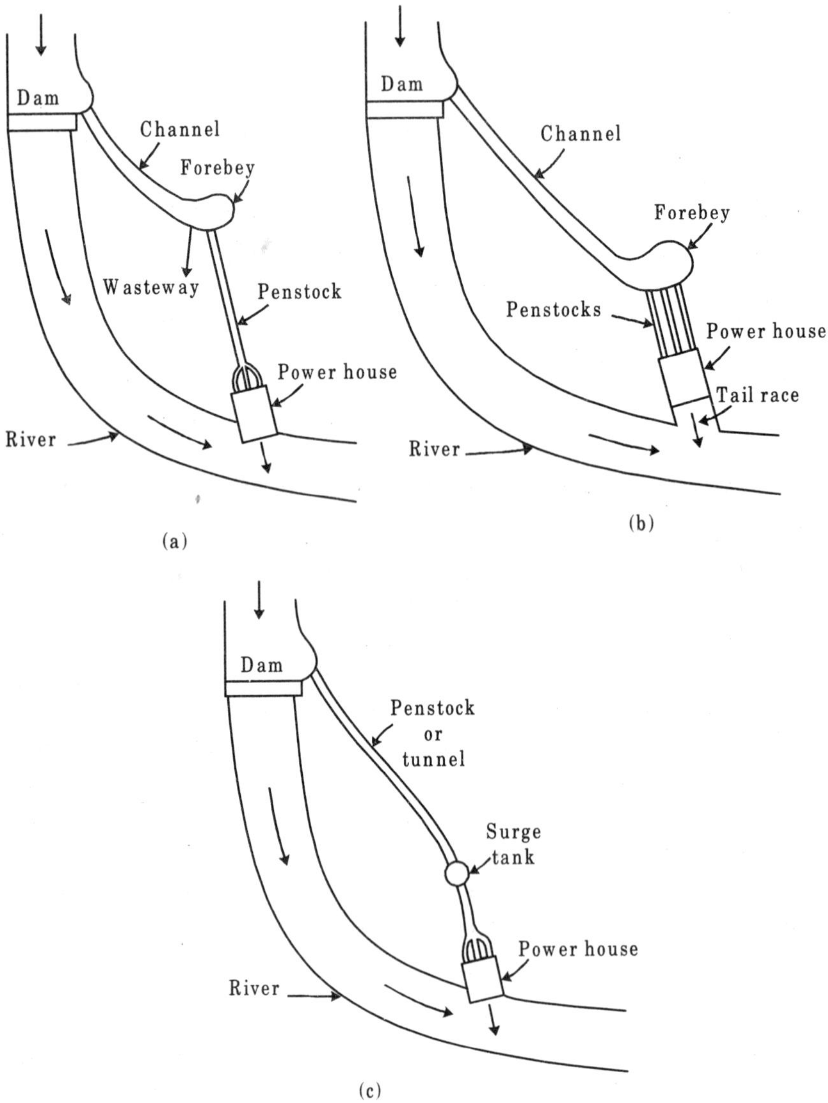

Fig. 24.5 *Divided-fall hydroelectric development*

24.14 COMPONENTS OF HYDROELECTRIC DEVELOPMENT

The various components of hydroelectric developments are described below.

Diversion Structures. The diversion structures commonly used are dams, weirs and barrages. A diversion structure is constructed across a river at suitable site to develop storage of water and to create head for the generation of power. In order to control the supply of water from the storage certain gates and valves are used. The different types of gates used are plain sliding gates, wheeled or roller gates, etc. Similarly the common types of valves used are butterfly and needle valves described in Appendix V.

Waterway. A waterway is passage through which water is carried from the storage reservoir to the power house. It may consist of tunnel, channel (or canal) or penstock. If a hill intervenes the reservoir and the power house, then a tunnel may be driven to provide the necessary waterway. The tunnel may be circular or horse-shoe shaped, lined or unlined, lined with concrete or reinforced concrete or steel depending on the nature of the rock through which it is driven. It may flow full as a pressure conduit or partly full as a channel.

Water may be conveyed from the reservoir to the turbines through penstocks or through a channel (called power channel or power canal) followed by penstocks. In the later case at the end of the channel a forebay is provided from which water is conveyed to the turbines through the penstocks.

Penstocks. Penstocks are the pipes of large diameter used for conveying water from the reservoir to the turbines. These are usually made of steel. However, reinforced concrete, cast iron and wooden penstocks are also used for low heads and relatively small developments. The thickness of the penstock is determined on the basis of the magnitude of stresses developed due to static pressures as well as water hammer pressures which may be developed due to sudden reduction in the flow caused by the governor when the load is suddenly reduced. Long penstocks are usually provided with a surge tank to absorb water hammer pressures and to provide water to meet sudden load increases.

A penstock is usually supported on piers, but when it is laid along slopes or encounters change in alignment, it should be supported on anchor blocks. As far as possible sharp bends in a penstock should be avoided because of the head loss and the large forces required to anchor the penstock. Air valves are provided at the sections of the penstocks where there are steep changes in the gradient.

In the case of long steel penstocks, expansion joints should be provided to take care of expansion and contraction due to changes in temperature. Long penstocks are usually branched at the lower end to serve several turbines [Fig. 24.5 (a) and (c)]. However, in the case of short penstocks, separate penstocks are ordinarily used for each turbine [Fig. 24.5 (b)].

Penstocks are usually provided with head gates which can be closed to permit repair of the penstock. An air-inlet valve and air duct connecting the penstock with the open air, should be provided immediately on the downstream of the gate. The air-inlet valve permits air to enter the penstock when the head gate is closed and the turbine gate is open, and thus prevent collapsing of the penstock which may occur due to sudden drainage of the penstock. A sufficient water depth should be provided above the penstock entrance to avoid formation of vortices which may carry air into the penstock and result in lowered turbine efficiency and undesirable pressure surges. This problem may more commonly arise in the case of penstocks taking off from the forebays. The entrance to the penstock should be properly designed to minimize the loss of head.

Forebay. A forebay is an enlarged body of water provided just in front of the penstocks. It is provided in the case of run-of-river plants and in the case of storage plants when the power house is located at a certain distance away from the dam and water is carried from the reservoir to the power house through a channel. However, if the power house is located close to the dam then since the penstocks directly take water from the reservoir, the reservoir itself will act as forebay. The main function of the forebay is to provide a small balancing storage upstream of the power house to store temporarily the water rejected by the plant when the load is reduced and to provide water to meet the instantaneous increased demand on account of increased load while the flow in the channel is being accelerated. The forebay may be developed by enlarging the channel just upstream of the intake for the penstocks leading water to the turbines in the power house. The forebay must be provided with a spillway or wasteway, so that excess water can be disposed of safely if the need arises.

Intake Structures. The water from the reservoir or forebay is let into the penstocks through intake structure. The main components of an intake structure are trash racks and gates. The trash rack (see Fig. 12.44) is provided to prevent the entry of debris into the water passage of the hydropower plant which may otherwise damage the wicket gates and turbine runners, or choke up the nozzle of impulse turbines. A debris cleaning device is usually fitted on the trash rack. Further if ice may get deposited on the trash rack, a heating element or some other ice removing equipment is also provided. The trash racks are provided ahead of the gates. The gates along with their hoisting arrangements are provided to control the entry of water into the penstocks.

Surge Tanks. A surge tank is a cylindrical open-topped storage tank which is connected to the penstock at a suitable point. Surge tanks are provided in the case of hydroelectric developments having long penstocks. These are provided to relieve the penstocks of excess pressure caused by water hammer and to provide additional supply of water when the turbines are in need of more water on account of increased load. A surge tank however, provides protection against water hammer pressure for only that portion of the penstock which lies on the upstream of it. As such surge tank should be provided as close to the power house as is possible according to the site conditions.

When the load on the turbine is steady and normal, a constant water surface will be maintained in the surge tank, which will be lower than the reservoir surface by an amount equal to the friction head loss in the portion of the penstock connecting the reservoir and the surge tank. When the load on the generator is reduced, turbine gates are closed by the governor to reduce the flow through the turbine and the water moving towards the turbine has to move backwards. The rejected water is then stored in the surge tank and water level in the surge tank rises. The retarding head so built up in the surge tank reduces the velocity of flow in the pipeline corresponding to the reduced discharge required by the turbine. When the load on the generator increases, the governor opens the turbine gates to increase the rate of flow entering the runner. The increased demand of water by the turbine is partly met by the water stored in the surge tank. As such the water level in the surge tank falls. In other words, the surge tank develops an accelerating head which increases the velocity of flow in the penstock to a value corresponding to the increased discharge required by the turbine.

In general the surge tanks may be classified as (i) Simple surge tanks and (ii) Differential surge tanks. A simple surge tank is a cylindrical open-topped tank connected to the penstock through a central vertical riser pipe or an orifice. A differential surge tank is also a cylindrical open-topped tank connected to the penstock through a central vertical riser pipe but in this case the riser pipe extends vertically into the tank upto considerable height and at the lower end of the riser pipe within the tank small ports or holes are provided. The main advantage of a differential surge tank is that for the same stabilising effect its capacity may be less than that of a simple surge tank. This is so because in a differential surge tank retarding and accelerating heads are developed more promptly than in a simple surge tank in which the heads only built up gradually as the tank fills.

The surge tanks must be high enough so that there is no spilling of water from the tank even with a full load change. This would, however, require a very high surge tank involving large cost if it is provided close to the power house. As such in order to reduce the height and hence the cost of the surge tank, it is generally provided at a point where the ground surface has a steep fall and the penstock drops rapidly to the power house. The height of a simple surge tank may also be reduced by providing an internal bellmouth spillway (with its crest upto the desired maximum water level in the tank) which permits the overflow of water from the tank and the same can be conveniently disposed of. However, the main drawback of this arrangement is that considerable amount of water is allowed to be wasted.

Power House. The power house of a hydroelectric development houses the various hydraulic and electric equipments. The various hydraulic equipments are turbines, gates or gate valves, governors etc. The various electrical equipments include generators, transformers, switching equipment, transmission lines and transmission structures, auxiliary electrical equipments etc. The power house consists of two main parts, a *substructure* to support the hydraulic and electrical equipment and a *superstructure* to house and protect this equipment. The generating units are always placed in a row at right angles to the direction of flow through the power house. An important feature of the superstructure is a travelling crane, spanning the width of the power house and of sufficient capacity to lift the heaviest single piece of equipment. This crane is needed to remove and carry turbine, generator or other equipment for servicing, repairs or replacement. A switch yard for the transmission of power is usually located outdoors near the power house.

Tailrace. The tailrace is the channel into which the water is discharged after passing through the turbines. If the power house is close to the stream, the outflow may be discharged directly into the stream. On the other hand if the power house is located away from the stream the tailrace is formed by constructing an artificial channel between the power house and the stream.

24.15 SELECTION OF SUITABLE TYPES OF TURBINES

The most commonly used turbines are Francis turbines, Kaplan turbines and Pelton wheel turbines. In addition to these three types of turbines, the other types of turbines which are also used though not very commonly are Deriaz (or Diagonal) turbines and Tubular turbines. The selection of a suitable type of turbines is usually governed by the following factors.

1. Head and Specific Speed. It has been found that there is a range of head and specific speed for which each type of turbine is most suitable which is given in Table 24.3.

TABLE 24.3

S.No.	*Head in metres*	*Type of turbine*	*Specific speed*
1	300 or more	Pelton wheel Single or Multiple jet	8.5 to 47 (in SI units) 10 to 55 (in metric units)
2	150 to 300	Pelton or Francis	30 to 85 (in SI units) 35 to 100 (in metric units)

Contd.

Contd.

3	60 to 150	Francis or Deriaz (or Diagonal)	85 to 188 (in SI units) 100 to 220 (in metric units)
4	Less than 60	Kaplan or Propeller or Deriaz or Tubular	188 to 860 (in SI units) 220 to 1000 (in metric units)

However, as a general rule, it may be stated that as far as possible a turbine with highest permissible specific speed should be chosen which will not only be cheapest in itself but its relatively small size and high rotational speed will reduce the size of the generator as well as power house. But the specific speed cannot be increased indefinitely because higher specific speed turbine is generally more liable to cavitation. However, the cavitation may be avoided by installing the turbine at a lower level with respect to the tail race.

(ii) **Part Load Operation**. The turbines may be required to work with considerable load variations. As the load deviates from the normal working load, the efficiency would also vary. At part load the performance of Kaplan and Pelton turbines is better in comparison to that of Francis and Propeller turbines. The variability of load will influence the choice of type of turbine if the head lies between 150 m to 300 m or lies below 30 m. For higher range of heads Pelton wheel is preferable for part load operation in comparison to Francis turbine, though the former involves higher initial cost. For heads below 30 m, Kaplan turbine is preferable for part load operation in comparison to Propeller turbine.

In addition to above mentioned factors there are certain other factors to be considered for the selection of the suitable type of a turbine. The *overall cost* which includes the initial cost and the running cost should also be considered. The *cavitation* characteristics of the turbine should also be considered since it affects the installation of reaction turbines viz., Francis and Kaplan turbines.

24.16 TYPICAL HYDROELECTRIC DEVELOPMENTS OF INDIA

Bhakra-Nangal Hydroelectric Project

Figure 24.6 shows a schematic section of Bhakra-Nangal hydroelectric project which is one of the major hydroelectric projects of India. It consists of Bhakra dam constructed across the river Sutlej and Nangal barrage across the river Sutlej about 13 km downstream of Bhakra dam. There are two power houses one on the left bank and the other on the right bank, located close to the dam. The left bank power house has 5 units of 90 MW each and the right bank power house has 5 units of 120 MW each. Francis turbines have been provided for these power plants which are classified as medium head plants and also they come under the category of storage or reservoir plants.

In addition to the above noted power houses, there are two more power houses in this project, which are located on a power channel taking off from the Nangal barrage. One of the power houses is located at Ganguwal which has in all 3 units out of which 2 units are of 24 MW each and 1 unit of 29 MW. The other power house is located at Kotla on the downstream of Ganguwal and has in all 3 units of capacities exactly same as at Ganguwal. The power plants at Ganguwal and Kotla may be classified as low head plants.

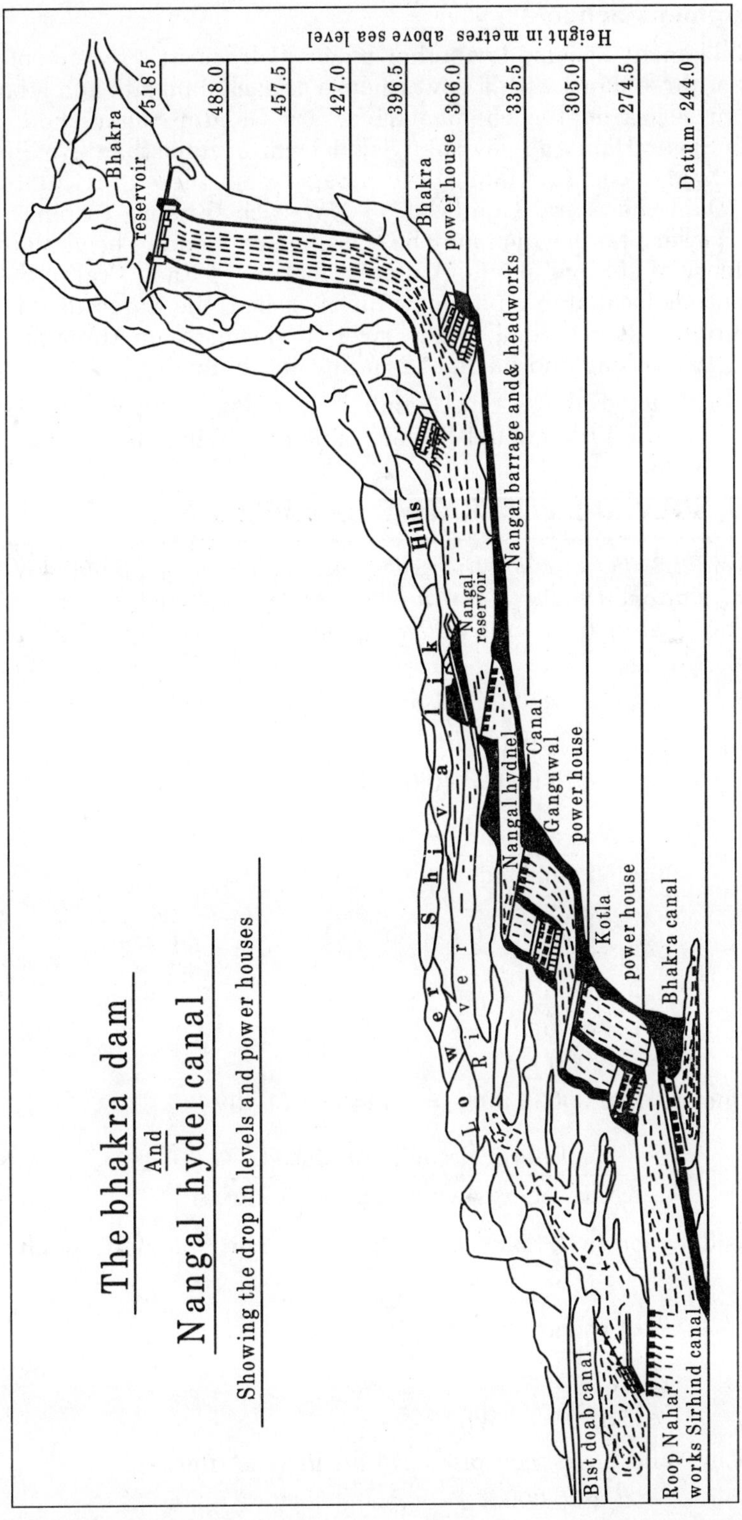

The bhakra dam
And
Nangal hydel canal
Showing the drop in levels and power houses
Height in metres above sea level
518.5
488.0
457.5
427.0
396.5
366.0
335.5
305.0
274.5
Datum- 244.0
Bhakra reservoir
Bhakra power house
Nangal barrage and& headworks
Hills
Lower Shivalik
River
Nangal reservoir
Nangal hydel Canal
Ganguwal power house
Kotla power house
Bhakra canal
Bist doab canal
Roop Nahar works Sirhind canal

Fig. 24.6

Chambal Valley Development Scheme

The Chambal Valley Development Scheme is another major hydroelectric scheme of India. The full development consists of three dams with a power house at each dam site and a barrage. The three dams constructed in series on the chambal river are Gandhi Sagar Dam in Madhya Pradesh and Rana Pratap Sagar Dam and Jawahar Sagar Dam in Rajasthan. At Gandhi Sagar Dam there are 5 units of 23 MW each. At Rana Pratap Sagar dam there are 4 units of 43 MW each. At Jawahar Sagar Dam there are 3 units of 33 MW each. Francis turbines have been provided for each of these power plants which may be classified as medium head plants and also they come under the category of storage plants. A unique feature of Rana Pratap Sagar project is that since the power house is located in a pit below the river bed, the water after issuing from the draft tubes of the turbines is returned to the river at downstream through a standard horseshoe shaped tunnel 1466 m long and having 12 m internal diameter.

A barrage has been constructed across the Chambal river near Kota city in Rajasthan to divert the water released from the Storage and Power Dams into irrigation canals.

SOLVED ILLUSTRATIVE EXAMPLES

Illustrative Example 24.1 *Two turbo-generators each of capacity 20,000 kW have been installed at a hydel power station. During a certain period the load on the hydel plant varies from 15 000 to 35 000 kW. Calculate (i) total installed capacity, (ii) load factor, (iii) plant factor, and (iv) utilization factor.*

Solution

(i) Total installed capacity of the hydel plant

$$= 2 \times 20\,000 = 40{,}000 \text{ kW}$$

(ii) Load factor $= \dfrac{\text{Average load}}{\text{Peak load}}$

Average load $= \left(\dfrac{15000 + 35000}{2}\right)$

$= 25{,}000$ kW

Peak load $= 35{,}000$ kW

$\therefore$ Load factor $=$ 25 00035 000 $= 0.7143$ or 71.43%

(iii) Plant factor $= \dfrac{\text{Energy actually produced}}{\text{Max. energy which can be produced}}$

If the time under consideration is T hours then the total energy actually produced during this time

$$= \left(\frac{15000 + 35000}{2}\right) T$$

$= 25{,}000\ T$ kWh.

Maximum energy which would have been produced during the time

$$= 2 \times 20{,}000 \times T$$

$$= 40{,}000\ T \text{ kWh}$$

$$\therefore \text{ Plant factor} = \frac{25{,}000T}{40{,}000T} = 0.625 \text{ or } 62.5\%$$

$$(iv) \text{ Utilization factor} = \frac{\text{Max. power developed}}{\text{Total power that can be developed}}$$

$$= \frac{35{,}000}{40{,}000} = 0.875 \text{ or } 87.5\%$$

Illustrative Example 24.2 *A run-of-river hydroelectric power station is proposed across a river at a site where a net head of 20 m is available on the turbine. The river carries a sustained minimum flow of 25 cumec in dry weather and behind the power station sufficient pondage is provided to supply daily peak load of demand with a load factor of 71%. Assuming the plant efficiency of 56%, determine the maximum generating capacity of the generator to be installed at the power house. If the daily load pattern indicates 21 hours average load and 3 hours of peak load, determine the volume of pondage to be provided to supply the daily demand.*

Solution

The average power developed corresponding to the dry weather flow is given by equation 24.1 as

$$P = 9.81Q\,(H - h_f)\,\eta$$

$$= 9.81 \times 25 \times 20 \times 0.56$$

$$= 2746.8 \text{ kW}$$

$$\text{Load factor} = \frac{\text{Average load}}{\text{Peak load}}$$

$$\therefore \quad \text{Peak load} = \frac{2746.8}{0.71} = 3868.7 \text{ kW}$$

$\therefore$ Maximum capacity of the generators to be installed

$$= 3868.7 \text{ kW}$$

During 3 hours of peak load the excess power to be developed from the pondage

$$= (3868.7 - 2746.8) = 1121.9 \text{ kW}$$

Discharge required for developing the excess power during the peak demand is

$$Q = \frac{1121.9}{9.81 \times 20 \times 0.56} = 10.21 \text{ m3/s}$$

$$\therefore \text{ Pondage} = (10.21 \times 3 \times 3600)$$

$$= 11.03 \times 10^4 \text{ m}^3$$

Illustrative Example 24.3 *The following data gives the average monthly flow in cumec in a dry year and it is intended to use it for designing a hydroelectric plant : July 15.0, Aug. 16.4, Sept. 15.6, Oct. 11.6, Nov. 11.3, Dec. 9.3, Jan. 7.9, Feb. 7.6, March 6.5, Apr. 9.8, May 10.7, June 12.2.*

Assuming the net head on the turbine as 100.5 m and efficiency of the plant 75% draw the power duration curve. It is intended to develop electric power at a firm rate of 8 000 kW either by providing a storage or by providing stand by diesel unit and no storage. In the two cases determine the minimum capacities of the reservoir or the diesel unit.

Solution

The following table shows the computations involved for the plotting of the power duration curve.

Month	*Flow (cumec)*	*Flow arranged in ascending order*	*Power* $[9.81Q(H-h_f)\times\eta]$ *(kW)*	*% of time of availability*
July	15.0	6.5	4 806	100.0
August	16.4	7.6	5 620	91.7
September	15.6	7.9	5 841	83.3
October	11.6	9.3	6 877	75.0
November	11.3	9.8	7 246	66.7
December	9.3	10.7	7 912	58.3
January	7.9	11.3	8 356	50.0
February	7.6	11.6	8 577	41.7
March	6.5	12.2	9 021	33.3
April	9.8	15.2	11 239	25.0
May	10.7	15.6	11 535	16.7
June	12.2	16.4	12 127	8.3

The power duration curve is plotted as shown in the accompanying figure.

As seen from the figure in order to obtain a firm power at the rate of 8 000 kW, the power corresponding to area *ABC* will be required to be developed from the storage. Further it is seen that for 55% of the time the power developed from the natural flow in the stream is in excess of the required power. But for the 45% of the time the natural flow will be required to be supplemented from the storage to develop the required power.

The area of the portion *ABC* is obtained by counting the squares as 675 mm^2 which is equivalent to

$$\frac{675\times1000\times10\times12}{10\times10\times100} = 8100 \text{ kW – month}$$

$$\text{The storage required} = \frac{8100\times30\times24\times3600}{9.81\times100.5\times0.75}$$

$$= 2.84\times10^7 \text{ m}^3 = 2{,}840 \text{ hectare – metre.}$$

With no storage the firm power available

$$= 4\ 806 \text{ kW}$$

∴ The capacity of the diesel unit required

$$= (8{,}000 - 4{,}806)$$

$$= 3{,}194 \text{ kW say } 3{,}200 \text{ kW.}$$

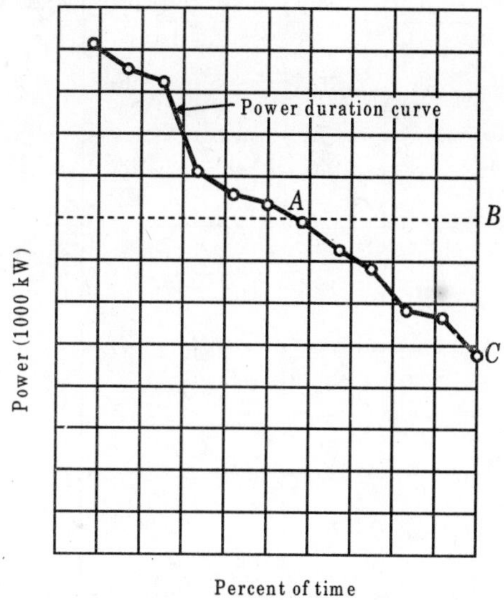

Fig. Ex. 24.3

Illustrative Example 24.4 *The water turbine of a hydel storage scheme has an output of 7500 kW, when working under a net head of 25 m and with an overall efficiency of 85%. The inflow in the reservoir during a year is as given below.*

Month	*J*	*F*	*M*	*A*	*M*	*J*	*J*	*A*	*S*	*O*	*N*	*D*
Inflow in 10^6 m^3	*105*	*111*	*93*	*84*	*75*	*90*	*123*	*93*	*96*	*102*	*103*	*99*

Find (i) the minimum capacity of reservoir to satisfy the uniform demand of water, (ii) the total quantity of water wasted during the year. Use graphical or analytical method. Assume that the reservoir is full at the beginning of January.

Solution

The problem is solved by analytical method as follows.

The discharge required for the turbines

$$= \frac{7500}{9.81 \times 25 \times 0.85} = 36 \text{ m}^3/\text{s}$$

∴ Flow required for the complete month

$$= (36 \times 30 \times 24 \times 3{,}600) \text{ m}^3$$

$$= 93 \times 10^6 \text{ m}^3$$

From the following table it is seen that the greatest reservoir depletion is 30 $\times 10^6$ m^3 and hence the minimum capacity of the reservoir required to satisfy the uniform demand of water

$$= 30 \times 10^6 \text{ m}^3$$

The total quantity of water wasted during the year

$$= 57 \times 10^6 \text{ m}^3$$

Month	*J*	*F*	*M*	*A*	*M*	*J*	*J*	*A*	*S*	*O*	*N*	*D*
Inflow in 10^6 m^3	105	111	93	84	75	90	120	90	96	102	108	99
Flow thro' turbine in 10^6 m^3	93	93	93	93	93	93	93	93	93	93	93	93
Draft from storage in 10^6 m^3	—	—	—	9	18	3	—	3	—	—	—	—
Storage refill in 10^6 m^3	—	—	—	—	—	—	27	—	3	3	—	—
Reservoir depletion in 10^6 m^3	—	—	—	9	27	30	3	6	3	—	—	—
Water wasted in 10^6 m^3	12	18	—	—	—	—	—	—	—	6	15	6

Illustrative Example 24.5 *During a low water week a river has an average daily flow of 32 m3/s with a fluctuation during the day requiring a pondage capacity of approximately 15% of the daily discharge. A hydroelectric plant is to be located on the river which will operate 5 days a week, 24 hours a day, but will supply power at a varying rate such that the daily load factor is 55%, corresponding to which the pondage required is equal to 0.2 times the mean flow to the turbine. On Saturdays and Sundays all the flow is ponded for use on rest of the days.*

If the effective head on the turbines when the pond is full is to be 25 m and the maximum allowable fluctuation in pond level is 1 m find (a) the surface area of the pond to satisfy all the operating conditions, (b) the weekly output at the switch board in kWh.

Assume turbine efficiency 55% and generator efficiency 92%.

Solution

Average daily flow in river = 32 m^3/s

$\therefore$ Total flow per day = $(32 \times 24 \times 3600)$

= 2.765×10^6 m^3

Total inflow on Saturday and Sunday

= $(2 \times 2.765 \times 10^6)$ m^3

= 5.53×10^6 m^3

This is to be used on the rest of the five days and hence the amount used daily from this pondage

$$= \frac{5.53 \times 10^6}{5} = 1.106 \times 10^6 \text{ m}^3$$

The corresponding additional flow to the turbine

$$= \frac{1.106 \times 10^6}{24 \times 3600} = 12.8 \text{ m}^3/\text{s}$$

∴ Total average flow to turbine

$$= (32 + 12.8) = 44.8 \text{ m}^3/\text{s}$$

It is given that the fluctuations in the flow of river require a pondage capacity of approximately 15% of the daily discharge. Hence additional pondage required on this account

$$= \left(\frac{32 \times 15}{100}\right) \times 24 \times 3600 = 0.415 \times 10^6 \text{ m}^3$$

The pondage required on account of the variation in the load

$$= (0.2 \times 44.8 \times 24 \times 3600) = 0.774 \times 10^6 \text{ m}^3$$

Hence the total pondage required

$$= (5.53 + 0.415 + 0.774) \times 10^6$$

$$= 6.719 \times 10^6 \text{ m}^3$$

Since the maximum allowable fluctuation in the pond level

$$= 1 \text{ m}$$

Surface area of pond $= 6.719 \times 10^6$ m^3

With a fluctuation of 1 m the average head on the turbine

$$= \left(\frac{25 + 24}{2}\right) = 24.5 \text{ m}$$

∴ Maximum power generated is

$$P = (9.81 \times 44.8 \times 24.5 \times 0.85 \times 0.92)$$

$$= 8.420 \times 10^3 \text{ kW}$$

The weekly output at the switch board is

$$E = [8.420 \times 10^3 \times (5 \times 24)] \times \frac{55}{100}$$

$$= 5.557 \times 10^5 \text{ kWh.}$$

Illustrative Example 24.6 *A run-of-river plant with an installed capacity of 12,000 kW operates at 25% load factor when it serves as a peak load station. What should be the minimum discharge in the stream so that it may serve as the base load station ? The plant efficiency may be taken as 85% when working under a head of 20 m. Also calculate the maximum load factor of the plant when the discharge in the stream is 30 m³/s.*

Solution

When the plant operates as peak load plant with a load factor of 25% the average power developed

$$= \frac{12000 \times 25}{100} = 3 \times 10^3 \text{ kW}$$

If Q is the minimum flow necessary for the plant to run as a base load plant, the power developed is

$$P = 9.81Q(H - h_f)\eta$$
$$= (9.81 \times Q \times 20 \times 0.85) = 166.77Q$$

Equating the two, we get

$$166.77Q = 3 \times 10^3$$

$$\therefore \quad Q = \frac{3 \times 10^3}{166.77} = 17.99 \text{ m}^3\text{/s}$$

If the stream flow is 30 m^3/s then the power developed is

$$P = (9.81 \times 30 \times 20 \times 0.85) = 5.003 \times 10^3 \text{ kW}$$

$$\therefore \text{ Maximum load factor } = \frac{5.003 \times 10^3}{12{,}000} = 0.4169 \text{ or } 41.69\%$$

Illustrative Example 24.7 *A run-of-river plant is installed on a river having a minimum flow of 12 m^3/s. If the plant is used as a peak load plant operating only for 6 hours a day, determine the firm capacity of the plant (a) without pondage, (b) with pondage but allowing 10% of the water to be lost in evaporation and other losses. Head at the plant is 15 m and the plant efficiency may be assumed as 80%.*

Solution

(a) Without pondage the firm capacity

$$= (9.81 \times 12 \times 15 \times 0.8) = 1.413 \times 10^3 \text{ kW}$$

(b) With pondage, the total volume of water stored during the 18 hours when the plant is not operating

$$= (12 \times 18 \times 3600) = 7.776 \times 10^5 \text{ m}^3$$

The loss of water due to evaporation etc.,

$$= (7.776 \times 10^5) \times \frac{10}{100}$$
$$= 0.7776 \times 10^5 \text{ m}^3$$

$\therefore$ Net amount of water available from the pondage for the generation of the power during 6 hours when the plant is operating

$$= (7.776 - 0.7776) \times 10^5$$
$$= 6.9984 \times 10^5 \text{ m}^3$$

The corresponding rate of flow available from the pondage

$$= \frac{6.9984 \times 10^5}{6 \times 3600} = 32.4 \text{ m}^3\text{/s}$$

$\therefore$ Total flow available for the power generation

$$= (12 + 32.4) = 44.4 \text{ m}^3\text{/s}$$

The firm capacity of the plant with pondage

$$= 9.81 \times 44.4 \times 15 \times 0.8$$

$$= 5.23 \times 10^3 \text{ kW}$$

REVIEW QUESTIONS

24.1 What are the different types of hydropower plants ? Describe each one briefly.

24.2 Write a brief note on the relative economics of a hydel plant and a thermal plant for power generation.

24.3 What do you understand by 'flow duration curve' ? How is it prepared?

24.4 Differentiate between storage and pondage. Support your answer with a neat sketch.

24.5 Describe with neat sketches pumped storage plant.

24.6 State briefly the functions of the following in a hydroelectric installation :

(i) Intake structure, (ii) Penstock, (iii) Anchor block, (iv) Surge tank.

24.7 How do you make the assessment of water power potential of a hydroelectric scheme ?

24.8 Explain the terms : (i) Load factor, (ii) Plant factor, (iii) Utilization factor. Derive a relation between these three factors.

24.9 Distinguish between a base-load power plant and a peak-load power plant.

24.10 Differentiate between firm power and secondary power.

24.11 Describe any one hydroelectric development scheme of India.

24.12 Write a note on the selection of a suitable type of turbine for a hydroelectric scheme.

24.13 Distinguish between a run-of-river plant and a storage plant.

24.14 Discuss briefly with neat sketches how high heads are obtained in power plants ?

24.15 Discuss the general classification of hydropower plants according to different considerations.

24.16 Draw a neat sketch of a hydropower plant and show the various elements.

24.17 Show that capacity factor is equal to the product of the load factor and utilization factor.

24.18 Explain the terms :

(i) Gross head, (ii) Net or effective head, (iii) Hydraulic efficiency, (iv) Installed capacity, (v) Overall efficiency, as applied to a hydropower plant.

Appendix–I
Khosla's Theory

APP. I.1 COMPOSITE FLOOR PROFILE

Figure App. I.1 shows a composite floor profile analysed by Khosla and his associates by using Schwarz–Christoffel transformation. As shown in Fig. App. I.1 the flow lines and equipotential lines of the actual section are no longer ellipses and hyperbolas. However, by using this transformation the distorted flownet of the actual section can be reduced to normal flownet consisting of elliptical and hyperbolic curves.

Let the plane of the composite floor profile be called z-plane (Fig. App. I.1 a), so that any point in this plane is given by

$$z = x + iy.$$

Further let the plane to which the transformation from the z-plane is carried out be called z-plane, so that any point in this plane is given by

$$\zeta = \xi + i\eta$$

If a polygon is located in the z-plane, then the Schwarz–Christoffel equation that maps it conformally on to z-plane is given by

$$\zeta = A\int \frac{d\zeta}{(\zeta-\zeta_1)^{\lambda_1}(\zeta-\zeta_2)^{\lambda_n}....(\zeta-\zeta_n)^{\lambda_n}} \qquad ...\text{(App. I. 1)}$$

where $\lambda_1\pi$, $\lambda_2\pi$,...$\lambda_n\pi$ are the changes in the angles at the vertices of the polygon in the z-plane as one transverses along the sides of the polygon in the positive sense ; ζ_1, ζ_2,...ζ_n are the coordinates of the points in the z-plane at which the vertices of the polygon of the z-plane map; and A is a complex constant.

When Eq. (A. I. 1) is applied to transform the polygon of the z-plane, the area bounded by the polygon transforms to the lower half of the ζ-plane bounded by real axis of the ζ-plane and all the vertices of the polygon of the ζ-plane map on to the corresponding points on the real axis of the ζ-plane. In other words the complicated boundary profile of the composite floor of the ζ-

plane is reduced to a straight line for which the solution of the Laplace's equation can be obtained and the flownet can be drawn. Moreover, as discussed below the uplift pressures at the key points of the composite floor may also be obtained.

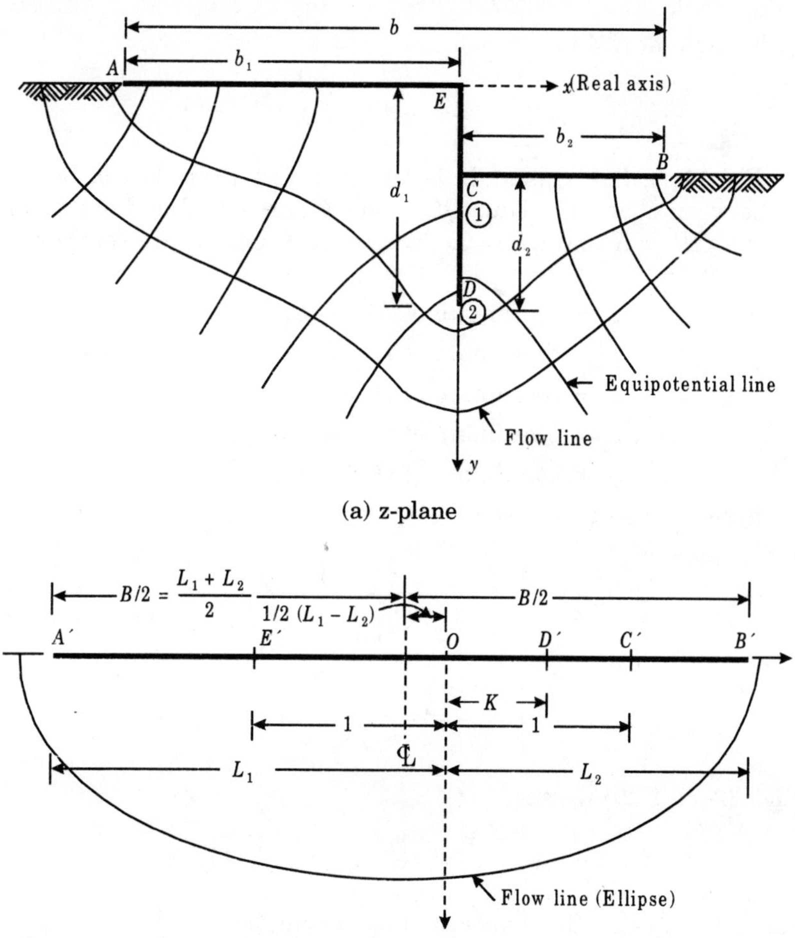

(a) z-plane

(b) ζ–plane

Fig. *App. I.1*

Step 1. Transformation Equation $z = f(\zeta)$

By applying Eq. (A. I. 1), the key points *A, E, D, C* and *B* of the composite floor profile in Fig. [App. I. 1 a] map on to the points *A′, E′, D′, C′* and *B′* respectively on the real axis of the ζ-plane in Fig. [App. I. 1 b]. It may be seen that the angles at the vertices *A, E, D, C* and *B* are π, π/2, 2π, π/2 and π respectively and these have changed to π at each of the points *A′, E′, D′, C′* and *B′* on transformation. Thus

$$\pi\lambda_1 = \text{change in angle at } A = (\pi - \pi) = 0$$

$$\pi\lambda_2 = \text{change in angle at } E = \left(\pi - \frac{\pi}{2}\right) = +\frac{\pi}{2}$$

$\pi\lambda_3$ = change in angle at $D = (\pi - 2\pi) = -\pi$

$\pi\lambda_4$ = change in angle at $C = \left(\pi - \frac{\pi}{2}\right) = +\frac{\pi}{2}$

$\pi\lambda_5$ = change in angle at $B = (\pi - \pi) = 0$

Hence $\lambda_1 = 0$; $\lambda_2 = \frac{1}{2}$; $\lambda_3 = -1$; $\lambda_4 = \frac{1}{2}$; and $\lambda_5 = 0$.

The origin in Fig. [App. I. 1 a] is at E, but that in Fig. [App. I. 1 b] has been chosen at O which is midway between E' and C'. Thus OE' and OC' are equal and are taken equal to unity. Further let $OA' = L_1$, $OD' = K$ and $OB' = L_2$, where L_1, K and L_2 are real numbers.

Thus,

$$\zeta_1 = \text{coordinate of } A' = -L_1$$
$$\zeta_2 = \text{coordinate of } E' = -1$$
$$\zeta_3 = \text{coordinate of } D' = +K$$
$$\zeta_4 = \text{coordinate of } C' = +1$$
$$\zeta_5 = \text{coordinate of } B' = +L_2$$

Substituting these values in Eq. [App. I. 1], we get

$$\therefore \quad \zeta = A\int \frac{d\zeta}{(\zeta+L_1)^0(\zeta+1)^{1/2}(\zeta-K)^{-1}(\zeta-1)^{1/2}(\zeta-L_2)^0}$$

or

$$\zeta = A\int \frac{(\zeta-K)\ d\zeta}{\sqrt{(\zeta^2-1)}} \qquad \text{...(App. I. 2)}$$

Integrating Eq. [App. I. 2], we get

$$\zeta = A\sqrt{\zeta^2-1} - AK\log\left(\zeta+\sqrt{\zeta^2-1}\right)+B \qquad \text{...(App. I. 3)}$$

where B is a constant of integration which may be a complex number.

Step 2. Determination of Constants A and B

The point C' where $\zeta = 1$, corresponds to the point C where $\zeta = i(d_1 - d_2)$.

Thus, substituting these values in Eq. [App. I. 3], we get

$$i(d_1 - d_2) = A \times 0 - AK\log(1+0) + B$$

or

$$B = i(d_1 - d_2) \qquad \text{...(App. I. 4)}$$

Similarly the point E' where $\zeta = -1$ corresponds to the point E where $\zeta = 0$. Thus, substituting these values in Eq. [App. I. 3], we get

$$0 = A \times 0 - AK\log(-1+0) + i(d_1 - d_2)$$

or

$$0 = -AK\log(-1) + i(d_1 - d_2)$$

Since $\log(-1) = i\pi$, we have

$$0 = -AK(i\pi) + i(d_1 - d_2)$$

or $$AK = \frac{d_1 - d_2}{\pi} \quad \text{...(App. I. 5)}$$

Step 3. Determination of Parameter K

The point D' where $\zeta = K < 1$ corresponds to the point D where $z = id_1$. Thus, substituting these values in Eq. [App. I. 3], we get

$$id_1 = A\sqrt{K^2 - 1} - AK\log\left(K + \sqrt{K^2 - 1}\right) + i(d_1 - d_2)$$

or $$id_1 = Ai\sqrt{1 - K^2} - AK\log\left(K + i\sqrt{1 - K^2}\right) + i(d_1 - d_2)$$

as $$K < 1$$

Let $K = \cos\theta$, then the above equation becomes

$$id_1 = A\,i\sin\theta - AK\log(\cos\theta + i\sin\theta) + i(d_1 - d_2)$$

or $$id_1 = A\,i\sin\theta - AK\log(e^{i\theta}) + i(d_1 - d_2)$$

or $$id_1 = A\,i\sin\theta - AK\,i\,\theta + i(d_1 - d_2)$$

or $$d_2 = A\sin\theta - AK\theta$$

Substituting the values of A and AK from Eq. [App. I 5] and since $K = \cos\theta$, we get

$$d_2 = \frac{d_1 - d_2}{\pi}\frac{\sin\theta}{\cos\theta} - \frac{d_1 - d_2}{\pi}\theta$$

or $$\frac{d_2}{d_1 - d_2}\pi = \tan\theta - \theta \quad \text{...(App. I. 6)}$$

Let $$\frac{d_2}{d_1 - d_2} = \delta, \text{ then}$$

$$\pi\delta = \tan\theta - \theta \quad \text{...(App. I. 7)}$$

or $$\pi\delta = \frac{\sqrt{1 - K^2}}{K} - \cos^{-1}K \quad \text{...(App. I. 7 a)}$$

Thus, from Eq. [App. I 7 a] the values of K can be obtained for different values of d.

Step 4. Final Transformation Equation

Substituting the values of A, AK and B in Eq. [App. I 3], we get the final transformation equation as

$$z = \frac{d_1 - d_2}{\pi K}\sqrt{\zeta^2 - 1} - \frac{d_1 - d_2}{\pi}\log\left(\zeta + \sqrt{\zeta^2 - 1}\right) + i(d_1 - d_2) \quad \text{...(App. I. 8)}$$

in which K is a known parameter the value of which may be obtained as indicated above.

Step 5. Determination of the Coordinates of A′ and B′ (*i.e.*, values of L_1 and L_2)

The point A' where $\zeta = -L_1$, corresponds to point A where $z = -b_1$. Thus, substituting these values in Eq. [App. I 8], we get

$$-b_1 = \frac{d_1 - d_2}{\pi K}\sqrt{L_1^2 - 1} - \frac{d_1 - d_2}{\pi}\log\left(-L_1 + \sqrt{L_1^2 - 1}\right) + i(d_1 - d_2)$$

or $$-b_1 = \frac{d_1 - d_2}{\pi K}\sqrt{L_1^2 - 1} - \frac{d_1 - d_2}{\pi}\log\left(L_1 - \sqrt{L^2 - 1}\right) - \frac{d_1 - d_2}{\pi}\log(-1) + i(d_1 - d_2)$$

or $$-b_1 = \frac{d_1 - d_2}{\pi K}\sqrt{L_1^2 - 1} - \frac{d_1 - d_2}{\pi}\log\left(L_1 - \sqrt{L_1^2 - 1}\right) \quad \text{...(App. I. 9)}$$

Similarly the point B' where $\zeta = -L_2$, corresponds to the point B where $z = \{b_2 + i(d_1 - d_2)\}$. Thus, substituting these values in Eq. [App. I 8], we get

$$b_2 + i(d_1 - d_2) = \frac{d_1 - d_2}{\pi}\sqrt{L_2^2 - 1} - \frac{d_1 - d_2}{\pi}\log\left(L_2 + \sqrt{L_2^2 - 1}\right) + i(d_1 - d_2)$$

or $$b_2 = \frac{d_1 - d_2}{\pi K}\sqrt{L_2^2 - 1} - \frac{d_1 - d_2}{\pi}\log\left(L_2 + \sqrt{L_2^2 - 1}\right) \quad \text{...(App. I. 10)}$$

Assuming $L_1 = \cosh \gamma_1$ and $L_2 = \cosh \gamma_2$, and substituting these values in Eq. [App. I 9 and App. I 10], we get

$$-b_1 = \frac{d_1 - d_2}{\pi K}\sinh \gamma_1 + \frac{d_1 - d_2}{\pi}\gamma_1$$

or $$\sinh \gamma_1 + K\gamma_1 = -\frac{\pi K b_1}{d_1 - d_2}$$

or $$\sinh \gamma_1 + K\gamma 1 = -pK\delta_1 \quad \text{...(App. I. 11)}$$

where $\delta_1 = \dfrac{b_1}{d_1 - d_2}$

Also $$b_2 = \frac{d_1 - d_2}{\pi K}\sinh \gamma_2 - \frac{d_1 - d_2}{\pi}\gamma_2$$

or $$\sinh \gamma_2 - K\gamma_2 = \frac{\pi K b_2}{d_1 - d_2}$$

or $$\sinh \gamma_2 - K\gamma_2 = pK\delta_2 \quad \text{...(App. I. 12)}$$

where $\delta_2 = \dfrac{b_2}{d_1 - d_2}$

Thus, knowing the values of b_1, b_2, d_1 and d_2, the values of δ_1, δ_2 and K can be computed, and using Eq. [App. I 11] and Eq. [App. I 12], γ_1 and γ_2 can be found. Knowing γ_1 and γ_2, $L_1 = \cosh \gamma_1$ and $L_2 = \cosh \gamma_2$ are known.

Thus, Eq. [App. I 8] along with Eqs [App. I 7, 9, 10, 11] and Eq. [App. I 12] give a complete solution for the transformation of the general case of Fig. [App. I. 1 a] into Fig. [App. I. 1 b].

Step 6. Pressure Distribution below the Floor

In section 13.14 it has been indicated that the flow field (comprising streamlines and equipotential lines) for the case of a horizontal floor (Fig. 13.9) may be represented by the following expression

$$z = \frac{b}{2} \cosh w$$

where $z = x + iy$; $w = u + iv$; x and y are the coordinates of any point P in the flow field; and u and v are the stream function and potential function (or pressure function) at P.

The above noted expression may be applied to the transformed floor of Fig. [App. I. 1 b] in which case $\frac{b}{2}$ = half the width of floor = $\frac{1}{2}(L_1 + L_2)$ and $z = \zeta\frac{1}{2}(L_1 - L_2)$; since the origin in Fig. 13.9 is at the middle of the floor but in Fig. [App. I. 1 b] it is at a distance of $\frac{1}{2}(L_1 + L_2)$ from the middle point. The equation for this case will thus become

$$\zeta + \frac{1}{2}(L_1 + L_2) = \frac{1}{2}(L_1 + L_2) \cosh w \qquad \text{...(App. I. 13)}$$

Putting $\frac{L_1 - L_2}{2} = \lambda$ and $\frac{L_1 - L_2}{2} = \lambda_1$, we get

$$\zeta = \lambda \cosh w - \lambda_1 \qquad \text{...(App. I. 13 a)}$$

Substituting this in Eq. [App. I 8], we get

$$z = \frac{d_1 - d_2}{\pi K}\sqrt{(\lambda \cosh w - \lambda_1)^2 - 1} - \frac{d_1 - d_2}{\pi} \log \{(\lambda \cosh w - \lambda_1) + \sqrt{(\lambda \cosh w - \lambda_1)^2 - 1}\} + i\,(d_1 - d_2) \qquad \text{...(App. I. 14)}$$

For the purpose of design one is mainly concerned with the pressure distribution along the base profile of the work, and the pressure gradient mainly at the exit end. For the streamline which hugs the base of the work, we have $u = 0$ so that

$$w = iv$$

Also

$$z = x + iy$$

Substituting these in Eq. [App. I 8], we get

$$x + iy = \frac{d_1 - d_2}{\pi K}\sqrt{(\lambda \cos v - \lambda_1)^2 - 1} - \frac{d_1 - d_2}{\pi} \log \{(\lambda \cos v - \lambda_1) + \sqrt{(\lambda \cos v - \lambda_1)^2 - 1}\} + i\,(d_1 - d_2) \qquad \text{...(15)}$$

Equation [App. I. 15] gives the pressure function v for any point whose coordinate is $(x + iy)$. This equation may be applied to determine the pressure distribution along different portions of the floor.

Step 7. Pressure Distribution for Length AE

For points along *AE* Fig. [App. I. 1 a], $y = 0$ and x varies from $-b_1$ to 0. Also a new variable β_1' is introduced which is defined by the equation

$$(\lambda \cos v - \lambda_1) = -\cosh \beta_1' \quad \text{...(App. I. 16)}$$

Substituting these in Eq. App. I (15), we get

$$x = \frac{d_1 - d_2}{\pi K} \sinh \beta_1' - \frac{d_1 - d_2}{\pi} \log(-\cosh \beta_1' + \sinh \beta_1') + i(d_1 - d_2)$$

or
$$x = \frac{d_1 - d_2}{\pi K} \sinh \beta_1' + \frac{d_1 - d_2}{\pi} \beta_1' - \frac{d_1 - d_2}{\pi} \log(-1) + i(d_1 - d_2)$$

or
$$x = \frac{d_1 - d_2}{\pi K} \sinh \beta_1' + \frac{d_1 - d_2}{\pi} \beta_1'$$

or
$$\frac{x}{d_1 - d_2} \pi K = \delta x \pi K = \sinh \beta_1' + K \beta_1' \quad \text{...(App. I. 17)}$$

where $\delta_x = \dfrac{x}{d_1 - d_2}$

From Eq. [App. I 17], corresponding to any point under the floor *AE* the value of β_1' can be calculated, which when substituted in Eq. [App. I 16] will give the pressure at that point. Hence along the floor *AE*

$$v = \cos{-1} \left(\frac{\lambda_1 - \cosh \beta_1'}{\lambda} \right)$$

But
$$v = \pi\phi = \pi \frac{P}{H}$$

∴
$$P = \frac{H}{\pi} \cos^{-1} \left(\frac{\lambda_1 - \cos \beta_1'}{\lambda} \right) \quad \text{...(App. I. 18)}$$

or
$$\phi = \frac{P}{H} = \frac{1}{\pi} \cos^{-1} \left(\frac{\lambda_1 - \cos \beta_1'}{\lambda} \right) \quad \text{...(App. I. 18 a)}$$

Equation [App. I 18] gives the pressure distribution along the floor *AE*.

Step 8. Pressure Distribution for Length CB

For points along *CB*, $y = (d_1 - d_2)$ and x varies from 0 to b_2. Also a new variable β_2' is introduced which is defined by the equation

$$(\lambda \cos v - \lambda_1) = \cosh \beta_2' \quad \text{...(App. I. 19)}$$

Substituting these in Eq. [App. I 15], we get

$$x + i(d_1 - d_2) = \frac{d_1 - d_2}{\pi K} \sinh \beta_2' - \frac{d_1 - d_2}{\pi} \log(\cosh \beta_2' + \sinh \beta_2') + i(d_1 - d_2)$$

or $$x = \frac{d_1 - d_2}{\pi K} \sinh \beta'_2 - \frac{d_1 - d_2}{\pi} \beta'_2$$

or $$\frac{x}{d_1 - d_2} \pi K = \delta_x \pi K = \sinh \beta'_2 - K \beta'_2 \quad \text{...(App. I. 20)}$$

where $\delta_x = \dfrac{x}{d_1 - d_2}$

From Eq. [App. I 20], corresponding to any point under the floor *CB* the value of β'_2 can be calculated, which when substituted in Eq. [App. I 19] will give the pressure at the point. Hence along the floor *CB*

$$v = \cos^{-1}\left(\frac{\lambda_1 + \cosh \beta'_2}{\lambda}\right) = \pi \frac{P}{H}$$

$\therefore$ $$P = \frac{H}{\pi} \cos^{-1}\left(\frac{\lambda_1 + \cosh \beta'_2}{\lambda}\right) \quad \text{...(App. I. 21)}$$

or $$\phi = \frac{P}{H} = \frac{1}{\pi} \cos^{-1}\left(\frac{\lambda_1 + \cosh \beta'_2}{\lambda}\right) \quad \text{...(App. I. 21 a)}$$

Equating Eq. [App. I 21] gives the pressure distribution along the floor *CB*.

Step 9. Pressure Distribution along ED and DC

For both the *u/s* and the *d/s* faces *ED* and *DC* of the pile $x = 0$, but for the *u/s* face *ED*, y varies from 0 to d_1 and for the *d/s* face *DC*, y varies from $(d_1 - d_2)$ to d_1. Also a new variable β is introduced which is defined by the equation

$$(\lambda \cos v - \lambda_1) = \cos \beta \quad \text{...(App. I. 22)}$$

Substituting these in Eq. [App. I 15], we get

$$iy = \frac{d_1 - d_2}{\pi K} i \sin \beta - \frac{d_1 - d_2}{\pi} \log (\cos \beta + i \sin \beta) + i(d_1 - d_2)$$

or $$iy = \frac{d_1 - d_2}{\pi K} i \sin \beta - \frac{d_1 - d_2}{\pi} i\beta + i(d_1 - d_2)$$

or $$\left(\frac{y}{d_1 - d_2} - 1\right) \pi K = \sin \beta - K\beta \quad \text{...(App. I. 23)}$$

From Eq. [App. I 23], corresponding to any point along the pile the value of β can be calculated which when substituted in Eq. [App. I 22] will give the pressure at that point. Hence along the pile

$$v = \cos^{-1}\left(\frac{\lambda_1 + \cos \beta}{\lambda}\right) = \pi \frac{P}{H}$$

$$\therefore \quad P = \frac{H}{\pi}\cos^{-1}\left(\frac{\lambda_1 + \cos\beta}{\lambda}\right) \quad \text{...(App. I. 24)}$$

or
$$\phi = \frac{P}{H} = \frac{1}{\pi}\cos^{-1}\left(\frac{\lambda_1 + \cos\beta}{\lambda}\right) \quad \text{...(App. I. 24 a)}$$

Equation [App. I 24] gives the pressure distribution along the *u/s* and *d/s* faces *ED* and *DC* of the pile.

Step 10. Pressures at Key Points

For design, it is generally enough to know the pressure at key points viz., E, *D* and *C*, and the same may be determined as indicated below.

(i) Pressure at E

From Eq. [App. I 13 a], we have

$$\zeta = \lambda \cosh w - \lambda_1$$

The point E' where $\zeta = -1$, corresponds to the point E where $u = 0$, so that $w = iv$. Thus, substituting these in the above equation, we get

$$-1 = \lambda \cos v - \lambda_1$$

or
$$v = \cos^{-1}\left(\frac{\lambda_1 - 1}{\lambda}\right)$$

For point E

$$v = v_E = \pi\frac{P_E}{H} = \pi\phi_E$$

$$\therefore \quad v_E = \cos^{-1}\left(\frac{\lambda_1 - 1}{\lambda}\right)$$

or
$$P_E = \frac{H}{\pi}\cos^{-1}\left(\frac{\lambda_1 - 1}{\lambda}\right) \quad \text{...(App. I. 25)}$$

or
$$\phi_E = \frac{1}{\pi}\cos^{-1}\left(\frac{\lambda_1 - 1}{\lambda}\right) \quad \text{...(App. I. 25 a)}$$

(ii) Pressure at C

From Eq. [App. I 13 a], we have

$$\zeta = \lambda \cosh w - \lambda_1$$

The point C' where $\zeta = +1$, corresponds to the point C where $u = 0$, so that $w = iv$. Thus, substituting these values in the above equation, we get

$$1 = \lambda \cos v - \lambda_1$$

or
$$v = \cos^{-1}\left(\frac{\lambda_1 + 1}{\lambda}\right)$$

For point C

$$v = v_C = \pi\frac{P_C}{H} = \pi\phi_C$$

$\therefore$
$$v_C = \cos^{-1}\left(\frac{\lambda_1+1}{\lambda}\right)$$

or
$$P_C = \frac{H}{\pi}\cos^{-1}\left(\frac{\lambda_1+1}{\lambda}\right) \qquad \text{...(App. I. 26)}$$

or
$$\phi_C = \frac{1}{\pi}\cos^{-1}\left(\frac{\lambda_1+1}{\lambda}\right) \qquad \text{...(App. I. 26 a)}$$

(iii) Pressure at D

From Eq. [App. I 13 a], we have

$$\zeta = \lambda \cosh w - \lambda_1$$

The point D' where $\zeta = K$, corresponds to the point D where $u = 0$, so that $w = iv$. Thus, substituting these values in the above equation, we get

$$K = \lambda \cos v - \lambda_1$$

or
$$v = \cos^{-1}\left(\frac{\lambda_1+K}{\lambda}\right)$$

For point D

$$v = v_D = \pi\frac{P_D}{H} = \pi\phi_D$$

$\therefore$
$$v_D = \cos^{-1}\left(\frac{\lambda_1+K}{\lambda}\right)$$

or
$$P_D = \frac{H}{\pi}\cos^{-1}\left(\frac{\lambda_1+K}{\lambda}\right) \qquad \text{...(App. I. 27)}$$

or
$$\phi_D = \frac{1}{\pi}\cos^{-1}\left(\frac{\lambda_1+K}{\lambda}\right) \qquad \text{...(App. I. 27 a)}$$

APP. I.1.2 Particular Cases

Some of the particular cases of the general case considered earlier are dealt with below. As the main concern is with the values of the pressures at the key points E, D and C and the value of the exit gradient, the value of only these quantities are obtained.

Case I. Floor with Pile Line at Downstream End with Step

As shown in Fig. [App. I. 2], in this case $CB = b_2 = 0$. Thus,

$$\delta_2 = \frac{b_2}{d_1 - d_2} = 0$$

and from Eq. [App. I 12], we have

$$\sinh \gamma_2 - K\gamma_2 = 0$$

or

$$\gamma_2 = 0$$

so that

$$L_2 = \cosh 0 = 1$$

Equation [App. I 15] will give the distribution of pressure along the base of the floor and pile line.

For this case

$$\lambda = \frac{L_1 + L_2}{2} = \frac{L_1 + 1}{2}$$

and

$$\lambda_1 = \frac{L_1 - L_2}{2} = \frac{L_1 - 1}{2}$$

or

$$\lambda_1 = \lambda - 1$$

$\therefore$

$$\lambda_1 - 1 = \frac{L_1 - 3}{2}$$

and

$$\lambda_1 + 1 = \frac{L_1 + 1}{2} = \lambda$$

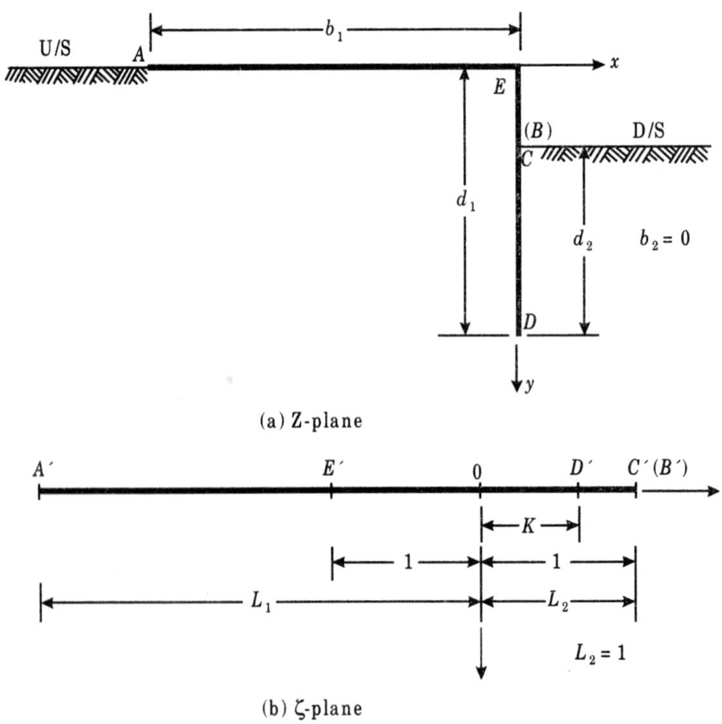

Fig. *App. I. 2*

Hence from Eq. [App. I 25, 26] and Eq. [App. I 27], we have

$$P_E = \frac{H}{\pi} \cos^{-1} \left(\frac{\lambda_1 - 1}{\lambda} \right)$$

or

$$P_E = \frac{H}{\pi} \cos^{-1} \left(\frac{\lambda - 2}{\lambda} \right)$$

or

$$P_E = \frac{H}{\pi} \cos^{-1} \left(\frac{L_1 - 3}{L_1 + 1} \right) \quad \text{...(App. I. 28 a)}$$

$$P_C = \frac{H}{\pi} \cos^{-1} \left(\frac{\lambda_1 + 1}{\lambda} \right)$$

or

$$P_C = \frac{H}{\pi} \cos^{-1} (1)$$

or

$$P_C = 0 \quad \text{...(App. I. 28 b)}$$

$$P_D = \frac{H}{\pi} \cos^{-1} \left(\frac{\lambda_1 + K}{\lambda} \right)$$

or

$$P_D = \frac{H}{\pi} \cos^{-1} \left(\frac{\lambda - 1 + K}{\lambda} \right)$$

or

$$P_D = \frac{H}{\pi} \cos^{-1} \left(\frac{2K + L_1 - 1}{L_1 + 1} \right) \quad \text{...(App. I. 28 c)}$$

Exit Gradient

Along the pile line from Eq. [App. I 23], we have

$$\left(\frac{y}{d_1 - d_2} - 1 \right) \pi K = \sin \beta - K\beta$$

or

$$y = \sin \beta - \frac{d_1 - d_2}{\pi} \beta + (d_1 - d_2)$$

To find the exit gradient at C, differentiate the above expression with respect to y, thus we have

$$\frac{dy}{dy} = 1 = \frac{d_1 - d_2}{\pi K} \cos \beta \frac{d\beta}{dy} - \frac{d_1 - d_2}{\pi} \frac{d\beta}{dy}$$

From Eq. [App. I 22], we have

$$\lambda \cos \upsilon - \lambda_1 = \cos \beta$$

Differentiating the above equation with respect to y, we get

$$\lambda \sin v \frac{dy}{dy} = \sin \beta \frac{d\beta}{dy}$$

or

$$\frac{d\beta}{dy} = \lambda \frac{\sin v}{\sin \beta} \frac{dv}{dy} \quad \text{...(App. I. 29)}$$

Substituting this, we get

$$1 = \frac{d_1 - d_2}{\pi K} \cos \beta \frac{\lambda \sin v}{\sin \beta} \frac{dv}{dy} - \frac{d_1 - d_2}{\pi} \frac{\lambda \sin v}{\sin \beta} \frac{dv}{dy} \quad \text{...(App. I. 30)}$$

At the exit end $v = 0$, thus from Eq. [App. I. 22], we have

$$\cos \beta = \lambda \cos v - \lambda_1$$

or

$$\cos \beta = \lambda - \lambda_1 = 1$$

$$\therefore \quad \beta = 0$$

Hence when $v = 0$, $\beta = 0$

and

$$\frac{\sin v}{\sin \beta} = \frac{0}{0}$$

which is indeterminate and it is essential to determine the limit of $\left(\frac{\sin v}{\sin \beta}\right)_{v=0,\ \beta=0}$

$$\text{Lt}\left(\frac{\sin v}{\sin \beta}\right)_{v=\beta=0} = \underset{\substack{v=0\\ \beta=0}}{\text{Lt}} \frac{\frac{d}{dy}(\sin v)}{\frac{d}{dy}(\sin \beta)}$$

$$= \underset{\substack{v=0\\ \beta=0}}{\text{Lt}} \frac{\cos v \frac{dv}{dy}}{\cos \beta \frac{d\beta}{dy}}$$

Substituting from Eq. [App. I. 29], we get

$$\text{Lt}\left(\frac{\sin v}{\sin \beta}\right)_{v=0,\beta=0} = \underset{\substack{v=0\\ \beta=0}}{\text{Lt}} \left\{\frac{\cos v}{\cos \beta} \frac{\sin \beta}{\lambda \sin v}\right\}$$

$$= \underset{\substack{v=0\\ \beta=0}}{\text{Lt}} \frac{\sin \beta}{\lambda \sin v}$$

$$\therefore \quad \left[\text{Lt}\left(\frac{\sin v}{\sin \beta}\right)\right]^2_{\substack{v=0\\ \beta=0}} = \frac{1}{\lambda}$$

or

$$\underset{\substack{v=0\\ \beta=0}}{\text{Lt}}\left(\frac{\sin v}{\sin \beta}\right) = \frac{1}{\sqrt{\lambda}} \qquad \text{...(App. I. 31)}$$

Substituting this in Eq. [App. I 30], we get

$$1 = \frac{d_1 - d_2}{\pi K}\frac{\lambda}{\sqrt{\lambda}}\left(\frac{dv}{dy}\right)_{v=0} - \frac{d_1 - d_2}{\pi}\frac{\lambda}{\sqrt{\lambda}}\left(\frac{dv}{dy}\right)_{v=0}$$

$$\therefore \quad \left(\frac{dv}{dy}\right)_{v=0} = \frac{\pi K}{(d_1 - d_2)\sqrt{\lambda}(1-K)}$$

Exit gradient

$$G_E = \left(\frac{dP}{dy}\right)_{v=0} - \frac{H}{\pi}\left(\frac{dv}{dy}\right)_{v=0}$$

or

$$G_E = \frac{H}{\pi}\frac{\pi K}{(d_1 - d_2)\sqrt{\lambda}(1-K)}$$

or

$$G_E = \frac{H}{d_1 - d_2}\frac{K}{1-K}\frac{1}{\sqrt{\lambda}} \qquad \text{...(App. I. 32)}$$

or

$$G_E = \frac{H}{d_2}\frac{d_2}{d_1 - d_2}\frac{K}{1-K}\frac{1}{\sqrt{\lambda}} \qquad \text{...(App. I. 32 a)}$$

Case II. Single Pile Line without Floor

As shown in Fig. [App. I. 3] for this case $b_1 = b_2 = 0$ and $d_1 = d_2 = d$.

For the cases where $d_1 = d_2$, $K = 0$, because the point D is equidistant from the points E and C and hence the corresponding point D' is midway between the points E' and C'.

$$\therefore \quad \frac{K}{d_1 - d_2} = \frac{0}{0}$$

which is indeterminate and hence it is essential to determine the limit of $\left(\dfrac{K}{d_1 - d_2}\right)_{d_1 = d_2}$

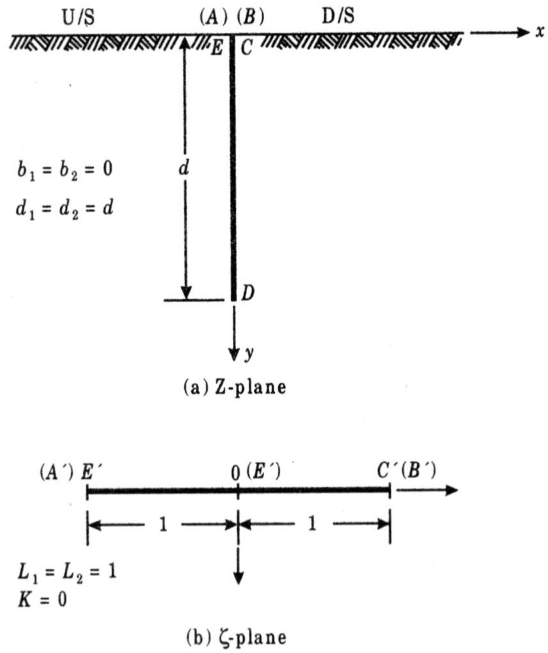

Fig. *App. I. 3*

From Eq. [App. I 7], we have

$$\tan\theta - \theta = \pi\delta = \frac{\pi d_2}{d_1 - d_2}$$

also $\cos\theta = K$.

Multiplying the left side by cos q and the right side by K, we get

$$\sin\theta - \theta\cos\theta = \pi d_2 \frac{K}{(d_1 - d_2)}$$

For $$K = 0, \theta = \frac{\pi}{2}$$

Hence the above equation becomes

$$1 - 0 = \pi d_2 \left(\frac{K}{d_1 - d_2}\right)_{\text{Lt } d_1 = d_2}$$

$$\therefore \quad \underset{d_1 = d_2}{\text{Lt}} \left(\frac{K}{d_1 - d_2}\right) = \frac{1}{\pi d_2} = \frac{1}{\pi d} \qquad \text{...(App. I. 33)}$$

For no floor, i.e., $b_1 = b_2 = 0$, since points A' and E', and C' and B' will be coincident $L_1 = L_2 = 1$, and hence

$$\lambda = \frac{L_1 + L_2}{2} = 1$$

$$\lambda_1 = \frac{L_1 - L_2}{2} = 0$$

Equation [App. I 15] for giving pressure distribution along the faces of pile line become for $x = 0$

$$iy = \frac{d_1 - d_2}{\pi K}\sqrt{\cos^2 v - 1} - \frac{d_1 - d_2}{\pi} - \log(\cos v + i \sin v) + i(d_1 - d_2)$$

or $$iy = \frac{d_1 - d_2}{\pi K} i \sin v - \frac{d_1 - d_2}{\pi} iv + i(d_1 - d_2)$$

or $$y = \frac{\pi d}{\pi} \sin v \qquad (\text{as } d_1 = d_2 = d)$$

or $$y = d \sin v \qquad \text{...(App. I. 34)}$$

or $$v = \sin^{-1}\left(\frac{y}{d}\right)$$

As $$v = \pi\phi = \frac{P}{H}$$

$\therefore$ $$P = \frac{H}{\pi} v$$

or $$P = \frac{H}{\pi} \sin^{-1}\left(\frac{y}{d}\right) \qquad \text{...(App. I. 35)}$$

or $$\phi = \sin^{-1}\left(\frac{y}{d}\right) \qquad \text{...(App. I. 35 a)}$$

Equation [App. I 35] gives the pressure distribution along the faces of the pile line.

For the point D, $y = d$ and hence from Eq. [App. I 35], we get

$$P_D = \frac{H}{\pi} \sin^{-1}(1)$$

or $$P_D = \frac{H}{2} \qquad \text{...(App. I. 35 b)}$$

From Eq. [App. I 32] exit gradient is given as

$$G_E = \frac{H}{1-K} \frac{K}{d_1 - d_2} \frac{1}{\sqrt{\lambda}}$$

or $$G_E = \frac{H}{1}\frac{1}{\pi}\frac{1}{\sqrt{1}}$$

or $$G_E = \frac{H}{\pi} \qquad \text{...(App. I. 36)}$$

Case III. Floor with Pile Line not at End (or Intermediate Pile Line)

As shown in Fig. [App. I. 4] for this case $d_1 = d_2 = d$. The solution for this case is however, obtained *ab initio* by using Schwarz–Christoffel transformation. Thus by applying Eq. [App. I 1], the key points *A, E, D, C* and *B* of the floor profile, in Fig. [App. I. 4 a] map on to the points *A′, E′, D′, C′* and *B′* respectively on the real axis of the ζ-plane in Fig. [App. I. 4 b]. As indicated earlier in this case also $\lambda_1 = 0$; $\lambda_2 = \frac{1}{2}$; $\lambda_3 = -1$; $\lambda_4 = \frac{1}{2}$ and $\lambda_5 = 0$. Further the origin in Fig. [App. I. 4 a] is at the points *E* and *C* which are coincident but that in Fig. [App. I. 4 b] has been chosen at *O* which is midway between *E′* and *C′*. Thus *OE′* and *OC′* are equal and taken equal to unity. Since the point *D* in the ζ plane is equidistant from the points *E* and *C*, the corresponding point *D′* in the ζ-plane will also be equidistant from the points *E′* and *C′*, and hence *D′* coincides with *O*. In other words in this case $K = 0$. Again in this case also let $OA' = L_1$ and $OB' = L_2$, where L_1 and L_2 are real numbers. Thus, in this case $\zeta_1 = -L_1$; $\zeta_2 = -1$; $\zeta_3 = 0$; $\zeta_4 = +1$ and $\zeta_5 = +L_2$.

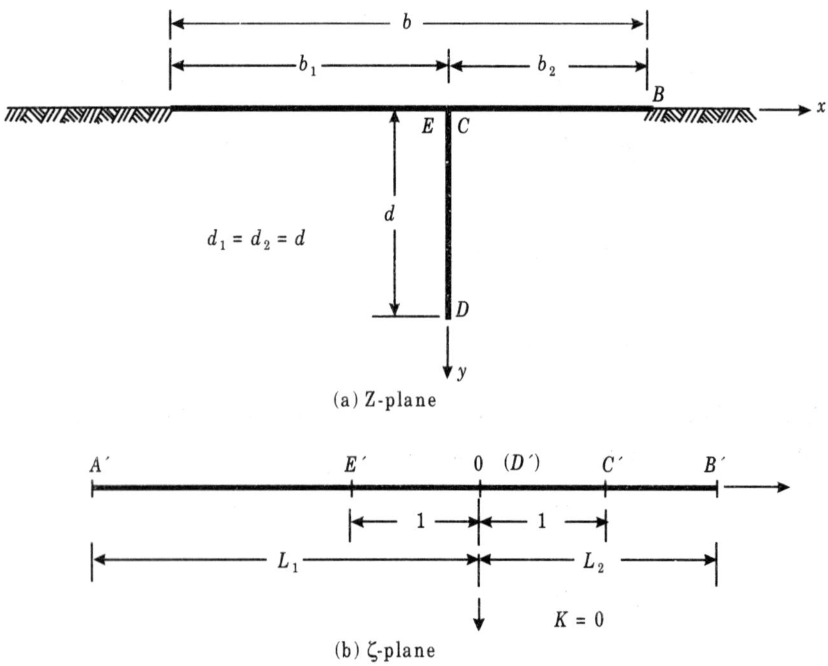

Fig. *App. I.4*

Substituting these values in 1, we get

$$z = A\int \frac{d\zeta}{(\zeta+L_1)^0(\zeta+1)^{1/2}(\zeta-0)^{-1}(\zeta-1)^{1/2}(\zeta-L_2)^0} \quad ...(\text{App. I. 37})$$

or

$$z = \int \frac{\zeta d\zeta}{(\zeta^2-1)^{1/2}} \quad ...(\text{App. I. 37})$$

Integrating Eq. [App. I 37], we get

$$z = A\sqrt{\zeta^2-1}+B \quad ...(\text{App. I. 38})$$

In order to determine the constant B, consider the coordinates of points C and C', thus

$$z = 0 \text{ and } \zeta = 1$$

Hence from Eq. [App. I 38], we get

$$B = 0$$

In order to determine the constant A, consider the coordinates of points D and D', thus

$$z = id \text{ and } \zeta = 0$$

Hence from Eq. [App. I 38], we get

$$id = A\sqrt{-1}+0 = Ai$$

$$\therefore \quad A = d$$

Hence the transformation equation becomes

$$z = d\sqrt{z^2-1} \quad ...(\text{App. I. 39})$$

Again for the horizontal floor of the ζ-plane, we have

$$\zeta + \frac{1}{2}(L_1 - L_2) = \frac{1}{2}(L_1 + L_2)\cosh w$$

Putting $\frac{L_1+L_2}{2} = \lambda$ and $\frac{L_1-L_2}{2} = \lambda_1$, we get

$$\zeta = \lambda \cosh w - \lambda_1$$

Substituting this in Eq. [App. I 39], we get

$$z = d\sqrt{(\lambda \cosh w - \lambda_1)^2 - 1} \quad ...(\text{App. I. 40})$$

Equation [App. I 40], may also be obtained directly from Eq. [App. I 14] by considering $d_1 = d_2 = d$; $K = 0$; and $\text{Lt}\left(\frac{K}{d_1-d_2}\right) = \frac{1}{\pi d}$.

For the streamline which hugs the base of the floor and the pile line $u = 0$, so that

$$w = iv$$

Also

$$z = x + iy$$

Substituting these in Eq. [App. I 40], we get

$$(x + iy) = d\sqrt{(\lambda \cos \upsilon - \lambda_1)^2 - 1} \quad \text{...(App. I. 41)}$$

Along the base of the floor and the pile line pressure may be determined by Eq. [App. I 41]. From Eq. [App. I 9], we have

$$-b_1 = \frac{d_1 - d_2}{\pi K}\sqrt{L_1^2 - 1}$$

or

$$-b_1 = d\sqrt{L_1^2 - 1}$$

Similarly from Eq. [App. I 10], we have

$$b_2 = \frac{d_1 - d_2}{\pi K}\sqrt{L_2^2 - 1}$$

or

$$b_2 = d\sqrt{L_2^2 - 1}$$

From the above expressions, we have

$$L_1 = \sqrt{\left(\frac{b_1}{d}\right)^2 + 1}$$

and

$$L_2 = \sqrt{\left(\frac{b_2}{d}\right)^2 + 1}$$

Let $\frac{b_1}{d} = \alpha_1$ and $\frac{b_2}{d} = \alpha_2$, then

$$L_1 = \sqrt{1 + \alpha_1^2} \quad \text{...(App. I. 42)}$$

$$L_2 = \sqrt{1 + \alpha_2^2} \quad \text{...(App. I. 43)}$$

Also

$$\lambda = \frac{L_1 + L_2}{2}$$

or

$$\lambda = \frac{\sqrt{1 + \alpha_1^2} + \sqrt{1 + \alpha_2^2}}{2} \quad \text{...(App. I. 44)}$$

and

$$\lambda_1 = \frac{L_1 - L_2}{2}$$

or

$$\lambda_1 = \frac{\sqrt{1 + \alpha_1^2} - \sqrt{1 + \alpha_2^2}}{2} \quad \text{...(App. I. 45)}$$

Equation [App. I 40] can be written as

$$(\lambda \cosh w - \lambda_1)^2 = 1 + \left(\frac{z}{d}\right)^2$$

or
$$(\lambda \cosh w - \lambda 1)^2 = 1+\left(\frac{x+iy}{d}\right)^2$$

or
$$(\lambda \cosh w - \lambda_1)^2 = \frac{x^2-y^2}{d^2} + \frac{2ixy}{d} \quad \text{...(App. I. 46)}$$

This is the general equation which will give streamlines and equipotential lines for this case.

Along the base $u = 0$, $y = 0$ and hence Eq. [App. I 40] becomes

$$\lambda \cos v - \lambda_1 = \pm\sqrt{\frac{d^2+x^2}{d^2}}$$

or
$$\cos v = \frac{\lambda_1 d \pm \sqrt{d^2+x^2}}{\lambda d}$$

or
$$v = \cos^{-1}\left(\frac{\lambda_1 d \pm \sqrt{d^2+x^2}}{\lambda d}\right)$$

As
$$v = \pi\phi = \pi\frac{P}{H}$$

or
$$P = \frac{H}{\pi} v$$

or
$$P = \frac{H}{\pi}\cos^{-1}\left(\frac{\lambda_1 d \pm \sqrt{d^2+x^2}}{\lambda d}\right) \quad \text{...(App. I. 47)}$$

Along the pile line $u = 0$ and $x = 0$, and hence Eq. [App. I 40] becomes

$$\lambda \cos v - \lambda_1 = \pm\sqrt{\frac{d^2-y^2}{d^2}}$$

or
$$\cos v = \frac{\lambda_1 d \pm \sqrt{d^2-y^2}}{\lambda d}$$

or
$$v = \cos^{-1}\left(\frac{\lambda_1 d \pm \sqrt{d^2-y^2}}{\lambda d}\right)$$

or $$P = \frac{H}{\pi}\cos^{-1}\left(\frac{\lambda_1 d \pm \sqrt{d^2 - y^2}}{\lambda d}\right) \quad \text{...(App. I. 48)}$$

In Eqs. [App. I 47 and 48] the values of pressure on the upstream of the pile will be given by the negative sign of the radical (i.e., under root term) and those on the downstream side by the positive sign.

For the points E and C, $x = 0$, and substituting this in Eq. [App. I 47], we get

$$P_E = \cos^{-1}\left(\frac{\lambda_1 - 1}{\lambda}\right)$$

and $$P_C = \cos^{-1}\left(\frac{\lambda_1 + 1}{\lambda}\right)$$

These are same as Eqs [App. I 25 and 26] derived earlier.

For the point D at the bottom of the pile $y = d$, so that from Eq. [App. I. 48], we get

$$P_D = \frac{H}{\pi}\cos^{-1}\left(\frac{\lambda_1}{\lambda}\right)$$

The curves of Fig. [App. I 13.13 a] Plate I are based on these equations.

Case IV. Floor With Pile Line at the Downstream End

As shown in Fig. [App. I 5] for this case $d_1 = d_2 = d$; $b_1 = b$; and $b_2 = 0$. The solution for this case is also obtained *ab initio* by using Schwarz Christoffel transformation. Thus, by applying equation 1, the key points A, E, D and C of the floor profile in Fig. [App. I 5 a] map on to the points A', E', D' and C' respectively on the real axis of the z-plane in Fig. [App. I 5 b]. For this case Schwarz–Christoffel equation becomes

$$z = A\int \frac{d\zeta}{(\zeta + L_1)^0(\zeta+1)^{1/2}(\zeta - 0)^{-1}(\zeta - 1)^{1/2}} \quad \text{...(App. I. 49)}$$

or $$z = A\int \frac{\zeta d\zeta}{(\zeta^2 - 1)^{1/2}} \quad \text{...(App. I. 50)}$$

Integrating Eq. [App. I 49], we get

$$z = A\sqrt{\zeta^2 - 1} + B$$

For the points C and C'

$$z = 0 \text{ and } \zeta = 1$$

$$\therefore \quad B = 0$$

For the points D and D'

$$z = id \text{ and } \zeta = 0$$

$$\therefore \quad id = A\sqrt{-1} + 0 = Ai$$

or $A = d$

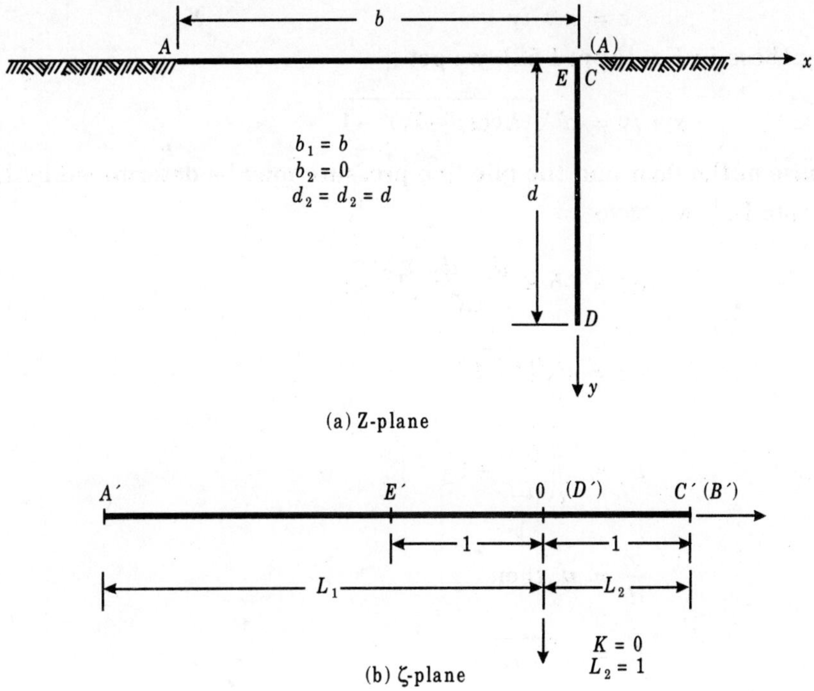

Fig. *App. I. 5*

Hence the transformation equation becomes

$$z = d\sqrt{\zeta^2 - 1} \qquad \text{...(App. I. 51)}$$

For the horizontal floor of the ζ-plane

$$z + \frac{1}{2}(L_1 - L_2) = \frac{1}{2}(L_1 + L_2)\cosh w$$

Putting $\frac{L_1 + L_2}{2} = \lambda$ and $\frac{L_1 - L_2}{2} = \lambda_1$, we get

$$\zeta = \lambda \cosh w - \lambda_1$$

Substituting this in Eq. [App. I 51], we get

$$z = d\sqrt{(\lambda \cosh \ w - \lambda_1)^2 - 1} \qquad \text{...(App. I. 52)}$$

Equation App.I (52) may also be obtained directly from Eq. App.I (14) by considering $d_1 = d_2 = d$; $K = 0$; and Lt $\left(\frac{K}{d_1 - d_2}\right) = \frac{1}{\pi d}$.

For the streamline which hugs the base of the floor and the pile line $u = 0$, so that

$$w = iv$$

Also $$z = x + iy$$

Substituting these in Eq. [App. I 52], we get

$$x + iy = d\sqrt{(\lambda \cos v - \lambda_1)^2 - 1} \quad \text{...(App. I. 53)}$$

Along the base of the floor and the pile line pressure may be determined by Eq. [App. I 53]. From Eq. [App. I 9], we have

$$-b1 = -b = \frac{d_1 - d_2}{\pi K}\sqrt{L_1^2 - 1}$$

or $$-b = d\sqrt{L_1^2 - 1}$$

or $$L_1 = \sqrt{\left(\frac{b}{d}\right)^2 + 1}$$

Let $$\frac{b}{d} = \alpha, \text{ then}$$

$$L_1 = \sqrt{1 + \alpha^2}$$

Similarly from Eq. [App. I 10], we have

$$b_2 = 0 = \frac{d_1 - d_2}{\pi K}\sqrt{L_2^2 - 1}$$

or $$L_2 = 1$$

which is also indicated by Fig. [App. I. 5 b].

$$\lambda = \frac{L_1 + L_2}{2}$$

or $$\lambda = \frac{\sqrt{1+\alpha^2} + 1}{2} \quad \text{...(App. I. 54)}$$

$$\lambda_1 = \frac{L_1 - L_2}{2}$$

or $$\lambda = \frac{\sqrt{1+\alpha^2} - 1}{2} \quad \text{...(App. I. 55)}$$

Equation [App. I 52] may be expressed as

$$(\lambda \cosh w - \lambda_1)^2 = 1 + \frac{x^2 - y^2}{d^2} + \frac{2ixy}{d^2} \quad \text{...(App. I. 56)}$$

This is the general equation which will give streamlines and equipotential lines for this case. Along the base of the floor $u = 0$, $y = 0$, and hence Eq. [App. I 53] becomes

$$\lambda \cos v - \lambda_1 = \pm\sqrt{\frac{d^2 + x^2}{d^2}}$$

or
$$\cos v = \frac{\lambda_1 d \pm \sqrt{d^2 + x^2}}{\lambda d}$$

or
$$v = \cos^{-1}\left(\frac{\lambda_1 d \pm \sqrt{d^2 + x^2}}{\lambda d}\right)$$

As
$$v = \pi\phi = \pi\frac{P}{H}$$

$\therefore$
$$P = \frac{H}{\pi} v$$

or
$$P = \frac{H}{\pi}\cos^{-1}\left(\frac{\lambda_1 d \pm \sqrt{d^2 + x^2}}{\lambda d}\right) \quad \text{...(App. I. 57)}$$

Along the pile line $u = 0$ and $x = 0$, and hence Eq. [App. I 53] becomes

$$\lambda \cos v - \lambda_1 = \pm\sqrt{\frac{d^2 - y^2}{d^2}}$$

or
$$\cos v = \frac{\lambda_1 d \pm \sqrt{d^2 - y^2}}{\lambda d}$$

or
$$v = \cos^{-1}\left(\frac{\lambda_1 d \pm \sqrt{d^2 - y^2}}{\lambda d}\right)$$

or
$$P = \frac{H}{\pi}\cos^{-1}\left(\frac{\lambda_1 d \pm \sqrt{d^2 - y^2}}{\lambda d}\right) \quad \text{...(58)}$$

Again in Eqs [App. I 57 and 58] the negative sign is to be used for determining the pressure on the upstream of the pile and the positive sign for those on the downstream side.

For the point E, $x = 0$ and hence from Eq. [App. I 57], we get

$$P_E = \frac{H}{\pi}\cos^{-1}\left(\frac{\lambda_1 - 1}{\lambda}\right)$$

Since $$\lambda - \lambda_1 = L_2 = 1,$$
$$\lambda_1 = \lambda - 1$$

$\therefore$ $$P_E = \frac{H}{\pi}\cos^{-1}\left(\frac{\lambda-2}{\lambda}\right) \qquad \text{...(App. I. 59)}$$

For the point C, $x = 0$ and hence from Eq. [App. I 57], we get

$$P_C = \frac{H}{\pi}\cos^{-1}\left(\frac{\lambda_1+1}{\lambda}\right)$$

or $$P_C = \frac{H}{\pi}\cos^{-1}\left(\frac{\lambda}{\lambda}\right) = 0 \qquad \text{...(App. I. 60)}$$

For the point D at the bottom of the pile $y = d$, and hence from Eq. [App. I 58], we get

$$P_D = \frac{H}{\pi}\cos^{-1}\left(\frac{\lambda_1}{\lambda}\right)$$

or $$P_D = \frac{H}{\pi}\cos^{-1}\left(\frac{\lambda-1}{\lambda}\right) \qquad \text{...(App. I. 61)}$$

The curves of Fig. [App. I. 13.13 b] Plate I are based on these equations.

Exit Gradient

Along the pile line $x = 0$, and hence from Eq. [App. I 53], we have

$$iy = d\sqrt{(\lambda\cos v - \lambda_1)^2 - 1}$$

Let $$(\lambda \cos v - \lambda_1) = \cos\beta \qquad \text{...(App. I. 62)}$$

$\therefore$ $$iy = d\sqrt{\cos^2\beta - 1}$$

or $$iy = id \sin\beta$$

or $$y = d\sin\beta$$

Differentiating it with respect to y, we get

$$1 = d\cos\beta\,\frac{d\beta}{dy} \qquad \text{...(App. I. 63)}$$

Differentiating Eq. [App. I 62] with respect to y, we get

$$\lambda \sin v\,\frac{dv}{dy} = \sin\beta\,\frac{d\beta}{dy} \qquad \text{...(App. I. 64)}$$

Substituting the value of $\left(\frac{d\beta}{dy}\right)$ from Eq. [App. I 63] in Eq. [App. I 64], we get

$$\lambda \sin v \frac{dv}{dy} = \frac{1}{d}\frac{\sin\beta}{\cos\beta}$$

or $$\frac{dv}{dy} = \frac{1}{\lambda d}\frac{1}{\cos\beta}\frac{\sin\beta}{\sin v} \quad \text{...(App. I. 65)}$$

At the exit end $v = 0$, thus from Eq. [App. I 62], we have

$$\cos\beta = (\lambda \cos 0 - \lambda_1)$$

or $$\cos\beta = (\lambda - \lambda_1) = 1$$

$\therefore$ $$\beta = 0$$

Hence when $$v = 0, \ \beta = 0,$$

and $$\left(\frac{\sin v}{\sin\beta}\right) = \frac{0}{0}$$

which is indeterminate and it is essential to determine the limit of $\left(\dfrac{\sin v}{\sin\beta}\right)_{v=0,\,\beta=0}$

As indicated earlier in the Case I

$$\underset{\substack{v=0\\ \beta=0}}{\text{Lt}}\left(\frac{\sin v}{\sin\beta}\right) = \frac{1}{\sqrt{\lambda}}$$

Substituting this in Eq. [App. I 65], for $v = 0$ and b = 0, we get

$$\left(\frac{dv}{dy}\right)_{v=0} = \frac{1}{d\sqrt{\lambda}}$$

Exit gradient

$$G_E = \left(\frac{dP}{dy}\right)_{v=0} \frac{H}{\pi}\left(\frac{dv}{dy}\right)_{v=0}$$

or $$G_E = \frac{H}{\pi}\left(\frac{1}{d\sqrt{\lambda}}\right)$$

or $$G_E = \frac{H}{d}\frac{1}{\pi\sqrt{\lambda}} \quad \text{...(App. I. 66)}$$

The curve of Fig. [App. I. 13.13 c] Plate II is based on this equation.

Case V. Floor with Pile Line at the Upstream End

As shown in Fig. [App. I. 6] for this case $d_1 = d_2 = d$; $b_1 = 0$; and $b_2 = b$. The solution for this case is also obtained *ab initio* by using Schwarz–Christoffel transformation. Thus, applying Eq. [App. I 1], the key points *E, D, C* and *B* of the floor profile in Fig. [App. I. 6 a] map on to the points E', D', C' and B' respectively on the real axis of the ζ-plane in Fig. [App. I. 6 b]. For this case Schwarz-Christoffel equation becomes

$$z = A\int \frac{d\zeta}{(\zeta+1)^{1/2}(\zeta-0)^{-1}(\zeta-1)^{1/2}(\zeta-L_2)^0}$$

or $$z = A\int \frac{\zeta d\zeta}{(\zeta^2-1)^{1/2}} \qquad \text{...(App. I. 67)}$$

Integrating Eq. [App. I 67], we get

$$z = A\sqrt{\zeta^2-1}+B \qquad \text{...(App. I. 68)}$$

For the points C and C'

$$z = 0 \text{ and } \zeta = 1$$

∴ $$B = 0$$

For points D and D'

$$z = id \text{ and } \zeta = 0$$

∴ $$id = A\sqrt{-1}+0 = Ai$$

or $$A = d$$

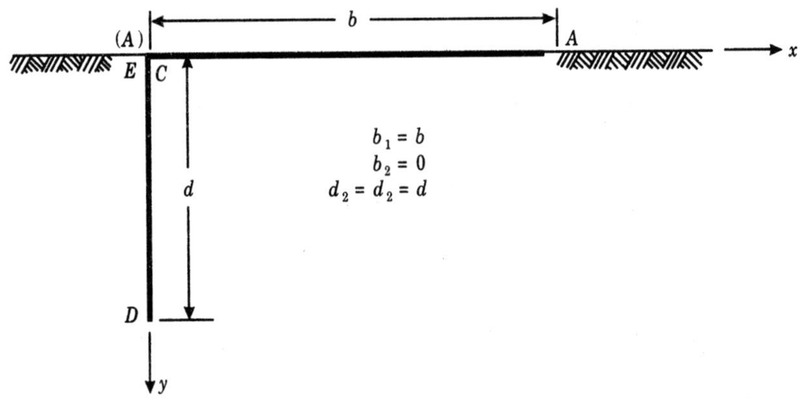

(a) Z-plane

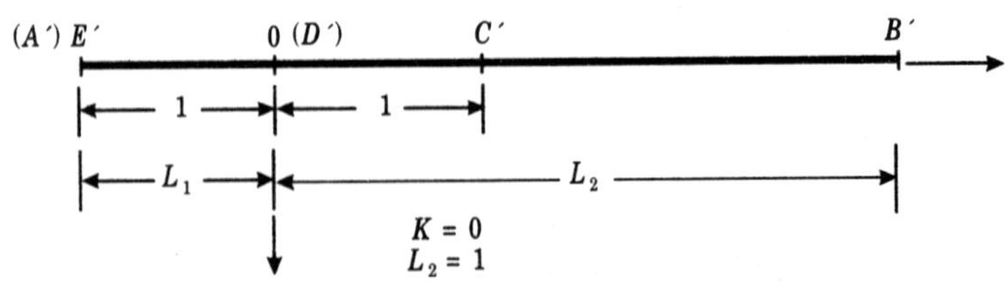

(b) ζ-plane

Fig. *App. I. 6*

Hence the transformation equation becomes

$$z = d\sqrt{\zeta^2 - 1} \quad \text{...(App. I. 69)}$$

For the horizontal floor of the ζ-plane

$$\zeta + \frac{1}{2}(L_1 - L_2) = \frac{1}{2}(L_1 + L_2)\cosh w$$

Putting $\frac{L_1 + L_2}{2} = 1$ and $\frac{L_1 - L_2}{2} = \lambda_1$, we get

$$\zeta = \lambda \cosh w - \lambda_1$$

Substituting this in Eq. [App. I 69], we get

$$z = d\sqrt{(\lambda \cosh w - \lambda_1)^2 - 1} \quad \text{...(App. I. 70)}$$

Equation [App. I 70] may also be obtained directly from Eq. [App. I 14] by considering

$$d_1 = d_2 = d\,;\, K = 0\,; \text{ and } \mathrm{Lt}\left(\frac{K}{d_1 - d_2}\right) = \frac{1}{\pi d}$$

For the streamline which hugs the base of the floor and the pile line $u = 0$, so that

$$w = iv$$

Also $z = x + iy$

Substituting these in Eq. [App. I 70], we get

$$x + iy = d\sqrt{(\lambda \cos v - \lambda_1)^2 - 1} \quad \text{...(71)}$$

Along the base of the floor and the pile line pressure may be determined by Eq. [App. I 71]. From Eq. [App. I 9], we have

$$-b_1 = 0 = \frac{d_1 - d_2}{\pi K}\sqrt{L_1^2 - 1}$$

or $L_1 = 1$

which is also indicated by Fig. [App. I. 6 b]

Similarly from Eq. [App. I 10], we have

$$b_2 = b = \frac{d_1 - d_2}{\pi K}\sqrt{L_2^2 - 1}$$

or

$$b = d\sqrt{L_2^2 - 1}$$

or

$$L_2 = \sqrt{\left(\frac{b}{d}\right)^2 + 1}$$

Let $$\frac{b}{d} = \alpha, \text{ then}$$

$$L_2 = \sqrt{1+\alpha^2}$$

$$\lambda = \frac{L_1 + L_2}{2}$$

or $$\lambda = \frac{1+\sqrt{1+\alpha^2}}{2} \quad \text{...(App. I. 72)}$$

$$\lambda_1 = \frac{L_1 - L_2}{2}$$

or $$\lambda_1 = \frac{1-\sqrt{1+\alpha^2}}{2} \quad \text{...(App. I. 73)}$$

Equation [App. I 70] may be expressed as

$$(\lambda \cosh w - \lambda_1)^2 = 1+\frac{x^2 - y^2}{d^2} + \frac{2ixy}{d} \quad \text{...(App. I. 74)}$$

This is the general equation which will give streamlines and equipotential lines for this case.

Along the base of the floor $u = 0$. $y = 0$, and hence Eq. [App. I 70] becomes

$$\lambda \cos v - \lambda_1 = \pm\sqrt{\frac{d^2 + x^2}{d^2}}$$

or $$\cos v = \frac{\lambda_1 d + \sqrt{d^2 + x^2}}{\lambda d}$$

or $$v = \cos^{-1}\left(\frac{\lambda_1 d + \sqrt{d^2 + x^2}}{\lambda d}\right)$$

As $$v = \pi\phi = \pi\frac{P}{H}$$

∴ $$P = \frac{H}{\pi} v$$

or $$P = \frac{H}{\pi}\cos^{-1}\left(\frac{\lambda_1 d \pm \sqrt{d^2 + x^2}}{\lambda d}\right) \quad \text{...(App. I. 75)}$$

Along the pile line $u = 0$ and $x = 0$, and hence Eq. [App. I 70] becomes

$$\lambda \cos v - \lambda_1 = \pm \sqrt{\frac{d^2 - y^2}{d^2}}$$

or
$$\cos v = \frac{\lambda_1 d \pm \sqrt{d^2 - y^2}}{\lambda d}$$

or
$$v = \cos^{-1}\left(\frac{\lambda_1 d \pm \sqrt{d^2 - y^2}}{\lambda d}\right)$$

or
$$P = \frac{H}{\pi}\cos^{-1}\left(\frac{\lambda_1 d \pm \sqrt{d^2 - y^2}}{\lambda d}\right) \quad \text{...(App. I. 76)}$$

Again in Eqs. [App. I 75 and 76] the negative sign is to be used for determining the pressure on the upstream of the pile and the positive sign for those on the downstream side.

For the point E, $x = 0$, and hence from Eq. [App. I 75], we get

$$P_E = \frac{H}{\pi}\cos^{-1}\left(\frac{\lambda_1 - 1}{\lambda}\right)$$

Since
$$\lambda = \frac{1 + L_2}{2}; \text{ and } \lambda_1 = \frac{1 - L_2}{2},$$

$$\lambda + \lambda_1 = 1$$

or
$$\lambda_1 = 1 - \lambda$$

$\therefore$
$$P_E = \frac{H}{\pi}\cos^{-1}\left(-\frac{\lambda}{\lambda}\right)$$

or
$$P_E = \frac{H}{\pi}\cos^{-1}(-1) = H \quad \text{...(App. I. 77)}$$

For the point C, $x = 0$ and hence from Eq. [App. I 75], we get

$$P_C = \frac{H}{\pi}\cos^{-1}\left(\frac{\lambda_1 + 1}{\lambda}\right)$$

or
$$P_C = \frac{H}{\pi}\cos^{-1}\left(\frac{2 - \lambda}{\lambda}\right) \quad \text{...(App. I. 78)}$$

From the point D at the bottom of the pile $y = d$, and hence from Eq. [App. I 76], we get

$$P_D = \frac{H}{\pi}\cos^{-1}\left(\frac{\lambda_1}{\lambda}\right)$$

or
$$P_D = \frac{H}{\pi}\cos^{-1}\left(\frac{1-\lambda}{\lambda}\right) \qquad \text{...(App. I. 79)}$$

It may be seen that for the same length of the floor when the pile line is at the upstream end instead of the downstream end, the pressure at C in the former case if it is called $P_{C_5} = H - P_{E_4}$; where P_{E_4} is the pressure at E in the Case 4. This is so because, if in Eq. [App. I 59], $\cos^{-1}\left(\frac{\lambda-2}{\lambda}\right) = \theta$, then $\cos^{-1}\left(\frac{2-\lambda}{\lambda}\right) = \pi - \theta$. Thus,

$$P_{E_4} = \frac{H}{\pi}\theta$$

and
$$P_{C_5} = \frac{H}{\pi}(\pi-\theta) = H - \frac{H}{\pi}\theta = H - P_{E_4}$$

Similarly it can also be shown that the pressure at D in the case 5 will be equal to (H – pressure at D in the case 4), i.e.,

$$P_{D_5} = H - P_{D_4}$$

SOVED ILLUSTRATIVE EXAMPLES

Illustrative Example App. I. 1 *A horizontal impervious floor of length 25 m is provided with a pile of depth 5 m placed at a distance of 10 m from the upstream end as shown in the Fig. App. I. 1 Determine analytically the uplift pressures at the points E, D and C.*

Solution

Given :
$$b = 25 \text{ m} : b_1 = 10 \text{ m};$$
$$b_2 = (25 - 10) = 15 \text{ m}; d = 5 \text{ m}$$

∴
$$\alpha_1 = \frac{b_1}{d} = \frac{10}{5} = 2$$

and
$$\alpha_2 = \frac{b_2}{d} = \frac{15}{5} = 3$$

$$\lambda = \frac{\sqrt{1+\alpha_1^2}+\sqrt{1+\alpha_2^2}}{2}$$

or
$$\lambda = \frac{\sqrt{1+(2)^2}+\sqrt{1+(3)^2}}{2} = 2.7$$

TABLE. App. I.1 List of equations derived by Khosls's Theory

Case No. (1)	*Description* (2)	*Figure No.* (3)	*Expression for pressure*					*Expression for Exit Gradient*
			P_A (4)	P_E (5)	P_C (6)	P_D (7)	P_B (8)	G_E (9)
General	Composite floor-Stepped floor with a pile line at the step	1	H	$\frac{H}{\pi}\cos^{-1}\left(\frac{\lambda_1-1}{\lambda}\right)$	$\frac{H}{\pi}\cos^{-1}\left(\frac{\lambda_1+1}{\lambda}\right)$	$\frac{H}{\pi}\cos^{-1}\left(\frac{\lambda_1+K}{\lambda}\right)$	0	—
Case I	Floor with pile line at downstream end with step	2	H	$\frac{H}{\pi}\cos^{-1}\left(\frac{\lambda}{\lambda}\right)$	0		0	
Case II	Single pile line without floor	3	H	H	0		0	
Case III	Floor with an intermediate pile line	4	H				0	—
Case IV	Floor with pile line at downstream end	5	H		0		0	$\frac{H}{d}\frac{1}{\pi\sqrt{\lambda}}$
Case V	Floor with pile at upstream	6	H	H			0	

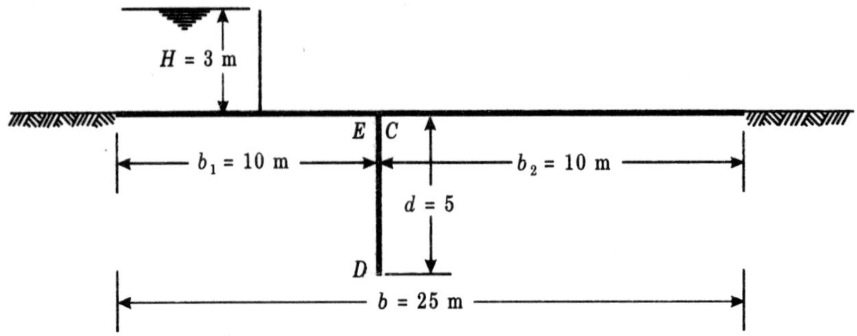

Fig. *Ex. App.I. 1*

$$\lambda_1 = \frac{\sqrt{1+\alpha_1^2} - \sqrt{1+\alpha_2^2}}{2}$$

or $$\lambda_1 = \frac{\sqrt{1+(2)^2} - \sqrt{1+(3)^2}}{2} = -0.46$$

$$P_E = \frac{H}{\pi}\cos^{-1}\left(\frac{\lambda_1 - 1}{\lambda}\right)$$

or $$P_E = \frac{3}{\pi}\cos^{-1}\left(\frac{-0.46-1}{2.7}\right)$$

or $$P_E = \frac{3}{\pi}(2.142) = 2.045 \text{ m}$$

$$P_C = \frac{H}{\pi}\cos^{-1}\left(\frac{\lambda_1 + 1}{\lambda}\right)$$

or $$P_C = \frac{3}{\pi}\cos^{-1}\left(\frac{-0.46+1}{2.7}\right)$$

or $$PC = \frac{3}{\pi}(1.369) = 1.308 \text{ m}$$

$$P_D = \frac{H}{\pi}\cos^{-1}\left(\frac{\lambda_1}{\lambda}\right)$$

or $$P_D = \frac{3}{\pi}\cos^{-1}\left(\frac{-0.46}{2.7}\right)$$

or $$P_D = \frac{3}{\pi}(1.742) = 1.663 \text{ m}$$

Illustrative Example App. I. 2 *A horizontal impervious floor of length 25 m is provided with a cutoff of 5 m depth at its downstream end as shown in the Fig. Ex. App. I. 2. Determine analytically the uplift pressure at the points E and D and also the exit gradient, if the head causing seepage is 3 m.*

Solution

Given: $b = 25\text{ m}\ ;\ d = 5\text{ m}\ ;\ H = 3\text{ m}$

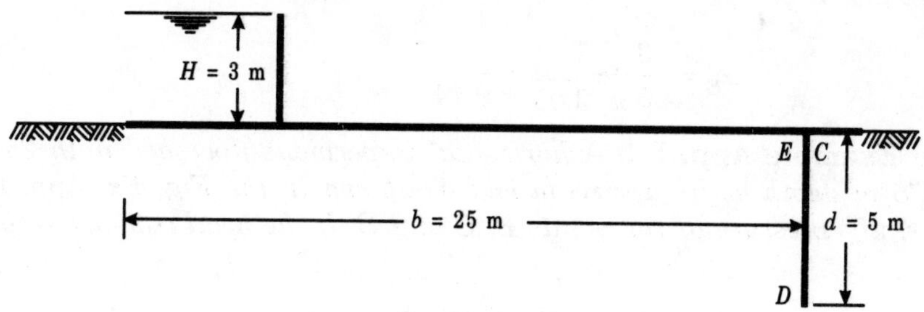

Fig. *Ex. App. I. 2*

Pressure at the point E is given by

$$P_E = \frac{H}{\pi}\cos^{-1}\left(\frac{\lambda-2}{\lambda}\right)$$

$$\lambda = \frac{1+\sqrt{1+\alpha^2}}{2}$$

$$\alpha = \frac{b}{d} = \frac{25}{5} = 5$$

$$\therefore \quad \lambda = \frac{1+\sqrt{1+(5)^2}}{2} = 3.05$$

$$\therefore \quad P_E = \frac{3}{\pi}\cos^{-1}\left(\frac{3.05-2}{3.05}\right)$$

or $$P_E = \frac{3}{\pi}(1.219) = 1.164\text{ m}$$

Pressure at the point D is given by

$$P_D = \frac{H}{\pi}\cos^{-1}\left(\frac{\lambda-1}{\lambda}\right)$$

or $$P_D = \frac{3}{\pi}\cos^{-1}\left(\frac{3.05-1}{3.05}\right)$$

or $$P_D = \frac{3}{\pi}(0.834) = 0.796 \text{ m}$$

Exit gradient is given by

$$G_E = \frac{H}{d}\frac{1}{\pi\sqrt{\lambda}}$$

or $$G_E = \frac{3}{5}\frac{1}{\pi\sqrt{3.05}} = \frac{1}{9.14}$$

Illustrative Example App. I. 3 *A horizontal impervious floor of length 25 m is provided with a cutoff of 5 m depth at its upstream end as shown in the Fig. Ex. App. I. 3. Determine analytically the uplift pressure at the points E, C and D, if the head causing seepage is 3 m.*

Solution

Given: $b = 25 \text{ m} ; d = 5 \text{ m} ; H = 3 \text{ m}$

Pressure at the point E is given by

$$P_E = H = 3 \text{ m}$$

Pressure at the point C is given by

$$P_C = \frac{H}{\pi}\cos^{-1}\left(\frac{2-\lambda}{\lambda}\right)$$

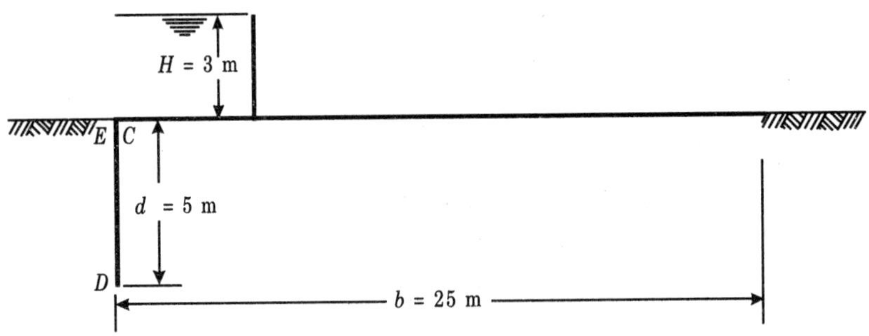

Fig. *Ex. App. I. 3*

$$\lambda = \frac{1+\sqrt{1+\alpha^2}}{2}$$

$$\alpha = \frac{b}{d} = \frac{25}{5} = 5$$

$$\therefore \quad \lambda = \frac{1+\sqrt{1+(5)^2}}{2} = 3.05$$

$$\therefore \qquad P_C = \frac{3}{\pi}\cos^{-1}\left(\frac{2-3.05}{3.05}\right)$$

or
$$P_C = \frac{3}{\pi}(1.922) = 1.836 \text{ m}$$

Pressure at the point D is given by

$$P_D = \frac{H}{\pi}\cos^{-1}\left(\frac{1-\lambda}{\lambda}\right)$$

or
$$P_D = \frac{3}{\pi}\cos^{-1}\left(\frac{1-3.05}{3.05}\right)$$

or
$$P_D = \frac{3}{\pi}(2.308) = 2.204 \text{ m.}$$

Appendix II
Multiple Choice Questions

1. Rabi crops are sown and harvested in the months of
   (a) April and September (b) October and March
   (c) February and June (d) January and May.
2. Kharif crops are sown and harvested in the months of
   (a) October and March (b) February and June
   (c) April and September (d) January and May.
3. Duty of water is defined as
   (a) the area of land in hectares which can be irrigated for growing any crop if one cumec of water is supplied continuously to the land for the entire base period of the crop.
   (b) the total depth of water over the irrigated land required by a crop grown on it during the entire base period of the crop.
   (c) the cumec of water required to be applied to grow any crop on 100 hectares of land.
   (d) the area of land in hectares which can be irrigated for growing any crop if one cumec of water is supplied continuously to the land for the entire crop period.
4. If a crop requires a total depth of 1 m of water for a base period of 150 days then the duty of water is
   (a) 1596 hectares/cumec (b) 1496 hectares/cumec
   (c) 1396 hectares/cumec (d) 1296 hectares/cumec.
5. If the duty for a base period of 120 days is 1500 hectares/cumec, then the delta for the crop is approximately
   (a) 690 mm (b) 790 mm
   (c) 890 mm (d) 990 mm.
6. Field capacity is defined as
   (a) the maximum amount of moisture which can be held by a soil against gravity
   (b) the maximum amount of moisture which can be held by a soil under gravity

(c) the maximum water holding capacity of the soil

(d) the minimum water holding capacity of the soil.

7. Hygroscopic water is defined as

(a) the readily available water for the use of plants

(b) the water which is adsorbed by the particles of the dry soil from the atmosphere

(c) the total water content of the soil when all the pores are filled with water

(d) the water held by the soil under capillary action.

8. Mixed cropping is defined as

(a) two or more crops grown during any year

(b) two or more crops grown during the same crop season in different fields

(c) two or more crop seasons of the year

(d) growing of two or more crops together in the same field during the same crop season.

9. Sodium adsorption ratio (SAR) is expressed as

(a) $\dfrac{Na^+}{\left[\frac{1}{2}\left(Ca^{++}+Mg^{++}\right)\right]^{1/2}}$

(b) $\dfrac{Na^+}{\frac{1}{2}\left(Ca^{++}+Mg^{++}\right)}$

(c) $\dfrac{Na^+}{\frac{1}{2}\left(Ca^{++}+Mg^{++}\right)^{1/2}}$

(d) $\dfrac{Na^+}{\left[2\left(Ca^{++}+Mg^{++}\right)\right]^{1/2}}$

10. If the pan evaporation indicates a total yearly evaporation of 1.5 m, then the annual water loss by evaporation from a reservoir of water surface area 100 km^2 will be

(a) 1.5×10^6 ha-m
(b) 1.5×10^4 ha-m
(c) 1.5×10^5 ha-m
(d) 1.5×10^8 ha-m

11. If the peak of a 3 hour unit hydrograph is 1000 cumec, the flood hydrograph peak for a net rainfall of 150 mm and a base flow of 200 cumec is

(a) 14 800 cumec
(b) 12 000 cumec
(c) 15 200 cumec
(d) 18 000 cumec

12. Infiltration process is

(a) movement of water through the soil

(b) absorption of water by surface of the soil

(c) saturation of the soil

(d) capillary rise of water in the soil.

13. If the area of drainage basin is doubled, other factors remaining the same the peak discharge according to Dicken's formula will be approximately

(a) 2 times the original discharge

(b) 1.68 times the original discharge

(c) 1.59 times the original discharge

(d) 0.5 times the original discharge.

14. If the area of drainage basin is doubled, other factors remaining the same the peak discharge according to Ryve's formula will be approximately
    (a) 1.41 times the original discharge
    (b) 2 times the original discharge
    (c) 1.59 times the original discharge
    (d) 1.68 times the original discharge.
15. The recurrence interval T is given by Gumbels method as (with usual notation)
    (a) $\frac{n}{m}$
    (b) $\frac{2n}{2m-1}$
    (c) $\frac{2n}{2m+C-1}$
    (d) $\frac{n}{m+C-1}$
16. Water enters the soil at capacity rate when
    (a) intensity of rainfall is more than the infiltration capacity of the soil
    (b) intensity of rainfall is less than the infiltration capacity of the soil
    (c) intensity of rainfall is just equal to the infiltration capacity of the soil
    (d) intensity of rainfall is only slightly less than the infiltration capacity of the soil.
17. The area below a direct runoff hydrograph is
    (a) always greater than that below the corresponding rainfall histogram
    (b) always less than that below the corresponding rainfall histogram
    (c) always equal to that below the corresponding rainfall histogram
    (d) sometimes greater and sometimes less than that below the corresponding rainfall histogram.
18. If the area of drainage basin becomes 4 times, other factors remaining the same the peak discharge according to Inglis formula will be approximately
    (a) 4 times the original discharge
    (b) 3 times the original discharge
    (c) 2 times the original discharge
    (d) 1.41 times the original discharge.
19. A confined aquifer is one in which
    (a) water surface under the ground is at atmospheric pressure
    (b) water is confined under pressure less than atmospheric pressure between impermeable strata
    (c) water is confined at atmospheric pressure between impermeable strata
    (d) water is confined under pressure greater than atmospheric pressure between impermeable strata.
20. An unconfined aquifer is one in which
    (a) water surface under the ground is at atmospheric pressure
    (b) water is under pressure greater than atmospheric pressure within two impervious strata
    (c) water is at pressure below atmospheric pressure within two impervious strata
    (d) water is either below or above atmospheric pressure within two impervious strata.
21. Discharge from a well penetrating an unconfined aquifer is given by (with usual notation)

(a) $Q = \dfrac{2\pi kb(H-h)}{\log_e(R/r)}$ (b) $Q = \dfrac{\pi kb(H-h)}{\log_e(R/r)}$

(c) $Q = \dfrac{\pi k(H^2-h^2)}{\log_e(R/r)}$ (d) $Q = \dfrac{2\pi k(H^2-h^2)}{\log_e(R/r)}$

22. Discharge from a well penetrating a confined aquifer is given by (with usual notation)

(a) $Q = \dfrac{\pi k(H^2-h^2)}{\log_e(R/r)}$ (b) $Q = \dfrac{2\pi kb(H-h)}{\log_e(R/r)}$

(c) $Q = \dfrac{2\pi k(H^2h^2)}{\log_e(R/r)}$ (d) $Q = \dfrac{\pi kb(H-h)}{\log_e(R/r}$

23. For a well penetrating an unconfined aquifer having permeability $k = 4 \times 10^{-4}$ m/s, the radius of influence for a drawdown of 5 m is approximately

(a) 100 m (b) 200 m

(c) 300 m (d) 400 m.

24. A well of radius 100 mm penetrates fully a confined aquifer. If the radius of the well is doubled and the radius of influence is 300 m in both the cases the percentage increase in the discharge is approximately

(a) 25.5% (b) 20.5%

(c) 15.5% (d) 9.5%.

25. The specific capacity of a well is defined as

(a) drawdown per unit discharge (b) discharge per unit drawdown

(c) drawdown per certain given discharge (d) discharge per certain given drawdown.

26. With increase in quantity of water supplied the yield of most crops

(a) continuously increases

(b) continuously decreases

(c) increases upto a certain limit and then falls

(d) decreases upto a certain limit and then increases.

27. Specific retention of a soil is defined as

(a) the ratio of the volume of water it will retain after saturation against the force of gravity to its own volume

(b) the ratio of the volume of water that after saturation, can be drained by gravity to its own volume

(c) the ratio of volume of interstices to the total volume of the soil

(d) the sum of porosity and specific yield.

28. Multipurpose reservoir is the one which is constructed

(*a*) to be used in combination of other reservoirs

(b) to serve one main purpose and other purposes are served incidentally

(c) to serve more than one purpose

(d) to supply water for more than one year.

29. For no tension at any point in a gravity dam the criterion to be satisfied is
    (a) the resultant of all the forces must always pass through the mid-point of the base of the dam.
    (b) the resultant force for all conditions of loading must pass through the middle third of the base .
    (c) the resultant of all forces must always pass through the upstream extremity of the middle third of the base.
    (d) the resultant of all the forces must always pass through the downstream extremity of the middle third of the base.
30. For a triangular dam section of height H for just no tension under the action of water pressure, self weight and uplift pressure the minimum base width required with usual notation is

    (a) $\dfrac{H}{S-1}$ (b) $\dfrac{H}{(S-1)^2}$

    (c) $\dfrac{H}{\sqrt{S}}$ (d) $\dfrac{H}{\sqrt{S-1}}$
31. Shear friction factor is expressed as (with usual notation)

    (a) $\dfrac{\mu \Sigma V + bC}{\Sigma H}$ (b) $\dfrac{\mu[\Sigma V + bC]}{\Sigma H}$

    (c) $\dfrac{\mu \Sigma V - bC}{\Sigma H}$ (d) $\dfrac{\mu[\Sigma V - bC]}{\Sigma H}$.
32. Maximum height of a dam upto which it may be designated as low dam is given by (with usual notation)

    (a) $H = \dfrac{f}{w\sqrt{(s-c+1)}}$ (b) $H = \dfrac{f}{w(s-c)}$

    (c) $H = \dfrac{f}{w(s-c+1)}$ (d) $H = \dfrac{f}{w\sqrt{s-c}}$
33. Outlets which maintain a constant discharge irrespective of the fluctuations in the water level of supply channel and water course are known as
    (a) Nonmodular outlets (b) Rigid modules
    (c) Semimodular outlets (d) Kennedy's gauge outlets.
34. The freeboard in a dam depends on the height of the wave which in turn depends on
    (a) wind velocity (b) fetch
    (c) depth of water in the reservoir (d) both wind velocity and fetch.
35. According to Creager for a gravity dam the crest width is most economical if it is about
    (a) 14% of the height of the dam (b) 20% of the height of the dam
    (c) 28% of the height of the dam (d) 30% of the height of the dam.

36. The slope of the upstream face of a flat-slab type butress dam
(a) is always less than 35°
(b) is always more than 45°
(c) usually varies from 35° to 45°
(d) is always equal to 45°.

37. According to thin cylinder theory the volume of concrete required for an arch dam would be minimum if the central angle is
(a) 130° 34′
(b) 133° 34′
(c) 136° 34′
(d) 139° 34′.

38. The rate of seepage through an earth dam obtained from a flow net is given by (with usual notation)
(a) $q = Kh\,(N_d \times N_f)$
(b) $q = Kh\,(N_d / N_f)$
(c) $q = kh\left(\sqrt{N_f N_d}\right)$
(d) $q = kh\,(N_f / N_d)$.

39. Due to valley storage the peak discharge of an outflow hydrograph is
(a) increased
(b) reduced
(c) not affected
(d) sometimes increased and sometimes reduced.

40. If for a gravity dam the coefficient of friction μ is 0.60 and the value of the sliding factor as obtained from the calculations is 0.68, the factor of safety against sliding is
(a) 0.882
(b) 1.133
(c) 0.408
(d) 0.639.

41. A detention basin for flood control is the one which is provided with
(a) uncontrolled outlets and spillway
(b) uncontrolled outlets but controlled spillway
(c) controlled outlets but uncontrolled spillway
(d) controlled outlets and spillway.

42. A retarding reservoir for flood control is the one which is provided with
(a) uncontrolled outlets and spillway
(b) uncontrolled outlets but controlled spillway
(c) controlled outlets but uncontrolled spillway
(d) controlled outlets and spillway.

43. The ordinate of a mass curve at any time is equal to
(a) the ordinate of the flood hydrograph at that time
(b) the sum of the ordinates of the flood hydrograph upto that time
(c) the area under the flood hydrograph upto that time
(d) the ordinate of the demand curve at that time.

44. No storage is required if the
(a) slope of the demand curve is always greater than that of the mass curve
(b) slope of the demand curve is always smaller than that of the mass curve
(c) slope of the demand curve is sometimes smaller and sometimes greater than that of the mass curve
(d) slope of the demand curve is sometimes equal to and sometimes greater than that of the mass curve.

45. For the upstream slope of an earth dam the most critical condition is
    (a) sudden drawdown condition
    (b) steady seepage condition
    (c) neither sudden drawdown nor steady seepage condition
    (d) during construction when the reservoir is allowed to be filled.

46. Seepage endangers the stability of an earth dam built on pervious foundation because of piping which depends on
    (a) height of the dam
    (b) quantity of seepage flow
    (c) value of the exit gradient
    (d) total storage capacity of the reservoir.

47. According to Khosla's theory for a hydraulic structure built on pervious foundation a cutoff is quite essential at
    (a) the upstream end
    (b) the mid-section
    (c) some intermediate section
    (d) the downstream end.

48. The undersluices in a diversion headworks are provided with a crest level
    (a) same as the rest of the weir
    (b) lower than the rest of the weir
    (c) higher than the rest of the weir
    (d) same as the crest of the head regulator.

49. According to Khosla's theory exit gradient is given by (with usual notation)
    (a) $G_E = \frac{H}{d}\frac{1}{\pi\sqrt{\lambda}}$
    (b) $G_E = \frac{H}{d}\pi\sqrt{\lambda}$
    (c) $G_E = \frac{d}{H}\frac{1}{\pi\sqrt{\lambda}}$
    (d) $G_E = \frac{H}{d}\pi\sqrt{\lambda}$

50. If a river in alluvial plain has a dominant discharge of 1600 cumec, the waterway for a bridge on this river should be approximately
    (a) 250 m
    (b) 240 m
    (c) 190 m
    (d) 180 m.

51. Gibb's module is a type of an outlet which
    (a) ensures a constant discharge irrespective of the fluctuations in the water level of supply channel and water course
    (b) ensures a constant discharge irrespective of the fluctuations in the water level of water course
    (c) does not ensure constant discharge irrespective of the fluctuations in water level of supply channel and water course
    (d) ensures a constant discharge irrespective of the fluctuations in the water level of supply channel.

52. If a river in alluvial plain has a dominant discharge of 2500 cumec, the meander length M_l is
    (a) 2580.5 m
    (b) 2680.5 m
    (c) 2780.5 m
    (d) 2880.5 m.

53. A land is known as waterlogged when
    (a) gravity drainage has ceased
    (b) permanent wilting point is reached

(c) the soil becomes completely saturated
(d) capillary fringe reaches the root zone of the plants.

54. If the regime velocity of flow in a Lacey's channel having hydraulic mean radius 1 m is 0.4 m/s, the silt factor is
(a) 0.4 (b) 0.5
(c) 0.6 (d) 0.7

55. For 1 m depth of flow in a canal with CVR equal to 0.90, according to Kennedy's theory the velocity of flow will be
(a) 0.395 m/s (b) 0.495 m/s
(c) 0.595 m/s (d) 0.695 m/s.

56. Lining of irrigation channels
(a) increases the possibility of area to be waterlogged
(b) does not affect the phenomenon of waterlogging
(c) is an effective antiwaterlogging measure
(d) increases the maintenance cost of the channels.

57. Height of capillary fringe
(a) is more in fine grained soils
(b) is more in coarse grained soils
(c) does not depend upon soil characteristics
(d) depends on the quantity of water in the soil.

58. Closed drains are provided to drain
(a) only surface water (b) only ground water
(c) both surface water and ground water (d) neither surface water nor ground water.

59. In Mitra's method of design of channel transitions the depth of flow is assumed to be
(a) varying in the transitions and the trough
(b) varying only in the contraction transition
(c) varying only in the expansion transition
(d) constant in the transitions as well as in the trough.

60. The upstream face of the earth dam is considered as
(a) equipotential line
(b) streamline
(c) neither streamline nor equipotential line
(d) sometimes equipotential line and sometimes streamline.

61. Syphon aqueduct is a cross drainage work provided to carry canal over a natural drain when
(a) canal bed is well above the H.F.L., of the natural drain
(b) canal bed is at the same level as the bed of the natural drain
(c) canal bed is below the H.F.L., of the natural drain
(d) canal bed is below the bed of the natural drain.

62. An outlet is a hydraulic structure which
(a) is a device at the head of water course connecting it with distributing channel

(b) regulates the supply of water from river to canal

(c) control the entry of silt into distributary

(d) regulates the supply of water from the main canal to a branch canal or distributary.

63. For safety against uplift in a structure where a hydraulic jump is forming on a sloping floor, the conditions to be considered are

(a) no flow with maximum ponding upstream

(b) with maximum discharge passing over the work

(c) condition (a) as well as condition (b) for the jump trough

(d) with any discharge less than maximum discharge passing over the work.

64. Falls are required in canals when the general ground slope is

(a) smaller than the slope of the canal

(b) greater than the slope of the canal

(c) equal to the slope of the canal

(d) sometimes equal to and sometimes smaller than the slope of the canal.

65. A canal is generally lined

(a) when it is in cutting

(b) when it is in embankment

(c) when it is partly in cutting and partly in embankment

(d) when bed and banks of the canal consist of highly impervious soil.

66. Silt excluder is provided

(a) in the canal on the downstream of the head regulator

(b) in the river on the downstream of the weir

(c) in the river adjacent to the head regulator

(d) in the river far off from the weir on the upstream side.

67. A run-of-river plant is the one which

(a) has a limited pondage (b) has no pondage at all

(c) is a pumped-storage plant (d) is a high head plant.

68. If a stream having a discharge of 500 cumec has a fall of 75 m, the power potential of the stream is

(a) 2.68×10^5 kW (b) 4.68×10^5 kW

(a) 5.68×10^5 kW (b) 3.68×10^5 kW

69. A pumped storage plant is the one which

(a) is used as base load plant (b) is same as run-of-river plant

(c) is used as peak load plant (d) is a high head plant.

70. Silt ejector is provided

(a) in the canal on the downstream of the head regulator

(b) in the river on the downstream of the weir

(c) in the river adjacent to the head regulator

(d) in the river far off from the weir on the upstream side.

71. An outlet is said to be proportional when flexibility is
(a) less than one
(b) greater than one
(c) equal to one
(d) sometimes less and sometimes greater than one.

72. The intermediate sheet pile is ineffective if it is
(a) larger in length than the outer ones
(b) smaller in length than the outer ones
(c) equal in length to the outer ones
(d) far off from the outer ones.

73. Lacey's equations can be used for the design of
(a) unlined channels only
(b) lined channels only
(c) both lined and unlined channels
(d) neither lined nor unlined channels.

74. An attracting groyne is the one which is
(a) inclined upstream
(b) inclined downstream
(c) normal to the bank
(d) same as a repelling groyne.

75. A repelling groyne is the one which is
(a) inclined upstream
(b) inclined downstream
(c) normal to the bank
(d) same as attracting groyne.

76. An aqueduct is a cross drainage work provided to carry canal over a natural drain when
(a) canal bed is at the same level as the bed of the natural drain
(b) canal bed is below the H.F.L., of the natural drain
(c) canal bed is well above the H.F.L., of the natural drain
(d) canal bed is below the bed of the natural drain.

77. A super passage is a cross drainage work provided to carry a natural drain over a canal when
(a) bed of the natural drain is at the same level as the canal bed
(b) bed of the natural drain is well above the canal F.S.L.
(c) bed of the natural drain is below the canal F.S.L.
(d) bed of the natural drain is below the canal bed.

78. If a natural stream carries a maximum discharge of 1.408×10^4 cumec and the mean particle size of the bed material is 1 mm, Lacey's regime scour depth is
(a) 9.4 m
(b) 8.4 m
(c) 10.4 m
(d) 7.4 m.

79. If the yearly delta for sugarcane is 2.5 m, the quantity of water needed for a 1 000 hectare sugarcane farm with water application efficiency 80% is
(a) 2 000 ha-m
(b) 2 250 ha-m
(c) 2 500 ha-m
(d) 3 125 ha-m.

80. The capacity of undersluices in a river barrage is usually equal to
(a) two times the maximum discharge of the offtaking canal
(b) maximum winter discharge
(c) 10 to 15% of the maximum flood discharge
(d) maximum of (a), (b) and (c).

81. Vertical drop falls are satisfactory for drops upto

(a) 0.5 m (b) 1.5 m

(c) 3.5 m (d) 5.0 m

82. The cistern in a vertical drop fall is provided with

(a) four rows of friction blocks (b) three rows of friction blocks

(c) two rows of friction blocks (d) one row of friction blocks.

83. The cistern of a flumed glacis falls is provided with

(a) four rows of friction blocks (b) three rows of friction blocks

(c) two rows of friction blocks (d) one row of friction blocks.

84. River training works are seldom required in

(a) rocky stage of a river (b) boulder stage of a river

(c) trough stage of a river (d) deltaic stage of a river.

85. The relation between duty D in hectare/cumec, depth of water Δ in metres and base period B in days is given by

(a) $\Delta = \dfrac{8.64B}{D}$ (b) $\Delta = \dfrac{8.64D}{B}$

(c) $\Delta = 8.64$ BD (d) $\Delta = \dfrac{8.64}{BD}$

86. The outlet factor is the duty at the head of the

(a) main canal (b) branch canal

(c) distributary (d) field channel or water course.

87. The ratio of the quantity of water stored in the root zone of the crops to the quantity of water actually delivered in the field is known as

(a) water conveyance efficiency (b) water application efficiency

(c) water use efficiency (d) water storage efficiency.

88. The water which may be utilized by the plants is available in the soil mainly in the form of

(a) gravitational water (b) capillary water

(c) hygroscopic water (d) chemical water.

89. The amount of irrigation water required to meet the evapotranspiration needs of the crop during its full growth is

(a) effective rainfall (b) consumptive use

(c) consumptive irrigation requirement (d) net irrigation requirement.

90. With the increase in the quantity of water supplied, the yield of most crops

(a) increases continuously

(b) decreases continuously

(c) increases upto a certain limit and then decreases

(d) increases upto a certain limit and then becomes constant.

91. The area between the isohyets 450 mm and 550 mm is 100 sq.km and between 550 mm and 650 mm is 150 sq. km. The average depth of annual precipitation over the above basin of 250 sq. km will be

(a) 500 mm (b) 530 mm

(c) 560 mm (d) 600 mm.

92. The best unit duration of storm for a unit hydrograph is

(a) 1 hour (b) one-fourth of basin lag

(c) one-half of basin lag (d) equal to basin lag.

93. The normal annual precipitation at stations *X, A, B* and *C* are 700 mm, 1000 mm, 900 mm and 800 mm respectively. If the storm precipitations at three stations *A, B* and *C* were 100 mm, 90 mm and 80 mm respectively, then the storm precipitation for station *X* will be

(a) 70 mm (b) 80 mm

(c) 90 mm (d) 100 mm

94. According to Dicken's formula for estimating floods the peak discharge is proportional to

(a) A (b) $A^{1/2}$

(c) $A^{2/3}$ (d) $A^{3/4}$

where *A* is catchment area in sq. km.

95. According to Ryve's formula for estimating floods, the peak discharge is proportional to

(a) A (b) $A^{1/2}$

(c) $A^{2/3}$ (d) $A^{3/4}$

where *A* is catchment area in sq. km.

96. The relation between probability *P* and recurrence interval *T* is given by

(a) $PT = 1$ (b) $PT^2 = 1$

(c) $PT^{-1} = 1$ (d) $PT^{-2} = 1$

97. If *b* is the aquifer thickness then the relationship between coefficient of permeability *k* and coefficient of transmissibility *T* is given by

(a) $T = \left(\frac{k}{b}\right)$ (b) $T = bk$

(c) $T = \sqrt{bk}$ (d) $T = \left(\frac{k}{b}\right)$

98. The water stored in the reservoir below the minimum pool level is called

(a) useful storage (b) surcharge storage

(c) dead storage (d) valley storage.

99. Trap efficiency of a reservoir is a function of

(a) Capacity/inflow ratio (b) Capacity/outflow ratio

(c) Outflow/inflow ratio (d) Only outflow and inflow.

100. The total capacity of a reservoir is 30 million cubic metre and dead storage is 5 million cubic metre. If the average volume of sediment deposited per year is 0.1 million cubic metre, then the usefulness of the reservoir will start reducing after

(a) 50 years (b) 100 years

(c) 150 years (d) 200 years.

101. Total force due to wave action on a gravity dam acts above the reservoir surface at a height of

(a) $\frac{h_w}{2}$ (b) $\frac{h_w}{3}$

(c) $\frac{3h_w}{8}$ (d) $\frac{2h_w}{3}$

where h_w is the height of wave.

102. When the reservoir is full the maximum compressive stress in a gravity dam is produced

(a) at the heel (b) at the toe

(c) within the middle third of the base (d) at the centre of the base.

103. The maximum permissible eccentricity for no tension at the base of a gravity dam is

(a) $B/2$ (b) $B/3$

(c) $B/4$ (d) $B/6$

where B is the base width.

104. According to Khosla's theory the exit gradient in the absence of a downstream cutoff is

(a) zero (b) infinity

(c) unity (d) very large.

105. A watershed canal

(a) irrigates only on one side

(b) avoids cross drainage works

(c) is generally aligned parallel to the contours of the area

(d) is most suitable for hilly areas.

106. A canal which is aligned at right angles to the contours is called

(a) contour canal (b) water shed canal

(c) side slope canal (d) branch canal.

107. Lacey's regime scour depth R is given by

(a) $1.35\left(\frac{q}{f}\right)^{1/3}$ (b) $1.35\left(\frac{q^2}{f}\right)^{1/3}$

(c) $1.35\left(\frac{q}{f}\right)^{1/6}$ (d) $1.35\left(\frac{q^2}{f}\right)^{1/6}$

where q is discharge per unit width and f is silt factor.

108. According to Lacey the bed slope S for a regime channel is given by

(a) $\frac{f^{4/3}}{3340Q^{1/2}}$ (b) $\frac{f^{2/3}}{3340Q^{1/2}}$

(c) $\dfrac{f^{5/3}}{3340Q^{1/6}}$ (d) $\dfrac{f^{1/3}}{3340Q^{1/6}}$

where f is silt factor and Q is discharge in cumec.

109. Wetted perimeter of a regime channel for a discharge of 64 cumec as per Lacey's theory will be
(a) 19 m (b) 38 m
(c) 57 m (d) 76 m

110. The sensitivity of a rigid module is
(a) zero (b) between zero and 1
(c) 1 (d) greater than 1.

111. In a Sarda type fall rectangular crest is used for discharge upto
(a) 6 cumec (b) 10 cumec
(c) 14 cumec (d) 20 cumec.

112. If the R.L's of canal bed level and high flood level of drainage are 212.0 m and 210.0 m respectively, then cross drainage work will be
(a) aqueduct (b) superpassage
(c) syphon (d) syphon aqueduct.

113. Levees are constructed
(a) parallel to the river flow
(b) transverse to the river flow
(c) at some inclination to river flow
(d) sometimes parallel and sometimes transverse to river flow.

114. The ratio of the quantity of water stored in the root zone to the quantity of water needed to bring the moisture content of the soil to the field capacity is known as
(a) water use efficiency (b) water storage efficiency
(c) water application efficiency (d) water conveyance efficiency.

115. The unit hydrograph may be obtained by dividing the ordinates of the direct runoff hydrograph of a storm by :
(a) duration of unit hydrograph (b) storm duration
(c) total runoff volume (d) direct runoff volume.

116. To determine the reservoir storage capacity for a given uniform demand, one of the following data is most useful :
(a) Unit hydrograph of the basin.
(b) Mass curve of the flow volumes for several consecutive years.
(c) Daily flow-duration curve of the stream at the dam site.
(d) Stage-discharge relation for the stream at the reservoir site.

117. Base width of the elementary profile of a gravity dam of 15 m height, built of material of sp. gr. 2.25 and resisting only hydrostatic force is
(a) 15 m (b) 10 m
(c) 7.5 m (d) 5 m

118. An irrigation channel is carrying $2 m^3/s$ to a culturable commanded area of 1500 ha with an intensity of irrigation of 50%. If the base period of the crop is 140 days the duty of channel water in $ha/m^3/s$ is

(a) 300 (b) 325

(c) 350 (d) 375

119. For a basin of area 500 ha, unit hydrograph is of a triangular shape with base period 20 hours. The peak discharge in m^3/hour is

(a) 4000 (b) 5000

(c) 6000 (d) 7000

120. Hydrograph is a plot of

(a) rainfall intensity against time (b) discharge against time

(c) cumulative rainfall against time (d) cumulative runoff against time.

121. *Assertion* (*A*): Duty for rice is more than that for wheat.

Reason (*R*) : Rice requires more water than wheat.

(a) Both *A* and *R* are incorrect (b) Both *A* and *R* are correct

(c) *A* is correct but *R* is incorrect (d) *A* is incorrect but *R* is correct.

122. *Assertion* (A): Duty of water is more at the head of a canal.

Reason (*R*): Large quantity of water is released at the head of the canal.

(a) Both *A* and *R* are correct (b) Both *A* and *R* are incorrect

(c) *A* is correct but *R* is incorrect (d) *A* is incorrect but *R* is correct.

123. *Assertion* (*A*): Mean annual rainfall worked out from 35 years of rainfall record gives almost true mean value.

Reason (*R*): In this case the error in the mean annual rainfall is less than 2%.

(a) *A* is correct but *R* is incorrect (b) *A* is incorrect but *R* is correct

(c) Both *A* and *R* are correct (d) Both *A* and *R* are incorrect.

124. *Assertion* (*A*): For an earth dam the u/s side slope is usually flatter than the d/s side slope.

Reason (*R*): This is done to ensure the safety of the dam under sudden drawdown condition.

(a) Both *A* and *R* are correct (b) Both *A* and *R* are incorrect

(c) A is correct and *R* is incorrect (d) *A* is incorrect and *R* is correct.

125. An 8h storm had 100 mm of rainfall and the resulting runoff was 50 mm. If the ϕ-index remains at the same value, a rainfall of 150 mm in a 12h storm produces a runoff in this catchment of

(a) 45 mm (b) 65 mm

(c) 75 mm (d) 85 mm

126. A 6h unit hydrograph of a catchment of area 126 km^2 is triangular in shape with a base width of 70h. A direct runoff hydrograph (DRH) of this catchment due to 2 cm of effective rainfall in 6 hours, will have a peak flow rate, in m^3/s, of

(a) 12.5 (b) 20.0

(c) 37.5 (d) 60.0

127. For a catchment with an area of 540 km^2, the equilibrium discharge of an *S*-curve obtained by 3h unit hydrograph summations is

(a) 200 m^3/s (b) 300 m^3/s

(c) 400 m^3/s (d) 500 m^3/s

128. In a gravity dam a maximum wave height of 1.2 m along with a wind set up of 0.2 m is expected at the maximum reservoir level. The minimum freeboard to be provided in this case is

(a) 1.4 m (b) 1.6 m

(c) 1.8 m (d) 2.0 m

129. At the base of a gravity dam section, the vertical stress at the toe was found to be 3 MPa. If the downstream face of the dam has a slope of 0.707 horizontal to 1 vertical and if there is no tailwater, the major principal stress at the toe in MPa is

(a) 3.00 (b) 4.50

(c) 6.00 (d) 7.50

130. Provision of a filter and a rock toe in an earthen dam is done to

(a) prevent the piping action in the dam section

(b) collect and drain out the seeping water

(c) reduce the seepage quantity

(d) have the resultant in the middle third of the base.

131. In a saddle-siphon spillway, an air vent is provided at the level of the full reservoir surface to

(a) break the siphonic action at that level

(b) initiate the siphonic action at that level

(c) prevent cavitation

(d) maintain ventilation inside the siphon.

132. In a solid-roller bucket type energy dissipator, the energy dissipation is

(a) due to the formation of a hydraulic jump

(b) partly due to the interaction of the free jet with air, and partly due to impact on the downstream channel bed

(c) due to interaction of two complementary rollers

(d) partly due to lateral spreading of the jet, and partly due to interaction of two rollers.

133. A salt concentration of 3200 ppm in a sample of water is equivalent to the electrical conductivity (EC) in milliomhos /cm of value

(a) 0.032 (b) 3.2

(c) 5.0 (d) 20.0

134. Sodium chloride salt concentration of 936 ppm, when expressed in milliequivalent per litre, has a value of

(a) 26.0 (b) 16.0

(c) 11.7 (d) 1.46

135. An irrigation water has the ion concentrations of Na^+, Ca^{++} and Mg^{++} as 20, 10 and 5 milliequivalents per litre, respectively. The SAR of this water is

(a) 3.65 (b) 5.16

(c) 2.67 (d) 7.30

136. A soil has an ESP = 20% and EC of the soil extract is 6 mmhos/cm, and the pH is 8.5. This soil can be classified as

(a) black alkali soil (b) saline-alkali soil
(c) saline soil (d) non-alkaline soil

137. A soil has an ESP = 18% and *EC* of the soil extract is 3 mmhos/cm. If the pH of the soil is 9.0, it can be classified as

(a) alkali soil (b) saline soil
(c) saline-alkali soil (d) acidic soil

138. A soil has an ESP = 12%, pH = 7.5 and *EC* = 6 mmhos/cm. This soil is classified as

(a) black alkali soil (b) saline-alkali soil
(c) white alkali soil (d) acidic soil

139. Alkali soils are reclaimed by

(a) leaching only (b) addition of gypsum and leaching
(c) addition of gypsum only (d) provision of drainage

140. Acidic soils are reclaimed by

(a) using limestone as a soil amendment (b) using gypsum as a soil amendment
(c) leaching of the soil (d) provision of drainage

141. For a meandering alluvial river in flood plain, the meander width is K times the width of the river, where the value of K is usually about

(a) 8 (b) 17
(c) 27 (d) 54

142. The meander length of an alluvial river is a function of the dominant discharge Q and it varies as

(a) Q (b) $Q^{-0.5}$
(c) $Q^{1/6}$ (d) $Q^{0.5}$

143. Meander ratio of an alluvial river in flood plain has a value of about

(a) 0.5 (b) 1.0
(c) 3.0 (d) 6.0

144. The meander ratio is the ratio of

(a) the length of the river channel to the axial length of the river
(b) the main curved channel length to the width of the meander belt
(c) the width of the meander belt to the meander length
(d) length of the meandering river to the number of loops in a stretch.

145. In a meandering river the deepest part of the river in a reach is found at

(a) the outer bank near the apex of the curve
(b) the crossings
(c) the mid-point in a cross-section
(d) the inner bank of the curve.

146. A guide bank is to be built for an alluvial river having a maximum flood discharge of 1000 m^3/s. The thickness of stone pitching required on the river side of this bund is

(a) 0.3 m (b) 0.6 m
(c) 0.9 m (d) 1.2 m

147. A bridge across an alluvial river has a length of 400 m between the abutments. The lengths of guide bank upstream and downstream of the bridge axis for this case would respectively, be
(a) 500 m and 100 m (b) 100 m and 500 m
(c) 800 m and 200 m (d) 200m and 800 m

148. In a launching apron, the total quantity of stone required for final position is initially laid in a horizontal length of
(a) 1.5 times the design depth of scour below HFL
(b) 1.5 times the design depth of scour below the floor level
(c) 2.0 times the design depth of scour below HFL
(d) 2.0 times the design depth of scour below the floor level.

149. A launching apron has a horizontal apron length of 1.5 times the depth of possible scour. If the apron is expected to be launched ultimately to a slope of 2 horizontal: 1 vertical with a thickness of 0.75 m, the average thickness of stone to be provided in the launching apron is
(a) 0.75 m (b) 1.00 m
(c) 1.12 m (d) 1.25 m

150. The flexibility of an outlet passing a discharge q from a distributary of discharge Q is defined as
(a) $(dq/q) / (dQ/Q)$ (b) $(dQ/Q) / (dq/q)$
(c) $(dq/q) . (dQ/Q)$ (d) $(dq/dQ) . (q/Q)$

151. An irrigation outlet is proportional when its
(a) sensitivity = 1 (b) flexibility = 1
(c) channel index = 1 (d) outlet index = 1

152. The sensitivity of a rigid module is
(a) zero (b) < 1 and > 0
(c) 1 (d) > 1

153. By considering the channel index as 5/3, the setting of an orifice type irrigation outlet to have proportionality is
(a) 0.15 (b) 0.30
(c) 0.67 (d) 0.90

154. The ratio of the rate of change of discharge in an irrigation outlet to the rate of change of discharge of the distributary is known as
(a) sensitivity (b) setting
(c) flexibility (d) minimum modular head ratio.

155. Gibbs module is
(a) a flexible module (b) an adjustable proportional module
(c) rigid module (d) non-modular outlet.

156. A submerged pipe outlet is
(a) a semi-modular outlet (b) a rigid module
(c) an adjustable proportional module (d) a non-modular outlet

157. The adjustable proportional module (APM) is
(a) an open flume outlet
(b) a modular outlet
(c) a non-modular outlet
(d) a semi-modular outlet.

158. The outlet factor is
(a) the duty based on the discharge of the outlet
(b) the ratio of mean discharge of the outlet to the full supply discharge of the distributary
(c) the inverse of the duty of the outlet
(d) the ratio of the number of days the outlet is actually open to the number of days it was supposed to be open as per design.

159. The open flume outlet is an example of
(a) a semi-modular outlet
(b) a non-modular outlet
(c) a rigid module
(d) an adjustable proportional module.

160. In a tail cluster
(a) the depth of water over the crests in all the outlets is kept the same
(b) the widths of all the outlets are kept the same
(c) the discharge in all the outlets is the same
(d) adjustable proportional modules are used at all the outlets.

161. In designing hydraulic structures in alluvial rivers, the equation that is used to calculate the normal depth of scour R for a discharge intensity of q m^3/s/m is
(a) $R = 4.75\ (q)^{1/2}$
(b) $R = 1.35\ (q/f)^{1/3}$
(c) $R = 1.35\ (q^2/f)^{1/3}$
(d) $R = 1.35\ (q/f^2)^{1/3}$

162. A cross regulator is built
(a) below the head of an offtaking canal at a suitable downstream location
(b) in a main canal downstream of an offtake canal or escape
(c) along the axis of the head of the offtaking canal
(d) at the upstream of an offtake canal

163. Cross regulators are utilized to
(a) increase the water level upstream of the structure when a main canal is running with low supplies
(b) control the sediment entering an offtake canal
(c) dispose excess of supplies from the main canal into a drainage
(d) control the silt load in the main canal.

164. In a Sarda type canal fall, the energy dissipation is by
(a) a free hydraulic jump
(b) a water cushion
(c) friction blocks only
(d) a hydraulic jump assisted by baffles.

165. In straight glacis type canal fall the energy dissipation is essentially due to
(a) water cushion
(b) baffle and friction blocks only
(c) fluming action
(d) hydraulic jump assisted by baffle blocks.

166. In a Sarda type canal fall, maximum uplift pressure on the floor is experienced when the upstream water surface is
   (a) at the crest level of the fall and the downstream water level is at the brim of the cistern
   (b) at the crest level of the fall and there is no water in the downstream channel
   (c) at the bed level of the upstream channel and the downstream water is at the brim of the cistern
   (d) at the full supply level and there is no flow in the downstream channel.

167. A trapezoidal notch fall maintains the normal depth in the upstream channel at
   (a) one discharge only (b) two discharges only
   (c) all discharges (d) no discharge at all.

168. A cross drainage work is termed as a *syphon* if it carries the canal
   (a) above the drainage with the drainage flowing under pressure
   (b) below the drainage with the canal flowing under pressure
   (c) below the drainage with the canal flowing at atmospheric pressure under the works
   (d) above the drainage with the drainage flowing at atmospheric pressure under the works.

169. A canal carries a discharge of 20 m^3/s at a depth of 1.5 m and has its bed 3.5 m higher than the bed of a drainage it has to cross. If the drainage has a high flood depth of 2.5 m, the type of cross drainage works appropriate to this site is
   (a) aqueduct (b) syphon aqueduct
   (c) syphon (d) super passage

170. The following data are available for a cross drainage project:

Items	*Canal*	*Drainage*
FSL/HFL	105.00 m	104.00 m
Bed level	100.00 m	102.00 m
Discharge	80 m^3/s	15 m^3/s

The most appropriate cross drainage work for this situation is
   (a) aqueduct (b) syphon aqueduct
   (c) syphon (d) super passage

171. A canal with a bed width of 15 m has a discharge of 35 m^3/s at a full supply depth of 2.0 m. It has to cross a 50 m wide drainage whose bed level is 1.9 m above the bed of the channel. At high flood, the flow depth in the drainage is 1.5 m. A suitable cross drainage work at this location would be
   (a) an aqueduct (b) a syphon aqueduct
   (c) a syphon (d) a super passage

172. A level crossing type of cross drainage work consists of
   (a) one regulator only (b) two weirs only
   (c) one weir and two regulators (d) two weirs and two regulators.

173. The structure which serves the purpose of a safety valve for a canal is
   (a) head regulator (b) cross regulator
   (c) canal escape (d) canal fall

174. Lacey in his regime equations has adjusted his silt factor to be unity for the standard silt used by Kennedy. It is known that Kennedy used Bari-Doab alluvium as the standard silt in his equation. Then the median size, in mm, of Bari-Doab alluvium and the critical velocity ratio for this silt are, respectively

(a) 0.32 and 1.00 (b) 0.57 an 0.90

(c) 1.00 and 1.00 (d) 1.00 and 1.76

175. An alluvium with a median size of 0.32 mm has Lacey's silt factor f of value

(a) 0.57 (b) 0.80

(c) 1.00 (d) 1.76

176. A regime canal has a discharge of 100 m^3/s. It will have a perimeter of

(a) 4.75 m (b) 10 m

(c) 47.5 m (d) 475 m

177. A regime canal has a width of 22.2 m and depth of flow of 1.7 m. The discharge in the canal is about

(a) 68 m^3/s (b) 30 m^3/s

(c) 7 m^3/s (d) 3 m^3/s

178. When an alluvial channel attains its regime it will have side slopes of value

(a) equal to the angle of repose of the alluvium

(b) equal to the angle of repose of the alluvium under water

(c) 0.5 vertical to 1 horizontal

(d) 0.5 horizontal to 1 vertical.

179. In a Lacey's regime channel the discharge is 100 m^3/s and the longitudinal slope is 0.0002. The silt factor for this channel is about

(a) 0.72 (b) 1.24

(c) 1.76 (d) 1.83

180. In a wide alluvial river the normal scour depth by Lacey's equation is 7.40 m for a discharge intensity of 10 m^3/s/m width. The normal scour depth in that river for a discharge intensity of 20 m^3/s/m width is

(a) 11.75 (b) 17.63

(c) 23.50 (d) 26.44

181. The minimum freeboard recommended for lined canals carrying a discharge greater than 10 m^3/s, is

(a) 0.30 m (b) 0.50 m

(c) 0.60 m (d) 0.75 m

182. The generally adopted value for the transmission loss from unlined canals in alluvial soils in the country is

(a) 1.8 cumec per million m^2 of wetted perimeter

(b) 2.0 cumec per million m^2 of wetted perimeter

(c) 2.5 cumec per million m^2 of wetted perimeter

(d) 3.0 cumec per million m^2 of wetted perimeter

183. If at the time of construction, a berm width of half the depth of cutting is left for a slope of : 1 in filling and 1 : 1 in cutting in a regime channel of flow depth D, the final berm width will be
(a) $D/4$ (b) $D/2$
(c) D (d) $3D/2$

184. A counter berm is provided
(a) at the water side of the slope in filling to provide protection against rain water
(b) at the outer side of the filling to provide adequate cover over the phreatic line
(c) in the cutting to provide quick silting of the sides to form a berm
(d) in the water side at the ground level to make bank line and bed line parallel.

185. The generally adopted minimum radius for an unlined alluvial canal carrying 150 m^3/s of discharge is
(a) 1500 m (b) 900 m
(c) 600 m (d) 300 m

186. A *bed bar* in an earthen irrigation channel is
(a) a borrow pit dug in the bed of the channel
(b) a masonry block in the channel bed to serve as a permanent mark of reference
(c) a deposit of sand in the channel bed
(d) a masonry wall on the bed of the channel for the purpose of silt ejection.

187. At a certain point in the floor of a weir the uplift pressure head due to seepage is 3.3 m. If the relative density of concrete is 2.5, the minimum thickness of floor required at this point to counteract the uplift pressure is
(a) 2.20 m (b) 2.00 m
(c) 1.80 m (d) 1.5 m

188. A barrage on a major river has been designed for a flood discharge of 7000 m^3/s and has been provided with a waterway of 358 m. The looseness factor of this barrage is
(a) 0.8 (b) 0.9
(c) 1.1 (d) 1.2

189. A weir consists of 32 m long horizontal floor with two sheet piles of 6 m and 8 m depth at the upstream and downstream ends of the floor respectively. Under an impounded depth of 4 m above the floor and with no tail water, the uplift pressure head at the mid-point of the floor by Bligh's creep theory is
(a) 2.50 m (b) 2.13 m
(c) 2.00 m (d) 1.87 m

190. A weir consists of 32 m long horizontal floor with two sheet piles of 6 m and 8 m depth at the upstream and downstream ends of the floor respectively. Under an impounded depth of 4 m above the floor and with no tail water the uplift pressure head at the mid-point of the floor by Lane's weighted creep theory is
(a) 2.50 m (b) 2.21 m
(c) 2.00 m (d) 1.79 m

191. A weir across an alluvial river has a horizontal floor of length 60 m and retains 6 m of water under full pond condition. If the downstream sheet pile is driven to a depth of 6 m below the average bed level, the exit gradient for this case is

(a) 0.100 (b) 0.135

(c) 0.200 (d) 0.318

192. A non-cohesive soil has a porosity of 30% and the relative density of soil particles is 2.70. The value of critical exit gradient for this soil is

(a) 0.81 (b) 1.00

(c) 1.19 (d) 1.89

193. The factor of safety normally adopted for the critical exit gradient in fine sand is

(a) 2 to 3 (b) 3 to 4

(c) 4 to 5 (d) 6 to 7

194. A graded inverted filter, used in a diversion structure, consists of

(a) graded sand with coarse part in contact with the foundation and finest part on the top

(b) graded sand with fine part in contact with the foundation material and coarsest part on the top

(c) a blanket of uniform size sand set inclined to the direction of seepage and loaded by masonry blocks

(d) a sand blanket covered by graded heavy stones on top.

195. Surcharge storage in a reservoir is the volume of storage between

(a) minimum pool level and normal pool level

(b) minimum pool level and average pool level

(c) normal pool level and maximum pool level

(d) dead storage level and maximum reservoir level.

196. The dead storage capacity of a storage reservoir is usually decided

(a) on the basis of expected life and sedimentation characteristics of the proposed reservoir

(b) by using the mass curve of inflow and demand

(c) by using the optimization techniques

(d) by economic analysis to get the minimization of the cost of the reservoir.

197. In a reservoir having a fetch of 40 km the height of wave in metres generated by a wind of 100 km/h could be expected to be about

(a) 1.5 (b) 2.0

(c) 2.5 (d) 3.0

198. An earthquake acceleration of 0.1 g acting vertically downward causes in a gravity dam

(a) an increase in the weight of dam by 10%

(b) a reduction in the unit weight of concrete only by 10%

(c) a decrease in the unit weights of concrete and water by 10%

(d) increase in the uplift pressure by 10%

199. The centre of pressure of wave pressure due to wave of height h_w acting on a gravity dam will be at a height above the maximum still water level of

(a) $\frac{h_w}{3}$ (b) $\frac{3h_w}{8}$

(c) $\frac{h_w}{2}$ (d) $\frac{2h_w}{3}$

200. In gravity dam design the horizontal 'silt and water' pressure is assumed as equivatent to that of a fluid with a mass density, in kg/m^3 of

(a) 360 (b) 1000

(c) 1360 (d) 1925

201. In gravity dam design the vertical 'silt and water' pressure is determined by assuming that the silt and water together have a density, in kg/m^3, of

(a) 360 (b) 1000

(c) 1360 (d) 1925

202. In gravity dam design ice pressure is normally accounted for by providing a pressure of value X kN/m^2 applied to the face of the dam over the anticipated area of contact of contact of ice with the face of the dam. The value of X is

(a) 250 (b) 25

(c) 2.5 (d) 0.5

203. In a gravity dam a maximum wave height of 1.2 m along with a wind set-up of 0.3 m is expected at the maximum water level. The minimum freeboard to be provided in this case is

(a) 1.5 m (b) 1.6 m

(c) 1.8 m (d) 1.9 m

204. The axis of a dam is the

(a) line joining the mid-point of the base

(b) centre line of the top width of the dam

(c) horizontal trace of the downstream edge of the top of the dam

(d) horizontal trace of the upstream edge of the top of the dam

205. At the base of a gravity dam section, the vertical stress at the toe is found to be 2×10^6 N/m^2. If the downstream face of the dam has a slope of 0.707 horizontal to 1 vertical and if there is no tail water, the maximum principal stress at the toe in N/m^2 is

(a) 3.00×10^6 (b) 2.83×10^6

(c) 2.50×10^6 (d) 2×10^6

206. Uplift pressure is considered in the analysis of gravity dams

(a) only when there is a drainage gallery in the dam

(b) only when there is tail water

(c) only when the reservoir is empty

(d) in all situations having water in the reservoir.

207. When a drainage gallery with its drainage pipe system is provided in a gravity dam, the uplift pressure is reduced

(a) at all levels below the drainage gallery

(b) at the foundation level only

(c) at all levels below the upstream level

(d) only when there is no tail water.

208. In a gravity dam a grout curtain is usually provided
   (a) near the toe to reduce the exit gradient
   (b) in the middle of the base to reduce seepage
   (c) near the heel to reduce uplift and seepage
   (d) at the middle third of the base nearest to the toe to reduce the reaction pressure.

209. Which one of the following defines aridity index (AI)?

(a) $AI = \frac{PET - AET}{PET}$ (b) $AI = \frac{AET - PET}{PET} \times 100$

(c) $AI = \frac{AET}{PET} \times 100$ (d) $AI = \frac{PET}{AET} \times 100$

where AET = Actual Evapotranspiration and
PET = Potential Evapotranspiration

210. Which one of the following statements is correct?
In snyder's method of synthetic unit hydrograph development, basin lag is taken as
   (a) the time interval between centroid of the rainfall excess and surface runoff
   (b) the time interval from mid point of the unit rainfall excess to the peak of the unit hydrograph
   (c) independent of rainfall duration
   (d) independent of catchment characteristics

211. A strom with 14 cm precipitation produces a direct runoff of 8 cm. The time distribution of the storm is as shown in the Table below:

Time from start (h)	*Incremental rain fall (cm/h)*
1	1.0
2	2.0
3	2.8
4	3.3
5	2.5
6	1.8
7	0.6

What is the value of ϕ-index of the storm?
   (a) 0.5 cm/h (b) 0.7 cm/h
   (c) 0.8 cm/h (d) 0.9 cm/h

212. Which one of the following is the correct sequence in the increasing order of the Froude number of flow assumed by the bed form of an alluvial stream with movable bed material?
   (a) Ripple— Plane bed — Dune — Plane bed— Antidune
   (b) Dune— Ripple— Plane bed — Antidune— Plane bed
   (c) Plane bed— Ripple–Dune— Plane bed — Anti dune
   (d) Plane bed— Ripple— Antidune—Dune— Plane bed

213. Match List-I with List-II and select the correct answer using the Codes given below the lists:

List-I	*List-II*
A. Rising limb of a hydrograph	1. Depends on intensity of rainfall
B. Falling limb of a hydrograph	2. Function of total channel length
C. Peak rate of flow	3. Function of catchment slope
D. Drainage density	4. Function of storage characteristics

Codes :

	A	B	C	D
(a)	3	4	1	2
(b)	1	4	3	2
(c)	3	2	1	4
(d)	1	2	3	4

214. A bridge has an expected life of 50 years and is designed for a flood magnitude of return period 100 years. What is the risk associated with this hydrologic design?

(a) $1 - (0.99)^{50}$ (b) $(0.5)^{50}$

(c) $(0.99)^{50}$ (d) $(0.99)^{100}$

215. Match List-I with List-II and select the correct answer using the codes given below the lists:

List-I	*List-II*
A. Unit hydrograph	1. Design flood
B. Synthetic unit hydrograph	2. Permeability
C. Darcy's Law	3. Ungauged basin
D. Rational method	4. 1 cm runoff

Codes :

	A	B	C	D
(a)	2	3	4	1
(b)	2	1	4	3
(c)	4	3	2	1
(d)	4	1	2	3

216. For a culturable commanded area of 1000 hectare with intensity of irrigation 50% the duty on field for a certain crop is 2000 hectares/cumec. What is the discharge required at head of water course with 25% losses of water?

(a) $\frac{3}{16}$ cumec (b) $\frac{1}{4}$ cumec

(c) $\frac{1}{3}$ cumec (d) $\frac{1}{2}$ cumec

217. What is the moisture depth available for evapotranspiration in root zone of 1 m depth of soil, if dry weight of soil is 1.5 gm/cc, wilting point is 10%?

(a) 450 mm (b) 300 mm

(c) 200 mm (d) 150 mm

218. A catchment area of 90 hectares has a runoff coefficient of 0.4. A storm of duration large than the time of concentration of the catchment and of intensity 4.5 cm/h creates a peak discharge rate of

(a) 11.3 m^3/s (b) 0.45 m^3/s

(c) 450 m^3/s (d) 4.5 m^3/s

219. Match List-I with List-II and select the correct answer using the codes given below the lists:

List-I	*List-II*
A. Standard Project Flood (SPF)	1. Includes catastrophic floods
B. maximum Probable Flood (MPF)	2. Includes floods of severs codition
C. Design flood	3. Peak flow obtained from observed data
D. Maximum flood	4. Flood of desired recurrence interval

Codes :

	A	B	C	D
(a)	2	1	4	3
(b)	1	2	3	4
(c)	2	1	3	4
(d)	1	2	4	3

220. Consider the follwoing statements:

The function of a cutoff in an earth dam is to

1. Reduce uplift pressures on the dam
2. Prevent undermining of foundation
3. Reduce loss of stored water
4. Support the dam

(a) 1 and 3 (b) 2 and 4

(c) 2 and 3 (d) 3 and 4

221. Which one of the following equations represents the downstream curve of the 'Ogee' spillway (where x and y are the coordinates of the crest profile measured from the apex of the crest, and H is the design head)?

(a) $x^{1.85} = 2\ H^{0.85}\ y$ (b) $x = 2\ H^{1.85}\ y^{0.85}$

(c) $x^{0.85} = 2\ H^{1.85}\ y$ (d) $x = 2\ H^{0.85}\ y^{1.85}$

222. For a saddle siphon, the maximum operative head is 6.25 m. The width and the height of the throat of the siphon are 4 m and 2 m respectively. The coeficient of dischrage is 0.90. How many units are required to pass a flood of 300 cumec?

(a) One (b) Two

(c) Three (d) Four

223. Match List-I (Well Hydraulics Parameters) with List-II (Definition) and select the correct answer using the Codes given below the lists:

List-I	*List-II*
A. Specific yield	1. Discharge per unit drawdown of well
B. Safe yield	2. Safe as specific retention

C. Specific capacity — 3. Measure of water that can be removed by pumping

D. Field capacity — 4. Limit of withdrawal from well without depletion of the aquifer

5. Water bearing capacity of aquifer

Codes :

	A	B	C	D
(a)	4	3	2	5
(b)	3	4	1	2
(c)	4	3	1	2
(d)	3	4	2	5

224. Leaching is a process
   (a) By which alkali salts present in the soil are dissolved and drained away
   (b) By which alkali salts in soil come up with water
   (c) Of draining excess water of irrigation
   (d) Which controls water logging

225. The ideal condition for energy dissipation in the design of spillways is the one when the tail water rating curve
   (a) Lies above jump rating curve at all discharges
   (b) Coincides with the jump rating cyrve at all dischrages
   (c) Lies below the jump rating curve at all dischrages
   (d) Lies either above or below the jump rating curve depending upon discharges

226. The worst condition of uplift on the floor of a siphon aqueduct occurs when there is
   (a) High flood flow in the drainage with canal dry
   (b) Full supply flow in the canal with drainage dry
   (c) High flood flow in the drainage with canal running full
   (d) Water level is at drainage bed and canal is dry

227. A check dam is a
   (a) Flood control structure
   (b) Soil conservation structure
   (c) River training structure
   (d) Water storage structure

228. The total number of independent equations that form Lacey's regime theory is
   (a) 2
   (b) 3
   (c) 4
   (d) 6

229. Which of the following pairs are correctly matched?
   1. Silt factor : Average size of silt particles
   2. Silt load : Volume of suspended sediment flowing with water in unit time
   3. Silt charge : Weight of silt per unit volume of water
   4. Silt grade : Gradation between different silt particles

Select the correct answer using the Codes given below:

Codes :

(a) 1 and 4 (b) 3 and 4

(c) 2 and 3 (d) 1 and 2

230. *Assertion (A)* : Canal escape serves as a safety valve for a canal.

Reason (R) : Canal escape discharges the channel due to sudden closure of outlets by the farmers.

(a) Both *A* and *R* are correct (b) Both *A* and *R* are incorrect

(c) *A* is incorrect but *R* is correct (d) *A* is incorrect but *R* is correct

231. The unit hydrograph theory is based on the assumption of

(a) Nonlinear response and time invariance

(b) Linear response and nonlinear time variance

(c) Time invariance and linear response

(d) Nonlinear response and nonlinear time variance

232. Balanced depth of cutting of canal is

(a) half the total depth of a canal

(b) half of full supply depth

(c) the maximum cut that an excavator can take

(d) where volume of cutting is equal to valume of filling

233. At a certain point in the floor of weir, the uplift pressure head due to seepage is 4.5 m. If the specific gravity of concrete is 2.5, the minimum thickness of floor required at this point to counteract the uplift pressure is

(a) 1 m (b) 2 m

(c) 3 m (d) 4 m

234. The Penman's avapo-transpiration equation is based on

(a) water budget method

(b) energy balance method

(c) mass transfer method

(d) energy balance and mass transfer approach

235. *Assertion A*: Sprinkler method of irrigation has a higher water application/use efficiency.

Reason R : Sprinkler system causes less interference in cultivation and other farming operations.

(a) Both *A* and *R* are correct (b) Both *A* and *R* are incorrect

(c) *A* is correct but R is incorrect (d) *A* is incorrect but *R* is correct

236. A constant angle arch dam when compared to a constant radius arch dam utilizes concrete quantity of about

(a) 33% (b) 43%

(c) 73% (d) 143%

237. A canal was designed to supply the irrigation needs of 1000 ha of land growing rice of 140 days base period and having a delta of 130 cm. If the canal water is used to irrigate wheat of base period 119 days and having a delta of 50 cm. The area that can be irrigated is

(a) 452 ha (b) 904 ha
(c) 1105 ha (d) 2210 ha

238. A well development
1. involves reversal of flow through the well screen
2. increases permeability towards the well
3. decreases permeability towards the well
4. is continued till sand/silt free water is pumped out

which of these statements is/are correct?
(a) 1, 3 and 4 (b) 1, 2 and 4
(c) 3 only (d) 1 and 4 variance

239. Modular limit of a canal outlet is the ratio of
(a) rate of change of discharge of out let to that of distributary
(b) water depth above outlet crest to the full supply depth of the channel
(c) water depth above the crest on downstream to that on upstream of outlet
(d) rate of change of discharge of an outlet to the rate of change of water level of the channel.

240. The performance of a well is measured by its
(a) specific capacity (b) specific yield
(c) storage coefficient (d) permeability coefficient

241. A two hour strom hydrograph has 5 units of direct runoff. The two hour unit hydrograph for this storm can be obtained by dividing the ordinates of the storm hydrograph by

(a) 2 (b) $\frac{2}{5}$
(c) 5 (d) $\frac{5}{2}$

242. Consider the following statements:
1. An ephemeral stream is the one which has a base flow contribution
2. Flow characteristics of a stream depend upon rainfall and also the climatic factors which influence evaptranspiration
3. Sequent peak algorithm is used for estimating runoff from rainfall

which of these statements is/are correct?
(a) 1, 2 and 3 (b) 1 and 3
(c) 2 and 3 (d) 2 alone

243. Match List-I with List-II and select the correct answer using the codes given below the lists:

List-I	*List-II*
A. ϕ-index	1. Used for meansurement of evapotran spiration for given vegetation
B. Lysimeter	2. Used for flow measurement
C. Dilution technique	3. Average rainfall above which the rainfall volume is equal to the runoff volume
D. Snyder's equation	4. Relates the basin log to the basin characteristics

Codes :

	A	B	C	D
(a)	3	1	2	4
(b)	4	2	1	3
(c)	3	2	1	4
(d)	4	1	2	3

244. Which one of the following statements is correct?

In a river, silt excluder and silt ejector are constructed

(a) at a location after the head regulator and at the head of the canal respectively

(b) at the head of the canal and at a location after the head regulator respectively

(c) at the same location

(d) at specific locations depending upon diverse factors and their locations do not follow a set pattern

245. Ten m^3/s of water is diverted to a 2 hectare field for 4 hours. Soil providing after irrigation showed that 0.3 m of water had been stored in the root zone. Water application efficiency in this case would be

(a) 96% (b) 66.67%

(c) 48% (d) 24%

246. The Lacey's silt factor for a particular alluvium is 2.0. This alluvium would comprise

(a) medium sand of size 0.5 mm (b) coarse sand of size 0.75 mm

(c) medium bajri of size 1.3 mm (d) coarse bajri of size 2.4 mm

247. The following data pertain to a natural drain crossing an irrigation canal:

Item	*Canal data*	*Drainage data*
Flow (m^3/s)	5	500
Bed level (m)	120	116
Depth of flow (m)	0.8	10

Which one of the following types of cross drainage should be recommended in this case?

(a) Aqueduct (b) Syphon aqueduct

(c) Syphon (d) Super passage

248. Gibb's module is a type of outlet which ensures

(a) constant discharge even if the water levels in the supply channel and water course fluctuate

(b) variable discharge as per the need

(c) constant discharge into the water course when the water levels in the supply channel vary

(d) constant discharge for varying water levels in the water course for a given water level in the supply channel.

249. In a barrage on pervious foundation, sheet piles are provided both upstream and downstream of the barrage to reduce uplift pressure and to prevent piping which one of the following statements is true in this regard?

(a) Compared to downstream sheet pile, the upstream sheet pile is more effective in reducing uplift and piping

(b) Compared to upstream sheet pile, the downstream sheet pile is more effective in reducing uplift and piping

(c) Downstream sheet pile is more effective in reducing uplift while the upstream sheet pile is more effective in reducing piping

(d) Upstream sheet pile is more effective in reducing uplift while the downstream sheet pile is more effective in reducing piping.

250. The probability that a 100 year flood is equalled or exceeded, at least once in 100 years is

(a) 99% (b) 64%

(c) 36% (d) 1%

251. The lift of a reservoir is determined by its capacity (*c*), volume of annual inflow into the reservoir (*I*) and concentration of sediment in the incoming flow (C_s). Life will be more if

(a) *C*, *I* and C_s are high (b) *C* and *I* are high but C_s is low

(c) C is high but *I* and C_s are low (d) *C*, *I* and C_s are low

252. Which of the following principles related to a unit hydrograph?

1. The hydrographs of direct runoff due to effective rainfall of equal duration have the same time base.
2. Effective rainfall is not uniformaly distributed within its duration.
3. Effective rainfall is uniformaly distributed throughout the whole area of drainage basin.
4. Hydrograph of direct runoff from a basin due to a given period of effective rainfall reflects the combination of all the physical characteristics of the basin

Select the correct answer using the codes given below:

Codes :

(a) 1, 2 and 3 (b) 1, 2, and 4

(c) 2, 3 and 4 (d) 1, 3, and 4

253. If the base period of a 6 hour unit hydrograph of a basin is 84 hours, then a 12 hour unit hydrograph derived from this 6 hour unit hydrograph will have a base period of

(a) 72 hours (b) 78 hours

(c) 84 hours (d) 90 hours

254. In curved reach of a meandering river, both deposition of sediments and erosion of bank occur. Which one of the following statements is true in this regard?

(a) Deposition of sediment occurs in the inner bank while the outer bank is subjected to erosion

(b) Deposition of sediment occurs in the outer bank while the inner bank is subjected to erosion.

(c) In the direction of flow, the outer bank undergoes erosion first and the sediment is later deposited on the outer bank downstream.

(d) In the direction of flow, the inner bank undergoes erosion first and the sediment is later deposited on the inner bank downstream.

255. Match List-I (Energy dissipation) with List-II (Water level and slope condition) and select the correct answer using the Codes given below the lists:

List-I	*List-II*
A. Roller bucket	1. TWL (tail channel water level) is slightly above JWL jump height water level) and the slope of the channel is mild.
B. Ski-jump bucket	2. TWL is considerably above JWL and the slope of the channel is mild
C. Standing wave basin with depressed floor	3. TWL is slightly below JWL and the slope of the channel as mild
D. Standing wave basin with raised floor	4. TWL is considerably less than JWL and the slope of the channel is steep

Codes :

	A	B	C	D
(a)	4	2	1	3
(b)	2	4	1	3
(c)	2	4	3	1
(d)	4	2	3	1

256. Which one of the following sets is used to control the seepage through the foundations of an earth dam:

(a) Chimney drain, upstream blanket and cut-off trench

(b) Cut-off sheet piles, upstream blanket and cut-off trench

(c) Upstream blanket, cut-off sheet piles and chimney drain

(d) Relief wells, upstream blanket and chimney drain.

257. Which one of the following statements is true?

(a) Khosla's theory can be correcty applied only in alluvium of finite depth.

(b) Length of floor has lesser effect on exit gradient than the depth of piles.

(c) In designing a hydraulic structure on permeable foundation by Khosla's theory, the slope correction is applicable to both the key points E and C of apile fixed at the end of the sloping floor.

(d) According to the Khola's theory the provision of a cut-off at the downstream end of the floor is optional.

258. Consider the following statements:

1. A 100 year flood discharge is greater than a 50 year flood discharge.
2. 90% dependeable flow is greater than 50% dependable flow.
3. Evaporation from salt-water surface is less than that from fresh-water surface.

Which of these statements are correct?

(a) 1 and 2 (b) 2 and 3

(c) 1 and 3 (d) 1, 2 and 3

259. The trap efficiency of a reservoir is a function of

(a) inflow into the reservoir (b) ratio of inflow to storage capacity

(c) ratio of reservoir capacity to inflow (d) reservoir capacity

260. The spacing of tile drains to relieve water logged land is directly proportional to the
(a) depth of drain below the ground surface
(b) depth of impervious strata from the drain
(c) depth of drain below the water level
(d) coefficient of permeability of the soil to be drained.

261. Probability of a 10-year flood to occur at least once in the next 4 years is
(a) 25% (b) 35%
(c) 50% (d) 65%

262. The probability P of a flood of recurrence interval T_r to occur at least once in the next N years is given by
(a) $P = 1 - \left(1 - \frac{1}{T_r}\right)^N$ (b) $P = 1 - \left(1 - \frac{1}{N}\right)^{T_r}$
(c) $P = \left(1 - \frac{1}{T_r}\right)^N$ (d) $P = \left(1 - \frac{1}{N}\right)^{T_r}$

263. The following data were recorded from an irrigated field:
1. Field capacity : 20%
2. Permanent wilting point : 10%
3. Permissible depletion of available soil moisture : 50%
4. Dry unit weight of soil: 14.715 kN/m^3 (1500 kg(f)/m^3)
5. Effective rainfall : 25 mm

Based on these data, the net irrigation requirement per metre depth of soil will be
(a) 75 mm (b) 125 mm
(c) 50 mm (d) 25 mm

264. If a 4-hour unit hydrograph of a certain basin has a peak ordinate of 80 m^3/s the peak ordinate of a 2 hour unit hydrograph for the same basin will be
(a) equal to 80 m^3/s (b) greater than 80 m^3/s
(c) less than 80 m^3/s (d) between 40 and 80 m^3/s

265. If one wants to be 90% sure that the design flood in a a dam project will not occur during the design life period of 100 years, the recurrence interval for such a flood would be
(a) about 90 years (b) equal to 100 years
(c) about 110 years (d) roughly 1000 years

266. A channel designed by Lacey's theory has a mean velocity of one m/s. The silt factor is unity. The hydraulic mean radius will be
(a) 2.5 m (b) 2 m
(c) 1m (d) 0.5 m

267. If D = scour depth measured below high flood level and d = depth of scour below the original bed level, then the width of the launching apron at the toe of the guide bank will generally be equal to

(a) 1.5 D (b) 1.5 d

(c) 2.0 D (d) 2.0 d

268. A 4-hour rainfall in a catchment of 250 km^2 produces rainfall depths of 6.2 cm and 5.0 cm in successive 2 hour unit periods. Assuming the ϕ index of the soil to be 1.2 cm/hour, the run-off volume in ha-m will be

(a) 16 (b) 22

(c) 1600 (d) 2200

269. An isochrone is a line on the basin map

(*a*) joining rain gauge stations with equal rainfall duration

(*b*) joining points having equal standard time

(*c*) connecting points having equal time of travel of surface runoff to the catchment outlet

(*d*) connecting points of equal rainfall depth in a given time interval

270. Match List-I with List-II and select the correct answer using the Codes given below the lists:

List-I	*List-II* (*Flood formula*)
A. Dicken's formula	1. $Q = CA^{0.8}$ (1+ 0.8 log T)
B. Ryve's formula	2. $Q = CA^{3/4}$
C. Indlis formula	3. $Q = CA^{2/3}$
D. Fuller's formula	4. $Q = \dfrac{124A}{\sqrt{A+10.4}}$

Codes :

	A	B	C	D
(a)	1	3	2	4
(b)	2	3	4	1
(c)	3	4	1	2
(d)	4	1	2	3

271. Sinking fund method is useful in

(a) Depreciation (b) Obsolescence

(c) Liquidation (d) Scrap value

272. A construction equipment has a useful life of 5 years after which it is to bereplaced by a new one. If the interest rate is 4%, the sinking fund factor will be

(a) 0.033 (b) 0.184

(c) 0.224 (d) 0.232

273. Capital recovery factor at 15% p.a. discrete compounding for 4 years is 0.35 Rs. 10,000 deposited now at 15% p.a. discrete compounding will yield an amount X at the end of each 4 years period in perpetuity. The value of X is

(a) Rs. 7,500 (b) Rs. $6{,}666\frac{2}{3}$

(c) Rs. 6,000 (d) Rs. 5,000

274. With the usual notations, sinking fund factor is given by

(a) $(1 + i)^n$

(b) $\dfrac{i}{(1+i)^n - 1}$

(c) $\dfrac{i}{(1+i)^n}$

(d) $\dfrac{i(1+i)^n}{(1+i)^n - 1}$

275. Sinking fund is

(a) the fund for rebuilding a structure when its economic life is over

(b) raised to meet maintenance cost

(c) the total sum to be paid to the municipal authorities by the tenants

(d) a part of the money kept in reserve for providing additional structures and structural modifications

276. For supplying water to a city, two alternate design concepts are being considered with 8% p.a. discrete compounding as the criterion. A rock tunnel of indefinitely long life will cost Rs. 30 Lakhs now, and will need Rs. 60,000 per year for operations. The alternative is by a pipe line system, with an expected life of 20 years, at a first cost of Rs. 20 lakhs now, and will need annual operating expenses of Rs. 92,000. In terms of annualised costs the tunnel system will have a relative advantage of (given C.R.F at 8% p.a. discrete compounding for 20 years = 0.10185) nearly

(a) Rs. 14,000

(b) Rs. 11,000

(c) Rs. 8,200

(d) Rs. 4,300

277. The outgo of investment in a project is estimated at 2000 units of money at year zero. The following four alternatives of revenue earning schemes are considered as possibilities each at 6% ROR

1. 600 units of money at the end of the years 1st to 5th.
2. 800 units at the end year 1, reducing at the rate of 100 units each year, ending with 400 units at the end of year 5.
3. 400 units at the end of year 1, increasing at the rate of 100 units each year, ending with 800 units at the end of year 5.
4. 400 units at the end of year 1, end year 2; with 600 units at the end of year 3; and with 800 units at the end of each of year 4 and year 5.

From the investor's point of view, the correct sequence in decreasing order of preference will be

(a) 2, 1, 3, 4

(b) 2, 4, 3, 1

(c) 3, 4, 2, 1

(d) 3, 1, 2, 4

ANSWERS

1. (b) 2. (c) 3. (a) 4. (d) 5. (a) 6. (a)
7. (b) 8. (d) 9. (a) 10. (b) 11. (c) 12. (a)
13. (b) 14. (c) 15. (d) 16. (a) 17. (b) 18. (c)
19. (d) 20. (a) 21. (c) 22. (b) 23. (c) 24. (d)
25. (b) 26. (c) 27. (a) 28. (c) 29. (b) 30. (d)
31. (a) 32. (c) 33. (b) 34. (d) 35. (a) 36. (c)
37. (b) 38. (d) 39. (b) 40. (a) 41. (d) 42. (a)
43. (c) 44. (b) 45. (a) 46. (c) 47. (d) 48. (b)
49. (a) 50. (c) 51. (a) 52. (b) 53. (d) 54. (a)
55. (b) 56. (c) 57. (a) 58. (b) 59. (d) 60. (a)
61. (c) 62. (a) 63. (c) 64. (b) 65. (b) 66. (c)
67. (a) 68. (d) 69. (c) 70. (a) 71. (c) 72. (b)
73. (a) 74. (b) 75. (a) 76. (c) 77. (b) 78. (a)
79. (d) 80. (d) 81. (b) 82. (c) 83. (a) 84. (a)
85. (a) 86. (d) 87. (b) 88. (b) 89. (b) 90. (c)
91. (c) 92. (c) 93. (a) 94. (d) 95. (c) 96. (a)
97. (b) 98. (c) 99. (a) 100. (a) 101. (c) 102. (b)
103. (d) 104. (b) 105. (b) 106. (c) 107. (b) 108. (c)
109. (b) 110. (a) 111. (c) 112. (a) 113. (a) 114. (b)
115. (d) 116. (b) 117. (b) 118. (d) 119. (b) 120. (b)
121. (d) 122. (d) 123. (c) 124. (a) 125. (c) 126. (b)
127. (d) 128. (c) 129. (b) 130. (b) 131. (a) 132. (c)
133. (c) 134. (b) 135. (d) 136. (b) 137. (a) 138. (c)
139. (b) 140. (a) 141. (b) 142. (d) 143. (c) 144. (c)
145. (a) 146. (b) 147. (a) 148. (b) 149. (c) 150. (a)
151. (b) 152. (a) 153. (b) 154. (c) 155. (c) 156. (d)
157. (d) 158. (a) 159. (a) 160. (a) 161. (c) 162. (b)
163. (a) 164. (b) 165. (d) 166. (b) 167. (b) 168. (b)
169. (a) 170. (b) 171. (c) 172. (c) 173. (c) 174. (a)
175. (c) 176. (c) 177. (b) 178. (d) 179. (b) 180. (a)
181. (d) 182. (c) 183. (c) 184. (b) 185. (a) 186. (b)
187. (a) 188. (b) 189. (b) 190. (b) 191. (b) 192. (c)
193. (d) 194. (b) 195. (c) 196. (a) 197. (b) 198. (c)
199. (b) 200. (c) 201. (d) 202. (a) 203. (d) 204. (d)
205. (a) 206. (d) 207. (c) 208. (c) 209. (a) 210. (b)
211. (d) 212. (c) 213. (a) 214. (a) 215. (c) 216. (c)
217. (b) 218. (d) 219. (d) 220. (c) 221. (a) 222. (d)
223. (b) 224. (a) 225. (b) 226. (a) 227. (b) 228. (c)
229. (a) 230. (c) 231. (c) 232. (d) 233. (c) 234. (d)

235. (c)	236. (c)	237. (d)	238. (b)	239. (c)	240. (a)
241. (c)	242. (d)	243. (a)	244. (b)	245. (b)	246. (c)
247. (b)	248. (a)	249. (d)	250. (b)	251. (d)	252. (d)
253. (d)	254. (c)	255. (c)	256. (b)	257. (b)	258. (c)
259. (c)	260. (d)	261. (b)	262. (a)	263. (c)	264. (b)
265. (d)	266. (a)	267. (b)	268. (d)	269. (c)	270. (b)
271. (a)	272. (b)	273. (d)	274. (b)	275. (a)	276. (d)
277. (a)					

Appendix III
Cause of Earthquake and Earthquake Intensity Scales

App III. 1. CAUSES OF EARTHQUAKE

An earthquake is caused by the snapping of the rocky layer inside the crust due to strain, resulting in a 'fault'. The moment the rock snaps, the earthquake is felt and when the fault slips due to friction, a series of elastic waves are set in motion which can be recorded photographically by seismographs. The first initial disturbance recorded on the seismograph due to an earthquake is called *primary wave* (or *P* wave or 'push' wave) which is a compression wave longitudinal in propagation and hence the fastest. The primary wave is followed by *secondary wave* (or *S* wave or 'shake' wave) which is a shear wave that travels transversely and hence slower than primary wave. The primary wave has an average velocity of about 8 km/s and that of secondary wave is about 4.5 km/s. By measuring the time interval between the recording of these two waves (to a fraction of a second if possible), it is possible to find out the arc distance at which the earthquake has occurred. If the time intervals between secondary and primary waves are available for atleast three separate seismic stations with a fairly wide geographic distribution, it is possible to determine the epicentre of the earthquake by large circles on the surface of the globe with the stations as centres and the respective arc distance as radii. The common point of intersection of the three circles is the epicentre. If data from more seismic stations are taken, it will be noticed that all these circles pass through the epicentre.

A new theory of earthquakes known as *plate tectonics* which was first proposed about three decades ago has now been accepted by geologists and scientists. The theory has grown out of the 'sea-floor spreading' advanced by Prof. Harry Hess of Princeton University in 1960.

The concept of plate tectonics states that the lithosphere or the outermost layer of the earth's surface is divided into seven or eight major rigid almost spherical plates, each about 100 km thick. The theory or concept of plate tectonics is fundamentally based on the theory of Continental Drift first proposed by Prof. Alfred Wegner in 1912.

These 100 km thick plates glide over a partially melted rock, a lubricating layer, so to say, called asthenosphere, which is the upper portion of the 'mantle' which is itself about 1600 km thick. The plates distinguished are: the European-Asian plate, the North American plate, the

African plate, the Indian plate, the Antarctic plate all plates corresponding to the continents and in addition, the Pacific plate and some other minor plates like the Philippine, the Nazca, the Cocos and the Carribean plates. The movement of the plates is so insignificant that it does not exceed about a centimetre per year.

Three types of boundaries between the plates have been recognized, viz., where the plates diverge from each other when new sea floor is formed along the mid-oceanic ridges, convergent junctures where plates overlap and thirdly where the plates glide past one another along narrow zones along the plate boundaries. The boundary layers of these plates are the earthquake zones. When the plates collide earthquake is caused. However, these plates seldom have head-on collisions but collide obliquely.

The European-Asian plate has shown a clear tendency to move towards the Equator along with the land masses of which it consists. This movement has brought it into collision with the Indian and South east Asian and Indian plats, clashing over Turkey, Iran and Afghanistan. This clash is the cause of earthquakes in Turkey and Iran. Similarly the destructive Gautemala earthquake of February 4, 1976 (7.3 on Richter scale) was caused by the rupture ultimately caused by the clash between the Cocos plate and Caribbean plates.

APP. III. 2 EARTHQUAKE INTENSITY SCALES

For classifying the earthquakes a number of earthquake intensity scales have been developed in which earthquakes have been classified on the basis of their destructiveness. These scales are used to rate and compare the intensity of earthquakes. Two such earthquake intensity scales which are commonly used are Mercalli scale and Richter scale. Both these scales are briefly described below.

1. ***Mercalli Scale.*** In Mercalli scale earthquakes are described and rated in terms of human reactions and observations, and the earthquakes have been classified into 12 classes which are assigned numbers from I to XII as indicated below.

Class of Earthquake	*Characteristics / Effects*
I	Not felt except by a very few under specially favourable circumstances.
II	Felt only by a few persons at rest, specially on upper floors of buildings; and delicately suspended objects may swing.
III	Felt quite noticeably indoors, specially on upper floors of buildings but many people do not recognize it as an earthquake; standing motor cars may rock slightly; and vibration may be felt like the passing of a truck.
IV	During the day felt indoors by many, outdoors by a few, at night some awakened; dishes, windows, doors disturbed; walls make creaking sound, sensation like heavy truck striking the building; and standing motor cars rocked noticeably.
V	Felt by nearly every one; many awakened; some dishes, windows, etc., broken; a few instances of cracked plaster; unstable objects overturned; disturbance of trees, poles and other tall objects noticed sometimes; and pendulum clocks may stop.

VI	Felt by all, many people frightened and run outdoors; some heavy furniture moved; a few instances of fallen plaster or damaged chimneys; and in general slight damage.
VII	Everybody runs outdoors, damage negligible in buildings of good design and construction; slight to moderate in well built ordinary structures; considerable in poorly built or badly designed structures; and some chimneys broken, noticed by persons driving motor cars.
VIII	Damage slight in specially designed structures; considerable in ordinary but substantial buildings with partial collapse; very heavy in poorly built structures; panel walls thrown out of framed structures; falling of chimney, factory stakes, columns, monuments and walls; heavy furniture overturned, sand and mud ejected in small amounts; changes in well water; and disturbs persons driving motor cars.
IX	Damage considerable in specially designed structures; well designed framed structures thrown out of plumb; very heavy in substantial buildings with partial collapse; buildings shifted off foundations; ground cracked conspicuously; and underground pipes broken.
X	Some well built wooden structures destroyed; most masonry and framed structures with foundations destroyed; ground badly cracked; rails bent; landslides considerable from river banks and steep slopes; shifted sand and mud, and water splashed over banks.
XI	Few, if any, masonry structures remain standing; bridges destroyed; broad fissures in ground, underground pipelines completely out of service; earth slumps and landslips in soft ground; and rails bent greatly.
XII	Total damage; waves seen on ground surface; lines of slight and levels distorted; and objects thrown upward into the air.

2. ***Richter Scale.*** In Richter scale magnitude of earthquake at its epicentre is indicated in terms of rating which indicates the strength or severity of earthquake. The rating scale is logarithmic and hence each increase of 1 on the scale represents an earthquake ten times as powerful to its preceding lower magnitude. The Richter scale is indicated below.

Rating	*Characteristics / Effect*
2.5	Generally not felt but recorded on seismograph
3.5	Felt by many people but rarely cause damage
4.5	Some local damage may occur, specially to poorly constructed buildings over small regions.
6.0	Can be destructive in areas upto about 100 km across

	where people live.
7.0	Major earthquake, can cause serious damage over layer areas.
8.0 and above	Great earthquake, can cause serious damage in areas several hundred kilometres across.

It may be noted that Mercalli scale is extremely subjective and the intensity of tremers varies from one place to another during an earthquake and hence the ratings in this case will be different for the same earthquake. On the other hand since Richter scale is based on the measurement of the magnitude of an earthquake at its epicentre, an earthquake has only one measurement on this scale. Thus, Richter scale is objective and quantitative in which the magnitude (gravity) of earthquake is calculated according to the amount of energy released during its happening.

APP. III. 3 VALUES OF BASIC SEISMIC COEFFICINT α_0 FOR SOME IMPORTANT TOWNS

City	*Zone*	α_0	*City*	*Zone*	α_0
Agra	III	0.04	Ambala	IV	0.05
Ahmedabad	III	0.04	Amritsar	IV	0.05
Ajmer	I	0.01	Asansol	III	0.04
Allahabad	II	0.02	Aurangabad	I	0.01
Almora	IV	0.05	Bahraich	IV	0.05
Bangalore	I	0.01	Kurnool	I	0.01
Barauni	IV	0.05	Lucknow	III	0.04
Bareilly	III	0.04	Ludhiana	IV	0.05
Bhatinda	III	0.04	Madurai	II	0.02
Bhilai	I	0.01	Mandi	V	0.08
Bhopal	II	0.02	Mangalore	III	0.04
Bhubaneswar	III	0.04	Monghyr	IV	0.05
Bhuj	V	0.08	Moradabad	IV	0.05
Bikaner	III	0.04	Mysore	I	0.01
Bakaro	III	0.04	Nagpur	II	0.02
Mumbai	III	0.04	Nainital	IV	0.05
Burdwan	III	0.04	Nasik	III	0.04
Calicut	III	0.04	Nellore	II	0.02
Chandigarh	IV	0.05	Panjim	III	0.04
Chennai	II	0.02	Patiala	III	0.04
Chitorgarh	I	0.01	Patna	IV	0.05
Coimbatore	III	0.04	Pilibhit	IV	0.05

Contd.

Contd.

City	*Zone*	α_0	*City*	*Zone*	α_0
Cuttack	III	0.04	Pondicherry	II	0.02
Darbhanga	V	0.08	Pune	III	0.04
Darjeeling	IV	0.05	Raipur	I	0.01
Dehradoon	IV	0.05	Rajkot	III	0.04
Delhi	IV	0.05	Ranchi	II	0.02
Durgapur	III	0.04	Roorkee	IV	0.05
Gangtok	IV	0.05	Rourkela	I	0.01
Gauhati	V	0.08	Sadiya	V	0.08
Gaya	III	0.04	Shimla	IV	0.05
Gorakhpur	IV	0.05	Sironj	I	0.01
Hyderabad	I	0.01	Srinagar	V	0.08
Imphal	V	0.08	Surat	III	0.04
Jabalpur	III	0.04	Tezpur	V	0.08
Jaipur	II	0.02	Thanjavur	II	0.02
Jamshedpur	II	0.02	Tiruchirapalli	II	0.02
Jhansi	I	0.01	Trivandrum	III	0.04
Jodhpur	I	0.01	Udaipur	II	0.02
Jorhat	V	0.08	Vadodara	III	0.04
Kanpur	III	0.04	Varanasi	III	0.04
Kathmandu	V	0.08	Vijayawada	III	0.04
Kohima	V	0.08	Visakhapatnam	II	0.02
Kolkata	III	0.04			

Appendix IV
National Water Policy

1. The Need for a National Water Policy

1.1 Water is a prime natural resource, a basic human need and a precious national asset. Planning and development of water resources need to be governed by national perspectives.

1.2 It has been estimated that out of the total precipitation of around 400 million hectare metres in the country, the surface water availability is about 178 million hectare metres. Out of this about 50% can be put to beneficial use because of topographical and other constraints. In addition there is a ground water potential of about 42 million hectare metres. The availability of water is highly uneven in both space and time. Precipitation is confined to only about three or four months in the year and varies from 10 cm in the western parts of Rajasthan to over 1000 cm at Cherrapunji in Meghalaya. Further, water does not respect state boundaries. Water as a resource is one and indivisible : rainfall, river waters, surface ponds and lakes and ground water are all part of one system; water is also a part of a larger ecological system.

1.3 Floods and droughts affect vast areas of the country, transcending state boundaries. About one third of the country is drought-prone. Floods affect an average area of around 9 million hectares per year. According to the National Commission on floods, the area susceptible to floods is around 40 million hectares. The approach to the management of drought and floods has to be coordinated and guided at the national level.

1.4 Even the planning and implementation of individual irrigation or multi-purpose projects, though done at the state level, involve a number of aspects and issues such as environmental protection, rehabilitation of project-affected people and livestock, public health consequences of water impoundment, dam safety, etc. On these matters common approaches and guidelines are necessary. Moreover, certain problems and weaknesses have affected a large number of projects all over the country. There have been substantial time and cost overruns on project. In some irrigation commands, problems of water-logging and soil salinity have emerged, leading to the degradation of good agricultural land. There are also complex problems of equity and social justice in regard to water distribution. The development and exploitation of the country's groundwater resources also give rise to questions of judicious and scientific resource management and conservation. All these questions need to be tackled on the basis of common policies and strategies.

1.5 The growth process and the expansion of economic activities inevitably lead to increasing demands for water for diverse purposes : domestic, industrial, agricultural, hydro-power, navigation, recreation, etc. So far the principal consumptive use of water has been for irrigation. While the irrigation potential is estimated to have increased from 19.5 million hectares at the time of Independence to about 68 million hectares at the end of the Sixth Plan, further development of a substantial order is necessary if the food and fibre needs of a growing population are to be met. The country's population which is over 750 million at present is expected to reach a level of around 1000 million by the turn of the century.

1.6 The production of foodgrains has increased from around 50 million tonnes in the fifties to about 150 million tonnes at present, but this will have to be raised to around 240 million tonnes by the year 2000 A.D. The drinking water needs of people and livestock have also to be met. In keeping with the objectives of the International Drinking Water Supply and Sanitation Decade Programme (1981–1991), adequate drinking water facilities have to be provided to the entire population in both urban and rural areas and sanitation facilities to 80% of the urban population and 25% of the rural population by the end of the decade. Domestic and industrial water needs have largely been concentrated in or near the principal cities, but the demand from rural society is expected to increase sharply as the development programmes improve economic conditions in the rural areas. The demand for water for hydro and thermal power generation and for other industrial uses is also likely to increase substantially. As a result water which is already a scarce resource will become even scarcer in future. This underscores the need for the utmost efficiency in water utilization and a public awareness of the importance of its conservation.

1.7 Another important aspect is water quality. Improvements in existing strategies and the innovation of new techniques resting on a strong science and technology base will be needed to eliminate the pollution of surface and ground water resources, to improve water quality and to step up the recycling and re-use of water. Science and technology and training have also important roles to play in water resources development in general.

1.8 Water is one of the most crucial elements in developmental planning. As the country prepares itself to enter the 21st century, efforts to develop, conserve, utilise and manage this important resource have to be guided by national perspectives. The need for a national water policy is thus abundantly clear : water is a scarce and precious national resource to be planned, developed and conserved as such, and on an integrated and environmentally sound basis, keeping in view the needs of the states concerned.

2. Information System

2.1 The prime requisite for resource planning is a well-developed information system. A standardised national information system should be established with a network of data banks and data bases, integrating and strengthening the existing central and state level agencies and improving the quality of data and the processing capabilities. There should be free exchange of data among the various agencies and duplication in data collection should be avoided. Apart from the data regarding water availability and actual water use, the system should also include comprehensive and reasonably reliable projections of future demands for water for diverse purposes.

3. Maximising Availability

3.1 The water resources available to the country should be brought within the category of utilizable resources to the maximum possible extent. The resources should be conserved and the availability augmented by measures for maximising retention and minimising losses.

3.2 Resource planning in the case of water has to be done for a hydrological unit such as a drainage basin as a whole, or for a sub-basin. All individual developmental projects and proposals should be formulated by the states and considered within the framework of such an overall plan for a basin or sub-basin, so that the best possible combination of options can be made.

3.3 Appropriate organisations should be established for the planned development and management of a river basin as a whole. Special multi-disciplinary units should be set up in each state to prepare comprehensive plans taking into account not only the needs of irrigation but also harmonising various other water uses, so that the available water resources are determined and put to optimum use having regard to subsisting agreements or awards of Tribunals under the relevant laws.

3.4 Water should be made available to water short areas by transfer from other areas including transfers from one river basin to another, based on a national perspective, after taking into account the requirements of the areas/basins.

3.5 Recycling and re-use of water should be an integral part of water resources development.

4. Project Planning

4.1 Water resource development projects should as far as possible be planned and developed as multipurpose projects. Provision for drinking water should be a primary consideration. The projects should provide for irrigation, flood mitigation, hydro-electric power generation, navigation, pisciculture and recreation wherever possible.

4.2 The study of the impact of a project during construction and later on human lives, settlements, occupations, economic and other aspects should be an essential component of project planning.

4.3 In the planning, implementation and operation of projects, the preservation of the quality of environment and the ecological balance should be a primary consideration. The adverse impact, if any, on the environment should be minimised and should be off-set by adequate compensatory measures.

4.4 There should be an integrated and multi-disciplinary approach to the planning, formulation, clearance and implementation of projects, including catchment treatment and management, environmental and ecological aspects, the rehabilitation of affected people and command area development.

4.5 Special efforts should be made to investigate and formulate projects either in, or for the benefit of, areas inhabited by tribal or other specially disadvantaged groups such as Scheduled Castes and Scheduled Tribes. In other areas also, project planning should pay special attention to the needs of Scheduled Castes and Scheduled Tribes and other weaker sections of society.

4.6 The planning of projects in hilly areas should take into account the need to provide assured drinking water, possibilities of hydro-power development and the proper approach to irrigation in such areas, in the context of physical features and constraints such as steep slopes, rapid run-off and incidence of soil erosion. The economic evaluation of projects in such areas should also take these factors into account.

4.7 Time and cost overruns and deficient realization of benefits characterising most irrigation projects should be overcome by upgrading the quality of project preparation and management. The under-funding of projects should be obviated by an optimal allocation of resources, having regard to the early completion of on-going projects as well as the need to reduce regional imbalances.

5. Maintenance and Modernisation

5.1 Structures and systems created through massive investments should be properly maintained in good health. Appropriate annual provisions should be made for this purpose in the budgets.

5.2 There should be a regular monitoring of structures and systems and necessary rehabilitation and modernisation programmes should be undertaken.

6. Safety of Structures

6.1 There should be proper organisational arrangements at the national and state levels for ensuring the safety of storage dams and other water-related structures. The Central guidelines on the subject should be kept under constant review and periodically updated and reformulated. There should be a system of continuous surveillance and regular visits by experts.

7. Ground Water Development

7.1 There should be a periodical reassessment on a scientific basis of the ground water potential, taking into consideration the quality of the water available and economic viability.

7.2 Exploitation of ground water resources should be so regulated as not to exceed the recharging possibilities, as also to ensure social equity. Ground water recharge projects should be developed and implemented for augmenting the available supplies.

7.3 Integrated and coordinated development of surface water and ground water and their conjunctive use, should be envisaged right from the project planning stage and should form an essential part of the project.

7.4 Over exploitation of ground water should be avoided near the coast to prevent ingress of sea water into sweet water aquifers.

8. Water Allocation Priorities

8.1 In the planning and operation of systems, water allocation priorities should be broadly as follows :

- Drinking water
- Irrigation
- Hydro-power
- Navigation
- Industrial and other uses.

However, these priorities might be modified if necessary in particular regions with reference to area specific considerations.

9. Drinking Water

9.1 Adequate drinking water facilities should be provided to the entire population both in urban and in rural areas by 1991. Irrigation and multipurpose projects should invariably include a drinking water component, wherever there is no alternative source of drinking water. Drinking water needs of human beings and animals should be the first charge on any available water.

10. Irrigation

10.1 Irrigation planning either in an individual project or in a basin as a whole should take into account the irrigability of land, cost effective irrigation options possible from all available sources of water and appropriate irrigation techniques. The irrigation intensity should be such as to extend the benefits of irrigation to as large a number of farm families as possible, keeping in view the need to maximise production.

10.2 There should be a close integration of water-use and land-use policies.

10.3 Water allocation in an irrigation system should be done with due regard to equity and social justice. Disparities in the availability of water between head-reach and tail-end farms and between large and small farms should be obviated by adoption of a rotational water distribution system and supply of water on a volumetric basis subject to certain ceilings.

10.4 Concerted efforts should be made to ensure that the irrigation potential created is fully utilised and the gap between the potential created and its utilisation is removed. For this purpose, the command area development approach should be adopted in all irrigation projects.

11. Water Rates

11.1 Water rates should be such as to convey the scarcity value of the resource to the users and to foster the motivation for economy in water-use. They should be adequate to cover the annual maintenance and operation charges and a part of the fixed costs. Efforts should be made to reach this ideal over a period, while ensuring the assured and timely supplies of irrigation water. The water rates for surface water and ground water should be rationalised with due regard to the interests of small and marginal farmers.

12. Participation of Farmers and Voluntary Agencies

12.1 Efforts should be made to involve farmers progressively in various aspects of management of irrigation systems, particularly in water distribution and collection of water rates. Assistance of voluntary agencies should be enlisted in educating the farmers in efficient water use and water management.

13. Water Quality

13.1 Both surface water and ground water should be regularly monitored for quality. A phased programme should be undertaken for improvements in water quality.

14. Water Zoning

14.1 Economic development and activities including agricultural, industrial and urban development, should be planned with due regard to the constraints imposed by the configuration of water availability. There should be a water zoning of the country and the economic activities should be guided and regulated in accordance with such zoning.

15. Conservation of Water

15.1 The efficiency of utilisation in all the diverse uses of water should be improved and an awareness of water as a scarce resource should be fostered. Conservation consciousness should be promoted through education, regulation, incentives and disincentives.

16. Flood Control and Management

16.1 There should be a master plan for flood control and management for each flood prone basin. Sound watershed management through extensive soil conservation, catchment-area treatment, preservation of forests and increasing the forest area and the construction of check-dams should be promoted to reduce the intensity of floods. Adequate flood-cushion should be provided in water storage projects wherever feasible to facilitate better flood management. An extensive network for flood forecasting should be established for timely warning to the settlements in the flood plains, along with the regulation of settlements and economic activity in the flood plain zones, to minimise the loss of the life and property on account of floods. While physical flood protection works like embankments and dykes will continue to be necessary, the emphasis should be on non-structural measures for the minimisation of losses, such as flood forecasting and warning and flood plain zoning, so as to reduce the recurring expenditure on flood relief.

17. Land Erosion by Sea or River

17.1 The erosion of land, whether by the sea in coastal areas or by river waters inland, should be minimised by suitable cost-effective measures. The States and Union territories should also undertake all requisite steps to ensure that indiscriminate occupation and exploitation of coastal strips of land are discouraged and that the location of economic activities in areas adjacent to the sea is regulated.

18. Drought Management

18.1 Drought prone areas should be made less vulnerable to drought-associated problems through soil-moisture conservation measures, water harvesting practices, the minimisation of evaporation losses, the development of the ground water potential and the transfer of surface water from surplus areas where feasible and appropriate. Pastures, forestry or other modes of development which are relatively less water-demanding should be encouraged. In planning water resource development projects, the needs of drought-prone areas should be given priority.

18.2 Relief works undertaken for providing employment to drought-stricken populations should preferably be for drought proofing.

19. Science and Technology

19.1 For effective and economical management of our water resources, the frontiers of knowledge need to be pushed forward in several directions by intensifying research efforts in various areas, including the following :

- hydrometeorology ;
- assessment of water resources ;
- snow and lake hydrology ;
- ground water hydrology and recharge ;
- prevention of salinity ingress ;

- water-harvesting ;
- evaporation and seepage losses ;
- economical designs for water resource projects ;
- crops and cropping systems ;
- sedimentation of reservoirs ;
- the safety and longevity of water related structures;
- river morphology and hydraulics ;
- soils and material research ;
- better water management practices and improvements in operational technology ;
- recycling and re-use ;
- use of sea water resources.

20. Training

20.1 A perspective plan for standardised training should be an integral part of water resource development. It should cover training in information systems, sectoral planning, project planning and formulation, project management, operation of projects and their physical structures and systems and the management of the water distribution systems. The training should extend to all the categories of personnel involved in these activities as also the farmers.

21. Conclusion

21.1 In view of the vital importance of water for human and animal life, for maintaining ecological balance and for economic and developmental activities of all kinds, and considering its increasing scarcity, the planning and management of this resource and its optimal, economical and equitable use has become a matter of the utmost urgency. The success of the national water policy will depend entirely on the development and maintenance of a national consensus and commitments to its underlying principles and objectives.

Appendix V
Control Valves for Pipes

Valves are required to be introduced in the pipelines for one or all of the following purposes :

(i) To stop the flow of liquid altogether

(ii) To regulate the rate of flow of liquid

(iii) To regulate the head of pressure in the pipelines.

According to the motion of the valve element, the control valves may be classified in the following three categories :

(a) Valve element moves transversely across the waterway

(b) Valve element rotates about an axis transverse in the waterway

(c) Valve element slides axially or longitudinally.

(a) **With transverse motion.** In this category of valves the following two types of valves may be classified.

(i) *Full-way or Sluice or Gate Valve.* This is very commonly used valve in which the moving element has the form of a disc or circular gate, wedge-shaped in cross-section, having a

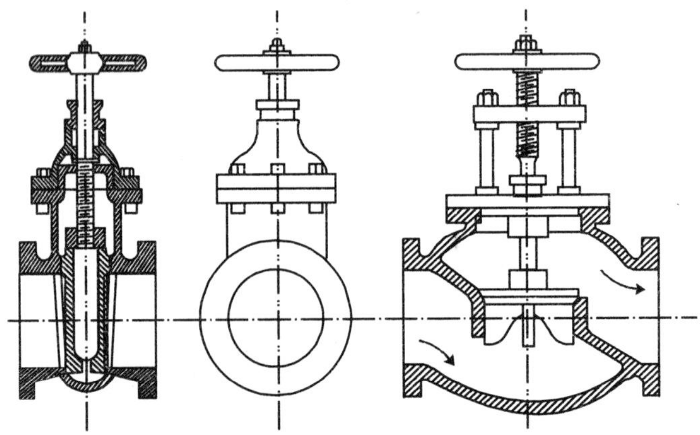

Fig. *App. V. 1*

nut which engage with the thread of the operating spindle as shown in the Fig. App. V. 1. In the fully-open position, the gate is withdrawn clear of the waterway and hence this valve is designated as full-way valve.

(ii) *Globe Valve or Stop Valve.* As shown in the accompanying figure, in this case the valve element is wing-guided mitre or disc, which in the closed position of the valve is forced on its circular seat by the screwed spindle. When this spindle releases the valve element, the pressure of the incoming liquid is sufficient to lift it, thus opening the valve.

(b) *With Rotary Motion.* The following two types of valves may be considered in this category.

(i) *Butterfly Valve.* In this valve the cylindrical valve body forms virtually a continuation of the pipe itself, and it houses a moving element of elliptical shape. Rotation of the valve spindle brings the valve element either into the closed position, or into the fully open position in which it is parallel with the pipe axis as shown in the Fig. App. V. 3.

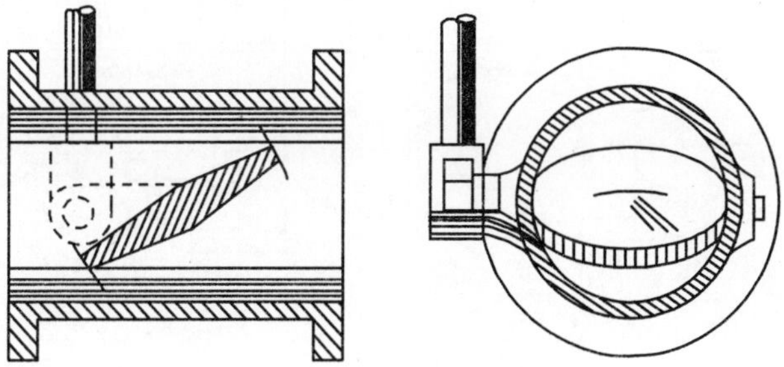

Butterfly valve

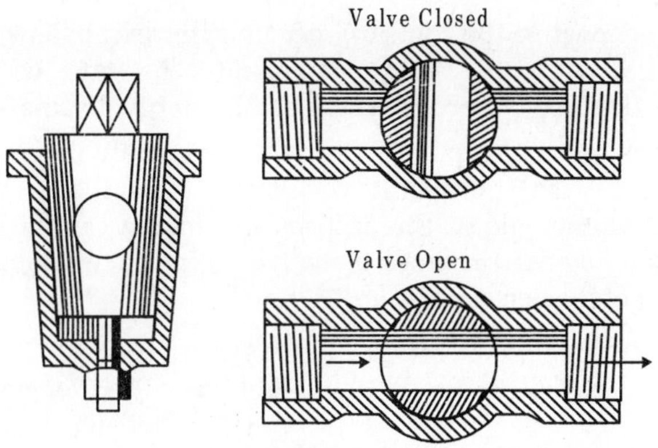

Plug Valve

Fig. *App. V. 3*

(ii) *Plug Valve.* A conical plug has a port which in the open position of the valve is in line with the pipe. After being turned through 90°, the plug completely closes the waterway as shown in the sectional view in the Fig. App. V. 3.

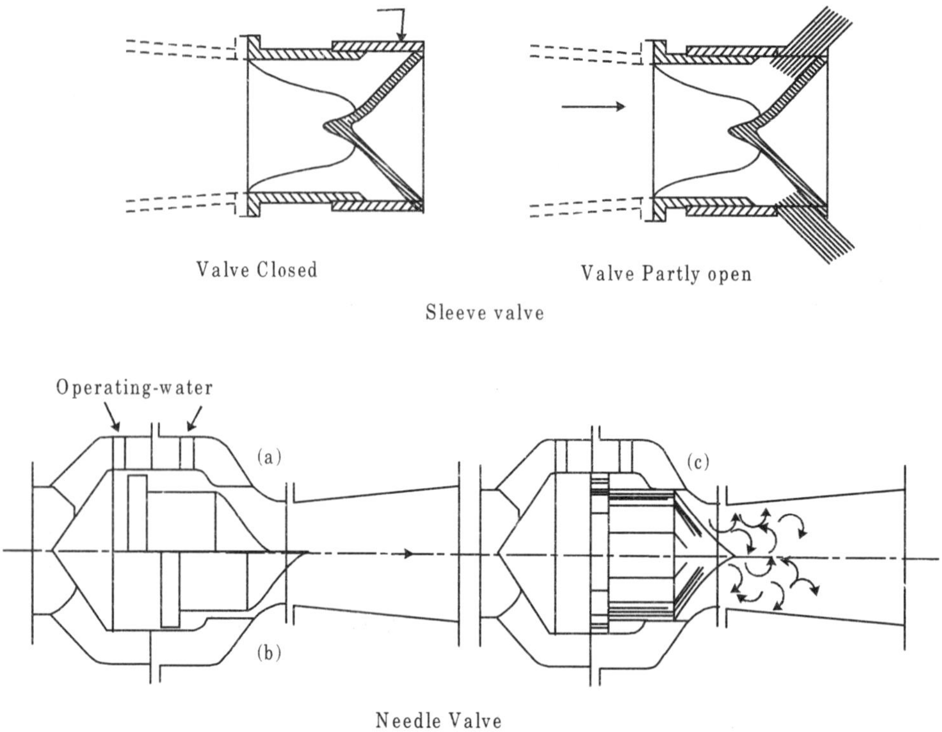

Fig. *App. V. 4*

(c) *With Axial Motion.* In this category the following two types of valves may be included.

(i) *Sleeve Valve.* As shown in the Fig. App. V. 4 a cylindrical sleeve surrounding the body of the valve can be moved longitudinally so as to open or close the water way. Such valves are usually fixed at the outlet end of the pipe, where it serves as an energy dissipator.

(ii) *Needle Valve.* It is another type of valve with an axially moving element. The operating piston is formed integrally with the needle which regulates the area of the waterway. High pressure water admitted to one side or the other of the piston causes the needle to take up the desired position. Accompanying figure shows the valve (a) in the fully open position, (b) in the fully closed position, (c) in the partly closed position.

The needle valve resists cavitation damage quite successfully, because as shown in the figure in the case of needle valve the zone of intense eddying lies near the axis of the waterway, far away from the pipe walls. The sluice valve is however, more susceptible to cavitation damage, particularly when used continuously in partly open position. However, in spite of this drawback the sluice valve is the most commonly used control valve, being suitable for the pipes of very small as well as very large diameters. The globe valve is most widely used in the domestic water tap. However, the globe valves are also used for the pipelines of very large diameters. The butterfly valves are generally used for regulating the flow of water in penstocks supplying water to the turbines installed in hydropower stations. Similarly the needle valves are also used for regulating the flow of liquid through large diameter pipes such as penstocks.

Appendix VI
Devices for Lifting Water

[A] DEVICES FOR LIFTING WATER

Lift irrigation requires the lifting of water from its source to the land to be irrigated. The various devices used for lifting water may be broadly classified as

1. Indigenous water lifting devices.
2. Mechanical water lifting devices.

(1) **Indigenous water lifting devices**

Several types of indigenous water lifting devices are used in India which are either manually-operated or animal operated. Some of these devices are briefly described below.

(i) *Swing basket.* Swing basket is one of the most ancient water lifting device. It consists of a basket or shovel like scoop to which four ropes are attached. Two persons stand facing each other and swing the basket to fill in water. The basket is raised and discharged into the field channel.

(ii) *Doon.* The doon is a manually operated boat-shaped trough, closed at one end and open at the other. The closed end of the trough is tied with a rope to a long wooden pole which is pivoted as a lever on a post. A counter-weight, either a large stone or a ball of dried mud, is fixed to the other end of the lever. At its centre, the trough is supported on a horizontal rod on which it rocks. Alternatively the open end of the trough is hinged at the discharge point. The trough is dipped into the water by applying the body weight and the force of the operator. The operator then releases the trough so that it rocks due to turning moment caused by the counter-weight and discharges the water from its open end into the field channel.

(iii) *Archimedian screw.* This device is also manually operated and consists of a hollow wooden or metal cylinder fitted with a helicoid screw inside it. The lower end of the screw ends in a shaft which is held in a bearing and supported on a vertical post. The upper end of the screw ends in a shaft with handle at its end and it is also held in a bearing and supported on a vertical post. The cylinder is placed at an angle of less than 30°, with its lower end immersed in water. When the handle is turned round and round, the water moves up through the cylinder and discharges from the upper end of the cylinder into the field channel.

(iv) *Water wheel.* The water wheel consists of small paddles mounted radially on a horizontal shaft. The wheel is fixed on a close-fitting concave trough. The shaft is driven by a bullock-wheel drive similar to the drive used in a Persian wheel. The wheel, when rotated, pushes the water to the field channel through the trough.

All the above noted devices may be used only for low lifts upto about 1.2 m.

(v) *Persian wheel.* The persian wheel consists of a chain and buckets or pots mounted on an open spoked wooden wheel of big diameter fixed in vertical position on the top of well and provided with a suitable driving mechanism. The capacity of the bucket or pot ranges from 7 to 14 litres. Two parallel loops of chains, joined by spacing bars and having metal buckets or earthen-ware pots attached to them at intervals, pass over the wheel with a portion of the chain dipping into the water in the well. A horizontal shaft extends from the central axis of the wheel to a small vertical cog-wheel or gear. The teeth of this vertical gear mesh with the teeth of a large horizontal wheel gear. To the shaft of the horizontal gear a long horizontal beam is fitted to which the animals are hitched.

When the Persian wheel is operated, the bucket-wheel revolves, causing the buckets or pots at the lower end of the chain to get filled with water and carried to the top with their open mouth upward. When the buckets or pots pass over the top of the wheel they spill their content into a trough, from where it is conveyed to the field channel. Persian wheel is useful for lifts upto 30 m.

(vi) *Chain pump.* The chain pump consists of an endless chain which is provided with leather discs or washers spaced at intervals of about 0.25 m. The chain passes over a notched wheel mounted on a suitable platform fixed on the top of the well with a portion of the chain dipping into the water in the well. On one side of the chain there is a pipe of about 100 mm diameter, having a flared opening at its bottom and connected to a trough at the top. The bottom of the pipe is submerged about 0.6 to 0.9 m below the surface of water. The discs have the same diameter as the inside of the pipe. When the wheel is turned each disc brings certain volume of water.

Chain pumps may be either animal operated or manually operated. In case of animal driven chain pumps, the arrangement of the driving mechanism is the same as in a Persian wheel. In manually operated chain pumps, a handle is fixed directly to the wheel shaft over which the chain works.

(vii) *Churus* or *Ring-and-Bucket lift.* It consists of a bucket or bag having a capacity of about 150 to 200 litres and made of leather or galvanized iron sheet. The bucket is fixed on a stout iron ring with an iron frame work at the top and is tied to a long rope. The rope passes over a pulley set on a wooden frame-work fixed on the top of the well. A pair of bullocks, hitched to the other end of the rope, provide the power to lift the bucket. The bullocks, while pulling up the full bucket walk down, on an earthen ramp sloped at an angle of 5 to 10 degrees.

(2) **Mechanical water lifting devices.** The various types of mechanical water lifting devices are as follows

(i) *Wind mill.* The wind mill utilises the natural power from wind to lift water. it consists of a large diameter vane wheel or rotor fixed on the top of a high steel or wooden tower and a reciprocating pump at its bottom. The upper end of the piston rod of the pump is fixed eccentrically to the vane wheel. The pump cylinder or at least the lower part of the suction pipe is always immersed in water in the well. When a strong wind blows, it rotates the vane wheel, which in turn works the pump and water is lifted up. Usually the delivery pipe rises to a reservoir where the water is stored to be used whenever required. The discharge from a windmill is usually intermittent and seasonal. Its capacity depends on wind velocity and frequency. The device is suitable to lift water from shallow and moderately deep wells, depending on the wind velocity.

(ii) *Hydraulic ram.* The hydraulic ram is a contrivance which utilizes the hydraulic energy of large amount of water falling through small height to lift a small quantity of this water to greater height. It consists of an inclined supply pipe connecting the source of water to a valve chamber which contains two automatic valves ; a waste valve opening downward and a delivery valve opening upward. Above the delivery valve there is an air vessel to the base of which the delivery pipe is attached.

Water flows from the source through the supply pipe to the valve chamber. The waste valve being open, water freely escapes through it and the flow is set up along the supply pipe. As the discharge through the waste valve increases, the velocity of flow of water in the supply pipe increases until the dynamic pressure on the lower face of the waste valve becomes sufficiently great to over-come its weight. The waste valve then closes rapidly and brings the water in the supply pipe suddenly to rest which gives rise to a rapid increase of pressure in the valve chamber which in turn causes the delivery valve to open. Water than escapes through the delivery valve into the air vessel, thus compresses the air and flows out through the delivery pipe to the discharge point. The flow through the delivery valve continues until the pressure in the valve chamber is reduced, the delivery valve then closes and the waste valve opens, thus again causing the water to flow through the waste valve. The same cycle of operations is then repeated. The air vessel serves the same purpose as it does on a reciprocating pump, *i.e.,* it reduces the fluctuations in flow through the delivery pipe.

The hydraulic ram involves considerable wastage of water, but it may be advantageously employed in situations such as hilly areas where no other outside source of power is available. The ratio of wasted water to pumped water for a well-designed hydraulic ram will vary from 6 : 1 to 2 : 1 depending upon the supply head, the lift, and other factors.

(iii) *Pumps.* There are many different types of pumps used for lifting water, some of which are as indicated below.

(a) *Centrifugal pumps.* A centrifugal pump has a rotating element called impeller which imparts energy to the water and lifts it to the required height. The impeller may be shaped to force water outward in a direction at right angles to its axis (*radial flow*), to give the water an axial as well as radial velocity (*mixed flow*), or to force the water in the axial direction alone (*axial-flow*). Radial-flow and mixed flow machines are commonly known as *centrifugal pumps* while axial-flow machines are called *axial-flow pumps.* Radial- and mixed flow impellers may be either open or closed. The open impeller consists of a hub to which the vanes are attached, while the closed impeller has plates or shrouds on each side of the vanes.

The casing of radial flow pumps may be either the volute type or the turbine type. A volute casing is designed to produce an equal velocity of flow around the circumference of the impeller and to reduce the velocity of the water as it enters the discharge pipe. In the turbine-type pump the impeller is surrounded by stationary guide vanes which reduce the velocity of the water and convert velocity head to pressure head. The casing surrounding the guide vanes is usually circular and concentric with the impeller.

A pump which has only one impeller is known as *single-stage pump.* On the other hand a pump is known as *multistage pump* if it has two or more impellers arranged in such a way that the discharge from one impeller enters the eye of the next impeller. The multistage pumps are used for higher lifts.

A *deep-well turbine pump* has several impellers on a vertical shaft suspended from the prime mover at the surface. This type of pump is instaled in a well casing of limited size and hence the entire assembly must be of smaller diameter. These pumps are usually of submersible type. A *submersible pump* is one in which pump and electric motor are placed below the water surface of a well. Delivery of water to the surface is through a riser pipe on which the assembly is suspended. The advantage of the submersible pump is the elimination of the long drive shaft from the ground surface to the pump. This reduces bearing friction and provides unobstructed pipe for delivery of water to the surface.

(b) *Reciprocating pump.* A reciprocating pump consists of a piston or plunger which moves to and fro in a close fitting cylinder. The cylinder is connected to suction and delivery pipes each of which is provided with a non-return or one-way valve called suction valve and delivery valve respectively. The function of non-return or one way valve is to admit water in one direction only. Thus the suction valve allows water only to enter the cylinder and the delivery valve permits only its discharge from the cylinder. During one stroke of the piston or plunger water is drawn into the cylinder and in the next stroke water is forced out from the cylinder. These strokes are known as suction and delivery strokes respectively. These pumps are therefore classified as displacement pumps.

A reciprocating pump may be single-acting or double-acting. A single-acting pump is the one in which discharge occurs only on alternate strokes of the piston or plunger. A double-acting pump discharges with each stroke and thus provide a more uniform flow. Many reciprocating pumps are provided with an air vessel on the discharge side of the pump. The air in the vessel compresses and expands on each stroke to cause a more uniform flow in the discharge pipe.

The simplest type of reciprocating pump which is commonly used is the hand-operated well pump which may be used for suction lifts upto about 6 m. However, the use of reciprocating pumps for lifting water through higher lifts is not common as compared to that of centrifugal pumps. This is so because the electric motor and centrifugal pump provide a compact pumping unit. Moreover, the centrifugal pumps are better suited to pump water containing solids than the reciprocating pumps.

(c) *Air-lift pumps.* An air-lift pump uses compressed air to deliver water from wells. The air is forced into the well through a small air pipe and discharged into the eduction pipe or rising main at the bottom of the well. The resulting mixture of air and water in the eduction pipe is lighter than the water outside the pipe and hence is forced upward by hydrostatic pressure. This type of pumps has been used for lifts as great as 150 m, but its efficiency is usually only between 25 and 50 percent.

(d) *Jet pumps.* In the case of *jet pump or ejector pump* compressed air or water is forced through a pipe and discharges from a nozzle in the throat of a vertical riser pipe at high velocity. The resulting jet creates a suction which draws the water up the riser pipe. The efficiency of jet pumps is rarely over 25 percent. Jet pumps are widely used for pumping water from small wells.

(e) *Rotary pumps.* The rotary pump consists of two cams or gears which mesh together and rotate in opposite directions. The rotating elements fit the casing closely, and the water trapped between them and the casing is forced through the pump as they rotate. This creates a partial vacuum and brings in a replacement supply of water along the inlet side. Water containing grit is very injurious to a rotary pump as the wear will destroy the seal between the cams and casing. The rotary pumps are, however, not commonly used for irrigation pumping.

[B] ADVANTAGES AND DISADVANTAGES OF WELL IRRIGATION OVER CANAL IRRIGATION

Advantages

The various advantages of well irrigation over canal irrigation are as follows.

(i) The well is under the direct control of the owner and hence it may be sunk and equipped as required.

(ii) The cultivator can lift water and use it as and when required by him. In canal irrigation he has to await his turn.

(iii) Isolated patches of land which cannot be supplied with canal water can be easily irrigated by wells. Wells may be located to command any desired land.

(iv) The entire system of well irrigation need not be constructed at once. Wells may be sunk gradually as irrigation develops. This leads to distribution of expenditure and avoids blocking up of capital and accumulation of interest before revenue starts coming in. In a canal project this is not possible as the expenditure on headworks and network of canals must be incurred before any irrigation can be done.

(v) The supply from a well can be maintained fairly constant. Water can be turned off at any moment, taking the advantage of rainfall.

(vi) Since the well is generally centrally located, the loss in transit is very much reduced. In the case of well irrigation the losses take place in water courses only. However, in a canal system the losses take place in the Main canal, the Branch and the Distributary before water reaches the water course. As such the duty of water in well irrigation is generally higher than that in the case of canal irrigation. It is generally found that the depth (or delta) required to mature a crop on well irrigation is about 2/3rd of the depth required on canal irrigation.

(vii) In the case of well irrigation volumetric assessment of water charges is possible resulting in economy of water due to more careful utilization and correct timing of watering.

(viii) Well pumping is an effective anti-waterlogging measure as it helps in lowering the subsoil water level and in draining of irrigated land which might become waterlogged. As such well irrigation helps to prevent waterlogging as well as to relieve the land already affected by waterlogging. On the other hand careless canal irrigation may lead to waterlogging.

(ix) Unless draught continues for several years, well irrigation does not fail in draught seasons while a canal irrigation may fail.

(x) With the help of well irrigation, it is possible to accomplish multiple cropping which essentially means intensive land use to take 2, 3 or even 4 crops a year.

(xi) The well water, which is warmer in cold weather and cooler in hot weather, is more agreeable to crops.

Disadvantages

The various disadvantages of well irrigation over canal irrigation are as follows.

(i) The well irrigation is about 3 to 5 times costlier than canal irrigation. This is so because high working expenses due to energy required for pumping and higher establishment charges, and depreciation due to limited life of tube well lead to high cost of well irrigation.

(ii) The well water being clear and free from silt does not help to improve the fertility of the soil. On the other hand the fine silt carried by the canal water to the fields helps to improve the soil fertility.

(iii) Due to mechanical defects in the pump or due to failure of power, water may not be available to the crop at a crucial time.

(iv) There may be difficulty of maintaining mechanical equipment in isolated areas.

It may, however, be stated that a combination of canal and well irrigation would be ideal where feasible.

Appendix VII
Indian Standards for use in Design of Irrigation and Hydraulic Structures

1.	IS : 457—1957	Code of practice for general construction of plain and reinforced concrete for dams and other massive structures
2.	IS : 1191—1971	Glossary of terms and symbols used in connection with the measurement of liquid flow with a free surface (first revision).
3.	IS : 1192—1981	Velocity-area methods for measurement of flow of water in open channels (first revision)
4.	IS : 1893—1984	Criteria for earthquake resistant design of structures (fourth revision).
5.	IS : 2800	Code of practice for tube well
	(i) Part 1—1991	Construction
	(ii) Part 2 — 1979	Testing
6.	IS : 3860—1966	Precast cement concrete slabs for canal lining. (Superseded by IS: 10646)
7.	IS : 3872—1966	Code of practice for lining of canals with burnt clay tiles.
8.	IS : 3873—1978	Code of practice for laying in situ cement concrete lining on canals (first revision)
9.	IS : 3910—1992	Requirements for water flow measurement in open channels-rotating elements current meters (first revision)
10.	IS : 3911—1966	Specification for surface floats
11.	IS : 3912—1966	Specification for sounding rods
12.	IS : 3918—1966	Code of practice for use of current meter (cup type) for water flow measurement

13. IS : 4410	Glossary of terms relating to river valley projects
(i) Part 1—1991	Irrigation practice
(ii) Part 2—1967	Project planning
(iii) Part 3—1988	River and river training
(iv) Part 5—1982	Canals (first revision)
(v) Part 6—1983	Reservoirs (first revision)
(vi) Part 7—1982	Engineering geology (first revision)
(vii) Part 8—1992	Dams and dam sections
(viii) Part 9—1982	Spillways and siphons (first revision)
(ix) Part 10—1988	Hydroelectric power station including water conductor system
(x) Part 11—1972–85	Hydrology
(xi) Part 12—1973	Diversion works
(xii) Part 13—1985	Operation, maintenance and repairs of river valley projects
(xiii) Part 14—1977	Soil conservation and reclamation
(xiv) Part 15—1973–92	Canal structures
(xv) Part 16—1977–81	Gates and valves
(xvi) Part 17—1977	Water requirements of crops
(xvii) Part 18—1983	Energy dissipation devices
(xviii) Part 20—1983	Tunnels
(xix) Part 21—1987	Flood control
14. IS : 4515—1967	Code of practice for boulder lining for canals. (Reaffirmed 1990)
15. IS : 4558—1983	Code of practice for under-drainage of lined canals (first revision)
16. IS : 4622—1992	Recommendations for structural design of fixed wheel gates (second revision).
17. IS : 4623—1984	Recommendations for structural design of radial gates (second revision). (Reaffirmed 1990)
18. IS : 4701—1982	Code of practice for earthwork on canals. (Reaffirmed 1990)
19. IS : 4839	Code of practice for maintenance of canals
(i) Part 1—1992	Unlined canals (second revision)
(ii) Part 2—1992	Lined canals (second revision)
(iii) Part 3—1992	Canals structures, drains, outlets, jungle clearance, plantation and regulation (second revision)
20. IS : 4877—1968	Guide for preparation of estimate for river valley projects
21. IS : 4880	Code of practice for design of tunnels conveying water
(i) Part 1—1987	General design
(ii) Part 2—1976	Geometric design (first revision). (Reaffirmed 1990)
(iii) Part 3—1976	Hydraulic design (first revision). (Reaffirmed 1990)
(iv) Part 4—1971	Structural design of concrete lining in rock. (Reaffirmed 1990)

(v) Part 5—1972	Structural design of concrete lining in soft strata and soils. (Reaffirmed 1990)
(vi) Part 6—1971	Tunnel supports. (Reaffirmed 1990)
(vii) Part 7—1975	Structural design of steel lining. (Reaffirmed 1990)
22. IS : 4997—1968	Criteria for design of hydraulic jump type stilling basins with horizontal and sloping apron. (Reaffirmed 1991)
23. IS : 5256—1992	Code of practice for sealing joints in concrete lining on canals. (first revision)
24. IS : 5477	Methods for fixing the capacities of reservoirs
(i) Part 1—1969	General requirements. (Reaffirmed 1990)
(ii) Part 2—1969	Dead storage. (Reaffirmed 1990)
(iii) Part 3—1969	Live storage. (Reaffirmed 1990)
(iv) Part 4—1971	Flood storage. (Reaffirmed 1990)
25. IS : 5497—1983	Guide for topographical surveys for river valley projects (first revision). (Reaffirmed 1990)
26 IS : 5542—1969	Guide for storm analysis. (Reaffirmed 1991)
27. IS : 5620—1985	Recommendations for structural design criteria for low head slide gates (second revision)
28. IS : 5968—1987	Guide for planning and layout of canal system for irrigation (first revision) (Reaffirmed 1992)
29. IS : 5973—1970	Pan evaporimeter (Amendment No. 3) (Reaffirmed 1990)
30 IS : 6059—1971	Recommendation for liquid flow measurement in open channels by weirs and flumes—weirs of finite crest width for free discharge. (Reaffirmed 1990)
31. IS : 6062—1971	Method of measurement of flow of water in open channels using standing wave flume fall. (Reaffirmed 1989)
32. IS : 6063—1971	Method of measurement of flow of water in open channels using standing wave flume. (Reaffirmed 1989)
33. IS : 6066—1984	Recommendations for pressure grouting of rock foundations in river valley projects (first revision). (Reaffirmed 1990)
34. IS : 6512—1984	Criteria for design of solid gravity dams (first revision). (Reaffirmed 1990)
35. IS : 6518—1992	Code of practice for control of sediment in reservoirs. (first revision)
36. IS : 6522—1972	Criteria for design of silt vanes for sediment control in offtaking canals. (Reaffirmed 1990)
37. IS : 6531—1992	Criteria for design of canal head regulators. (first revision)
38. IS : 6532—1972	Code of practice for design, installation, observation and maintenance of uplift pressure pipes for hydraulic structures on permeable foundations. (Reaffirmed 1991)

39.	IS : 6926—1973	Code of practice for diamond core drilling for site investigation for river valley projects. (Reaffirmed 1990)
40.	IS : 6934—1973	Recommendations for hydraulic design of high ogee overflow spillways. (Reaffirmed 1991)
41.	IS : 6936—1992	Guide for location, selection and hydraulic design of canal escapes. (first revision)
42.	IS : 6939—1992	Methods for determination of evaporation from reservoirs. (first revision)
43.	IS : 6955—1973	Code of practice for sub-surface exploration for earth and rockfill dams. (Reaffirmed 1990)
44.	IS : 6966—1989	Criteria for hydraulic design of barrages and weirs: Part 1 Alluvial Reaches (first revision)
45.	IS : 7092	Aluminium alloy tubes for irrigation purposes
	(i) Part 1—1992	Welded tubes (second revision)
	(ii) Part 2—1987	Extruded tubes (second revision). (Reaffirmed 1992)
46.	IS : 7112—1973	Criteria for design of cross-section for unlined canals in alluvial soil. (Reaffirmed 1990)
47.	IS : 7113—1973	Code of practice for soil-cement lining for canals. (Reaffirmed 1990)
48.	IS : 7114—1973	Criteria for hydraulic design of cross regulators for canals. (Reaffirmed 1990)
49.	IS : 7331—1981	Code of practice for inspection and maintenance of cross drainage works (first revision). (Reaffirmed 1992)
50.	IS : 7349—1989	Code of practice for operation and maintenance of barrages and weirs (first revision)
51.	IS : 7356	Code of practice for installation, maintenance and observation of instruments for pore pressure measurements on earth dams
	(i) Part 1—1992	Porous tube piezometers
	(ii) Part 2—1976	Twin tube hydraulic piezometers. (Reaffirmed 1991)
52	IS : 7365—1985	Criteria for hydraulic design of bucket type energy dissipators (first revision). (Reaffirmed 1990)
53.	IS : 7436	Guide for types of measurements for structures in river valley projects and criteria for choice and location of measuring instruments
	(i) Part 1—1974	For earth and rockfill dams. (Reaffirmed 1990)
	(ii) Part 2—1976	For concrete and masonry dams. (Reaffirmed 1990)
54.	IS : 7500—1974	Code of practice for installation and observation of crossarms for measurement of internal vertical movement in earth dams. (Reaffirmed 1991)
55.	IS : 7720—1991	Criteria for investigation, planning and layout of barrages and weirs (first revision)

56. IS : 7784	Code of practice for design of cross drainage works
(i) Part 1—1975	General features. (Reaffirmed 1987)
(ii) Part 2—1983	Specific requirements. (Reaffirmed 1992). Section 1. Aqueducts
(iii) Part 2—1980	Specific requirements. (Reaffirmed 1992) Section 2. Super passage
(iv) Part 2—1981	Specific requirements. (Reaffirmed 1992). Section 3. Canal syphons
(v) Part 2—1980	Specific requirements. (Reaffirmed 1992). Section 4. Level crossings
(vi) Part 2—1980	Specific requirements. (Reaffirmed 1992). Section 5. Syphon Aqueducts
57. IS : 7873—1975	Code of practice for lime concrete lining for canals. (Reaffirmed 1990)
58. IS : 7894—1975	Code of practice for stability analysis of earth dams. (Reaffirmed 1991)
59. IS : 7986—1976	Code of practice for canal outlets. (Reaffirmed 1990)
60. IS : 8389—1983	Code of practice for installation and use of raingauges, recording (first revision). (Reaffirmed 1990)
61. IS : 8408—1976	Criteria for river training works for barrages and weirs in alluvium. (Reaffirmed 1990)
62. IS : 8414—1977	Guidelines for design of under seepage control measures for earth and rockfill dams. (Reaffirmed 1988)
63 IS : 8826—1978	Guidelines for design of large earth and rockfill dams. (Reaffirmed 1991)
64. IS : 8835—1978	Guidelines for planning and design of surface drains. (Reaffirmed 1990)
65. IS : 9012—1978	Recommended practice for shotcreting. (Reaffirmed 1992)
66. IS : 9097—1979	Guidelines for laying lining of canals with hot bitumen or bituminous felts. (Reaffirmed 1990)
67. IS : 9447—1980	Guidelines for assessment of seepages losses from canals by analytical method. (Reaffirmed 1991)
68. IS : 9451—1985	Guidelines for lining of canals in expansive soils. (first revision). (Reaffirmed 1991).
69. IS : 9452	Code of practice for measurement of seepage losses from canals.
(i) Part 1—1980	Ponding method. (Reaffirmed 1991)
(ii) Part 2—1980	Inflow-outflow method. (Reaffirmed 1990)
(iii) Part 3—1988	Seepage meter method. (Reaffirmed 1990)
70. IS : 9696—1980	Code of practice for installation of farm drainage tile or pipe system. (Reaffirmed 1992)
71. IS : 9698—1980	Code of practice for lining of canals with low density polyethylene film. (Reaffirmed 1991)

72.	IS : 9913—1981	Code of practice for construction of cross drainage works. (Reaffirmed 1992)
73.	IS : 10084	Criteria for design of diversion works
	(*i*) Part 1—1982	Coffer dam. (Reaffirmed 1991)
74.	IS : 10135—1985	Code of practice for drainage system for gravity dams, their foundations and abutments (first revision)
75.	IS : 10137—1982	Guidelines for selection of spillways and energy dissipators. (Reaffirmed 1990)
76.	IS : 10430—1982	Criteria for design of lined canals and guidelines for selection of type of lining. (Reaffirmed 1991)
77.	IS : 10635—1983	Guidelines for freeboard requirements in embankment dams. (Reaffirmed 1990)
78.	IS : 10646—1991	Canal linings-cement concrete tiles (first revision)
79.	IS : 10751—1983	Criteria for design of guide banks for alluvial rivers, (Reaffirmed 1992)
80.	IS : 11130—1984	Criteria for structural design of barrages and weirs. (Reaffirmed 1990)
81.	IS : 11155—1984	Code of practice for construction of spillways and similar overflow structures. (Reaffirmed 1990)
82.	IS : 11189—1983	Methods of tube well development. (Reaffirmed 1990)
83.	IS : 11223—1985	Guidelines for fixing spillway capacity. (Reaffirmed 1990)
84.	IS : 11293	Guidelines for the design of grout curtains
	(i) Part 1—1985	For earth and rockfill dams. (Reaffirmed 1990)
85.	IS : 11385—1985	Code of practice for subsurface exploration for canals and cross drainage works. (Reaffirmed 1990)
86.	IS : 11388—1985	Recommendations for design of trash rack for intakes. (Reaffirmed 1991)
87.	IS : 11399	Guidelines for estimating output norms of items of work in construction of river valley project
	(i) Part 1—1985	Earthwork excavation. (Reaffirmed 1990)
88.	IS : 11485—1985	Criteria for hydraulic design of sluices in concrete and masonry dams. (Reaffirmed 1990)
89.	IS : 11532—1985	Guidelines for construction of river embankments (levees) (Reaffirmed 1991)
90.	IS : 11570—1985	Criteria for hydraulic design of irrigation intake structures (Reaffirmed 1991)
91.	IS : 11624—1986	Guidelines for the quality of irrigation water. (Reaffirmed 1992)
92.	IS : 11625—1986	Criteria for hydraulic design of penstocks. (Reaffirmed 1991)
93.	IS : 12028—1987	Guidelines for maintenance of river embankments (levees). (Reaffirmed 1992)

94. IS : 12094—1987	Guidelines for planning and design of river embankments (levees) (Reaffirmed 1992)
95. IS : 12138—1987	Vocabulary for basic types of earthmoving machinery. (Reaffirmed 1992)
96. IS : 12169—1987	Criteria for design of small embankment dams. (Reaffirmed 1992)
97. IS : 12182—1987	Guidelines for determination of effects of sedimentation in planning and performance of reservoirs. (Reaffirmed 1992)
98. IS : 12200—1987	Code of practice for provision of waterstops at transverse contraction joints in masonry and concrete dams. (Reaffirmed 1992)
99. IS : 12232	Rotating sprinkler
(i) Part 1—1987	Design and operational requirement. (Reaffirmed 1992)
(ii) Part 2—1987	Test method for uniformity of distribution. (Reaffirmed 1992)
100. IS: 12379—1988	Code of practice for lining water courses and field channels (Reaffirmed 1992)
101. IS : 12629—1989	Water well drilling-rotary hose for rotary drilling-recommended practice for care and use
102. IS : 12720—1989	Criteria for structural design of spillway training walls and divide walls
103. IS : 12731—1989	Hydraulic design of impact type energy dissipators-recommendations
104. IS : 12785-1989	Irrigation equipment-strainer type filters. (Amendment 1)
105. IS : 12786—1989	Irrigation equipment-Polyethylene pipes for irrigation laterals. (Amendment 1)
106. IS : 12804—1989	Criteria for estimation of aeration demand for spillways and outlet structures
107. IS : 12966	Code of practice for galleries and other openings in dams
(i) Part 1—1992	General requirements
(ii) Part 2—1990	Structural design
108. IS : 13028—1991	Guidelines for overall planning of river basin
109. IS : 13056—1991	Deepwell handpumps. (Amendment 1)
110. IS : 13062—1991	Irrigation equipment and systems—evaluation of field irrigation efficiencies—guidelines
111. IS : 13142—1991	Proforma for reporting progress of benefits created by river valley projects. (Amendment 1)
112. IS : 13143—1991	Joints in concrete lining of canals-sealing compound
113. IS : 13144—1991	Recommendations for provision of facilities outside the dams
114. IS : 13218	Proforma for reporting progress during construction for river valley projects

(i) Part 1—1992	Irrigation works
(ii) Part 2—1992	Hydel works
(iii) Part 3—1992	Flood control
115. IS : 13487—1992	Irrigation equipment-emitters
116. IS : 13488—1992	Irrigation equipment-emitting pipe system
117. IS : 13551—1992	Structural design of spillway pier and crest criteria
118. IS : 13623—1993	Criteria for choice of gates and hoists
119. IS : 13645—1993	Guniting the upstream face of masonry dams-guidelines

Index

D